SIGNAL PROCESSING AND OPTIMIZATION FOR TRANSCEIVER SYSTEMS

Presenting the first complete treatment of MIMO transceiver optimization, this self-contained book provides all the mathematical information needed to understand transceiver optimization in a single volume. It begins with a review of digital communication fundamentals, and then moves on to a detailed study of joint transceiver optimization, starting from simple single-input single-output channels all the way to minimum bit error rate transceivers for MIMO channels. Crucial background material is covered, such as Schur-convex functions, matrix calculus, and constrained optimization, together with eight appendices providing further background material on topics such as matrix theory, random processes, and sampling theory. A final ninth appendix provides a grand summary of all the optimization results. With 360 illustrations, over 70 worked examples, and numerous summary tables provided to aid understanding of key concepts, this book is ideal for graduate students, practitioners, and researchers in the fields of communications and signal processing.

P. P. VAIDYANATHAN is a Professor of Electrical Engineering at the California Institute of Technology, where he has been a faculty member since 1983. He is an IEEE Fellow and has co-authored over 400 technical papers and two previous books in the area of signal processing. He has received numerous awards, including four awards for journal papers, the Award for Excellence in Teaching at the California Institute of Technology three times, and the Technical Achievement Award of the IEEE Signal Processing Society.

SEE-MAY PHOONG is a Professor in the Graduate Institute of Communication Engineering and the Department of Electrical Engineering at the National Taiwan University. He is a recipient of the Charles H. Wilts Prize for outstanding independent doctoral research at the California Institute of Technology and the Chinese Institute of Electrical Engineering's Outstanding Youth Electrical Engineer Award.

YUAN-PEI LIN is a Professor in Electrical Engineering at the National Chiao Tung University, Taiwan. She is a recipient of the Ta-You Wu Memorial Award, the Chinese Institute of Electrical Engineering's Outstanding Youth Electrical Engineer Award, and of the Chinese Automatic Control Society's Young Engineer in Automatic Control Award.

SIGNAL PROCESSING AND OPTIMIZATION FOR TRANSCEIVER SYSTEMS

P. P. VAIDYANATHAN

California Institute of Technology

SEE-MAY PHOONG

National Taiwan University

YUAN-PEI LIN

National Chiao Tung University, Taiwan

CAMBRIDGE
UNIVERSITY PRESS

CAMBRIDGE UNIVERSITY PRESS

Shaftesbury Road, Cambridge CB2 8EA, United Kingdom

One Liberty Plaza, 20th Floor, New York, NY 10006, USA

477 Williamstown Road, Port Melbourne, VIC 3207, Australia

314–321, 3rd Floor, Plot 3, Splendor Forum, Jasola District Centre, New Delhi – 110025, India

103 Penang Road, #05–06/07, Visioncrest Commercial, Singapore 238467

Cambridge University Press is part of Cambridge University Press & Assessment, a department of the University of Cambridge.

We share the University's mission to contribute to society through the pursuit of education, learning and research at the highest international levels of excellence.

www.cambridge.org
Information on this title: www.cambridge.org/9780521760799

First published 2010

A catalogue record for this publication is available from the British Library

ISBN 978-0-521-76079-9 Hardback

To Usha, Vikram, Sagar, and my parents
— P. P. Vaidyanathan

To our families
— See-May Phoong and Yuan-Pei Lin

Contents at a glance

Contents

Part 1: Communication fundamentals

Part 3: Mathematical background

Part 4: Appendices

Preface

Digital communication systems have been studied for many decades, and they have become an integral part of the technological world we live in. Many excellent books in recent years have told the story of this communication revolution, and have explained in considerable depth the theory and applications. Since the late 1990s particularly, there have been a number of significant contributions to digital communications from the signal processing community. This book presents a number of these recent developments, with emphasis on the use of filter bank precoders and equalizers. Optimization of these systems will be one of the main themes in this book. Both multiple-input multiple-output (MIMO) systems and single-input single-output (SISO) systems will be considered.

The book is divided into four parts. Part 1 contains introductory material on digital communication systems and signal processing aspects. In Part 2 we discuss the optimization of transceivers, with emphasis on MIMO channels. Part 3 provides mathematical background material for optimization of transceivers. This part can be used as a reference, and will be useful for readers wishing to pursue more detailed literature on optimization. Part 4 contains eight appendices on commonly used material such as matrix theory, Wiener filtering, and so forth. Thus, while it is assumed that the reader has some exposure to digital communications and signal processing at the introductory level, there is plenty of review material at the introductory level (Part 1) and at the advanced level (Parts 3 and 4). The material in Parts 2 and 3 will be useful for students wishing to pursue advanced work in the field, which is still a very active area for research. A detailed outline of the book can be found in Sec. 1.5 of Chap. 1.

Some of the material herein has been tested in the classroom, and a considerable part has benefited students at an advanced level. While many of the results in Part 2 can be regarded as results which appeared since the late 1990s, the mathematical foundation for this material is much older. Starting with the days of Shannon and Nyquist, there have been many giants in the field who contributed to this strong foundation since the 1940s. However, because of technological advances and the directions in which applications evolved, such as wireless communication and DSL technology, some of the theoretical problems have been revisited and some new problems solved by researchers in recent years. This freshness and novelty in the midst of old grandeur can clearly been seen from the combination of topics covered in Parts 1, 2, and 3 of the book.

We have endeavored to come up with a text that will be useful in the classroom, and which will serve as a research reference for advanced students. The writing style is in the form of an easy-to-read text book with detailed theory, plenty of examples, discussions, and homework problems. It is self-contained for students with an introductory background in signal processing and communications.

Acknowledgements

The wonderful environment provided by the California Institute of Technology, and the generous support from the Office of Naval Research and the National Science Foundation, have been crucial in developing the material covered in this book. As mentioned in the introductory and historical review sections of this book, many great minds have been involved in making the fields of communication and signal processing what they are today. Without their fundamental contributions this book would have been impossible.

Many graduate students have participated in extensive discussions relating to the material in this book. It is my pleasure to thank them here, and especially acknowledge the extensive discussions I have had with Chun-Yang Chen and Ching-Chih Weng regarding the material in Part 2 of the book.

For a project of this magnitude, long hours of hard work and concentration are absolutely essential. I have to thank Usha for creating the peaceful atmosphere which is crucial for the success of such a project. She has shown infinite patience during the long evenings and weekends of my absorption in this book. Her total unconditional love and sincere support, and the enthusiasm and love from Vikram and Sagar, are much appreciated!

P. P. Vaidyanathan
California Institute of Technology, Pasadena, CA

It is our pleasure to acknowledge the support of National Taiwan University and National Chiao-Tung University. We would also like to thank the National Science Council, Taiwan, for continued support of our research.

See-May Phoong
National Taiwan University, Taipei, Taiwan

and

Yuan-Pei Lin
National Chiao Tung University, Hsinchu, Taiwan

Part 1 Communication fundamentals

1 Introduction

1.1 Introduction

Digital communication systems have been studied for many decades, and they have become an integral part of the technological world we live in. Many excellent books in recent years have told the story of this communication revolution, and have explained in considerable depth the theory and applications. Since the late 1990s particularly, there have been a number of significant contributions to digital communications from the signal processing community. This book presents a number of these recent developments, with emphasis on the use of filter bank precoders and equalizers. Optimization of these systems will be one of the main themes in this book. Both multiple-input multiple-output (MIMO) systems and single-input single-output (SISO) systems will be considered. It is assumed that the reader has had some exposure to digital communications and signal processing at the introductory level. Many text books cover this prerequisite, and some are mentioned at the beginning of Sec. 1.5.

Before we describe the contents of the book we first give an introductory description of analog and digital communication systems in the next few sections. The scope and outline of the book will be described in Sec. 1.5.

1.2 Communications systems

Figure 1.1(a) shows the schematic of a simple analog communication system. Here we have a message signal $s(t)$ which is transmitted over a channel to produce the signal $y(t)$ at the receiver end. In many practical systems the channel can be modeled as a linear time invariant, or **LTI**, system followed by an additive noise source $q(t)$. This is shown in Fig. 1.1(b), where the channel impulse response is indicated as $h(t)$.

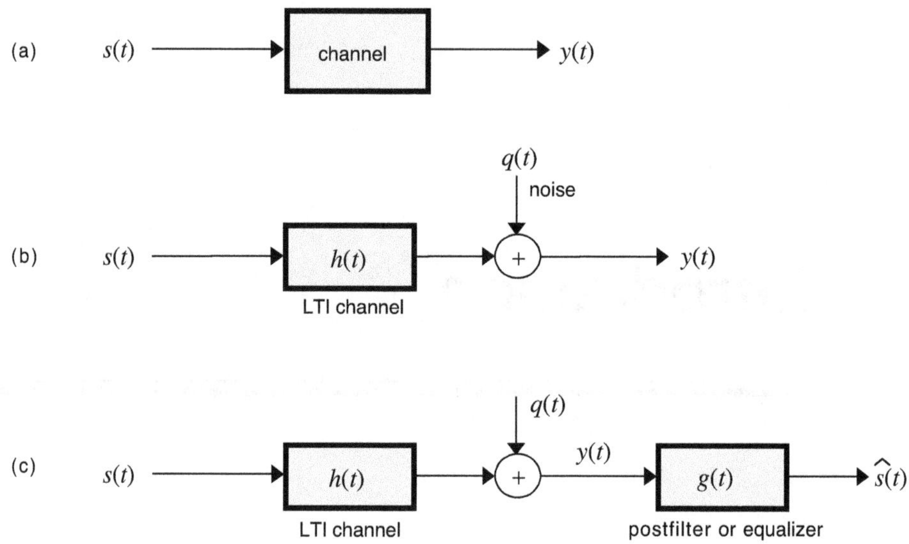

Figure 1.1. An analog communication system. (a) Channel with input $s(t)$ and output $y(t)$. (b) The channel modeled as a linear time invariant system followed by an additive noise source. (c) The channel followed by a postfilter or equalizer at the receiver.

The received signal $y(t)$ in Fig. 1.1(b) can be expressed in the form

$$y(t) = \int_{-\infty}^{\infty} h(\tau)s(t-\tau)d\tau \; + q(t). \tag{1.1}$$

The first term above represents a convolution integral. In practice the channel is *causal* so that $h(t)$ is zero for $t < 0$. In this case the lower limit of the integral can be taken as 0 rather than $-\infty$.

Next consider Fig. 1.1(c), where the received signal $y(t)$ is processed using an LTI system called the **equalizer** or the **postfilter.** The purpose of an equalizer is to compensate for the distortion caused by the convolution with the channel $h(t)$, and to reduce the effect of the channel noise. The equalizer should be designed by taking into account the knowledge of $h(t)$ and whatever knowledge might be available about the statistics of the noise $q(t)$. The reconstructed signal $\widehat{s}(t)$ then serves as an approximation of $s(t)$. The reconstruction error is given by

$$e(t) = \widehat{s}(t) - s(t). \tag{1.2}$$

Figure 1.2 shows a further enhancement at the transmitter end. The message signal $s(t)$ is first passed through an LTI system called the **prefilter** or **precoder**. This system has impulse response $f(t)$. The prefilter "shapes" the message $s(t)$ appropriately.

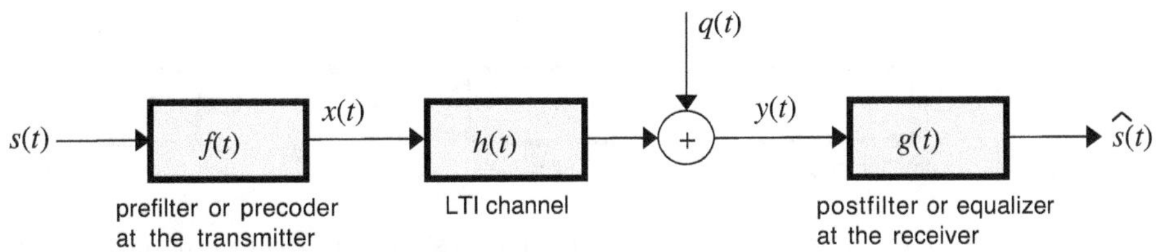

Figure 1.2. An analog communication system with a linear precoder at the transmitter and a linear equalizer at the receiver.

Given our knowledge of the channel and the noise statistics, it is possible to choose the prefilter (jointly with the equalizer) such that $\widehat{s}(t)$ approximates $s(t)$ as best as possible. For this we have to specify some error criterion such as, for example, the mean square error. Also, appropriate constraints on the transmitted power have to be specified. The precoder at the transmitter and the equalizer at the receiver are together referred to as a **transceiver**. Thus, we often talk of optimal design of a transceiver $\{f(t), g(t)\}$ for a given channel.[1] In later chapters we will consider many recent variations of this classical problem, especially in the context of *digital* communications. The multi-input multi-output (MIMO) version of this problem is especially important as we shall see. It is often convenient to use a frequency domain representation for the communication system. Thus, let $H(j\omega)$ denote the Fourier transform of $h(t)$, that is,

$$H(j\omega) = \int_{-\infty}^{\infty} h(t)e^{-j\omega t}dt. \tag{1.3}$$

This is called the **frequency response** of the channel. Similarly, let $F(j\omega)$ and $G(j\omega)$ represent the frequency responses of the precoder $f(t)$ and equalizer $g(t)$.

Figure 1.3 shows a redrawing of Fig. 1.2 in terms of these Fourier transform quantities. In terms of this notation the effective channel from $s(t)$ to $\widehat{s}(t)$ is

$$H_{eff}(j\omega) = G(j\omega)H(j\omega)F(j\omega). \tag{1.4}$$

With $(s * f)(t)$ denoting the **convolution** of $s(t)$ with $f(t)$, the effective channel impulse response can be written in the form

$$h_{eff}(t) = (g * h * f)(t). \tag{1.5}$$

There are many variations of the above channel model. In some situations, like mobile communications, the channel is modeled as a slowly time varying system rather than an LTI system. In some scenarios the channel impulse response $h(t)$ is regarded as a random variable drawn from a known distribution. The noise source $q(t)$ is often modeled as a Gaussian random process with known power spectrum. We shall come to the details later.

[1]Sometimes the entire system in the figure is loosely referred to as the transceiver.

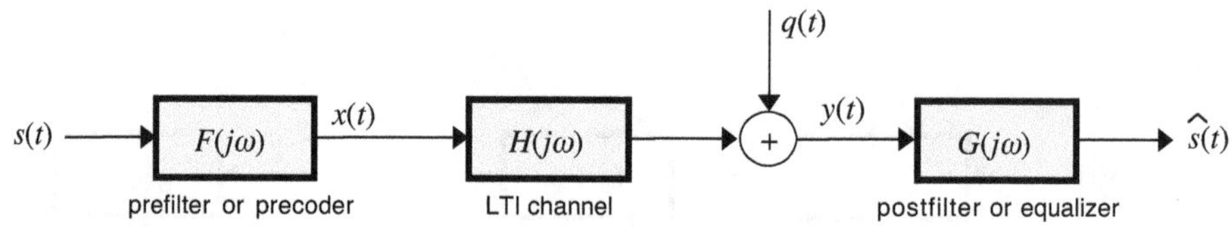

Figure 1.3. The analog communication system represented in terms of frequency responses.

1.3 Digitial communication systems

In the preceding section, the signal $s(t)$ was regarded as a continuous time signal with continuous (unquantized) amplitude. In a digital communication system, the messages are *quantized* amplitudes, transmitted in *discrete* time. Figure 1.4 shows the schematic of a digital communication system. Here we have a *discrete-time* message or signal $s(n)$ which we wish to transmit over a *continuous-time* channel. The amplitudes of $s(n)$ are "digitized," that is they come from a finite set of **symbols**. This collection of symbols is called a **constellation**. We shall come to details of digitization later.[2] Since $s(n)$ is a discrete-time signal and the channel is continuous-time, the signal is first converted into a continuous-time signal $x(t)$ as indicated in the figure.

The conversion from $s(n)$ to $x(t)$ can be described schematically in two steps. The building block indicated as D/C is a *discrete-to-continuous-time converter*, and it converts $s(n)$ to a signal $s_c(t)$ given by

$$s_c(t) = \sum_{n=-\infty}^{\infty} s(n)\delta_c(t - nT). \tag{1.6}$$

Here $\delta_c(t)$ is the *impulse* or *Dirac delta* function [Oppenheim and Willsky, 1997]. Thus, the sample $s(n)$ is converted into an impulse positioned at time nT. The sample spacing T determines the speed with which the message samples are conveyed. Since we have $1/T$ symbols per second, the **symbol rate** is given by

$$f_s = \frac{1}{T} \text{ Hz}. \tag{1.7}$$

The prefilter $F(j\omega)$ at the transmitter performs a convolution to produce the output

$$x(t) = \sum_{n=-\infty}^{\infty} s(n)f(t - nT), \tag{1.8}$$

[2]Examples of constellations include PAM and QAM systems to be described in Sec. 2.2.

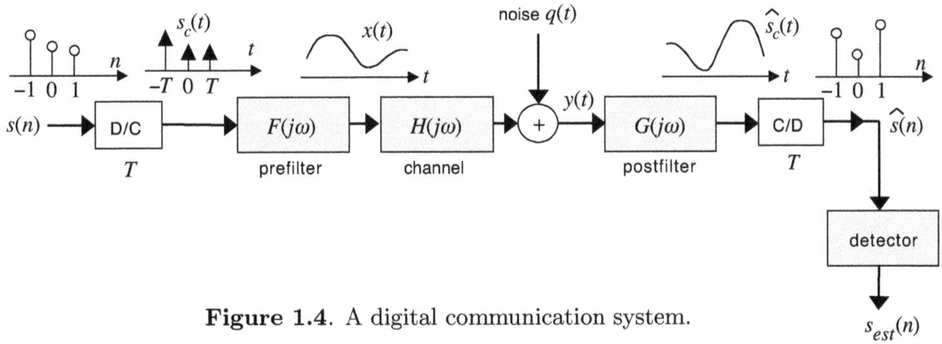

Figure 1.4. A digital communication system.

where $f(t)$ is the impulse response of $F(j\omega)$. Typically $f(t)$ is a smooth, finite duration function, as demonstrated in Fig. 1.5(a). In practice, $f(t)$ is **causal**, that is, it is zero for negative time. In the figure it is shown to be noncausal for generality. Note that $x(t)$ is a weighted sum of uniformly shifted versions of the impulse response $f(t)$. The weight on the nth shifted version $f(t - nT)$ is the nth message sample $s(n)$. This construction of $x(t)$ from $s(n)$ is demonstrated in Fig. 1.5(b).

The signal $x(t)$ is then transmitted over the continuous-time channel $H(j\omega)$, which also adds noise $q(t)$:

$$y(t) = \int_{-\infty}^{\infty} h(\tau)x(t-\tau)d\tau \quad + q(t). \tag{1.9}$$

This is the signal that will be observed at the receiver. The goal at the receiver is to reconstruct the original discrete signal $s(n)$ from this noisy and distorted continuous-time signal $y(t)$. First, the postfilter $G(j\omega)$ at the receiver processes $y(t)$ to produce $\widehat{s}_c(t)$, which is then **sampled at the rate** $f_s = 1/T$ to obtain a reconstructed version of $s(n)$:

$$\widehat{s}(n) = \widehat{s}_c(nT). \tag{1.10}$$

The box labeled C/D is the *continuous-to-discrete-time converter*, and performs the sampling operation (1.10). The reconstruction error is given by

$$e(n) = \widehat{s}(n) - s(n). \tag{1.11}$$

Given the knowledge of the channel $H(j\omega)$ and the noise statistics, it is possible to design the filters $F(j\omega)$ and $G(j\omega)$ to minimize an appropriate measure of reconstruction error. A simple example of an "appropriate measure" is the mean square error (i.e., the average value of $|e(n)|^2$).

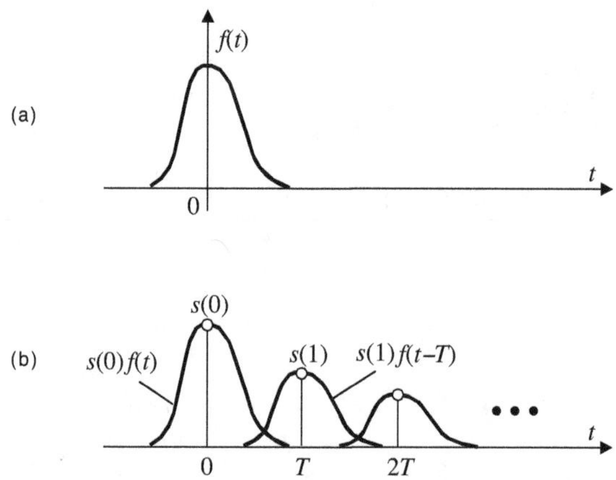

Figure 1.5. (a) Impulse response $f(t)$ of the prefilter $F(j\omega)$, and (b) the signal $x(t)$ generated from the samples $s(n)$ by interpolation with the function $f(t)$.

Note that, while $s(n)$ belongs to the signal constellation, the quantity $\widehat{s}(n)$ does not. In practice there is a device called the **detector** at the receiver (Fig. 1.4), which obtains an *estimated constellation symbol* $s_{est}(n)$ from the quantity $\widehat{s}(n)$. The *probability of symbol error* is defined to be the probability that $s_{est}(n)$ differs from $s(n)$. The minimization of this probability is another important optimization problem. Such optimizations and several generalizations will be discussed in appropriate sections of the book.

Bandlimiting. In practice the bandwidth allowed for a transceiver is limited. This bandlimiting is enforced by using lowpass filters at the transmitter and receiver. These filters can be incorporated as parts of $F(j\omega)$ and $G(j\omega)$. The transmitted signal $x(t)$ therefore occupies a fairly narrow bandwidth of the form

$$-\sigma < \omega < \sigma, \tag{1.12}$$

and is called the **baseband** signal. The bandwidth can be a few kHz to MHz, depending on application. The signal $x(t)$ is actually used to modulate a high-frequency carrier, and the modulated signal is transmitted either wirelessly using antennas or on wirelines. A discussion of carrier modulation is included in Sec. 2.4. The channel model discussed above is called the *baseband model*, as it does not show the carrier explicitly. Similarly the continuous-time system described in Sec. 1.2 also represents a baseband model. $\triangledown\triangledown\triangledown$

1.3.1 Discrete-time equivalent

We will see in later sections that the problem of designing the digital communication system of Fig. 1.4 can be reformulated entirely in terms of discrete-time transfer functions as in Fig. 1.6.

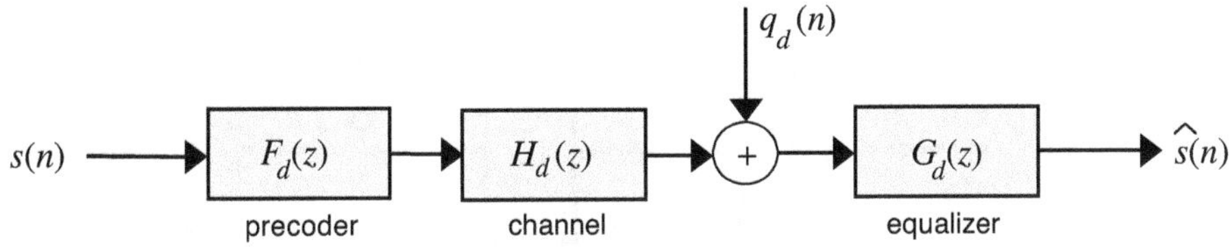

Figure 1.6. An all-discrete equivalent of the digital communication system.

Here $H_d(z)$ is the transfer function of an equivalent *discrete-time* channel. It is the z-transform of an equivalent digital channel impulse response $h_d(n)$, that is

$$H_d(z) = \sum_{n=-\infty}^{\infty} h_d(n)z^{-n}. \tag{1.13}$$

Similarly, $F_d(z)$ and $G_d(z)$ are the transfer functions of the discrete-time precoder and equalizer. The subscript d (for "discrete"), which is just for clarity, is usually dropped. In practice $H_d(z)$ is causal and can be approximated by a finite impulse response, or **FIR**, system so that

$$H_d(z) = \sum_{n=0}^{L} h_d(n)z^{-n}. \tag{1.14}$$

The problem of optimizing the precoder $F_d(z)$ and equalizer $G_d(z)$ for fixed channel $H_d(z)$ and fixed noise statistics will be addressed in later chapters.

1.4 MIMO channels

The transceivers described so far have one input signal $s(n)$ and a corresponding output $\widehat{s}(n)$. These are called *single-input single-output*, or **SISO**, transceivers. An important communication system that comes up frequently in this book is the *multi-input multi-output*, or **MIMO**, channel. Figure 1.7 shows a MIMO channel assumed to be linear and time-invariant with a transfer function matrix $\mathbf{H}(z)$, usually an FIR system:

$$\mathbf{H}(z) = \sum_{n=0}^{L} \mathbf{h}(n)z^{-n}. \tag{1.15}$$

The sequence $\mathbf{h}(n)$, called the MIMO impulse response, is a sequence of matrices. If the channel has P inputs and J outputs then $\mathbf{H}(z)$ has size $J \times P$, and so does each of the matrices $\mathbf{h}(n)$. The MIMO communication channel is used to transmit a vector signal $\mathbf{s}(n)$ with M components:

$$\mathbf{s}(n) = \begin{bmatrix} s_0(n) & s_1(n) & \ldots & s_{M-1}(n) \end{bmatrix}^T. \tag{1.16}$$

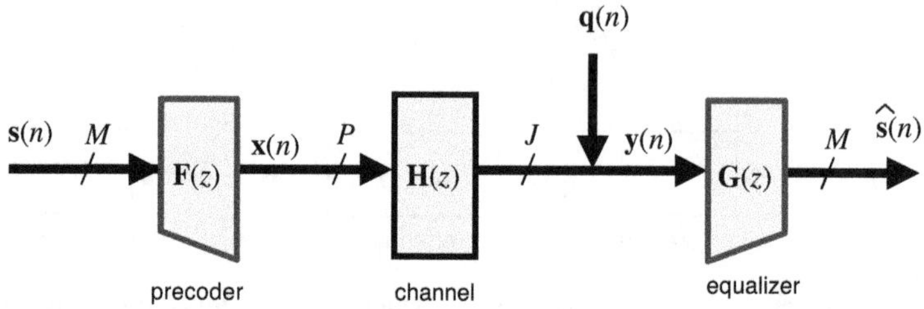

Figure 1.7. A digital communication system.

The precoder $\mathbf{F}(z)$ transforms this sequence $\mathbf{s}(n)$ into another sequence $\mathbf{x}(n)$. We will see that the choice of $\mathbf{F}(z)$ plays an important role in the performance of the communication system. The channel produces the inevitable distortion represented by the transfer function $\mathbf{H}(z)$ and the noise vector $\mathbf{q}(n)$. Thus the signal obtained at the receiver is

$$\mathbf{y}(n) = \sum_{k=0}^{L} \mathbf{h}(k)\mathbf{x}(n-k) \; + \; \mathbf{q}(n). \tag{1.17}$$

The equalizer $\mathbf{G}(z)$ seeks to reconstruct $\mathbf{s}(n)$ from this distorted version:

$$\widehat{\mathbf{s}}(n) = \sum_{k} \mathbf{g}(k)\mathbf{y}(n-k). \tag{1.18}$$

The joint design of the transceiver $\{\mathbf{F}(z), \mathbf{G}(z)\}$ is an important problem in modern digital communications. The MIMO transceiver shown in the figure can be used to transmit messages

$$s_k(n), \; 0 \leq k \leq M-1, \tag{1.19}$$

from M separate users. It can also be used to transmit information from one user by representing the message $s(n)$ from the user in the form of a vector $\mathbf{s}(n)$; such systems are called block-based transceivers for SISO channels. They have many advantages as we shall see. MIMO channels also arise from the use of multiple antennas for single users; a detailed discussion of how MIMO channels arise will be given in Sec. 4.5.

A special case of the MIMO system arises when the channel is **memoryless**, that is, the transfer function $\mathbf{H}(z)$ is just a constant $\mathbf{H}$. This corresponds to the situation where $L = 0$ in Eq. (1.15). Optimization of transceivers for memoryless MIMO channels will be the focus of some of the chapters in this book.

1.5 Scope and outline

The reader is assumed to have some familiarity with introductory topics in communications and signal processing. References for such background material include Proakis [1995], Oppenheim and Willsky [1997], Oppenheim and Schafer [1999], Lathi [1998], Haykin [2001], Mitra [2001], and Antoniou [2006], among other excellent texts. Advanced material related to the topics in this book can also be found in Ding and Li [2001], and Giannakis, *et al.* [2001]. Books on the very important related areas of wireless and multiuser communications include Rappaport [1996], Verdu [1998], Goldsmith [2005], Haykin and Moher [2005], and Tse and Viswanath [2005].

The book is divided into four parts. Part 1 contains introductory material on digital communication systems and signal processing aspects. In part 2 we discuss the optimization of transceivers, with emphasis on MIMO channels. Part 3 provides mathematical background material for optimization of transceivers. This part can be used as a reference, and will be very useful for readers wishing to pursue more detailed literature on optimization. Part 4 contains eight appendices on commonly used material such as matrix theory, Wiener filtering, and so forth.

The history of digital communication theory is fascinating. It is a humbling experience to look back and reflect on the tremendous insights and accomplishments of the communications and signal processing society in the last six decades. Needless to say, much of the recent research is built upon six to seven decades of this solid foundation. A detour into history will be provided in Chap. 9, where we present a historical perspective of transceiver design, equalization, and optimization, all of which originated in the early 1960s, and have continued to this day to be research topics. All references to literature will be given in the specific chapters as appropriate. An extensive reference list is given at the end of the book. In what follows we briefly describe the four parts of the book.

Part 1: Communication fundamentals

Part 1 consists of Chapters 1 to 8. In Chap. 2 we review basic topics in digital communication systems, such as signal constellations, carrier modulation, and so forth. Formulas for probabilities of error in symbol detection are derived. Matched filtering, which is used in some receiver systems, is discussed in some detail. In Chap. 3 we describe digital communication systems using the language of multirate filter banks. Such a representation is very useful for transceivers with or without redundancy, and has many applications as we shall see throughout the book.

In Chap. 4 we describe digital communication systems using discrete-time language. This chapter also introduces symbol spaced equalizers (SSE) and fractionally spaced equalizers (FSE). The minimum mean square error (MMSE) equalizer is also introduced in this chapter. Chapter 5 discusses a number of fundamental techniques that are commonly used in digital communications. First a detailed discussion of the matched filter is provided. Then we discuss optimal sequence estimators, such as the maximum likelihood (ML) detector and the Viterbi alogrithm. Nonlinear methods, such as the decision feedback equalizer and nonlinear precoders, are introduced. Chapter 6 is a brief discussion of channel capacity with emphasis on MIMO channels.

Chapter 7 introduces *redundant precoders*, including *zero-padded* and *cyclic-*

prefixed precoders. The redundant precoder is an integral part of many of the transceiver designs today. For example, cyclic prefix systems are employed in *orthogonal frequency division multiplexing* (OFDM) systems and discrete multitone (DMT) systems, used in *digital subscriber loop* (DSL) technology. The introduction of redundancy allows us to compensate or equalize the effects of a linear channel very efficiently – for example, an FIR channel can be equalized without the use of IIR equalizers. In Chap. 8 we discuss zero-padded systems in greater detail and introduce zero-forcing FIR equalizers, which can perfectly equalize FIR channels by exploiting the redundancy in the transmitted symbol stream. A number of properties of such equalizers are studied.

Part 2: Transceiver optimization

Part 2 consists of Chapters 9 to 19. Chapter 9 gives a brief historical introduction to transceiver optimization, and provides a detailed outline for Chapters 10 to 19. Briefly, Chap. 10 discusses the optimization of transceivers for scalar channels, and Chap. 11 discusses the optimization of transceivers for MIMO diagonal channels.

Chapters 12 and 13 discuss the minimization of mean square error in transceivers (MMSE transceivers) for general (nondiagonal) channels with and without the so-called *zero-forcing* constraint. Chapter 14 discusses the minimization of transmitted power for fixed performance criteria (such as error probability). This chapter also shows how one can perform *bit allocation* among the symbol streams optimally.

Chapter 15 discusses transceiver optimization for the special case where the precoder at the transmitter is constrained to be orthogonal. In Chap. 16 we consider the minimization of symbol error rates or bit error rates (BER), which are more directly related to practical performance than mean square errors. There is a close connection between MMSE transceivers and minimum-BER transceivers as we shall see in that chapter. The results of transceiver optimization are applied in Chaps. 17 and 18 to the case of cyclic-prefix systems and zero-padded systems, respectively. These are single-input single-output (SISO) channels turned into multi-input multi-output (MIMO) channels by introducing redundancy as described in Chap. 7. Chapter 19 discusses the decision feedback equalizer for MIMO channels. The joint optimization of transceiver matrices with decision feedback is discussed in detail.

Part 3: Mathematical background

Part 3 consists of Chapters 20 to 22. Some of the mathematical background needed for the optimization chapters is given in these chapters. This includes matrix calculus, Schur convex functions, and nonlinear optimization tools. Matrix calculus is a less commonly reviewed topic, so Chap. 20 offers a detailed review. Schur convex functions have played a major role in transceiver optimization in recent years, and the review in Chap. 21 will be useful to readers wishing to pursue the literature in depth. Chapter 22 is a review of constrained optimization theory, which is useful in some of the chapters on transceiver optimization.

Part 4: Appendices

There are eight appendices at the end of the book. They contain short discussions on useful topics from inequalities, matrix theory, singular value decompositions, random processes, Wiener filtering, sampling theory, and so forth. In addition, there are appendices at the ends of some individual chapters, which contain useful material relevant to those chapters. Book appendices are numbered as Appendix A, Appendix B, and so forth. Chapter appendices are numbered as Appendix 2.A (App. A at the end of Chap. 2), and so forth. Appendix I at the end of the book gives a summary of the main optimization results in Part 2 of the book, with each major result summarized in one page.

1.6 Commonly used notations

Bold-faced letters, such as $\mathbf{A}$ and $\mathbf{v}$, indicate matrices and vectors. Superscript $T, *$, and $\dagger$, as in $\mathbf{A}^T, \mathbf{A}^*$, and $\mathbf{A}^\dagger$ denote, respectively, the transpose, conjugate, and transpose-conjugate of a matrix. The determinant of a square matrix $\mathbf{A}$ is denoted as $\det(\mathbf{A})$, and the trace as $\mathrm{Tr}(\mathbf{A})$, with brackets omitted when redundant. Given two Hermitian matrices $\mathbf{A}$ and $\mathbf{B}$, the notation $\mathbf{A} \geq \mathbf{B}$ means that $\mathbf{A} - \mathbf{B}$ is positive semidefinite, and $\mathbf{A} > \mathbf{B}$ means that $\mathbf{A} - \mathbf{B}$ is positive definite (Appendix B). For a continuous-time function $h(t)$ the *Laplace transform* is denoted as $H(s)$ and the *Fourier transform* as $H(j\omega)$. The frequency variable $f = \omega/2\pi$ is also sometimes used. For a discrete-time function $g(n)$ the *z-transform* is denoted as $G(z)$ and the Fourier transform as $G(e^{j\omega})$. The tilde notation on a function of z is defined as follows:

$$\widetilde{\mathbf{H}}(z) = \mathbf{H}^\dagger(1/z^*).$$

Thus,

$$\mathbf{H}(z) = \sum_n \mathbf{h}(n) z^{-n} \Rightarrow \widetilde{\mathbf{H}}(z) = \sum_n \mathbf{h}^\dagger(n) z^n,$$

so that the tilde notation effectively replaces all coefficients with the transpose conjugates, and replaces z with $1/z$. For example,

$$H(z) = h(0) + h(1)z^{-1} \Rightarrow \widetilde{H}(z) = h^*(0) + h^*(1)z,$$

and

$$H(z) = \frac{a_0 + a_1 z^{-1}}{1 + b_1 z^{-1}} \Rightarrow \widetilde{H}(z) = \frac{a_0^* + a_1^* z}{1 + b_1^* z}$$

Note that $\widetilde{\mathbf{H}}(e^{j\omega}) = \mathbf{H}^\dagger(e^{j\omega})$. That is, the tilde notation reduces to transpose conjugation on the unit circle.

2

Review of basic ideas from digital communication

2.1 Introduction

In this chapter we briefly review introductory material from the theory and practice of digital communication. The reader familiar with such introductory material can use this chapter primarily as a reference for later chapters. For a more detailed treatement one should consult standard communication texts such as Proakis [1995], Lathi [1998], or Haykin [2001].

2.1.1 Chapter overview

The schematic representation of digital communication systems described earlier in Sec. 1.3 is reproduced in Fig. 2.1. As mentioned in that section, the input symbol stream has sample values $s(n)$ chosen from a finite set of values called the symbol alphabet, constellation, or code words – we shall consistently use the term *constellation*. In Sec. 2.2 we shall describe two commonly used constellations called the PAM (pulse amplitude modulation) and the QAM (quadrature PAM) constellations.

Given the symbol stream $s(n)$ the transmitter generates the baseband waveform

$$x(t) = \sum_{n=-\infty}^{\infty} s(n)f(t - nT),$$

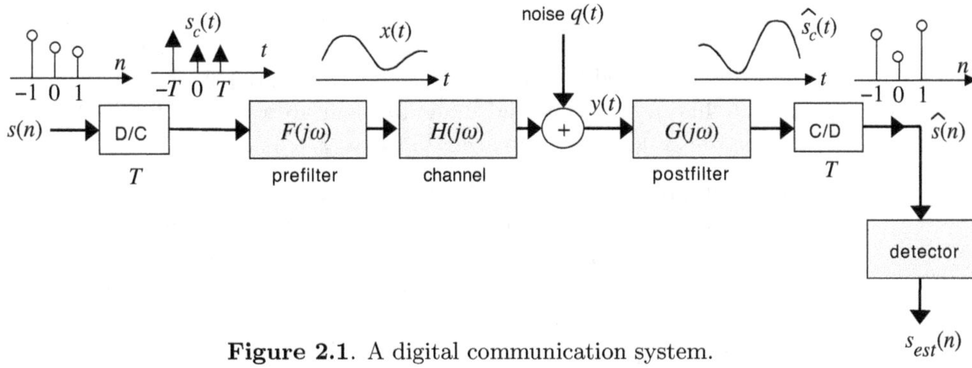

Figure 2.1. A digital communication system.

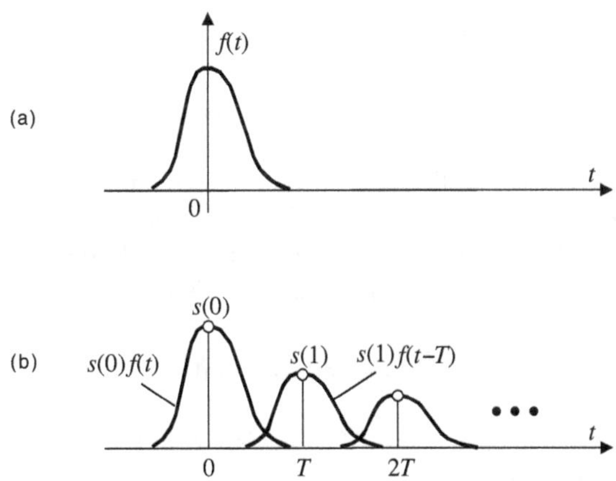

Figure 2.2. (a) Impulse response $f(t)$ of the prefilter $F(j\omega)$, and (b) the signal $x(t)$ generated from the samples $s(n)$ by interpolation with the function $f(t)$.

where $f(t)$ is a prefilter waveform. We also say that $f(t)$ is the transmitted *pulse* (and sometimes we use the notation $p(t)$). This waveform is usually time-limited, and most of its energy is confined to a narrow band of frequencies such as $|\omega| < \sigma$ called the **baseband** (sometimes written as base band). So $f(t)$ can be considered to be nearly **bandlimited**. A typical $f(t)$ is shown in Fig. 2.2(a), and a pictorial description of the waveform $x(t)$ is reproduced in Fig. 2.2(b) (from Sec. 1.3). The signal $x(t)$ is filtered by the channel $H(j\omega)$, and the receiver receives a noisy version of this signal. The received signal is further filtered by $G(j\omega)$ and then sampled at the symbol rate $1/T$. From the sampled version $\widehat{s}(n)$ the receiver has to identify or **detect** the symbol $s(n)$ which was actually transmitted. Since the channel introduces distortions (due to $H(j\omega)$ and noise $q(t)$), the detected symbol $s_{est}(n)$ can sometimes be different from $s(n)$, resulting in an error. Under appropriate assumptions on the statistics of the reconstruction error it is possible to compute the **probability of error** in symbol identification. This is done in Sec. 2.3. In practice the baseband signal $x(t)$ is used to modulate a high-frequency carrier signal, and the modulated signal is transmitted. At the receiver this signal is demodulated to extract the (noisy version of) the baseband signal. The details of this modulation and demodulation are different for PAM and QAM constellations, and will be presented in Sec. 2.4.

Returning to the postfilter $G(j\omega)$, we mentioned in Chap. 1 that it is called the equalizer, and compensates for the channel distortions. In practice, this equalization is often performed with a digital filter *after* the sampling process. The analog filter $G(j\omega)$ then plays a different role called **matched filtering**. The purpose of this filter is to maximize the signal-to-noise ratio at the sample locations, and its optimal choice depends on the channel $H(j\omega)$ as well as on the power spectrum of the noise $q(t)$. Matched filtering is reviewed in Sec. 2.5, and some practical details are discussed in Sec. 2.6.

2.2 Signal constellations

Two of the popular constellations widely used today are the **PAM** (pulse amplitude modulation) and the **QAM** (quadrature amplitude modulation) constellations. The PAM constellation has real-valued numbers for $s(n)$, whereas the QAM constellation has complex numbers. A b-bit PAM or QAM constellation has $M = 2^b$ allowed numbers, called **constellation symbols** or **codewords** (sometimes written as code words), and is also called an M-PAM or M-QAM constellation.

Figure 2.3 shows a PAM constellation with M codewords. Note that M is even (a power of 2), and adjacent codewords are separated by a fixed amount $2A$. No codeword has value zero. The positive number A can be chosen to adjust the average energy per codeword as we shall see later. Figure 2.4 shows a QAM constellation with $M = 16$ codewords. This is a 4-bit constellation. Once again, no codeword has value zero. Note that any 4-bit QAM code word has the form

$$z = x + jy,$$

where x and y are 2-bit PAM words. The components x and y are sometimes called the **in-phase** and **quadrature** components. A b-bit QAM constellation has real and imaginary parts coming from $(b/2)$-bit PAM constellations. The QAM constellations we use here are called **square** QAM constellations. More generally they can be **rectangular** constellations with b_1 bits for x and b_2 bits for y. There are more general types of QAM constellations including circular ones [Proakis, 1995], but we shall not elaborate on those here. Unless mentioned otherwise, we always imply square constellations when we use the term QAM. So the number of bits b is an even integer. As in PAM, any word in the square QAM constellation has nearest neighbors at distance $2A$.

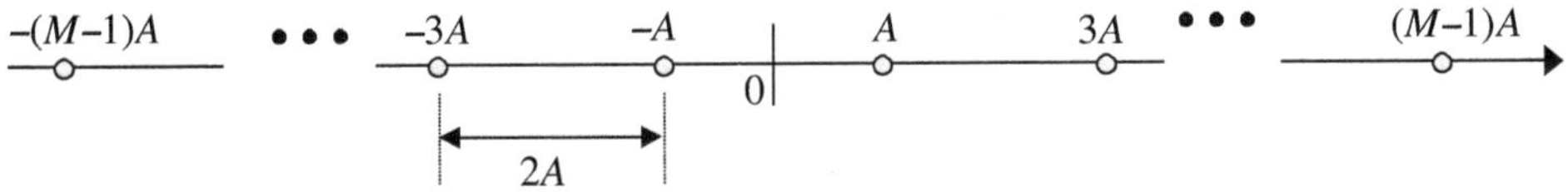

Figure 2.3. A *PAM* constellation with M codewords.

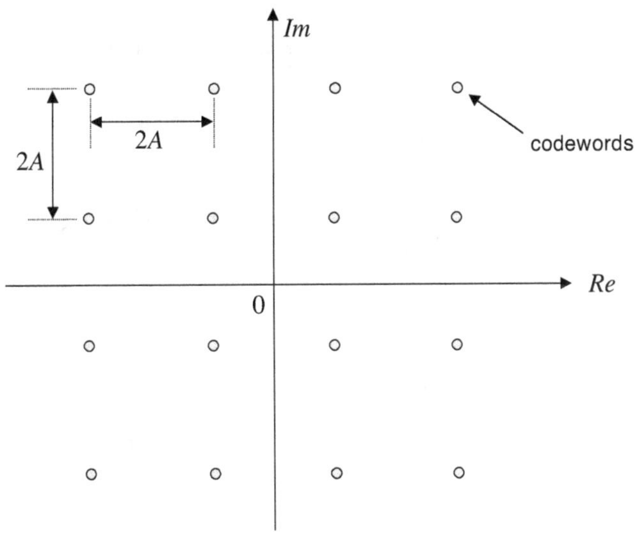

Figure 2.4. A 4-bit *QAM* constellation with $2^4 = 16$ codewords.

2.2.1 Average energy in a PAM constellation

It is useful to have an expression for the average energy in the symbols. In a PAM constellation, the words are real numbers of the form $(2n+1)A$, where n is an integer. The energy of a symbol is simply $(2n+1)^2A^2$. Note that for every positive codeword there is a corresponding negative codeword with the same magnitude. So the average energy of a b-bit PAM constellation (with $M = 2^b$ words) is

$$E_{ave,PAM} = \frac{2A^2}{M}\Big(1^2 + 3^2 + 5^2 + \ldots + (M-1)^2\Big). \tag{2.1}$$

If the M codewords are equally likely, then this represents the average energy per sample in the sequence $s(n)$. Using the fact (see below) that

$$1^2 + 3^2 + 5^2 + \ldots + (M-1)^2 = \frac{M(M^2-1)}{6}, \tag{2.2}$$

we therefore have

$$E_{ave,PAM} = \frac{(M^2-1)A^2}{3} = \frac{(2^{2b}-1)A^2}{3}, \tag{2.3}$$

so that

$$A = \sqrt{\frac{3E_{ave,PAM}}{2^{2b}-1}}.$$

Since $b = \log_2 M$, the quantity

$$E_{b,PAM} = \frac{E_{ave,PAM}}{\log_2 M} = \frac{(M^2-1)A^2}{3\log_2 M}, \tag{2.4}$$

is called the *energy per bit.*

Derivation of Eq. (2.2). It is well known that $\sum_{k=1}^{n} k^2 = n(n+1)(2n+1)/6$. Now, for even M we can write

$$\begin{aligned}
&1 + 3^2 + \ldots + (M-1)^2 \\
&= \sum_{k=1}^{M} k^2 - (2^2 + 4^2 + \ldots + M^2) \\
&= \sum_{k=1}^{M} k^2 - 4\Big(1 + 2^2 + \ldots + (\frac{M}{2})^2\Big) \\
&= \frac{M(M+1)(2M+1)}{6} - \frac{1}{6} \times \frac{4M}{2}\Big(\frac{M}{2}+1\Big)(M+1) \\
&= \frac{M(M+1)}{6}\Big(2M+1-(M+2)\Big) = \frac{M(M^2-1)}{6},
\end{aligned}$$

which proves Eq. (2.2). ▽ ▽ ▽

2.2.2 Average energy in a QAM constellation

Now consider a b-bit QAM constellation. There are $M = 2^b$ codewords of the form

$$z = x + jy,$$

where x and y belong to $(b/2)$-bit constellations with $\sqrt{M}$ words each. Thus the average energy in the real part is as in Eq. (2.3) with M replaced by $\sqrt{M} = 2^{b/2}$:

$$E_{ave,x} = \frac{(M-1)A^2}{3}.$$

The same is true for the imaginary part y. So the average energy of a b-bit QAM constellation (with $M = 2^b$ words) is

$$E_{ave,QAM} = \frac{2(M-1)A^2}{3} = \frac{2(2^b-1)A^2}{3}. \tag{2.5}$$

The energy per bit is therefore

$$E_{b,QAM} = \frac{E_{ave,QAM}}{\log_2 M} = \frac{2(M-1)A^2}{3\log_2 M}. \tag{2.6}$$

2.3 Error probability

In a digital communication system the receiver constructs an approximation $\widehat{s}(n)$ of the transmitted symbol stream $s(n)$. The reconstructed version $\widehat{s}(n)$ differs from $s(n)$ because of errors introduced by the channel $H(j\omega)$ and the noise $q(t)$. We can write

$$\widehat{s}(n) = s(n) + e(n), \tag{2.7}$$

where $e(n)$ is the reconstruction error, which can often be modeled as random noise. Thus, even though $s(n)$ is a codeword belonging to a constellation such as a PAM constellation, the reconstructed number $\widehat{s}(n)$ is not. In practice $\widehat{s}(n)$ is processed by a decision making device, called the **detector**, which takes the signal $\widehat{s}(n)$ and estimates the transmitted codeword $s(n)$ (Fig. 2.5). There is a nonzero probability that this estimated codeword $s_{est}(n)$ is different from the original codeword $s(n)$. This error probability depends on the statistics of the error term $e(n)$ in Eq. (2.7). Figure 2.6 shows how the received symbols $\widehat{s}(n)$ get spread out into a "cloud" owing to the noise $e(n)$ in a QAM constellation. The received signal can be anywhere in the shaded areas. If the shaded area for a symbol overlaps with the corresponding shaded area of an adjacent symbol, there is nonzero probability of symbol error. In this section we derive mathematical expressions for error probabilities.

2.3.1 Error probability for PAM signals

First consider the PAM constellation shown in Fig. 2.7 for 3 bits. We have indicated small vertical lines called decision boundaries. These are placed exactly midway between every pair of symbols. If $\widehat{s}(n)$ falls within a pair of decision boundaries, then the unique codeword within those boundaries is assumed to be transmitted because it is the closest codeword. This is the estimated symbol $s_{est}(n)$ corresponding to the transmitted symbol $s(n)$. To demonstrate, Fig. 2.8(a) shows the threshold detector characteristics for 1-bit PAM. This figure says that the 1-bit PAM detector estimates the symbol according to the rule

$$s_{est}(n) = \begin{cases} A & \text{if } \widehat{s}(n) \geq 0 \\ -A & \text{if } \widehat{s}(n) < 0. \end{cases} \tag{2.8}$$

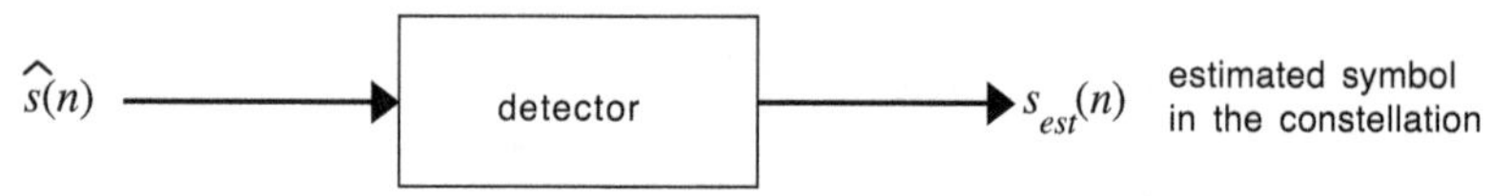

Figure 2.5. The detector, or decision device, at the receiver takes the reconstructed symbol $\widehat{s}(n)$ and maps it into a symbol $s_{est}(n)$ in the constellation, which is regarded as an estimate of the transmitted symbol $s(n)$.

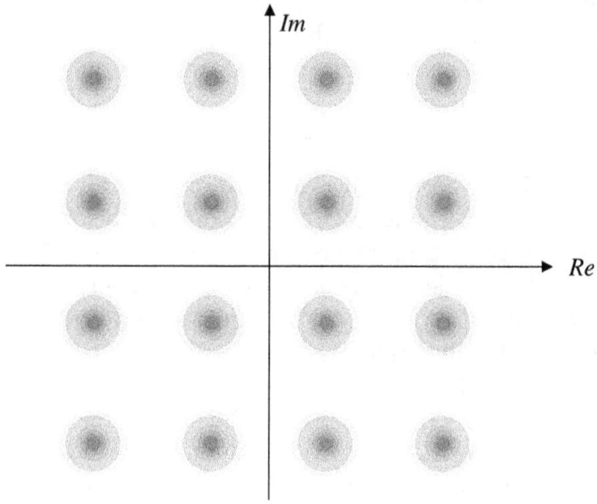

Figure 2.6. Noise cloud associated with a 4-bit QAM constellation.

Figure 2.8(b) shows the threshold detector characteristics for an arbitrary number of bits. As a specific example, assume that the transmitted symbol at time n was $s(n) = 3A$, as highlighted in Fig. 2.7. If the error term $e(n)$ (which is a random variable) is such that

$$|e(n)| < A \tag{2.9}$$

then $\widehat{s}(n)$ is within the shaded box shown in Fig. 2.7, and the symbol estimation is correct, that is, $s_{est}(n) = s(n)$. If $e(n)$ has magnitude larger than A then the symbol is estimated to be A (for $e(n) < -A$) or $5A$ (for $e(n) > A$), and there is an error.

The probability of error in the decision can be calculated if we know the **probability density function** (pdf) of the additive error term $e(n)$ in Eq. (2.7). Figure 2.7 shows an example of this pdf, denoted as $f_E(e)$.

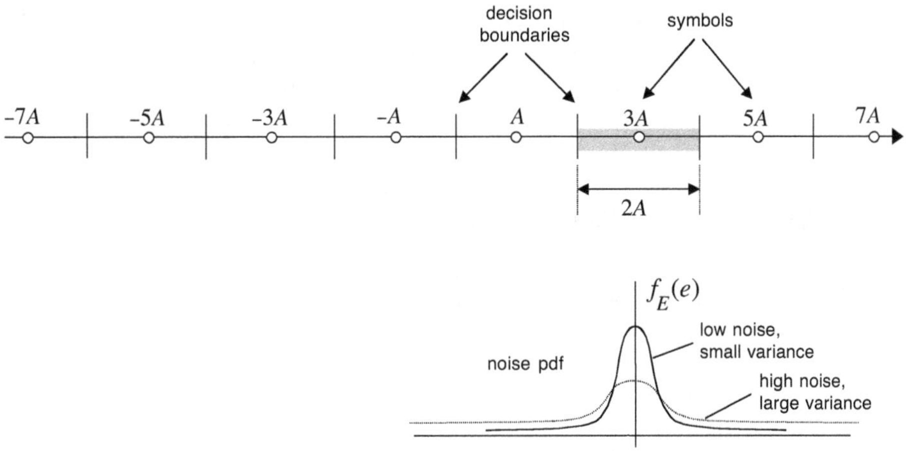

Figure 2.7. Explanation of how the detector works, and how decision errors occur in a PAM constellation. See text.

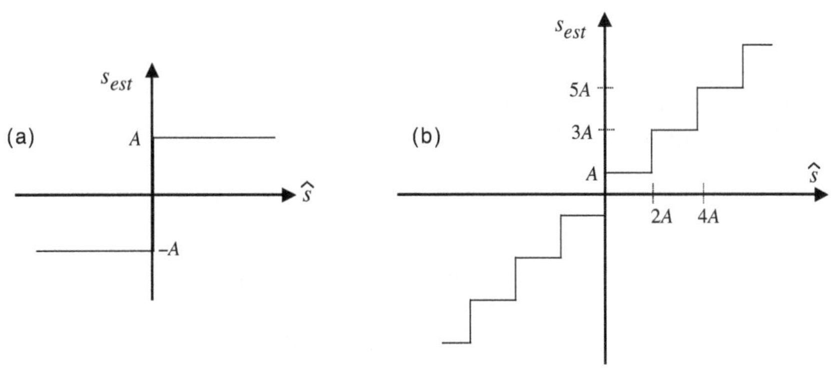

Figure 2.8. Threshold detector characteristics (a) for 1-bit PAM and (b) for a general PAM.

The probability of incorrectly deciding that $5A$ was transmitted instead of $3A$ is the probability that $e(n) > A$, that is,

$$\mathcal{P}(3A \text{ detected as } 5A) = \int_{A}^{\infty} f_E(e)de. \tag{2.10}$$

The probability of incorrectly deciding that A was transmitted instead of $3A$ is the probability that $e(n) < -A$, that is,

$$\mathcal{P}(3A \text{ detected as } A) = \int_{-\infty}^{-A} f_E(e)de. \tag{2.11}$$

The probability of an incorrect decision is therefore the sum of the above two integrals.[3] For a *boundary symbol* such as $7A$ or $-7A$, the contribution to error comes only from one integral. For example, since there are no symbols past $7A$, the probability of error is just the integral in (2.11), and similarly for the transmitted symbol $-7A$, the probability of error is the integral in (2.10). Only for the $M-2$ *interior symbols* is the error probability the sum of the integrals in (2.10) and (2.11).

2.3.1.A Case of Gaussian noise

In many communication systems it is reasonable to assume that the error $e(n)$ is a *Gaussian* random variable with zero mean, that is

$$f_E(e) = \frac{1}{\sqrt{2\pi\sigma_e^2}} e^{-(e^2/2\sigma_e^2)}, \tag{2.12}$$

where σ_e^2 is the variance of $e(n)$. Since $e(n)$ is usually also assumed to be white, we say that it is additive white Gaussian noise, or AWGN. Figure 2.9 demonstrates this pdf for two values of the variance σ_e^2. Since $f_E(e)$ is symmetric, the integrals (2.10) and (2.11) are identical in this case. In fact this integral can be expressed elegantly in terms of the so-called **$\mathcal{Q}$-function**. This function is defined as the integral

$$\mathcal{Q}(v) = \int_v^\infty f_N(u)du, \tag{2.13}$$

where $f_N(u)$ is the *normal density* $e^{-u^2/2}/\sqrt{2\pi}$, that is, Gaussian with zero mean and unit variance. Thus

$$\mathcal{Q}(v) = \frac{1}{\sqrt{2\pi}} \int_v^\infty c^{-u^2/2} du. \tag{2.14}$$

The $\mathcal{Q}$-function is related to the *complementary error function* $\text{erfc}(x)$ by

$$\mathcal{Q}(x) = 0.5\,\text{erfc}(x/\sqrt{2}), \tag{2.15}$$

or equivalently $\text{erfc}(x) = 2\mathcal{Q}(\sqrt{2}x)$. A plot is shown in Fig. 2.10. Notice how it decreases monotonically. In Sec. 21.2.3 we will see that $\mathcal{Q}(x)$ is a convex function. Using the $\mathcal{Q}$-function we can express the integral (2.10) as

$$\begin{aligned}\int_A^\infty f_E(e)de &= \frac{1}{\sqrt{2\pi\sigma_e^2}} \int_A^\infty e^{-(e^2/2\sigma_e^2)} de \\ &= \frac{1}{\sqrt{2\pi}} \int_{A/\sigma_e}^\infty e^{-(u^2/2)} du = \mathcal{Q}(A/\sigma_e),\end{aligned}$$

where we have used the change of variables $u = e/\sigma_e$.

[3]Note that if the error pdf $f_E(e)$ is confined to the range $-A < e < A$ then the symbol error probability is zero. That is, the receiver can perfectly identify the transmitted symbol even though there is channel noise.

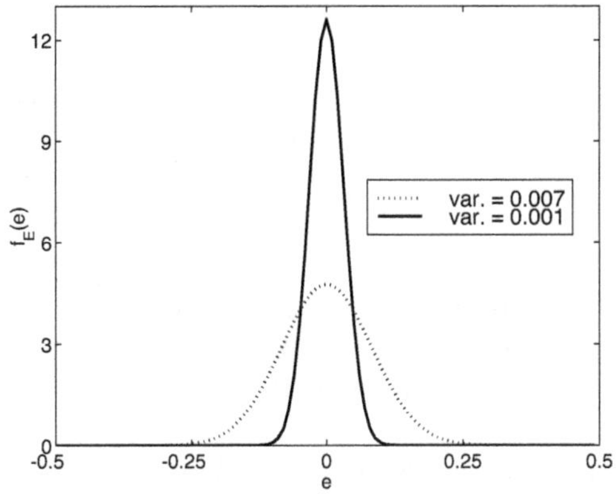

Figure 2.9. Examples of the Gaussian density function for zero mean and two values of the variance σ_e^2. For large variance, the plot is more spread out. For small variance the plot is taller. The area under each curve is unity.

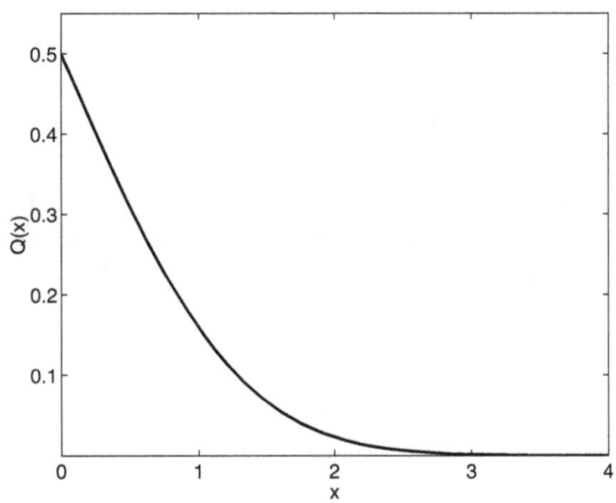

Figure 2.10. A plot of $\mathcal{Q}(x) = \int_x^\infty e^{-u^2/2} du/\sqrt{2\pi}$, for $x \geq 0$.

Thus the probability of error for any interior symbol is $2\mathcal{Q}(A/\sigma_e)$ whereas the probability of error for each boundary symbol is just $\mathcal{Q}(A/\sigma_e)$. Assuming all codewords are equally likely, the average error probability is therefore given by

$$\mathcal{P}_{e,PAM} = \frac{2(M-2)\mathcal{Q}(A/\sigma_e) + \mathcal{Q}(A/\sigma_e) + \mathcal{Q}(A/\sigma_e)}{M} = \frac{2\mathcal{Q}(A/\sigma_e)(M-1)}{M}.$$

Substituting $M = 2^b$, where b is the number of bits, we therefore obtain

$$\mathcal{P}_{e,PAM} = 2(1 - 2^{-b})\mathcal{Q}(A/\sigma_e). \tag{2.16}$$

Using the expression for the average energy (2.3) in a QAM constellation, we can rewrite this as

$$\mathcal{P}_{e,PAM} = 2(1 - 2^{-b})\mathcal{Q}\Big(\sqrt{\frac{3E_{ave}}{(2^{2b} - 1)\sigma_e^2}}\Big), \tag{2.17}$$

where the subscript PAM on E_{ave} has been deleted for simplicity. The error probabilities are also called the **symbol error rates** (SER) because they tell us what fraction of symbols are expected to be in error, given a long symbol stream.

Summary. Let the detector input have the form $\widehat{s}(n) = s(n) + e(n)$, where $s(n)$ is a b-bit PAM symbol and $e(n)$ is zero-mean Gaussian with variance σ_e^2, and let the average energy of the PAM constellation be E_{ave}. Then the average error probability in detecting $s(n)$ is given by Eq. (2.17). The error probability can also be written as (2.16), where A is the amplitude of the smallest codeword (Fig. 2.3). In these expressions, $\mathcal{Q}(.)$ is the integral defined in Eq. (2.14).

Example 2.1: One-bit PAM

Consider the case of 1-bit PAM, also known as a binary antipodal or PSK (phase-shift keying) or BPSK (binary phase-shift keying) constellation. This is shown in Fig. 2.11. Setting $b = 1$ in Eq. (2.17), the average error probability is given by

$$\mathcal{P}_{e,PSK} = \mathcal{Q}(A/\sigma_e). \tag{2.18}$$

We can also use (2.17) to get the equivalent expression

$$\mathcal{P}_{e,PSK} = \mathcal{Q}\Big(\sqrt{\frac{E_{ave}}{\sigma_e^2}}\Big). \tag{2.19}$$

Figure 2.12 shows a typical error pdf $f_E(e)$. Also shown is the pdf of the received signal when the symbol $-A$ is transmitted for two different noise variances. The shaded area, which represents the probability of error (probability that a $-A$ is judged as an A), is smaller when the noise variance is smaller.

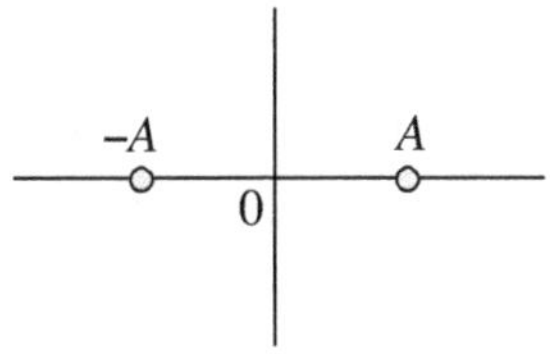

Figure 2.11. The 1-bit PAM constellation.

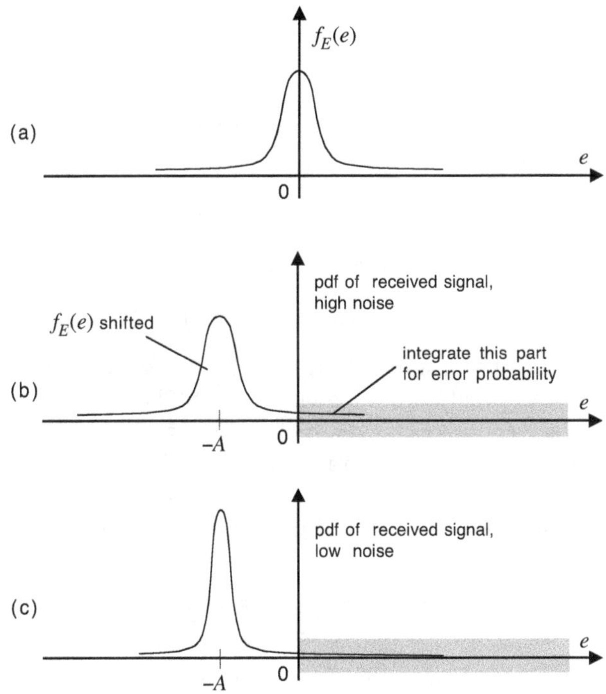

Figure 2.12. (a) The pdf of the error e (assumed Gaussian). (b) and (c) The pdf of the received signal $\widehat{s}(n)$ when the transmitted symbol is $-A$ and the noise has pdf $f_E(e)$. (b) Large noise variance, and (c) small noise variance.

2.3.2 Error probability for QAM signals

Suppose the symbols $s(n)$ in Fig. 2.1 are drawn from a QAM constellation such as the one in Fig. 2.4. Once again the receiver constructs an approximation $\widehat{s}(n)$

of the transmitted symbol stream $s(n)$, of the form

$$\widehat{s}(n) = s(n) + e(n), \tag{2.20}$$

where $e(n)$ is the reconstruction error. Recall here that $s(n)$, and hence $\widehat{s}(n)$, are complex numbers. It is usually reasonable to assume that the error $e(n)$ is complex and has the form

$$e(n) = e_r(n) + je_i(n),$$

where $e_r(n)$ and $e_i(n)$ are independent zero-mean Gaussian random variables with identical variance $0.5\sigma_e^2$. In this case the total variance of the complex error $e(n)$ is

$$0.5\sigma_e^2 + 0.5\sigma_e^2 = \sigma_e^2.$$

The joint pdf of the variables $[e_r(n), e_i(n)]$ is given by

$$f_E(e_r, e_i) = \frac{e^{-e_r^2/\sigma_e^2}}{\sqrt{\pi\sigma_e^2}} \times \frac{e^{-e_i^2/\sigma_e^2}}{\sqrt{\pi\sigma_e^2}} = \frac{e^{-(e_r^2+e_i^2)/\sigma_e^2}}{\pi\sigma_e^2}. \tag{2.21}$$

Figure 2.13 demonstrates this for $\sigma_e^2 = 0.01$. This is a special case of a so-called **circularly symmetric** complex Gaussian random variable.[4] Next, the complex symbol $s(n)$ is of the form

$$s_r(n) + js_i(n),$$

where $s_r(n)$ and $s_i(n)$ are PAM symbols. Since

$$|s(n)|^2 = s_r^2(n) + s_i^2(n),$$

it follows that the average energy of the constellation is the sum of the average energies of the real and imaginary parts. Thus, for a b-bit QAM constellation with average energy E_{ave}, the real and imaginary parts are $(b/2)$-bit PAM constellations with average energy $E_{ave}/2$, and each of these PAM constellations sees an error source with variance $\sigma_e^2/2$. For the real-part PAM the probability of error can be obtained from Eq. (2.17) by replacing b, E_{ave}, and σ_e^2 with half their values:

$$\mathcal{P}_{e,re} = 2(1 - 2^{-b/2})\mathcal{Q}\Big(\sqrt{\frac{3E_{ave}}{(2^b-1)\sigma_e^2}}\Big). \tag{2.22}$$

Since the factor of one-half cancels out in the ratio E_{ave}/σ_e^2, this is nothing but the error probability for a $(b/2)$-bit PAM constellation with energy E_{ave} and noise variance σ_e^2. Similarly for the imaginary part

$$\mathcal{P}_{e,im} = 2(1 - 2^{-b/2})\mathcal{Q}\Big(\sqrt{\frac{3E_{ave}}{(2^b-1)\sigma_e^2}}\Big). \tag{2.23}$$

The QAM symbol is detected *correctly* if the real part and imaginary part are *both detected correctly.* The probability for this is

$$(1 - \mathcal{P}_{e,re})^2.$$

[4] A detailed discussion of circularly symmetric complex random variables can be found in Sec. 6.6.

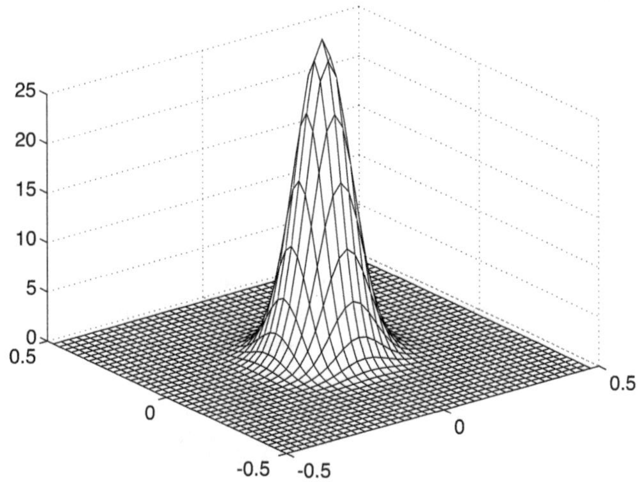

Figure 2.13. The pdf of Eq. (2.21) plotted for $\sigma_e^2 = 0.01$.

The probability of error in detection of the QAM symbol is therefore

$$\mathcal{P}_{e,QAM}(b) = 1 - (1 - \mathcal{P}_{e,PAM}(b/2))^2, \tag{2.24}$$

where we have used the functional arguments to indicate the number of bits. Thus $\mathcal{P}_{e,PAM}(b/2)$ is the error probability for a $(b/2)$-bit PAM constellation with energy E_{ave} and noise variance σ_e^2. For small errors the preceding equation can be approximated as

$$\mathcal{P}_{e,QAM}(b) = 1 - \Big(1 - 2\mathcal{P}_{e,PAM}(b/2) + \mathcal{P}^2_{e,PAM}(b/2)\Big) \approx 2\mathcal{P}_{e,PAM}(b/2), \tag{2.25}$$

where we have neglected $\mathcal{P}^2_{e,PAM}(b/2)$. This approximation is quite reasonable in practice. For example, even if $\mathcal{P}_{e,PAM}(b/2) = 10^{-3}$ (a rather large value), its square is 10^{-6}, which can be neglected. Thus the error probability for the QAM constellation can be approximated by

$$\mathcal{P}_{e,QAM}(b) \approx 2\mathcal{P}_{e,PAM}(b/2) = 4(1 - 2^{-b/2})\mathcal{Q}\Big(\sqrt{\frac{3E_{ave}}{(2^b - 1)\sigma_e^2}}\Big), \tag{2.26}$$

where b is the number of bits, E_{ave} is the average energy of the constellation, and σ_e^2 is the variance of the complex Gaussian error term $e(n)$ at the input of the detector. For comparsion, recall that a b-bit PAM constellation with the same energy E_{ave} and noise variance σ_e^2 would have error probability

$$\mathcal{P}_{e,PAM}(b) = 2(1 - 2^{-b})\mathcal{Q}\Big(\sqrt{\frac{3E_{ave}}{(2^{2b} - 1)\sigma_e^2}}\Big). \tag{2.27}$$

Example 2.2: Two-bit QAM or QPSK

Consider the case of 2-bit QAM, also known as a QPSK (quadrature phase-shift keying) constellation. This is shown in Fig. 2.14. Setting $b = 2$ in Eq. (2.26), the average error probability is given by

$$\mathcal{P}_{e,QPSK} = 2\mathcal{Q}\Big(\sqrt{\frac{E_{ave}}{\sigma_e^2}}\Big). \quad (2.28)$$

For both the PAM and QAM systems, note that the error probability depends on the ratio E_{ave}/σ_e^2, rather than the individual values of the energy E_{ave} and the error variance σ_e^2. This ratio is called the signal-to-error ratio or signal-to-noise ratio **SNR** *at the input of the detector:*

$$SNR = \frac{E_{ave}}{\sigma_e^2} \quad (2.29)$$

Figure 2.15 shows plots of the symbol error probability $\mathcal{P}_{e,PAM}$ as a function of this SNR for PAM systems, for various values of the number of bits b. Figure 2.16 shows similar plots for QAM systems. To compare PAM and QAM systems, it is useful to introduce the bit error rates (BER), which are related to the symbol error rates. Before doing this we have to introduce a binary coding system called the Gray code.

2.3.3 Gray codes

In digital communication systems we are often required to transmit binary streams. These streams can be converted to PAM or QAM symbols by appropriately grouping the bits. This is called the **symbol modulation** process. For example, if the binary stream is divided into blocks of size 3:

$$\ldots\ 010\ 001\ 111\ 011\ 100\ \ldots$$

then each 3-bit block can be turned into a 3-bit PAM symbol. More generally, a b-bit block can be translated into an M-word constellation with $M = 2^b$. There are many ways to define the mapping from the binary representation to the constellation words. Figure 2.17 shows an example of such a representation for a 3-bit PAM constellation. Thus the preceding binary sequence is converted to

$$\ldots - A, -5A, 3A, -3A, 7A, \ldots$$

In this example the binary words are assigned such that adjacent symbols in the constellation (Fig. 2.17) differ only in *one* of the bit locations. Such a representation is called a **Gray code** [Proakis, 1995].

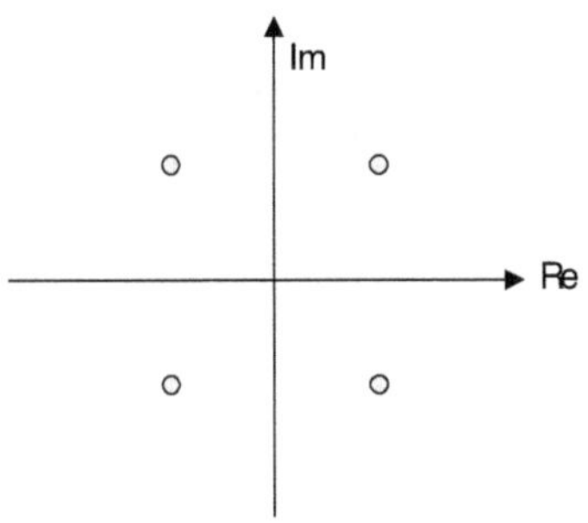

Figure 2.14. The 2-bit QAM constellation, also known as a QPSK constellation.

An important property of Gray codes is that, for reasonably high SNR, the symbol error rate can be related to the bit error rate in a simple manner. Thus, assume the SNR at the input of the detector is large enough, so that, when there is a symbol error, the estimated symbol is an adjacent symbol (rather than a symbol that is far away). In this case, only one bit is in error. Thus, in a symbol stream with N symbols, if there are K symbol errors then there are K bit errors as well. The **symbol error rate** (SER) is K/N, but since N symbols have Nb bits, the **bit error rate** (BER) is K/Nb. Thus, for a Gray coded system with b bits,

$$BER = \frac{SER}{b}. \tag{2.30}$$

From Eq. (2.25) we know that the symbol error probabilities for the PAM and QAM systems (for fixed SNR) are related approximately by

$$P_{e,QAM}(b) \approx 2P_{e,PAM}(b/2). \tag{2.31}$$

Using this, a simple relation between the bit error rates can be derived. Thus, dividing both sides of the preceding equation by b we get

$$\frac{P_{e,QAM}(b)}{b} \approx \frac{2P_{e,PAM}(b/2)}{b} = \frac{P_{e,PAM}(b/2)}{b/2},$$

which shows that the *BER for b-bit QAM is identical to BER for b/2 bit PAM:*

$$BER_{QAM}(b) \approx BER_{PAM}(b/2). \tag{2.32}$$

Thus, for a given error rate, the QAM system can transmit twice as many bits. However, QAM also requires twice as much bandwidth compared to PAM, as we shall see in Sec. 2.4.3.

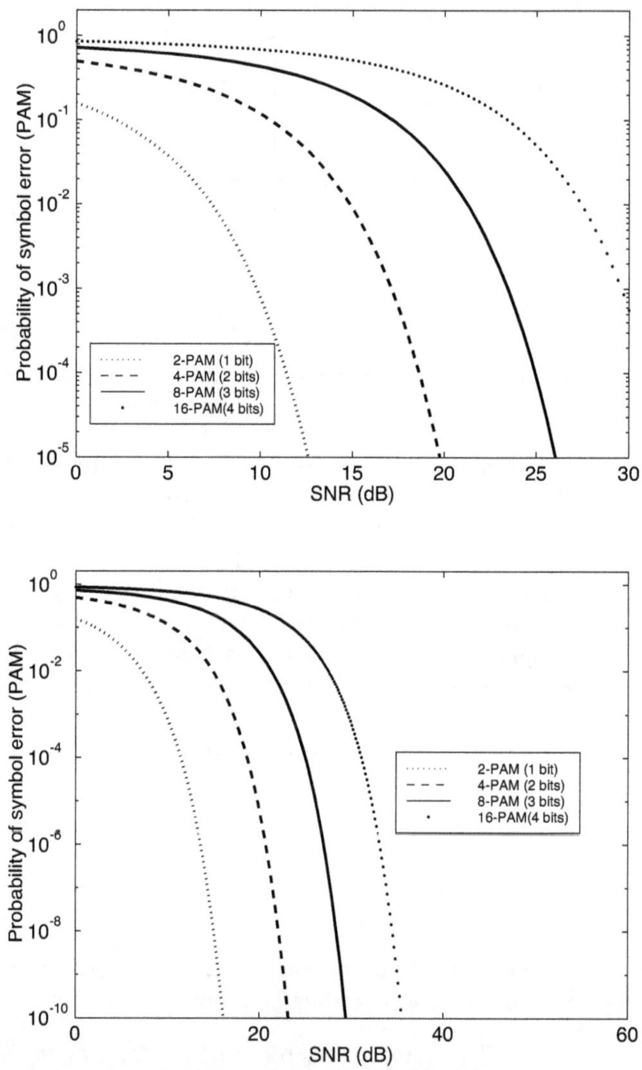

Figure 2.15. The symbol error probability for PAM. The top plot shows an SNR range from 0 to 30 dB; the bottom plot shows a wider range of SNR.

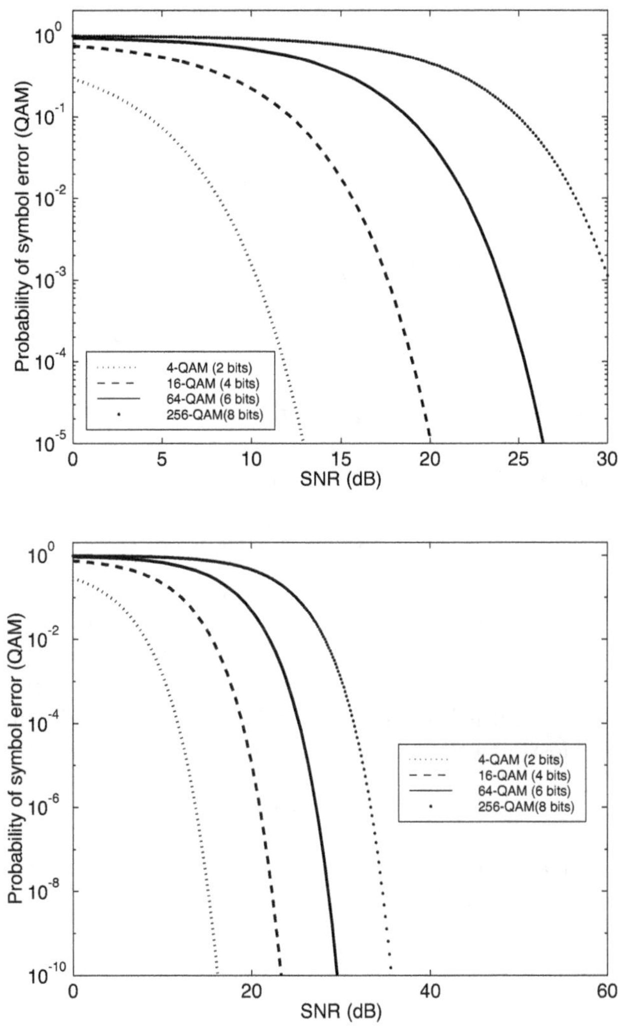

Figure 2.16. The symbol error probability for QAM. The top plot shows an SNR range from 0 to 30 dB; the bottom plot shows a wider range of SNR.

As a reminder, note that (2.32) holds under two assumptions: (a) the Gray code has been used for the binary representation, and (b) the SNR at the detector input is large enough to justify the approximation (2.25). Both of these are very reasonable assumptions. In simulations and examples that arise in the book, we shall always show the *symbol* error rates for consistency. The plots can readily be converted to BER plots by using the relation (2.30). As the reader can see, the expression for BER involves a few assumptions which may not hold accurately in practice. More exact expressions for BER tend to be very complicated. The reader wishing to investigate this topic further should read Cho and Yoon [2002] and references therein.

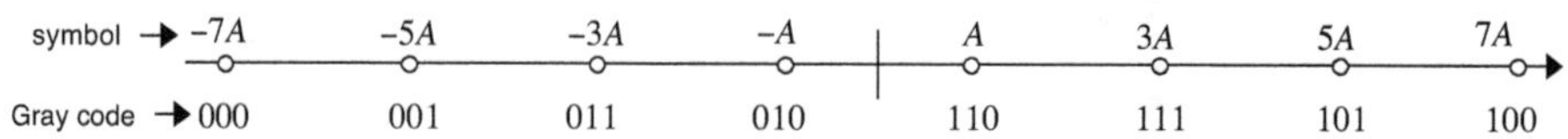

Figure 2.17. The 3-bit PAM constellation and the corresponding Gray code.

2.4 Carrier-frequency modulation

A PAM signal or waveform has the form

$$x(t) = \sum_n s(n)p(t - nT), \tag{2.33}$$

where the symbol stream $s(n)$ comes from the PAM constellation, and $p(t)$ is the baseband pulse. This signal can be generated by setting the prefilter $f(t) = p(t)$ in Fig. 2.1. The pulse $p(t)$ is usually a *time-limited* signal and can also be regarded as approximately *bandlimited*, with its Fourier transform confined mostly to the baseband region

$$-\sigma < \omega < \sigma \tag{2.34}$$

as demonstrated in Fig. 2.18(a). In practice the signal $x(t)$ is used to modulate a high-frequency **carrier** with frequency

$$\omega_c >> \sigma,$$

and the modulated signal is transmitted [Lathi, 1998]. For example, the modulated signal

$$x_{DSB}(t) = x(t) \cos \omega_c t$$

is said to be the double side band (DSB) modulated version of $x(t)$. The energy distribution in frequency is as demonstrated in Fig. 2.18(b). A variation of this called single side band (SSB) modulation is also shown in the figure and will be explained below. But first we have to explain some intricacies relating to these plots:

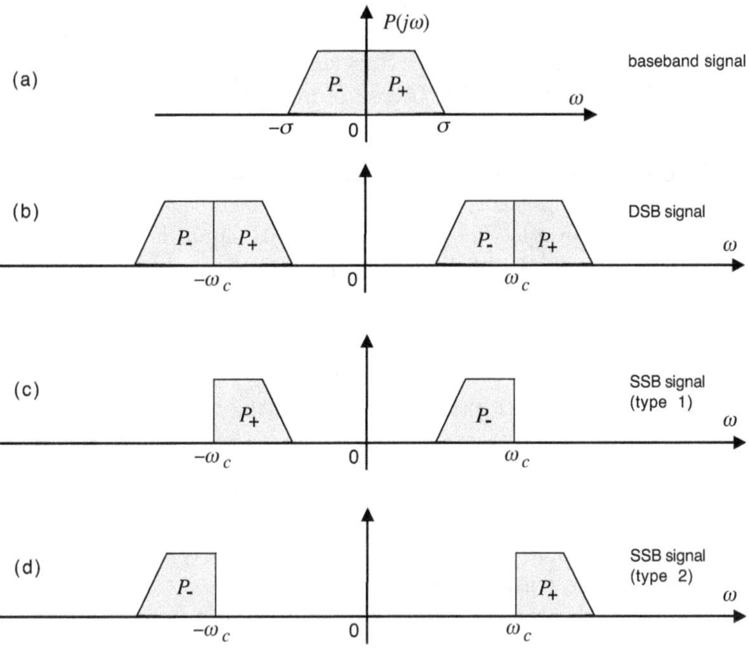

Figure 2.18. Fourier transforms of (a) the baseband signal $p(t)$, (b) the double side band (DSB) modulated version, and (c), (d) single side band (SSB) modulated versions.

Intricacies. If $s(n)$ is a wide sense stationary random process with power spectrum $S_{ss}(e^{j\omega})$, then $x(t)$ in Eq. (2.33) is a cyclo WSS process (Sec. E.3 in Appendix E) with average power spectrum

$$S_{xx}(j\omega) = \frac{1}{T} S_{ss}(e^{j\omega T})|P(j\omega)|^2. \tag{2.35}$$

Since $S_{ss}(e^{j\omega T})$ is periodic with period $2\pi/T$, the bandwidth of $x(t)$ is determined essentially by $P(j\omega)$. In the following discussion, which involves frequency bands, we shall therefore consider the modulation of the pulse $p(t)$ even though, in actual practice, it is $x(t)$ that is modulated. $\bigtriangledown \bigtriangledown \bigtriangledown$

The cosine-modulated pulse waveform has the form

$$p_{_{DSB}}(t) = p(t) \cos \omega_c t. \tag{2.36}$$

Thus, assuming $p(t)$ has an approximately bandlimited Fourier transform $P(j\omega)$ as demonstrated in Fig. 2.18(a), the modulated version $p_{DSB}(t)$ has the Fourier transform shown in Fig. 2.18(b). This occupies twice as much bandwidth as the baseband signal. A more economic version of the modulated signal can be obtained by retaining only one side band for the positive baseband frequency and one for negative baseband frequency.[5] There are two ways to do this, and these are shown in Figs. 2.18(c) and 2.18(d). These are referred to as *single side band* or **SSB** modulations. The PAM/SSB signal occupies half as much bandwidth as the PAM/DSB signal, and is therefore more desirable.

2.4.1 Generation of single side band signals

A conceptually simple way to obtain an SSB signal is to first process the baseband signal using a **Hilbert transformer**. This is an ideal filter with frequency response

$$H_{hi}(j\omega) = \begin{cases} j & \text{for } \omega < 0 \\ -j & \text{for } \omega > 0, \end{cases} \tag{2.37}$$

as shown in Fig. 2.19. If $p(t)$ is the input to the filter $H_{hi}(j\omega)$ then the output is called the Hilbert transform of $p(t)$, which we temporarily denote as $\widehat{p}(t)$. Thus

$$\widehat{P}(j\omega) = H_{hi}(j\omega)P(j\omega).$$

The Fourier transforms of $p(t)$ and $\widehat{p}(t)$ are clearly related as demonstrated in Fig. 2.20. There exist standard design techniques to approximate the Hilbert transformer with practical filters [Oppenheim and Schafer, 1999].

It is now clear that the positive and negative halves of the Fourier transform $P(j\omega)$ can be separated by combining $p(t)$ and $\widehat{p}(t)$ appropriately. Thus

$$p(t) + j\widehat{p}(t)$$

has the Fourier transform $P(j\omega) + j\widehat{P}(j\omega)$, which is (twice) the right half of $P(j\omega)$ as shown in Fig. 2.21(a). Similarly the left half $P_-(j\omega)$ can be obtained from

$$p(t) - j\widehat{p}(t)$$

(Fig. 2.21(b)). The SSB versions shown in Fig. 2.18 can be obtained by modulating these signals and combining them. Thus the signal

$$0.5(p(t) + j\widehat{p}(t))e^{-j\omega_c t} + 0.5(p(t) - j\widehat{p}(t))e^{j\omega_c t}$$

represents the SSB version in Fig. 2.18(c), whereas

$$0.5(p(t) + j\widehat{p}(t))e^{j\omega_c t} + 0.5(p(t) - j\widehat{p}(t))e^{-j\omega_c t}$$

gives the SSB version in Fig. 2.18(d). Thus the first type of SSB modulated signal has the form

$$p_{SSB,1}(t) = p(t)\cos\omega_c t + \widehat{p}(t)\sin\omega_c t \quad \text{(Type 1 SSB)},$$

and the second type of SSB modulated signal has the form

$$p_{SSB,2}(t) = p(t)\cos\omega_c t - \widehat{p}(t)\sin\omega_c t \quad \text{(Type 2 SSB)}.$$

[5] Since $p(t)$ is real, $P(-j\omega) = P^*(j\omega)$, so the information about the other half is not lost.

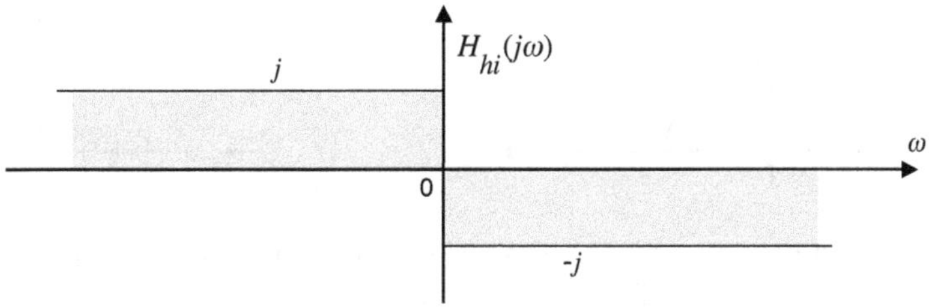

Figure 2.19. Frequency response of an ideal Hilbert transformer.

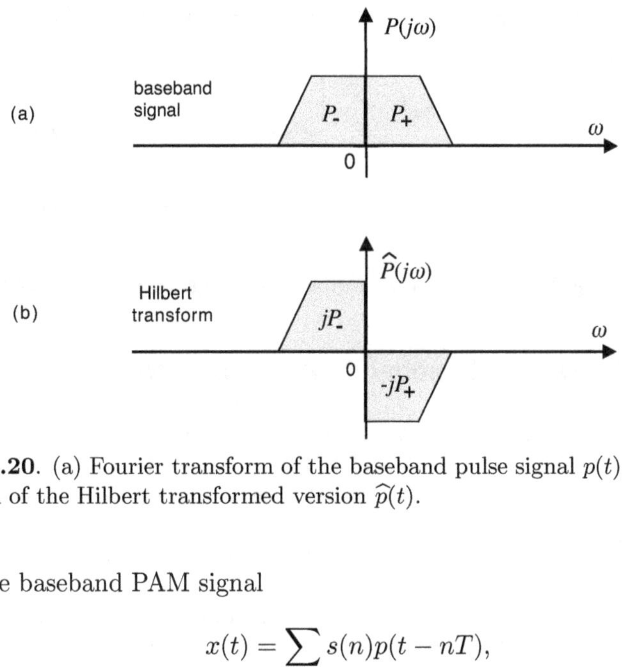

Figure 2.20. (a) Fourier transform of the baseband pulse signal $p(t)$, and (b) Fourier transform of the Hilbert transformed version $\widehat{p}(t)$.

Given the baseband PAM signal

$$x(t) = \sum_n s(n)p(t - nT),$$

the SSB modulated signal can therefore be generated by using

$$x_{SSB,1}(t) = x(t)\cos\omega_c t + \widehat{x}(t)\sin\omega_c t \quad \text{(Type 1 SSB)}, \tag{2.38}$$

$$x_{SSB,2}(t) = x(t)\cos\omega_c t - \widehat{x}(t)\sin\omega_c t \quad \text{(Type 2 SSB)}, \tag{2.39}$$

as shown in Fig. 2.22. Note that the pulse $p(t)$ is real, and so is $s(n)$ in the case of PAM. So the SSB signals in the preceding equation are real-valued for all t.

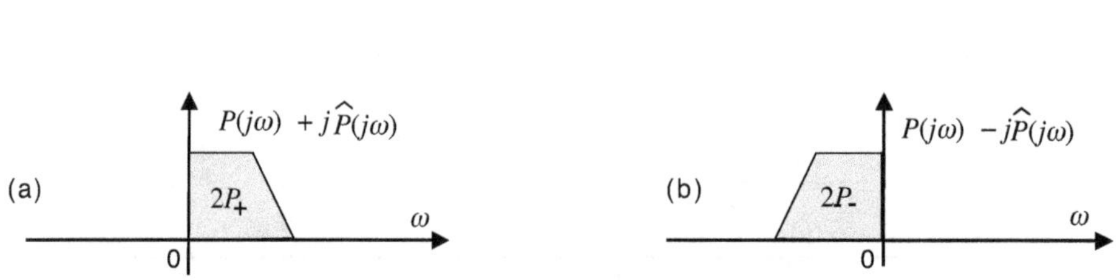

Figure 2.21. (a), (b) The right half and left half of the Fourier transform $P(j\omega)$, separated using the Hilbert transform.

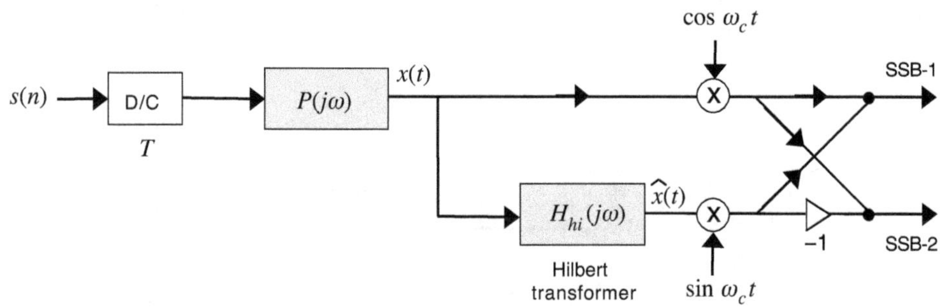

Figure 2.22. Block diagram for generating the two types of SSB signals.

2.4.2 Extracting baseband PAM from modulated versions

Given the modulated PAM signal in DSB form, or in SSB form of either type, we can extract the baseband signal $x(t)$ at the receiver by using the structure shown in Fig. 2.23. The input to this system is the modulated signal

$$x_{mod}(t) = x(t)\cos\omega_c t + \alpha\widehat{x}(t)\sin\omega_c t,$$

where α is 0 for DSB, 1 for Type 1 SSB, and -1 for Type 2 SSB. This signal is multiplied by the locally generated carrier $2\cos\omega_c t$ to yield the output[6]

$$\begin{aligned}2x_{mod}(t)\cos\omega_c t &= 2x(t)\cos^2\omega_c t + 2\alpha\widehat{x}(t)\sin\omega_c t\cos\omega_c t\\ &= x(t)(1+\cos 2\omega_c t) + \alpha\widehat{x}(t)\sin 2\omega_c t\\ &= x(t) + x(t)\cos 2\omega_c t + \alpha\widehat{x}(t)\sin 2\omega_c t.\end{aligned}$$

The second and third terms in the last line above are high frequency components, and their energy is contained mostly around the frequency $2\omega_c$. The first term

[6]This assumes that the receiver knows the carrier frequency ω_c and furthermore that there is no unknown phase offset. (If $\cos\omega_c t$ is received as $\cos(\omega_c t+\phi)$, where ϕ is *unknown*, we say there is an unknown phase offset.)

is the baseband term confined to $-\sigma < \omega < \sigma$. The lowpass filter in Fig. 2.23 retains the first term, and rejects the high-frequency terms. Thus the baseband PAM signal can be extracted from any of the three modulated versions, by using this simple structure.

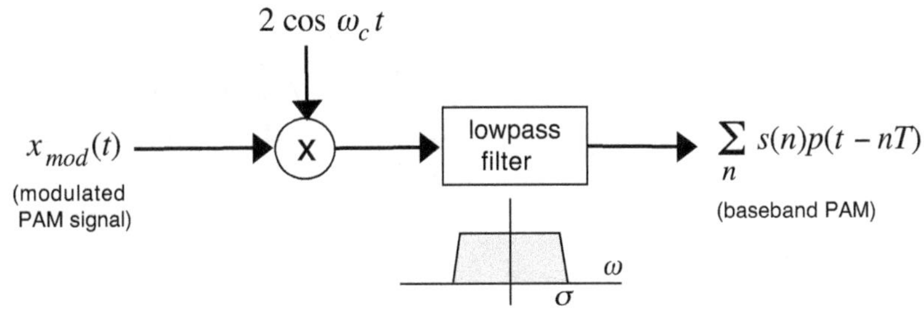

Figure 2.23. Receiver structure which extracts the baseband PAM signal by demodulating the modulated version $x_{mod}(t)$. Here $x_{mod}(t)$ can be a DSB or an SSB version.

2.4.3 Modulating the QAM symbol

Recall that the QAM symbol is complex, with the real and imaginary parts representing separate PAM constellations. But the modulated QAM signal can still be represented as a *real* signal of the form [Lathi, 1998]

$$x(t) = \sum_n s_c(n)p(t - nT)\cos\omega_c t - \sum_n s_s(n)p(t - nT)\sin\omega_c t, \tag{2.40}$$

where $s_c(n)$ and $s_s(n)$ are real signals (in-phase and quadrature components) drawn from PAM constellations.[7] The complex signal $s_c(n) + js_s(n)$, belonging to the QAM constellation, can actually be transmitted by transmitting the real signal $x(t)$ given above because the components $s_c(n)$ and $s_s(n)$ can readily be extracted from $x(t)$ as we shall see (Sec. 2.4.4).

[7]The notation $x_{QAM}(t)$ would have been appropriate for (2.40), but for simplicity we use $x(t)$ in this section.

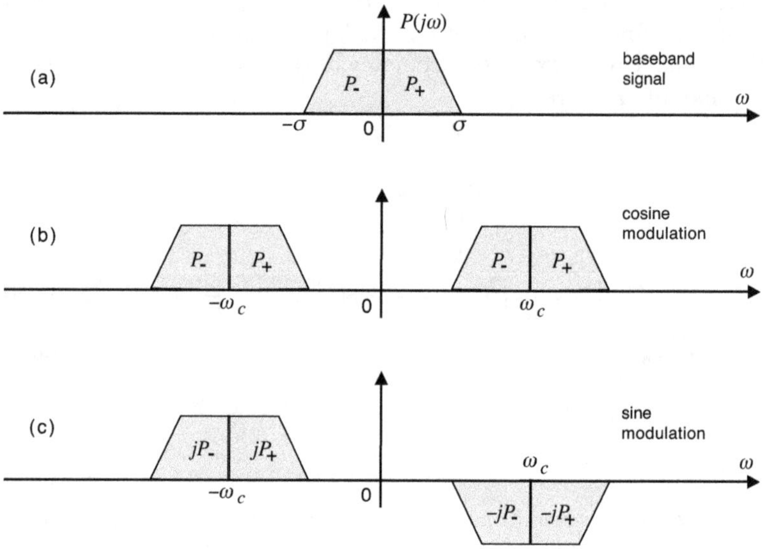

Figure 2.24. Fourier transforms of (a) the baseband pulse signal $p(t)$, (b) its cosine modulated version, and (c) its sine modulated version.

Assume the real baseband signal $p(t)$ has a bandlimited Fourier transform as in Fig. 2.24(a). Since

$$\cos \omega_c t = \frac{e^{j\omega_c t} + e^{-j\omega_c t}}{2} \quad \text{and} \quad \sin \omega_c t = \frac{e^{j\omega_c t} - e^{-j\omega_c t}}{2j},$$

the cosine modulated version has the Fourier transform

$$\frac{P(j(\omega - \omega_c)) + P(j(\omega + \omega_c))}{2},$$

and the sine modulated version has the Fourier transform

$$\frac{P(j(\omega - \omega_c)) - P(j(\omega + \omega_c))}{2j} = \frac{-jP(j(\omega - \omega_c)) + jP(j(\omega + \omega_c))}{2},$$

as indicated in Figs. 2.24(b) and 2.24(c). Since the cosine and sine modulated versions carry different messages $\{s_c(n)\}$ and $\{s_s(n)\}$, we cannot eliminate any of the side bands. Thus we need to retain both side bands, and the QAM signal (2.40) is a DSB signal. The QAM signal therefore requires **twice as much bandwidth** as a PAM/SSB signal. But, since the QAM signal carries two messages $s_c(n)$ and $s_s(n)$ at the same rate, the bandwidth used per real symbol is identical for the two systems. One advantage of a QAM system is that it does not require the Hilbert transformer. It is very easy to generate the QAM signal by using Eq. (2.40), as shown schematically in Fig. 2.25.

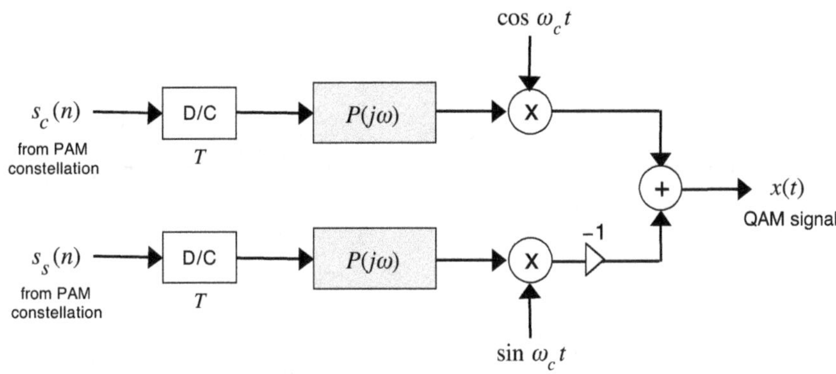

Figure 2.25. Generating the real QAM signal $x(t)$ from $s_c(n)$ and $s_s(n)$, which are the real and imaginary parts of the complex QAM symbol $s_c(n) + js_s(n)$.

2.4.4 Extracting the complex QAM signal from the real version

We now explain how the components $s_c(n)$ and $s_s(n)$ can be extracted from the real QAM signal shown in Eq. (2.40). For this, refer to Fig. 2.26 where the real QAM signal $x(t)$ is multiplied with $2\cos\omega_c t$ and $2\sin\omega_c t$ separately. Using Eq. (2.40) we see that

$$\begin{aligned} &2x(t)\cos\omega_c t \\ &\quad = 2\sum_n s_c(n)p(t-nT)\cos^2\omega_c t - 2\sum_n s_s(n)p(t-nT)\sin\omega_c t\cos\omega_c t \\ &\quad = \sum_n s_c(n)p(t-nT)(1+\cos 2\omega_c t) - \sum_n s_s(n)p(t-nT)\sin 2\omega_c t. \end{aligned}$$

The term $p(t-nT)\cos 2\omega_c t$ has its energy mostly concentrated around the high frequency $2\omega_c$ (since $p(t)$ is a baseband signal with bandwidth $2\sigma << 2\omega_c$). Similarly, the term $p(t-nT)\sin 2\omega_c t$ has its energy mostly concentrated around $2\omega_c$. The lowpass filter in the figure therefore retains only the term

$$\sum_n s_c(n)p(t-nT), \tag{2.41}$$

which contains the real part of the QAM symbol $s_c(n) + js_s(n)$. Similarly, the lowpass filter in the lower branch extracts the signal

$$\sum_n s_s(n)p(t-nT), \tag{2.42}$$

which contains the imaginary part of the QAM symbol $s_c(n) + js_s(n)$.

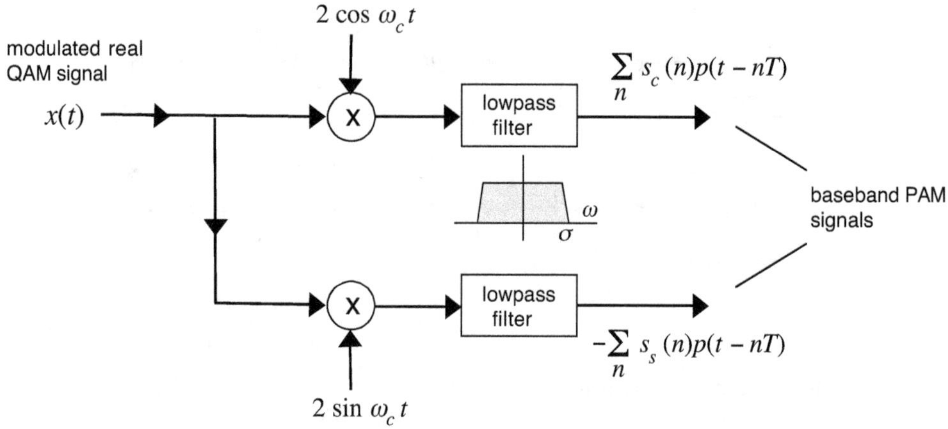

Figure 2.26. Demodulation of a real QAM signal $x(t)$ to obtain the baseband PAM components $\sum_n s_c(n)p(t-nT)$ and $-\sum_n s_s(n)p(t-nT)$.

Thus the complex QAM symbol can be extracted from the real QAM signal $x(t)$ by using the structure of Fig. 2.26 at the receiver. In practice, only a noisy and distorted version of $x(t)$ is received at the receiver, because of the effects of the channel. So, the outputs of Fig. 2.26 are also distorted. The symbols $s_c(n)$ and $s_s(n)$ can be extracted from these versions with certain predictable error probabilities, as explained in Sec. 2.3.2.

Complex baseband equivalent. As mentioned at the beginning of Sec. 2.4.3, the acutal QAM signal $x(t)$ is real. When a real bandpass signal is transmitted over a real channel, it is often convenient to represent this with an equivalent lowpass signal and a lowpass channel. Such equivalent signal and equivalent channel are, in general, complex. Appendix 2.A at the end of this chapter describes this in detail. This complex baseband equivalent is commonly used in all discussions and is in fact very useful when we discuss digital implementations of the transmitter and receiver. $\bigtriangledown\bigtriangledown\bigtriangledown$

2.5 Matched filtering

We now review the concept of matched filtering, which lies at the heart of receiver designs in digital communication systems. Matched filters were first introduced by North [1963], and have since been used widely in communication and radar systems. The basic idea will be described here, and some deeper details will be postponed to Chap. 5 (Sec. 5.2).

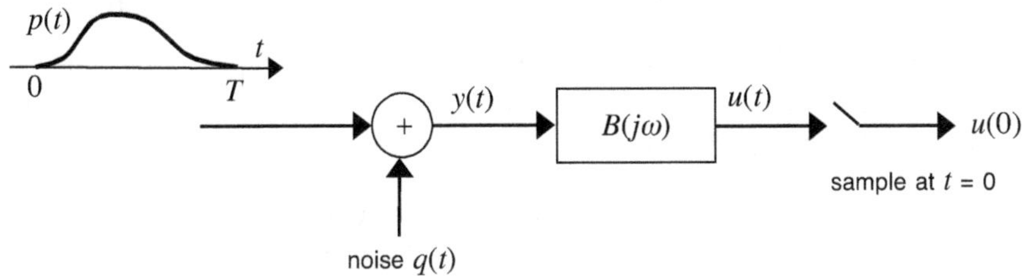

Figure 2.27. Filtering the baseband pulse plus $p(t)$ contaminated with noise.

Consider Fig. 2.27, which shows a baseband pulse $p(t)$ on the left. Imagine that we have a noisy version $y(t)$ of this pulse, where the noise $q(t)$ is assumed to be a zero-mean wide sense stationary (WSS) process with power spectrum $S_{qq}(j\omega)$ (Appendix E). This noisy version is passed through a filter with frequency response $B(j\omega)$. We would like to choose the filter $B(j\omega)$ such that the signal-to-noise ratio in the waveform $u(t)$ is maximized at time $t = 0$. In what follows we shall indicate[8] the impulse response of $B(j\omega)$ as $\beta(t)$.

To formulate this problem mathematically, observe that $y(t) = p(t) + q(t)$ so that

$$u(t) = \underbrace{(\beta * p)(t)}_{\text{signal part } w(t)} + \underbrace{(\beta * q)(t)}_{\text{noise part } e(t)} . \tag{2.43}$$

Here $*$ denotes convolution. The signal component in the sample $u(0)$ is[9]

$$w(0) = (\beta * p)(0) = \int_{-\infty}^{\infty} B(j\omega)P(j\omega)\frac{d\omega}{2\pi}. \tag{2.44}$$

The noise component $e(t)$ has power spectrum $S_{ee}(j\omega) = |B(j\omega)|^2 S_{qq}(j\omega)$. Its mean square value (which is independent of t owing to the WSS assumption) is therefore given by

$$\sigma_e^2 = \int_{-\infty}^{\infty} S_{ee}(j\omega)\frac{d\omega}{2\pi} = \int_{-\infty}^{\infty} |B(j\omega)|^2 S_{qq}(j\omega)\frac{d\omega}{2\pi} \tag{2.45}$$

The signal-to-noise ratio at time $t = 0$ is therefore

$$\frac{|w(0)|^2}{\sigma_e^2} = \frac{\left| \int_{-\infty}^{\infty} B(j\omega)P(j\omega)\frac{d\omega}{2\pi} \right|^2}{\int_{-\infty}^{\infty} |B(j\omega)|^2 S_{qq}(j\omega)\frac{d\omega}{2\pi}} \tag{2.46}$$

[8]Note that we have used $B(j\omega)$ instead of $G(j\omega)$ because $G(j\omega)$ usually includes the matched filter part and an equalizer part. The equalizer part is a transversal filter that can be implemented as a digital filter, as we shall see in later chapters.

[9]We use the fact that, for any Fourier transformable signal $x(t)$, the relation $x(0) = \int_{-\infty}^{\infty} X(j\omega)d\omega/2\pi$ holds.

We now prove the following result, which assumes $S_{qq}(j\omega) \neq 0$ for all ω. In practice this assumption is needed only over a finite bandwidth, since the transmitter and receiver have bandlimiting filters to eliminate unwanted interferences.

♠**Lemma 2.1.** *Matched filtering.* The filter $B(j\omega)$, which maximizes the signal-to-noise ratio (2.46) for fixed $P(j\omega)$ and $S_{qq}(j\omega)$ (i.e., fixed pulse shape $p(t)$ and noise spectrum), is of the form

$$B(j\omega) = \frac{cP^*(j\omega)}{S_{qq}(j\omega)} \tag{2.47}$$

where c is an arbitrary nonzero constant. The maximized value of the signal-to-noise ratio (2.46) is given by

$$\frac{|w(0)|^2}{\sigma_e^2} = \int_{-\infty}^{\infty} \frac{|P(j\omega)|^2}{S_{qq}(j\omega)} \frac{d\omega}{2\pi}. \tag{2.48}$$

The filter (2.47) is said to be the *matched filter* for the pulse shape $p(t)$ and noise spectrum $S_{qq}(j\omega)$. ◇

Proof. Consider the numerator of Eq. (2.46). We have

$$\left| \int_{-\infty}^{\infty} B(j\omega)P(j\omega)\frac{d\omega}{2\pi} \right|^2 = \left| \int_{-\infty}^{\infty} B(j\omega)\sqrt{S_{qq}(j\omega)} \frac{P(j\omega)}{\sqrt{S_{qq}(j\omega)}} \frac{d\omega}{2\pi} \right|^2,$$

where it is assumed that $S_{qq}(j\omega) \neq 0$. By using Cauchy-Schwartz inequality (Appendix A) on the right-hand side we therefore obtain

$$\left| \int_{-\infty}^{\infty} B(j\omega)P(j\omega)\frac{d\omega}{2\pi} \right|^2 \leq \int_{-\infty}^{\infty} |B(j\omega)|^2 S_{qq}(j\omega)\frac{d\omega}{2\pi} \int_{-\infty}^{\infty} \frac{|P(j\omega)|^2}{S_{qq}(j\omega)} \frac{d\omega}{2\pi}.$$

This implies

$$\frac{|w(0)|^2}{\sigma_e^2} \leq \int_{-\infty}^{\infty} \frac{|P(j\omega)|^2}{S_{qq}(j\omega)} \frac{d\omega}{2\pi}, \tag{2.49}$$

and equality is achieved if and only if there is equality in Cauchy-Schwartz inequality, that is (Appendix A),

$$B(j\omega)\sqrt{S_{qq}(j\omega)} = c\frac{P^*(j\omega)}{\sqrt{S_{qq}(j\omega)}} \tag{2.50}$$

for some $c \neq 0$. The optimal filter $B(j\omega)$ therefore has the form (2.47). The maximized SNR is the right-hand side of (2.49), which equals (2.48). ▽▽▽

2.5.1 Case of white noise

For the special case where the noise is white, we have

$$S_{qq}(j\omega) = \frac{\mathcal{N}_0}{2} \tag{2.51}$$

for all ω, as shown in Fig. 2.28. Since c is arbitrary, we can take the optimal filter to be

$$B(j\omega) = P^*(j\omega). \tag{2.52}$$

Taking the inverse Fourier transform, the impulse response of the filter is

$$\beta(t) = p^*(-t). \tag{2.53}$$

For white noise, the matched filter $\beta(t)$ is therefore the *time-reversed conjugate* of the pulse $p(t)$. Figure 2.29 shows an example of a real pulse $p(t)$ and its matched filter $p(-t)$. Now, the signal component $w(t)$ at the output of the filter $\beta(t)$ is

$$w(t) = (\beta * p)(t) = \int_{-\infty}^{\infty} p(\tau)p^*(\tau - t)d\tau, \tag{2.54}$$

which is nothing but the *deterministic autocorrelation* of $p(t)$. This is demonstrated in Fig. 2.29(c). Note in particular that

$$w(0) = \int_{-\infty}^{\infty} |p(\tau)|^2 d\tau = \text{energy of } p(t). \tag{2.55}$$

Thus, when the filter $\beta(t)$ is matched to the pulse, the output at time zero is the pulse energy. Moreover the maximized signal-to-noise ratio (2.48) becomes

$$\frac{|w(0)|^2}{\sigma_e^2} = \frac{2}{\mathcal{N}_0}\int_{-\infty}^{\infty} |P(j\omega)|^2 \frac{d\omega}{2\pi} = \frac{2\int_{-\infty}^{\infty} |p(t)|^2 dt}{\mathcal{N}_0} = \frac{2E_p}{\mathcal{N}_0}, \tag{2.56}$$

where E_p is the energy in the pulse $p(t)$.

Remarks

1. ***Autocorrelation of baseband pulse.*** For finite-duration $p(t)$, even though the duration of $R_{pp}(\tau)$ is twice as long, the energy in the autocorrelation $R_{pp}(\tau)$ is more narrowly concentrated, as demonstrated in Fig. 2.29(c). This is the advantage of using a matched filter: it concentrates the signal energy at the point where the detector samples the signal. For a given pulse duration $p(t)$, the achievable narrowness of the autocorrelation depends on the **bandwidth** of the pulse; the larger the bandwith, the narrower the autocorrelation can be. Thus, a baseband pulse $p(t)$ with *large energy* and *large bandwidth* results in better performace at the output of the matched filter, and, furthermore, adjacent pulses can be spaced closer (because of the narrowness of the autocorrelation). Of course, bandwidth and energy are valuable resources which cannot be increased arbitrarily.

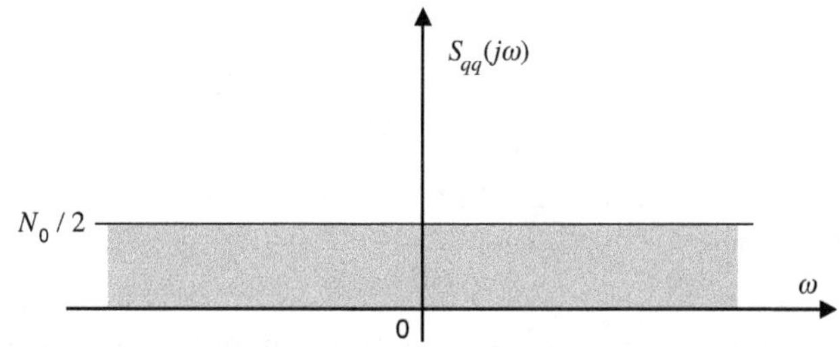

Figure 2.28. White noise spectrum.

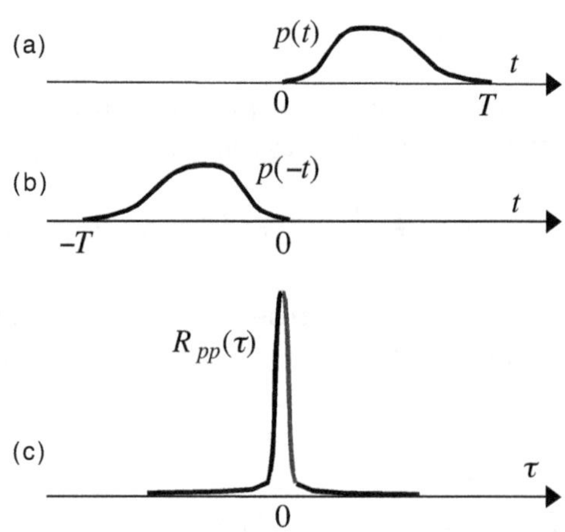

Figure 2.29. (a) Example of a pulse $p(t)$, (b) the matched filter for $p(t)$ when the noise is white, and (c) autocorrelation of $p(t)$, exhibiting a sharp peak at $\tau = 0$.

2. ***Causality***. For a causal pulse $p(t)$ with duration T, the matched filter $p^*(-t)$ is noncausal. In practice the shifted version $p^*(T-t)$, which is causal, can be used. The filter output is now the shifted version of the autocorrelation, that is, $R_{pp}(\tau - T)$. Instead of sampling this at $\tau = 0$, the receiver has to sample it at $\tau = T$.

3. ***Pulse compression.*** An interesting question here is the following. Given a pulse duration T and pulse bandwidth[10] σ (as in Fig. 2.18(a)), what

[10] A time-limited signal cannot also be bandlimited. What is meant in this context is that "most" of the energy of the Fourier transform is confined to a finite region.

is the shape of the pulse $p(t)$ that makes the autocorrelation $R_{pp}(\tau)$ narrowest (in some well defined mathematical sense)? This is called the *pulse compression* problem because $p(t)$ is "compressed" into $R_{pp}(\tau)$. The pulse compression ratio, defined in the next section, is the extent to which the duration has been compressed when going from $p(t)$ to $R_{pp}(\tau)$. The reader interested in this important topic should pursue references such as Van Trees [2001] and Levanon and Mozeson [2004].

2.5.2 The chirp or LFM signal

A commonly used waveform in radar applications is the **chirp** waveform

$$p(t) = e^{j\pi Kt^2}. \tag{2.57}$$

Here K is a positive constant called the chirp parameter. It has the dimension $(\text{Hz})^2$ so that Kt^2 is dimensionless. The above signal is called the *complex chirp* to distinguish it from the real version $\cos(\pi Kt^2)$. The finite-duration version of this waveform provides an excellent compression ratio as we shall see. For a preview of what the waveform looks like, see Figs. 2.32 and 2.35. We can regard $p(t)$ as a signal with "instantaneous frequency" which increases linearly with time. For the waveform $e^{j\omega_0 t}$ the fixed frequency ω_0 is the derivative of the phase $\omega_0 t$. Similarly for the chirp waveform the "instantaneous frequency" is the derivative

$$\frac{d(\pi Kt^2)}{dt} = 2\pi Kt \quad \text{rad/s}.$$

Dividing by 2π we get the instantaneous frequency in hertz:

$$f_t = Kt \quad \text{Hz}. \tag{2.58}$$

Since this is linearly increasing with time, $p(t)$ is also called a linear frequency modulated (or **LFM**) waveform. It can be shown that the Fourier transform of $p(t)$ is given by [Papoulis, 1968][11]

$$P(j\omega) = \frac{1}{\sqrt{K}} e^{j\pi/4} e^{-j\omega^2/4\pi K} = \frac{1}{\sqrt{K}} e^{j\pi/4} e^{-j\pi f^2/K}, \tag{2.59}$$

where we have used $\omega = 2\pi f$. Note that $|P(j\omega)| = 1/\sqrt{K}$ for all ω, and $|p(t)| = 1$ for all t. Thus, both the function *and* its Fourier transform are spread out uniformly! The real part of the complex chirp is the real-chirp function

$$p_{re}(t) = \cos(\pi Kt^2), \tag{2.60}$$

and this has the Fourier transform (Problem 2.10)

$$P_{re}(j\omega) = \frac{1}{\sqrt{K}} \cos\left(\frac{\omega^2}{4\pi K} - \frac{\pi}{4}\right). \tag{2.61}$$

[11] See Problem 2.10. Note that $p(t)$ is not absolutely integrable, nor square integrable. Its Fourier transform exists only if it is interpreted as a Cauchy principal value [Papoulis, 1968], that is, as the limit of $\int_{-a}^{a} p(t)e^{-j2\pi ft}dt$ as $a \to \infty$.

2.5.2.A Finite-duration chirp

In practice the chirp is truncated to be a finite duration pulse. Thus the truncated complex chirp is

$$p_T(t) = \begin{cases} e^{j\pi K t^2} & -\frac{T}{2} \le t \le \frac{T}{2} \\ 0 & \text{otherwise.} \end{cases} \tag{2.62}$$

The subscript T is temporary, for added clarity. The instantaneous frequency $f_t = Kt$ changes from $-KT/2$ to $KT/2$, so the total frequency sweep, denoted as F_m, is given by

$$F_m = KT. \tag{2.63}$$

The product

$$D \triangleq TF_m = KT^2 \tag{2.64}$$

is dimensionless and is called the **time-bandwidth product** of the chirp. Note that $p_T(t)$ is an even function, that is, $p_T(t) = p_T(-t)$. The Fourier transform therefore satisfies

$$P_T(j\omega) = P_T(-j\omega). \tag{2.65}$$

In particular $|P_T(j\omega)|^2$ is (real and) even, which shows that the autocorrelation of the complex chirp $p_T(t)$ is real and even:

$$R(\tau) = R(-\tau). \tag{2.66}$$

2.5.2.B Fourier transform of finite-duration chirp

The Fourier transform $P_T(j\omega)$ is a convolution of $P(j\omega)$ in Eq. (2.59) with an appropriate sinc function. There is no simple closed form for this, but it has been shown [Klauder *et al.*, 1960] that for large values of the time-bandwidth product D, most of the energy of the Fourier transform is in the region

$$-\frac{F_m}{2} \le f \le \frac{F_m}{2}, \tag{2.67}$$

as demonstrated in Fig. 2.30. (Actually there are ripples in the plot which we have not shown. See Fig. 2.34 for a preview.) The magnitude at $f = F_m/2$ is about 6 dB below the zero-frequency value. It has been claimed [Klauder *et al.*, 1960] that, when $D = 10$, about 95% of the energy is contained in $|f| \le F_m/2$, whereas, for $D = 100$, about 99% of the energy is in $|f| \le F_m/2$. Thus the time-limited chirp is *approximately bandlimited* with total bandwidth $\approx F_m$.

2.5.2.C Autocorrelation of finite-duration chirp

The autocorrelation of $p_T(t)$ has been shown to be

$$R(\tau) = \begin{cases} T\left(1 - \frac{|t|}{T}\right)\left[\dfrac{\sin \pi F_m t\left(1 - \frac{|t|}{T}\right)}{\pi F_m t\left(1 - \frac{|t|}{T}\right)}\right] & -T \le t \le T \\ 0 & \text{otherwise,} \end{cases} \tag{2.68}$$

and zero outside, which is approximately a sinc function for small t:

$$R(\tau) \approx \frac{T \sin(\pi F_m t)}{\pi F_m t}, \quad |t| << T. \tag{2.69}$$

The most important point is that the energy of this function is concentrated in a much narrower duration of time than the original pulse $p_T(t)$, which had duration T (Fig. 2.31). Since the first zero crossing of the sinc is at $t = 1/F_m$, we see that the duration of the main lobe is

$$\Delta\tau = \frac{2}{F_m}, \tag{2.70}$$

where F_m is the total frequency sweep ($\approx$ bandwidth) of $p_T(t)$. Since $F_m = KT$, we see that the significant duration of the autocorrelation is

$$\Delta\tau = \frac{2}{KT}, \tag{2.71}$$

whereas the pulse $p_T(t)$ itself has duration T. The **pulse compression ratio**, defined as

$$\mathcal{C} = \frac{\text{duration of } p_T(t)}{\text{duration of } R(\tau)}, \tag{2.72}$$

is therefore

$$\mathcal{C} = \frac{KT^2}{2} = \frac{F_m T}{2}. \tag{2.73}$$

Thus the *pulse compression ratio is proportional to the time-bandwidth product.* Summarizing, the complex chirp or LFM signal $p_T(t)$ of duration T and chirp parameter K has the following properties:

1. The total frequency sweep is given by $F_m = KT$.

2. The bandwidth is approximately $F_m = KT$ for large time-bandwidth product TF_m.

3. The pulse compression ratio is $F_m T/2$, or equivalently, $KT^2/2$, and is therefore proportional to the time-bandwidth product D given by $TF_m = KT^2$.

4. The autocorrelation has main-lobe width $2/F_m$.

Thus the compression ratio is achieved at the expense of bandwidth. The more we spread the energy in the frequency domain, the larger the compression ratio. Note also that the time-domain spreading of the pulse helps to increase the total pulse energy for fixed peak power (equivalently reduce the peak power for fixed energy). It should be mentioned finally that, even though a narrow autocorrelation is very attractive from an SNR viewpoint because all energy is concentrated in a narrow region, this also increases the sensitivity to sampling-time errors.

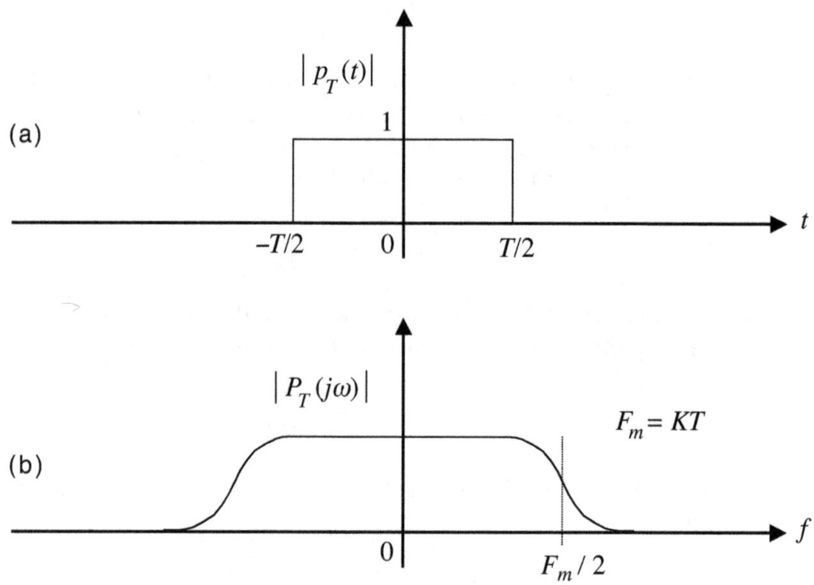

Figure 2.30. (a) Magnitude of the complex chirp signal, and (b) approximate sketch of its Fourier transform magnitude. There are ripples in the plot which we have not shown. See [Klauder *et al.*, 1960] for exact plots.

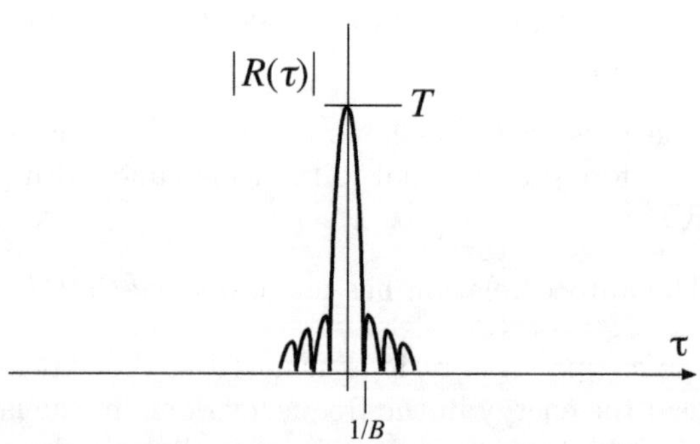

Figure 2.31. Approximate autocorrelation magnitude for the chirp signal.

Example 2.3: The chirp signal

Consider an example of the chirp signal with duration $T = 600\,\mu$s, and chirp parameter $K = 0.0001$ in units of $(\text{MHz})^2$. Figure 2.32 shows the real and imaginary parts of $p(t) = e^{j\pi Kt^2}$ and the autocorrelation $R(\tau)$. Note how sharp the autocorrelation is, compared to the pulse duration. Figure 2.33 shows the autocorrelation of the chirp together with the autocorrelation of a rectangular pulse of duration $600\,\mu$s. The height of $p(t)$ is scaled such that the autocorrelation has a peak value of unity.

In this example the bandwidth is $F_m = KT = 60$ kHz. The magnitude of the Fourier transform $P(j\omega)$ is shown in Fig. 2.34 for $0 \leq f \leq 500$ kHz (with passband normalized to have a maximum of 0 dB). Note that the passband extends from 0 to about $F_m/2 = 30$ kHz, as expected from the theory. The time-bandwidth product in this example is

$$F_m T = 36.$$

So the pulse compression ratio is $F_m T/2 = 18$. This means that the duration of the main lobe of the autocorrelation plot is about 18 times smaller than the pulse width of $600\,\mu$s.

As a practical matter, it is desirable to use the real causal signal

$$p(t) = \begin{cases} \cos(\pi K t^2) & 0 \leq t \leq T \\ 0 & \text{otherwise} \end{cases} \tag{2.74}$$

as the chirp. In this case the autocorrelation continues to be very sharp, as in the above example. This is demonstrated in Fig. 2.35, which shows the pulse and the autocorrelation ($p(t)$ is scaled such that the autocorrelation has a peak value of unity). In this example we once again have chosen $T = 600\mu$s, and chirp parameter $K = 0.0001\ (\text{MHz})^2$. Note, however, that the autocorrelation plot in Fig. 2.35 is even sharper than the one in Fig. 2.32. This is intuitively expected from the fact that the bandwidth is wider, as seen from the plot of $|P(j2\pi f)|$. To see why the bandwidth is higher, write

$$\cos(\pi K t^2) = 0.5e^{j\pi K t^2} + 0.5e^{-j\pi K t^2}.$$

The first term has a frequency sweep from 0 to KT, and the second term has a sweep from 0 to $-KT$. So there are dominant frequency components in the range $-KT$ to KT, that is, -60 to 60 kHz in this case. In the previous example of a complex chirp it was only from -30 to 30 kHz.

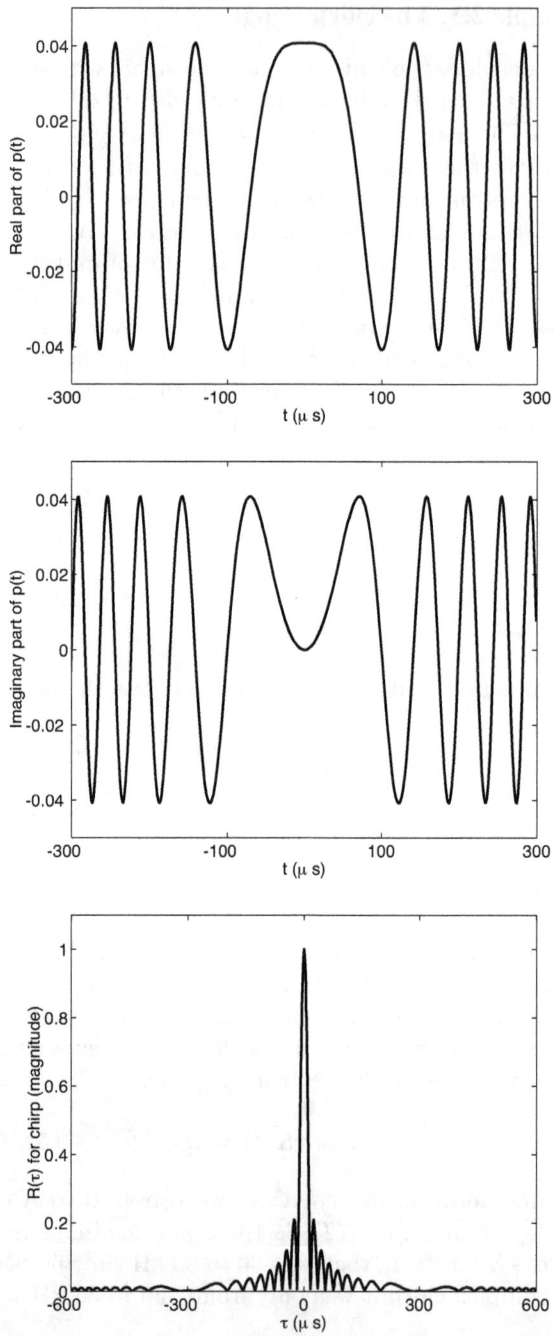

Figure 2.32. Example 2.3: A complex chirp signal. Real part (top); imaginary part (middle); autocorrelation magnitude (bottom).

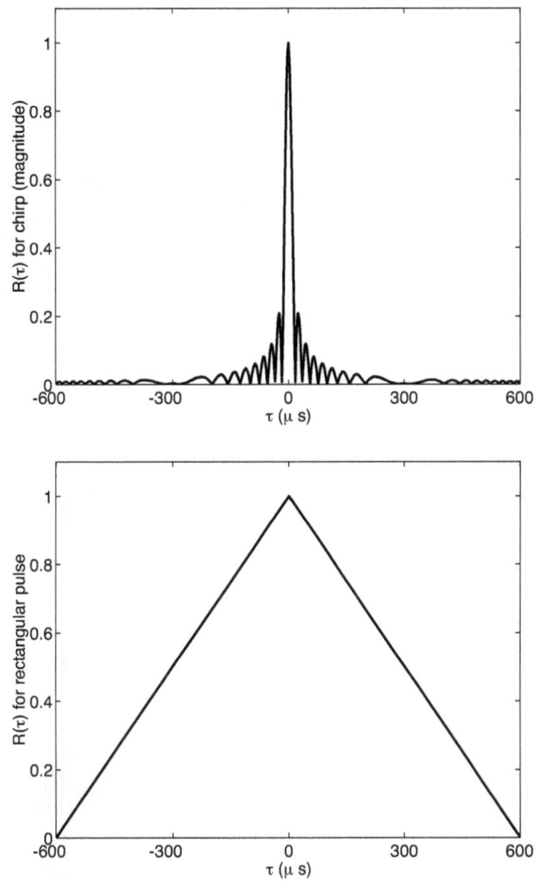

Figure 2.33. Autocorrelation of the chirp signal in Example 2.3 (top); autocorrelation of a rectangular pulse of the same duration (bottom).

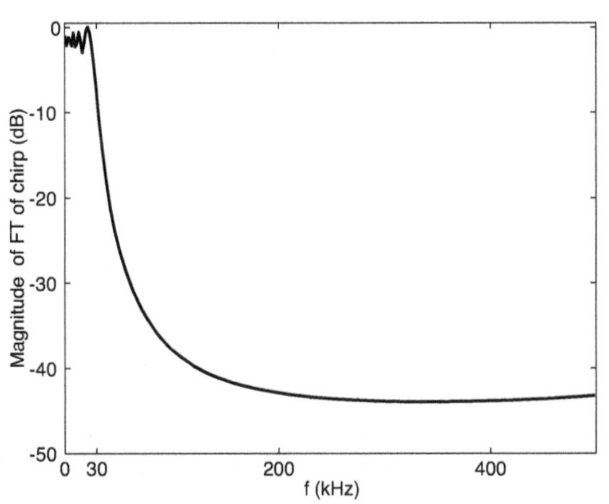

Figure 2.34. Fourier transform magnitude $|P(j2\pi f)|$ for the chirp signal of Example 2.3, plotted on a dB scale.

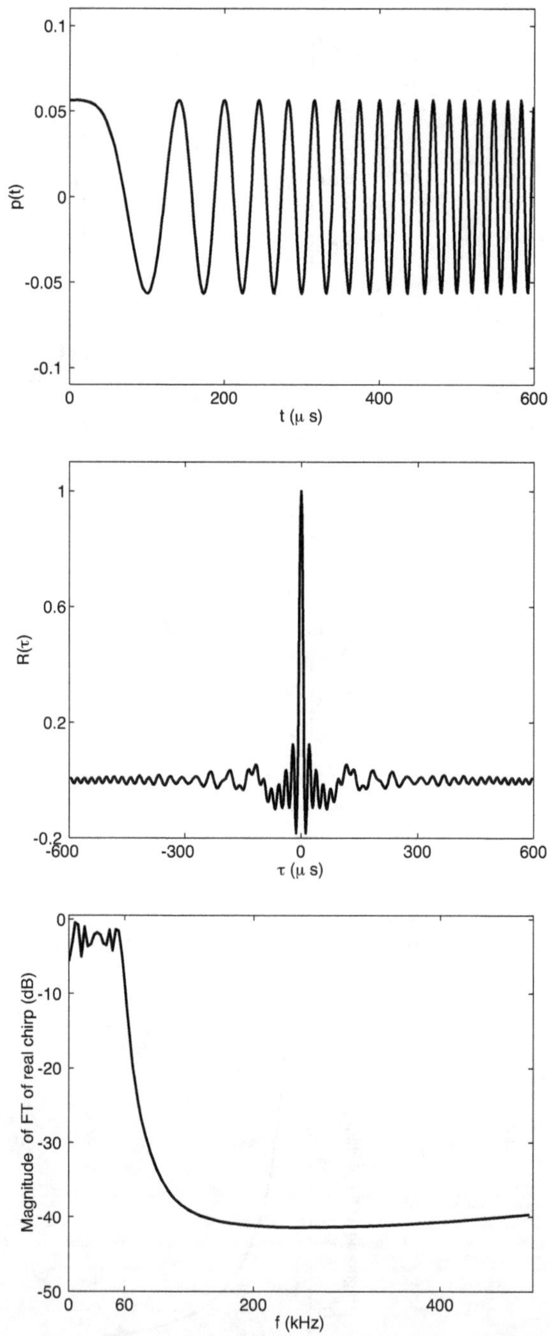

Figure 2.35. Example of a real chirp signal $p(t)$ (top), its autocorrelation (middle), and Fourier transform magnitude $|P(j2\pi f)|$ on a dB scale (bottom).

2.5.3 SNR Improvement due to matched filtering

As explained in Sec. 1.3, the transmitter and receiver have bandlimiting filters to eliminate unwanted interferences. These filters are bandlimited to the baseband $-\sigma < \omega < \sigma$. The integral in Eq. (2.48) can therefore be replaced with

$$\frac{|w(0)|^2}{\sigma_e^2} = \int_{-\sigma}^{\sigma} \frac{|P(j\omega)|^2}{S_{qq}(j\omega)} \frac{d\omega}{2\pi}. \tag{2.75}$$

If we do not perform matched filtering at the receiver but merely use a bandlimiting filter

$$B(j\omega) = \begin{cases} 1 & -\sigma < \omega < \sigma \\ 0 & \text{otherwise}, \end{cases} \tag{2.76}$$

then the signal component of interest would be

$$w(T_p) = \int_{-\sigma}^{\sigma} e^{j\omega T_p} P(j\omega) \frac{d\omega}{2\pi}. \tag{2.77}$$

Here T_p denotes the time at which the *signal component* at the output of $\beta(t)$ is maximum (in response to input $p(t)$). Assuming $p(t)$ is itself approximately bandlimited to $-\sigma < \omega < \sigma$ (Fig. 2.18(a)), the sample $w(T_p)$ is essentially the peak value of $p(t)$. The noise variance at the output of the bandlimiting filter is

$$\sigma_e^2 = \int_{-\sigma}^{\sigma} S_{qq}(j\omega) \frac{d\omega}{2\pi} \tag{2.78}$$

so that

$$\frac{|w(T_p)|^2}{\sigma_e^2} = \frac{|\int_{-\sigma}^{\sigma} e^{j\omega T_p} P(j\omega) \frac{d\omega}{2\pi}|^2}{\int_{-\sigma}^{\sigma} S_{qq}(j\omega) \frac{d\omega}{2\pi}}. \tag{2.79}$$

The SNR improvement γ due to matched filtering is therefore obtained by taking the ratio of (2.75) to (2.79):

$$\gamma = \frac{\displaystyle\int_{-\sigma}^{\sigma} S_{qq}(j\omega) \frac{d\omega}{2\pi}}{\left| \displaystyle\int_{-\sigma}^{\sigma} e^{j\omega T_p} P(j\omega) \frac{d\omega}{2\pi} \right|^2} \int_{-\sigma}^{\sigma} \frac{|P(j\omega)|^2}{S_{qq}(j\omega)} \frac{d\omega}{2\pi}. \tag{2.80}$$

Example 2.4: SNR improvement for white noise

To understand how the expression (2.80) works, it is useful to consider the special case of white noise again. In this case $S_{qq}(j\omega) = \mathcal{N}_0/2$ so that

$$\gamma = \frac{2\sigma \displaystyle\int_{-\sigma}^{\sigma} |P(j\omega)|^2 \frac{d\omega}{2\pi}}{\left| \displaystyle\int_{-\sigma}^{\sigma} e^{j\omega T_p} P(j\omega) \frac{d\omega}{2\pi} \right|^2}.$$

Since the time-limited pulse $p(t)$ is assumed to be approximately bandlimited to $-\sigma < \omega < \sigma$, it follows that

$$\int_{-\sigma}^{\sigma} |P(j\omega)|^2 \frac{d\omega}{2\pi} \approx \int_0^T p^2(t)dt,$$

which is the pulse energy. Similarly, $\int_{-\sigma}^{\sigma} e^{j\omega T_p} P(j\omega)\frac{d\omega}{2\pi} \approx p(T_p)$ so that

$$\gamma \approx \frac{2\sigma \int_0^T p^2(t)dt}{|p(T_p)|^2}. \tag{2.81}$$

Thus the improvement factor is proportional to the ratio of the energy to the peak magnitude-squared of the pulse. It also depends on the bandwidth 2σ, which is the extent to which the energy is spread out in frequency.

2.5.4 Matched filtering and detector performance

Returning to Fig. 2.27, the sampled signal at time zero is

$$u(0) = w(0) + e(0). \tag{2.82}$$

With $B(j\omega)$ chosen as a matched filter (2.47) (with $c = 1$), we have

$$w(0) = \int_{-\infty}^{\infty} B(j\omega)P(j\omega)\frac{d\omega}{2\pi} = \int_{-\infty}^{\infty} \frac{|P(j\omega)|^2}{S_{qq}(j\omega)} \frac{d\omega}{2\pi}, \tag{2.83}$$

and the variance of the noise term $e(0)$ is

$$\sigma_e^2 = \int_{-\infty}^{\infty} S_{qq}(j\omega)|B(j\omega)|^2 \frac{d\omega}{2\pi} = \int_{-\infty}^{\infty} \frac{|P(j\omega)|^2}{S_{qq}(j\omega)} \frac{d\omega}{2\pi}. \tag{2.84}$$

Assuming a PAM constellation is used by the transmitter, the actual transmitted pulse is a scaled version $s(0)p(t)$, where $s(0)$ is a PAM symbol such as $\pm 1, \pm 3$, and so forth.

From the noisy sample $u(0)$ the detector at the receiver has to estimate the transmitted symbol $s(0)$. The error probability in a PAM constellation has the form (2.16), where $2A$ is the distance between the words. The quantity $w(0)$ therefore plays the role of A, and the PAM constellation as seen by the detector is as in Fig. 2.36. The decision boundaries are midway between the symbols as shown. By applying Eq. (2.16) we therefore get

$$\mathcal{P}_{e,PAM} = 2(1 - 2^{-b})\mathcal{Q}\Big(\frac{w(0)}{\sigma_e}\Big). \tag{2.85}$$

Here the argument $w(0)/\sigma_e$ of the $\mathcal{Q}$-function is determined by (2.83) and (2.84) (or equivalently (2.48)), and reduces to (2.56) for white noise. Thus, with a matched filter and white noise, the PAM symbol error probability is

$$\mathcal{P}_{e,PAM} = 2(1 - 2^{-b})\mathcal{Q}\Big(\sqrt{\frac{2E_p}{\mathcal{N}_0}}\Big), \tag{2.86}$$

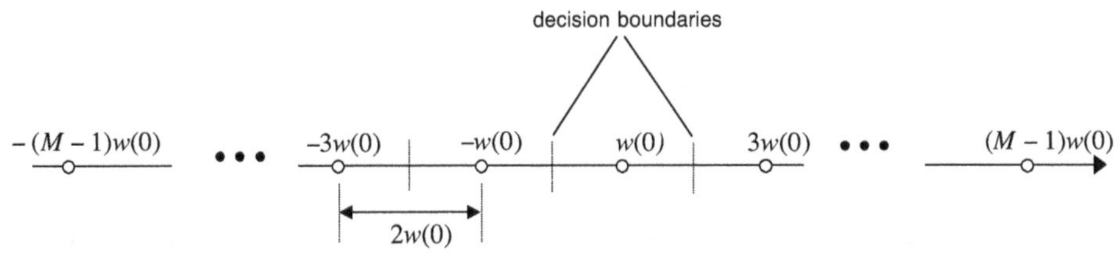

Figure 2.36. The PAM constellation as seen by the receiver.

where

$$E_p = \int_{-\infty}^{\infty} |p(t)|^2 dt \tag{2.87}$$

is the pulse energy. Assuming that the PAM symbols are transmitted with identical probabilties, the average energy of the *transmitted* constellation has the form (2.3) with A^2 replaced by E_p. Thus

$$E_{ave} = \frac{(2^{2b} - 1)E_p}{3} \tag{2.88}$$

Substituting this into Eq. (2.86) we obtain

$$\mathcal{P}_{e,PAM} = 2(1 - 2^{-b})\mathcal{Q}\Big(\sqrt{\frac{6E_{ave}}{(2^{2b} - 1)\mathcal{N}_0}}\Big). \tag{2.89}$$

Summarizing, Eq. (2.89) is the average error probability in a b-bit PAM system with average transmitted energy E_{ave}, if a matched filter is used at the receiver. This holds under the assumption that the noise is white with power spectrum $\mathcal{N}_0/2$. For example, in the 1-bit PAM case we have

$$\mathcal{P}_{e,PAM} = \mathcal{Q}\Big(\sqrt{\frac{2E_{ave}}{\mathcal{N}_0}}\Big). \tag{2.90}$$

2.5.5 Other sources of errors in detection

In Fig. 2.36, if the receiver does not identify the decision thresholds correctly, then there can be serious error in the symbol detection. This can happen, for example, if the detector's knowledge of $w(0)$ is off by a scale factor. The same problem can arise for QAM systems as well. For example, consider Fig. 2.37. On the left is shown a QAM constellation with the decision thresholds (vertical and horizontal lines) correctly placed. For this constellation, if the decision theresholds are estimated to be as in the figure on the right, then the probability of error is clearly very large.

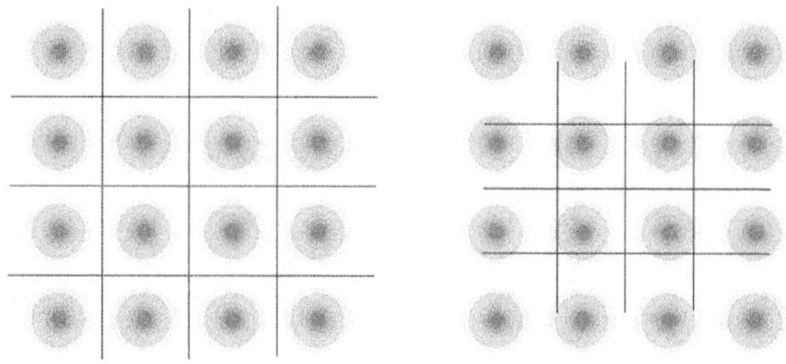

Figure 2.37. A QAM constellation with decision thresholds (vertical and horizontal lines) correctly placed (left) and incorrectly placed (right).

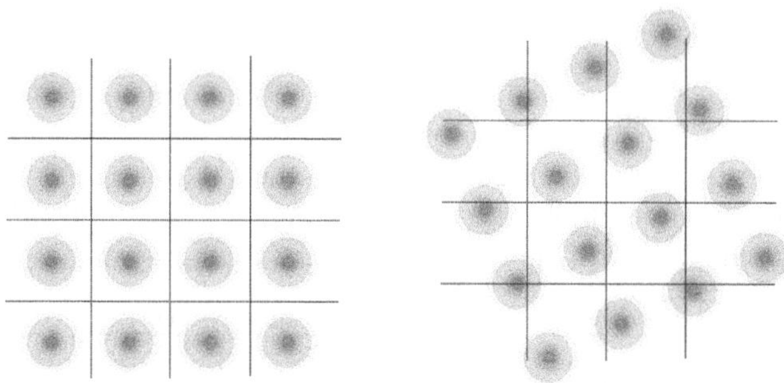

Figure 2.38. A QAM constellation with threshold lines correctly placed (left), and the rotated QAM constellation with threshold lines unchanged.

A similar type of error occurs when the symbol stream is multiplied by an unknown complex constant. This happens if the channel impulse response can only be estimated within a scalar ambiguity as is sometimes the case. The unknown complex constant has the effect of rotating the received constellation with respect to the transmitted one, as shown on the right in Fig. 2.38. Once again, if decision thresholds are chosen by ignoring this rotation, there is serious error in symbol detection. In practice, the scale factor ambiguity is resolved by transmitting "pilot symbols" to the receiver.

Yet another source of error is the so-called **bias** introduced by the equalizer at the receiver. Such a situation often arises in minimum mean square error (MMSE) equalizers, which have the property that the error does not have zero mean (Sec. 16.3). In this case the symbols in the received constellation are shifted by the nonzero mean (on the average), which creates serious error in the detection. In practice, whenever MMSE equalizers are used, the bias is estimated and removed before symbol detection (Sec. 16.3).

2.6 Practical considerations in matched filtering

In the previous section we assumed that a single pulse $p(t)$ was transmitted (Fig.

2.27), and designed a matched filter for it. The single-pulse case, often referred to as one-shot transmission [Lucky, 1973], helps us to understand the basic ideas in a simple way. But in practice the transmitted signal $x(t)$ is a series of pulses spaced apart uniformly by T seconds, as demonstrated in Fig. 2.39(b). The received signal is therefore of the form[12]

$$y(t) = \underbrace{\sum_n s(n)p(t-nT)}_{x(t)} + q(t).$$

The receiver has to detect the transmitted symbols $s(n)$ from this noisy baseband signal. In the figure it is assumed that the adjacent pulses do not overlap. This assumption is not necessary in the following discussion, as we shall see later. Figure 2.40 shows the baseband signal processing system. The noisy signal $y(t)$ is filtered by the matched filter $B(j\omega)$, and then sampled at the symbol rate $1/T$. The resulting signal $\widehat{s}(n)$ is then tested by the threshold detector in order to make the decision as to what symbol was transmitted at discrete time n. Assuming white noise, the matched filter is $p(-t)$ for real $p(t)$. Figure 2.41 shows the signal $x(t)$, the causal matched filter $p(T-t)$ (which is $p(-t)$ shifted by T), and the signal component of the matched filter output

$$w(t) = \int \beta(\tau)x(t-\tau)d\tau.$$

This output is nothing but the sum of convolutions of the matched filter $p(T-t)$ with the individual pulses in Fig. 2.41(a). Since each of these convolutions is the autocorrelation of $p(t)$ (shifted by a multiple of T), we obtain the plot shown in Fig. 2.41(c). Since $p(t)$ is designed such that its autocorrelation is a narrow pulse (as demonstrated by Fig. 2.29(c) and Fig. 2.35, middle), the matched filter output is a uniformly spaced series of pulses, and the heights of the pulses depend on the PAM symbol stream transmitted. Thus, since Fig. 2.41(a) shows that the symbol stream is $1, 3, -1, 1, -1, \ldots$, the pulses in Fig. 2.41(c) have heights proportional to $1, 3, -1, 1, -1, \ldots$ (but delayed by T seconds because the matched filter is causal). The scale factor α depends on the pulse $p(t)$.

The detector can identify the transmitted symbol stream by sampling Fig. 2.41(c) at the locations nT, and using decision boundaries as in Fig. 2.36. Note that the decision boundaries depend on the scale factor α, so the detector should know what α is. This scale factor, which depends on $p(t)$, is easy to identify. In practice, the matched filter input is a noisy version of $x(t)$, since the channel noise $q(t)$ is added to it. The output of the matched filter is therefore not as clean as it is shown in Fig. 2.41(c). But, since the matched filter is designed to maximize the signal-to-noise ratio (i.e., maximize the heights of the peaks in Fig. 2.41(c) compared to σ_e, where σ_e^2 is the noise vairance), the correct symbols $s(n)$ can still be identified in practice, with a small probability of error.

[12]The signal also has an additional distortion due to the filtering effects of the channel transfer function. We assume here that this has been compensated for, or "equalized." Channel equalization is a major topic and will be discussed in later chapters.

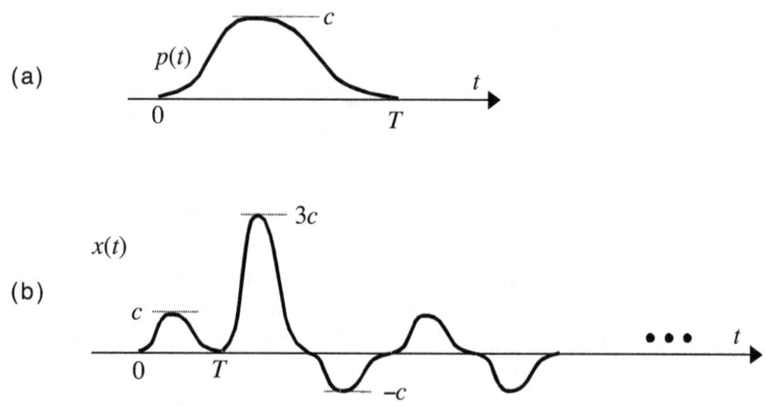

Figure 2.39. (a) Example of a baseband pulse $p(t)$, and (b) example of a baseband signal $x(t) = \sum_n s(n)p(t - nT)$, where $s(n)$ comes from the PAM constellation.

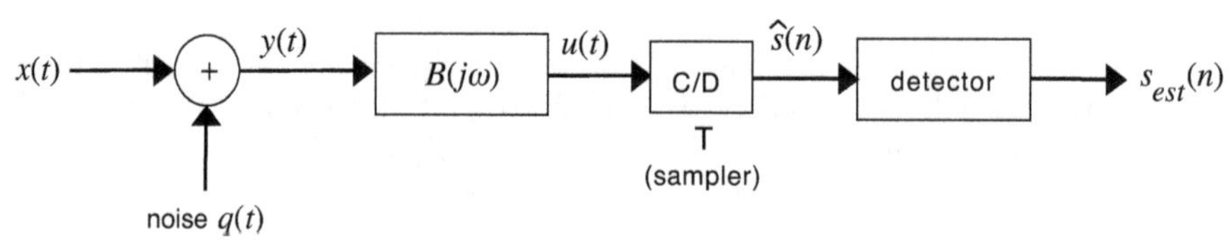

Figure 2.40. The received signal $y(t)$ filtered by a matched filter $B(j\omega)$ before sampling and symbol detection.

Remarks

1. ***Overlapping pulses.*** It is not necessary to assume that the adjacent pulses in Fig. 2.41(a) do not overlap. If the pulse duration exceeds the symbol spacing T, then there is overlap, as demonstrated in Fig. 2.42. This is quite acceptable, as long as the pulses at the output of the matched filter in Fig. 2.41(c) are still well separated. If these pulses are not well separated then there is interference between the symbols, and equalization is necessary. Nevertheless, in Chap. 5 we will see that matched filtering before sampling continues to be the appropriate operation. It turns out that equalization can always be performed digitally after sampling (Sec. 5.2).

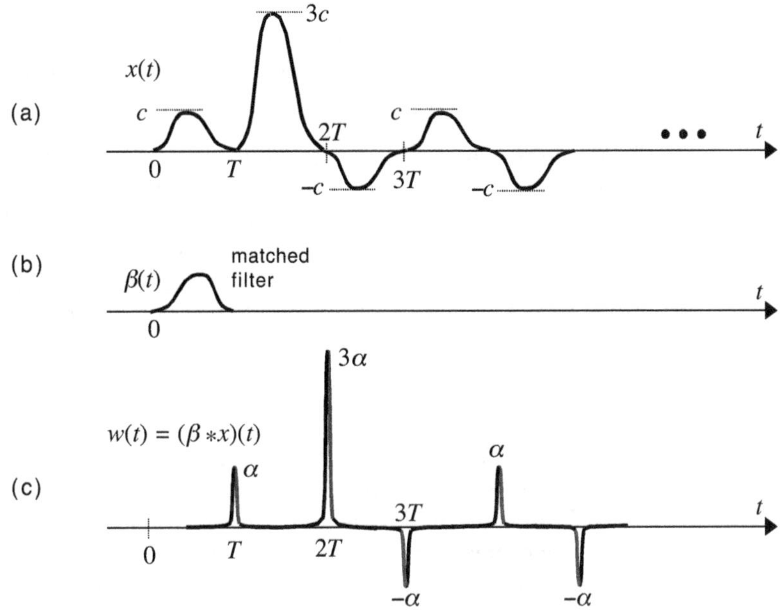

Figure 2.41. (a) The signal $x(t)$ at the receiver input, (b) matched filter impulse response, and (c) matched filter output (signal component).

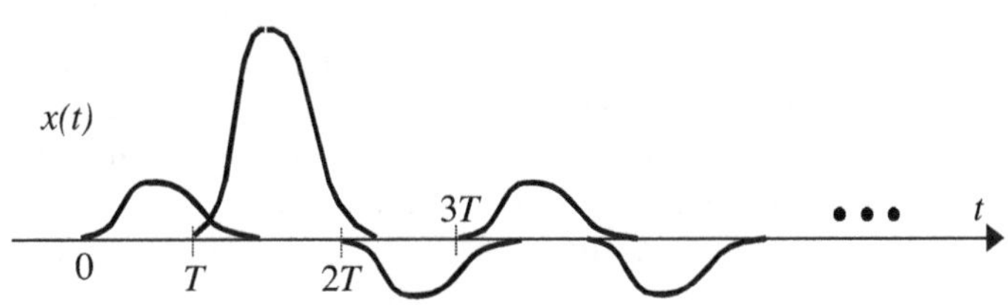

Figure 2.42. The transmitted signal $x(t)$ with overlapping pulses $p(t - nT)$.

2. ***Channel impulse response.*** In the preceding discussions we assumed that the received signal is equal to the transmitted signal $x(t)$ plus noise. In practice the transmitted signal is also distorted by the channel frequency response $H(j\omega)$. Thus the effective pulse is the convolution of $p(t)$ with the channel impulse response $h(t)$, and the channel output has the form

$$\sum_n s(n)(h * p)(t - nT), \tag{2.91}$$

where $*$ denotes convolution. The matched filter in this case is

$$\beta(t) = p_{eff}^*(-t), \tag{2.92}$$

where $p_{eff}(t) = (h * p)(t)$ is the "effective pulse." Since the duration of $p_{eff}(t)$ is the sum of durations of $h(t)$ and $p(t)$, it usually exceeds the symbol duration T.

3. ***Synchronization.*** Referring again to Fig. 2.41(c), which is the signal component at the output of the matched filter, we see that it is crucial for the receiver to sample it at the peak locations. A slight offset in sampling can decrease the signal-to-noise ratio significantly. If T_d denotes the duration of the effective pulse $p_{eff}(t)$ (which could be longer than the symbol spacing T), then the ideal sample locations have the form

$$t_n = T_d + nT$$

for integer n, because the convolution of $p_{eff}(t)$ with the causal matched filter $p_{eff}^*(T_d - t)$ has its peak value at time T_d. As long as the receiver knows T_d exactly (and, of course, knows the symbol spacing T), it can perform the sampling at the ideal locations, and we say that the receiver is **synchronized** to the transmitter. Since the effective pulse $p_{eff}(t)$ is the transmitted pulse $p(t)$ convolved with the channel $h(t)$, it is therefore important to estimate the duration of the channel impulse response correctly. We shall not discuss synchronization in detail. The reader is referred to standard texts such as Proakis [1995] or Lathi [1998], and references therein.

4. ***Decision boundaries.*** As explained earlier, the PAM constellation which is presented to the detector is as in Fig. 2.36. The symbol spacing is $2w(0)$, where $w(0)$ is the peak at the matched filter output, in response to the baseband pulse $p(t)$ sent by the transmitter. Since the effective pulse is $(p * h)(t)$, the above peak is proportional to the quantity $\sqrt{E_p}$, where E_p is the energy in the effective pulse. This quantity should be known to the receiver accurately, in order to form the decision boundaries indicated in the figure. Otherwise the decisions made by the detector can be severely in error. In a mathematical sense this is analogous to the synchronization problem (amplitude synchronization rather than time synchronization).

2.7 Concluding remarks

Claude Shannon [1948, 1949] laid the foundation for information theory and digital communications. Harry Nyquist [1928], who is regarded as one of the pioneers of digital communications, worked with telegraphic signals. He showed that the signaling speed which can be handled by a communication system is proportional to the bandwidth available. Nyquist also considered distortion-free transmission and derived the zero-forcing condition, today known as the Nyquist condition. He demonstrated a wide variety of pulse shapes, besides the sinc function, which have the zero-forcing property. In this chapter we reviewed many basic ideas from the theory of digital communications. Further advanced considerations relating to matched filtering will be discussed in Chap. 5, where we review more detailed material on transceiver design techniques.

Appendix to Chapter 2

2.A The baseband signal and the passband signal

The basic message $x(t)$ transmitted by a communication system is usually a lowpass signal called the **baseband signal** (Fig. 2.43(a)). When this is modulated with a carrier the result becomes a bandpass signal. For example, the PAM/DSB signal has the form $x(t)\cos\omega_c t$, and its Fourier transform magnitude is as in Fig. 2.43(b).

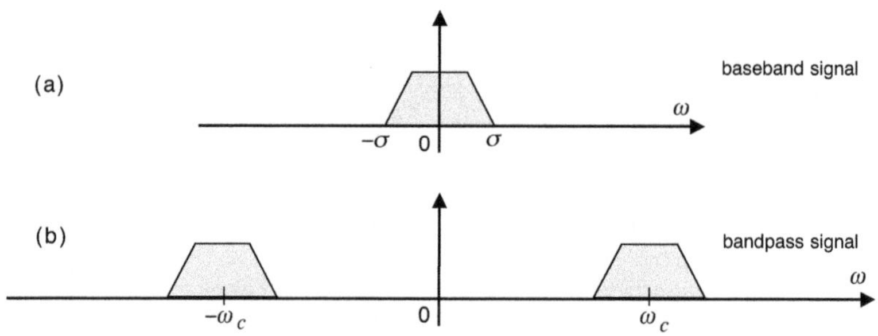

Figure 2.43. Fourier transform magnitudes of (a) a baseband signal, and (b) the corresponding bandpass signal obtained by modulation.

We always assume $\sigma < \omega_c$ so that each copy of the baseband signal shown in Fig. 2.43(b) is confined entirely to one side of the frequency axis. Note that if $x(t)$ is real, its Fourier transform has Hermitian symmetry around $\omega = 0$ (i.e., $X(j\omega) = X^*(-j\omega)$) so the Fourier transform of $x(t)\cos\omega_c t$ has such symmetry in the neighborhood of ω_c (and $-\omega_c$). Imagine such a signal is communicated over a channel with some frequency response $H(j\omega)$. Assuming the channel impulse repsonse is real, $H(j\omega)$ has Hermitian symmetry around $\omega = 0$, but of course, not around ω_c in general (see Fig. 2.44(b)). So, the channel output, which is still bandpass, does not usually have Hermitian symmetry in the neighborhood of ω_c (Fig. 2.44(c)), even if the channel input does. So, in communication theory, a **bandpass signal** typically signifies a real signal with passbands centered around $\pm\omega_c$ and with $\sigma < \omega_c$, but the passbands do not necessarily exhibit any symmetry around ω_c or $-\omega_c$.[13]

[13] Another example is a signal of the form $x(t)\sin\omega_c t$ (which is one of the two terms in the QAM signal $x(t)\cos\omega_c t - y(t)\sin\omega_c t$). This has Fourier transform $-0.5jX(j(\omega-\omega_c)) + 0.5jX(j(\omega+\omega_c))$. So the copy of the Fourier transform around ω_c is phase-shifted by $-\pi/2$, which means the phase response has no antisymmetry around ω_c (and similarly around $-\omega_c$). So the Hermitian symmetry of $X(j\omega)$ (around $\omega = 0$) is not present in the modulated version (around $\omega = \omega_c$ or $\omega = -\omega_c$).

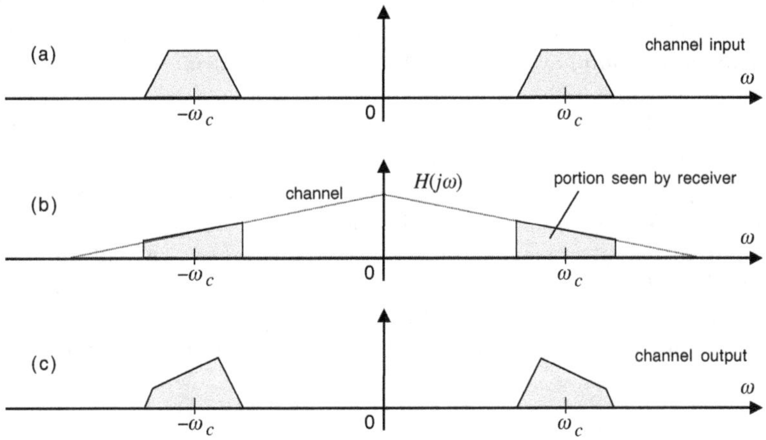

Figure 2.44. Examples of Fourier transforms of (a) the channel input, (b) the channel, and (c) channel output which does not have Hermitian symmetry in the neighborhood ω_c or $-\omega_c$.

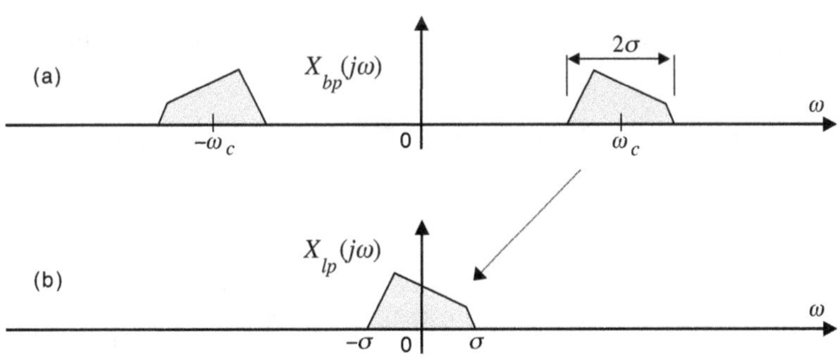

Figure 2.45. Fourier transforms of (a) a *bandpass* signal, and (b) the corresponding (lowpass) *baseband* signal.

Given any real bandpass signal $x_{bp}(t)$ with $\omega_c > \sigma$ as in Fig. 2.45(a), consider the lowpass signal $x_{lp}(t)$ constructed as shown in Fig. 2.45(b). Thus, the band around ω_c is moved to zero frequency, and the band around $-\omega_c$ is discarded. We say that $x_{lp}(t)$ is the (lowpass) **baseband equivalent** of $x_{bp}(t)$ (with respect to carrier frequency ω_c). Similarly, $x_{bp}(t)$ is the **bandpass equivalent** of $x_{lp}(t)$, and can be expressed analytically as

$$x_{bp}(t) = \mathrm{Re}\Big(x_{lp}(t)e^{j\omega_c t}\Big). \tag{2.93}$$

Sometimes a factor of 2 is included on the right-hand side, but we shall keep it simple. Since the frequency bands of the real signal $x_{bp}(t)$ do not in general exhibit symmetry around ω_c, the baseband equivalent $X_{lp}(j\omega)$ has no Hermitian symmetry around $\omega = 0$. Thus *the baseband equivalent* $x_{lp}(t)$ *is not necessarily real,* even though the bandpass signal $x_{bp}(t)$ is real.

We will see that it is extremely useful to have the definition of a baseband equivalent signal. Among other things, it allows us to perform efficient digital signal processing in the complex domain, both at the transmitter and at the receiver. Assume now that a real bandpass signal $x_{bp}(t)$ is passed through a real bandpass channel $h_{bp}(t)$ as shown in Fig. 2.44. Then the output is the bandpass signal $y_{bp}(t) = (h_{bp} * x_{bp})(t)$ with Fourier transform $Y_{bp}(j\omega) = H_{bp}(j\omega)X_{bp}(j\omega)$. So we can define the baseband equivalent $y_{lp}(t)$ of the output signal exactly as we did in Fig. 2.45. The baseband equivalents are then related as

$$Y_{lp}(j\omega) = H_{lp}(j\omega)X_{lp}(j\omega). \tag{2.94}$$

Thus, the communication of the modulated signal $x_{bp}(t)$ over a bandpass channel $h_{bp}(t)$ is mathematically equivalent to the communication of the lowpass baseband signal $x_{lp}(t)$ over the lowpass baseband channel $h_{lp}(t)$. Notice again that *the baseband channel* $h_{lp}(t)$ *is in general complex*, even though the physical channel $h_{bp}(t)$ and the signal $x_{lp}(t)$ may be real.

2.A.1 Baseband channel model for QAM

In QAM transmission the real bandpass signal transmitted has the form

$$\begin{aligned} x_{bp}(t) &= \sum_k s_c(k)p(t-kT)\cos\omega_c t - \sum_k s_s(k)p(t-kT)\sin\omega_c t \\ &= \text{Re}\sum_k \Big(s_c(k) + js_s(k)\Big)p(t-kT)e^{j\omega_c t}, \end{aligned}$$

where $s_c(k), s_s(k)$, and $p(t)$ are real. Here $p(t)$ is the baseband waveform and $s_c(k) + js_s(k)$ represents the complex QAM symbol. The complex baseband equivalent is therefore

$$x_{lp}(t) = \sum_k \Big(s_c(k) + js_s(k)\Big)p(t-kT). \tag{2.95}$$

Assume this is transmitted on a real bandpass channel $h_{bp}(t)$ with lowpass equivalent $h_{lp}(t)$:

$$h_{bp}(t) = \text{Re}\Big(h_{lp}(t)e^{j\omega_c t}\Big). \tag{2.96}$$

Then the channel output (ignoring noise) is:

$$\begin{aligned} y_{bp}(t) &= \Big(h_{bp} * x_{bp}\Big)(t) \\ &= \text{Re}\Big((x_{lp} * h_{lp})(t)e^{j\omega_c t}\Big) \end{aligned}$$

$$= \text{Re}\left[\left(\sum_k \Big(s_c(k) + js_s(k)\Big)p(t - kT) * h_{lp}(t)\right)e^{j\omega_c t}\right]$$

$$= \text{Re}\left[\left(\sum_k \Big(s_c(k) + js_s(k)\Big)h_{lp,e}(t - kT)\right)e^{j\omega_c t}\right]$$

$$= \text{Re}\Big[y_{lp}(t)e^{j\omega_c t}\Big],$$

where

$$h_{lp,e} = (h_{lp} * p)(t),$$

and $y_{lp}(t)$ denotes the complex signal

$$y_{lp}(t) = \sum_k \Big(s_c(k) + js_s(k)\Big)h_{lp,e}(t - kT) = y_r(t) + jy_i(t).$$

So

$$y_{bp}(t) = \text{Re}\Big[y_{lp}(t)e^{j\omega_c t}\Big] = y_r(t)\cos\omega_c t - y_i(t)\sin\omega_c t. \tag{2.97}$$

The QAM demodulator (Sec. 2.4.4) simply separates out $y_r(t)$ and $y_i(t)$. But since $h_{lp,e}(t)$ is complex, $y_r(t)$ is a linear combination of $s_c(k)$ as well as the interference component $s_s(k)$. Similarly $y_i(t)$ is a linear combination of $s_s(k)$ and $s_c(k)$. With $y_r(t)$ and $y_i(t)$ sampled at the output of the QAM demodulator, we can say that the complex signal $y_{lp}(nT)$ is available for digital signal processing. Define the complex discrete-time equivalent channel

$$h_d(n) = h_{lp,e}(nT), \tag{2.98}$$

and the sampled signal $y_d(n) = y_{lp}(nT)$. We then have

$$y_d(n) = \sum_k \Big(s_c(k) + js_s(k)\Big)h_d(n - k) + q(n), \tag{2.99}$$

where we have added the term $q(n)$ to represent channel noise (which we ignored until now). The digital signal processor sees only this signal $y_d(n)$. Summarizing, the complex QAM symbol $s_c(k) + js_s(k)$ is effectively filtered through the complex discrete-time equivalent channel $h_d(n)$. In later chapters, we will find it convenient to use this equivalent complex channel $h_d(n)$ in all discussions. Thus:

1. The digital transmitter performs complex signal processing on the complex sequence $s_c(k) + js_s(k)$ (such as precoding, cyclic prefixing, and so forth, as explained in later chapters).

2. The digital receiver performs complex signal processing on the complex signal $y_d(n)$ (in fact a noisy version of it), such as channel equalization, to recover the original symbol stream $s_c(k) + js_s(k)$.

Thus, even though a pair of real symbol streams $s_c(k)$ and $s_s(k)$ are modulated and transmitted on the real bandpass channel, the equivalent lowpass baseband channel is complex, and the digital signal processor performs all operations in the complex domain. So we use complex mathematics freely at the transmitter and receiver.

2.A.2 Baseband channel model for PAM

In the PAM/DSB case all equations are as above but $s_s(k) = 0$. Thus, after demodulation and sampling of the channel output, the digital equivalent is

$$y_d(n) = \sum_k s_c(k) h_d(n-k). \tag{2.100}$$

Note that $h_d(n)$ is complex though $s_c(n)$ is real. That is, the real input $s_c(n)$ leaks into the imaginary part of $y_d(n)$. The receiver usually compensates for the distortion caused by the filter $h_d(n)$ using a digital equalizer, sometimes called a time-domain equalizer (TDE). The cascade of $h_d(n)$ and this equalizer is an equivalent digital channel with a shorter, real, impulse response. In fact we can take the real part of the output of the equalizer to remove the imaginary part (which only has noise). Thus, after time-domain equalization (and taking real parts),

$$y'_d(n) = \sum_k s_c(k) h'_d(n-k) + q(n), \tag{2.101}$$

where $h'_d(n)$ is real, and $q(n)$ is real-valued noise coming from the channel noise.

2.A.2.A PAM/SSB case

In the PAM/SSB case the details are different. Recall that the PAM/SSB signal has the form

$$x_{bp}(t) = s(t)\cos\omega_c t - \widehat{s}(t)\sin\omega_c t \tag{2.102}$$

for Type 2 SSB (and for Type 1 SSB, we have + sign instead of −). Here $\widehat{s}(t)$ is the Hilbert transform of the real signal $s(t)$:

$$s(t) = \sum_k s_c(k)p(t-kT), \quad \widehat{s}(t) = \sum_k s_c(k)\widehat{p}(t-kT).$$

So

$$x_{bp}(t) = \text{Re}\left[\left(\sum_k s_c(k)\Big(p(t-kT) + j\widehat{p}(t-kT)\Big)\right)e^{j\omega_c t}\right]. \tag{2.103}$$

So the lowpass baseband equivalent is

$$x_{lp}(t) = \left(\sum_k s_c(k)\Big(p(t-kT) + j\widehat{p}(t-kT)\Big)\right). \tag{2.104}$$

Now, let the real bandpass channel be represented as in Eq. (2.96), that is, $h_{bp}(t) = \text{Re}[h_{lp}(t)e^{j\omega_c t}]$. The received signal is

$$\begin{aligned} y_{bp}(t) &= (x_{bp} * h_{bp})(t) \\ &= \text{Re}\Big((x_{lp} * h_{lp})(t)e^{j\omega_c t}\Big) \\ &= \text{Re}\left[\left(\sum_k s_c(k)\Big(p(t-kT) + j\widehat{p}(t-kT)\Big) * h_{lp}(t)\right)e^{j\omega_c t}\right]. \end{aligned}$$

This can be rewritten as

$$y_{bp}(t) = \text{Re}\left[\Big(\sum_k s_c(k)h_{lp,e}(t-kT)\Big)e^{j\omega_c t}\right], \qquad (2.105)$$

where

$$h_{lp,e}(t) = \Big(p * h_{lp} + j\widehat{p} * h_{lp}\Big)(t) = h_r(t) + jh_i(t). \qquad (2.106)$$

Here $s_c(k)$, $p(t)$, and $\widehat{p}(t)$ are real, but $h_{lp}(t)$ is complex. So the real and imaginary parts $p(t-kT)$ and $\widehat{p}(t-kT)$ get mixed up by convolution with $h_{lp}(t)$. Normally the PAM/SSB demodulator just multiplies the received signal by $\cos\omega_c t$ and performs lowpass filtering. This removes $\widehat{p}(t)$ and recovers $p(t)$. But, since $h_{lp}(t)$ mixes up $p(t)$ and $\widehat{p}(t)$, the receiver now has to be smarter, and must perform a demodulation with $\cos\omega_c t$ and one with $\sin\omega_c t$ (like a QAM demodulator). This recovers the real and imaginary parts of

$$\sum_k s_c(k)\Big(h_r(t-kT) + jh_i(t-kT)\Big).$$

So the received digital signal after sampling has the simple form

$$y_d(n) = \sum_k s_c(k)h_d(n-k), \qquad (2.107)$$

where

$$h_d(n) = h_r(nT) + jh_i(nT).$$

One may argue that we can simply keep the real part of $y_d(n)$ and minimize complication. But this is like wasting the power of the signal in the imaginary part. For example, if $h_d(n)$ happens to be purely imaginary, this is a disaster because the real part of $y_d(n)$ has no information about the signal. So we take the entire complex signal $y_d(n)$, perform "time-domain equalization" (TDE), and take the real part to get

$$y'_d(n) = \sum_k s_c(k)h'_d(n-k) + q(n), \qquad (2.108)$$

where $h'_d(n)$ is real and $q(n)$ is real-valued noise. Thus at this point the digital signal processor simply sees a real PAM signal $s_c(k)$ and a real channel $h'_d(n)$.

Problems

2.1. For a 2-bit QAM system suppose the SNR is such that the symbol error probability is $\mathcal{P}_e = 10^{-5}$. If the desired $\mathcal{P}_e$ has to be decreased to 10^{-6}, then by how much should the SNR (in dB) be increased? Repeat for 6-bit QAM.

2.2. Repeat Problem 2.1 for 2-bit PAM and 6-bit PAM.

2.3. Consider Fig. 2.7, which explains how decision errors arise due to additive noise. In the figure the noise probability density function $f_E(e)$ resembles a Gaussian. Suppose instead that $f_E(e)$ is a uniform pdf:

$$f_E(e) = \begin{cases} \dfrac{1}{2B} & \text{for } -B < e < B \\ 0 & \text{otherwise.} \end{cases}$$

1. What is the probability of error for an interior symbol, in terms of A and B?
2. Is it possible for this probability of error to be zero? If so, under what conditions?
3. Find an example of a non-Gaussian noise pdf for which the probability of error can never be zero no matter how small the (nonzero) noise variance is.

2.4. Consider a signal $x(t) = \sum_{n=-\infty}^{\infty} s(n)f(t-nT)$. This is of the same form as the signals transmitted by digital communication systems (Sec. 2.1). Assume that we know $f(t)$ and we know that $x(t)$ has this form.

1. Let $x_d(n)$ be the uniformly sampled version $x_d(n) = x(nT)$, and let $f_d(n) = f(nT)$. Assume that $f_d(n)$ has z-transform $F_d(z)$ in some appropriate region of convergence. Show that $s(n)$ can be recovered from the samples $x_d(n)$ by using a filtering operation of the form

$$s(n) = \sum_{k=-\infty}^{\infty} g(k)x_d(n-k).$$

How do you identify the impulse response $g(k)$?

2. What are the conditions on $F_d(z)$ so that $G(z) = \sum_k g(k)z^{-k}$ is stable?
3. Given the samples $x(nT)$, explain how you can reconstruct the signal $x(t)$ for all t?

Note that the signal $x(t)$ is in general not bandlimited unless $f(t)$ also happens to be bandlimited. This problem shows that even a non-bandlimited signal can sometimes be reconstructed from its uniform samples if we know the model which generates the signal.

2.5. Show how the 3-bit Gray code in Fig. 2.17 can be extended to 4-bit PAM. *Hint.* You only have to make sure that the 16 codewords are distinct and that adjacent codewords differ exactly in one bit location.

2.6. Consider the system shown in the figure, where $H(j\omega)$ is a linear time-invariant system.

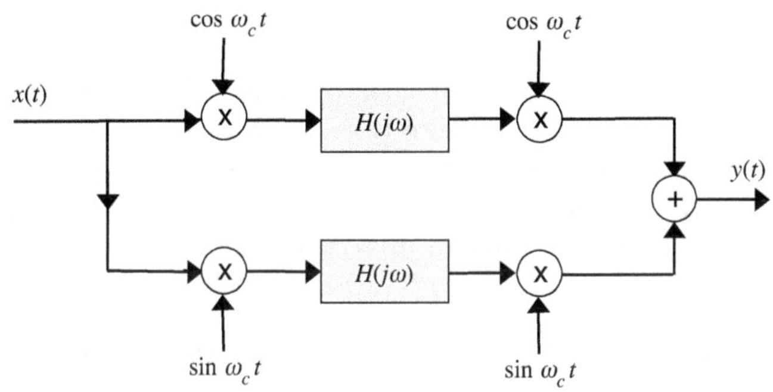

Figure P2.6.

1. Show that the modulator $\cos\omega_c t$ is a linear time varying device, that is, the operation $y_1(t) = x(t)\cos\omega_c t$ is linear time varying. Similarly the modulator $\sin\omega_c t$ is time varying.
2. Show, however, that the overall system from $x(t)$ to $y(t)$ in the figure is linear and time invariant (LTI). What is its transfer function and what is the impulse response? Thus the system is an example of an LTI system, with time varying building blocks in it.

2.7. Consider the system shown in Fig. P2.7, where $H(j\omega)$ is a linear time invariant system. This is a modified version of the system in Problem 2.6. Is this an LTI system?

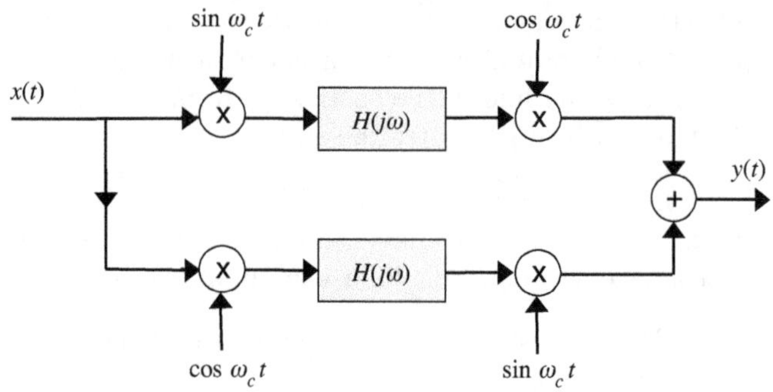

Figure P2.7.

2.8. For the finite-duration chirp $p_T(t)$ given in Eq. (2.62) let $K = 0.001$ (MHz)^2 and $T = 1$ ms. Compute the approximate bandwidth F_m and the pulse compression ratio $\mathcal{C}$. Plot the real and imaginary parts of $p_T(t)$, and the Fourier transform magnitude $|P_T(j2\pi f)|$ in dB. Also plot the autocorrelation of $p_T(t)$. You can use computer aided methods such as Matlab for this.

2.9. For the complex chirp $p(t) = e^{j\pi K t^2}$ the Fourier transform is given in Eq. (2.59). Assuming the truth of this, show that the real chirp $p_{re}(t) = \cos(\pi K t^2)$ has the Fourier transform

$$P_{re}(j\omega) = \frac{1}{\sqrt{K}} \cos\Big(\frac{\omega^2}{4\pi K} - \frac{\pi}{4}\Big).$$

For the real chirp $q_{re}(t) = \sin(\pi K t^2)$ show that the Fourier transform is

$$Q_{re}(j\omega) = \frac{1}{\sqrt{K}} \sin\Big(\frac{\pi}{4} - \frac{\omega^2}{4\pi K}\Big).$$

2.10. We now sketch out a derivation of the formula for the Fourier transform (FT) (2.59) of the complex chirp. The proof is tricky because $p(t)$ is not absolutely integrable, and the FT exists only as a Cauchy principal value. Assuming its existence we now verify that it is of the form (2.59).

1. If $P(j\omega)$ is the FT of $p(t)$ then we know that the FT of $dp(t)/dt$ is $j\omega P(j\omega)$ and the FT of $tp(t)$ is $jdP(j\omega)/d\omega$. Using these show that the FT of the complex chirp in Eq. (2.57) satisfies the differential equation

$$\frac{dP(j\omega)}{d\omega} = -\frac{j\omega}{2\pi K} P(j\omega).$$

2. Solve this differential equation and show that the solution has the form

$$P(j\omega) = P(0) e^{-j\omega^2/(4\pi K)}.$$

3. Show that $P(0) = (1/\sqrt{K}) \int_{-\infty}^{\infty} e^{j\pi t^2} dt$. Using the not-so-obvious fact [Abramowitz and Stegun, 1965, page 300-301] that

$$\int_0^{\infty} \cos(\pi t^2/2)dt = \int_0^{\infty} \sin(\pi t^2/2)dt = 0.5$$

verify indeed that $P(0) = e^{j\pi/4}/\sqrt{K}$.

3

Digital communication systems and filter banks

3.1 Introduction

It has been known for a number of years that digital filter banks play a useful role in the representation and design of digital communication systems. Examples where a filter bank framework is attractive include transmultiplexers, redundant precoders, and fractionally spaced equalizers, to name a few.

In this chapter we introduce basic concepts relating to multirate filter bank theory such as decimation, interpolation, polyphase representations, and transmultiplexing. Section 3.2 introduces multirate building blocks, polyphase representations, and basic multirate identities. In Secs. 3.3 and 3.4 we introduce decimation and interpolation filters. Operations such as blocking and parsing are described in Secs. 3.5 and 3.6 using multirate building blocks. Polyphase representations of decimators and interpolators are described in Sec. 3.7. The transmultiplexer is described in detail in Sec. 3.8. In Sec. 3.9 we show how the scalar channel in the transmultiplexer system can be represented in blocked form using an algebraic structure called the pseudocirculant matrix.

Notations are as described in Sec. 1.6. In this chapter the notation $C(z)$ is used for channel transfer functions, as $H_k(z)$ and $F_k(z)$ are used to represent filters.

3.2 Multirate building blocks

The decimator and expander are two fundamental building blocks in multirate signal processing systems. The M-fold **decimator**, shown in Fig. 3.1(a), is

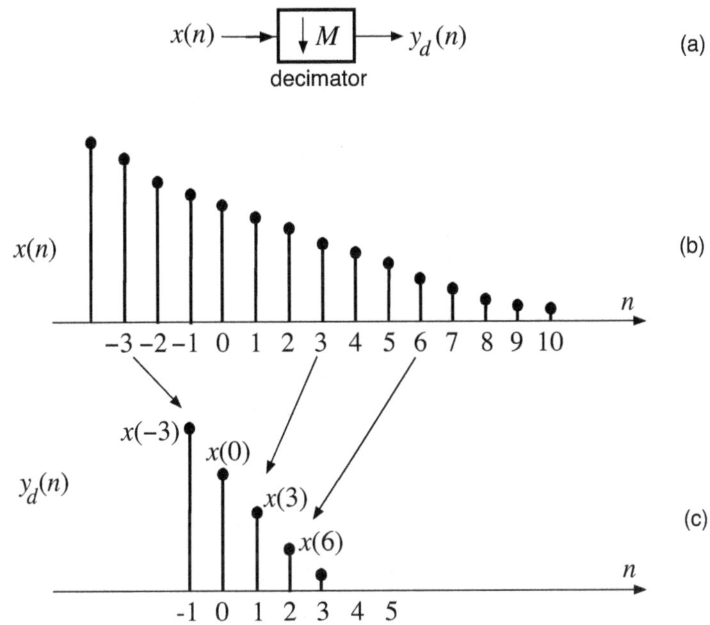

Figure 3.1. (a) The M-fold decimator or downsampler, and (b),(c) examples of input and output signals with $M = 3$.

defined by the input-output relation

$$y_d(n) = x(Mn), \tag{3.1}$$

which means that every Mth sample of the input is retained. This is demonstrated in the figure for $M = 3$. Note that the samples are renumbered by the decimator such that

$$y_d(0) = x(0), \quad y_d(1) = x(3), \quad y_d(2) = x(6), \ \ldots,$$

and so forth. The decimator is also known as the **downsampler**. In general, decimation results in loss of information. But if a signal is bandlimited in a certain way then it can be decimated without loss of information, as we shall see in Sec. 3.2.4. The M-fold **expander** or **upsampler**, shown in Fig. 3.2(a), is defined in the time domain by the input-output relation

$$y_e(n) = \begin{cases} x(n/M) & n = \text{multiple of } M \\ 0 & \text{otherwise.} \end{cases} \tag{3.2}$$

Thus the expander merely inserts zero-valued samples in a systematic manner, namely $M - 1$ zeros are inserted between adjacent samples of $x(n)$. This is demonstrated in parts (b) and (c) of the figure for $M = 3$. Evidently there is no loss of information due to the expanding operation. From the definition of the expander we see that

$$y_e(Mn) = x(n),$$

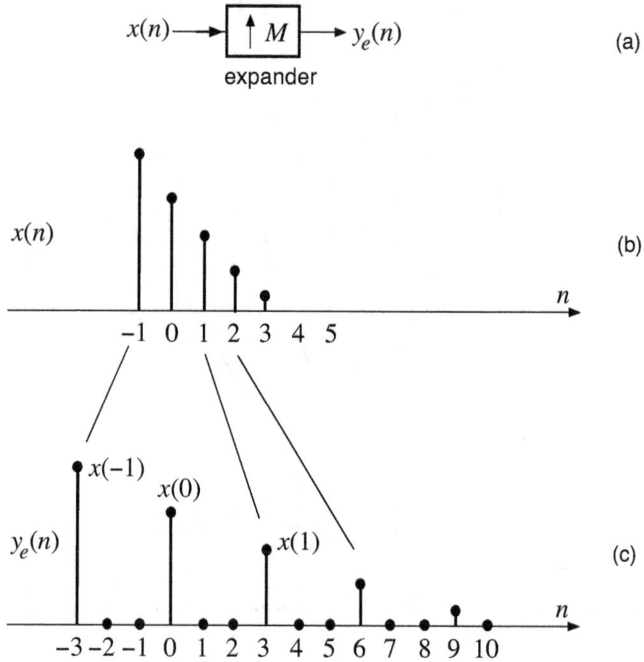

Figure 3.2. (a) The M-fold expander or upsampler, and (b),(c) examples of input and output signals with $M = 3$.

which shows that we can recover the input $x(n)$ from $y_e(n)$ simply by M-fold decimation. The expander is usually followed by a digital filter called an interpolation filter, which makes it more interesting (Sec. 3.4).

3.2.1 Polyphase decomposition

Consider a signal $x(n)$ with z-transform $X(z) = \sum_{n=-\infty}^{\infty} x(n)z^{-n}$. This can always be rewritten in the form

$$X(z) = \sum_{k=0}^{M-1} \sum_{n=-\infty}^{\infty} x(Mn+k)z^{-(Mn+k)} = \sum_{k=0}^{M-1} z^{-k} \sum_{n=-\infty}^{\infty} x(Mn+k)z^{-Mn}$$

for any positive integer M. Equivalently,

$$X(z) = \sum_{k=0}^{M-1} z^{-k} X_k(z^M) \quad \text{(Type 1 polyphase)}, \tag{3.3}$$

where $X_k(z)$ is the z-transform of the kth subsequence

$$x_k(n) = x(Mn + k), \quad 0 \leq k \leq M - 1. \tag{3.4}$$

Equation (3.3) is called the Type 1 polyphase decomposition (or representation) of $X(z)$. The M quantities $X_k(z)$ are said to be the **polyphase components** of $X(z)$ [equivalently $x_k(n)$ are the M polyphase components of $x(n)$].[1] Note that the 0th polyphase component $x_0(n)$ is nothing but the decimated version of $x(n)$, that is,

$$x_0(n) = x(Mn).$$

It is also possible to define the polyphase decomposition

$$X(z) = \sum_{k=0}^{M-1} z^k \widehat{X}_k(z^M) \quad \text{(Type 2 polyphase)}, \tag{3.5}$$

where the Type 2 polyphase components $\widehat{X}_k(z)$ are z-transforms of

$$\widehat{x}_k(n) = x(Mn - k)$$

instead of $x(Mn + k)$. Figure 3.3 shows how the polyphase components can be represented in terms of delays and decimators. A time-domain example for Type 1 polyphase decomposition is shown in Fig. 3.4 for $M = 3$; note how the samples of the polyphase components are numbered with consecutive integers. By applying the polyphase decomposition to a transfer function $H(z)$ we can write it in the form

$$H(z) = \begin{cases} \sum_{k=0}^{M-1} z^{-k} R_k(z^M) & \text{(Type 1)} \\ \sum_{k=0}^{M-1} z^{k} E_k(z^M) & \text{(Type 2)}. \end{cases} \tag{3.6}$$

The usefulness of polyphase decompositions will become clear later. Polyphase representations were introduced in the seventies by Bellanger and Daguet [1974].

3.2.2 Transform domain formulas

In the z-domain the input-output relation for the expander can be written as

$$Y_e(z) = X(z^M) \qquad \text{(expander)}. \tag{3.7}$$

To see this note that, since $y_e(n) = 0$ when n is not a multiple of M, we get

$$Y_e(z) = \sum_{n=-\infty}^{\infty} y_e(Mn) z^{-Mn} = \sum_{n=-\infty}^{\infty} x(n) z^{-Mn} \quad = X(z^M),$$

[1] Clearly $X_k(z)$ depends on M, so a notation such as $X_k^{(M)}(z)$ would have been more appropriate; but the quantity M is almost always clear from the context. So we shall use the simpler notation $X_k(z)$.

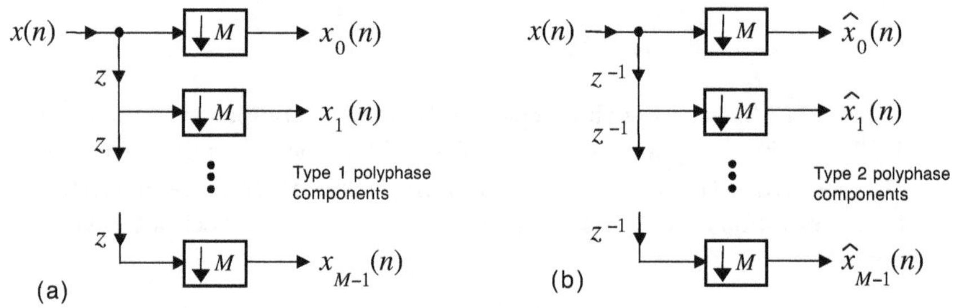

Figure 3.3. A simple block diagram interpretation of polyphase components; (a) Type 1 polyphase components $x_k(n)$, and (b) Type 2 polyphase components $\widehat{x}_k(n)$.

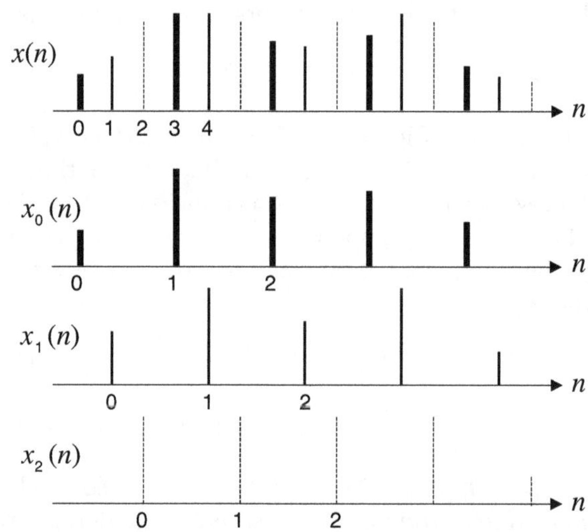

Figure 3.4. A signal $x(n)$ and its three polyphase components $x_k(n)$, for $M = 3$. Note carefully how the samples are renumbered.

where we have used the fact that $y_e(Mn) = x(n)$. Next, for the decimator we claim that the input-output relation in the z-domain can be written as

$$Y_d(z) = \frac{1}{M} \sum_{\ell=0}^{M-1} X(z^{1/M} W^{\ell}) \qquad \text{(decimator)}, \tag{3.8}$$

where the quantity W is defined as

$$W = e^{-j2\pi/M}. \tag{3.9}$$

Note that W is an Mth root of unity, that is,

$$W^M = 1.$$

It is easily verified that W satisfies the property

$$\sum_{m=0}^{M-1} W^{-km} = \begin{cases} M & \text{for } k = 0 \\ 0 & \text{for } 1 \leq k \leq M-1. \end{cases} \tag{3.10}$$

For $k = 0$ this is obvious; when $1 \leq k \leq M-1$, rewrite the left-hand side as $(1 - W^{-kM})/(1 - W^{-k})$. Since $W^M = 1$ and $W^k \neq 1$, this is zero indeed.

Proof of Eq. (3.8). Replacing z with zW^ℓ in the polyphase representation (3.3) we get

$$X(zW^\ell) = \sum_{k=0}^{M-1} z^{-k} W^{-k\ell} X_k(z^M).$$

Thus

$$\sum_{\ell=0}^{M-1} X(zW^\ell) = \sum_{k=0}^{M-1} z^{-k} X_k(z^M) \sum_{\ell=0}^{M-1} W^{-k\ell} = M X_0(z^M), \tag{3.11}$$

where we have used Eq. (3.10). Thus $MX_0(z) = \sum_{\ell=0}^{M-1} X(z^{1/M} W^\ell)$; since $Y_d(z) = X_0(z)$, the proof is complete. ▽ ▽ ▽

3.2.2.A *Frequency-domain viewpoint*

It is important to understand Eqs. (3.7) and (3.8) in terms of the frequency variable ω. Substituting $z = e^{j\omega}$ we have

$$Y_e(e^{j\omega}) = X(e^{j\omega M}) \quad \text{(expander)} \tag{3.12}$$

and

$$Y_d(e^{j\omega}) = \frac{1}{M} \sum_{\ell=0}^{M-1} X(e^{j(\omega - 2\pi\ell)/M}) \quad \text{(decimator)}. \tag{3.13}$$

Thus, for the expander, $Y_e(e^{j\omega})$ is obtained by **squeezing**, or shrinking, the plot of $X(e^{j\omega})$ by a factor of M (Fig. 3.5). For the decimator, the output $Y_d(e^{j\omega})$ is a sum of M terms. The 0th term contains $X(e^{j\omega/M})$, which is nothing but an M-fold **stretched version** of M. The 1st term is

$$X(e^{j(\omega - 2\pi)/M})$$

which is the stretched version shifted by 2π. More generally, the ℓth term given by $X(e^{j(\omega - 2\pi\ell)/M})$ is the stretched version shifted by $2\pi\ell$. Figure 3.6 demonstrates this for $M = 3$.

In Eq. (3.13) the shifted versions in general overlap with the original stretched version. This is called the **aliasing effect** and is similar to aliasing created by undersampling a bandlimited signal [Oppenheim and Willsky, 1997].

3.2.2.B Notations for decimation and expansion

Since the transform-domain expressions for decimation are rather complicated, we often abbreviate them as follows:

$$X(e^{j\omega})\Big|_{\downarrow M} \quad \text{or} \quad \Big[X(e^{j\omega})\Big]_{\downarrow M} \stackrel{\Delta}{=} \frac{1}{M}\sum_{\ell=0}^{M-1} X(e^{j(\omega-2\pi\ell)/M}), \tag{3.14}$$

$$X(z)\Big|_{\downarrow M} \quad \text{or} \quad \Big[X(z)\Big]_{\downarrow M} \stackrel{\Delta}{=} \frac{1}{M}\sum_{\ell=0}^{M-1} X(z^{1/M}W^{\ell}). \tag{3.15}$$

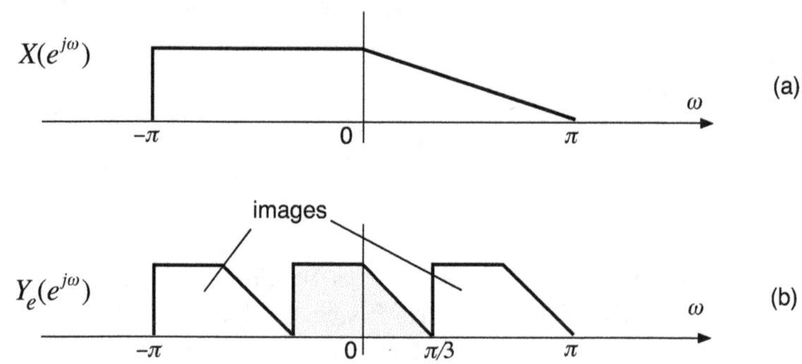

Figure 3.5. Fourier transforms of (a) input to the expander and (b) output of the expander for $M = 3$. In general there are $M - 1$ images.

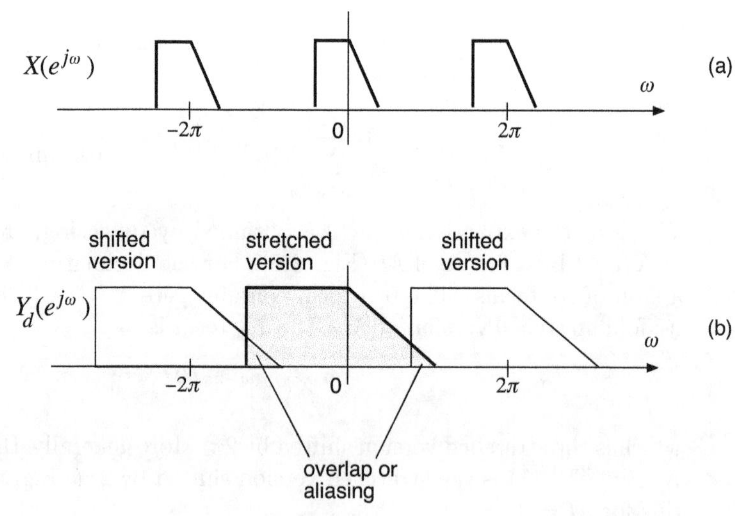

Figure 3.6. Fourier transforms of (a) input to the decimator and (b) output of the decimator for $M = 3$. In general there are $M - 1$ shifted versions of the stretched version.

Similarly, the notation $[x(n)]_{\downarrow M}$ is often used in the time domain instead of $x(Mn)$. For the expander we use similar notations, i.e., $[x(n)]_{\uparrow M}$ and $[X(z)]_{\uparrow M}$. These notations are especially useful in the simplification of complicated mathematical expressions containing multirate operators.

3.2.3 Multirate identities

In many applications decimators and expanders are used in cascade with linear time invariant systems (digital filters). Since the decimator and expander are

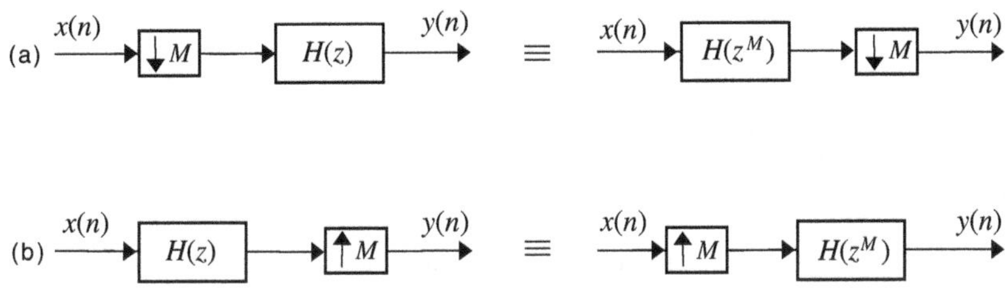

Figure 3.7. The Noble identities. (a) Rule for interchanging decimators with transfer functions, and (b) rule for interchanging expanders with transfer functions.

linear but time-varying building blocks [Vaidyanathan, 1993], they cannot in general be interchanged with other systems in the cascade. However, under some conditions a restricted amount of movement is possible. For example, the two systems shown in Fig. 3.7(a) are equivalent and so are the two systems shown in Fig. 3.7(b). These are known as **noble identities**. For convenience we shall refer to them as the first and the second noble identity, respectively.

Proving the noble identities. The filter $H(z^M)$ on the right in Fig. 3.7(a) has the output $X(z)H(z^M)$. The output is therefore the decimated version

$$Y(z) = [X(z)H(z^M)]_{\downarrow M} = \frac{1}{M}\sum_{k=0}^{M-1} X(z^{1/M}W^k)H\Big((z^{1/M}W^k)^M\Big),$$

where $W = e^{-j2\pi/M}$. By using the fact that $W^M = 1$ we find

$$Y(z) = \frac{H(z)}{M}\sum_{k=0}^{M-1} X(z^{1/M}W^k) = H(z)\Big(X(z)\Big)_{\downarrow M}.$$

The above right hand side is precisely the output of the system on the left side of Fig. 3.7(a). This proves the first identity. For the second identity note that the output of the system on the left side of Fig. 3.7(b) is

$$Y(z) = \Big(X(z)H(z)\Big)_{\uparrow M} = X(z^M)H(z^M),$$

which is precisely the output of the system on the right side of Fig. 3.7(b). This establishes the second identity. $\bigtriangledown \bigtriangledown \bigtriangledown$

Note that the first noble identity can also be written in the form

$$\Big(H(z^M)X(z)\Big)_{\downarrow M} = H(z)\Big(X(z)\Big)_{\downarrow M}. \tag{3.16}$$

As this is valid for any two z-transforms $X(z)$ and $H(z)$, it is useful to write this as

$$\Big(X_1(z^M)X_2(z)\Big)_{\downarrow M} = X_1(z)\Big(X_2(z)\Big)_{\downarrow M}. \tag{3.17}$$

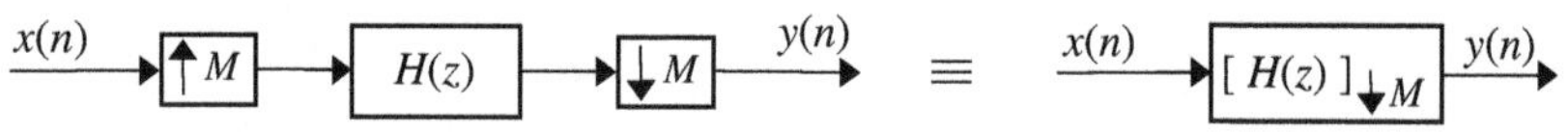

Figure 3.8. The polyphase identity.

Next consider the cascade shown on the left in Fig. 3.8. Here a transfer function $H(z)$ is sandwiched between an expander and a decimator. The output can be written as

$$Y(z) = \Big(H(z)X(z^M)\Big)_{\downarrow M} = X(z)\Big(H(z)\Big)_{\downarrow M}, \tag{3.18}$$

where we have used Eq. (3.17). Thus the system is equivalent to a simple transfer function $[H(z)]_{\downarrow M}$, which represents an LTI system with the decimated impulse response $h(Mn)$. This proves the identity in Fig. 3.8, which is called the **polyphase identity.** Note that it is nothing but the first noble identity with the roles of $X(z)$ and $H(z)$ interchanged.

3.2.4 Alias-free decimation

Consider again the frequency-domain relation (3.13) for a decimator. This was demonstrated in Fig. 3.6. Note that the stretched version $X(e^{j\omega/M})$ in general overlaps with some of the shifted versions. Because of this the original input $x(n)$ cannot in general be recovered from the decimated output $y_d(n) = x(Mn)$. This overlap is commonly referred to as **aliasing**. It is clear that, if the original signal $x(n)$ is **bandlimited** to the region

$$-\pi/M \leq \omega < \pi/M,$$

then there is no aliasing because the shifted versions do not overlap with the stretched version. In this case we can recover the higher rate signal $x(n)$ from its decimated version $y_d(n)$.

A signal $x(n)$ need not be a bandlimited lowpass signal in order to be recoverable from its M-fold decimated version. For example, consider the **band-pass** signal with Fourier transform shown in Fig. 3.9. The total bandwidth is

$2\pi/M$. The reader can verify that if this is decimated then the stretched version $X(e^{j\omega/M})$ does not overlap with the shifted versions $X(e^{j(\omega-2\pi\ell)/M})$. We can therefore recover $x(n)$ from $x(Mn)$ by using an ideal bandpass filter with passband region coinciding with the nonzero part of the original signal.

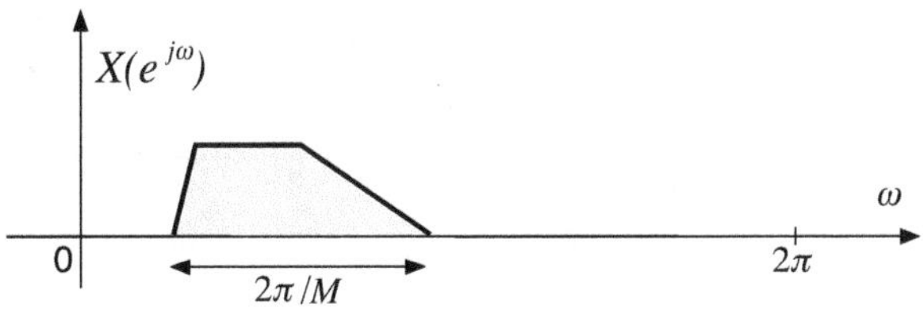

Figure 3.9. A bandpass signal with total bandwidth $2\pi/M$.

3.2.5 Decimation of multiband signals

Consider a signal $x(n)$ with Fourier transform as shown in Fig. 3.10(a). This is a two-band signal. A special case of this arises when $x(n)$ is a **real bandpass signal**, in which case the left half is the mirror image of the right half. What happens if we decimate such signals? The fact that the total width of the signal band is less than $2\pi/M$ leads to the hope that there will be no aliasing when we decimate such signals by M. This, however, is not always the case. Let us decompose $x(n)$ as

$$x(n) = x_L(n) + x_R(n),$$

where $x_L(n)$ is a complex signal whose Fourier transform is the left half shown in Fig. 3.10(a), and similarly $x_R(n)$ represents the right half. The decimated version of the original signal $x(n)$ is simply the sum of the two decimated versions

$$x(Mn) = x_L(Mn) + x_R(Mn).$$

When $x_L(n)$ is decimated by M there is no aliasing.[2] That is, the shifted copies $X_L(e^{j(\omega-2\pi\ell)/M})$ do not overlap for distinct ℓ in $0 \le \ell \le M-1$. The same is true when $x_R(n)$ is decimated by M. However, since $x(Mn)$ is the sum of the two decimated versions we have

$$X(e^{j\omega})\big|_{\downarrow M} = X_L(e^{j\omega})\big|_{\downarrow M} + X_R(e^{j\omega})\big|_{\downarrow M}.$$

It is possible that a shifted copy

$$X_L(e^{j(\omega-2\pi\ell)/M})$$

from the first term overlaps with a shifted copy $X_R(e^{j(\omega-2\pi k)/M})$ from the second term. So a real bandpass signal, and more generally a multiband signal (with

[2] In fact, we could decimate it by $2M$.

possibly more than two bands), cannot in general be decimated by M without aliasing even if the total width of the signal band is less than $2\pi/M$.

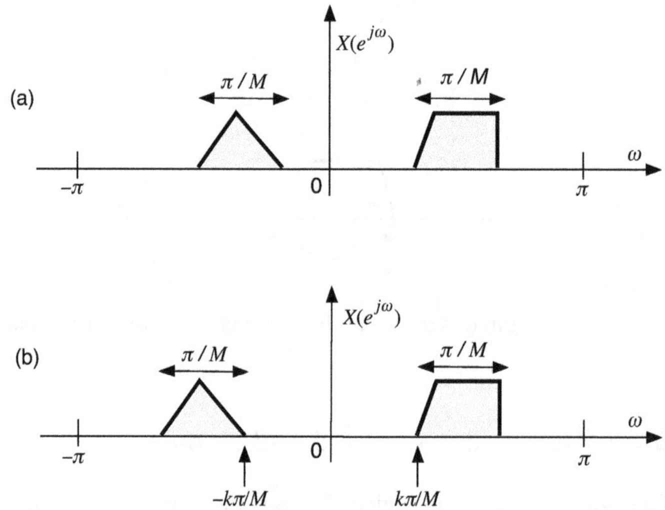

Figure 3.10. Two-band signals with total bandwidth $2\pi/M$. (a) An arbitrary example, and (b) band location which allows decimation by M without aliasing.

Bandpass sampling theorem. For the case of **two-band** signals it is possible to develop conditions under which M-fold decimation can be done without aliasing (i.e., conditions under which the overlap described above does not occur). Suppose the total bandwidth is less than $2\pi/M$ and the bands are located as in Fig. 3.10(b). That is, the band edges are integer multiples of π/M. In this case it can be shown (Problem 3.3) that M-fold decimation causes no aliasing. This is sometimes referred to as the bandpass sampling theorem. $\bigtriangledown\bigtriangledown\bigtriangledown$

3.2.5.A Alias-free(M) regions

More generally, if a region $\mathcal{S}$ of frequencies for which $X(e^{j\omega})$ is nonzero is such that the shifted versions

$$\mathcal{S}+\frac{2\pi\ell}{M},\quad 0\le\ell\le M-1,$$

are disjoint, then the terms $X(e^{j(\omega-(2\pi\ell/M))})$ do not overlap for different values of ℓ. Since a change of variables ($\omega\to\omega/M$) does not affect this property, we conclude that the quantities $X(e^{j(\omega-2\pi\ell)/M})$ in the summation which represents $[X(e^{j\omega})]_{\downarrow M}$ do not overlap. Such a signal $x(n)$ can be decimated by M without causing aliasing. Such a region $\mathcal{S}$ is called an *alias-free(M)* region.

3.3 Decimation filters

From Sec. 3.2.4 we know that if a signal is lowpass with effective bandwidth less than $2\pi/M$ then it can be decimated by M without aliasing. The same is true of bandpass signals and multiband signals of various kinds as summarized earlier. Given an arbitrary signal $x(n)$ we have to pass it through an appropriate discrete-time filter $H(z)$ before it satisfies the alias-free(M) condition. This is shown in Fig. 3.11(a). The preconditioning filter $H(z)$ is called the **decimation filter**. It can be a lowpass, bandpass or highpass filter, as demonstrated in parts (b)–(d) of the figure for $M = 3$. In fact the filter could be a multiband filter with alias-free support as demonstrated in Fig. 3.12 for $M = 3$. The filter $H(e^{j\omega})$ and the shifted versions $H(e^{j(\omega-2\pi/3)})$ and $H(e^{j(\omega-4\pi/3)})$ do not overlap as seen from the figure. In practice, of course, the filter cannot have the ideal responses shown. There can only be *finite attenuation* in the stopbands. But the effect of aliasing can be made arbitrarily small by making this attenuation large.

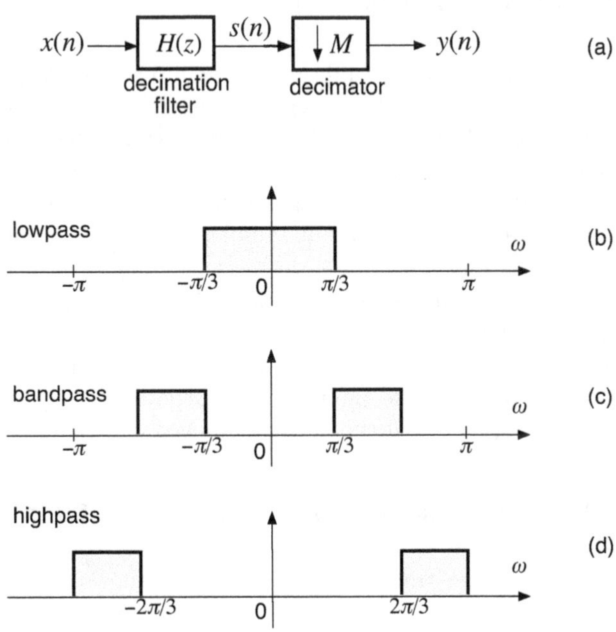

Figure 3.11. (a) The decimation filter, and (b)–(d) examples of alias-free(3) frequency responses for $H(z)$.

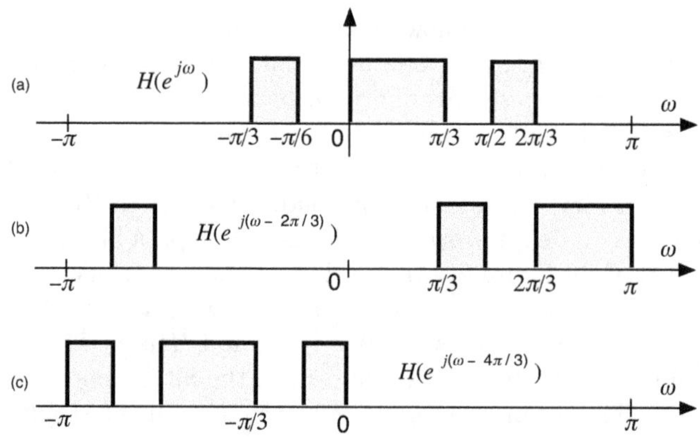

Figure 3.12. (a) A multiband alias-free(3) decimation filter response $H(e^{j\omega})$. (b), (c) The shifted versions $H(e^{j(\omega-2\pi/3)})$ and $H(e^{j(\omega-4\pi/3)})$ as seen in the region $-\pi \leq \omega < \pi$.

3.4 Interpolation filters

An interpolation filter is a discrete-time filter used at the output of an expander as shown in Fig. 3.13(a). This combination of an expander followed by an interpolation filter is said to be a discrete-time **interpolator**. From its definition we know that the expander inserts zero-valued samples between adjacent samples of the input. The purpose of the interpolation filter is to replace these zeros with weighted averages of the input samples, as we shall explain later.

First consider what happens in the frequency domain. Recall that the expander has M squeezed copies (images) of the input Fourier transform $X(e^{j\omega})$. The interpolation filter retains one out of these M copies of the Fourier transform. This is demonstrated in Fig. 3.13 for the case when the filter is lowpass and $M = 3$. Thus the interpolated output $Y(e^{j\omega})$ is nothing but a squeezed version of $X(e^{j\omega})$ with the images removed.

3.4.1 Time-domain view of interpolation filter

Returning to the time domain, Fig. 3.14 shows the various signals involved. The output of the interpolation filter is the convolution of its input $s(n)$ with the impulse response $h(n)$, that is,

$$y(n) = \sum_{k=-\infty}^{\infty} s(k)h(n-k) = \sum_{k=-\infty}^{\infty} s(kM)h(n-kM),$$

where the second equality follows because $s(k) = 0$ unless k is a multiple of M.

Since $s(kM) = x(k)$ by definition, we therefore have

$$y(n) = \sum_{k=-\infty}^{\infty} x(k)h(n - kM). \tag{3.19}$$

In our example, since $H(e^{j\omega})$ is ideal lowpass, the impulse response $h(n)$ is the sinc function

$$h(n) = \frac{\sin(\pi n/M)}{\pi n},$$

where $M = 3$ in our case. The convolution $(s * h)(n)$ results in the replacement of the zero-valued samples of $s(n)$ with **interpolated values** as demonstrated in the figure. The result $y(n)$ has a smooth lowpass appearance because of the filtering.

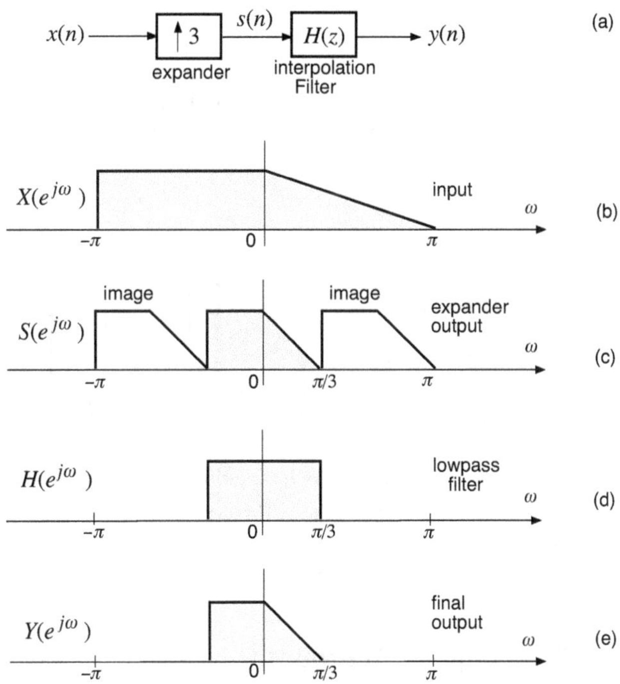

Figure 3.13. (a) The discrete-time interpolator, and (b)–(e) explanation of its operation in the frequency domain for $M = 3$.

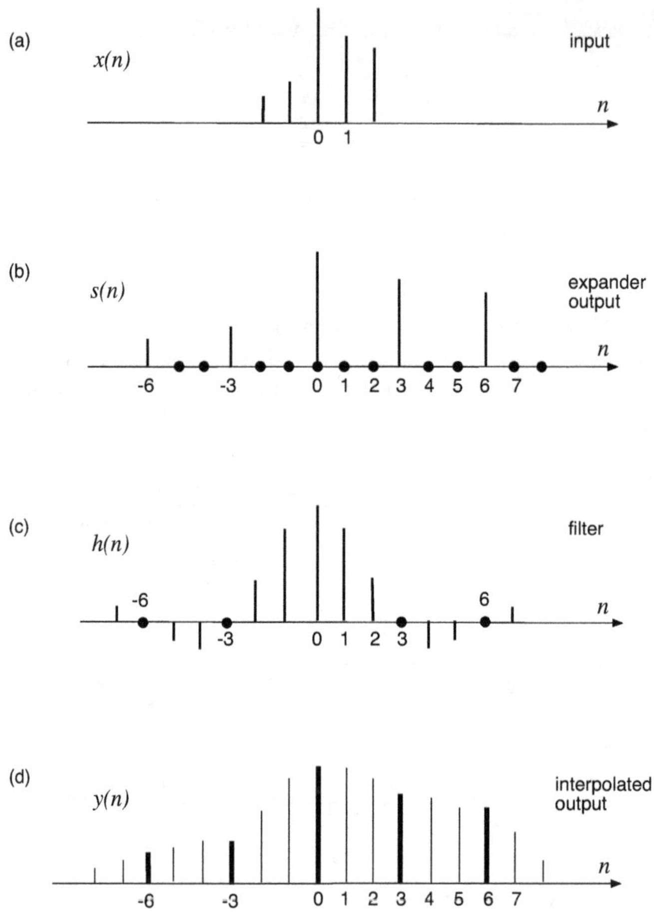

Figure 3.14. Operation of the interpolation filter in the time domain. (a) Input signal, (b) output of the expander, (c) impulse response of the interpolation filter, and (d) the final interpolated output.

3.4.2 The Nyquist(M) property

Note that the sinc impulse response satisfies the property

$$h(Mn) = \delta(n) \qquad \text{(Nyquist}(M)\text{ property).} \tag{3.20}$$

That is, it has regular zero crossings at nonzero multiples of M, and moreover $h(0) = 1$. A filter $H(z)$ satisfying this property is called a Nyquist(M) filter or just a **Nyquist filter**. We will show that this property ensures that the original samples $x(n)$ of the input are retained by the interpolation filter without distortion, as shown by the thick lines in Fig. 3.14(d). More precisely, we will show

$$y(Mn) = x(n).$$

Thus an M-fold decimated version of the interpolated signal $y(n)$ returns the original signal $x(n)$ without distortion.

Proof. The fact that the Nyquist(M) condition ensures $y(Mn) = x(n)$ can be verified directly from Eq. (3.19). Thus, setting $n = \ell M$, this equation can be rewritten as

$$y(\ell M) = \sum_{k=-\infty}^{\infty} x(k)h(\ell M - kM) = \sum_{k=-\infty}^{\infty} x(k)g(\ell - k),$$

where $g(n) \stackrel{\Delta}{=} h(Mn)$. In view of the Nyquist(M) property we have $g(n) = \delta(n)$ so that $y(\ell M) = x(\ell)$ indeed. $\bigtriangledown \bigtriangledown \bigtriangledown$

We have proved that $y(nM) = (x * g)(n)$, where $*$ represents convolution. So it is clear that $y(nM) = x(n)$ for all inputs $x(n)$ **if and only if** the Nyquist condition $h(Mn) = \delta(n)$ is satisfied. In practice the interpolation filter is a realizable system with *finite attenuation* in stopbands. The images are therefore not completely suppressed. It is, however, still possible to satisfy the Nyquist property exactly, thus ensuring that original samples $x(n)$ are preserved.

3.4.3 Interpolation filters in communications

In all the above discussions the interpolation filter is assumed lowpass. It is also possible to have *bandpass or highpass interpolation filters.* These will be especially useful in applications which involve filter banks (Sec. 3.8). Figure 3.15 shows an interpolation filter used as a transmitting filter in a communication system. When the filter has lowpass characteristics, the interpolated signal is as in part (b). The filter can also be a bandpass filter as shown in part (c). Part (d) shows an example of the interpolating impulse response $f(n)$ when $F(z)$ is a real-coefficient bandpass filter as in part (e).

We will see in later chapters that the interpolation filters in communication applications are often not required to be good approximations to ideal bandpass filters. Other considerations come into play. Simple FIR filters with only about 13 dB stopband attenuation are sometimes used.

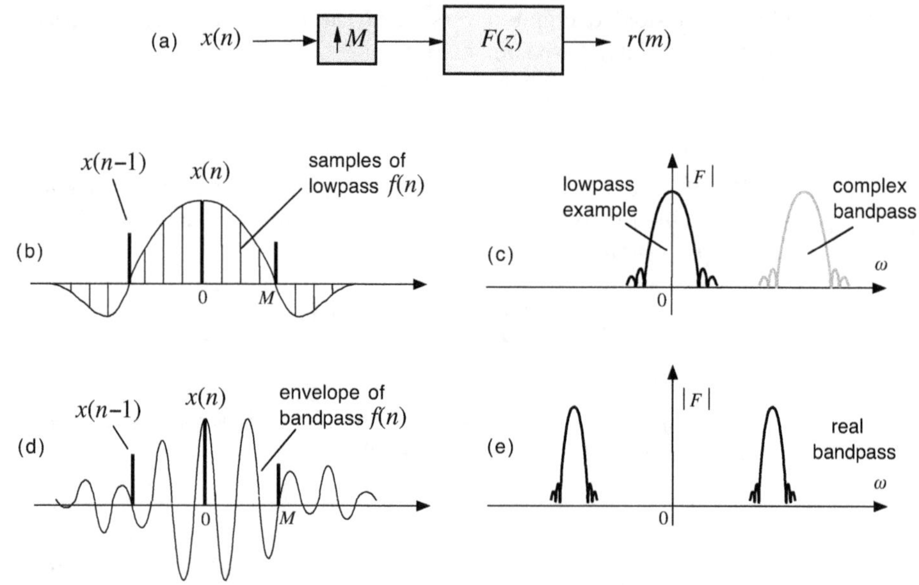

Figure 3.15. (a) An interpolation filter used as a transmitting filter in a communication system. (b) Example where $F(z)$ is lowpass; (c) examples of lowpass and bandpass interpolation filters; (d) example where $F(z)$ is a real-coefficient bandpass filter; and (e) response of a real bandpass filter.

3.5 Blocking and unblocking

The multirate building blocks introduced in Sec. 3.2 can be used to describe schematically two basic operations called blocking and unblocking. Thus consider Fig. 3.16(a). Here the vector $\mathbf{x}(n)$ is made from successive samples of $x(n)$. Note also that the vectors $\mathbf{x}(n)$ and $\mathbf{x}(n+1)$ have no overlapping samples. Thus the samples of the vector $\mathbf{x}(n)$ represent successive nonoverlapping blocks of the input stream $x(n)$ (Fig. 3.16(c)). We therefore say that $\mathbf{x}(n)$ is a P-blocked version or just a **blocked version** of $x(n)$. The inverse operation is shown in Fig. 3.16(b). Here a set of P signals is **interleaved** by the combination of expanders and the delay chain. We say that $x(n)$ is the **unblocked** version of the set $\{x_k(n)\}$. We also say that the signals $x_k(n)$ are **time-multiplexed** to get $x(n)$. If the original samples $x(n)$ are spaced apart by τ seconds, then the samples of $\mathbf{x}(n)$ are spaced apart by $P\tau$ seconds, as shown in the figure. Blocking and unblocking can also be regarded as **serial-to-parallel conversion** of data and vice versa.

If the two systems in parts (a) and (b) are connected in cascade (in either order) it is clear that we get identity. This is shown in Figs. 3.17(a) and 3.17(b). This *trivial multirate identity* is surprisingly useful as we shall see.

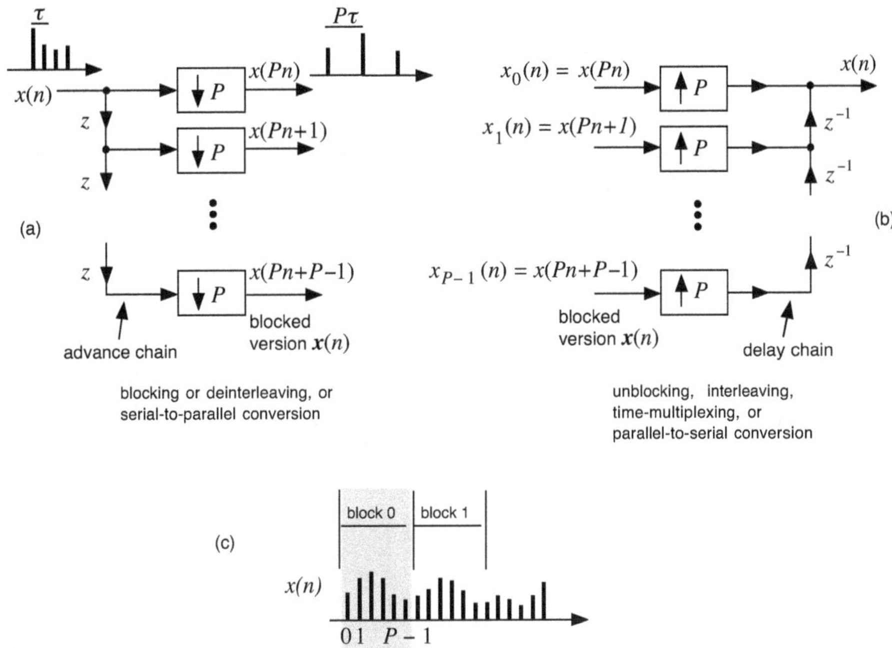

Figure 3.16. (a) Blocking a scalar signal $x(n)$, (b) the unblocking or interleaving structure, and (c) blocking demonstrated in the time domain.

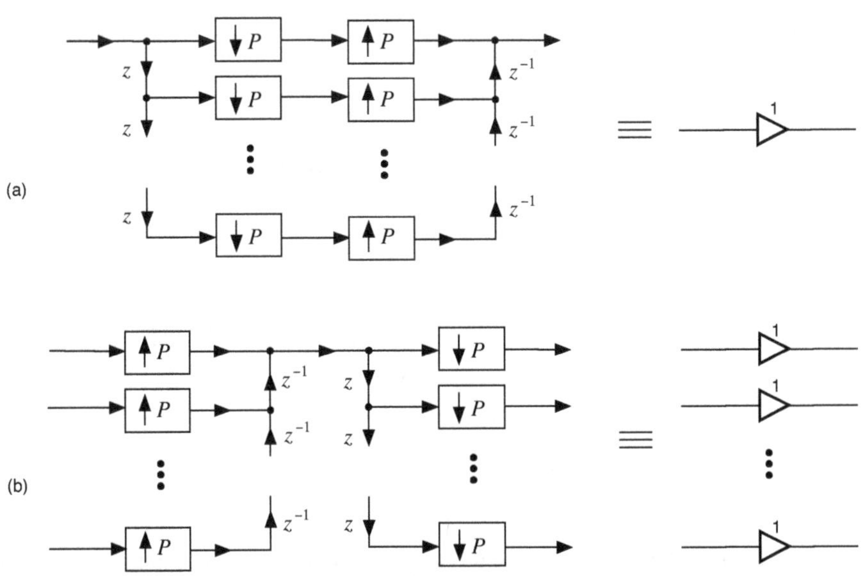

Figure 3.17. Two systems that are equivalent to identity systems. (a) Blocking followed by unblocking, and (b) unblocking followed by blocking.

3.6 Parsing a scalar signal into a vector signal

In Sec. 3.5 we explained how a signal $x(n)$ can be converted into a vector signal $\mathbf{x}(n)$ by a process called blocking. There is another way to convert a scalar sequence into a vector sequence, which is commonly used in some communication systems such as the DMT system (Chap. 7). This is called **parsing** and is usually applied on a binary sequence.

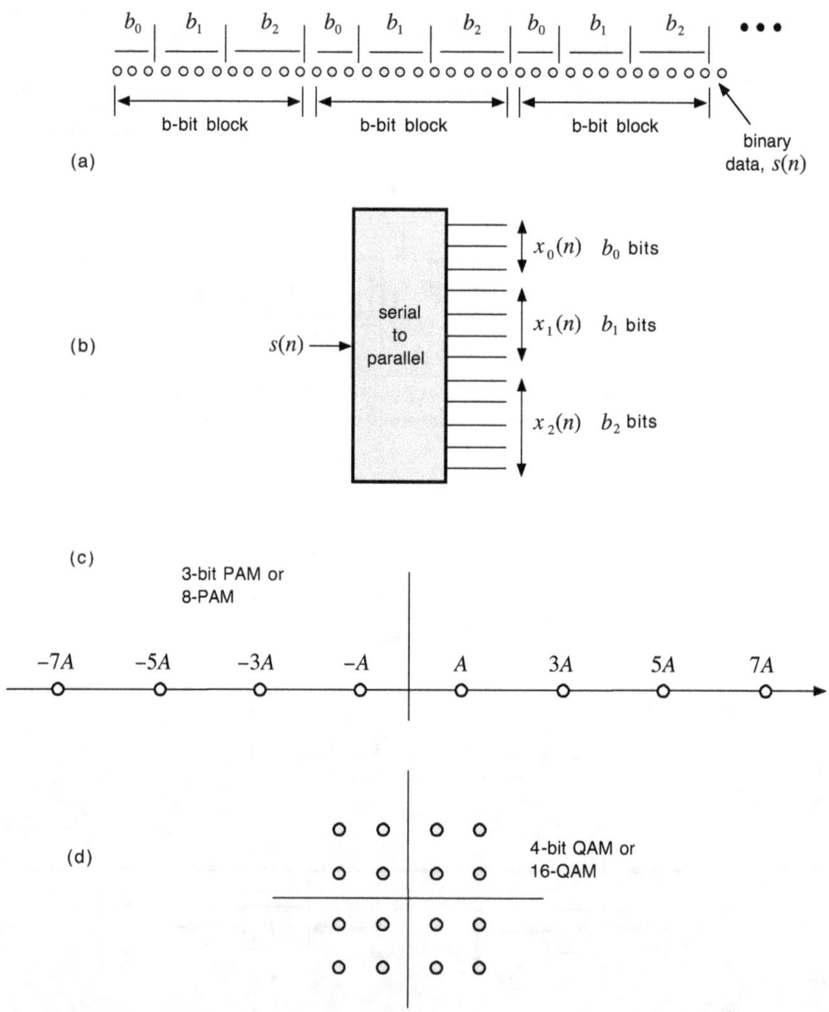

Figure 3.18. Parsing a binary sequence to create M constellations. (a) A binary data stream divided into blocks of size b; (b) each block divided into M groups with the kth group converted into a b_k-bit sequence $x_k(n)$; (c), (d) example of PAM and QAM constellations.

Thus consider Fig. 3.18(a), which shows a binary sequence. This is first divided into nonoverlapping blocks of size b. Then the b bits are divided into M groups with b_k bits in the kth group. Thus

$$b = \sum_{k=0}^{M-1} b_k.$$

The kth group of b_k bits is then converted into a signal in a standard modulation format such as a PAM or a QAM constellation (Sec. 2.2). For example, if $b_1 = 3$ then it can be converted into a codeword in the 3-bit PAM constellation shown in Fig. 3.18(c). Similarly, if $b_1 = 4$, it can be converted into a 4-bit PAM or a 4-bit QAM signal. Thus the set of M signals in Fig. 3.18(b) are finite-bit signals belonging to standard modulation formats.

3.7 Decimation and interpolation in polyphase form

An important application of the polyphase decomposition is in the implementation of decimation and interpolation filters. We describe this in this section.

3.7.1 Decimation and interpolation filters

First consider the decimation filter reproduced in Fig. 3.19(a). Notice that one out of every M output samples is discarded. It is therefore wasteful to compute all the outputs of $H(z)$. The structure for $H(z)$ can be rearranged in such a way that none of the discarded samples is computed in the first place. For this we use the polyphase decomposition. By using the Type 2 polyphase decomposition

$$H(z) = \sum_{k=0}^{M-1} z^k E_k(z^M)$$

we can redraw Fig. 3.19(a) to get Fig. 3.19(b). The first noble identity (Fig. 3.7) can now be applied to redraw this in the form of Fig. 3.19(c). This is called the **polyphase implementation** of the decimation filter. Note that none of the computed results is discarded, and furthermore all the computations (filtering with $E_k(z)$) are implemented at the *lower* rate (i.e., after decimation). The polyphase implementation is therefore computationally very efficient. Note from the figure that the *Type 2* polyphase components $E_k(z)$ of the filter are operating individually on the *Type 1* polyphase components $x_k(n)$ of the input $x(n)$ (compare Fig. 3.3(a)) before the results are added to produce $y(n)$. Equivalently the polyphase components $E_k(z)$ operate on the blocked version $\mathbf{x}(n)$ of the input $x(n)$.

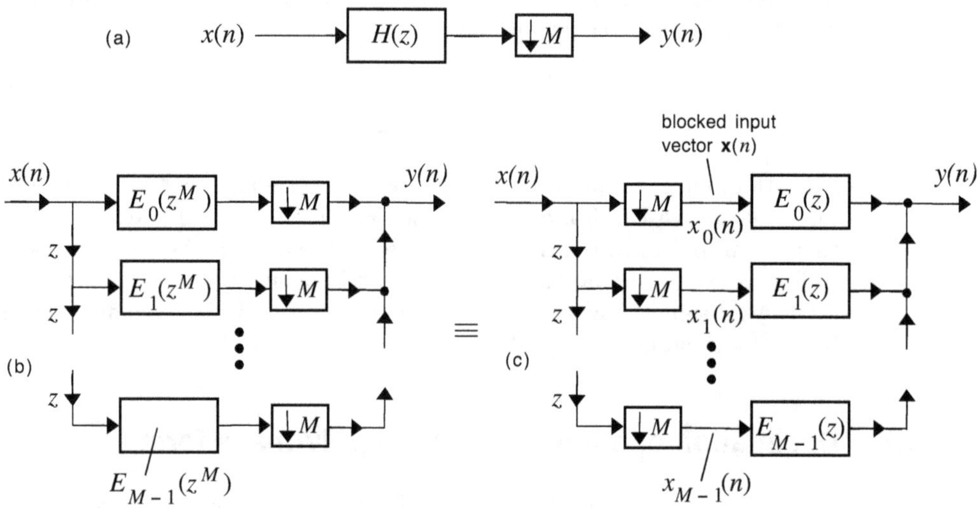

Figure 3.19. (a) A decimation filter, (b) Type 2 polyphase representation, and (c) simplification with the help of noble identity 1.

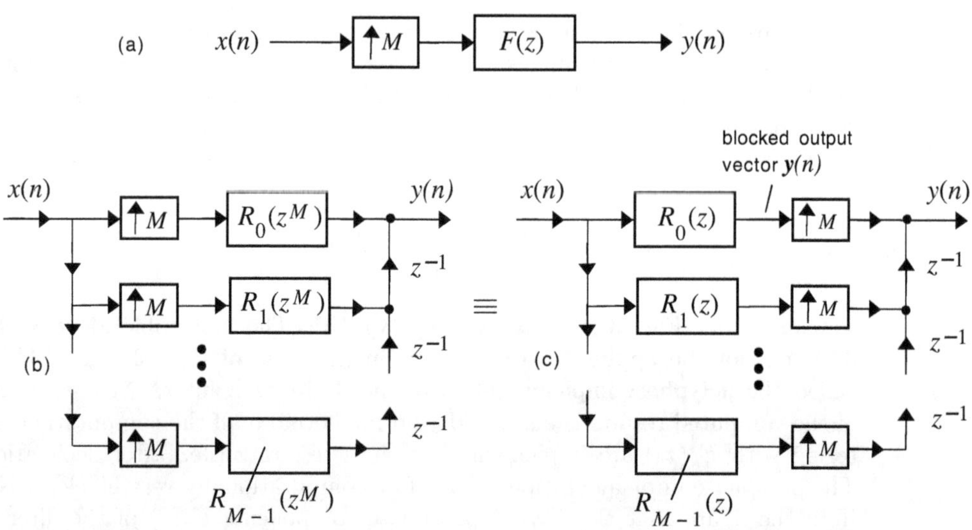

Figure 3.20. (a) An interpolation filter, (b) Type 1 polyphase representation, and (c) simplification with the help of the second noble identity.

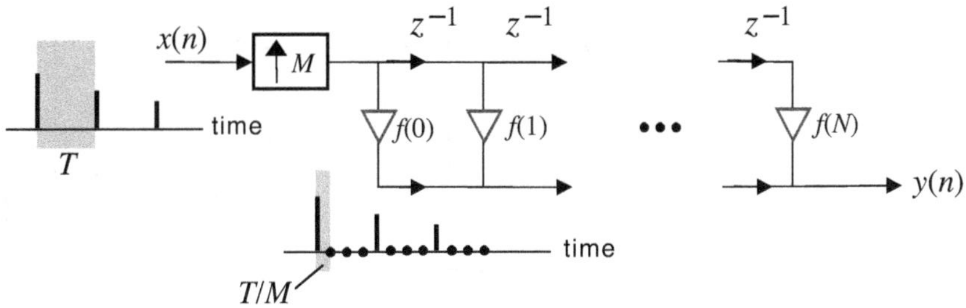

Figure 3.21. An FIR interpolation filter implemented directly.

Next consider the interpolation filter shown in Fig. 3.20(a). This is demonstrated in Fig. 3.21 for the FIR case; in this implementation the zero-valued samples inserted by the expander enter the multipliers $f(k)$ and result in wasted computation. Furthermore the multipliers operate at the higher rate (they only have T/M seconds per computation), where T is the separation between input samples). To avoid such inefficient computations we represent the filter in its Type 1 polyphase form:

$$F(z) = \sum_{k=0}^{M-1} z^{-k} R_k(z^M).$$

Then the system can be redrawn as in Fig. 3.20(b). By using the second noble identity (Fig. 3.7) this can be simplified to the form shown in Fig. 3.20(c). In this implementation there are no zero-valued samples entering the filters $R_k(z)$. Furthermore all the computations are taking place at the lower rate (i.e., before the expanders). Observe that the output $y(n)$ is the *interleaved version* of the outputs of $R_k(z)$. Thus each Type-1 polyphase component $R_k(z)$ of the filter computes the corresponding Type 1 polyphase component of the output $y(n)$, and these results are interleaved to get $y(n)$. In order to maximize the efficiency of computation in multirate systems we *move all decimators as far to the left as possible*, and *move all expanders as far to the right as possible.* In this way, the computational building blocks operate at the lowest possible rate.

3.7.2 Transmitting filter banks or synthesis filter banks

We next consider the system shown in Fig. 3.22(a). In this system, there are M interpolation filters with inputs $s_k(n)$, and their outputs are added. The interpolation ratio P is in general different from M (usually $P > M$ in communication applications). The system is called a **synthesis filter bank** because it combines a set of signals $s_k(n)$ into a single signal $x(n)$. In digital communications, such a filter bank is used as a **transmitting filter bank** or transmitter filter bank. This system will be described in greater detail in Sec. 3.8. We now show how this filter bank can be expressed in polyphase form.

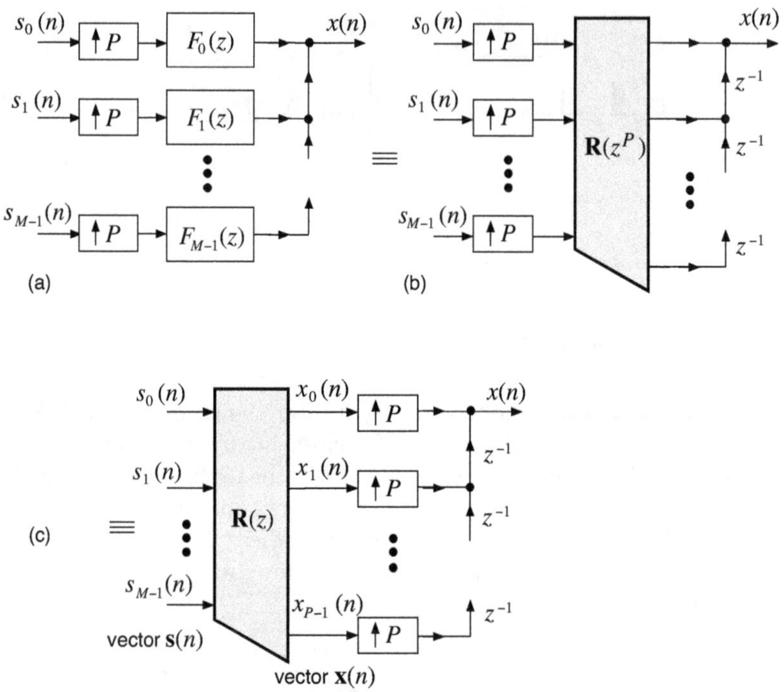

Figure 3.22. (a) A transmitting filter bank, also known as a synthesis filter bank, (b) polyphase version, and (c) simplification with the use of the second noble identity.

Assume that each filter $F_m(z)$ is expressed in Type 1 polyphase form:

$$F_m(z) = \sum_{k=0}^{P-1} z^{-k} R_{km}(z^P). \tag{3.21}$$

Expressing the bank of M filters as a row vector we therefore get

$$[F_0(z) \quad F_1(z) \quad \dots \quad F_{M-1}(z)] = [1 \quad z^{-1} \quad \dots \quad z^{-(P-1)}]\,\mathbf{R}(z^P), \tag{3.22}$$

where $\mathbf{R}(z)$ is the $P \times M$ matrix of polyphase components:

$$\mathbf{R}(z) = \begin{bmatrix} R_{00}(z) & R_{01}(z) & \dots & R_{0,M-1}(z) \\ R_{10}(z) & R_{11}(z) & \dots & R_{1,M-1}(z) \\ \vdots & \vdots & \ddots & \vdots \\ R_{P-1,0}(z) & R_{P-1,1}(z) & \dots & R_{P-1,M-1}(z) \end{bmatrix}. \tag{3.23}$$

We say that $\mathbf{R}(z)$ is the **polyphase matrix** of the transmitter filter bank. Equation (3.22) is said to be the polyphase representation of the filter bank, and can be drawn as in Fig. 3.22(b). With the help of the second noble identity, this polyphase representation can be redrawn as shown in Fig. 3.22(c). In this representation the input vector $\mathbf{s}(n)$ is filtered by the MIMO system $\mathbf{R}(z)$ to produce the output vector $\mathbf{x}(n)$, whose components are interleaved (unblocked) to produce the scalar output $x(n)$.

3.7.3 Receiving filter banks or analysis filter banks

Now consider the system shown in Fig. 3.23(a). In this system, there are M decimation filters with a common input $x(n)$. The decimation ratio P is in general different from M (usually $P > M$ in communication receivers and $P \leq M$ in data compression analysis banks). Such a system is called an **analysis filter bank** because it splits a signal $x(n)$ into M components. In digital communication systems, such a filter bank is used as a **receiving filter bank** or receiver filter bank, and this will be described in greater detail in Sec. 3.8. We now show how this filter bank can be expressed in polyphase form.

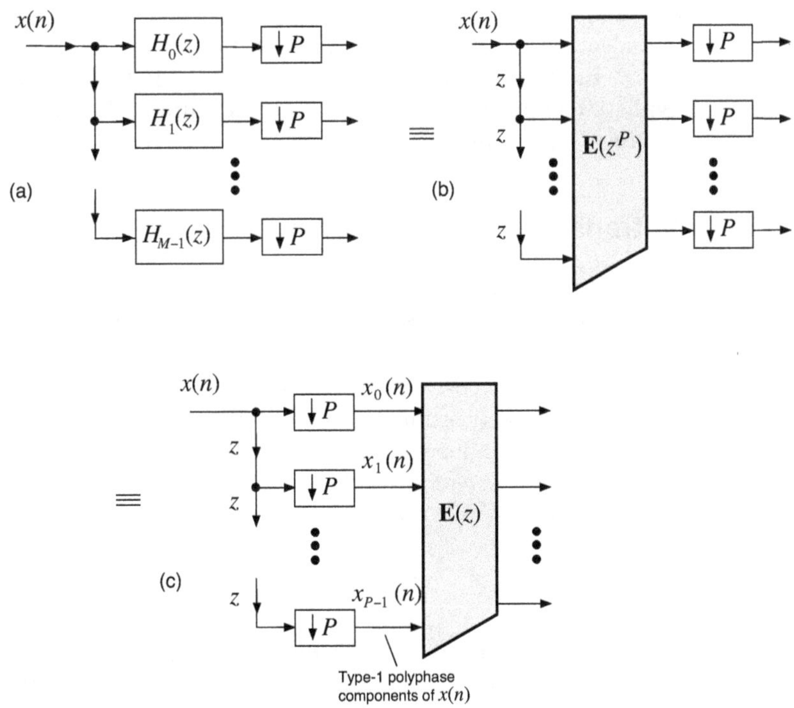

Figure 3.23. (a) A receiving filter bank, also known as analysis filter bank, (b) Type-2 polyphase version, and (c) simplification with the use of the first noble identity.

Assume that each filter $H_m(z)$ is expressed in Type 2 polyphase form:

$$H_m(z) = \sum_{k=0}^{P-1} z^k E_{mk}(z^P). \tag{3.24}$$

Expressing the bank of M filters as a column vector we therefore get

$$\begin{bmatrix} H_0(z) \\ H_1(z) \\ \vdots \\ H_{M-1}(z) \end{bmatrix} = \mathbf{E}(z^P) \begin{bmatrix} 1 \\ z \\ \vdots \\ z^{(P-1)} \end{bmatrix}, \tag{3.25}$$

where $\mathbf{E}(z)$ is the $M \times P$ matrix of polyphase components:

$$\mathbf{E}(z) = \begin{bmatrix} E_{00}(z) & E_{01}(z) & \dots & E_{0,P-1}(z) \\ E_{10}(z) & E_{11}(z) & \dots & R_{1,P-1}(z) \\ \vdots & \vdots & \ddots & \vdots \\ E_{M-1,0}(z) & E_{M-1,1}(z) & \dots & E_{M-1,P-1}(z) \end{bmatrix}. \tag{3.26}$$

We say that $\mathbf{E}(z)$ is the **polyphase matrix** of the receiving filter bank (or analysis filter bank). Equation (3.25) is said to be the polyphase representation of the filter bank, and can be represented as in Fig. 3.23(b). With the help of the first noble identity, this polyphase representation can be redrawn as shown in Fig. 3.23(c). In this representation the input $x(n)$ is first blocked into its Type 1 polyphase components $x_k(n)$ and these are filtered by the MIMO system $\mathbf{E}(z)$ to produce the output vector.

3.8 The transmultiplexer system

Consider Fig. 3.24, which shows a system called the transmultiplexer or **filter bank transceiver**. This system and its variations will be central to many of our discussions in this book. The signals $s_k(n)$ are symbol streams (such as PAM or QAM signals; see Sec. 2.2).[3] These could be symbols generated by different users who wish to transmit messages over a communication channel. Or they could be different independent parts of the signals generated by one user (Sec. 3.6). The symbol streams $s_k(n)$ are passed through the interpolation filters or transmitting filters $F_k(z)$ to produce the signals

$$x_k(n) = \sum_i s_k(i) f_k(n - iM). \tag{3.27}$$

The filters $F_k(z)$ are also called **pulse-shaping filters** because they take each sample of $s_k(n)$ and "put a pulse $f_k(n)$ around it" (see Fig. 3.15). The sum $x(n)$ of the signals $x_k(n)$ is then transmitted over a common channel.

---

[3] Even though we have shown only the discrete-time signals $s_k(n)$ and $\widehat{s}_k(n)$, in practice there will be D/C and C/D converters, as described in Sec. 1.3, to interface with real continuous-time channels.

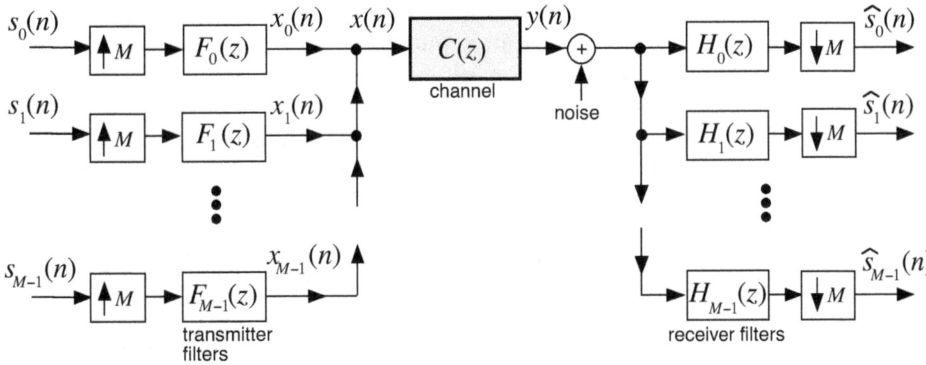

Figure 3.24. The M-user transmultiplexer or transceiver system.

The channel is described by a linear time invariant filter $C(z)$ followed by additive noise. At the receiver end, the filters $H_k(z)$ have the task of separating the signals and reducing them to the original rates by decimation. The transmitted symbols $s_k(n)$ are then identified from the signals $\widehat{s}_k(n)$, which in general are distorted versions of the symbols $s_k(n)$ because of the combined effects of channel, noise, and the filters. The general goal is to identify the symbols $s_k(n)$ accurately in the presence of these distortions. The choice of filters $\{F_k(z), H_m(z)\}$ depends on the details of the specific instances of application. We will discuss details of this throughout the book.

3.8.1 The multiplexing operation

We can always regard the set of signals $s_k(n)$ as the polyphase components of a hypothetical signal $s(n)$, as demonstrated in Fig. 3.25(a) for $M = 3$. That is, we can regard $s_k(n)$ as the components of a **time-multiplexed** signal $s(n)$. On the other hand, with the filters $F_k(z)$ chosen as a contiguous set of bandpass filters (Fig. 3.25(b)), we can regard the transmitted signal $x(n)$ as a **frequency-multiplexed** version of the set of signals $s_k(n)$. To see this consider an arbitrary $s_1(n)$ with a Fourier transform as demonstrated in Fig. 3.25(c). The expander squeezes this Fourier transform by a factor of M, and the interpolation filter $F_1(z)$ retains only one copy (Fig. 3.25(d)). Thus the output $x(n)$ of the synthesis bank has a Fourier transform as shown in Fig. 3.25(e); it is just a concatenation of squeezed versions of $S_k(e^{j\omega})$ for $0 \leq k \leq M-1$. Thus $x(n)$ is the frequency-multiplexed version of $\{s_k(n)\}$.

Historical remarks. The earliest applications of the transmultiplexer were indeed conversions from time-domain multiplexing (TDM) to frequency-domain multiplexing (FDM) of telephone signals prior to their transmission [Bellanger, 1982].

The received signal is then converted back from FDM to TDM format. So transmultiplexers are often referred to as TDM to FDM (and vice versa) converters. The same structure is used today in many different configurations. Some examples are discrete multitone systems (DMT), and filter bank precoders (Chap. 7). In fact the transmitting filter bank has been frequently referred to as the *filter bank precoder* in more recent literature. In these examples a variation called the redundant transmultiplexer is used and is described in Sec. 3.8.2.

The reader interested in early literature on transmultiplexers will enjoy reading Bellanger and Daguet [1974], Narasimha and Peterson [1979], Ansari and Liu [1982], and Bellanger [1982]. The study of transmultiplexers in the context of formal filter bank theory was pioneered in Vetterli [1986] and further developed in Koilpillai, Nguyen, and Vaidyanathan [1991]. Also see Akansu *et al.* [1998], Vaidyanathan [2001], Vaidyanathan, *et al.* [2002], and Vaidyanathan and Vrcelj [2004]. The advantages of redundant filter banks have been established by more recent work of many authors. We will elaborate on this in many of the chapters to follow. $\bigtriangledown\bigtriangledown\bigtriangledown$

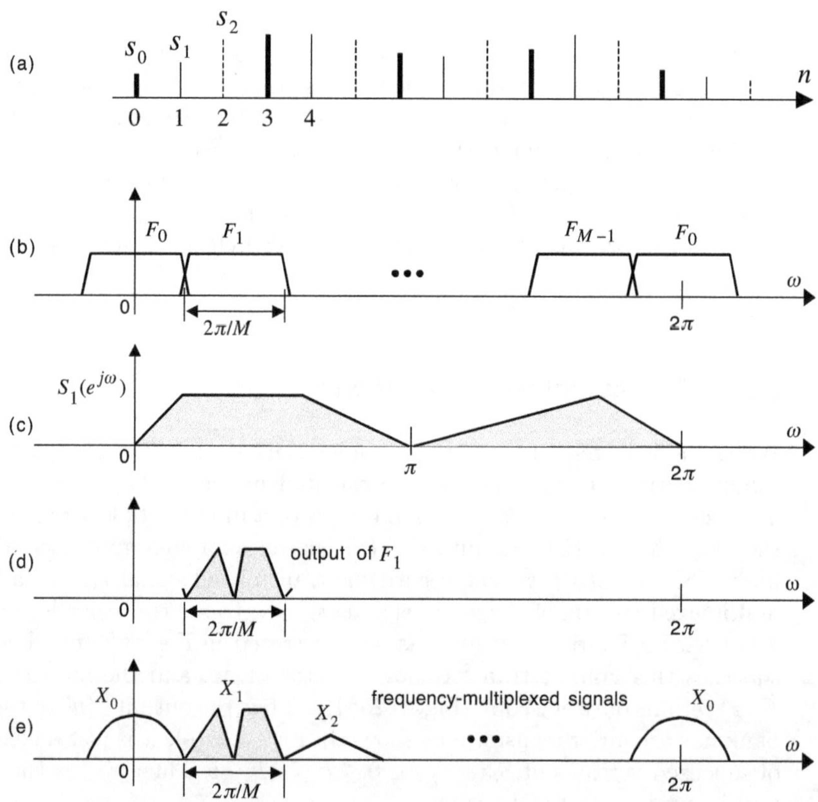

Figure 3.25. The frequency-multiplexing operation performed by the transmultiplexer. (a) Original time-domain-multiplexed (TDM) signal; (b) the filter bank response; (c) Fourier transform of a signal $s_1(n)$; (d) output of the corresponding filter $F_1(z)$; (e) frequency-multiplexed (FDM) signal $X(e^{j\omega})$, which is input to the channel.

3.8.2 Redundancy in transmultiplexers

The reader will notice that, if the filters $F_k(z)$ in Fig. 3.25(b) are not perfectly bandlimited to a width of $2\pi/M$, then the Fourier transforms $X_k(e^{j\omega})$ will contain a little bit of leakage because of the multiple images created by the expanders. Thus the spectra $X_k(e^{j\omega})$ overlap for adjacent k, and cannot in general be separated perfectly. A simple way to solve this problem is to introduce redundancy into the symbol stream. To understand how this can be done consider Fig. 3.26, where the expander $\uparrow P$ is allowed to be different from M. If $P > M$ we say that the system has redundancy. The spectra $S_k(e^{j\omega})$ are now squeezed by a larger amount P, and the M filters $F_k(e^{j\omega})$ have a smaller bandwidth $2\pi/P$, which allows a gap between the filters called **guard bands**; see Fig. 3.27(a). The outputs of the filters in this case are also sketched in Fig. 3.27(b). Note that there are M filters with uniformly spaced center frequencies, and there are $P > M$ copies of spectra created by each expander. The portion that is filtered by $F_k(e^{j\omega})$ therefore comes from more than one copy, but it has all the information about $X_k(e^{j\omega})$. There is a circular shift within the band, as demonstrated for $X_1(e^{j\omega})$ in Fig. 3.27(b) (compare with Fig. 3.25).

Transmultiplexers with $P = M$ are called **minimal** transmultiplexers, and those with $P > M$ are called **redundant** transmultiplexers. With redundancy, the signals $s_k(n)$ are easier to separate at the receiver. In practice the filters are not ideal bandlimiting filters; in fact filters of fairly small orders are used. In such systems the introduction of redundancy is still useful but it takes various special forms such as **zero padding** and **cyclic prefixing**, as we shall explain in Chap. 7. Such methods allow us to eliminate the interference between $s_k(n)$ for different k. It is possible to avoid redundancy altogether in a method based on the theory of **biorthogonal** or perfect reconstruction filter banks. As we shall explain later such a method is not very practical, so a certain amount of redundancy is almost always desirable. Redundancy is used in many practical systems, such as orthogonal frequency division multiplexing (OFDM) systems and discrete multitone (DMT) systems to be described later. Redundancy is useful not only in channel equalization, but also in channel identification [Scaglione, Giannakis, and Barbarossa, 1999a, 1999b].

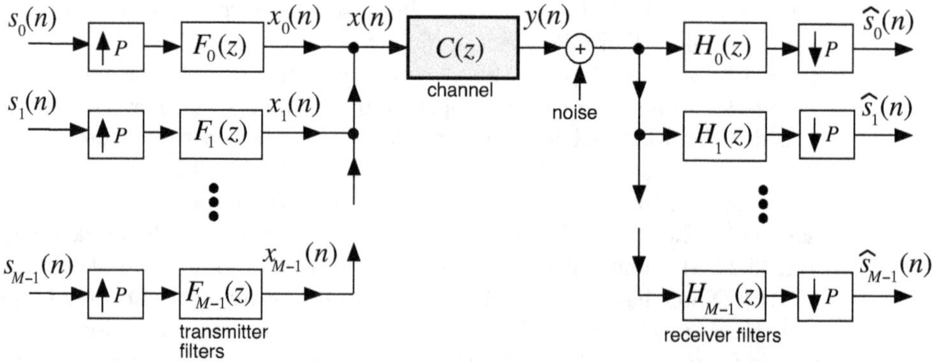

Figure 3.26. The M-user transmultiplexer system with redundancy ($P > M$).

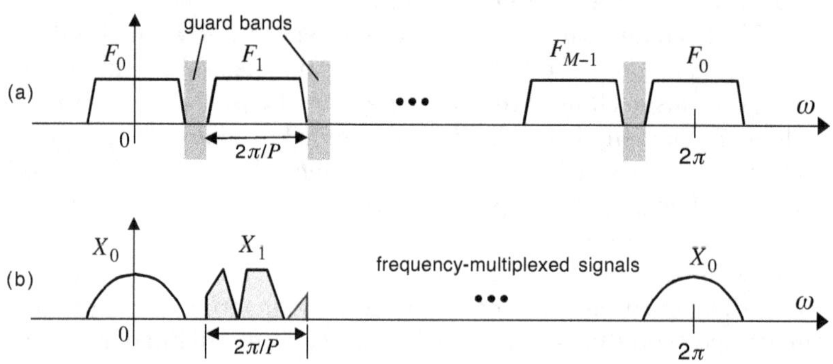

Figure 3.27. (a) Examples of frequency responses of filters in the transmultiplexer with redundancy, and (b) examples of outputs of the filters. See text.

3.8.3 Types of distortion in transmultiplexers

The received signals $\widehat{s}_k(n)$ in general are different from $s_k(n)$ for several reasons.

1. First, there is *interchannel* interference. This means that $\widehat{s}_k(n)$ is affected not only by $s_k(n)$, but also by $s_m(n), m \neq k$. This is also sometimes called multiuser interference, or **MUI**. The term "multiuser" interference arises from the fact that historically the transmultiplexer was used to multiplex several users in TDM format into an FDM format [Bellanger and Daguet, 1974]. Today, the same mathematical structure can be used to describe a wider range of communication systems as we shall see.

2. Second, for any fixed k, the signal $\widehat{s}_k(n)$ depends also on $s_k(n-m)$ for $m \neq 0$. This is due to the effect of filtering created by the channel and the various filters. This is called intersymbol interference, or **ISI**, and can also be referred to as *intrachannel* interference.

3. Finally there is additive channel noise.

The task at the receiver is to minimize the effects of these distortions so that the transmitted symbols $s_k(n)$ can be detected from $\widehat{s}_k(n)$ with acceptably low probabilities of error. We will see in later chapters that this task is easier when there is redundancy (i.e., $P > M$). We shall present a quantitative study of ISI and MUI in the following section.

3.9 Analysis of the transmultiplexer system

In this section we show that the transmultiplexer can be redrawn in a form that does not require the use of multirate building blocks. Such a redrawing represents the channel $C(z)$ in the form of a $P \times P$ transfer matrix with a special form called the **pseudocirculant** form. This mathematical form is very useful in the theoretical study of communication systems.

3.9.1 Blocked version of the transceiver system

Consider the path from $s_m(n)$ to $\widehat{s}_k(n)$. In the absence of channel noise this path is simply a transfer function sandwiched between an expander and a decimator as shown in Fig. 3.28(a). By using the polyphase identity (Fig. 3.8) we see that this path has the transfer function

$$T_{km}(z) = [H_k(z)C(z)F_m(z)]_{\downarrow P}. \tag{3.28}$$

Thus, ignoring the channel noise for a moment, the transmultiplexer can be described as a simple transfer matrix $\mathbf{T}(z)$. The nondiagonal elements of the matrix, namely $T_{km}(z), k \neq m$, represent the path from the mth to the kth user. If the matrix is diagonal, that is, if we have

$$T_{km}(z) = 0, \quad k \neq m, \tag{3.29}$$

then there is *no multiuser interference.* With MUI eliminated each diagonal element $T_{kk}(z)$ represents the transfer function from $s_k(n)$ to $\widehat{s}_k(n)$. If

$$T_{kk}(z) = 1 \tag{3.30}$$

(more generally cz^{-N} for some N and $c \neq 0$), then $\widehat{s}_k(n)$ is affected by only a single sample of $s_k(n)$ (e.g., $s_k(n-1)$ does not affect it). That is, there is *no intersymbol interference* or ISI in the kth path.

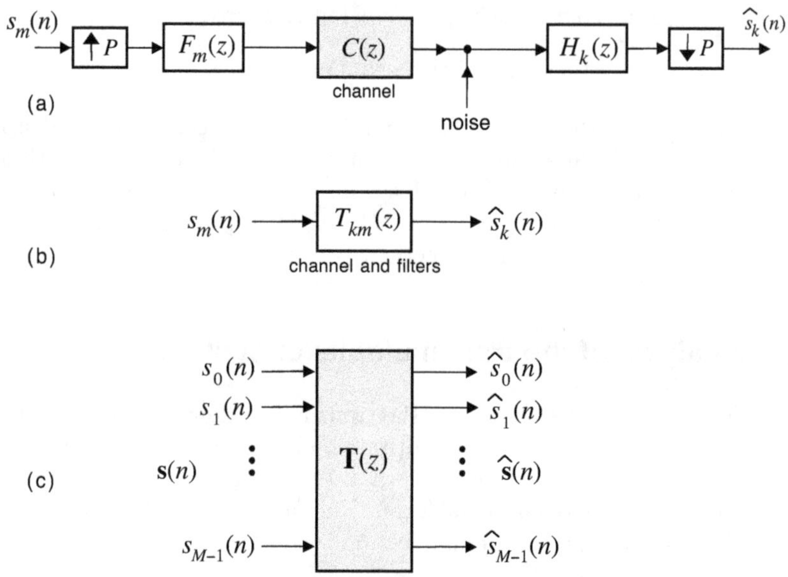

Figure 3.28. (a) The path from the mth input to the kth output in the transmultiplexer; (b) equivalent system with noise ignored; (c) matrix representation of the transmultiplexer in part (a), with noise ignored.

Thus, *nonzero nondiagonal* elements of $\mathbf{T}(z)$ represent MUI, whereas *nonconstant diagonal* elements represent ISI. If the system is free from both MUI and ISI then

$$\mathbf{T}(z) = \mathbf{I}, \tag{3.31}$$

and we see that there is **perfect symbol recovery** or **PR property**

$$\widehat{s}_k(n) = s_k(n) \tag{3.32}$$

in the absence of noise. More generally in a PR transmultiplexer, $\mathbf{T}(z)$ is allowed to be a diagonal matrix with diagonal elements $c_k z^{-N_k}$, so that

$$\widehat{s}_k(n) = c_k s_k(n - N_k). \tag{3.33}$$

In such a transmultiplexer the channel noise is the only source of imperfection which makes $\widehat{s}_k(n)$ different from $s_k(n)$. Noise was ignored in the preceding discussion, but we shall handle it with great care in later chapters where we discuss different types of receivers, such as the zero-forcing receiver, the MMSE receiver, and the minimum BER receiver.

3.9.2 The pseudocirculant channel matrix

It is useful to develop a more detailed matrix representation which shows the effects of the channel and the transmitting and receiving filters separately. To develop such a representation consider Fig. 3.29, where the transmitter and receiver filter bank are shown in polyphase form (Secs. 3.7.2, 3.7.3). Consider the system shown in the gray box in Fig. 3.29. We will see that this is an LTI system with a transfer matrix $\mathbf{C}_b(z)$ which only depends on the scalar transfer function $C(z)$ and the integer P. Thus the entire transceiver system can be drawn in the form of a cascade of multiple-input multiple-output, or **MIMO** systems as shown in Fig. 3.30. The transmitter polyphase matrix $\mathbf{R}(z)$ is also known as the **precoder matrix** and the receiver polyphase matrix $\mathbf{E}(z)$ is also known as the **equalizer matrix**.

Since the matrix $\mathbf{C}_b(z)$ operates on the blocked version of the channel input $x(n)$ to produce the blocked version of the channel output $y(n)$, we say that $\mathbf{C}_b(z)$ is the **blocked version of the channel** $C_b(z)$. We will see that the transfer matrix $\mathbf{C}_b(z)$ has a special form called the pseudocirculant form. Pseudocirculant matrices were originally introduced in the filter bank literature in the context of alias-free filter banks and blocked LTI systems (see Vaidyanathan and Mitra [1988] and Vaidyanathan [1993]). In our context, their importance arises from the fact that the blocked channel matrix $\mathbf{C}_b(z)$ is a pseudocirculant matrix (Theorem 3.1). To describe what a pseudocirculant is, let the Type 1 polyphase decomposition of the channel be given by

$$C(z) = C_0(z^P) + z^{-1}C_1(z^P) + \ldots + z^{-(P-1)}C_{P-1}(z^P). \tag{3.34}$$

Consider an example with $P = 5$. Then it can be shown (see Theorem 3.1 below) that the matrix $\mathbf{C}_b(z)$ has the form

$$\begin{bmatrix} C_0(z) & z^{-1}C_4(z) & z^{-1}C_3(z) & z^{-1}C_2(z) & z^{-1}C_1(z) \\ C_1(z) & C_0(z) & z^{-1}C_4(z) & z^{-1}C_3(z) & z^{-1}C_2(z) \\ C_2(z) & C_1(z) & C_0(z) & z^{-1}C_4(z) & z^{-1}C_3(z) \\ C_3(z) & C_2(z) & C_1(z) & C_0(z) & z^{-1}C_4(z) \\ C_4(z) & C_3(z) & C_2(z) & C_1(z) & C_0(z) \end{bmatrix}. \tag{3.35}$$

Note that, along any line parallel to the main diagonal, all elements are identical. Such a matrix is called a **Toeplitz** matrix. Furthermore, any column can be obtained from the previous column by shifting it downwards by one, circulating the element that spills over, and attaching a z^{-1} element to the circulated element. If there were no extra z^{-1} to be attached like this, the matrix would be a **circulant** (Appendix D). The presence of the extra z^{-1} elements gives it the name **pseudocirculant**. The matrix $\mathbf{C}_b(z)$ is also often denoted as $\mathbf{C}_{ps}(z)$, where the subscript ps is a reminder of "pseudocirculant."

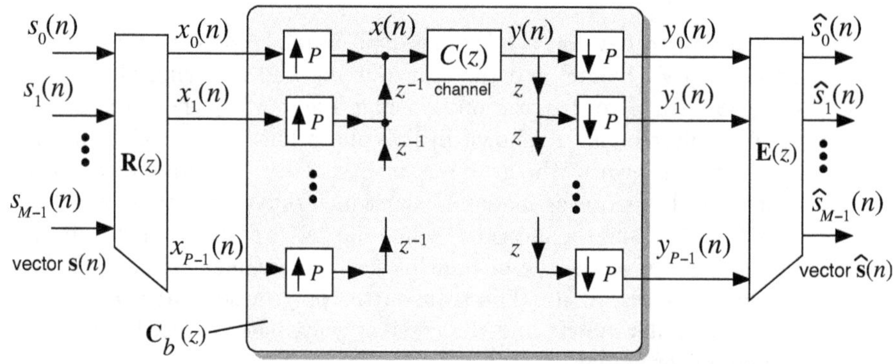

Figure 3.29. The M-user transmultiplexer or transceiver system in polyphase form.

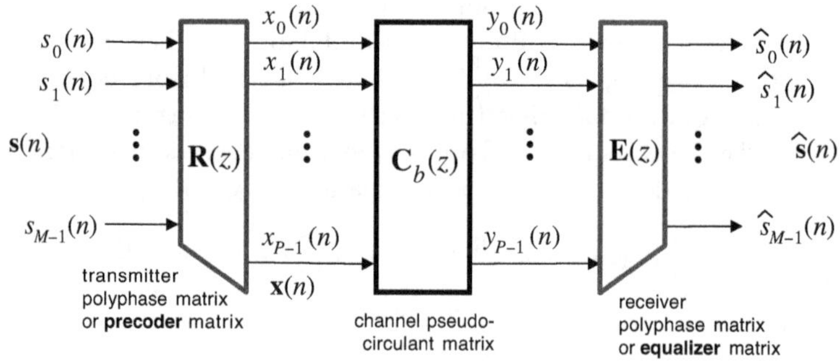

Figure 3.30. The M-user transmultiplexer or transceiver system in matrix form with transmitter and receiver polyphase matrices indicated.

Here then is the formal definition of a pseudocirculant matrix:

♠**Definition 3.1.** *Pseudocirculant matrices.* A $P \times P$ matrix $\mathbf{C}_b(z)$ is said to be pseudocirculant if any column (except the leftmost column which is arbitrary) is obtained from the preceding column by performing the following operations: (a) shift down by one element, (b) recirculate the spilled element to the top, and (c) multiply the recirculated element by the delay z^{-1}. ◇

A number of simple observations should now be made.

1. If we set $z = 1$ the pseudocirculant becomes an ordinary *circulant* matrix (Appendix D).

2. The leftmost column of the matrix completely determines all the other columns. In fact any row or column contains all the elements $C_k(z)$ in some order, with a possible z^{-1} factor.

3. If any row is shifted to the *right* and the spilled element recirculated and multiplied by z, we get the next row. Similarly, if any column is shifted *up* and the spilled element recirculated to the bottom and multiplied by z, we get the previous column, and so forth. The reader can discover more such properties.

4. Other minor variations are possible in the definition of pseudocirculants (for example, we can shift a column *up* and then circulate the spilled element with z^{-1}). Throughout this book, Definition 3.1 will be used.

♠**Theorem 3.1.** *Blocked versions and pseudocirculants.* Let $\mathbf{C}_b(z)$ be the $P \times P$ blocked version of a channel $C(z)$ as defined in Fig. 3.29. Then $\mathbf{C}_b(z)$ is a pseudocirculant. ◇

Proof. The system with input $x_m(n)$ and output $y_k(n)$ in Fig. 3.29 is nothing but $z^{(k-m)}C(z)$ sandwiched between the expander $\uparrow P$ and the decimator $\downarrow P$. According to the polyphase identity this is therefore an LTI system with transfer function

$$[\mathbf{C}_b(z)]_{km} = \Big(z^{(k-m)}C(z)\Big)_{\downarrow P}. \tag{3.36}$$

Substituting from Eq. (3.34) this becomes

$$[\mathbf{C}_b(z)]_{km} = \left(z^{(k-m)}\Big[C_0(z^P) + z^{-1}C_1(z^P) + \ldots + z^{-(P-1)}C_{P-1}(z^P)\Big]\right)_{\downarrow P}$$

For example, the elements of the 0th column of $\mathbf{C}_b(z)$ are obtained from this by setting $m = 0$. Thus

$$[\mathbf{C}_b(z)]_{k,0} = \left(z^{k}\Big[C_0(z^P) + z^{-1}C_1(z^P) + \ldots + z^{-(P-1)}C_{P-1}(z^P)\Big]\right)_{\downarrow P}, \tag{3.37}$$

for $0 \le k \le P-1$. From elementary multirate identities (see Eq. (3.17)) we know that

$$[z^k D(z^P)]_{\downarrow P} = D(z)[z^k]_{\downarrow P}.$$

Now observe that

$$[z^k]_{\downarrow P} = \begin{cases} z^{k/P} & \text{if } k \text{ is a multiple of } P \\ 0 & \text{otherwise.} \end{cases}$$

By using these identities in Eq. (3.37) it follows that $[\mathbf{C}_b(z)]_{k,0} = C_k(z)$. Thus the 0th column of $\mathbf{C}_b(z)$ has the elements $C_k(z)$ in increasing order, as demonstrated in Eq. (3.35). Next consider the first column ($m = 1$). We have

$$[\mathbf{C}_b(z)]_{k,1} = \left(z^{k-1}\Big[C_0(z^P) + z^{-1}C_1(z^P) + \ldots + z^{-(P-1)}C_{P-1}(z^P)\Big]\right)_{\downarrow P}.$$

When $k = 0$ the last term is $[z^{-P}C_{P-1}(z^P)]_{\downarrow P} = z^{-1}C_{P-1}(z)$. All other terms are zero. Similarly, when $k = 1$ the only nonzero term is $[C_0(z^P)]_{\downarrow P} = C_0(z)$. Proceeding thus, the entire matrix $\mathbf{C}_b(z)$ can be constructed and has the pseudocirculant form demonstrated in Eq. (3.35). $\bigtriangledown\bigtriangledown\bigtriangledown$

For example, assume that the channel is FIR with order L:

$$C(z) = \sum_{n=0}^{L} c(n)z^{-n}. \tag{3.38}$$

If we choose $P > L$ then the polyphase components $C_k(z)$ are constants, and we get a simple form of the pseudocirculant matrix. For example, suppose $L = 2$ and $P = 5$. Then $\mathbf{C}_b(z)$ is the 5×5 matrix given by

$$\mathbf{C}_b(z) = \begin{array}{c} \begin{array}{ccc|cc} & P-L & & L & \end{array} \\ \left(\begin{array}{ccc|cc} c(0) & 0 & 0 & z^{-1}c(2) & z^{-1}c(1) \\ c(1) & c(0) & 0 & 0 & z^{-1}c(2) \\ c(2) & c(1) & c(0) & 0 & 0 \\ 0 & c(2) & c(1) & c(0) & 0 \\ 0 & 0 & c(2) & c(1) & c(0) \end{array}\right) \end{array}. \tag{3.39}$$

On the other hand if $L = P = 5$ we get

$$\mathbf{C}_b(z) =$$

$$\begin{array}{c} \begin{array}{ccc|cc} & P-L & & L & \end{array} \\ \left(\begin{array}{ccc|cc} c(0)+z^{-1}c(5) & z^{-1}c(4) & z^{-1}c(3) & z^{-1}c(2) & z^{-1}c(1) \\ c(1) & c(0)+z^{-1}c(5) & z^{-1}c(4) & z^{-1}c(3) & z^{-1}c(2) \\ c(2) & c(1) & c(0)+z^{-1}c(5) & z^{-1}c(4) & z^{-1}c(3) \\ c(3) & c(2) & c(1) & c(0)+z^{-1}c(5) & z^{-1}c(4) \\ c(4) & c(3) & c(2) & c(1) & c(0)+z^{-1}c(5) \end{array}\right) \end{array}$$

Note in this example that the diagonal elements are no longer constants: they are FIR filters of order one.

3.9.3 Perfect symbol recovery

From Fig. 3.30 we see that the received symbol vector $\widehat{\mathbf{s}}(n)$ is the output of the MIMO system

$$\mathbf{T}(z) = \mathbf{E}(z)\mathbf{C}_b(z)\mathbf{R}(z)$$

in response to the input vector sequence $\mathbf{s}(n)$. From this we can come to a number of conclusions:

1. MUI is eliminated if and only if $\mathbf{E}(z)\mathbf{C}_b(z)\mathbf{R}(z)$ is diagonal.
2. With MUI eliminated, ISI is also eliminated if and only if the condition $\mathbf{E}(z)\mathbf{C}_b(z)\mathbf{R}(z) = \mathbf{T}_0$ holds (i.e., the left-hand side is a constant matrix).

3. There is perfect symbol recovery or PR property (in absence of channel noise) if and only if $\mathbf{E}(z)\mathbf{C}_b(z)\mathbf{R}(z) = \mathbf{I}$. This condition is equivalent to $\widehat{\mathbf{s}}(n) = \mathbf{s}(n)$. More generally, since $\widehat{s}_k(n) = c_k s_k(n - N_k)$ is acceptable for nonzero c_k, the product
$$\mathbf{E}(z)\mathbf{C}_b(z)\mathbf{R}(z)$$
just has to be a diagonal matrix with nonzero diagonal delay elements.

It turns out that these conditions are easier to satisfy with redundant transmultiplexers ($P > M$) as we shall elaborate later (Chap. 7). For example, if $P = M$ and the channel is FIR then the PR property cannot be satisfied with FIR transmitter and receiver filters. We need IIR filters, which may or may not be stable depending on the channel transfer function $C(z)$.

3.10 Concluding remarks

We will find in the next few chapters that many of the recent developments on transceiver theory and design can be efficiently represented in terms of the multirate filter bank language introduced in this chapter. The introduction given here is brief but sufficient for the purposes of our discussions. Readers interested in a more detailed treatement of multirate systems and filter banks can find it in Crochiere and Rabiner [1983], Vaidyanathan [1990, 1993], Fliege [1994], Vetterli and Kovačević [1995], Strang and Nguyen [1996], and Harris [2004]. There is an elegant mathematical formulation to describe filter banks and transmultiplexers in a unified way using the theory of **frames**. This description accommodates both redundant and non-redundant filter banks, and can be applied both to the continuous- and the discrete-time cases. Background material and details on this very interesting topic can be found in a number of references, including Heil and Walnut [1989], Daubechies [1992], Daubechies, Landau, and Landau [1995], Janssen [1997], Burrus, Gopinath, and Guo [1998], Mallat [1998], Bölcskei [1999], and Vaidyanathan and Djokovic [2000].

Problems

3.1. The decimator and expander in cascade are in general not interchangeable. That is, the two systems in Fig. P3.1 are in general not equivalent.

Figure P3.1.

1. For $M = L \neq 1$, prove that the systems are not equivalent by constructing an input $x(n)$ such that $y_1(n) \neq y_2(n)$.
2. For $M \neq L$, prove that the two systems are equivalent *if* M and L are coprime (i.e., they do not have any common integer factors other than unity).
3. For $M \neq L$, prove that the two systems are equivalent *only if* M and L are coprime.

3.2. Consider the multirate system shown in part (a) of Fig. P3.2. Assume that M and L are coprime integers. In view of Euclid's theorem this implies that there exist integers N and K such that $NL + MK = 1$. Show then that this system can be drawn as in part (b). With z replaced in part (a) by z^{-k} for arbitrary integer k, how does the system in part (b) change?

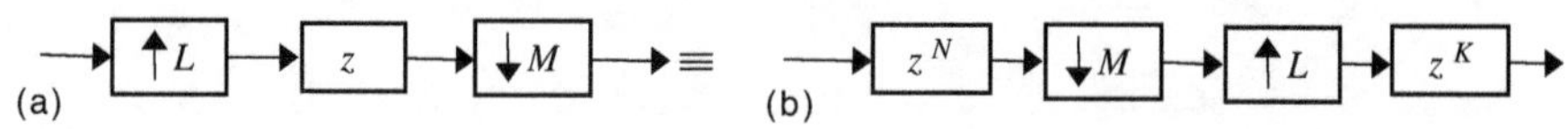

Figure P3.2.

3.3. *Bandpass sampling.* For the signal shown in Fig. P3.3, the total bandwidth (sum of passband widths) is less than $2\pi/M$ and the bands are located as shown, that is, the band edges are integer multiples of π/M. Show that M-fold decimation causes no aliasing. This result is called the bandpass sampling theorem.

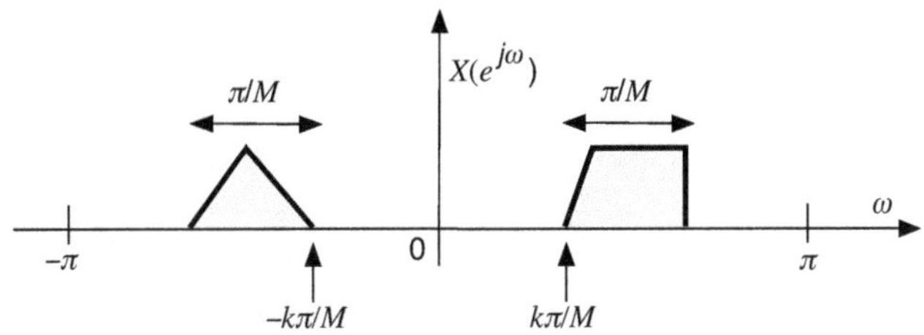

Figure P3.3.

3.4. *Fractional decimation.* Consider Fig. P3.4(a), which shows a discrete-time signal bandlimited to $-2\pi/3 < \omega < 2\pi/3$. Since the total bandwidth is greater than π we cannot decimate this signal by two, or any integer factor, without causing aliasing. But since only two-thirds of the region $[-\pi, \pi)$ is occupied by the signal, it appears that we ought to be able to "decimate by a fraction" such as 3/2, so that the Fourier transform stretches by 3/2 and fills the entire range $[-\pi, \pi)$, as shown in Fig. P3.4(b). This indeed is the case, but we have to be careful because building blocks such as $\downarrow M$ and $\uparrow L$ are defined only for integer factors. It is possible to combine such integer building blocks with filters to achieve "fractional decimation." Such a structure is shown in Fig. P3.4(c). Assume $L = 2$ and $M = 3$ in the following.

1. With $X(e^{j\omega})$ as in Fig. P3.4(a), sketch the Fourier transform of $s(n)$.
2. Make a choice of the filter $H(e^{j\omega})$ such that only the lowpass image in $S(e^{j\omega})$ is retained. Plot the Fourier transform of the filter output $r(n)$.
3. Show that the Fourier transform of the decimated version $y(n)$ is indeed the fractionally stretched version shown in Fig. P3.4(b).
4. Is the filter in Part 2 above unique? What is the largest possible transition bandwidth?

The signal $y(n)$ constructed as above can be regarded as the fractionally decimated version of $x(n)$ with decimation ratio 3/2 (more generally M/L). In practice the ideal filter $H(e^{j\omega})$ can only be approximated. But since its transition band is allowed to be large, very good approximations are possible with low filter orders.

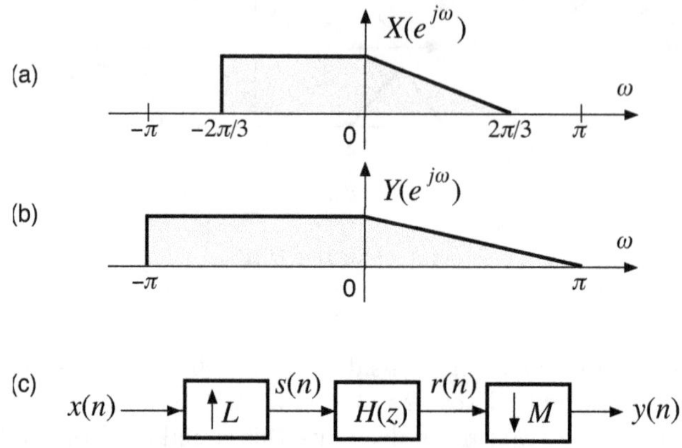

Figure P3.4.

3.5. *Nyquist property and power complementarity.* Consider a filter $F(z)$ written in polyphase form $F(z) = \sum_{k=0}^{M-1} z^{-k} R_k(z^M)$.

1. Show that

$$[\widetilde{F}(z)F(z)]_{\downarrow M} = \sum_{k=0}^{M-1} \widetilde{R}_k(z)R_k(z),$$

where the tilde notation $\widetilde{F}(z)$ is as defined in Sec. 1.6.

2. Let $G(e^{j\omega}) \stackrel{\Delta}{=} |F(e^{j\omega})|^2$. Show that $G(z)$ has the Nyquist(M) property, that is,

$$g(Mn) = \delta(n),$$

if and only if the polyphase components of $F(z)$ are power complementary, that is, they satisfy the property

$$\sum_{k=0}^{M-1} |R_k(e^{j\omega})|^2 = 1 \qquad \text{(P3.5)}$$

for all ω.

3.6. *Sum of allpass filters.* Consider two filters $H_0(z)$ and $H_1(z)$ defined as follows:

$$H_0(z) = \frac{A_0(z) + A_1(z)}{2}, \quad H_1(z) = \frac{A_0(z) - A_1(z)}{2},$$

where $A_0(z)$ and $A_1(z)$ are allpass filters, that is, $|A_k(e^{j\omega})| = 1$ for all ω. Show that $H_0(z)$ and $H_1(z)$ are power complementary, that is, $|H_0(e^{j\omega})|^2 + |H_1(e^{j\omega})|^2 = 1$ for all ω. Note also that $H_0(z)$ and $H_1(z)$ are allpass complementary, that is, $H_0(z) + H_1(z) = \text{allpass}$.

3.7. *Sum of allpass filters: generalization.* Consider a filter bank with M filters $H_k(z)$ given by

$$H_k(z) = \frac{1}{M}\sum_{m=0}^{M-1} W^{km} A_m(z),$$

where $A_m(z)$ are allpass ($|A_m(e^{j\omega})| = 1$ for all ω). Show that $H_k(z)$ are power complementary, that is,

$$\sum_{k=0}^{M-1} |H_k(e^{j\omega})|^2 = 1 \tag{P3.7}$$

for all ω. Also show that $H_k(z)$ are allpass complementary, that is,

$$\sum_{k=0}^{M-1} H_k(z) = \text{allpass}.$$

3.8. *Biorthogonality.* Consider the transmultiplexer system of Fig. 3.26 and assume $P = M$ (no redundancy). Let $\mathbf{R}(z)$ and $\mathbf{E}(z)$ represent the $M \times M$ polyphase matrix of the synthesis and analysis bank, respectively, as in Secs. 3.7.2–3.7.3.

1. Assuming that the channel $C(z) = 1$ and that there is no noise, show that the necessary and sufficient condition for perfect symbol recovery ($\widehat{s}_k(n) = s_k(n)$ for all k) is given by $\mathbf{E}(z)\mathbf{R}(z) = \mathbf{I}$.
2. Now consider the analysis/synthesis system shown in Fig. P3.8. Here a single signal $s(n)$ is first analyzed using the filter bank $\{H_k(z)\}$, and this is followed by the synthesis bank $\{F_k(z)\}$. There are M filters in each bank, and the decimation ratios are equal to M as well. This is called a maximally decimated analysis/synthesis bank and finds application in data compression. Show that the necessary and sufficient condition for perfect reconstruction ($\widehat{s}(n) = s(n)$) is given by $\mathbf{R}(z)\mathbf{E}(z) = \mathbf{I}$. For square matrices, since $\mathbf{R}(z)\mathbf{E}(z) = \mathbf{I}$ is satisfied if and only if $\mathbf{E}(z)\mathbf{R}(z) = \mathbf{I}$, we see that the condition for perfect reconstruction in the transmultiplexer system is identical to the corresponding condition in the analysis/synthesis system.
3. For the analysis/synthesis system shown in Fig. P3.8, show that the condition for perfect reconstruction ($\widehat{s}(n) = s(n)$) can also be written as

$$[F_m(z)H_k(z)]_{\downarrow M} = \delta(k-m). \tag{P3.8}$$

Filter banks satisfying the condition (P3.8) are called biorthogonal filter banks. The biorthogonality condition (P3.8) is therefore equivalent to perfect reconstruction for both the analysis/synthesis system and the transmultiplexer system when $P = M$.

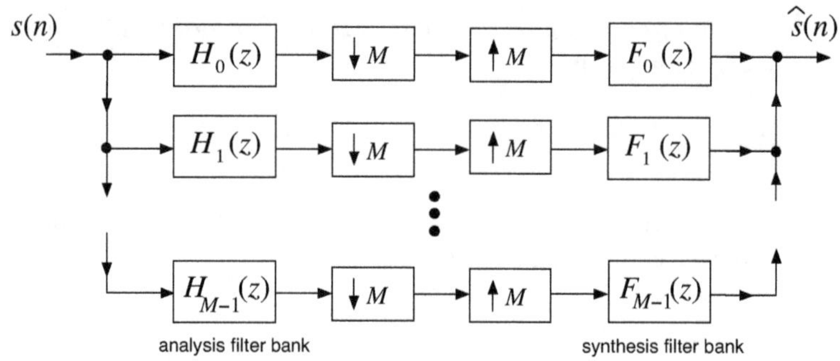

Figure P3.8.

3.9. For the FIR channel $C_2(z) = 1 + 2z^{-1} + 3z^{-2}$, what is the 5×5 pseudocirculant matrix $\mathbf{C}_b(z)$ which results from blocking the channel by $P = 5$? Repeat for the channels $C_3(z) = 1 + 2z^{-1} + 3z^{-2} + 4z^{-3}$ and $C_5(z) = 1 + 2z^{-1} + 3z^{-2} + 4z^{-3} + 5z^{-4} + 6z^{-5}$.

3.10. It can be shown that the product of two pseudocirculant matrices is a pseudocirculant.

1. Verify this for the 3×3 case by explicitly calculating $\mathbf{P}_1(z)\mathbf{P}_2(z)$, where
$$\mathbf{P}_k(z) = \begin{bmatrix} a_k(z) & z^{-1}c_k(z) & z^{-1}b_k(z) \\ b_k(z) & a_k(z) & z^{-1}c_k(z) \\ c_k(z) & b_k(z) & a_k(z) \end{bmatrix}.$$
2. It can also be shown that products of pseudocirculants commute, that is, $\mathbf{P}_1(z)\mathbf{P}_2(z) = \mathbf{P}_2(z)\mathbf{P}_1(z)$. Verify this for the above 3×3 example by explicitly calculating $\mathbf{P}_2(z)\mathbf{P}_1(z)$.

The same properties can readily be verified for circulant matrices as well.

3.11. Unlike pseudocirculants, the product of two Toeplitz matrices is not necessarily Toeplitz.

1. Show this by explicitly multiplying out
$$\mathbf{T}_1 = \begin{bmatrix} a_1 & b_1 & c_1 \\ d_1 & a_1 & b_1 \\ e_1 & d_1 & a_1 \end{bmatrix} \quad \text{and} \quad \mathbf{T}_2 = \begin{bmatrix} a_2 & b_2 & c_2 \\ d_2 & a_2 & b_2 \\ e_2 & d_2 & a_2 \end{bmatrix}.$$
2. Show also that $\mathbf{T}_1\mathbf{T}_2 \neq \mathbf{T}_2\mathbf{T}_1$, that is, Toeplitz matrices do not in general commute in a product.

3.12. Let $\mathbf{T}_1$ and $\mathbf{T}_2$ be two lower triangular Toeplitz matrices as shown below:

$$\mathbf{T}_1 = \begin{bmatrix} a_1 & 0 & 0 \\ d_1 & a_1 & 0 \\ e_1 & d_1 & a_1 \end{bmatrix} \quad \text{and} \quad \mathbf{T}_2 = \begin{bmatrix} a_2 & 0 & 0 \\ d_2 & a_2 & 0 \\ e_2 & d_2 & a_2 \end{bmatrix}.$$

1. Compute the product $\mathbf{T}_1\mathbf{T}_2$. Is it Toeplitz?
2. Compute the product $\mathbf{T}_2\mathbf{T}_1$. Is it the same as $\mathbf{T}_1\mathbf{T}_2$?
3. More generally, if $\mathbf{T}_1$ and $\mathbf{T}_2$ are $N \times N$ Toeplitz, can you argue that $\mathbf{T}_1\mathbf{T}_2$ is Toeplitz and that $\mathbf{T}_1\mathbf{T}_2 = \mathbf{T}_2\mathbf{T}_1$? *Hint.* The matrix vector multiplication $\mathbf{Tx}$, where $\mathbf{T}$ is lower triangular Toeplitz, is related to convoltuion of a causal input with a causal impulse response.

3.13. Consider the system shown in Fig. P3.13 with input $x(n)$ and output $\widehat{x}(n)$. The delay chain and advance chain are made from z^N rather than z, where N and M are coprime (i.e., they have no common factor other than unity). Show that this system has the perfect reconstruction property, that is, $\widehat{x}(n) = x(n)$. Feel free to use the results of Problem 3.8.

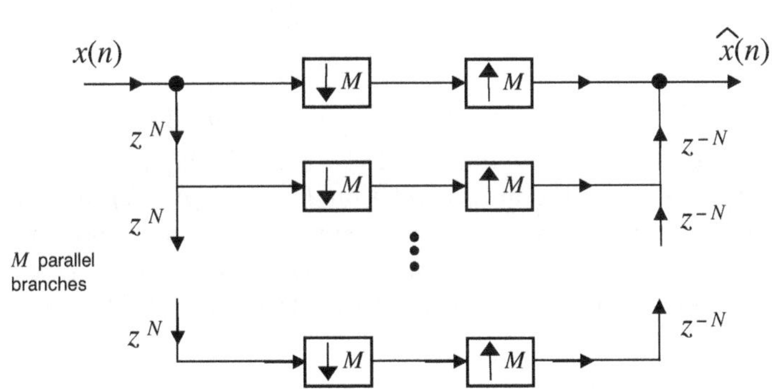

Figure P3.13.

3.14. Consider the system shown in Fig. P3.14 with input $x(n)$ and output $\widehat{x}(n)$. The delay chain and advance chain are made from z^N rather than z, where N and M are coprime (i.e., they have no common factor other than unity). Show that this system has the perfect reconstruction property, that is, $\widehat{x}(n) = x(n)$, if and only if $T_k(z) = 1$ for all k. Use the results of Problem 3.8 if necessary.

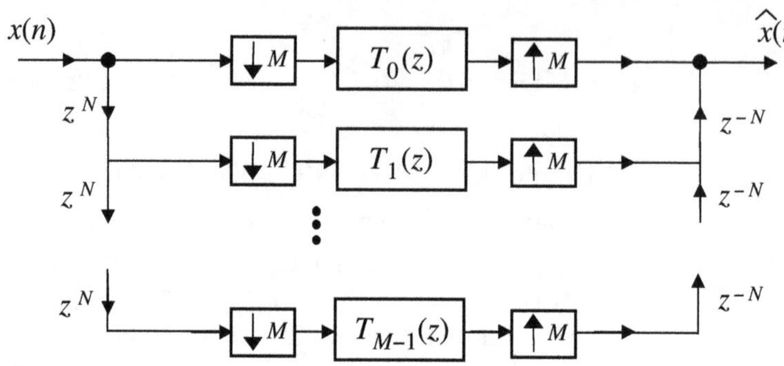

Figure P3.14.

3.15. For this problem, first you have to review Sec. E.3 in Appendix E. Let $s(n)$ be a discrete-time WSS process with power spectrum

$$S_{ss}(e^{j\omega}) = \frac{1}{1+\rho^2+2\rho\cos\omega}$$

Define

$$y(n) = \begin{cases} s(n/2) & n \text{ even}, \\ 0 & n \text{ odd}. \end{cases}$$

That is, $y(n)$ is the output of the two-fold expander ($\uparrow 2$) in response to input $s(n)$. Is the autocorrelation $E[y(n)y^*(n-k)]$ independent of n? If not, can you define an average autocorrelation (analogous to what we do in Sec. E.3 of Appendix E)? If this is indeed possible, then express the average power spectrum of $y(n)$ in terms of $S_{ss}(e^{j\omega})$.

4

Discrete-time representations

4.1 Introduction

The digital communication system described in Chap. 2 is reproduced in Fig. 4.1. This system contains both continuous-time and discrete-time quantities. The signal $x_c(t)$ entering the channel is a continuous-time interpolated version of the sequence $s(n)$, whereas the signal $\widehat{s}(n)$ at the receiver is a sampled version of the filtered received signal. It is often convenient to represent the communication system entirely in terms of equivalent discrete-time quantities such as digital filters. This is indeed possible based on the relationship between continuous-time signals and their uniformly sampled versions. In this chapter we first describe this interconnection (Secs. 4.2 and 4.3). This is done both for SISO channels and MIMO channels. The raised cosine function, which arises in the context of pulse shaping, is described in Sec. 4.4. The multiuser communication system is briefly described in Sec. 4.5. In Secs. 4.6–4.8 we discuss equalization of the digitized channel using digital filters, under the so-called zero-forcing constraint. Section 4.9 contains further remarks on the digital design of pre and postfilters. Digital equalizers without zero forcing (so-called minimum mean square error equalizers) will be introduced in Sec. 4.10.

Notations. Whenever it is necessary to distinguish continuous-time quantities from discrete-time we use the subscript c (continuous) or d (discrete). Where no subscript is used, the distinction will be quite clear from the context. The notation $\delta_c(t)$ denotes the continuous-time Dirac delta function, and $\delta(n)$ denotes the discrete-time impulse.

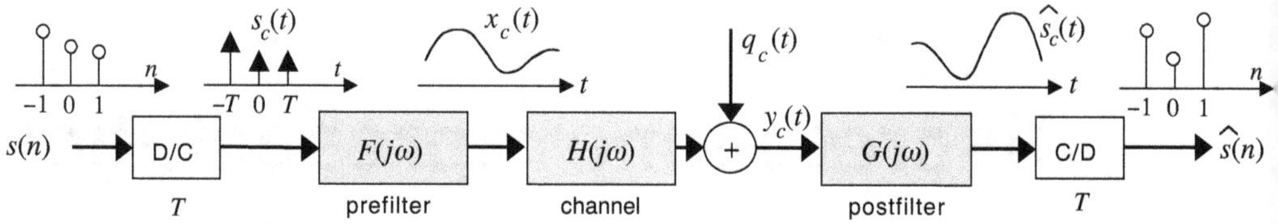

Figure 4.1. A discrete-time symbol stream $s(n)$ transmitted through a noisy continuous-time channel. The reconstructed symbol stream is $\widehat{s}(n)$.

4.2 Conversion between continuous and discrete time

The C/D and D/C building blocks make an appearance in the digital communication system shown in Fig. 4.1. The C/D (continuous-to-discrete) converter is shown separately in Fig. 4.2(a) and represents uniform sampling. Thus the input-output relation is

$$y(n) = y_c(nT) \qquad \text{(C/D conversion).}$$

In terms of Fourier transforms the relation is therefore [Oppenheim and Willsky, 1997]

$$Y(e^{j\omega}) = \frac{1}{T}\sum_{k=-\infty}^{\infty} Y_c\Big(j\Big(\frac{\omega+2\pi k}{T}\Big)\Big) \qquad \text{(C/D conversion).} \tag{4.1}$$

The sample spacing T is usually indicated under the C/D box as shown in the figure. The D/C (discrete-to-continuous) converter is shown in Fig. 4.2(b) and has the input-output relation given by

$$s_c(t) = \sum_{n=-\infty}^{\infty} s(n)\delta_c(t-nT) \qquad \text{(D/C conversion),} \tag{4.2}$$

where $\delta_c(t)$ is the Dirac delta function. Thus, the sample $s(n)$ is converted into an impulse positioned at time nT. The D/C converter is therefore merely a conceptual device which converts "lollipops into delta functions," and assigns a time dimension to the samples (i.e., assigns $T, 2T, 3T, \ldots$ etc.). Taking Fourier transforms, Eq. (4.2) yields

$$S_c(j\omega) = S(e^{j\omega T}) \qquad \text{(D/C conversion),} \tag{4.3}$$

where $S(e^{j\omega})$ is the Fourier transform of $s(n)$.

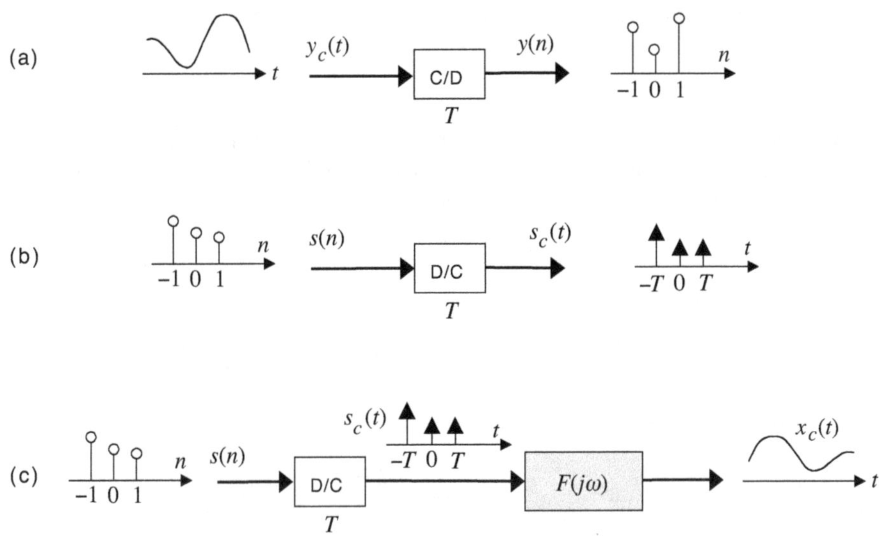

Figure 4.2. (a) The C/D (continuous-to-discrete) converter, (b) the D/C (discrete-to-continuous) converter, and (c) the D/C converter followed by filtering.

The usefulness of the D/C converter arises when we combine it with a continuous-time filter $F(j\omega)$ as shown in Fig. 4.2(c). The filter performs a convolution to produce the continuous-time output

$$x_c(t) = \sum_{n=-\infty}^{\infty} s(n)f(t-nT) \qquad \text{(D/C converter with filter)}, \tag{4.4}$$

where $f(t)$ is the impulse response of $F(j\omega)$. Thus $x_c(t)$ can be regarded as an interpolated version of $s(n)$. Taking Fourier transforms on both sides of (4.4) we obtain

$$X_c(j\omega) = \sum_{n=-\infty}^{\infty} s(n)e^{-j\omega nT}F(j\omega),$$

which can be rewritten as

$$X_c(j\omega) = S(e^{j\omega T})F(j\omega) \qquad \text{(D/C with filter)}. \tag{4.5}$$

4.2.1 The sampling identity

Next we derive a simple identity pertaining to the C/D and D/C operators. Figure 4.3(a) shows a continuous-time filter $H_c(j\omega)$ sandwiched between the D/C and C/D operators. Since the output of $H_c(j\omega)$ is given by

$$y_c(t) = \sum_k x(k)h_c(t-kT), \tag{4.6}$$

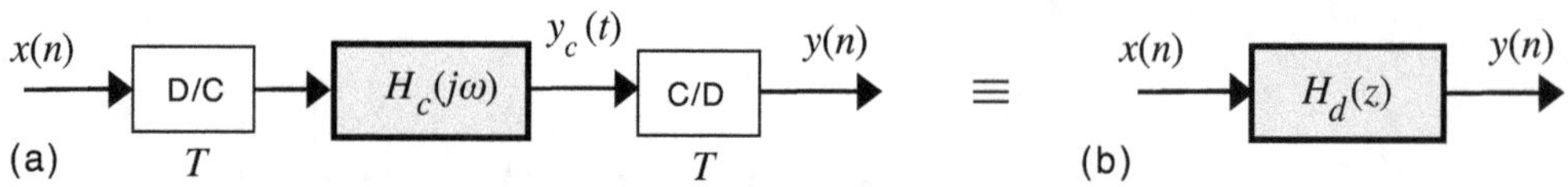

Figure 4.3. The sampling identity. (a) A continuous-time filter $H_c(j\omega)$ sandwiched between the D/C and C/D building blocks, and (b) equivalent digital filter system. The impulse response of $H_d(z)$ is $h_d(n) = h_c(nT)$.

it follows that

$$y(n) = y_c(nT) = \Big[\sum_k x(k)h_c(t - kT)\Big]_{t=nT} = \sum_k x(k)h_c(nT - kT).$$

This can be written as a discrete-time convolution

$$y(n) = \sum_k x(k)h_d(n - k), \tag{4.7}$$

where

$$h_d(n) \triangleq h_c(nT). \tag{4.8}$$

Thus a continuous-time LTI system sandwiched between the C/D and D/C operators as in Fig. 4.3(a) is equivalent to a discrete-time LTI system (Fig. 4.3(b)) with transfer function

$$H_d(z) = \sum_{n=-\infty}^{\infty} h_d(n)z^{-n} = \sum_{n=-\infty}^{\infty} h_c(nT)z^{-n}. \tag{4.9}$$

This result will be referred to as the *sampling identity*, and is similar to the polyphase identity of Sec. 3.2.3.

4.3 Discrete-time representations of channels

In this section we explain how digital communication systems can be represented entirely in terms of discrete-time transfer functions. Such representations also give rise to optimal transceiver problem formulations, with the prefilter and equalizer defined entirely in the discrete-time domain.

4.3.1 Digital communication over a SISO channel

The digital communication system is reproduced in Fig. 4.4. Ignoring noise for a moment, the system from $s(n)$ to $\widehat{s}(n)$ can be regarded as an LTI system (sampling identity, Fig. 4.3). Its transfer function has the form

$$H_d(z) = \sum_n h_d(n)z^{-n}, \tag{4.10}$$

where $h_d(n)$ is the sampled version of the impulse response of the cascasde $G(j\omega)H(j\omega)F(j\omega)$, that is,

$$h_d(n) = h_c(nT). \tag{4.11}$$

Here

$$h_c(t) = (g * h * f)(t), \tag{4.12}$$

with $*$ representing convolution. The system of Fig. 4.4 can therefore be represented as in Fig. 4.5, where the additive noise $q(n)$ is the sampled version of $q_c(t)$ filtered by $G(j\omega)$, that is,

$$q(n) = (q_c * g)(nT). \tag{4.13}$$

The prefilter $F(j\omega)$ and equalizer $G(j\omega)$ can ideally be chosen in such a way that

$$h_c(nT) = \delta(n), \tag{4.14}$$

or equivalently $H_d(z) = 1$. This is called the **zero-forcing** (ZF) condition. In this case we have a system with no intersymbol interference or **ISI**. That is, the sample $\widehat{s}(n)$ at the receiver is not affected by $s(n-k), k \neq 0$. The only nonideality in this case is the channel noise $q(n)$. Note that the product

$$H_c(j\omega) = G(j\omega)H(j\omega)F(j\omega) \tag{4.15}$$

need not be identity for the zero-forcing condition to be satisifed. From the sampling theorem we see that the zero-forcing or ISI-free condition (4.14) is equivalent to

$$\frac{1}{T}\sum_{k=-\infty}^{\infty} H_c\Big(j\Big(\omega + \frac{2\pi k}{T}\Big)\Big) = 1. \tag{4.16}$$

Figure 4.6(a) shows a typical example of $H_c(j\omega)$ and its shifted versions taking part in the summation above. If the bandwidth of $H_c(j\omega)$ is less than $2\pi/T$ it is clear that the ISI-free condition (4.16) cannot be satisfied.[1] In a **minimum-bandwidth** system, $H_c(j\omega)$ has total bandwidth exactly equal to $2\pi/T$. In this case the terms in Eq. (4.16) do not overlap, and the only way to satisfy the ISI-free condition is to force $H_c(j\omega)$ to have the ideal response

$$H_c(j\omega) = \begin{cases} T & -\pi/T \leq \omega < \pi/T \\ 0 & \text{otherwise,} \end{cases} \tag{4.17}$$

[1]The "bandwidth" includes negative and positive frequencies. Thus the region $|\omega| \leq \pi/T$ corresponds to a bandwidth of $2\pi/T$.

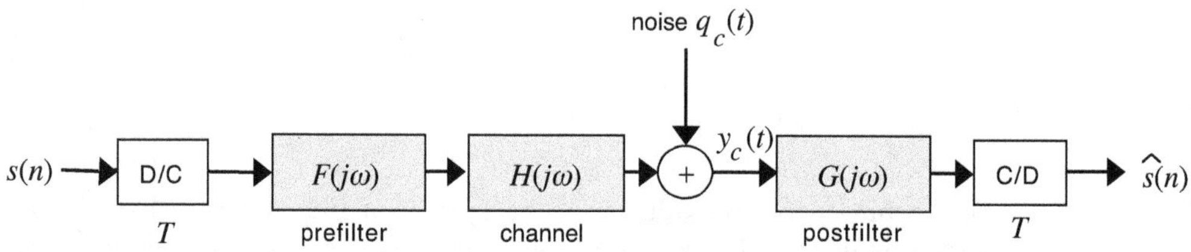

Figure 4.4. The digital communication system.

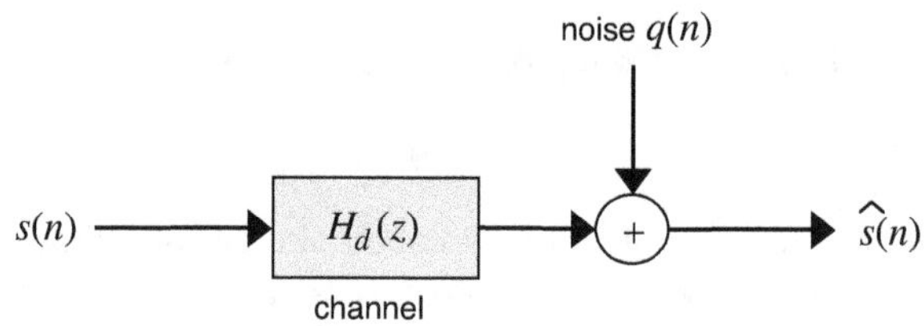

Figure 4.5. The all-discrete-time equivalent of the digital communication system of Fig. 4.4.

so that $h_c(t)$ is the sinc function:

$$h_c(t) = \frac{\sin(\pi t/T)}{\pi t/T}. \tag{4.18}$$

The ideal response and its shifted versions are demonstrated in Fig. 4.6(b).

4.3.1.A Minimum bandwidth versus excess bandwidth

As mentioned above, $H_c(j\omega)$ should have a minimum bandwidth of $2\pi/T$ if the symbol spacing is T seconds. In an *excess-bandwidth* system the bandwidth of $H_c(j\omega)$ exceeds $2\pi/T$ as shown in Fig. 4.6(a). Then there is more flexibility in the choice of the exact shape of $H_c(j\omega)$, and the design of the filters $F(j\omega)$ and $G(j\omega)$ (for a given channel $H(j\omega)$) becomes easier. Thus, if we want to enforce ISI cancellation, then we have two choices: either (a) use minimum bandwidth, in which case the only choice of $H_c(j\omega)$ is the ideal response corresponding to a sinc, or (b) use excess bandwidth, in which case $H_c(j\omega)$ is more flexible and the designs of $F(j\omega)$ and $G(j\omega)$ are easier. For example, if $H_c(j\omega)$ is taken to be the Fourier transform of the so-called **raised-cosine** function (Sec. 4.4), then the condition $h_c(nT) = \delta(n)$ prevails, though there is excess bandwidth.

In practice, given a channel $H(j\omega)$, the filters $F(j\omega)$ and $G(j\omega)$ can only be designed to satisfy the zero-forcing condition (4.16) approximately. So, the discrete-time system $H_d(z)$ is not identity, though it can often be reasonably

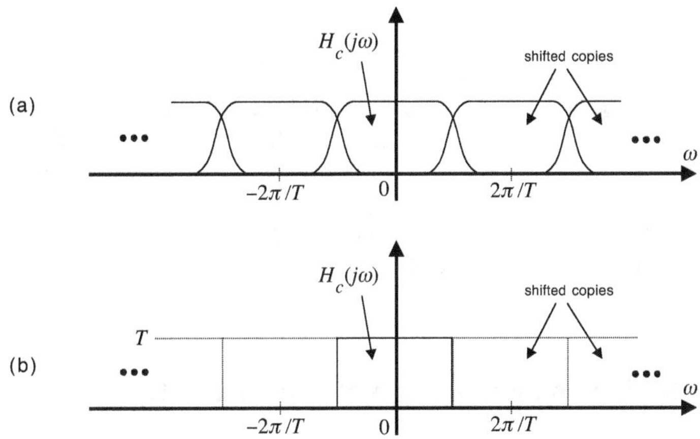

Figure 4.6. The effective channel $H_c(j\omega)$ and its shifted versions in Eq. (4.16). (a) Case where $H_c(j\omega)$ has excess bandwidth, and (b) case where $H_c(j\omega)$ has minimum bandwidth.

approximated by an FIR transfer function

$$H_d(z) = \sum_{n=0}^{L} h_d(n) z^{-n}. \tag{4.19}$$

There are also designs where a **controlled amount of ISI** is allowed by letting $h_c(nT)$ be equal to a specified finite-duration sequence such as the **duobinary** pulse (Sec. 5.9). (Recall that $h_c(t)$ can be controlled by designing $G(j\omega)$ and $F(j\omega)$.) In this case there is no need for excess bandwidth.

4.3.1.B The digital transceiver

As explained above, the equivalent digital channel $H_d(z)$ is in general not identity. Whenever this is the case, $H_d(z)$ can be compensated digitally. There are many ways to do this. For example, we can insert a digital precoder $F_d(z)$ and a digital equalizer $G_d(z)$ and jointly optimize the pair $\{F_d(z), G_d(z)\}$ such that

$$G_d(z) H_d(z) F_d(z) = 1, \tag{4.20}$$

and furthermore an objective function such as the *mean square reconstruction error* due to channel noise, or *symbol error probability*, is minimized. The condition (4.20) is again the zero-forcing constraint, this time enforced using the digital filters $F_d(z)$ and $G_d(z)$. We will return to such optimization problems in later chapters. Once again, remember that the discrete-time model given in Fig. 4.5 does not assume that the continuous-time channel and filters in Fig. 4.4 are bandlimited in any way, though there is some bandlimiting in practice (Chap. 2).

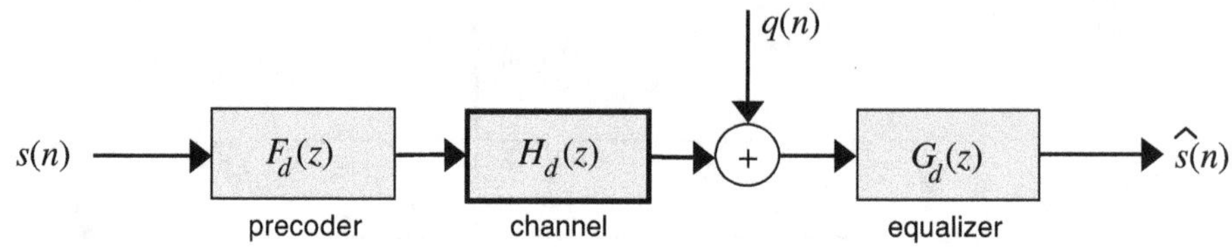

Figure 4.7. The all-discrete-time equivalent diagram of the transceiver, with a digital precoder $F_d(z)$ and an equalizer $G_d(z)$.

4.3.2 Digital communication over a MIMO channel

Figure 4.8 shows the case where M symbol streams $s_k(n), 0 \leq k \leq M-1$, have to be transmitted over a communication medium with transfer matrix $\mathbf{H}(j\omega)$ and additive noise sources $q_k(t)$. The reconstructed signal $\widehat{s}_m(n)$ depends not only on $s_m(n)$, but also on $s_k(\ell)$, where $k \neq m$ and/or $n \neq \ell$. The interference between $s_m(n)$ and $s_k(n)$ for the same n is called **intrablock interference**, and the interference between $s_m(n)$ and $s_k(\ell)$ for $\ell \neq n$ is called **interblock interference**. These can be regarded as two forms of ISI. When we say that the transceiver is free from ISI, or equivalently satisfies *zero forcing*, we imply that both interblock and intrablock interferences have been eliminated.

The MIMO channel can often be approximated by a transfer matrix $\mathbf{H}(j\omega)$. The (m,k)th element $H_{mk}(j\omega)$ of this matrix represents the channel connecting the output of $F_k(j\omega)$ to the input of $G_m(j\omega)$. This transceiver can be made more general by replacing the scalar prefilters and postfilters with matrix transfer functions $\mathbf{F}(j\omega)$ and $\mathbf{G}(j\omega)$ as in Fig. 4.9. This implies that collaboration is allowed between different "users" $s_k(n)$. Such a situation arises, for example, in DMT systems where a single user is "divided" into different frequency bands (Chap. 7). The situation also arises to a limited extent in the so-called broadcast channels (Sec. 4.5). Denoting the cascaded system by $\mathbf{H}_c(j\omega)$, we have

$$\mathbf{H}_c(j\omega) = \mathbf{G}(j\omega)\mathbf{H}(j\omega)\mathbf{F}(j\omega). \tag{4.21}$$

Let $\mathbf{h}_c(t)$ denote the impulse response of this MIMO system, that is,

$$\mathbf{h}_c(t) = \int_{-\infty}^{\infty} \mathbf{H}_c(j\omega)e^{j\omega t}\frac{d\omega}{2\pi}. \tag{4.22}$$

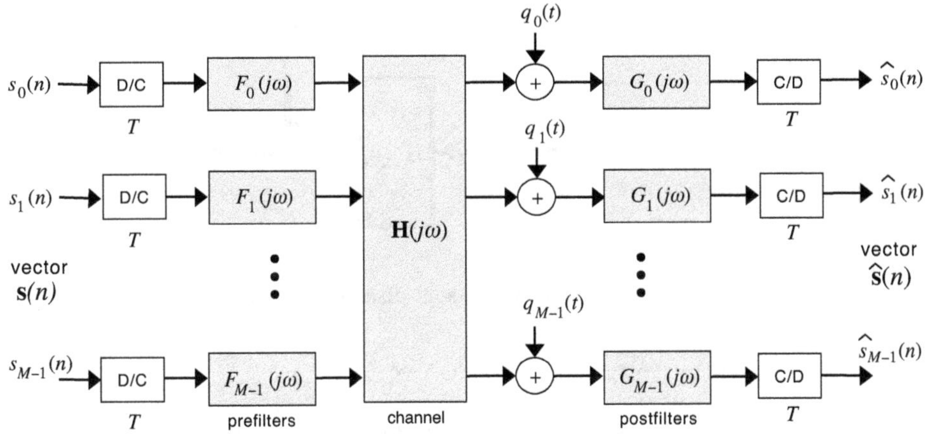

Figure 4.8. A MIMO channel with multiple transmitted messages $s_k(t)$.

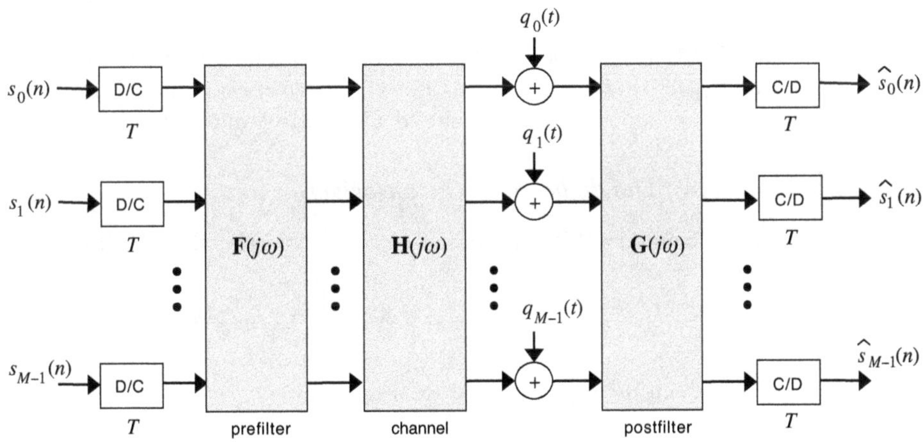

Figure 4.9. The MIMO channel with a MIMO prefilter and a MIMO postfilter or equalizer.

Using the results of Sec. 4.3.1 again, we see that the system from $s_k(n)$ to $\widehat{s}_m(n)$ is a discrete-time LTI system whose impulse response is the sampled version $[\mathbf{h}_c(nT)]_{mk}$. Thus, the MIMO system from the vector $\mathbf{s}(n)$ to the vector $\widehat{\mathbf{s}}(n)$ can be modeled as an LTI system with impulse response

$$\mathbf{h}_d(n) = \mathbf{h}_c(nT) \tag{4.23}$$

and transfer function

$$\mathbf{H}_d(z) = \sum_n \mathbf{h}_d(n) z^{-n}. \tag{4.24}$$

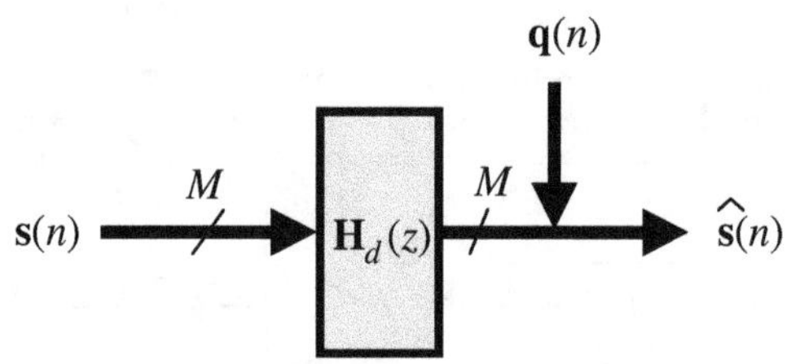

Figure 4.10. The all-discrete-time MIMO channel.

Defining the continuous-time noise vector

$$\mathbf{q}_c(t) = [\,q_0(t) \quad q_1(t) \quad \ldots \quad q_{M-1}(t)\,]^T \tag{4.25}$$

and the sampled noise vector

$$\mathbf{q}(n) = (\mathbf{g} * \mathbf{q}_c)(nT) = \int_{-\infty}^{\infty} \mathbf{g}(\tau)\mathbf{q}_c(nT - \tau)d\tau, \tag{4.26}$$

we see that the discrete-time equivalent can be represented entirely using $\mathbf{H}_d(z)$ and $\mathbf{q}(n)$ (Fig. 4.10). Note that the discrete-time model does not assume that the continuous-time channel and filters are bandlimited.

4.3.2.A The digital MIMO transceiver

If $\mathbf{F}(j\omega)$ and $\mathbf{G}(j\omega)$ are designed appropriately, then $\mathbf{H}_d(z)$ is (approximately) FIR:

$$\mathbf{H}_d(z) = \sum_{n=0}^{L} \mathbf{h}_d(n)z^{-n}. \tag{4.27}$$

If this can be approximated by a constant:

$$\mathbf{H}_d(z) = \mathbf{h}_d(0), \tag{4.28}$$

then the equivalent digital MIMO channel becomes **memoryless**. More generally $\mathbf{H}_d(z)$ has memory, and can be compensated for by using a digital precoder $\mathbf{F}_d(z)$ and a digital equalizer $\mathbf{G}_d(z)$, as shown in Fig. 4.11.

The optimal choice of $\mathbf{F}_d(z)$ and $\mathbf{G}_d(z)$ is an important problem. The details of this optimization problem depend on the objective function that we desire to optimize. One example is the minimization of the total mean square error,

$$\sum_{k=0}^{M-1} E[|\widehat{s}_k(n) - s_k(n)|^2], \tag{4.29}$$

subject to the constraint

$$\mathbf{G}_d(z)\mathbf{H}_d(z)\mathbf{F}_d(z) = \mathbf{I}. \tag{4.30}$$

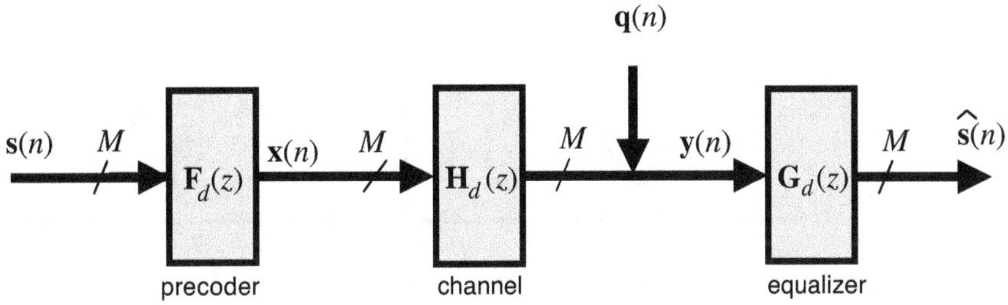

Figure 4.11. The all-discrete-time MIMO transceiver with digital precoder $\mathbf{F}_d(z)$ and equalizer $\mathbf{G}_d(z)$.

This is simply the *zero-forcing* constraint rewritten for the MIMO case. Under this constraint all intrablock and interblock interferences are eliminated, and the only distortion is due to channel noise. We shall address a number of optimization problems of the above type in Chaps. 10–19.

4.4 The raised-cosine function

In Sec. 4.3.1 we considered the digital communication system and constrained the product of $F(j\omega), G(j\omega)$, and $H(j\omega)$ to be a Nyquist(T) filter $H_c(j\omega)$, where T is the symbol spacing. The Nyquist(T) constraint ensures that the discrete-time system from $s(n)$ to $\widehat{s}(n)$ is free from intersymbol interference, and suffers only from channel noise.

The function $H_c(j\omega)$, in addition to being Nyquist(T), should also be a bandlimited function (Fig. 4.6). The bandwidth of $H_c(j\omega)$ should at least be equal to the symbol rate $\omega_s = 2\pi/T$. Usually the bandwidth is higher than this. Having the excess bandwidth allows more freedom to design the Nyquist filter; a filter with total bandwidth exactly $2\pi/T$ would have to be an ideal lowpass filter in order to be also Nyquist(T) (see the discussion of Fig. 4.6). So, as a matter of practical convenience, the bandwidth of the effective channel is usually larger than the symbol rate.

In this section we introduce a bandlimited function $p(t)$ which also has the Nyquist(T) property. The bandwidth of the function is not restricted to be $2\pi/T$. It can be chosen to be anything between $2\pi/T$ and $4\pi/T$, depending on the value of a parameter called β. The Fourier transform of this family of functions is shown in Figure 4.12. This is called a **raised-cosine** function [Lucky, Salz, and Weldon, 1968], sometimes abbreviated as *rcos* [Assalini and Tonello, 2004].

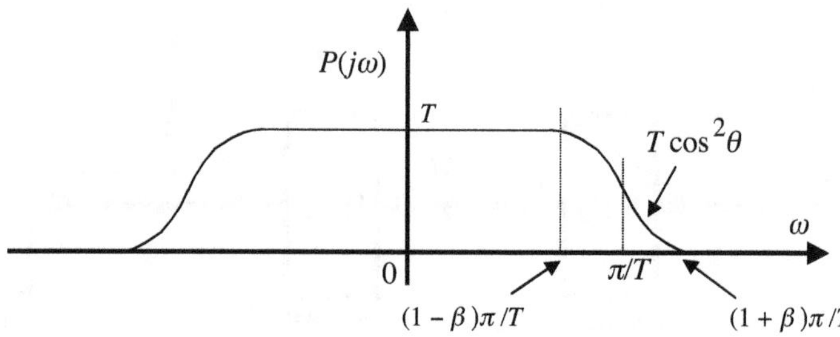

Figure 4.12. The raised-cosine function $P(j\omega)$.

The function $P(j\omega)$ can be mathematically expressed as follows:

$$P(j\omega) = \begin{cases} T & \text{for } |\omega| \leq (1-\beta)\pi/T \\ T\cos^2\left(\dfrac{T}{4\beta}\left(|\omega| - (1-\beta)\dfrac{\pi}{T}\right)\right) & \text{for } (1-\beta)\pi/T < |\omega| < (1+\beta)\pi/T \\ 0 & \text{for } |\omega| \geq (1+\beta)\pi/T. \end{cases} \tag{4.31}$$

The parameter β is chosen to be in the range $0 \leq \beta \leq 1$. Figure 4.13 shows the function $P(j\omega)$ for the two extreme cases $\beta = 0$ and $\beta = 1$.

1. For $\beta = 0$, the function reduces to an ideal lowpass filter with cutoff frequency π/T.
2. For $\beta = 1$, the function has twice as much bandwidth, and is given by
$$P(j\omega) = \begin{cases} \cos^2(\omega T/4) & \text{for } |\omega| \leq 2\pi/T \\ 0 & \text{otherwise.} \end{cases} \tag{4.32}$$
3. For $0 < \beta < 1$, the function has intermediate bandwidth as seen from Fig. 4.12.

The name **raised cosine** comes from the fact that $\cos^2\theta$ can be written as

$$\cos^2\theta = \frac{1+\cos 2\theta}{2} \tag{4.33}$$

A plot of this would be the cosine function raised vertically by one, and then divided by two.

4.4.1 Nyquist property of the raised cosine

The function $P(j\omega)$ has the additional property that

$$\sum_{k=-\infty}^{\infty} P(j(\omega + k\omega_s)) = T, \quad \text{for all } \omega, \tag{4.34}$$

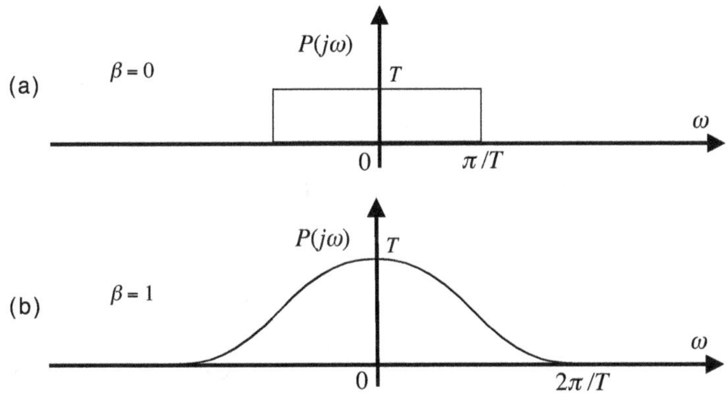

Figure 4.13. Two special cases of the raised-cosine function $P(j\omega)$: (a) $\beta = 0$ and (b) $\beta = 1$.

where $\omega_s = 2\pi/T$. Thus, uniformly shifted versions of the raised-cosine function add exactly to produce a constant response for all ω. While this property is obvious for the special case with $\beta = 0$ (Fig. 4.13(a)), it is not so obvious for the general case. From sampling theory (Appendix G) we know that Eq. (4.34) is equivalent to the Nyquist property

$$p(nT) = \delta(n), \tag{4.35}$$

where $p(t)$ is the inverse Fourier transform of $P(j\omega)$. The Nyquist property is proved by inspection of the inverse Fourier transform of $P(j\omega)$ given by (Problem 4.3)

$$p(t) = \frac{\sin(\pi t/T)}{(\pi t/T)} \left(\frac{\cos(\beta\pi t/T)}{1 - (2\beta t/T)^2} \right), \quad 0 \leq \beta \leq 1. \tag{4.36}$$

The sinc factor in Eq. (4.36) ensures the Nyquist(T) property (4.35) regardless of what β is. The function $p(t)$ is plotted in Fig. 4.14 for $\beta = 0$ and $\beta = 1$ (top) and for $\beta = 0.2$ and $\beta = 0.8$ (bottom). For $\beta = 0$, the function $p(t)$ becomes the sinc function

$$p(t) = \frac{\sin(\pi t/T)}{(\pi t/T)}$$

whereas for $\beta = 1$ this becomes

$$p(t) = \frac{\sin(2\pi t/T)}{(2\pi t/T)} \left(\frac{1}{1 - (2t/T)^2} \right). \tag{4.37}$$

At $t = T/2$, even though $\sin(2\pi t/T) = 0$, the denominator term $1 - (2t/T)^2$ is also zero. Taking limits as $t \to T/2$ we can show that $p(T/2) = 0.5$.

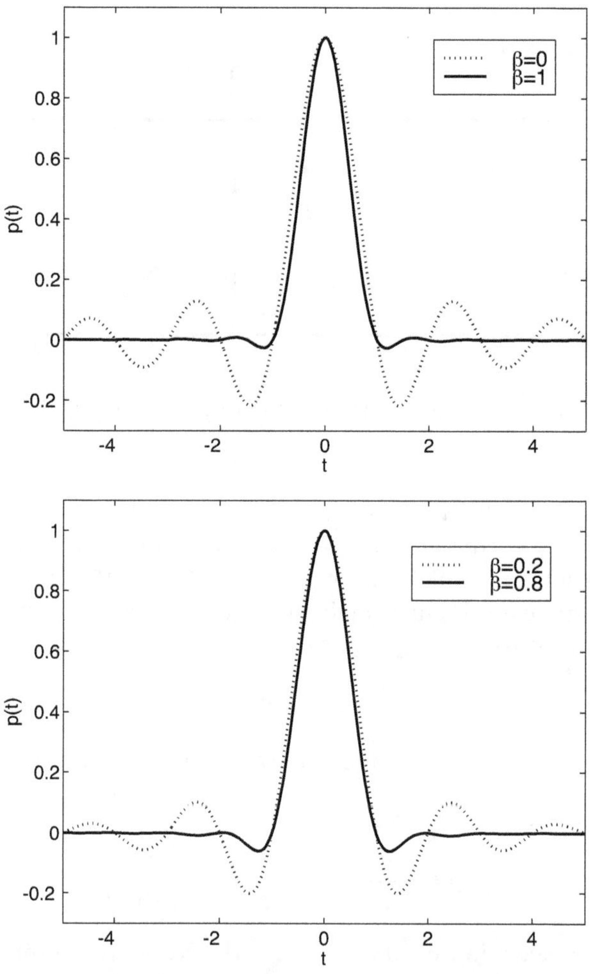

Figure 4.14. The raised-cosine pulse for various values of β. Here $T = 1$.

4.4.2 Generalizations

Generalized versions of the raised cosine have been reported by a number of researchers, and the topic is still of some research interest. These modified versions have some advantages, such as reduced timing estimation error, and reduced error probability for fixed timing errors in the receiver. The reader should consult references such as Beaulieu, Tan, and Damen [2001], Assalini and Tonello [2004], Tan and Beaulieu [2004], and Lee and Beaulieu [2008].

4.5 MIMO systems and multiuser systems

MIMO channels can arise from a single user in a number of ways. For example, the symbol stream $s(n)$ can be blocked and zero-padded or cyclic-prefixed to obtain a matrix equivalent $\mathbf{H}$ of the scalar channel (Chap. 7). Another context is in multiantenna systems. The symbol stream from a single user is first converted from serial to parallel format (by blocking) and each component transmitted through a separate antenna. The antennas are spread apart in space, for example in the form of a uniform linear array. This is shown in Fig. 4.15. The transmitter has M antennas and the receiver also has M antennas. Such multiantenna systems are beneficial because they bring in the additional dimension of spatial diversity [Proakis and Salehi, 2008]. The transmission system between these antennas is a MIMO channel described by an $M \times M$ channel matrix $\mathbf{H}$. Since all the transmitted components $s_k(n)$ are simultaneously available at the transmitter, there can be **collaboration** between them. For example, we can use a $P \times M$ linear precoder $\mathbf{F}_d$ (or a nonlinear precoder) followed by P transmitting antennas. The receiver can use J receiving antennas and an $M \times J$ receiver matrix $\mathbf{G}_d$ (or a nonlinear receiver such as a decision feedback device) to reconstruct the transmitted vector $\mathbf{s}(n)$. In this case the channel is a $J \times P$ transfer matrix. The discrete-time equivalent, derived as in Sec. 4.3, then takes the form in Fig. 4.16, where a linear precoder and a linear equalizer are also shown. We can also employ *power* and *bit allocation* at the transmitter.

The preceding discussion considers a *single-user* system. In Fig. 4.15 it is also possible that the messages $s_k(n)$ correspond to different users. In a *multiuser* system like this, collaboration at the transmitter and receiver may not be possible. There are two typical scenarios which admit a limited amount of collaboration:

1. In a *broadcast system* (also called a downlink system) the multiple messages are transmitted from a central station (e.g., a base station or a satellite). In this case, collaboration is possible at the transmitter. So a precoder $\mathbf{F}_d$ can be used. However, the users at the receiving ends are in different locations and do not collaborate. Each receiver therefore has a single scalar equalizer $G_m(j\omega)$. The received signal usually has interference components from the other users, and these have to be filtered out.

2. In a *multiple-access* system (also called an uplink) several users transmit messages towards a central station. In this case it is possible for the received signal components to collaborate, so a receiver matrix $\mathbf{G}_d$ can be used. Each transmitter, however, can only have a scalar precoder. No power or bit allocation is possible. Many types of multiple access are possible, such as time-division, frequency-division, and code-division multiple access systems, respectively, abbreviated as TDMA, FDMA, and CDMA systems.

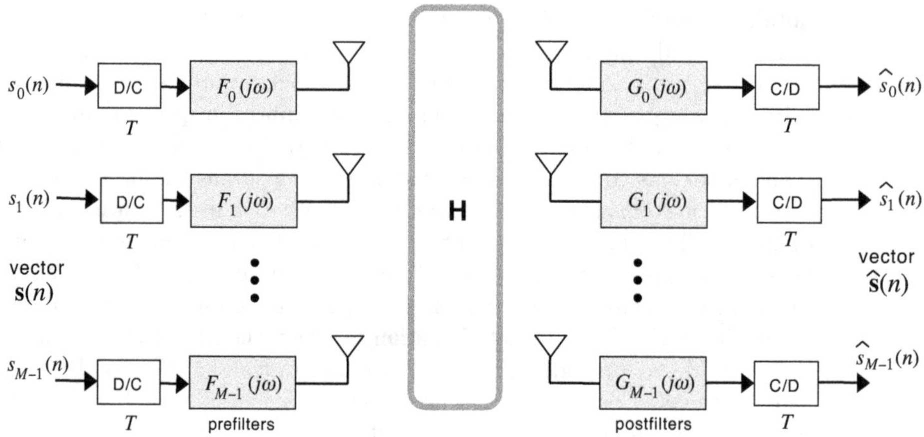

Figure 4.15. A multiple-antenna system for communicating the message from a single user. The effective channel is a MIMO system.

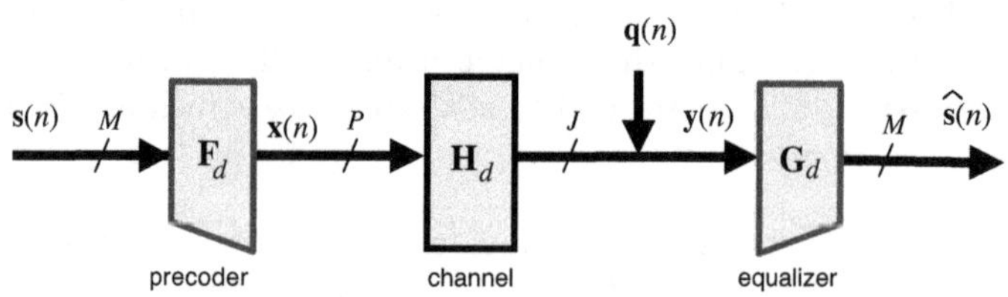

Figure 4.16. Digital equivalent of the MIMO transceiver obtained from the multiple-antenna system for single-user communication.

It is also possible that each user in a multiuser system uses multiple antennas at the transmitter and/or receiver. In this case each user can employ matrix precoders and equalizers, in addition to the limited collaboration between users mentioned above. Many other types of multiuser systems exist, such as two-way systems and store-and-forward networks. The interested reader can consult Proakis and Salehi [2008] and references therein.

4.6 Digital equalization

As explained in Sec. 4.3.1 the digital communication system has the equivalent digital filter model given by Fig. 4.5. Here, $H_d(z)$ has impulse response

$$h_d(n) = h_c(nT),$$

where $h_c(t)$ is the impulse response of the equivalent continuous-time channel

$$H_c(j\omega) = F(j\omega)H(j\omega)G(j\omega)$$

(Fig. 4.4). Ideally, if the pre- and post-filters $F(j\omega)$ and $G(j\omega)$ are such that the product $H_c(j\omega)$ satisfies the zero-crossing constraint $h_c(nT) = \delta(n)$, then $H_d(z) = 1$. In practice it is difficult to design continuous-time filters $F(j\omega)$ and $G(j\omega)$ to satisfy this condition exactly. So the digital filter $H_d(z)$ is not identity. But it can be approximated well with an FIR filter. A simple way to eliminate the channel distortion is to use the inverse digital filter $1/H_d(z)$ at the receiver, after sampling (i.e., after the C/D converter). Then the equivalent digital system is as shown in Fig. 4.17(a). The filter $1/H_d(z)$ is called an **equalizer** because it equalizes or eliminates the effect of $H_d(z)$. It is also called a symbol spaced equalizer, or **SSE**, to distinguish it from the so-called fractionally spaced equalizer (Sec. 4.8). The use of $1/H_d(z)$ assumes that the channel transfer function $H_d(z)$ is known to the receiver. If $H_d(z)$ is an FIR channel, then $1/H_d(z)$ is IIR, and it is required to be a stable transfer function (i.e., all its poles should be inside the unit circle).

The equalizer $1/H_d(z)$ is a **zero-forcing** digital equalizer because its cascade with $H_d(z)$ is identity. An important observation here is that $1/H_d(z)$ is *not necessarily the best equalizer* (even assuming that it is stable), because there is channel noise in practice. The noise manifests in the form of a reconstruction error $e_1(n)$ as shown in Fig. 4.17(a). Given a signal $s(n)$ with additive noise component $e_1(n)$, it is possible to design a filter $W(z)$, called the *Wiener filter* (Fig. 4.17(b)), such that the output $\widehat{s}(n)$ of this filter in response to $s(n)+e_1(n)$ is closer to $s(n)$. The cascaded system

$$G_d(z) = \frac{1}{H_d(z)} \times W(z)$$

is therefore a better equalizer filter than $1/H_d(z)$. We shall study this in greater detail in Sec. 4.10.

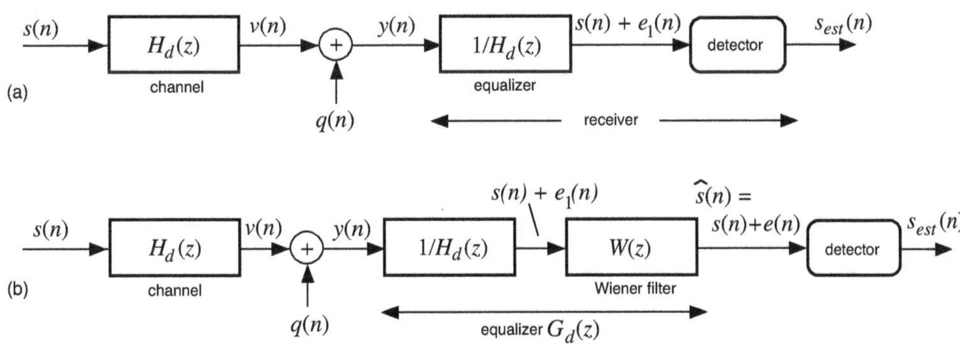

Figure 4.17. (a) The digital communication system with a receiver which eliminates the channel distortion with the help of an equalizer. (b) More general structure for equalization.

We know that the transmitted symbol $s(n)$ belongs to a constellation such as, for example, the PAM or QAM constellation (Sec. 2.2). The detector shown in the figure obtains an estimate $s_{est}(n)$ of this constellation symbol based on the noisy received symbol $\widehat{s}(n)$.

4.7 Oversampling the received signal

Consider Fig. 4.18(a), where the received and filtered signals are sampled with sample spacing T/L, where L is an integer. The sampling rate at the receiver is L times higher than the symbol rate $1/T$. In this case the equivalent digital model is different from Fig. 4.5. We will see in Sec. 4.8 that the digital model is an interpolation filter model (see Fig. 4.21 for a preview). One advantage of oversampling at the receiver is that the digital equalizer, indicated as $G_d(e^{j\omega})$ in Fig. 4.18(a), becomes simpler. For example, FIR channels can be equalized with FIR equalizers, as we shall see. The equalizers are therefore guaranteed to be stable. However, additional bandwidth is required in order to take advantage of oversampling, as we shall explain (Sec. 4.8.5).

One remark is in order here. In principle, *a receiver with oversampling can always be replaced with a receiver with no oversampling.* To show this we use the so-called "noble identity" shown in Fig. 4.19(a) (proved in Sec. G.2 of Appendix G). Using this we see that the digital equalizer $G_d(e^{j\omega})$ can be moved to the left of the C/D block, as shown in Fig. 4.18(b). Defining

$$\widehat{G}(j\omega) = G(j\omega)G_d(e^{j\omega T/L}),$$

and using the simple identity shown in Fig. 4.19(b), we can redraw this as in Fig. 4.18(c). Thus the digital equalizer $G_d(e^{j\omega})$ has been absorbed into the equalizer $\widehat{G}(j\omega)$. This shows that a receiver with oversampling can be replaced with an equivalent receiver with no oversampling.

But there are some practical advantages to oversampling. For example, we will see that FIR equalizers can be used to equalize FIR channels (Sec. 4.8). There are also important advantages when we take into account errors in timing phase at the receiver; see Gitlin and Weinstein [1981], Proakis [1995], and references therein.

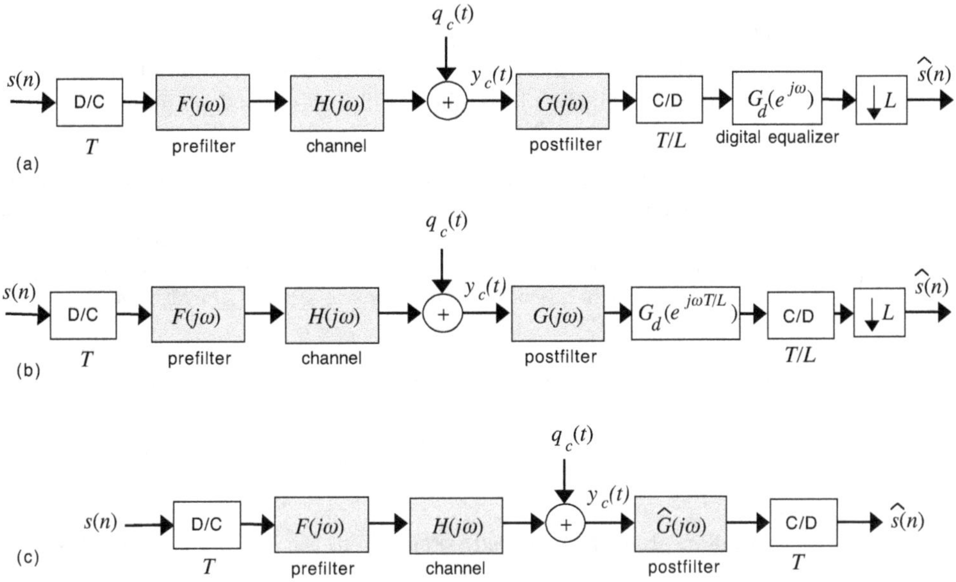

Figure 4.18. (a) The digital communication system with oversampling at the receiver followed by a digital equalizer, (b) equivalent diagram obtained by using a noble identity, and (c) simplified equivalent system with no oversampling.

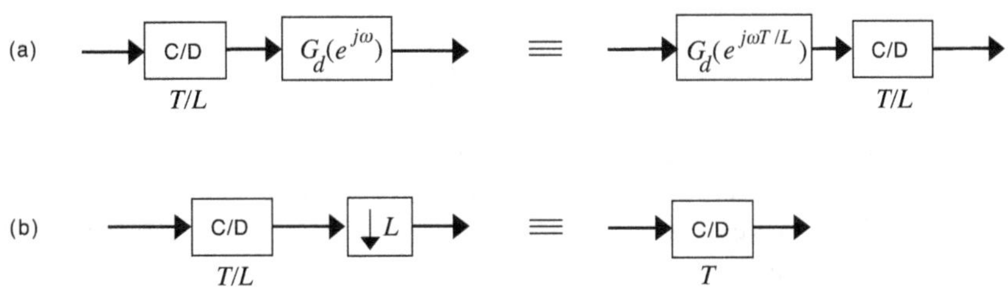

Figure 4.19. (a) and (b) Identities used in the simplification of Fig. 4.18.

4.8 Fractionally spaced equalizers

Consider Fig. 4.20, where the received signal $y_c(t)$ is sampled at a rate L times higher than the symbol rate $1/T$. In this section we analyze this system in some depth. We begin by obtaining the digital equivalent model.

> ***Notations.*** In previous sections we represented the digital channel using the notation $H_d(z)$ (as in Fig. 4.5). We shall find that this notation is not sufficient because the digital channel depends on the oversampling factor L. So we use the notation $C_L(z)$ here (defined below). For the special case of $L = 1$, this reduces to $C_1(z)$, which is nothing but the system $H_d(z)$ of Fig. 4.5. ▽ ▽ ▽

For the oversampled system of Fig. 4.20 the channel output is given by

$$y_c(t) = \sum_{k=-\infty}^{\infty} s(k)h_c(t - kT) + q_c(t) \tag{4.38}$$

as before, but the sampled version is $y_L(n) = y_c(nT/L)$, that is,

$$y_L(n) = \sum_{k=-\infty}^{\infty} s(k)h_c\Big(\frac{nT}{L} - kT\Big) + q_c\Big(\frac{nT}{L}\Big), \tag{4.39}$$

where the subscript L is a reminder that we have oversampled by a factor L. Similarly, let $c_L(n)$ denote the oversampled version of the effective channel $h_c(t)$, and let $q_L(n)$ denote the oversampled noise sequence, that is,

$$c_L(n) = h_c\Big(\frac{nT}{L}\Big), \quad q_L(n) = q_c\Big(\frac{nT}{L}\Big). \tag{4.40}$$

Then

$$y_L(n) = \sum_{k=-\infty}^{\infty} s(k)c_L(n - kL) + q_L(n). \tag{4.41}$$

The first term simply represents an *interpolation filter* operating on the input $s(n)$ (see Sec. 3.4). In terms of multirate signal processing notations (Chap. 3), the preceding equation therefore has the beautiful interpretation shown in Fig. 4.21, where

$$C_L(z) = \sum_n c_L(n)z^{-n}. \tag{4.42}$$

The obvious way to equalize the effect of the channel in the oversampled system is to use the inverse of $C_L(z)$ as the equalizer, as shown in Fig. 4.22(a). Then the reconstruction error $e(n)$ is simply $q_L(n)$ passed through $1/C_L(z)$ and decimated by L. As pointed out earlier, the inverse $1/C_L(z)$ may be unstable, or it may have poles very close to the unit circle. An important observation here is that, if we want to cancel out the effect of $C_L(z)$ completely, it is *not necessary* to use $1/C_L(z)$ as the equalizer because the output is decimated by L anyway.

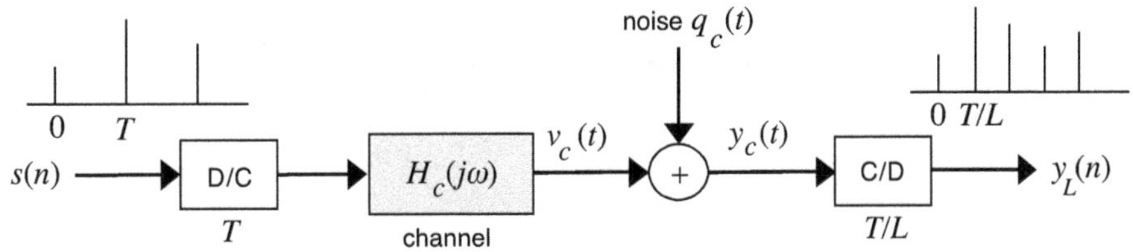

Figure 4.20. A digital communication system with oversampling at the receiver; the oversampling ratio is L.

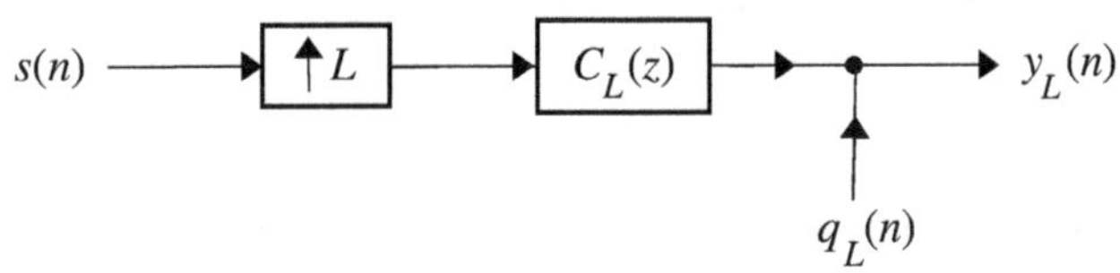

Figure 4.21. Discrete-time equivalent of a digital communication system with oversampling at the receiver. Here L is the oversampling ratio, and the received samples are represented as $y_L(n)$.

To see this, consider Fig. 4.22(b). The system from $s(n)$ to the output $\widehat{s}(n)$ is an LTI system, and its transfer function can be computed by using the polyphase identity (Sec. 3.2.3):

$$T(z) = [C_L(z)G_L(z)]_{\downarrow L}\,, \tag{4.43}$$

which is the *distortion function* in the absence of noise. In general there exist many choices of $G_L(z)$ which make this equal to unity. In fact if the channel $C_L(z)$ is FIR, we will see (Theorem 4.1) that there almost surely exists an FIR solution $G_L(z)$ satisfying

$$[C_L(z)G_L(z)]_{\downarrow L} = 1. \tag{4.44}$$

Since the equalizer $G_L(z)$ operates at the higher sampling rate, it is called a *fractionally spaced equalizer*, abbreviated as **FSE**.

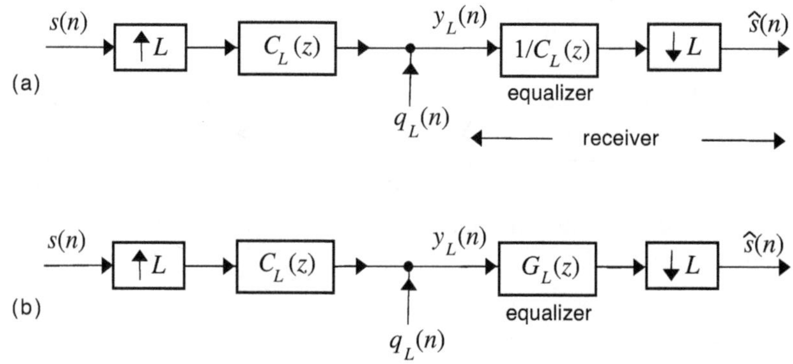

Figure 4.22. Examples of *fractionally spaced equalizers.* (a) The channel inverse $1/C_L(z)$ is directly used, and (b) a different filter $G_L(z)$ is used.

Example 4.1: Fractionally spaced FIR equalizer

Let $L = 2$ and assume that the channel is

$$C_L(z) = 1 + 2z^{-1} + 4z^{-2} + z^{-3} + z^{-5}.$$

Suppose we choose the equalizer to be the FIR filter

$$G_L(z) = \frac{16z - 3 - 4z^{-2}}{29}.$$

(We will soon explain how to *find* such equalizers.) Then

$$C_L(z)G_L(z) = \frac{16z + 29 + 58z^{-1} - 11z^{-3} - 7z^{-5} - 4z^{-7}}{29}$$

so that

$$T(z) = [C_L(z)G_L(z)]_{\downarrow 2} = 1.$$

In practice the advance operator in $G_L(z)$ can be eliminated by using $z^{-2}G_L(z)$. If we do this then

$$T(z) = [z^{-2}C_L(z)G_L(z)]_{\downarrow 2} = z^{-1},$$

and the only distortion is the unit delay.

Note that the FIR channel in the preceding example can be perfectly equalized with the FIR equalizer $G_L(z)$. So, what is the general theory behind this? What are the conditions on the channel which allow this to be done, and how do we solve for the FIR equalizer in general? The answers are contained in the next theorem. In what follows we express the channel $C_L(z)$ and equalizer $G_L(z)$ in polyphase forms (Sec. 3.2.1):

$$C_L(z) = \sum_{k=0}^{L-1} z^{-k} R_k(z^L), \tag{4.45}$$

and

$$G_L(z) = \sum_{k=0}^{L-1} z^{k} E_k(z^L). \tag{4.46}$$

♠**Theorem 4.1.** *FIR fractionally spaced equalizers.* Given an FIR channel $C_L(z) = \sum_{n=0}^{N} c_L(n) z^{-n}$, there exists an FIR $G_L(z)$ such that

$$T(z) \stackrel{\Delta}{=} [C_L(z)G_L(z)]_{\downarrow L} = 1 \tag{4.47}$$

if and only if there is no factor of the form $(1-\alpha z^{-1}), \alpha \neq 0$, common to all polyphase components $R_k(z)$ of $C_L(z)$. ◇

Proof. With $C_L(z)$ and $G_L(z)$ expressed in polyphase form, we have

$$\begin{aligned} C_L(z)G_L(z) &= \sum_{k=0}^{L-1} z^{-k} R_k(z^L) \sum_{\ell=0}^{L-1} z^{\ell} E_\ell(z^L) \\ &= \sum_{m=-L+1}^{L-1} z^m A_m(z^L), \end{aligned}$$

where $A_m(z)$ are appropriate sums of products of $R_k(z)$ and $E_\ell(z)$. Since

$$[z^m A_m(z^L)]_{\downarrow L} = \begin{cases} A_m(z) & \text{if } m = 0 \\ 0 & \text{if } m \text{ is not a multiple of } L, \end{cases}$$

we see that

$$T(z) = [C_L(z)G_L(z)]_{\downarrow L} = A_0(z) = \sum_{k=0}^{L-1} R_k(z)E_k(z)$$

where $R_k(z)$ are FIR filters. If the polyphase components $\{R_k(z)\}$ had a common factor of the form $(1-\alpha z^{-1}), \alpha \neq 0$, then it would be a factor of the above right-hand side. So there cannot be FIR filters $E_k(z)$ such that the right-hand side is unity. On the other hand, if $\{R_k(z)\}$ do not have such a factor, then, from Euclid's theorem (Appendix H), it follows that there exist causal FIR filters $E_k(z)$ such that

$$\sum_{k=0}^{L-1} R_k(z)E_k(z) = 1,$$

as long as there is no factor of the form z^{-K} common to all $R_k(z)$. Even if there is such a factor z^{-K}, we can still find FIR solutions $E_k(z)$ with the noncausal factor z^K in them. An FIR filter $G_L(z)$ can therefore be identified such that $[C_L(z)G_L(z)]_{\downarrow L} = 1$. $\bigtriangledown\bigtriangledown\bigtriangledown$

Theorem 4.1 says that there should be no common zero shared by all the L polyphase components of $C_L(z)$, that is, their greatest common factor should be of the form

$$\gcd\{R_k(z)\} = z^{-K} \tag{4.48}$$

for some integer K. If the conditions of the theorem are satisfied, then the FIR components $E_k(z)$ of the equalizer satisfying

$$\sum_{k=0}^{L-1} E_k(z)R_k(z) = 1 \tag{4.49}$$

can be found using Euclid's algorithm (Appendix H). If the solution $G_L(z)$ is not causal, a causal solution $z^{-JL}G_L(z)$ can be obtained by using large enough J, and results in an extra delay z^{-J} at the output.

As a practical matter, it is sufficient to choose the oversampling factor to be $L = 2$. Then the oversampled channel and equalizer have the form

$$C_2(z) = R_0(z^2) + z^{-1}R_1(z^2), \quad G_L(z) = E_0(z^2) + zE_1(z^2). \tag{4.50}$$

Given an arbitrary FIR channel $C_2(z)$ it is unlikely that its polyphase components $R_0(z)$ and $R_1(z)$ have a common factor. So, in nearly all practical situations we can find FIR $E_0(z)$ and $E_1(z)$ to equalize the channel.

Example 4.2: Fractionally spaced FIR equalizer: polyphase view

Now it is easy to see how the equalizer in Ex. 4.1 was found. We have $C_2(z) = 1 + 2z^{-1} + 4z^{-2} + z^{-3} + z^{-5}$ so that

$$R_0(z) = 1 + 4z^{-1} \quad \text{and} \quad R_1(z) = 2 + z^{-1} + z^{-2},$$

which have no common factor. If we apply Euclid's algorithm summarized in Appendix H, we find (Problem 4.4)

$$E_0(z) = \frac{-3 - 4z^{-1}}{29}, \quad E_1(z) = \frac{16}{29}$$

Indeed, we can verify by explicit calculation that

$$R_0(z)E_0(z) + R_1(z)E_1(z) = 1.$$

The FIR equalizer is then

$$G_2(z) = E_0(z^2) + zE_1(z^2) = (16z - 3 - 4z^{-2})/29$$

as claimed earlier.

4.8.1 FSE structure in polyphase form

With $C_L(z)$ and $G_L(z)$ expressed in the polyphase forms (4.45) and (4.46), the system in Fig. 4.23(a) can be redrawn as in Fig. 4.23(b). The system within the shaded box is the identity system, as proved earlier in Chap. 3 (Fig. 3.17). So the system simplifies to the one shown in Fig. 4.23(c). Note that the channel has been converted into an equivalent *single-input multi-ouput system* and the equalizer is an equivalent *multi-input single-output system.* In short this is equivalent to an analysis filter bank followed by a synthesis filter bank without any decimators or expanders.

Biorthogonal partners. If two transfer functions $C_L(z)$ and $G_L(z)$ are such that $[C_L(z)G_L(z)]_{\downarrow L} = 1$, we say that they are biorthogonal partners with respect to the integer L. With the transfer functions expressed in polyphase forms (4.45) and (4.46), this is equivalent to the condition

$$\sum_{k=0}^{L-1} R_k(z)E_k(z) = 1, \tag{4.51}$$

as shown by the proof of Theorem 4.1. Given the transfer function $C_L(z)$, its inverse $1/C_L(z)$ serves as a biorthogonal partner. As shown by Example 4.1 the partner is not unique. In fact, whenever $C_L(z)$ is FIR, we can find an FIR partner (but not an FIR inverse unless $C_L(z) = cz^{-m}$) as long as the polyphase components of $C_L(z)$ do not share a common factor $1-\alpha z^{-1}, \alpha \neq 0$, as shown by Theorem 4.1. Further details about the theory and applications of biorthogonal partners can be found in Vaidyanathan and Vrcelj [2001]. The concept has also been extended to the case of matrix transfer functions, and is applicable to the equalization of MIMO channels [Vrcelj and Vaidyanathan, 2002]. ▽ ▽ ▽

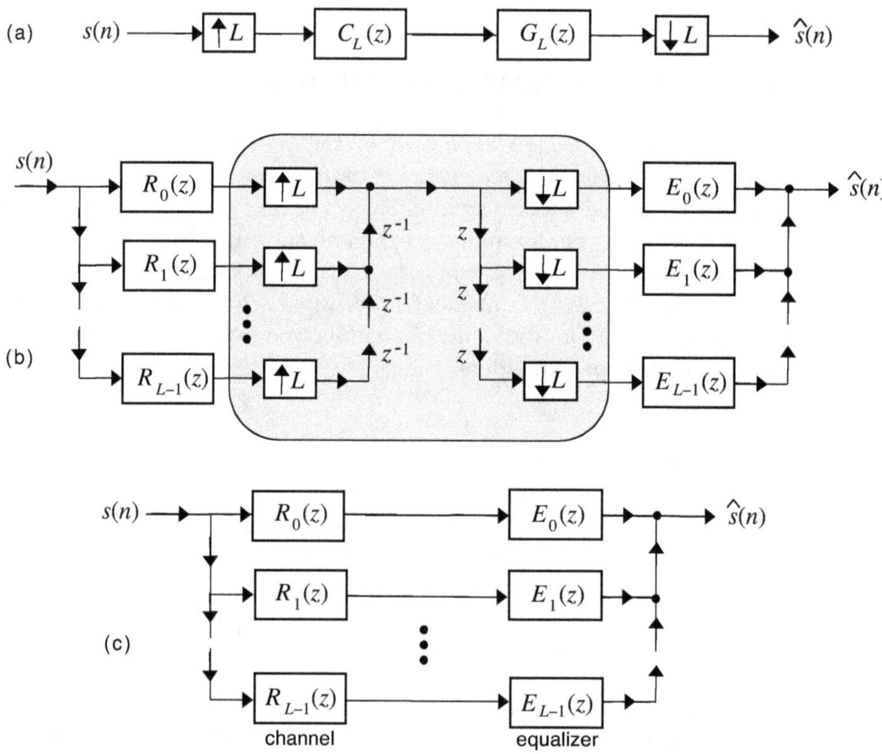

Figure 4.23 (a) The oversampled channel $C_L(z)$ and fractionally spaced equalizer $G_L(z)$, (b) polyphase representation, and (c) simplified equivalent.

4.8.2 Comparing SSE and FSE with examples

For the special case where the fractionally spaced equalizer uses an oversampling rate $L = 2$, Fig. 4.24(a) shows the equalizer system and Fig. 4.24(b) shows the corresponding polyphase representation.

The noise sources $q_{2,0}(n)$ and $q_{2,1}(n)$ are the even and odd samples of the oversampled channel noise $q_2(n)$. Consider a second-order FIR channel

$$C_1(z) = 1 - 2\rho\cos\theta z^{-1} + \rho^2 z^{-2}$$

with a complex conjugate pair of zeros at $\rho e^{\pm j\theta}$, and assume

$$\rho = 0.999, \quad \theta = 0.2\pi.$$

So $1/C_1(z)$ has poles $0.999e^{\pm 0.2j\pi}$ very close to the unit circle. Assume the channel noise $q(n)$ is white Gaussian with zero mean and variance

$$\sigma_q^2 = 10^{-5}.$$

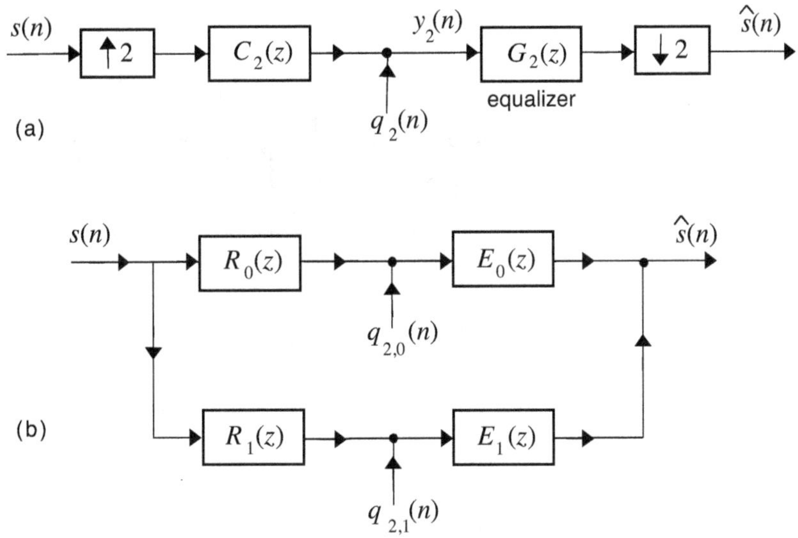

Figure 4.24. Special case with oversampling factor $L = 2$. (a) The oversampled channel $C_2(z)$ and fractionally spaced equalizer $G_2(z)$; (b) equivalent representation using polyphase components.

The input signal $s(n)$ to be transmitted on the channel is as shown in the top plot of Fig. 4.25 (upper trace). This is only a "toy" signal (not a sequence of symbols from a realistic constellation), as the purpose here is only to demonstrate the basic difference between SSE and FSE. The noise is also shown in the top plot (bottom trace; shifted down by 0.5 for clarity). If we use a symbol spaced equalizer (SSE) $1/C_1(z)$, then the reconstructed signal $\widehat{s}(n)$ is as shown in the middle plot. Note the large amplification of error created by $1/C_1(z)$, which has poles very close to the unit circle. If we use a fractionally spaced equalizer as in Fig. 4.24(b), with oversampling factor $L = 2$, then there is dramatic improvement in performance (bottom plot).

4.8.2.A Solving for the FIR FSE

In this example the original and oversampled channels are

$$C_1(z) = 1.0000 - 1.6164z^{-1} + 0.9980z^{-2} \tag{4.52}$$

$$C_2(z) = 1.0000 - 0.3082z^{-1} - 1.6164z^{-2} - 0.3092z^{-3} + 0.9980z^{-4}. \tag{4.53}$$

The polyphase compoments of $C_2(z)$ are

$$R_0(z) = 1.0000 - 1.6164z^{-1} + 0.9980z^{-2}, \quad R_1(z) = -0.3082 - 0.3092z^{-1}.$$

Solving for the polyphase components $E_0(z)$ and $E_1(z)$ of the equalizer, satisfying

$$R_0(z)E_0(z) + R_1(z)E_1(z) = 1,$$

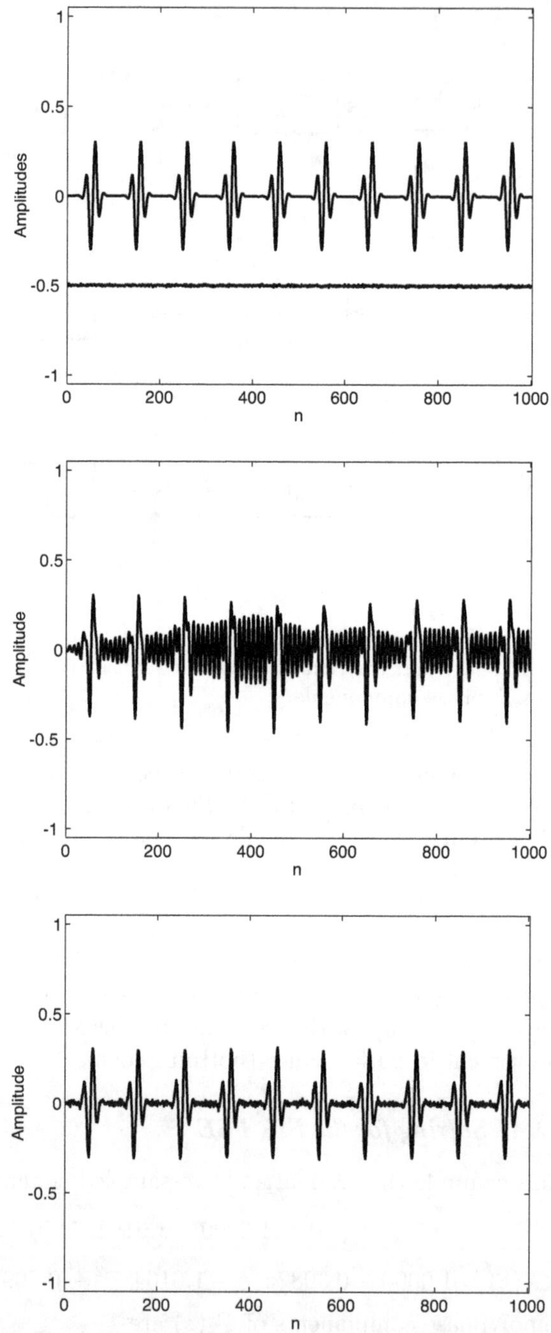

Figure 4.25. Comparison of symbol spaced equalizer (SSE) and fractionally spaced equalizer (FSE). Top: transmitted signal $s(n)$ (upper trace) and channel noise $q_1(n)$ (lower trace). Middle: reconstructed signal for the SSE. Bottom: reconstructed signal for the FSE. The channel $C_1(z)$ is second-order FIR with zeros at $0.999e^{\pm 0.2j\pi}$. So $1/C_1(z)$ is stable, but has poles very close to the unit circle.

we find

$$E_0(z) = 0.2776z^{-1}, \quad E_1(z) = -2.3440z^{-1} + 0.8959z^{-2}.$$

Here we have inserted a delay z^{-1} into each polyphase component so that the FSE transfer function is causal:

$$G_2(z) = E_0(z^2) + zE_1(z^2) = -2.3440z^{-1} + 0.2776z^{-2} + 0.8959z^{-3}. \quad (4.54)$$

The reader can verify by explicit computation that

$$T(z) = [C_2(z)G_2(z)]_{\downarrow 2} = z^{-1}$$

and similarly $R_0(z)E_0(z) + R_1(z)E_1(z) = z^{-1}$. Thus $\widehat{s}(n) = s(n-1)$ in the absence of noise. If we had not inserted delays into $E_k(z)$ we would have the noncausal equalizer $G_2(z) = -2.3440z + 0.2776 + 0.8959z^{-1}$.

4.8.2.B Channel with zeros outside the unit circle

A variation of the example is obtained by changing ρ from 0.999 to 1.01. In this case, the channel has zeros outside the unit circle. So $1/C_1(z)$ is unstable, and the symbol spaced equalizer does not work at all. That is, the plot of $\widehat{s}(n)$ grows unboundedly because the tiny channel noise is amplified by the unstable system $1/C_1(z)$. The fractionally spaced equalizer, however, works wonderfully, as seen from the reconstructed signal shown in Fig. 4.26 (bottom trace).

In this example we have

$$C_1(z) = 1.0000 - 1.6342z^{-1} + 1.0201z^{-2}$$

and

$$C_2(z) = 1.0000 - 0.3171z^{-1} - 1.6342z^{-2} - 0.3071z^{-3} + 1.0201z^{-4}.$$

The polyphase components of the channel are

$$R_0(z) = 1.0000 - 1.6342z^{-1} + 1.0201z^{-2}, \quad R_1(z) = -0.3171 - 0.3071z^{-1}.$$

Solving for the polyphase components $E_0(z)$ and $E_1(z)$ of the equalizer, we find

$$E_0(z) = 0.2649z^{-1}, \quad E_1(z) = -2.3183z^{-1} + 0.8799z^{-2}.$$

Again, we have inserted a delay z^{-1} into each polyphase component so that the FSE transfer function is causal:

$$G_2(z) = E_0(z^2) + zE_1(z^2) = -2.3183z^{-1} + 0.2649z^{-2} + 0.8799z^{-3}.$$

The reader can verify by explicit calculation that $T(z) = [C_2(z)G_2(z)]_{\downarrow 2} = z^{-1}$.

4.8.2.C Scatter diagrams

For the SSE and FSE systems in the above example, assume now that the input $s(n)$ is actually a symbol stream from a 6-bit QAM constellation. This

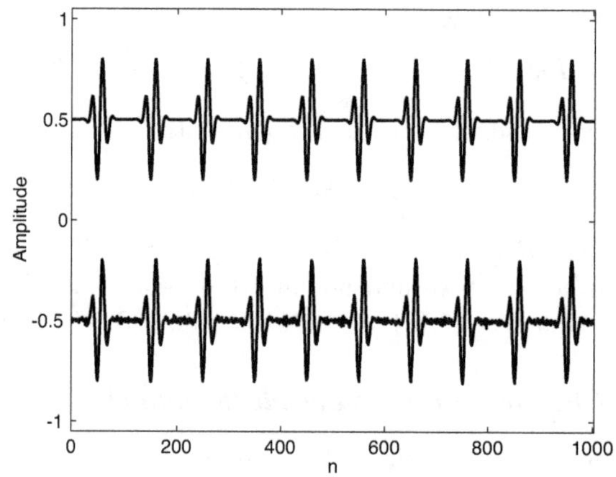

Figure 4.26. Performance of the FSE when the channel has zeros *outside* the unit circle. The transmitted and reconstructed signals are shown (upper and lower traces, respectively; vertical shifting is for clarity). The channel $C_1(z)$ is second-order FIR with zeros at $1.01e^{\pm 0.2j\pi}$, so $1/C_1(z)$ is *unstable*. The SSE method yields unbounded results for $\widehat{s}(n)$, but the FSE method works well, as the above plot shows.

constellation has 64 symbols, as shown in Fig. 4.27 (top). If such a symbol stream is transmitted over the noisy channel then the received signal $\widehat{s}(n)$ after equalization will contain distorted versions of the constellation symbols because some of the channel noise still gets through the equalizers.

Assume that $s(n)$ is a symbol stream with uncorrelated symbols. If we plot the locations of the received noisy symbols in the complex plane for a large number of received samples, we get a plot like the one shown in Fig. 4.27 (middle). In this example, the channel is as in Eq. (4.52) and the equalizer $1/C_1(z)$ (SSE) is assumed to be used. The noise is assumed to be zero-mean complex with independent real and imaginary parts, each having a variance of 0.00004. We see that the constellation symbols are spread out, or scattered, due to the random noise from the channel. The plot in the figure is called a **scatter diagram**. It is similar to the noise cloud diagram shown in Fig. 2.6, Chap. 2. From the scatter diagram we see that, if the noise is large enough, a received symbol can be mistaken for a neighboring symbol.

Instead of using the SSE, if we use the fractionally spaced equalizer (4.54), then the scatter diagram is as shown in the bottom plot of Fig. 4.27. The scatter diagram for the FSE is clearly much better than that for the SSE.

For reference, Fig. 4.28 shows what happens to the constellation if the channel noise is added directly to the input constellation (bottom). Clearly, the effect of noise acting directly on the constellation is negligible. But when the noise is amplified by the equalizers, its effect on the received symbols is more noticeable, as seen from Fig. 4.27. Still, the use of FSE produces a much better scatter diagram than SSE in this example.

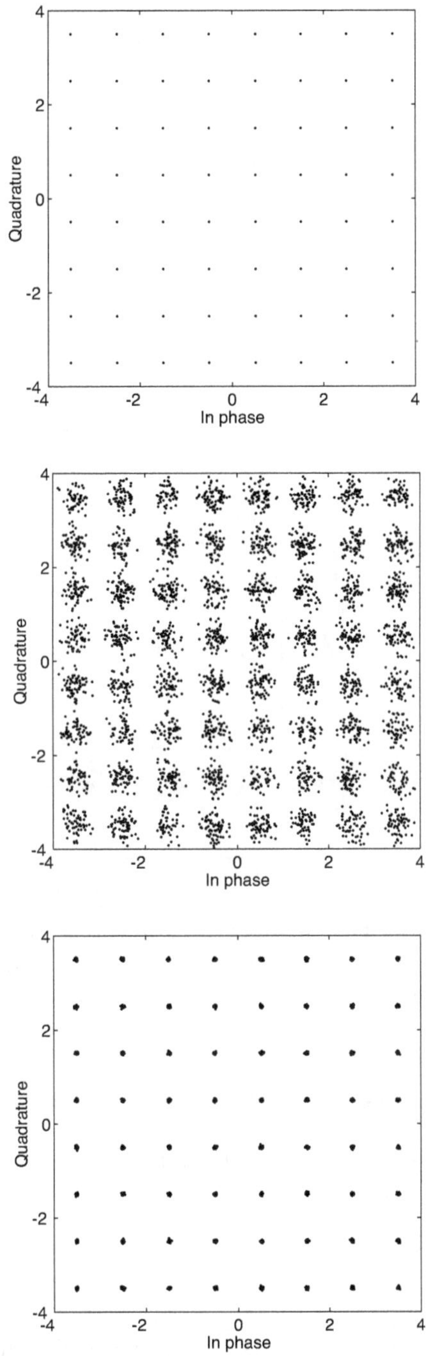

Figure 4.27. The input QAM constellation (top), scatter diagram for the equalized symbol stream with SSE (middle), and scatter diagram for the equalized symbol stread with FSE (bottom). The channel is as in Eq. (4.52), the SSE is $1/C_1(z)$, and the FSE is as in (4.54).

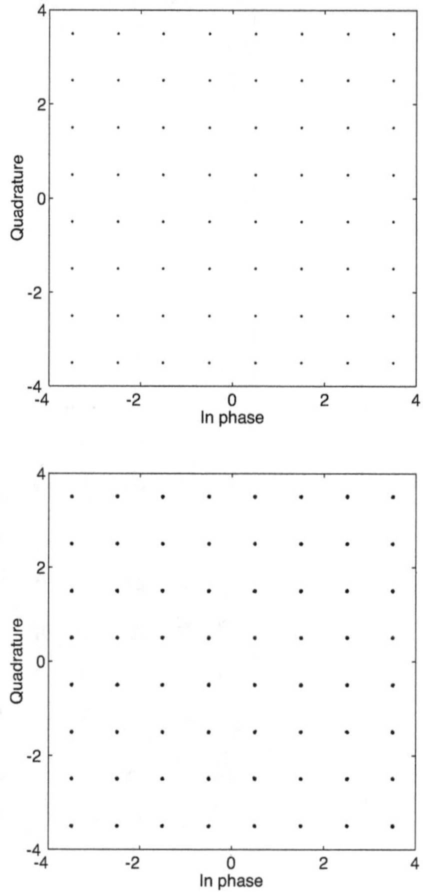

Figure 4.28. The input QAM constellation (top), and the scatter diagram obtained by adding the channel noise directly to the input (bottom). The amount of noise is the same as the channel noise, which produces Fig. 4.27 after equalization.

4.8.3 The zero-forcing property of the equalizer

The condition for perfect symbol recovery (in the absence of noise) in Fig. 4.23(a) is that the transfer function defined as

$$D(z) = C_L(z)G_L(z)$$

satisfies the property $[D(z)]_{\downarrow L} = 1$. From Chap. 3 (Sec. 3.4.2) we know that this is nothing but the Nyquist(L) property of an interpolation filter. This property ensures that there is no intersymbol interference or **ISI**. That is, the nth sample $\widehat{s}(n)$ depends only on the nth sample $s(n)$, and not on $s(n-k), k \neq 0$. Figure 4.29(a) shows an example of the impulse response of $D(z)$ satisfying the property $[D(z)]_{\downarrow L} = 1$ for $L = 3$.

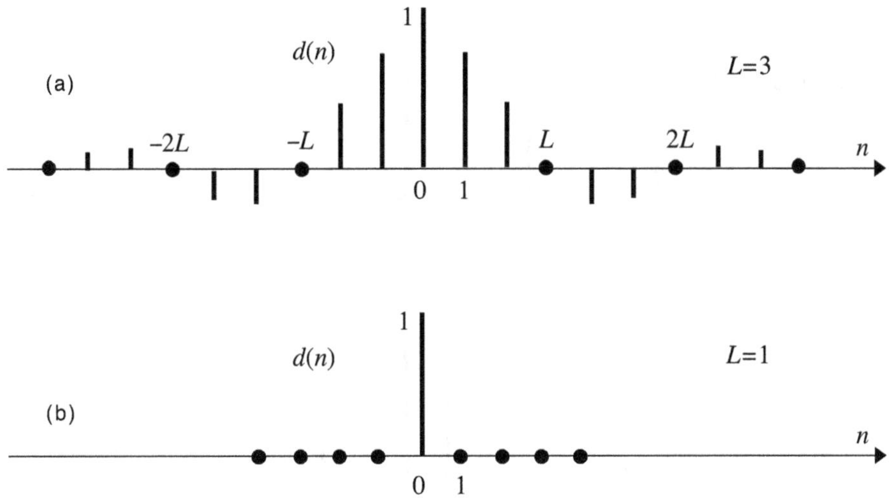

Figure 4.29. Examples of impulse response of the product $D(z) = C_L(z)G_L(z)$ satisfying the Nyquist(L) condition. (a) Case where $L = 3$ and (b) case where $L = 1$.

Notice that the condition $[D(z)]_{\downarrow L} = 1$ is equivalent to the statement that

$$d(n) = \begin{cases} 1 & \text{for } n = 0 \\ 0 & \text{for } n = \text{nonzero multiples of } L, \end{cases}$$

or, equivalently,

$$d(Ln) = \delta(n).$$

That is, $d(n)$ has **zero crossings** at nonzero multiples of L.

Figure 4.30 shows an example of the input symbol stream and the impulse response $d(n)$, with real-time dimensions indicated. Since $G_L(z)$ is such that these zero crossings are forced on the impulse response of the product $D(z) = C_L(z)G_L(z)$, it is a **zero-forcing** equalizer (ZFE). For $L = 1$ (SSE case) this zero-crossing property says that $d(n)$ is as in Fig. 4.29(b), that is, $D(z) = 1$. That is, the zero-forcing symbol spaced equalizer is just the channel inverse $G_L(z) = 1/C_L(z)$. But for the FSE case there is a lot more freedom because the values of $d(n)$ when n is not a multiple of L are allowed to be arbitrary.

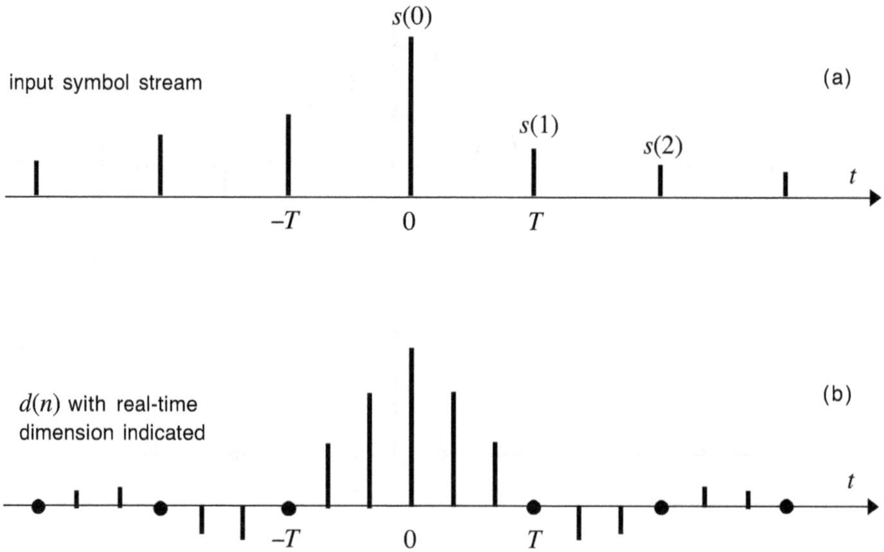

Figure 4.30. (a) The input symbol stream with real-time dimensions, and (b) the impulse response of $D(z) = C_L(z)G_L(z)$ with real-time dimensions, which results in perfect symbol recovery in absence of noise.

4.8.4 Non-uniqueness of FSE

The fact that the biorthogonal partner is not unique means that the FSE is not unique for a given channel. Even though all FSEs are equivalent in the absence of channel noise, this is not the case when there is noise. In practice it is therefore wise to choose the FSE such that the effect of noise is minimized. To see how this can be accomplished, consider the example where $L = 2$. In this case the channel $C_2(z)$ and the FSE function $G_2(z)$ have the polyphase form

$$C_2(z) = R_0(z^2) + z^{-1}R_1(z^2), \quad G_2(z) = E_0(z^2) + zE_1(z^2), \tag{4.55}$$

and the condition for perfect equalization is

$$R_0(z)E_0(z) + R_1(z)E_1(z) = 1. \tag{4.56}$$

Note now that, given any transfer function $A(z)$, the following identity holds trivially:

$$R_0(z)R_1(z)A(z) - R_1(z)R_0(z)A(z) = 0. \tag{4.57}$$

This means that, if we define a new transfer function $\widehat{G}_2(z)$ with polyphase components

$$\widehat{E}_0(z) = E_0(z) + R_1(z)A(z), \quad \widehat{E}_1(z) = E_1(z) - R_0(z)A(z), \tag{4.58}$$

then

$$\begin{aligned} T(z) &= R_0(z)\widehat{E}_0(z) + R_1(z)\widehat{E}_1(z) \\ &= R_0(z)\Big(E_0(z) + R_1(z)A(z)\Big) + R_1(z)\Big(E_1(z) - R_0(z)A(z)\Big) \\ &= R_0(z)E_0(z) + R_1(z)E_1(z) \\ &= 1, \end{aligned}$$

so that $\widehat{G}_2(z)$ is a valid zero-forcing FSE as well. Since this is true for any $A(z)$, we can optimize it to minimize the effect of the channel noise in the reconstructed symbol stream. Figure 4.31 demonstrates for $L = 2$ how the FSE scheme can be implemented with $A(z)$ in place. In practice we typically fix the order of $A(z)$ (so that the cost is fixed) and optimize its coefficients based on our knowledge of the statistics of noise. Then the performance of the FSE becomes even better than what we have seen in the examples of Figs. 4.25 and 4.26. Further details of this idea can be found in Vrcelj and Vaidyanathan [2003].

4.8.5 Need for excess BW

Suppose the continuous-time channel $H_c(j\omega)$ is bandlimited to $|\omega| < \pi/T$, so that it does not have any excess bandwidth (see Sec. 4.3.1.A), as shown in Fig. 4.32(a). In this case, the oversampled digital channel $C_L(e^{j\omega})$ is bandlimited to $|\omega| < \pi/L$, as demonstrated in Fig. 4.32(b). Now consider the zero-forcing condition (4.44), reproduced below:

$$[C_L(e^{j\omega})G_L(e^{j\omega})]_{\downarrow L} = 1. \tag{4.59}$$

Since $C_L(e^{j\omega})$ is bandlimited to $|\omega| < \pi/L$, the decimated expression on the left hand side is nothing but the stretched version (Sec. 3.2.2)

$$\frac{C_L(e^{j\omega/L})G_L(e^{j\omega/L})}{L}, \quad \text{for } 0 \le |\omega| < \pi. \tag{4.60}$$

This repeats periodically with period 2π. So the zero-forcing condition can be written as

$$C_L(e^{j\omega})G_L(e^{j\omega}) = \begin{cases} L, & |\omega| < \pi/L \\ 0 & \text{otherwise.} \end{cases} \tag{4.61}$$

Thus the zero-forcing equalizer can be taken as

$$G_L(e^{j\omega}) = \begin{cases} L/C_L(e^{j\omega}), & |\omega| < \pi/L \\ 0 & \text{otherwise.} \end{cases} \tag{4.62}$$

This means that we are essentially inverting the channel as in the SSE case, so there is no advantage to oversampling, as far as equalizer design is concerned.

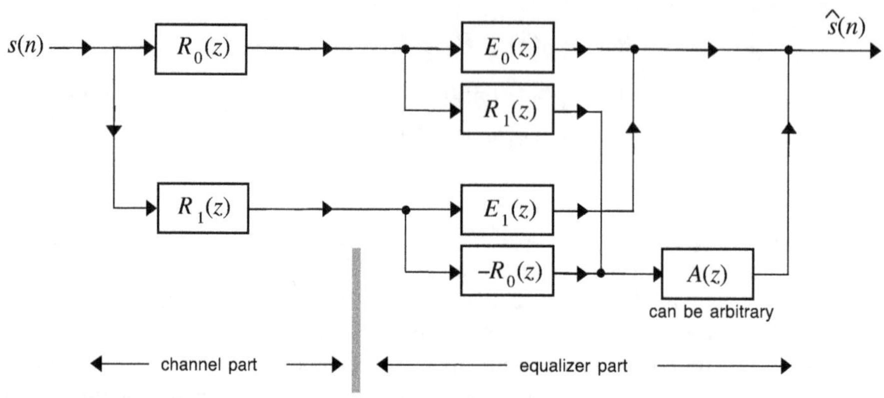

Figure 4.31. The FSE schematic for $L = 2$, with the redundant transfer function $A(z)$ included. In the absence of channel noise the ouput $\widehat{s}(n)$ is unaffected by the choice of $A(z)$. Thus in the presence of noise, $A(z)$ offers an extra degree of freedom that can be optimized such that $\widehat{s}(n)$ is as close to $s(n)$ as possible.

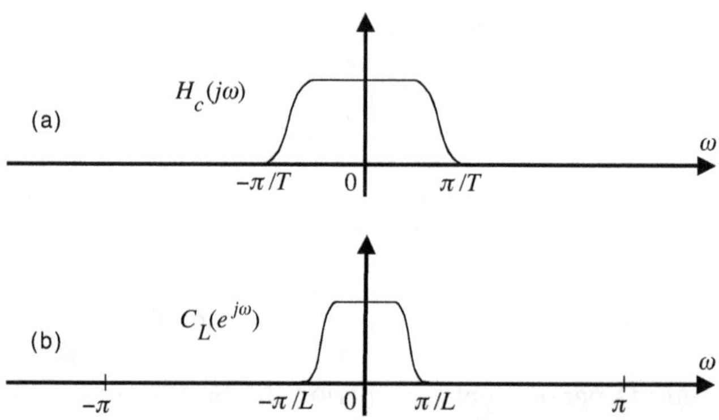

Figure 4.32. (a) A continuous-time channel without excess bandwidth, and (b) the oversampled version with oversampling ratio L.

Summarizing, when there is no excess bandwidth, (i.e., $H_c(j\omega)$ is bandlimited to $\omega| < \pi/T$), there is no benefit to oversampling. The benefits of FSE arise when there is excess bandwidth, which is usually the case in practice. In Chap. 10 we shall briefly revisit this issue in the context of optimal pre- and post-filtering in transceivers.

4.9 Noble identities and digital design of filters

In a digital communication system, such as the one in Fig. 4.1, the prefilter $F(j\omega)$ and postfilter $G(j\omega)$ are designed based on a number of considerations including the minimization of reconstruction error (Chap. 10). These filters often turn out to be complicated functions of frequency. As such, it is not easy to design and implement them as continuous-time filters. It turns out, however, that these filters can be approximated quite well by using *digital filters*, which operate in the sampled data domain. To explain this idea we first introduce a couple of identities based on the sampling theorem. Shown in Fig. 4.33, these are called **noble identities** and they are proved in Appendix G (Sec. G.2). The first identity says that a digital filter $P(z)$ *following* a C/D converter can be moved to the left of the C/D converter, as long as ω in $P(e^{j\omega})$ is replaced with ωT. The second identity says that a digital filter *preceding* a D/C converter can be moved to the right by using the identity shown in Fig. 4.33(b). Now consider Fig. 4.34(a), which shows the digital communication system with a digital prefilter $F_d(e^{j\omega})$ and a continuous-time prefilter $F_c(j\omega)$. Similarly there are two postfilters as shown. (Notice the slight difference in notation with respect to Fig. 4.18.) By using the two noble identities this system can be redrawn as in Fig. 4.34(b) where the combined prefilter is

$$F(j\omega) = F_c(j\omega)F_d(e^{j\omega T}), \tag{4.63}$$

and similarly the combined postfilter is

$$G(j\omega) = G_c(j\omega)G_d(e^{j\omega T}). \tag{4.64}$$

This shows that the digital filter part $F_d(e^{j\omega T})$ (and similarly $G_d(e^{j\omega T})$) can be completely absorbed into the continuous-time filter. At this point the reader might wonder why it is necessary to use a digital filter at all. There are some advantages to this. While the continuous-time filter cannot approximate *arbitrary* shapes easily, the design of *digital* filters with nearly arbitrary frequency response shapes is quite straightforward [Oppenheim and Schafer, 1999]. In the continuous-time case, *lowpass* filters are easy to design. It is therefore practicable to use $G_c(j\omega)$ to approximate a good lowpass filter in $|\omega| < \pi/T$, and $G_d(e^{j\omega})$ to achieve arbitrary shapes within $|\omega| < \pi/T$. This is demonstrated in Fig. 4.35. Part (a) shows an arbitrary digital filter response $G_d(e^{j\omega})$ and part (b) shows $G_d(e^{j\omega T})$, which is exactly the digital filter response, appropriately scaled. So this is a periodic function with period $2\pi/T$. If this is multiplied by the ideal lowpass filter in part (c), then the result is $G_d(e^{j\omega T})$, restricted to the first period $|\omega| < \pi/T$.

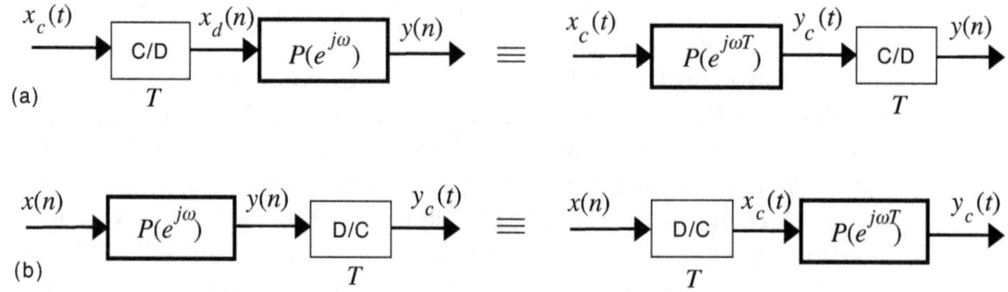

Figure 4.33. The noble identities, (a) for C/D converters, and (b) for D/C converters.

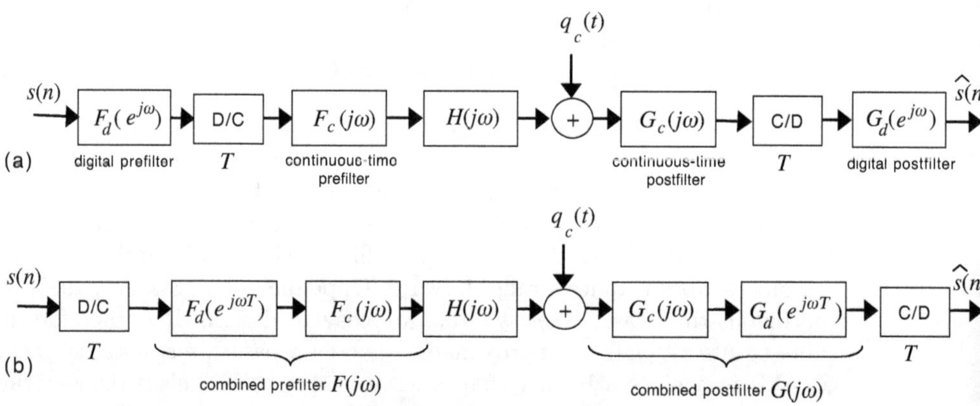

Figure 4.34. (a) The digital communication system with a pair of analog pre- and postfilters, and a pair of digital pre- and postfilters. (b) Equivalent redrawing.

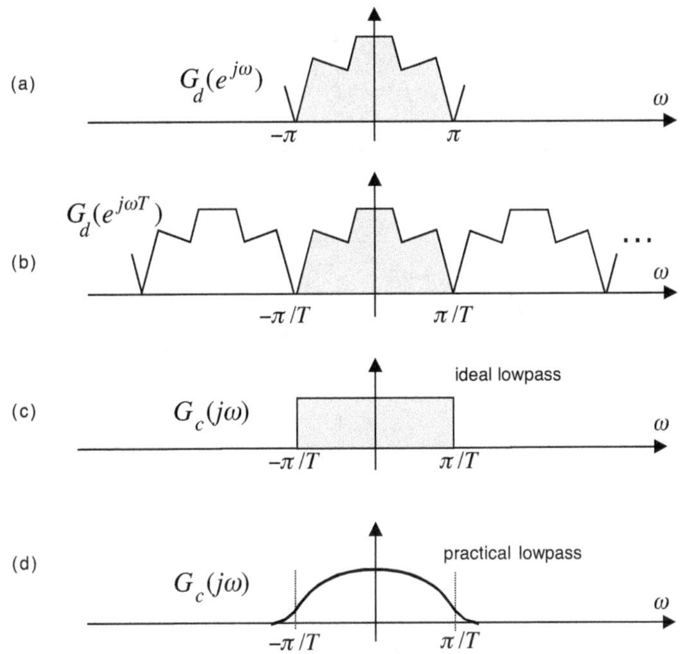

Figure 4.35. (a) Example of a digital postfilter response $G_d(e^{j\omega})$; (b) the scaled response $G_d(e^{j\omega T})$; (c) example of an ideal lowpass postfilter $G_c(j\omega)$; (d) example of a nonideal lowpass postfilter $G_c(j\omega)$.

In practice, the continuous-time filter can only approximate the ideal lowpass response, for example, as in Fig. 4.35(d). In this case the product $G_d(e^{j\omega T})G_c(j\omega)$ can still approximate any desired shape in $|\omega| < \pi/T$ because the digital filter can be adjusted in $|\omega| < \pi$. However, in the region $|\omega| > \pi/T$, the filter $G_d(e^{j\omega T})$ can only repeat itself. The response $G_c(j\omega)$ attentuates these extra copies with an efficacy that depends on how good a lowpass filter it is. If the receiver uses *oversampling*, then there are some additional advantages in the design of the filters. Equation (4.64) is now replaced with

$$G(j\omega) = G_c(j\omega)G_d(e^{j\omega T/L}), \tag{4.65}$$

where L is the oversampling factor. In this case the response $G_d(e^{j\omega T/L})$ has a longer period $2\pi L/T$. For sufficiently large L, the first period of $G_d(e^{j\omega T/L})$ will be able to cover the entire region where $G_c(j\omega)$ is significant. See Fig. 4.36. Thus, even in a situation where a postfilter $G(j\omega)$ has to be designed to have *excess bandwidth*, we can achieve this by appropriate design of the digital filter $G_d(e^{j\omega})$, as long as the receiver uses a large enough oversampling factor L. In fact, if L is sufficiently large, then there is a significant gap between $L\pi/T$ and σ, as indicated in the figure. This means that the digital filter can be designed with a transition band, or "don't care" band, as indicated in Fig. 4.36(d), which makes the filter design problem easier.

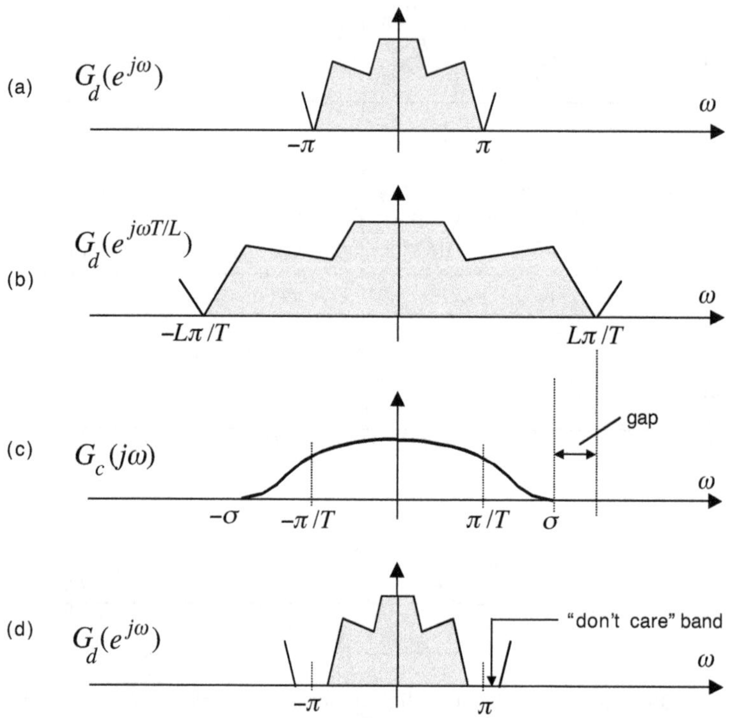

Figure 4.36. (a) Example of a digital postfilter response $G_d(e^{j\omega})$; (b) the scaled response $G_d(e^{j\omega T/L})$, with oversampling ratio L; (c) example of a nonideal lowpass postfilter $G_c(j\omega)$ with excess bandwidth; (d) example of a digital postfilter $G_d(e^{j\omega})$ with a "don't care" region.

4.10 MMSE equalization

The equalizers introduced in Secs. 4.6 and 4.8 work under the zero-forcing constraint. This constraint guarantees that the transfer function between the transmitted signal $s(n)$ and the received signal $\widehat{s}(n)$ is identity. It is possible to relax this constraint, and design the equalizer such that the mean square error between $s(n)$ and $\widehat{s}(n)$,

$$\mathcal{E} = E[|\widehat{s}(n) - s(n)|^2], \tag{4.66}$$

is minimized. Such an equalizer is called a minimum mean square error equalizer, or **MMSE**, equalizer. The theory of such equalizers comes from the theory of **Wiener** filtering (Appendix F). MMSE equalizers offer smaller mean square reconstruction error than zero-forcing equalizers. With the detector appropriately designed, this eventually results in smaller error probability, as we shall repeatedly observe in later chapters.

In this section we describe how MMSE equalizers are designed. We will first address the case of a MIMO channel and then consider the special case of channels with single input and output. Thus, we begin with the equalization of a discrete-time channel (Fig. 4.37) with transfer matrix $\mathbf{C}(z)$, possibly rectangular. This matrix is presumably obtained by discretizing a continuous-time channel as in Sec. 4.3 (the notation $\mathbf{H}_d(z)$ is replaced with the more convenient $\mathbf{C}(z)$ here). The equalizer $\mathbf{G}(z)$ is assumed to have the form

$$\mathbf{G}(z) = \sum_{m=-\infty}^{\infty} \mathbf{g}(m) z^{-m}, \tag{4.67}$$

even though in practice it is desirable to make it a causal and FIR system. We make the assumption that the input $\mathbf{s}(n)$ and the noise $\mathbf{q}(n)$ are jointly wide sense stationary processes (Appendix E). So all the signals involved are WSS, including the reconstruction error

$$\mathbf{e}(n) = \widehat{\mathbf{s}}(n) - \mathbf{s}(n). \tag{4.68}$$

Readers not familiar with Wiener filtering may want to review Appendix F at this time, especially the orthogonality principle (Sec. F.2 in Appendix F). As for other background material, the discussions in this section depend heavily on the language of random processes (power spectra, autocorrelations, and so forth) reviewed in Appendix E.

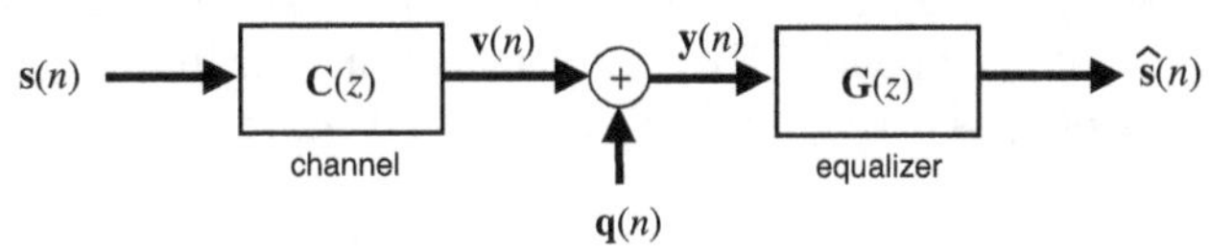

Figure 4.37. A channel with transfer function $\mathbf{C}(z)$ and additive noise $\mathbf{q}(n)$,

4.10.1 MMSE equalizers based on Wiener filtering

The reconstructed signal $\widehat{\mathbf{s}}(n)$ is a linear combination of samples of the received noisy signal $\mathbf{y}(n)$, that is,

$$\widehat{\mathbf{s}}(n) = \sum_{m=-\infty}^{\infty} \mathbf{g}(m)\mathbf{y}(n-m), \tag{4.69}$$

where the received signal is

$$\mathbf{y}(n) = \sum_{m=-\infty}^{\infty} \mathbf{c}(m)\mathbf{s}(n-m) + \mathbf{q}(n). \tag{4.70}$$

In MMSE designs, the goal is to choose the coefficients $\mathbf{g}(m)$ of the linear equalizer such that the reconstruction error has minimum mean square value, that is,

$$\mathcal{E} = E[\mathbf{e}^\dagger(n)\mathbf{e}(n)] \tag{4.71}$$

is minimized. This is the optimal equalizer, also known as the **MMSE** equalizer or the **Wiener** equalizer. The theorem below gives a closed-form expression for the MMSE equalizer in terms of various power spectra. In what follows, the notations described in Appendix E are used. Thus $\mathbf{S_{uv}}(z)$ represents the cross power spectrum between two jointly WSS processes $\mathbf{u}(n)$ and $\mathbf{v}(n)$. Similarly $\mathbf{R_{uv}}(k)$ represents the cross correlation sequence between $\mathbf{u}(n)$ and $\mathbf{v}(n)$. In the following discussion the channel $\mathbf{C}(z)$ is not restricted to be a square matrix.

♠**Theorem 4.2.** *MMSE equalizer.* The filter $\mathbf{G}(z)$ which minimizes the mean square reconstruction error is

$$\mathbf{G}(z) = \mathbf{S_{sy}}(z)\mathbf{S}^{-1}_{\mathbf{yy}}(z), \tag{4.72}$$

assuming the indicated inverse exists. This can also be rewritten as

$$\mathbf{G}(z) = \Big(\mathbf{S_{ss}}(z)\widetilde{\mathbf{C}}(z) + \mathbf{S_{sq}}(z)\Big)\mathbf{S}^{-1}_{\mathbf{yy}}(z), \tag{4.73}$$

where $\widetilde{\mathbf{C}}(z) = [\mathbf{C}(1/z^*)]^\dagger$. ◇

Proof. The reconstructed signal $\widehat{s}(n)$ in Eq. (4.69) is a linear combination of samples of the observed random process $\mathbf{y}(.)$. From the orthogonality principle (Appendix F) we know that the optimal $\{\mathbf{g}(m)\}$ is such that the error $\mathbf{e}(n)$ is orthogonal to $\mathbf{y}(m)$ for all m, that is,

$$E[\mathbf{e}(n)\mathbf{y}^\dagger(n-k)] = \mathbf{0} \quad \text{for all } k. \tag{4.74}$$

The left-hand side can be rewritten as

$$E\Big[\Big(\widehat{\mathbf{s}}(n) - \mathbf{s}(n)\Big)\mathbf{y}^\dagger(n-k)\Big] = E\Big[\Big(\sum_{m=-\infty}^{\infty} \mathbf{g}(m)\mathbf{y}(n-m) - \mathbf{s}(n)\Big)\mathbf{y}^\dagger(n-k)\Big].$$

So Eq. (4.74) can be written in the form

$$\sum_{m=-\infty}^{\infty} \mathbf{g}(m)\mathbf{R_{yy}}(k-m) = \mathbf{R_{sy}}(k). \tag{4.75}$$

Taking z-transforms, we therefore arrive at $\mathbf{G}(z)\mathbf{S_{yy}}(z) = \mathbf{S_{sy}}(z)$. Equation (4.72) follows from this. To prove (4.73), observe that (4.70) implies

$$\mathbf{S_{ys}}(z) = \mathbf{C}(z)\mathbf{S_{ss}}(z) + \mathbf{S_{qs}}(z)$$

(use Sec. E.2.4, Appendix E). Since $\mathbf{S_{sy}}(z) = \widetilde{\mathbf{S}}_{\mathbf{ys}}(z)$ and $\mathbf{S_{ss}}(z) = \widetilde{\mathbf{S}}_{\mathbf{ss}}(z)$, Eq. (4.72) reduces to the expression (4.73) claimed in the theorem. ▽ ▽ ▽

4.10.1.A *Memoryless case*

An important special case arises when the channel and the equalizer are memoryless, that is, constant matrices:

$$\mathbf{C}(z) = \mathbf{C}, \quad \mathbf{G}(z) = \mathbf{G}. \tag{4.76}$$

In this case, if we start from the orthogonality principle and proceed as above, we can show (Problem 4.9) that the optimal $\mathbf{G}$ is

$$\mathbf{G} = \Big(\mathbf{R_{ss}}\mathbf{C}^\dagger + \mathbf{R_{sq}}\Big)\mathbf{R}_{\mathbf{yy}}^{-1}, \tag{4.77}$$

where the $\mathbf{R}$-matrices are autocorrelations defined by

$$\mathbf{R_{ss}} = E[\mathbf{s}(n)\mathbf{s}^\dagger(n)], \quad \mathbf{R_{yy}} = E[\mathbf{y}(n)\mathbf{y}^\dagger(n)], \quad \mathbf{R_{sq}} = E[\mathbf{s}(n)\mathbf{q}^\dagger(n)].$$

4.10.2 Uncorrelated noise

For the special case where $\mathbf{s}(n)$ and $\mathbf{q}(n)$ are zero-mean uncorrelated processes we have

$$E[\mathbf{s}(n)\mathbf{q}^\dagger(n-k)] = \mathbf{0}, \tag{4.78}$$

or equivalently $\mathbf{S_{sq}}(z) = \mathbf{0}$. This is an important special case, because it usually arises in practice. For this case it is useful to derive an expression for the power spectrum of the error.

♠**Theorem 4.3.** *MMSE equalizers for uncorrelated noise.* When $\mathbf{s}(n)$ and $\mathbf{q}(n)$ are zero-mean uncorrelated processes the MMSE equalizer is

$$\mathbf{G}(z) = \mathbf{S_{ss}}(z)\widetilde{\mathbf{C}}(z)\Big(\mathbf{C}(z)\mathbf{S_{ss}}(z)\widetilde{\mathbf{C}}(z) + \mathbf{S_{qq}}(z)\Big)^{-1}. \tag{4.79}$$

The power spectrum of the error is

$$\mathbf{S_{ee}}(z) = \Big(\mathbf{S}_{\mathbf{ss}}^{-1}(z) + \widetilde{\mathbf{C}}(z)\mathbf{S}_{\mathbf{qq}}^{-1}(z)\mathbf{C}(z)\Big)^{-1}. \tag{4.80}$$

Here the channel $\mathbf{C}(z)$ is possibly a rectangular matrix. ◇

Proof. From Eq. (4.70) we have

$$\mathbf{S_{yy}}(z) = \mathbf{C}(z)\mathbf{S_{ss}}(z)\widetilde{\mathbf{C}}(z) + \mathbf{S_{qq}}(z).$$

Substituting this into (4.73) and using $\mathbf{S_{sq}}(z) = \mathbf{0}$ yields Eq. (4.79). The error autocorrelation is

$$\mathbf{R_{ee}}(k) = E[\mathbf{e}(n)\mathbf{e}^\dagger(n-k)] = E\Big[\mathbf{e}(n)\Big(\widehat{\mathbf{s}}(n-k) - \mathbf{s}(n-k)\Big)^\dagger\Big].$$

Since $\mathbf{e}(n)$ is orthogonal to $\mathbf{y}(m)$ it is also orthogonal to $\widehat{\mathbf{s}}(n)$ (which is the linear combination (4.69)). Thus $E[\mathbf{e}(n)\widehat{\mathbf{s}}^\dagger(n-k)] = \mathbf{0}$ so that

$$\mathbf{R}_{\mathbf{ee}}(k) = -E\Big[\mathbf{e}(n)\mathbf{s}^\dagger(n-k)\Big] = -E\Big[\widehat{\mathbf{s}}(n)\mathbf{s}^\dagger(n-k)\Big] + E\Big[\mathbf{s}(n)\mathbf{s}^\dagger(n-k)\Big].$$

Taking z-transforms this is equivalent to

$$\mathbf{S}_{\mathbf{ee}}(z) = -\mathbf{S}_{\widehat{\mathbf{s}}\mathbf{s}}(z) + \mathbf{S}_{\mathbf{ss}}(z). \tag{4.81}$$

But

$$\widehat{\mathbf{s}}(n) = (\mathbf{g} * \mathbf{c} * \mathbf{s})(n) + (\mathbf{g} * \mathbf{q})(n),$$

where $*$ denotes convolution. This yields

$$\mathbf{S}_{\widehat{\mathbf{s}}\mathbf{s}}(z) = \mathbf{G}(z)\mathbf{C}(z)\mathbf{S}_{\mathbf{ss}}(z) + \mathbf{G}(z)\mathbf{S}_{\mathbf{qs}}(z) = \mathbf{G}(z)\mathbf{C}(z)\mathbf{S}_{\mathbf{ss}}(z),$$

where we have used $\mathbf{S}_{\mathbf{sq}}(z) = \mathbf{0}$. So the error spectrum (4.81) takes the form

$$\mathbf{S}_{\mathbf{ee}}(z) = \Big(\mathbf{I} - \mathbf{G}(z)\mathbf{C}(z)\Big)\mathbf{S}_{\mathbf{ss}}(z).$$

Substituting from Eq. (4.79) we then have

$$\mathbf{S}_{\mathbf{ee}}(z) = \mathbf{S}_{\mathbf{ss}}(z) - \mathbf{S}_{\mathbf{ss}}(z)\widetilde{\mathbf{C}}(z)\Big(\mathbf{C}(z)\mathbf{S}_{\mathbf{ss}}(z)\widetilde{\mathbf{C}}(z) + \mathbf{S}_{\mathbf{qq}}(z)\Big)^{-1}\mathbf{C}(z)\mathbf{S}_{\mathbf{ss}}(z).$$

We now use the matrix inversion lemma (Appendix B, Sec. B.4.2), reproduced below:

$$(\mathbf{P} + \mathbf{QRS})^{-1} = \mathbf{P}^{-1} - \mathbf{P}^{-1}\mathbf{Q}(\mathbf{SP}^{-1}\mathbf{Q} + \mathbf{R}^{-1})^{-1}\mathbf{SP}^{-1}. \tag{4.82}$$

This holds for square nonsingular $\mathbf{P}$ and $\mathbf{R}$ even if $\mathbf{Q}$ and $\mathbf{S}$ are not square. Identifying

$$\mathbf{P}^{-1} = \mathbf{S}_{\mathbf{ss}}(z), \quad \mathbf{R}^{-1} = \mathbf{S}_{\mathbf{qq}}(z), \quad \mathbf{Q} = \widetilde{\mathbf{C}}(z), \quad \text{and} \quad \mathbf{S} = \mathbf{C}(z),$$

we therefore obtain Eq. (4.80). $\bigtriangledown \bigtriangledown \bigtriangledown$

Some special cases are worth noting:

1. ***Identity channel.*** For $\mathbf{C}(z) = \mathbf{I}$ the error spectrum (4.80) reduces to

$$\mathbf{S}_{\mathbf{ee}}(z) = \Big(\mathbf{S}_{\mathbf{ss}}^{-1}(z) + \mathbf{S}_{\mathbf{qq}}^{-1}(z)\Big)^{-1}.$$

   This is an interesting expression. It represents the inverse of the sum of the inverses of the spectra of $\mathbf{s}(n)$ and $\mathbf{q}(n)$.

2. ***White signal and noise.*** Suppose $\mathbf{s}(n)$ is zero-mean and white, so that $E[\mathbf{s}(n)\mathbf{s}^\dagger(m)] = \mathbf{0}$ for $n \neq m$. Assume further that $E[\mathbf{s}(n)\mathbf{s}^\dagger(n)] = \sigma_s^2\mathbf{I}$, that is, the components $s_k(n)$ are uncorrelated for different k, and have identical mean square value σ_s^2. Thus $\mathbf{S}_{\mathbf{ss}} = \sigma_s^2\mathbf{I}$. Assume similarly that $\mathbf{S}_{\mathbf{qq}} = \sigma_q^2\mathbf{I}$. Then

$$\mathbf{S}_{\mathbf{ee}}(z) = \sigma_q^2\left(\frac{\sigma_q^2}{\sigma_s^2}\mathbf{I} + \widetilde{\mathbf{C}}(z)\mathbf{C}(z)\right)^{-1} = \sigma_s^2\left(\mathbf{I} + \frac{\sigma_s^2}{\sigma_q^2}\widetilde{\mathbf{C}}(z)\mathbf{C}(z)\right)^{-1}.$$

3. ***Memoryless case.*** If the channel and equalizer are memoryless, as in Eq. (4.76), then, by proceeding as in the proof of Theorem 4.3, we can show (Problem 4.9) that the autocorrelation matrix of the minimized error is

$$\mathbf{R_{ee}} = \left(\mathbf{R_{ss}^{-1}} + \mathbf{C}^\dagger\mathbf{R_{qq}^{-1}}\mathbf{C}\right)^{-1}, \tag{4.83}$$

where $\mathbf{R_{ss}} = E[\mathbf{s}(n)\mathbf{s}^\dagger(n)]$ and $\mathbf{R_{qq}} = E[\mathbf{q}(n)\mathbf{q}^\dagger(n)]$.

Example 4.3: MMSE equalizer for scalar channel

Consider the special case of scalar channels, where all matrices become scalars. In this case (4.79) yields

$$G(e^{j\omega}) = \frac{S_{ss}(e^{j\omega})C^*(e^{j\omega})}{S_{qq}(e^{j\omega}) + S_{ss}(e^{j\omega})|C(e^{j\omega})|^2} \tag{4.84}$$

The noise spectrum $S_{qq}(e^{j\omega})$ is usually nonzero for all ω in the band of interest. So, even if $C(e^{j\omega}) = 0$ for some frequencies, the equalizer is still defined for all ω, unlike in the zero-forcing case. If $C(e^{j\omega}) \neq 0$ for all ω, then the equalizer can also be written as

$$G(e^{j\omega}) = \frac{1/C(e^{j\omega})}{1 + \dfrac{S_{qq}(e^{j\omega})}{S_{ss}(e^{j\omega})|C(e^{j\omega})|^2}} \tag{4.85}$$

The numerator $1/C(e^{j\omega})$ is the zero-forcing part. The rest of the right-hand side is the "Wiener part," or correction factor, which makes the filter statistically optimal. From (4.80) the power spectrum of the error is given by

$$S_{ee}(e^{j\omega}) = \frac{1}{\dfrac{1}{S_{ss}(e^{j\omega})} + \dfrac{|C(e^{j\omega})|^2}{S_{qq}(e^{j\omega})}} \tag{4.86}$$

which can be rewritten as

$$S_{ee}(e^{j\omega}) = \frac{S_{ss}(e^{j\omega})}{1 + \dfrac{S_{ss}(e^{j\omega})|C(e^{j\omega})|^2}{S_{qq}(e^{j\omega})}} \tag{4.87}$$

The Wiener filter (4.85) is in general *not realizable* as a causal stable filter even if the channel is a rational FIR or IIR filter; it can only be approximated. See Problems 4.13–4.15. A number of remarks are now in order.

1. ***When is zero-forcing the best?*** Observe first that $G(e^{j\omega_0}) = 1/C(e^{j\omega_0})$ if and only if the noise spectrum $S_{qq}(e^{j\omega_0}) = 0$ at the frequency ω_0. Thus, where the noise is zero, the best equalizer is certainly the "common sense solution" $1/C(e^{j\omega_0})$. If the channel is noiseless ($q(n) = 0$) then $S_{qq}(e^{j\omega}) = 0$ for all ω, and the best equalizer is the zero-forcing equalizer.

2. ***Handling channel-nulls.*** When the noise is nonzero then the equalizer depends on the strength of noise relative to the signal. At some frequency, if the effective noise is much stronger than the signal, that is, if

$$\frac{S_{qq}(e^{j\omega_0})}{|C(e^{j\omega_0})|^2} >> S_{ss}(e^{j\omega_0}),$$

then $G(e^{j\omega_0}) \approx 0$ there. Frequencies where the channel gain $C(e^{j\omega}) = 0$ are called **channel nulls**. At such frequencies, we see that the optimal filter is zero, and the noise does not get through! *And neither does the signal.* Note that, if the equalizer were the zero-forcing equalizer $1/C(e^{j\omega})$ for all ω, then the error spectrum would be

$$S_{ee}(e^{j\omega}) = \frac{S_{qq}(e^{j\omega})}{|C(e^{j\omega})|^2} \tag{4.88}$$

This can become arbitrarily large at the channel nulls. The *optimal MMSE solution prevents such a catastrophe* by suppressing the output where there are channel nulls: indeed, from Eq. (4.86), we see that, in the neighborhood of channel nulls, the error spectrum becomes

$$S_{ee}(e^{j\omega_0}) \approx S_{ss}(e^{j\omega_0}).$$

This is consistent with the fact that the output of $G(e^{j\omega})$ is zero, so the reconstruction error is the input itself.

Summarizing this discussion, we note that for very low and very high SNRs the *optimum equalizer* takes the following approximate forms:

$$G(e^{j\omega}) \approx \begin{cases} 0 & \text{very low } SNR \\ 1/C(e^{j\omega}) & \text{very high } SNR, \end{cases} \tag{4.89}$$

where the term *SNR* is used here temporarily to denote the ratio

$$\frac{S_{qq}(e^{j\omega})}{S_{ss}(e^{j\omega})|C(e^{j\omega})|^2}$$

The corresponding reconstruction error spectrum is

$$S_{ee}(e^{j\omega}) \approx \begin{cases} S_{ss}(e^{j\omega}) & \text{very low } SNR \\ S_{qq}(e^{j\omega})/|C(e^{j\omega})|^2 & \text{very high } SNR. \end{cases} \tag{4.90}$$

4.10.2.A Special case of white signal and white noise

Let us assume that the transmitted signal and channel noise are both white, with power spectra

$$S_{ss}(e^{j\omega}) = \sigma_s^2, \quad S_{qq}(e^{j\omega}) = \sigma_q^2.$$

From Eqs. (4.85) and (4.87) the optimal equalizer $G(e^{j\omega})$ and the corresponding error spectrum are

$$G(e^{j\omega}) = \frac{1/C(e^{j\omega})}{1 + \dfrac{\sigma_q^2}{\sigma_s^2|C(e^{j\omega})|^2}} \tag{4.91}$$

and

$$S_{ee,mmse}(e^{j\omega}) = \frac{\sigma_s^2}{1 + \dfrac{\sigma_s^2 |C(e^{j\omega})|^2}{\sigma_q^2}} \tag{4.92}$$

where the extended subscript is to distinguish this from the zero-forcing case. In the latter case the equalizer is $G(e^{j\omega}) = 1/C(e^{j\omega})$ and the error spectrum is

$$S_{ee,ZF}(e^{j\omega}) = \frac{\sigma_q^2}{|C(e^{j\omega})|^2} \tag{4.93}$$

Note that $S_{ee,mmse}(e^{j\omega}) \leq S_{ee,ZF}(e^{j\omega})$ for each frequency ω. Furthermore, $S_{ee,mmse}(e^{j\omega})$ is bounded above by σ_s^2, whereas $S_{ee,ZF}(e^{j\omega})$ can get arbitrarily large because the channel response $C(e^{j\omega})$ can be close to zero at some frequencies. The mean square value of the reconstruction error can be calculated as the integral

$$\mathcal{E} = E\,|\widehat{s}(n) - s(n)|^2 = \frac{1}{2\pi}\int_0^{2\pi} S_{ee}(e^{j\omega})d\omega,$$

where $S_{ee}(e^{j\omega})$ is as in Eq. (4.92) or Eq. (4.93).

Example 4.4: Zero-forcing versus MMSE equalizers

To consider a specific example with numbers, let $\sigma_s^2 = 1.0, \quad \sigma_q^2 = 0.1$, and

$$C(z) = \prod_{k=1}^{2}\Big(1 - 2\rho_k \cos\theta_k z^{-1} + \rho_k^2 z^{-2}\Big),$$

with

$$\rho_1 = 0.94,\ \rho_2 = 0.96,\ \theta_1 = 0.2\pi,\text{ and } \theta_2 = 0.32\pi.$$

Figure 4.38 (top) shows the magnitude response $|C(e^{j\omega})|$. The channel is FIR, so the inverse $1/C(z)$ is IIR. It has poles inside the unit circle, but they are rather close to the unit circle. This makes $1/|C(e^{j\omega})|$ rather large as ω approaches θ_k. The error spectrum $S_{ee,ZF}(e^{j\omega})$ is therefore large. Figure 4.38 shows the plots of the channel response $|C(e^{j\omega})|$, the error spectrum for the *zero-forcing equalizer*, and the error spectrum for the *MMSE equalizer* (Eqs. (4.92),(4.93)). Note that the peak error spectrum for the zero-forcing equalizer is about 90 times larger than the error spectrum for the MMSE equalizer. The integrals of these spectra yield the mean square reconstruction errors:

$$E|\widehat{s}(n) - s(n)|^2 = \begin{cases} 7.7872 & \text{zero forcing} \\ 0.3351 & \text{MMSE.} \end{cases}$$

Replacing a zero-forcing equalizer with an MMSE equalizer therefore results in a reduction in MSE by a factor of about 23 (about 13.6 dB).

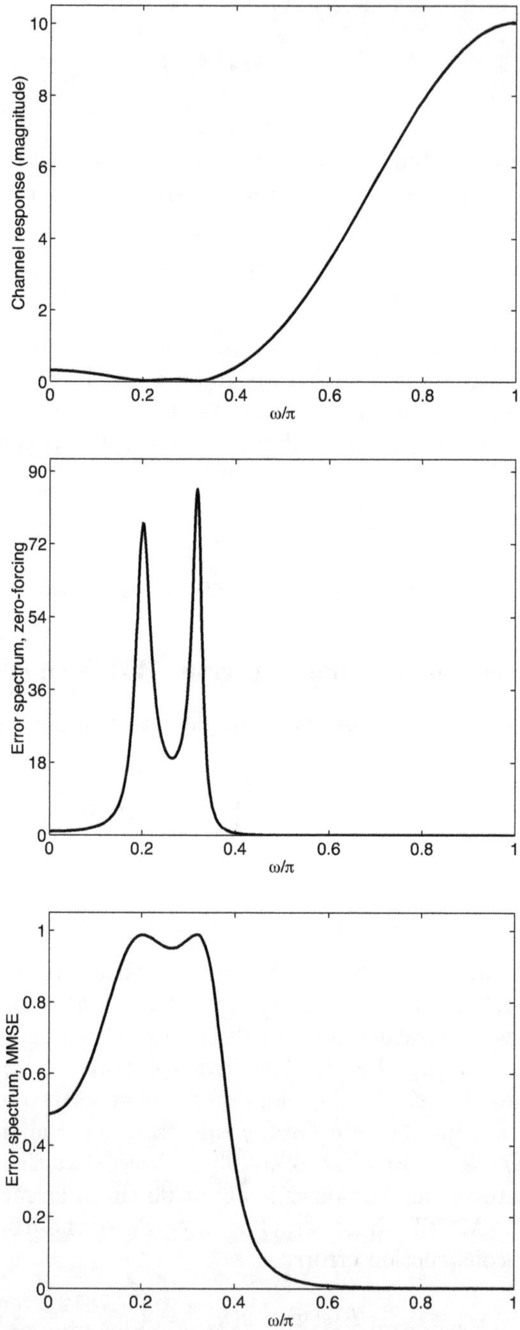

Figure 4.38. Comparison of zero-forcing and MMSE equalizers Top: magnitude response of the channel. Middle: error spectrum with zero-forcing equalizer. Bottom: error spectrum with MMSE equalizer.

4.11 Concluding remarks

In this chapter we developed discrete-time representations for digital communication channels. These representations allow us to analyze the system as if it is an all-digital system. For example, the problem of designing an optimal transceiver, that is, *jointly* designing the pre- and postfilters to minimize the mean square error, can be viewed entirely in the discrete-time domain. We shall return to such optimization problems in Chaps. 9–19, and treat systems with and without zero forcing in great detail. In this chapter we also discussed oversampled receivers and fractionally spaced equalizers. Fractionally spaced equalizers have been studied extensively in the literature, and the reader wishing to pursue the history should refer to Proakis [1995]. Very interesting discussions on the design issues can be found in Triechler, Fijalkow, and Johnson [1996], and advanced research material can be found in Tong, Xu, and Kailath [1994], Borah *et al.* [2001], Vaidyanathan and Vrcelj [2001, 2002], Vrcelj and Vaidyanathan [2002, 2003], and references therein. The use of fractionally spaced equalizers in jointly optimal transceivers for MIMO channels is described in Weng and Vaidyanathan [2008]. Fractionally spaced equalization allows us to equalize an FIR channel with FIR rather than IIR filters. Another way to achieve the same goal would be to introduce redundancy at the transmitter, instead of oversampling at the receiver. This idea will be introduced in Chap. 7.

Problems

4.1. Referring to Fig. 4.3, let $H_c(s) = 1/(s^2 + 2\alpha s + \alpha^2 + \beta^2)$ represent the transfer function of a causal continuous-time channel, where α and β are real. Find the impulse response $h_d(n)$ of the discrete-time equivalent channel. For what values of α is $h_d(n)$ stable?

4.2. Consider a causal MIMO channel with transfer matrix

$$\mathbf{H}_c(s) = \begin{bmatrix} \dfrac{1}{s+a} & \dfrac{1}{s+b} \\ \\ \dfrac{1}{s+c} & \dfrac{1}{(s+a)(s+b)} \end{bmatrix},$$

where $a, b, c > 0$. With the discrete-time equivalent channel defined as in Eq. (4.23), compute the impulse response matrix $\mathbf{h}_d(n)$. Find the transfer function $\mathbf{H}_d(z)$ and express its entries in rational form. Does $\mathbf{H}_d(z)$ represent a stable channel (i.e., does it have all poles in $|z| < 1$)?

4.3. Show that the inverse Fourier transform of the raised-cosine function, defined in Eq. (4.31), is given by Eq. (4.36).

4.4. Use Euclid's algorithm and derive the expressions for $E_0(z)$ and $E_1(z)$ claimed in Example 4.2.

4.5. We now consider examples of FIR filter pairs satisfying Eq. (4.44).

1. For the FIR filters $C_L(z) = 1 + z^{-1} + z^{-2}$ and $G_L(z) = z$, show that Eq. (4.44) is satisfied with $L = 2$.
2. In the above example $G_L(z)$ is not causal. What is the left-hand side of Eq. (4.44) if $G_L(z)$ is replaced with the causal filter $G_L(z) = z^{-1}$?
3. Construct your own example of causal FIR filters satisfying the condition $C_L(z)G_L(z)]_{\downarrow 2} = z^{-K}$ for some K, but this time make sure that $C_L(z)$ has order ≥ 3.

4.6. *Fractional oversampling.* Consider the digital communication system shown in Fig. P4.6(a), where the receiver uses oversampling as in Secs. 4.7–4.8. But now the oversampling ratio is a rational number L/M (ratio of integers L and M) rather than an integer L. Without loss of generality, we can assume that M and L are coprime. This is called a fractionally oversampled system and results in a rational FSE rather than an integer FSE. Assume the oversampling ratio $L/M > 1$.

1. Show that the sampled output can be expressed as

$$y(n) = \sum_{k=-\infty}^{\infty} s(k)c_L(Mn - kL) + q_c(MTn/L), \qquad \text{(P4.6)}$$

where $c_L(n) = c_c(nT/L)$ (channel $c_c(t)$ oversampled by L).

2. Show that the system described by Eq. (P4.6) can be represented using multirate building blocks as shown in Fig. P4.6(b), where $q(n) = q_c(MTn/L)$.

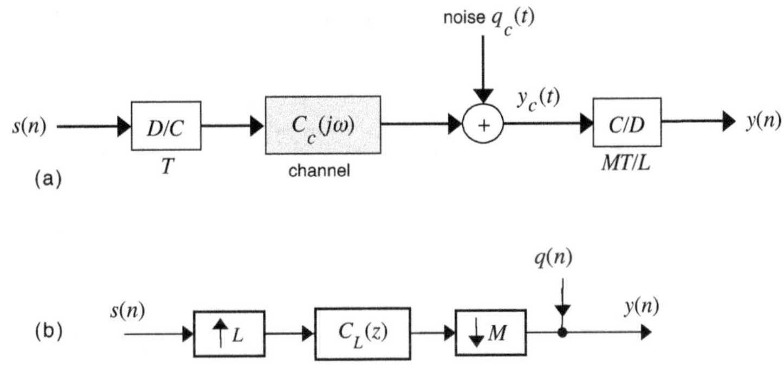

Figure P4.6.

4.7. We now consider the problem of designing a zero-forcing equalizer for the fractionally oversampled system of Problem 4.6. Since the equivalent channel in Fig. P4.6 is a multirate system, so is the equalizer as indicated in Fig. P4.7(a). The noise has not been shown, as it is irrelevant to the zero-forcing condition. The goal in this problem is to show how to choose $H(z)$ (for fixed $C_L(z)$, L, and M) such that the zero-forcing condition $\widehat{s}(n) = s(n)$ is satisfied.

1. Let $C_L(z)$ and $H(z)$ be represented in the polyphase forms

$$C_L(z) = \sum_{k=0}^{M-1} z^k A_k(z^M), \quad H(z) = \sum_{k=0}^{M-1} z^{-k} B_k(z^M).$$

Assume as in Problem 4.6 that M and L are coprime. Then there exist integers N and K such that $NL + KM = 1$. Using this, along with noble identities (Sec. 3.2.3) and the fact that $\uparrow L$ and $\downarrow M$ can be interchanged when L and M are coprime (Problem 3.1), show that the system can be redrawn as in Fig. P4.7(b). The system within the gray box is an LTI system with transfer matrix $\mathbf{T}(z)$ whose elements are $T_{mk}(z) = z^{(k-m)K} A_k(z)B_m(z)$.

2. In Problem 3.13 we showed that the system in Fig. P3.13 has perfect reconstruction when N and M in that figure are coprime. Using this

show that the system in Fig. P4.7(b) has perfect reconstruction (i.e., $\widehat{s}(n) = s(n)$) if

$$\left[z^{(k-m)K} A_k(z) B_m(z)\right]_{\downarrow L} = \delta(k-m), \quad 0 \leq k, m \leq M-1. \quad \text{(P4.7)}$$

So there are M^2 conditions in (P4.7) (where $M < L$ because the oversampling ratio $L/M > 1$). For the case where $M = 1$ (integer oversampling) this reduces to the single condition $[A_0(z)B_0(z)]_{\downarrow L} = 1$, which is equivalent to $[C_L(z)H(z)]_{\downarrow L} = 1$ as in Sec. 4.8.

For the curious reader: In fact it can be shown that the condition (P4.7) is both necessary and sufficient for perfect reconstruction.

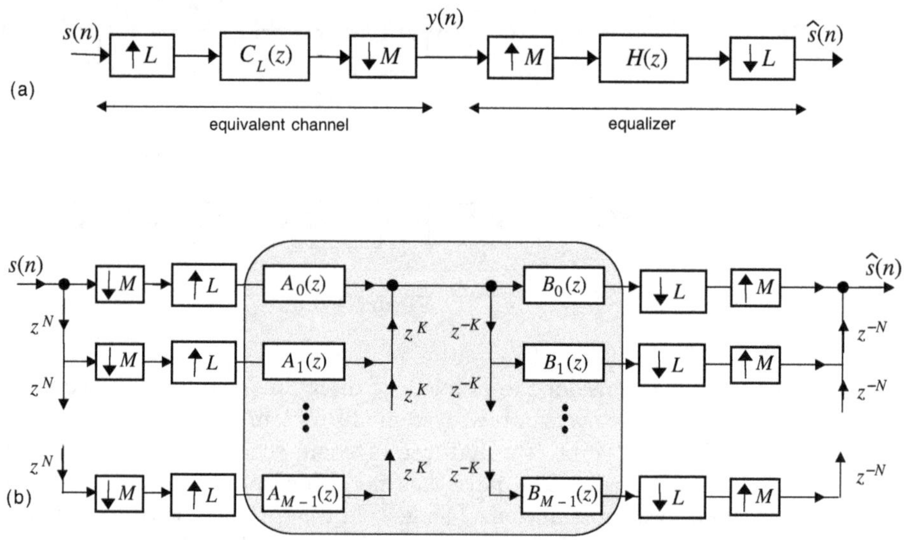

Figure P4.7.

4.8. Returning to the rational FSE system of Problem 4.7 we now construct an example with $M = 2, L = 3$.

1. Find the smallest pair of integers N and K such that $NL + KM = 1$, and write down the four conditions (P4.7) explicitly.
2. Construct an FIR example by selecting some FIR channel $C_L(z)$ (with at least two nonzero coefficients, to avoid trivialities), and finding FIR $B_0(z)$ and $B_1(z)$ to satisfy the four conditions in (P4.7).
3. What is the resulting FIR equalizer $H(z)$ of Fig. 4.7(a)?

4.9. Under the assumptions given in Sec. 4.10.1 consider the memoryless case where Eq. (4.76) holds. Show indeed that the optimal $\mathbf{G}$ is given by Eq. (4.77). Furthermore, when the noise is uncorrelated as in Sec. 4.10.2, show that the autocorrelation of minimized error in the memoryless case is indeed given by (4.83).

4.10. Consider Fig. 4.37 and assume we have a memoryless MIMO channel with transfer function matrix

$$\mathbf{C} = \begin{bmatrix} 1 & 0 \\ 2 & \epsilon \end{bmatrix},$$

where $\epsilon > 0$. Assume the signal and noise are zero-mean processes with constant power spectra $\mathbf{S_{ss}}(z) = \mathbf{I}$ and $\mathbf{S_{qq}}(z) = \sigma_q^2\mathbf{I}$.

1. Find an expression for the MMSE equalizer filter $\mathbf{G}$, the error autocorrelation $\mathbf{R_{ee}}$, and the total mean square error $\mathcal{E}$ (trace of $\mathbf{R_{ee}}$).
2. Simplify the expression for $\mathcal{E}$ when σ_q^2 becomes arbitrarily small for fixed ϵ.
3. Simplify the expression for $\mathcal{E}$ when ϵ becomes arbitrarily small for fixed σ_q^2.

4.11. In Problem 4.10 suppose we use a zero-forcing equalizer $\mathbf{G}$ instead of an MMSE equalizer.

1. Find an expression for $\mathbf{G}$, the error autocorrelation $\mathbf{R_{ee}}$, and the total mean square error $\mathcal{E}$ (trace of $\mathbf{R_{ee}}$).
2. For $\sigma_q^2 = 0.01$ and $\epsilon = 0.1$, compute the total mean square error $\mathcal{E}$. How does this compare with the MMSE value obtained from Problem 4.10?
3. For $\sigma_q^2 = 0.01$ and $\epsilon = 0.001$, compute the total mean square error $\mathcal{E}$. How does this compare with the MMSE value obtained from Problem 4.10?

4.12. In Fig. 4.37 assume the MIMO channel is FIR with transfer function matrix

$$\mathbf{C}(z) = \frac{1}{2}\begin{bmatrix} 1+z^{-1} & 1-z^{-1} \\ 1-z^{-1} & 1+z^{-1} \end{bmatrix}.$$

Assume the signal and noise are zero-mean processes with constant power spectra $\mathbf{S_{ss}}(z) = \mathbf{I}$ and $\mathbf{S_{qq}}(z) = \sigma_q^2\mathbf{I}$, with $\sigma_q^2 \neq 0$.

1. Find expressions for the MMSE equalizer filter $\mathbf{G}(z)$ and the error power spectrum $\mathbf{S_{ee}}(z)$.
2. Assume we use a zero-forcing equalizer $\mathbf{G}(z)$ instead. Find expressions for $\mathbf{G}(z)$ and $\mathbf{S_{ee}}(z)$ (i.e., find all the elements $G_{km}(z)$ and $S_{km}(z)$ explicitly).
3. By comparing the expressions for the error power spectra $\mathbf{S_{ee}}(z)$ in the above two cases, can you argue explicitly that the MSE for the MMSE solution is smaller than the MSE for the ZF solution (which of course should be the case by the definition of MMSE)?

4.13. In Fig. 4.37 assume $C(z)$ is a scalar FIR channel with transfer function $C(z) = 1 + \rho z^{-1}$. Assume the signal and noise are zero-mean processes with constant power spectra $S_{ss}(z) = 1$ and $S_{qq}(z) = \sigma_q^2$.

1. Find a closed-form expression for the MMSE equalizer $G(z)$.
2. Is there a causal stable impulse response $g(n)$ for this equalizer when $\rho \neq 0$? Explain your answer.

4.14. Repeat Problem 4.13 when the channel is IIR with transfer function $C(z) = 1/(1 + \rho z^{-1})$.

4.15. Repeat Problems 4.13 and 4.14 for the case where the equalizer is zero-forcing instead of MMSE.

5

Classical transceiver techniques

5.1 Introduction

In this chapter we review a number of well established ideas in digital communication. Section 5.2 revisits the matched filter introduced in Sec. 2.5 and discusses it from the viewpoint of information sufficiency and reconstructibility. We establish the generality of matched filtering as a fundamental front-end tool in receiver design. It is often necessary to make sure that the noise sequence at the input of the detector is white. The digital filter that ensures such a property, the so-called *sampled noise whitening filter,* is described in Sec. 5.3. A vector space interpretation of matched filtering is presented in Sec. 5.4, and offers a very useful viewpoint.

Estimation of the transmitted symbol stream from the received noisy stream is one of the basic tasks performed in any communication receiver. The foundation for this comes from optimal sequence estimation theory, which is briefly discussed in Sec. 5.5. This includes a review of maximum likelihood and maximum a posteriori methods. The Viterbi algorithm for sequence estimation is described in Sec. 5.6. While this is one of the most well known algorithms, simpler suboptimal methods, such as decision feedback equalizers (DFE), have also become popular. The motivation for DFE, which is a nonlinear equalizer, is explained in Sec. 5.7. A nonlinear precoder called the Tomlinson-Harashima precoder can be used as an alternative to the nonlinear equalizer at the receiver, and is described in Sec. 5.8. Other precoding and pulse-shaping methods such as duobinary methods are described in Sec. 5.9, which is an introduction to partial response signaling. Such techniques allow us to deal with intersymbol interference satisfactorily without taking up excess bandwidth.

5.2 Matched filtering and reconstructibility

In most digital communication systems the continuous-time received signal is filtered with a matched filter before sampling. Regardless of how the sampled signal is processed later, the use of the matched filter does not result in any loss of optimality (as long as the filter is matched to the convolution of the transmitted pulse and the channel). This has to do with the fact that a matched filter provides sufficient statistics [Forney, 1972], [Lucky, 1973], [Proakis and Salehi, 2008]. This property of the matched filter in fact allows us to perform all further optimization digitally, in the sampled domain. In practice the output of the matched filter is often modified by a noise whitening filter [Forney, 1972], [Andersen, 1973], which simplifies further digital processing. In this and the next few sections we review some of these ideas.

Consider Fig. 5.1(a), which shows a digital communication system with channel $H(j\omega)$ and receiver filter $G(j\omega)$. We have not shown the prefilter or pulse-shaping filter $F(j\omega)$, but these can be assumed to be part of $H(j\omega)$. We will establish an important result on the relationship between the noise-free channel output $y(t)$ and the signal component of the sampled signal $r(n)$. Since we are interested in the signal component at the moment, we concentrate on the noise-free channel shown in Fig. 5.1(b) for the rest of this section.

5.2.1 Reconstructibility of y(t)

The first question we address is this: from the sampled version $r(n)$ of the filtered signal in Fig. 5.1(b), can we recover the received signal $y(t)$? That is, does $r(n)$ contain all the information in $y(t)$? In general the answer depends on how $G(j\omega)$ is chosen, so we would like to find the conditions on $G(j\omega)$ which allow such reconstruction. In short we are looking for something analogous to the **sampling theorem**, which allows one to reconstruct a signal from bandlimited samples. We will also address the reconstructibility of $s(n)$ from $r(n)$. Throughout this section "sampling" refers to uniform sampling at the rate

$$\omega_s = 2\pi/T,$$

where T is the symbol spacing for $s(n)$. The notation $[X(j\omega)]_{\downarrow T}$ stands for the Fourier transform of the sampled version $x_d(n) = x(nT)$ (see Appendix G at the end of the book):

$$X_d(e^{j\omega}) = [X(j\omega)]_{\downarrow T} = \frac{1}{T}\sum_{k=-\infty}^{\infty} X\Big(\frac{j(\omega+2\pi k)}{T}\Big). \tag{5.1}$$

First we make some preliminary observations.

1. ***Bandlimited case***. If $y(t)$ is bandlimited to an alias-free(T) band $\mathcal{A}$ (Appendix G, Sec. G.3), which happens when the channel $H(j\omega)$ is bandlimited to $\mathcal{A}$, then so is the output of $G(j\omega)$. So the output of $G(j\omega)$ can be recovered from its sampled version by using an ideal bandpass filter with passband restricted to $\mathcal{A}$. The signal $y(t)$ can then be recovered as long as $G(j\omega)$ is nonzero in the band $\mathcal{A}$. The question becomes more interesting when $y(t)$ is not an ideally bandlimited signal.

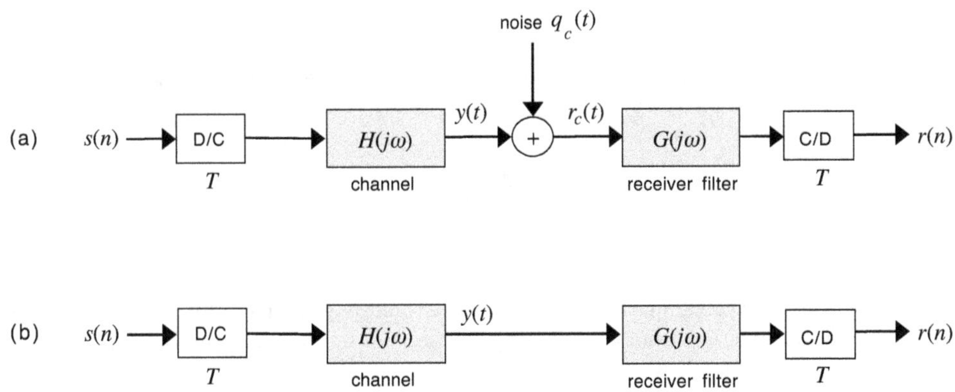

Figure 5.1. (a) The digital communication system, and (b) system with channel noise ignored.

2. ***Need for filtering.*** If the receiver filter $G(j\omega) = 1$ (i.e., we use no filtering at all), then it is very easy to find examples such that $y(t)$ cannot be reconstructed from $r(n)$. For example, suppose the channel $H(j\omega)$ is such that $h(nT) = 0$ for all n. An example is shown in Fig. 5.2. Since the transfer function from $s(n)$ to $r(n)$ is $H_d(z) = \sum_n h(nT)z^{-n}$, it follows that $H_d(z) = 0$, that is, $r(n)$ is zero for all n, and $y(t)$ (which is not necessarily zero) cannot be recovered.

Thus some filtering is in general necessary in order for the samples $r(n)$ to preserve the information about $y(t)$. Interestingly enough, we will see that there **always exists a filter** $G(j\omega)$ such that $y(t)$ can be reconstructed from $r(n)$. Whenever the receiver filter has this property (i.e., the signal $r(n)$ in Fig. 5.1(b) can be used to reconstruct $y(t)$) we say that it is **information lossless**. Consider the following simple choice for the receiver filter:

$$G(j\omega) = H^*(j\omega), \tag{5.2}$$

or equivalently

$$g(t) = h^*(-t). \tag{5.3}$$

Note that (5.3) can be interpreted as the matched filter for $h(t)$ in the white-noise case (Sec. 2.5.1). We now prove the following result:

♠**Lemma 5.1**. *Reconstructibility of y(t).* Given any channel $H(j\omega)$, if we choose the receiver filter as in (5.2), then the received signal $y(t)$ can be recovered from the samples $r(n)$. ◇

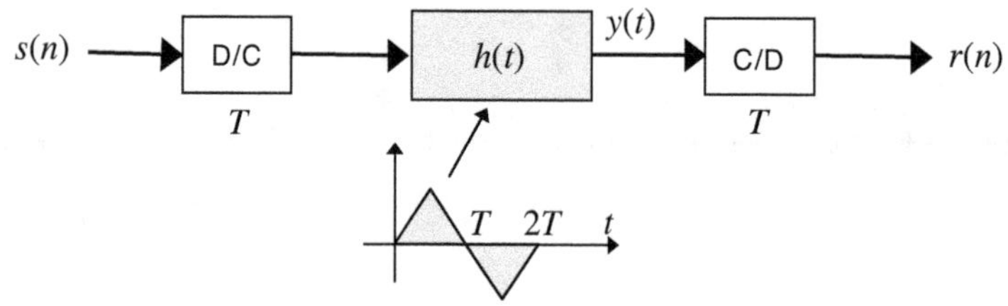

Figure 5.2. Example of a channel $h(t)$ without any receiver filter $g(t)$. In this case the transfer function from $s(n)$ to $r(n)$ is zero, and no information about $y(t)$ is retained after sampling.

Proof. The Fourier transform at the output of $G(j\omega)$ in Fig. 5.1(b) is

$$S(e^{j\omega T})H(j\omega)G(j\omega) = S(e^{j\omega T})|H(j\omega)|^2.$$

The Fourier transform of the sampled version $r(n)$ is therefore

$$R(e^{j\omega}) = \frac{S(e^{j\omega})}{T} \sum_{k=-\infty}^{\infty} \left| H\Big(\frac{j(\omega + 2\pi k)}{T}\Big)\right|^2. \tag{5.4}$$

If the summation above is nonzero for all ω, define

$$A^2(e^{j\omega T}) = \Big(\frac{1}{T}\sum_{k=-\infty}^{\infty} \Big|H\Big(j\Big(\omega + \frac{2\pi k}{T}\Big)\Big)\Big|^2\Big)^{-1}. \tag{5.5a}$$

We can then reconstruct $s(n)$ simply by passing $r(n)$ through the inverse filter $A^2(e^{j\omega})$ (Fig. 5.3). From $s(n)$ we can then obtain $y(t)$ by using its very definition as shown in the figure.

More generally we have to allow for the possibility that the sum in (5.4) can be zero for some ω. So let us define

$$A^2(e^{j\omega T}) = \begin{cases} \Big(\frac{1}{T}\sum_{k=-\infty}^{\infty} \Big|H\Big(j\Big(\omega + \frac{2\pi k}{T}\Big)\Big)\Big|^2\Big)^{-1} & \text{when } H(j\omega) \neq 0 \\ 0 & \text{otherwise.} \end{cases} \tag{5.5b}$$

When $H(j\omega) \neq 0$ for some ω the sum in (5.4) is positive and the inverse indicated in (5.5b) exists for that ω. With $A^2(e^{j\omega T})$ as above, (5.4) implies

$$R(e^{j\omega T}) = \frac{S(e^{j\omega T})}{A^2(e^{j\omega T})} \quad \text{when } H(j\omega) \neq 0.$$

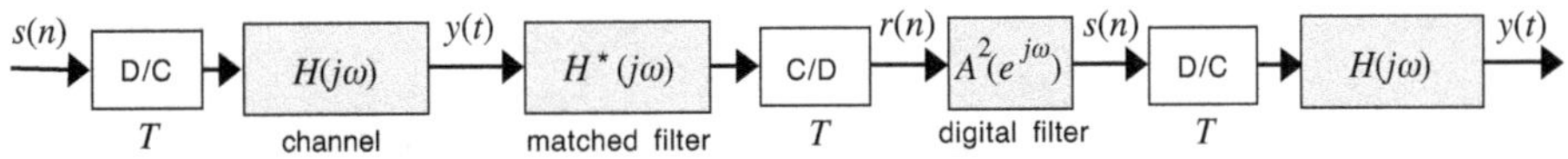

Figure 5.3. Reconstruction of $y(t)$ from the sampled output $r(n)$ of the matched filter $H^*(j\omega)$ when the channel $H(j\omega)$ is such that $(|H(j\omega)|^2)_{\downarrow T} > 0$ everywhere.

Defining the reconstruction filter $B(j\omega) = A^2(e^{j\omega T})H(j\omega)$, the reconstructed signal (Fig. 5.4(a)) is:

$$\begin{aligned} Y_1(j\omega) &= R(e^{j\omega T})A^2(e^{j\omega T})H(j\omega) \\ &= \begin{cases} \dfrac{S(e^{j\omega T})}{A^2(e^{j\omega T})}A^2(e^{j\omega T})H(j\omega) & \text{for } H(j\omega) \neq 0 \\ 0 & \text{otherwise} \end{cases} \\ &= S(e^{j\omega T})H(j\omega) = Y(j\omega). \end{aligned}$$

That is, $y_1(t) = y(t)$ indeed. $\nabla \nabla \nabla$

By using the appropriate noble identity from Sec. G.2 in Appendix G, we can redraw the reconstruction system of Fig. 5.3 in various equivalent forms, as in Fig. 5.4. Thus the periodic filter (or transversal filter) $A^2(e^{j\omega T})$ can be lumped either with the matched filter $H^*(j\omega)$ (Fig. 5.4(b)), or with the final reconstruction filter $H(j\omega)$ on the right (Fig. 5.4(a)). It can also be split equally into two factors $A(e^{j\omega T}) \geq 0$ as shown in Fig. 5.4(c).

Pulse matching is not sufficient

Now consider Fig. 5.5, where we have shown a transmitting pulse $p(t)$ and a channel $h(t)$. The effective channel seen by the receiver is the convolution $(p*h)(t)$. As shown in Lemma 5.1, if the receiver filter is matched to $(p*h)(t)$ then the information about the channel output $y(t)$ is preserved in the samples $r(n)$. But in the system shown, the receiver filter is matched only to the transmitted pulse $p(t)$. Is this also sufficient to preserve information about $y(t)$?

The answer in general is *no*. There might be loss of information. To see this observe that the impulse response of the cascade is the convolution of $p(t)$, $h(t)$, and $p^*(-t)$, which is equal to the convolution of $h(t)$ with $R_{pp}(t)$ (autocorrelation of the pulse). The impulse response $h_d(n)$ of the discrete-time LTI system from $s(n)$ to $r(n)$ is the sampled version of this convolution:

$$h_d(n) = \int_{-\infty}^{\infty} h(t)R_{pp}(nT - t)dt. \tag{5.6}$$

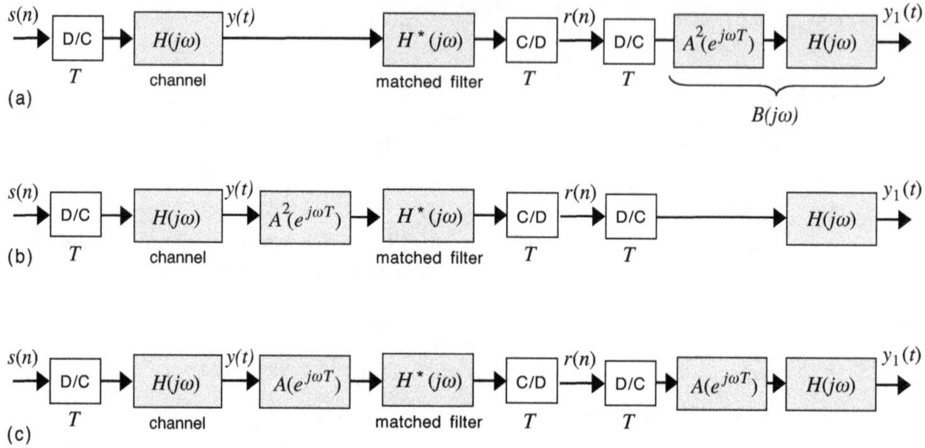

Figure 5.4. Reconstruction of $y(t)$ for the case of arbitrary channel $H(j\omega)$. (a)–(c) Three equivalent arrangements.

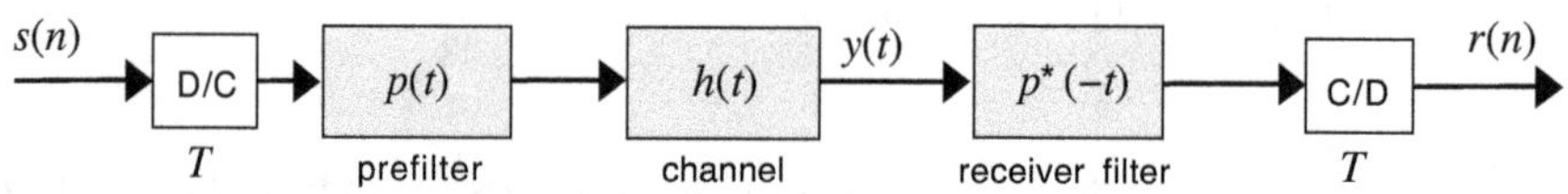

Figure 5.5. Situation where the receiver filter $g(t)$ matches only the transmitted pulse $p(t)$ and ignores the channel.

Consider the example of the transmitted pulse $p(t)$ and channel shown in Fig. 5.6. The autocorrelation $R_{pp}(t)$ is also shown. In this case it can be verified (Problem 5.4) that $h_d(n)$ given by (5.6) is zero for all n. This means that the sampled output $r(n)$ is always zero, and the information about the channel output $y(t)$ is completely destroyed!

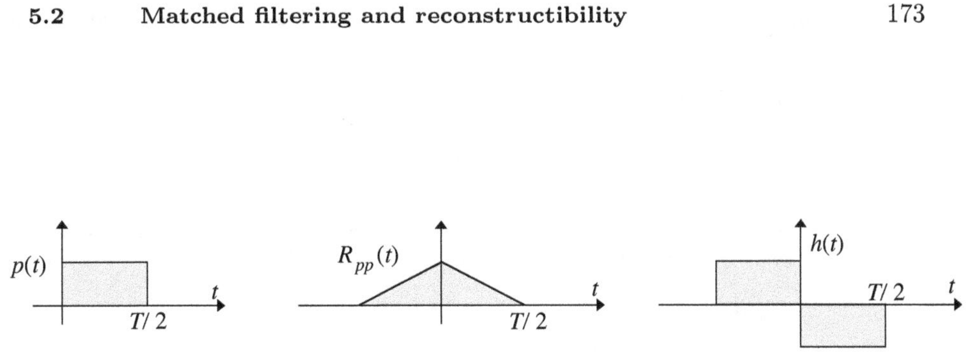

Figure 5.6. An example of a pulse shape $p(t)$, its autocorrelation $R_{pp}(t)$, and the channel $h(t)$. In this example $h_d(n) = 0$, so the sampled output $r(n)$ of the receiver filter does not preserve information about the channel output $y(t)$.

5.2.2 Reconstructibility of s(n)

In practice, the sampled signal $r(n)$ is buried in additive noise, and it has to be processed to get an estimate of the transmitted symbol stream. This is the symbol detection stage and there are many methods for this.

The topic of recovery of $s(n)$ from $r(n)$ under the zero-noise assumption, which we now discuss, is of interest in the design of the so-called zero-forcing receivers. Note that reconstructibility of $y(t)$ does not imply that of $s(n)$. As shown in the proof of Lemma 5.1, the signal $s(n)$ is easily reconstructed from $r(n)$ if

$$\frac{1}{T}\sum_{k=-\infty}^{\infty}\left|H\Big(\frac{j(\omega+2\pi k)}{T}\Big)\right|^2 > 0 \quad \text{for all } \omega. \tag{5.7}$$

In Sec. G.3 of Appendix G we define the idea of alias-free(T) bands which the reader may want to review at this time. Examples of such bands include lowpass regions, such as $|\omega| \leq \pi/T$, and bandpass regions such as $\omega_a \leq \omega \leq \omega_a + 2\pi/T$. More general alias-free(T) bands with multiple passbands are possible, as demonstrated in Appendix G. It is also clear from that Appendix that the condition (5.7) is equivalent to the statement that there exists an alias-free(T) region $\mathcal{A}$ such that $H(j\omega)$ is nonzero everywhere in $\mathcal{A}$. We now prove the following result:

♠**Lemma 5.2**. *Reconstructibility of $s(n)$.* Referring to Fig. 5.1(b), there exists a receiver filter $G(j\omega)$ such that $s(n)$ can be recovered from the sampled output $r(n)$ of $G(j\omega)$, if and only if there exists an alias-free(T) band $\mathcal{A}$, where the channel $H(j\omega)$ is nonzero for all ω. ◇

Proof. First assume there is an alias-free(T) band $\mathcal{A}$ where $H(j\omega) \neq 0$ everywhere, that is, (5.7) holds. Choosing $G(j\omega) = H^*(j\omega)$, we then have (5.4). Defining $A^2(e^{j\omega})$ as in (5.5a) and using Fig. 5.3 we can reconstruct $s(n)$ indeed.

Conversely, assume that there is no alias-free(T) band $\mathcal{A}$ such that $H(j\omega) \neq 0$ everywhere in $\mathcal{A}$. This means that there are regions $\mathcal{R}$ of frequency such that if $\omega_0 \in \mathcal{R}$ then

$$H\Big(j\Big(\omega_0 + k\frac{2\pi}{T}\Big)\Big) = 0$$

for *all* integers k. If this is the case then, for any receiver filter $G(j\omega)$, the frequency response from $s(n)$ to $r(n)$, namely

$$H_d(e^{j\omega})$$

$$= [H(j\omega)G(j\omega)]_{\downarrow T} = \frac{1}{T}\sum_{k=-\infty}^{\infty} H\Big(j\Big(\frac{\omega+2\pi k}{T}\Big)\Big)G\Big(j\Big(\frac{\omega+2\pi k}{T}\Big)\Big),$$

will be zero at $\omega = \omega_0 T$. That is, there will be frequency regions $\mathcal{R}_d$ in the digital filter $H_d(e^{j\omega})$ where the response is zero. This means that $s(n)$ cannot entirely be reconstructed from $r(n)$. Only those parts of $S(e^{j\omega})$ which do not fall in $\mathcal{R}_d$ can be reconstructed. $\bigtriangledown \bigtriangledown \bigtriangledown$

Under the conditions of the above lemma, the quantity $A^2(e^{j\omega T})$ defined in Eq. (5.5b) simplifies to

$$A^2(e^{j\omega T}) = \Big(\frac{1}{T}\sum_{k=-\infty}^{\infty}\Big|H\Big(j\Big(\omega + \frac{2\pi k}{T}\Big)\Big)\Big|^2\Big)^{-1}, \tag{5.8}$$

or equivalently

$$A^2(e^{j\omega}) = \Big(\Big[|H(j\omega)|^2\Big]_{\downarrow T}\Big)^{-1}. \tag{5.9}$$

The following result is an immediate consequence of the above lemma.

♠**Corollary 5.1**. *Reconstruction of $s(n)$.* If $s(n)$ is recoverable from $r(n)$ (i.e., Eq. (5.7) holds), it can be recovered by $A^2(e^{j\omega})$, as indicated in Fig. 5.7. $\diamondsuit$

The reconstruction filters might turn out to be ideal filters. Their practical realizability is not of concern at the moment; in practice one can implement approximations of these ideal filters.

5.2.3 Relation to linear independence

The positivity condition in Eq. (5.7) has made repeated appearances in our discussions. It is necessary for the reconstructibility of $s(n)$ in Fig. 5.1(b), but not for the reconstructibility of $y(t)$. This condition can be interpreted in a number of ways. In particular it is equivalent to linear independence of the functions $h(t - nT)$.

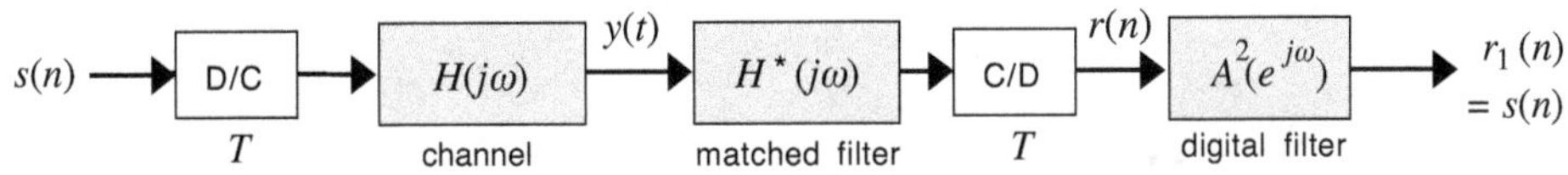

Figure 5.7. Reconstruction of $s(n)$ from $r(n)$.

Since these relationships will be encountered frequently, we shall summarize them as a lemma. A deeper discussion in the setting of ℓ_2 and L_2 spaces (Appendix A) is given in Sec. 5.4.

♠**Lemma 5.3**. Given a function $h(t)$ with Fourier transform $H(j\omega)$, the following statements are equivalent:

1. The set of functions $\{h(t-nT)\}$ is linearly independent, that is, there does not exist a nonzero sequence $c(n) \in \ell_2$ such that $\sum_n c(n)h(t-nT) = 0$ for all t.
2. $\sum_{k=-\infty}^{\infty} \left|H\Big(j\Big(\omega + \frac{2\pi k}{T}\Big)\Big)\right|^2 > 0$ for all ω.
3. There exists an alias-free(T) band $\mathcal{A}$ such that $H(j\omega)$ is nonzero everywhere in it. That is, the channel $H(j\omega)$ has *sufficient passband width* (though not necessarily as a single contiguous piece) for sending symbols at the rate $1/T$.
4. There does *not* exist a region $\mathcal{R}$ of frequencies $\theta_a < \omega < \theta_b$ such that $H(j\omega)$ vanishes everywhere in $\mathcal{R}$ and in *all its shifted versions* $\mathcal{R} + 2\pi k/T$ (see Fig. 5.8 for a demonstration). $\diamondsuit$

Proof. Assume the functions $\{h(t - nT)\}$ are linearly dependent. Then there exists a nonzero sequence $c(n) \in \ell_2$ such that

$$\sum c(n)h(t - nT) = 0$$

for all t. Taking Fourier transforms, this implies $C(e^{j\omega T})H(j\omega) = 0$ for all ω, which implies

$$\left|C(e^{j\omega T})\right|^2 \sum_{k=-\infty}^{\infty} \left|H\Big(j\Big(\omega + \frac{2\pi k}{T}\Big)\Big)\right|^2 = 0 \tag{5.10}$$

for all ω. Since $c(n) \in \ell_2$ is a nonzero sequence, $C(e^{j\omega T})$ is nonzero in some region $\omega_a < \omega < \omega_b$. In this region, therefore, the summation in the preceding left-hand side must vanish identically.

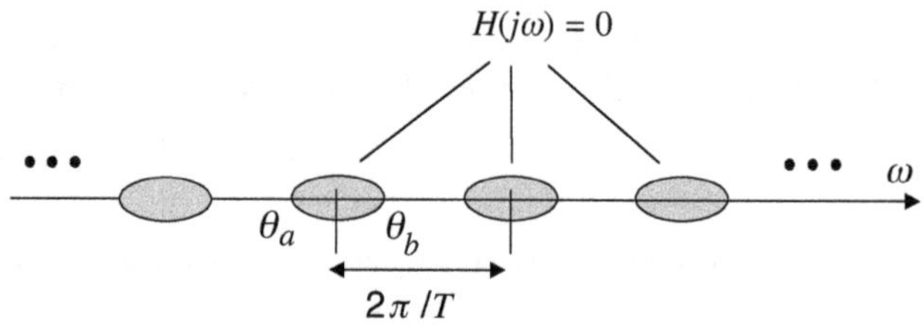

Figure 5.8. A situation where $H(j\omega)$ is identically zero in a region $\mathcal{R}$ and all shifted regions $\mathcal{R}+2\pi k/T$. In this situation the set of functions $\{h(t-nT)\}$ is not linearly independent. See Lemma 5.3.

So, if the summation has to be nonzero everywhere then $\{h(t-nT)\}$ have to be linearly independent. This proves that statement 2 implies statement 1. Conversely, assume linear independence. Then, as long as $C(e^{j\omega})$ is nonzero somewhere, the left-hand side of (5.10) is nonzero somewhere. By concentrating $C(e^{j\omega})$ in an arbitrary narrow band of frequencies we can then argue that statement 2 has to be true. So statements 1 and 2 are equivalent. The equivalence of statements 2 and 3 follows essentially from the definition of alias-free(T) bands (Appendix G). Now consider statement 4. If it is true, then, for any ω_0, at least one of the terms in the summation in statement 2 is nonzero, so statement 4 implies statement 2. If statement 4 fails, that is, if there is $\mathcal{R}$ such that $H(j\omega)$ vanishes on $\mathcal{R}+2\pi k/T$ for all integer k, then clearly the summation in statement 2 is zero on $\mathcal{R}$. So, failure of statement 4 implies that of statement 2, or equivalently statement 2 implies statement 4. So we have proved that all four statements are equivalent. $\bigtriangledown \bigtriangledown \bigtriangledown$

5.2.4 Generality of matched filtering

The preceding results show that matched filters can always be used at the receiver prior to sampling, without loss of generality. Thus, if $s(n)$ is reconstructible, then it can be reconstructed by first using the matched filter $G(j\omega) = H^*(j\omega)$, sampling its output, and filtering it with an appropriate digital filter (Fig. 5.7). Even if $s(n)$ is not reconstructible, we can still always reconstruct $y(t)$ from the signal component of $r(n)$ by first using the matched filter $G(j\omega) = H^*(j\omega)$, sampling its output, and filtering it with an appropriate filter as in Fig. 5.4(a). That is, $G(j\omega) = H^*(j\omega)$ is information lossless. Some further points are summarized below.

1. ***Sufficient statistics.*** The fact that a matched filter followed by a sampler is information lossless can be regarded as an extended version of the *sampling theorem* for the receiver structure. In a statistical setting one says that $r(n)$

constitutes a *sufficient statistic* for $y(t)$ [Forney, 1972], [Proakis, 1995]. For a more rigorous discussion of this viewpoint, see Blahut [1990] and references therein. A good discussion of the concept of sufficient statistics can be found in Sec. 10.5 of Moon and Stirling [2000].

2. ***More general receiver filter.*** Even though we have proved that $G(j\omega) = H^*(j\omega)$ is information lossless, there are more general filters which have such a property. For example, the choice

$$G(j\omega) = H^*(j\omega)D(j\omega) \tag{5.11}$$

is information lossless when $D(j\omega) > 0$ for all ω (see Appendix 5.A at the end of this chapter). There is, however, one positive advantage to be gained by using the matched filter $H^*(j\omega)$, namely the signal-to-noise ratio after sampling is maximized (for the case of white noise $q_c(t)$).

3. ***Jointly optimal transceivers with ZF.*** In Sec. 10.2.1 we will show how a prefilter and equalizer should be jointly optimized to minimize the mean square error under a power constraint and the zero-forcing constraint. We will see that the expression for the optimal receiver filter $G(j\omega)$, however, is *not* in the form of a matched filter (i.e., not matched to $H(j\omega)$). This does not mean that there is loss of information because the matched filter (and the more general form (5.11)) has only been shown to be *sufficient,* but not necessary, for information losslessness. In fact, since the zero-forcing condition is enforced in Sec. 10.2.1 it will automatically force information losslessness.

More examples of information lossless receiver filters

We conclude by mentioning some of the well known examples where the receiver filter indeed has the general form (5.11) and is therefore information lossless:

1. ***Matched filter for colored noise.*** In Sec. 2.5 we showed that, when the noise is not white but has a power spectrum $S_{qq}(j\omega)$, the receiver filter

$$G(j\omega) = \frac{H^*(j\omega)}{S_{qq}(j\omega)} \tag{5.12}$$

serves as the matched filter, that is, it maximizes SNR after sampling. This expression indeed has the form (5.11) with $D(j\omega) > 0$. Thus the above matched filter is also information lossless.

2. ***Noise-whitening receiver filter.*** In some applications the receiver filter is chosen as (5.11) where $D(j\omega)$ is such that the noise component after filtering and sampling is white. Such "whitening matched filters," discussed in Sec. 5.3, are also information lossless.

5.3 The sampled-noise whitening receiver filter

Referring again to Fig. 5.1(a), we now derive the condition on the filter $G(j\omega)$ which ensures that the sampled version of the noise (i.e., the noise component in $r(n)$) is white [Forney, 1972], [Andersen, 1973]. This is of importance when the sequence $r(n)$ is processed by a Viterbi decoder to estimate the transmitted symbol stream (Sec. 5.6). The figure is reproduced in Fig. 5.9. Assume that $q_c(t)$ is WSS with power spectrum $S_{qq,c}(j\omega)$, and let $q(n)$ be the noise after sampling (i.e., at the output of the C/D converter). The noise spectrum at the output of $G(j\omega)$ is $S_{qq,c}(j\omega)|G(j\omega)|^2$. The power spectrum of the noise component of $r(n)$ is the aliased version of this, that is,

$$S_{qq}(e^{j\omega}) = \Big[S_{qq,c}(j\omega)|G(j\omega)|^2\Big]_{\downarrow T}, \tag{5.13}$$

where the notation $[X(j\omega)]_{\downarrow T}$ is defined by Eq. (5.1). Assuming $S_{qq,c}(j\omega)$ is constant with value $\mathcal{N}_0$ in the band of interest[1] (i.e., $q_c(t)$ is white in the band of interest), we therefore have

$$S_{qq}(e^{j\omega}) = \mathcal{N}_0\Big[|G(j\omega)|^2\Big]_{\downarrow T}. \tag{5.14}$$

The noise $q(n)$ is therefore white if and only if

$$\Big[|G(j\omega)|^2\Big]_{\downarrow T} = c \tag{5.15}$$

for all ω, where $c > 0$ is a constant. In particular, the magnitude-square $|G(j\omega)|^2$ should be an alias-free(T) filter. More explicitly, Eq. (5.15) means

$$\frac{1}{T}\sum_{k=-\infty}^{\infty}\Big|G\Big(j\Big(\frac{\omega+2\pi k}{T}\Big)\Big)\Big|^2 = c. \tag{5.16}$$

With $g(t)$ denoting the impulse response $g(t)$ of the filter $G(j\omega)$, Eq. (5.16) can be rewritten in the time domain as

$$\int_{-\infty}^{\infty} g(t)g^*(t-kT)dt = c\delta(k). \tag{5.17}$$

This property is equivalent to

$$\int_{-\infty}^{\infty} g(t-mT)g^*(t-kT)dt = c\delta(m-k) \tag{5.18}$$

for all integers k, m. That is, the shifted version $g(t-mT)$ is **orthogonal** to the shifted version $g(t-kT)$ whenever integers k and m are distinct.

[1] If the channel noise $q_c(t)$ is not white to begin with, we can introduce a filter $1/\sqrt{S_{qq,c}(j\omega)}$ in front of $G(j\omega)$ to whiten it first.

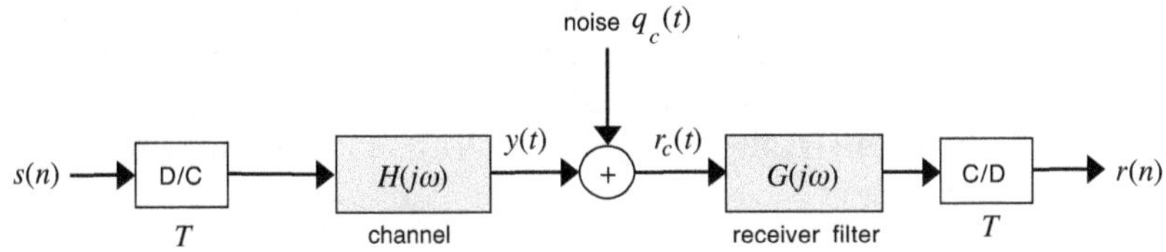

Figure 5.9. The digital communication system.

So we say that the set of shifted functions

$$\{g(t-kT)\} \tag{5.19}$$

forms an orthogonal set. It is an orthogonal basis for the space of functions of the form

$$x(t) = \sum_k c_k g(t-kT). \tag{5.20}$$

Restricting c_k to be sequences with finite energy (i.e., $c_k \in \ell_2$), we can show that this space is a subspace of L_2 (i.e., finite-energy functions, see Appendix A).

Orthonormalization. Given an arbitrary filter $G(j\omega)$, we can convert it into a filter satisfying orthogonality (5.18) under some mild conditions. For this consider the sum

$$S_g(e^{j\omega T}) = \frac{1}{T}\sum_{k=-\infty}^{\infty}\left|G\left(j\left(\omega+\frac{2\pi k}{T}\right)\right)\right|^2.$$

The notation on the left is justified by the fact that the right-hand side has period $\omega_s = 2\pi/T$ in the variable ω. Assuming that $G(j\omega)$ is nonzero everywhere in some alias-free(T) band, we see that

$$S_g(e^{j\omega T}) > 0 \tag{5.21}$$

for all ω. (This is related to linear independence and bandwidth sufficiency; see Lemma 5.3.) Assuming therefore that Eq. (5.21) holds, define a new filter

$$G_{new}(j\omega) = \frac{G(j\omega)}{\sqrt{S_g(e^{j\omega T})}} \tag{5.22}$$

Then

$$\left[|G_{new}(j\omega)|^2\right]_{\downarrow T} = \frac{\left[|G(j\omega)|^2\right]_{\downarrow T}}{S_g(e^{j\omega})} = \frac{S_g(e^{j\omega})}{S_g(e^{j\omega})} = 1.$$

for all ω. Thus, in the time domain,

$$\int_{-\infty}^{\infty} g_{new}(t)g^*_{new}(t-kT)dt = \delta(k), \tag{5.23}$$

and the basis $\{g_{new}(t-kT)\}$ is orthonormal. $\bigtriangledown\bigtriangledown\bigtriangledown$

The filter $G_{new}(j\omega)$ which produces white noise samples has been referred to as the whitened-matched filter [Forney, 1972] and the sample-whitened matched filter [Andersen, 1973] in the literature because, in those papers, the filter $G(j\omega)$ was matched to $H(j\omega)$. But since the whitening procedure is the same whether $G(j\omega)$ is matched to $H(j\omega)$ or not, we say that $G_{new}(j\omega)$ is a **sampled-noise whitening receiver filter**. Figure 5.10(a) shows an implementation of the filter $G_{new}(j\omega)$, where

$$G_d(e^{j\omega}) = \frac{1}{\sqrt{S_g(e^{j\omega})}}$$

Note that the orthonormalizing filter $G_d(e^{j\omega T})$ is a **transversal filter**, that is, it has the form

$$G_d(e^{j\omega T}) = \sum_n g_d(n) e^{-j\omega T n}, \tag{5.24}$$

where $g_d(n)$ is a discrete-time sequence. Using the appropriate noble identity (see Sec. G.2 in Appendix G) this can be moved to the right of the sampler and implemented as a digital filter $G_d(z)$, as shown in Fig. 5.10(b).

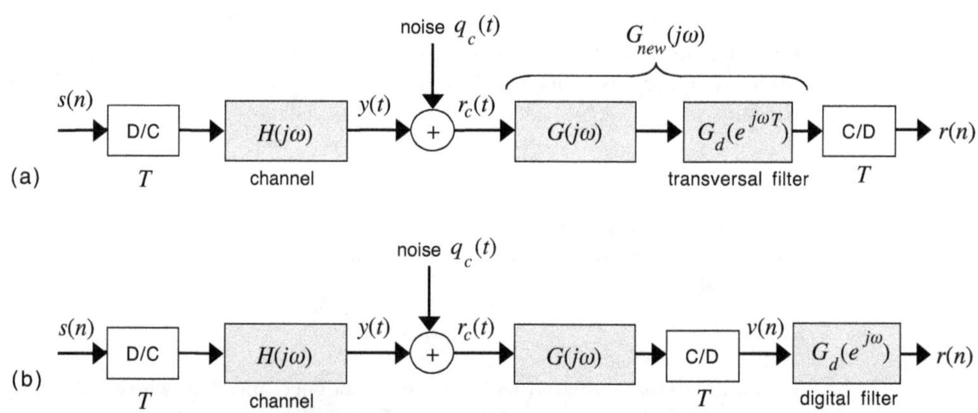

Figure 5.10. (a) Transceiver with sampled-noise whitening receiver filter, and (b) digital implementation of the transversal part.

5.4 Vector space interpretation of matched filtering

We now give an interesting interpretation of matched filtering based on the language of vector spaces. For this discussion the reader may first want to review Appendix A on ℓ_2-spaces, L_2-spaces, etc. We begin by observing that the noise-free channel output can be expressed as

$$y(t) = \sum_{n=-\infty}^{\infty} s(n)h(t - nT), \tag{5.25}$$

and is a linear combination of $h(t)$ and shifted versions $h(t - nT)$, where the shifts are in integer multiples of T. So we say that $y(t)$ belongs to a space V_h of signals spanned by the basis functions

$$\ldots\ h(t+T), h(t), h(t-T), h(t-2T), \ldots \tag{5.26}$$

The term "basis" is used here under the assumption that $\{h(t-nT)\}$ are linearly independent. Assuming that $s(n) \in \ell_2$ (space of finite energy sequences), the space V_h is a subspace of L_2 (space of finite energy functions), and we write $V_h \subset L_2$. Under the condition that

$$S_h(e^{j\omega}) \stackrel{\Delta}{=} \frac{1}{T} \sum_{k=-\infty}^{\infty} \left| H\Big(j\Big(\frac{\omega + 2\pi k}{T}\Big)\Big)\right|^2 > 0 \tag{5.27}$$

for all ω, we can construct an orthonormal basis for the space V_h, as shown in Sec. 5.3.[2] For this we define

$$H_{new}(j\omega) = \frac{H(j\omega)}{\sqrt{S_h(e^{j\omega T})}} \tag{5.28}$$

As we have already shown, $h_{new}(t - nT)$ is an orthonormal set:

$$\int_{-\infty}^{\infty} h_{new}(t - mT)h^*_{new}(t - kT)dt = c\delta(m - k). \tag{5.29}$$

Observe further that

$$h_{new}(t) = \sum_n c(n)h(t - nT), \tag{5.30}$$

where $c(n)$ is the inverse transform of $1/\sqrt{S_h(e^{j\omega})}$. Thus $h_{new}(t - kT) \in V_h$ for all k. Similarly

$$h(t) = \sum_n d(n)h_{new}(t - nT), \tag{5.31}$$

where $d(n)$ is the inverse transform of $\sqrt{S_h(e^{j\omega})}$, which means that $h(t-kT)$ is in the space spanned by $\{h_{new}(t-kT)\}$. Thus $\{h_{new}(t-nT)\}$ and $\{h(t-nT)\}$ span

[2]This positivity condition is related to linear independence and bandwidth sufficiency; see Lemma 5.3.

the same space V_h, the only difference being that the former is an orthonormal basis. Thus, given any function $y(t) \in V_h$, we can write

$$y(t) = \sum_n e(n) h_{new}(t - nT), \tag{5.32}$$

where $e(m)$ are the coefficients of expansion of $y(t)$ using the basis $\{h_{new}(t - nT)\}$. Thus we have

$$\begin{aligned} \int_{-\infty}^{\infty} y(t) h_{new}^*(t - mT) dt &= \sum_n e(n) \int_{-\infty}^{\infty} h_{new}(t - nT) h_{new}^*(t - mT) dt \\ &= \sum_n e(n) \delta(m - n) \quad \text{(using Eq. (5.29))} \end{aligned}$$

so that

$$\int_{-\infty}^{\infty} y(t) h_{new}^*(t - mT) dt = e(m). \tag{5.33}$$

Defining the function

$$g(t) = h_{new}^*(-t) \tag{5.34}$$

we see that

$$\int y(t) h_{new}^*(t - mT) dt = \int y(t) g(mT - t) dt,$$

so that the preceding equation can be rewritten as

$$\int_{-\infty}^{\infty} y(t) g(mT - t) dt = e(m). \tag{5.35}$$

Note that $\{h(t - nT)\}$ and $\{g^*(nT - t)\}$ span the same space V_h. The process of extracting the coefficients $e(m)$ from $y(t)$ can therefore be interpreted as filtering followed by sampling, as shown in Fig. 5.11.

Summarizing, given the output $y(t)$ of the channel $h(t)$ as in Fig. 5.12(a), it can be expressed in the form (5.32) using the orthonormal basis $h_{new}(t - nT)$. The coefficients of expansion $e(n)$ can be obtained by filtering $y(t)$ with the receiver filter $g(t)$ having frequency response $H_{new}^*(j\omega)$, that is,

$$G(j\omega) = \frac{H^*(j\omega)}{\sqrt{S_h(e^{j\omega T})}}, \tag{5.36}$$

as indicated in Fig. 5.12(a) on the right. This is simply the normalized matched filter, whose sampled output has its noise component whitened. These samples $e(m)$ contain the full information about $y(t)$ in view of Eq. (5.32). Using a noble identity we can redraw the system as in Fig. 5.12(b). Since $e(m)$ contains all the information about $y(t)$, the sampled version $r(n)$ of the output of the matched filter $H^*(j\omega)$ also contains this information.

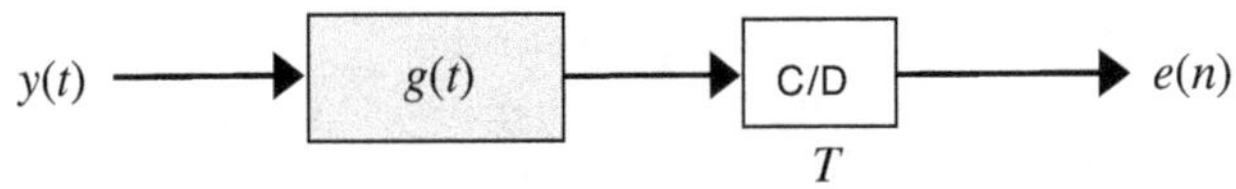

Figure 5.11. Filtering interpretation of the computation of the expansion coefficients $e(n)$.

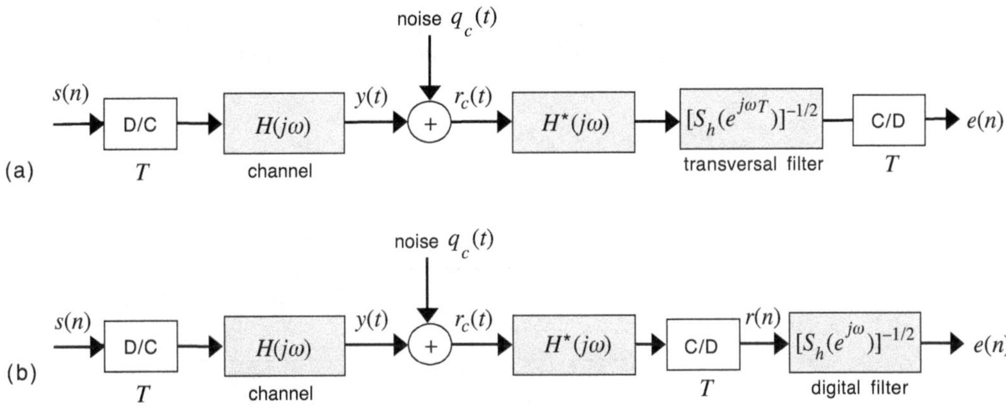

Figure 5.12. The matched filter idea rederived from a subspace approach. (a) Matched filter followed by a transversal filter, and (b) digital implementation of the transversal part.

A number of interesting conclusions can now be drawn.

1. ***Receiver filter space should include channel.*** Refer again to the basic system reproduced at the top of Fig. 5.13. Imagine that we have a fixed receiver filter $g(t)$ with $\{g^*(nT-t)\}$ spanning a certain space V_g. Let $h(t)$ be a channel not contained entirely in V_g, though it is in L_2. We can always write

$$h(t) = h_0(t) + h_1(t), \tag{5.37}$$

where $h_0(t) \in V_g$ and $h_1(t) \in V_g^{\perp}$, where $V_g^{\perp}$ is the orthogonal complement of V_g. Then $h_1(t)$ is orthogonal to $g^*(nT-t)$ for any n, that is,

$$\int_{-\infty}^{\infty} h_1(t)g(nT-t)dt = 0. \tag{5.38}$$

Thus

$$\int_{-\infty}^{\infty} h(t)g(nT-t)dt = \int_{-\infty}^{\infty} h_0(t)g(nT-t)dt. \tag{5.39}$$

That is, the sampled output $r(n)$ of the receiver filter contains information only about the $h_0(t)$ part. The $h_1(t)$ part, which is in the orthogonal complement $V_g^\perp$, is lost. An example is shown in Fig. 5.13. In this case we can reconstruct from the sampled signal $r(n)$ only the part $\sum_n s(n)h_0(t-nT)$, which is the orthogonal projection of $\sum_n s(n)h(t-nT)$ onto the subspace V_g. Summarizing, if $\{g^*(nT-t)\}$ spans a certain space V_g, then the sampled output $r(n)$ of the receiver filter $g(t)$ contains all the information about $y(t)$ (the output of $h(t)$) if and only if $h(t) \in V_g$, or equivalently $V_h \subset V_g$, where V_h is the span of $\{h(t-nT)\}$.

2. ***Receiver filter space should minimize noise***. Given any $h(t) \in V_h$ it is clear that the space spanned by $\{g^*(nT-t)\}$ should include V_h, but should it be exactly identical to V_h? The answer is, this is the optimal choice when there is noise and interference. In general any interfering signal that enters $g(t)$ has components in V_h and $V_h^\perp$. Choosing $g(t)$ such that $\{g^*(nT-t)\}$ is a basis for V_h implies that the portion of interference present in $V_h^\perp$ is suppressed.

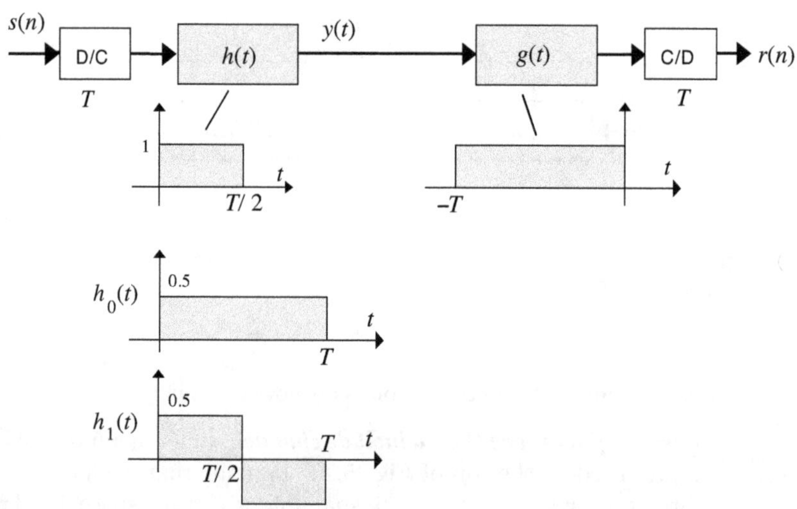

Figure 5.13. Example of a situation where the space spanned by $\{g^*(nT-t)\}$ includes part of $h(t)$ but not all of it. With $h(t)$ written as $h_0(t) + h_1(t)$, the part $h_1(t)$ is suppressed by the receiver filter and sampler.

5.5 Optimal estimates of symbols and sequences

The process of estimating a transmitted symbol stream from received noisy samples is central to the successful operation of digital communication receivers. In this section we discuss different types of estimates such as the maximal likelihood (ML) estimates, the maximum a posterior (MAP) estimates, and so forth.

5.5.1 Estimates of a symbol based on the received sample

The probability that the transmitted symbol $s(n)$ at time n is s_k, given that $r(n)$ was received, is denoted by the standard notation[3]

$$p(s_k|r). \tag{5.40}$$

This is called the *a posterior probability* that s_k was transmitted, given that the received signal is r. Here the time argument (n) has been deleted for simplicity. Given the received signal r, if we choose the estimate s_{est} of the transmitted symbol such that

$$p(s_{est}|r) \geq p(s_k|r) \tag{5.41}$$

for all possible choices of s_k, then the symbol estimate s_{est} is said to be the **maximum a posterior** or **MAP** estimate.

Next, the conditional pdf of the received signal r, given that s_k was transmitted at time n, is denoted by

$$f(r|s_k). \tag{5.42}$$

This is also called the **likelihood function**, as it indicates the likelihood that a certain symbol was transmitted, given that the received signal was such and such. Note that we have used the letter $f(.)$ to denote pdfs and the letter $p(.)$ to denote probabilities. Given the received signal r, if we choose the estimate s_{est} of the transmitted symbol such that

$$f(r|s_{est}) \geq f(r|s_k) \tag{5.43}$$

for all possible choices of s_k, then the symbol estimate s_{est} is said to be the **maximum likelihood** or **ML** estimate. To relate the MAP and ML estimates of the transmitted symbol, we use Bayes' rule from probability theory. According to this rule [Papoulis, 1965], [Peebles,1987], we have

$$p(s_k|r) = \frac{f(r|s_k)p(s_k)}{f(r)}, \tag{5.44}$$

where $p(s_k)$ is the a priori probability that s_k was transmitted at time n. Note that $f(r)$ can be written as a sum over all the possible values of the transmitted symbol:

$$f(r) = \sum_{k=0}^{M-1} f(r|s_k)p(s_k). \tag{5.45}$$

[3]The conditional probability of x given y is denoted as $p(x|y)$ [Papoulis, 1965], [Peebles,1987]. More elaborate notations such as $p_{X|Y}(x|y)$ [Peebles, 1987] are avoided for simplicity.

Since this is a sum over all k, it does not depend on the transmitted symbol s_k. If we make the assumption that *all transmitted symbols have identical probabilities,* that is, $p(s_k) = 1/M$ for all k, then

$$p(s_k|r) = cf(r|s_k) \tag{5.46}$$

for constant c. This implies that symbol estimates based on the **MAP and ML criteria are identical,** whenever $p(s_k)$ is the same for all symbols.

5.5.2 Symbol estimate based on a sequence of received samples

In general there is intersymbol interference, which means that the received signal $r(n)$ depends not only on the present value of the transmitted symbol $s(n)$, but also on some of the past values $s(n-m)$. Assuming that the discrete-time channel $H_d(z)$ (i.e., the sampled version of the cascade $F(j\omega)H(j\omega)G(j\omega)$) is FIR with order L, the received signal $r(n)$ depends on $s(n)$ and the past values

$$s(n-1), s(n-2), \ldots, s(n-L). \tag{5.47}$$

Thus a received sample $r(n)$ has information about the present and L past symbols. Putting it another way, the estimate of the present symbol $s(n)$ will benefit from the present and *future* observations

$$[\,r(n) \quad r(n+1) \quad \ldots \quad r(n+L)\,].$$

Note also that the past received samples $r(n-k), k > 0$, can be correlated to the current symbol $s(n)$ because of memory in the sequence $s(.)$ itself. Thus, assuming an arbitrary starting time $n = 0$, an estimate of $s(n)$ based on the vector of present, past, and L future observations

$$\mathbf{r} = [\,r(0) \quad r(1) \quad \ldots \quad r(n+L)\,] \tag{5.48}$$

will be more accurate than estimating it from the sample $r(n)$ alone. Similar to what we did in Sec. 5.5.1 we can now define the quantity

$$p(s_k|\mathbf{r}). \tag{5.49}$$

This is the *a posterior probability* that s_k was transmitted at time n, given the received samples $\mathbf{r}$ in Eq. (5.48). Given the received signal samples $\mathbf{r}$, if we choose the estimate s_{est} of the transmitted symbol such that

$$p(s_{est}|\mathbf{r}) \geq p(s_k|\mathbf{r}) \tag{5.50}$$

for all possible choices of s_k, then the symbol estimate s_{est} is the *MAP estimate* based on the sequence of observations (5.48). Next, the conditional pdf that the received signal samples are $\mathbf{r}$, given that s_k was transmitted at time n, is denoted by

$$f(\mathbf{r}|s_k). \tag{5.51}$$

This is the likelihood function. Given the received signal samples $\mathbf{r}$, if we choose the estimate s_{est} of the transmitted symbol $s(n)$ such that

$$f(\mathbf{r}|s_{est}) \geq f(\mathbf{r}|s_k) \tag{5.52}$$

for all possible choices of s_k, then the symbol estimate s_{est} is the *ML estimate* of the symbol $s(n)$ based on the sequence of observations $\mathbf{r}$ in Eq. (5.48). By appealing to Bayes' rule we can again conclude that the MAP and ML estimates of $s(n)$ yield the same answer whenever the probabilities $p(s_k)$ of the symbol values are identical.

Sequence estimates based on sequences. The preceding discussions are readily generalized to the case where an entire segment of the transmitted sequence is estimated based on a received sequence. In this case the MAP estimate is defined by the property

$$p(\mathbf{s}_{est}|\mathbf{r}) \geq p(\mathbf{s}_k|\mathbf{r}), \tag{5.53}$$

and the ML estimate is defined by the property

$$f(\mathbf{r}|\mathbf{s}_{est}) \geq f(\mathbf{r}|\mathbf{s}_k). \tag{5.54}$$

In these equations, $\mathbf{r}$ is a measurement (e.g., a received sequence of noisy samples), $\mathbf{s}_{est}$ is an estimate of a vector of transmitted symbols, and $\mathbf{s}_k$ is any valid choice of such a vector from the constellation. $\bigtriangledown\bigtriangledown\bigtriangledown$

Thus there are many possibilities: we can estimate (a) a symbol based on the observation of a sample, (b) a symbol based on the observation of a *sequence* of samples, or (c) a *sequence* of symbols based on the observation of a *sequence* of samples.

5.5.3 MAP or ML estimates and error probabilities

Let $\mathbf{s}_{est}$ be the estimated value of a vector symbol based on some observation $\mathbf{r}$. If this is an MAP estimate then the average probability of error is minimized. To prove this assume that the correct value of the estimate is $\mathbf{s}_k$. Then the error probability is

$$\mathcal{P}_e(\mathbf{s}_k) = 1 - \int_{\mathcal{R}_k} f(\mathbf{r}|\mathbf{s}_k)d\mathbf{r},$$

where $\mathcal{R}_k$ is the region such that if $\mathbf{r} \in \mathcal{R}_k$ the decision is correct. Thus the error probability averaged over all symbols is

$$\begin{aligned}\mathcal{P}_e = \sum_{k=0}^{M-1} p(\mathbf{s}_k)\mathcal{P}_e(\mathbf{s}_k) &= \sum_{k=0}^{M-1} p(\mathbf{s}_k) - \sum_{k=0}^{M-1} p(\mathbf{s}_k)\int_{\mathcal{R}_k} f(\mathbf{r}|\mathbf{s}_k)d\mathbf{r} \\ &= 1 - \sum_{k=0}^{M-1} p(\mathbf{s}_k)\int_{\mathcal{R}_k} f(\mathbf{r}|\mathbf{s}_k)d\mathbf{r}\end{aligned}$$

$$= 1 - \sum_{k=0}^{M-1} \int_{\mathcal{R}_k} p(\mathbf{s}_k|\mathbf{r}) f(\mathbf{r}) d\mathbf{r}$$

$$= 1 - \int_{\mathcal{R}} p(\mathbf{s}_k|\mathbf{r}) f(\mathbf{r}) d\mathbf{r},$$

where $\mathcal{R}$ is the union of the disjoint regions $\mathcal{R}_k$. The third equality is obtained by using Bayes' rule, that is, $f(\mathbf{r}|\mathbf{s}_k)p(\mathbf{s}_k) = p(\mathbf{s}_k|\mathbf{r})f(\mathbf{r})$. So we have proved that the average error probability is

$$\mathcal{P}_e = 1 - \int_{\mathcal{R}} p(\mathbf{s}_k|\mathbf{r}) f(\mathbf{r}) d\mathbf{r}. \tag{5.55}$$

It is now clear that this is minimized by maximizing $p(\mathbf{s}_k|\mathbf{r})$ for each k. So the *MAP estimate also minimizes average error probabilities.* Since the ML estimate agrees with the MAP estimate when all $p(\mathbf{s}_k)$ are equal, the *ML estimate also minimizes error probabilty when all* $\mathbf{s}_k$ *have identical probabilities.*

5.5.4 The ML estimate in the Gaussian case

The maximum likelihood (ML) method finds an estimate s_{est} from the measurement r such that the conditional probability $f(r|s_{est})$ is maximized. Now assume that we have an additive white Gaussian noise or **AWGN** channel, that is,

$$r(n) = s(n) + q(n), \tag{5.56}$$

where $q(n)$ is zero-mean white Gaussian noise with variance σ_q^2. For a fixed transmitted symbol $s(n) = s_k$, we see that $r(n)$ is a Gaussian random variable with mean s_k. Its density function is therefore

$$f(r|s_k) = \frac{1}{\sqrt{2\pi\sigma_q^2}} \exp\Big(\frac{-(r - s_k)^2}{2\sigma_q^2}\Big), \tag{5.57}$$

where the time argument (n) has been omitted for simplicity. Maximizing this quantity is therefore equivalent to minimizing

$$D^2(r, s_k) = (r - s_k)^2. \tag{5.58}$$

That is, given the received sample $r(n)$ at time n, the best estimate of the symbol $s(n)$ at time n would be that value s_k in the constellation which **minimizes the distance** $D(r, s_k)$. Similarly, suppose we have received a sequence of K samples

$$\mathbf{r} = [\, r(0) \quad r(1) \quad \ldots \quad r(K-1) \,] \tag{5.59}$$

and want to estimate the first K symbols

$$\mathbf{s} = [\, s(0) \quad s(1) \quad \ldots \quad s(K-1) \,] \tag{5.60}$$

such that the conditional pdf

$$f(\mathbf{r}|\mathbf{s}_k) \tag{5.61}$$

is maximized. Note that if $s(n)$ comes from a constellation having M symbols, then $\mathbf{s}$ can have M^K possible values. The notation $\mathbf{s}_k$ represents one such value. The result of maximizing (5.61) would be the ML estimate $\mathbf{s}_k$ of the transmitted symbol sequence $\mathbf{s}$ based on the received noisy sequence $\mathbf{r}$.

We now use the assumption that the Gaussian noise $q(n)$ is white. The **white-noise assumption** means that the components of the Gaussian vector $\mathbf{r}$ become statistically independent, and the pdf becomes a product of components pdfs:

$$f(\mathbf{r}|\mathbf{s}_k) = \left(\frac{1}{\sqrt{2\pi\sigma_q^2}}\right)^K \prod_{n=0}^{K-1} \exp\Big(\frac{-(r(n)-s_k(n))^2}{2\sigma_q^2}\Big)$$

$$= \left(\frac{1}{\sqrt{2\pi\sigma_q^2}}\right)^K \exp\Big(\frac{-\sum_{n=0}^{K-1}(r(n)-s_k(n))^2}{2\sigma_q^2}\Big) \tag{5.62}$$

The ML estimate $\mathbf{s}_k$ is therefore such that

$$D^2(\mathbf{r}, \mathbf{s}_k) = \sum_{n=0}^{K-1}(r(n)-s_k(n))^2 \tag{5.63}$$

is minimized. That is, given the received sequence $r(n), 0 \leq n \leq K-1$, the best estimate of the transmitted sequence $s(n), 0 \leq n \leq K-1$ (with elements $s(n)$ in the constellation), is the one which **minimizes the distance** $D(\mathbf{r}, \mathbf{s}_k)$. We conclude with a couple of observations.

1. ***Sequence estimate versus symbol estimate***. Note that the ML estimate of an entire sequence produces a result that can be different from the results obtained from the ML estimation of each individual symbol. This is because the samples $s(n)$ may have some dependencies. For example, imagine that $s(n)$ is the output of a filter $F(z)$ in response to some original symbol stream $s_{in}(n)$ with independent samples. Then there is dependency amongst the samples $s(n)$ owing to ISI created by $F(z)$. The filter $F(z)$ could be a channel, or it could even be a precoding filter, as in a duobinary system (Sec. 5.9).

2. ***Relation between MAP, ML, and minimum-distance methods.*** As pointed out in Sec. 5.5.1 the MAP estimate is identical to the ML estimate when all allowed values of the symbol have identical probability. Furthermore, as shown above, when the additive noise is white-Gaussian the ML estimate $\mathbf{s}_k$ is the one which has minimum distance from the received vector $\mathbf{r}$. Thus when the symbol values have identical probability and the noise is white-Gaussian, the MAP, ML, and minimum-distance methods yield the same result.

5.6 Viterbi algorithm for channel equalization

The idea of using an entire sequence of received samples to obtain an MAP estimate of a transmitted symbol was perhaps first introduced in a paper by Chang and Hancock [1966]. It was later noticed that the method in the paper does not minimize symbol error probability, but rather minimizes the error probability for an entire symbol sequence viewed as one vector. The method was later modified by other authors and culminated in a version credited to Abend and Fritchman [1970], which achieves minimum average symbol error probability. The disadvangtage of these early algorithms is their computational complexity.

A fundamentally different method for optimal estimation of an entire sequence was developed by Viterbi [1967]. This is based on the representation of the FIR channel using a finite state machine called a trellis, and offers great advantages in terms of complexity. Viterbi's algorithm has therefore become the most widely known method today. It was originally introduced for the decoding of convolutional codes [Viterbi, 1967]. It was found later, indpendently by three researchers (namely, by Forney, Kobayashi, and Omura ; see Lucky [1973]), that this algorithm can be used for the ML equalization of FIR channels. That is, it can be used for the ML detection of sequences which have been subject to intersymbol interference because of an FIR channel. This ISI can be either directly caused by the channel transfer function $H(z)$ or indirectly present due to the design of the transmit waveform (e.g., duobinary waveforms used in partial response signaling, Sec. 5.9). In either case the algorithm is equally applicable. We now give a brief description of the algorithm along with examples. Detailed discussions on the Viterbi algorithm can be found in many references, e.g., Forney [1972, 1973], [Blahut, 1990], and [Proakis, 1995]. A fascinating history can be found in Lucky [1973].

5.6.1 The trellis diagram

Figure 5.14 shows the discrete-time equivalent of a digital communication channel. Here the FIR filter

$$H(z) = \sum_{n=0}^{L} h(n)z^{-n} \tag{5.64}$$

has impulse response $h(n) = h_c(nT)$, where $h_c(t)$ is the inverse Fourier transform of the equivalent channel $H_c(j\omega) = G(j\omega)H(j\omega)F(j\omega)$ (Sec. 4.3.1). The figure shows an example where this FIR filter has order $L = 2$. The noise $q(n)$ can be assumed to be white if the matched filter prior to sampling is appropriately designed as explained in Sec. 5.3. The signal $r(n)$ suffers from intersymbol interference due to the FIR filter $H(z)$, and there is additive noise as well. Note that, at any time n, the vector of signals stored in the delay elements

$$\mathbf{x}(n) = [\, s(n-1) \quad s(n-2)\,] \tag{5.65}$$

is the **state vector**. The elements of this state vector are the transmitted symbols, which can only take a discrete set of values from the constellation. The FIR filter can be viewed as a **finite state machine** or FSM, whose states $\mathbf{x}(n)$ can take only a finite set of values. Since each sample can take M possible values (size of the symbol constellation), there can be M^L possible values for the state vector $\mathbf{x}(n)$.

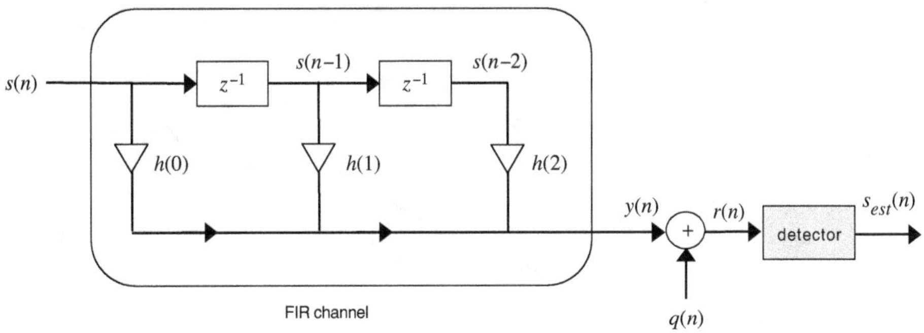

Figure 5.14. The discrete-time equivalent of the digital communication channel.

To explain the Viterbi algorithm we first need to understand something called the **trellis** diagram. This diagram is useful to keep track of the states as a function of the input sequence. It has one stage per instant of time. In each stage, each of the M^L possible values of the state is assigned a node. For example, when $M = 2$ and $L = 2$ there are $M^L = 4$ possible states, as shown in Fig. 5.15(a). Assume that the finite state machine is in a particular state $\mathbf{x}(n)$ at time n. When the next input $s(n)$ comes in, the values of the state variables change, and the trellis diagram shows all the M state transitions that are possible (one for each possible value of $s(n)$).

The trellis diagram shown in Fig. 5.15(a) is for the second-order FIR system of Fig. 5.14 with one-bit PAM input. So the input constellation is $\{1, -1\}$. Thus the four possible states of the system are $(1, 1), (-1, 1), (1, -1)$, and $(-1, -1)$. Assume for example that the system is initially in the state $(1, 1)$, and an input $s(0) = 1$ comes along. Then after one unit of time this input appears at the output of the first delay, and the state remains $(1, 1)$. On the other hand if the input is $s(0) = -1$, then the state changes to $(-1, 1)$. These two choices are indicated in the figure with a **heavy arrow** for $s(0) = 1$ and a **light arrow** for $s(0) = -1$. So the state at time $n = 1$ can be one of the two states indicated. From each of these two states two transitions are possible, depending on whether $s(1) = 1$ or -1. The trellis diagram therefore grows as indicated in Fig. 5.15(a).

The figure starts with some initial state $(1, 1)$ and then considers every possible choice of the input sample at every succeeding instant of time. After each state has been reached at least once, a steady condition is reached, and the trellis diagram becomes a **periodic** pattern, with the same set of criss-crosses repeated for ever. In fact, right from the beginning we can consider all possible states and draw a "full trellis."

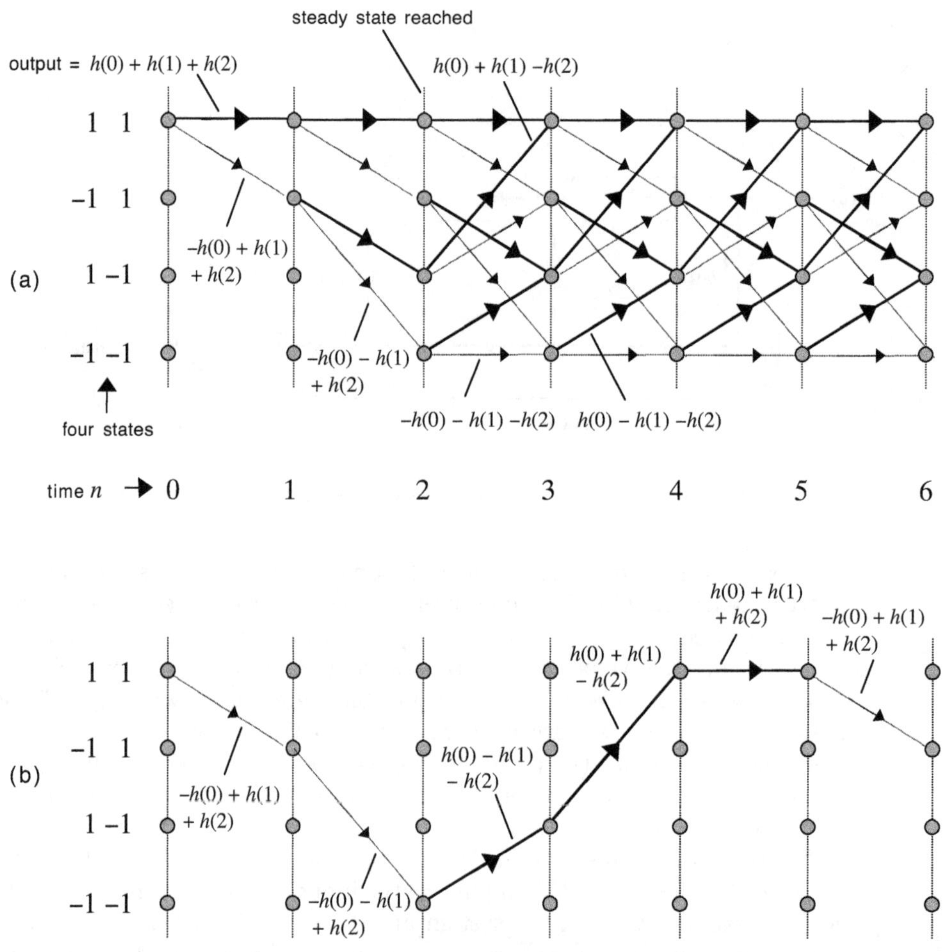

Figure 5.15. (a) The trellis diagram for the channel $H(z) = h(0)+h(1)z^{-1}+h(2)z^{-2}$, with one-bit PAM constellation (symbol values ± 1). (b) A specific path through the trellis corresponding to the input sequence $[-1 - 1 \ 1 \ 1 \ 1 - 1]$. In the figure, heavy arrows represent input symbol 1, and the light arrows represent input symbol -1. The labels on the edges represent filter outputs corresponding to the present state and present input. For example if the present input is -1 and the state is $[-1, 1]$, then the filter output is $-h(0) - h(1) + h(2)$.

Showing the noise-free outputs

Given an input $s(n)$ and a state vector $\mathbf{x}(n)$ at time n, the inputs to all the multipliers $h(k)$ in Fig. 5.14 are determined. This determines the output $y(n)$ completely. All the branches in the trellis are labeled with the output value

$y(n)$ that would result from the combination of $s(n)$ and $\mathbf{x}(n)$. For example, if $s(0) = 1$ and $\mathbf{x}(0) = [1\ 1]$ then $y(0) = h(0) + h(1) + h(2)$, and this is indicated as a branch gain in Fig. 5.15. Similarly if $s(0) = -1$ and $\mathbf{x}(0) = [1\ \ 1]$, then $y(0) = -h(0) + h(1) + h(2)$ as indicated. As another example, consider the state $\mathbf{x}(1) = [-1\ \ 1]$. If the input is $s(1) = -1$ then $y(1) = -h(0) - h(1) + h(2)$ as indicated.

Thus, given an input symbol stream $s(0), s(1), s(2), \ldots$ and an initial state $\mathbf{x}(0)$, we can trace a path through the trellis and obtain all the states through which the machine passes, as well as all the outputs that have been generated. For example, if the first six input symbols are

$$[\,s(0)\quad s(1)\quad s(2)\quad s(3)\quad s(4)\quad s(5)\,] = [\,-1\quad -1\quad 1\quad 1\quad 1\quad -1\,]$$

then the path traced on the trellis is as shown in Fig. 5.15(b), assuming that the initial state is $\mathbf{x}(0) = [1\ \ 1]$. The states traversed are clearly seen from this path. The output samples $y(n)$ can be figured out from the state and input information, and are indicated in Fig. 5.15(b).

Computing the output $y(n)$ starting from the input $s(n)$ is an easy task. The Viterbi algorithm seeks to perform the inverse operation, namely, to estimate the input sequence $s(n)$ from the output sequence $y(n)$, or, rather, a noisy version $r(n)$ of the output sequence. One way to do this would be to implement the inverse filter $1/H(z)$, but this can amplify noise. Instead of a linear filtering approach the Viterbi algorithm takes a nonlinear approach, and searches the trellis for the "most likely path" that would lead to the observed noisy output $r(n)$.

5.6.2 Finding the closest path

Assume that the first K samples of the noisy received signal are

$$r(0), r(1), \ldots, r(K-1), \tag{5.66}$$

so the trellis has K transitions. We can first make a list of all possible paths in this trellis starting from any intial state $\mathbf{x}(0)$ to any final state $\mathbf{x}(K)$. The kth path defines a potential output sequence

$$y_k(0), y_k(1), \ldots, y_k(K-1). \tag{5.67}$$

We can now define the distance $D(r, y_k)$ between the sequences $r(n)$ and $y_k(n)$ as follows:

$$D^2(r, y_k) = \sum_{n=0}^{K-1} |r(n) - y_k(n)|^2 \tag{5.68}$$

and choose the path $y_k(n)$ that yields the smallest value for this distance. This is the path through the trellis that yields an output sequence $y_k(n)$ closest to the observed sequence $r(n)$. The quantity $D(r, y_k)$ can be regarded as the Euclidean distance between the sequences $r(n)$ and $y_k(n)$. Thus, we can find the noiseless output sequence $y_k(n)$ which is closest to the output $r(n)$, and then identify the corresponding input from the trellis path that has been identified. Remember here that the path has input information: if a branch in the path has a heavy arrow then the input symbol is 1, and if it is a light arrow the input symbol is -1. Thus, having found the closest path, we can identify the transmitted symbol stream which generates this path. Of course, we have not argued that the distance measure based on the Euclidean distance (5.68) is optimal in any way, for example in the sense of minimizing error probabilities, but we shall return to that in Sec. 5.6.4.

The disadvantage of the "brute force" method described above is that we have too many paths to search, and the computational load becomes unresonable very quickly. However, there is great deal of structure in the trellis, and Viterbi's algorithm is an ingenious way to exploit this structure and reduce the computational load dramatically. We describe this next.

5.6.3 The Viterbi algorithm

The Viterbi algorithm is best described with an example. Consider an FIR channel with the transfer function

$$H(z) = 1 + 0.5z^{-1} \tag{5.69}$$

as in Fig. 5.16(a). Let the input $s(n)$ be drawn from a 1-bit PAM (BPSK) constellation so that $s(n) = \pm 1$. The state vector is just a scalar:

$$x(n) = s(n-1). \tag{5.70}$$

Figure 5.16(b) shows the basic trellis module for an input sequence of length one. The system has two possible states, namely 1 and -1. If $s(n) = 1$ then the next state becomes a 1, whereas if $s(n) = -1$, the next state becomes a -1. The output at any time n is $y(n) = s(n) + 0.5s(n-1)$, that is,

$$y(n) = s(n) + 0.5x(n). \tag{5.71}$$

Recall that the trellis carries the following information:

1. The state transition is shown using a thick arrow if the input is 1 and a thin arrow if the input is -1.
2. The output calculated using (5.71) is indicated alongside the transitioning arrow.

For example, if the system is in state 1 and the input is a -1, then the next state is -1 and the output is $-1 + 0.5 = -0.5$. This is indicated by the thin down going arrow in Fig. 5.16(b) labeled -0.5. The other arrows in the figure are obtained similarly. To work out a specific example, we now assume that the first few samples of the noisy output are

$$\begin{array}{cccccc} n & 0 & 1 & 2 & 3 & 4 \\ r(n) & 1 & -0.4 & -0.8 & 0.1 & 1.1 \end{array} \tag{5.72}$$

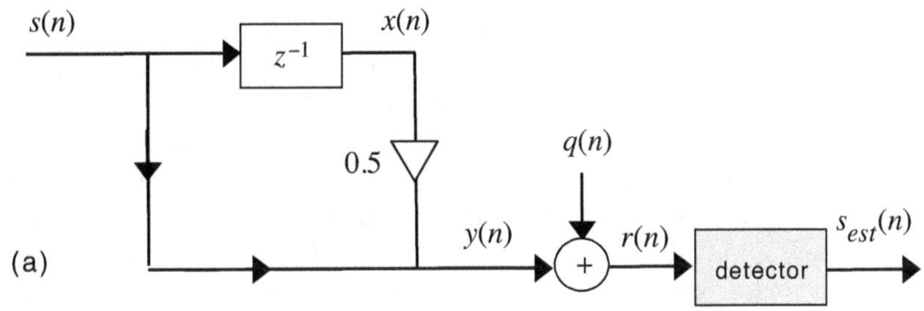

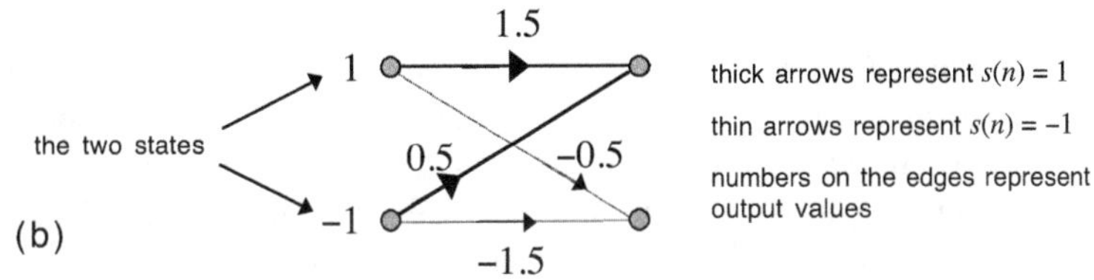

Figure 5.16. (a) A first-order FIR channel, and (b) the trellis module to be used in Viterbi's algorithm. The input is assumed to be from a 1-bit PAM constellation.

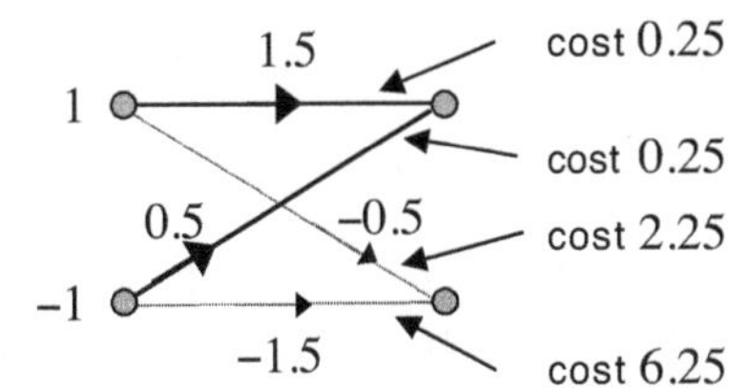

Figure 5.17. The first stage of the Viterbi algorithm. With output $r(0) = 1$, the costs of all possible state transitions are indicated.

Starting from this output sequence, and *without* knowledge of the noise variance σ_q^2, we show how to estimate the closest path in the trellis which would have given rise to this output. This process will automatically yield estimates of the state sequence $x(n)$, the noiseless output sequence $y(n)$, and, most importantly, the input sequence $s(n)$.

To initialize the procedure, note that if the initial state $x(0)$ is 1 and the input $s(0)$ is also 1, then the noise-free output would be $y(0) = 1.5$, as shown by the top horizontal path in Fig. 5.17. We define the cost associated with this path to be

$$\text{cost} = |r(0) - y(0)|^2 = 0.25.$$

This is nothing but the square of the distance between the quantities $r(0)$ and $y(0)$. The costs corresponding to all four combinations of the initial state and input are indicated in the figure. The minimum-cost path to reach the state -1 is the thin downward arrow, whereas the minimum-cost path to reach the state 1 can be taken to be either of the two choices. This results in the trellis shown at the top of Fig. 5.18 (stage 0). In this figure, the minimum cost of reaching each node (state) is indicated in bold. *This idea of computing all possible paths to a state and retaining only the minimum-cost path is central to the algorithm.* It is continued with each successive output sample $r(n)$ one after another. Thus the trellis grows, and in the process we can estimate an initial segment of the transmitted symbol stream as demonstrated next.

1. ***Stage 1 (time n = 1).*** Consider the output $y(1)$. The four possible ways to generate this output are shown at stage 1 in the figure, and the four possible values of $y(1)$ are indicated alongside the four arrows. Since the actual (noisy) output is $r(1) = -0.4$, the costs $|r(1) - y(1)|^2$ associated with the two choices to reach state 1 are

$$|-0.4 - 1.5|^2 = 3.61 \quad \text{or} \quad |-0.4 - 0.5|^2 = 0.81.$$

   These costs must be added to the earlier cost of reaching the state from which these paths originate. Thus the total cost to reach state 1 at time $n = 1$ is given by

$$\text{either} \quad 0.25 + |-0.4 - 1.5|^2 = 3.86 \quad \text{or} \quad 2.25 + |-0.4 - 0.5|^2 = 3.06.$$

   Similarly, at time $n = 1$ the total cost to reach state -1 is given by

$$\text{either} \quad 0.25 + |-0.4 + .5|^2 = 0.26 \quad \text{or} \quad 2.25 + |-0.4 + 1.5|^2 = 3.46.$$

   Retaining only the path to each state that has minimum cost we obtain the pruned trellis shown in stage 2, with the costs to reach the states indicated as 3.06 and 0.26.

2. ***Stage 2 (time n = 2).*** We now add one more trellis stage and compute all the costs corresponding to the next noisy output $r(2) = -0.8$. These are indicated in the figure. Retaining one path per state with minimum cost, we get the pruned tree shown in stage 3. At this point we notice something interesting. Namely, the initial segments of length two in the minimum-cost paths to the two states have **merged** into a *common initial sub-path.* This means that we are ready to make a decision about the two transmitted symbols $s(0)$ and $s(1)$. Recalling here that thin arrows correspond to $s(n) = -1$ and thick arrows to $s(n) = 1$, we see that the symbol estimates so far are

$$s_{est}(0) = 1 \quad \text{and} \quad s_{est}(1) = -1.$$

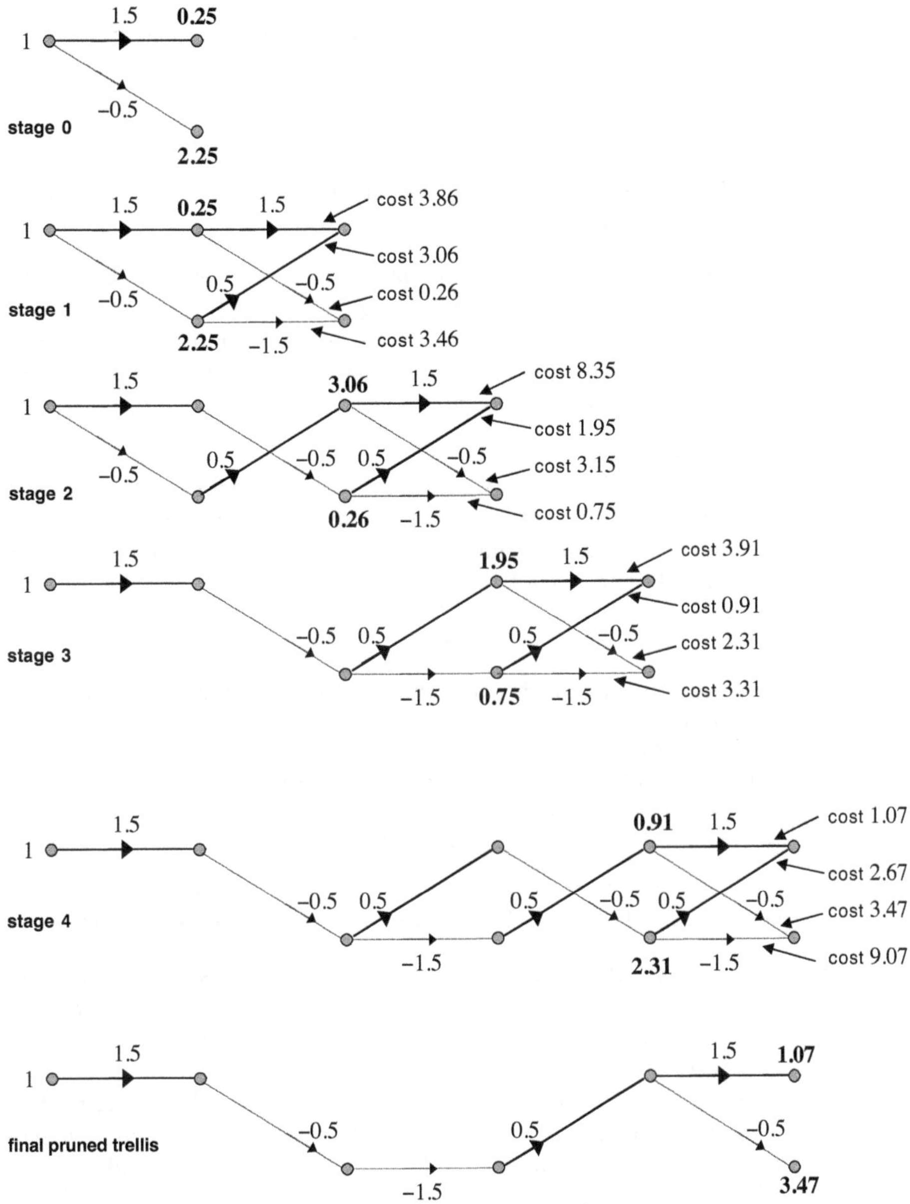

Figure 5.18. Successive stages in the development of the Viterbi algorithm. The noisy channel output is $1, -0.4, -0.8, 0.1, 1.1, \ldots$ From these the first four symbols of the transmitted signal are estimated as $1, -1, -1, 1$.

The estimated state sequence so far is:

$$x_{est}(0) = 1, x_{est}(1) = 1 \quad \text{and} \quad x_{est}(2) = -1,$$

and the estimated noiseless output is:

$$y_{est}(0) = 1.5 \quad \text{and} \quad y_{est}(1) = -0.5.$$

Since the estimate of $s(n)$ is of greater interest, we henceforth pay attention only to that.

3. ***Stage 3 (time n = 3).*** Continuing with the construction of the trellis we find that the next noisy output is $r(3) = 0.1$. Adding the trellis and computing all the costs, we obtain the four numbers indicated on the right in stage 3. Retaining only the path to each state that has minimum cost we obtain the pruned trellis shown in stage 4.

4. ***Stage 4 (time n = 4).*** Adding the next trellis corresponding to $r(4) = 1.1$ and retaining paths with lowest cost, we finally obtain the pruned trellis shown at the bottom. At this point we again notice that there arises a longer common initial path shared by the two minimum-cost paths. This allows us to make the decisions

   $$s_{est}(2) = -1 \quad \text{and} \quad s_{est}(3) = 1.$$

   We cannot continue any further without obtaining more samples of the output.

Summarizing, the noisy observed output sequence

$$r(n) = [\,1 \quad -0.4 \quad -0.8 \quad 0.1 \quad 1.1\,]$$

has been used to estimate the first four samples of the transmitted sequence to be $s_{est}(n) = [\,1 \quad -1 \quad -1 \quad 1\,]$. A number of points should now be noted.

1. ***Decoding delay.*** In the above example we first estimated the inputs $s(0)$ and $s(1)$ from the outputs $r(0), r(1)$, and $r(2)$. When the new output sample $r(3)$ arrived, it was not possible to estimate the next input symbol $s(2)$ because the "initial common path" or "merger" did not get longer, so there was still uncertainty about the closest path. When the next sample $r(4)$ arrived and the trellis grew, suddenly the initial common path grew by *two* units, and we were able to estimate both $s(2)$ and $s(3)$ at this stage. Thus, no matter how many outputs have been received (i.e., how much the trellis has grown), it is not possible to make new decisions on input symbols unless the "initial common path" grows. In general therefore, having received the first K noisy output samples

   $$r(0), r(1), \ldots, r(K-1), \tag{5.73}$$

   the Viterbi algorithm can only complete an estimate of $K - D$ symbols

   $$s(0), s(1), \ldots, s(K-D-1), \tag{5.74}$$

where D is called the *decision delay* or *decoding delay*. The quantity D changes as the algorithm progresses, and it can increase or decrease (Problem 5.7). It is not possible to find an upper bound on this delay, though it appears to get worse as the channel order L increases. In practice it has been found that we can safely assume $D \leq 5L$ (see discussions in Forney [1972] and Proakis [1995]). Typically, the longer the channel impulse response, the longer is the decoding delay.

2. ***Complexity***. For each new output sample $r(n)$ the Viterbi algorithm computes the cost associated with every possible way of reaching every possible next state and compares them. Since there are M^L states in general, the computational complexity grows rapidly with L. Since the growth is exponential with the channel order L, this becomes impractical for large L. However, for small L and small constellation sizes it is quite practical.

3. ***Other related algorithms.*** Viterbi's algorithm can be related to well known algorithms for finding the shortest path in a graph, such as Dijkstra's algorithm. See McEliece, Ash, and Ash [1989] for details. The algorithm can also be related to dynamic programming [Forney, 1973]. A good discussion of these connections can be found in Chap. 19 of Moon and Stirling [2000].

5.6.4 Viterbi algorithm and maximum likelihood estimates

The Viterbi algorithm described above identifies a path through the trellis which results in an output $y_k(n)$ closest to the noisy output $r(n)$. Under some assumptions this results in an optimal (maximum likelihood) estimation of the input sequence $s(n)$. More importantly this also minimizes the probability of error in the estimation of the input, viewed as a sequence. We now elaborate on these remarks. Given the first K noisy output samples

$$\widehat{\mathbf{s}}_K = [\, r(0) \quad r(1) \quad \ldots \quad r(K-1)\,] \tag{5.75}$$

the Viterbi algorithm obtains an estimate $\mathbf{y}_{est}$ of $K-D$ noise free outputs

$$\mathbf{y} = [\, y(0) \quad y(1) \quad \ldots \quad y(K-D-1)\,], \tag{5.76}$$

where D is the decoding delay. This estimate is obtained by minimizing $\sum_{n=0}^{K-1} |r(n) - y_{est}(n)|^2$, that is,

$$\sum_{n=0}^{K-1} |r(n) - y_{est}(n)|^2 \leq \sum_{n=0}^{K-1} |r(n) - y_k(n)|^2, \tag{5.77}$$

where $y_k(n)$ is the noise-free output for any path along the trellis. Note that we are minimizing a distance measure over K samples even though only $K - D$ samples are estimated. For $K >> D$ it is reasonable to approximate this statement by saying that

$$\sum_{n=0}^{K-D-1} |r(n) - y_{est}(n)|^2 \tag{5.78}$$

is minimized. Defining

$$\mathbf{r} = [\, r(0) \quad r(1) \quad \ldots \quad r(K-D-1) \,]$$

we can therefore say that $\mathbf{y}_{est}$ is a minimum distance estimate of $\mathbf{y}$ based on $\mathbf{r}$. When the noise is AWGN, this estimate is therefore an ML estimate (Sec. 5.5.4). Thus

$$f(\mathbf{r}|\mathbf{y}_{est}) \geq f(\mathbf{r}|\mathbf{y}) \tag{5.79}$$

for any other feasible noise-free output vector $\mathbf{y}$. Observe now that the vector $\mathbf{y}$ is related to the transmitted symbol vector $\mathbf{s}$ as follows:

$$\underbrace{\begin{bmatrix} y(0) \\ y(1) \\ \vdots \\ y(N) \end{bmatrix}}_{\mathbf{y}} = \underbrace{\begin{bmatrix} h(0) & 0 & \ldots & 0 \\ h(1) & h(0) & \ldots & 0 \\ \vdots & \vdots & \ddots & \vdots \\ h(N) & h(N-1) & \ldots & h(0) \end{bmatrix}}_{\mathbf{A}} \underbrace{\begin{bmatrix} s(0) \\ s(1) \\ \vdots \\ s(N) \end{bmatrix}}_{\mathbf{s}}, \tag{5.80}$$

where $N = K - D - 1$. In fact Eq. (5.80) is true for any $N \geq 0$. Since $s(n)$ belong to a constellation with M possible discrete values, the vector $\mathbf{s}$ can take M^{N+1} discrete values. So the output vector $\mathbf{y}$ can take at most M^{N+1} discrete values. In fact it takes exactly M^{N+1} distinct values assuming that the matrix above is nonsingular, that is, $h(0) \neq 0$.[4] Thus, even though each sample $y(n)$ can in principle have many more than M possible values (because it is a linear combination of the samples $s(n-k)$), the vector $\mathbf{y}$ comes from a set with precisely M^{N+1} discrete values. A number of points should now be noted.

1. ***ML property.*** Since each $\mathbf{s}$ from the discrete set maps to a unique $\mathbf{y}$ and vice versa, we see that (5.79) also implies

$$f_1(\mathbf{r}|\mathbf{s}_{est}) \geq f_1(\mathbf{r}|\mathbf{s}), \tag{5.81}$$

   where $f_1(.|.)$ represent the conditional pdf for $\mathbf{r}$ given $\mathbf{s}$. Thus, the fact that $\mathbf{y}_{est}$ is an ML estimate of $\mathbf{y}$ implies that $\mathbf{s}_{est}$ is an ML estimate of $\mathbf{s}$ (based on the received vector $\mathbf{r}$).

2. ***MAP property.*** Next, assume that the symbols $s(n)$ are independent and identically distributed, with identical probabilities for all symbols in the constellation. Then the M^{N+1} discrete values of $\mathbf{s}$ have identical probabilities. Thus the ML property of the estimate $\mathbf{s}_{est}$ also implies that it is an MAP estimate (Sec. 5.5.1).

3. ***Error-event probability.*** From the discussion of Sec. 5.5.3 it therefore follows that this estimated vector $\mathbf{s}_{est}$ has the minimum error probability property. We simply say that the error-event probability has been minimized [Forney, 1972]. That is, the probability of error in the estimation of the vector, viewed as one entity, is minimized.

[4] If $\mathbf{s}_1$ and $\mathbf{s}_2$ are two distinct values of $\mathbf{s}$ then $\mathbf{y}_1 - \mathbf{y}_2 = \mathbf{A}(\mathbf{s}_1 - \mathbf{s}_2)$. If $\mathbf{s}_1 - \mathbf{s}_2 \neq \mathbf{0}$ and $\mathbf{A}$ is nonsingular, it follows that $\mathbf{y}_1 - \mathbf{y}_2 \neq \mathbf{0}$. So $\mathbf{y}$ takes M^{N+1} distinct values like $\mathbf{s}$.

Here is a summary of the assumptions used in the above discussions on optimality of the estimates: (a) the noise $q(n)$ is AWGN, (b) the symbols $s(n)$ are independent and identically distributed, with identical probabilities for all symbols in the constellation, and (c) the decoding delay D is small compared to the sequence length.

Discussions

1. In the AWGN case it can also be argued that the algorithm finds the *state sequence* $\mathbf{x}(n)$ that is "most likely" to have generated the noisy output sequence $r(n)$.
2. Even though the error-event probability has been minimized, the algorithm does not necessarily minimize the *symbol error probability.* However, the symbol by symbol error probability can be bounded above and below using elegant formulas [Forney, 1972], [Proakis, 1995].
3. As a practical matter, the symbol error probabilities obtained using the Viterbi algorithm are typically much smaller than the results obtained using suboptimal methods such as decision feedback equalization [Proakis, 1995].
4. It should also be remarked here that in the non-Gaussian case it is possible to modify the path metrics appropriately (i.e., the definition of distances between paths in the trellis) so that the ML property is retained. In fact the algorithm was originally developed for the optimal decoding of convolutional codes where the path metrics are based on Hamming distances rather than Euclidean distances [Viterbi, 1967], [Proakis, 1995].

5.7 Decision feedback equalizers

Given an FIR channel

$$H(z) = h(0) + h(1)z^{-1} + \ldots + h(L)z^{-L}, \tag{5.82}$$

with input $s(n)$ and output $y(n)$, suppose we equalize the channel at the receiver by using the inverse

$$\frac{1}{H(z)} = \frac{1/h(0)}{1 + [h(1)/h(0)]z^{-1} + \ldots + [h(L)/h(0)]z^{-L}} \tag{5.83}$$

With $r(n)$ denoting the received signal, this inverse can be implemented using the following difference equation:

$$\widehat{s}(n) = -[h(1)/h(0)]\widehat{s}(n-1) \ \ldots - [h(L)/h(0)]\widehat{s}(n-L) + r(n). \tag{5.84}$$

This is shown in Fig. 5.19 for $L = 3$. The equalizer output $\widehat{s}(n)$ is used as the input to the detector, which produces the final symbol estimates $s_{est}(n)$.

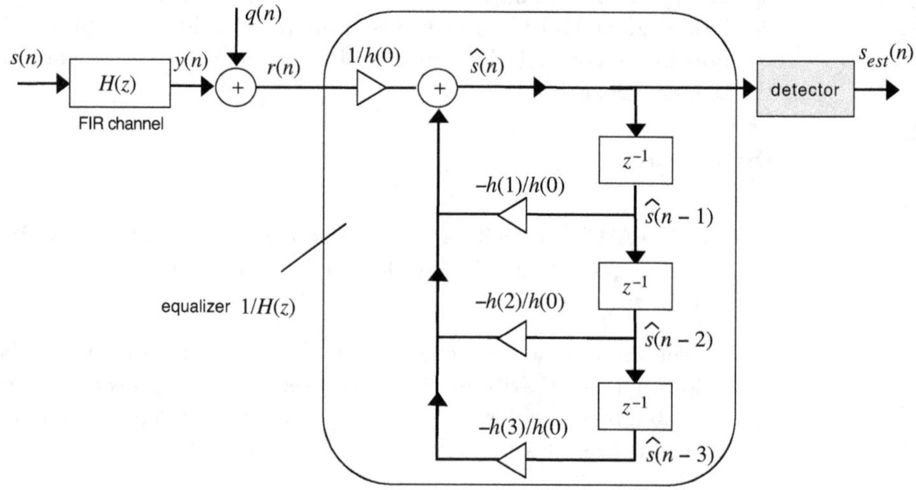

Figure 5.19. An IIR equalizer for the FIR channel $H(z)$.

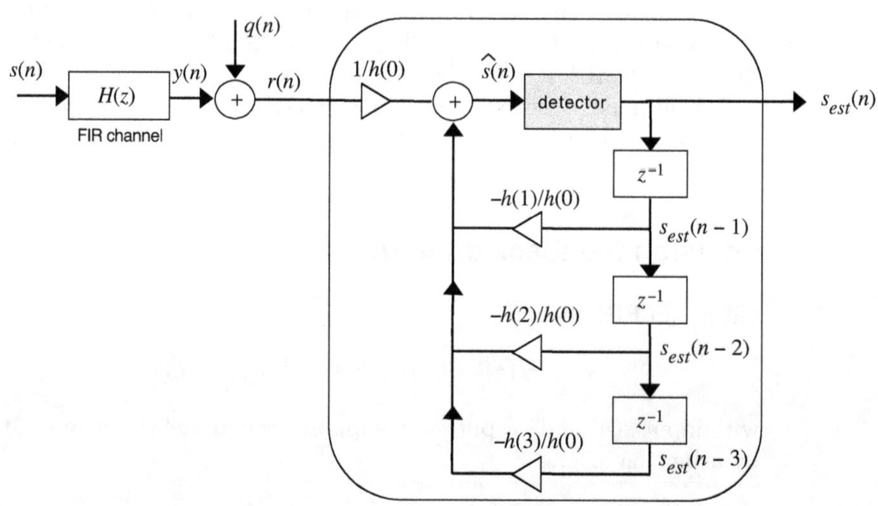

Figure 5.20. Moving the detector into the feedback loop. In this system the past decisions $s_{est}(n-k)$ are fed back.

This receiver with an IIR equalizer can pose some difficulties in practice. First, if $H(z)$ has zeros on or outside the unit circle then the equalizer $1/H(z)$ is unstable, and $\widehat{s}(n)$ can become unbounded because of the channel noise. Second, even if $1/H(z)$ is stable, it can have poles close to the unit circle, amplifying the noise significantly; if $h(0)$ is small, the multiplier $1/h(0)$ can be very large, again amplifying noise. A simple trick which avoids unboundedness is shown in Fig. 5.20, where the detector is moved into the feedback loop.

Since the detector output has a finite value taken from the signal constellation, all the signals in the modified structure are bounded. In this structure, the L past decisions

$$s_{est}(n-1), s_{est}(n-2), \ldots, s_{est}(n-L) \tag{5.85}$$

are used, along with the current received signal $r(n)$, to make the present decision $s_{est}(n)$. This decision is then fed back through the delay chain. This gives rise to the beautiful idea of a *decision feedback equalizer*, abbreviated as **DFE**. The DFE is a nonlinear equalizer because of the nonlinear operator (detector) in the feedback loop. In practice the performance of the DFE system is much better than that of the linear equalizer [Lucky, 1973], [Salz, 1973], [Proakis, 1995]. The feedback multipliers are not usually taken to be $-h(n)/h(0)$; rather, their values are optimized for performance. Figure 5.21 shows a generalized version of the DFE system. Here

$$B(z) = b_1 z^{-1} + b_2 z^{-2} + \ldots + b_N z^{-N} \tag{5.86}$$

and

$$A(z) = a_0 + a_1 z^{-1} + \ldots + a_M z^{-M}. \tag{5.87}$$

The filter $B(z)$ plays a similar role, as explained in Fig. 5.20, feeding back past decisions. The filter $A(z)$ performs some FIR equalization before the decision feedback loop takes over.

Decision feedback methods have an interesting history. The idea of using past decisions to correct the present can be traced back to the work of Milnor [1929]. The patent by MacColl [1936] contains the DFE idea for arbitrary channel length, applied for telegraph reception. The first serious mathematical work on DFE appears to be a detailed report by Austin [1967] from the MIT Lincoln Laboratories. Early optimization of receiver filters (feedforward and feedback) in a DFE system was done by Monsen [1971]; the feedforward filter is shown to separate beautifully into a matched filter and a digital filter. The difficult problem of joint optimization of the transmitting and receiving filters in a DFE system was addressed by Price [1972] under a zero-forcing consraint. Salz [1973] worked out the jointly optimum MMSE-DFE solution; see also Falconer and Foschini [1973]. The research on this topic has continued for a few decades (e.g., see Vandendorpe *et al.*, [1998] and references therein). The idea of feeding back past decisions to improve future ones can also be applied in the spatial domain (i.e., from component to component of a vector signal in MIMO communication), as discussed in great detail in Chap. 19.

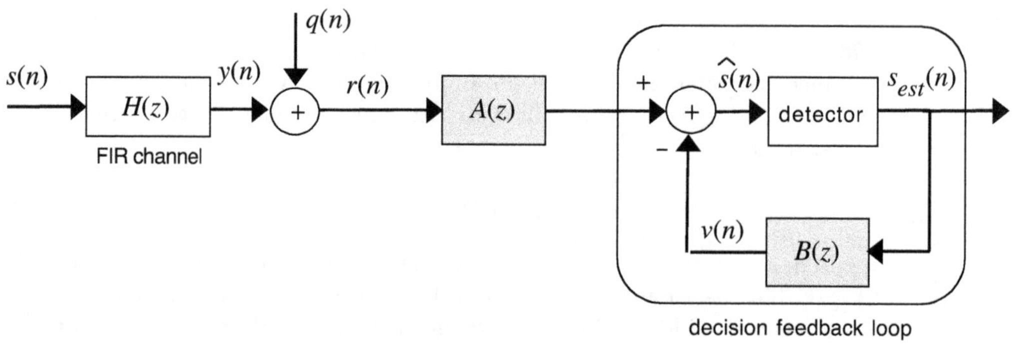

Figure 5.21. Final form of the decision feedback equalizer.

5.8 Precoders for pre-equalization of channels

Consider an FIR channel

$$H(z) = 1 + h(1)z^{-1} + \ldots + h(L)z^{-L}, \tag{5.88}$$

where we have assumed $h(0) = 1$ for simplicity. We can equalize this channel by using the inverse $1/H(z)$ at the transmitter instead of at the receiver. This is called pre-equalization or equalization with a prefilter. It is also sometimes referred to as *precoding*. This is shown in Fig. 5.22, where

$$B(z) = h(1)z^{-1} + \ldots + h(L)z^{-L} \tag{5.89}$$

so that

$$\frac{X(z)}{S(z)} = \frac{1}{1 + B(z)} = \frac{1}{H(z)} \tag{5.90}$$

as desired.

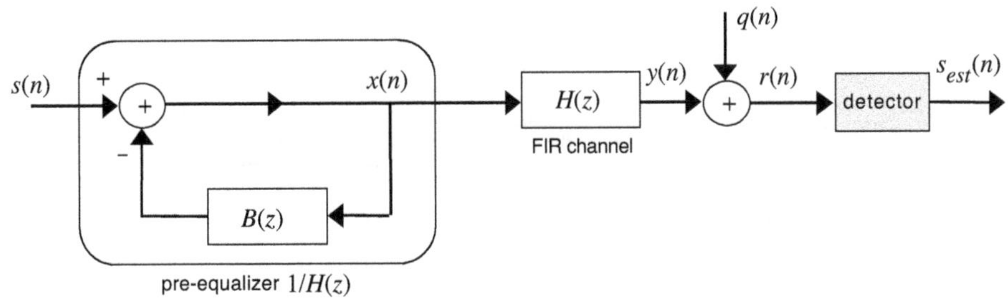

Figure 5.22. An IIR pre-equalizer for the FIR channel $H(z)$, inserted at the transmitter end.

Note that the receiver uses only a detector, and no equalizers. In Sec. 5.7 we explained that if $1/H(z)$ is used at the receiver then the channel noise can get severely amplified. This problem is no longer present in the pre-equalized transceiver. However, a dual problem arises at the transmitter. Namely, the amplitudes of the samples $x(n)$ which enter the channel can get very large (especially if $1/H(z)$ has poles close to or outside the unit circle), increasing the channel input power. In the case of equalization at the receiver we overcame this problem by moving the detector, so that it sits inside the feedback loop. There is a somewhat similar trick we can perform for the pre-equalizer. Namely, we insert an amplitude-limiting operator inside the feedback loop, indicated as "mod V" (read as modulo V) in Fig. 5.23. The description of this operator is given below; we will see that it is a nonlinear device, and this makes the pre-equalizer a nonlinear equalizer. The mod V operator is used at the receiver also, just prior to detection. We will show that, in spite of the presence of nonlinearly, the pre-equalization operation perfectly equalizes the channel (i.e., $\widehat{s}(n) = s(n)$ in the absence of noise).

The mod V operator is a memoryless device with input-output characteristics described by Fig. 5.24. Given any real number u, it produces a number v in the range[5]

$$-0.5V \leq v < 0.5V \tag{5.91}$$

such that $u - v = 0 \bmod V$ (i.e., $u - v$ is a multiple of V). That is, given an arbitrary real u, the operator simply subtracts a multiple of V from it so that the result is in the range $[-0.5V, 0.5V)$. Thus, imagine that the real axis is divided into intervals of length V. If a number u does not fall in the fundamental period given by (5.91), it is simply brought back to the appropriate position within this period. This is demonstrated in Fig. 5.25. The numbers marked $\times$, which are outside the range $[-0.5V, 0.5V)$, are mapped into the numbers marked by little circles falling within the desired range $[-0.5V, 0.5V)$.

[5] For complex constellations such as QAM, the definition of the modulo V operator can be appropriately extended.

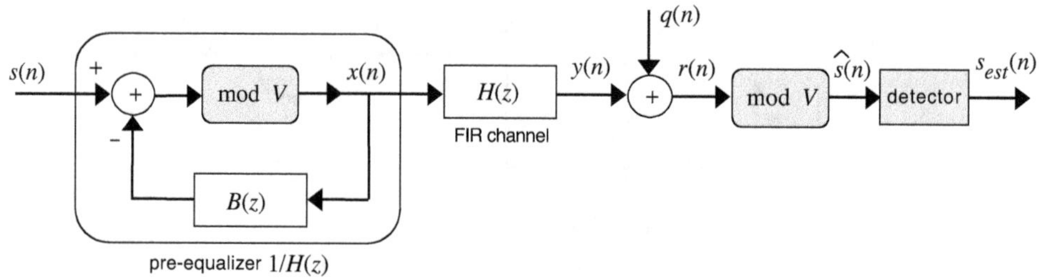

Figure 5.23. Insertion of modulo operators at the transmitter and receiver of the pre-equalizer system.

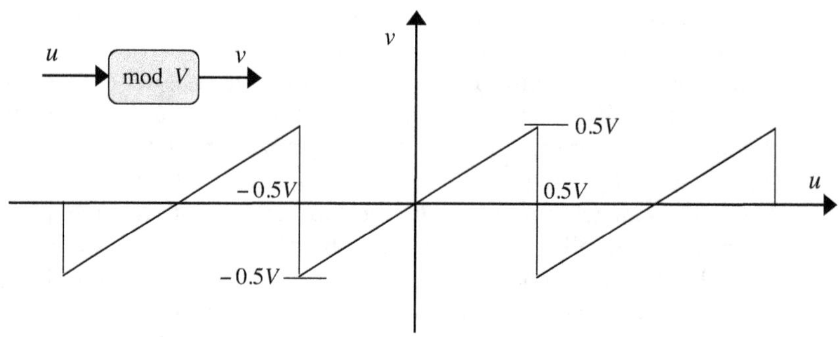

Figure 5.24. Definition of the mod V operator.

In the above example, we have indicated the PAM words for a 2-bit constellation for reference. The fundamental interval $[-0.5V, 0.5V)$ is chosen such that the PAM words are included in it. This is a crucial requirement for successful operation of the equalizer. Note that in Fig. 5.23, even though the signal $s(n)$ is from a PAM constellation, the computed output $x(n)$ is not necessarily a PAM word (as demonstrated by the little circles in Fig. 5.25). Thus the transmitted signal $x(n)$ is not in the constellation, although it satisfies $-0.5V \leq x(n) < 0.5V$. Moreover, it is possible for $x(n)$ to be larger than the largest PAM word because there is some room between $3A$ and $0.5V$ (and similarly betweem $-3A$ and $-0.5V$). Of course, $0.5V$ can be made arbitrarily close to (but larger than) $3A$ by choosing V.

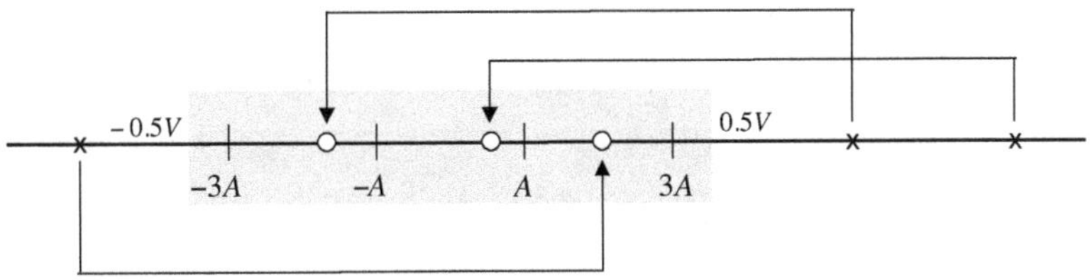

Figure 5.25. Examples showing the application of the mod V operator. The points $\pm A$ and $\pm 3A$ correspond to the words in a 2-bit PAM constellation.

What makes the pre-equalization system successful is an important property of the modulo operator. Given two real numbers u_1 and u_2, it is readily verified that

$$(u_1 + u_2)\text{mod } V = \Big(u_1 \text{ mod } V \ + \ u_2 \text{ mod } V\Big) \text{ mod } V. \tag{5.92}$$

Using this we will show that for the transceiver in Fig. 5.23, as long as $-0.5V \leq s(n) < 0.5V$ for all n, the signal $\widehat{s}(n)$ at the receiver is exactly equal to $s(n)$ in absence of channel noise:

$$\widehat{s}(n) = s(n). \tag{5.93}$$

Thus, even though $x(n)$ may have any value in the range $-0.5V \leq x(n) < 0.5V$, after it passes through the channel $H(z)$ and the modulo operator again, the result is precisely the transmitted symbol $s(n)$ from the constellation. When there is channel noise, this is modified to

$$\widehat{s}(n) = [s(n) + q(n)] \text{ mod } V = \Big(s(n) + [q(n) \text{ mod } V]\Big) \text{ mod } V. \tag{5.94}$$

Thus, even though the modulo operator is nonlinear, the transceiver reproduces $s(n)$ perfectly at the receiver, except for channel noise. If $q(n)$ is small compared to the room between the largest PAM symbol and the boundary $0.5V$, Eq. (5.94) can be written as $\widehat{s}(n) = s(n) + q(n)$. In this case it is as though there is no modulo operation at all, and the channel has been perfectly pre-equalized!

Proof of Eq. (5.93). Since $B(z)$ is as in Eq. (5.89), we see from Fig. 5.23 that

$$x(n) = \Big(-h(1)x(n-1) - \ldots - h(L)x(n-L) + s(n)\Big)\text{mod } V,$$

which implies

$$\Big(x(n) + h(1)x(n-1) + \ldots + h(L)x(n-L)\Big)\text{mod } V = s(n). \tag{5.95}$$

But the output of the channel $H(z)$ in Fig. 5.23 is

$$y(n) = x(n) + h(1)x(n-1) + \ldots + h(L)x(n-L).$$

Thus

$$\begin{aligned}
\widehat{s}(n) &= r(n) \bmod V \\
&= [y(n) + q(n)] \bmod V \\
&= [x(n) + h(1)x(n-1) + \ldots + h(L)x(n-L) + q(n)] \bmod V \\
&= \Big([x(n) + h(1)x(n-1) + \ldots + h(L)x(n-L)] \bmod V \; + q(n) \bmod V\Big) \bmod V \\
&= [s(n) + q(n)] \bmod V
\end{aligned}$$

where we have used Eq. (5.95). This proves Eq. (5.94), which reduces to Eq. (5.93) when there is no noise. $\bigtriangledown \bigtriangledown \bigtriangledown$

The idea of introducing a pre-equalizer with modulo arithmetic was introduced by Tomlinson [1971] and Harashima and Miyakawa [1972]. It is therefore referred to as the Tomlinson–Harashima–Miyakawa precoding system. Apparently it has its origin in the work of Gerrish and Howson [1967]. The MIMO version of the Tomlinson–Harashima–Miyakawa precoder has important applications in multiuser systems operating in the broadcast mode [Proakis and Salehi, 2008].

5.9 Controlled ISI and partial-response signals

Returning again to the digital communication system of Fig. 4.1 with prefilter $F(j\omega)$, equalizer $G(j\omega)$, and channel $H(j\omega)$, let us re-examine the product

$$H_c(j\omega) = F(j\omega)H(j\omega)G(j\omega). \tag{5.96}$$

The channel $H(j\omega)$ is given, and the filters $F(j\omega)$ and $G(j\omega)$ are designed based on some requirements. The more flexibility we have in the shape of the product $H_c(j\omega)$, the easier it is to design the analog filters $F(j\omega)$ and $G(j\omega)$. Since the product $H_c(j\omega)$ determines the equivalent digital communication channel with impulse response $h_d(n) = h_c(nT)$, its shape determines the properties of the digital channel

$$H_d(z) = \sum_n h_d(n) z^{-n}. \tag{5.97}$$

If $H_c(j\omega)$ is required to be such that ISI is eliminated (i.e., the zero-forcing condition is satisfied), then

$$h_d(n) = \delta(n),$$

or equivalently

$$\frac{1}{T} \sum_{k=-\infty}^{\infty} H_c\Big(j\Big(\omega + \frac{2\pi k}{T}\Big)\Big) = 1. \tag{5.98}$$

Figure 5.26(a) shows a typical example of $H_c(j\omega)$ and its shifted versions taking part in the summation above. If the bandwidth of $H_c(j\omega)$ is less than $2\pi/T$ it is clear that the ISI-free condition (5.98) cannot be satisfied. In a **minimum-bandwidth** system, $H_c(j\omega)$ has total bandwidth exactly equal to $2\pi/T$. In this case the terms in Eq. (5.98) do not overlap, and the only way to satisfy the ISI-free condition is to force $H_c(j\omega)$ to have the ideal response

$$H_c(j\omega) = \begin{cases} T & -\pi/T \leq \omega < \pi/T \\ 0 & \text{otherwise,} \end{cases} \tag{5.99}$$

so that $h_c(t)$ is the sinc function:

$$h_c(t) = \frac{\sin(\pi t/T)}{\pi t/T}. \tag{5.100}$$

The ideal response and its shifted versions are demonstrated in Fig. 5.26(b). In an **excess bandwidth** system the bandwidth of $H_c(j\omega)$ exceeds $2\pi/T$, as shown in Fig. 5.26(a). Then there is more flexibility in the choice of the exact shape of $H_c(j\omega)$, and the design of the filters $F(j\omega)$ and $G(j\omega)$ (for a given channel $H(j\omega)$) becomes easier. Thus, if we want to enforce ISI cancellation, then we have two choices: either (a) use minimum bandwidth, in which case the only choice of $H_c(j\omega)$ is the ideal response corresponding to a sinc, or (b) use excess bandwidth, in which case $H_c(j\omega)$ is more flexible and the designs of $F(j\omega)$ and $G(j\omega)$ are easier. The raised cosine function (Sec. 4.4) provides a family of examples with excess bandwidth.

In short, if we wish to have minimum bandwidth and a more practical $H_c(j\omega)$ (instead of the ideal lowpass response), then we cannot eliminate ISI. Fortunately, a brilliant trick was introduced in the early literature to overcome this dilemma [Lender, 1963]. Namely, accept a certain controlled amount of ISI, and then equalize it digitally. For example, we may want to accept a $H_c(j\omega)$ such that $h_d(n) \stackrel{\Delta}{=} h_c(nT)$ has the form

$$h_d(n) = \delta(n) + \delta(n-1), \tag{5.101}$$

as demonstrated in Fig. 5.27(a). We then say that $h_c(t)$ is a **duobinary** pulse. The equivalent digital channel is now

$$H_d(z) = 1 + z^{-1}. \tag{5.102}$$

The expression for the minimum-bandwidth duobinary pulse $h_c(t)$ can be found by finding the bandlimited signal whose sampled version is $h_d(n)$ shown above. Such an $h_c(t)$ is simply the lowpass filtered (or sinc-interpolated) version of the impulse train

$$\delta_c(t) + \delta_c(t - T)$$

where $\delta_c(t)$ is the continuous-time impulse [Oppenheim and Willsky, 1997]. That is,

$$h_c(t) = \frac{\sin(\pi t/T)}{\pi t/T} + \frac{\sin(\pi(t-T)/T)}{\pi(t-T)/T} \tag{5.103}$$

so that

$$H_c(j\omega) = \begin{cases} T(1 + e^{-j\omega T}) & \text{for } -\pi/T \leq \omega < \pi/T \\ 0 & \text{otherwise.} \end{cases}$$

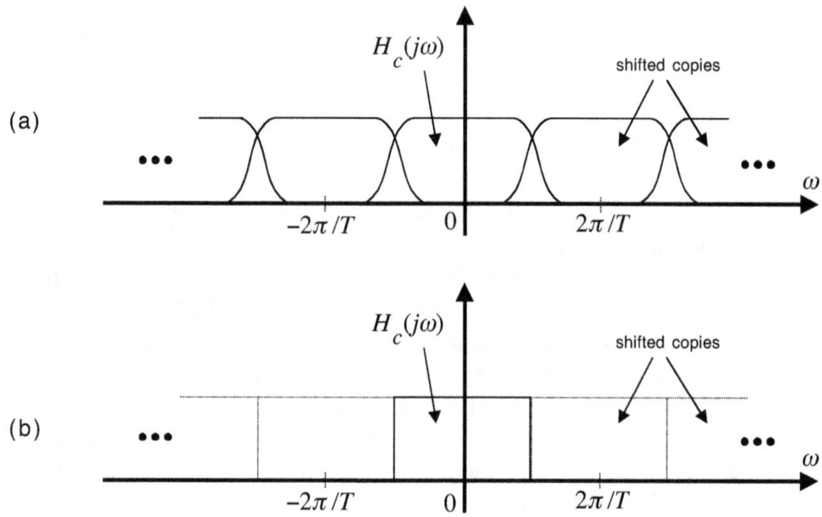

Figure 5.26. The effective channel $H_c(j\omega)$ and its shifted versions in Eq. (5.98). (a) Case where $H_c(j\omega)$ has excess bandwidth, and (b) case where $H_c(j\omega)$ has minimum bandwidth.

which simplifies to

$$H_c(j\omega) = \begin{cases} 2Te^{-j\omega T/2}\cos(\omega T/2) & \text{for } -\pi/T \leq \omega < \pi/T \\ 0 & \text{otherwise.} \end{cases} \tag{5.104}$$

From Eq. (5.103) it is clear that $h_c(nT) = 0$ except when $n = 0, 1$. Figure 5.27 shows $h_d(n), h_c(t)$, and the magnitude of $H_c(j\omega)$ for the duobinary pulse.

The duobinary system achieves minimum bandwidth without requiring that $H_c(j\omega)$ be an ideal brickwall lowpass filter. The price paid for this is that the digital channel $H_d(z)$ is not unity, but instead $H_d(z) = 1 + z^{-1}$. This equivalent digital channel can be equalized in one of many ways: we can use Viterbi's algorithm (Sec. 5.6), or DFE (Sec. 5.7), or the Tomlinson–Harashima–Miyakawa pre-equalizer as demonstrated in Fig. 5.28 (see Sec. 5.8 for details).

Another example of a minimum-bandwidth pulse $h_c(t)$ which gives rise to controlled ISI is the **modified duobinary** pulse. In this case we have

$$h_d(n) = \delta(n+1) - \delta(n-1). \tag{5.105}$$

The digital channel

$$H_d(z) = z - z^{-1} \tag{5.106}$$

can be equalized using any standard technique. The expression for the modified duobinary pulse $h_c(t)$ with minimum bandwidth, and its Fourier transform can be derived by proceeding as in the duobinary case (Problem 5.9).

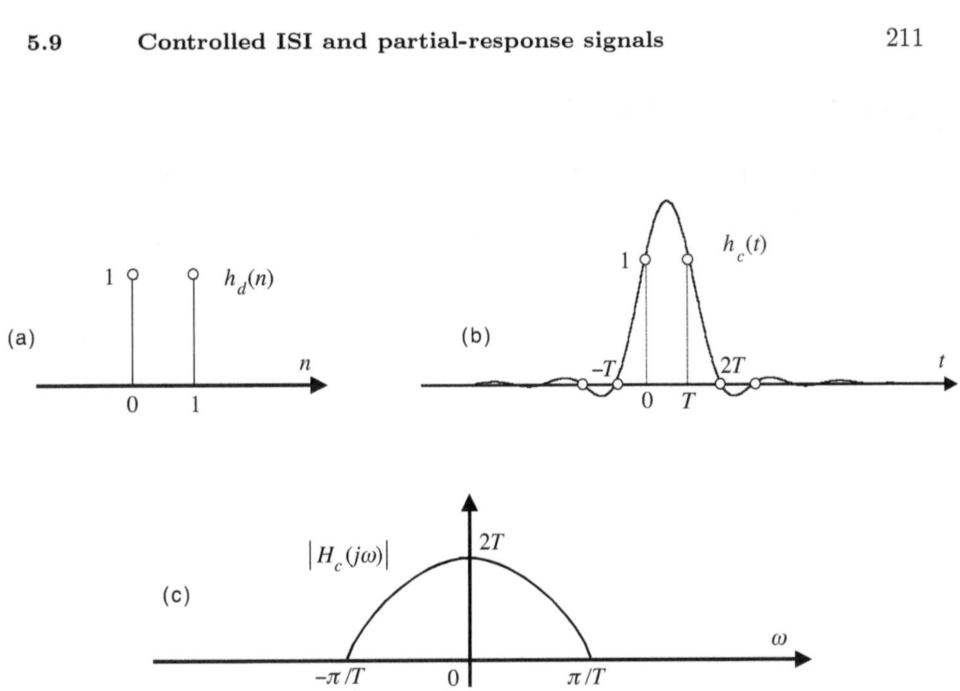

Figure 5.27. The duobinary-pulse system. (a) Impulse response of the equivalent digital channel $H_d(z)$, (b) the duobinary pulse $h_c(t)$ which gives rise to such a digital channel, and (c) its Fourier transform magnitude.

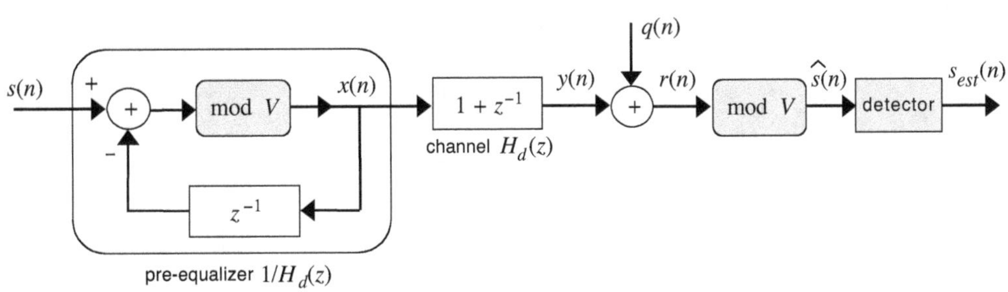

Figure 5.28. Pre-equalization of the digital channel $H_d(z)$ corresponding to the duobinary pulse $h_c(t)$. See Sec. 5.8 for details.

The idea of duobinary signals can be extended to obtain more general pulse shapes such as the **polybinary pulse**, and so forth. The general name for the class of pulses $h_c(t)$ which allow a controlled amount of ISI is **partial-response signals**. The original classical references on this topic are Lender [1964], Howson [1965], and Kretzmer [1966], and a discussion can be found in Lucky, *et al.* [1968].

5.10 Concluding remarks

The material described in this chapter is fundamental to most digital communication systems and is dealt with in great detail in a number of references. Material for further reading includes Forney [1972, 1973], Lucky [1973], Haykin [1988], Blahut [1990], and Proakis and Salehi [2008].

Appendix to Chapter 5

5.A General form of receiver filter

We now prove the claim made in Sec. 5.2.4, namely that if the receiver filter $G(j\omega) = H^*(j\omega)D(j\omega),\ D(j\omega) > 0$ is used before sampling, then (the signal component of) $r(n)$ retains all the information about the noise-free channel output $y(t)$. The Fourier transform at the output of $G(j\omega)$ is

$$S(e^{j\omega T})H(j\omega)G(j\omega) = S(e^{j\omega T})|H(j\omega)|^2 D(j\omega).$$

The Fourier transform of the sampled version $r(n)$ is therefore

$$R(e^{j\omega}) = \frac{S(e^{j\omega})}{T}\sum_{k=-\infty}^{\infty}\left|H\Big(\frac{j(\omega+2\pi k)}{T}\Big)\right|^2 D\Big(\frac{j(\omega+2\pi k)}{T}\Big). \tag{5.107}$$

Now define $A^2(e^{j\omega T})$ as

$$\begin{cases}\left(\dfrac{1}{T}\displaystyle\sum_{k=-\infty}^{\infty}\left|H\Big(j\Big(\omega+\frac{2\pi k}{T}\Big)\Big)\right|^2 D\Big(\frac{j(\omega+2\pi k)}{T}\Big)\right)^{-1} & \text{when } H(j\omega)\neq 0\\ 0 & \text{otherwise.}\end{cases} \tag{5.108}$$

When $H(j\omega) \neq 0$ for some ω the sum in Eq. (5.107) is positive (since $D(ju) > 0$ for all u) and the inverse indicated in Eq. (5.108) exists for that ω. With $A^2(e^{j\omega T})$ as above, Eq. (5.107) implies

$$R(e^{j\omega T}) = \frac{S(e^{j\omega T})}{A^2(e^{j\omega T})} \qquad \text{when } H(j\omega)\neq 0.$$

Defining the reconstruction filter $B(j\omega) = A^2(e^{j\omega T})H(j\omega)$, the reconstructed signal (Fig. 5.4(a)) is

$$\begin{aligned} Y_1(j\omega) &= R(e^{j\omega T})A^2(e^{j\omega T})H(j\omega)\\ &= \begin{cases}\dfrac{S(e^{j\omega T})}{A^2(e^{j\omega T})}A^2(e^{j\omega T})H(j\omega) & \text{for } H(j\omega)\neq 0\\ 0 & \text{otherwise}\end{cases}\\ &= S(e^{j\omega T})H(j\omega) = Y(j\omega).\end{aligned}$$

That is, $y_1(t) = y(t)$ indeed.

Problems

5.1. Consider the receiver filter

$$G(j\omega) = \begin{cases} 1 & 0 \le |\omega| < \pi/T \\ 0.1 & \pi/T \le |\omega| \le 2\pi/T \\ 0 & \text{otherwise.} \end{cases}$$

Does this satisfy the orthogonality property (5.18)? If not, can you orthonormalize it and find $G_{new}(j\omega)$ satisfying this property?

5.2. Consider the receiver filter

$$G(j\omega) = \begin{cases} 1 & 0 \le |\omega| < 0.9\pi/T \\ 0 & \text{otherwise.} \end{cases}$$

Does this satisfy the orthogonality property (5.18)? If not, can you orthonormalize it and find $G_{new}(j\omega)$ satisfying this property?

5.3. Consider the receiver filter $G(j\omega) = 1/(1 + j\omega)$. Does this satisfy the orthogonality property (5.18)?

5.4. Consider Fig. 5.5 again, where $p(t)$ and $h(t)$ are as in Fig. 5.6. Verify that the impulse response from $s(n)$ to $r(n)$ is zero for all n.

5.5. Consider the system in Fig. P5.5, where we have a channel $h(t)$ and a receiving filter $g(t)$. Let

$$g(t) = \begin{cases} \sin(2\pi t/T) & 0 \le t \le T \\ 0 & \text{otherwise.} \end{cases}$$

Find an example of $h(t)$ such that $r(n)$ does not contain enough information to reconstruct $y(t)$.

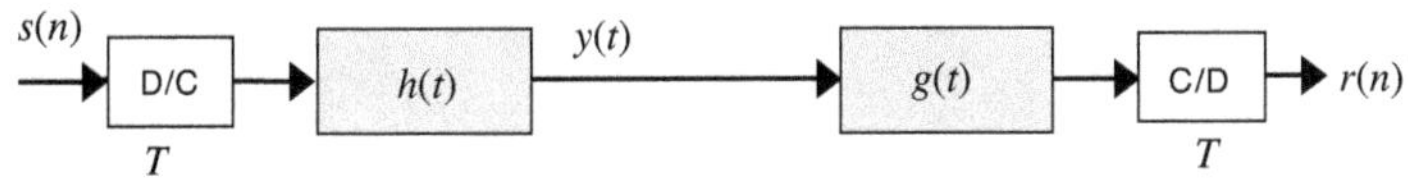

Figure P5.5.

5.6. Consider an FIR channel $H(z) = 1 + 0.5z^{-1} + 0.25z^{-2}$ with PAM constellation $\{1, -1\}$. Draw the complete trellis module for the Viterbi algorithm analogous to what we did in Fig. 5.16(b).

5.7. In Sec. 5.6.3 we presented an example of how the Viterbi algorithm works. In that example the received noisy sequence $r(n)$ was as in Eq. (5.72). Instead of this suppose $r(n)$ is as follows:

$$\begin{array}{cccccccc} n & 0 & 1 & 2 & 3 & 4 & 5 & 6 \\ r(n) & 1 & -0.4 & -0.8 & 0.1 & 0.1 & 0.1 & 1.1 \end{array} \quad \text{(P5.7)}$$

1. Rerun the Viterbi algorithm. Which of the symbols $s(n)$ in the range $0 \leq n \leq 6$ are you able to identify, and what are the estimated values of these symbols?
2. What is the decoding delay for $s(2)$ in this example? (The decoding delay for $s(2)$ in Sec. 5.6.3 was two, because we needed to use received samples up until $r(4)$.)

5.8. Consider the simple DFE receiver example shown in Fig. 5.20. This works for an FIR channel of order 3. Now consider the IIR channel

$$H(z) = \frac{a_0 + a_1 z^{-1} + a_2 z^{-2} + a_3 z^{-3}}{1 + d_1 z^{-1} + d_2 z^{-2} + d_3 z^{-3}}$$

How would you modify the receiver to perform equalization with decision feedback?

5.9. For the modified duobinary pulse mentioned at the end of Sec. 5.9, derive the pulse shape $h_c(t)$ and its Fourier transform $H_c(j\omega)$ having minimum bandwidth. Plot $h_c(t)$ and the magnitude $|H_c(j\omega)|$.

5.10. Referring to Sec. 5.9, suppose we wish to design a polybinary pulse with minimum bandwidth and $h_d(n) = c_0\delta(n) + c_1\delta(n-1) + c_2\delta(n-2)$. What is the pulse shape $h_c(t)$ and what is its Fourier transform $H_c(j\omega)$?

6

Channel capacity

6.1 Introduction

The fundamentals of digital communications were developed many decades ago in the ground breaking work of Shannon [1948, 1949]. Central to the capability of a communication channel is the channel capacity, which is studied in great detail in many communication texts [Gallager, 1968], [McEliece, 1977], [Cover and Thomas, 1991], [Proakis, 1995]. The channel capacity is the upper bound on the rate at which information can be communicated through a noisy channel with arbitrarily small error probability. The capacity can be achieved, in principle, by appropriate channel coding. We summarize here some of the results that are central to our discussions. For details the reader should consult one of the above references.

6.2 Ideal lowpass channel

Figure 6.1(a) shows a channel with transfer function $H(f)$ and additive Gaussian noise $q(t)$. (For convenience we have used f instead of ω for frequency.) Assume that $H(f)$ is an ideal bandlimited channel with total bandwidth $2B$ Hz (one-sided bandwidth of B):

$$H(f) = \begin{cases} 1 & -B \leq f < B \\ 0 & \text{otherwise.} \end{cases} \tag{6.1}$$

Since the receiver filter can be restricted to this band, the noise is also bandlimited. Assume the noise has a flat spectrum within this band as shown in Fig. 6.1(b). Let the total signal power be p_0. Assume this is uniformly distributed within the band as shown in Fig. 6.1(c). The total noise power in the channel band is the integral of the noise spectrum:

$$\sigma_q^2 = \mathcal{N}_0 B. \tag{6.2}$$

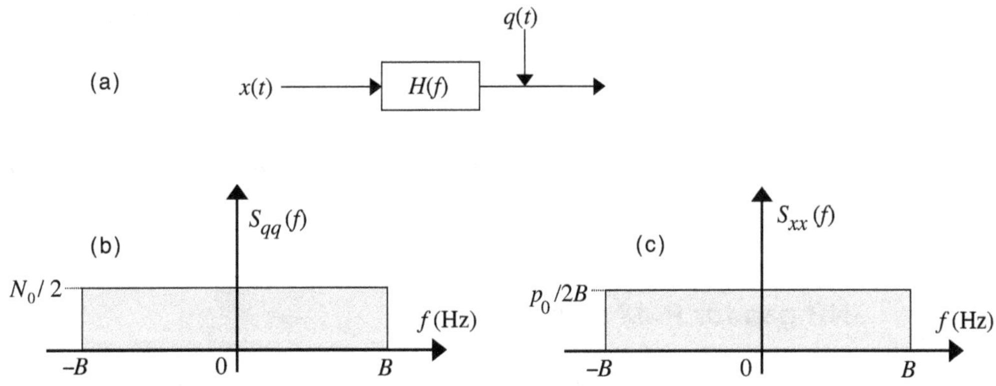

Figure 6.1. (a) A communication channel; (b), (c) examples of the noise and signal power spectra.

Assuming that $x(t)$ and $q(t)$ are real,[1] the channel capacity per use is [Cover and Thomas, 1991], [Proakis, 1995]

$$C_{once} = 0.5 \log_2\left(1 + \frac{p_0}{\sigma_q^2}\right), \tag{6.3}$$

which can be written as

$$C_{once} = 0.5 \log_2\left(1 + \frac{p_0}{\mathcal{N}_0 B}\right) \qquad \text{bits per use.} \tag{6.4}$$

Evidently the capacity is infinity when there is no noise ($\mathcal{N}_0 = 0$), for any power $p_0 > 0$.

How many times can we use the channel per second? The answer depends on the bandwidth. Since the channel impulse response is the sinc function,

$$h(t) = \frac{\sin(2\pi Bt)}{\pi t}, \tag{6.5}$$

the zero crossings are spaced apart by $1/2B$ seconds, so we can use the channel $2B$ times per second without intersymbol interference. Thus the capacity for a (one-sided) bandwidth of B Hz is $(2BC_{once})$ per second, that is,

$$\mathcal{C} = B \log_2\left(1 + \frac{p_0}{\mathcal{N}_0 B}\right) \quad \text{bits per second.} \tag{6.6}$$

We can find the limiting value of this capacity as the bandwidth tends to ∞. For this we use the approximation

$$\ln(1 + x) \approx x, \quad |x| << 1,$$

[1]The case of complex and MIMO channels is discussed in Sec. 6.7.

so that

$$\mathcal{C} = (B \log_2 e) \ \ln\Big(1 + \frac{p_0}{\mathcal{N}_0 B}\Big) \approx (B \log_2 e) \ \frac{p_0}{\mathcal{N}_0 B} \approx \frac{1.44 p_0}{\mathcal{N}_0},$$

where we have used $\log_2 e \approx 1.44$. Thus, for bandwidth tending to ∞, the capacity becomes

$$\mathcal{C}_\infty \approx \frac{1.44 p_0}{\mathcal{N}_0} \quad \text{bits per second.} \tag{6.7}$$

This capacity increases *linearly* with total power p_0.

6.3 SNR gap for PAM signals

In Sec. 2.3.1 we derived an expression for the probability of error in the detection of a PAM symbol. When the number of bits b is large enough, we can write $1 - 2^{-b} \approx 1$ and then invert this expression to obtain

$$b = 0.5 \log_2 \left(1 + \frac{3}{[\mathcal{Q}^{-1}(\mathcal{P}_e/2)]^2} \frac{\mathcal{E}_{ave}}{\sigma_e^2} \right), \tag{6.8}$$

where $\mathcal{E}_{ave}$ and σ_e^2 are the signal power and noise variance at the input of the detector, respectively. Now, according to information theory, the capacity of an additive Gaussian noise channel with input power $\mathcal{E}_{ave}$ and noise variance σ_e^2 is given by

$$\mathcal{C} = 0.5 \log_2 \left(1 + \frac{\mathcal{E}_{ave}}{\sigma_e^2} \right) \quad \text{bits per use.} \tag{6.9}$$

This is the upper bound on the number of bits that can be transmitted with arbitrarily small probability of error. According to Shannon, this upper bound can be achieved by channel coding. On the other hand Eq. (6.8) gives the number of PAM bits that can safely be used without channel coding, and without exceeding a specified error probability $\mathcal{P}_e$, when the signal-to-noise ratio at the detector is $\mathcal{E}_{ave}/\sigma_e^2$. By comparing (6.8) and (6.9) we see that, for the same transmission rate ($b = \mathcal{C}$) and small enough $\mathcal{P}_e$, uncoded PAM requires a higher signal-to-noise ratio because of the factor $3/[\mathcal{Q}^{-1}(\mathcal{P}_e/2)]^2$, when compared with the SNR for ideally channel-coded transmission. Thus

$$SNR_{pam} = SNR_{ideal} \times \frac{[\mathcal{Q}^{-1}(\mathcal{P}_e/2)]^2}{3}. \tag{6.10}$$

The excess SNR depends only on the error probability. For example, if $\mathcal{P}_e = 10^{-7}$ we have $[\mathcal{Q}^{-1}(\mathcal{P}_e/2)]^2 = (5.33)^2$, and the excess factor is

$$\frac{[\mathcal{Q}^{-1}(\mathcal{P}_e/2)]^2}{3} = 9.47, \tag{6.11}$$

which corresponds to $10 \log_{10}(9.47) = 9.76$ dB. This is the **SNR gap** of PAM transmission with $\mathcal{P}_e = 10^{-7}$. Summarizing, *uncoded PAM requires an excess SNR of 9.76 dB for achieving an error probability of* 10^{-7}. This is the "excess" compared to what would be required for transmission with ideal channel coding, which would in fact achieve error probability arbitrarily close to zero.

6.4 Capacity of frequency-dependent channel

Imagine we have a channel with transfer function $H(f)$ and additive Gaussian noise with a power spectrum $S_{qq}(f)$, not necessarily flat. Let $\mathcal{F}$ denote the frequency band where $H(f)$ is nonzero. With $S_{xx}(f)$ denoting the input power spectrum, the signal-to-noise ratio in a very narrow band of width Δf centered around a frequency f_k is given by

$$\frac{S_{xx}(f_k)|H(f_k)|^2\Delta f}{S_{qq}(f_k)\Delta f}. \tag{6.12}$$

If we only had this narrow band available for transmission, then the channel capacity would be

$$C_{once} = \log_2\left(1 + \frac{S_{xx}(f_k)|H(f_k)|^2}{S_{qq}(f_k)}\right). \tag{6.13}$$

This is essentially obtained by using Eq. (6.3). However there is one difference. It is assumed now that the signal and noise, limited to the bandwidth Δf, can be complex. In this case we have to add the capacities from the real and imaginary paths, so the factor of 0.5 disappears.[2] Since the channel can be used Δf times per second, the capacity for a **subband** of bandwidth Δf is

$$C_{\Delta f} = \Delta f\ \log_2\left(1 + \frac{S_{xx}(f_k)|H(f_k)|^2}{S_{qq}(f_k)}\right) \quad \text{bits per second.} \tag{6.14}$$

By adding up these capacities of the individual subbands we get the total capacity. The sum becomes an integral in the limit, so the total capacity is

$$\mathcal{C} = \int_{\mathcal{F}} \log_2\left(1 + \frac{S_{xx}(f)|H(f)|^2}{S_{qq}(f)}\right) df \quad \text{bits per second,} \tag{6.15}$$

where $\mathcal{F}$ is the frequency band where $H(f)$ is nonzero. The total channel input power is

$$p_0 = \int_{\mathcal{F}} S_{xx}(f)df. \tag{6.16}$$

For fixed p_0 we can allocate the power across the frequency band intelligently by shaping the power spectrum $S_{xx}(f)$. The problem therefore is to maximize the integral (6.15) subject to the power constraint (6.16). This can be done by setting up an appropriate Lagrangian and setting its derivative to zero.

[2] Actually, in the complex case the Gaussian noise is assumed to be zero-mean circularly symmetric, and the signal achieving capacity is also zero-mean circularly symmetric Gaussian. We will discuss these details in Secs. 6.6 and 6.7.

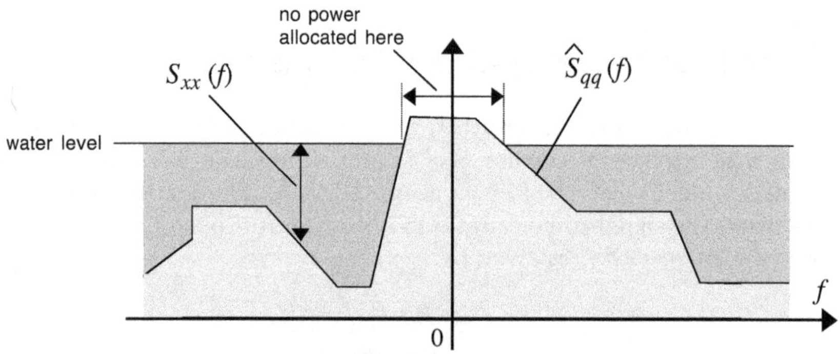

Figure 6.2. Demonstration of the water-filling power allocation.

The solution is well known from Shannon's work (see Sec. 12-2-1 of Proakis [1995]), and is similar to the water-filling solution discussed in Sec. 22.3 later: we simply take the input power spectrum to be

$$S_{xx}(f) = \begin{cases} \dfrac{1}{\lambda} - \dfrac{S_{qq}(f)}{|H(f)|^2} & \text{if this is non-negative} \\ 0 & \text{otherwise.} \end{cases} \tag{6.17}$$

This allocation is demonstrated in Fig. 6.2. Note that the effective channel noise spectrum which enters this crucial equation is

$$\widehat{S}_{qq}(f) = \frac{S_{qq}(f)}{|H(f)|^2} \tag{6.18}$$

that is, the spectrum of the noise $q(t)$ shaped by $1/H(f)$. Here λ is a constant that arises from setting up a Lagrangian for the constrained optimization problem. Its value can be found by using the total power constraint (6.16). While the frequency shaping of power may not be practicable as prescribed by (6.17), a similar *power allocation* can readily be made by splitting the channel into a finite number of subbands as done in discrete multitone (DMT) systems (Chap. 7).

6.5 Splitting the channel into subbands

Consider the additive Gaussian noise channel of Fig. 6.1(a) with an ideal lowpass filter $H(f)$ as in Eq. (6.1), but with noise spectrum as shown in Fig. 6.3(b). Unlike in Fig. 6.1(b) the noise spectrum is different in different parts of the passband. If the signal power is uniformly distributed as in Fig. 6.3(c) then the total signal power is p_0 and noise power is

$$\sigma_q^2 = \frac{(\mathcal{N}_0 + \mathcal{N}_1)B}{2}. \tag{6.19}$$

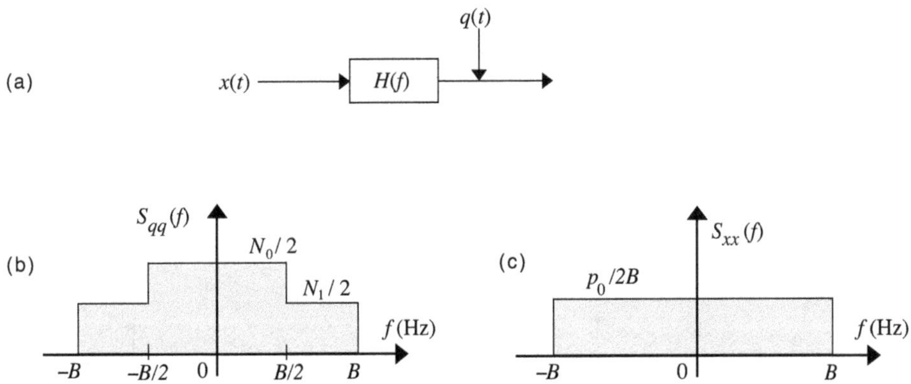

Figure 6.3. (a) An additive Gaussian noise channel, (b) noise spectrum, and (c) signal spectrum.

So the capacity of the channel with uniform power allocation is a slightly modified version of (6.6):

$$\mathcal{C}_{unsplit} = B \log_2\left(1 + \frac{2p_0}{(\mathcal{N}_0 + \mathcal{N}_1)B}\right) \quad \text{bits per second.} \tag{6.20}$$

The meaning of the subscript will become clear later. Equation (6.20) reduces to (6.6) when $\mathcal{N}_0 = \mathcal{N}_1$. If one part of the channel is bad, that is, $\mathcal{N}_0 >> \mathcal{N}_1$, then the above capacity becomes very small, no matter how good the other part of the channel is. If the quality of the channel varies across the band in this way, then it is useful to split the channel into subbands and use the subchannels independently. In this way, parts of the channel that are very poor do not ruin the performance of the good parts. Thus, imagine that the channel can be split into two bands as in Fig. 6.4(a), where subband 0 has total bandwidth B with noise and signal spectra as in Fig. 6.4(b), and subband 1 has total bandwidth B with noise and signal spectra as in Fig. 6.4(c).

Assume for the moment that there is no power allocation, that is, let

$$p_{00} = p_{01} = p_0. \tag{6.21}$$

Then the total signal and noise power in subband 0 are $p_0/2$ and $\mathcal{N}_0 B/2$, whereas those in subband 1 are $p_0/2$ and $\mathcal{N}_1 B/2$. By modifiying (6.6) appropriately we see that the capacities of the individual channels are

$$\mathcal{C}_0 = 0.5B \log_2\left(1 + \frac{p_0}{\mathcal{N}_0 B}\right), \quad \mathcal{C}_1 = 0.5B \log_2\left(1 + \frac{p_0}{\mathcal{N}_1 B}\right), \tag{6.22}$$

so that the total capacity after subband splitting is

$$\mathcal{C}_{split} = 0.5B \log_2\left(1 + \frac{p_0}{\mathcal{N}_0 B}\right) + 0.5B \log_2\left(1 + \frac{p_0}{\mathcal{N}_1 B}\right) \tag{6.23}$$

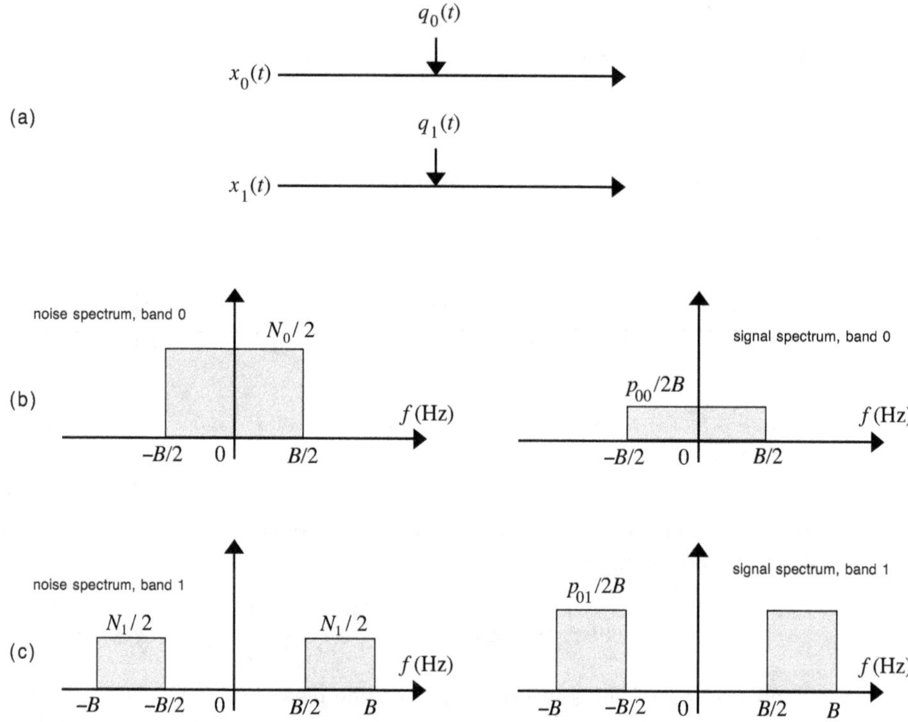

Figure 6.4. (a) Splitting the channel into subbands, (b) noise and signal spectra in subband 0, and (c) noise and signal spectra in subband 1.

bits per second. We can rewrite the capacity of the unsplit channel (6.20) in the form

$$\mathcal{C}_{unsplit} = B\log_2\Big(1+\frac{p_0}{(0.5\mathcal{N}_0+0.5\mathcal{N}_1)B}\Big). \tag{6.24}$$

Now, it turns out that the function

$$\psi(x) = \log_2\Big(1+\frac{A}{x}\Big) \tag{6.25}$$

is strictly convex in x for $A, x > 0$ (because $\partial^2\psi/\partial x^2 > 0$; see Sec. 21.2 in Chap. 21). This implies that

$$\log_2\Big(1+\frac{A}{\alpha x_0+(1-\alpha)x_1}\Big) \le \alpha\log_2\Big(1+\frac{A}{x_0}\Big)+(1-\alpha)\log_2\Big(1+\frac{A}{x_1}\Big) \tag{6.26}$$

for any α in $0 \le \alpha \le 1$, with strict inequality unless $x_0 = x_1$, or $\alpha = 0$ or 1. Using this with $\alpha = 0.5$ we see immediately that

$$C_{unsplit} < C_{split} \tag{6.27}$$

unless $\mathcal{N}_0 = \mathcal{N}_1$. Thus, even without any power allocation we obtain improvement in capacity as long as the different parts of the channel are used independently. If we allocate different powers to the two subbands then the expression for capacity becomes

$$\mathcal{C}_{alloc} = 0.5B\log_2\Big(1+\frac{p_{00}}{\mathcal{N}_0 B}\Big) + 0.5B\log_2\Big(1+\frac{p_{01}}{\mathcal{N}_1 B}\Big). \tag{6.28}$$

Since the total power is p_0 we have the constraint

$$\frac{p_{00}+p_{01}}{2} = p_0. \tag{6.29}$$

For fixed bandwidth, noise densities $\mathcal{N}_0$ and $\mathcal{N}_1$, and total power p_0 we can maximize the capacity (6.28) by optimally allocating the numbers p_{00} and p_{01}. This is similar to the parallel channel problem discussed in Sec. 22.3 later, and the solution is again the water-filling allocation:

$$p_{0k} = \begin{cases} \dfrac{1}{\lambda} - \mathcal{N}_k B & \text{if this is non-negative} \\ 0 & \text{otherwise.} \end{cases} \tag{6.30}$$

Here λ is a constant that arises from setting up a Lagrangian for the constrained optimization problem. It can be found from the power constraint (6.29).

Splitting the channel into a finite number of subbands is a practical way to approximate the ideal power allocation formula (6.17). A simple way to mechanize the splitting of a channel into subbands is indicated in Fig. 6.5(a). Here the transmitter has two ideal filters bandlimited to the two halves of the channel bandwidth (Fig. 6.5(b)). The same pair of ideal filters is used at the receiver to split the received signal into subbands. The effect is the creation of two independent parallel channels with half the original bandwidth, as shown in Fig. 6.5(c). The idea of power allocation also arises in many other forms in digital communication, not necessarily in the context of channel capacity.

Example 6.1: Splitting the channel into subbands

Assume $H(f)$ is the ideal lowpass channel in Eq. (6.1). In Fig. 6.3, let $B = 1$ MHz (total bandwidth 2 MHz), and let

$$p_0 = 1\,\text{mW}, \quad \mathcal{N}_0 = 2\times10^{-6}\,\text{mW/Hz}, \quad \text{and} \quad \mathcal{N}_1 = 2\times10^{-8}\,\text{mW/Hz}.$$

So, subband 1 is 100 times less noisier than subband 0. Using the above numbers we can calculate the capacity of the unsplit channel [Eq. (6.20)], the capacity of the split channel with equal power in the subbands [Eq. (6.23)], and the capacity of the split channel with optimum power allocation [Eq. (6.28)]. The results in kilobits per second are:

$$C_{unsplit} = 993\,\text{kb/s}, \quad C_{split} = 3{,}129\,\text{kb/s}, \quad \text{and} \quad \mathcal{C}_{alloc} = 3{,}329\,\text{kb/s}.$$

The optimal power allocation obtained from Eq. (6.30) is such that $p_{00} = 0.01$ and $p_{01} = 1.99$, so the powers allocated to the subbands are

$$p_{00}/2 = 0.005\,\text{mW} \quad \text{and} \quad p_{01}/2 = 0.995\,\text{mW}.$$

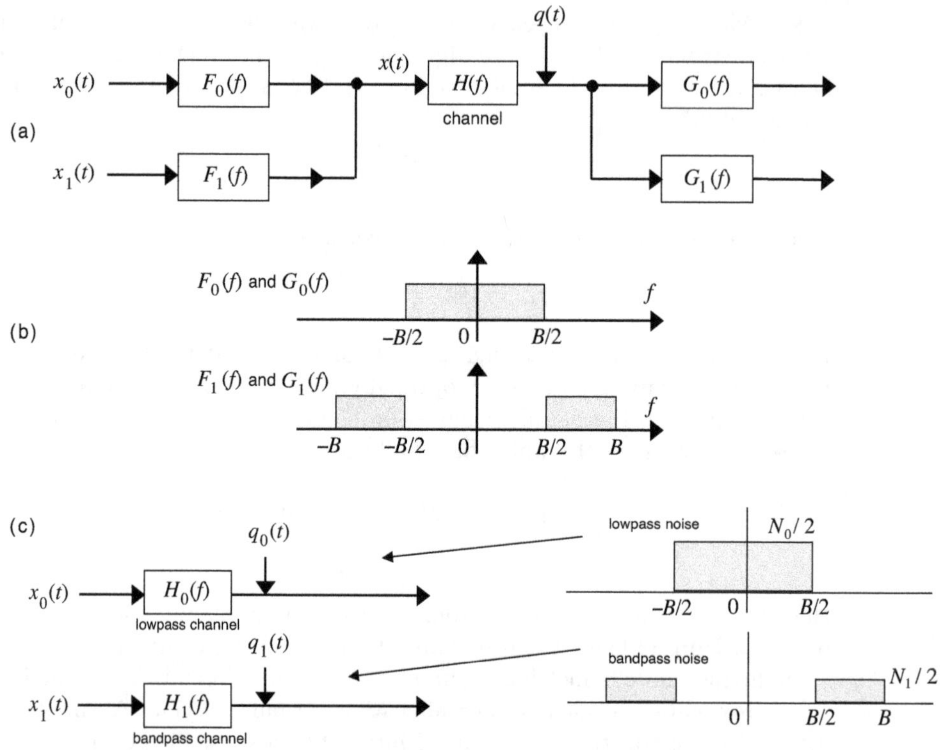

Figure 6.5. (a) A bank of filters at the transmitter and at the receiver, to split the channel into subbands, (b) frequency responses of the filters, and (c) equivalent parallel pair of channels.

Thus, when power is optimally allocated, most of it goes to subband 1, and nearly nothing goes to subband 0. Note that there is a significant improvement in capacity owing to band spitting, and further improvement owing to power allocation.

6.6 Circularly symmetric complex random vectors

When random vectors arising in digital communication systems are complex, they often satisfy a property called circular symmetry. We now describe this property, which plays a crucial role in the derivation of channel capacity for the complex case. Let

$$\mathbf{x} = \mathbf{x}_{re} + j\mathbf{x}_{im} \tag{6.31}$$

be a complex random vector with autocorrelation

$$\mathbf{R}_{\mathbf{xx}} \stackrel{\Delta}{=} E[\mathbf{x}\mathbf{x}^{\dagger}] = \mathbf{P} + j\mathbf{Q} \tag{6.32}$$

so that

$$\mathbf{R_{xx}} = \Big(E[\mathbf{x}_{re}\mathbf{x}_{re}^T] + E[\mathbf{x}_{im}\mathbf{x}_{im}^T]\Big) + j\Big(E[\mathbf{x}_{im}\mathbf{x}_{re}^T] - E[\mathbf{x}_{re}\mathbf{x}_{im}^T]\Big). \tag{6.33}$$

Clearly the real part $\mathbf{P}$ is symmetric and the imaginary part $\mathbf{Q}$ is antisymmetric. Note that the individual terms

$$E[\mathbf{x}_{re}\mathbf{x}_{re}^T], \quad E[\mathbf{x}_{im}\mathbf{x}_{im}^T], \quad E[\mathbf{x}_{re}\mathbf{x}_{im}^T] \tag{6.34}$$

cannot be obtained from a knowledge of the autocorrelation, so $E[\mathbf{x}\mathbf{x}^\dagger]$ only has limited information about correlations. Suppose the real and imaginary parts are related such that

$$E[\mathbf{x}_{re}\mathbf{x}_{re}^T] = E[\mathbf{x}_{im}\mathbf{x}_{im}^T] \tag{6.35}$$

and

$$E[\mathbf{x}_{re}\mathbf{x}_{im}^T] = -E[\mathbf{x}_{im}\mathbf{x}_{re}^T], \tag{6.36}$$

so that

$$\mathbf{R_{xx}} = 2E[\mathbf{x}_{re}\mathbf{x}_{re}^T] + 2jE[\mathbf{x}_{im}\mathbf{x}_{re}^T]. \tag{6.37}$$

That is, the real and imaginary parts have identical autocorrelation, and furthermore, the cross correlation between them is antisymmetric. We then say that the complex random vector is **circularly symmetric**. Such complex random vectors arise frequently in digital communication models. For such vectors, $\mathbf{R_{xx}}$ indeed contains all information about correlation components.

6.6.1 Properties of circularly symmetric random vectors

Equations (6.35) and (6.36) are the defining equations for circular symmetry of a complex random vector. In what follows we will first show that this property can be restated in a number of other ways.

1. ***Symmetry in correlation.*** Given the complex random vector $\mathbf{x} = \mathbf{x}_{re} + j\mathbf{x}_{im}$, define the real vector

$$\mathbf{u} = \begin{bmatrix} \mathbf{x}_{re} \\ \mathbf{x}_{im} \end{bmatrix}. \tag{6.38}$$

The autocorrelation of $\mathbf{u}$ is

$$\mathbf{R_{uu}} = \begin{bmatrix} E[\mathbf{x}_{re}\mathbf{x}_{re}^T] & E[\mathbf{x}_{re}\mathbf{x}_{im}^T] \\ E[\mathbf{x}_{im}\mathbf{x}_{re}^T] & E[\mathbf{x}_{im}\mathbf{x}_{im}^T] \end{bmatrix} \tag{6.39}$$

and contains all the individual information listed in Eq. (6.34). It is clear that this reduces to the form

$$\mathbf{R_{uu}} = 0.5 \begin{bmatrix} \mathbf{P} & -\mathbf{Q} \\ \mathbf{Q} & \mathbf{P} \end{bmatrix} \tag{6.40}$$

if and only if both Eq. (6.35) and (6.36) are true. So Eq. (6.40) is often used as an equivalent definition for circular symmetry [Telatar, 1999].

2. ***Pseudocorrelation.*** Given the complex random vector $\mathbf{x} = \mathbf{x}_{re} + j\mathbf{x}_{im}$, the quantity $E[\mathbf{x}\mathbf{x}^T]$ is sometimes referred to as the *pseudocorrelation* matrix [Tse and Viswanath, 2005]. It is not the correlation matrix, because of the appearance of $\mathbf{x}^T$ instead of $\mathbf{x}^\dagger$. We have

$$E[\mathbf{x}\mathbf{x}^T] = \Big(E[\mathbf{x}_{re}\mathbf{x}_{re}^T] - E[\mathbf{x}_{im}\mathbf{x}_{im}^T]\Big) + j\Big(E[\mathbf{x}_{im}\mathbf{x}_{re}^T] + E[\mathbf{x}_{re}\mathbf{x}_{im}^T]\Big). \quad (6.41)$$

This is zero if and only if both Eqs. (6.35) and (6.36) are true. Thus, circular symmetry can equivalently be stated as the following property:

$$E[\mathbf{x}\mathbf{x}^T] = \mathbf{0}. \quad (6.42)$$

That is, the pseudocorrelation is zero.

3. ***Rotational invariance.*** Given the complex random vector $\mathbf{x} = \mathbf{x}_{re} + j\mathbf{x}_{im}$, define $\mathbf{y} = e^{j\theta}\mathbf{x}$. Thus the real and imaginary parts of each component of $\mathbf{y}$ are related to the corresponding component of $\mathbf{x}$ as follows:

$$\begin{bmatrix} y_{k,re} \\ y_{k,im} \end{bmatrix} = \begin{bmatrix} \cos\theta & -\sin\theta \\ \sin\theta & \cos\theta \end{bmatrix} \begin{bmatrix} x_{k,re} \\ x_{k,im} \end{bmatrix}.$$

The vector on the left is a rotated version of the vector on the right (counter-clockwise rotation by θ). So we say that $\mathbf{y}$ is a rotated version of $\mathbf{x}$. Clearly $\mathbf{y}$ has the same autocorrelation as $\mathbf{x}$, but its pseudocorrelation is different:

$$E[\mathbf{y}\mathbf{y}^T] = e^{2j\theta}E[\mathbf{x}\mathbf{x}^T]. \quad (6.43)$$

This is equal to the pseudocorrelation of $\mathbf{x}$ for all possible choices of θ if and only if the pseudocorrelation $E[\mathbf{x}\mathbf{x}^T] = \mathbf{0}$. Thus circular symmetry of $\mathbf{x}$ is equivalent to the property that the pseudocorrelation be invariant to multiplication by $e^{j\theta}$ for any real θ. We express this by saying that pseudocorrelation is rotation invariant for circularly symmetric $\mathbf{x}$.

Summarizing, the complex random vector $\mathbf{x}$ is called circularly symmetric if any one of the four equivalent conditions are satisfied, namely, (a) Eqs. (6.35) and (6.36) both hold, (b) Eq. (6.40) holds, (c) Eq. (6.42) holds, or (d) the pseudocorrelation of $e^{j\theta}\mathbf{x}$ is the same for all θ. Further discussions on such random vectors can be found in Therrien [1992] and Picinbono [1994]. We now show Telatar [1999] that certain common operations arising in communication theory preserve circular symmetry:

♠**Lemma 6.1.** *Linear transformation.* If $\mathbf{x}$ is circularly symmetric then so is $\mathrm{y} = \mathbf{A}\mathbf{x}$ for any matrix $\mathbf{A}$. ◊

Proof. Circular symmetry of $\mathbf{x}$ implies that the pseudocorrelation $E[\mathbf{x}\mathbf{x}^T] = \mathbf{0}$. So $E[\mathbf{y}\mathbf{y}^T] = \mathbf{A}E[\mathbf{x}\mathbf{x}^T]\mathbf{A}^T = \mathbf{0}$, that is, $\mathbf{y}$ is circularly symmetric. ▽▽▽

♠**Lemma 6.2.** *Superposition.* Assume $\mathbf{x}$ and $\mathbf{q}$ are circularly symmetric and independent, and at least one of them has zero mean. Then $\mathbf{y} = \mathbf{x} + \mathbf{q}$ is circularly symmetric. ◊

Proof. The pseudocorrelation of $\mathbf{y}$ is

$$E[\mathbf{y}\mathbf{y}^T] = E[\mathbf{x}\mathbf{x}^T] + E[\mathbf{q}\mathbf{q}^T] + E[\mathbf{x}\mathbf{q}^T] + E[\mathbf{q}\mathbf{x}^T].$$

Since $\mathbf{x}$ and $\mathbf{q}$ are circularly symmetric, the first two terms on the right-hand side are zero. Next, the cross-term can be written as $E[\mathbf{x}\mathbf{q}^T] = E[\mathbf{x}]E[\mathbf{q}^T]$ because of statistical independence. Using the further fact that $E[\mathbf{x}]$ or $E[\mathbf{q}]$ is zero, we therefore get $E[\mathbf{y}\mathbf{y}^T] = \mathbf{0}$, proving that $\mathbf{y}$ is circularly symmetric indeed. $\bigtriangledown\bigtriangledown\bigtriangledown$

♠**Lemma 6.3.** *Algebraic identities from circular symmetry.* Let $\mathbf{C}_{\mathbf{xx}} = \mathbf{P}+j\mathbf{Q}$ have the inverse

$$\mathbf{C}_{\mathbf{xx}}^{-1} = \widehat{\mathbf{P}} + j\widehat{\mathbf{Q}}. \tag{6.44}$$

Then the inverse of

$$\mathbf{C}_{\mathbf{uu}} \stackrel{\Delta}{=} 0.5 \begin{bmatrix} \mathbf{P} & -\mathbf{Q} \\ \mathbf{Q} & \mathbf{P} \end{bmatrix}$$

is given by

$$\mathbf{C}_{\mathbf{uu}}^{-1} = 2 \begin{bmatrix} \widehat{\mathbf{P}} & -\widehat{\mathbf{Q}} \\ \widehat{\mathbf{Q}} & \widehat{\mathbf{P}} \end{bmatrix}. \tag{6.45}$$

Furthermore

$$\det\left[2\mathbf{C}_{\mathbf{uu}}\right] = \left[\det \mathbf{C}_{\mathbf{xx}}\right]^2 \tag{6.46}$$

for any covariance $\mathbf{C}_{\mathbf{xx}} = \mathbf{P} + j\mathbf{Q}$. $\diamondsuit$

Proof. To prove Eq. (6.45) observe that the equation $(\mathbf{P}+j\mathbf{Q})(\widehat{\mathbf{P}}+j\widehat{\mathbf{Q}}) = \mathbf{I}$ implies

$$\mathbf{P}\widehat{\mathbf{P}} - \mathbf{Q}\widehat{\mathbf{Q}} = \mathbf{I} \quad \text{and} \quad \mathbf{P}\widehat{\mathbf{Q}} + \mathbf{Q}\widehat{\mathbf{P}} = \mathbf{0}.$$

Using this it follows that

$$\begin{bmatrix} \mathbf{P} & -\mathbf{Q} \\ \mathbf{Q} & \mathbf{P} \end{bmatrix} \begin{bmatrix} \widehat{\mathbf{P}} & -\widehat{\mathbf{Q}} \\ \widehat{\mathbf{Q}} & \widehat{\mathbf{P}} \end{bmatrix} = \mathbf{I},$$

which proves Eq. (6.45). To prove Eq. (6.46) first observe the identity

$$\begin{bmatrix} \mathbf{I} & j\mathbf{I} \\ \mathbf{0} & \mathbf{I} \end{bmatrix} \begin{bmatrix} \mathbf{P} & -\mathbf{Q} \\ \mathbf{Q} & \mathbf{P} \end{bmatrix} \begin{bmatrix} \mathbf{I} & -j\mathbf{I} \\ \mathbf{0} & \mathbf{I} \end{bmatrix} = \begin{bmatrix} \mathbf{C}_{\mathbf{xx}} & \mathbf{0} \\ \mathbf{Q} & \mathbf{C}_{\mathbf{xx}}^* \end{bmatrix}.$$

Using the fact that $[\det \mathbf{C}_{\mathbf{xx}}]$ is real (because $\mathbf{C}_{\mathbf{xx}}$ is Hermitian), and the fact that

$$\det \begin{bmatrix} \mathbf{A} & \mathbf{0} \\ \mathbf{C} & \mathbf{D} \end{bmatrix} = \det \mathbf{A} \det \mathbf{D}$$

for square matrices $\mathbf{A}$ and $\mathbf{D}$, Eq. (6.46) follows. $\bigtriangledown\bigtriangledown\bigtriangledown$

Example 6.2: Scalar complex random variable with circular symmetry

Consider a zero-mean scalar complex random variable $x = x_{re} + jx_{im}$. We have

$$C_{xx} = \sigma_x^2 = \sigma_{x_{re}}^2 + \sigma_{x_{im}}^2.$$

This shows that $P = \sigma_x^2$ and $Q = 0$ in the scalar case. Define

$$\mathbf{u} = \begin{bmatrix} x_{re} \\ x_{im} \end{bmatrix}$$

as usual. Then the covariance of $\mathbf{u}$ is

$$\mathbf{C_{uu}} = \begin{bmatrix} E[x_{re}^2] & E[x_{re}x_{im}] \\ E[x_{re}x_{im}] & E[x_{im}^2] \end{bmatrix}.$$

If x is circularly symmetric then from the defining equations (6.35) and (6.36) we get

$$E[x_{re}^2] = E[x_{im}^2] \quad \text{and} \quad E[x_{re}x_{im}] = 0. \tag{6.47}$$

Thus the real and imaginary parts are orthogonal, with identical variances:

$$\sigma_{x_{re}}^2 = \sigma_{x_{im}}^2 = \sigma_x^2/2. \tag{6.48}$$

The covariance matrix $\mathbf{C_{uu}}$ becomes

$$\mathbf{C_{uu}} = 0.5 \begin{bmatrix} \sigma_x^2 & 0 \\ 0 & \sigma_x^2 \end{bmatrix}.$$

6.6.2 Circularly symmetric variables with specific covariances

Given an arbitrary covariance matrix (i.e., any positive semidefinite matrix) we can generate a circularly symmetric complex random vector $\mathbf{x}$ with this specific covariance. For this, first consider a zero-mean circularly symmetric complex random vector $\mathbf{x}$ such that

$$\mathbf{R_{xx}} = \sigma_x^2\mathbf{I}. \tag{6.49}$$

It follows by comparing Eqs. (6.35)–(6.37) that

$$E[\mathbf{x}_{re}\mathbf{x}_{re}^T] = E[\mathbf{x}_{im}\mathbf{x}_{im}^T] = 0.5\sigma_x^2\mathbf{I}, \tag{6.50}$$

and

$$E[\mathbf{x}_{im}\mathbf{x}_{re}^T] = \mathbf{0}. \tag{6.51}$$

Thus, a circularly symmetric complex random vector with covariance $\sigma_x^2\mathbf{I}$ has real and imaginary parts restricted as above; the individual components of the real part are uncorrelated random variables, and so are the components of the imaginary part. All these components have (zero mean and) identical variance $0.5\sigma_x^2$. Furthermore each component of the real part is uncorrelated to each component of the imaginary part. A complex random vector with the above properties is readily generated by generating $2N$ real independent zero-mean random variables with variance $\sigma_x^2/2$ and defining the $N \times 1$ real part $\mathbf{x}_{re}$ using N of these variables and the $N \times 1$ imaginary part $\mathbf{x}_{im}$ using the remaining N variables.

Arbitrary covariances. Given an arbitrary covariance matrix $\mathbf{R}$ (i.e., any positive semidefinite matrix), we can generate a circularly symmetric complex random vector $\mathbf{x}$ with this covariance as follows: first generate a circularly symmetric complex random vector $\mathbf{s}$ with covariance $\mathbf{I}$, and then let $\mathbf{x} = \mathbf{A}\mathbf{s}$, where $\mathbf{R} = \mathbf{A}\mathbf{A}^\dagger$. (Such an $\mathbf{A}$ exists since $\mathbf{R}$ is positive semidefinite.) Then $\mathbf{x}$ has the desired covariance $\mathbf{R}$, and is circularly symmetric by Lemma 6.1. $\bigtriangledown \bigtriangledown \bigtriangledown$

6.6.3 Gaussian circularly symmetric random vectors

Our discussion of circularly symmetric complex random vectors so far has not been confined to any specific distribution. We now turn our attention to the Gaussian case because it arises frequently in MIMO communications. First of all, an $N \times 1$ complex random vector

$$\mathbf{x} = \mathbf{x}_{re} + j\mathbf{x}_{im} \tag{6.52}$$

is said to be Gaussian if the real vector

$$\mathbf{u} = \begin{bmatrix} \mathbf{x}_{re} \\ \mathbf{x}_{im} \end{bmatrix} \tag{6.53}$$

is Gaussian, that is, it has the pdf

$$f_{\mathbf{u}}(\mathbf{u}) = \frac{1}{\sqrt{\det(2\pi\mathbf{C}_{\mathbf{uu}})}} e^{-\frac{1}{2}(\mathbf{u}-\mathbf{m})^T \mathbf{C}_{\mathbf{uu}}^{-1}(\mathbf{u}-\mathbf{m})}, \tag{6.54}$$

where $\mathbf{m}$ and $\mathbf{C}_{\mathbf{uu}}$ are real (mean and covariance, respectively, of $\mathbf{u}$). For the special case where $\mathbf{x} - \mathbf{m}_{\mathbf{x}}$ has circular symmetry, we know that

$$\mathbf{C}_{\mathbf{uu}} = 0.5 \begin{bmatrix} \mathbf{P} & -\mathbf{Q} \\ \mathbf{Q} & \mathbf{P} \end{bmatrix}, \tag{6.55}$$

where the complex vector $\mathbf{x}$ has covariance

$$\mathbf{C}_{\mathbf{xx}} = \mathbf{P} + j\mathbf{Q}. \tag{6.56}$$

In this case we will show that the above pdf can be rewritten in the form

$$f_{\mathbf{u}}(\mathbf{u}) = f_{\mathbf{x}}(\mathbf{x}) = \frac{1}{\det(\pi\mathbf{C}_{\mathbf{xx}})} e^{-(\mathbf{x}-\mathbf{m}_{\mathbf{x}})^\dagger \mathbf{C}_{\mathbf{xx}}^{-1}(\mathbf{x}-\mathbf{m}_{\mathbf{x}})}. \tag{6.57}$$

Note that $\mathbf{C}_{\mathbf{xx}}$ is complex with size $N \times N$, whereas $\mathbf{C}_{\mathbf{uu}}$ is real with size $2N \times 2N$. If the complex vector $\mathbf{x}$ is Gaussian (i.e., $\mathbf{u}$ is Gaussian) but $\mathbf{x} - \mathbf{m}_{\mathbf{x}}$ is not circularly symmetric, then the expression on the right-hand side of Eq. (6.57) does not describe its pdf; we can only use the original form (6.54).

Derivation of Eq. (6.57). We now show that Eq. (6.54) can be rewritten as Eq. (6.57) when $\mathbf{x}-\mathbf{m_x}$ is circularly symmetric. Let $\mathbf{A} = \mathbf{A}_{re} + j\mathbf{A}_{im}$ be an arbitrary $N \times N$ matrix and define

$$\mathbf{B} = \begin{bmatrix} \mathbf{A}_{re} & -\mathbf{A}_{im} \\ \mathbf{A}_{im} & \mathbf{A}_{re} \end{bmatrix}.$$

With the $2N \times 1$ real vector $\mathbf{u}$ defined in terms of the complex vector $\mathbf{x}$ as in Eq. (6.53), we can verify the following identity by direct substitution:

$$\mathbf{u}^T\mathbf{B}\mathbf{u} = \text{Re}\,[\mathbf{x}^\dagger \mathbf{A}\mathbf{x}]. \tag{6.58}$$

Identifying $\mathbf{B}$ with $0.5\mathbf{C}_{\mathbf{uu}}^{-1}$ and using Eqs. (6.44) and (6.45), it therefore follows that

$$0.5\mathbf{u}^T\mathbf{C}_{\mathbf{uu}}^{-1}\mathbf{u} = \mathbf{x}^\dagger \mathbf{C}_{\mathbf{xx}}^{-1}\mathbf{x}. \tag{6.59}$$

The same holds if we replace $\mathbf{u}$ and $\mathbf{x}$ by the zero-mean versions $\mathbf{u}-\mathbf{m}$ and $\mathbf{x}-\mathbf{m_x}$. Next, from Eq. (6.46) we have $\det\,[2\mathbf{C}_{\mathbf{uu}}] = [\det \mathbf{C}_{\mathbf{xx}}]^2$. Using this and Eq. (6.59) in Eq. (6.54) we obtain Eq. (6.57). $\bigtriangledown\bigtriangledown\bigtriangledown$

What we have shown is that if $\mathbf{x}$ is Gaussian (i.e., $\mathbf{u}$ is Gaussian as in Eq. (6.54)) and if $\mathbf{x}-\mathbf{m_x}$ is circularly symmetric, then the pdf of $\mathbf{x}$ can be expressed as in Eq. (6.57). Conversely, suppose we have a complex random vector $\mathbf{x}$ whose pdf can be written as in Eq. (6.57). Does this mean that $\mathbf{x}-\mathbf{m_x}$ is circularly symmetric Gaussian? Yes, indeed; the expression (6.57) defines $\mathbf{C}_{\mathbf{xx}}$, and hence defines $\mathbf{P}$ and $\mathbf{Q}$ via (6.56). Using these, define a matrix $\mathbf{C}_{\mathbf{uu}}$ as in Eq. (6.55). Then Eqs. (6.59) and (6.46) are true (because these are mere algebraic identities), so Eq. (6.57) can be written as in Eq. (6.54) proving that $\mathbf{u}$ is real Gaussian with covariance (6.55), i.e., $\mathbf{x}-\mathbf{m_x}$ is circularly symmetric!

Summarizing, *the pdf of a complex random vector* $\mathbf{x}$ *can be described by (6.57) if and only if* $\mathbf{x}-\mathbf{m_x}$ *is circularly symmetric Gaussian.* If $\mathbf{x}-\mathbf{m_x}$ is Gaussian but not circularly symmetric we have to use the original form (6.54) to describe the pdf.

Example 6.3: Complex scalar Gaussian

Consider again a zero-mean scalar complex random variable $x = x_{re} + jx_{im}$ as in Ex. 6.2. We say x is Gaussian if $f_{\mathbf{u}}(\mathbf{u})$ is as in Eq. (6.54). When x is circularly symmetric *and* Gaussian, Eq. (6.54) reduces to (6.57):

$$f_{\mathbf{u}}(\mathbf{u}) = f_x(x) = \frac{1}{\pi\sigma_x^2}e^{-\frac{|x|^2}{\sigma_x^2}} \tag{6.60}$$

and furthermore

$$\sigma_{x_{re}}^2 = \sigma_{x_{im}}^2 = \sigma_x^2/2 \tag{6.61}$$

from Ex. 6.2. So $f_{\mathbf{u}}(\mathbf{u})$ can be rewritten as

$$f_{\mathbf{u}}(\mathbf{u}) = \frac{1}{\sqrt{2\pi\sigma_x^2/2}} e^{-x_{re}^2/2(\sigma_x^2/2)} \times \frac{1}{\sqrt{2\pi\sigma_x^2/2}} e^{-x_{im}^2/2(\sigma_x^2/2)}. \tag{6.62}$$

Since $\sigma_{x_{re}}^2 = \sigma_{x_{im}}^2 = \sigma_x^2/2$, this is nothing but the product of the individual pdfs of x_{re} and x_{im}. This happens because, in the Gaussian case, the uncorrelatedness of x_{re} and x_{im} (induced by the circular symmetry of x) implies that they are statistically independent as well.

Indeed, given any pdf of the form (6.60), it can always be rewritten in the form (6.62) which shows that such an x has uncorrelated real and imaginary parts with identical variances, showing that the expression (6.60) always represents a circularly symmetric complex Gaussian x. The pdf of a more general complex zero-mean Gaussian x would have to be expressed in the form

$$f(x_{re}, x_{im}) = \frac{1}{\sqrt{\det(2\pi\mathbf{C}_{\mathbf{uu}})}} e^{-\frac{1}{2}[x_{re} \quad x_{im}]\mathbf{C}_{\mathbf{uu}}^{-1}\begin{bmatrix} x_{re} \\ x_{im} \end{bmatrix}}, \tag{6.63}$$

where

$$\mathbf{C}_{\mathbf{uu}} = \begin{bmatrix} \sigma_{x_{re}}^2 & \rho \\ \rho & \sigma_{x_{im}}^2 \end{bmatrix},$$

with $\rho = E[x_{re}x_{im}]$.

6.6.4 Entropy of Gaussian random vectors

The differential entropy [Cover and Thomas, 1991] of a real random vector $\mathbf{u}$ with pdf $f_{\mathbf{u}}(\mathbf{u})$ is defined as

$$\mathcal{H}(f_{\mathbf{u}}) = -\int f_{\mathbf{u}}(\mathbf{u}) \ln f_{\mathbf{u}}(\mathbf{u}) d\mathbf{u}, \tag{6.64}$$

where the integration is over all the components of $\mathbf{u}$. When there is no confusion of notations we indicate $\mathcal{H}(f_{\mathbf{u}})$ by the simpler notation $\mathcal{H}(\mathbf{u})$.

If $\mathbf{x}$ is a complex Gaussian vector, then its pdf can be expressed as in Eq. (6.54) where $\mathbf{u}$ is as in Eq. (6.53) and $\mathbf{C}_{\mathbf{uu}}$ is its covariance. Then the entropy evaluated using Eq. (6.64) is given by

$$\mathcal{H}(f_{\mathbf{u}}) = \ln \sqrt{\det\,(2\pi e\mathbf{C}_{\mathbf{uu}})}. \tag{6.65}$$

Note that the mean value of the random vector plays no role in this expression. For the special case where $\mathbf{x} - \mathbf{m}_{\mathbf{x}}$ is circularly symmetric, Eq. (6.46) holds. So, when a complex Gaussian vector $\mathbf{x}$ is such that $\mathbf{x} - \mathbf{m}_{\mathbf{x}}$ is circularly symmetric, the differential entropy is

$$\mathcal{H}(f_{\mathbf{x}}) = \ln \det\,(\pi e\mathbf{C}_{\mathbf{xx}}). \tag{6.66}$$

For the special case of a scalar complex Gaussian x, if $x - m_x$ is circularly symmetric, the entropy of x is therefore

$$\mathcal{H}(f_x) = \ln(\pi e \sigma_x^2), \tag{6.67}$$

where $\sigma_x^2 = E[|x - m_x|^2]$. In this case $x = x_{re} + jx_{im}$, where the real and imaginary parts are independent Gaussians with variance $\sigma_x^2/2$. The real and imaginary parts each have an entropy [Cover and Thomas, 1991]

$$\frac{1}{2}\ln(2\pi e \sigma_{re}^2) = \frac{1}{2}\ln(\pi e \sigma_x^2). \tag{6.68}$$

Thus the entropy of the complex circularly symmetric Gaussian x is the sum of the entropies of the real and imaginary parts. We can also use $\log_2$ instead of ln in the above expressions. If we use $\log_2$ the entropy is in bits, whereas if we use ln then the entropy is in nats.

6.6.5 Relation to other definitions

The definition for circularly symmetric complex random vectors is given by Eqs. (6.35)–(6.36). Let us refer to this as *Definition 1.* This is different from the definition given in Tse and Viswanath [2005] which says that $\mathbf{x}$ is circularly symmetric if $e^{j\theta}\mathbf{x}$ has the same pdf as $\mathbf{x}$; let us refer to this as *Definition 2.* We now make a number of observations.

1. Definition 2 implies in particular that $E[e^{j\theta}\mathbf{x}] = E[\mathbf{x}]$ for all θ. That is, $e^{j\theta}E[\mathbf{x}] = E[\mathbf{x}]$ for all θ, which implies $E[\mathbf{x}] = \mathbf{0}$. But Definition 1 (used in this book) does not imply zero mean, as seen from Example 6.4 below. So the two definitions are not equivalent. Even for zero-mean random vectors, the definitions are not equivalent because Definition 2 restricts the entire pdf instead of just the second-order moment.

2. If Definition 2 holds then so does Definition 1 because if the pdf of $\mathbf{x}$ is identical to that of $e^{j\theta}\mathbf{x}$ then the pseudocorrelations ought to be unchanged, that is,
$$E[\mathbf{x}\mathbf{x}^T] = e^{2j\theta}E[\mathbf{x}\mathbf{x}^T] \tag{6.69}$$
for any θ. This implies $E[\mathbf{x}\mathbf{x}^T] = \mathbf{0}$, which in turn is equivalent to Definition 1 (Sec. 6.6.1). Thus Definition 2 implies Definition 1.

3. For the case of zero-mean Gaussian $\mathbf{x}$, it can be shown that the converse is also true, that is, Definition 1 does imply Definition 2. This is because a zero-mean Gaussian satisfying Definition 1 has a pdf of the form
$$f(\mathbf{x}) = \frac{1}{\det(\pi\mathbf{C}_{\mathbf{xx}})}e^{-\mathbf{x}^\dagger\mathbf{C}_{\mathbf{xx}}^{-1}\mathbf{x}} \tag{6.70}$$
as explained in Sec. 6.6.3. If we replace $\mathbf{x}$ with $e^{j\theta}\mathbf{x}$ then the covariance $\mathbf{C}_{\mathbf{xx}}$ does not change, nor does the quadratic form $\mathbf{x}^\dagger\mathbf{C}_{\mathbf{xx}}^{-1}\mathbf{x}$. So $f(\mathbf{x})$ is unchanged as demanded by Definition 2. So in the zero-mean Gaussian case the two definitions are equivalent.

Unless mentioned otherwise, the term circular symmetry in our discussions always means Eqs. (6.35) and (6.36), which corresponds to Definition 1.

6.6.5.A Zero mean not implied by Definition 1

For the special case of a scalar random variable, the definition of circular symmetry given in Eqs. (6.35) and (6.36) is equivalent to

$$E[x_{re}^2] = E[x_{im}^2], \quad E[x_{re}x_{im}] = 0. \tag{6.71}$$

The condition $E[x_{re}x_{im}] = 0$ is just orthogonality. So, given *any two* real orthogonal random variables x_{re} and x_{im} with identical mean square values, the complex random variable $x = x_{re} + jx_{im}$ is circularly symmetric. Since x_{re} and x_{im} can have arbitrary mean, x need not have zero mean.

Example 6.4: Circularly symmetric variables with nonzero mean

Let x_1 and x_2 be real random variables with

$$E[x_1] = E[x_2] = m \neq 0, \quad E[x_1^2] = E[x_2^2] = 1, \quad \text{and} \quad E[x_1x_2] = \rho,$$

with $0 < \rho < 1$. Define

$$y = x_1 - \frac{x_2}{\rho}$$

so that

$$E[yx_1] = 0,$$

and $E[y] = m(1 - \rho^{-1}) \neq 0$. Let

$$x_{re} = \beta y = \beta x_1 - \frac{\beta x_2}{\rho} \quad \text{and} \quad x_{im} = x_1,$$

where $\beta > 0$ is a constant such that $E[x_{re}^2] = 1$. Then $E[x_{re}^2] = E[x_{im}^2] = 1$, and

$$E[x_{re}x_{im}] = 0,$$

by construction. Furthermore, $E[x_{re}]$ and $E[x_{im}]$ are both nonzero. Thus

$$x \stackrel{\Delta}{=} x_{re} + jx_{im}$$

is circularly symmetric with nonzero mean.

Example 6.5: Circularly symmetric Gaussian with nonzero mean

In Ex. 6.4, note that, by construction,

$$\begin{bmatrix} x_{re} \\ x_{im} \end{bmatrix} = \begin{bmatrix} \beta & -\beta/\rho \\ 1 & 0 \end{bmatrix} \begin{bmatrix} x_1 \\ x_2 \end{bmatrix}. \tag{6.72}$$

Assume the original real random variables x_1 and x_2 from which we started are jointly Gaussian. That is, let the real vector $[x_1 \quad x_2]^T$ be Gaussian.

Then so is the linearly transformed real vector $[x_{re} \quad x_{im}]^T$ (Appendix E, Sec. E.4.1). This is equivalent to saying that $x = x_{re} + jx_{im}$ is Gaussian. In short, the random variable x created in this example is a complex circularly symmetric Gaussian, with nonzero mean.

6.6.5.B Restriction on shifts

Let $\mathbf{x}$ be circularly symmetric with mean $\mathbf{m}$. Define

$$\mathbf{y} = \mathbf{x} + \mathbf{c}, \tag{6.73}$$

where $\mathbf{c}$ is a constant. Note that $\mathbf{y}$ is the random vector $\mathbf{x}$ with its mean shifted to $\mathbf{m} + \mathbf{c}$. We claim that $\mathbf{y}$ is circularly symmetric if and only if either $\mathbf{c} = \mathbf{0}$ or $\mathbf{c} = -2\mathbf{m}$. That is, circular symmetry is preserved if and only if the mean is shifted to $-\mathbf{m}$. No other nonzero shift of the mean will work.

Proof. Recall that circular symmetry of $\mathbf{y}$ is equivalent to $E[\mathbf{y}\mathbf{y}^T] = \mathbf{0}$, which can be rewritten as

$$E[\mathbf{x}\mathbf{x}^T] + \mathbf{c}\mathbf{c}^T + E[\mathbf{x}]\mathbf{c}^T + \mathbf{c}E[\mathbf{x}^T] = \mathbf{0}.$$

Since $\mathbf{x}$ is circularly symmetric, $E[\mathbf{x}\mathbf{x}^T] = \mathbf{0}$. So the preceding can be rewritten as $\mathbf{c}\mathbf{c}^T + \mathbf{m}\mathbf{c}^T + \mathbf{c}\mathbf{m}^T = \mathbf{0}$, or equivalently,

$$\mathbf{m}\mathbf{c}^T = -\mathbf{c}\Big(\mathbf{m}^T + \mathbf{c}^T\Big). \tag{6.74}$$

The matrices on both sides of this equation have rank at most one. The matrix on the left has all columns proportional to $\mathbf{m}$ and the matrix on the right has all columns proportional to $\mathbf{c}$. So we must have

$$\mathbf{c} = \alpha\mathbf{m} \tag{6.75}$$

for some scalar α. Thus, the shift $\mathbf{c}$ should be parallel to $\mathbf{m}$. If $\mathbf{m} = \mathbf{0}$ then $\mathbf{c} = \mathbf{0}$ is the only choice. For $\mathbf{m} \neq \mathbf{0}$ we can use Eq. (6.75) to rewrite Eq. (6.74) as

$$\alpha(2 + \alpha)\mathbf{m}\mathbf{m}^T = \mathbf{0}. \tag{6.76}$$

So, either $\alpha = 0$, in which case we have $\mathbf{c} = \mathbf{0}$ from Eq. (6.75), or $\alpha = -2$, in which case $\mathbf{c} = -2\mathbf{m}$. $\bigtriangledown\bigtriangledown\bigtriangledown$

6.7 Capacity for MIMO and complex channels

An important result in the theory of information is that the complex random vector $\mathbf{x}$ that has maximum entropy for fixed covariance $\mathbf{C_{xx}}$ is such that $\mathbf{x} - \mathbf{m_x}$ is circularly symmetric Gaussian. We shall not prove it here, but the proof can be found in Telatar [1999], and Telatar and Tse [2000]. Based on this it can be shown that the so-called mutual information between the input and output of a "standard complex channel" (defined below) under a total power constraint is maximized by choosing the input to be a zero-mean circularly symmetric complex Gaussian vector. This maximum mutual information is the capacity of such a channel. We now elaborate on some of these details.

6.7.1 Mutual information

Consider a channel described by

$$\mathbf{y} = \mathbf{H}\mathbf{x} + \mathbf{q}, \tag{6.77}$$

where $\mathbf{H}$ is a known complex channel matrix, possibly rectangular with size $J \times P$ (see Fig. 6.6). Assume that the signal $\mathbf{x}$ and the noise $\mathbf{q}$ are independent. Assume further that $\mathbf{q}$ is a zero-mean circularly symmetric complex Gaussian random vector (Sec. 6.6) with covariance $\mathbf{C}_{\mathbf{qq}} = \sigma_q^2\mathbf{I}$. We shall refer to such a channel as the "standard complex channel," as it makes frequent appearances in our discussions. The assumption on $\mathbf{q}$ means in particular (Sec. 6.6.2) that the components q_k of $\mathbf{q}$ have independent real and imaginary parts with identical variances, that is,

$$E[(q_{k,re})^2] = E[(q_{k,im})^2] = \sigma_q^2/2. \tag{6.78}$$

Readers familiar with fundamentals of information theory [Gallager, 1968], [Cover and Thomas, 1991], will recall that the **capacity** of the channel (under appropriate constraints) is the maximum possible value of the **mutual information** $\mathcal{I}(\mathbf{x};\mathbf{y})$ that can be achieved by choosing the statistics of $\mathbf{x}$. We will consider the constraint to be the power constraint $E[\mathbf{x}^\dagger\mathbf{x}] = p_0$, that is,

$$\text{Tr}\, E[\mathbf{x}\mathbf{x}^\dagger] = p_0. \tag{6.79}$$

For the above MIMO channel, we now show that the choice of $\mathbf{x}$ which maximizes the mutual information is a zero-mean **circularly symmetric complex Gaussian** [Telatar, 1999], and furthermore, with $\mathbf{x}$ so restricted the expression for mutual information takes the following form:

$$\mathcal{I}(\mathbf{x};\mathbf{y}) = \log\det\Big(\mathbf{I} + \frac{1}{\sigma_q^2}\mathbf{H}\mathbf{C}_{\mathbf{xx}}\mathbf{H}^\dagger\Big), \tag{6.80}$$

where $\mathbf{C}_{\mathbf{xx}}$ is the covariance of $\mathbf{x}$. The **capacity** of the channel is the maximum value of this quantity as $\mathbf{C}_{\mathbf{xx}}$ is varied over all possible positive semidefinite matrices satisfying the power constraint $\text{Tr}\,\mathbf{C}_{\mathbf{xx}} = p_0$.

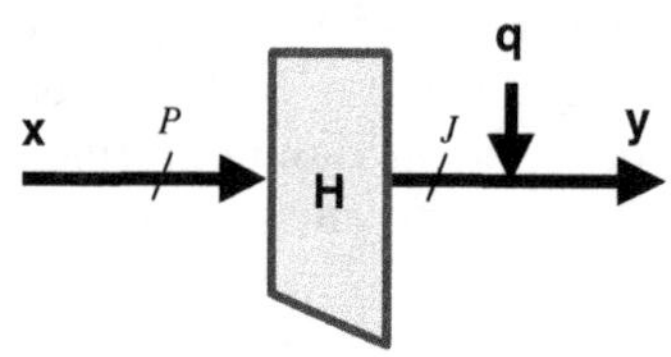

Figure 6.6. A MIMO channel $\mathbf{H}$ with additive noise $\mathbf{q}$.

Derivation of Eq. (6.80). For the standard channel described above the mutual information can be expressed as follows:

$$\begin{aligned}\mathcal{I}(\mathbf{x};\mathbf{y}) &= \mathcal{H}(\mathbf{y}) - \mathcal{H}(\mathbf{y}|\mathbf{x}) \\ &= \mathcal{H}(\mathbf{y}) - \mathcal{H}(\mathbf{Hx}+\mathbf{q}|\mathbf{x}) \\ &= \mathcal{H}(\mathbf{y}) - \mathcal{H}(\mathbf{q}|\mathbf{x}) \\ &= \mathcal{H}(\mathbf{y}) - \mathcal{H}(\mathbf{q}).\end{aligned}$$

The first equality follows essentially from the definition of mutual information; here $\mathcal{H}(\mathbf{y})$ denotes the entropy of $\mathbf{y}$ and $\mathcal{H}(\mathbf{y}|\mathbf{x})$ denotes the entropy of $\mathbf{y}$ given $\mathbf{x}$. The third equality arises because, for fixed $\mathbf{x}$, the random vector $\mathbf{Hx}+\mathbf{q}$ is simply the random vector $\mathbf{q}$ with mean value changed by the constant $\mathbf{Hx}$. The last equality follows from the statistical independence of $\mathbf{q}$ and $\mathbf{x}$. So we have proved that

$$\mathcal{I}(\mathbf{x};\mathbf{y}) = \mathcal{H}(\mathbf{y}) - \mathcal{H}(\mathbf{q}). \tag{6.81}$$

Now consider maximizing this mutual information by adjusting the input covariance. The input and output covariances are related as

$$\mathbf{C_{yy}} = \mathbf{HC_{xx}H}^\dagger + \mathbf{C_{qq}} \tag{6.82}$$

because of the independence of $\mathbf{x}$ and $\mathbf{q}$. We now use the result [Telatar, 1999] that $\mathcal{H}(\mathbf{y})$ is maximized by making $\mathbf{y}-\mathbf{m_y}$ a circularly symmetric complex Gaussian vector. In view of Lemmas 6.1 and 6.2 this is accomplished simply by making $\mathbf{x}-\mathbf{m_x}$ a circularly symmetric complex Gaussian vector. For fixed covariance $\mathbf{C_{xx}}$, since the autocorrelation is

$$\mathbf{R_{xx}} = \mathbf{C_{xx}} + \mathbf{m_x}\mathbf{m_x}^\dagger,$$

the power in $\mathbf{x}$ (trace of the autocorrelation) can only decrease by taking the mean to be zero, so we take $\mathbf{x}$ to be zero-mean. Summarizing, the mutual information $\mathcal{I}(\mathbf{x};\mathbf{y})$ for the "standard channel" is maximized for fixed power by taking $\mathbf{x}$ to be a (zero-mean) circularly symmetric complex Gaussian vector. With $\mathbf{x}$ so chosen, $\mathbf{y}$ is a circularly symmetric complex Gaussian vector with covariance

$$\mathbf{C_{yy}} = \mathbf{HC}_{xx}\mathbf{H}^\dagger + \sigma_q^2\mathbf{I}. \tag{6.83}$$

From Eq. (6.66) it therefore follows that

$$\mathcal{H}(\mathbf{y}) = \ln\det\left(\pi e \mathbf{C_{yy}}\right) = \ln\det\left(\pi e(\mathbf{H}\mathbf{C}_{xx}\mathbf{H}^\dagger + \sigma_q^2\mathbf{I})\right) \tag{6.84}$$

and, furthermore,

$$\mathcal{H}(\mathbf{q}) = \ln\det\left(\pi e \sigma_q^2 \mathbf{I}\right). \tag{6.85}$$

Substituting into Eq. (6.81), the result (6.80) follows. $\bigtriangledown\bigtriangledown\bigtriangledown$

The problem of maximizing the mutual information (6.80) can be reformulated by considering Fig. 6.7. Here the vector $\mathbf{s}$ is restricted to be a zero-mean circularly symmetric complex Gaussian with covariance matrix $\mathbf{C_{ss}} = \mathbf{I}$. The vector $\mathbf{x}$, which is the output of the linear transformation $\mathbf{F}$, is also circularly symmetric complex Gaussian for any $\mathbf{F}$ (Lemma 6.1, Sec. 6.6). Since the covariance of $\mathbf{x}$ is

$$\mathbf{C_{xx}} = \mathbf{F}\mathbf{F}^\dagger, \tag{6.86}$$

we can realize any covariance matrix by appropriate choice of $\mathbf{F}$. So the problem of maximizing the mutual information $\mathcal{I}(\mathbf{x}, \mathbf{y})$ between $\mathbf{x}$ and $\mathbf{y}$ can be solved by fixing $\mathbf{s}$ to be a zero-mean circularly symmetric complex Gaussian with covariance $\mathbf{I}$, and optimizing $\mathbf{F}$. Since the mutual information can now be written as

$$\mathcal{I}(\mathbf{x}; \mathbf{y}) = \log\det\left(\mathbf{I} + \frac{1}{\sigma_q^2}\mathbf{H}\mathbf{F}\mathbf{F}^\dagger\mathbf{H}^\dagger\right), \tag{6.87}$$

we only have to maximize this by optimizing $\mathbf{F}$ subject to the power constraint which now becomes

$$\text{Tr}\left(\mathbf{F}\mathbf{F}^\dagger\right) = p_0. \tag{6.88}$$

The same optimization problem also arises in a different context, namely that of optimizing the precoder in a decision feedback transceiver without the zero-forcing constraint (Sec. 19.4).

6.7.2 Solution to the maximum mutual information problem

At this point it is convenient to represent the channel $\mathbf{H}$ and the precoder $\mathbf{F}$ using their singular value decompositions (Appendix C):

$$\mathbf{F} = \mathbf{U}_f\mathbf{\Sigma}_f\mathbf{V}_f^\dagger \quad \text{and} \quad \mathbf{H} = \mathbf{U}_h\mathbf{\Sigma}_h\mathbf{V}_h^\dagger, \tag{6.89}$$

where $\mathbf{U}_f, \mathbf{V}_f, \mathbf{U}_h$, and $\mathbf{V}_h$ are unitary matrices, and $\mathbf{\Sigma}_f$ and $\mathbf{\Sigma}_h$ are diagonal matrices with non-negative diagonal elements (singular values of $\mathbf{F}$ and $\mathbf{H}$). Note that $\mathbf{H}$ and $\mathbf{\Sigma}_h$ are rectangular matrices; all other matrices are square. Since $\mathbf{F}\mathbf{F}^\dagger = \mathbf{U}_f\mathbf{\Sigma}_f^2\mathbf{U}_f^\dagger$ the power constraint (6.88) becomes

$$\sum_{k=0}^{P-1} \sigma_{f,k}^2 = p_0, \tag{6.90}$$

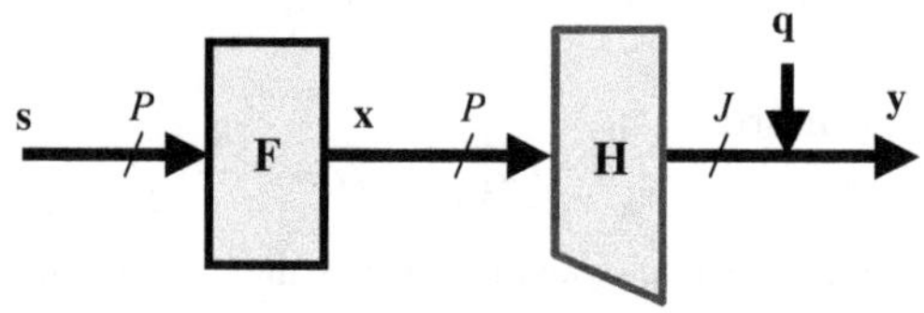

Figure 6.7. A MIMO channel $\mathbf{H}$ with precoder $\mathbf{F}$.

and furthermore

$$\begin{aligned}\mathcal{I}(\mathbf{x};\mathbf{y}) &= \log\det\Big(\mathbf{I}+\frac{1}{\sigma_q^2}\mathbf{H}\mathbf{F}\mathbf{F}^\dagger\mathbf{H}^\dagger\Big)\\ &= \log\det\Big(\mathbf{I}+\frac{1}{\sigma_q^2}\mathbf{U}_h\mathbf{\Sigma}_h\mathbf{V}_h^\dagger\mathbf{U}_f\mathbf{\Sigma}_f^2\mathbf{U}_f^\dagger\mathbf{V}_h\mathbf{\Sigma}_h^\dagger\mathbf{U}_h^\dagger\Big)\\ &= \log\det\mathbf{U}_h\Big(\mathbf{I}+\frac{1}{\sigma_q^2}\mathbf{\Sigma}_h\mathbf{V}_h^\dagger\mathbf{U}_f\mathbf{\Sigma}_f^2\mathbf{U}_f^\dagger\mathbf{V}_h\mathbf{\Sigma}_h^\dagger\Big)\mathbf{U}_h^\dagger\\ &= \log\det\Big(\mathbf{I}+\frac{1}{\sigma_q^2}\mathbf{\Sigma}_h\mathbf{V}_h^\dagger\mathbf{U}_f\mathbf{\Sigma}_f^2\mathbf{U}_f^\dagger\mathbf{V}_h\mathbf{\Sigma}_h^\dagger\Big)\\ &= \log\det\Big(\mathbf{I}+\frac{1}{\sigma_q^2}\mathbf{\Sigma}_h\mathbf{U}\mathbf{\Sigma}_f^2\mathbf{U}^\dagger\mathbf{\Sigma}_h^\dagger\Big),\end{aligned}$$

where $\mathbf{U}=\mathbf{V}_h^\dagger\mathbf{U}_f$. Since $\mathbf{V}_h$ is invertible and $\mathbf{U}_f$ is free to choose, we can regard $\mathbf{U}$ as a unitary matrix that can be freely chosen in the maximization process. So the goal is to choose the $P\times P$ unitary matrix $\mathbf{U}$ and the diagonal matrix $\mathbf{\Sigma}_f$ with diagonal elements $\sigma_{f,k}\geq 0$ such that the preceding expression is maximized. Defining the $P\times P$ matrix

$$\mathbf{P}=\mathbf{U}\mathbf{\Sigma}_f^2\mathbf{U}^\dagger, \tag{6.91}$$

we have

$$\mathcal{I}(\mathbf{x};\mathbf{y})=\log\det\Big(\mathbf{I}+\frac{1}{\sigma_q^2}\mathbf{\Sigma}_h\mathbf{P}\mathbf{\Sigma}_h^\dagger\Big). \tag{6.92}$$

Since the matrix within the brackets is positive definite, we have, from Hadamard's inequality (Appendix B, Sec. B.6)

$$\det\Big(\mathbf{I}+\frac{1}{\sigma_q^2}\mathbf{\Sigma}_h\mathbf{P}\mathbf{\Sigma}_h^\dagger\Big)\leq\prod_{k=0}^{P-1}\Big(1+\frac{1}{\sigma_q^2}\sigma_{h,k}^2P_{kk}\Big) \tag{6.93}$$

with equality if and only if the matrix is diagonal, that is, if and only if $\mathbf{P}$ is diagonal. That is, given any positive semidefinite $\mathbf{P}$, if we replace it with a diagonal matrix having the diagonal elements P_{kk}, then the above determinant can only increase (while the power constraint continues to be satisfied). We can therefore carry out the maximization of $\mathcal{I}(\mathbf{x};\mathbf{y})$ under the constraint that $\mathbf{P}$ be diagonal. That is, we can choose the unitary matrix $\mathbf{U} = \mathbf{I}$, or equivalently

$$\mathbf{U}_f = \mathbf{V}_h \tag{6.94}$$

so that

$$\mathcal{I}(\mathbf{x};\mathbf{y}) = \log \prod_{k=0}^{P-1} \Big(1 + \frac{1}{\sigma_q^2}\sigma_{h,k}^2 P_{kk}\Big) = \sum_{k=0}^{P-1} \log\Big(1 + \frac{1}{\sigma_q^2}\sigma_{h,k}^2\sigma_{f,k}^2\Big).$$

For a fixed set of numbers $\sigma_{h,k}^2/\sigma_q^2$, the set of numbers $\{\sigma_{f,k}^2\}$ should therefore be optimized to maximize

$$\mathcal{I}(\mathbf{x};\mathbf{y}) = \sum_{k=0}^{P-1} \log\Big(1 + \frac{1}{\sigma_q^2}\sigma_{h,k}^2\sigma_{f,k}^2\Big) \tag{6.95}$$

subject to the power constraint

$$\sum_{k=0}^{P-1} \sigma_{f,k}^2 = p_0. \tag{6.96}$$

This is exactly the problem we solve in Chap. 22 (Sec. 22.3) when we demonstrate the applications of the KKT technique for constrained optimization. We show there that the optimum solution is a water-filling solution, taking the form

$$\sigma_{f,k}^2 = \begin{cases} \dfrac{1}{\lambda} - \dfrac{\sigma_q^2}{\sigma_{h,k}^2} & \text{if this is non-negative} \\ 0 & \text{otherwise,} \end{cases} \tag{6.97}$$

where λ depends on the total available power p_0.

Discussion

1. The above is precisely also the solution we shall obtain in Sec. 19.4 for the optimum precoder when the receiver uses an optimum DFE equalizer without zero forcing.
2. Note that the matrix $\mathbf{V}_f$ in the optimal $\mathbf{F}$ can be taken to be identity. Only $\mathbf{U}_f$ and $\mathbf{\Sigma}_f$ matter as far as maximizing the mutual information is concerned.
3. The maximized mutual information (6.95) depends only on channel singular values and transmitted power, and not on the matrices $\mathbf{U}_h$ and $\mathbf{V}_h$.
4. Note in particular that if the channel is diagonal, that is

   $$\mathbf{H} = \mathbf{\Sigma}_h,$$

   so that $\mathbf{U}_h = \mathbf{I}$ and $\mathbf{V}_h = \mathbf{I}$, the optimal precoder $\mathbf{F}$ can also be taken to be diagonal.

5. There is a factor of 0.5 in (6.3) because we considered a real channel (real signals, real noise). For complex channels with circularly symmetric Gaussian noise, the diagonalized channel has P paths, and each path is a complex channel. If a complex path has circular symmetric Gaussian noise with variance σ_q^2 and circularly symmetric Gaussian input with variance σ_s^2, then it is equivalent to two parallel channels with real Gaussian inputs (variance $\sigma_s^2/2$) and real noise variance $(\sigma_q^2/2)$. The capacity of each real channel is therefore

$$\mathcal{C}_{real} = 0.5 \log_2\Big(1 + \frac{\sigma_s^2}{\sigma_q^2}\Big) \quad \text{bits per use,} \tag{6.98}$$

as in Sec. 6.2 so that the complex channel has twice as much capacity:

$$\mathcal{C}_{cmplx} = \log_2\Big(1 + \frac{\sigma_s^2}{\sigma_q^2}\Big) \quad \text{bits per use.} \tag{6.99}$$

This explains why there is no factor of 0.5 in Eq. (6.95).

6.7.3 Arbitrary noise covariance

In all our discussions we have assumed that the channel noise has covariance $\sigma_q^2\mathbf{I}$. This is not a loss of generality. For, if the noise is circulary symmetric Gaussian with a positive definite covariance $\mathbf{C_{qq}}$, we can write

$$\mathbf{C_{qq}} = \mathbf{A}\mathbf{A}^\dagger \tag{6.100}$$

for appropriate $\mathbf{A}$ and redraw the system as in Fig. 6.8. In this system the noise $\mathbf{w}$ has correlation $\mathbf{I}$ and continues to be circularly symmetric. The nonsingular matrix $\mathbf{A}$ can be assumed to be part of the receiver and does not affect capacity. So the problem reduces to the form considered earlier with the channel matrix $\mathbf{H}$ replaced by $\mathbf{A}^{-1}\mathbf{H}$.

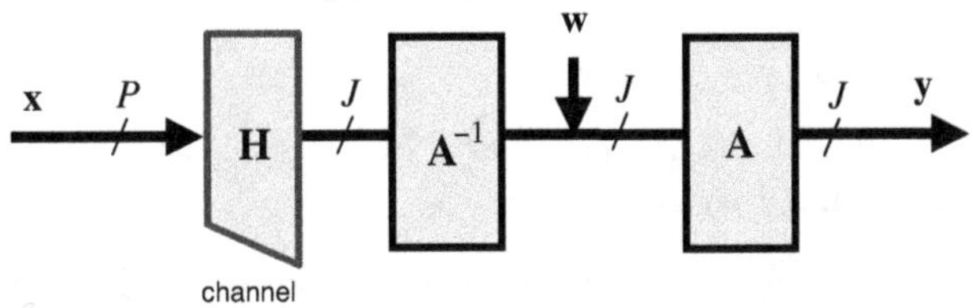

Figure 6.8. Redrawing the channel so that the noise $\mathbf{q}$ with arbitrary covariance is replaced with $\mathbf{w}$ which has covariance $\mathbf{I}$.

6.8 Concluding remarks

All discussions in this chapter assumed that the channel transfer function is deterministic. However, there are situations where the channel varies with time. There are also cases where the channel can only be described statistically, with a certain probability distribution for each of its coefficients. In these situations the derivation of the capacity formula becomes very interesting, though more involved. The interested reader should study Telatar [1999].

Problems

6.1. Sketch a plot of the capacity $\mathcal{C}$ in Eq. (6.6) as a function of B for $0 \leq B \leq \infty$, for fixed $p_0/\mathcal{N}_0$.

6.2. Note that $p_0/\mathcal{N}_0$ in Eq. (6.6) should be specified in hertz so that $p_0/(\mathcal{N}_0 B)$ is dimensionless. For a channel with bandwidth $B = 100$ Hz what is the capacity if the ratio $p_0/\mathcal{N}_0$ is 100 Hz? If the bandwidth is increased to 1 MHz (with $p_0/\mathcal{N}_0$ at 100 Hz) then what is the capacity?

6.3. In Problem 6.2 is it possible to increase the bandwidth so that the capacity is twice its value at the bandwidth of 1 MHz? If so, what is the new bandwidth?

6.4. For $B = 1$ MHz and $p_0/\mathcal{N}_0 = 100$ Hz, let $\mathcal{C}_1$ be the value of the capacity calculated from Eq. (6.6).

1. Suppose we want to increase the power p_0 (with all other quantities fixed) so that the capacity is doubled. What is the new ratio $p_0/\mathcal{N}_0$?
2. Suppose we want to increase the power p_0 (with all other quantities fixed) so that the capacity is ten times higher than $\mathcal{C}_1$. What is the new ratio $p_0/\mathcal{N}_0$?

6.5. Does the capacity $\mathcal{C}$ in Eq. (6.6) have any convexity property? That is, is it a convex (or a concave) function of B? Justify your answer. (A review of convex functions can be found in Sec. 21.2.)

6.6. In Sec. 6.3 we showed that for PAM with error probability $\mathcal{P}_e = 10^{-7}$ the SNR gap from capacity is 9.76 dB. What is the SNR gap with $P_e = 10^{-5}$? How about with $P_e = 10^{-9}$?

6.7. Consider again Ex. 6.1, where we had $\mathcal{C}_{split}/\mathcal{C}_{unsplit} > 1$. Is there any upper bound on this ratio? If not, find an example of $\mathcal{N}_0$ and $\mathcal{N}_1$ such that this ratio is arbitrarily large, say 10.0.

6.8. Compute the mutual information (6.80) for the following channel:

$$\mathbf{H} = \begin{bmatrix} 1 & 0 \\ 1 & \epsilon \end{bmatrix},$$

assuming $\mathbf{C}_{\mathbf{xx}} = \mathbf{I}$ and $\sigma_q^2 = 1$. Plot this as a function of $\epsilon > 0$. How large can you make this if you are free to choose ϵ?

6.9. Consider the problem of maximizing the mutual information (6.95) subject to the power constraint (6.96). Assume $P = 3, \sigma_q^2 = 1$, and channel singular values $\sigma_{h,k}$ such that

$$\sigma_{h,0}^2 = 1,\ \sigma_{h,1}^2 = 0.1,\ \text{and}\ \sigma_{h,2}^2 = 0.01.$$

Find an example of total power p_0 such that exactly two out of the three scalars $\sigma_{f,k}$ in the water-filling formula (6.97) are nonzero. What is the maximized mutual information?

6.10. With all other quantities as in Problem 6.9, find an example of total power p_0 such that $\sigma_{f,k}$ is nonzero for all three k. What is the maximized mutual information in your example?

7

Channel equalization with transmitter redundancy

7.1 Introduction

In this chapter we consider the effect of introducing redundancy into the symbol stream at the transmitter of a digital communication channel. The introduction of a sufficient amount of redundancy helps to equalize the channels more easily. For example FIR channels can be equalized with FIR filters without the need for oversampling at the receiver as in Secs. 4.7–4.8. Two types of redundancies will be discussed here. The first one, discussed in Sec. 7.2, is zero padding (ZP), where a block of zeros is inserted between adjacent blocks of input samples. The second one, studied in Sec. 7.3, is called cyclic prefixing (CP), where a subset of input samples is repeated in each block. In Sec. 7.4 we show how the CP system can be represented in terms of a circulant matrix. Important variations of cyclic-prefix systems such as single-carrier (SC-CP) systems and multicarrier systems, also called orthogonal frequency division multiplexing (OFDM) systems, are discussed in Sec. 7.5. Cyclic prefixing is commonly employed in OFDM systems, and in discrete multitone (DMT) systems used in DSL technology. Some details about the DMT system are discussed in Sec. 7.6.

7.2 Zero padding

Figure 7.1 explains the zero-padding operation on a discrete-time signal $s(n)$ (the symbol stream to be transmitted). We divide the signal into blocks of length M and insert L zeros at the end of each block to obtain the zero-padded result $x(n)$, which is then sent over the channel. By convention the kth block begins at $n = kP$ and ends at $n = kP + P - 1$, where

$$P = M + L. \tag{7.1}$$

Figure 7.2 shows how the zero-padding operation can be represented using the decimator and expander building blocks (Sec. 3.2). First we block the signal $s(n)$ using the advance chain and decimators ($\downarrow M$), and then add the block of L zeros. We then interleave these $M+L$ signals using the expander ($\uparrow P$) and the delay chain. The signal $x(n)$ is then transmitted over the channel. The received signal $y(n)$ is distorted because of the channel transfer function $C(z)$ and the additive noise $q(n)$. The goal at the receiver is to extract an estimate of $s(n)$ from this received signal.

The number of zeros L inserted per block will be called the **redundancy**. If L is large enough it allows us to eliminate inter-block interference, as we shall explain. Assuming that the spacing between the samples in $s(n)$ and $x(n)$ is identical, we see that the number of useful original samples we can send has been reduced to M samples per block of size P. We express this by defining the **redundancy ratio**

$$\gamma = \frac{P}{M} = \frac{M+L}{M}. \tag{7.2}$$

Note that if the zero-padded signal $x(n)$ has to preserve the same information rate as $s(n)$, then the samples of $x(n)$ would have to be spaced closer by a factor γ (see Fig. 7.3). This means the channel bandwidth has to be increased by γ. For this reason γ is also known as the **bandwidth expansion factor.**

We will assume that the channel $C(z)$ is causal FIR with transfer function

$$C(z) = \sum_{n=0}^{L} c(n)z^{-n}. \tag{7.3}$$

In this expression $c(L)$ could possibly be zero. The main point is that the redundancy L (number of zeros padded per block) is chosen to be *at least as large as the channel order.* We will assume that

$$c(0) \neq 0. \tag{7.4}$$

This is not a loss of generality, as we can always delay $x(n)$ appropriately and redefine the coefficients $c(n)$ to make this happen.

In App. 2.A of Chap. 2 we saw that the *baseband equivalent* of a continuous time channel is in general complex, whether the transmitted symbol stream is PAM or QAM. The effective discrete-time equivalent system is therefore a complex channel $C(z)$ with complex input in the case of QAM, and a complex channel with real input in the case of PAM. Since the system can be represented mathematically in this way using the complex baseband model, the discrete-time channel $C(z)$ in this chapter can be regarded as this baseband channel. Operations such as cyclic prefixing, zero padding, precoding, and equalizing, are performed on this all-discrete equivalent. Furthermore, in the case of PAM we indicated (Sec. 2.A.2 in App. 2.A of Chap. 2) that by using time domain equalization this complex channel can be replaced with a (shorter) real channel. So in the PAM case $C(z)$ can be assumed to be real, even though this assumption is not necessary for the purpose of this chapter.

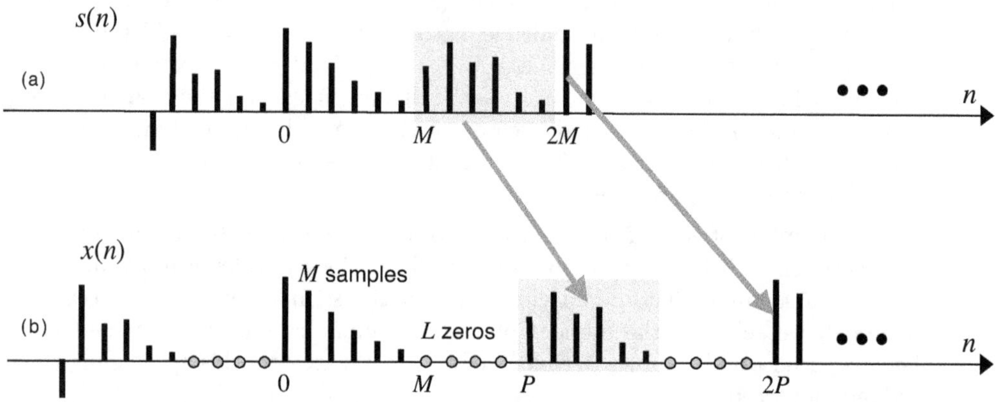

Figure 7.1. (a) A signal $s(n)$ and (b) its zero-padded version $x(n)$. Here, L zeros are padded at the end of each block of M samples.

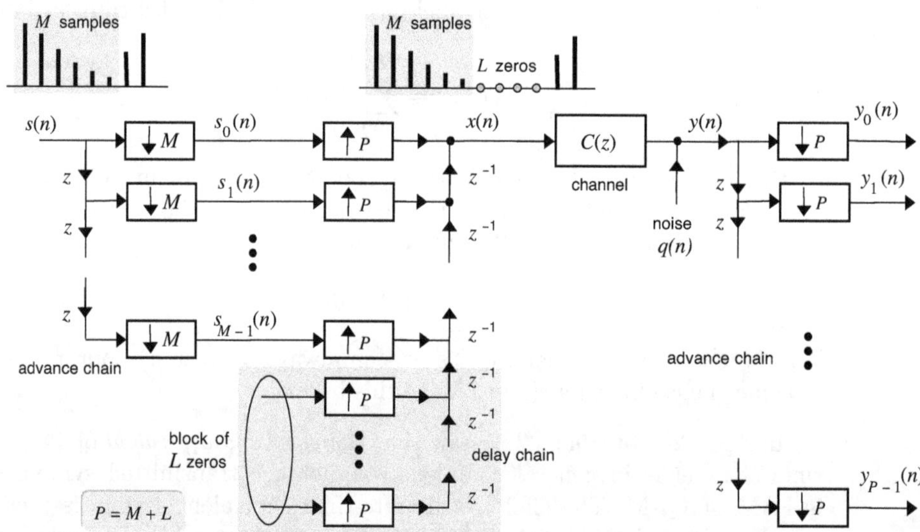

Figure 7.2. The zero-padding transmitter and corresponding receiver represented using multirate building block notation.

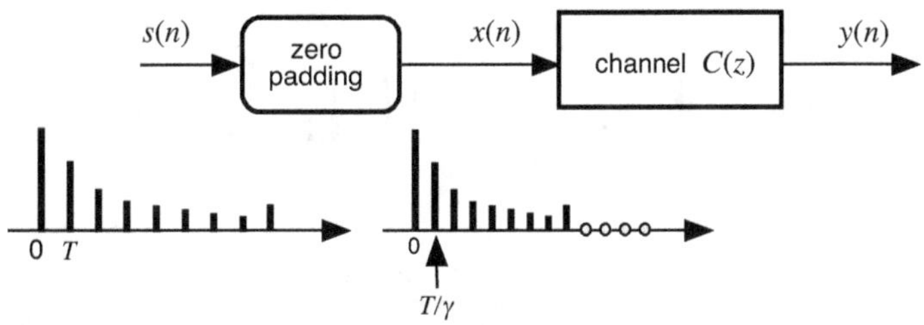

Figure 7.3. The zero-padding system with time scale for $x(n)$ squeezed such that the number of symbols transmitted per unit time is not changed.

7.2.1 Elimination of interblock interference

To see how interblock interference, or **IBI**, is eliminated by zero padding, consider the 0th block first. Ignore noise for a moment. When the M nonzero samples of the input are convolved with the channel impulse response, the result has length $P = M + L$, that is, the channel output due to the 0th block is

$$y(0), y(1), \ldots, y(P-1).$$

See Fig. 7.4. This does not therefore extend into the next block. The same is true for any block located at

$$kP \leq n < (k+1)P.$$

Thus, zero padding ensures that the channel output in response to each input block of length M is unaffected by the previous input block, and IBI has been eliminated.

With IBI eliminated, we now argue that the input block of M samples can actually be recovered from the corresponding output block of P samples (in fact, M output samples from the block suffice). Since the output $y(n)$ is the convolution of $x(n)$ with $c(n)$ we can write for the kth block

$$\begin{bmatrix} y(kP) \\ y(kP+1) \\ \\ \vdots \\ \\ \\ y(kP+P-1) \end{bmatrix} = \begin{bmatrix} c(0) & 0 & \ldots & 0 \\ c(1) & c(0) & \ldots & 0 \\ \vdots & \vdots & \ddots & \vdots \\ c(L) & & & \\ 0 & c(L) & & \\ \vdots & & \ddots & \vdots \\ 0 & 0 & \ldots & c(L) \end{bmatrix} \begin{bmatrix} x(kP) \\ x(kP+1) \\ \\ \vdots \\ \\ \\ x(kP+M-1) \end{bmatrix}. \quad (7.5)$$

The $P \times M$ matrix in Eq. (7.5) is a **banded Toeplitz matrix**.[1]

[1] A Toeplitz matrix is such that all the elements on the diagonal are identical and, further-

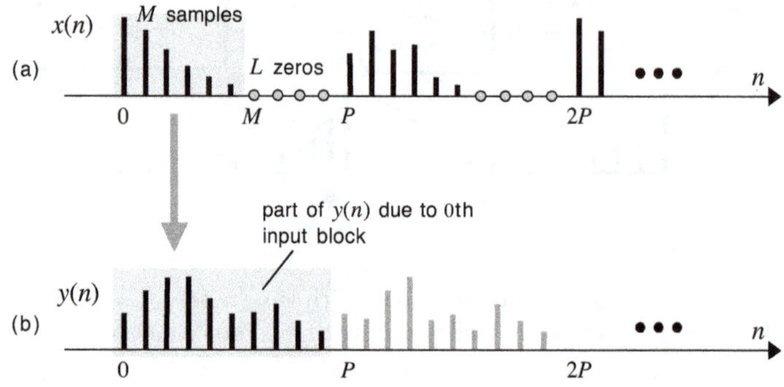

Figure 7.4. (a) The channel input padded with zeros, and (b) one block of the channel output. See text.

To recover the M input samples we only need to consider the first M equations above, and write

$$\begin{bmatrix} y(kP) \\ y(kP+1) \\ \vdots \\ y(kP+M-1) \end{bmatrix} = \begin{bmatrix} c(0) & 0 & \dots & 0 \\ c(1) & c(0) & \dots & 0 \\ \vdots & \vdots & \ddots & \vdots \\ c(M-1) & c(M-2) & \dots & c(0) \end{bmatrix} \begin{bmatrix} x(kP) \\ x(kP+1) \\ \vdots \\ x(kP+M-1) \end{bmatrix}. \tag{7.6}$$

Since $c(0) \neq 0$ the $M \times M$ matrix is *nonsingular* and can be inverted to yield the M input samples $x(n)$ which in turn equal the M transmitted symbols $s(kM), \ldots, s(kM+M-1)$. In practice there is channel noise, and the right-hand sides of Eqs. (7.5) and (7.6) have a noise term as well, so the recovery of $x(n)$ is not perfect. We will see later that, with noise present, inversion of the matrix in (7.6) is not the best thing to do. *All the P samples of $y(n)$* in Eq. (7.5) should be used to estimate $x(n)$ (equivalently $s(n)$) with minimum error, using the ideas of statistical optimal filtering (Sec. 4.10).

Note that the price paid for the suppression of interblock interference is the bandwidth expansion factor, or redundancy ratio γ, defined in Eq. (7.2). For a given channel order L we can choose M to be *arbitrarily large*, and make $\gamma \to 1$. While this appears to be a simple way to avoid bandwidth expansion, there are some practical issues to keep in mind. For example, the *computational complexity* involved in the inversion of (7.5) grows rapidly with M. Secondly, there is amplification of the ever-present channel noise in the inversion process. This amplification increases as M is increased, as we shall see in future chapters. Before we get into these details, let us see what happens if we choose M as *small* as possible, namely $M = 1$.

---

more, that all the elements along any line parallel to the diagonal are identical. Since the $L+1$ nonzero elements in each column form a "band," we say the matrix in Eq. (7.5) is a full banded Toeplitz matrix. The matrix in Eq. (7.6) is not full banded because some of the elements $c(k)$ drop out as we move to the right.

Example 7.1: Case where $M = 1$

When $M = 1$ the channel input and output are as in Fig. 7.5. Each block of the channel output is just a scaled copy of the channel impulse response $c(n)$. Thus, the input samples $x(n)$ can readily be identified since the channel is known. For example we can identify $x(0)$ in many ways:

$$x(0) = y(0)/c(0),\ x(0) = y(1)/c(1),\ \ldots,$$

and so forth. In practice there is noise, and we can filter out the noise to some extent by averaging out the estimates of $x(n)$ obtained from different output samples in a block. In any case, with $M = 1$, the process of equalization is very simple and the computational complexity is almost trivial. What is the price paid for this simplicity? With $M = 1$ the bandwidth expansion ratio becomes

$$\gamma = L + 1,$$

which is as bad as it can get! We send only one useful sample per $L + 1$ channel input samples.

7.2.2 Equalization at the zero-prefix receiver

Consider again Eq. (7.6), which shows the relation between the M transmitted samples in one block and the received samples in the same block. Since the M samples of $y(n)$ in the block are obtained by convolution of the M samples of $x(n)$ in the same block with a causal FIR filter $c(n)$, it is clear that we can recover these M samples of $x(n)$ from the M samples of $y(n)$ by using the inverse filter

$$H(z) \stackrel{\Delta}{=} \frac{1}{C(z)} = \sum_{n=0}^{\infty} h(n) z^{-n}. \tag{7.7}$$

This is an IIR filter if the channel $C(z)$ is FIR. The inversion process can be expressed by the following equations:

$$\begin{bmatrix} x(kP) \\ x(kP+1) \\ \vdots \\ x(kP+M-1) \end{bmatrix} = \begin{bmatrix} h(0) & 0 & \ldots & 0 \\ h(1) & h(0) & \ldots & 0 \\ \vdots & \vdots & \ddots & \vdots \\ h(M-1) & h(M-2) & \ldots & h(0) \end{bmatrix} \begin{bmatrix} y(kP) \\ y(kP+1) \\ \vdots \\ y(kP+M-1) \end{bmatrix}. \tag{7.8}$$

This result also follows by inverting the triangular matrix in (7.6). Since the matrix is Toeplitz and lower triangular, the inverse has the same property, and defines $h(n)$. Thus, even though no IIR filtering is involved in the equalization, the recovery of the M input samples block by block is equivalent to computing the M outputs of an IIR filter in response to a causal signal of length M in each block. Zero padding merely decouples the computations so that each block operates independently of the others. In fact, since only the first M outputs of the IIR filter $H(z)$ are computed, we can replace it with the FIR filter

$$H_M(z) = \sum_{n=0}^{M-1} h(n)z^{-n}$$

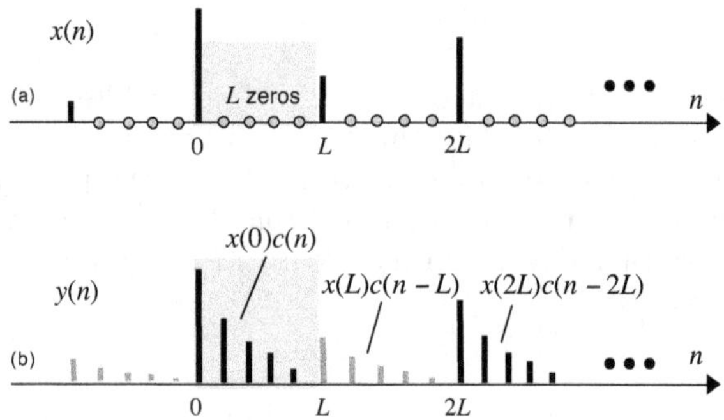

Figure 7.5. The case of $M = 1$. (a) The channel input and (b) the channel output. See text.

without affecting the computations in any way! Note that the computation of the M samples of $x(n)$ requires

$$1 + 2 + \ldots + M = \frac{M(M+1)}{2} \tag{7.9}$$

multiplications in general. Thus the equalizer has a computational complexity of about $M/2$ *multiplications per output sample.* Large block size therefore implies more computations.

7.2.2.A Effect of zeros of the channel

The discussions so far have not involved the locations of the zeros of the FIR channel $C(z)$. This, however, is important in the actual implementation of the computation (7.8), as zeros of $C(z)$ outside the unit circle imply unstable poles of $1/C(z)$. This is demonstrated in the next example.

Example 7.2: Non minimum-phase Channels

A minimum-phase FIR system has all its zeros in $|z| < 1$. In this case $1/C(z)$ is a causal stable filter, and the coefficients $h(n)$ are bounded. If $C(z)$ has some zeros on or outside the unit circle, then it is said to be a *non minimum-phase channel.* In this case the causal filter $1/C(z)$ is unstable and the truncated version $H_M(z)$ tends to have large coefficients. This implies a large noise gain for the channel noise. For example, let $C(z) = 1 - 2z^{-1}$, which has a zero outside the unit circle at $z = 2$. Then since

$$\frac{1}{1-2z^{-1}} = 1 + 2z^{-1} + 4z^{-2} + 8z^{-3} + \ldots$$

we see that

$$H_M(z) = \sum_{n=0}^{M-1} 2^n z^{-n}$$

and the coefficients $h(n)$ grow exponentially. For example if $M = 21$ then $h(M-1) = 1{,}048{,}576$ which exceeds a million! Thus, for non minimum-phase channels we cannot choose M to be arbitrarily large.

Equation (7.8) and the filtering interpretation with an FIR filter $H_M(z)$ resulted from inverting the channel equation (7.6). This corresponds to the case where only the first M samples from the received block of size P are retained. In practice we can avoid the problem described in the above example if we retain all the P samples in each block. We shall return to this in Sec. 8.3. We will see that if the equalizer is appropriately designed then the noise amplification is actually independent of whether the channel zeros are inside or outside the unit circle!

The general schematic for equalization is to use a matrix $\mathbf{T}$ at the receiver end as shown on the right side of Fig. 7.6(a). The matrix $\mathbf{T}$ is chosen depending on the channel $C(z)$, and our knowledge of the second-order statistics of the channel noise $q(n)$. Interleaving (or unblocking) the signals $\widehat{s}_k(n)$ as shown in Fig. 7.6(b) yields the reconstructed version $\widehat{s}(n)$ of the signal $s(n)$.

7.2.2.B Summary (zero padding)

Here is a summary of the main points about zero-padding precoders:

1. ***Elimination of interblock interference.*** Zero padding helps to avoid interblock interference (IBI) and allows us to recover each block of the input from the corresponding block of the channel output. The method works as long as the channel order is no larger than L, the number of zeros added per block.

2. ***Bandwidth expansion.*** If the block length M of the input stream $s(n)$ is chosen to be large ($M >> L$) then there is no significant bandwidth expansion, that is, $\gamma \approx 1$ in Eq. (7.2).

3. ***Complexity.*** However, large block size M implies more computation per recovered input symbol (see Eq. (7.9)). It takes about $0.5M$ multiplications to recover each symbol $s(n)$ if we directly use the FIR filter $H_M(z)$.

7.2.3 Generalization of the precoder matrix

A further generalization of the zero-padding precoder allows a matrix transformation $\mathbf{R}_1$ in the transmitter end as shown in Fig. 7.6. If we set $\mathbf{R}_1 = \mathbf{I}$ we get the precoder discussed so far (Fig. 7.2). If we *jointly optimize* $\mathbf{R}_1$ and $\mathbf{T}$ based on our knowledge of channel noise statistics, it results in the best transceiver design. This topic is studied in greater detail in Chaps. 12 and 13. Another generalization is in the introduction of memory in the precoder and receiver matrices, as shown in Fig. 7.7.

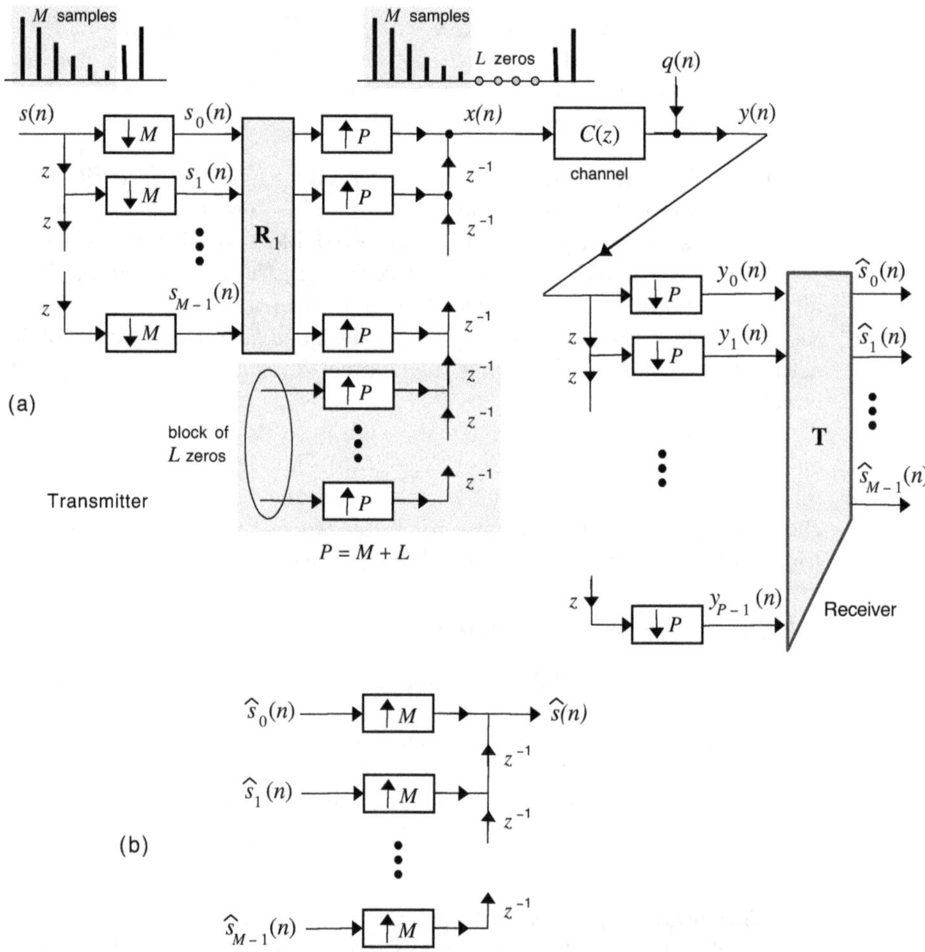

Figure 7.6. (a) Insertion of a linear precoder matrix $\mathbf{R}_1$ in a zero-padding transmitter. The receiver uses a postcoder matrix $\mathbf{T}$ to compensate for $\mathbf{R}_1$. (b) Interleaving (or unblocking) the signals $\widehat{s}_k(n)$, we obtain $\widehat{s}(n)$, the reconstructed version of $s(n)$.

This is a generalization of Fig. 7.6 in two ways. First, instead of padding zeros, we allow a general transformation from a vector of size M to a vector of size P. Second, the matrices $\mathbf{R}(z)$ and $\mathbf{E}(z)$ have memory, that is, they are general z-transforms rather than constants. These are typically chosen as polynomials (MIMO FIR filters). In fact, if we look at the system between the inputs $s_k(n)$ and outputs $\widehat{s}_k(n)$ in Fig. 7.7, this is nothing but the redundant transmultiplexer introduced in Chap. 3, drawn in polyphase form (see Fig. 3.29). Interleaving (or unblocking) the signals $\widehat{s}_k(n)$ as shown in Fig. 7.6(b) yields the reconstructed version $\widehat{s}(n)$ of the signal $s(n)$. In Problem 7.6 we revisit this system and show that there are two ways to look at it from the point of view of filter banks.

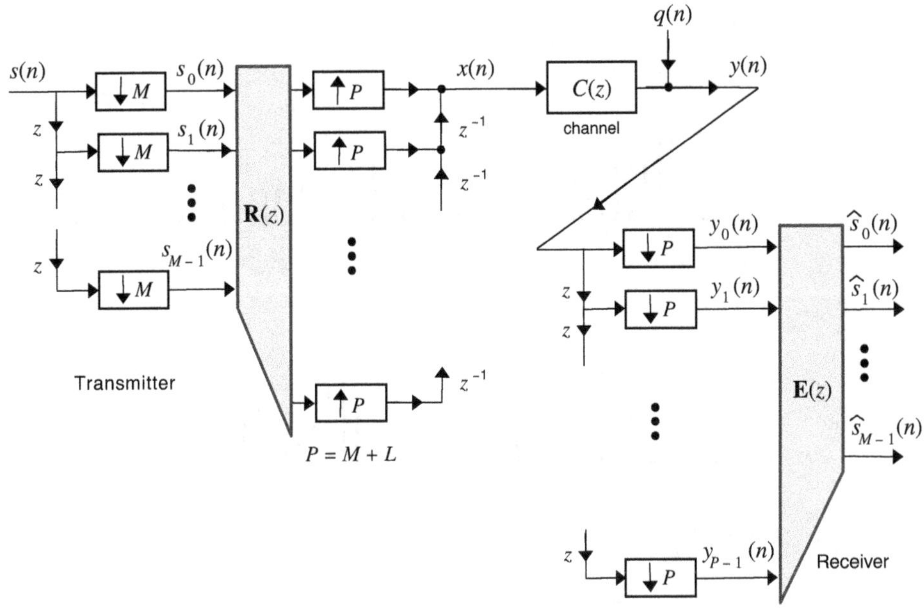

Figure 7.7. Insertion of a more general linear precoder matrix $\mathbf{R}(z)$ in the transmitter. The receiver uses a postcoder or equalizer matrix $\mathbf{E}(z)$. This system is equivalent, in polyphase form, to the general transmultiplexer introduced in Sec. 3.8. Interleaving the signals $\widehat{s}_k(n)$, we obtain $\widehat{s}(n)$, the reconstructed version of $s(n)$.

7.3 Introduction of the cyclic prefix

In Sec. 7.2 we divided the symbol stream $s(n)$ into blocks of length M and inserted L zeros at the end of each block. There is another way to introduce redundancy: namely, we copy the L samples at the end of each block into the L locations at the beginning. This scheme is called cyclic prefixing, and was first introduced by Peled and Ruiz [1980]. It has now become a standard method, commonly used in many communication systems that employ OFDM or DMT techniques.

The idea of cyclic prefixing is explained in Fig. 7.8. The construction implicitly assumes $M \geq L$. We shall make the stronger assumption that

$$M > L \tag{7.10}$$

for reasons of convenience, that will be clear later. The new symbol stream $x(n)$ can be considered to have blocks of length

$$P = M + L,$$

with the first L samples identical to the last L samples. The set of L samples at the beginning is referred to as a **cyclic prefix** for the original block of M samples.

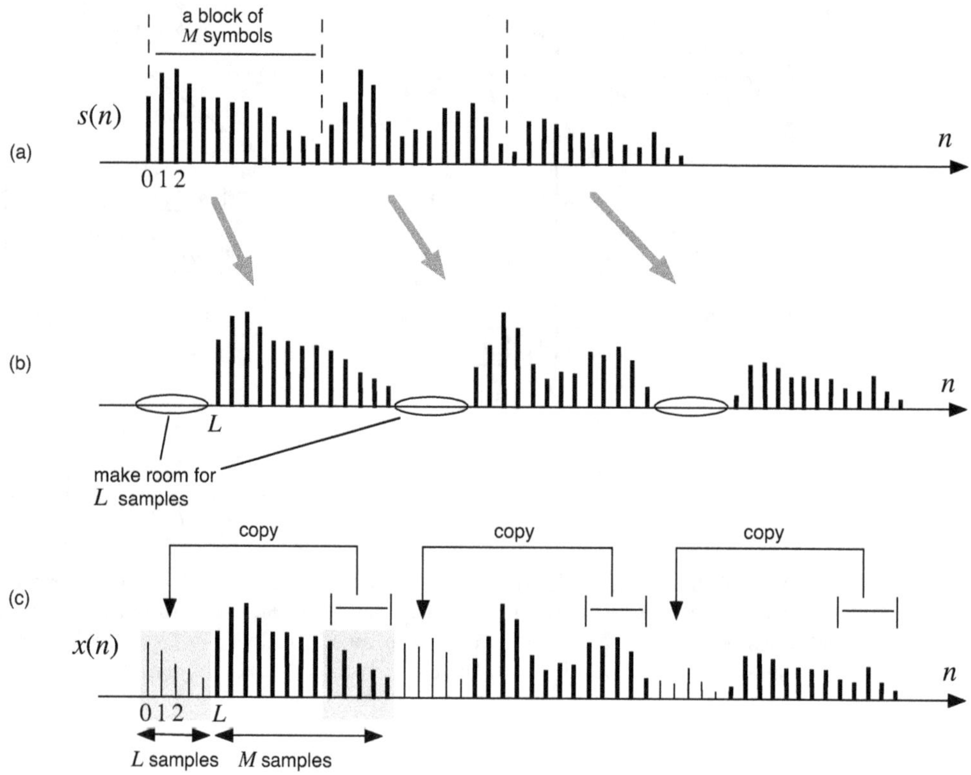

Figure 7.8. (a)–(c) Description of how cyclic prefixing is carried out.

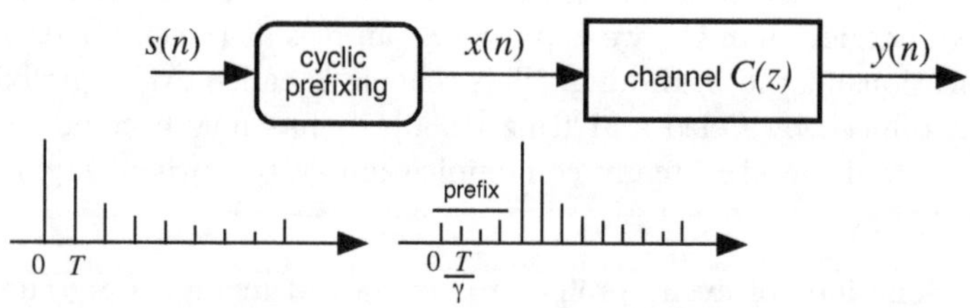

Figure 7.9. Schematic of channel with cyclic-prefixed symbol stream.

As in zero padding, we therefore have a redundancy of L symbols per M original symbols and the bandwidth expansion factor is

$$\gamma = \frac{P}{M} = \frac{M+L}{M} \tag{7.11}$$

as before. Figure 7.9 shows the cyclic-prefixing system in the form of a block diagram. If we wish the original data rate to be preserved in spite of introduction of the redundancy then the samples of $x(n)$ have to be spaced closer than those of $s(n)$ by a factor γ. We will show that if this new symbol stream $x(n)$ is transmitted through the channel $C(z)$, then, from its output $y(n)$, we can recover $s(n)$ perfectly (ignoring noise, of course), with **no IIR filtering**. Thus cyclic prefixing serves the same purpose as zero padding, but channel equalization now uses frequency-domain computations, as we shall show.

7.3.1 Working principle of the cyclic-prefix system

The signal $x(n)$ containing the cyclic prefix is convolved with the channel impulse response $c(n)$, which has length $L+1$. Figure 7.10(b) shows how the samples of $c(n)$ enter the computation of the output sample $y(L)$. The samples involved in the computation of $y(M+L-1)$ are also shown. Note that causality ensures that none of the output blocks is affected by future input blocks. Moreover, the prefix of length L ensures that the last M samples in the mth block depend only on the input samples in that block but not on the previous block. So, once again, *interblock interference is eliminated* as in zero padding. Since the first L input samples are identical to those at the end of the block, we see that the last M samples of the output are computed as if we are performing a **circular convolution** or cyclic convolution of $x(n)$ with $c(n)$ [Oppenheim and Schafer, 1999]. That is, if we denote the last M samples of the mth block temporarily as

$$a(n) = x(Pm + L + n),\ b(n) = y(Pm + L + n), \quad \text{for } 0 \leq n \leq M-1, \tag{7.12}$$

then

$$b(n) = \sum_{\ell=0}^{M-1} c(\ell)a((n-\ell)) \ \text{(cyclic convolution)}, \tag{7.13}$$

where $a((i))$ denotes the periodic extension of $a(i)$ with period M. Defining the M-point DFTs

$$A[k] = \sum_{n=0}^{M-1} a(n)W^{kn}, \quad B[k] = \sum_{n=0}^{M-1} b(n)W^{kn}, \tag{7.14}$$

where

$$W = e^{-j2\pi/M}, \tag{7.15}$$

we then have, from the circular convolution theorem [Oppenheim and Schafer, 1999]

$$B[k] = C[k]A[k], \tag{7.16}$$

where $C[k]$ is the M-point DFT of the channel, that is,

$$C[k] = \sum_{n=0}^{L} c(n)W^{kn} \tag{7.17}$$

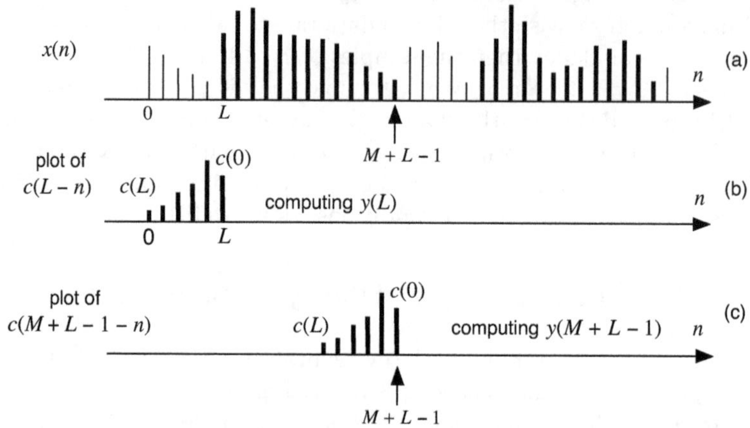

Figure 7.10. Description of how the channel output samples are computed when the input symbol stream has a cyclic prefix.

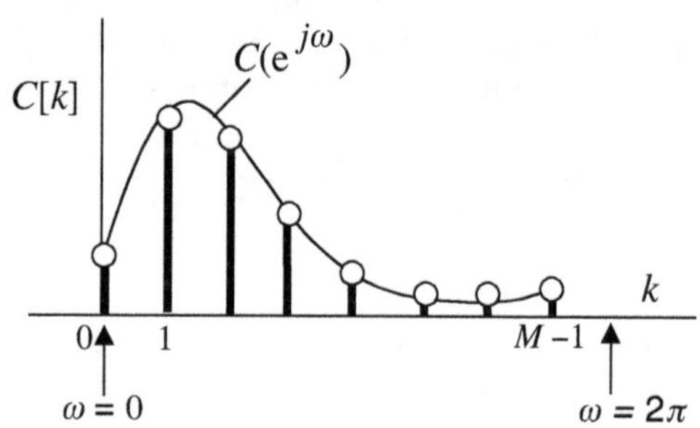

Figure 7.11. Samples $C[k]$ of the channel frequency response $C(e^{j\omega})$, taken at uniformly spaced frequencies.

where we have used the fact that the channel order $L < M$. Observe here that $C[k]$ is nothing but the channel frequency response $C(e^{j\omega})$ sampled uniformly:

$$C[k] = C(e^{j2\pi k/M}). \tag{7.18}$$

See Fig. 7.11. Summarizing, *the linear convolution of $s(n)$ with the channel $c(n)$ has been turned into a circular convolution by the process of cyclic prefixing.* Furthermore, there is no interference between blocks. This scheme works as long as the cyclic-prefix length $L \geq$ channel order, and $M > L$.

7.3.2 The cyclic-prefix receiver

Figure 7.12 shows the cyclic-prefix system in block diagram form using multirate notations. The symbol stream $s(n)$ is blocked with block size M and then the last L samples are copied at the top to make the block size $P = M + L$. These samples are then interleaved to obtain $x(n)$, the channel input. The blocked version of the channel output $y(n)$ has P components $y_k(n)$, as shown in the figure, on the right. For convenience we have numbered the last M components as $\widehat{y}_k(n)$, that is,

$$\widehat{y}_k(n) = y_{P-M+k}(n), \quad 0 \le k \le M-1.$$

Recall that $a(n)$ and $b(n)$ in Eq. (7.12) are the last M samples of $x(n)$ and $y(n)$ in each block. In terms of the notations in Fig. 7.12 we can write

$$\begin{bmatrix} s_0(n) \\ s_1(n) \\ \vdots \\ s_{M-1}(n) \end{bmatrix} = \begin{bmatrix} a(0) \\ a(1) \\ \vdots \\ a(M-1) \end{bmatrix}, \quad \begin{bmatrix} \widehat{y}_0(n) \\ \widehat{y}_1(n) \\ \vdots \\ \widehat{y}_{M-1}(n) \end{bmatrix} = \begin{bmatrix} b(0) \\ b(1) \\ \vdots \\ b(M-1) \end{bmatrix},$$

With the DFTs $A[k]$ and $B[k]$ defined[2] as in (7.14) we see that

$$\begin{bmatrix} B[0] \\ B[1] \\ \vdots \\ B[M-1] \end{bmatrix} = \mathbf{W} \begin{bmatrix} \widehat{y}_0(n) \\ \widehat{y}_1(n) \\ \vdots \\ \widehat{y}_{M-1}(n) \end{bmatrix}, \text{ and } \begin{bmatrix} A[0] \\ A[1] \\ \vdots \\ A[M-1] \end{bmatrix} = \mathbf{W} \begin{bmatrix} s_0(n) \\ s_1(n) \\ \vdots \\ s_{M-1}(n) \end{bmatrix},$$

where $\mathbf{W}$ is the $M \times M$ DFT matrix with elements

$$[\mathbf{W}]_{k\ell} = W^{k\ell} = e^{-j2\pi k\ell/M}.$$

[2] A brief review of DFT can be found in Problem 7.11 and problems following it.

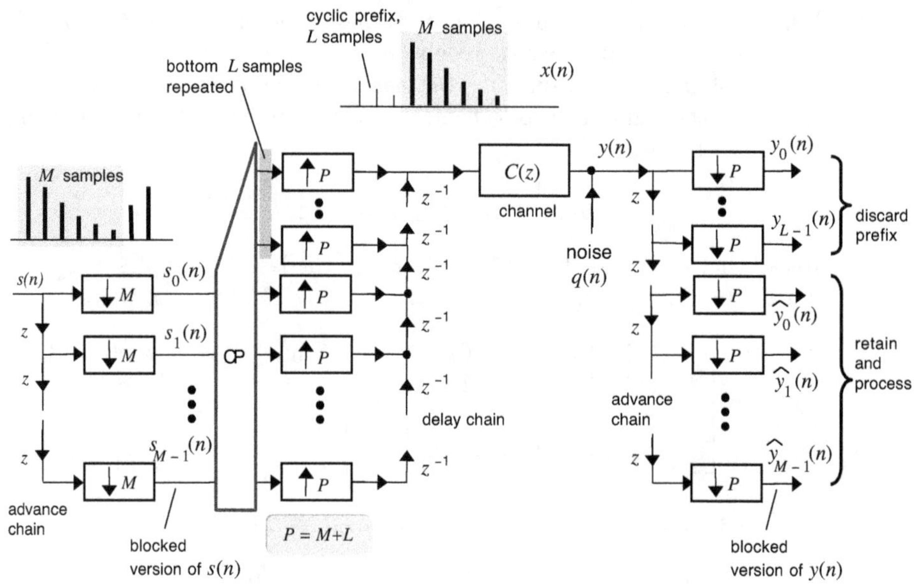

Figure 7.12. The cyclic-prefix transceiver shown in terms of multirate building block notation.

Since $B[k] = C[k]A[k]$ (ignoring noise for a moment), it then follows that

$$\mathbf{W}\begin{bmatrix} \widehat{y}_0(n) \\ \widehat{y}_1(n) \\ \vdots \\ \widehat{y}_{M-1}(n) \end{bmatrix} = \underbrace{\begin{bmatrix} C[0] & 0 & \dots & 0 \\ 0 & C[1] & \dots & 0 \\ \vdots & \vdots & \ddots & \vdots \\ 0 & 0 & \dots & C[M-1] \end{bmatrix}}_{\text{call this } \mathbf{\Lambda}_c} \mathbf{W}\begin{bmatrix} s_0(n) \\ s_1(n) \\ \vdots \\ s_{M-1}(n) \end{bmatrix}, \tag{7.19}$$

which shows that

$$\begin{bmatrix} \widehat{y}_0(n) \\ \widehat{y}_1(n) \\ \vdots \\ \widehat{y}_{M-1}(n) \end{bmatrix} = \mathbf{W}^{-1}\mathbf{\Lambda}_c\mathbf{W}\begin{bmatrix} s_0(n) \\ s_1(n) \\ \vdots \\ s_{M-1}(n) \end{bmatrix}. \tag{7.20}$$

Thus the channel output in block number n depends only on the symbol stream transmitted at block number n. This is because interblock interference has been elimimated, as we have already seen. Assuming

$$C[k] \neq 0, \tag{7.21}$$

for all k in the preceding equation we obtain

$$\begin{bmatrix} s_0(n) \\ s_1(n) \\ \vdots \\ s_{M-1}(n) \end{bmatrix} = \mathbf{W}^{-1}\mathbf{\Lambda}_c^{-1}\mathbf{W} \begin{bmatrix} \widehat{y}_0(n) \\ \widehat{y}_1(n) \\ \vdots \\ \widehat{y}_{M-1}(n) \end{bmatrix}. \tag{7.22}$$

This is a crucial equation. It shows that the receiver structure shown in Fig. 7.13 recovers the transmitted signal $s(n)$ perfectly, that is, $\widehat{s}(n) = s(n)$ in absence of noise. *Summarizing*, this is how the **cyclic-prefix receiver** works:

1. ***Discard prefix.*** Divide the received signal $y(n)$ into blocks of size $P = L+M$. Discard the cyclic-prefix part (first L samples of $y(n)$ in each block).

2. ***Perform DFT***. Perform a DFT of the remaining M samples $\widehat{y}_0(n), \ldots, \widehat{y}_{M-1}(n)$ in the received block of $y(n)$.

3. ***Perform frequency-domain equalization***. Multiply these DFT coefficients with $1/C[k]$. This is the frequency-domain equalization step. Thus the inversion of the channel is performed in the frequency-domain as shown in Fig. 7.13 (recall that $C[k]$ are samples of the channel frequency response). This equalization should be regarded as a **zero-forcing** method, because the transfer function from the transmitted signal $s(n)$ to the received signal $\widehat{s}(n)$ is identity. Other equalization methods such as the MMSE method are also possible, and are preferred because of the presence of noise (Sec. 4.10).

4. ***Perform IDFT***. Then perform the inverse DFT to return to the time domain, and unblock the result to obtain the signal $\widehat{s}(n)$. Ideally, $\widehat{s}(n) = s(n)$ in absence of channel noise $q(n)$.

In practice, the noise $q(n)$ is filtered through the DFT and IDFT matrices and the multipliers $1/C[k]$, and shows up at the output $\widehat{s}(n)$. The reconstruction error due to noise,

$$e(n) = \widehat{s}(n) - s(n),$$

will be analyzed in later chapters, where we develop techniques to minimize the mean square value of $e(n)$. We will see that the optimal receiver in the presence of noise does not use the multipliers $1/C[k]$ explicitly, and the assumption that $C[k] \neq 0$ becomes unnecessary.

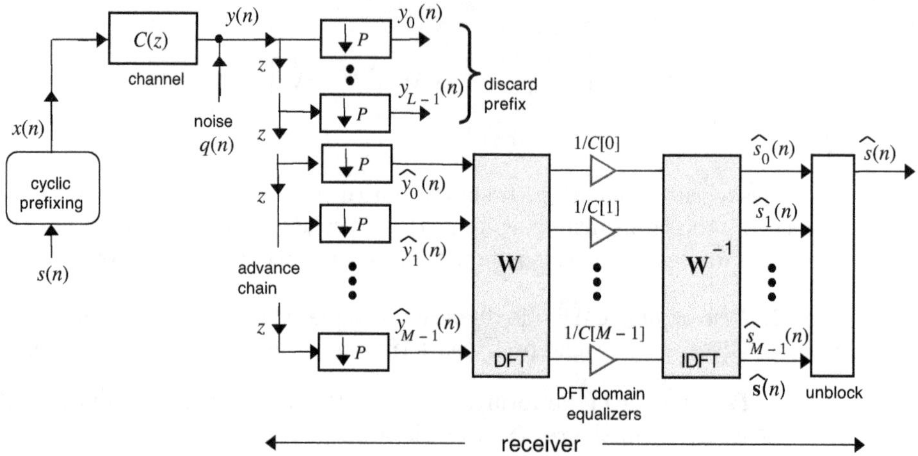

Figure 7.13. The cyclic-prefix receiver.

Example 7.3: Cyclic prefix for a non minimum-phase channel

For channels with zeros outside the unit circle, we saw in Sec. 7.2 that the computations involved in the zero-padded equalizer could become unstable. This was in part because the coefficients of the truncated FIR filter $H_M(z)$ could be very large, as demonstrated in Ex. 7.2 with $C(z) = 1 - 2z^{-1}$. Now, in the cyclic-prefix based equalizer, only the reciprocals of the DFT coefficients $C[k]$ are involved. For any fixed M we have

$$C[k] = C(e^{j2\pi k/M}) = \sum_{n=0}^{L} c(n)e^{-j2\pi kn/M}.$$

As long as $C(z)$ has no zeros on the unit circle, $C[k] \neq 0$. Thus it is possible that $1/C[k]$ are not very large, even if $C(z)$ has zeros outside the unit circle. For the case where $C(z) = 1 - 2z^{-1}$, there is a zero outside the unit circle. But we have

$$\left|C[k]\right| = \left|1 - 2e^{-j2\pi k/M}\right| \geq 1,$$

which shows that $|1/C[k]| \leq 1$. Thus the cyclic-prefix system does not unduly amplify channel noise, unlike the zero-padded system of Ex. 7.2.

Note, however, that if a channel has unit circle zeros close to the DFT frequencies $2\pi k/M$, then $C[k]$ are very small, and the *cyclic-prefix system also amplifies noise significantly.* The use of MMSE equalizers rather than zero-forcing equalizers $1/C[k]$ can take care of this problem, as we shall see later. Another important point is that if the zero-padded system makes use of all the $P = M + L$ samples at the receiver to perform equalization, then the existence of channel zeros on or outside the unit circle ceases to be an issue, as explained in Sec. 8.5.

If we introduce additional redundancy in an appropriate way ($2L$ instead of L samples per block) it is possible to show that the equalizers are the reciprocals of

$$C_1[k] = \sum_{n=0}^{L} c(n)e^{-j2\pi k/(M+L)}, \quad 0 \leq k \leq M+L-1,$$

which are samples of the channel frequency response taken at a closer spacing $2\pi/(M+L)$. If appropriately designed, it can be shown that only M of these $M+L$ samples need to be inverted at the receiver. Choosing the M largest samples we ensure that none of them is zero (because the Lth order FIR channel can have at most L zeros on the unit circle). This clever idea has been used to design receivers that are **resistant to channel** nulls [Liang and Tran, 2002]. See Problem 7.23.

Use of FFT in equalization. The cyclic-prefix transceiver uses DFT and inverse DFT operations, both of which can be performed efficiently using the FFT algorithm. For this it is desirable to choose the block length M to be a power of two, so that radix-2 FFT can be employed. The number of complex multipliers in an M-point radix-2 FFT is equal to [Oppenheim and Schafer, 1999]

$$\frac{M}{2}\log_2 M - \frac{3M}{2} + 2.$$

Since $\mathbf{W}^{-1} = \mathbf{W}^*/M$, the complexity of the inverse DFT operation is the same.[3] Counting the M multipliers $1/C[k]$ in Fig. 7.13, we have a total of

$$M\log_2 M - 2M + 4 \approx M\log_2 M$$

multiplications. From Sec. 7.2 we know that the complexity of the channel equalizer based on *zero padding* is nearly $M^2/2$ per block, which is much larger for large M (unless we use fast convolution methods in the implementation of the receiver). Recall from Sec. 7.2 that large M is desirable if we want to keep the bandwidth expansion ratio $\gamma = (M+L)/M$ small. $\bigtriangledown \bigtriangledown \bigtriangledown$

7.4 The circulant matrix representation

Equation (7.20), which was crucial in the design of the cyclic-prefix receiver, is a beautiful equation. To appreciate the significance of this in a different light, recall that cyclic prefixing with $M > L$ ensures that the operation of the channel is turned into a circular convolution. More specifically, the last M samples of $y(n)$ in each block can be regarded as the circular convolution of $c(n)$ with the M samples of $s(n)$ in a block. Written in the form of an equation we have

[3] The extra M multiplications due to the scale factor $1/M$ are binary shifts if M is a power of two.

$$\underbrace{\begin{bmatrix} \widehat{y}_0(n) \\ \widehat{y}_1(n) \\ \vdots \\ \widehat{y}_{M-1}(n) \end{bmatrix}}_{\widehat{\mathbf{y}}(n)} = \mathbf{C} \underbrace{\begin{bmatrix} s_0(n) \\ s_1(n) \\ \vdots \\ s_{M-1}(n) \end{bmatrix}}_{\mathbf{s}(n)}, \tag{7.23}$$

where $\mathbf{C}$ is a **circulant matrix** (see Appendix D). For example when $L = 3$ and $M = 4$ we have

$$\begin{bmatrix} \widehat{y}_0(n) \\ \widehat{y}_1(n) \\ \widehat{y}_2(n) \\ \widehat{y}_3(n) \end{bmatrix} = \underbrace{\begin{pmatrix} c(0) & c(3) & c(2) & c(1) \\ c(1) & c(0) & c(3) & c(2) \\ c(2) & c(1) & c(0) & c(3) \\ c(3) & c(2) & c(1) & c(0) \end{pmatrix}}_{\mathbf{C}} \begin{bmatrix} s_0(n) \\ s_1(n) \\ s_2(n) \\ s_3(n) \end{bmatrix},$$

and when $L = 3$ and $M = 6$ we have

$$\begin{bmatrix} \widehat{y}_0(n) \\ \widehat{y}_1(n) \\ \widehat{y}_2(n) \\ \widehat{y}_3(n) \\ \widehat{y}_4(n) \\ \widehat{y}_5(n) \end{bmatrix} = \underbrace{\begin{pmatrix} c(0) & 0 & 0 & c(3) & c(2) & c(1) \\ c(1) & c(0) & 0 & 0 & c(3) & c(2) \\ c(2) & c(1) & c(0) & 0 & 0 & c(3) \\ c(3) & c(2) & c(1) & c(0) & 0 & 0 \\ 0 & c(3) & c(2) & c(1) & c(0) & 0 \\ 0 & 0 & c(3) & c(2) & c(1) & c(0) \end{pmatrix}}_{\mathbf{C}} \begin{bmatrix} s_0(n) \\ s_1(n) \\ s_2(n) \\ s_3(n) \\ s_4(n) \\ s_5(n) \end{bmatrix}.$$

Note finally that if $L = M = 3$ we get

$$\begin{bmatrix} \widehat{y}_0(n) \\ \widehat{y}_1(n) \\ \widehat{y}_2(n) \end{bmatrix} = \underbrace{\begin{pmatrix} c(0)+c(3) & c(2) & c(1) \\ c(1) & c(0)+c(3) & c(2) \\ c(2) & c(1) & c(0)+c(3) \end{pmatrix}}_{\mathbf{C}} \begin{bmatrix} s_0(n) \\ s_1(n) \\ s_2(n) \end{bmatrix},$$

which is still circulant but the elements $c(0)$ and $c(3)$ are no longer separate. The restriction $M > L$ mentioned at the beginning of this section is required to avoid this.

Now, it turns out that *any circulant matrix can be diagonalized by the DFT matrix*, that is,

$$\mathbf{W}\mathbf{C}\mathbf{W}^{-1} = \mathbf{\Lambda}_c, \tag{7.24a}$$

or equivalently

$$\mathbf{C} = \mathbf{W}^{-1}\mathbf{\Lambda}_c\mathbf{W}, \tag{7.24b}$$

where

$$\mathbf{\Lambda}_c = \begin{bmatrix} C[0] & 0 & \ldots & 0 \\ 0 & C[1] & \ldots & 0 \\ \vdots & \vdots & \ddots & \vdots \\ 0 & 0 & \ldots & C[M-1], \end{bmatrix}$$

with $C[k] = \sum_{n=0}^{M-1} c(n)W^{kn}$ denoting the M-point DFT of $c(n)$. This diagonalization result is proved in Appendix D. Substituting this into Eq. (7.23) we obtain

$$\begin{bmatrix} \widehat{y}_0(n) \\ \widehat{y}_1(n) \\ \vdots \\ \widehat{y}_{M-1}(n) \end{bmatrix} = \mathbf{W}^{-1}\mathbf{\Lambda}_c\mathbf{W} \begin{bmatrix} s_0(n) \\ s_1(n) \\ \vdots \\ s_{M-1}(n) \end{bmatrix}, \tag{7.25}$$

which upon inversion yields Eq. (7.20) again! Thus, the receiver structure shown in Fig. 7.13 is a manifestation of the fact that *the channel appears to be a circulant matrix* from the viewpoint of the vectors $\mathbf{s}(n)$ and $\widehat{\mathbf{y}}(n)$. For a preview, see Fig. 7.15(a), which we shall discuss again later.

From pseudocirculants to circulants

Yet another way to see this is from Sec. 3.9. We proved therein that the channel as seen by a transmultiplexer system is a pseudocirculant. For example, when $L = 3$ and $M = 4$ we get

$$\mathbf{C}_b(z) = \begin{array}{c} \begin{array}{ccc|cccc} & L & & & & M & \end{array} \\ \left(\begin{array}{ccc|cccc} c(0) & 0 & 0 & 0 & z^{-1}c(3) & z^{-1}c(2) & z^{-1}c(1) \\ c(1) & c(0) & 0 & 0 & 0 & z^{-1}c(3) & z^{-1}c(2) \\ c(2) & c(1) & c(0) & 0 & 0 & 0 & z^{-1}c(3) \\ c(3) & c(2) & c(1) & c(0) & 0 & 0 & 0 \\ 0 & c(3) & c(2) & c(1) & c(0) & 0 & 0 \\ 0 & 0 & c(3) & c(2) & c(1) & c(0) & 0 \\ 0 & 0 & 0 & c(3) & c(2) & c(1) & c(0) \end{array}\right) \end{array}.$$

Since the receiver drops the prefix part (first L samples in each block) it effectively sees only the matrix

$$[\,\mathbf{0} \quad \mathbf{I}_M\,]\,\mathbf{C}_b(z) = \begin{array}{c} \begin{array}{ccc|cccc} & L & & & M & & \end{array} \\ \left(\begin{array}{ccc|cccc} c(3) & c(2) & c(1) & c(0) & 0 & 0 & 0 \\ 0 & c(3) & c(2) & c(1) & c(0) & 0 & 0 \\ 0 & 0 & c(3) & c(2) & c(1) & c(0) & 0 \\ 0 & 0 & 0 & c(3) & c(2) & c(1) & c(0) \end{array}\right) \end{array}.$$

That is, the received vector is

$$\begin{bmatrix} \widehat{y}_0(n) \\ \widehat{y}_1(n) \\ \widehat{y}_2(n) \\ \widehat{y}_3(n) \end{bmatrix} = \begin{array}{c} \begin{array}{ccc|cccc} & L & & & M & & \end{array} \\ \left(\begin{array}{ccc|cccc} c(3) & c(2) & c(1) & c(0) & 0 & 0 & 0 \\ 0 & c(3) & c(2) & c(1) & c(0) & 0 & 0 \\ 0 & 0 & c(3) & c(2) & c(1) & c(0) & 0 \\ 0 & 0 & 0 & c(3) & c(2) & c(1) & c(0) \end{array}\right) \end{array} \begin{bmatrix} s_1(n) \\ s_2(n) \\ s_3(n) \\ \hdashline s_0(n) \\ s_1(n) \\ s_2(n) \\ s_3(n) \end{bmatrix},$$

where the first L samples of vector of $s_k(n)$'s on the right are identical to the last L samples because of the cyclic prefixing at the transmitter. Thus we can add the first L columns of the matrix of $c(n)$'s to the last L columns and write the preceding equation more compactly as follows:

$$\begin{bmatrix} \widehat{y}_0(n) \\ \widehat{y}_1(n) \\ \widehat{y}_2(n) \\ \widehat{y}_3(n) \end{bmatrix} = \underbrace{\begin{pmatrix} c(0) & c(3) & c(2) & c(1) \\ c(1) & c(0) & c(3) & c(2) \\ c(2) & c(1) & c(0) & c(3) \\ c(3) & c(2) & c(1) & c(0) \end{pmatrix}}_{\text{circulant } \mathbf{C}} \begin{bmatrix} s_0(n) \\ s_1(n) \\ s_2(n) \\ s_3(n) \end{bmatrix}.$$

This explains in yet another way why the channel appears to be a circulant when we use a cyclic prefix of appropriate length.

Summary (cyclic-prefix systems)

1. If we divide the transmitted symbol stream $s(n)$ into blocks of length M and insert a cyclic prefix of length L at the beginning of each block, where $L \geq$ channel order and $M > L$, the effect of the channel appears to be a circular convolution, and interblock interference is eliminated.

2. Equivalently the vector $\widehat{\mathbf{y}}(n)$ is the vector $\mathbf{s}(n)$ multiplied with a circulant channel matrix $\mathbf{C}$.

3. Since circulants can be diagonalized by the DFT matrix, as in Eq. (7.24a) (or equivalently since circular convolution implies pointwise multiplication of DFT coefficients), the receiver structure can be implemented as in Fig. 7.13 with the help of a DFT, an IDFT, and frequency-domain equalizers.

Some variations of the cyclic-prefix system will be presented in the next section. Improved receivers which take into account the effect of channel noise will be presented in later chapters.

7.5 Variations of the cyclic-prefix system

We now derive some variations of the cyclic-prefix system. It will be convenient to use a diagram where blocked versions of signals are used (the reader should review Sec. 3.5 on blocking and unblocking conventions).

Figure 7.14 shows the cyclic-prefix scheme in block diagram form. In the figure, $\mathbf{p}(n)$ represents the cyclic-prefix part, which is merely a repetition of the last L entries of $\mathbf{s}(n)$. The channel noise is not shown, but will be analyzed in later chapters. The matrix $\mathbf{\Lambda}_e$ on the receiver side is a diagonal matrix representing the zero-forcing equalizer coefficients in the DFT domain, that is,

$$\mathbf{\Lambda}_e = \begin{bmatrix} 1/C[0] & 0 & \dots & 0 \\ 0 & 1/C[1] & \dots & 0 \\ \vdots & \vdots & \ddots & \vdots \\ 0 & 0 & \dots & 1/C[M-1] \end{bmatrix}.$$

We now show how to obtain other equivalent forms of the cyclic-prefix equalizer. In Fig. 7.15(a) we have shown a portion of the system that is relevant for further manipulations. This representation follows from the relation $\widehat{\mathbf{y}}(m) = \mathbf{Cs}(m)$ proved earlier (see Eq. (7.23)). The entire system in Fig. 7.15(a) is an identity system (since the channel noise is ignored). Since $\mathbf{AB} = \mathbf{I}$ implies $\mathbf{BA} = \mathbf{I}$ for two square matrices $\mathbf{A}$ and $\mathbf{B}$, it follows that the system shown in Fig. 7.15(b) is also identity. Thus the inverse DFT can be moved to the transmitter side! By the same reasoning we see that the system shown in Fig. 7.15(c) is identity as well. This shows that the entire computation can be moved to the transmitter end. This is practicable as long as the channel coefficients $C[k]$ are known to the transmitter.

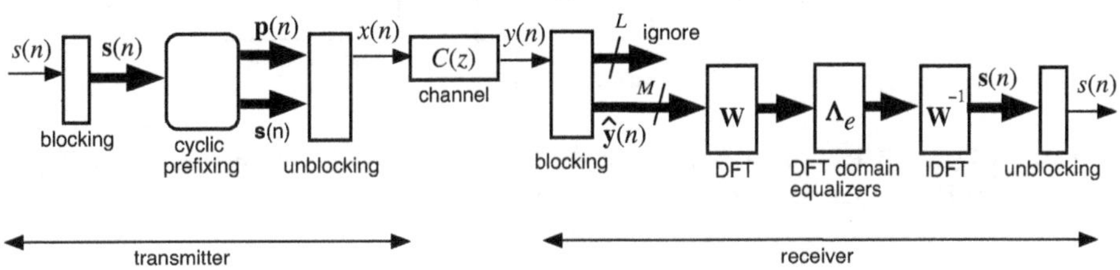

Figure 7.14. The DFT-based cyclic-prefix system in block diagram form.

The original cyclic-prefix system of Fig. 7.14 and the two modified cyclic-prefix systems resulting from these arguments are summarized in Fig. 7.16. Note that we have indicated the channel input as $x_1(n)$ in Fig. 7.16(b) and as $x_2(n)$ in Fig. 7.16(c). Similarly we have used notations like $s_1(n)$, $p_1(n)$, $s_2(n)$, $p_2(n)$ and so forth to emphasize that the internal signals are not identical in the three figures. For a given symbol stream $s(n)$, the actual signal $x(n)$ that enters the communication channel is different in the three systems shown in Figs. 7.16(a), 7.16(b), and 7.16(c). However, the working principle behind channel equalization is the same in all three cases, namely introduction of a cyclic prefix followed by *subsequent cancellation of the channel* by insertion of $1/C[k]$ somewhere.

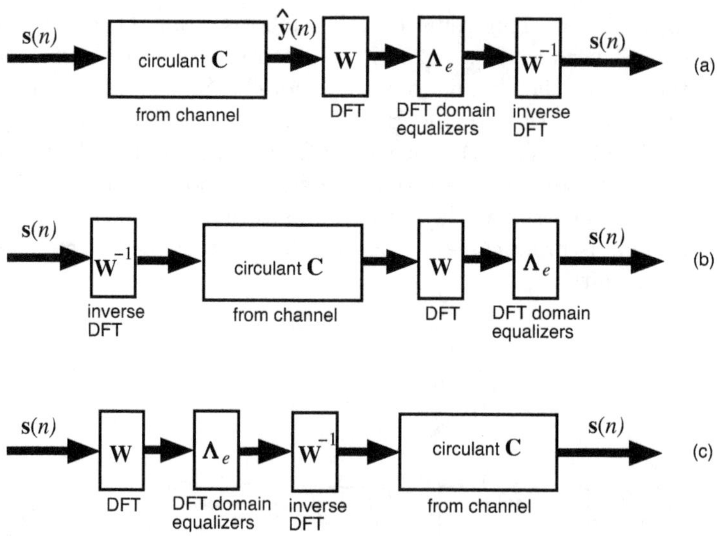

Figure 7.15. (a) A portion of the DFT based cyclic prefix system in Fig. 7.14. (b), (c) Rearrangements of the system shown in part (a)

Remarks

1. ***The multicarrier (MC) system.*** The configuration in Fig. 7.16(b) is called a multicarrier-cyclic-prefix (MC-CP) system. It is usually referred to as the discrete multitone or **DMT** system when the transmitter allows bit allocation among the components of $\mathbf{s}(n)$. Bit allocation is considered in greater detail in Chap. 14. The DMT system is widely used in DSL and ADSL services (high-speed internet via telephone lines) [Starr, Cioffi, and Silverman, 1999]. The multicarrier system is typically referred to as an orthogonal frequency division multiplexing, or **OFDM**, system when there is no bit allocation at the transmitter. The name "multicarrier" comes from the fact that the components of the vector $\mathbf{s}_1(n)$ can be regarded as bandpass signals because the IDFT system at the transmitter can be interpreted as a filter bank (Sec. 7.6).

2. ***The single-carrier (SC) system.*** By contrast, the configuration of Fig. 7.16(a) is called a *single-tone* or a *single-carrier* cyclic-prefix (SC-CP) system because there are no bandpass filters in the transmitter. These systems have considerable importance [Falconer *et al.*, 2002], [Wang, Ma, and Giannakis, 2004], and we will discuss them again in Sec. 17.6.

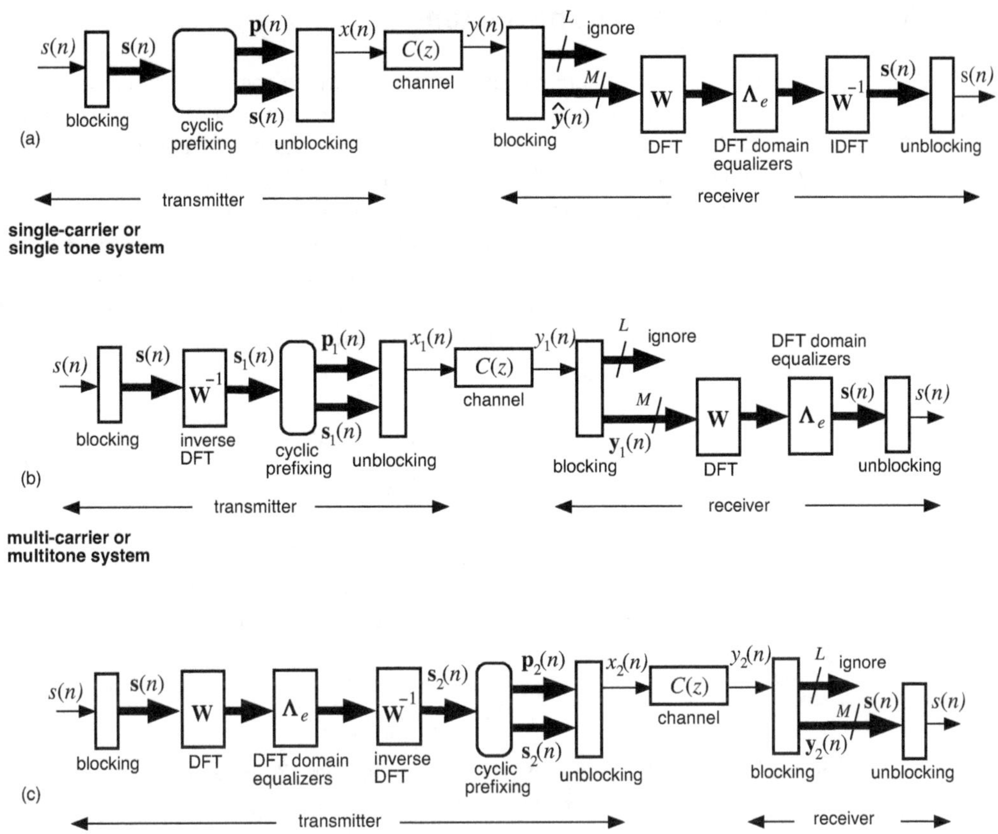

Figure 7.16. Modified forms of the cyclic-prefix system. (a) Original system (single tone system) (b) inverse DFT moved to transmitter end (multitone system), and (c) all computations moved to the transmitter side. In (c) the receiver simply retains the last M samples $\mathbf{y}_2(n)$ out of every successive set of L samples.

3. ***The peak power problem.*** The presence of the IDFT matrix $\mathbf{W}^{-1}$ in the transmitter of the DMT system introduces a new problem. The transmitted symbol stream $s(n)$ usually has a certain average power and peak power. When $s(n)$ is transformed into $x_1(n)$ using the IDFT, the maximum possible magnitude of $x_1(n)$ increases to very large values for large M. That is, the peak power input to the channel increases. This is called the peak to average power ratio problem (PAPR) and creates difficulties in the design of electronic equipment at the transmitter end. There are a number of papers which study the peak power problem and offer solutions to it by careful design of the signal constellation from which the symbol stream $s(n)$ is drawn. See, for example, Müller and Huber [1997], Jones [1999], and Kou, Lu, and Antoniou [2002]. Note that the single-tone system of Fig. 7.16(a) transmits the symbol stream $s(n)$ with prefix. Thus the *single-tone system does not suffer from the peak power problem.*

7.6 The discrete multitone system

The three transceivers shown in Fig. 7.16 can be regarded as communicating either a scalar symbol stream $s(n)$ or a vector signal $\mathbf{s}(n)$ whose components $s_k(n)$ come from different users. Let us examine again the DMT system of Fig. 7.16(b), shown in Fig. 7.17 in greater detail. The subscripts on the signals shown earlier in Fig. 7.16(b) have been dropped for simplicity.

Further details on the transmitter and receiver end are shown in Fig. 7.18, where the blocking and unblocking operations are shown explicitly using multirate signal processing language (Chap. 3). Here the matrix or precoder $\mathbf{R}$ in the transmitter performs the inverse DFT and prefixing, that is,

$$\mathbf{R} = \begin{array}{c} \\ L \\ M \end{array} \overset{M}{\begin{pmatrix} \mathbf{R}_{cyc} \\ \mathbf{W}^{-1} \end{pmatrix}}, \tag{7.26}$$

where $\mathbf{R}_{cyc}$ is a copy of the last L rows of $\mathbf{W}^{-1}$. We now see that the transceiver can be regarded as a redundant transmultiplexer in polyphase form (Sec. 3.9) with transmitting polyphase matrix $\mathbf{R}$ as above and receiving polyphase matrix

$$\mathbf{E} = M \overset{\;\;L \quad\; M}{\begin{pmatrix} \mathbf{0} & \mathbf{W} \end{pmatrix}}. \tag{7.27}$$

This system can therefore be redrawn as in Fig. 7.19, where the transmitting and receiving filters are, respectively,

$$[F_0(z) \quad F_1(z) \quad \ldots \quad F_{M-1}(z)] = \mathbf{e}_P^T(z) \begin{bmatrix} \mathbf{R}_{cyc} \\ \mathbf{W}^{-1} \end{bmatrix} \tag{7.28}$$

and

$$\begin{bmatrix} H_0(z) \\ H_1(z) \\ \vdots \\ H_{M-1}(z) \end{bmatrix} = \mathbf{E}\mathbf{e}_P(z^{-1}) = z^L \mathbf{W}\mathbf{e}_M(z^{-1}), \tag{7.29}$$

where $\mathbf{e}_K(z)$ is the delay chain vector, that is,

$$\mathbf{e}_K^T(z) = [1 \quad z^{-1} \quad \ldots \quad z^{-(K-1)}].$$

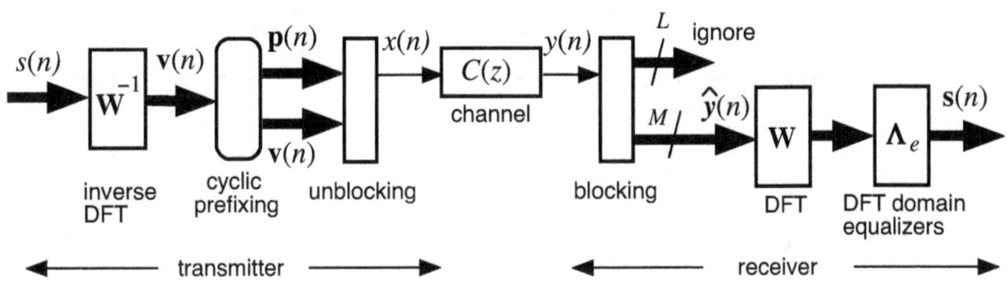

Figure 7.17. The DMT system for communicating a vector $\mathbf{s}(n)$ on a scalar channel $C(z)$.

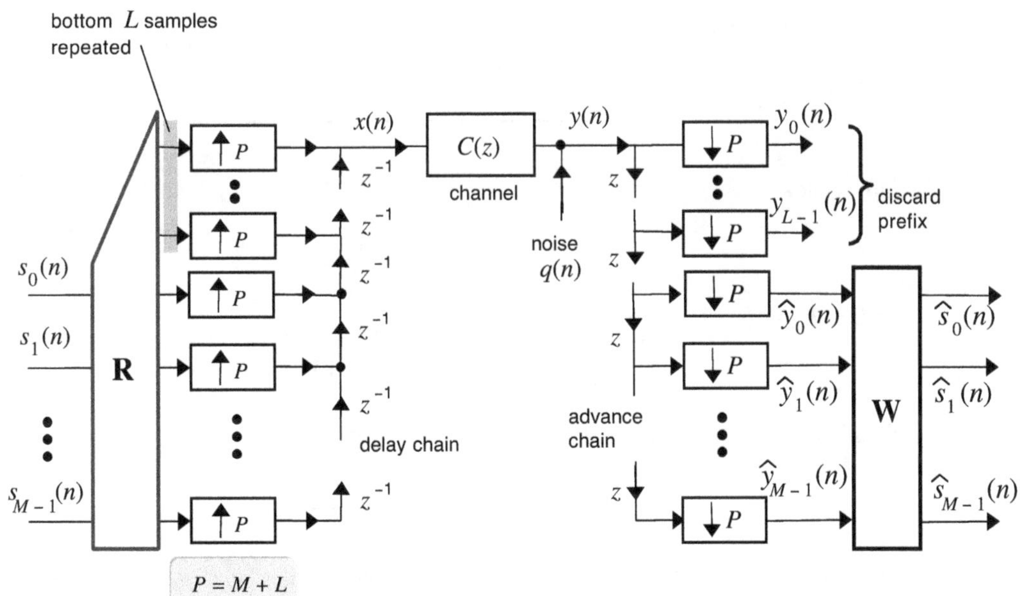

Figure 7.18. The DMT system redrawn in terms of multirate building blocks.

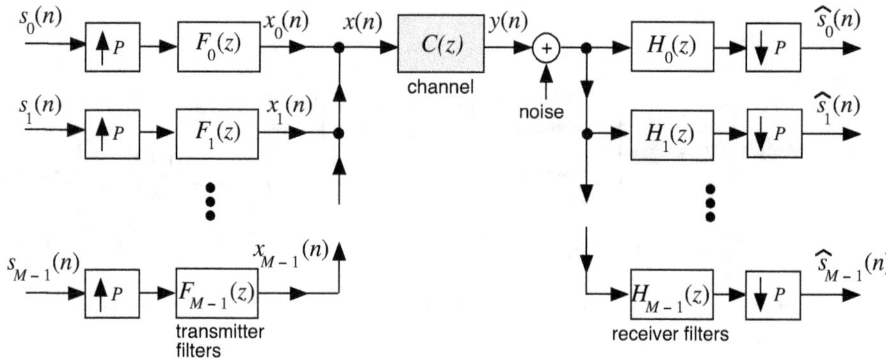

Figure 7.19. The DMT system redrawn in terms of a transmitting filter bank and a receiving filter bank.

We can write the receiving filters more explicitly as follows:

$$\begin{bmatrix} H_0(z) \\ H_1(z) \\ H_2(z) \\ \vdots \\ H_{M-1}(z) \end{bmatrix} = z^L \underbrace{\begin{bmatrix} 1 & 1 & 1 & \dots & 1 \\ 1 & W & W^2 & \dots & W^{M-1} \\ 1 & W^2 & W^4 & \dots & W^{2(M-1)} \\ \vdots & \vdots & \vdots & \ddots & \vdots \\ 1 & W^{M-1} & W^{2(M-1)} & \dots & W^{(M-1)^2} \end{bmatrix}}_{\mathbf{W}} \begin{bmatrix} 1 \\ z \\ z^2 \\ \vdots \\ z^{(M-1)} \end{bmatrix}.$$

This yields

$$H_k(z) = z^L \sum_{n=0}^{M-1} z^n W^{kn} \tag{7.30}$$

for $0 \le k \le M-1$. Note in particular that

$$H_0(z) = z^L(1 + z + z^2 + \dots z^{M-1}).$$

This is a linear phase filter with frequency response

$$H_0(e^{j\omega}) = e^{j\omega K}\frac{\sin(\omega M/2)}{\sin(\omega/2)},$$

where $K = L + 0.5(M-1)$. The magnitude is shown in Fig. 7.20(a) and we see that this is a very simple lowpass filter with about 13 dB difference between the passband maximum and first stopband peak. From (7.30) we see that

$$|H_k(e^{j\omega})| = |H_0(e^{j(\omega - \frac{2\pi k}{M})})|,$$

which shows that the filter responses are uniformly shifted versions of the **prototype** response $|H_0(e^{j\omega})|$. This is demonstrated in Fig. 7.20(b) for $H_1(z)$, and in Fig. 7.20(c) for all M filters (for $M = 8$).

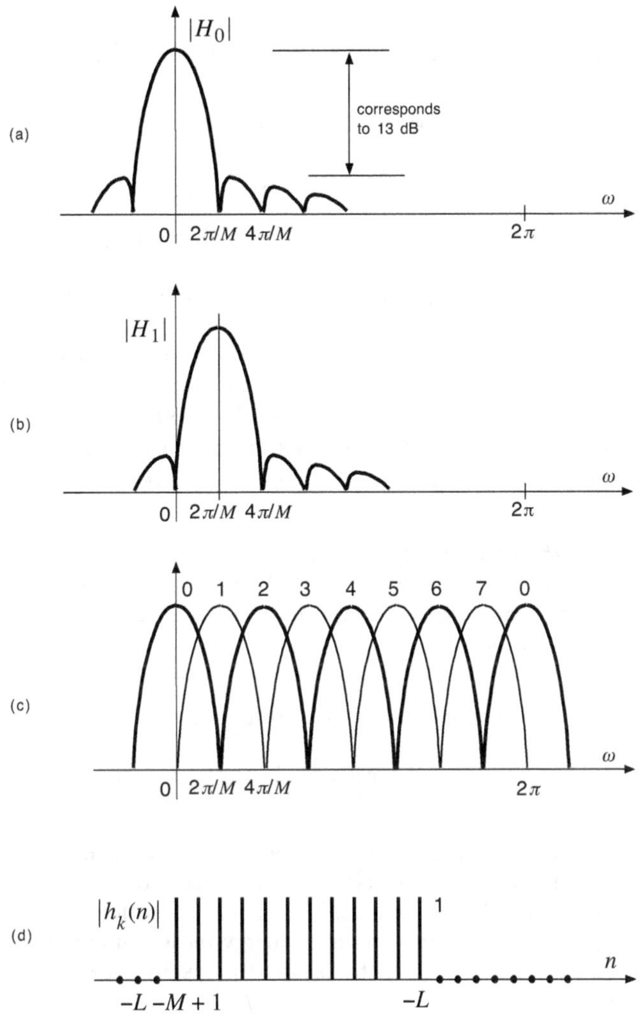

Figure 7.20. Responses of the receiving filters in the DMT system. (a) Prototype magnitude $|H_0(e^{j\omega})|$, (b) magnitude of $H_1(e^{j\omega})$, (c) magnitudes of all filters with sidelobes suppressed for clarity, and (d) impulse response magnitudes.

Because of the uniformly shifted nature of the filters, this system is called the **uniform DFT filter bank**. All the receiving filters $H_k(z)$ are FIR with length M, and their impulse responses $h_k(n)$ are nonzero only in

$$-L - M + 1 \leq n \leq -L.$$

All the impulse responses have unit magnitude in this region as shown in Fig. 7.20(d). With proper delays inserted, the filters can be adjusted to the causal form

$$z^{-(L+M-1)}H_k(z) = W^{-k}\sum_{n=0}^{M-1} W^{-kn}z^{-n}.$$

The impulse responses of these causal forms are given by

$$h_k^c(n) = W^{-k}W^{-kn}, \quad 0 \le n \le M-1.$$

Thus, all the filters are related to the 0th filter $h_0^c(n)$ according to the relation

$$h_k^c(n) = W^{-k}h_0^c(n)W^{-kn} = W^{-k}h_0^c(n)e^{j2\pi kn/M}.$$

That is, the filters are simple modulated versions of the prototype $H_0(z)$. Even though the figure shows that $h_k(n)$ have constant magnitudes in their regions of support, these are bandpass filters. This can be better appreciated by looking at the real parts of $W^k h_k^c(n)$, demonstrated in Fig. 7.21 for $k = 0, 1, 2$. These are cosines of increasing frequencies as k increases.

Why the name multitone? The transmitting filters are also FIR but they have order P as seen from Eq. (7.28). If $M >> L$ then it can be verified that the kth transmitting and receiving filters have approximately the same magnitude response, that is,

$$|F_k(e^{j\omega})| \approx |H_k(e^{j\omega})|. \tag{7.31}$$

Each transmitting filter is an interpolation filter and produces the output

$$x_k(n) = \sum_m s_k(m)f_k(n - mP). \tag{7.32}$$

Thus the symbols $x_k(n)$ are interpolated with the pulse shapes $f_k(n)$. Each transmitting filter imposes its own pulse stream onto its input $s_k(n)$. The energy of the pulse $f_k(n)$ is dominantly in the kth **frequency** band. Each pulse $f_k(n)$ can be regarded as a *tone*, with its energy mostly in the neighborhood of a single frequency $\omega_k = k2\pi/M$. This is why the system is called a *multitone system* [Kalet, 1989].

The DFT matrix has the property that its rows are **orthogonal**, that is,

$$\mathbf{W}\mathbf{W}^\dagger = M\mathbf{I}. \tag{7.33}$$

Similarly, the columns of $\mathbf{W}^{-1}$ are mutually orthogonal. Notice from (7.28) that the columns of $\mathbf{W}^{-1}$ are the coefficients $f_k(n)$ of the transmitting filters $F_k(z)$ (ignoring the prefix part for a moment). Thus

$$\sum_n f_k(n)f_m^*(n) \approx M\delta(k-m),$$

where the approximate nature of the relation arises from the fact that the prefix part has been ignored. To the extent of this approximation we can say that the symbol streams $s_k(n)$ are modulated by signals $f_k(n)$ which are orthogonal to each other. Since the kth orthogonal component is concentrated around the kth frequency $2\pi k/M$, the DMT system therefore falls under the category of systems called *orthogonal frequency division multiplexing*, or **OFDM**, systems.

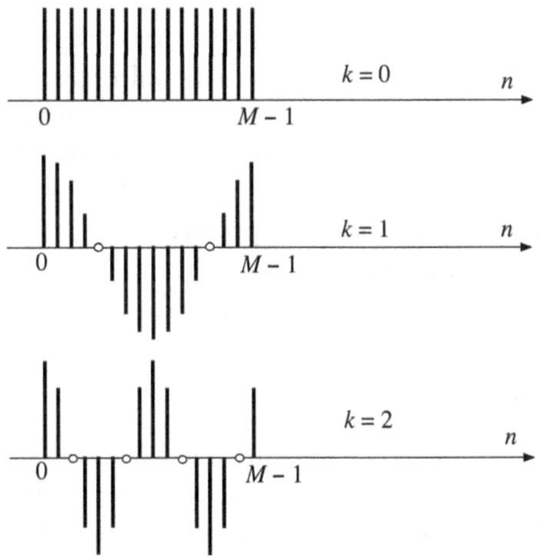

Figure 7.21. Real parts of $W^k h_k^c(n)$ for $k = 0, 1, 2$.

Historically, the early versions of the multitone system used continuous-time modulators (multiplications with $\cos\omega_k t$ and $\sin\omega_k t$) to produce multiple carriers. But in digital multitone systems these are performed in discrete time using the IDFT matrix at the transceiver, and then converting the single multitone signal into a continuous-time signal (using D/C converters). The receiver samples the output of the continuous-time channel, and then uses the DFT matrix. All the signal processing is therefore digital. $\bigtriangledown \bigtriangledown \bigtriangledown$

7.7 Concluding remarks

In this chapter we studied the idea of introducing redundancy at the transmitter of a digital communication system. We showed that this gives rise to a block-based system with no inter-block interference. Summarizing some of the main points of the chapter, if a channel $C(z)$ is representable by an Lth order FIR filter we can equalize it by using $1/C(z)$, which is an IIR equalizer. To avoid an IIR filter (which can be unstable) we can use either a zero-padding or a cyclic-prefix based system. The zero-padding scheme requires receiver complexity proportional to M^2, where M is the block length. The cyclic-prefix scheme can be implemented using FFT, and requires complexity proportional to $M \log_2 M$ [Oppenheim and Schafer, 1999]. The zero-padding receiver can also be implemented by using an FFT based fast convolution algorithm for Eq. (7.8). These fast implementations are especially important for large M, which is desirable in order to reduce the bandwidth expansion ratio $(M + L)/M$.

At the end of Sec. 7.2 we saw that the zero-padding transceiver system is a special case of the redundant transmultiplexer system. This is clear from a comparison of Fig. 7.6 with the transmultiplexer in polyphase form (Fig. 7.7). The cyclic-prefix system is also a special case of the redundant transmultiplexer, as seen from Fig. 7.18. The channel-equalization methods introduced in this chapter are therefore both special cases of the transmultiplexer configuration.

In Chap. 4 we described communication systems which use oversampling at the receiver. Such receivers use **fractionally spaced equalizers**, and their advantages were explained in Sec. 4.8. A natural question then would be this: can we use oversampling receivers for the case where the transmitter introduces redundancy as we have done in this chapter? This is indeed possible. For example, the use of **oversampling in cyclic-prefix receivers** results in certain simplifications. One of these is that the inverse $1/C[k]$ no longer appears at the receiver. For more details the interested reader is referred to Problems 7.25 and 7.26.

Historical remarks

With our background on transmultiplexers in Secs. 3.8–3.9 and DMT systems in Sec. 7.6, this is a nice place to recall some of the history behind OFDM systems. From Sec. 3.8 we know that we can multiplex several users by separating them in the frequency domain using filter banks with good filter responses. This requires guard bands, as shown earlier in Fig. 3.27, and therefore results in inefficient use of bandwidth. In 1966, Chang introduced the idea of **orthogonal** frequency division multiplexing (OFDM) whereby one can introduce orthogonality between modulating signals. This was then used to separate the different messages; there is no need for the messages to be overlap-free in frequency [Chang, 1966]. This idea was further developed by other authors, and Cimini [1985] showed application in digital mobile communications; he used the DFT and its fast computation using the FFT to perform the orthogonal multiplexing efficiently. The orthogonality property is reflected in the unitarity of the DFT matrix shown in Eq. (7.33).

The cyclic-prefix idea was introduced by Peled and Ruiz [1980]. The multitone channel was introduced by Kalet [1989], who showed that if the frequency range of the channel can be partitioned into subbands, with each subband carrying a different constellation with different power, this gives added flexibility for power and bit allocation. Later on, Sari, Karam and Jeanclaude [1994] analyzed the OFDM system for mobile radio and showed that the single-carrier cyclic-prefix (SC-CP) system actually outperforms the multicarrier (OFDM) system from the point of view of bit error rate. In Chaps. 16 and 17 we shall study the theory of this and provide a formal justification. The discrete multitone (DMT) transceiver with bit allocation, for DSL applications on telephone lines, was proposed in 1991 in the landmark paper by Chow, Tu, and Cioffi [1991]. Many interesting variations and extensions have since been proposed [Rizos, Proakis, and Nguyen, 1994], [Tzannes, et al., 1994], [Sandberg and Tzannes, 1995]. The success of DSL systems and their variations is now history. Optimal transceivers with bit allocation will be discussed in Chap. 14.

A variation of OFDM systems called OFDM/OQAM (OFDM with offset-QAM) has been of great interest in recent years. To understand what OQAM means, recall from Sec. 2.4.3 that a QAM signal has the representation

$$x_{QAM}(t) = \sum_n s_c(n)p(t-nT)\cos\omega_c t - \sum_n s_s(n)p(t-nT)\sin\omega_c t, \qquad (7.34)$$

or equivalently

$$x_{QAM}(t) = \text{Re}\Big[\sum_n \Big(s_c(n) + js_s(n)\Big)p(t - nT)e^{j\omega_c t}\Big], \tag{7.35}$$

where $s_c(n)$ and $s_s(n)$ are the real symbol streams defining the complex QAM stream $s_c(n) + js_s(n)$, and $p(t)$ is the real baseband pulse. Thus, the samples of the complex symbol stream

$$s(n) = s_c(n) + js_s(n) \tag{7.36}$$

are separated by T seconds, and each sample has the baseband waveform $p(t - nT)$ around it. In a staggered QAM, or **SQAM**, system, the real and imaginary parts are staggered in time, and the signal has the form

$$x_{SQAM}(t) = \sum_n s_c(n)p(t - nT)\cos\omega_c t - \sum_n s_s(n)p(t - nT - t_0)\sin\omega_c t \tag{7.37}$$

that is,

$$x_{SQAM}(t) = \text{Re}\Big[\sum_n \Big(s_c(n)p(t - nT) + js_s(n)p(t - nT - t_0)\Big)e^{j\omega_c t}\Big], \tag{7.38}$$

where t_0 is the amount of stagger, with $0 \le t_0 < T$. Thus, the real and imaginary parts $s_c(n)$ and $s_s(n)$ are not overlapping in time. Instead, they are staggered by an amount $t_0 < T$ (Fig. 7.22). The special case where $t_0 = T/2$ is usually known as the offset QAM system and abbreviated as **OQAM**.[4] The history of OQAM systems is fairly long, some of the pioneering papers being Saltzberg [1967], Gitlin and Ho [1975], and Hirosaki [1980]. Gitlin and Ho have shown that the receiver in an OQAM system is more robust to error in the carrier phase.

When the OQAM signal is passed through a precoding filter, the real and imaginary samples get mixed up in general, and the output is a sequence of complex numbers spaced apart by $T/2$ seconds. This increases the channel bandwidth required to communicate the signal (the signal rate being still only $1/T$ complex samples per second). This bandwidth inefficiency is avoided in the OFDM version of the OQAM [Hirosaki, 1980], [Hirosaki, 1981], [Vangelista and Laurenti, 2001]. The main advantage of OFDM/OQAM systems as compared to OFDM/QAM systems is that the former provide more freedom for shaping the transmitter pulse. Optimization of the transmitter pulse is crucial in time varying channels where it is desirable to concentrate the pulse in sufficiently narrow time-frequency bins. See Le Floch, Alard, and Berrou [1995], Vahlin and Holte [1996], Haas and Belfiore [1997], and references therein. Optimum decision feedback equalizers for OQAM systems are discussed in Tu [1993]. Another important advantage of OFDM/OQAM is that it does not require a guard interval like cyclic prefix, so there is no bandwidth expansion unlike in cyclic-prefixed OFDM/QAM. See Lin *et al.* [2009] and references therein.

---

[4]Sometimes OQAM is interpreted as *orthogonal* QAM.

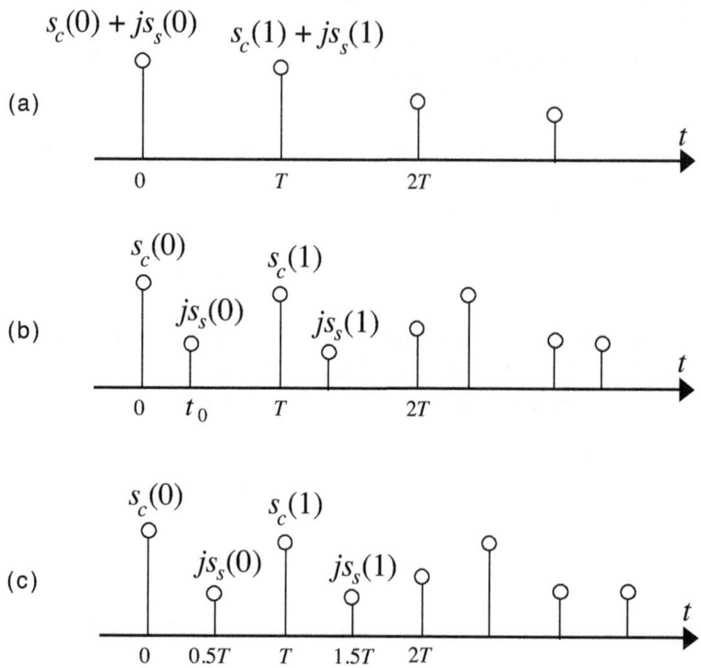

Figure 7.22. (a) The conventional QAM symbol stream, (b) the symbol stream for the SQAM signal, and (c) the symbol stream for the OQAM signal.

There is a fundamental relation between OFDM/OQAM systems and perfect reconstruction filter banks [Bölcskei, Duhamel, and Hleiss, 1999], [Siohan, Siclet, and Lacaille, 2002]. It has been shown by Siohan *et al.* [2002] that an M band OFDM/OQAM system can be represented using a cosine modulated filter bank in the transmultiplexer configuration [Vaidyanathan, 1993], and that the perfect equalization constraint can be related mathematically to the perfect reconstruction condition in a maximally decimated cosine modulated filter bank.

Problems

7.1. Let the channel transfer function be $C(z) = 1 + 2z^{-1}$, so we have $L = 1$. For a zero-padding system with block size $M = 4$, let $\mathbf{A}$ be the 5×4 matrix of the coefficients $c(n)$ in Eq. (7.5), and let $\mathbf{C}$ be the 4×4 matrix in Eq. (7.6). Write down these matrices explicitly.

7.2. In Problem 7.1 what is the bandwidth expansion ratio created by zero padding? What are the values of this ratio if (a) M is increased to 16, and (b) M is increased to 128.

7.3. Let $C(z)$ be as in Problem 7.1. For a zero-padding system with block size $M = 8$, let $\mathbf{A}$ be the 9×8 matrix of the coefficients $c(n)$ in Eq. (7.5), and let $\mathbf{C}$ be the 8×8 matrix in Eq. (7.6). Use a computer to do the following:

1. Compute the inverse $\mathbf{C}^{-1}$. What is the largest magnitude among the elements $[\mathbf{C}^{-1}]_{km}$ in the inverse?
2. Compute the left-inverse $\mathbf{A}^{\#}$ (Sec. C.2, Appendix C) of the matrix $\mathbf{A}$. What is the largest magnitude among the elements $[\mathbf{A}^{\#}]_{km}$ in the inverse?

So we see that, by adding one more row to the matrix $\mathbf{C}$ to form $\mathbf{A}$, we have significantly reduced the size of the largest-magnitude element in the inverse! This reduces the amplification of channel noise, as we shall see in Chap. 8.

7.4. Let the channel transfer function be $C(z) = 2 - 3z^{-1}$.

1. Write down the 4×4 circulant matrix $\mathbf{C}$ which represents the equivalent channel in a cyclic-prefix system with block size $M = 4$.
2. What are the eigenvalues of the matrix $\mathbf{C}$?
3. Write down a set of M linearly independent eigenvectors for the matrix $\mathbf{C}$.

7.5. Consider again the channel in Problem 7.3 and consider cyclic prefixing with block size $M = 8$. Now let $\mathbf{C}$ denote the 8×8 circulant matrix representing the channel. Using a computer, calculate the inverse $\mathbf{C}^{-1}$ and identify the largest magnitude among the elements $[\mathbf{C}^{-1}]_{km}$ in the inverse. How does this compare with the largest magnitudes computed in Problem 7.3?

7.6. *Filter bank representations of transceivers.* Consider Fig. 7.7 and assume $\widehat{s}_k(n)$ are interleaved to produce $\widehat{s}(n)$ as in Fig. 7.6(b).

1. Show that this system can be drawn using filter bank notation as in Fig. P7.6(a). The coefficients of the filters $\{A_k(z)\}$ can be regarded as a code for the symbols $s(n)$ before they are transmitted.
2. Show that the system can also be redrawn in the form shown in Fig. P7.6(b). Now the coefficients of $\{F_k(z)\}$ are regarded as the code for the symbols $s(n)$.

Thus, a redundant transceiver system with bandwidth expansion factor P/M can be represented using either of these two configurations.

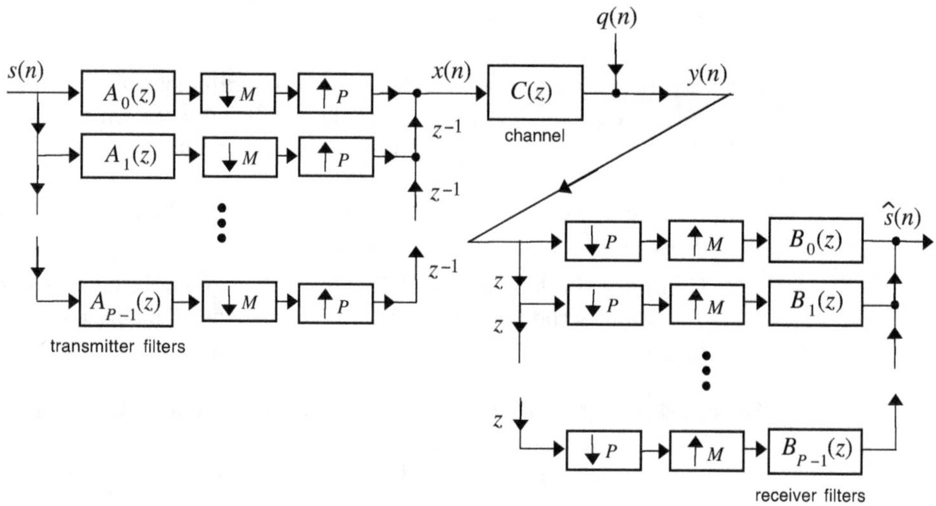

Figure P7.6(a).

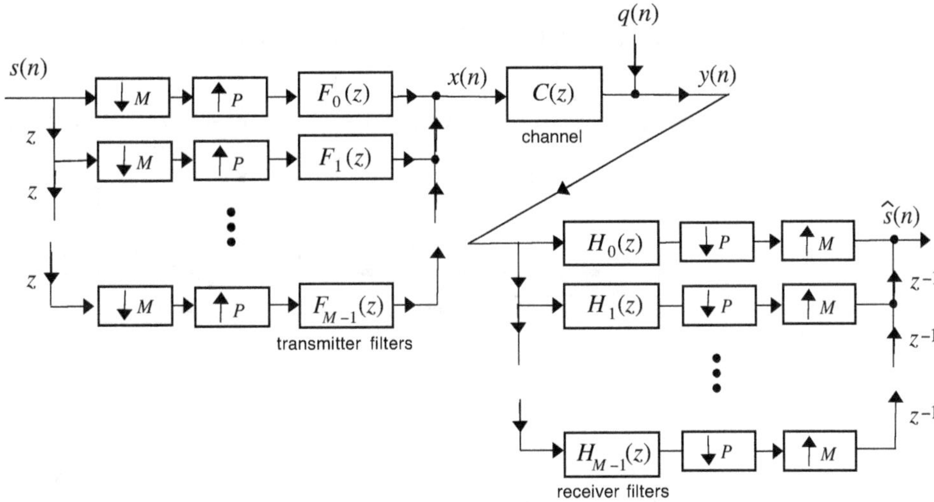

Figure P7.6(b).

7.7. *Multiuser systems.* Figure P7.7 shows a multiuser system where the kth user sends a signal $s_k(n)$ (modulated by the transmitting filter $F_k(z)$) through a channel $C_k(z)$. You can imagine that each user uses a single antenna to transmit the waveform. The signals from all M users are collected by a receiver with a single antenna. The receiver uses M filters $H_k(z)$ to separate out the different users. Let the filters be expressed in the polyphase forms

$$F_m(z) = \sum_{i=0}^{P-1} z^{-i} R_{i,m}(z^P), \; H_k(z) = \sum_{i=0}^{P-1} z^{i} E_{k,i}(z^P). \tag{P7.7}$$

Using the pseudocirculant theory of Sec. 3.9, show that the path from $s_m(n)$ to $\widehat{s}_k(n)$ is an LTI system with transfer function

$$T_{km}(z) = \begin{bmatrix} E_{k,0}(z) & E_{k,1}(z) & \ldots & E_{k,P-1}(z) \end{bmatrix} \mathbf{C}_m(z) \begin{bmatrix} R_{0,m}(z) \\ R_{1,m}(z) \\ \vdots \\ R_{P-1,m}(z) \end{bmatrix},$$

where $\mathbf{C}_m(z)$ is the $P \times P$ pseudocirculant (blocked version) associated with $C_m(z)$. From this expression we see that the system in general suffers from *multiuser interference* (MUI) because $\widehat{s}_k(n)$ can be affected by $s_m(\ell), m \neq k$. There is also *intersymbol interference* (ISI) because $\widehat{s}_k(n)$ can be affected by $s_k(\ell), \ell \neq n$ (more generally by $s_m(\ell), \ell \neq n$).

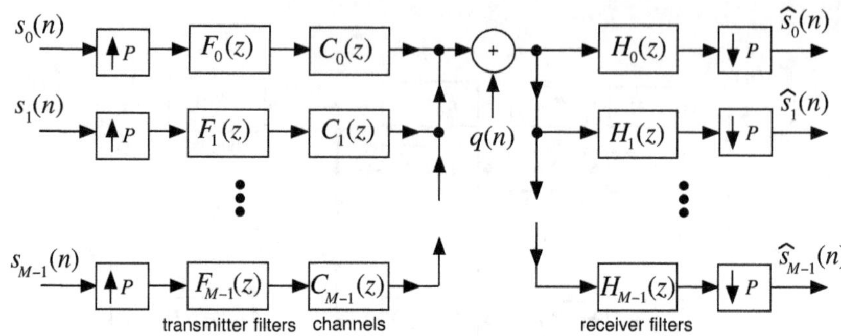

Figure P7.7.

7.8. In Problem 7.7 assume that the channels are FIR:

$$C_m(z) = \sum_{n=0}^{L} c_m(n) z^{-n}.$$

You will show how ISI can be eliminated by appropriate choice of the integer P, and the transmitting and receiving filters. Let $P = M + L$, and assume the filters are expressed in polyphase form (P7.7).

1. Show that if the last L polyphase components of $F_m(z)$ are equal to zero, then the transfer function $T_{km}(z)$ of Problem 7.7 reduces to

$$T_{km}(z) = [\, E_{k,0}(z) \quad E_{k,1}(z) \quad \dots \quad E_{k,P-1}(z) \,] \, \mathbf{A}_m \begin{bmatrix} R_{0,m}(z) \\ R_{1,m}(z) \\ \vdots \\ R_{P-L-1,m}(z) \end{bmatrix},$$

where $\mathbf{A}_m$ is a $P \times M$ full banded Toeplitz matrix:

$$\mathbf{A}_m = \begin{bmatrix} c_m(0) & 0 & \dots & 0 \\ c_m(1) & c_m(0) & \dots & 0 \\ \vdots & \vdots & \ddots & \vdots \\ c_m(L) & & & \\ 0 & c_m(L) & & \\ \vdots & & \ddots & \vdots \\ 0 & 0 & \dots & c_m(L) \end{bmatrix}.$$

Restricting the polyphase components like this is equivalent to *zero padding* at the transmitter, as demonstrated in Fig. P7.8.

2. Furthermore suppose that the filters $F_m(z)$ and $H_k(z)$ are causal and anticausal FIR respectively, with lengths $\leq P$. Show that *ISI is eliminated*, that is, $\widehat{s}_k(n)$ does not depend on $s_m(\ell), \ell \neq n$.

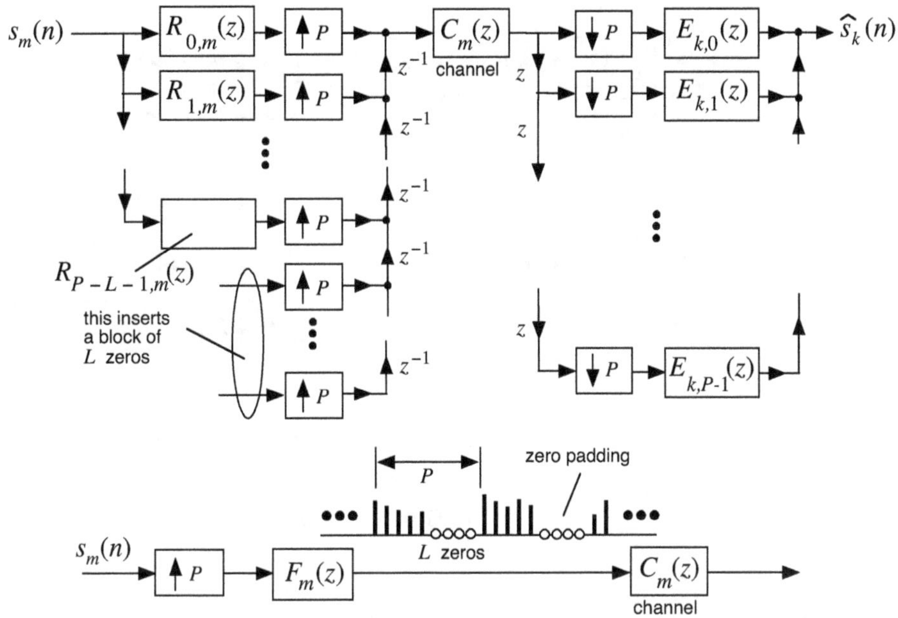

Figure P7.8.

7.9. Zero jamming. In Problem 7.8 we eliminated ISI by restricting the last L polyphase components of $F_m(z)$ to be zero (zero padding). Instead of doing this, a second approach would be to restrict the first L polyphase components of $H_k(z)$ to be zero. Show that this *eliminates ISI* if, furthermore, the filters $F_m(z)$ and $H_k(z)$ are causal and anticausal FIR respectively, with lengths $\leq P$. This is called the *zero-jamming* (rather than the zero-padding) approach. It can be implemented by using the structure shown in Fig. P7.9.

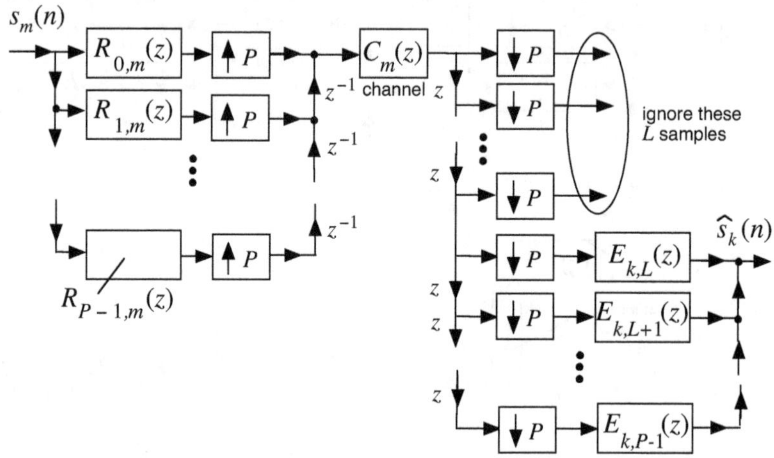

Figure P7.9.

7.10. In Problem 7.8 we showed how ISI can be eliminated in a multiuser system by zero padding when the filter lengths are restricted to be $\leq P$, where $P = M+L$ and L is the order of the FIR channel $C_m(z)$. We now show how these restricted filters can be designed such that *multiuser interference* is also eliminated. Let $\mathbf{A}_m$ be the $P \times M$ banded Toeplitz matrix given in Problem 7.8.

1. For any nonzero number ρ_k, show that

$$\begin{bmatrix} 1 & \rho_k^{-1} & \ldots & \rho_k^{-(P-1)} \end{bmatrix} \mathbf{A}_m = C_m(\rho_k) \begin{bmatrix} 1 & \rho_k^{-1} & \ldots & \rho_k^{-(M-1)} \end{bmatrix},$$

where $C_m(z) = \sum_{n=0}^{L} c_m(n) z^{-n}$. In other words, any exponential row vector is like a left-eigenvector of $\mathbf{A}_m$ (except that the vectors on the two sides have different sizes).

2. Using this, show that if the transmitting and receiving filters are

$$F_m(z) = r_{0,m} + r_{1,m} z^{-1} + \ldots + r_{M-1,m} z^{-(M-1)},$$

$$H_k(z) = a_k(1 + \rho_k^{-1} z + \rho_k^{-2} z^2 + \ldots + \rho_k^{-(P-1)} z^{(P-1)}),$$

then the transfer function from $s_m(n)$ to $\widehat{s}_k(n)$ is given by

$$T_{km}(z) = a_k C_m(\rho_k) F_m(\rho_k). \qquad \text{(P7.10}a\text{)}$$

3. Given M distinct numbers $\rho_k, 0 \leq k \leq M-1$, write a set of M linear equations from which we can solve for the M nonzero coefficients of $F_m(z)$ such that

$$F_m(\rho_k) = \delta(k-m), \quad 0 \leq k \leq M-1.$$

This will eliminate multiuser interference by forcing $T_{km} = 0$ for $k \neq m$.

4. Show that the M equations in part 3 are linearly independent.

With the above design, we have $T_{kk}(z) = a_k C_k(\rho_k)$, which is independent of z. So ISI is also eliminated. If the numbers ρ_k are chosen such that $C_k(\rho_k) \neq 0$, we can force $T_{kk}(z) = 1$. Thus, by choosing an arbitrary set of M distinct nonzero numbers ρ_k (one for each user) with the only restriction that $C_k(\rho_k) \neq 0$, then defining the receiver filters $H_k(z)$ using these ρ_k, and setting

$$a_k = \frac{1}{C_k(\rho_k)} \tag{P7.10b}$$

and choosing the transmitting filters $F_m(z)$ as described above, we can completely eliminate multiuser interference and ISI. Notice that cancellation of MUI does not even require a knowledge of the channels! This method was pioneered in a paper by Giannakis *et al.* [2000].

7.11. *DFT matrices.* In Sec. 7.3.2 we came across the DFT matrix, which is central to the implementation of cyclic-prefix based systems. In this and the following several problems we study the DFT and its properties. (All matrix notations are as in Sec. 1.6.) As mentioned in the chapter, the $M \times M$ DFT matrix $\mathbf{W}$ has elements $[\mathbf{W}]_{k\ell} = W^{k\ell}$, where $W = e^{-j2\pi/M}$. Note that $W^M = 1$, so W is called an Mth root of unity. There are M distinct Mth roots of unity, namely $W^k, 0 \leq k \leq M-1$.

1. Verify that $\mathbf{W}^T = \mathbf{W}$ (i.e., $\mathbf{W}^\dagger = \mathbf{W}^*$).
2. Show that

$$\sum_{m=0}^{M-1} W^{km} = \begin{cases} M & \text{if } k \text{ is a multiple of } M \\ 0 & \text{otherwise.} \end{cases}$$

3. Using this, show that $\mathbf{W}^\dagger\mathbf{W} = M\mathbf{I}$, that is, $\mathbf{W}/\sqrt{M}$ is a unitary matrix.
4. Show that $\mathbf{W}^{-1} = \mathbf{W}^*/M$.

7.12. Let $\mathbf{W}$ denote the $M \times M$ DFT matrix. In this problem we consider integer powers of this matrix.

1. Show that the squared matrix $\mathbf{W}^2$ has the simple form

$$\mathbf{W}^2 = M \times \begin{bmatrix} 1 & \mathbf{0} \\ \mathbf{0} & \mathbf{J}_{M-1} \end{bmatrix},$$

where $\mathbf{J}_{M-1}$ is the $(M-1) \times (M-1)$ reversal matrix. For example, if $M = 4$,

$$\mathbf{W}^2 = 4 \begin{bmatrix} 1 & 0 & 0 & 0 \\ 0 & 0 & 0 & 1 \\ 0 & 0 & 1 & 0 \\ 0 & 1 & 0 & 0 \end{bmatrix}.$$

2. Show that $\mathbf{W}^4 = M^2\mathbf{I}$. Based on this, argue that if λ is an eigenvalue of $\mathbf{W}/\sqrt{M}$, then it has to be one of the following four numbers: $1, -1, j$, or $-j$.
3. Given an M-point sequence, suppose we compute the DFT of the DFT of the DFT of the DFT of $x(n)$. What is the answer?

7.13. Let $X[k]$ be the M-point DFT of $x(n)$. Show that

$$\sum_{k=0}^{M-1} |X[k]|^2 = M \sum_{n=0}^{M-1} |x(n)|^2.$$

This is called the Parseval relation for DFT.

7.14. Let $x(n)$ be an M-point signal ($M > 1$) such that it is its own DFT, that is, $X[n] = x(n)$ for $0 \le n \le M-1$. Then what can you say about $x(n)$?

7.15. In DFT theory the notation $((n))_M$ is often used. This means that n should be interpreted modulo M. So $((n))_M$ is the remainder of n divided by M, and has values in the range $[0, M-1]$. The subscript is deleted if it is clear from the context. For example, if $M = 4$ then $((4)) = 0, ((5)) = 1, ((-1)) = 3$, and so on. In what follows $X[k]$ is the M-point DFT of $x(n)$. We say that $x(n)$ is even if $x((-n)) = x(n)$ (and odd if $x((-n)) = -x(n)$). Similarly $X[k]$ is even if $X[((-k))] = X[k]$. Prove the following:

1. $X[((-k))]$ is the M-point DFT of $x((-n))$.
2. If $x(n)$ is real, then $X[k] = X^*[((-k))]$.
3. If $x(n) = x^*((-n))$ then $X[k]$ is real.
4. If $x(n)$ is real and even, then $X[k]$ is real and even.

7.16. *Circular convolutions.* Given two sequences $x(n)$ and $h(n)$ for $0 \le n \le M-1$, their circular convolution is an M-point sequence defined by

$$y(n) = (x \otimes h)(n) = \sum_{m=0}^{M-1} x(m)h((n-m)), \qquad 0 \le n \le M-1,$$

where $((n))$ is as defined in Problem 7.15. Show that $Y[k] = X[k]H[k]$. Thus circular convolution in the time domain is equivalent to multiplication in the DFT domain. From this it also follows that circular convolution is commutative, that is, $(x \otimes h)(n) = (h \otimes x)(n)$.

7.17. Let $x(n) = \{0, 1, 2, 3\}$ and $h(n) = \{2, 4, 5, 1\}$. Compute the circular convolution of $x(n)$ and $h(n)$.

7.18. *The fast Fourier transform.* Computation of the DFT

$$X[k] = \sum_{n=0}^{M-1} x(n)W^{nk}$$

requires M multiplications per output sample, so that a total of M^2 multiplications are required. However, owing to the properties of the number W, there is a great deal of structure in the definition, and we can reduce the number of multiplications significantly. The fast Fourier transform (FFT) does precisely this. If M is a power of two, then the number of computations is reduced to about $cM \log_2 M$, where c is a constant. The idea behind the FFT is very simple, as explained in this problem.

1. If M is even, show that we can write

$$X[k] = \sum_{n=0}^{\frac{M}{2}-1} x(2n)W^{2nk} + W^k \sum_{n=0}^{\frac{M}{2}-1} x(2n+1)W^{2nk}, 0 \le k \le \frac{M}{2} - 1 \tag{P7.18a}$$

$$X[k+\frac{M}{2}] = \sum_{n=0}^{\frac{M}{2}-1} x(2n)W^{2nk} - W^k \sum_{n=0}^{\frac{M}{2}-1} x(2n+1)W^{2nk}, 0 \le k \le \frac{M}{2} - 1 \tag{P7.18b}$$

   Thus, $X[k]$ can be computed for all M values of k by computing the $M/2$-point DFT of the even samples $x(2n)$ and the odd samples $x(2n+1)$ and combining the answers.

2. Show that the preceding computation takes only $(M/2)^2 + M$ multiplications. For example, if $M = 2^{10}$, how does this compare with M^2?

If $M/2$ is even then the $(M/2)$-point DFTs can also be computed by further decomposing them into $(M/4)$-point DFTs. The reader can see that if M is a power of two then this process can be repeated until there are only two-point DFTs to compute. A two-point DFT requires no multiplications (because $W = -1$ for $M = 2$). So all the multiplications come from the factors W^k (called twiddle factors) in equations of the form (P7.18a,b) at every stage of decomposition. The number of multiplications becomes proportional to $M \log_2 M$.

7.19. In the FFT algorithm, since some of the powers of W are trivial multipliers (e.g., $W^0 = 1, W^{M/2} = -1, W^{M/4} = -j$, and $W^{-M/4} = j$), the exact number of multiplications when M is a power of two is smaller than what is outlined in Problem 7.18. Assuming $x(n)$ is possibly complex, show that the precise number of complex multiplications is only

$$\frac{M}{2} \log_2 M - \frac{3M}{2} + 2. \tag{P7.19}$$

How many complex additions are required in the above algorithm?

7.20. Using Eq. (P7.19) evaluate the number of multiplications required to compute (a) a 2-point FFT, (b) a 4-point FFT, (c) an 8-point FFT, and (d) a 2^{10}-point FFT. For $M = 2^{10}$ a direct computation of the DFT (without FFT) would require about $(1{,}024)^2$ (i.e., about a million) multiplications.

7.21. We now consider a special case of Problem 7.10 where all the M user channels are identical: $C_m(z) = C(z)$. Choose $\rho_k = W^{-k}$, where $W \stackrel{\Delta}{=} e^{-j2\pi/M}$ so that $H_k(z) = a_k \sum_{n=0}^{P-1} z^n W^{nk}$ with $a_k = (1/C(W^{-k}))$.

1. Show that the condition $F_m(\rho_k) = \delta(k-m)$ yields the transmitting filter expression $F_m(z) = \sum_{n=0}^{M-1} W^{-mn} z^{-n}/M$.
2. If all users are in one place, show that the system can be drawn as shown in Fig. P7.21, where $\mathbf{W}$ is the $M \times M$ DFT matrix and $\mathbf{W}_1$ is the submatrix of $\mathbf{W}$ obtained by retaining the first $L = (P-M)$ columns. The equalizers

$$a_k = 1/C(W^{-k}) = 1/C(e^{j2\pi k/M})$$

invert the channel frequency response, sampled on the DFT grid (i.e., at $\omega = 2\pi k/M$). These equalizers are similar to what we obtain in a cyclic-prefix system.

In the multiuser scenario, the symbol streams $s_m(n)$ are separate independent users. But sometimes they can collaborate as in Fig. P7.21. This happens, for example, when $s_m(n)$ are obtained from a single user by blocking, parsing, or by use of a multiple antenna system.

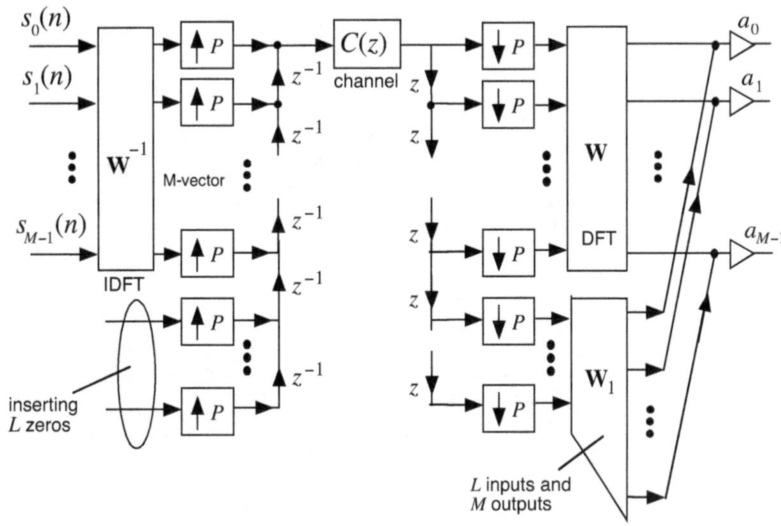

Figure P7.21.

7.22. *Bandwidth expansion factor.* Consider again the multiuser system of Problem 7.7. Suppose we take each user stream $s_m(n)$ and perform K-fold blocking (Sec. 3.5). This gives a new representation for the mth user, as shown in Fig. P7.22. The K "subusers" $s_{m,i}(n)$ are merely substreams of $s_m(n)$. We have used the new expanders $\uparrow \widehat{P}$ instead of $\uparrow P$. This figure is therefore similar to the one in Problem 7.7, but we now have MK "users" (though the MK channels are not all different). The relation $P = M + L$ is now replaced with $\widehat{P} = MK + L$. By repeating the ideas in the earlier problems above, we can show that multiuser interference and ISI can be cancelled as before.

1. In Problem 7.7 we have M users multiplexed with an interpolation factor of P. So the bandwidth expansion factor was $\gamma = P/M = (M + L)/M$. Since the number of users M is fixed, γ is fixed. With the system redrawn as in Fig. P7.22, where $\widehat{P} = MK + L$, what is the bandwidth expansion factor now?
2. What happens to this bandwidth expansion factor as the free integer parameter K is made arbitrarily large?

The fact that the bandwidth expansion factor in the multiuser system can be reduced to any desired value by using this simple trick was pointed out by Giannakis *et al.* [2000].

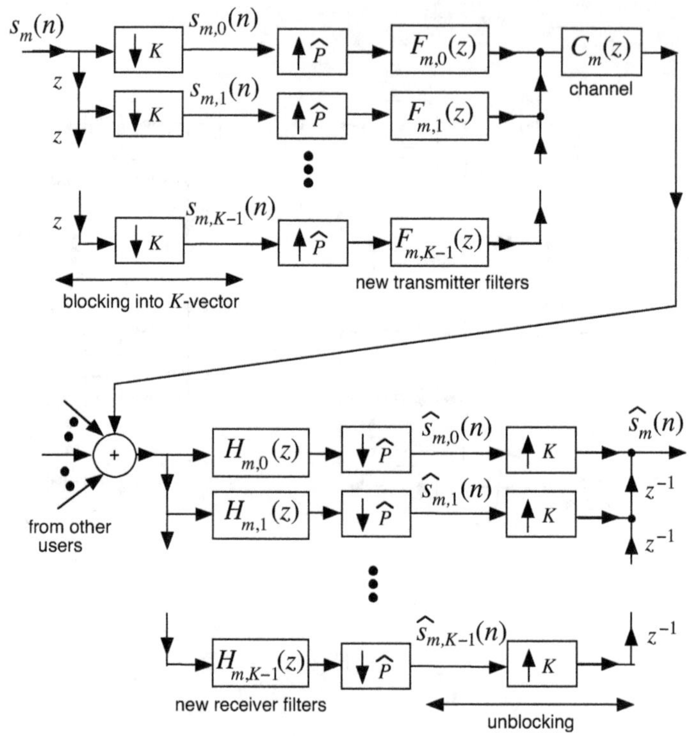

Figure P7.22.

7.23. *Null resistance.* The cyclic-prefix system described in Sec. 7.3.2 uses the equalizer multipliers $1/C[k], 0 \le k \le M-1$, at the receiver. Here $C[k]$ are the M DFT coefficients of the FIR channel (assumed to be of order L). In general, some of these can be very small, or even equal to zero. This creates practical difficulties in the implementation, as the noise can be severely amplified by $1/C[k]$. There are two practical ways to overcome this problem. One is to replace the zero-forcing equalizers $1/C[k]$ with MMSE equalizers. The other is to use extra redundancy at the transmitter to incorporate what is called *null resistance.* In this problem we describe the second approach [Liang and Tran, 2002]. Consider the modified cyclic-prefix system shown in Fig. P7.23. Here the blocked version $\mathbf{s}(n)$ of size M is transformed into a vector $\mathbf{r}(n)$ of size $N = M + L$ using a matrix $\mathbf{Q}$. The vector $\mathbf{r}(n)$ is treated as if this is to be transmitted. Thus, we perform N-point IDFT, and insert cyclic prefix, and so forth, as shown.

1. Express the $N \times N$ transfer matrix from $\mathbf{s}_1(n)$ to $\mathbf{y}_1(n)$ in terms of the channel coefficients $c(n)$.
2. Show that the transfer function from $\mathbf{r}(n)$ to $\mathbf{v}(n)$ is a diagonal matrix with elements $C(e^{j2\pi k/N}), 0 \le k \le N-1$. These are the N-point DFT coefficients of the channel.

3. Since the channel has order L, at most L of its N-point DFT coefficients are zero. So there exist at least M nonzero coefficients, since $N = M + L$. Let

$$C(e^{j2\pi k_i/N}), \quad 0 \leq i \leq M-1,$$

be the M nonzero coefficients with the largest magnitude. The "selector" at the receiver picks these M components, that is, $u_i(n) = v_{k_i}(n)$. What are the elements of the diagonal equalizer $\mathbf{\Lambda}_e$ which ensure that $\widehat{r}_i(n) = r_{k_i}(n), 0 \leq i \leq M-1$? Are these diagonal elements finite and nonzero?

4. Assume that the $N \times M$ matrix $\mathbf{Q}$ is chosen such that any set of M rows is linearly independent. Then show that regardless of what the M integers $\{k_i\}$ above are, there exists a matrix $\mathbf{R}$ such that its output is the original vector $\mathbf{s}(n)$ (in absence of channel noise, of course).

Thus the method provides perfect equalization even if the channel has nulls in frequency. The use of the integer $N = M + L$ simply avoids these nulls in the equalization process. The system is therefore referred to as a null-resistant system. The matrix $\mathbf{Q}$ with the property that any set of M rows is linearly independent is called a valid null-resistance matrix.

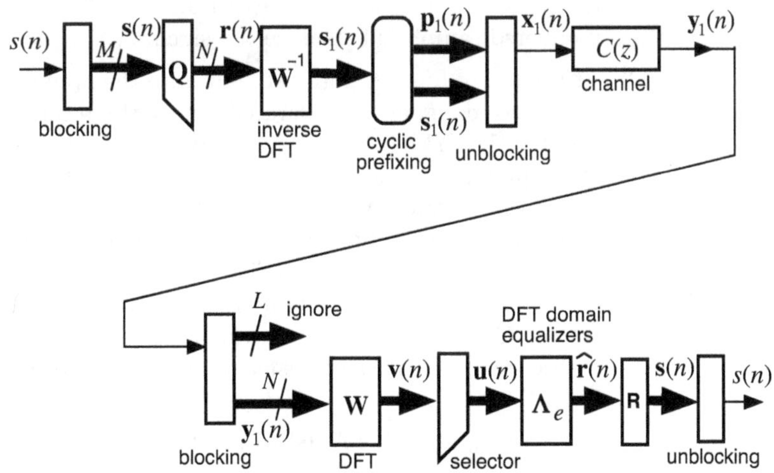

Figure P7.23.

7.24. Let $\mathbf{Q}$ be an $N \times M$ matrix with $N > M$ such that the kth row is

$$[1 \quad a_k \quad a_k^2 \quad \dots \quad a_k^{M-1}].$$

Assume that $a_k \neq a_m$ for $0 \leq k < m \leq N-1$. Show then that $\mathbf{Q}$ is a valid null-resistance matrix, that is, any set of M rows is linearly independent. For example, if we take the $N \times N$ DFT matrix and form $\mathbf{Q}$ by retaining the first M columns, we get a valid null-resistance matrix.

7.25. *Oversampled cyclic-prefix receivers.* In Chap. 4 (Secs. 4.7 and 4.8) we considered digital communication systems with oversampling at the receiver. This gave rise to the so-called fractionally spaced equalizer (FSE). In this Problem we develop the FSE version of the cyclic-prefix system [Vaidyanathan and Vrcelj, 2002], [Tepedelenlioĝlu and Challagulla, 2004]. Consider Fig. P7.25(a) where the transmitter performs cyclic prefixing as usual and $C_c(s)$ is a continuous-time channel with finite duration impulse repsonse $c_c(t)$. We have a D/C converter with sample spacing T at the transmitter and a C/D converter at the receiver with sample spacing $T/2$. So the receiver oversamples by a factor of two.

1. With $h(n) = c_c(nT/2)$ and $H(z) = \sum_n h(n)z^{-n}$, let $H(z) = R_0(z^2) + z^{-1}R_1(z^2)$ be the polyphase representation of $H(z)$. Show that the path from $x(n)$ to $y(n)$, shown separately on the left in Fig. P7.25(b), can be represented by the multirate model shown on the right of the same figure. Note that $y_0(n)$ and $y_1(n)$ are the even and odd numbered samples of the received signal $y(n)$.
2. At the top of Fig. P7.25(c) we show how the even and odd samples of the received signal are separated out at the receiver. The even part $y_0(n)$ is then P-blocked, and the first L samples discarded, as in the cyclic-prefix receiver. The vector consisting of the last M samples in the block is named $\mathbf{y}_0(n)$. Starting from $y_1(n)$, we define an $M \times 1$ vector $\mathbf{y}_1(n)$ similarly. With $\mathbf{s}(n)$ denoting the $M \times 1$ blocked version of the symbol stream $s(n)$ (Fig. P7.25(a)), show that the transfer matrix $\mathbf{C}_0$ from $\mathbf{s}(n)$ to $\mathbf{y}_0(n)$ is a circulant, and so is the transfer matrix $\mathbf{C}_1$ from $\mathbf{s}(n)$ to $\mathbf{y}_1(n)$, as long as $R_0(z)$and $R_1(z)$ are FIR with order $\leq L$. These transfer matrices are shown at the bottom of Fig. P7.25(c).
3. How are $\mathbf{C}_0$ and $\mathbf{C}_1$ related to $H(z) = R_0(z^2) + z^{-1}R_1(z^2)$?

Thus the discrete-time equivalent of the cyclic-prefix transceiver with two-fold oversampling can be represented by a pair of circulant matrices which come from the two polyphase components $\{R_0(z), R_1(z)\}$ of the oversampled discrete-time channel $H(z)$.

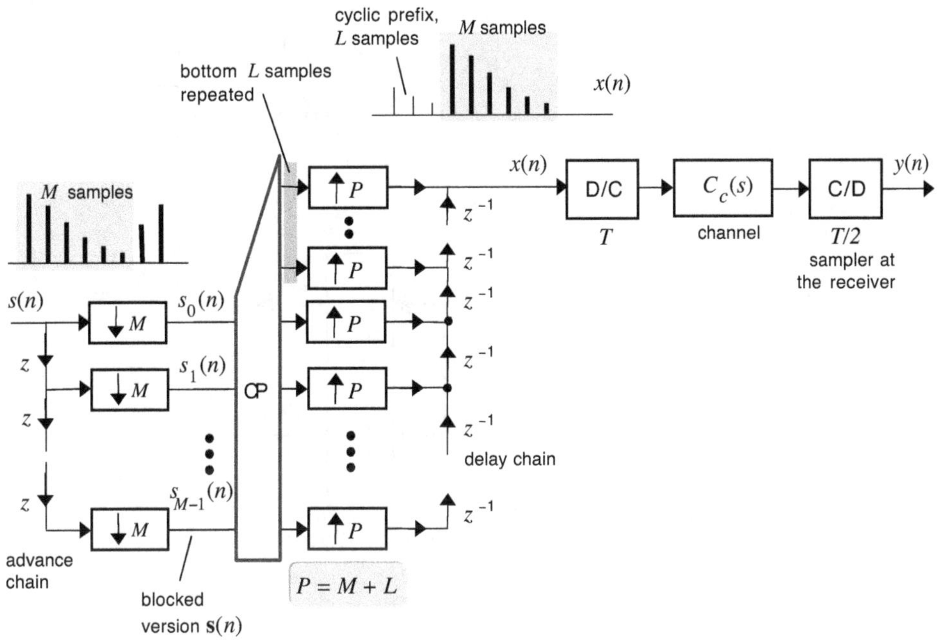

Figure P7.25(a).

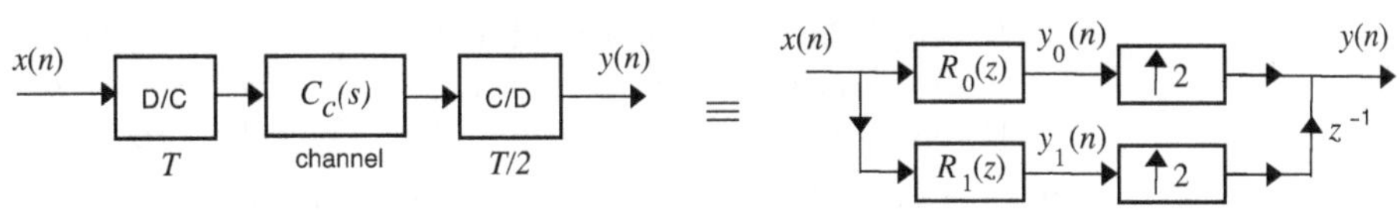

Figure P7.25(b).

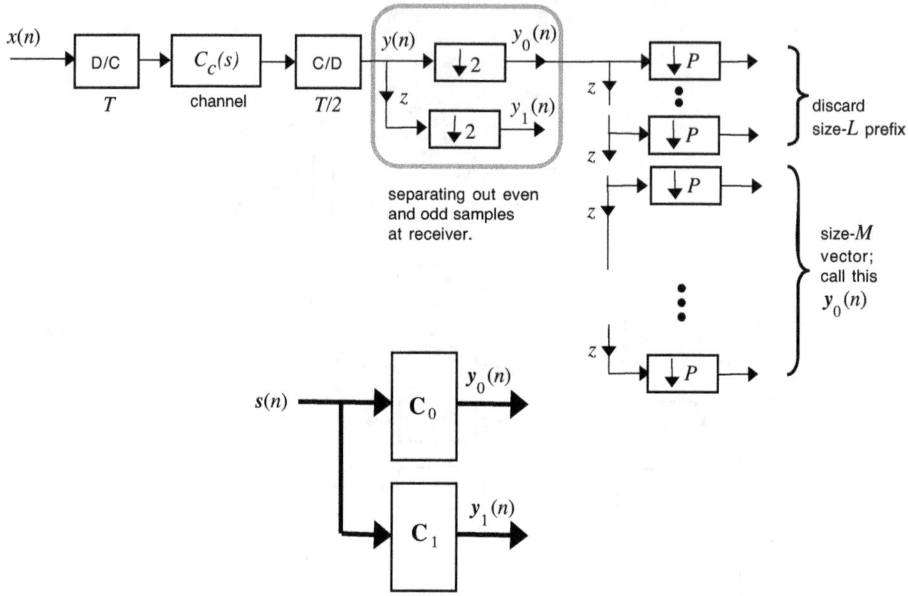

Figure P7.25(c).

7.26. Referring again to Problem 7.25, we now consider the problem of reconstructing $\mathbf{s}(n)$ (equivalently the scalar symbol stream $s(n)$) from the vectors $\mathbf{y}_0(n)$ and $\mathbf{y}_1(n)$. Consider Fig. 7.26(a), which shows a receiver structure with $M \times M$ DFT matrices $\mathbf{W}_M$ and IDFT matrices, and diagonal matrices $\mathbf{\Lambda}_{e,0}$ and $\mathbf{\Lambda}_{e,1}$ (equalizer coefficients).

1. Figure 7.26(b) shows a modified version of Fig. 7.26(a). If the first figure has the zero-forcing property (i.e., $\widehat{\mathbf{s}}(n) = \mathbf{s}(n)$ ignoring noise), then show that the second figure also has this property.
2. Now consider the problem of choosing the diagonal matrices $\mathbf{\Lambda}_{e,0}$ and $\mathbf{\Lambda}_{e,1}$ (equalizers) at the receiver such that the zero-forcing constraint $\widehat{\mathbf{s}}(n) = \mathbf{s}(n)$ is satisfied. With $\mathbf{C}_0$ and $\mathbf{C}_1$ representing $M \times M$ circulant matrices, we know that they can be written as

$$\mathbf{C}_0 = \mathbf{W}_M^{-1}\mathbf{\Lambda}_0\mathbf{W}_M \quad \text{and} \quad \mathbf{C}_1 = \mathbf{W}_M^{-1}\mathbf{\Lambda}_1\mathbf{W}_M,$$

where $\mathbf{\Lambda}_0$ and $\mathbf{\Lambda}_1$ are diagonal matrices. Show that the ZF condition is satisfied if and only if

$$\mathbf{\Lambda}_0\mathbf{\Lambda}_{e,0} + \mathbf{\Lambda}_1\mathbf{\Lambda}_{e,1} = \mathbf{I}_M. \qquad \text{(P7.26}a\text{)}$$

3. Argue that the diagonal elements

$$C_0[k] \triangleq [\mathbf{\Lambda}_0]_{kk}, \quad C_1[k] \triangleq [\mathbf{\Lambda}_1]_{kk},$$

are proportional to the M uniformly spaced samples of the polyphase components $R_0(e^{j\omega})$ and $R_1(e^{j\omega})$, i.e., the M-point DFTs of $r_0(n) = c_c(nT)$ and $r_1(n) = c_c(nT + 0.5T)$. Thus, instead of inverting the M DFT coefficients of the channel, as is normally done in a cyclic-prefixed ZF system, the ZF condition in the cyclic-prefixed FSE system does something different, namely it combines the M DFT coefficients of $r_0(n)$ and $r_1(n)$ to force

$$C_0[k]E_0[k] + C_1[k]E_1[k] = 1, \quad 0 \leq k \leq M - 1, \qquad \text{(P7.26}b\text{)}$$

where $E_i[k]$ is the kth diagonal element of $\mathbf{\Lambda}_{e,i}$.

Thus, unless $C_0[k]$ and $C_1[k]$ are both very small for a certain k, the equalizer coefficients $E_0[k]$ and $E_1[k]$ do not get unduly large. Furthermore, for a given pair of numbers $C_0[k]$ and $C_1[k]$, the numbers $E_0[k]$ and $E_1[k]$ satisfying Eq. (P7.26b) are not unique. For example if $E_0[k], E_1[k]$ satisfies Eq. (P7.26b), then so does the pair

$$\Big(E_0[k] + \alpha(k)C_1[k]\Big), \quad \Big(E_1[k] - \alpha(k)C_0[k]\Big), \qquad \text{(P7.26}c\text{)}$$

for any set of M numbers $\alpha(k)$. This freedom offered by $\alpha[k]$ can be used [Vaidyanathan and Vrcelj, 2002] to minimize reconstruction error due to channel noise (which has been ignored in this problem).

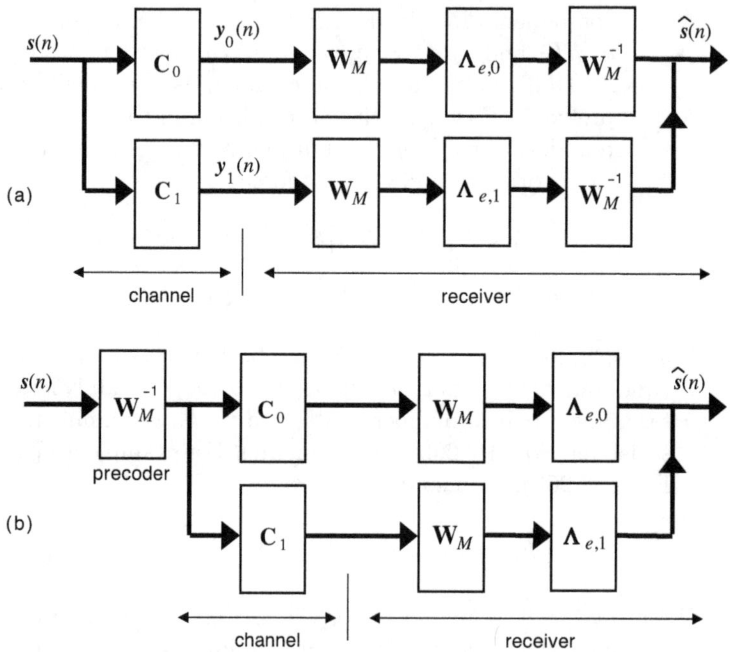
s(n)
C0
y0(n)
WM
Λe,0
WM-1
ŝ(n)
C1
y1(n)
WM
Λe,1
WM-1
(a)
channel
receiver
s(n)
WM-1
precoder
C0
WM
Λe,0
ŝ(n)
C1
WM
Λe,1
(b)
channel
receiver

Figure P7.26.

8

The lazy precoder with a zero-forcing equalizer

8.1 Introduction

The zero-padded precoder described in Chap. 7 is reproduced in Fig. 8.1. In this system, $C(z)$ is a scalar FIR channel:

$$C(z) = \sum_{n=0}^{L} c(n)z^{-n}.$$

The input $s(n)$ is transmitted in blocks of length M, with L zeros padded to each block. The precoder at the transmitter is represented by the matrix $\mathbf{R}_1$ and the equalizer matrix is $\mathbf{T}$. The special choice of the precoder given by

$$\mathbf{R}_1 = \mathbf{I}$$

is called the lazy precoder, and is shown in Fig. 8.2. Thus, the transmitter merely pads zeros, but does not perform any transformation. All collaboration between components happens at the receiver. As pointed out in Chap. 4, such a scenario is applicable in multiuser transceivers operating in the multiple access or uplink mode (Sec. 4.5). In Chap. 7 we showed (see Eq. (7.5)) that the system from the vector $\mathbf{s}(n)$ to the vector $\mathbf{y}(n)$ is a linear system with a constant transfer matrix. Including the additive noise contributed by the channel, this equation takes the form

$$\mathbf{y}(n) = \mathbf{A}\mathbf{s}(n) + \mathbf{q}(n), \tag{8.1}$$

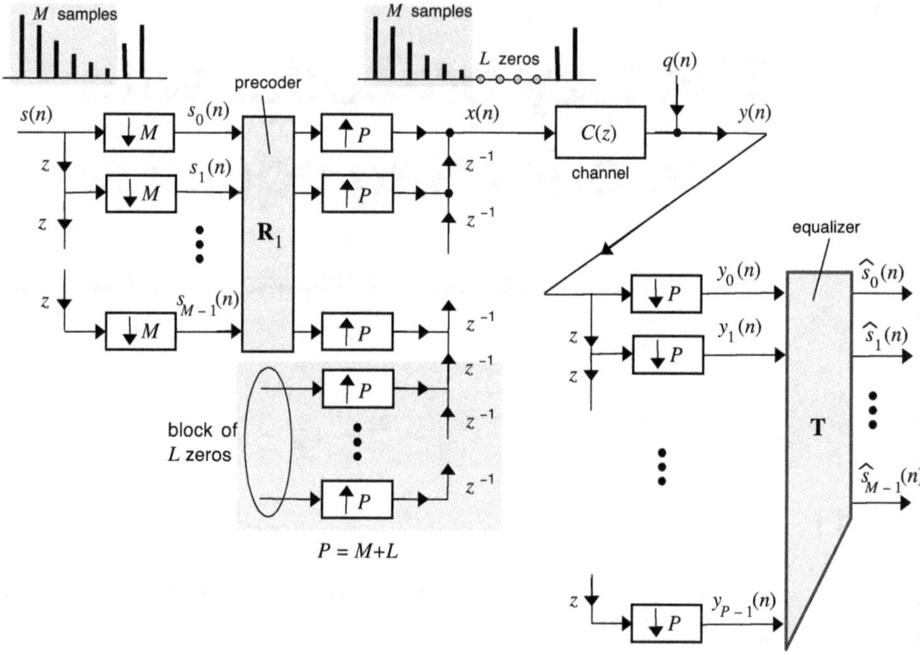

Figure 8.1. Transceiver with a zero-padding precoder.

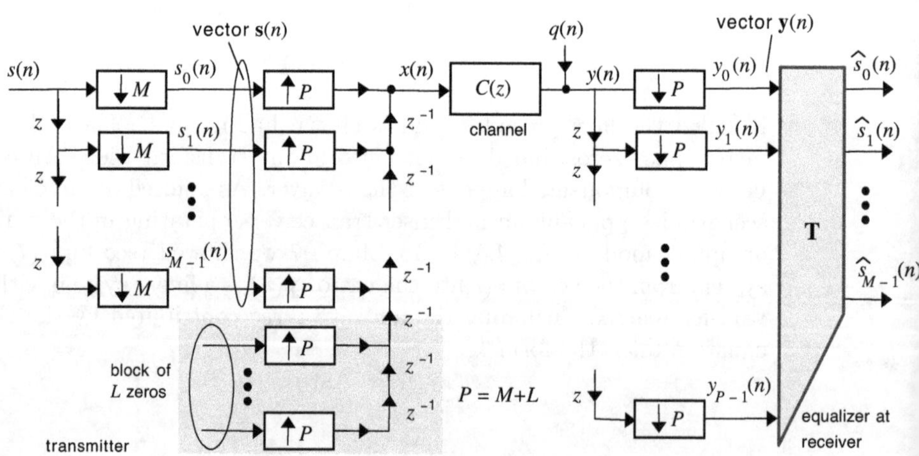

Figure 8.2. Transceiver with a lazy precoder.

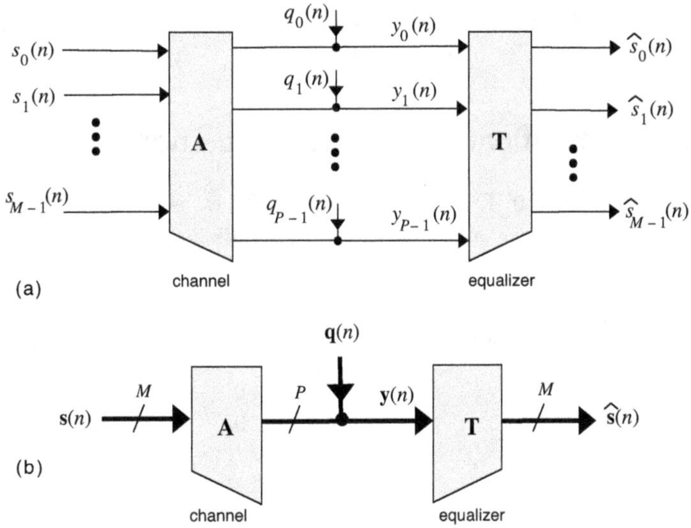

Figure 8.3. (a) The lazy precoder in vector notation. (b) Further simplified notation.

where $\mathbf{A}$ is the $P \times M$ full banded Toeplitz matrix of channel coefficients:

$$\mathbf{A} = \begin{bmatrix} c(0) & 0 & \ldots & 0 \\ c(1) & c(0) & \ldots & 0 \\ \vdots & \vdots & \ddots & \vdots \\ c(L) & & & \\ 0 & c(L) & & \\ \vdots & & \ddots & \vdots \\ 0 & 0 & \ldots & c(L) \end{bmatrix}. \tag{8.2}$$

The system from the input vector $\mathbf{s}(n)$ to the reconstructed vector $\widehat{\mathbf{s}}(n)$ can therefore be drawn in the form shown in Fig. 8.3(a). The simplified vector notation is shown in Fig. 8.3(b). Here $\mathbf{A}$ is the $P \times M$ matrix representing the channel, and $\mathbf{T}$ is the equalizer.

In this chapter we study the lazy precoder with a zero-forcing equalizer. In Sec. 8.2 we will find an expression for the average reconstruction error in terms of the matrix $\mathbf{A}$. We will see that the error is determined by the Frobenius norm of the left inverse of $\mathbf{A}$. The received signal in each block has $P = M+L$ samples, and at least M samples are required to get a reconstruction of the transmitted block. Even though it is preferable to use all P samples of the received block $\mathbf{y}(n)$, it is in principle possible to use $K < P$ samples. In Sec. 8.3–8.4 we will study the effect of increasing K progressively from M to P. In Sec. 8.5 we will show that the matrix $\mathbf{A}^\dagger\mathbf{A}$ is determined entirely by the autocorrelation of the channel $c(n)$. The practical consequences of this property are also elaborated.

Recall from Sec. 7.2 that increasing the block size M has the advantage of decreasing the bandwidth expansion factor γ. We will show in Sec. 8.6 that increasing M has a disadvantage, namely the average reconstruction error per symbol necessarily increases [Ohno, 2006], [Vaidyanathan, 2007].

8.2 Noise amplification and Frobenius norm

The matrix $\mathbf{A}$ in Eq. (8.2) has full rank M as long as the channel $C(z)$ is not identically zero for all z. We can prove this as follows:

Proof. If $C(z)$ is not identically zero for all z, then $c(0), c(1), \ldots, c(L)$ cannot all be zero. Let $c(K)$ be the first nonzero value. Then the $P \times M$ matrix $\mathbf{A}$ has the form

$$\mathbf{A} = \begin{bmatrix} \mathbf{0} & \mathbf{0} & \mathbf{0} & \ldots \\ c(K) & 0 & 0 & \\ c(K+1) & c(K) & 0 & \\ c(K+2) & c(K+1) & c(K) & \\ \vdots & \vdots & & \ddots \end{bmatrix}.$$

Since $c(K) \neq 0$, the M columns are independent, that is, the rank is M. $\bigtriangledown \bigtriangledown \bigtriangledown$

Since $\mathbf{A}$ has rank M, it has a left inverse, that is, an $M \times P$ matrix $\mathbf{A}^{\#}$ which satisfies

$$\mathbf{A}^{\#}\mathbf{A} = \mathbf{I}_M.$$

Note that since $P > M$ this inverse is not unique. This left inverse can be used as the equalizer

$$\mathbf{T} = \mathbf{A}^{\#} \tag{8.3}$$

in Fig. 8.3. Premultiplying both sides of Eq. (8.1) by the left inverse we obtain[1]

$$\mathbf{A}^{\#}\mathbf{y}(n) = \mathbf{s}(n) + \mathbf{A}^{\#}\mathbf{q}(n).$$

The quantity on the left, which can be computed from received noisy data, therefore represents an estimate $\widehat{\mathbf{s}}(n)$ of the nth block $\mathbf{s}(n)$ of the transmitted symbol stream $s(n)$. The error in the estimation of $\mathbf{s}(n)$ is clearly the second term

$$\mathbf{e}(n) \stackrel{\Delta}{=} \mathbf{A}^{\#}\mathbf{q}(n). \tag{8.4}$$

The mean square error of reconstruction can be written as

$$\mathcal{E}_{reco} = E\Big[(\mathbf{A}^{\#}\mathbf{q}(n))^{\dagger}\mathbf{A}^{\#}\mathbf{q}(n)\Big] = \text{Tr}\left(\mathbf{A}^{\#}E[\mathbf{q}(n)\mathbf{q}^{\dagger}(n)](\mathbf{A}^{\#})^{\dagger}\right). \tag{8.5}$$

Since the channel noise $q(n)$ is assumed to be white with variance σ_q^2, the blocked version $\mathbf{q}(n)$ satisfies

$$E[\mathbf{q}(n)\mathbf{q}^{\dagger}(n)] = \sigma_q^2\mathbf{I}.$$

[1]The reader not familiar with left inverses and matrix norms should review Appendix C at this time.

Thus

$$\mathcal{E}_{reco} = \sigma_q^2 \text{Tr}\left(\mathbf{A}^{\#}(\mathbf{A}^{\#})^{\dagger}\right) = \sigma_q^2 \|\mathbf{A}^{\#}\|^2, \tag{8.6}$$

where $\|\mathbf{T}\|$ denotes the **Frobenius norm** of $\mathbf{T}$, defined such that

$$\|\mathbf{T}\|^2 = \sum_k \sum_m |T_{km}|^2. \tag{8.7}$$

In Eq. (8.6) we have used the fact (Appendix C) that

$$\|\mathbf{T}\|^2 = \text{Tr}\,(\mathbf{T}^{\dagger}\mathbf{T}) = \text{Tr}\,(\mathbf{T}\mathbf{T}^{\dagger}). \tag{8.8}$$

Thus the channel noise amplification factor is precisely the square of the Frobenius norm of the left inverse $\mathbf{A}^{\#}$. Therefore the importance of choosing the inverse to be the *minimum-norm inverse* is amply clear! From Appendix C we know that the minimum-norm left inverse of $\mathbf{A}$ has the closed form expression

$$\mathbf{A}^{\#} = (\mathbf{A}^{\dagger}\mathbf{A})^{-1}\mathbf{A}^{\dagger} \tag{8.9}$$

and can therefore be readily calculated. The quantity $\|\mathbf{A}^{\#}\|^2$ can be found using Eq. (8.8) and is given by

$$\|\mathbf{A}^{\#}\|^2 = \text{Tr}\left(\mathbf{A}^{\#}(\mathbf{A}^{\#})^{\dagger}\right) = \text{Tr}\,(\mathbf{A}^{\dagger}\mathbf{A})^{-1},$$

where we have used the form (8.9). In practice, since there are M symbols in each block (i.e., $\mathbf{e}(n)$ has M components) we divide Eq. (8.6) by M to get the average reconstruction error variance

$$\mathcal{E} = \frac{\sigma_q^2 \|\mathbf{A}^{\#}\|^2}{M}. \tag{8.10}$$

From Appendix C we know that if σ_k denotes the **singular values** of $\mathbf{A}$ (i.e., σ_k^2 are eigenvalues of $\mathbf{A}^{\dagger}\mathbf{A}$) then

$$\|\mathbf{A}\|^2 = \sum_{k=0}^{M-1} \sigma_k^2 \quad \text{and} \quad \|\mathbf{A}^{\#}\|^2 = \sum_{k=0}^{M-1} \frac{1}{\sigma_k^2}$$

Thus, the reconstruction error (8.10) can be expressed as follows:

$$\mathcal{E} = \frac{\sigma_q^2 \|\mathbf{A}^{\#}\|^2}{M} = \frac{\sigma_q^2}{M} \sum_{k=0}^{M-1} \frac{1}{\sigma_k^2} \tag{8.11}$$

where σ_q^2 is the variance of the channel noise $q(n)$, and σ_k are the singular values of the channel matrix $\mathbf{A}$. The noise gain due to equalization is therefore

$$\text{noise gain} = \frac{\|\mathbf{A}^{\#}\|^2}{M} = \frac{1}{M} \sum_{k=0}^{M-1} \frac{1}{\sigma_k^2} \tag{8.12}$$

8.3 Frobenius norm of left inverse as A grows taller

We now observe an important property which has direct application in channel equalization. Let $\mathbf{A}$ be $P \times M$ with $P \geq M$ and assume its rank is M. Define the taller matrix

$$\mathbf{B} = \begin{bmatrix} \mathbf{A} \\ \mathbf{a} \end{bmatrix}, \tag{8.13}$$

where $\mathbf{a} \neq \mathbf{0}$ is a row vector. Since $\|\mathbf{B}\|^2$ is the total energy of the elements of $\mathbf{B}$ (Eq. (8.7)), it is obvious that

$$\|\mathbf{B}\| > \|\mathbf{A}\|. \tag{8.14}$$

Let $\mathbf{A}^{\#}$ be the unique minimum-norm left inverse of $\mathbf{A}$ and let $\mathbf{B}^{\#}$ be the unique minimum-norm left inverse of $\mathbf{B}$, so that

$$\mathbf{A}^{\#}\mathbf{A} = \mathbf{I}_M, \quad \mathbf{B}^{\#}\mathbf{B} = \mathbf{I}_M.$$

We now prove the following, which does not require that $\mathbf{A}$ be Toeplitz.

♠**Lemma 8.1.** *Frobenius norm of the minimum-norm left inverse.* In the above set up,

$$\|\mathbf{B}^{\#}\| \leq \|\mathbf{A}^{\#}\|. \tag{8.15}$$

That is, even though $\mathbf{B}^{\#}$ has more columns than $\mathbf{A}^{\#}$ (because $\mathbf{B}$ has more rows than $\mathbf{A}$), the norm of $\mathbf{B}^{\#}$ cannot be larger than that of $\mathbf{A}^{\#}$. ◇

Proof. Observe first that the matrix

$$[\,\mathbf{A}^{\#} \quad \mathbf{0}\,] \tag{8.16}$$

is a valid left inverse for $\mathbf{B}$ because

$$[\,\mathbf{A}^{\#} \quad \mathbf{0}\,]\,\mathbf{B} = [\,\mathbf{A}^{\#} \quad \mathbf{0}\,]\begin{bmatrix} \mathbf{A} \\ \mathbf{a} \end{bmatrix} = \mathbf{A}^{\#}\mathbf{A} = \mathbf{I}_M. \tag{8.17}$$

The left inverse (8.16) clearly has the same norm as $\mathbf{A}^{\#}$ (because the extra column of zeros does not change the energy in the elements). This shows that there exists at least one solution to the left inverse of $\mathbf{B}$ which has norm identical to that of $\mathbf{A}^{\#}$. So the minimum-norm left inverse $\mathbf{B}^{\#}$, by its very definition, satisfies Eq. (8.15). ▽▽▽

8.4 Application in equalization

Consider again the equation for the received block of data

$$\mathbf{y}(n) = \mathbf{A}\mathbf{s}(n) + \mathbf{q}(n), \tag{8.18}$$

where $\mathbf{A}$ is the full banded Toeplitz matrix of channel coefficients given in Eq. (8.2). This is a $P \times M$ matrix, where $P = M + L$.

We now define certain **submatrices** of $\mathbf{A}$ which are important for further discussions. Let $\mathbf{A}_K$ denote the submatrix obtained by keeping the top K rows of $\mathbf{A}$. For $K \geq M$ the matrix $\mathbf{A}_K$ has full rank M (assuming $c(0) \neq 0$). For example, with $M = 3$ and $L = 2$ we have

$$\mathbf{A}_3 = \begin{bmatrix} c(0) & 0 & 0 \\ c(1) & c(0) & 0 \\ c(2) & c(1) & c(0) \end{bmatrix}, \mathbf{A}_4 = \begin{bmatrix} c(0) & 0 & 0 \\ c(1) & c(0) & 0 \\ c(2) & c(1) & c(0) \\ 0 & c(2) & c(1) \end{bmatrix}, \mathbf{A}_5 = \begin{bmatrix} c(0) & 0 & 0 \\ c(1) & c(0) & 0 \\ c(2) & c(1) & c(0) \\ 0 & c(2) & c(1) \\ 0 & 0 & c(2) \end{bmatrix}. \tag{8.19}$$

Note the following properties of these matrices:

1. $\mathbf{A}_K$ is lower triangular and Toeplitz for all K.
2. For $K = M$ the matrix $\mathbf{A}_K$ is also a *square* matrix.
3. For $K = P$ the matrix $\mathbf{A}_K$ is *full banded* Toeplitz.

Note from Eq. (8.18) that we also have

$$\mathbf{y}_K(n) = \mathbf{A}_K\mathbf{s}(n) + \mathbf{q}_K(n), \tag{8.20}$$

where $\mathbf{y}_K(n)$ contains the first K components of $\mathbf{y}(n)$, and $\mathbf{q}_K(n)$ contains the first K components of $\mathbf{q}(n)$. For $K \geq M$, since $\mathbf{A}_K$ has full rank, it has a left inverse. Letting $\mathbf{A}_K^{\#}$ denote the minimum-norm left inverse as usual, we have

$$\mathbf{A}_K^{\#}\mathbf{y}_K(n) = \mathbf{s}(n) + \mathbf{A}_K^{\#}\mathbf{q}_K(n). \tag{8.21}$$

Proceeding as in the derivation of Eq. (8.6) we conclude again that

$$\mathcal{E}_{reco,K} = \sigma_q^2 \|\mathbf{A}_K^{\#}\|^2, \tag{8.22}$$

where the subscript K is a reminder that we have retained K components of the output vector $\mathbf{y}(n)$ in the reconstruction of $\mathbf{s}(n)$. Since we can write

$$\mathbf{A}_{K+1} = \begin{bmatrix} \mathbf{A}_K \\ \times \end{bmatrix}$$

it follows that the minimum-norm left inverse of $\mathbf{A}_{K+1}$ has a smaller Frobenius norm than $\mathbf{A}_K$ (Lemma 8.1). This shows that

$$\mathcal{E}_{reco,K+1} \leq \mathcal{E}_{reco,K}$$

for fixed M. That is, the reconstruction error (8.22) can only improve as K increases. As we make $\mathbf{A}_K$ taller and taller, that is, *as we use more and more output samples from the block* $\mathbf{y}(n)$, the effect of channel noise becomes smaller. This is intuitively obvious as well.

We will demonstrate this with numerical examples. For typical channels, the improvement that we obtain as K increases from M to P is significant, especially when the channel $C(z)$ has zeros outside the unit circle, as we shall demonstrate. Thus, even though it is sufficient to retain M samples of the output per block (of size $P = M + L$), it is always wiser to use all the P received data samples in the estimation of $\mathbf{s}(n)$.

Example 8.1: Advantage of making the A-matrix taller

Consider the example of the fourth-order channel

$$C(z) = 1 + 2z^{-1} + 5z^{-2} + 10z^{-3} - z^{-4}$$

and let $M = 8$. Since $P = M + L = 12$ we see that the matrix $\mathbf{A}_K$ has 8 columns, and the number of rows K can be 8, 9, 10, 11, or 12. For each of these cases we have calculated $\|\mathbf{A}_K^{\#}\|^2/M$. The results are as follows:

K (no. of rows)	$\|\mathbf{A}_K^{\#}\|^2/M$	$\|\mathbf{A}_K^{\#}\|^2$ dB (normalized)
8	$2.4360 \times 10^{+3}$	0
9	$1.0201 \times 10^{+3}$	−3.78
10	$2.8898 \times 10^{+2}$	−9.26
11	1.0181×10^{-2}	−53.79
12	1.0168×10^{-2}	−53.79

Notice how the norm decreases dramatically as the number of rows is increased from 10 to 11. Thus the channel noise amplification is improved by about 45 dB if we keep eleven rows of $\mathbf{A}^{\#}$ instead of ten! In some cases, there is similarly a jump in quality as the number of rows increases from M to $M + 1$. For example, try the channel in Problem 8.1.

8.5 The autocorrelation property

As seen in the previous sections, the noise gain depends on the eigenvalues of the matrix $\mathbf{A}^{\dagger}\mathbf{A}$. Consider the full size ($P \times M$) channel matrix (8.2). If we compute the product $\mathbf{R} = \mathbf{A}^{\dagger}\mathbf{A}$ explicitly, we will find that it is a *Hermitian, positive definite, and Toeplitz matrix.* That is, it is a valid autocorrelation matrix for a fictitious wide sense stationary (WSS) random process. For example, if $M = 4, L = 2$, and $C(z) = 1 + 2z^{-1} + 4z^{-2}$ then

$$\mathbf{A} = \begin{bmatrix} 1 & 0 & 0 & 0 \\ 2 & 1 & 0 & 0 \\ 4 & 2 & 1 & 0 \\ 0 & 4 & 2 & 1 \\ 0 & 0 & 4 & 2 \\ 0 & 0 & 0 & 4 \end{bmatrix} \quad \text{and} \quad \mathbf{A}^\dagger\mathbf{A} = \begin{bmatrix} 21 & 10 & 4 & 0 \\ 10 & 21 & 10 & 4 \\ 4 & 10 & 21 & 10 \\ 0 & 4 & 10 & 21 \end{bmatrix}.$$

This result holds for any M and L, and is a consequence of the full banded Toeplitz property of $\mathbf{A}$, as we shall show. Imagine $\mathbf{A}_K$ is only a partial matrix obtained by dropping rows from $\mathbf{A}$. For example if

$$\mathbf{A}_K = \begin{bmatrix} 1 & 0 & 0 & 0 \\ 2 & 1 & 0 & 0 \\ 4 & 2 & 1 & 0 \\ 0 & 4 & 2 & 1 \\ 0 & 0 & 4 & 2 \end{bmatrix} \quad \text{then} \quad \mathbf{A}_K^\dagger\mathbf{A}_K = \begin{bmatrix} 21 & 10 & 4 & 0 \\ 10 & 21 & 10 & 4 \\ 4 & 10 & 21 & 10 \\ 0 & 4 & 10 & 5 \end{bmatrix}.$$

The Toeplitz property is violated by $\mathbf{A}_K^\dagger\mathbf{A}_K$ because of the last diagonal element. So $\mathbf{A}_K^\dagger\mathbf{A}_K$ is Toeplitz for $K = M + L$ but not necessarily so for smaller K.

Proof of Toeplitz property. To prove that $\mathbf{A}^\dagger\mathbf{A}$ is Toeplitz when $\mathbf{A}$ is full banded Toeplitz, note that the mth column of $\mathbf{A}$ is the full impulse response $c(n)$ shifted down by m. Thus the (k, m) element of $\mathbf{A}^\dagger\mathbf{A}$ is

$$[\mathbf{A}^\dagger\mathbf{A}]_{km} = \sum_n c^*(n-k)c(n-m) = \sum_{n=0}^{L} c(n)c^*(n-k+m) = r(k-m),$$

where

$$r(\ell) \stackrel{\Delta}{=} \sum_n c(n)c^*(n-\ell)$$

is the autocorrelation of the deterministic sequence $c(n)$. Using the fact that $r(\ell) = r^*(-\ell)$ we see that $\mathbf{A}^\dagger\mathbf{A}$ has the form demonstrated below for $M = 4$.

$$\mathbf{R} \stackrel{\Delta}{=} \mathbf{A}^\dagger\mathbf{A} = \begin{bmatrix} r(0) & r^*(1) & r^*(2) & r^*(3) \\ r(1) & r(0) & r^*(1) & r^*(2) \\ r(2) & r(1) & r(0) & r^*(1) \\ r(3) & r(2) & r(1) & r(0) \end{bmatrix}. \tag{8.23}$$

This matrix is clearly Hermitian and Toeplitz. $\bigtriangledown \bigtriangledown \bigtriangledown$

Here are some consequences of the fact that $\mathbf{A}^\dagger\mathbf{A}$ has the form (8.23):

1. ***Insensitivity to channel phase.*** Given an FIR channel

$$C(z) = \sum_{n=0}^{L} c(n)z^{-n} = c(0)\prod_{k=1}^{L}(1 - z^{-1}z_k),$$

suppose we create a new channel

$$C_{new}(z) = \sum_{n=0}^{L} c_{new}(n)z^{-n} = c(0)\prod_{k=1}^{L}(1 - z^{-1}z_k) \times \frac{z_m^* - z^{-1}}{1 - z^{-1}z_m}$$

This is an FIR channel with the mth zero z_m replaced by $1/z_m^*$. Since the factor

$$\frac{z_m^* - z^{-1}}{1 - z^{-1}z_m}$$

is an allpass filter, the magnitude response is unchanged:

$$|C_{new}(e^{j\omega})| = |C(e^{j\omega})|.$$

Only the phase response of the channel is changed by this. Since the autocorrelation $r(k)$ is the inverse Fourier transform of $|C(e^{j\omega})|^2$, it follows that $c(n)$ and $c_{new}(n)$ have the same autocorrelation. Thus, even though the full banded Toeplitz matrix $\mathbf{A}$ is different for $C(z)$ and $C_{new}(z)$, the matrix $\mathbf{A}^\dagger\mathbf{A}$ is identical for both channels.

2. ***Zero locations of channel, and noise gain.*** Since $\mathbf{A}^\dagger\mathbf{A}$ is the same for $C(z)$ and $C_{new}(z)$, it follows that $||\mathbf{A}^{\#}||$ (which depends only on $\mathbf{A}^\dagger\mathbf{A}$) is also unchanged. Since the reconstruction error at the receiver (due to channel noise) has the amplification factor $||\mathbf{A}^{\#}||^2/M$ (Eq. (8.10)) it then follows that the channel noise amplification is insensitive to whether the zeros of the channel are inside or outside the unit circle. This is a surprising result and follows from the fact that the receiver uses all P noisy samples in every block for the identification of the transmitted symbols. By contrast if the receiver had used only M of the received samples, then the equalization would be tantamount to inverting a square lower triangular Toeplitz matrix, like $\mathbf{A}_3$ in Eq. (8.19). In this case, zeros of $C(z)$ outside the unit circle can create a large noise gain as demonstrated in Ex. 8.1 (where $C(z)$ has zeros with magnitudes 2.2611, 2.2611, 2.0529, and 0.0953).

3. ***Channel with unit circle zeros.*** If an FIR channel has unit circle zeros, then the inverse $1/C(z)$ is unstable (even if we are willing to accept noncausal inverses). Thus there is no stable equalizer at all (if there is no redundancy like zero padding), and the channel noise is amplified in an unbounded manner by $1/C(z)$. But in a zero-padded system, the equalization works perfectly well: the full banded matrix $\mathbf{A}$ still has full rank, so $\sigma_i^2 > 0$ for all i, and the noise gain $||\mathbf{A}^{\#}||^2/M$ is finite.

4. ***Norm of A.*** Even though only the norm of the inverse $\mathbf{A}^{\#}$ is involved in the discussion of error, we would like to point out an interesting property satisfied by the norm of $\mathbf{A}$ itself. We have $||\mathbf{A}||^2 = \mathrm{Tr}\,(\mathbf{A}^\dagger\mathbf{A})$. Since the diagonal elements of $\mathbf{A}^\dagger\mathbf{A}$ are all equal to $r(0)$, it follows that

$$||\mathbf{A}||^2 = \mathrm{Tr}\,(\mathbf{A}^\dagger\mathbf{A}) = M\,r(0),$$

where $r(0) = \sum_n |c(n)|^2$ = energy of the channel. Note therefore that the quantity

$$\frac{\|\mathbf{A}\|^2}{M} = r(0) = \sum_{n=0}^{L} |c(n)|^2 \tag{8.24}$$

and depends only on the channel energy $r(0)$ and not on M.

Example 8.2: Channels with zeros outside the unit circle

Consider the channel

$$C(z) = 1 + z^{-1} + 0.31z^{-2} + 0.03z^{-3}$$

This has order $L = 3$ and the three zeros are *inside* the unit circle: $z_1 = -0.3, z_2 = -0.5$, and $z_3 = -0.2$. Choose $M = 8$ so that $P = M + L = 11$. Then the size of $\mathbf{A}_K$ can be $K = 8, 9, 10$, or 11. Calculations show:

K	$\|\mathbf{A}_K^{\#}\|^2/M$
8	2.37
9	2.05
10	2.02
11	2.02

The noise gain therefore decreases only slightly as we increase the size of $\mathbf{A}_K$. Now consider the channel

$$C_{rev}(z) = 0.03 + 0.31z^{-1} + z^{-2} + z^{-3},$$

which is the time-reversed version of $C(z)$. This has all the zeros *outside* the unit circle. Calculations show:

K	$\|\mathbf{A}_K^{\#}\|^2/M$
8	2.03×10^{13}
9	1.30×10^{8}
10	3.03×10^{3}
11	2.02

For $C_{rev}(z)$, since the zeros are outside the unit circle, $\|\mathbf{A}_K^{\#}\|^2$ is very large for $K = M, M+1$, and $M+2$. But for the full size matrix $\mathbf{A}_K$ with $K = 11$, the quantity $\|\mathbf{A}_K^{\#}\|^2$ is identical for $C(z)$ and $C_{rev}(z)$ as expected.

Example 8.3: Channels with zeros on the unit circle

Consider the channel $C(z) = \sum_{n=0}^{7} z^{-n}$, which has all seven zeros on the unit circle at the points

$$z_k = e^{-j2\pi k/7}, \quad 0 \leq k \leq 6.$$

We have $L = 7$, and choosing $M = 8$ we have $P = M + L = 15$. Calculations show:

K	$\|\mathbf{A}_K^{\#}\|^2/M$
8	1.87
9	1.75
10	1.59
11	1.45
12	1.31
13	1.18
14	1.05
15	0.89

Thus as K increases the quantity $\|\mathbf{A}_K^{\#}\|^2/M$ gets smaller, though not as dramatically as the case where the zeros of $C(z)$ are outside the unit circle. For unit circle zeros with higher multiplicity, the choice of large K becomes very crucial. For example let

$$C(z) = 1 + 3z^{-1} + 3z^{-2} + z^{-3},$$

which has three zeros at $z = -1$. With $M = 8$ so that $P = M + L = 11$, calculations show:

K	$\|\mathbf{A}_K^{\#}\|^2/M$
8	734.25
9	21.03
10	5.03
11	3.81

In this example there is a major improvement as K increases from M to $M + 1$. Again, the noise gain is least when $K = M + L = 11$.

In practice of course, an arbitrary channel can have zeros inside, on or outside the unit circle. The above examples show that, as long as we choose $K = P$ (i.e., use all the P samples in the output block at the receiver), the quantity $\|\mathbf{A}_K^{\#}\|^2$ is insensitive to these details, and the equalization can be done without undue noise amplification.

8.6 Effect of increasing the block size

In all examples so far, the block size M was held constant. We found that, as long as we keep all the P samples in the block, the noise gain is only mildly sensitive to zero locations of the channel. In this section we assume this, and study the effect of increasing M. The advantage of large M is that the bandwidth expansion ratio

$$\gamma = \frac{M+L}{M}$$

gets smaller. But how about the performance? For the lazy precoder with zero-forcing equalizer, how does the noise gain given by Eq. (8.12) change as M increases? It has been shown in the literature that for any channel this quantity increases monotonically with M [Ohno, 2006]. The formal proof of monotonicity is included in Appendix 8.A at the end of this chapter. In what follows, we show some examples of this increasing behavior. In all examples, the channels are normalized to have unit energy, that is $\sum_n |c(n)|^2 = 1$.

The top plot in Fig. 8.4 shows the behavior of $\|\mathbf{A}^{\#}\|^2/M$, for a fifth-order FIR channel with coefficients $c(0), c(1), \ldots, c(5)$ randomly chosen:

$$\mathbf{c}_1 = \begin{bmatrix} -0.1797 & -0.6920 & 0.0521 & 0.1195 & -0.4764 & 0.4948 \end{bmatrix}.$$

The noise gain increases monotonically and tends to settle down at a steady value for large M. The middle plot in the figure is for the channel

$$\mathbf{c}_2 = \begin{bmatrix} 0.5074 & -0.4347 & 0.6957 & 0.0265 & 0.0386 & 0.2595 \end{bmatrix}.$$

Unlike the first channel, the gain seems to be increasing steadily in the form of a straight line. It appears that there is something fundamentally different between the first and the second channels. Calculating the zeros of the channels we find that the absolute values of the zeros of the channels are:

$$3.8207, 1.2626, 0.8853, 0.8853, 0.7282 \quad (\text{for } \mathbf{c}_1);$$

$$1.0000, 1.0000, 0.9233, 0.9233, 0.5999 \quad (\text{for } \mathbf{c}_2).$$

Thus the second channel has a pair of unit circle zeros (a complex conjugate pair $0.5500 \pm j0.8352$). In fact, experiments have shown repeatedly that whenever the channel has some zeros on the unit circle with multiplicity one, the noise gain tends to have a straight line growth. This is demonstrated in the bottom plot of Fig. 8.4 for the simple channel

$$C_3(z) = c(0)(1 + z^{-1})$$

which has a unit circle zero of multiplicity one at $z = -1$. (In all examples the constant $c(0)$ is adjusted so that the channel has unit energy.) Figure 8.5 shows what happens when we have double zeros at a point on the unit circle. The plot is for the channel

$$C_4(z) = c(0)(1 + z^{-1})^2$$

and appears to be parabolic. Even for a reasonably small value of M the noise gain is quite large. Note that the scale for the vertical axis is very different when compared with Fig. 8.4. If the zeros are moved a little bit away from the unit circle then the growth is not as drastic. This is shown by the middle and bottom plots in Fig. 8.5 for $C_5(z) = c(0)(1 + 0.95z^{-1})^2$ and $C_6(z) = c(0)(1 + 0.9z^{-1})^2$, respectively. In these cases the growth slows down and tends to saturate. Finally Fig. 8.6 shows the behavior for the channel

$$C_7(z) = c(0)(1 + z^{-1})^3,$$

which has a triple zero on the unit circle. The growth of noise gain is unacceptable indeed. Again notice that the vertical scale is significantly different from the earlier plots. A couple of remarks are now in order:

1. Even though it is unlikely that a practical channel will have multiple zeros on the unit circle, it is possible that the channel gain is quite small in some frequency regions, thereby approximating the behavior of unit circle zeros. So the observations made here have some practical relevance.

2. While the monotone increasing behavior of $\|\mathbf{A}^{\#}\|^2/M$ as M increases has been proved (Appendix 8.A), the reason for the specific nature of the growth outlined above (for unit circle zeros) has not been theoretically studied.

8.7 Concluding remarks

The lazy precoder with a zero-forcing equalizer is admittedly a restricted special case of the more general transceiver with precoder $\mathbf{R}$ and equalizer $\mathbf{T}$. The noise gain can be reduced considerably by optimization. For example we can follow up $\mathbf{T}$ with a Wiener filter (Sec. 4.10). In fact we can reduce the reconstruction error even further by jointly optimizing $\mathbf{R}$ and $\mathbf{T}$ for a given channel and noise statistics (Chaps. 12 and 13). Another sophisticated method is called bit allocation, whereby fewer bits are transmitted on parts of the channel with poor frequency response (Sec. 14); this has the effect of selectively eliminating some parts of the channel which have very low gain.

Even though optimization helps a lot, the zero-padded optimal transceiver still exhibits a monotonically increasing reconstruction error as M increases! This will be elaborated in Chap. 18. As we shall see in Chap. 17, the situation is quite different for cyclic-prefix sytems (single as well as multicarrier).

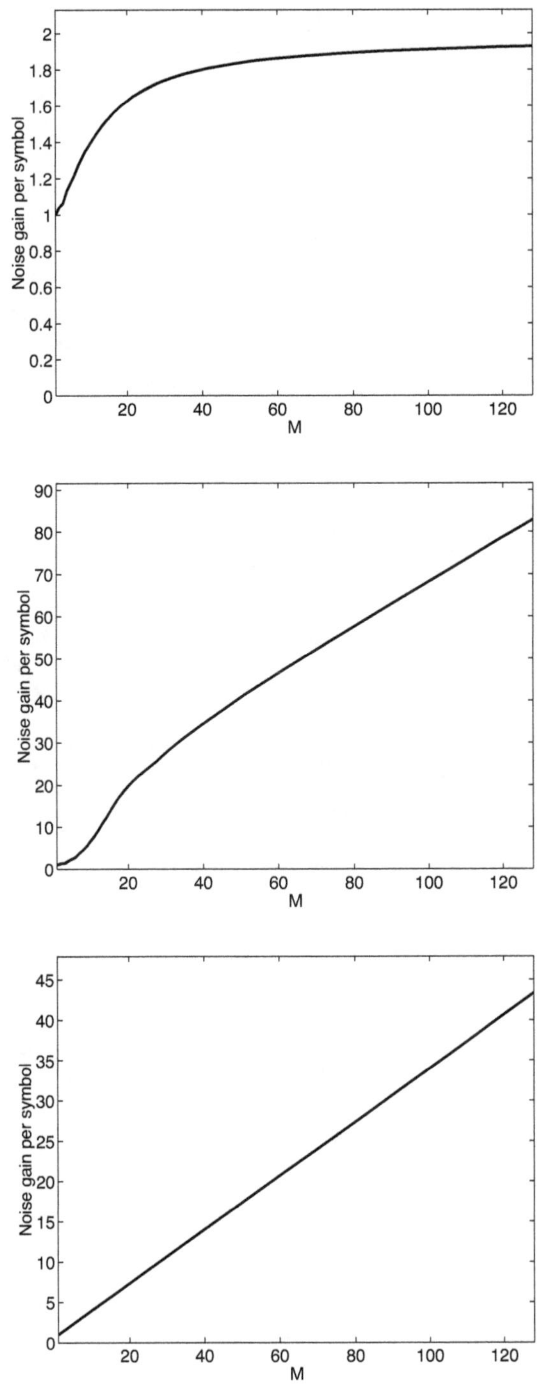

Figure 8.4. The behavior of lazy precoders (with zero-forcing equalizers). The noise gain is shown as a function of block size M. Random channel (top), semi-random channel with a complex conjugate pair of unit-circle zeros (middle), and channel proportional to $(1 + z^{-1})$ (bottom).

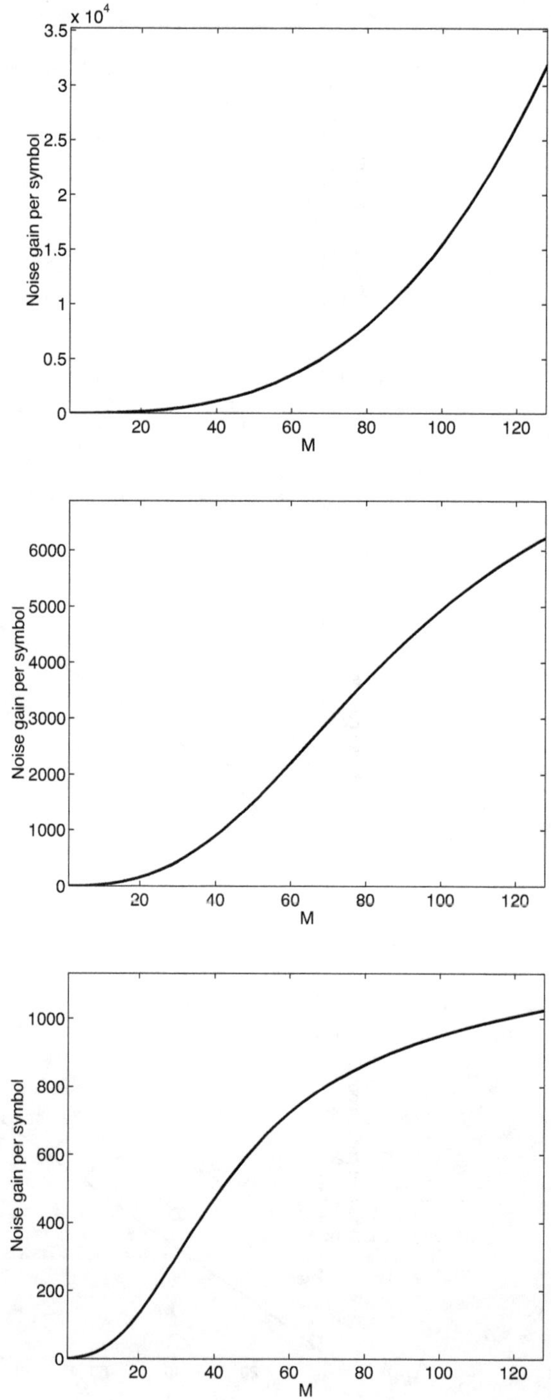

Figure 8.5. The behavior of lazy precoders (with zero-forcing equalizers). The noise gain is shown as a function of block size M. The channels are proportional to $(1+z^{-1})^2$ (top), $(1+.95z^{-1})^2$ (middle), and $(1+.9z^{-1})^2$ (bottom).

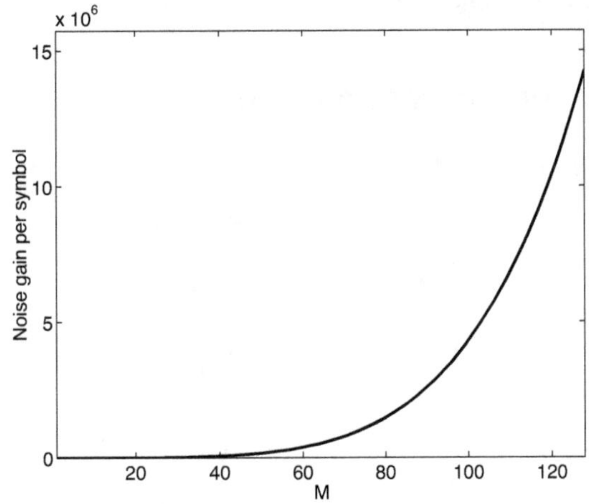

Figure 8.6. The behavior of lazy precoders (with zero-forcing equalizers). The noise gain is shown as a function of block size M for channel proportional to $(1 + z^{-1})^3$.

Appendix to Chapter 8

8.A Monotonicity of noise gain

With $\mathbf{A}_M$ denoting the $(M+L) \times M$ banded Toeplitz matrix (8.2), denote

$$\mathbf{R}_M \stackrel{\Delta}{=} \mathbf{A}_M^\dagger \mathbf{A}_M. \tag{8.25}$$

Using Eq. (8.8) with $\mathbf{T} = \mathbf{A}_M^\#$, we have

$$\|\mathbf{A}_M^\#\|^2 = \text{Tr}\Big(\mathbf{A}_M^\#(\mathbf{A}_M^\#)^\dagger\Big) = \text{Tr}\,(\mathbf{A}_M^\dagger \mathbf{A}_M)^{-1} \tag{8.26}$$

so that

$$\text{noise gain } = \frac{1}{M}\|\mathbf{A}_M^\#\|^2 = \frac{1}{M}\text{Tr}\left(\mathbf{R}_M^{-1}\right). \tag{8.27}$$

Proving that the noise gain is monotonically increasing is therefore equivalent to proving that

$$\frac{1}{M+1}\text{Tr}\left(\mathbf{R}_{M+1}^{-1}\right) \geq \frac{1}{M}\text{Tr}\left(\mathbf{R}_M^{-1}\right). \tag{8.28}$$

It should be recalled here that $\mathbf{R}_M$ is positive definite and Toeplitz (see Eq. (8.23)). We begin by proving the following lemma. The results in this appendix are based on the work of Ohno [2006].

♠**Lemma 8.2.** Let $\mathbf{R}_{M+1}$ be $(M+1) \times (M+1)$ positive definite. Partition it as

$$\mathbf{R}_{M+1} = \begin{bmatrix} \mathbf{R}_M & \mathbf{v} \\ \mathbf{v}^\dagger & c \end{bmatrix}, \tag{8.29}$$

where $\mathbf{R}_M$ is $M \times M$. Then

$$[\mathbf{R}_{M+1}^{-1}]_{kk} \geq [\mathbf{R}_M^{-1}]_{kk} \tag{8.30}$$

for $0 \leq k \leq M-1$. Similarly, suppose $\mathbf{S}_M$ is the lower $M \times M$ principal submatrix of $\mathbf{R}_{M+1}$, that is,

$$\mathbf{R}_{M+1} = \begin{bmatrix} d & \mathbf{u}^\dagger \\ \mathbf{u} & \mathbf{S}_M \end{bmatrix}. \tag{8.31}$$

Then we have

$$[\mathbf{R}_{M+1}^{-1}]_{k+1,k+1} \geq [\mathbf{S}_M^{-1}]_{kk} \tag{8.32}$$

for $0 \leq k \leq M-1$. ◇

Note that since $\mathbf{R}_{M+1}$ is positive definite, $\mathbf{R}_M$ and $\mathbf{S}_M$ are positive definite, and $c, d > 0$. The Lemma does not assume that the matrices are Toeplitz.

Proof. Using the formula for inverses of matrices in partitioned form (Sec. B.4.3, Appendix B) we get

$$\mathbf{R}_{M+1}^{-1} = \begin{bmatrix} \mathbf{I} & -\mathbf{R}_M^{-1}\mathbf{v} \\ \mathbf{0} & 1 \end{bmatrix} \begin{bmatrix} \mathbf{R}_M^{-1} & 0 \\ \mathbf{0} & \dfrac{1}{c - \mathbf{v}^\dagger \mathbf{R}_M^{-1}\mathbf{v}} \end{bmatrix} \begin{bmatrix} \mathbf{I} & 0 \\ -\mathbf{v}^\dagger \mathbf{R}_M^{-1} & 1 \end{bmatrix}.$$

$$= \begin{bmatrix} \mathbf{R}_M^{-1} + \dfrac{\mathbf{R}_M^{-1}\mathbf{v}\mathbf{v}^\dagger\mathbf{R}_M^{-1}}{c - \mathbf{v}^\dagger\mathbf{R}_M^{-1}\mathbf{v}} & \dfrac{-\mathbf{R}_M^{-1}\mathbf{v}}{c - \mathbf{v}^\dagger\mathbf{R}_M^{-1}\mathbf{v}} \\ \dfrac{-\mathbf{v}^\dagger\mathbf{R}_M^{-1}}{c - \mathbf{v}^\dagger\mathbf{R}_M^{-1}\mathbf{v}} & \dfrac{1}{c - \mathbf{v}^\dagger\mathbf{R}_M^{-1}\mathbf{v}} \end{bmatrix}$$

Since $\mathbf{R}_{M+1}^{-1}$ is also positive definite, all its diagonal elements are positive, so in particular $c - \mathbf{v}^\dagger\mathbf{R}_M^{-1}\mathbf{v} > 0$. Since $\mathbf{R}_M^{-1}\mathbf{v}\mathbf{v}^\dagger\mathbf{R}_M^{-1}$ is positive semidefinite, $[\mathbf{R}_M^{-1}\mathbf{v}\mathbf{v}^\dagger\mathbf{R}_M^{-1}]_{kk} \geq 0$. This proves Eq. (8.30). Next, from Eq. (8.31) we have

$$\mathbf{R}_{M+1}^{-1} = \begin{bmatrix} 1 & \mathbf{0} \\ -\mathbf{S}_M^{-1}\mathbf{u} & \mathbf{I} \end{bmatrix} \begin{bmatrix} \dfrac{1}{d - \mathbf{u}^\dagger\mathbf{S}_M^{-1}\mathbf{u}} & \mathbf{0} \\ \mathbf{0} & \mathbf{S}_M^{-1} \end{bmatrix} \begin{bmatrix} 1 & -\mathbf{u}^\dagger\mathbf{S}_M^{-1} \\ \mathbf{0} & \mathbf{I} \end{bmatrix}$$

$$= \begin{bmatrix} \dfrac{1}{d - \mathbf{u}^\dagger\mathbf{S}_M^{-1}\mathbf{u}} & \dfrac{-\mathbf{u}^\dagger\mathbf{S}_M^{-1}}{d - \mathbf{u}^\dagger\mathbf{S}_M^{-1}\mathbf{u}} \\ \dfrac{-\mathbf{S}_M^{-1}\mathbf{u}}{d - \mathbf{u}^\dagger\mathbf{S}_M^{-1}\mathbf{u}} & \mathbf{S}_M^{-1} + \dfrac{\mathbf{S}_M^{-1}\mathbf{u}\mathbf{u}^\dagger\mathbf{S}_M^{-1}}{d - \mathbf{u}^\dagger\mathbf{S}_M^{-1}\mathbf{u}} \end{bmatrix},$$

where we have used one of the two inversion formulas in Sec. B.4.3 (Appendix B). Using the same argument as before, Eq. (8.32) follows immediately. $\bigtriangledown\bigtriangledown\bigtriangledown$

♠**Lemma 8.3.** Consider $M+1$ real numbers $b_0, b_1 \ldots, b_M$ and M real numbers $c_0, c_1 \ldots, c_{M-1}$ such that

$$b_k \geq \begin{cases} c_k & 0 \leq k \leq M-1 \\ c_{k-1} & 1 \leq k \leq M. \end{cases} \tag{8.33}$$

Then

$$\frac{1}{M+1}\sum_{m=0}^{M} b_m \geq \frac{1}{M}\sum_{m=0}^{M-1} c_m. \tag{8.34}$$

That is, the arithmetic mean of $\{b_m\}$ is at least as large as the arithmetic mean of $\{c_m\}$. $\diamondsuit$

Proof. Equation (8.33) implies $\sum_{m=0}^{k-1} b_m \geq \sum_{m=0}^{k-1} c_m$ for $1 \leq k \leq M$, and $\sum_{m=\ell+1}^{M} b_m \geq \sum_{m=\ell}^{M-1} c_m$ for $0 \leq \ell \leq M-1$. Setting $\ell = k$ and adding these two inequalities we get $\sum_{m=0}^{k-1} b_m + \sum_{m=k+1}^{M} b_m \geq \sum_{m=0}^{M-1} c_m$ or equivalently

$$\sum_{m=0}^{M} b_m - \sum_{m=0}^{M-1} c_m \geq b_k \quad \text{for} \quad 1 \leq k \leq M-1. \tag{8.35}$$

For $k = M$ also this is true because of Eq. (8.33). Adding this for all k in $1 \leq k \leq M$ we have

$$M \sum_{m=0}^{M} b_m - M \sum_{m=0}^{M-1} c_m \geq \sum_{k=1}^{M} b_k \geq \sum_{k=0}^{M-1} c_k \quad (\text{using } b_k \geq c_{k-1} \text{ for } k \geq 1). \quad (8.36)$$

This is equivalent to $M \sum_{m=0}^{M} b_m - (M+1) \sum_{m=0}^{M-1} c_m \geq 0$, so the proof of Eq. (8.34) is complete. $\bigtriangledown \bigtriangledown \bigtriangledown$

Since the matrix $\mathbf{R}_M$ has the form (8.23), it follows that

$$\mathbf{R}_{M+1} = \begin{bmatrix} \mathbf{R}_M & \mathbf{v} \\ \mathbf{v}^\dagger & R(0) \end{bmatrix} = \begin{bmatrix} R(0) & \mathbf{u}^\dagger \\ \mathbf{u} & \mathbf{R}_M \end{bmatrix}. \quad (8.37)$$

It is therefore clear from Lemma 8.2 that

$$[\mathbf{R}_{M+1}^{-1}]_{kk} \geq [\mathbf{R}_M^{-1}]_{kk}, \quad \text{and} \quad [\mathbf{R}_{M+1}^{-1}]_{k+1,k+1} \geq [\mathbf{R}_M^{-1}]_{kk}. \quad (8.38)$$

Using Lemma 8.3 we therefore conclude that

$$\frac{1}{M+1} \sum_{k=0}^{M} [\mathbf{R}_{M+1}^{-1}]_{kk} \geq \frac{1}{M} \sum_{k=0}^{M-1} [\mathbf{R}_M^{-1}]_{kk},$$

which is equivalent to the desired result (8.28). Even though the Toeplitz property was not required in Lemma 8.2, it is easily verified that if Eq. (8.37) holds for all M then the matrix $\mathbf{R}_{M+1}$ has to be Toeplitz anyway.

Problems

8.1. Consider Ex. 8.1 again. With M and L as in that example, suppose we use the channel $C(z) = 4 - 20z^{-1} + 33z^{-2} - 20z^{-3} + 4z^{-4}$. With notations as in Sec. 8.5 compute $\|\mathbf{A}_K^{\#}\|^2/M$ for $8 \leq K \leq 12$, and make a table as in Ex. 8.1. Observe what happens when K increases from M to $M + 1$.

8.2. Consider the channel $C(z) = 1+0.5z^{-1}$. Let the block size be $M = 8$. With notations as in Sec. 8.5 compute $\|\mathbf{A}_K^{\#}\|^2/M$ for $K = M$ and $K = M + 1$. Repeat for the time-reversed version $C_{rev} = 0.5 + z^{-1}$.

8.3. In this problem the block size is fixed at $M = 64$. Compute the noise gain (8.12) (where $\mathbf{A}$ is the $P \times M$ full banded Toeplitz matrix), for the three channels

$$C_1(z) = c_1(0)(1 + z^{-1}),\ C_2(z) = c_2(0)(1 + z^{-1})^2,\ C_3(z) = c_3(0)(1 + z^{-1})^3,$$

where $c_k(0)$ are scale factors such that the channel energy $\sum_n |c_k(n)|^2$ is unity in each case.

8.4. In Problem 8.3 the channels have one, two, and three zeros at the same point on the unit circle. Repeat the problem with the channel zeros moved a little bit inside the unit circle:

$$C_1(z) = c_1(0)(1+\rho z^{-1}),\ C_2(z) = c_2(0)(1+\rho z^{-1})^2,\ C_3(z) = c_3(0)(1+\rho z^{-1})^3,$$

where $\rho = 0.9$. Again, make sure $c_k(0)$ are such that the channel energy $\sum_n |c_k(n)|^2$ is unity in each case.

8.5. Consider the inequality (8.28), which was proved assuming that $\mathbf{R}_M$ is a positive definite Toeplitz matrix. In this problem we emphasize the fact that the Toeplitz property is crucial. In particular the fact that $\mathbf{R}_M$ is a leading principal submatrix of $\mathbf{R}_{M+1}$, that is,

$$\mathbf{R}_{M+1} = \begin{bmatrix} \mathbf{R}_M & \mathbf{v} \\ \mathbf{v}^{\dagger} & R(0) \end{bmatrix}, \qquad (P8.5)$$

is not enough. Thus, let

$$\mathbf{R}_2 = \begin{bmatrix} 2 & 1 \\ 1 & 2 \end{bmatrix}$$

and

$$\mathbf{R}_3 = \begin{bmatrix} 2 & 1 & 0 \\ 1 & 2 & a \\ 0 & a & 2 \end{bmatrix},$$

where a is real, so Eq. (P8.5) holds for $M = 2$.

(a) What are the eigenvalues λ_1, λ_0 of $\mathbf{R}_2$?

(b) Express the eigenvalues η_2, η_1, η_0 of $\mathbf{R}_3$ in terms of a.

(c) Show that

$$\frac{1}{2}\mathrm{Tr}\left(\mathbf{R}_2^{-1}\right) = \frac{2}{3}, \qquad \frac{1}{3}\mathrm{Tr}\left(\mathbf{R}_3^{-1}\right) = \frac{11 - a^2}{6(3 - a^2)}$$

You can use the fact that the trace is the sum of eigenvalues.

(d) Show that $\mathbf{R}_2$ is positive definite, and that $\mathbf{R}_3$ is positive definite if and only if $a^2 < 3$.

(e) For $0 < a^2 < 1/3$ show that

$$\frac{1}{3}\mathrm{Tr}\left(\mathbf{R}_3^{-1}\right) < \frac{1}{2}\mathrm{Tr}\left(\mathbf{R}_2^{-1}\right),$$

though $\mathbf{R}_3$ and $\mathbf{R}_2$ are both positive definite.

Thus, even though the positive definite matrix $\mathbf{R}_3$ has $\mathbf{R}_2$ as its leading principal submatrix, the inequality (8.28) is violated.

Part 2 Transceiver optimization

9 History and outline

9.1 Introduction

In this part we consider the optimization of the transceiver system shown in Fig. 9.1. Here $\mathbf{H}(z)$ is a MIMO channel characterized by a $J \times P$ transfer matrix. The transmitter has a $P \times M$ precoder $\mathbf{F}(z)$ and the receiver has an $M \times J$ equalizer $\mathbf{G}(z)$. The goal is to optimize $\mathbf{F}(z)$ and $\mathbf{G}(z)$ subject to appropriate constraints such that some measure of performance is minimized. This simple system leads to a multitude of interesting optimization problems depending upon what is being optimized and what the constraints are. For example one might wish to minimize the bit error rate under a power constraint. Or one might want to minimize the transmitted power under a certain bit error rate constraint. Another example might be to minimize the mean square reconstruction error for fixed power. The optimization can be performed with or without the **zero forcing** or **ZF** constraint

$$\mathbf{G}(z)\mathbf{H}(z)\mathbf{F}(z) = \mathbf{I}. \tag{9.1}$$

And in some situations one might impose the constraint that the precoder be very simple (e.g., $\mathbf{F}(z) = \mathbf{I}$, or the IDFT matrix) so that the transmitter does not require any knowledge of the channel. Most of our attention in this book will be restricted to the case where the channel $\mathbf{H}(z)$ is memoryless, and is represented by the constant matrix $\mathbf{H}$. In this case the precoder and equalizer are also constant matrices $\mathbf{F}$ and $\mathbf{G}$.

Notations. Notations are as described in Sec. 1.6. In all the optimization chapters, we use the notations $\mathbf{F}, \mathbf{H}$, and $\mathbf{G}$ for the precoder, channel, and equalizer. This is unlike in some of the earlier chapters where we used $C(z)$, $\mathbf{C}$, etc. for channels. Since the notations within a chapter are fixed and consistent, this should cause no confusion to the reader. $\bigtriangledown \bigtriangledown \bigtriangledown$

The results derived for the general transceiver in Fig. 9.1 can be applied to special cases such as zero-padded transceivers (Sec. 7.2) and cyclic prefix trans-

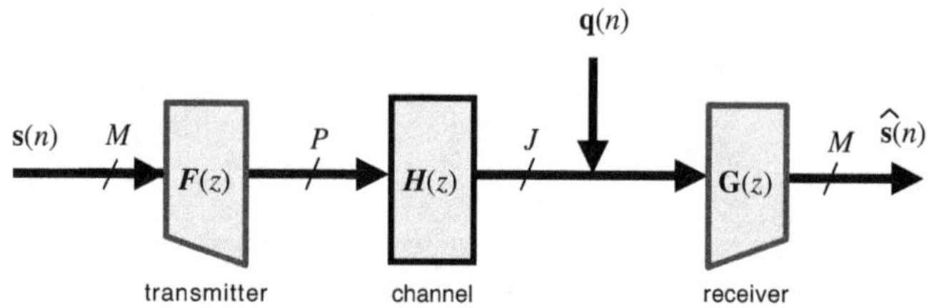

Figure 9.1. The general form of a transceiver sytem with channel $\mathbf{H}(z)$, precoder $\mathbf{F}(z)$, and equalizer $\mathbf{G}(z)$.

ceivers. Such special cases are useful to convert scalar channels into MIMO channels without interblock interference; examples include OFDM and DMT transceivers (Sec. 7.3). This will be elaborated further in Chaps. 17 and 18. For the most part we will assume that the signal $\mathbf{s}(n)$ and noise $\mathbf{q}(n)$ are zero mean, uncorrelated WSS processes, with covariances $\sigma_s^2\mathbf{I}$ and $\sigma_q^2\mathbf{I}$. There is nearly no loss of generality in this, as explained in Chap. 12. In Chap. 14, which considers bit allocation as in DMT systems, the users $s_k(n)$ are allowed to have different powers, so the covariance of $\mathbf{s}(n)$ is allowed to be a diagonal matrix which may not be identity.

9.2 A brief history of transceiver optimization

Transceiver optimization has had a long history starting from the late 1950s. In the context of digital communications, research in this area has been especially intense starting from the 1990s, thanks to the technological breakthroughs that led to DSL, MIMO, and wireless communication systems. However, since much of the recent work has its roots in the rich history of transceivers, it is important to appreciate the historical perspective. With this in mind we now give a brief overview of the excellent body of research in this field. Many good papers had to be omitted in the interest of brevity. The discussion of the papers mentioned will only be brief. However, the essence of many of these papers will be covered in detail in later chapters, as outlined later in Sec. 9.3.

9.2.1 General remarks on early work on equalization

Work on the optimization of equalizers in digital communications started in the early 1960s. Tufts considered the problem with and without zero-forcing constraints [Tufts, 1960, 1965]. The joint optimization of the transmitter and receiver was considered by Smith [1965]. Besides zero-forcing, the case of partial zero forcing (duobinary) was also considered. Lucky [1965] showed how an FIR equalizer can actually be designed in practice by taking into account the knowledge of the channel impulse response. The maximum possible intersymbol interference at the receiver was shown to be proportional to the ℓ_1 norm of the impulse response of the cascade of the channel and equalizer. This peak distortion, rather than a mean square error, was minimized. With appropriate normalization, Lucky was able to formulate the problem of finding optimal equalizer coefficients as a convex problem. The optimal filter coefficients are identified as part of a training mode. In this approach the objective to be minimized was independent of the symbol stream and the noise. As long as the distortion caused by the unequalized channel was sufficiently small (i.e., the channel does not completely close the so-called eye-diagram), the optimal equalizer was shown to be a zero-forcing equalizer. The mathematics of the adaptive FIR equalizer which works continuously in real time was soon developed [Lucky, 1966] as an extension of this fundamental work. In this work Lucky developed a decision directed method for maximum likelihood estimation. The adaptive FIR equalizer based on the minimization of mean square error rather than peak error was soon developed [Lucky and Rudin, 1966], and an extensive mathematical study, including convergence analysis, was published by Gersho [1969]. An excellent survey of early literature on equalization can be found in Lucky [1973].

9.2.2 Joint optimization of continuous-time filters

One of the oldest papers to address a joint optimization problem similar to the $(\mathbf{F}, \mathbf{G})$ optimization was the 1952 paper by Costas from the General Electric Co. In that paper, published in the proceedings of the IRE, a single-input single-output (SISO) analog communication channel was considered [Costas, 1952]. This is shown in Fig. 9.2. For such a system the author addressed the problem of optimizing the prefilter $F(j\omega)$ and the postfilter (or equalizer) $G(j\omega)$ (also known as pre-emphasis and de-emphasis networks) such that the continuous-time signal $s(t)$ was reconstructed with minimum mean square error, that is,

$$\mathcal{E}_{reco} \stackrel{\Delta}{=} E|\widehat{s}(t) - s(t)|^2 \tag{9.2}$$

was minimized.[1] The channel transfer function was assumed to be identity, but the noise $q(t)$ was allowed to have arbitrary power spectrum. The input $s(t)$ and the additive channel noise were assumed to be jointly WSS and uncorrelated, but with arbitrary power spectra $S_{ss}(j\omega)$ and $S_{qq}(j\omega)$. Under these assumptions it was shown that the optimal receiver filter $G(j\omega)$ is given by

$$G(j\omega) = \frac{F^*(j\omega)S_{ss}(j\omega)}{|F(j\omega)|^2 S_{ss}(j\omega) + S_{qq}(j\omega)} \tag{9.3}$$

The reader will recognize this as the **Wiener** filter which extracts $s(t)$ from the noisy signal $y(t)$ (see Sec. F.3.1 in Appendix F). This is the best $G(j\omega)$ to use for any fixed $F(j\omega)$. With this the mean square reconstruction error becomes

$$\mathcal{E}_{reco} = \int_{-\infty}^{\infty} \frac{S_{qq}(j\omega)S_{ss}(j\omega)}{|F(j\omega)|^2 S_{ss}(j\omega) + S_{qq}(j\omega)} \frac{d\omega}{2\pi} \tag{9.4}$$

[1]Note that this is an analog objective function which does not restrict the amplitudes of $s(t)$ (as would a digital communication system). So there is no consideration of minimizing error probability at this point.

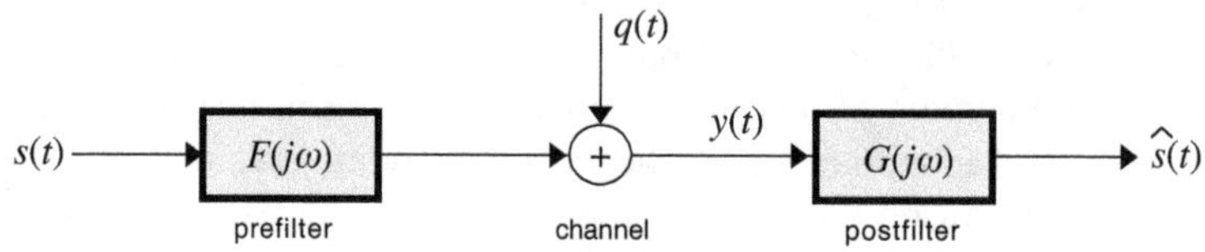

Figure 9.2. A continuous-time transceiver system.

This error was then minimized with respect to the prefilter $F(e^{j\omega})$ subject to a power constraint

$$\int_{-\infty}^{\infty} S_{ss}(j\omega)|F(j\omega)|^2 \frac{d\omega}{2\pi} = p_0. \tag{9.5}$$

This resulted in the optimum solution

$$|F(j\omega)|^2 = \begin{cases} \gamma\sqrt{\frac{S_{qq}(j\omega)}{S_{ss}(j\omega)}} - \frac{S_{qq}(j\omega)}{S_{ss}(j\omega)} & \text{if this is non-negative,} \\ 0 & \text{otherwise.} \end{cases} \tag{9.6}$$

Here the constant γ is such that the power constraint is satisfied. Intuitively, if the noise $S_{qq}(j\omega)$ is very strong compared to the signal $S_{ss}(j\omega)$ at a certain frequency, then the second term dominates, making the right-hand side negative. This forces $|F(j\omega)|^2 = 0$. The wisdom is therfore to choose the prefilter to be zero at frequencies where the channel is bad; $F(j\omega)$ can be taken to be any spectral factor of the above expression. For practical reasons, of course, one would take the spectral factor which yields a stable filter. The preceding solution can readily be derived by applying the **Karush–Kuhn–Tucker** (KKT) theory for constrained optimization (Chap. 22).

The joint optimization problem under power constraint, formulated by Costas, is fundamental. As technology progressed from analog communication systems to digital systems and then to MIMO systems and wireless systems, the problem and its solution have made repeated appearances in appropriately modified forms since the 1950s. We will therefore see the basic mathematical formulation and solutions of the form (9.6) repeatedly. For example, when the analog communication channel is replaced with a digital system, a similar problem arises. When the SISO digital communication system is generalized to a MIMO system, a matrix version of the problem arises (Chaps. 12 and 13). Other variations of the MIMO digital communication problem include those where a zero-forcing constraint is applied (Chap. 12), those where optimal bit allocation is possible among users (Chap. 14), and those where bit error rate is minimized (Chap. 16).

9.2.3 Optimization with zero forcing, for TV signals

In 1964 R. A. Bruce from the Bell Telephone Laboratories considered a joint optimization problem involving a prefilter and a postfilter. The application was in the transmission of television signals after sampling and quantization using pulse code modulation (PCM) [Bruce, 1964]. A condition similar to the zero forcing condition was imposed, and the filters were optimized using a standard power constraint. Simplifying the solution to its essentials, one of the main conclusions was that the prefilter magnitude has the form

$$|F(j\omega)| = \gamma\Big(\frac{S_{qq}(j\omega)}{|H(j\omega)|^2 S_{ss}(j\omega)}\Big)^{1/4}. \tag{9.7}$$

This dependence of prefilters on the fourth root of the *noise-to-signal ratio* is again a fundamental feature of systems which use zero forcing. We will see a similar feature in all the advanced and modern versions of this result whenever zero forcing is in play (e.g., Chaps. 11 and 12). In a zero-forcing system the prefilter has larger gain in regions where the channel is bad (i.e., $S_{qq}/|H(j\omega)|^2$ is large). In a MMSE system which does not impose zero forcing (as in Costas' work), this is not true; in fact, the prefilter gain is zero when the noise power is excessively large (Eq. (9.6)).

Whether zero forcing is imposed or not, note that the prefilter and postfilter share the task of optimal filtering. Setting the prefilter to unity and peforming all equalization at the receiver is a loss of optimality.

9.2.4 Joint optimization of filters for discrete-time symbol streams

The problem of transmitting a discrete-time sequence $s(n)$ over a continuous-time channel $H(j\omega)$ also involves optimization of continuous-time transmitting and receiving filters. But this is more complicated because continuous and discrete time are involved together in the structure. To discuss this, consider again the simple schematic diagram of the digital communication system, reproduced in Fig. 9.3. The schematic building block D/C (discrete-to-continuous-time converter) simply converts $s(n)$ to an impulse train of samples spaced apart by T seconds. The prefilter $F(j\omega)$ at the transmitter therefore produces the continuous-time signal

$$x(t) = \sum_{n=-\infty}^{\infty} s(n)f(t-nT), \tag{9.8}$$

where $f(t)$ is the impulse response of $F(j\omega)$. This signal is transmitted over the channel which is described by a transfer function $H(j\omega)$ and additive noise $q(t)$. The received signal $y(t)$ is then filtered by the receiver postfilter $G(j\omega)$ and sampled by the C/D (continuous-to-discrete) converter to obtain $\widehat{s}(n) = \widehat{s}_c(nT)$. The symbol reconstruction error is given by

$$e(n) = \widehat{s}(n) - s(n). \tag{9.9}$$

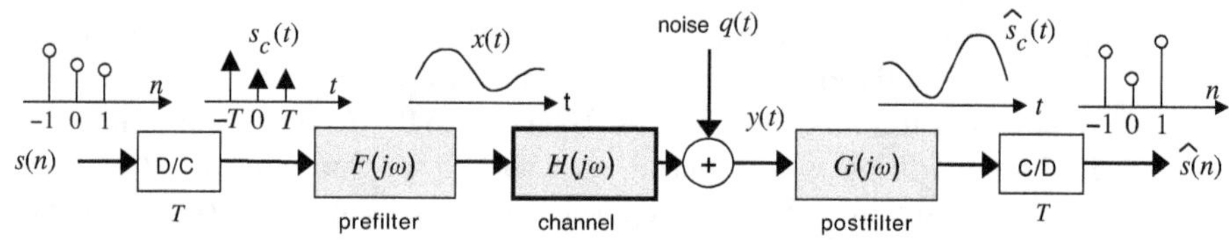

Figure 9.3. A discrete-time symbol stream $s(n)$ transmitted through a noisy channel. The reconstructed symbol stream is $\widehat{s}(n)$. See text.

Under appropriate statistical assumptions, we can formulate the mean squared reconstruction error as

$$\mathcal{E}_{mse} = E|\widehat{s}(n) - s(n)|^2 \tag{9.10}$$

and minimize it subject to a power constraint similar to (9.5). For example, optimization of the two filters for minimum mean square error of the symbol stream was considered in the fundamental work by Berger and Tufts [1967]. Therein it is once again found that the receiver filter is Wiener-like and the transmitting filter has a form similar in flavor to the Costas filter (9.6), appropriately modified to account for the fact that the transmitted signal is not a continuous waveform but a pulse stream. Classic work on automatic equalization of digital communication channels appeared in the mid sixties [Becker *et al.*, 1965], [Lucky, 1965, 1966].

In 1971, Chan and Donaldson from the University of British Columbia again addressed the problem of joint optimal design of pre- and postfilters for a SISO communication channel [Chan and Donaldson, 1971], but in a more general way. The usual power constraint was imposed, but the zero-forcing constraint and the statistical assumptions on the signal and noise components were relaxed (e.g., there could be correlation between signal and noise). Furthermore the authors worked out a general theory which was able to cover sampling and quantization as in digital communication systems. Several types of modulations, such as PAM, PCM, and DPCM, emerged as special cases. The PAM scenario envisioned by the authors (Fig. 3 in Chan and Donaldson [1971]) in fact resembles the modern digital transmultiplexer system with M inputs and M outputs. The general solution is similar to Costas' solution — Wiener-like filter $G(j\omega)$ at the receiver, and a transmitter filter $F(j\omega)$ solved using the KKT theory of constrained optimization. The design of filters for the case of bandlimited channels, under zero intersymbol interference constraint, was addressed later [Chevillat and Ungerboeck, 1982]. The theory of digital communication systems with decision feedback equalization was developed in the early 1970s [Falconer and Foschini, 1973], [Salz, 1973], [Witsenhausen, 1975].

9.2.5 Turning the hybrid system into an all-discrete system

In Chap. 4 we showed how a discrete-time communication system (with a continuous-time channel) can be redrawn as an all discrete-time system. We found that in the equivalent representation the digital channel is the sampled version of the convolution of the continuous-time channel, precoder, and equalizer. So, the three components are not separately shown in the all-discrete equivalent. It was shown by Ericson in the early 1970s that the problem of jointly optimizing $F(j\omega)$ and $G(j\omega)$ in Fig. 9.3 can be reduced to the problem of optimizing digital filters $F_d(z)$ and $G_d(z)$ in an equivalent system of the form shown in Fig. 9.4 [Ericson, 1971, 1973]. This is a purely discrete-time system. The channel transfer function $H_d(z)$ and the noise $q_d(n)$ are related to $H(j\omega)$ and $q(t)$ in a rather subtle way, as described in Chap. 10 (Sec. 10.4). One optimizes $F_d(z)$ and $G_d(z)$ and then identifies the original filters $F(j\omega)$ and $G(j\omega)$ using a simple relation. Ericson's observation of the equivalence between the hybrid system and the discrete-time system was motivated by the fact that the optimum filters in earlier work by many people always turned out to be "bandlimited" in a generalized sense.

9.2.6 MIMO channels

Multi-input multi-output, or MIMO, channels (Fig. 9.1) were soon realized to be of great interest [Brandenburg and Wyner, 1974]. The optimization of the precoder was considered as early as 1976 [Lee and Petersen, 1976]. During the early and mid 1980s the optimization of MIMO channels started receiving more attention. In 1984, Amitay and Salz considered minimum MSE equalization in 2×2 channels [Amitay and Salz, 1984]. In 1985, Salz considered the problem of optimizing continuous-time MIMO filters for the case where a discrete-time vector sequence $\mathbf{s}(n)$ had to be transmitted over a continuous-time channel $\mathbf{H}(j\omega)$. So the system is similar to Fig. 9.3 except that $\mathbf{F}(j\omega), \mathbf{H}(j\omega)$, and $\mathbf{G}(j\omega)$ are square matrices. This can be regarded as the MIMO version of Costas' original problem in discrete time, and the solution indeed has similarities. Salz showed that the equalizer can be derived from Wiener-like equations and the prefilter identified using KKT conditions. A nontrivial step in the MIMO case is to prove that the optimal solution can be obtained by diagonalizing the channel using singular value decomposition (SVD), and deriving optimal filters for the diagonal channel.[2] A simple and ingenious theorem due to Witsenhausen, based on the idea of **Schur-convex** functions, was used by Salz [1985] to reduce the problem to that of optimizing a diagonal matrix. After this, the simple KKT theory of constrained optimization does the rest. The details of this theory, generalized for the case of rectangular channels, prefilters, and equalizers, will be presented in Chap. 13.

[2]It should be noted that the optimal prefilter and equalizer matrices for diagonal channels turn out to be diagonal only for certain objective funtions, e.g., MMSE systems with average power constraint. Other types of power constraints have been considered as well; see [Weng and Vaidyanathan, 2010] and references therein. For objective functions, such as average BER, this is not the case, as we shall see in Chap. 11.

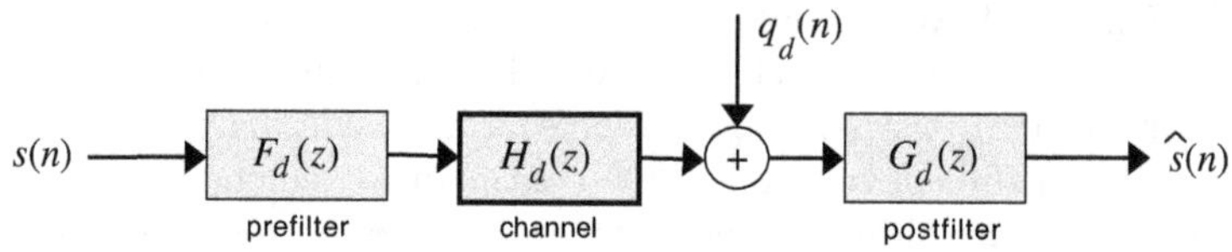

Figure 9.4. Discrete-time equivalent of the hybrid system in Fig. 9.3.

In 1988 Malvar and Staelin from the Massachusetts Institute of Technology addressed the joint optimization of the prefilter and equalizer matrices for the case where they are allowed to be rectangular constant matrices [Malvar and Staelin, 1988b].[3] Furthermore, instead of imposing a total power constraint as in all earlier work, they imposed a constraint on individual components. The solution therefore had a different flavor. Namely, there were extra unitary matrix factors in **F** and **G**, e.g., **Hadamard** matrix factors. These are similar in vein to the Hadamard factors that were shown to arise (much later in history) when BER rather than MSE is minimized (Secs. 11.5 and 16.2). The joint optimization problem was also revisited by Yang and Roy [1994a, 1994b]; this paper also provides a good historical perspective.

The equalization of scalar channels with memory can be converted to that of MIMO channels without memory in a number of ways. Two of the most common techniques for this are the **zero-padding** method and the **cyclic-prefixing** method. Such conversion allows for bit allocation and/or power allocation for different subchannels generated from one user. In a pioneering paper in 1997, Xia considered the equalization of scalar channels with redundant precoders, emphasizing zero-padded systems [Xia, 1997]. In a series of papers in 1999, Scaglione and colleagues considered the equalization as well as blind identification of scalar channels with redundant precoders, especially zero-padded systems [Scaglione *et al.*, 1999a, 1999b]. They also addressed the problem of optimizing the precoder and equalizer for various criteria, including mutual information [Scaglione *et al.*, 2000]. In the same year, Li and Ding also considered the joint optimization of communication systems with redundant filter banks under power constraints [Li and Ding, 1999]. Their derivation of the optimal solution closely parallels that of Salz [1985], including the use of Witsenhausen's result to justify diagonal solutions.

Another class of equalization systems which converts scalar channels into memoryless MIMO channels is based on the use of cyclic-prefix redundancy [Peled and Ruiz, 1980]. These have lead to **OFDM** and **DMT** systems, which are discussed in detail in Chaps. 7 and 17. DMT systems use optimal bit allocation, and have traditionally used DFT and inverse DFT matrices for the equalizer and precoder.

[3] In earlier work, Malvar and Staelin also considered similar problems in the context of decimation and interpolation of signals [Malvar and Staelin, 1988a].

In 2000, Lin and Phoong removed the a priori DFT constraint, and showed how to optimize an orthonormal precoder and an equalizer in a zero-padded DMT system [Lin and Phoong, 2000]. They showed how the optimal **bit allocation** problem can be decoupled from the optimization of the precoder. Their work assumed the zero-forcing constraint. In spite of this constraint there is still some flexibility in optimizing the equalizer because of the zero-padding redundancy at the transmitter. Lin and Phoong also showed how to optimize the redundant part. In a later paper they showed that, under appropriate assumptions, the orthonormality restriction on the precoder was not a loss of generality in a bit-allocated system [Lin and Phoong, 2001b]. The same is not true in other types of systems (e.g., OFDM with transmitter and receiver jointly optimized for MSE, and systems optimized for BER), as we shall see. The optimization of MIMO transceivers with bit allocation will be considered in its general form in Chap. 14.

It is sometimes desirable to optimize a DMT system with specific quality of service (QoS) constraints for each of the users. The above work of Lin and Phoong can readily be extended to this case as shown by Pandharipande and Dasgupta [2003] who used Schur convex functions to establish some interesting results for this problem. Optimization with QoS constraints has also been addressed by other authors [Palomar, Lagunas and Cioffi, 2004], [Yasotharan, 2006]. Transceiver optimization in a multiuser context has been addressed by others as well, e.g., see Serbetli and Yener [2004].

The important problem of optimizing transceivers for **bit error rate** was considered by a number of authors. In 2000, Wang and colleagues considered a number of important aspects of this problem, and optimized a multiuser detector for BPSK modulation [Wang, Lu, and Antoniou, 2000]. In 2003, Lin and Phoong addressed theoretical optimality formally, and proved a number of important results for the case of BER optimized OFDM systems. For example, they showed for "reasonable SNR" that the **single-carrier cyclic-prefix** system has optimum BER performance, whereas the conventional (multicarrier) OFDM system has the worst performance (among the orthonormal-precoder class with or without zero forcing). At the same time Ding and colleagues also worked on a similar problem; they addressed both zero-padded precoders and cyclic-prefix precoders under the zero-forcing constraint. Many identical conclusions were presented in these two papers [Ding *et al.*, 2003], [Lin and Phoong, 2003]. Joint optimization of transceivers for minimum BER in binary modulation systems was considered by Hjϕrungnes, Diniz, and de Campos [2004]. The optimization of MIMO transceivers for BER will be presented in its general form in Chap. 16. We will see that zero-padded and cyclic-prefix systems follow as special cases of this general derivation. The problem of bit rate maximization has also been considered by a number of authors (see Arslan, Evans, and Kiaei [2001], and references therein).

By now the reader should have realized that the transceiver optimization problem is in fact a collection of many different problems, depending upon the objective funtion to be minimized and the constraints to be used. The objective function can be the mean square error or the bit error rate or transmitted power;

there may or may not be a zero-forcing constraint; there may or may not be an opportunity to allocate bits optimally. And so on and so forth. Historically, depending on the combination of such requirements, a different problem was formulated and solved each time. In 2003 Palomar and colleagues showed how to unify many of these problems into an elegant setting based on convex optimization [Palomar, Cioffi, and Lagunas, 2003], [Palomar and Barbarosa, 2005], [Palomar and Jiang, 2006]. Also see Guess [2003]. This formulation covers a wide range of objective functions and constraints for the case of MIMO channels. Palomar and colleagues showed that the nature of the solution can be divided into different classes depending upon whether the objective function is **Schur convex** or **Schur concave.** Not surprisingly many of the earlier results can be derived as special cases of this grand scheme. The trend of literature in this field took a definite change after these publications.

Most of the results described above for MIMO channels assume the channel, precoder, and equalizer to be constant matrices, possibly rectangular. Some of these earlier papers allowed frequency dependence [Salz, 1985], [Yang and Roy, 1994a, 1994b], [Li and Ding, 1999], but ended up with solutions that require ideal unrealizable filters. More recently some authors have included precoder matrices with finite memory and have shown how to *numerically optimize* these filters for practical applications [Farhang-Boroujeny and Lin, 2003], [Mertins, 2003], [Phoong, Chang, and Chen, 2005], [Vijaya Krishna and Hari, 2006]. For example, Mertins allowed precoders to be FIR matrices. In that work a scalar channel with memory was converted into a MIMO channel by using redundancy as in earlier papers. But the author made allowances for the possibility that in practice the redundancy may not be sufficient to render the MIMO equivalent channel a memoryless constant. To allow for the frequency dependence of the MIMO equivalent, the author therefore allowed the precoder to be an FIR filter matrix rather than a constant matrix; the equalizer in general can be IIR. The work by Phoong, Chang, and Chen [2005] restrained both the precoder and equalizer to be FIR filters. The paper by Vijaya Krishna and Hari [2006] performed the optimization for the case of minimum-redundancy precoders, a concept which originated from earlier work by Lin and Phoong [2002].[4] Also see [Ribeiro, de Campos, and Diniz, 2002].

As mentioned in the historical remarks at the end of Chap. 7, the OFDM idea was pioneered by Chang [1966, 1970], and further developed by Cimini [1985]. The cyclic-prefix idea was first introduced by Peled and Ruiz [1980]. One of the pioneers who introduced the idea of the multitone channel was Kalet [1989] who showed that if the frequency range of the channel can be partitioned into subbands, with each subband carrying a different constellation with different power, this gives added flexibility for power and bit allocation. Also see Bingham [1990]. Later on, Sari *et al.* [1994] analyzed the OFDM system for mobile radio and showed that the single-carrier cyclic-prefix (SC-CP) system actually

[4] For a scalar FIR channel of order L, the minimum redundancy required in a zero-padded or cyclic-prefix precoder system had previously been thought to be L samples per block. Lin and Phoong [2002] showed that this is not necessary – the minimum required redundancy depends on the properties of the Smith form decomposition of the blocked channel, and can be as small as one sample per block!

outperforms the multicarrier (OFDM) system from the point of view of bit error rate. The discrete multitone (DMT) transceiver with bit allocation, for DSL applications on telephone lines, was proposed in 1991 in the landmark paper by Chow *et al.* [1991].

In Sec. 5.7 we described decision feedback equalizers, also known as **DFE**s. These equalizers use past decisions on symbols to refine the decision on the present symbol. This idea can be extended to decision feedback equalization in memoryless MIMO channels. Here, past decisions within a block are used to improve future decisions in that block, at a given block-time instant. This makes the transceiver a nonlinear system. Transceivers without decision feedback are often referred to as **linear transceivers**. The decision feedback transceiver has a precoder $\mathbf{F}$, feedforward equalizer $\mathbf{G}$, and a feedback matrix $\mathbf{B}$. These matrices can be jointly optimized to minimize the bit error rate, with or without the zero-forcing constraint. Assuming there is no bit allocation, the performance of optimized DFE transceivers has been shown to be strictly better than that of linear transceivers. (When bit allocation is allowed, the situation is different [Weng, Chen, and Vaidyanathan, 2010a].) It has also been shown that optimal DFE systems without zero forcing also maximize the mutual information between the input and output of the channel. Recent fundamental work on this optimization was done by Zhang *et al.* [2002], Zhang, Kavčić, and Wong [2005], Jiang, Hager, and Li [2005a], Jiang, Li, and Hager, [2005b, 2005c], and Xu *et al.* [2006]. Further important developments based on Schur convexity and majorization theory can be found in Jiang, Palomar, and Varanasi [2007], Jiang, Hager, and Li [2008], Shenouda and Davidson [2008], and Weng, Chen, and Vaidyanathan [2010a].[5] In many of these references the channel $\mathbf{H}$ is assumed to be known. There are many methods to estimate channels, although this topic will not be addressed in this book. The interested reader should consult references listed in Tong, Xu, and Kailath [1994], Tong *et al.* [1995], Giannakis [1997], Scaglione *et al.* [1999b], and Su and Vaidyanathan [2007a, 2007b].

Such is the brief history of transceiver optimization, as it evolved over many decades. Unavoidably we have been able to mention only a limited number of papers which directly relate to the story that is going to unfold in the next few chapters. As mentioned above, a unifying approach to many optimization problems in transceiver theory has been proposed by Palomar *et al.* [2003], based on the theory of Schur-convex funtions. While this approach is very elegant and effective, we shall not use it here. We will use the notion of Schur convex functions only for the problem of bit error rate minimization. For the other optimization problems we shall take a conventional approach, which uses traditional tools from variational calculus and linear algebra.

The reason for our choice of this approach is that it is always valuable to have a first exposure to these problems via traditional tools. After this the student will be well prepared to appreciate many of the excellent papers in the literature that take other approaches.

[5] These ideas have also found applications in transform coding theory [Weng, Chen, and Vaidyanathan, 2010b].

9.3 Outline for Part 2

We now provide a brief outline of the topics covered in various chapters in this part. Chapters 20, 21, and 22 in Part 3 provide the mathematical background that might be useful to the reader during the study of these chapters. The material in Chaps. 20 and 21 (matrix calculus and Schur-convex functions) contains much greater detail than what is necessary to follow the chapters in this book. The reason for including these extensive chapters is that they will serve as useful references for the reader who wishes to pursue the literature in greater detail.

Chapter 10: Optimal SISO transceivers

The optimizations of the scalar prefilter and the scalar equalizer for a SISO channel are considered in this chapter. This problem is considered in a number of contexts. First, there are *continuous-time* filters (prefilter and equalizer) to be optimized in the *digital* communication system. Then there are digital filters to be optimized in the equivalent all-digital system, as we shall explain. The relation between continuous-time and sampled-time optimization problems is treated in some detail here.

Chapter 11: Optimal transceivers for diagonal channels

In this chapter we consider the simple case of diagonal channels. That is, in the notation of Fig. 9.1, $M = P = J$ and $\mathbf{H}$ is a diagonal matrix. Given $\mathbf{H}$, we optimize the diagonal matrices $\mathbf{F}$ and $\mathbf{G}$ to minimize the mean square error in reconstruction, with and without the zero-forcing (ZF) constraint. These are known, respectively, as the **ZF-MMSE** problem and the **pure-MMSE** problem. While the diagonal channel offers a relatively simple set of problems to solve, the solutions are quite illuminating. They place in evidence the main difference between MMSE systems with and without the zero-forcing constraint. During these exercises, the student is also introduced to the KKT conditions (Karush–Kuhn–Tucker conditions) which are widely used in constrained optimization problems (also reviewed in Chap. 22).

One important result in this chapter is that, when the precoder and equalizer are jointly optimized, the result is such that the channel equalization is split between the transmitter and the receiver. When the channel information is available at the transmitter, it is therefore a loss of generality to perform all equalization at the receiver.

This chapter also shows how to minimize the average bit error rate (BER) by using extra unitary matrices at the transmitter and receiver. We will see that, even though $\mathbf{H}$ is diagonal, the optimal precoder and equalizer which minimize the bit error rate are *not* diagonal. To arrive at this result we use the theory of Schur-convex functions (reviewed in Chap. 21). For the case where the objective is just the MSE with or without the ZF constraint, it turns out that the optimal $\mathbf{F}$ and $\mathbf{G}$ are indeed diagonal; the proof follows as a special case of the results proved in the following chapters.

Chapter 12: MMSE transceivers with zero-forcing equalizers

In this chapter we consider the joint optimization of the precoder $\mathbf{F}$ and equalizer $\mathbf{G}$ for nondiagonal channels (with P, J, and M possibly different integers). More specifically, we consider minimization of the mean square reconstruction error under the zero-forcing constraint (the ZF-MMSE problem). We prove that the optimal solution is such that the channel is "diagonalized," that is, $\mathbf{F}$ and $\mathbf{G}$ have unitary factors in them such that the cascade of $\mathbf{F}, \mathbf{H}$, and $\mathbf{G}$ resembles the diagonal system addressed in Chap. 11. For rectangular matrices this is a nontrivial result, and requires some elegant theorems from matrix theory. The chapter makes use of matrix calculus with complex matrices; the requisite background for this is reviewed in Chap. 20. The results of this chapter will be applied to OFDM as well as zero-padded transmitters in Chaps. 17 and 18.

One interesting outcome is that the optimal precoder can be assumed to have orthogonal columns, that is $\mathbf{F}^\dagger\mathbf{F}$ is a diagonal matrix, even though it is not necessarily identity. That is, the precoder matrix is orthogonal, though not orthonormal. Thus, precoder orthogonality is not a loss of generality (under the conditions and assumptions stated in the chapter).

Chapter 13: MMSE transceivers without zero-forcing equalizers

In this chapter we reconsider the joint optimization of the precoder $\mathbf{F}$ and equalizer $\mathbf{G}$ for nondiagonal channels without the zero-forcing constraint (the pure-MMSE problem). While the mathematics of derivation is different from that of Chap. 12, the final form of the optimal solution is again such that the channel is diagonalized. The details of the solution differ from Chap. 12 in essentially the same way that the ZF-MMSE solution and pure-MMSE solution for diagonal channels differ (Chap. 11). The results of this chapter will be applied to OFDM as well as zero-padded transmitters in Chaps. 17 and 18.

As in the ZF-MMSE case, the pure-MMSE solution is also such that the optimal precoder matrix $\mathbf{F}$ is orthogonal, though not orthonormal.

Chapter 14: Bit allocation and power minimization

In this chapter we optimize $\mathbf{G}$ and $\mathbf{F}$ jointly to minimize the transmitter power under some specified constraints on the bit error rates for the symbols $s_k(n)$. Zero forcing is assumed. The optimization of $\mathbf{F}$ and $\mathbf{G}$ is performed jointly with the optimization of bits allocated to the symbol streams $s_k(n)$. As in Chaps. 12 and 13 we again find that the optimum solution "diagonalizes" the channel. The results of this chapter are applicable to discrete multitone or DMT systems (Chap. 17).

As in the preceding chapters, the solution is such that the precoder matrix $\mathbf{F}$ is orthogonal. In fact we will see that it can be assumed to be orthonormal in this chapter without loss of generality.

Chapter 15: Transceivers with orthonormal precoders

In this chapter we consider transceivers in which the precoder is restricted to be orthonormal. Subject to this constraint we show how the precoder and equalizer can be jointly optimized with or without the zero-forcing constraint. For the case where $\mathbf{F}$ is rectangular (i.e., $P > M$) we will show that the best $\mathbf{F}$ is again such that the channel is "diagonalized." Special cases of this system include the single-carrier and multicarrier cyclic-prefix (SC-CP and MC-CP) systems that are normally used in OFDM and DMT applications.

Chapter 16: Minimization of error probability in transceivers

In this chapter we consider the problem of minimizing the symbol error rate. It will be shown that linear transceivers which have been optimized to minimize the mean square error constraint can be converted to transceivers with minimum symbol error rate by the introduction of a pair of unitary matrices, one at the transmitter and one at the receiver. So there is a very simple connection between MMSE transceivers and transceivers with minimum error rate. This connection holds true whether there is zero forcing or not.

Chapter 17: Optimization of cyclic-prefix transceivers

This chapter is an application of the results of Chaps. 12–16 for the case of cyclic prefix systems which convert a scalar channel into a MIMO memoryless channel. The results covered in this chapter therefore include OFDM systems optimized for MMSE and BER with or without zero forcing. Also covered are optimal DMT systems with bit allocation. Many numerical examples of optimal OFDM and DMT systems are presented to demonstrate the key principles.

Chapter 18: Optimization of zero-padded transceivers

This chapter is an application of the results of Chaps. 12–16 for the case of zero padded (ZP) systems, which convert a scalar channel into a rectangular MIMO memoryless channel. This chapter also analyzes the so-called **lazy precoder** system in some detail. Several numerical examples are presented to demonstrate the ideas of the chapter. One outcome of the examples is the observation that the BER performance of optimal ZP systems degrades with increasing block size, especially when the channel has zeros close to the unit circle. For the case of OFDM systems, however, the BER performance is not a monotonically increasing or decreasing function of the block size – it all depends on the relative positioning of the channel nulls with respect to the DFT frequencies, which depend on the block size.

Chapter 19: Optimization of DFE transceivers

This is the final chapter for Part 2. We consider MIMO transceivers with decision feedback and optimize the transceiver matrices jointly to minimize the mean square reconstruction error. This is a very detailed chapter, in a way climaxing the earlier results on optimal transceivers. Cases with and without zero forcing are considered separately in the same chapter. It is shown in this chapter that MMSE DFE transceivers also automatically have minimum symbol error probability. It is also shown that, for the case where zero forcing is not imposed, the optimal DFE transceiver also maximizes the mutual information between the channel input and output.

Optimization summary

Appendix I gives a summary of the main optimization results in Part 2 of the book, with each major result summarized on one page.

10

Single-input single-output transceiver optimization

10.1 Introduction

In this chapter we consider the optimization of scalar filters for single-input single-output (SISO) channels. A number of optimization problems which arise in different contexts will be considered. In Sec. 10.2 we begin with the digital communication system of Fig. 10.1 for a fixed channel $H(j\omega)$. We consider the optimization of the continuous-time prefilter (transmitted pulse shape) and postfilter (equalizer) to minimize the mean square reconstruction error under the zero-forcing condition on the product $F(j\omega)H(j\omega)G(j\omega)$. The zero-forcing condition does not uniquely determine the above product. It will be shown that the optimal product (under the zero-forcing condition) is the so-called optimal compaction filter of the channel (Sec. 10.2.3). Usually the filters that result from the above optimization problem are ideal, unrealizable, filters and can only be approximated. The equivalent digital channel therefore requires further equalization. In Sec. 10.3 we consider the problem of jointly optimizing a *digital* prefilter–postfilter pair to minimize the mean square error. Both the zero-forcing and the non-ZF situation are considered.

Section 10.4 revisits Fig. 10.1 for an arbitrary channel $H(j\omega)$ from a more general viewpoint and formulates some general conditions on the filters $F(j\omega)$ and $G(j\omega)$ for optimality. The most general forms of the postfilter and prefilter for optimality are established. These forms were first derived by Ericson [1971, 1973]. Using these results we can argue that the optimization of the continuous-time filters in Fig. 10.1 can always be reformulated as the optimization of a digital prefilter–postfilter pair.

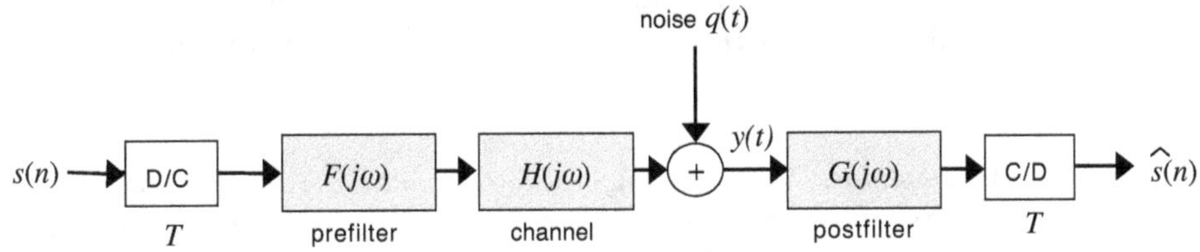

Figure 10.1. A digital communication system to transmit messages through a continuous-time channel.

However, some standard continuous-time filtering is required first: namely a matched filter at the receiver, and a so-called "optimum compaction filter" at the transmitter. Historically, Ericson's work unifies many earlier papers in transceiver literature which solved the SISO transceiver optimization problems under various settings.

In Sec. 10.5 we revisit SISO channels with oversampling at the receiver, and some simple examples of optimization are considered. In Sec. 10.6 we briefly mention a different kind of problem, namely the optimal pulse-shaping problem for the case where a *single* pulse is transmitted over a continuous-time channel. The criterion to be optimized is the signal-to-noise ratio at the receiver. This "single shot" scenario creates a different solution compared to the earlier pulse-shaping solutions because there is no intersymbol interference. Such pulse design problems arise in pulsed radar systems.

10.2 Optimization of the SISO communication system

Consider the digital communication system shown in Fig. 10.1 which transmits a sequence of symbols $s(n)$ over a continuous-time channel. This system was described in detail in earlier chapters. We now consider the problem of optimizing the prefilter $F(j\omega)$ and equalizer $G(j\omega)$ to minimize the mean square error.

10.2.1 Minimizing MSE under the product constraint

Let us assume that the transfer function $G(j\omega)H(j\omega)F(j\omega)$ is constrained to be a fixed function of frequency:

$$G(j\omega)H(j\omega)F(j\omega) = H_c(j\omega) \quad \text{(product constraint)}. \tag{10.1}$$

Under this condition we will optimize $G(j\omega)$ and $F(j\omega)$ such that the mean square error is minimized. If we wish to have the zero-forcing property, then $H_c(j\omega)$ should be such that its impulse response is Nyquist(T), that is,

$$h_c(nT) = \delta(n) \quad \text{(ZF constraint)}. \tag{10.2}$$

We will derive expressions for the optimal pair of filters $\{F(j\omega), G(j\omega)\}$ for fixed $H_c(j\omega)$ satisfying the ZF constraint, and fixed transmitted power. The exact shape of $H_c(j\omega)$ does not matter in the formulation, as the solutions will be expressed in terms of $H_c(j\omega)$. The shape of $H_c(j\omega)$ will itself be optimized later in Sec. 10.2.2.

Assumptions. We will assume that the noise $q(t)$ is WSS with zero mean and power spectrum $S_{qq}(j\omega)$. The signal $s(n)$ is assumed to be a random process with independent samples having zero mean and variance σ_s^2. We also assume that $s(n)$ and $q(t)$ are statistically independent for all integer n and real t. $\bigtriangledown\bigtriangledown\bigtriangledown$

From Appendix E (Sec. E.3) we know that the output of $F(j\omega)$ is a CWSS(T) process with power spectrum $\sigma_s^2|F(j\omega)|^2/T$. So the transmitted power (average variance at the output of $F(j\omega)$) is given by

$$p_0 = \frac{\sigma_s^2}{T}\int_{-\infty}^{\infty}|F(j\omega)|^2\frac{d\omega}{2\pi} \tag{10.3}$$

For fixed power p_0 and fixed $H_c(j\omega)$ satisfying the ZF constraint (10.2), we shall minimize the mean square error

$$\mathcal{E}_{mse} = E|e(n)|^2,$$

where

$$e(n) = \widehat{s}(n) - s(n).$$

In view of the zero-forcing constraint, the reconstruction error is entirely due to the noise $q(t)$ filtering through $G(j\omega)$. Thus

$$e(n) = \int_{-\infty}^{\infty} g(\tau)q(nT-\tau)d\tau.$$

Since $q(t)$ is WSS, the mean square value of $e(n)$ is independent of n and is given by

$$\mathcal{E}_{mse} = \int_{-\infty}^{\infty} S_{qq}(j\omega)|G(j\omega)|^2\frac{d\omega}{2\pi}$$

In view of the product constraint (10.1) we can rewrite this as

$$\mathcal{E}_{mse} = \int_{-\infty}^{\infty} S_{qq}(j\omega)\frac{|H_c(j\omega)|^2}{|F(j\omega)|^2|H(j\omega)|^2}\frac{d\omega}{2\pi} \tag{10.4}$$

Thus the product constraint (10.1) has been eliminated, and the goal is to optimize $F(j\omega)$ such that Eq. (10.4) is minimized subject to the constraint (10.3).

♠**Theorem 10.1.** *Optimal SISO transceiver.* Consider the digital communication system shown in Fig. 10.1. Under the power constraint (10.3) and the product constraint (10.1), where $H_c(j\omega)$ satisfies the ZF constraint (10.2), the optimal combination of the filters $F(j\omega)$ and $G(j\omega)$ is given by

$$F(j\omega) = \beta e^{j\theta_f(\omega)}\left|\frac{H_c(j\omega)}{H(j\omega)}\right|^{1/2} S_{qq}^{1/4}(j\omega) \tag{10.5}$$

and

$$G(j\omega) = \frac{e^{-j[\theta(\omega)+\theta_f(\omega)]}}{\beta}\left|\frac{H_c(j\omega)}{H(j\omega)}\right|^{1/2} S_{qq}^{-1/4}(j\omega). \tag{10.6}$$

The minimized mean square error per symbol is given by

$$\mathcal{E}_{mmse} = \frac{\sigma_s^2}{p_0 T}\left(\int_{-\infty}^{\infty} \frac{S_{qq}^{1/2}(j\omega)}{|H(j\omega)|}|H_c(j\omega)|\,\frac{d\omega}{2\pi}\right)^2. \tag{10.7}$$

In these expressions $\theta_f(\omega)$ is arbitrary (e.g., it can be taken to be zero for simplicity), $\theta(\omega)$ is the phase of $H(j\omega)/H_c(j\omega)$, and the constant β is computed using the power constraint (10.3). $\diamondsuit$

Proof. From Cauchy-Schwartz inequality (Appendix A) we have

$$\int_{-\infty}^{\infty} S_{qq}(j\omega)\frac{|H_c(j\omega)|^2}{|F(j\omega)|^2|H(j\omega)|^2}\frac{d\omega}{2\pi}\int_{-\infty}^{\infty}|F(j\omega)|^2\frac{d\omega}{2\pi}$$
$$\geq \left(\int_{-\infty}^{\infty} S_{qq}^{1/2}(j\omega)\left|\frac{H_c(j\omega)}{H(j\omega)}\right|\frac{d\omega}{2\pi}\right)^2. \tag{10.8}$$

Substituting from Eq. (10.3) and rearranging this we get

$$\int_{-\infty}^{\infty} S_{qq}(j\omega)\frac{|H_c(j\omega)|^2}{|F(j\omega)|^2|H(j\omega)|^2}\frac{d\omega}{2\pi}$$
$$\geq \frac{\sigma_s^2\left(\int_{-\infty}^{\infty} S_{qq}^{1/2}(j\omega)|H_c(j\omega)/H(j\omega)|d\omega/2\pi\right)^2}{p_0 T} \tag{10.9}$$

Equality is achieved when the two integrands on the left-hand side of Eq. (10.8) are equal up to a scale factor. So we can take the optimal prefilter solution to be as in Eq. (10.5), where $\theta_f(\omega)$ is an arbitrary phase factor and the constant β is such that the power constraint (10.3) is satisfied. The optimum equalizer $G(j\omega)$ can then be found from the product constraint (10.1), and yields Eq. (10.6). The minimized MSE is given by the right hand side of Eq. (10.9), which proves Eq. (10.7). $\triangledown\triangledown\triangledown$

Remarks

1. ***Bandlimiting.*** The optimal solution assumes that the channel $H(j\omega)$ and the noise spectrum $S_{qq}(j\omega)$ are nonzero for all ω (because these appear in certain denominators). The error expression (10.7) also requires the assumption $H(j\omega) \neq 0$. In practice, the channel is bandlimited and so are the filters $G(j\omega)$ and $F(j\omega)$. So these assumptions are required to hold only in this band. All the infinite integrals in the preceding results are then replaced with finite integrals limited to the baseband.

2. ***Realizable approximations.*** In general the optimal solutions (10.5) and (10.6) represent unrealizable filters. In practice these can only be approximated. We also have the further requirement of causality of these filters. Causality can be achieved by introducing a sufficient amount of delay into the approximations of the ideal solutions.

3. ***Matched filter interpretation.*** From Eqs. (10.5) and (10.6) we see that the optimal filters $G(j\omega)$ and $F(j\omega)$ are related as follows:

$$G(j\omega) = \frac{e^{-j\theta(\omega)} F^*(j\omega)}{\beta^2 S_{qq}^{1/2}(j\omega)} \tag{10.10}$$

For the special case where (a) the noise is white ($S_{qq}(j\omega) = \sigma_q^2$), and (b) $\theta(\omega) = 0$ (phase of $H(j\omega)/H_c(j\omega)$ is zero), this reduces to

$$G(j\omega) = cF^*(j\omega) \tag{10.11}$$

for constant c. This expression is precisely the matched filter expression derived in Chap. 2 (Sec. 2.5), where the receiver filter was matched to the transmitted pulse shape. Note however, that the more general relation (10.10) satisfied by the optimal pair does not resemble the matched filter, which actually has the form

$$G_m(j\omega) = \frac{cF^*(j\omega)H^*(j\omega)}{S_{qq}(j\omega)} \tag{10.12}$$

as shown in Sec. 2.5.

10.2.2 Optimal choice of the product filter

The optimization in Sec. 10.2.1 was performed for a fixed product filter $H_c(j\omega)$ in Eq. (10.1). This product is arbitrary except for the zero-forcing constraint (10.2). The minimized mean square error (10.7) clearly depends on our choice of $H_c(j\omega)$. This raises the following question: what is the optimal choice of $H_c(j\omega)$ that minimizes (10.7) subject to the zero-forcing constraint (10.2)? We will address this question in this section. The main result to be proved is that $H_c(j\omega)$ should be chosen as an **optimal compaction filter** for the effective channel defined as

$$H_{eff}(j\omega) = \frac{H(j\omega)}{S_{qq}^{1/2}(j\omega)} \tag{10.13}$$

The meaning of "optimal compaction filter" will be clear as the discussion progresses. It is useful to note here that the Nyquist condition $h_c(nT) = \delta(n)$ is equivalent to

$$\frac{1}{T}\sum_{k=-\infty}^{\infty} H_c(j(\omega + k\omega_s)) = 1 \quad \text{(Nyquist constraint)} \tag{10.14}$$

for all ω, where

$$\omega_s = \frac{2\pi}{T}$$

See the review of sampling theory given in Appendix G.

10.2.2.A Formulating the optimization problem

The integral in Eq. (10.7) can be written as $\int A(\omega)B(\omega)d\omega/2\pi$, where

$$A(\omega) \stackrel{\Delta}{=} \frac{S_{qq}^{1/2}(j\omega)}{|H(j\omega)|}, \qquad B(\omega) \stackrel{\Delta}{=} \frac{|H_c(j\omega)|}{T} \tag{10.15}$$

are non-negative quantities. Given $A(\omega) \geq 0$, find a function $B(\omega) \geq 0$ such that

$$\psi \stackrel{\Delta}{=} \int_{-\infty}^{\infty} A(\omega)B(\omega)\frac{d\omega}{2\pi} \tag{10.16}$$

is minimized under the constraint

$$\sum_{k=-\infty}^{\infty} B(\omega + k\omega_s) = 1 \tag{10.17}$$

for all ω. Since the left-hand side is periodic, it is sufficient to ensure the condition for $0 \leq \omega < \omega_s$. For convenience we rewrite the objective function as

$$\psi = \int_0^{\omega_s} \sum_{k=-\infty}^{\infty} A(\omega + k\omega_s)B(\omega + k\omega_s)\frac{d\omega}{2\pi} \tag{10.18}$$

Since $A(\omega)$ is non-negative, the non-negative $B(\omega)$ which minimizes Eq. (10.18) should be such that, for each frequency ω_0 in $0 \leq \omega_0 < \omega_s$, the sum

$$\sum_{k=-\infty}^{\infty} A(\omega_0 + k\omega_s)B(\omega_0 + k\omega_s) \tag{10.19}$$

is minimized under the constraint (10.17). Let k_0 be an integer such that

$$A(\omega_0 + k_0\omega_s) \leq A(\omega_0 + k\omega_s)$$

for all integer k.

♠**Lemma 10.1**. If $B(j\omega)$ is chosen such that

$$B(\omega_0 + k\omega_s) = \begin{cases} 1 & \text{for } k = k_0 \\ 0 & \text{otherwise,} \end{cases} \tag{10.20}$$

then Eq. (10.19) is minimized, and the constraint (10.17) is satisfied as well. ◇

Proof. Let a_k be a fixed set of non-negative numbers arranged such that

$$a_0 \leq a_1 \leq a_2 \ldots \tag{10.21}$$

We are required to find $b_k \geq 0$ such that $\sum_k b_k = 1$ and $\sum_k a_k b_k$ is minimized. Now,

$$\begin{aligned}\sum_k a_k b_k &\geq a_0 \sum_k b_k \quad \text{(using Eq. (10.21))} \\ &= a_0 \quad (\text{using } \textstyle\sum_k b_k = 1).\end{aligned}$$

Thus the minimum value of $\sum_k a_k b_k$ is a_0, and it is readily achieved by choosing $b_0 = 1$ and $b_k = 0$ otherwise. Identifying a_k and b_k with $A(\omega_0+k\omega_s)$ and $B(\omega_0 + k\omega_s)$ (appropriately reordered), the result follows. $\bigtriangledown \bigtriangledown \bigtriangledown$

While the solution (10.20) is not unique because there could be more than one k_0 with minimum value $A(\omega_0+k_0\omega_s)$, it is clear that one optimal solution can readily be identified this way. Since $A(j\omega)$ and $B(j\omega)$ are defined as in Eq. (10.15), we can restate this result as follows. In the following statement $H_{eff}(j\omega) = H(j\omega)/S_{qq}^{1/2}(j\omega)$ denotes the effective channel (see Eq. (10.13)).

♠**Theorem 10.2**. *Optimal product filter.* For every ω_0 in $0 \leq \omega_0 < \omega_s$ assign

$$H_c(j(\omega_0 + k\omega_s)) = \begin{cases} T & \text{for } k = k_0 \\ 0 & \text{otherwise,} \end{cases} \tag{10.22}$$

where k_0 is an integer such that

$$|H_{eff}(j(\omega_0 + k_0\omega_s))| \geq |H_{eff}(j(\omega_0 + k\omega_s))| \tag{10.23}$$

for all k, thus defining $H_c(j\omega)$ for all ω. This $H_c(j\omega)$ is such that the mean square error (10.7) is minimized subject to the Nyquist constraint (10.14). $\diamondsuit$

In practice, the channel $H(j\omega)$ is bandlimited, which means that the test (10.23) needs to be conducted only for a finite number of integers k (see Example 10.1). Even though the "number" of frequencies is uncountably infinite, the filters can often be graphically constructed, as we shall demonstrate later with examples.

10.2.3 Optimum compaction filters

The Nyquist function $H_c(j\omega)$ constructed in Theorem 10.2 is called the **optimal compaction filter** for the function $H_{eff}(j\omega)$. For every frequency ω_0, this filter passes $H_{eff}(j(\omega_0 + k\omega_s))$ for one value of k (where this is maximum) and rejects the rest. In some sense the compaction filter extracts the best part, or rather, "the best alias-free(T) band"[1] of the channel $H_{eff}(j\omega)$. The theorem essentially says that using this "best part" minimizes the reconstruction error. Compaction filters also arise naturally in optimal subband coding problems [Unser, 1993], [Tsatsanis and Giannakis, 1995], [Kirac and Vaidyanathan, 1998], [Vaidyanathan, 1998]. Note that, by construction, the optimum compaction filter $H_c(j\omega)$ has the following properties.

[1] Alias-free bands are discussed in Appendix G.

1. ***Unrealizability.*** $H_c(j\omega)$ is an ideal filter with passband gain equal to T and stopband gain equal to zero. So the filter can only be approximated in practice.

2. ***Multiband property.*** $H_c(j\omega)$ may have more than one passband, but the total bandwidth, counting all passbands, is $\omega_s = 2\pi/T$.

3. ***Nyquist(T) property.*** $H_c(j\omega)$ satisfies the Nyquist(T) property, that is,

$$h_c(nT) = \delta(n), \tag{10.24}$$

or equivalently

$$\sum_{k=-\infty}^{\infty} H_c(j(\omega + k\omega_s)) = T. \tag{10.25}$$

4. ***Alias-free property.*** Since no two terms in Eq. (10.25) overlap (owing to the defining equation (10.22)), the filter $H_c(j\omega)$ has the alias-free property. So, a signal $y(t)$ which is bandlimited to the passband of $H_c(j\omega)$ can be reconstructed from its samples $y(nT)$ by passing the sampled version through the filter $H_c(j\omega)$. The passband region of $H_c(j\omega)$ is an example of alias-free(T) regions described in Appendix G.

Example 10.1: Optimum compaction filters

Figure 10.2(a) shows an example of the effective channel $H_{eff}(j\omega)$. In this example $h_{eff}(t)$ is complex, since $|H_{eff}(j\omega)|$ is not symmetric. The symbol frequency $\omega_s = 2\pi/T$ is indicated in the figure. The total bandwidth of $H_{eff}(j\omega)$ is $\sigma + 3\omega_s$, and is considerably larger than ω_s. To construct the filter $H_c(j\omega)$ described in Theorem 10.2, imagine that the frequency axis is divided into consecutive intervals of length ω_s. Four such intervals overlap with the passband of $H_{eff}(j\omega)$ as indicated in the figure. To construct the compaction filter we proceed as follows: given any ω_0 in $[0, \omega_s]$ we compare the magnitudes of $H_{eff}(j\omega)$ at the frequencies $\omega_0 + k\omega_s$ for integer values of k. For fixed ω_0, one such frequency falls inside each region in Fig. 10.2(a). We simply choose one of the samples with maximum magnitude (i.e., pick one integer k which maximizes the magnitude of $H_{eff}(j(\omega_0 + k\omega_s))$), and discard the rest. In this way, we isolate the dominant part of $H_{eff}(j\omega)$, as demonstrated in Fig. 10.2(b). The optimal compaction filter $H_c(j\omega)$ by definition is an ideal filter whose passband coincides with this chosen band, as shown in Fig. 10.2(c). The magnitudes of optimal pre- and postfilters (10.5) and (10.6) (up to scale) are shown in Fig. 10.2(d), assuming $S_{qq}(j\omega)$ is constant.

If the effective channel has real impulse response $h_{eff}(t)$ then a compaction filter can be constructed with real impulse response $h_c(t)$. This is because when $h_{eff}(t)$ is real, $|H_{eff}(j\omega)|$ is symmetric, and $H_c(j\omega)$ will turn out to be symmetric too. Figure 10.3(a) shows an example. Proceeding as in the preceding example, we arrive at the optimal compaction filter shown in Fig. 10.3(c).

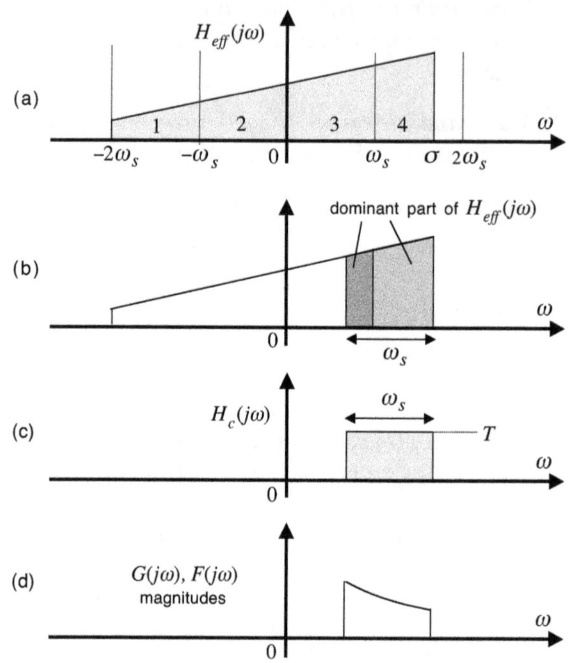

Figure 10.2. Example of construction of an optimum compaction filter. Shape of the effective channel, (b) dominant aliasfree(T) band of the channel, the optimum compaction filter, and (d) magnitudes of optimal pre- and postfilt (up to scale) assuming $S_{qq}(j\omega)$ is constant.

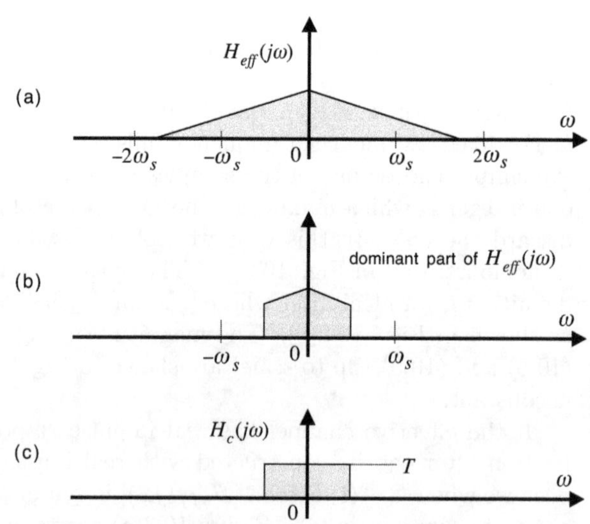

Figure 10.3. Example of construction of an optimum compaction filter with impulse response. (a) Shape of the effective channel, (b) dominant alias-free band of the channel, and (c) the optimum compaction filter.

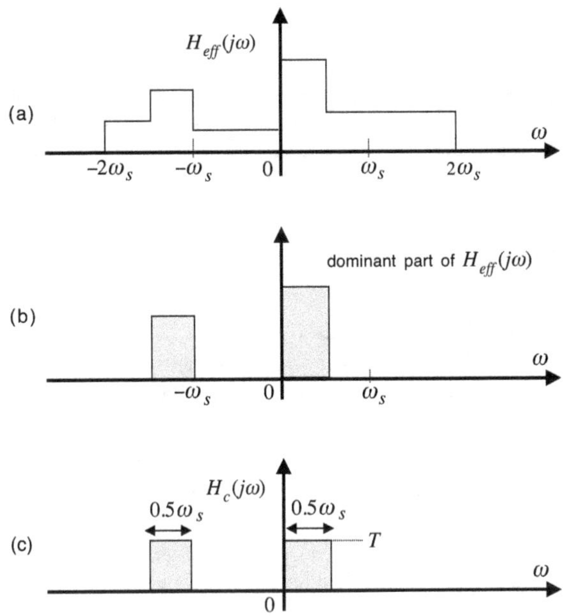

Figure 10.4. Another example of construction of an optimum compaction filter. (a) Shape of the effective channel, (b) dominant alias-free(T) band of the channel, and (c) the optimum compaction filter.

A third example of $H_{eff}(j\omega)$ is shown in Fig. 10.4(a), and the corresponding compaction filter is shown in Fig. 10.4(c). In this case the compaction filter has multiple passbands. Note that the total bandwidth of the compaction filter (counting all passbands) is ω_s.

10.3 The all-discrete SISO channel

From Chap. 4 we know that the system of Fig. 10.1 is equivalent to a purely discrete-time system with transfer function

$$H_d(z) = \sum_n h_c(nT)z^{-n},$$

with some additive noise $q_d(n)$. If the product $H_c(j\omega) = F(j\omega)H(j\omega)G(j\omega)$ does not satisfy the Nyquist(T) property exactly, then the equivalent digital channel $H_d(z)$ is not identity. In this case we can further equalize the channel using digital filters. This is one example of a situation where digital signal processing can be used to further improve the performance. Consider Fig. 10.5

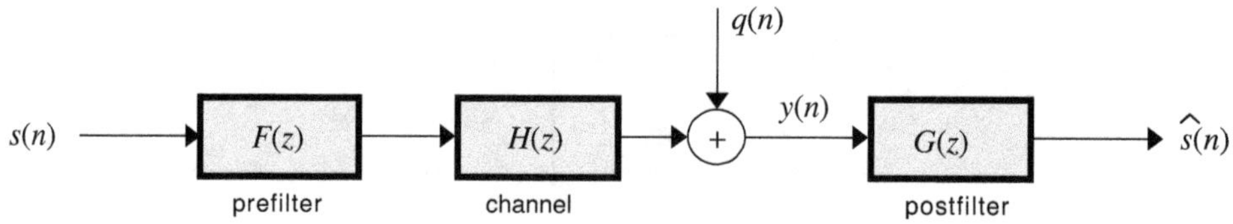

Figure 10.5. A single-input single-output all-digital transceiver.

which shows a single-input single-output transceiver in discrete time (with the subscript d dropped for simplicity). Here the sequence $s(n)$ is prefiltered using a digital filter $F(z)$ and transmitted on the channel $H(z)$, which also adds noise $q(n)$. The received noisy signal $y(n)$ is equalized with the digital postfilter $G(z)$. Assume that $s(n)$ and $q(n)$ are zero-mean, uncorrelated, jointly WSS processes with power spectra $S_{ss}(e^{j\omega})$ and $S_{qq}(e^{j\omega})$. The reconstruction error

$$e(n) = \widehat{s}(n) - s(n) \tag{10.26}$$

has mean square value

$$\mathcal{E}_{mse} \stackrel{\Delta}{=} E|\widehat{s}(n) - s(n)|^2. \tag{10.27}$$

In this section we minimize this quantity by optimizing the digital prefilter $F(z)$ and postfilter $G(z)$. As described in Chap. 9 the foundation for such optimization was laid in the early work by Costas [1952] and Bruce [1964], who considered continuous-time filters and signals.

10.3.1 MMSE transceiver without zero forcing (pure MMSE)

We first assume that the filters are not constrained to satisfy the zero-forcing property $F(z)H(z)G(z) = 1$. For fixed prefilter $F(z)$ and channel $H(z)$, we can appeal to Wiener filter theory (Appendix F) and show that the postfilter which minimizes $\mathcal{E}_{mse}$ is

$$G(e^{j\omega}) = \frac{H^*(e^{j\omega})F^*(e^{j\omega})S_{ss}(e^{j\omega})}{S_{ss}(e^{j\omega})|H(e^{j\omega})F(e^{j\omega})|^2 + S_{qq}(e^{j\omega})} \tag{10.28}$$

With the postfilter chosen as this, the reconstruction error has power spectrum

$$S_{ee}(e^{j\omega}) = \frac{S_{ss}(e^{j\omega})S_{qq}(e^{j\omega})}{S_{ss}(e^{j\omega})|H(e^{j\omega})F(e^{j\omega})|^2 + S_{qq}(e^{j\omega})} \tag{10.29}$$

so that

$$\mathcal{E}_{mse} = \int_{-\pi}^{\pi} \frac{S_{ss}(e^{j\omega})S_{qq}(e^{j\omega})}{S_{ss}(e^{j\omega})|H(e^{j\omega})F(e^{j\omega})|^2 + S_{qq}(e^{j\omega})} \frac{d\omega}{2\pi} \tag{10.30}$$

Justifications. These expressions can readily be obtained from Wiener filter theory (Appendix F), which was also used in Chap. 4. More specifically, we only have to use Eqs. (4.79) and (4.80) with appropriate substitutions, e.g., replace the matrix $\mathbf{C}(z)$ with the scalar $F(z)H(z)$, the matrix $\mathbf{S}_{\mathbf{xx}}(z)$ with the scalar $S_{ss}(z)$, and so forth. The postfilter $\mathbf{H}(z)\mathbf{C}^{-1}(z)$ of that section corresponds to $G(z)$ above. $\triangledown \triangledown \triangledown$

With $G(e^{j\omega})$ thus eliminated, the error in Eq. (10.30) is only a function of the prefilter $F(e^{j\omega})$. We can minimize this subject to the power constraint

$$\int_{-\pi}^{\pi} S_{ss}(e^{j\omega})|F(e^{j\omega})|^2 \frac{d\omega}{2\pi} = p_0. \tag{10.31}$$

For this we form the Lagrangian function

$$\psi(\alpha) = \int_{-\pi}^{\pi} \frac{S_{ss}(e^{j\omega})S_{qq}(e^{j\omega})}{\alpha S_{ss}(e^{j\omega})|H(e^{j\omega})|^2 + S_{qq}(e^{j\omega})} \frac{d\omega}{2\pi} + \lambda\Big(\int_{-\pi}^{\pi} \alpha S_{ss}(e^{j\omega})\frac{d\omega}{2\pi} - p_0\Big), \tag{10.32}$$

where we have used the notation

$$\alpha = |F(e^{j\omega})|^2.$$

We can find the optimal α by writing the necessary conditions for a local optimum, namely $\partial\psi(\alpha)/\partial\alpha = 0$. The result is

$$\frac{S_{qq}(e^{j\omega})S_{ss}(e^{j\omega})|H(e^{j\omega})|^2}{\Big(\alpha S_{ss}(e^{j\omega})|H(e^{j\omega})|^2 + S_{qq}(e^{j\omega})\Big)^2} = \lambda,$$

from which we can solve for α to obtain

$$|F(e^{j\omega})|^2 = \gamma\sqrt{\frac{S_{qq}(e^{j\omega})}{|H(e^{j\omega})|^2 S_{ss}(e^{j\omega})}} - \frac{S_{qq}(e^{j\omega})}{|H(e^{j\omega})|^2 S_{ss}(e^{j\omega})}$$

where γ is a constant. In the preceding derivation we ignored the constraint $\alpha \geq 0$. If this is appropriately incorporated using KKT theory (Chap. 22) then the resulting optimum $|F(e^{j\omega})|^2$ takes the form

$$|F(e^{j\omega})|^2 = \begin{cases} \gamma\sqrt{\dfrac{S_{qq}(e^{j\omega})}{|H(e^{j\omega})|^2 S_{ss}(e^{j\omega})}} - \dfrac{S_{qq}(e^{j\omega})}{|H(e^{j\omega})|^2 S_{ss}(e^{j\omega})} & \text{if this is} \geq 0 \\ 0 & \text{otherwise.} \end{cases} \tag{10.33}$$

The prefilter $F(e^{j\omega})$ is taken to be the stable spectral factor of this expression.[2] The constant γ is chosen to satisfy the power constraint (10.31). The reader should compare Eq. (10.33) with the water-filling formula given in Eq. (6.17)

[2] Note that if $|H(e^{j\omega})|$ is smaller than a threshold, then $F(e^{j\omega}) = 0$, so it is not necessary to assume $H(e^{j\omega}) \neq 0$.

for optimizing channel capacity. The formulas are different because the first term in Eq. (10.33) is not a constant. The construction in Eq. (10.33) can at best be regarded as "pseudo" water filling.

Let $\mathcal{F}$ be the set of frequencies in $[-\pi, \pi]$ for which $|F(e^{j\omega})|^2$ is nonzero, and let $\mathcal{F}^c$ be the complementary set. From Eqs. (10.31) and (10.33) we have

$$p_0 = \gamma \int_{\mathcal{F}} \sqrt{\frac{S_{qq}(e^{j\omega})S_{ss}(e^{j\omega})}{|H(e^{j\omega})|^2}} \frac{d\omega}{2\pi} - \int_{\mathcal{F}} \frac{S_{qq}(e^{j\omega})}{|H(e^{j\omega})|^2} \frac{d\omega}{2\pi}$$

so that

$$\gamma = \frac{p_0 + \displaystyle\int_{\mathcal{F}} \frac{S_{qq}(e^{j\omega})}{|H(e^{j\omega})|^2} \frac{d\omega}{2\pi}}{\displaystyle\int_{\mathcal{F}} \sqrt{\frac{S_{qq}(e^{j\omega})S_{ss}(e^{j\omega})}{|H(e^{j\omega})|^2}} \frac{d\omega}{2\pi}}, \qquad (10.34)$$

An expression for the minimized error $\mathcal{E}_{mse}$ can be obtained by substituting Eq. (10.33) into Eq. (10.30) which yields

$$\mathcal{E}_{mmse} = \frac{1}{\gamma} \int_{\mathcal{F}} \sqrt{\frac{S_{qq}(e^{j\omega})S_{ss}(e^{j\omega})}{|H(e^{j\omega})|^2}} \frac{d\omega}{2\pi} + \int_{\mathcal{F}^c} S_{ss}(e^{j\omega}) \frac{d\omega}{2\pi},$$

where the second term arises from the fact that when $F(e^{j\omega}) = 0$, the integrand in Eq. (10.30) reduces to $S_{ss}(e^{j\omega})$. Substituting for γ from Eq. (10.34) yields the final expression

$$\mathcal{E}_{mmse} = \frac{\left(\displaystyle\int_{\mathcal{F}} \sqrt{\frac{S_{qq}(e^{j\omega})S_{ss}(e^{j\omega})}{|H(e^{j\omega})|^2}} \frac{d\omega}{2\pi}\right)^2}{p_0 + \displaystyle\int_{\mathcal{F}} \frac{S_{qq}(e^{j\omega})}{|H(e^{j\omega})|^2} \frac{d\omega}{2\pi}} + \int_{\mathcal{F}^c} S_{ss}(e^{j\omega}) \frac{d\omega}{2\pi} \qquad (10.35)$$

Since the mean squared error has been minimized without the zero-forcing constraint, this is called the **pure-MMSE** solution.

10.3.2 MMSE transceiver with zero forcing (ZF-MMSE)

The zero-forcing constraint in Fig. 10.5 implies

$$F(z)H(z)G(z) = 1. \qquad (10.36)$$

In this case the reconstruction error is simply the noise $q(n)$ filtered by $G(e^{j\omega})$. The error spectrum is therefore

$$S_{ee}(e^{j\omega}) = S_{qq}(e^{j\omega})|G(e^{j\omega})|^2 = \frac{S_{qq}(e^{j\omega})}{|F(e^{j\omega})|^2|H(e^{j\omega})|^2}$$

The mean squared reconstruction error is

$$\mathcal{E}_{mse} = \int_{-\pi}^{\pi} \frac{S_{qq}(e^{j\omega})}{|F(e^{j\omega})|^2|H(e^{j\omega})|^2} \frac{d\omega}{2\pi} \qquad (10.37)$$

We can use the Cauchy-Schwartz inequality as in Sec. 10.2.1, but, just to demonstrate a different technique, we will use the Lagrange multiplier approach here. To minimize Eq. (10.37) under the power constraint (10.31) we form the Lagrangian

$$\psi(\alpha) = \int_{-\pi}^{\pi} \frac{S_{qq}(e^{j\omega})}{\alpha|H(e^{j\omega})|^2}\frac{d\omega}{2\pi} + \lambda\Big(\int_{-\pi}^{\pi} \alpha S_{ss}(e^{j\omega})\frac{d\omega}{2\pi} - p_0\Big), \tag{10.38}$$

where we have used the notation $\alpha = |F(e^{j\omega})|^2$. Setting $\partial\psi(\alpha)/\partial\alpha = 0$ we obtain the necessary conditions for an interior optimum:

$$\frac{S_{qq}(e^{j\omega})}{\alpha^2|H(e^{j\omega})|^2} = \lambda S_{ss}(e^{j\omega}).$$

Solving for $\alpha = |F(e^{j\omega})|^2$ we get

$$|F(e^{j\omega})|^2 = \gamma\sqrt{\frac{S_{qq}(e^{j\omega})}{|H(e^{j\omega})|^2 S_{ss}(e^{j\omega})}}, \tag{10.39}$$

where γ is chosen to satisfy the power constraint (10.31). Substituting into Eq. (10.37), the minimized mean square error is

$$\mathcal{E}_{ZF,mmse} = \frac{1}{\gamma}\int_{-\pi}^{\pi}\sqrt{\frac{S_{qq}(e^{j\omega})S_{ss}(e^{j\omega})}{|H(e^{j\omega})|^2}}\frac{d\omega}{2\pi} \tag{10.40}$$

From the power constraint (10.31) we have

$$p_0 = \gamma\int_{-\pi}^{\pi} S_{ss}(e^{j\omega})\sqrt{\frac{S_{qq}(e^{j\omega})}{|H(e^{j\omega})|^2 S_{ss}(e^{j\omega})}}\frac{d\omega}{2\pi} = \gamma\int_{-\pi}^{\pi}\sqrt{\frac{S_{ss}(e^{j\omega})S_{qq}(e^{j\omega})}{|H(e^{j\omega})|^2}}\frac{d\omega}{2\pi}$$

Solving for γ and substituting into Eq. (10.40) we obtain

$$\mathcal{E}_{ZF,mmse} = \frac{1}{p_0}\left(\int_{-\pi}^{\pi}\sqrt{\frac{S_{qq}(e^{j\omega})S_{ss}(e^{j\omega})}{|H(e^{j\omega})|^2}}\frac{d\omega}{2\pi}\right)^2. \tag{10.41}$$

The prefilter $F(e^{j\omega})$ is taken to be the stable spectral factor of Eq. (10.39). The optimal equalizer is then obtained from the zero-forcing condition Eq. (10.36). Since the mean square error has been minimized under the ZF condition, this is called the **ZF-MMSE** solution. Notice that unlike in the pure MMSE case, we have to assume $H(e^{j\omega}) \neq 0$ in the zero-forcing case. Otherwise we cannot satisfy Eq. (10.36). Under this assumption Eq. (10.39) is well defined.

Summary

The minimized mean-squared errors with and without zero forcing are summarized below:

$$\mathcal{E}_{mmse} = \begin{cases} \dfrac{1}{p_0}\left(\displaystyle\int_{-\pi}^{\pi}\sqrt{\dfrac{S_{qq}(e^{j\omega})S_{ss}(e^{j\omega})}{|H(e^{j\omega})|^2}}\dfrac{d\omega}{2\pi}\right)^2 & \text{(with ZF)} \\ \dfrac{\left(\displaystyle\int_{\mathcal{F}}\sqrt{\dfrac{S_{qq}(e^{j\omega})S_{ss}(e^{j\omega})}{|H(e^{j\omega})|^2}}\dfrac{d\omega}{2\pi}\right)^2}{p_0+\displaystyle\int_{\mathcal{F}}\dfrac{S_{qq}(e^{j\omega})}{|H(e^{j\omega})|^2}\dfrac{d\omega}{2\pi}} + \displaystyle\int_{\mathcal{F}^c} S_{ss}(e^{j\omega})\dfrac{d\omega}{2\pi} & \text{(no ZF).} \end{cases} \tag{10.42}$$

Discussion

A number of points should be observed:

1. ***Effective noise spectrum.*** The channel transfer function $H(z)$ and the noise spectrum $S_{qq}(e^{j\omega})$ appear together in these expressions, in the form $S_{qq}(e^{j\omega})/|H(e^{j\omega})|^2$. They never appear separately. The ratio given by $S_{qq}(e^{j\omega})/|H(e^{j\omega})|^2$ can be regarded as the effective noise spectrum.

2. ***ZF-MMSE error can be unbounded.*** As seen from Eq. (10.39), for the ZF-MMSE solution the prefilter magnitude $|F(e^{j\omega})|$ is large where the channel is bad (i.e., $S_{qq}(e^{j\omega})/|H(e^{j\omega})|^2$ is large). So the ZF-MMSE solution pumps more power in these bad regions. As a result, the minimized mean square error (10.41) can be very large for bad channels. For example, if $|H(e^{j\omega})| \to 0$ in some frequency regions, then the error is unbounded.

3. ***Pure-MMSE error is bounded.*** The pure-MMSE solution behaves differently. First, the filter magnitude $|F(e^{j\omega})|$ does not keep increasing with $S_{qq}(e^{j\omega})/|H(e^{j\omega})|^2$. In regions where $S_{qq}(e^{j\omega})/|H(e^{j\omega})|^2$ is larger than a threshold, the filter becomes zero, as seen from Eq. (10.33). In such regions, the reconstruction error is obtained from the second term in Eq. (10.35) and is therefore bounded. In fact it is readily verified using Cauchy-Schwartz inequality that the first term in Eq. (10.35) is bounded above by $\int_{\mathcal{F}} S_{ss}(e^{j\omega})d\omega/2\pi$:

$$\begin{aligned} \frac{\left(\displaystyle\int_{\mathcal{F}}\sqrt{\dfrac{S_{qq}(e^{j\omega})S_{ss}(e^{j\omega})}{|H(e^{j\omega})|^2}}\dfrac{d\omega}{2\pi}\right)^2}{p_0+\displaystyle\int_{\mathcal{F}}\dfrac{S_{qq}(e^{j\omega})}{|H(e^{j\omega})|^2}\dfrac{d\omega}{2\pi}} &\le \frac{\displaystyle\int_{\mathcal{F}}\dfrac{S_{qq}(e^{j\omega})}{|H(e^{j\omega})|^2}\dfrac{d\omega}{2\pi}\int_{\mathcal{F}}S_{ss}(e^{j\omega})\dfrac{d\omega}{2\pi}}{p_0+\displaystyle\int_{\mathcal{F}}\dfrac{S_{qq}(e^{j\omega})}{|H(e^{j\omega})|^2}\dfrac{d\omega}{2\pi}} \\ &\le \int_{\mathcal{F}}S_{ss}(e^{j\omega})\frac{d\omega}{2\pi} \end{aligned}$$

   Consequently,

$$\mathcal{E}_{mmse} \le \int_{\mathcal{F}} S_{ss}(e^{j\omega})\frac{d\omega}{2\pi} + \int_{\mathcal{F}^c} S_{ss}(e^{j\omega})\frac{d\omega}{2\pi} = \int_{-\pi}^{\pi} S_{ss}(e^{j\omega})\frac{d\omega}{2\pi} = \sigma_s^2.$$

So, in the pure-MMSE case the mean square error is bounded above by the signal variance σ_s^2.

4. ***MIMO channels.*** We will see that the basic mathematical formulation and solution for the optimal transceiver given in this section makes repeated appearances in more complicated scenarios. For example, when the SISO digital communication system becomes a MIMO system, we will see similar results (Chaps. 12 and 13). Instead of integrals involving $|H(e^{j\omega})|$, the expressions will have summations involving the *singular values* of a MIMO memoryless channel.

5. ***Error probabilities.*** Even though we have minimized mean square errors in this section, the more important quantity in practice is the symbol error probability. It can be shown that the pure-MMSE system has smaller symbol error probability than the ZF-MMSE system (see Appendix 16.D of Chap. 16).

10.4 General forms of optimal filters

In Sec. 10.2 we considered the optimization of the transceiver of Fig. 10.6 under the zero-forcing constraint. With the zero-forcing constraint removed the problem becomes more general, though more difficult. A number of such generalized results were reported in the early literature. In the early 1970s Ericson observed that in many of these problems the solutions have some common properties, and showed that the optimal combination of the prefilter $F(j\omega)$ and postfilter $G(j\omega)$ in Fig. 10.6 can be restricted in certain ways without loss of generality [Ericson, 1971, 1973]. More specifically, for any objective function which depends only on the noise spectrum at the detector input and the amount of transmitted power, there is a certain standard form for the filters $G(j\omega)$ and $F(j\omega)$. In this section we shall elaborate on this. One consequence of this observation is that the optimal design of $F(j\omega)$ and $G(j\omega)$ can be converted into an optimal design problem involving an all-digital system like Fig. 10.5, as we shall explain. Historically, Ericson's work unifies many earlier papers in transceiver literature which obtained rather similar solutions for different optimization problems.

For simplicity we restrict our discussions to the case where the objective to be minimized is the mean square error, subject to the average power constraint (10.3). But these observations also hold for "any reasonable objective function" as Ericson puts it. The contents of this section can be used primarily as a reference. For this section, the reader should review Appendix G at the end of the book, especially notations such as $\downarrow T$, jargon such as "alias-free(T) bands," and the noble identities.

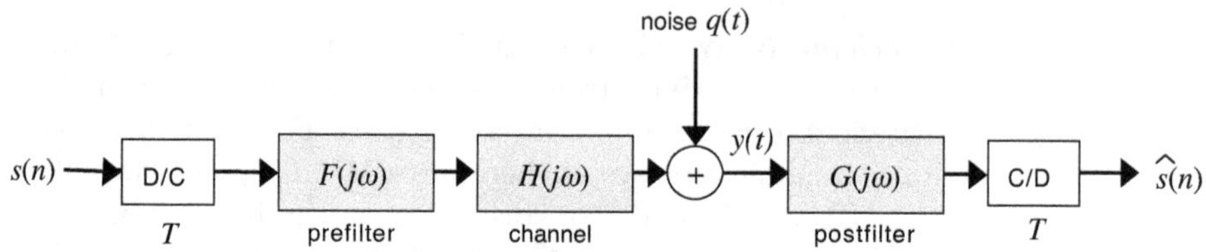

Figure 10.6. A digital communication system to transmit messages through a continuous-time channel.

10.4.1 The receiver filter G

We begin by considering the receiver filter $G(j\omega)$. The first observation is as follows:

♠**Lemma 10.2.** *Form of* $G(j\omega)$. In Fig. 10.6 assume that the noise power spectrum is $S_{qq}(j\omega) = 1$ for all ω. Given some combination of filters $F(j\omega)$ and $G(j\omega)$, suppose we replace $G(j\omega)$ with

$$\widehat{G}(j\omega) = P(e^{j\omega T})F^*(j\omega)H^*(j\omega), \tag{10.43}$$

where

$$P(e^{j\omega}) = \frac{\Big[F(j\omega)H(j\omega)G(j\omega)\Big]_{\downarrow T}}{\Big[|F(j\omega)H(j\omega)|^2\Big]_{\downarrow T}} \tag{10.44}$$

This replacement does not alter the transfer function of the discrete-time system from $s(n)$ to $\widehat{s}(n)$, and it does not increase the noise power spectrum at the output of the C/D converter for any frequency ω. $\Diamond$

Thus, any "reasonable" objective function which depends only on the sampled noise spectrum can only get smaller (or remain unchanged) when $G(j\omega)$ is replaced with $\widehat{G}(j\omega)$. As a result the optimum $G(j\omega)$ can be assumed, without loss of generality, to be of the form (10.43). Note that Eq. (10.43) resembles a **matched filter** for $F(j\omega)H(j\omega)$ except for the **transversal filter** part $P(e^{j\omega T})$. This part can be implemented as a digital filter as we shall see.

Proof of Lemma 10.2. Observe first that

$$\begin{aligned}\Big[F(j\omega)H(j\omega)\widehat{G}(j\omega)\Big]_{\downarrow T} &= \Big[|F(j\omega)H(j\omega)|^2 P(e^{j\omega T})\Big]_{\downarrow T} \\ &= \Big[|F(j\omega)H(j\omega)|^2\Big]_{\downarrow T} P(e^{j\omega}) \\ &= \Big[F(j\omega)H(j\omega)G(j\omega)\Big]_{\downarrow T},\end{aligned}$$

where we have substituted for $P(e^{j\omega})$ using Eq. (10.44). The second equality is obtained by using the identity (G.6) from Appendix G. Since the frequency response function from $s(n)$ to $\widehat{s}(n)$ is given by

$$[F(j\omega)H(j\omega)G(j\omega)]_{\downarrow T}$$

it therefore follows that this transfer function is unchanged when $\widehat{G}(j\omega)$ replaces $G(j\omega)$. Now consider the noise power spectrum which is given by $|G(j\omega)|^2$ at the output of $G(j\omega)$. At the output of the C/D converter this becomes

$$\Big[|G(j\omega)|^2\Big]_{\downarrow T}.$$

With $G(j\omega)$ replaced by $\widehat{G}(j\omega)$ we have

$$\begin{aligned}
&\Big[|\widehat{G}(j\omega)|^2\Big]_{\downarrow T} \\
&\quad = \Big[|P(e^{j\omega T})F^*(j\omega)H^*(j\omega)|^2\Big]_{\downarrow T} \\
&\quad = \Big[|F(j\omega)H(j\omega)|^2\Big]_{\downarrow T}|P(e^{j\omega})|^2 \text{ (from Eq. (G.6), Appendix G)} \\
&\quad = \frac{\Big|[F(j\omega)H(j\omega)G(j\omega)]_{\downarrow T}\Big|^2}{\Big[|F(j\omega)H(j\omega)|^2\Big]_{\downarrow T}}.
\end{aligned}$$

Observe now that the last numerator can be expanded as follows:

$$\begin{aligned}
&\Big|\frac{1}{T}\sum_k F(j(\omega+2\pi k)/T)H(j(\omega+2\pi k)/T)G(j(\omega+2\pi k)/T)\Big|^2 \\
&\le \frac{1}{T}\sum_k\Big|F(j(\omega+2\pi k)/T)H(j(\omega+2\pi k)/T)\Big|^2\frac{1}{T}\sum_k\Big|G(j(\omega+2\pi k)/T)\Big|^2 \\
&= \Big[|F(j\omega)H(j\omega)|^2\Big]_{\downarrow T}\Big[|G(j\omega)|^2\Big]_{\downarrow T},
\end{aligned}$$

where $T = 2\pi/\omega_s$. Here the second line follows from the Cauchy-Schwartz inequality. This proves that

$$\Big[|\widehat{G}(j\omega)|^2\Big]_{\downarrow T} \le \Big[|G(j\omega)|^2\Big]_{\downarrow T}.$$

That is, the noise power spectrum (at the output of the C/D converter) cannot increase at any frequency ω, when we replace the receiver filter $G(j\omega)$ with $\widehat{G}(j\omega)$. $\triangledown\triangledown\triangledown$

10.4.1.A More general noise power spectrum

Lemma 10.2 assumed that $S_{qq}(j\omega) = 1$. Now imagine that $S_{qq}(j\omega)$ is arbitrary but nonzero for all ω. Suppose we rearrange Fig. 10.6 as in Fig. 10.7(a), where the noise $\widehat{q}(t)$ has power spectrum equal to unity as in Lemma 10.2.

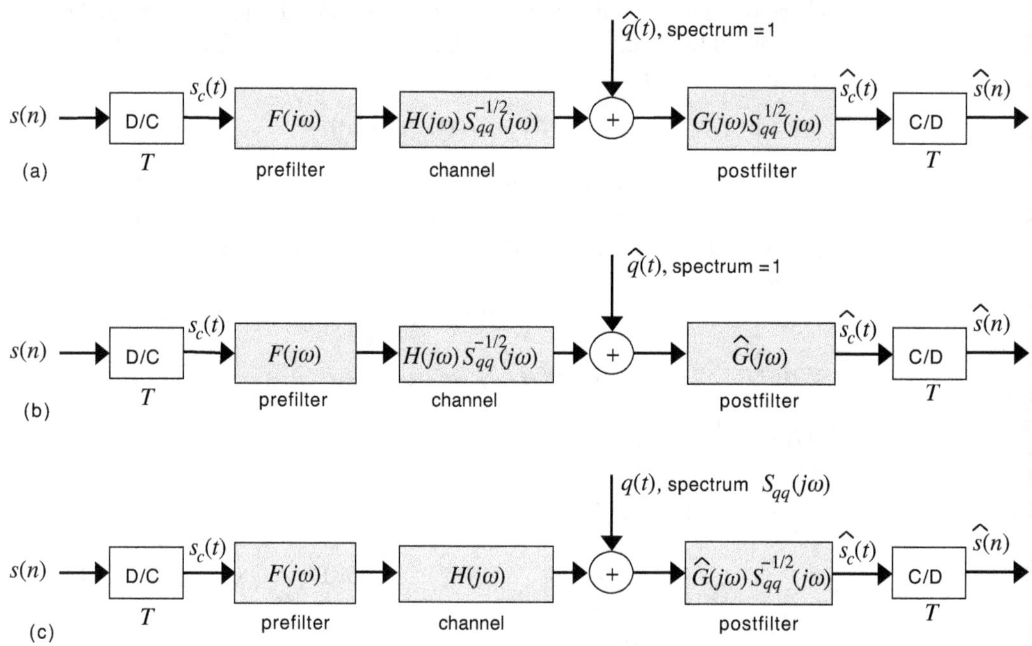

Figure 10.7. (a)–(c) Pertaining to the generalization of Lemma 10.2 for arbitrary channel noise spectrum $S_{qq}(j\omega) > 0$.

Then the signal path from $s_c(t)$ to $\widehat{s}_c(t)$ is unchanged. Furthermore the noise power spectrum at the input of the C/D converter is also unchanged. So this system is indistinguishable from the original system. From Lemma 10.2 we know that the filter $G(j\omega)S_{qq}^{1/2}(j\omega)$ in this figure can be replaced with the filter

$$\widehat{G}(j\omega) = P(e^{j\omega T})F^*(j\omega)H^*(j\omega)S_{qq}^{-1/2}(j\omega), \tag{10.45}$$

where

$$\begin{aligned} P(e^{j\omega}) &= \frac{[F(j\omega)H(j\omega)S_{qq}^{-1/2}(j\omega)G(j\omega)S_{qq}^{1/2}(j\omega)]_{\downarrow T}}{[|F(j\omega)H(j\omega)|^2 S_{qq}^{-1}(j\omega)]_{\downarrow T}} \\ &= \frac{[F(j\omega)H(j\omega)G(j\omega)]_{\downarrow T}}{[|F(j\omega)H(j\omega)|^2 S_{qq}^{-1}(j\omega)]_{\downarrow T}} \end{aligned}$$

without increasing the reconstruction error. This system is shown in Fig. 10.7(b). This can further be rearranged as in Fig. 10.7(c) without changing the overall transfer function or the noise spectrum at the input of the C/D converter. The final receiver filter is therefore

$$G_1(e^{j\omega}) = P(e^{j\omega T})F^*(j\omega)H^*(j\omega)S_{qq}^{-1}(j\omega). \tag{10.46}$$

Summarizing, we have shown the following.

♠**Lemma 10.3.** *Form of* $G(j\omega)$. In Fig. 10.6 assume that the noise power spectrum is $S_{qq}(j\omega) > 0$ for all ω. Given some combination of filters $F(j\omega)$ and $G(j\omega)$, suppose we replace $G(j\omega)$ with

$$G_1(j\omega) = \frac{P(e^{j\omega T})F^*(j\omega)H^*(j\omega)}{S_{qq}(j\omega)}, \tag{10.47}$$

where

$$P(e^{j\omega}) = \frac{\Big[F(j\omega)H(j\omega)G(j\omega)\Big]_{\downarrow T}}{\Big[|F(j\omega)|^2|H(j\omega)|^2/S_{qq}(j\omega)\Big]_{\downarrow T}}. \tag{10.48}$$

Then the signal component of $\widehat{s}(n)$ is unchanged, and the noise power spectrum at the output of the C/D converter does not increase for any ω. ◊

Remarks

1. ***Parts of the optimal filter.*** Thus the signal component at the output of Fig. 10.8(c) is identical to that in Fig. 10.8(a), and the noise spectrum does not increase at any frequency. The filter (10.47) has the form

   $$G_1(j\omega) = B(j\omega)P(e^{j\omega T}).$$

   This is product of a transversal part $P(e^{j\omega T})$ (periodic in ω) and a non-transversal part $B(j\omega)$.

2. ***Digital filter part.*** In view of the noble identity shown in Fig. G.1 of Appendix G, the transversal part $P(e^{j\omega T})$ can be realized by implementing a digital filter $P(z)$ with frequency response $P(e^{j\omega})$ after the C/D unit (i.e., after sampling at the receiver). See Fig. 10.8(c).

3. ***Matched filter interpretation.*** With $P(e^{j\omega T})$ so moved, the optimum filter $G_1(j\omega)$ has the remaining factor

   $$B(j\omega) = \frac{F^*(j\omega)H^*(j\omega)}{S_{qq}(j\omega)}. \tag{10.49}$$

   This can be regarded as a matched filter corresponding to the cascaded filter $F(j\omega)H(j\omega)$ with additive noise spectrum $S_{qq}(j\omega)$.

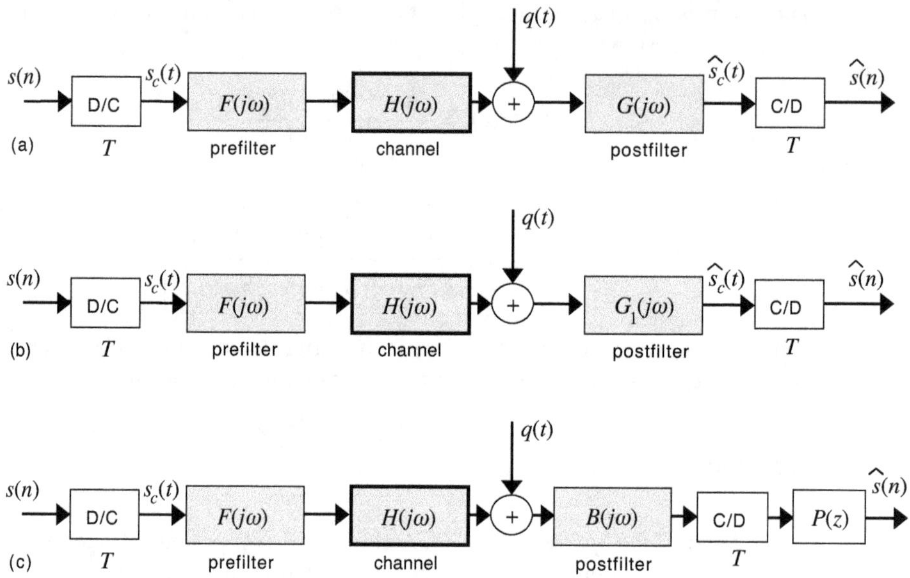

Figure 10.8. (a)–(c) Pertaining to Lemma 10.3. See text.

10.4.2 Form of the optimal prefilter

We now consider the system between the D/C and C/D converter shown in Fig. 10.8(c), reproduced in Fig. 10.9. Here $B(e^{j\omega})$ is as in Eq. (10.49). Since

$$F(j\omega)H(j\omega)B(j\omega) = \frac{|F(j\omega)H(j\omega)|^2}{S_{qq}(j\omega)},$$

the frequency response function between $s(n)$ and $\widehat{s}_1(n)$ is (using "sampling identity" from Sec. 4.2.1)

$$\mathcal{H}(e^{j\omega}) = \frac{1}{T}\sum_k \left(\frac{\left|F\Big(j\Big(\frac{\omega+2\pi k}{T}\Big)\Big)H\Big(j\Big(\frac{\omega+2\pi k}{T}\Big)\Big)\right|^2}{S_{qq}\Big(j\Big(\frac{\omega+2\pi k}{T}\Big)\Big)} \right). \tag{10.50}$$

The noise power spectrum at the input of the C/D converter is

$$S_{qq}(j\omega)|B(j\omega)|^2 = \frac{S_{qq}(j\omega)|F(j\omega)H(j\omega)|^2}{S_{qq}^2(j\omega)} = \frac{|F(j\omega)H(j\omega)|^2}{S_{qq}(j\omega)}$$

so that the noise psd at the output of the C/D converter is also given by Eq. (10.50). We now show that the transfer function (10.50) remains unchanged if $F(j\omega)$ is replaced by an appropriately bandlimited filter $F_a(j\omega)$ (and $B(j\omega)$ is adjusted using Eq. (10.49)). This will be useful to convert the filter optimization problem into an all-discrete-time problem as we shall see.

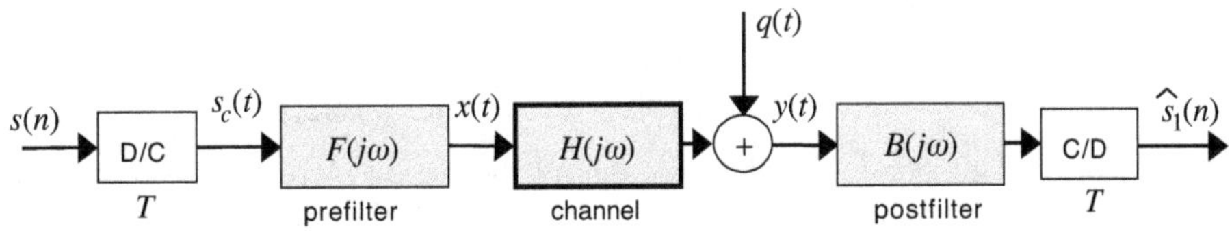

Figure 10.9. Pertaining to the optimal form of $F(j\omega)$.

10.4.2.A Introducing the alias-free prefilter $F_a(j\omega)$

At this point the reader should review the concept of alias-free(T) bands and filters from Appendix G. Given a prefilter $F(j\omega)$, suppose we define $F_a(j\omega)$ such that

$$|F_a(j\omega)|^2 = \begin{cases} \dfrac{S_{qq}(j\omega)}{|H(j\omega)|^2} \displaystyle\sum_k \frac{\Big|F(j(\omega + k\omega_s))H(j(\omega + k\omega_s))\Big|^2}{S_{qq}(j(\omega + k\omega_s))} & \text{for } \omega \in \mathcal{A} \\ 0 & \text{otherwise,} \end{cases}$$

where $\mathcal{A}$ is an alias-free(T) band (with $T = 2\pi/\omega_s$). Clearly $F_a(j\omega)$ is an alias-free(T) filter, that is, its output can be sampled uniformly with spacing T without aliasing. We can rewrite

$$|F_a(j\omega)|^2 \frac{|H(j\omega)|^2}{S_{qq}(j\omega)} = \begin{cases} \displaystyle\sum_k \frac{\Big|F(j(\omega + k\omega_s))H(j(\omega + k\omega_s))\Big|^2}{S_{qq}(j(\omega + k\omega_s))} & \text{for } \omega \in \mathcal{A} \\ 0 & \text{otherwise.} \end{cases}$$

Since $|F_a(j\omega)|^2$ is alias-free(T), so is the preceding left-hand side. It therefore follows from Lemma G.1 in Appendix G that

$$\sum_k \frac{\Big|F_a(j(\omega + k\omega_s))H(j(\omega + k\omega_s))\Big|^2}{S_{qq}(j(\omega + k\omega_s))} = \sum_k \frac{\Big|F(j(\omega + k\omega_s))H(j(\omega + k\omega_s))\Big|^2}{S_{qq}(j(\omega + k\omega_s))}. \tag{10.51}$$

Thus the right-hand side of Eq. (10.50) is unchanged when we replace $F(j\omega)$ with the alias-free filter $F_a(j\omega)$ and replace $B(j\omega)$ with

$$B_a(j\omega) = \frac{F_a^*(j\omega)H^*(j\omega)}{S_{qq}(j\omega)}. \tag{10.52}$$

Since the right-hand side of (10.50) represents the transfer function from $s(n)$ to $\widehat{s}_1(n)$ as well as the noise spectrum at the output of the C/D converter, it follows from Eq. (10.51) that these two quantities are unchanged by the replacement of $F(j\omega)$ and $B(j\omega)$ with $F_a(j\omega)$ and $B_a(j\omega)$. That is, the signal and noise components of the output $\widehat{s}_1(n)$ are unchanged.

10.4.2.B Choice of the alias-free(T) prefilter $F_a(j\omega)$

The preceding results hold for any choice of the alias-free(T) region $\mathcal{A}$. In particular, suppose $\mathcal{A}$ is chosen such that if $\omega \in \mathcal{A}$ then

$$\frac{|H(j\omega)|^2}{S_{qq}(j\omega)} \geq \frac{|H(j(\omega + k\omega_s))|^2}{S_{qq}(j(\omega + k\omega_s))} \tag{10.53}$$

for any integer k. That is, we define the region $\mathcal{A}$ to be the best alias-free(T) band of the channel (i.e., the portion where the channel gain divided by the noise spectrum is maximized). Then replacement of $F(j\omega)$ with $F_a(j\omega)$ *does not increase* the transmitted power!

Proof. The transmitted power is given by Eq. (E.42) in Appendix E. With $F(j\omega)$ replaced by $F_a(j\omega)$ this power becomes

$$\begin{aligned}
&\frac{1}{T}\int_{-\infty}^{\infty} \Big|F_a(j\omega)\Big|^2 S_{ss}(e^{j\omega T})\frac{d\omega}{2\pi} \\
&= \int_{\mathcal{A}} \frac{S_{qq}(j\omega)}{|H(j\omega)|^2} \sum_k \frac{\Big|F(j(\omega + k\omega_s))H(j(\omega + k\omega_s))\Big|^2}{S_{qq}(j(\omega + k\omega_s))} S_{ss}(e^{j\omega T})\frac{d\omega}{2\pi} \\
&= \int_{\mathcal{A}} \sum_k \frac{\Big|F(j(\omega + k\omega_s))H(j(\omega + k\omega_s))\Big|^2}{S_{qq}(j(\omega + k\omega_s))} \times \frac{S_{qq}(j\omega)}{|H(j\omega)|^2} S_{ss}(e^{j\omega T})\frac{d\omega}{2\pi} \\
&\leq \int_{\mathcal{A}} \sum_k \Big|F(j(\omega + k\omega_s))\Big|^2 S_{ss}(e^{j\omega T})\frac{d\omega}{2\pi} \quad \text{(from Eq. (10.53))} \\
&= \int_{\mathcal{A}} \sum_k \Big|F(j(\omega + k\omega_s))\Big|^2 S_{ss}(e^{j(\omega + k\omega_s)T})\frac{d\omega}{2\pi} \quad (\text{since } \omega_s T = 2\pi) \\
&= \int_{-\infty}^{\infty} |F(j(\omega)|^2 S_{ss}(e^{j\omega T})\frac{d\omega}{2\pi}.
\end{aligned}$$

The last equality follows from the alias-free($2\pi/\omega_s$) property of $\mathcal{A}$ (which ensures that the regions $\{\mathcal{A}+k\omega_s\}$ are nonoverlapping for different values of k, and cover the entire frequency axis as k varies over all integers). $\bigtriangledown \bigtriangledown \bigtriangledown$

Thus, the optimal system can be assumed to be such that the prefilter has the form $F_a(j\omega)$ and the equalizer has the form $B_a(j\omega)$ (followed by the digital equalizer $P(z)$, which is located to the right of the C/D converter).

10.4.3 The all-digital equivalent

The transceiver system with anti-alias filters $F_a(j\omega)$ and $B_a(j\omega)$ is shown in Fig. 10.10. Since $F_a(j\omega)$ and $B_a(j\omega)$ are restricted to the alias-free(T) band $\mathcal{A}$ we can write

$$F_a(j\omega) = H_{id}(j\omega)F_d(e^{j\omega T}), \quad B_a(j\omega) = H_{id}(j\omega)B_d(e^{j\omega T})$$

for some transversal filters $F_d(e^{j\omega T})$ and $B_d(e^{j\omega T})$, where $H_{id}(j\omega)$ is an ideal filter defined by

$$H_{id}(j\omega) = \begin{cases} 1 & \omega \in \mathcal{A} \\ 0 & \text{otherwise.} \end{cases}$$

The filter $H_{id}(j\omega)$ is also called the *indicator function* for the set $\mathcal{A}$. By using the noble identities in Figs. G.1 and G.2 (see Appendix G), we can therefore redraw Fig. 10.10(a) as in Fig. 10.10(b). The system

$$H_{ess} = H_{id}(j\omega)H(j\omega)$$

is the "essence of the channel," that is, the channel transfer function restricted to its *best* alias-free(T) band $\mathcal{A}$ (as defined by Eq. (10.53)). Since the ideal filter $H_{id}(j\omega)$ extracts the "best part" of $|H(j\omega)|^2/S_{qq}(j\omega)$, it is the **optimum compaction filter** for $|H(j\omega)|^2/S_{qq}(j\omega)$ (Sec. 10.2.3). By using the identity shown in Fig. 4.3 this system can be redrawn as the all-discrete system shown in Fig. 10.10(c), where $H_d(z)$ is the digital channel with impulse response

$$h_d(n) = h_{ess}(nT)$$

and $q_d(n)$ is a noise source whose power spectrum is the "aliased" version of $S_{qq}(j\omega)H_{id}(j\omega)$, that is,

$$S_{qq,d}(e^{j\omega}) = \frac{1}{T}\sum_k S_{qq}\Big(j\Big(\frac{\omega + 2\pi k}{T}\Big)\Big) H_{id}\Big(j\Big(\frac{\omega + 2\pi k}{T}\Big)\Big).$$

With the digital filter $P(z)$ also included as shown earlier in Fig. 10.8(c), the original system, reproduced in Fig. 10.11(a), takes the final equivalent form shown in Fig. 10.11(b), where $G_d(z) = B_d(z)P(z)$. In practice most of the optimization problems minimize some function which depends on the noise spectrum at the input of the detector, for a fixed transmitted power. In these cases therefore, the joint optimization of the continuous-time filters $F(j\omega)$ and $G(j\omega)$ is equivalent to the optimization of the digital filters $F_d(z)$ and $G_d(z)$.

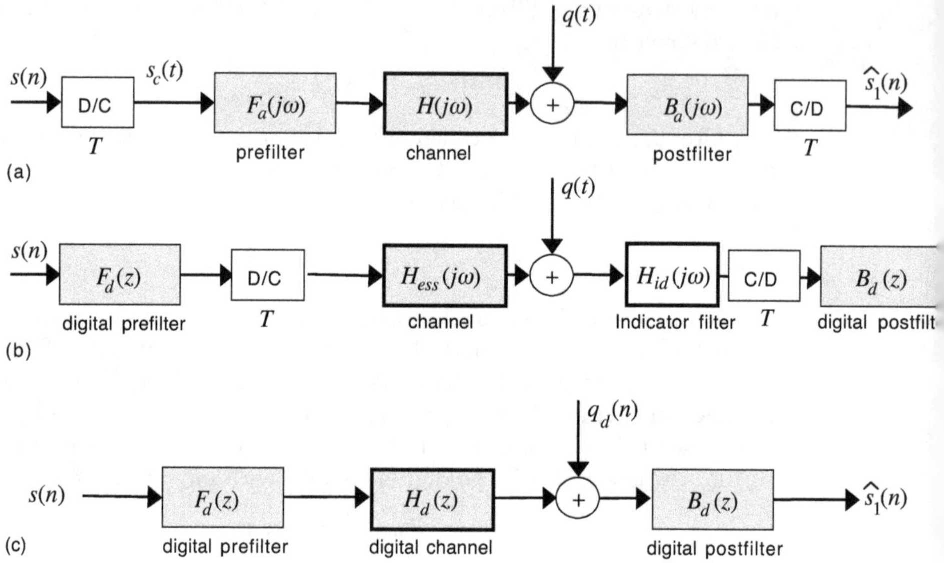

Figure 10.10. (a) The transceiver with alias-free filters $F_a(j\omega)$ and $B_a(j\omega)$, (b) rearrangement, and (c) final discrete-time form.

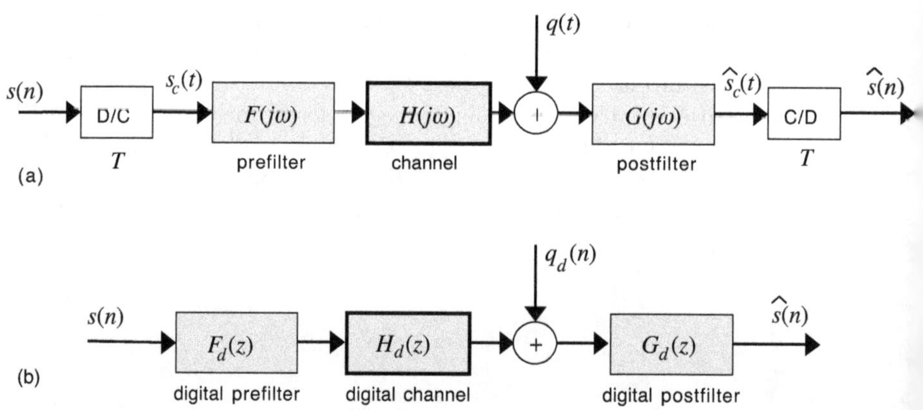

Figure 10.11. (a) The original communication system and (b) the discrete time equivalent.

10.5 Excess bandwidth and oversampling

In Sec. 4.7 we found that a transceiver with oversampling at the receiver can be reduced to an equivalent system with minimal sampling. The performance of the optimal transceiver in Sec. 10.2, which minimizes the mean square error under the zero-forcing constraint, cannot therefore be bettered by any oversampled system, as long as ideal unrealizable filters are allowed in the solution. As explained in Secs. 4.7 and 4.8, the advantage of oversampling arises when we consider practical (FIR) implementations.

In this section we will continue the use of ideal filters. We will work through a specific example of an **excess bandwidth** system. Thus, consider Fig. 10.12, where the channel $H(j\omega)$ is ideal lowpass:

$$H(j\omega) = \begin{cases} 1 & \text{for } |\omega| < 2\pi/T \\ 0 & \text{otherwise.} \end{cases} \tag{10.54}$$

The total bandwidth is $4\pi/T$, which is twice the minimum bandwidth of $2\pi/T$. In this example the noise power spectrum is assumed to be given by

$$S_{qq}(j\omega) = \begin{cases} A & \text{for } |\omega| < \pi/T \\ \epsilon & \pi/T < |\omega| < 2\pi/T, \end{cases} \tag{10.55}$$

and zero otherwise, as shown in the figure. We choose the prefilter $F(j\omega)$ to be

$$F(j\omega) = \begin{cases} f_0 & \text{for } |\omega| < \pi/T \\ f_1 & \text{for } \pi/T < |\omega| < 2\pi/T, \end{cases} \tag{10.56}$$

where

$$f_0^2 + f_1^2 = 1.$$

In the two frequency bands where the noise has two different levels, the channel input powers are proportional to f_0^2 and f_1^2, respectively. We will explore the optimal way to distribute the power (i.e., choose f_0 and f_1). We choose the receiver filter $G(j\omega)$ to be of the form

$$G(j\omega) = \begin{cases} 1/f_0 & \text{for } |\omega| < \pi/T \\ 1/f_1 & \text{for } \pi/T < |\omega| < 2\pi/T, \end{cases} \tag{10.57}$$

so that the product $H_c(j\omega) = G(j\omega)H(j\omega)F(j\omega)$ has the simple form

$$H_c(j\omega) = \begin{cases} 1 & \text{for } |\omega| < 2\pi/T \\ 0 & \text{otherwise,} \end{cases} \tag{10.58}$$

just like $H(j\omega)$. For this system with excess bandwidth we consider the **oversampled** receiver with oversampling factor $L = 2$. Oversampling is indicated by the label $T/2$ under the C/D box.

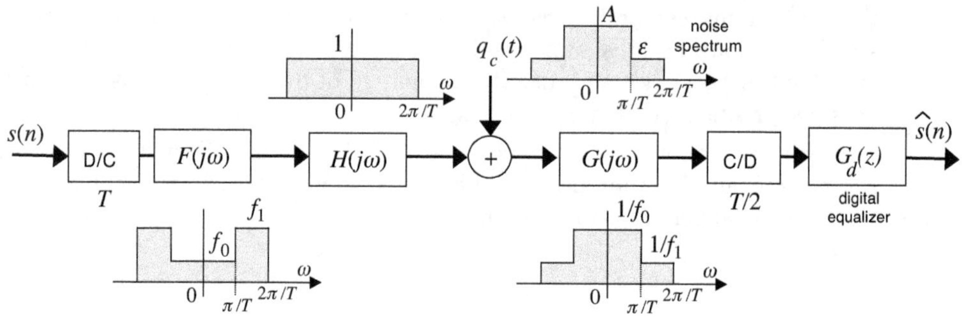

Figure 10.12. A channel with excess bandwidth and colored noise. The precoder $F(j\omega)$ is chosen to have two levels of gain, and the receiver filter $G(j\omega)$ is adjusted such that $F(j\omega)H(j\omega)G(j\omega)$ is constant in the passband.

To analyze this system we first replace it with the equivalent system shown in Fig. 10.13(a). Here $\widehat{q}(t)$ is the noise at the output of $G(j\omega)$, and has spectrum

$$\widehat{S}_{qq}(j\omega) = \begin{cases} \dfrac{A}{f_0^2} & \text{for } |\omega| < \pi/T \\ \\ \dfrac{\epsilon}{f_1^2} & \text{for } \pi/T < |\omega| < 2\pi/T, \end{cases}$$

and zero otherwise. The equivalent digital channel can be obtained as described in Sec. 4.8, and is shown in Fig. 10.13(b). Here $H_d(e^{j\omega})$ has the impulse response $h_d(n) = h_c(nT/2)$. In fact, $H_d(e^{j\omega})$ and the noise spectrum $S_d(e^{j\omega})$ are simply the aliased versions of $H_c(j\omega)$ and $\widehat{S}_{qq}(j\omega)$ at the sampling rate $2\pi L/T$ (with $L = 2$).

The digital equalizer $G_d(z)$, which is a **fractionally spaced equalizer** (FSE, see Sec. 4.8), is chosen to have the two-level response shown in Fig. 10.13(b). We will optimize α and β to minimize the mean square reconstruction error. So the quantities to be optimized are the power allocation at the transmitter (determined by f_0 and f_1) and the relative equalizer gains in the two bands, α and β. With the digital filter $G_d(z)$ chosen as shown, the zero-forcing condition is (see Problem 10.7)

$$\beta + \alpha = 1. \tag{10.59}$$

The noise variance at the output of $G_d(z)$ is also the mean square reconstruction error because of zero forcing. So the mean square error is

$$\mathcal{E}_{FSE} = T\Big(\frac{A\beta^2}{f_0^2} + \frac{\epsilon\alpha^2}{f_1^2}\Big). \tag{10.60}$$

(a)

(b)

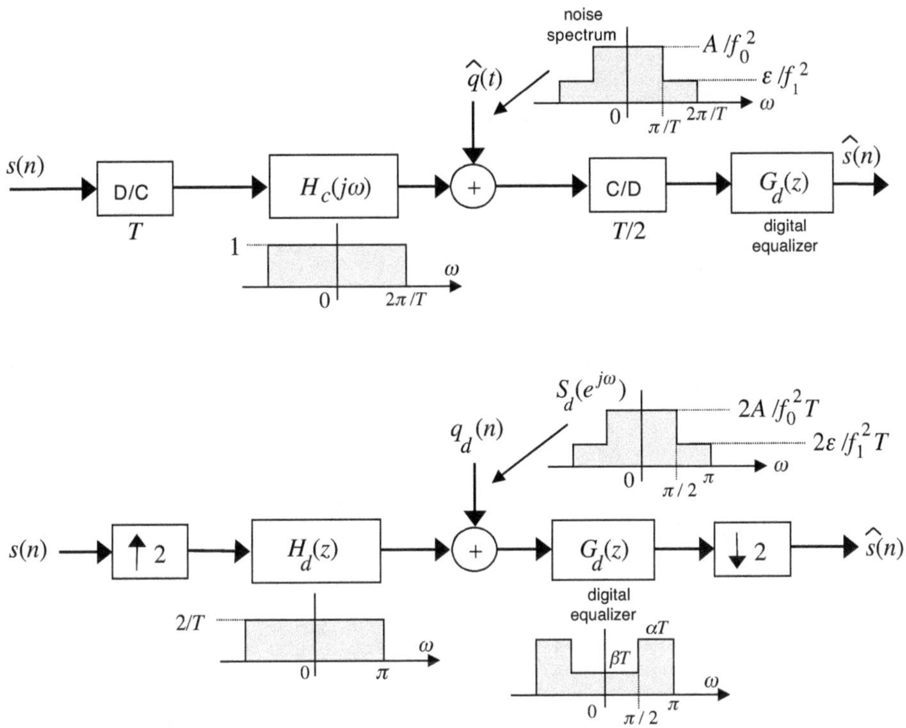

Figure 10.13. (a) The digital communication system with oversampling at the receiver, and power allocation at the transmitter. (b) Equivalent digital channel with FSE $G_d(z)$.

Denoting $D = f_0^2$ for convenience and using the constraints $f_1^2 = 1 - f_0^2$ and $\alpha = 1 - \beta$, we therefore have

$$\mathcal{E}_{FSE} = T\Big(\frac{A\beta^2}{D} + \frac{\epsilon(1-\beta)^2}{1-D}\Big). \tag{10.61}$$

For fixed D, we can find the optimal β by setting $\partial\mathcal{E}_{FSE}/\partial\beta = 0$. The result is

$$\beta = \frac{\epsilon D}{\epsilon D + A(1-D)} \tag{10.62}$$

Substituting into the preceding expression for $\mathcal{E}_{FSE}$, we find the minimized MSE is given by

$$\mathcal{E}_{FSE} = \frac{A\epsilon T}{A + D(\epsilon - A)} \tag{10.63}$$

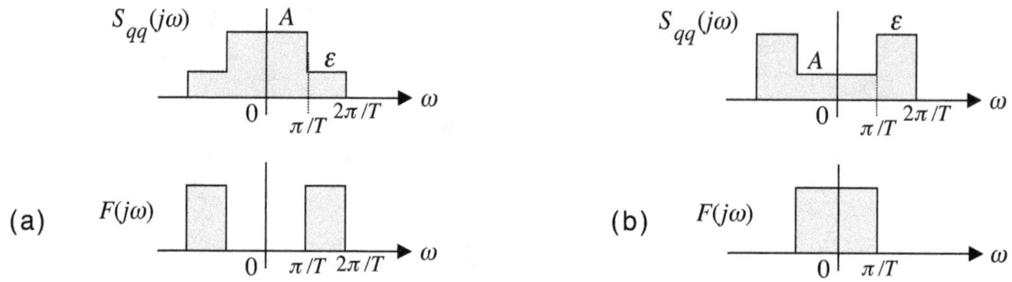

Figure 10.14. (a), (b). The noise power spectrum and the optimal choice of the prefilter for the two cases.

This expression can further be minimized by choosing D (which determines power allocation at the transmitter). Since $0 \leq D \leq 1$, it is clear that the optimal D can be taken to be

$$D = \begin{cases} 1 & \epsilon > A \\ 0 & \epsilon \leq A. \end{cases} \tag{10.64}$$

That is, the filter $F(j\omega)$ should be such that[3] $f_0 = 1$ and $f_1 = 0$ if $\epsilon > A$ and $f_0 = 0$ and $f_1 = 1$ if $\epsilon \leq A$. This is demonstrated in Fig. 10.14. This shows that the optimal FSE system in this example is such that the total bandwidth, where the product $H_c(j\omega)$ is nonzero, is only $2\pi/T$ rather than $4\pi/T$. The passbands of the filters are such that the best part of the channel (the part with least noise) is used by the transceiver. This is similar to the use of an optimal compaction filter as described in Sec. 10.2.3. The minimized error $\mathcal{E}_{FSE}$ now reduces to

$$\mathcal{E}_{FSE} = \begin{cases} AT & \text{when } \epsilon > A \\ \epsilon T & \text{when } \epsilon \leq A. \end{cases} \tag{10.65}$$

It is easily verified (see Problem 10.10) that this is also the mean square error achieved by the optimal transceiver of Sec. 10.2, which does not use oversampling (assuming the energy of $F(j\omega)$ is fixed in this comparison). This demonstrates that if we ignore the fact that the filters are ideal and unrealizable, the oversampled system shows no advantage in terms of mean square error.

10.6 Optimal pulse shape in single-pulse case

In Sec. 2.5 we introduced the matched filter, but did not consider the problem of optimizing the transmitted pulse. Consider a noisy channel with a *single* input pulse $p(t)$ of duration T, as shown in Fig. 10.15. Here $H(j\omega)$ is the channel

[3]Note that if $f_1 = 0$, then $1/f_1$ in the filter $G(j\omega)$ is undefined. But since the transmitter sends zero power in this band, we simply take $G(j\omega)$ to be zero here.

transfer function and $q(t)$ is additive noise with zero mean and power spectrum $S_{qq}(j\omega)$. Observe that

$$u(t) = \underbrace{(b * h * p)(t)}_{\text{signal part } w(t)} + \underbrace{(b * q)(t)}_{\text{noise part } e(t)} \tag{10.66}$$

where $*$ denotes convolution. The signal component in the sample $u(0)$ is

$$w(0) = (b * h * p)(0) = \int_{-\infty}^{\infty} B(j\omega)H(j\omega)P(j\omega)\frac{d\omega}{2\pi}, \tag{10.67}$$

The noise component $e(t)$ has mean square value

$$\sigma_e^2 = \int_{-\infty}^{\infty} S_{ee}(j\omega)\frac{d\omega}{2\pi} = \int_{-\infty}^{\infty} |B(j\omega)|^2 S_{qq}(j\omega)\frac{d\omega}{2\pi}, \tag{10.68}$$

so that the signal-to-noise ratio at time $t = 0$ is

$$\frac{|w(0)|^2}{\sigma_e^2} = \frac{\left|\int_{-\infty}^{\infty} B(j\omega)H(j\omega)P(j\omega)\frac{d\omega}{2\pi}\right|^2}{\int_{-\infty}^{\infty} |B(j\omega)|^2 S_{qq}(j\omega)\frac{d\omega}{2\pi}}. \tag{10.69}$$

Before we proceed to optimize the filter $B(j\omega)$ and the pulse $P(j\omega)$, some remarks are in order. The preceding expression assumes that a single pulse is transmitted over the channel. Such a "one-shot" transmission is a good approximation to a real system if the pulses are sufficiently separated as in pulsed radar applications [Skolnik, 2001]. Here each transmitted pulse is reflected from a target and received back at the antenna. The "channel" impulse response $h(t)$ arises from the so-called **range spread** of the target [Bell, 1993], [Van Trees, 2001]. Successive transmitted pulses are well separated, and adjacent reflected pulses from a given target do not overlap. The model under discussion is applicable in these cases. However, pulses reflected from targets at different ranges can overlap if the targets are too close. The pulse width therefore determines range resolution.[4]

[4]The width of the pulse autocorrelation, which is smaller than the width of the pulse, determines the range resolution when the receiver uses a matched filter.

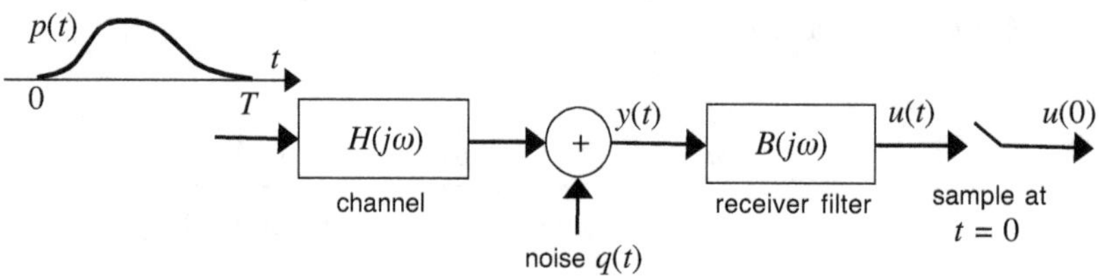

Figure 10.15. A pulse transmitted over a noisy channel.

10.6.1 Optimum receiver filter

Proceeding as in Sec. 2.5 we can show that the receiver filter $B(j\omega)$, which maximizes the signal-to-noise ratio (10.69) is

$$B(j\omega) = \frac{cP^*(j\omega)H^*(j\omega)}{S_{qq}(j\omega)}, \tag{10.70}$$

where c is an arbitrary nonzero constant, and $S_{qq}(j\omega)$ is assumed to be nonzero. The maximized value of the signal-to-noise ratio (10.69) is given by

$$\frac{|w(0)|^2}{\sigma_e^2} = \int_{-\infty}^{\infty} |P(j\omega)|^2 \frac{|H(j\omega)|^2}{S_{qq}(j\omega)} \frac{d\omega}{2\pi} \tag{10.71}$$

The filter (10.70) is the *matched filter* for the combination of pulse shape $p(t)$, channel $H(j\omega)$, and noise spectrum $S_{qq}(j\omega)$.

Details of the derivation. Consider the numerator of Eq. (10.69). We have

$$\Big|\int_{-\infty}^{\infty} B(j\omega)H(j\omega)P(j\omega)\frac{d\omega}{2\pi}\Big|^2 = \Big|\int_{-\infty}^{\infty} B(j\omega)\sqrt{S_{qq}(j\omega)}\frac{P(j\omega)H(j\omega)}{\sqrt{S_{qq}(j\omega)}}\frac{d\omega}{2\pi}\Big|^2.$$

So Cauchy-Schwartz inequality (Appendix A) yields

$$\Big|\int_{-\infty}^{\infty} BHP\frac{d\omega}{2\pi}\Big|^2 \le \int_{-\infty}^{\infty} |B|^2 S_{qq}\frac{d\omega}{2\pi}\int_{-\infty}^{\infty}\frac{|HP|^2}{S_{qq}}\frac{d\omega}{2\pi},$$

where the argument $(j\omega)$ is deleted for simplicity. Thus Eq. (10.69) yields

$$\frac{|w(0)|^2}{\sigma_e^2} \le \int_{-\infty}^{\infty} |P(j\omega)|^2 \frac{|H(j\omega)|^2}{S_{qq}(j\omega)} \frac{d\omega}{2\pi} \tag{10.72}$$

Equality is achieved if and only if there is equality in Cauchy-Schwartz inequality, that is (Appendix A),

$$B(j\omega)\sqrt{S_{qq}(j\omega)} = \frac{cP^*(j\omega)H^*(j\omega)}{\sqrt{S_{qq}(j\omega)}} \tag{10.73}$$

for some $c \neq 0$. The optimal filter $B(j\omega)$ therefore has the form (10.70). The maximized SNR is the right-hand side of Eq. (10.72), which indeed justifies Eq. (10.71). $\bigtriangledown\bigtriangledown\bigtriangledown$

10.6.2 Optimum pulse

For any given pulse shape $p(t)$ the maximum signal-to-noise ratio, achievable by choice of the receiver filter $B(j\omega)$, is given by Eq. (10.71). The next question is, what is the optimum pulse shape $p(t)$ (equivalently $P(j\omega)$) which maximizes the SNR? With no constraints on $p(t)$, Eq. (10.71) can be arbitrarily large. In practice, the SNR (10.71) is maximized under the energy constraint

$$\int_0^T |p(t)|^2 dt = E_p. \tag{10.74}$$

Before presenting the solution, it is useful to make a few definitions. Let

$$S_{hh}(j\omega) = \frac{|H(j\omega)|^2}{S_{qq}(j\omega)}. \tag{10.75}$$

This can be regarded as the "spectrum" (magnitude square) of the effective channel

$$H_{eff}(j\omega) = \frac{H(j\omega)}{\sqrt{S_{qq}(j\omega)}}. \tag{10.76}$$

Thus the SNR to be optimized becomes

$$\frac{|w(0)|^2}{\sigma_e^2} = \int_{-\infty}^{\infty} |P(j\omega)|^2 S_{hh}(j\omega)\frac{d\omega}{2\pi}. \tag{10.77}$$

Let $R_{hh}(\tau)$ denote the inverse FT of $S_{hh}(j\omega)$. This is nothing but the deterministic autocorrelation of the effective channel:

$$R_{hh}(\tau) = \int_{-\infty}^{\infty} h_{eff}(t)h_{eff}^*(t-\tau)dt, \tag{10.78}$$

where $h_{eff}(t)$ is the inverse FT of $H_{eff}(j\omega)$. Note that the inverse FT of $|P(j\omega)|^2$ is the autocorrelation of $p(t)$, that is,

$$R_{pp}(\tau) = \int_0^T p(t)p^*(t-\tau)dt. \tag{10.79}$$

♠**Theorem 10.3.** The optimum pulse $p(t)$ which maximizes Eq. (10.77) subject to Eq. (10.74) is the solution to the eigen equation

$$\int_0^T p(\tau)R_{hh}(t-\tau)d\tau = \lambda p(t), \quad 0 \leq t \leq T, \tag{10.80}$$

corresponding to the maximum eigenvalue λ. The maximized SNR has the form

$$\frac{|w(0)|^2}{\sigma_e^2} = \lambda E_p, \tag{10.81}$$

where E_p is the pulse energy (10.74). $\diamond$

Proof. Using Parseval's relation, Eq. (10.77) can be written as

$$\frac{|w(0)|^2}{\sigma_e^2} = \int_{-\infty}^{\infty} R_{pp}^*(\tau) R_{hh}(\tau) d\tau,$$

that is,

$$\frac{|w(0)|^2}{\sigma_e^2} = \int_0^T p^*(t) \Big(\int_{-\infty}^{\infty} p(t-\tau) R_{hh}(\tau) d\tau \Big)\, dt. \tag{10.82}$$

The pulse energy constraint can be written as

$$\int_0^T p^*(t) p(t) dt = E_p. \tag{10.83}$$

To maximize (10.82) under the constraint (10.83) we form the Lagrangian

$$\phi = \int_0^T p^*(t) \Big(\int_{-\infty}^{\infty} p(t-\tau) R_{hh}(\tau) d\tau \Big) dt - \lambda \Big(\int_0^T p^*(t) p(t) dt - E_p \Big),$$

where $-\lambda$ is the Lagrange multiplier. To obtain the optimal pulse, we set the gradient of this quantity, with respect to $p^*(t)$, to zero.[5] This yields the following necessary condition for optimality:

$$\int_{-\infty}^{\infty} p(t-\tau) R_{hh}(\tau) d\tau = \lambda p(t), \quad 0 \le t \le T. \tag{10.84}$$

With a change of variables we can rewrite the left-hand side as the integral $\int_{-\infty}^{\infty} p(\tau) R_{hh}(t-\tau) d\tau$. Since $p(t)$ is nonzero only in $0 \le t \le T$, Eq. (10.84) takes the final form (10.80). Substituting into Eq. (10.82) we get

$$\begin{aligned} \frac{|w(0)|^2}{\sigma_e^2} &= \int_0^T p^*(t) \Big(\int_{-\infty}^{\infty} p(t-\tau) R_{hh}(\tau) d\tau \Big) dt \\ &= \lambda \int_0^T p^*(t) p(t) dt = \lambda E_p \end{aligned}$$

using Eq. (10.84). This proves Eq. (10.81). $\bigtriangledown \bigtriangledown \bigtriangledown$

---

[5]Since $p(t)$ is complex, we have to use the complex gradient operator; review Sec. 20.3 at this time. The reader not familiar with calculus of variations for functionals should review appropriate reference material, e.g., Chap. 4 of Kirk [1970].

Discussions

1. ***Linear system representation.*** Notice that the integral (10.80) represents the convolution of the finite duration pulse $p(t)$ with the autocorrelation $R_{hh}(\tau)$. The result of this convolution can be arbitrarily long. Equation (10.80) says that this result, truncated to the region $0 \leq t \leq T$, matches $p(t)$ exactly except for the scale factor λ. This is schematically represented in Fig. 10.16 where we have an LTI system with impulse response $R_{hh}(t)$ (or frequency response $|H(j\omega)|^2/S_{qq}(j\omega)$) followed by a time domain truncator. The optimal solution $p(t)$ should be such that if $p(t)$ is input to this system, then the output is precisely $p(t)$ except for a scale factor λ. We say that $p(t)$ is an *eigenfunction* of the linear system defined by the cascade shown in the figure, and that λ is the *eigenvalue.*

2. ***Nature of eigenvalues.*** Readers familiar with Hermitian operators and quadratic forms [Helstrom, 1968] will recognize that Eq. (10.82) is a quadratic form involving a linear operator (an integral operator). If the integrals are regarded as summations this is especially easy to recognize. Since the original form (10.77) is nonnegative for any $p(t)$, we say that the operator is positive semidefinite. It can be shown that the linear operator under discussion has a *discrete* set of eigenvalues:

$$\lambda_0 \geq \lambda_1 \geq \lambda_2 \ldots$$

   and furthermore these are nonnegative because of positive semidefiniteness of the operator.

3. ***Prolate spheroidal functions.*** For the special case where $R_{hh}(j\omega)$ is a sinc function [that is, the quantity $|H(j\omega)|^2/S_{qq}(j\omega)$ is ideal lowpass], the solution $p(t)$ is called a *prolate spheroidal wave function*, and has been studied extensively [Slepian, 1978].

10.6.2.A MIMO radar

As explained at the beginning of this section, the analysis of the channel with a single input pulse is more appropriate in pulsed radar than in digital communications. In the radar application the inverse transform of $H(j\omega)$ is the target impulse response $h(t)$ (for a range-spread target). We find that the optimal $p(t)$ is the eigenfunction corresponding to the largest eigenvalue of a linear operator determined by the target impulse response $h(t)$.

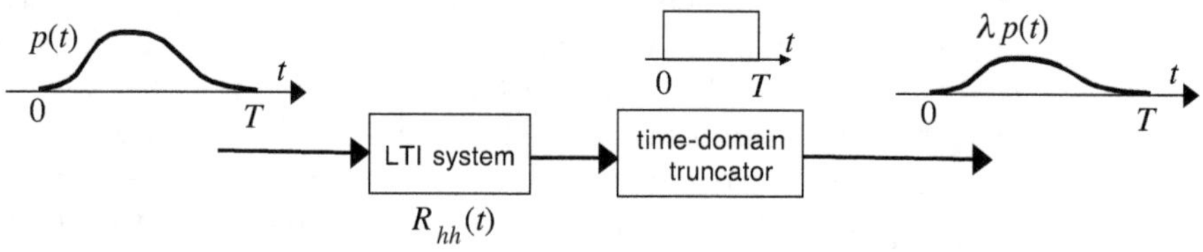

Figure 10.16. Eigenfunction property of the optimal pulse $p(t)$. See text.

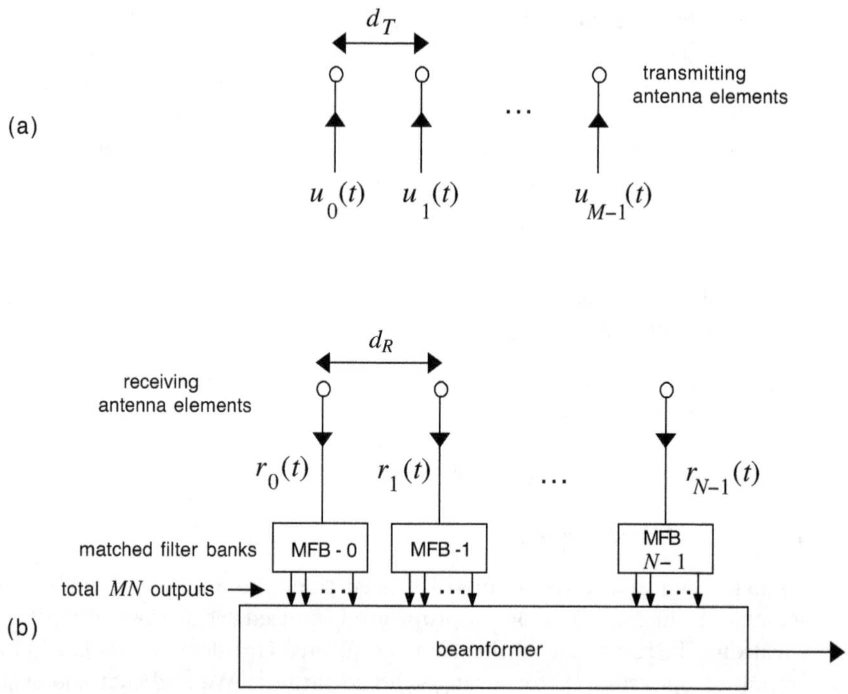

Figure 10.17. Elements of a MIMO radar system. (a) Transmitting antenna array with different waveforms fed into different antenna elements. (b) Receiving antenna array with matched filter banks and beamformer.

In recent years, a variation of the radar called the MIMO radar has become important [Bliss and Forsythe, 2003], [Robey *et al.*, 2004], [San Antonio, Fuhrmann, and Robey, 2006], [Li and Stoica, 2007], [Chen and Vaidyanathan, 2007a, 2007b, 2008, 2009]. In a MIMO radar we have a transmitting antenna array with M elements and a receiving array with N elements (see Fig. 10.17).

The transmitting elements transmit orthogonal waveforms $u_k(t)$. At the receiver, the signal from each of the N antenna elements is processed through a bank of M matched filters, each filter matched to one of the waveforms $u_k(t)$. In this way, MN independent signals can be generated at the receiver. If the M transmitting antennas and the N receiving antennas are spaced appropriately, then these MN signals can be used to define a **virtual array** with MN elements. This increases the number of degrees of freedom available at the receiver for beamforming. For example if we have ten transmitting and ten receiving antennas, the number of degrees of freedom can be as large as 100, even though there are only 20 physical antenna elements. The advantages and compromises involved in the design of MIMO radars are discussed extensively in the references cited above.

An important problem in the field of MIMO radars is the design of the waveforms $u_k(t)$. When the target has range spread (i.e., an impulse response $h(t)$ that is not a delta function), then the design of the waveforms should be such that the signal-to-noise ratios at the outputs of the MN matched filters are maximized. A challenging problem in such waveform optimization is to take into account the *signal-dependent clutter components* which are, unfortunately, always present in the received signal.

10.7 Concluding remarks

In this chapter we considered single-input single-output channels. Expressions for the optimal prefilter and postfilter were derived for various different choices of criteria. These results, which involve ideal unrealizable filters, are important from a theoretical point of view. In practice SISO channels are often equalized indirectly by first introducing a redundancy at the transmitter such as zero padding or cyclic prefixing (Chap. 7). Such redundancy converts the channel into a memoryless MIMO system, which is easier to equalize using practically realizable means such as matrix multiplications and so forth.

Problems

10.1. Consider a bandlimited channel

$$H(j\omega) = \begin{cases} 1 & 0 \le |\omega| < 0.5\pi/T \\ \epsilon & 0.5\pi/T \le |\omega| < \pi/T \\ 0 & \text{otherwise.} \end{cases}$$

Let the noise power spectrum be $S_{qq}(j\omega) = \sigma_q^2$ in $|\omega| \le \pi/T$. Assuming the power $p_0 = 1$ and $H_c(j\omega) = 1$, plot the magnitudes of a valid pair of optimal filters $F(j\omega)$ and $G(j\omega)$ as given in Theorem 10.1. Compute the minimized error $\mathcal{E}_{mmse}$. What happens to this error as $\epsilon \to 0$?

10.2. In Example 10.1 we considered a number of cases of the effective channel $H_{eff}(j\omega)$ which gave rise to compaction filters with one or two passbands. Construct an example of $H_{eff}(j\omega)$ such that the compaction filter has *four* passbands. Remember that the sum of all passband widths should be exactly $\omega_s = 2\pi/T$.

10.3. Referring to Sec. 10.2.3, construct an example of an effective channel $H_{eff}(j\omega)$ such that the optimum compaction filter has at least two *distinct* solutions $H_{c,1}(j\omega)$ and $H_{c,2}(j\omega)$. The example should be such that the product of the two solutions has total bandwidth strictly smaller than $2\pi/T$.

10.4. In Sec. 10.2 we came across continuous-time signals with Fourier transforms restricted to alias-free(T) regions. These can have disjoint bands in frequency, as shown by the example of Fig. 10.4(c), although the total width of all the bands is limited and depends on T. Find an example of an alias-free(T) Fourier transform $X(j\omega)$, with *four* distinct bands, where $X(j\omega)$ is nonzero.

10.5. In Sec. 10.3.2 we solved the MMSE transceiver problem under the zero-forcing constraint. In this problem we assume $S_{ss}(e^{j\omega}) = 1$ and $S_{qq}(e^{j\omega}) = \sigma_q^2$. Assume $H(e^{j\omega})$ is nonzero everywhere so $1/H(e^{j\omega})$ is bounded. Suppose we replace this optimal transceiver with $F(e^{j\omega}) = c$ and $G(e^{j\omega}) = [cH(e^{j\omega})]^{-1}$, where c is such that the power constraint is satisfied. This is called the lazy precoder system.

1. What is the mean square reconstruction error $\mathcal{E}_{mse}$?
2. What is the ratio $\mathcal{G} = \mathcal{E}_{mse}/\mathcal{E}_{ZF,mmse}$? Show that this can be expressed entirely in terms of $|H(e^{j\omega})|$. This is the gain obtained by going from the lazy precoder to the jointly optimal transceiver.
3. Using Cauchy-Schwartz inequality verify that $\mathcal{G} \ge 1$. When does equality arise?

10.6. In Problem 10.5 there is no upper bound on the gain $\mathcal{G}$.

1. Show this by constructing a channel $H(e^{j\omega})$ (nonzero everywhere so that $1/H(e^{j\omega})$ is bounded) such that the expression for the gain takes arbitrarily large values. *Hint.* Make $1/|H(e^{j\omega})|$ negligible everywhere compared to its value in a narrow band of width $\Delta\omega$, and make $\Delta\omega \to 0$.
2. Find an example of $H(e^{j\omega})$ such that the gain $\mathcal{G}$ is exactly 10.0.

10.7. For the digital equivalent of the oversampled system in Fig. 10.13(b), show that the zero-forcing condition is satisfied if and only if $\beta + \alpha = 1$.

10.8. In Sec. 10.2.3 we explained how to design an optimum compaction filter for any specified effective channel $H_{eff}(j\omega)$. Suppose

$$H_{eff}(j\omega) = \frac{H(j\omega)}{S_{qq}^{1/2}(j\omega)},$$

where $H(j\omega)$ is as in Fig. 10.12, and $S_{qq}(j\omega)$ is the noise spectrm shown in that figure. Find the optimum alias-free(T) compaction filter for this $H_{eff}(j\omega)$ for two cases: (a) when $A > \epsilon$, and (b) when $A < \epsilon$.

10.9. In Problem 10.8, assume $A = \epsilon$. What are the possible solutions to the optimum compaction filter? Sketch one solution.

10.10. Show that the mean square error (10.65) achieved by the optimal FSE system is the same as the mean square error achieved by the optimal transceiver of Sec. 10.2 without oversampling, as long as the filter $H_c(j\omega)$ in Sec. 10.2 is constrained to be the appropriate compaction filter. Assume that the energy $\int_0^{2\pi} |F(j\omega)|^2 d\omega/2\pi$ of the prefilter is fixed for the two systems being compared.

11

Optimal transceivers for diagonal channels

11.1 Introduction

Consider a transceiver system of the form shown in Fig. 11.1. For a fixed channel $\mathbf{H}$ and noise statistics, an important problem is the optimization of the precoder $\mathbf{F}$ and equalizer $\mathbf{G}$ in order to minimize an appopriate objective function such as the mean square reconstruction error or the bit error rate. In this chapter we consider this problem for the simple special case where the channel transfer matrix $\mathbf{H}$ is an $M \times M$ diagonal matrix

$$\mathbf{H} = \begin{bmatrix} H_0 & 0 & \dots & 0 \\ 0 & H_1 & \dots & 0 \\ \vdots & \vdots & \ddots & \vdots \\ 0 & 0 & \dots & H_{M-1} \end{bmatrix} \tag{11.1}$$

with independent additive Gaussian noise sources $q_k(n)$ as shown in Fig. 11.2. We can regard this as a single-user system, where $s_k(n)$ is obtained from a signal $s(n)$ by blocking or parsing (see Secs. 3.5 and 3.6); $s_k(n)$ could be the kth frequency bin in the DMT or OFDM system, for example. We can also regard this as an M-user communication system, provided it is practicable to collect the M signals in one transmitting location for common processing.

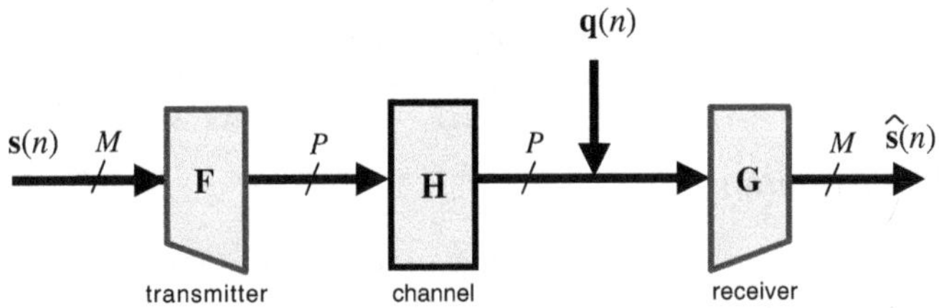

Figure 11.1. The general form of a transceiver sytem with channel **H**, precoder **F**, and equalizer **G**.

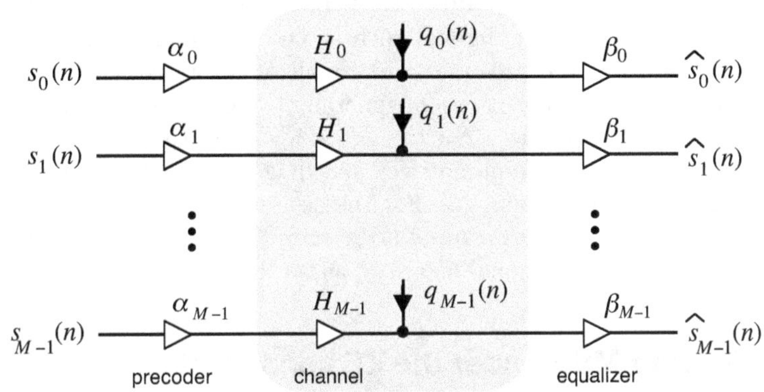

Figure 11.2. A simple diagonal channel made from a set of parallel independent channels. The precoder and equalizer are assumed to be diagonal matrices.

In Fig. 11.2 the transmitter is assumed to have an $M \times M$ diagonal precoder

$$\mathbf{F} = \begin{bmatrix} \alpha_0 & 0 & \ldots & 0 \\ 0 & \alpha_1 & \ldots & 0 \\ \vdots & \vdots & \ddots & \vdots \\ 0 & 0 & \ldots & \alpha_{M-1} \end{bmatrix}, \tag{11.2}$$

and the receiver has a $M \times M$ diagonal equalizer

$$\mathbf{G} = \begin{bmatrix} \beta_0 & 0 & \ldots & 0 \\ 0 & \beta_1 & \ldots & 0 \\ \vdots & \vdots & \ddots & \vdots \\ 0 & 0 & \ldots & \beta_{M-1} \end{bmatrix}. \tag{11.3}$$

With suitable assumptions on the signals $s_k(n)$ and noise sources $q_k(n)$ we shall consider a number of optimization problems. While the problems are considerably simpler than those involving nondiagonal channels (Chaps. 12–19), the diagonal case offers significant insights into the issues involved. Furthermore, the solutions form an integral part of the solutions for nondiagonal channels. Since $\mathbf{F}, \mathbf{H}$, and $\mathbf{G}$ are memoryless, the *time-argument* (n) on the signals will be dropped whenever convenient.

Chapter outline

Section 11.2 considers the case where the total reconstruction error is minimized (in the mean square sense) under the zero-forcing (ZF) constraint $\beta_k = 1/[H_k\alpha_k]$. This is called the **ZF-MMSE** transceiver. In Sec. 11.3 the ZF condition is removed and the multipliers α_k and β_k are *jointly* optimized, resulting in the MMSE transceiver, also called the **pure-MMSE** transceiver (to differentiate from ZF-MMSE). Both of these problems are solved under a fixed power constraint. In Sec. 11.4 we consider a totally different problem, namely that of *maximizing the capacity* of the set of parallel independent channels under a power constraint.

Even for a diagonal channel such as the one mentioned above, it is not in general true that the precoder and equalizer matrices can be restricted to be diagonal matrices without loss of generality. It depends on what objective function we want to optimize. For the *ZF-MMSE* and the *pure-MMSE* problems it turns out that the optimal matrices *are* diagonal as seen from the more general analysis of Chaps. 12 and 13. For the case where the objective function to be minimized is the average symbol error rate, the best precoder and equalizer for diagonal $\mathbf{H}$ are *not* diagonal matrices, as we shall see in Sec. 11.5 of this chapter.

11.2 Minimizing MSE under the ZF constraint

We now consider the system shown in Fig. 11.3 where

$$\beta_k = \frac{1}{H_k\alpha_k} \tag{11.4}$$

This choice of β_k makes it a zero-forcing (ZF) system. In this case the reconstruction error is entirely due to the noise sources $q_k(n)$ amplified through $1/H_k$ and $1/\alpha_k$. We assume that the signals $s_k(n)$ and noise sources $q_k(n)$ are zero-mean WSS processes. To state the statistical assumptions concisely, let $\mathbf{s}(n)$ and $\mathbf{q}(n)$ be column vectors whose components are $s_k(n)$ and $q_k(n)$. Then the covariance matrices are assumed to be as follows:

$$\mathbf{R_{ss}} = \sigma_s^2\mathbf{I}, \quad \mathbf{R_{qq}} = \text{diag}[\sigma_{q_0}^2, \sigma_{q_1}^2, \ldots, \sigma_{q_{M-1}}^2], \quad \text{and} \quad \mathbf{R_{sq}} = \mathbf{0}. \tag{11.5}$$

Note in particular that the signal variance is assumed identical for all k, that is,

$$\sigma_s^2 = E|s_k(n)|^2. \tag{11.6}$$

With these assumptions, the reconstruction error for the kth signal has variance

$$\mathcal{E}_k = \frac{\sigma_{q_k}^2}{|\alpha_k|^2|H_k|^2} \tag{11.7}$$

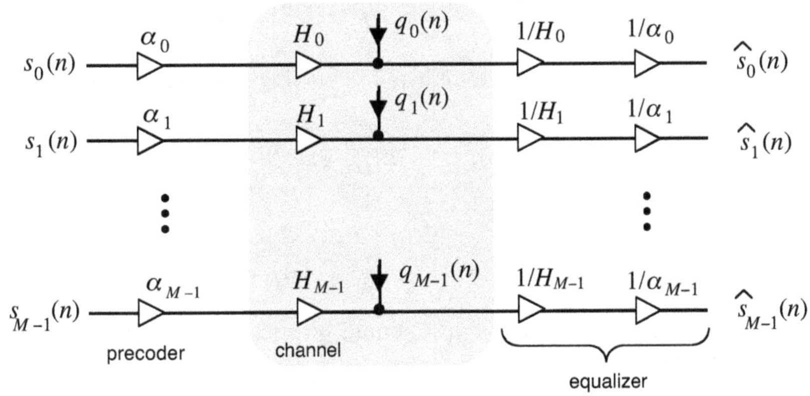

Figure 11.3. The diagonal channel with diagonal precoder and equalizer, satisfying the zero-forcing constraint.

Our goal is to optimize the multipliers α_k in order to minimize the sum of reconstruction errors:

$$\mathcal{E}_{mse} = \sum_{k=0}^{M-1} \frac{\sigma_{q_k}^2}{|\alpha_k|^2 |H_k|^2} \tag{11.8}$$

Since the mean square error is minimized under the zero-forcing constraint, this will be called the **ZF-MMSE** system. The minimization is done under an appropriate power constraint. Since the power input to the kth channel is $\sigma_s^2|\alpha_k|^2$, the power constraint has the form

$$\sigma_s^2 \sum_{k=0}^{M-1} |\alpha_k|^2 = p_0, \tag{11.9}$$

where p_0 is the total power input to the channel. The numbers α_k can be varied to achieve any power distribution, and the quantity σ_s^2 does not add extra flexibility. So we can set $\sigma_s^2 = 1$, *without loss of generality.* We shall however retain σ_s^2 in all equations, as it enables us to visualize some of the mathematical expressions more clearly.

11.2.1 Solving for the optimal multipliers

The notation

$$x_k = |\alpha_k|^2 \geq 0 \tag{11.10}$$

will be used temporarily for convenience. To solve the above constrained optimization problem define the Lagrangian function

$$\psi = \sum_{k=0}^{M-1} \frac{\sigma_{q_k}^2}{x_k|H_k|^2} + \lambda\Big(\sigma_s^2 \sum_{k=0}^{M-1} x_k - p_0\Big), \tag{11.11}$$

where λ is the Lagrange multiplier. Setting $\partial\psi/\partial x_k = 0$ we get

$$\frac{-\sigma_{q_k}^2}{x_k^2|H_k|^2} + \lambda\sigma_s^2 = 0,$$

which yields

$$x_k = \frac{c^2\sigma_{q_k}}{|H_k|\sigma_s}, \tag{11.12}$$

where the constant $c^2 \stackrel{\Delta}{=} 1/\sqrt{\lambda}$ should be such that the power constraint (11.9) is satisfied, that is,

$$c^2 = \frac{p_0}{\sigma_s}\left(\sum_{m=0}^{M-1} \frac{\sigma_{q_m}}{|H_m|}\right)^{-1}. \tag{11.13}$$

The optimum multipliers α_k therefore can be taken to be the nonnegative numbers defined by

$$\alpha_{k,opt} = c\left(\frac{\sigma_{q_k}}{|H_k|\sigma_s}\right)^{1/2}. \tag{11.14}$$

From Eq. (11.4) we have

$$\beta_{k,opt} = \frac{1}{H_k\alpha_{k,opt}} \tag{11.15}$$

which therefore yields

$$|\beta_{k,opt}| = \frac{1}{|H_k\alpha_{k,opt}|} = \frac{1}{c}\left(\frac{\sigma_s}{\sigma_{q_k}}\right)^{1/2} \times \frac{1}{|H_k|^{1/2}} \tag{11.16}$$

The minimized objective function $\mathcal{E}_{mse}$ is obtained by substituting the optimal α_k into Eq. (11.8). The result is

$$\mathcal{E}_{ZFmmse} = \frac{\sigma_s^2}{p_0}\left(\sum_{k=0}^{M-1} \frac{\sigma_{q_k}}{|H_k|}\right)^2. \tag{11.17}$$

A few remarks are in order:

1. ***Effective noise.*** Note that the quantity $\sigma_{q_k}^2/|H_k|^2$ appears in all the expressions. This can be regarded as the effective noise variance of the kth channel. This will become more clear when we discuss Fig. 11.4 below.
2. ***Power allocation.*** The signal power at the output of α_k is the power input to the kth channel H_k. This is given by

   $$P_k = \sigma_s^2\alpha_{k,opt}^2 = \frac{\gamma\sigma_{q_k}}{|H_k|}$$

   where γ is a temporary notation for the constant part. Thus channels with *large effective noise* $\sigma_{q_k}^2/|H_k|^2$ get *more power* in the optimal system.

3. ***Channels with poor gains create problems.*** For a fixed set of noise powers $\sigma_{q_k}^2$, the error can become arbitrarily large if some of the channel gains H_k get arbitrarily small. So, there is no channel-independent upper bound on the reconstruction error (11.17). This problem disappears if the zero-forcing constraint is removed as shown in Sec. 11.3.

4. ***No inequality constraint?*** In the optimization we did not explicitly incorporate the inequality constraint $x_k \geq 0$. The solution (11.12) however turned out to be positive, so all is well. Formally, if we incorporate $x_k \geq 0$ and use the Karush–Kuhn–Tucker conditions for optimality (Chap. 22), we obtain the same result as above.

5. ***Minimum or maximum?*** In general, setting the derivative of the Lagrangian to zero yields the necessary conditions for a local extremum, which can be a minimum or a maximum. In our problem, we know from Eq. (11.8) that $\mathcal{E}_{mse}$ has really no upper bound (no finite maximum). It can be made arbitrarily large by making α_n arbitrarily small for some n (for which $\sigma_{q_n}^2$ is nonzero) and adjusting the remaining α_k such that the power constraint holds. Since $\mathcal{E}_{mse}$ has no upper bound, the stationary point obtained from the Lagrange approach can only be a minimum and not a maximum. As the solution is unique, this minimum has to be global as well.

6. ***Noise-to-signal ratio.*** From Eq. (11.17) we see that the noise-to-signal ratio at the receiver (i.e., at the detector input) is

$$\frac{\mathcal{E}_{ZFmmse}}{\sigma_s^2} = \frac{1}{p_0}\left(\sum_{k=0}^{M-1} \frac{\sigma_{q_k}}{|H_k|}\right)^2.$$

   This depends on the channel input power p_0 and the quality of the channels $\sigma_{q_k}/|H_k|$, but not on σ_s^2, as expected.

7. ***Half-whitening.*** Consider Fig. 11.4, which shows a redrawing of the diagonal system of Fig. 11.3. The signals $\widehat{s}_k(n)$ are identical in both systems (for the same inputs $s_k(n)$). The equivalent noise sources are q_k/H_k. We see that

$$\widehat{s}_k(n) = s_k(n) + e_k(n),$$

   where $e_k(n)$ has the variance

$$\mathcal{E}_k = \frac{\sigma_{q_k}^2}{|H_k|^2} \times \frac{1}{\alpha_k^2}$$

   If α_k were chosen such that $\alpha_k^2 = \sigma_{q_k}^2/|H_k|^2$ then

$$\mathcal{E}_k = 1$$

   for all k, that is, the error variances would have been equalized. Such a choice of multipliers would be called a "whitening" system.[1] But the

[1]The term whitening comes about by imagining that k is a frequency index (as in OFDM), in which case the $\mathcal{E}_k$ are like the samples (in frequency) of an error spectrum.

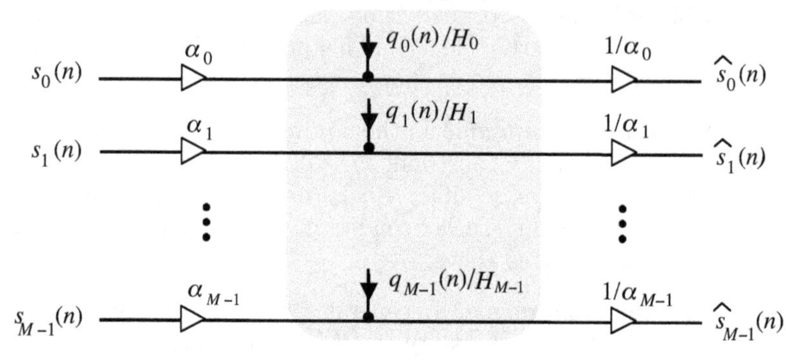

Figure 11.4. Redrawing the channel paths for further interpretation.

actual choice of α_k is as in Eq. (11.14), and this results in

$$\mathcal{E}_k = \gamma\sqrt{\frac{\sigma_{q_k}^2}{|H_k|^2}}$$

for some constant γ. That is, the error variances are not equal, but their original distribution $\sigma_{q_k}^2/|H_k|^2$ is replaced by the square root distribution. So the optimal choice of α_k is called a *half-whitening system.* In short, under the zero-forcing constraint, half-whitening minimizes reconstruction error. A similar phenomenon arises in certain data compression problems [Jayant and Noll, 1984].

11.3 Minimizing MSE without ZF constraint

We now reconsider the optimization problem with the zero-forcing constraint removed. This will lead to a smaller mean square error. Such systems are called pure-MMSE systems to distinguish them from ZF-MMSE systems. Unlike in a ZF-MMSE system, we will see that the pure-MMSE system forces an upper bound on the error, independent of the channels (i.e., regardless of how large $\sigma_{q_k}^2$ are, or how small H_k are).

The MMSE system is studied in considerable detail in Chap. 22 where we develop it as an example of the application of the Karush–Kuhn–Tucker (KKT) theory of optimization. In Chap. 22 we consider the example where $H_k = 1$ for all k (with $\sigma_{q_k}^2$ allowed to be different for different k). The results can be modified readily for the case of arbitrary H_k as elaborated next.

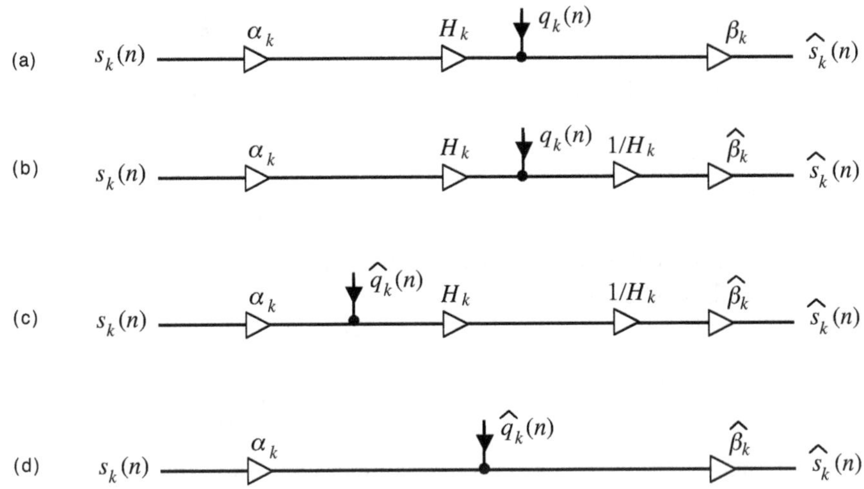

Figure 11.5. Redrawing the kth channel path for further simplification.

In Fig. 11.5(a) we show the kth branch of the system with channel multiplier H_k. Assuming $H_k \neq 0$ we can always write β_k in the form[2]

$$\beta_k = \frac{\widehat{\beta}_k}{H_k}, \tag{11.18}$$

as shown in Fig. 11.5(b). We now move the noise source to the left of H_k to obtain an equivalent noise source $\widehat{q}_k(n) = q_k(n)/H_k$. This is shown in Fig. 11.5(c). Note that $\widehat{q}_k(n)$ has variance

$$\widehat{\sigma}^2_{q_k} \stackrel{\Delta}{=} \frac{\sigma^2_{q_k}}{|H_k|^2} \tag{11.19}$$

The equivalent system of Fig. 11.5(c) can now be redrawn as in Fig. 11.5(d). In this system we have a channel with unit gain and a noise source with a certain variance $\widehat{\sigma}^2_{q_k}$. With each branch of Fig. 11.2 modified in this way, the results of Chap. 22 can be applied here. For example, for fixed α_k the optimal $\widehat{\beta}_k$ that minimizes the mean square error

$$\mathcal{E}_k = E[|\widehat{s}_k(n) - s_k(n)|^2] \tag{11.20}$$

[2]If $H_k = 0$ for some k, then no matter how we choose α_k and β_k, the quantity $\widehat{s}_k(n)$ would only be the noise amplified by β_k. So we take the solution to be $\alpha_k = \beta_k = 0$ when $H_k = 0$ for a particular k.

is the Wiener multiplier given by

$$\widehat{\beta}_k = \frac{\alpha_k^*}{|\alpha_k|^2 + \dfrac{\widehat{\sigma}_{q_k}^2}{\sigma_s^2}} = \frac{\alpha_k^*}{|\alpha_k|^2 + \dfrac{\sigma_{q_k}^2}{|H_k|^2\sigma_s^2}} \tag{11.21}$$

With this choice of $\widehat{\beta}_k$ the mean square error becomes

$$\mathcal{E}_k = \frac{\widehat{\sigma}_{q_k}^2}{|\alpha_k|^2 + \dfrac{\widehat{\sigma}_{q_k}^2}{\sigma_s^2}}, \tag{11.22}$$

as shown in the proof after Eq. (22.37) in Chap. 22. The total MSE summing over all M channels is therefore

$$\mathcal{E}_{mse} = \sum_{k=0}^{M-1} \frac{\widehat{\sigma}_{q_k}^2}{|\alpha_k|^2 + \dfrac{\widehat{\sigma}_{q_k}^2}{\sigma_s^2}} \tag{11.23}$$

Assuming that the noise variances are ordered such that

$$\widehat{\sigma}_{q_0}^2 \leq \widehat{\sigma}_{q_1}^2 \leq \ldots \leq \widehat{\sigma}_{q_{M-1}}^2, \tag{11.24}$$

the optimal values of the precoder multipliers α_k under the power constraint can be obtained from

$$|\alpha_k|^2 = \begin{cases} \dfrac{\widehat{\sigma}_{q_k}}{\sigma_s^2}\left(\dfrac{p_0 + \sum_{\ell=0}^{K-1}\widehat{\sigma}_{q_\ell}^2}{\sum_{\ell=0}^{K-1}\widehat{\sigma}_{q_\ell}} - \widehat{\sigma}_{q_k}\right), & 0 \leq k \leq K-1 \\ 0 & \text{otherwise.} \end{cases} \tag{11.25}$$

This follows from Eq. (22.48) in Chap. 22 with appropriate modification of notations. Thus, depending on the total available power, only a certain number K of multipliers can be nonzero, and K can be determined as explained in Sec. 22.4.2 of Chap. 22. The minimized mean square error is given by Eq. (22.49) in Chap. 22, with appropriate modification of notations:

$$\mathcal{E}_{mmse} = \frac{\sigma_s^2\left(\sum_{\ell=0}^{K-1}\widehat{\sigma}_{q_\ell}\right)^2}{p_0 + \sum_{\ell=0}^{K-1}\widehat{\sigma}_{q_\ell}^2} + (M-K)\sigma_s^2. \tag{11.26}$$

The second term $(M-K)\sigma_s^2$ arises because, whenever $\alpha_k = 0$, the mean square error is σ_s^2, as seen from Eq. (11.22). The final expression for the optimal system can be obtained by replacing $\widehat{\sigma}_{q_k}^2$ with Eq. (11.19) in all expressions. Thus the optimal premultipliers α_k are determined from

$$|\alpha_k|^2 = \begin{cases} \left(\dfrac{p_0 + \displaystyle\sum_{\ell=0}^{K-1}\frac{\sigma_{q_\ell}^2}{|H_\ell|^2}}{\displaystyle\sum_{\ell=0}^{K-1}\frac{\sigma_{q_\ell}}{|H_\ell|}} - \dfrac{\sigma_{q_k}}{|H_k|}\right)\dfrac{\sigma_{q_k}}{|H_k|\sigma_s^2}, & 0 \leq k \leq K-1 \\ 0 & \text{otherwise,} \end{cases} \tag{11.27}$$

and the optimal equalizers β_k are given by

$$\beta_k = \frac{1}{H_k}\left(\frac{\alpha_k^*}{|\alpha_k|^2 + \dfrac{\sigma_{q_k}^2}{|H_k|^2\sigma_s^2}}\right). \tag{11.28}$$

The minimized mean square error is

$$\mathcal{E}_{mmse} = \frac{\sigma_s^2\left(\sum_{\ell=0}^{K-1}\dfrac{\sigma_{q_\ell}}{|H_\ell|}\right)^2}{p_0 + \sum_{\ell=0}^{K-1}\dfrac{\sigma_{q_\ell}^2}{|H_\ell|^2}} + (M-K)\sigma_s^2. \tag{11.29}$$

Note that α_k can be taken as the non-negative square root of Eq. (11.27). In this case

$$\beta_k = \frac{1}{H_k}\left(\frac{\alpha_k}{\alpha_k^2 + \dfrac{\sigma_{q_k}^2}{|H_k|^2\sigma_s^2}}\right). \tag{11.30}$$

Thus, for real H_k the optimal α_k and β_k can be assumed to be real as expected. We conclude this section with a number of important remarks:

1. ***ZF MMSE versus pure MMSE***. From Eq. (11.25) we see that if the channel quality is below a threshold (effective noise $\sigma_{q_k}^2/|H_k|^2$ above a threshold), then the first expression in Eq. (11.25) becomes negative, which means the solution is $\alpha_k = 0$. So in that case, the channel is not used at all. This "blockout property" is not observed in the optimal solution (11.14) with the zero-forcing constraint. According to Eq. (11.14) the quantity $|\alpha_k|^2$ increases monotonically as the effective noise $\sigma_{q_k}^2/|H_k|^2$ increases. That is, the ZF solution is such that the "bad channels" are put to use by allocating more power to them. On the other hand, the pure-MMSE solution is such that "bad channels" are simply eliminated from consideration.

2. ***Threshold.*** The threshold value for blockout is not obvious from Eq. (11.27) because it is implicit. The reader should review Chap. 22 for a discussion of interesting details pertaining to the MMSE solution given above.

3. ***Non-monotonicity***. For the pure-MMSE system the quantity $|\alpha_k|^2$ is not necessarily a monotone function of $\sigma_{q_k}^2/|H_k|^2$. For example, see the plot in Sec. 22.4.3 of Chap. 22. Channels with very large or very small $\sigma_{q_k}^2/|H_k|^2$ tend to get very little power.

4. ***Signal-to-noise ratio.*** The average signal-to-noise ratio at the detector input, given by $\mathcal{E}_{mmse}/\sigma_s^2$, does not depend on σ_s^2 as seen from (11.29):

$$\frac{\mathcal{E}_{mmse}}{\sigma_s^2} = \frac{\left(\sum_{\ell=0}^{K-1}\dfrac{\sigma_{q_\ell}}{|H_\ell|}\right)^2}{p_0 + \sum_{\ell=0}^{K-1}\dfrac{\sigma_{q_\ell}^2}{|H_\ell|^2}} + (M-K).$$

This depends only on the channel and the power p_0.

11.4 Maximizing channel capacity

Given a diagonal channel characterized by the channel gains H_k and zero-mean additive Gaussian noise sources $q_k(n)$ as shown in Fig. 11.6(a), and given the power constraint

$$\sum_{k=0}^{M-1} \sigma_{s_k}^2 = p_0, \tag{11.31}$$

how do we distribute the powers $\sigma_{s_k}^2$ such that the channel capacity is maximized? In Chap. 22 we address this problem for the special case where $H_k = 1$ for all k. The results of Chap. 22 can readily be modified for the case of arbitrary H_k as we show in this section. In this section we assume for simplicity that the signals, noise, and the channel are real-valued.

The channel capacity [Cover and Thomas, 1991] depends only on the channel input powers and the noise powers. It gives an upper bound on the number of information bits per sample (or per second, when appropriately modified) that can be transmitted with arbitrarily small error. Bit rates arbitrarily close to the capacity can be achieved with sophisticated channel-coding techniques. The capacity, by definition, does not depend on the signal processing at the transmitter or the receiver.

For example, suppose we change the kth path in Fig. 11.6(a) as shown in Fig. 11.6(b) (left), by inserting a multiplier $1/H_k$ at the receiver (as in a zero-forcing system). Such an insertion clearly does not change the capacity of the kth channel. The kth path can then be redrawn as shown on the right-hand side of Fig. 11.6(b), where the noise source is

$$\widehat{q}_k(n) = \frac{q_k(n)}{H_k} \tag{11.32}$$

This equivalence follows because $\widehat{s}_k(n) = s_k(n) + q_k(n)/H_k$ for both of these systems. The capacity of the original system in Fig. 11.6(a) is therefore equal to the capacity of the system shown in Fig. 11.6(c). If we distribute the powers $\sigma_{s_k}^2$ to maximize the capacity in Fig. 11.6(c) subject to the power constraint (11.31), this is equivalent to maximizing the capacity of the original system in Fig. 11.6(a). In Chap. 22 we maximized the capacity of the system in Fig. 11.6(c). These results can therefore be applied here. The results are summarized next with appropriate change of notations. The capacity of the kth channel per use is given by

$$C_k = 0.5 \log_2\Big(1 + \frac{\sigma_{s_k}^2}{\widehat{\sigma}_{q_k}^2}\Big),$$

where

$$\widehat{\sigma}_{q_k}^2 = \frac{\sigma_{q_k}^2}{|H_k|^2}, \tag{11.33}$$

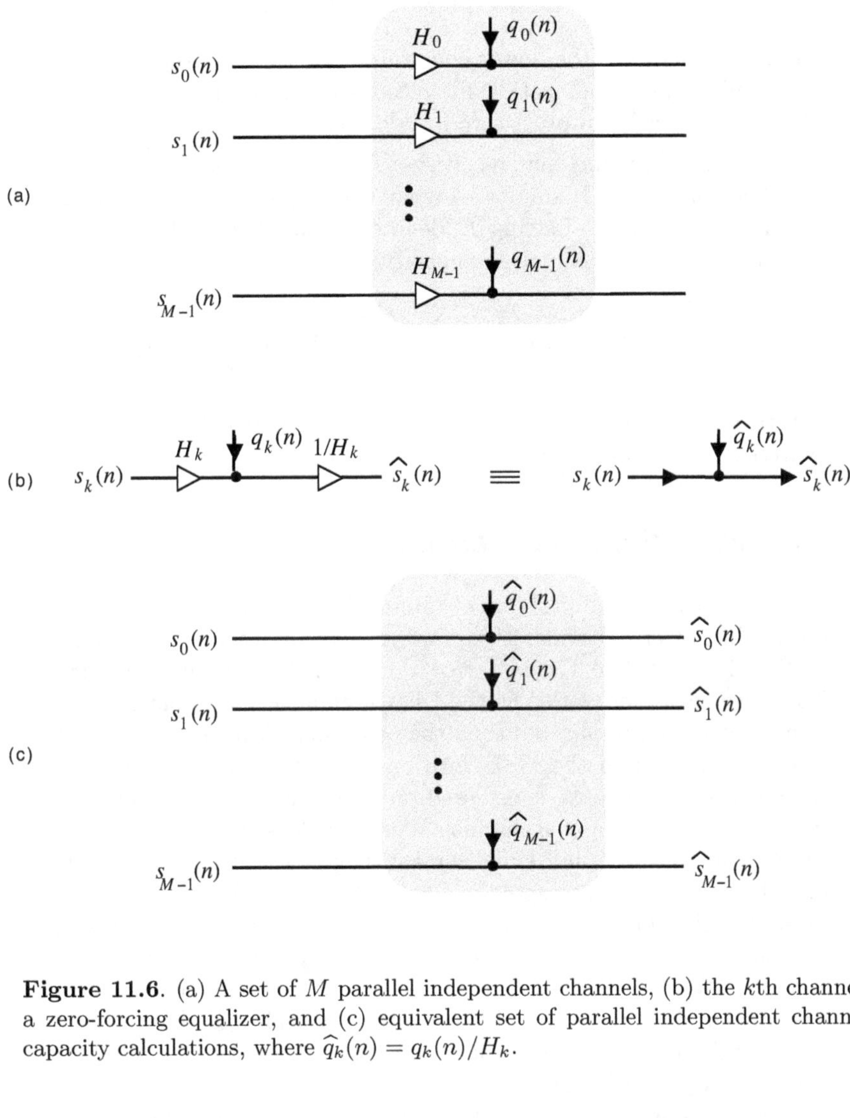

Figure 11.6. (a) A set of M parallel independent channels, (b) the kth channel with a zero-forcing equalizer, and (c) equivalent set of parallel independent channels for capacity calculations, where $\widehat{q}_k(n) = q_k(n)/H_k$.

from Eq. (11.32). $\mathcal{C}_k$ is the number of bits per sample that can be transmitted reliably on the kth channel if channel coding of arbitrary sophistication is available. The capacity of the parallel set of M channels is

$$\mathcal{C} = \sum_{k=0}^{M-1} \mathcal{C}_k = 0.5 \sum_{k=0}^{M-1} \log_2\Big(1 + \frac{\sigma_{s_k}^2}{\widehat{\sigma}_{q_k}^2}\Big). \tag{11.34}$$

From Chap. 22 (Sec. 22.3) we know that the power allocation which maximizes this capacity is given by

$$\sigma_{s_k}^2 = \begin{cases} \eta - \widehat{\sigma}_{q_k}^2 & \text{if this is} \geq 0 \\ 0 & \text{otherwise,} \end{cases} \tag{11.35}$$

where η is such that the power constraint (11.31) is satisfied. This is called the **water-filling** power allocation for reasons described in Chap. 22 (Sec. 22.3), which also describes a procedure for identifying η.

Note that a channel may receive zero power if the noise $\widehat{\sigma}_{q_k}^2$ is too large. This "blockout" property is similar to what we observed for the MMSE solution at the end of Sec. 11.3. Letting K denote the number of channels which receive nonzero power, the maximized capacity is

$$C_{water} = 0.5K \log_2 \left(\frac{p_0 + \sum_{k=0}^{K-1} \widehat{\sigma}_{q_k}^2}{K} \right) - 0.5 \sum_{k=0}^{K-1} \log_2 \widehat{\sigma}_{q_k}^2. \tag{11.36}$$

Substituting $\widehat{\sigma}_{q_k}^2 = \sigma_{q_k}^2/|H_k|^2$ in Eqs. (11.35) and (11.36) results in the final expressions for these quantities.

11.5 Minimizing the symbol error rate

We now consider the problem of minimizing the average symbol error rate (or probability of error) under the power constraint. Since the symbol error rate is directly related to the bit error rate (BER) for Gray coded systems (Sec. 2.3.3), we also refer to this as the BER minimization problem. This problem is somewhat more complicated than the problems addressed in Secs. 11.2 and 11.3 where we minimized the MSE with or without the zero-forcing constraint. In these earlier sections we considered only diagonal precoders and equalizers. It turns out that for diagonal channels there is no loss of generality in assuming that the MMSE precoder and equalizer are diagonal matrices. This follows from a more formal analysis presented in Chaps. 12 and 13. However, for the case where we minimize symbol error rate for a diagonal channel, it is a loss of generality to assume that the precoder and equalizer are diagonal, as we shall show in this section. We will demonstrate the existence of a nondiagonal precoder which has smaller BER than any diagonal precoder. Throughout the section we will assume that there is no bit allocation, that is, each $s_k(n)$ is represented by a fixed b-bit constellation.

First consider Fig. 11.7(a), where the precoder and equalizer are assumed to be diagonal matrices with diagonal elements α_k and β_k. Assume

$$\beta_k = \frac{1}{\alpha_k H_k}$$

for each k, so we have a *zero-forcing* system. Assume the signal and noise statistics to be precisely as described at the beginning of Sec. 11.2, and assume further that all noise sources have identical variance. Thus

$$\sigma_s^2 = E|s_k(n)|^2 \quad \text{and} \quad \sigma_q^2 = E|q_k(n)|^2. \tag{11.37}$$

The power constraint is

$$\sigma_s^2 \sum_{k=0}^{M-1} |\alpha_k|^2 = p_0. \tag{11.38}$$

Clearly the noise variance at the output of β_k is $\sigma_q^2|\beta_k|^2$, so that the signal-to-noise ratio at the input of the kth detector is

$$SNR_k = \frac{\sigma_s^2}{\sigma_q^2|\beta_k|^2}. \tag{11.39}$$

We assume that the samples $s_k(n)$ come from PAM or QAM constellations. If each $q_k(n)$ is Gaussian[3] with variance $\sigma_{q_k}^2$, the probability of error in detection of $s_k(n)$ is (Sec. 2.3)

$$\mathcal{P}_e(k) = c\mathcal{Q}\Big(\frac{A}{\sqrt{x_k}}\Big), \tag{11.40}$$

where

$$c = 2(1 - 2^{-b}), \quad A = \sqrt{\frac{3\sigma_s^2}{2^{2b} - 1}}, \tag{11.41}$$

in the PAM case, and

$$c = 4(1 - 2^{-b/2}), \quad A = \sqrt{\frac{3\sigma_s^2}{2^b - 1}}, \tag{11.42}$$

in the QAM case. Here x_k is the noise variance at the input of the kth detector:

$$x_k = \sigma_q^2|\beta_k|^2.$$

Thus the average probability of error has the form

$$\mathcal{P} = \frac{c}{M}\sum_{k=0}^{M-1} \mathcal{Q}\Big(\frac{A}{\sqrt{x_k}}\Big). \tag{11.43}$$

11.5.1 Introducing the unitary matrix U

Now consider Fig. 11.7(b), which shows a unitary matrix $\mathbf{U}^\dagger$ at the transmitter, and its inverse $\mathbf{U}$ at the receiver.[4] The noise power at the input of the kth detector changes to $[\mathbf{U}\boldsymbol{\Lambda}\mathbf{U}^\dagger]_{kk}$, where

$$\boldsymbol{\Lambda} = \sigma_q^2 \begin{bmatrix} |\beta_0|^2 & 0 & \ldots & 0 \\ 0 & |\beta_1|^2 & \ldots & 0 \\ \vdots & \vdots & \ddots & \vdots \\ 0 & 0 & \ldots & |\beta_{M-1}|^2 \end{bmatrix} \tag{11.44}$$

[3] In the QAM case, circularly symmetric complex Gaussian, as described in Sec. 2.3.2.

[4] The unitary property means $\mathbf{U}\mathbf{U}^\dagger = \mathbf{I}$, so $\mathbf{U}$ is the inverse of $\mathbf{U}^\dagger$.

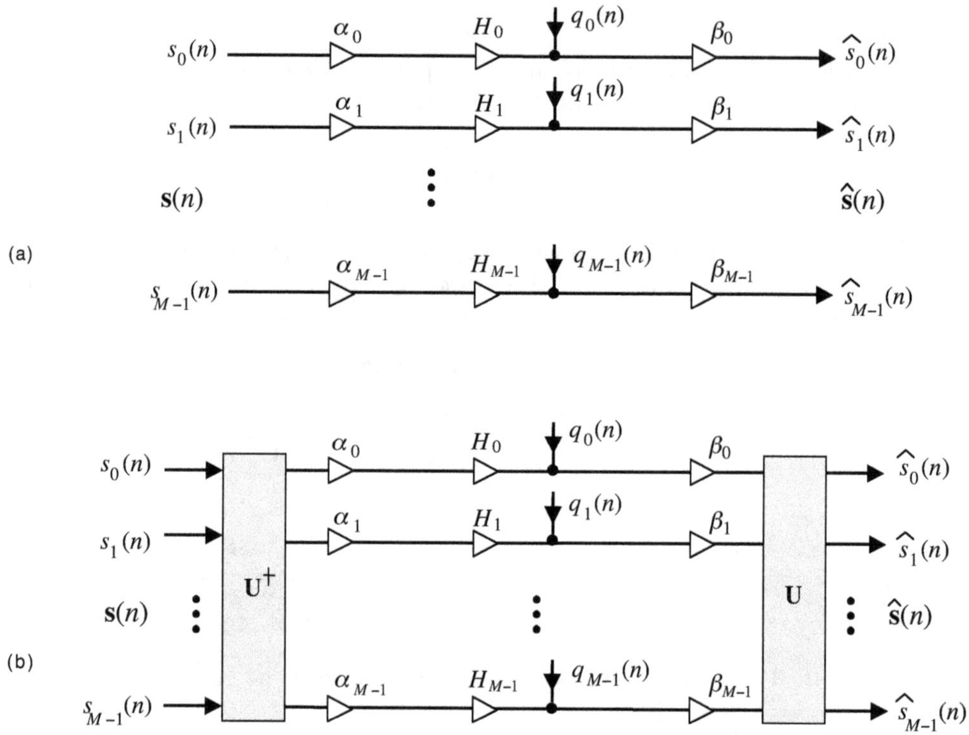

Figure 11.7. (a) A diagonal channel with diagonal precoder and equalizer. This is a zero-forcing system if $\beta_k = 1/\alpha_k H_k$. (b) Modified transceiver with extra unitary matrices $\mathbf{U}^\dagger$ and $\mathbf{U}$. This continues to be a ZF system as long as $\beta_k = 1/\alpha_k H_k$.

is the noise covariance at the input of $\mathbf{U}$. Note however the following points:

1. The *total* reconstruction error is

$$\mathcal{E}_{mse} = \text{Tr }(\mathbf{U}\boldsymbol{\Lambda}\mathbf{U}^\dagger) = \text{Tr }(\mathbf{U}^\dagger\mathbf{U}\boldsymbol{\Lambda}) = \text{Tr }\boldsymbol{\Lambda} = \sigma_q^2 \sum_k |\beta_k|^2, \qquad (11.45)$$

   and is independent of $\mathbf{U}$. Only the individual components $[\mathbf{U}\boldsymbol{\Lambda}\mathbf{U}^\dagger]_{kk}$ change with $\mathbf{U}$.

2. The channel input power is unchanged since the covariance matrix at the output of $\mathbf{U}^\dagger$ continues to be $\sigma_s^2\mathbf{I}$.

3. The system continues to enjoy the zero-forcing property, so the *signal component* of the power at the input of the kth detector is unchanged.

But the fact that the individual error variances $[\mathbf{U}\mathbf{\Lambda}\mathbf{U}^\dagger]_{kk}$ depend on $\mathbf{U}$ makes the average error probability dependent on $\mathbf{U}$:

$$\mathcal{P} = \frac{c}{M}\sum_{k=0}^{M-1} \mathcal{Q}\Big(\frac{A}{\sqrt{y_k}}\Big), \tag{11.46}$$

where

$$y_k = [\mathbf{U}\mathbf{\Lambda}\mathbf{U}^\dagger]_{kk} = k\text{th error variance.} \tag{11.47}$$

What is the best choice of unitary $\mathbf{U}$ that minimizes Eq. (11.46)? We can apply the theory of **Schur-convex** functions (Chap. 21) to answer this question. In Sec. 21.2.3 of Chap. 21 we show that

$$\mathcal{Q}(A/\sqrt{y_k}) \text{ is } \begin{cases} \text{convex in } y_k & \text{for } y_k < A^2/3 \\ \text{concave in } y_k & \text{for } y_k > A^2/3. \end{cases} \tag{11.48}$$

If the signal-to-noise ratios at the inputs of detectors are sufficiently large, then we can assume that this convexity condition is satisfied. From Sec. 21.4 in Chap. 21 we then conclude that the average probability of error (11.46) is a *Schur-convex function* of the vector

$$\mathbf{y} = [\, y_0 \quad y_1 \quad \ldots \quad y_{M-1} \,]. \tag{11.49}$$

Since

$$\sum_k y_k = \text{Tr}\,(\mathbf{U}\mathbf{\Lambda}\mathbf{U}^\dagger) = \sigma_q^2 \sum_k |\beta_k|^2 \tag{11.50}$$

is independent of $\mathbf{U}$, we are looking for the best set of positive numbers y_k with a fixed sum, such that the Schur-convex function (11.46) is minimized. From Sec. 21.5 in Chap. 21 we know that the solution is the set with identical y_k:

$$y_0 = y_1 = \ldots = y_{M-1}. \tag{11.51}$$

As long as there exists a unitary matrix $\mathbf{U}$ such that the diagonal elements of $\mathbf{U}\mathbf{\Lambda}\mathbf{U}^\dagger$ are identical, that serves as an optimal solution for $\mathbf{U}$. For example, if we choose $\mathbf{U}$ to be the normalized **DFT matrix**:

$$\mathbf{U} = \frac{\mathbf{W}}{\sqrt{M}}, \tag{11.52}$$

where $[\mathbf{W}]_{km} = e^{-j2\pi km/M}$ then $\mathbf{U}\mathbf{\Lambda}\mathbf{U}^\dagger$ is circulant because $\mathbf{\Lambda}$ is diagonal (see Appendix D). This ensures, indeed, that the y_k are all equal. More generally, it can be shown that any unitary $\mathbf{U}$ whose elements have identical magnitude:

$$|U_{km}| = \frac{1}{\sqrt{M}} \tag{11.53}$$

will work, as explained in Sec. 21.5.1.B of Chap. 21. Besides the DFT and IDFT, another example of such a unitary matrix would be the so-called **Hadamard matrix**. Hadamard matrices are very attractive because they do not have nontrivial multipliers. For example, the 2×2 and 4×4 (normalized) Hadamard matrices are

$$\mathbf{U}_2 = \frac{1}{\sqrt{2}}\begin{bmatrix} 1 & 1 \\ 1 & -1 \end{bmatrix} \quad \text{and} \quad \mathbf{U}_4 = \frac{1}{2}\begin{bmatrix} 1 & 1 & 1 & 1 \\ 1 & -1 & 1 & -1 \\ 1 & 1 & -1 & -1 \\ 1 & -1 & -1 & 1 \end{bmatrix}. \tag{11.54}$$

The 2×2 DFT matrix is identical to the Hadamard matrix, but for larger sizes, the DFT has complex entries. It can be implemented with the FFT, which makes it efficient, especially when M is a power of two. Note that the DFT matrix exists for any integer M, but the Hadamard matrix exists only for certain values of the block size M; it always exists when M is a power of two. If $\mathbf{U}_M$ is Hadamard, then so is

$$\frac{1}{\sqrt{2}}\begin{bmatrix} \mathbf{U}_M & \mathbf{U}_M \\ \mathbf{U}_M & -\mathbf{U}_M \end{bmatrix}.$$

11.5.2 Expression for minimized error rate

With $\mathbf{U}$ chosen as above we have y_k identical for all k, so, from Eq. (11.50),

$$y_k = \frac{\sigma_q^2}{M}\sum_{\ell=0}^{M-1} |\beta_\ell|^2 = \frac{\mathcal{E}_{mse}}{M}$$

for all k. The average symbol error rate Eq. (11.46) can therefore be written as

$$\mathcal{P} = c\mathcal{Q}\left(\frac{A}{\sqrt{\mathcal{E}_{mse}/M}}\right). \tag{11.55}$$

The quantity $\mathcal{E}_{mse}$ does not depend on $\mathbf{U}$ (see (11.45)) but it depends on how α_k is designed. Since $\mathcal{Q}$ is smaller for smaller $\mathcal{E}_{mse}$ it follows that α_k and β_k should be chosen to minimize this mean square error $\mathcal{E}_{mse}$. This means that the original system in Fig. 11.7(a) should be designed to have minimum mean square error property, and the matrix $\mathbf{U}$ should be designed as described above in order to minimize average error probability.

It is clear from here that *a system with minimum error probability also enjoys minimum MSE*, but an arbitrary MMSE system (e.g., the one with $\mathbf{U}$ replaced by identity) does not necessarily have minimum error probability.

Example 11.1: Effect of equalizing error components

Consider a transceiver with $M = 5$, where the mean square errors x_k are given by the elements of the following vector:

$$\mathbf{x} = \begin{bmatrix} 0.01 & 0.1 & 0.001 & 0.0001 & 0.0001 \end{bmatrix}.$$

Assume we have a 2-bit PAM system, and let $\sigma_s^2 = 1$. The average error probability calculated using Eq. (11.46) is then 0.0236. If we equalize these errors using a unitary matrix, then the resulting error vector is

$$\mathbf{y} = 0.0222\,[1 \quad 1 \quad 1 \quad 1 \quad 1]$$

and the average error probability computed using Eq. (11.55) becomes 0.0020. The gain in using the unitary matrix is therefore

$$\mathcal{G} = 0.0236/0.0020 = 11.8.$$

Thus the use of a unitary matrix to equalize the error components can be very powerful. Several examples of minimum error probability systems will be presented in Chaps. 17 and 18.

11.5.2.A How large should the SNR be?

For QAM symbols, since A is as in (11.42), the convexity condition $y_k < A^2/3$ can be written as

$$y_k < \frac{\sigma_s^2}{(2^b - 1)} \tag{11.56}$$

that is, the SNR σ_s^2/y_k at the input of the kth detector should be

$$SNR_k = \frac{\sigma_s^2}{y_k} \geq (2^b - 1). \tag{11.57}$$

Notice that as the number of bits b is increased the threshold $(2^b - 1)$ also increases. For example if $b = 6$ (64-QAM) then

$$SNR_k \geq 63, \tag{11.58}$$

which is about 18 dB. The reader may wonder whether this is an unusually high SNR. Observe however that this is the SNR *at the detector input.* For 64-QAM this SNR corresponds to a symbol error probability of 0.14. Assuming we have a Gray coded system, the bit error probabilty is approximately $0.14/6 = 0.023$. Except in situations where this error is acceptable, we can assume that the SNR is large enough to justify the convexity requirement on $\mathcal{P}_e(k)$. As another example, let $b = 2$ (4-QAM, or QPSK). The requirement now is

$$SNR_k \geq 3 \tag{11.59}$$

or about 4.77 dB, which corresponds to a symbol error probability of 0.0816. Again, in situations where the error probabilties are smaller than this, the convexity condition is satisfied.

Summary. For any chosen set of multipliers $\{\alpha_k\}$ and $\{\beta_k\}$ in Fig. 11.7(b) satisfying the zero-forcing condition, the unitary $\mathbf{U}$ that minimizes the average error probability is *any* unitary matrix such that

$$|U_{km}| = 1/\sqrt{M}.$$

With $\mathbf{U}$ so chosen, the average error probability is as in Eq. (11.55). This is further minimized by choosing $\{\alpha_k\}$ and $\{\beta_k\}$ such that the system of Fig. 11.7(a) is an MMSE system under the power constraint (11.38). Since the system of Fig. 11.7(b) has the same MMSE and channel input power as Fig. 11.7(a) for any unitary $\mathbf{U}$, this means that the system in Fig. 11.7(b) is also an MMSE system. Thus the minimum BER system is also an MMSE system, even though the converse is not true (for example, if $\mathbf{U}$ is replaced with identity, the MMSE property continues to hold but the BER property is in general lost).

Even though the above results were proved for the case where $\{\alpha_k\}$ and $\{\beta_k\}$ are constrained by the zero-forcing (ZF) property, the results can be extended for the pure-MMSE case as well. But we have to be careful about bias in the reconstruction error.

Bias removal. The pure-MMSE system has the property that the error $\widehat{s}_k(n) - s_k(n)$ is biased, that is, its mean value is not zero. This bias should be removed before the detector makes its decisions about the transmitted symbol. The exact expression for error probability therefore needs a correction. The analysis is somewhat subtle, and will be presented in careful detail in Sec. 16.4 of Chap. 16. For now, we merely mention that, with bias removed, the expression for error probability (11.55) should be replaced with

$$\mathcal{P} = c\mathcal{Q}\left(A\sqrt{\frac{1}{\mathcal{E}_{mse}/M} - \frac{1}{\sigma_s^2}}\right), \tag{11.60}$$

where σ_s^2 is the mean square value of $s_k(n)$. In the numerical example presented in Sec. 11.6, this is the expression used for the error probability in the pure-MMSE case. $\bigtriangledown \bigtriangledown \bigtriangledown$

11.6 Examples of optimal diagonal transceivers

In this section we present some examples which compare the performances of optimal diagonal transceivers with and without zero forcing.

Example 11.2

In this example we choose $M = 16$ and 2-bit PAM constellations for the signals $s_k(n)$. The signal variances are assumed to be $\sigma_s^2 = 1$. We first consider a channel with H_k/σ_{q_k}, as shown in Fig. 11.8 (top plot). The quantity H_k/σ_{q_k} is plotted because, in the performance formulas (such as MSE, BER, and so forth), the channel gains H_k and noise variances $\sigma_{q_k}^2$ do not make separate appearances. They always appear in the form of the ratio H_k/σ_{q_k}, which is therefore the effective channel quality.

In this example the effective channel quality tapers off to smaller and smaller values as k increases. The middle plot in the figure shows the MSE per symbol for two systems, namely the ZF-MMSE system and the pure-MMSE system. The average probability of error is also shown in the figure (bottom) for various values of the input power. This plot has been obtained by using the optimal unitary matrix $\mathbf{U}$ which minimizes the average BER. The plot shows both the ZF-MMSE system (Eq. (11.55)) and the pure-MMSE system (Eq. (11.60)). From these plots it is clear that the system without zero forcing always has better performance, though it is more noticeable for low values of power. For large powers, the performance of the system with zero forcing approaches that of the system without zero forcing.[5]

Figure 11.9 shows another channel (top). The quality of this channel is worse than the previous example, as it tapers off faster and has many subchannels with small gain. The middle plot in the figure shows the MSE per symbol for MMSE systems with and without zero forcing. The optimized probability of error is also shown in the bottom figure. The optimal performance seen in this figure is clearly much worse than the one in Fig. 11.8. This clearly demonstrates the effect of channel quality on the performance.

Example 11.3: Improvement due to BER optimization

In this example we compare the average error probability of the systems with and without the optimizing unitary matrix. The channel shown earlier in Fig. 11.8 is used again. So we have $M = 16$. As in earlier examples we assume $\sigma_s^2 = 1$. We consider MMSE systems with zero forcing in this example. Figure 11.10 shows the performances of the BER-optimized system and the system with $\mathbf{U} = \mathbf{I}$ for three input signal constellations, namely 2-bit PAM, 3-bit PAM, and 4-bit PAM. A number of points should be noted.

1. As the constellation size increases for a given power, the BER gets larger, as one would expect.
2. Next, the BER-optimized systems are significantly better than the corresponding unoptimized systems for input powers greater than a threshold. Note that the threshold power gets larger as the number of bits gets larger.
3. Notice finally that in all cases the region below the threshold has an impractically large error probability, and is of limited practical interest.

[5]The error probability plots are based on theoretical formulas such as Eq. (11.55) and Eq. (11.60). Plots based on Monte Carlo simulations are more realistic, and usually agree well will these theoretical plots for reasonably high signal-to-noise ratios. Such Monte Carlo simulations are included in Sec. 19.6 for some of the examples for completeness.

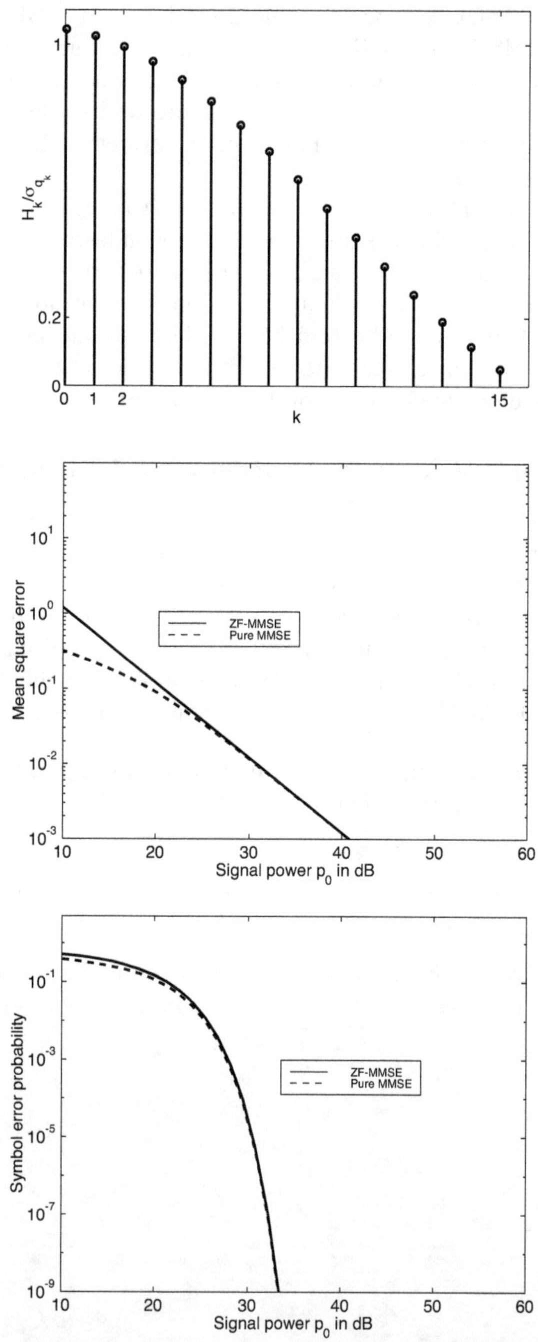

Figure 11.8. Performances of various transceivers in Example 11.2. The channel is diagonal. Channel quality H_k/σ_{q_k} (top), MSE plots (middle), and error probability plots (bottom). The error probability plots are for the minimum BER systems obtained using the optimum unitary matrix $\mathbf{U}$.

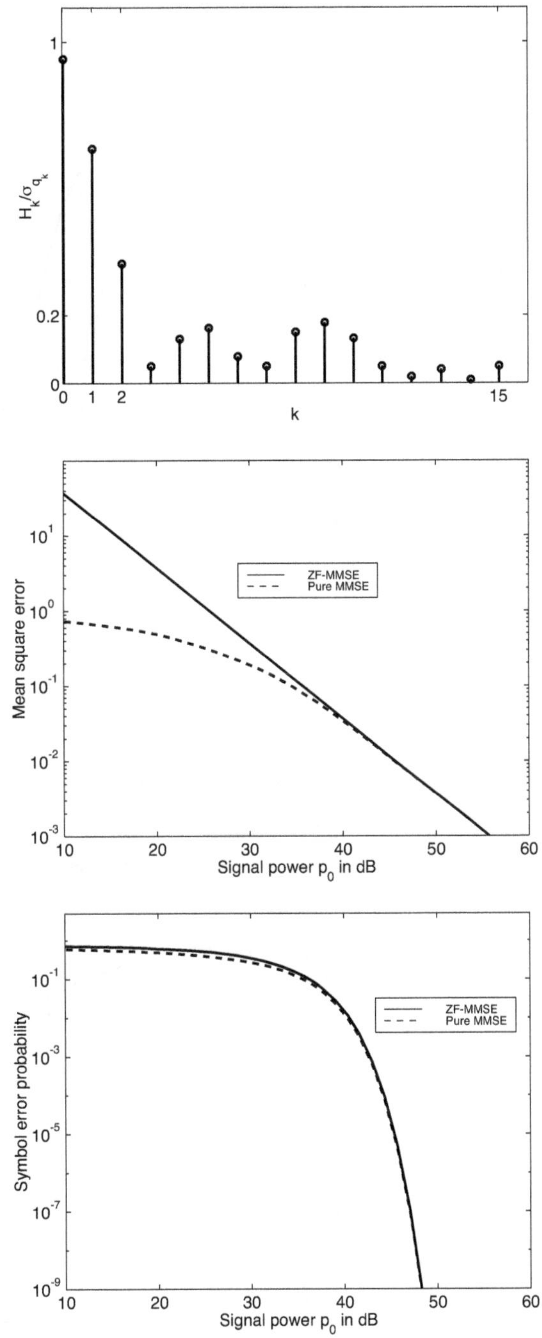

Figure 11.9. Performances of various transceivers for another diagonal channel in Example 11.2. Channel quality H_k/σ_{q_k} (top), MSE plots (middle), and error probability plots (bottom). The error probability plots are for the minimum BER systems obtained using the optimum unitary matrix $\mathbf{U}$.

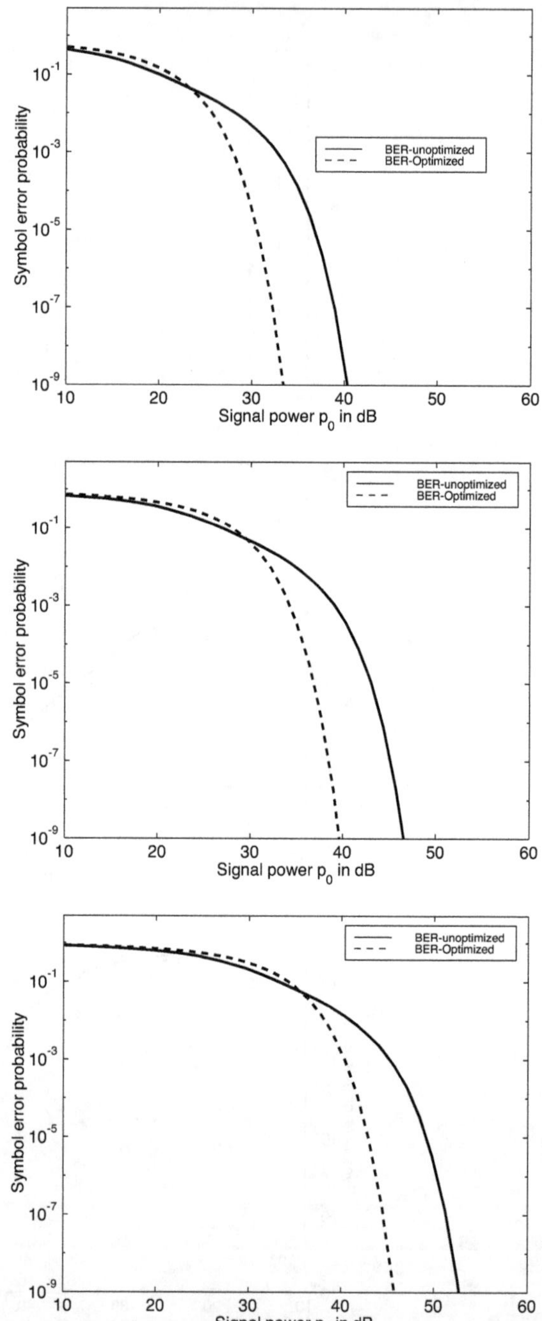

Figure 11.10. Average error probability for the ZF-MMSE transceiver with and without the unitary matrix $\mathbf{U}$ to minimize BER. The diagonal channel of Fig. 11.8 is used in all calculations. Three constellation sizes are used: 2-bit PAM (top), 3-bit PAM (middle), and 4-bit PAM (bottom).

11.6.1. Lazy precoder versus optimal precoder

In some situations the transmitter does not have the channel information (i.e., knowledge of the quantities $\sigma_{q_k}^2$ and H_k). In this case it is not possible to insert the optimal multipliers α_k at the transmitter, and one typically uses the system shown in Fig. 11.11. At the transmitter there is only one degree of freedom α for power control. If the receiver has the channel information, then we can design the multipliers β_k at the receiver to minimize the total mean square error. On the other hand, if the zero-forcing constraint is imposed, then there is nothing to optimize, and the receiver multipliers have the form

$$\beta_k = \frac{1}{H_k\alpha}.$$

The power constraint is

$$M\sigma_s^2\alpha^2 = p_0,$$

and the mean square reconstruction error is

$$\sum_{k=0}^{M-1}\sigma_{q_k}^2|\beta_k|^2 = \frac{1}{\alpha^2}\sum_{k=0}^{M-1}\frac{\sigma_{q_k}^2}{|H_k|^2} = \frac{M\sigma_s^2}{p_0}\sum_{k=0}^{M-1}\frac{\sigma_{q_k}^2}{|H_k|^2} \tag{11.61}$$

Without loss of generality we can set $\sigma_s^2 = 1$ as before. Once again, since only the ratio H_k/σ_{q_k} appears in this expression, it can be regarded as the effective channel in all discussions. Since the transmitter does not perform any operation except scaling, we call this a **lazy** zero-forcing (ZF) system. We can still insert the unitary matrices $\mathbf{U}$ and $\mathbf{U}^\dagger$ as in Sec. 11.5 to minimize the average error probability, because these matrices are channel independent. Any unitary matrix satisfying Eq. (11.53) will do.

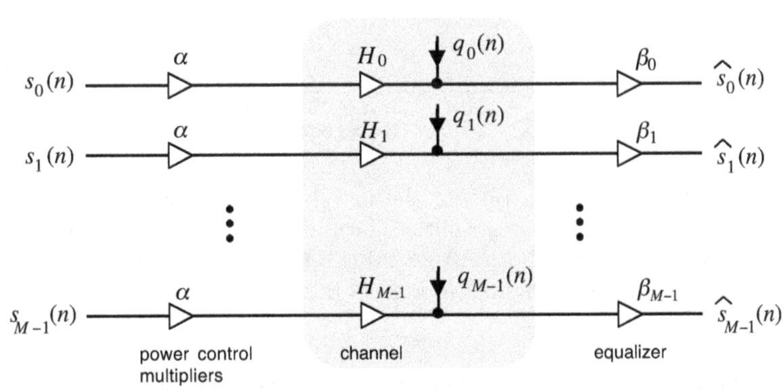

Figure 11.11. The transceiver with a lazy precoder, diagonal channel, and equalizer multipliers β_k.

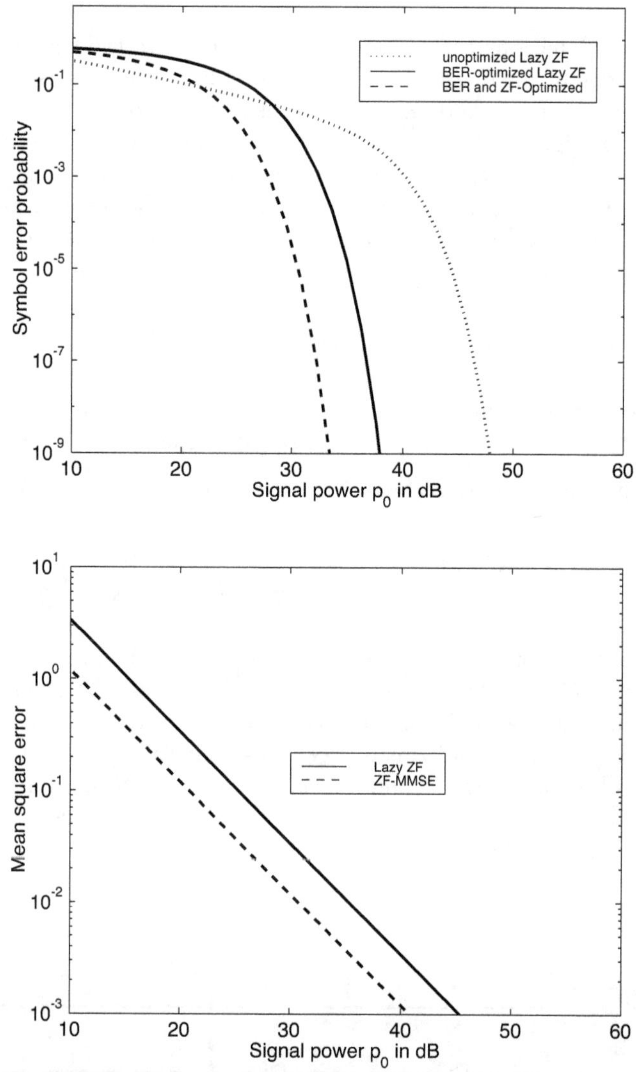

Figure 11.12. Comparison of the lazy precoder with the optimal precoder, both under the zero-forcing constraint. Shown are the average error probability plot (top) and MSE per symbol (bottom). A 2-bit PAM constellation is assumed. $M = 16$ and $\sigma_s^2 = 1$. The diagonal channel in Fig. 11.8 is used.

Example 11.4: Lazy precoder versus optimal precoder

Consider again the channel $\{H_k/\sigma_{q_k}\}$ shown in Fig. 11.8 (top). In this example we assume $M = 16$, $\sigma_s^2 = 1$, and a 2-bit PAM constellation. With a zero-forcing equalizer assumed in Fig. 11.11, the minimized probability of error (using the unitary matrix $\mathbf{U}$) is plotted in the top of Fig. 11.12 (solid curve). For comparison the plot also shows the performance of the ZF-MMSE system, which uses optimal multipliers at the transmitter (dashed curve). This shows that there is significant loss in performance if all equalization is performed at the receiver, without splitting it between the transmitter and the receiver as in an optimal system. For completeness, the top plot in the figure also shows the average symbol error rate when the lazy ZF system is used without the unitary Hadamard matrices that minimize error probability (dotted curve). A comparison of the dotted curve and the solid curve shows that the latter is superior except in the region of very high error probabilities. So, merely by inserting the *channel-independent* Hadamard matrices at the transmitter and receiver, we can obtain significant improvement of the lazy-precoder/ZF-equalizer combination! The bottom plot in Fig. 11.12 shows the mean square errors of the two systems (lazy ZF and ZF-MMSE). This plot is a comparison of Eqs. (11.61) and (11.17).

11.7 Concluding remarks

In this chapter we concentrated on the case of the diagonal channel. This simple special case has already revealed a number of important differences between the zero-forcing and pure-MMSE solutions. We have also seen that the symbol error rate can be reduced by appending an appropriate unitary matrix at the transmitter and its inverse at the receiver. In the following chapters these results will be generalized for the case of nondiagonal channels.

Problems

11.1. For the diagonal transceiver with zero forcing shown in Fig. 11.3, what is the minimized error $\mathcal{E}_{ZFmmse}$ if $\sigma_s^2 = 1$, $p_0 = 1$, and $\sigma_{q_k}/|H_k|$ has the following values (with $M = 4$): $\{0.01, 0.05, 0.1, 0.2\}$? If we remove the ZF constraint, then what is the error $\mathcal{E}_{mmse}$?

11.2. Consider the set of parallel independent Gaussian channels described in Sec. 11.4, and assume $M = 4$. Assume the values of $\widehat{\sigma}_{q_k}^2$ are

$$\{0.1, 0.05, 0.01, 0.005\}.$$

Choose four examples of the power p_0 such that the number of nonzero powers $\sigma_{s_k}^2$ in the water-filling allocation (11.35) are, respectively, one, two, three, and four. For each example you have chosen, compute the capacity (11.36).

11.3. Consider Problem 11.1 again, and assume we use 2-bit PAM. With optimal unitary matrices inserted as in Sec. 11.5.1, what is the minimized average error probability $\mathcal{P}$ for the MMSE systems with and without ZF? For the system without zero forcing, be sure to use the bias-removed formula (11.60). (For simplicity, go ahead and assume that the convexity condition described in Sec. 11.5.1 holds.)

11.4. Repeat Problem 11.3 for 2-bit QAM.

11.5. Repeat Problem 11.3 for 4-bit PAM and 4-bit QAM.

11.6. In Sec. 11.5.2 we showed that for 6-bit QAM the SNR at the detector input has to be at least 63 in order for $\mathcal{Q}(A/\sqrt{y_k})$ to be convex. What is this SNR threshold for the case of 4-bit QAM? What are the thresholds for $2, 3$, and 6 bit PAM?

12

MMSE transceivers with zero-forcing equalizers

12.1 Introduction

In this chapter we consider the general form of the transceiver shown in Fig. 12.1. The channel is a $P \times P$ constant matrix $\mathbf{H}$ representing a memoryless linear time invariant system.[1] The precoder $\mathbf{F}$ is a possibly rectangular matrix ($P \times M$ with $P \geq M$) and the equalizer $\mathbf{G}$ is $M \times P$. Figure 12.2 shows the same system in simplified vector notation. Under the zero-forcing (ZF) constraint

$$\mathbf{GHF} = \mathbf{I}_M \tag{12.1}$$

and an appropriate power constraint, we seek to find the optimal precoder and equalizer pair $\{\mathbf{F}, \mathbf{G}\}$ which minimizes the mean squared reconstruction error. The ZF condition implies in particular that the *precoder and equalizer matrices have full rank* M, and that the channel matrix has rank $\geq M$. The ZF constraint is a loss of generality, but it simplifies the optimization problem. In Chap. 13 we shall consider the problem of transceiver optimization without the ZF constraint.

Scope and outline

In Sec. 12.2 we explain the statistical assumptions on the signal and noise vectors, which will be used throughout the chapter. The formulation of the optimization problem will be done in Sec. 12.3, and some observations will be made about the form of the solution. The solution to the problem will be developed in Sec. 12.4. In Sec. 12.5 we discuss some generalizations of the

[1]The assumption that $\mathbf{H}$ is a square matrix simplifies notations a little. The results are readily generalized to rectangular channels, as shown in Appendix 12.C at the end of the chapter.

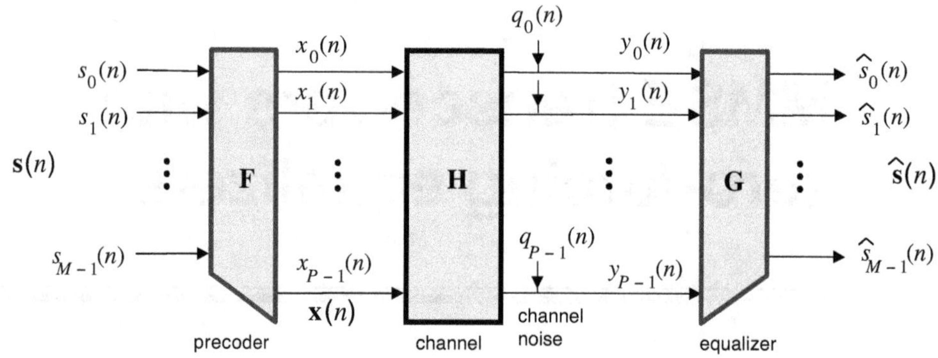

Figure 12.1. The general form of transceiver considered in this chapter.

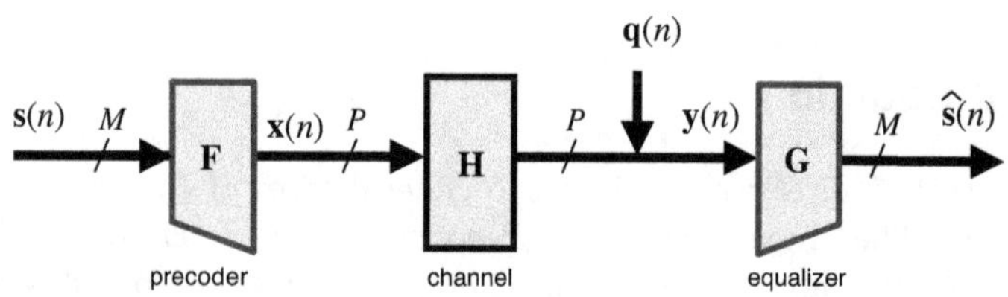

Figure 12.2. Vector notation for the transceiver considered in this chapter.

assumptions on the input covariance matrix. Some of the mathematical details pertaining to the optimal solutions are discussed in Appendices 12.A and 12.B at the end of the chapter. Finally, Appendix 12.C discusses the generalization of the transceiver optimization problem for the case of rectangular channels.

12.2 Assumptions on noise and signal statistics

We assume throughout this chapter that the signal $\mathbf{s}(n)$ and noise $\mathbf{q}(n)$ are zero-mean uncorrelated processes with correlation matrices

$$\mathbf{R_{ss}} = \sigma_s^2 \mathbf{I}_M \quad \text{and} \quad \mathbf{R_{qq}} = \sigma_q^2 \mathbf{I}_P. \tag{12.2}$$

Because of the zero-mean assumptions the *correlation* matrices are often referred to as *covariance* matrices. Equation (12.2) implies that the signals $s_k(n)$ are uncorrelated to each other and so are the noise sources $q_k(n)$. Furthermore, all the signals $s_k(n)$ have identical variance or power σ_s^2, and all the noise sources $q_k(n)$ have identical power σ_q^2. As pointed out at the end of Sec. 12.4.4, we can set $\sigma_s^2 = 1$ in all discussions without loss of generality, but we will retain σ_s^2 just for additional clarity of some of the equations. Some further remarks regarding the assumption (12.2) are now appropriate.

12.2.1 Noise covariance matrix

It turns out that the assumption $\mathbf{R_{qq}} = \sigma_q^2\mathbf{I}$ is not a loss of generality. To see this consider Fig. 12.3(a), which shows a transceiver with a more general noise covariance matrix $\mathbf{R_{qq}}$. Since this matrix is Hermitian and positive definite we can write it in the form (see Sec. B.6, Appendix B)

$$\mathbf{R_{qq}} = \mathbf{R}\mathbf{R}^{\dagger} \tag{12.3}$$

for appropriate $\mathbf{R}$. Such an $\mathbf{R}$ can be identified, for example, by using the Cholesky decomposition [Horn and Johnson, 1985]. Now define

$$\mathbf{q}_1(n) = \mathbf{R}^{-1}\mathbf{q}(n)$$

and consider the new system shown in Fig. 12.3(b), where the matrices $\mathbf{R}$ and $\mathbf{R}^{-1}$ have been inserted. In this system the noise vector at the input of $\mathbf{G}$ is

$$\mathbf{R}\mathbf{q}_1(n) = \mathbf{q}(n),$$

which is the same as in the original system of Fig. 12.3(a). Furthermore the signal path from $\mathbf{s}(n)$ to $\widehat{\mathbf{s}}(n)$ still has the transfer function $\mathbf{GHF}$ because $\mathbf{R}$ cancels $\mathbf{R}^{-1}$. Thus the signal $\widehat{\mathbf{s}}(n)$ in Fig. 12.3(b) is the same as the signal $\widehat{\mathbf{s}}(n)$ in Fig. 12.3(a). This system can further be redrawn as in Fig. 12.3(c), where

$$\mathbf{G}_1 = \mathbf{GR}, \quad \text{and} \quad \mathbf{H}_1 = \mathbf{R}^{-1}\mathbf{H}. \tag{12.4}$$

If we now find the optimum precoder $\mathbf{F}$ and equalizer $\mathbf{G}_1$ for the new channel $\mathbf{H}_1$ with noise covariance $\mathbf{I}$, we can compute the optimal matrix $\mathbf{G}$ for the original system of Fig. 12.3(a) simply by using $\mathbf{G} = \mathbf{G}_1\mathbf{R}^{-1}$.

This shows that it is always possible to draw the transceiver such that the noise covariance is identity. In practice it is slightly more convenient to assume $\mathbf{R_{qq}} = \sigma_q^2\mathbf{I}$ so that the effect of the noise variance σ_q^2 can be seen more explicitly in the expressions for various quantities such as the mean square error and optimal transceiver coefficients.

12.2.2 Signal covariance matrix

We will assume that the components $s_k(n)$ of the signal vector $\mathbf{s}(n)$ are uncorrelated and come from zero-mean constellations. In general these can have different powers, that is, the quantities

$$\sigma_{s_k}^2 = E|s_k(n)|^2 \tag{12.5}$$

need not be the same for all k. In this case the signal covariance is a diagonal

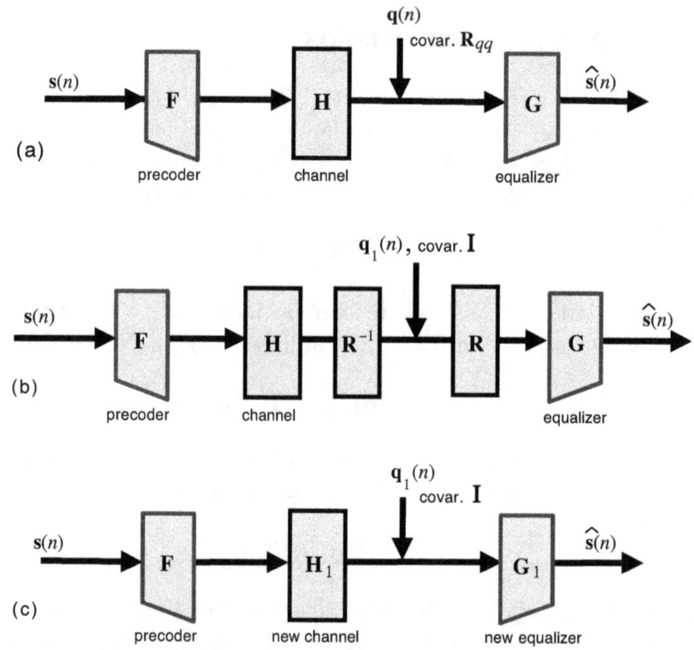

Figure 12.3. (a) The transceiver system with noise covariance $\mathbf{R}_{\mathbf{qq}}$. (b) An equivalent system with noise covariance $= \mathbf{I}$. (c) Simplification of part (b), where $\mathbf{H}_1 = \mathbf{R}^{-1}\mathbf{H}$ and $\mathbf{G}_1 = \mathbf{G}\mathbf{R}$.

matrix of the form

$$\mathbf{R}_{\mathbf{ss}} = \underbrace{\begin{bmatrix} \sigma_{s_0}^2 & 0 & \ldots & 0 \\ 0 & \sigma_{s_1}^2 & \ldots & 0 \\ \vdots & \vdots & \ddots & \vdots \\ 0 & 0 & \ldots & \sigma_{s_{M-1}}^2 \end{bmatrix}}_{\text{call this } \mathbf{\Lambda}_s}. \tag{12.6}$$

It is common that the M components $s_k(n), 0 \leq k \leq M-1$, arise from a *single* user. For example, $s_k(n)$ could be the M samples within one block of a block-based system such as the zero-padded (ZP) or cyclic-prefix (CP) system (Chap. 7). Since $s(n)$ is usually assumed to be an iid (independent identically distributed) sequence, it follows in these cases that $s_k(n)$ are not only uncorrelated, but have identical powers, that is, $\sigma_{s_k}^2 = \sigma_s^2$ for all k, so that

$$\mathbf{R}_{\mathbf{ss}} = \sigma_s^2\mathbf{I}. \tag{12.7}$$

The results derived in this and the following chapter are most appropriate for this situation. Another situation of interest is that the M signals $s_k(n)$ are derived from a single user stream $s(n)$ by *parsing* as in a DMT system, rather than by blocking (Secs. 3.5 and 3.6). In this case $s_k(n)$ would have different powers. This scenerio allows *bit allocation* to minimize transmitted power for fixed error probabilities, as we shall explain in detail in Chap. 14. Some further remarks for the case of arbitrary diagonal $\mathbf{R_{ss}}$ can be found at the end of this chapter (Sec. 12.5).

12.3 Problem formulation

Under the zero-forcing constraint the reconstruction error in Fig. 12.2 comes entirely from the channel noise $\mathbf{q}(n)$ processed through $\mathbf{G}$. Thus

$$\mathbf{e}(n) = \widehat{\mathbf{s}}(n) - \mathbf{s}(n) = \mathbf{Gq}(n) \tag{12.8}$$

so that the error covariance is

$$\mathbf{R_{ee}} = \mathbf{GR_{qq}G}^\dagger = \sigma_q^2\mathbf{GG}^\dagger, \tag{12.9}$$

where we have used the assumption $\mathbf{R_{qq}} = \sigma_q^2\mathbf{I}$. The total mean square error is therefore[2]

$$\mathcal{E}_{mse} = E[\mathbf{e}^\dagger(n)\mathbf{e}(n)] = \mathrm{Tr}(\mathbf{R_{ee}}) = \sigma_q^2\mathrm{Tr}(\mathbf{GG}^\dagger). \tag{12.10}$$

Now, the goal is to minimize $\mathcal{E}_{mse}$ under the power constraint

$$\sum_{k=0}^{M-1} E[\,|x_k(n)|^2] = p_0, \tag{12.11}$$

where $x_k(n)$ are the signals at the channel input (Fig. 12.1). We can express the left-hand side as

$$\sum_{k=0}^{M-1} E[\,|x_k(n)|^2] = \mathrm{Tr}(\mathbf{R_{xx}}) = \mathrm{Tr}\Big(\mathbf{FR_{ss}F}^\dagger\Big) = \sigma_s^2\mathrm{Tr}\Big(\mathbf{FF}^\dagger\Big),$$

where we have used the assumption $\mathbf{R_{ss}} = \sigma_s^2\mathbf{I}$. So the power constraint can be rewritten as $\sigma_s^2\mathrm{Tr}(\mathbf{FF}^\dagger) = p_0$. The optimization problem of interest is therefore as follows:

> ***Problem 1. The ZF-MMSE problem.*** Find the precoder $\mathbf{F}$ and equalizer $\mathbf{G}$ to minimize the total MSE
>
> $$\mathcal{E}_{mse} = \sigma_q^2\mathrm{Tr}(\mathbf{GG}^\dagger) \tag{12.12}$$
>
> subject to the power constraint
>
> $$\sigma_s^2\mathrm{Tr}\Big(\mathbf{FF}^\dagger\Big) = p_0 \tag{12.13}$$

[2] A more appropriate notation would have been $\mathcal{E}_{zfmse}$ where zf is a reminder of the zero-forcing constraint, but we keep it simple here.

and the zero-forcing constraint

$$\mathbf{GHF} = \mathbf{I}, \tag{12.14}$$

where the channel $\mathbf{H}$ and the quantities σ_q^2, σ_s^2, and p_0 are fixed. $\bigtriangledown \bigtriangledown \bigtriangledown$

Before proceeding to the solution we would like to give a flavor of what the results are going to look like. We will see (Sec. 12.4) that the minimized mean square error per symbol (i.e., $\mathcal{E}_{mse}/M$) with zero forcing takes the form

$$\mathcal{E}_{opt} = \frac{\sigma_s^2 \sigma_q^2}{M p_0} \Big(\sum_{k=0}^{M-1} \frac{1}{\sigma_{h,k}} \Big)^2, \tag{12.15}$$

where

$$\sigma_{h,0} \geq \sigma_{h,1} \geq \ldots \geq \sigma_{h,M-1} > 0 \tag{12.16}$$

are the first M dominant singular values of the channel. We now compare this with the mean square error obtained by using the lazy precoder and zero-forcing equalizer described in Chap. 8. The expression for this error is (from Eq. (8.11))

$$\mathcal{E}_{lazy} = \frac{\sigma_q^2}{M} \sum_{k=0}^{M-1} \frac{1}{\sigma_{h,k}^2} \tag{12.17}$$

For the lazy precoder, since $\mathbf{F} = \mathbf{I}$, the channel input power is $p_0 = \sigma_s^2 M$. By letting $\sigma_s^2 = 1$ and setting $p_0 = M$ we can compare the above two errors for identical powers:

$$\mathcal{E}_{opt} = \frac{\sigma_q^2}{M^2} \Big(\sum_{k=0}^{M-1} \frac{1}{\sigma_{h,k}} \Big)^2; \quad \mathcal{E}_{lazy} = \frac{\sigma_q^2}{M} \sum_{k=0}^{M-1} \frac{1}{\sigma_{h,k}^2}.$$

The gain obtained by going from the lazy precoder system to the jointly optimal system is therefore

$$\mathcal{G} = \frac{\mathcal{E}_{lazy}}{\mathcal{E}_{opt}} = \frac{M \sum_{k=0}^{M-1} 1/\sigma_{h,k}^2}{\Big(\sum_{k=0}^{M-1} 1/\sigma_{h,k} \Big)^2} \tag{12.18}$$

Example 12.1: ZF-MMSE versus lazy precoder

How large can the gain get? First it is readily verified that $\mathcal{G} \geq 1$. This follows from Cauchy-Schwartz inequality (Appendix A at the end of the book) which says that

$$\sum_k |a_k|^2 \sum_k |b_k|^2 \geq \Big| \sum_k a_k b_k^* \Big|^2$$

(Use $a_k = 1$ and $b_k = 1/\sigma_{h,k}$.) Furthermore, $\mathcal{G} \leq M$, as the following argument shows: since $\sigma_{h,k} \geq \sigma_{h,M-1}$, we have

$$\sum_{k=0}^{M-1} \frac{1}{\sigma_{h,k}^2} \leq \frac{1}{\sigma_{h,M-1}} \sum_{k=0}^{M-1} \frac{1}{\sigma_{h,k}},$$

so that

$$\mathcal{G} \leq \frac{M/\sigma_{h,M-1}}{\left(\sum_{k=0}^{M-1} 1/\sigma_{h,k}\right)} \leq \frac{M/\sigma_{h,M-1}}{1/\sigma_{h,M-1}} = M.$$

Thus the gain $\mathcal{G}$ obtained by going from the lazy precoder system to the jointly optimal system satisfies

$$1 \leq \mathcal{G} \leq M. \tag{12.19}$$

There is no gain when

$$\sigma_{h,0} = \sigma_{h,1} = \ldots = \sigma_{h,M-1}. \tag{12.20}$$

The gain approaches the upper bound M when the smallest singular value $\sigma_{h,M-1}$ becomes arbitrarily small compared to the others. Thus for large M, if the least significant singular value is very small, we get a large gain. For a numerical example, let $M = 2$ and

$$\sigma_{h,0} = 1, \ \sigma_{h,1} = 0.1.$$

Then the gain is

$$\mathcal{G} = \frac{2(1+100)}{(1+10)^2} = 1.67,$$

which is quite modest. On the other hand, if $M = 128$, $\sigma_{h,127} = 10^{-4}$, and $\sigma_{h,k} = 1$ for $k < 127$, then

$$\mathcal{G} = \frac{128(127+10^8)}{(127+10^4)^2} = 124.8,$$

a significant gain indeed!

Note that since $\sigma_{h,0} \geq \sigma_{h,k}$ for all k, we have $M/\sigma_{h,0} \leq \sum_k 1/\sigma_{h,k}$, that is,

$$\frac{1}{\sum_k 1/\sigma_{h,k}} \leq \frac{\sigma_{h,0}}{M}.$$

So, from the first inequality of the third equation in Ex. 12.1 it follows that

$$\mathcal{G} \leq \frac{M}{\sigma_{h,M-1}} \times \frac{\sigma_{h,0}}{M}$$

Thus

$$\mathcal{G} \leq \frac{\sigma_{h,0}}{\sigma_{h,M-1}},$$

which yields yet another bound besides $\mathcal{G} \le M$. This is a tighter bound when $\sigma_{h,0}/\sigma_{h,M-1} < M$.

12.3.1 Simplification using the ZF constraint

It is possible to absorb the ZF constraint into the objective function so that only the power constraint needs to be handled explicitly. For this observe that the ZF constraint (12.14) says that $\mathbf{G}$ has to be a left inverse of $\mathbf{HF}$. Let us assume that this is the *minimum norm* left inverse.[3] Then from Appendix C we have

$$\mathbf{G} = \Big((\mathbf{HF})^\dagger\mathbf{HF}\Big)^{-1}(\mathbf{HF})^\dagger = \Big(\mathbf{F}^\dagger\mathbf{H}^\dagger\mathbf{HF}\Big)^{-1}\mathbf{F}^\dagger\mathbf{H}^\dagger. \tag{12.21}$$

The output noise covariance matrix $\sigma_q^2\mathbf{GG}^\dagger$ therefore has the form

$$\sigma_q^2\mathbf{GG}^\dagger = \sigma_q^2\Big(\mathbf{F}^\dagger\mathbf{H}^\dagger\mathbf{HF}\Big)^{-1}\mathbf{F}^\dagger\mathbf{H}^\dagger\mathbf{HF}\Big(\mathbf{F}^\dagger\mathbf{H}^\dagger\mathbf{HF}\Big)^{-1} = \sigma_q^2\Big(\mathbf{F}^\dagger\mathbf{H}^\dagger\mathbf{HF}\Big)^{-1}.$$

Since the ZF constraint has been incorporated to eliminate $\mathbf{G}$ from the problem, the optimization problem reduces to the following:

Problem 2. Optimize the precoder $\mathbf{F}$ to minimize

$$\mathcal{E}_{mse} = \sigma_q^2 \ \mathrm{Tr}\Big(\mathbf{F}^\dagger\mathbf{H}^\dagger\mathbf{HF}\Big)^{-1} \tag{12.22}$$

subject to the power constraint

$$\sigma_s^2 \ \mathrm{Tr}[\mathbf{FF}^\dagger] = p_0, \tag{12.23}$$

where the channel $\mathbf{H}$ and the quantities σ_q^2, σ_s^2, and p_0 are fixed. $\bigtriangledown\bigtriangledown\bigtriangledown$

Remember throughout that $\mathbf{F}$ has rank M and $\mathbf{H}$ has rank $\ge M$ as required by the zero-forcing constraint.

12.3.2 General form of F

We now show that the matrix $\mathbf{F}$, which minimizes (12.22) subject to the constraint (12.23) can be assumed, without loss of generality, to be of the form

$$\mathbf{F} = \mathbf{U}_f\begin{bmatrix}\mathbf{\Sigma}_f \\ \mathbf{0}\end{bmatrix}, \tag{12.24}$$

where $\mathbf{U}_f$ is $P \times P$ unitary and $\mathbf{\Sigma}_f$ is $M \times M$ diagonal:

$$\mathbf{\Sigma}_f = \begin{bmatrix}\sigma_{f,0} & 0 & \ldots & 0 \\ 0 & \sigma_{f,1} & \ldots & 0 \\ \vdots & \vdots & \ddots & \vdots \\ 0 & 0 & \ldots & \sigma_{f,M-1}\end{bmatrix}, \tag{12.25}$$

[3] We argue in Appendix 12.A at the end of the chapter that there is no loss of generality in choosing this particular left inverse.

with $\sigma_{f,k} > 0$. That is, the singular value decomposition (SVD; see Appendix C) given by

$$\mathbf{F} = \mathbf{U}_f \mathbf{\Lambda}_f \mathbf{V}_f^\dagger$$

can be assumed to be such that $\mathbf{V}_f = \mathbf{I}_M$.

Proof. Substituting the SVD form $\mathbf{F} = \mathbf{U}_f \mathbf{\Lambda}_f \mathbf{V}_f^\dagger$ into Eq. (12.22) we see that

$$\begin{aligned}
\mathcal{E}_{mse} &= \text{Tr}(\mathbf{V}_f \mathbf{\Lambda}_f^\dagger \mathbf{U}_f^\dagger \mathbf{H}^\dagger \mathbf{H} \mathbf{U}_f \mathbf{\Lambda}_f \mathbf{V}_f^\dagger)^{-1} \\
&= \text{Tr}(\mathbf{V}_f^\dagger \mathbf{V}_f \mathbf{\Lambda}_f^\dagger \mathbf{U}_f^\dagger \mathbf{H}^\dagger \mathbf{H} \mathbf{U}_f \mathbf{\Lambda}_f)^{-1} \\
&= \text{Tr}(\mathbf{\Lambda}_f^\dagger \mathbf{U}_f^\dagger \mathbf{H}^\dagger \mathbf{H} \mathbf{U}_f \mathbf{\Lambda}_f)^{-1},
\end{aligned}$$

where the second equality is obtained by using $\text{Tr}(\mathbf{AB}) = \text{Tr}(\mathbf{BA})$ and the third equality is obtained by using $\mathbf{V}_f^\dagger \mathbf{V}_f = \mathbf{I}$. Thus the matrix $\mathbf{V}_f$ has no role in the expression for $\mathcal{E}_{mse}$. Similarly the matrix $\mathbf{V}_f$ drops out of the expression $\text{Tr}(\mathbf{F}\mathbf{F}^\dagger)$ in the power constraint. So we can ignore it and take $\mathbf{F}$ to be of the form $\mathbf{F} = \mathbf{U}_f \mathbf{\Lambda}_f$. Since $\mathbf{F}$ is assumed to have rank M, it follows that this can be written in the form (12.24). $\triangledown \triangledown \triangledown$

Some remarks are now in order:

1. ***Orthogonal codes are not a loss of generality***. With $\mathbf{F}$ expressed in the form (12.24), it follows that

$$\mathbf{F}^\dagger \mathbf{F} = [\,\mathbf{\Sigma}_f \quad \mathbf{0}\,]\, \mathbf{U}_f^\dagger \mathbf{U}_f \begin{bmatrix} \mathbf{\Sigma}_f \\ \mathbf{0} \end{bmatrix} = \mathbf{\Sigma}_f^2 = \text{ diagonal.} \qquad (12.26)$$

Thus, the optimal solution can be assumed, without loss of generality, to be such that $\mathbf{F}^\dagger \mathbf{F}$ is diagonal, that is, *the M columns of $\mathbf{F}$ are orthogonal.* Observe now that the channel input vector is given by

$$\mathbf{x} = \mathbf{F}\mathbf{s} = \sum_{k=0}^{M-1} s_k(n) \mathbf{f}_k,$$

where $\mathbf{f}_k$ is the kth column of $\mathbf{F}$ and $s_k(n)$ is the kth user signal. Thus the orthogonality of the columns $\mathbf{f}_k$ means that the channel input $\mathbf{x}$ is a linear combination of M orthogonal vectors, with the kth vector $\mathbf{f}_k$ representing the kth user signal $s_k(n)$. Since $\mathbf{f}_k$ can be regarded as the code for user k, we can say that *the use of orthogonal codes is not a loss of generality*, if the goal is to minimize MSE under the zero-forcing and power constraints. A similar result will be observed for several other types of optimality criteria as well, as described in later chapters.

2. ***Minimizing bit error rate***. Even though $\mathbf{V}_f$ has no role in the ZF-MMSE transceiver, we will see in Chap. 16 that it plays an important role in the minimization of the bit error rate.

12.3.3 Stationarity condition for optimality

We next use the Lagrangian approach to write down a necessary condition for optimality (the stationary condition, Sec. 20.3.5, Chap. 20). Thus we first set up a real-valued Lagrangian function

$$\psi(\mathbf{F}, \mathbf{F}^*) = \sigma_q^2 \operatorname{Tr}\Big(\mathbf{F}^\dagger \mathbf{H}^\dagger \mathbf{H} \mathbf{F}\Big)^{-1} + \mu\Big(\sigma_s^2 \operatorname{Tr}(\mathbf{F}\mathbf{F}^\dagger) - p_0\Big), \tag{12.27}$$

where μ is a real Lagrange multiplier. Since the variable of optimization is the complex matrix $\mathbf{F}$, we have to use matrix calculus for this (*complex* gradient method, Chap. 20). Thus, to write down the necessary condition for optimality (stationarity), we set the complex gradient with respect to $\mathbf{F}^*$ to zero:[4]

$$\frac{\partial \psi(\mathbf{F}, \mathbf{F}^*)}{\partial \mathbf{F}^*} = \mathbf{0}. \tag{12.28}$$

From Table 20.4 of Chap. 20 we know that

$$\frac{\partial \operatorname{Tr}(\mathbf{F}\mathbf{F}^\dagger)}{\partial \mathbf{F}^*} = \mathbf{F} \tag{12.29}$$

and

$$\frac{\partial \operatorname{Tr}\Big(\mathbf{F}^\dagger \mathbf{H}^\dagger \mathbf{H} \mathbf{F}\Big)^{-1}}{\partial \mathbf{F}^*} = -(\mathbf{H}^\dagger \mathbf{H})\mathbf{F}(\mathbf{F}^\dagger \mathbf{H}^\dagger \mathbf{H} \mathbf{F})^{-2}. \tag{12.30}$$

Thus the stationary condition (12.28) reduces to

$$\sigma_q^2 \mathbf{H}^\dagger \mathbf{H} \mathbf{F}(\mathbf{F}^\dagger \mathbf{H}^\dagger \mathbf{H} \mathbf{F})^{-2} = \mu \sigma_s^2 \mathbf{F}. \tag{12.31}$$

Premultiplying both sides by $\mathbf{F}^\dagger$ we find that the following is a necessary condition for optimality:

$$\sigma_q^2 (\mathbf{F}^\dagger \mathbf{H}^\dagger \mathbf{H} \mathbf{F})^{-1} = \mu \sigma_s^2 \mathbf{F}^\dagger \mathbf{F} \quad \text{(stationarity)}. \tag{12.32}$$

Using Eq. (12.24) we see that

$$\mathbf{F}^\dagger \mathbf{H}^\dagger \mathbf{H} \mathbf{F} = [\,\boldsymbol{\Sigma}_f \quad \mathbf{0}\,] \mathbf{U}_f^\dagger \mathbf{H}^\dagger \mathbf{H} \mathbf{U}_f \begin{bmatrix} \boldsymbol{\Sigma}_f \\ \mathbf{0} \end{bmatrix} = \boldsymbol{\Sigma}_f (\mathbf{U}_f^\dagger \mathbf{H}^\dagger \mathbf{H} \mathbf{U}_f)_M \boldsymbol{\Sigma}_f, \tag{12.33}$$

where the notation $(\mathbf{A})_M$ denotes the $M \times M$ leading principal matrix of $\mathbf{A}$ (Sec. B.2.1, Appendix B).

Two important points should now be observed. Since the right-hand side of Eq. (12.32) can be assumed to be diagonal without loss of generality (Eq. (12.26)), it follows that $(\mathbf{F}^\dagger \mathbf{H}^\dagger \mathbf{H} \mathbf{F})$ is diagonal as well! Since $\boldsymbol{\Sigma}_f$ is diagonal, Eq. (12.33) shows that $(\mathbf{U}_f^\dagger \mathbf{H}^\dagger \mathbf{H} \mathbf{U}_f)_M$ can be assumed to be diagonal as well. Summarizing, we have proved the following:

♠**Lemma 12.1.** *Form of the optimal precoder* $\mathbf{F}$. The optimal solution $\mathbf{F}$ which minimizes Eq. (12.22) subject to Eq. (12.23) can be assumed to be of the form (12.24), where $\mathbf{U}_f$ is unitary and $\boldsymbol{\Sigma}_f$ is diagonal with positive diagonal elements. Furthermore,

[4] At this point the reader may want to review Sec. 20.3 (Chap. 20) on the mathematics of differentiation with respect to complex matrices.

1. The optimal $\mathbf{F}$ has to satisfy the stationarity condition (12.32), where $\mathbf{H}$ is the channel matrix.
2. Consequently both $\mathbf{F}^\dagger\mathbf{H}^\dagger\mathbf{H}\mathbf{F}$ and $\mathbf{F}^\dagger\mathbf{F}$ are diagonal matrices.
3. As a result, $(\mathbf{U}_f^\dagger\mathbf{H}^\dagger\mathbf{H}\mathbf{U}_f)_M$ can be assumed to be diagonal as well, where $\mathbf{U}_f$ is the unitary matrix appearing in Eq. (12.24). $\diamondsuit$

12.4 Solution to the ZF-MMSE problem

With $\mathbf{F}$ restricted to be of the form (12.24), where $\mathbf{U}_f$ is unitary and $\boldsymbol{\Sigma}_f$ diagonal, the objective function (12.22) becomes

$$\begin{aligned}\mathcal{E}_{mse} &= \sigma_q^2 \operatorname{Tr}\Big(\mathbf{F}^\dagger\mathbf{H}^\dagger\mathbf{H}\mathbf{F}\Big)^{-1} \\ &= \sigma_q^2 \operatorname{Tr}\Big(\begin{bmatrix}\boldsymbol{\Sigma}_f & \mathbf{0}\end{bmatrix}\mathbf{U}_f^\dagger\mathbf{H}^\dagger\mathbf{H}\mathbf{U}_f\begin{bmatrix}\boldsymbol{\Sigma}_f \\ \mathbf{0}\end{bmatrix}\Big)^{-1} \\ &= \sigma_q^2 \operatorname{Tr}\Big(\boldsymbol{\Sigma}_f(\mathbf{U}_f^\dagger\mathbf{H}^\dagger\mathbf{H}\mathbf{U}_f)_M\boldsymbol{\Sigma}_f\Big)^{-1},\end{aligned}$$

where the notation $(\mathbf{A})_M$ denotes the $M \times M$ leading principal submatrix of $\mathbf{A}$ (Sec. B.2.1, Appendix B). The power constraint (12.23) can also be rewritten as

$$\begin{aligned}p_0 &= \sigma_s^2 \operatorname{Tr}(\mathbf{F}\mathbf{F}^\dagger) \\ &= \sigma_s^2 \operatorname{Tr}\Big(\mathbf{U}_f\begin{bmatrix}\boldsymbol{\Sigma}_f \\ \mathbf{0}\end{bmatrix}\begin{bmatrix}\boldsymbol{\Sigma}_f & \mathbf{0}\end{bmatrix}\mathbf{U}_f^\dagger\Big) \\ &= \sigma_s^2 \operatorname{Tr}\Big(\mathbf{U}_f^\dagger\mathbf{U}_f\begin{bmatrix}\boldsymbol{\Sigma}_f \\ \mathbf{0}\end{bmatrix}\begin{bmatrix}\boldsymbol{\Sigma}_f & \mathbf{0}\end{bmatrix}\Big) \\ &= \sigma_s^2 \operatorname{Tr}(\boldsymbol{\Sigma}_f^2),\end{aligned}$$

where we have used the trace identity $\operatorname{Tr}(\mathbf{AB}) = \operatorname{Tr}(\mathbf{BA})$ to get the third equality, and the unitarity of $\mathbf{U}_f$ to get the last equality. Thus the optimization problem described by Eqs. (12.22) and (12.23) reduces to the following:

Problem 3. Find a $P \times P$ unitary matrix $\mathbf{U}_f$ and an $M \times M$ diagonal matrix $\boldsymbol{\Sigma}_f$ with positive diagonal elements $\sigma_{f,k} > 0$, to minimize

$$\mathcal{E}_{mse} = \sigma_q^2 \operatorname{Tr}\Big(\boldsymbol{\Sigma}_f(\mathbf{U}_f^\dagger\mathbf{H}^\dagger\mathbf{H}\mathbf{U}_f)_M\boldsymbol{\Sigma}_f\Big)^{-1} \tag{12.34}$$

subject to the constraint

$$\sigma_s^2 \operatorname{Tr}(\boldsymbol{\Sigma}_f^2) = p_0, \tag{12.35}$$

where the channel $\mathbf{H}$ and the quantities σ_q^2, σ_s^2, and p_0 are fixed. $\nabla\nabla\nabla$

It turns out that the solution to the problem automatically satisfies the constraint $\sigma_{f,k} > 0$, so we shall ignore it for simplicity. The goal of optimization now is to choose $\mathbf{U}_f$ and $\boldsymbol{\Sigma}_f$. Note that $\mathbf{U}_f$ does not affect the power constraint (12.35). For any fixed $\boldsymbol{\Sigma}_f$ we will first show that the choice of unitary $\mathbf{U}_f$ which minimizes $\mathcal{E}_{mse}$ is independent of $\boldsymbol{\Sigma}_f$. Once we find the best $\mathbf{U}_f$, the error $\mathcal{E}_{mse}$ depends only on $\boldsymbol{\Sigma}_f$. This can be further minimized subject to the power constraint (12.35).

12.4.1 Optimal ordering of diagonal elements

From Lemma 12.1 we know that the optimal solution can be assumed to be such that the $M \times M$ submatrix $(\mathbf{U}_f^\dagger \mathbf{H}^\dagger \mathbf{H} \mathbf{U}_f)_M$ is diagonal, so

$$(\mathbf{U}_f^\dagger \mathbf{H}^\dagger \mathbf{H} \mathbf{U}_f)_M = \text{diag}\,[\,\mu_0 \quad \mu_1 \quad \ldots \quad \mu_{M-1}\,] \tag{12.36}$$

for appropriate μ_k. This means the $P \times P$ matrix $\mathbf{U}_f^\dagger \mathbf{H}^\dagger \mathbf{H} \mathbf{U}_f$ can be written in the form

$$\mathbf{U}_f^\dagger \mathbf{H}^\dagger \mathbf{H} \mathbf{U}_f = \begin{array}{c} \\ M \\ \\ \\ \\ P-M \end{array} \left(\begin{array}{ccccc|c} \mu_0 & 0 & \ldots & 0 & & \times \\ 0 & \mu_1 & \ldots & 0 & & \times \\ \vdots & \vdots & \ddots & \vdots & & \vdots \\ 0 & 0 & \ldots & \mu_{M-1} & & \times \\ \hline \times & \times & \ldots & \times & & \times \end{array}\right), \tag{12.37}$$

where $\times$ denotes entries that are possibly nonzero. Using Eq. (12.36) we can rewrite Eq. (12.34) as

$$\mathcal{E}_{mse} = \sigma_q^2 \sum_{k=0}^{M-1} \frac{1}{\mu_k \sigma_{f,k}^2} = \sigma_q^2 \sum_{k=0}^{M-1} \frac{a_k}{\mu_k}, \tag{12.38}$$

where $\sigma_{f,k}$ are the diagonal elements of $\boldsymbol{\Sigma}_f$ (Eq. (12.25)) and we have used the notation

$$a_k = \frac{1}{\sigma_{f,k}^2} \tag{12.39}$$

for simplicity. Assume for the moment that a_k is fixed in Eq. (12.38), and let these be numbered such that

$$a_0 \geq a_1 \geq \ldots \geq a_{M-1} > 0. \tag{12.40}$$

Then we can assume without loss of generality that

$$\mu_0 \geq \mu_1 \geq \ldots \geq \mu_{M-1} > 0. \tag{12.41}$$

Note that the numbers a_k come from the precoder and the numbers μ_k come from the channel.

Proof of Eq. (12.41). Assume the contrary. For example, suppose $\mu_0 < \mu_1$. Then

$$\Big(\frac{a_0}{\mu_0} + \frac{a_1}{\mu_1}\Big) - \Big(\frac{a_0}{\mu_1} + \frac{a_1}{\mu_0}\Big) = (a_0 - a_1)\Big(\frac{1}{\mu_0} - \frac{1}{\mu_1}\Big) > 0,$$

which shows that

$$\frac{a_0}{\mu_0} + \frac{a_1}{\mu_1} > \frac{a_0}{\mu_1} + \frac{a_1}{\mu_0}.$$

So an interchange of μ_0 and μ_1 reduces the objective function (12.38) further. From Eq. (12.36) we see that such an interchange can be effected by postmultiplying $\mathbf{U}_f$ with a permutation matrix (Sec. B.5.1 of Appendix B). Thus, define the new $\mathbf{U}_f$ as

$$\mathbf{U}_{new} = \mathbf{U}_f \begin{bmatrix} \mathbf{P} & \mathbf{0} \\ \mathbf{0} & \mathbf{I}_{P-M} \end{bmatrix}, \tag{12.42}$$

where $\mathbf{P}$ is a permutation. This is still unitary, and $(\mathbf{U}_{new}^\dagger \mathbf{H}^\dagger \mathbf{H} \mathbf{U}_{new})$ is still of the required form (12.37). Therefore we see that, under the numbering convention (12.40), the optimal $\mathbf{U}_f$ should be such that Eq. (12.41) is satisfied. $\bigtriangledown\bigtriangledown\bigtriangledown$

12.4.2 Finding the optimal $\mathbf{U}_f$

Note that the unitary matrix $\mathbf{U}_f$ need not diagonalize the $P \times P$ matrix $\mathbf{H}^\dagger\mathbf{H}$ completely. Only partial diagonalization of the $M \times M$ leading principal submatrix is called for, as shown in Eq. (12.37). We know however that since $\mathbf{H}^\dagger\mathbf{H}$ is Hermitian, there exists a unitary $\widehat{\mathbf{U}}_f$ diagonalizing it fully, that is,

$$\widehat{\mathbf{U}}_f^\dagger \mathbf{H}^\dagger \mathbf{H} \widehat{\mathbf{U}}_f = \begin{bmatrix} \sigma_{h,0}^2 & 0 & \dots & 0 \\ 0 & \sigma_{h,1}^2 & \dots & 0 \\ \vdots & \vdots & \ddots & \vdots \\ 0 & 0 & \dots & \sigma_{h,P-1}^2 \end{bmatrix}. \tag{12.43}$$

Here

$$\sigma_{h,0} \ge \sigma_{h,1} \ge \dots \ge \sigma_{h,P-1} \ge 0, \tag{12.44}$$

are the singular values of the channel $\mathbf{H}$. Setting $\mathbf{U}_f = \widehat{\mathbf{U}}_f$ certainly satisfies Eq. (12.37). We will prove the subtle fact that this choice of $\mathbf{U}_f$ serves as the optimal choice as well:

♠**Lemma 12.2.** *Optimal unitary* $\mathbf{U}_f$. The optimal unitary matrix $\mathbf{U}_f$, which minimizes Eq. (12.34) subject to Eq. (12.35) for fixed $\mathbf{\Sigma}_f$ and $\mathbf{H}$ can be assumed, without loss of generality, to be the unitary matrix which diagonalizes $\mathbf{H}^\dagger\mathbf{H}$. $\diamondsuit$

The idea of the proof is as follows: with $\mathbf{U}_f$ yielding a partial diagonalization (12.37), and under the numbering convention (12.41), we will show that

$$\sigma_{h,0}^2 \ge \mu_0,\ \sigma_{h,1}^2 \ge \mu_1,\ \dots,\sigma_{h,M-1}^2 \ge \mu_{M-1}. \tag{12.45}$$

Thus

$$\mathcal{E}_{mse} = \sigma_q^2 \sum_{k=0}^{M-1} \frac{a_k}{\mu_k} \ge \sigma_q^2 \sum_{k=0}^{M-1} \frac{a_k}{\sigma_{h,k}^2} \tag{12.46}$$

since $a_k > 0$. The lower bound on the right-hand side can therefore be achieved if we choose the optimal $\mathbf{U}_f$ to be the unitary matrix that diagonalizes $\mathbf{H}^\dagger\mathbf{H}$ completely and arranges diagonal elements as in Eq. (12.44). The proof of the above lemma is therefore complete once we prove Eq. (12.45):

Proof of Eq. (12.45). The proof is based on a beautiful theorem on Hermitian matrices which we first review (see Horn and Johnson [1985], p. 189). Recall first that all eigenvalues of a Hermitian matrix are real. Let $\mathbf{A}$ be a $P \times P$ Hermitian matrix and let $\mathbf{A}_M$ be any $M \times M$ principal submatrix, that is, a matrix obtained by deleting any $P-M$ rows and the corresponding columns from $\mathbf{A}$. Let the eigenvalues of $\mathbf{A}$ be arranged as

$$\lambda_0(\mathbf{A}) \geq \lambda_1(\mathbf{A}) \ldots \geq \lambda_{P-1}(\mathbf{A}) \tag{12.47}$$

and let the eigenvalues of $\mathbf{A}_M$ be arranged as

$$\lambda_0(\mathbf{A}_M) \geq \lambda_1(\mathbf{A}_M) \ldots \geq \lambda_{M-1}(\mathbf{A}_M). \tag{12.48}$$

Then the eigenvalues $\lambda_k(\mathbf{A}_M)$ are interleaved between the eigenvalues $\lambda_k(\mathbf{A})$ as follows:

$$\begin{aligned}
\lambda_0(\mathbf{A}) &\geq \lambda_0(\mathbf{A}_M) \geq \lambda_L(\mathbf{A}), \\
\lambda_1(\mathbf{A}) &\geq \lambda_1(\mathbf{A}_M) \geq \lambda_{L+1}(\mathbf{A}), \\
&\vdots \\
\lambda_{M-1}(\mathbf{A}) &\geq \lambda_{M-1}(\mathbf{A}_M) \geq \lambda_{P-1}(\mathbf{A}),
\end{aligned}$$

where $L = P - M$. This is demonstrated in Fig. 12.4. Now consider any unitary matrix $\mathbf{U}_f$ which has performed the partial diagonalization (12.37). The eigenvalues of the $M \times M$ leading principal submatrix of $\mathbf{U}_f^\dagger\mathbf{H}^\dagger\mathbf{H}\mathbf{U}_f$ are clearly μ_k, whereas the eigenvalues of $\mathbf{U}_f^\dagger\mathbf{H}^\dagger\mathbf{H}\mathbf{U}_f$ are just $\sigma_{h,k}^2$ (eigenvalues of $\mathbf{H}^\dagger\mathbf{H}$) regardless of $\mathbf{U}_f$. The interlacing property

$$\sigma_{h,k}^2 \geq \mu_k \geq \sigma_{h,k+L}^2, \quad 0 \leq k \leq M-1, \tag{12.49}$$

therefore holds. This implies in particular that Eq. (12.45) is true. $\bigtriangledown \bigtriangledown \bigtriangledown$

12.4.3 Finding the optimal $\mathbf{\Sigma}_f$

With the optimal unitary $\mathbf{U}_f$ shown to be the matrix that diagonalizes $\mathbf{H}^\dagger\mathbf{H}$, we now turn to the computation of the optimal coefficients a_k in Eq. (12.38) which determine the diagonal elements of $\mathbf{\Sigma}_f$. For this, recall again the necessary condition (12.32) for optimality derived earlier based on stationarity of the Lagrangian. This condition is reproduced below:

$$(\mathbf{F}^\dagger\mathbf{H}^\dagger\mathbf{H}\mathbf{F})^{-1} = \eta\mathbf{F}^\dagger\mathbf{F}, \tag{12.50}$$

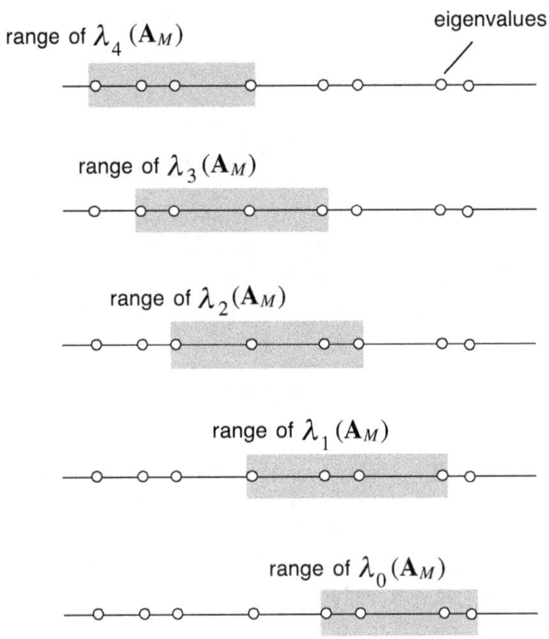

Figure 12.4. The interlacing property of the eigenvalues of a Hermitian matrix and its principal submatrix.

where $\eta = \mu\sigma_s^2/\sigma_q^2$. Now, in view of Eq. (12.24) we have the two equations

$$\mathbf{F}^\dagger\mathbf{H}^\dagger\mathbf{H}\mathbf{F} = \mathbf{\Sigma}_f\left(\mathbf{U}_f^\dagger\mathbf{H}^\dagger\mathbf{H}\mathbf{U}_f\right)_M\mathbf{\Sigma}_f \quad \text{and} \quad \mathbf{F}^\dagger\mathbf{F} = \mathbf{\Sigma}_f^2 \tag{12.51}$$

so that Eq. (12.50) simplifies to

$$\mathbf{\Sigma}_f^{-1}\left(\mathbf{U}_f^\dagger\mathbf{H}^\dagger\mathbf{H}\mathbf{U}_f\right)_M^{-1}\mathbf{\Sigma}_f^{-1} = \eta\mathbf{\Sigma}_f^2,$$

which shows that

$$\mathbf{\Sigma}_f^4 = \frac{1}{\eta}\left(\mathbf{U}_f^\dagger\mathbf{H}^\dagger\mathbf{H}\mathbf{U}_f\right)_M^{-1}. \tag{12.52}$$

Since the optimal $\mathbf{U}_f$ can be taken to be such that $\left(\mathbf{U}_f^\dagger\mathbf{H}^\dagger\mathbf{H}\mathbf{U}_f\right)$ is diagonal with diagonal elements $\sigma_{h,k}^2$, we conclude that the optimum diagonal elements in $\mathbf{\Sigma}_f$ have the form

$$\sigma_{f,k} = \frac{c}{\sigma_{h,k}^{1/2}}, \quad 0 \le k \le M-1. \tag{12.53}$$

Here $c > 0$ is such that the power constraint (12.35) is satisfied, that is,

$$\sigma_s^2\sum_{k=0}^{M-1}\sigma_{f,k}^2 = p_0.$$

This yields

$$c^2 = \frac{p_0}{\sigma_s^2 \sum_{k=0}^{M-1} \frac{1}{\sigma_{h,k}}} \tag{12.54}$$

Summarizing, the optimum $\mathbf{F}$ has the form

$$\mathbf{F} = \mathbf{U}_f \begin{bmatrix} \boldsymbol{\Sigma}_f \\ \mathbf{0} \end{bmatrix}, \tag{12.55}$$

where $\mathbf{U}_f$ is the $P \times P$ unitary matrix which diagonalizes $\mathbf{H}^\dagger\mathbf{H}$ and orders the diagonal elements in decreasing order, and $\boldsymbol{\Sigma}_f$ is an $M \times M$ diagonal matrix whose diagonal elements are given by Eq. (12.53). Substituting $\mu_k = \sigma_{h,k}^2$ and $\sigma_{f,k}$ from Eq. (12.53) into Eq. (12.38), the minimized mean square error simplifies to

$$\mathcal{E}_{mmse} = \frac{\sigma_s^2\sigma_q^2}{p_0}\left(\sum_{k=0}^{M-1} \frac{1}{\sigma_{h,k}}\right)^2. \tag{12.56}$$

Not surprisingly, the expressions (12.53) and (12.56) are similar to the expressions in Chap. 11 where we optimized a diagonal transceiver for a diagonal channel, subject to the ZF constraint. The fact that the Lagrangian stationary conditions eventually produce a *minimum* rather than a *maximum*, can be justified readily (see the discussion in Sec. 22.5.2).

12.4.4 Expressing the solution using the channel SVD

We know that the channel can always be represented using SVD in the form

$$\mathbf{H} = \mathbf{U}_h\boldsymbol{\Sigma}_h\mathbf{V}_h^\dagger, \tag{12.57}$$

where $\mathbf{U}_h$ and $\mathbf{V}_h$ are $P \times P$ unitary matrices and $\boldsymbol{\Sigma}_h$ is diagonal with diagonal elements $\sigma_{h,k} \geq 0$ (channel singular values). So

$$\mathbf{H}^\dagger\mathbf{H} = \mathbf{V}_h\boldsymbol{\Sigma}_h^2\mathbf{V}_h^\dagger, \quad \text{that is,} \quad \mathbf{V}_h^\dagger\mathbf{H}^\dagger\mathbf{H}\mathbf{V}_h = \boldsymbol{\Sigma}_h^2. \tag{12.58}$$

This shows that we can choose the unitary matrix $\mathbf{U}_f$ in the precoder $\mathbf{F}$ to be

$$\mathbf{U}_f = \mathbf{V}_h \tag{12.59}$$

The expression for the zero-forcing equalizer can be derived from Eq. (12.21), which is reproduced below:

$$\mathbf{G} = \Big((\mathbf{HF})^\dagger\mathbf{HF}\Big)^{-1}(\mathbf{HF})^\dagger. \tag{12.60}$$

Observe first that

$$\begin{aligned}
\mathbf{HF} &= \mathbf{U}_h\boldsymbol{\Sigma}_h\mathbf{V}_h^\dagger\mathbf{U}_f \begin{bmatrix} \boldsymbol{\Sigma}_f \\ \mathbf{0} \end{bmatrix} \\
&= \mathbf{U}_h\boldsymbol{\Sigma}_h\mathbf{V}_h^\dagger\mathbf{V}_h \begin{bmatrix} \boldsymbol{\Sigma}_f \\ \mathbf{0} \end{bmatrix} \\
&= \mathbf{U}_h\boldsymbol{\Sigma}_h \begin{bmatrix} \boldsymbol{\Sigma}_f \\ \mathbf{0} \end{bmatrix} \\
&= \mathbf{U}_h[\boldsymbol{\Sigma}_h]_{P\times M}\boldsymbol{\Sigma}_f
\end{aligned}$$

so that

$$(\mathbf{HF})^\dagger\mathbf{HF} = [\,\boldsymbol{\Sigma}_f \quad \mathbf{0}\,]\,\boldsymbol{\Sigma}_h\mathbf{U}_h^\dagger\mathbf{U}_h\boldsymbol{\Sigma}_h\begin{bmatrix}\boldsymbol{\Sigma}_f\\ \mathbf{0}\end{bmatrix} = \boldsymbol{\Sigma}_f^2[\boldsymbol{\Sigma}_h^2]_M,$$

where $[\boldsymbol{\Sigma}_h]_M$ is the $M \times M$ leading principal submatrix of $\boldsymbol{\Sigma}_h$. Thus the optimal equalizer is

$$\mathbf{G} = [\boldsymbol{\Sigma}_h]_M^{-2}\boldsymbol{\Sigma}_f^{-2}\,[\,\boldsymbol{\Sigma}_f \quad \mathbf{0}\,]\,\boldsymbol{\Sigma}_h\mathbf{U}_h^\dagger = \boldsymbol{\Sigma}_f^{-1}[\boldsymbol{\Sigma}_h]_M^{-1}[\mathbf{U}_h^\dagger]_{M\times P}. \tag{12.61}$$

The equalizer can be simplified a little bit by observing that

$$\left(\boldsymbol{\Sigma}_f^{-1}[\boldsymbol{\Sigma}_h]_M^{-1}\right)_{kk} = \frac{1}{\sigma_{f,k}\sigma_{h,k}} = \frac{1}{c\sigma_{h,k}^{1/2}}$$

where we have used Eq. (12.53). The equalizer matrix can therefore be written in the form

$$\mathbf{G} = [\,\boldsymbol{\Sigma}_g \quad \mathbf{0}\,]\,\mathbf{U}_h^\dagger, \tag{12.62}$$

where

$$\boldsymbol{\Sigma}_g = \frac{1}{c}\begin{bmatrix} \sigma_{h,0}^{-1/2} & 0 & \ldots & 0 \\ 0 & \sigma_{h,1}^{-1/2} & \ldots & 0 \\ \vdots & \vdots & \ddots & \vdots \\ 0 & 0 & \ldots & \sigma_{h,M-1}^{-1/2} \end{bmatrix} = \frac{1}{c}(\boldsymbol{\Sigma}_h)_M^{-1/2}. \tag{12.63}$$

As a reminder, the optimal precoder is

$$\mathbf{F} = \mathbf{V}_h\begin{bmatrix}\boldsymbol{\Sigma}_f\\ \mathbf{0}\end{bmatrix}, \tag{12.64}$$

where $\boldsymbol{\Sigma}_f$ is $M \times M$ diagonal with diagonal elements (12.53):

$$\boldsymbol{\Sigma}_f = c\begin{bmatrix} \sigma_{h,0}^{-1/2} & 0 & \ldots & 0 \\ 0 & \sigma_{h,1}^{-1/2} & \ldots & 0 \\ \vdots & \vdots & \ddots & \vdots \\ 0 & 0 & \ldots & \sigma_{h,M-1}^{-1/2} \end{bmatrix} = c(\boldsymbol{\Sigma}_h)_M^{-1/2}. \tag{12.65}$$

Summarizing the above results, we have proved the following:

♠**Theorem 12.1.** *Optimal ZF-MMSE transceiver.* The solution to the ZF-MMSE transceiver optimization problem described in Eqs. (12.12)–(12.14) can be summarized as follows:

1. ***Precoder.*** The optimal precoder has the form (12.64), where (a) $\mathbf{V}_h$ is the unitary matrix that occurs in the SVD of the channel Eq. (12.57) (i.e., it diagonalizes $\mathbf{H}^\dagger\mathbf{H}$ as in Eq. (12.58)), and (b) $\boldsymbol{\Sigma}_f$ is the diagonal matrix shown in (12.65), where $\sigma_{h,k}$ are the first M dominant singular values of the channel $\mathbf{H}$. The constant c is such that the power constraint is satisfied.

2. ***Equalizer.*** The optimal equalizer has the form (12.62), where $\mathbf{U}_h$ is the unitary matrix that occurs in the SVD of the channel Eq. (12.57), and $\mathbf{\Sigma}_g$ is the diagonal matrix shown in Eq. (12.63).

3. ***Minimized error.*** With the above choice of precoder and equalizer, the mean square error is minimized and has the value given by

$$\mathcal{E}_{mmse} = \frac{\sigma_s^2\sigma_q^2}{p_0}\left(\sum_{k=0}^{M-1}\frac{1}{\sigma_{h,k}}\right)^2, \tag{12.66}$$

as proved earlier in Eq. (12.56). This expression depends only on the M dominant singular values $\sigma_{h,k}$ of the channel. ◇

Recall from Chap. 10 that the ZF-MMSE SISO transceiver for the scalar channel $H(e^{j\omega})$ has MMSE given by Eq. (10.41). For the special case of white noise ($S_{qq}(e^{j\omega}) = \sigma_q^2$) and white signal ($S_{ss}(e^{j\omega}) = \sigma_s^2$), that expression reduces to

$$\mathcal{E}_{siso} = \frac{\sigma_q^2\sigma_s^2}{p_0}\left(\int_{-\pi}^{\pi}\frac{1}{|H(e^{j\omega})|}\frac{d\omega}{2\pi}\right)^2. \tag{12.67}$$

This is strikingly similar to Eq. (12.66) indeed! If we replace $|H(e^{j\omega})|$ with the singular values of the MIMO memoryless channel $\mathbf{H}$ (and replace integrals with summations) the result is precisely Eq. (12.66). Figure 12.5 summarizes the receiver structure. We conclude this section with some interesting remarks.

1. ***Optimal transceiver diagonalizes the channel.*** The unitary matrix $\mathbf{V}_h$ in the precoder and the unitary matrix $\mathbf{U}_h^\dagger$ in the equalizer together make the channel $\mathbf{H}$ look like a diagonal channel with diagonal elements $\sigma_{h,k}$ because

$$\mathbf{U}_h^\dagger\mathbf{H}\mathbf{V}_h = \begin{bmatrix}\sigma_{h,0} & 0 & \ldots & 0\\ 0 & \sigma_{h,1} & \ldots & 0\\ \vdots & \vdots & \ddots & \vdots\\ 0 & 0 & \ldots & \sigma_{h,P-1}\end{bmatrix}.$$

Thus the system of Fig. 12.5 is equivalent to Fig. 12.6, where

$$\sigma_{f,k} = \frac{c}{\sqrt{\sigma_{h,k}}}, \quad \sigma_{g,k} = \frac{1}{c\sqrt{\sigma_{h,k}}}$$

Note that only the M dominant singular values $\sigma_{h,k}$ of the channel matter because of the block of $P-M$ zeros entering $\mathbf{V}_h$ and the block of $P-M$ signals dropped at the output of $\mathbf{U}_h^\dagger$. The simple system of Fig. 12.6 was studied independently in Chap. 11 where we derived the optimal values of $\sigma_{f,k}$ and $\sigma_{g,k}$ under the zero-forcing condition. This explains why the optimal values Eq. (12.63) and Eq. (12.65) are similar to the expressions derived for the simple case of diagonal channels.

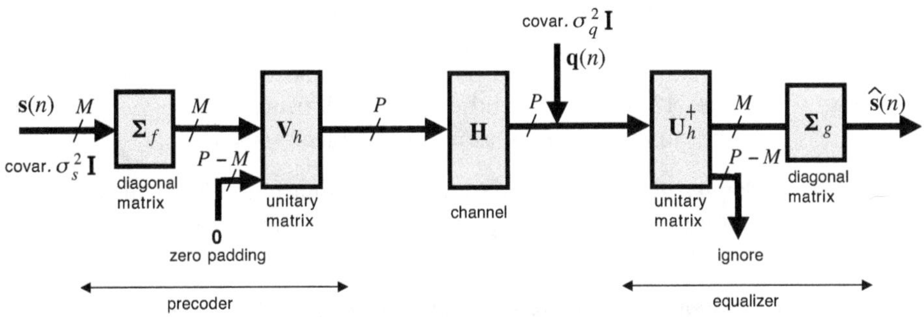

Figure 12.5. The structure of the MMSE transceiver with the zero-forcing constraint. Here $\mathbf{U}_h$ and $\mathbf{V}_h$ are $P \times P$ unitary matrices which diagonalized the channel, that is, $\mathbf{\Sigma} = \mathbf{U}_h^\dagger \mathbf{H} \mathbf{V}_h$. The diagonal matrices $\mathbf{\Sigma}_f$ and $\mathbf{\Sigma}_g$ are optimally chosen as described in the text.

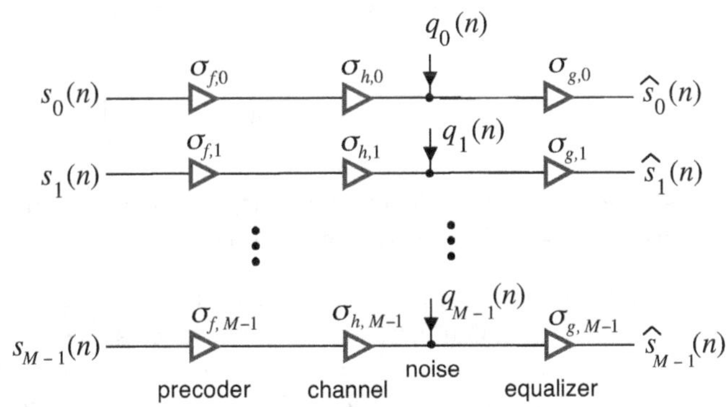

Figure 12.6. The equivalent diagonal-channel system.

2. ***Precoder is orthogonal.*** The optimal precoder is such that $\mathbf{F}^\dagger\mathbf{F} = \mathbf{\Sigma}_f^2$, which is diagonal (see Eq. (12.51)). That is, the M columns of the optimal solution $\mathbf{F}$ are orthogonal.

3. ***Half and a half.*** This diagonalized channel is equalized by the multipliers $\sigma_{h,k}^{-1/2}$ in the transmitter and receiver. Thus the optimal transceiver distributes the task of equalization *equally* between the transmitter and the receiver, rather than leaving all of the equalization to be performed at the receiver. This was also noticed in Chap. 11 where a diagonal transceiver was optimized for a diaognal channel. This "division of labor" is analogous to the idea of *half-whitening* known in data compression theory [Jayant and Noll, 1984].

4. ***Minimized error and channel singular values.*** Examination of Eq. (12.66)

shows that the minimized error can be large for channels with very small singular values $\sigma_{h,k}$. Indeed there is no upper bound on how large the error can get because the smallest nonzero singular value $\sigma_{h,M-1}$ can be arbitrarily small for very bad channels. A similar observation was also made in Chap. 11 where we considered the special case of diagonal channels. In Chap. 13, where we study the MMSE transceiver *without* the zero-forcing constraint, we will see that this problem goes away.

5. ***Minimized error and noise power.*** Equation (12.66) also shows that the minimized error is proportional to the channel noise variance σ_q^2, as one would expect.

6. ***Minimized error and channel input power.*** From Eq. (12.66) we see that $\mathcal{E}_{mmse}$ decreases with increasing channel input power p_0. This is intuitively appealing of course, and it comes about as follows: if we increase the channel power p_0 for fixed σ_s^2, then the multipliers $\sigma_{f,k} = c/\sqrt{\sigma_{h,k}}$ in the transmitter are larger (because c is scaled up). So the multipliers $\sigma_{g,k} = 1/c\sqrt{\sigma_{h,k}}$ at the receiver will get correspondingly smaller, which reduces the reconstruction error.

7. ***Signal-to-noise ratio.*** From Eq. (12.66) we see that

$$\frac{\mathcal{E}_{mmse}}{\sigma_s^2} = \frac{\sigma_q^2}{p_0}\left(\sum_{k=0}^{M-1}\frac{1}{\sigma_{h,k}}\right)^2.$$

   This ratio is clearly independent of σ_s^2. In fact, σ_s^2 can be set to unity in all discussions without loss of generality.

8. ***Diagonal channel.*** Suppose the channel $\mathbf{H}$ is a diagonal matrix with real non-negative diagonal elements. It then follows that the unitary matrices in its SVD are $\mathbf{U}_h = \mathbf{V}_h = \mathbf{I}$, so the optimal precoder and equalizer (essentially $\boldsymbol{\Sigma}_f$ and $\boldsymbol{\Sigma}_g$) are also diagonal matrices. (In Chap. 11 this was assumed without proof.) If $P > M$ then the first M dominant diagonal elements of $\mathbf{H}$ are used by the transceiver as seen in Fig. 12.5. If the diagonal channel $\mathbf{H}$ has complex diagonal entries then the matrices $\mathbf{U}_h$ and $\mathbf{V}_h$ are diagonal unitary matrices (i.e., their diagonal entries are of the form $e^{j\theta_k}$), and can be combined with diagonal matrices $\boldsymbol{\Sigma}_f$ and $\boldsymbol{\Sigma}_g$. The result is again an optimal transceiver with diagonal precoder and equalizer.

Example 12.2: ZF-MMSE transceiver

Consider the simple 2×2 channel

$$\mathbf{H} = \frac{1}{2\sqrt{2}}\begin{bmatrix}\sqrt{3}-(1/16) & 1+(\sqrt{3}/16)\\ \sqrt{3}+(1/16) & 1-(\sqrt{3}/16)\end{bmatrix}.$$

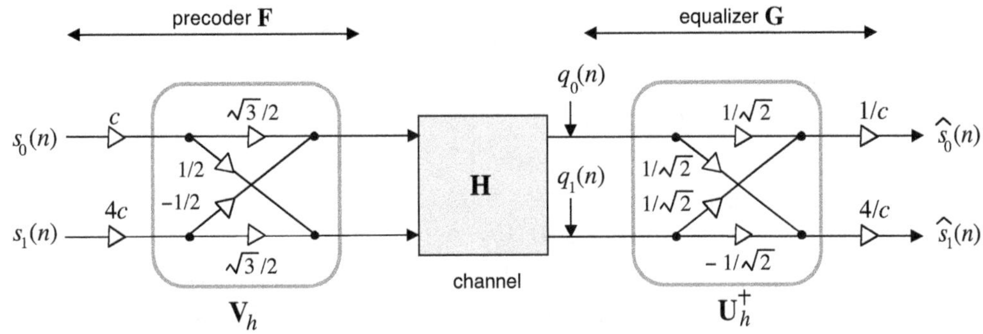

Figure 12.7. Example 12.2. The precoder and equalizer for the optimal (ZF-MMSE) transceiver for the channel $\mathbf{H}$.

The SVD of this channel is given by

$$\mathbf{H} = \underbrace{\frac{1}{\sqrt{2}}\begin{bmatrix}1 & 1\\ 1 & -1\end{bmatrix}}_{\mathbf{U}_h} \times \underbrace{\begin{bmatrix}1 & 0\\ 0 & (1/16)\end{bmatrix}}_{\boldsymbol{\Sigma}_h} \times \underbrace{\frac{1}{2}\begin{bmatrix}\sqrt{3} & 1\\ -1 & \sqrt{3}\end{bmatrix}}_{\mathbf{V}_h^\dagger}.$$

This determines the optimal linear transceiver completely, as shown in Fig. 12.7, except for the constant c which is determined from the power constraint. More extensive examples will be presented in Chaps. 17 and 18 where we compare several optimal transceivers on the basis of symbol error probability.

12.5 Optimizing the noise-to-signal ratio

In this chapter we have so far assumed a simple signal covariance matrix $\mathbf{R_{ss}} = \sigma_s^2\mathbf{I}$ and minimized the trace of $\mathbf{R_{ee}} = \sigma_q^2\mathbf{G}\mathbf{G}^\dagger$ subject to the power constraint $p_0 = \sigma_s^2\text{Tr}\,(\mathbf{F}\mathbf{F}^\dagger)$ and the zero-forcing constraint $\mathbf{GHF} = \mathbf{I}$. If $\mathbf{R_{ss}}$ is a more general diagonal matrix $\boldsymbol{\Lambda}_s$ of the form (12.6) it makes more sense to minimize the sum of error-to-signal ratios rather than the sum of errors. For, it is the error-to-signal ratios at the detectors that determine the performance of the receiver. Thus, it is more appropriate to minimize the trace of the matrix defined as

$$\mathbf{R}_{new} \stackrel{\Delta}{=} \sigma_q^2\boldsymbol{\Lambda}_s^{-1/2}\mathbf{G}\mathbf{G}^\dagger\boldsymbol{\Lambda}_s^{-1/2}.$$

With $\mathbf{R_{ss}} = \boldsymbol{\Lambda}_s$ the power constraint is modified to

$$p_0 = \sigma_s^2\text{Tr}\,(\mathbf{F}\boldsymbol{\Lambda}_s\mathbf{F}^\dagger). \tag{12.68}$$

Define

$$\mathbf{G}_1 = \boldsymbol{\Lambda}_s^{-1/2}\mathbf{G} \quad \text{and} \quad \mathbf{F}_1 = \mathbf{F}\boldsymbol{\Lambda}_s^{1/2}. \tag{12.69}$$

Then

$$\mathbf{R}_{new} = \sigma_q^2 \mathbf{G}_1 \mathbf{G}_1^\dagger \tag{12.70}$$

and the power constraint is

$$p_0 = \text{Tr } (\mathbf{F}_1 \mathbf{F}_1^\dagger). \tag{12.71}$$

Observe also that

$$\mathbf{GHF} = \mathbf{\Lambda}_s^{1/2} \mathbf{G}_1 \mathbf{H} \mathbf{F}_1 \mathbf{\Lambda}_s^{-1/2}. \tag{12.72}$$

Thus the condition $\mathbf{G}_1\mathbf{H}\mathbf{F}_1 = \mathbf{I}$ implies $\mathbf{GHF} = \mathbf{I}$ and vice versa. So, minimizing the trace of $\mathbf{R}_{new}$ under the power constraint and ZF constraint is equivalent to minimizing the trace of $\sigma_q^2 \mathbf{G}_1 \mathbf{G}_1^\dagger$ under the constraints

$$\mathbf{G}_1 \mathbf{H} \mathbf{F}_1 = \mathbf{I} \quad \text{and} \quad p_0 = \text{Tr } (\mathbf{F}_1 \mathbf{F}_1^\dagger).$$

This is the simplified problem we addressed in Eqs. (12.12)–(12.14) of Sec. 12.3. The solution in Sec. 12.4 can therefore be modified to the case where $\mathbf{\Lambda}_s \neq \sigma_s^2 \mathbf{I}$ by using the simple modification described above.

Structural interpretation. Let $\{\mathbf{F}_1^o, \mathbf{G}_1^o\}$ be the optimal solution for the simplified problem. Then from Eq. (12.69) we see that the optimal solution to the original problem with $\mathbf{R_{ss}} = \mathbf{\Lambda}_s$ is given by

$$\mathbf{F}^o = \mathbf{F}_1^o \mathbf{\Lambda}_s^{-1/2}, \quad \mathbf{G}^o = \mathbf{\Lambda}_s^{1/2} \mathbf{G}_1^o$$

So the optimal transceiver for the original system can be represented as in Fig. 12.8. Here the input to $\mathbf{F}_1^o$ has covariance matrix $\mathbf{I}$. Thus, the original input with covariance $\mathbf{\Lambda}_s$ has merely been transformed to a new input with covariance $\mathbf{I}$, to produce the simplified optimization problem involving $\{\mathbf{F}_1, \mathbf{G}_1\}$.

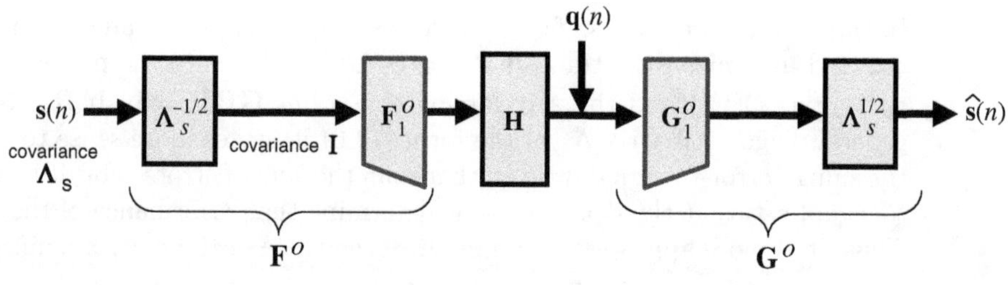

Figure 12.8. The simplified transceiver $\{\mathbf{F}_1^o, \mathbf{G}_1^o\}$ in terms of the original transceiver $\{\mathbf{F}^o, \mathbf{G}^o\}$.

12.6 Concluding remarks

In this chapter we considered the problem of optimizing a transceiver under the zero-forcing constraint. The solution to the problem is such that the channel $\mathbf{H}$ is effectively diagonalized, and the ZF-MMSE solution for the diagonal channel $\boldsymbol{\Sigma}_h$ is implemented implicitly. In the next chapter we will consider the optimization of transceivers without the zero-forcing constraint. The application of these optimization results for the case of zero-padded and cyclic-prefixed systems will be considered in later chapters. A one-page summary of the results of this chapter is included in Appendix I.

Appendices to Chapter 12

12.A Generality of minimum-norm inverse

The zero-forcing constraint is equivalent to $\mathbf{GHF} = \mathbf{I}$. That is, $\mathbf{G}$ is a left inverse of $\mathbf{HF}$. Since $\mathbf{HF}$ is in general rectangular, this left inverse is not unique. We now show that there is no loss of generality in restricting the left inverse $\mathbf{G}$ to be the *minimum-norm* left inverse (see Sec. C.4 in Appendix C).[5] The proof works even when the channel $\mathbf{H}$ is rectangular, so we assume $\mathbf{H}$ has size $K \times P$ for generality.

The matrix $\mathbf{HF}$ has size $K \times M$. We need $K \geq M$ so that the rank of $\mathbf{H}$ is at least as large as M (as required by the zero-forcing condition). Now, $\mathbf{HF}$ has a singular value decomposition (SVD) of the form

$$\mathbf{HF} = \underbrace{\mathbf{U}}_{K\times K} \begin{bmatrix} \boldsymbol{\Sigma} \\ \mathbf{0} \end{bmatrix} \underbrace{\mathbf{V}^\dagger}_{M\times M}, \tag{12.73}$$

where $\mathbf{U}$ and $\mathbf{V}$ are square unitary matrices. Assuming that $\mathbf{HF}$ has full rank M, the $M \times M$ diagonal matrix $\boldsymbol{\Sigma}$ is invertible. From Appendix C we know that the unique minimum-norm left inverse of $\mathbf{HF}$ has the form

$$\mathbf{G}_{MNLI} = \mathbf{V}\,[\,\boldsymbol{\Sigma}^{-1} \quad \mathbf{0}\,]\,\mathbf{U}^\dagger. \tag{12.74}$$

On the other hand, the most general left inverse (Appendix C) has the form

$$\mathbf{G} = \underbrace{\mathbf{V}}_{M\times M}\,[\,\boldsymbol{\Sigma}^{-1} \quad \mathbf{A}\,]\,\underbrace{\mathbf{U}^\dagger}_{K\times K}, \tag{12.75}$$

where $\mathbf{A}$ is an arbitrary matrix of size $M \times (K - M)$. Figure 12.9(a) shows the transceiver with the above receiver matrix indicated in detail, and Fig. 12.9(b) shows the receiver with noise $\mathbf{q}(n)$ only as its input. Let the output of $\mathbf{U}^\dagger$ in Fig. 12.9(b) be denoted as $\mathbf{w}(n)$, that is,

$$\mathbf{w}(n) = \begin{bmatrix} \mathbf{w}_1(n) \\ \mathbf{w}_2(n) \end{bmatrix}. \tag{12.76}$$

Since $\mathbf{q}(n)$ has correlation matrix $\sigma_q^2\mathbf{I}$, the correlation matrix of $\mathbf{w}(n)$ is

$$\mathbf{R}_{\mathbf{ww}} = \sigma_q^2\mathbf{U}^\dagger\mathbf{U} = \sigma_q^2\mathbf{I} = \begin{bmatrix} \mathbf{R}_{11} & \mathbf{R}_{12} \\ \mathbf{R}_{21} & \mathbf{R}_{22} \end{bmatrix}, \tag{12.77}$$

where $\mathbf{R}_{km} = E[\mathbf{w}_k(n)\mathbf{w}_m^\dagger(n)]$. Thus

$$\mathbf{R}_{11} = \sigma_q^2\mathbf{I}_M, \ \mathbf{R}_{22} = \sigma_q^2\mathbf{I}_{K-M}, \text{ and } \mathbf{R}_{12} = \mathbf{0}. \tag{12.78}$$

[5]That is, the MMSE with minimum-norm left inverse is no greater than the MMSE with the left inverse allowed to be arbitrary.

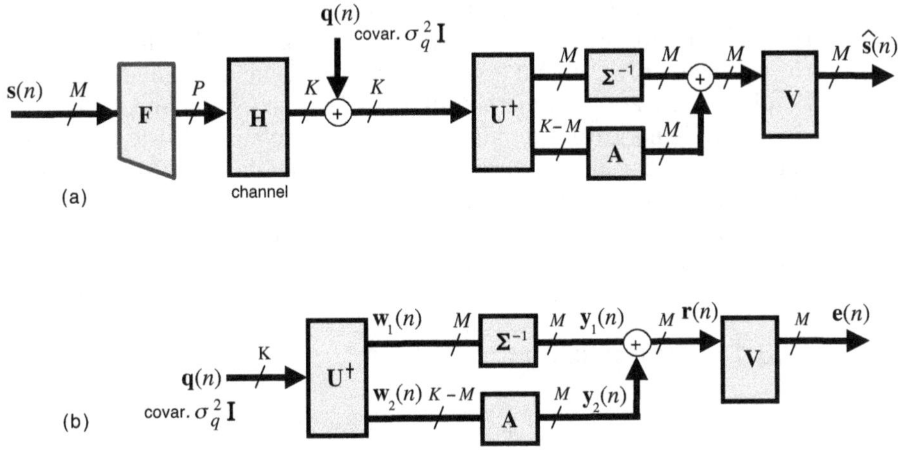

Figure 12.9. (a) The zero-forcing transceiver with receiver shown in SVD form, and (b) the receiver shown separately with channel noise as its only input.

The correlation matrices of $\mathbf{e}(n)$ and $\mathbf{r}(n)$ in Fig. 12.9(b) are related as follows:

$$\mathbf{R_{ee}} = \mathbf{V}\mathbf{R_{rr}}\mathbf{V}^\dagger.$$

From the unitarity of $\mathbf{V}$ it then follows that

$$\text{Tr}(\mathbf{R_{ee}}) = \text{Tr}(\mathbf{V}\mathbf{R_{rr}}\mathbf{V}^\dagger) = \text{Tr}(\mathbf{V}^\dagger\mathbf{V}\mathbf{R_{rr}}) = \text{Tr}(\mathbf{R_{rr}}).$$

To minimize the trace of $\mathbf{R_{ee}}$ we therefore minimize the trace of $\mathbf{R_{rr}}$. Since $\mathbf{w}_1(n)$ and $\mathbf{w}_2(n)$ in the figure have no correlation (i.e., $\mathbf{R}_{12} = \mathbf{0}$ from Eq. (12.78)), we have

$$\mathbf{R_{rr}} = E[\mathbf{y}_1\mathbf{y}_1^\dagger] + E[\mathbf{y}_2\mathbf{y}_2^\dagger] = \mathbf{\Sigma}^{-1}\mathbf{R}_{11}\mathbf{\Sigma}^{-1} + \mathbf{A}\mathbf{R}_{22}\mathbf{A}^\dagger = \sigma_q^2\mathbf{\Sigma}^{-2} + \sigma_q^2\mathbf{A}\mathbf{A}^\dagger$$

so that

$$\text{Tr}\;(\mathbf{R_{rr}}) = \sigma_q^2\text{Tr}(\mathbf{\Sigma}^{-2}) + \sigma_q^2\text{Tr}(\mathbf{A}\mathbf{A}^\dagger). \tag{12.79}$$

But the diagonal elements of $\mathbf{A}\mathbf{A}^\dagger$ are non-negative, which shows that the choice $\mathbf{A} = \mathbf{0}$ minimizes the right-hand side. Thus the best left inverse is the minimum-norm left inverse (12.74).

12.B Diagonalization approach to optimization

Refer again to Fig. 12.5, which represents the optimal transceiver derived in this chapter. The cascade of $\mathbf{V}_h, \mathbf{H}$, and $\mathbf{U}_h^\dagger$ has the transfer function

$$\mathbf{U}_h^\dagger\mathbf{H}\mathbf{V}_h = \begin{bmatrix} \sigma_{h,0} & 0 & \dots & 0 \\ 0 & \sigma_{h,1} & \dots & 0 \\ \vdots & \vdots & \ddots & \vdots \\ 0 & 0 & \dots & \sigma_{h,P-1} \end{bmatrix},$$

which is the diagonal matrix of singular values of the channel. Since the unitary matrices can always be inserted as in the figure to diagonalize the channel, the following question arises: if we first diagonalize the channel and then find the optimal precoder and equalizer of the diagonal channel, will this not be simpler than the direct method described in Secs. 12.3 and 12.4? It turns out that there is no great simplification because we still need to establish that the optimal precoder and equalizer for the diagonal channel are diagonal matrices. For the interested reader, we present in this appendix some of the details of such a "diagonalization approach." This approach also relies upon the concepts of majorization and Schur convexity reviewed in Chap. 21.

12.B.1 Diagonalizing the problem

Consider the transceiver system shown in Fig. 12.10(a). With the channel $\mathbf{H}$ represented in SVD form, $\mathbf{H} = \mathbf{U}_h\mathbf{\Sigma}_h\mathbf{V}_h^\dagger$, this can be redrawn as in Fig. 12.10(b). With the matrices $\mathbf{F}$ and $\mathbf{G}$ defined as $\mathbf{F} = \mathbf{V}_h^\dagger\mathbf{F}_1$ and $\mathbf{G} = \mathbf{G}_1\mathbf{U}_h$, the system can be redrawn as in Fig. 12.10(c). The original noise source $\mathbf{q}_1(n)$ is related to the new noise $\mathbf{q}(n)$ as follows:

$$\mathbf{q}(n) = \mathbf{U}_h^\dagger\mathbf{q}_1(n). \tag{12.80}$$

If $\mathbf{q}_1(n)$ has covariance $\sigma_q^2\mathbf{I}$, then so does $\mathbf{q}(n)$. A couple of points should now be noted (Problem 12.4):

1. The power at the output of $\mathbf{F}_1$ in Fig. 12.10(a) is equal to the power at the output of $\mathbf{F}$ in Fig. 12.10(c).

2. Assume $\{\mathbf{F}, \mathbf{G}\}$ is optimal for Fig. 12.10(c) in some sense (e.g., minimizing the mean square error), for a given power p_0 at the output of $\mathbf{F}$. Then $\{\mathbf{F}_1, \mathbf{G}_1\}$ is optimal for Fig. 12.10(a) for the same power p_0 at the output of $\mathbf{F}_1$.

Thus, if we solve for the optimal pair $\{\mathbf{F}, \mathbf{G}\}$ for the diagonal channel $\mathbf{\Sigma}_h$ in Fig. 12.10(c), we can then find the optimal $\{\mathbf{F}_1, \mathbf{G}_1\}$ readily using

$$\mathbf{F}_1 = \mathbf{V}_h\mathbf{F}, \quad \mathbf{G}_1 = \mathbf{G}\mathbf{U}_h^\dagger. \tag{12.81}$$

Since the matrix $\mathbf{\Sigma}_h$ is diagonal, the details of this optimization are somewhat simpler, though the simplification is not very significant as we shall see.

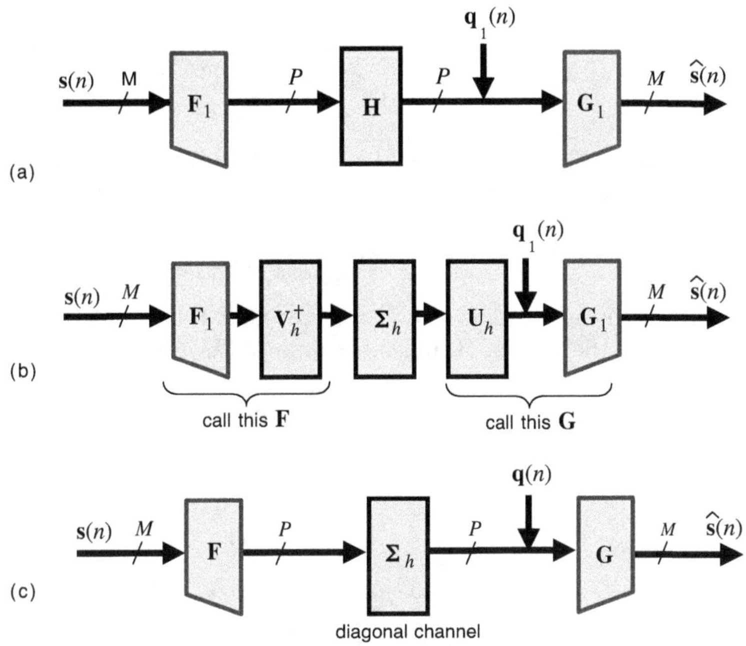

Figure 12.10. (a) Transceiver with channel $\mathbf{H}$, (b) SVD representation of the channel, and (c) equivalent transceiver with diagonal channel.

12.B.2 ZF-MMSE solution for the diagonal channel

For the system shown in Fig. 12.10(c) consider again the ZF-MMSE problem discussed in Sec. 12.3. We know that the problem can be reduced to the minimization of Eq. (12.22). That is, find the precoder $\mathbf{F}$ such that

$$\mathcal{E}_{mse} = \sigma_q^2 \text{Tr}\Big(\mathbf{F}^\dagger \mathbf{\Sigma}_h^2 \mathbf{F}\Big)^{-1} \tag{12.82}$$

is minimized. The optimal $\mathbf{F}$ cannot be assumed to be diagonal because we have not yet proved this to be the case. However, we know from Sec. 12.3.2 that $\mathbf{F}$ can always be assumed to be of the form

$$\mathbf{F} = \mathbf{U}_f \begin{bmatrix} \mathbf{\Sigma}_f \\ \mathbf{0} \end{bmatrix}, \tag{12.83}$$

where $\mathbf{U}_f$ is $P \times P$ unitary and $\mathbf{\Sigma}_f$ is $M \times M$ diagonal. Thus

$$\begin{aligned} \mathcal{E}_{mse} &= \sigma_q^2 \text{Tr}\Big(\mathbf{F}^\dagger \mathbf{\Sigma}_h^2 \mathbf{F}\Big)^{-1} \\ &= \sigma_q^2 \text{Tr}\Big([\,\mathbf{\Sigma}_f \quad \mathbf{0}\,]\,\mathbf{U}_f^\dagger \mathbf{\Sigma}_h^2 \mathbf{U}_f \begin{bmatrix} \mathbf{\Sigma}_f \\ \mathbf{0} \end{bmatrix}\Big)^{-1} \end{aligned}$$

$$= \sigma_q^2 \text{Tr}\Big(\mathbf{\Sigma}_f^2 (\mathbf{U}_f^\dagger \mathbf{\Sigma}_h^2 \mathbf{U}_f)_M\Big)^{-1}$$

$$= \sigma_q^2 \text{Tr}\Big[\Big((\mathbf{U}_f^\dagger \mathbf{\Sigma}_h^2 \mathbf{U}_f)_M\Big)^{-1} \mathbf{\Sigma}_f^{-2}\Big] = \sigma_q^2 \sum_{k=0}^{M-1} \frac{B_{kk}}{\sigma_{f,k}^2},$$

where $(\mathbf{A})_M$ denotes the $M \times M$ leading principal submatrix of $\mathbf{A}$ as usual. In the preceding, $B_{kk} > 0$ are the diagonal elements of the Hermitian positive definite matrix

$$\mathbf{B} \stackrel{\Delta}{=} \Big((\mathbf{U}_f^\dagger \mathbf{\Sigma}_h^2 \mathbf{U}_f)_M\Big)^{-1}. \tag{12.84}$$

Here it is assumed that $\sigma_{f,k} > 0$ (i.e., that $\mathbf{F}$ has rank M). Let us take a closer look at the matrix $\mathbf{B}$ and its relation to $\mathbf{U}_f$. Write the unitary matrix $\mathbf{U}_f$ in partitioned form:

$$\mathbf{U}_f = \begin{array}{c} \begin{array}{cc} M & P-M \end{array} \\ \begin{pmatrix} \mathbf{U}_1 & \mathbf{U}_2 \end{pmatrix} \end{array}.$$

If we replace $\mathbf{U}_f$ with

$$\mathbf{U}_{f,new} = \begin{array}{c} \begin{array}{cc} M & P-M \end{array} \\ \begin{pmatrix} \mathbf{U}_1\mathbf{U}_b & \mathbf{U}_2 \end{pmatrix} \end{array} \tag{12.85}$$

for some $M \times M$ unitary $\mathbf{U}_b$, then $\mathbf{B}$ in Eq. (12.84) is replaced with (Problem 12.5)

$$\mathbf{B}_{new} = \mathbf{U}_b^\dagger \mathbf{B} \mathbf{U}_b. \tag{12.86}$$

So, as far as $\mathbf{B}$ is concerned, the freedom to choose $\mathbf{U}_f$ is equivalent to the freedom to choose the unitary matrix $\mathbf{U}_b$.

12.B.3 Conversion to a majorization problem

To proceed further we will use the ideas of majorization and Schur convexity reviewed in Chap. 21. If we assume the ordering

$$\sigma_{f,0} \leq \sigma_{f,1} \leq \ldots \leq \sigma_{f,M-1} \tag{12.87}$$

then it can be shown that

$$B_{00} \leq B_{11} \leq \ldots \leq B_{M-1,M-1}. \tag{12.88}$$

If this is not the case then we can interchange pairs of elements B_{kk} and B_{mm} to reduce the MSE further (as we did in Sec. 12.4.1); we have the freedom to do that by interchanging the columns of the matrix $\mathbf{U}_b$. So without loss of generality we assume the above orderings. Now, from Ex. 1 in Sec. 21.4.1 (Chap. 21), we know that the sum

$$\sum_{k=0}^{M-1} a_k x_k \tag{12.89}$$

is *Schur-concave* in the region

$$x_0 \leq x_1 \leq \ldots \leq x_{M-1}$$

if we have $a_0 \geq a_1 \geq \ldots \geq a_{M-1} \geq 0$. Thus, $\mathcal{E}_{mse}$ is Schur-concave in the vector

$$\begin{bmatrix} B_{00} & B_{11} & \ldots B_{M-1,M-1} \end{bmatrix}. \tag{12.90}$$

12.B.4 Optimal choice of $\mathbf{U}_f$

Consider now the choice of the matrix $\mathbf{U}_b$. Given any $\mathbf{B}$ whatsoever, suppose we choose $\mathbf{U}_b$ such that $\mathbf{B}_{new}$ in Eq. (12.86) is diagonal. Then diag($\mathbf{B}_{new}$) majorizes diag($\mathbf{B}$) (Sec. 21.5.1). So, if we want to minimize a Schur-concave function of diag($\mathbf{B}$) the best thing to do would be to choose $\mathbf{U}_b$ (i.e., choose $\mathbf{U}_f$) such that $\mathbf{B}$ is diagonal. Having established that

$$\mathbf{B} = \left((\mathbf{U}_f^\dagger \boldsymbol{\Sigma}_h^2 \mathbf{U}_f)_M \right)^{-1}$$

has to be diagonal, we now use Lemma 12.2 to see that $\mathbf{U}_f$ can be chosen such that $\mathbf{U}_f^\dagger \boldsymbol{\Sigma}_h^2 \mathbf{U}_f$ itself is diagonal. Thus, we finally obtain

$$\mathcal{E}_{mse} = \sigma_q^2 \sum_{k=0}^{M-1} \frac{B_{kk}}{\sigma_{f,k}^2} = \sigma_q^2 \sum_{k=0}^{M-1} \frac{1}{\sigma_{h,k}^2 \sigma_{f,k}^2} \tag{12.91}$$

For any choice of $\sigma_{f,k}$ whatsoever it is clear that the M values of $\sigma_{h,k}$ should be the largest M singular values of the channel, otherwise $\mathcal{E}_{mse}$ can be reduced further by making such a choice. So it follows that the best $\mathbf{U}_f$ should have its columns ordered such that the largest singular values of the channel are the first M diagonal elements of $\boldsymbol{\Sigma}_h$. The optimal choice of $\sigma_{f,k}$ in Eq. (12.91) is done precisely as in Sec. 12.4.3.

12.C Rectangular channel

So far in this chapter we have assumed that the channel matrix $\mathbf{H}$ is a square matrix. For the case where $\mathbf{H}$ is a $K \times P$ (rectangular) matrix as shown in Fig. 12.11, the results extend readily as we shall show next. Thus consider Problem 1 described by Eqs. (12.12)–(12.14) again. The ZF constraint $\mathbf{GHF} = \mathbf{I}$ can again be eliminated by defining $\mathbf{G}$ to be the minimum-norm left inverse of $\mathbf{HF}$ as in (12.21). Then the problem reduces to the form shown in Eqs. (12.22) and (12.23). Thus the problem is again stated entirely in terms of the $P \times M$ matrix $\mathbf{F}$ and the $P \times P$ matrix $\mathbf{H}^\dagger \mathbf{H}$. The integer K has disappeared, and does not affect the optimum precoder $\mathbf{F}$, which therefore still has the form (12.55), that is,

$$\mathbf{F} = \mathbf{U}_f \begin{bmatrix} \boldsymbol{\Sigma}_f \\ \mathbf{0} \end{bmatrix},$$

where $\mathbf{U}_f$ is as in Eq. (12.59), that is,

$$\mathbf{U}_f = \mathbf{V}_h.$$

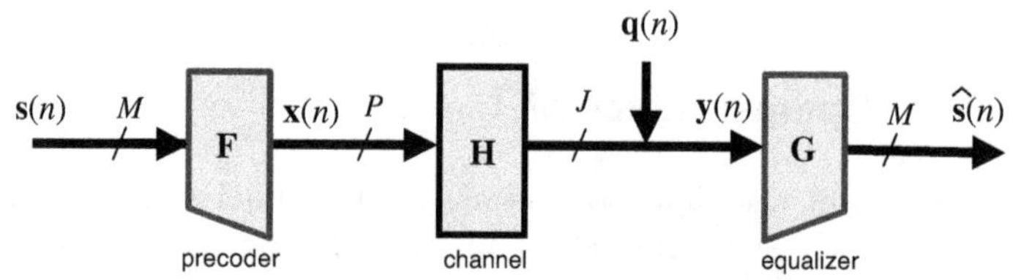

Figure 12.11. The transceiver with a rectangular channel.

To express the corresponding zero-forcing equalizer conveniently, consider again the channel in SVD form:

$$\mathbf{H} = \underbrace{\mathbf{U}_h}_{K\times K} \underbrace{\mathbf{\Sigma}_h}_{K\times P} \underbrace{\mathbf{V}_h^\dagger}_{P\times P}. \tag{12.92}$$

Note the sizes of the matrices; in particular $\mathbf{U}_h$ and $\mathbf{V}_h$ can be of different sizes. Proceeding as in Sec. 12.4.4 we have

$$\mathbf{HF} = \underbrace{\mathbf{U}_h}_{K\times K} \underbrace{\begin{bmatrix} [\mathbf{\Sigma}_h]_M \\ \mathbf{0} \end{bmatrix}}_{K\times M} \underbrace{\mathbf{\Sigma}_f}_{M\times M},$$

where $[\mathbf{\Sigma}_h]_M$ is the $M \times M$ leading principal submatrix of $\mathbf{\Sigma}_h$, containing the dominant singular values

$$\sigma_{h,0} \geq \sigma_{h,1} \geq \ldots \geq \sigma_{h,M-1}$$

Thus

$$(\mathbf{HF})^\dagger \mathbf{HF} = \mathbf{\Sigma}_f^2 [\mathbf{\Sigma}_h^2]_M.$$

The zero-forcing equalizer is therefore found to be

$$\begin{aligned} \mathbf{G} &= \Big((\mathbf{HF})^\dagger \mathbf{HF}\Big)^{-1} (\mathbf{HF})^\dagger \\ &= [\mathbf{\Sigma}_h]_M^{-2} \mathbf{\Sigma}_f^{-2} \mathbf{\Sigma}_f \left[[\mathbf{\Sigma}_h]_M \quad \mathbf{0} \right] \mathbf{U}_h^\dagger \\ &= [\mathbf{\Sigma}_h]_M^{-1} \mathbf{\Sigma}_f^{-1} [\mathbf{U}_h^\dagger]_{M\times K}, \end{aligned}$$

which is exactly as in Eq. (12.61), with the exception that $[\mathbf{U}_h^\dagger]_{M\times K}$ appears instead of $[\mathbf{U}_h^\dagger]_{M\times P}$. Thus the optimal transceiver structure has the form shown in Fig. 12.12, which is identical to the structure in Fig. 12.5 except for the sizes of $\mathbf{H}$ and $\mathbf{U}_h$. The diagonal matrices $\mathbf{\Sigma}_f$ and $\mathbf{\Sigma}_g$ are exactly as described in Theorem 12.1.

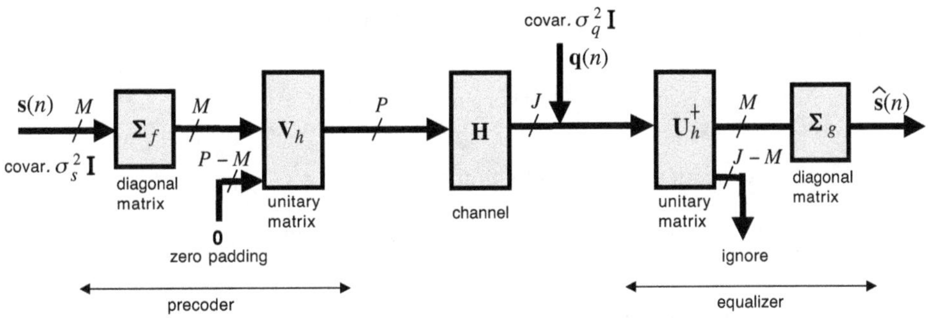

Figure 12.12. The optimal transceiver with rectangular channel. This is essentially identical to the optimal system shown in Fig. 12.5 for the square channel. The diagonal matrices $\boldsymbol{\Sigma}_f$ and $\boldsymbol{\Sigma}_g$ are as described in Theorem 12.1, and the unitary matrices $\mathbf{U}_h$ and $\mathbf{V}_h$ are from the channel SVD (12.92).

Problems

Note. Unless mentioned otherwise, assume $\sigma_s^2 = 1$.

12.1. Consider the MIMO channel

$$\mathbf{H} = \begin{bmatrix} 3 & 3 \\ 3 & 2 \end{bmatrix}$$

with noise variance $\sigma_q^2 = 0.01$.

1. Compute the singular values of the channel.
2. With the zero-forcing transceiver $\{\mathbf{F}, \mathbf{G}\}$ designed to minimize the mean square error as in Sec. 12.4, what is the mean square reconstuction error assuming the power is $p_0 = 2$?
3. Now assume $\mathbf{F}$ is the lazy precoder ($\mathbf{F} = \mathbf{I}$ so that $p_0 = 2$ as before), and $\mathbf{G}$ is the zero-forcing equalizer as in Sec. 12.3.1. What is the mean square reconstuction error?
4. What is the ratio of the MSE values obtained in the above two methods? This is the gain $\mathcal{G}$ obtained when going from a lazy precoder system to the optimal system.

12.2. Repeat Problem 12.1 for the channel

$$\mathbf{H} = \begin{bmatrix} 3 & 1 \\ 1 & 4 \end{bmatrix}.$$

Is the gain $\mathcal{G}$ higher or lower compared to Problem 12.1? Explain.

12.3. With all other quantities as in Problem 12.1, find an example of a channel $\mathbf{H}$ such that the gain $\mathcal{G}$ obtained when going from a lazy precoder system to the optimal system is exactly 1.5. Does there exist a channel such that this gain is 2.5 instead?

12.4. We showed that the transceiver system shown in Fig. 12.10(a) can be redrawn as in Fig. 12.10(c), where $\mathbf{F} = \mathbf{V}_h^\dagger \mathbf{F}_1$ and $\mathbf{G} = \mathbf{G}_1 \mathbf{U}_h$ $\mathbf{q}(n) = \mathbf{U}_h^\dagger \mathbf{q}_1(n)$.

1. Show that power at the output of $\mathbf{F}_1$ in Fig. 12.10(a) is equal to the power at the output of $\mathbf{F}$ in Fig. 12.10(c).
2. Assume $\{\mathbf{F}, \mathbf{G}\}$ is optimal for Fig. 12.10(c) in some sense (e.g., minimizing the mean square error), for a given power p_0 at the output of $\mathbf{F}$. Then show that $\{\mathbf{F}_1, \mathbf{G}_1\}$ is optimal for Fig. 12.10(a) for the same power p_0 at the output of $\mathbf{F}_1$.

12.5. Referring to Eq. (12.84), suppose we replace the unitary matrix $\mathbf{U}_f$ with $\mathbf{U}_{f,new}$ in Eq. (12.85) for some $M \times M$ unitary $\mathbf{U}_b$. Show that $\mathbf{U}_{f,new}$ remains unitary, and that $\mathbf{B}$ in Eq. (12.84) becomes $\mathbf{B}_{new} = \mathbf{U}_b^\dagger \mathbf{B} \mathbf{U}_b$.

12.6. Consider the transceiver of Fig. 12.2 with $P = M + L > M$. Assume $\mathbf{F}$ and $\mathbf{H}$ are fixed, with rank M. Let the SVD of the product $\mathbf{HF}$ be

$$\mathbf{HF} = \mathbf{U} \begin{bmatrix} \boldsymbol{\Sigma} \\ \mathbf{0} \end{bmatrix} \mathbf{V}^\dagger,$$

where $\mathbf{U}$ is $P \times P$ unitary, $\mathbf{V}$ is $M \times M$ unitary, and $\boldsymbol{\Sigma}$ is $M \times M$ diagonal with positive diagonal elements. Without loss of generality, write the equalizer $\mathbf{G}$ in the form $\mathbf{G} = \mathbf{G}_1\mathbf{U}^\dagger$, where $\mathbf{G}_1$ is $M \times P$. Suppose $\mathbf{G}_1$ is expressed in the partitioned form

$$\mathbf{G}_1 = \begin{bmatrix} \mathbf{G}_a & \mathbf{G}_b \end{bmatrix},$$

where $\mathbf{G}_a$ is $M \times M$ and $\mathbf{G}_b$ is $M \times L$.

1. Show that there is a unique choice of $\mathbf{G}_a$ which satisfies the zero-forcing condition. Find an expression for this $\mathbf{G}_a$.
2. Show that $\mathbf{G}_b$ can be chosen arbitrarily without affecting the zero-forcing property. So, $\mathbf{G}_b$ offers extra freedom in the design.
3. Let $\mathbf{R}_{\mathbf{qq}}$ denote the autocorrelation of the channel noise $\mathbf{q}$, and suppose we wish to minimize the mean square value of the reconstruction error (i.e., the trace of the autocorrelation of $\widehat{\mathbf{s}}(n) - \mathbf{s}(n)$) subject to the zero-forcing constraint. Then what is the optimal choice of the matrix $\mathbf{G}_b$? Use the orthogonality principle (Sec. F.2.1) cleverly. (A less elegant way would be to apply matrix calculus methods for noise canceling as in Ex. 20.19 in Chap. 20.)
4. For the special case where the channel noise satisfies $\mathbf{R}_{\mathbf{qq}} = \sigma_q^2\mathbf{I}$, show that the optimal $\mathbf{G}_b = \mathbf{0}$.

13

MMSE transceivers without zero forcing

13.1 Introduction

In Chap. 12 we assumed that the transceiver satisfies the zero-forcing (ZF) condition $\mathbf{GHF} = \mathbf{I}$. Under this condition we optimized the sum of mean square errors under an appropriate power constraint. The ZF condition however is a loss of generality. That is, if we minimize the error

$$\mathbf{e}(n) = \widehat{\mathbf{s}}(n) - \mathbf{s}(n) \tag{13.1}$$

in the mean square sense by optimizing the precoder $\mathbf{F}$ and the equalizer $\mathbf{G}$ jointly, then the solution is in general not such that $\mathbf{GHF} = \mathbf{I}$. For the special case of diagonal channels, the discussion in Sec. 11.3 (Chap. 11) already demonstrates this fact. In this chapter we study the optimal transceiver which minimizes the mean square error

$$\mathcal{E}_{reco} = E[\mathbf{e}^{\dagger}(n)\mathbf{e}(n)] \tag{13.2}$$

by jointly optimizing $\mathbf{F}$ and $\mathbf{G}$. This is called the **pure-MMSE** or simply the MMSE transceiver, to distinguish it from the ZF-MMSE transceiver of Chap. 12.

Chapter outline

The transceiver system is reproduced in Fig. 13.1. For generality we assume that the channel $\mathbf{H}$ is *rectangular*, with size $J \times P$. In Sec. 13.3 we derive the optimum equalizer $\mathbf{G}$ for any fixed pair of $\mathbf{F}$ and $\mathbf{H}$ matrices. This also yields an expression for the minimum value of the mean square error achievable for any fixed $\mathbf{F}$ and $\mathbf{H}$. In Secs. 13.4 and 13.5 we further optimize this error by minimizing

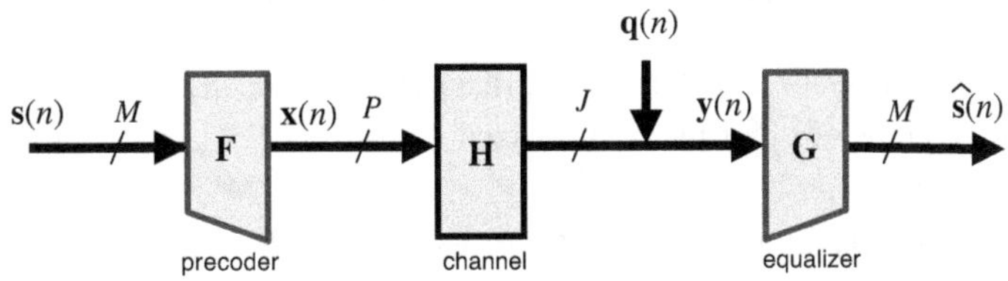

Figure 13.1. The channel $\mathbf{H}$ with precoder $\mathbf{F}$ and equalizer $\mathbf{G}$.

it with respect to $\mathbf{F}$. The complete MMSE trasnsceiver structure is summarized in Sec. 13.6 and compared with the ZF-MMSE transceiver of Chap. 12. The presentation in this chapter is based mostly on the clear exposition given by Li and Ding [1999]. Also see the pioneering work of Salz [1985].

Statistical assumptions As in earlier chapters, we assume that the signal $\mathbf{s}(n)$ and noise $\mathbf{q}(n)$ are jointly WSS, uncorrelated processes with zero mean. As explained in Sec. 12.2.1 the noise $\mathbf{q}(n)$ can be assumed, without loss of generality, to have the covariance

$$\mathbf{R_{qq}} = \sigma_q^2 \mathbf{I}_J. \tag{13.3}$$

The appropriate assumption one makes on the signal covariance $\mathbf{R_{ss}}$ depends on the application in hand. All the main results in this chapter assume

$$\mathbf{R_{ss}} = \sigma_s^2 \mathbf{I}_M. \tag{13.4}$$

In Sec. 13.3 we shall retain the general notations $\mathbf{R_{qq}}$ and $\mathbf{R_{ss}}$, but in all other sections the specific forms (13.3) and (13.4) are used in order to keep the discussions simple.

13.2 Formulation of the problem

The reconstruction error can be expressed as

$$\mathbf{e}(n) = \widehat{\mathbf{s}}(n) - \mathbf{s}(n) = \mathbf{Gy}(n) - \mathbf{s}(n) = \mathbf{G}\Big(\mathbf{HFs}(n) + \mathbf{q}(n)\Big) - \mathbf{s}(n)$$

so that

$$\mathbf{e}(n) = \Big(\mathbf{GHF} - \mathbf{I}\Big)\mathbf{s}(n) + \mathbf{Gq}(n). \tag{13.5}$$

Since $\mathbf{s}(n)$ and $\mathbf{q}(n)$ are uncorrelated zero-mean processes, the error covariance matrix is

$$\mathbf{R_{ee}} = (\mathbf{GHF} - \mathbf{I})\mathbf{R_{ss}}(\mathbf{GHF} - \mathbf{I})^\dagger + \mathbf{GR_{qq}G}^\dagger. \tag{13.6}$$

The mean square error $\mathcal{E}_{mse}$ is the trace of this:

$$\mathcal{E}_{mse} = \text{Tr}\,\mathbf{R_{ee}} = \text{Tr}\,\Big((\mathbf{GHF}-\mathbf{I})\mathbf{R_{ss}}(\mathbf{GHF}-\mathbf{I})^\dagger + \mathbf{GR_{qq}G}^\dagger\Big). \tag{13.7}$$

Substituting Eqs. (13.3) and (13.4) the optimization problem to be considered in this chapter can be summarized as follows:

Problem 1. For a fixed $J \times P$ channel $\mathbf{H}$ optimize the $P \times M$ precoder $\mathbf{F}$ and $M \times J$ equalizer $\mathbf{G}$ jointly such that

$$\mathcal{E}_{mse} = \text{Tr}\;\sigma_s^2\Big((\mathbf{GHF}-\mathbf{I})(\mathbf{GHF}-\mathbf{I})^\dagger + \frac{\sigma_q^2}{\sigma_s^2}\mathbf{GG}^\dagger\Big) \tag{13.8}$$

is minimized subject to the power constraint

$$\sigma_s^2\text{Tr}\,\Big(\mathbf{FF}^\dagger\Big) = p_0, \tag{13.9}$$

where σ_s^2, σ_q^2, and p_0 are fixed. $\bigtriangledown\bigtriangledown\bigtriangledown$

In the absence of noise ($\mathbf{R_{qq}} = \mathbf{0}$) it is obvious from Eq. (13.7) that the zero forcing condition $\mathbf{GHF} = \mathbf{I}$ is the best choice, as it leads to zero error in reconstruction. With noise present, the optimal $\mathbf{G}$ is different, and has the expression (13.14) derived in the next section.

13.3 MMSE equalizer for fixed precoder

In this section we will assume that the precoder $\mathbf{F}$ (and channel $\mathbf{H}$) are fixed and solve for the optimal equalizer $\mathbf{G}$, which minimizes the mean squared reconstruction error. The resulting mean square error then depends only on the precoder $\mathbf{F}$, which can be further optimized. The problem of optimizing $\mathbf{G}$ for fixed $\mathbf{F}$ can be regarded as a Wiener filtering problem (Appendix F). However, we provide a self-contained derivation here, to obtain direct insight into the problem.

In Fig. 13.2 the transceiver is redrawn in a slightly simplified form, with the matrix $\mathbf{C}$ defined as

$$\mathbf{C} = \mathbf{HF}. \tag{13.10}$$

So the immediate goal is to optimize $\mathbf{G}$ for fixed $\mathbf{C}$ so that

$$\mathcal{E}_{mse} = \text{Tr}\,\Big((\mathbf{GC}-\mathbf{I})\mathbf{R_{ss}}(\mathbf{GC}-\mathbf{I})^\dagger + \mathbf{GR_{qq}G}^\dagger\Big) \tag{13.11}$$

is minimized. For a fixed nonzero $\mathbf{R_{qq}}$, and a fixed set of matrices $\mathbf{C}$ and $\mathbf{R_{ss}}$, the best $\mathbf{G}$ can be obtained by using the *orthogonality principle* (Sec. F.2.1 of Appendix F) as we show below. A second approach based on variational techniques (matrix calculus) is given in Problem 13.8.

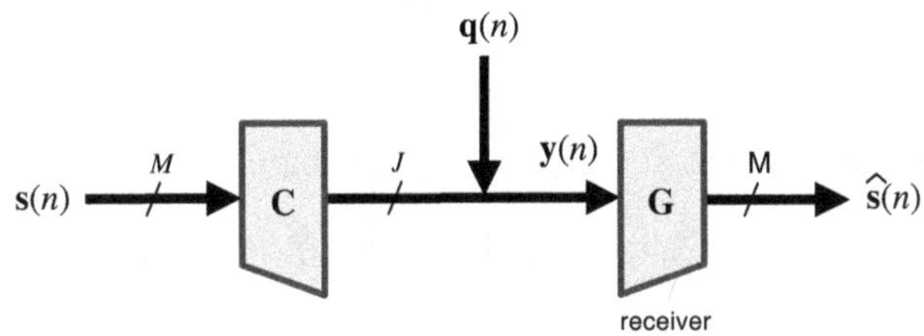

Figure 13.2. Redrawing of Fig. 13.1, where $\mathbf{C} = \mathbf{HF}$.

♠**Lemma 13.1.** *Optimal equalizer for fixed precoder.* The optimal equalizer $\mathbf{G}$ for fixed $\mathbf{C}$ is given by

$$\mathbf{G}_{mmse} = \mathbf{R_{ss}}\mathbf{C}^\dagger\left(\mathbf{C}\mathbf{R_{ss}}\mathbf{C}^\dagger + \mathbf{R_{qq}}\right)^{-1}, \tag{13.12}$$

and the corresponding mean square error is

$$\mathcal{E}_{mse} = \text{Tr}\left(\mathbf{R_{ss}} - \mathbf{R_{ss}}\mathbf{C}^\dagger\left(\mathbf{C}\mathbf{R_{ss}}\mathbf{C}^\dagger + \mathbf{R_{qq}}\right)^{-1}\mathbf{C}\mathbf{R_{ss}}\right), \tag{13.13}$$

whenever the indicated matrix inverse exists. ◇

The assumption on the existence of the indicated inverse is reasonable because $\mathbf{R_{qq}}$ can be assumed nonsingular (hence positive definite). Since $\mathbf{C}\mathbf{R_{ss}}\mathbf{C}^\dagger$ is positive (semi)definite, the sum $\mathbf{C}\mathbf{R_{ss}}\mathbf{C}^\dagger + \mathbf{R_{qq}}$ is positive definite, and its inverse exists. Substituting $\mathbf{C} = \mathbf{HF}$, the optimum solution given in the lemma takes the form

$$\mathbf{G}_{mmse} = \mathbf{R_{ss}}\mathbf{F}^\dagger\mathbf{H}^\dagger\left(\mathbf{H}\mathbf{F}\mathbf{R_{ss}}\mathbf{F}^\dagger\mathbf{H}^\dagger + \mathbf{R_{qq}}\right)^{-1}. \tag{13.14}$$

The corresponding minimized mean square error is

$$\mathcal{E}_{mse} = \text{Tr}\left(\mathbf{R_{ss}} - \mathbf{R_{ss}}\mathbf{F}^\dagger\mathbf{H}^\dagger\left(\mathbf{H}\mathbf{F}\mathbf{R_{ss}}\mathbf{F}^\dagger\mathbf{H}^\dagger + \mathbf{R_{qq}}\right)^{-1}\mathbf{H}\mathbf{F}\mathbf{R_{ss}}\right). \tag{13.15}$$

Proof of Lemma 13.1. Since the estimate $\widehat{\mathbf{s}}(n)$ of $\mathbf{s}(n)$ is a linear function of the observed signal $\mathbf{y}(n)$, we can apply the orthogonality principle (Appendix F, Sec. F.2.1). This says that the optimum $\mathbf{G}$ is such that the error $\mathbf{e}(n) = \widehat{\mathbf{s}}(n) - \mathbf{s}(n)$ is orthogonal to the observation $\mathbf{y}(n)$, that is,

$$E[\mathbf{y}(n)\mathbf{e}^\dagger(n)] = \mathbf{0}, \quad \text{for all } n. \tag{13.16}$$

Since the signals are WSS and the matrices $\mathbf{C}$ and $\mathbf{G}$ are constants with no memory, we will drop the argument (n) everywhere. Substituting $\mathbf{e} = (\widehat{\mathbf{s}} - \mathbf{s})$ and $\mathbf{y} = (\mathbf{Cs} + \mathbf{q})$ in Eq. (13.16) we get

$$E[(\mathbf{Cs} + \mathbf{q})\widehat{\mathbf{s}}^\dagger] = E[(\mathbf{Cs} + \mathbf{q})\mathbf{s}^\dagger].$$

Using the fact that $\widehat{\mathbf{s}} = \mathbf{G}(\mathbf{Cs} + \mathbf{q})$ this becomes

$$E[(\mathbf{Cs} + \mathbf{q})(\mathbf{s}^\dagger\mathbf{C}^\dagger\mathbf{G}^\dagger + \mathbf{q}^\dagger\mathbf{G}^\dagger)] = E[(\mathbf{Cs} + \mathbf{q})\mathbf{s}^\dagger].$$

Since $\mathbf{s}$ and $\mathbf{q}$ are zero-mean and uncorrelated vectors, this reduces to

$$\mathbf{C}\mathbf{R}_{\mathbf{ss}}\mathbf{C}^\dagger\mathbf{G}^\dagger + \mathbf{R}_{\mathbf{qq}}\mathbf{G}^\dagger = \mathbf{C}\mathbf{R}_{\mathbf{ss}}$$

from which it follows that $\mathbf{G}^\dagger = (\mathbf{C}\mathbf{R}_{\mathbf{ss}}\mathbf{C}^\dagger + \mathbf{R}_{\mathbf{qq}})^{-1}\mathbf{C}\mathbf{R}_{\mathbf{ss}}$. The optimal $\mathbf{G}$ is therefore as in (13.12) indeed. With $\mathbf{G}$ chosen thus, the error $\mathbf{e}$ is orthogonal to the observations $\mathbf{y}$. So it is also orthogonal to $\widehat{\mathbf{s}} = \mathbf{Gy}$. Thus $E[\mathbf{e}\widehat{\mathbf{s}}^\dagger] = \mathbf{0}$, and the error covariance is

$$\begin{aligned} E[\mathbf{e}\mathbf{e}^\dagger] = E[\mathbf{e}(\widehat{\mathbf{s}} - \mathbf{s})^\dagger] &= -E[\mathbf{e}\mathbf{s}^\dagger] \\ &= -E\Big((\mathbf{GC} - \mathbf{I})\mathbf{s} + \mathbf{Gq}\Big)\mathbf{s}^\dagger = -(\mathbf{GC} - \mathbf{I})\mathbf{R}_{\mathbf{ss}}, \end{aligned}$$

where we have also used the fact that $E[\mathbf{q}\mathbf{s}^\dagger] = \mathbf{0}$. Substituting from (13.12),

$$\mathbf{R}_{\mathbf{ee}} = E[\mathbf{e}\mathbf{e}^\dagger] = \mathbf{R}_{\mathbf{ss}} - \mathbf{R}_{\mathbf{ss}}\mathbf{C}^\dagger\Big(\mathbf{C}\mathbf{R}_{\mathbf{ss}}\mathbf{C}^\dagger + \mathbf{R}_{\mathbf{qq}}\Big)^{-1}\mathbf{C}\mathbf{R}_{\mathbf{ss}}. \qquad (13.17)$$

This proves Eq. (13.13). $\bigtriangledown\bigtriangledown\bigtriangledown$

13.4 Formulating the optimal precoder problem

For fixed precoder $\mathbf{F}$ and channel $\mathbf{H}$ we have shown that the smallest possible value of the reconstruction error is given by Eq. (13.15). The next step is to minimize this expression with respect to $\mathbf{F}$ subject to the power constraint. We will do this under the assumption that the signal and noise covariances have the simple form

$$\mathbf{R}_{\mathbf{ss}} = \sigma_s^2\mathbf{I}_M, \quad \mathbf{R}_{\mathbf{qq}} = \sigma_q^2\mathbf{I}_J. \qquad (13.18)$$

With this assumption, the MMSE equalizer in Eq. (13.12) becomes

$$\mathbf{G}_{mmse} = \mathbf{C}^\dagger\Big(\mathbf{C}\mathbf{C}^\dagger + \frac{\sigma_q^2}{\sigma_s^2}\mathbf{I}_J\Big)^{-1}, \qquad (13.19)$$

and the error covariance (13.17) becomes

$$\mathbf{R}_{\mathbf{ee}} = \sigma_s^2\bigg(\mathbf{I}_M - \mathbf{C}^\dagger\Big(\mathbf{C}\mathbf{C}^\dagger + \frac{\sigma_q^2}{\sigma_s^2}\mathbf{I}_J\Big)^{-1}\mathbf{C}\bigg). \qquad (13.20)$$

Substituting $\mathbf{C} = \mathbf{HF}$ we see that the problem we have to solve is as follows:

Problem 2. For a fixed channel $\mathbf{H}$ optimize the precoder $\mathbf{F}$ such that

$$\mathcal{E}_{mse} = \mathrm{Tr}\,\sigma_s^2\left(\mathbf{I}_M - \mathbf{F}^\dagger\mathbf{H}^\dagger\left(\mathbf{HFF}^\dagger\mathbf{H}^\dagger + \frac{\sigma_q^2}{\sigma_s^2}\mathbf{I}_J\right)^{-1}\mathbf{HF}\right) \tag{13.21}$$

is minimized subject to the power constraint

$$\sigma_s^2\mathrm{Tr}\left(\mathbf{FF}^\dagger\right) = p_0, \tag{13.22}$$

where σ_s^2, σ_q^2, and p_0 are fixed. $\bigtriangledown\bigtriangledown\bigtriangledown$

13.4.1 Rewriting the error covariance

Before proceeding further it is convenient to rearrange the expressions for $\mathbf{G}$ and $\mathbf{R_{ee}}$ in Eqs. (13.19) and (13.20). For this we first observe the identity

$$\left(\mathbf{C}^\dagger\mathbf{C} + \frac{\sigma_q^2}{\sigma_s^2}\mathbf{I}_M\right)\mathbf{C}^\dagger = \mathbf{C}^\dagger\left(\mathbf{CC}^\dagger + \frac{\sigma_q^2}{\sigma_s^2}\mathbf{I}_J\right), \tag{13.23}$$

which is verified just by multiplying out the matrices. From this it follows that

$$\mathbf{C}^\dagger\left(\mathbf{CC}^\dagger + \frac{\sigma_q^2}{\sigma_s^2}\mathbf{I}_J\right)^{-1} = \left(\mathbf{C}^\dagger\mathbf{C} + \frac{\sigma_q^2}{\sigma_s^2}\mathbf{I}_M\right)^{-1}\mathbf{C}^\dagger. \tag{13.24}$$

So we can rewrite the optimal equalizer (13.19) as

$$\mathbf{G}_{mmse} = \left(\mathbf{C}^\dagger\mathbf{C} + \frac{\sigma_q^2}{\sigma_s^2}\mathbf{I}_M\right)^{-1}\mathbf{C}^\dagger, \tag{13.25}$$

and furthermore the error covariance (13.20) can be rearranged as

$$\begin{aligned}\mathbf{R_{ee}} &= \sigma_s^2\left(\mathbf{I}_M - \mathbf{C}^\dagger\left(\mathbf{CC}^\dagger + \frac{\sigma_q^2}{\sigma_s^2}\mathbf{I}_J\right)^{-1}\mathbf{C}\right)\\ &= \sigma_s^2\left(\mathbf{I}_M - \left(\mathbf{C}^\dagger\mathbf{C} + \frac{\sigma_q^2}{\sigma_s^2}\mathbf{I}_M\right)^{-1}\mathbf{C}^\dagger\mathbf{C}\right)\\ &= \sigma_s^2\left(\mathbf{C}^\dagger\mathbf{C} + \frac{\sigma_q^2}{\sigma_s^2}\mathbf{I}_M\right)^{-1}\left(\mathbf{C}^\dagger\mathbf{C} + \frac{\sigma_q^2}{\sigma_s^2}\mathbf{I}_M - \mathbf{C}^\dagger\mathbf{C}\right).\end{aligned}$$

This shows that the error covariance with the MMSE equalizer can be written as

$$\mathbf{R_{ee}} = \sigma_q^2\left(\mathbf{C}^\dagger\mathbf{C} + \frac{\sigma_q^2}{\sigma_s^2}\mathbf{I}_M\right)^{-1} = \sigma_s^2\left(\mathbf{I}_M + \frac{\sigma_s^2}{\sigma_q^2}\mathbf{C}^\dagger\mathbf{C}\right)^{-1}$$

Putting back the meaning of $\mathbf{C}$, namely $\mathbf{C} = \mathbf{HF}$, we get

$$\mathbf{R_{ee}} = \sigma_s^2\Big(\mathbf{I}_M + \frac{\sigma_s^2}{\sigma_q^2}\mathbf{F}^\dagger\mathbf{H}^\dagger\mathbf{HF}\Big)^{-1}.$$

Since the $P\times P$ Hermitian matrix $\mathbf{H}^\dagger\mathbf{H}$ can be diagonalized by a unitary matrix $\mathbf{V}_h$ we can write

$$\mathbf{H}^\dagger\mathbf{H} = \mathbf{V}_h\boldsymbol{\Sigma}_h^2\mathbf{V}_h^\dagger, \tag{13.26}$$

where

$$\boldsymbol{\Sigma}_h^2 = \text{diag}\,[\,\sigma_{h,0}^2 \quad \sigma_{h,1}^2 \quad \ldots \quad \sigma_{h,P-1}^2\,], \tag{13.27}$$

with $\sigma_{h,k} \geq 0$ representing the singular values of the channel. Thus

$$\mathbf{R_{ee}} = \sigma_s^2\Big(\mathbf{I}_M + \frac{\sigma_s^2}{\sigma_q^2}\underbrace{\mathbf{F}^\dagger\mathbf{V}_h\boldsymbol{\Sigma}_h^2}_{M\times P}\underbrace{\mathbf{V}_h^\dagger\mathbf{F}}_{P\times M}\Big)^{-1}. \tag{13.28}$$

We now use the trace identity (see Appendix 13.A at the end of the chapter)

$$\text{Tr}\,(\mathbf{I}_M + \mathbf{BA})^{-1} = (M-P) + \text{Tr}\,(\mathbf{I}_P + \mathbf{AB})^{-1}, \tag{13.29}$$

which holds for any two matrices $\mathbf{A}$ and $\mathbf{B}$ with sizes $P\times M$ and $M\times P$, respectively, to obtain

$$\text{Tr}\,\mathbf{R_{ee}} = \sigma_s^2(M-P) + \sigma_s^2\,\text{Tr}\,\Big(\mathbf{I}_P + \frac{\sigma_s^2}{\sigma_q^2}\mathbf{V}_h^\dagger\mathbf{FF}^\dagger\mathbf{V}_h\boldsymbol{\Sigma}_h^2\Big)^{-1}. \tag{13.30}$$

For fixed $\mathbf{H}$ we have to minimize this by optimizing the precoder $\mathbf{F}$.

13.4.2 Rewriting the error covariance

The first term in Eq. (13.30) is just a constant. The trace in the second term can be written as

$$\phi = \text{Tr}\,\Big(\mathbf{I}_P + \mathbf{Q}\boldsymbol{\Lambda}\Big)^{-1},$$

where

$$\mathbf{Q} = \alpha\mathbf{V}_h^\dagger\mathbf{FF}^\dagger\mathbf{V}_h \quad \text{and} \quad \boldsymbol{\Lambda} = \boldsymbol{\Sigma}_h^2 \tag{13.31}$$

are $P\times P$ matrices, and

$$\alpha = \sigma_s^2/\sigma_q^2. \tag{13.32}$$

Since $\mathbf{V}_h$ depends only on the channel, the problem of optimizing the precoder $\mathbf{F}$ is the same as optimizing $\mathbf{Q}$. From this we can identify $\mathbf{FF}^\dagger$ using Eqs. (13.31). The optimal precoder $\mathbf{F}$ can then be identified as any $P\times M$ square root of $\mathbf{FF}^\dagger$. The power constraint (13.22) can be rewritten as as $\text{Tr}\,\mathbf{Q} = c$, where

$$c = \frac{\alpha p_0}{\sigma_q^2}. \tag{13.33}$$

Summarizing, the optimization problem to be solved in order to identify the optimal precoder is the following:

Problem 3. Given the $P \times P$ diagonal matrix $\mathbf{\Lambda}$ with non-negative diagonal elements, optimize the $P \times P$ positive semidefinite matrix $\mathbf{Q}$ such that

$$\phi = \text{Tr}\left(\mathbf{I}_P + \mathbf{Q\Lambda}\right)^{-1} \tag{13.34}$$

is minimized subject to the constraint

$$\text{Tr}\,\mathbf{Q} = c \tag{13.35}$$

for some fixed $c > 0$. $\bigtriangledown \bigtriangledown \bigtriangledown$

13.5 Solution to the optimal precoder problem

It only remains to solve the optimization problem described by Eqs. (13.34) and (13.35). We will first show that there is *no loss of generality in constraining the solution* $\mathbf{Q}$ *to be a diagonal matrix.* This is a crucial observation, as it simplifies the problem a great deal. It was first pointed out by Witsenhausen in an appendix of a classic paper by Salz on optimization of communication systems [Salz, 1985]. Witsenhausen's proof, given below, uses the idea of Schur convexity reviewed in detail in Chap. 21.

♠**Lemma 13.2.** *Optimal* $\mathbf{Q}$ *is diagonal.* In the optimization problem described by Eqs. (13.34) and (13.35), the solution $\mathbf{Q}$ can be assumed, without loss of generality, to be a diagonal matrix. $\diamondsuit$

Proof. Using $\text{Tr}\,(\mathbf{I} + \mathbf{BA}) = \text{Tr}\,(\mathbf{I} + \mathbf{AB})$ for square matrices $\mathbf{A}$ and $\mathbf{B}$ (Appendix 13.A) we can write Eq. (13.34) as

$$\phi = \text{Tr}\left(\mathbf{I}_P + \mathbf{Q\Lambda}\right)^{-1} = \text{Tr}\left(\mathbf{I}_P + \mathbf{\Lambda}^{1/2}\mathbf{Q}\mathbf{\Lambda}^{1/2}\right)^{-1}.$$

With η_i denoting the eigenvalues of $\mathbf{\Lambda}^{1/2}\mathbf{Q}\mathbf{\Lambda}^{1/2}$, the eigenvalues of $\mathbf{I}_P + \mathbf{\Lambda}^{1/2}\mathbf{Q}\mathbf{\Lambda}^{1/2}$ are $1+\eta_i$, so the eigenvalues of $(\mathbf{I}_P+\mathbf{\Lambda}^{1/2}\mathbf{Q}\mathbf{\Lambda}^{1/2})^{-1}$ are $1/(1+\eta_i)$. Since the trace of a matrix is the sum of its eigenvalues, it then follows that

$$\phi = \sum_{i=0}^{P-1} \frac{1}{1+\eta_i} \tag{13.36}$$

Using the diagonal elements of $\mathbf{Q}$ define the diagonal matrix

$$\mathbf{Q}_d = \text{diag}\left[\, q_{00} \quad q_{11} \quad \ldots \quad q_{P-1,P-1}\,\right]. \tag{13.37}$$

Recall now that the eigenvalues of a Hermitian matrix majorize its diagonal elements (Sec. 21.5 of Chap. 21). Since $\mathbf{\Lambda}^{1/2}\mathbf{Q}\mathbf{\Lambda}^{1/2}$ is Hermitian we therefore have

$$\left[\,\eta_0 \quad \eta_1 \quad \ldots \quad \eta_{P-1}\,\right] \succ \left[\,\lambda_0 q_{00} \quad \lambda_1 q_{11} \quad \ldots \quad \lambda_{P-1} q_{P-1,P-1}\,\right]. \tag{13.38}$$

Next, since the function $1/(1+x)$ is convex (i.e., its second derivative is positive), it follows from Theorem 21.3 (Chap. 21) that

$$\sum_{i=0}^{P-1} \frac{1}{1+x_i} \tag{13.39}$$

is Schur-convex in the vector $[x_0 \quad x_1 \quad \ldots \quad x_{P-1}]$. This shows that

$$\sum_{i=0}^{P-1} \frac{1}{1+\eta_i} \geq \sum_{i=0}^{P-1} \frac{1}{1+\lambda_i q_{ii}} \tag{13.40}$$

That is, given any $\mathbf{Q}$, replacing it with the diagonal matrix $\mathbf{Q}_d$ can only decrease the objective function ϕ. $\bigtriangledown \bigtriangledown \bigtriangledown$

Without loss of generality, we can therefore replace $\mathbf{Q}$ with a *diagonal* matrix and optimize its diagonal elements. With $\mathbf{Q}$ restricted to be a diagonal matrix, the problem described by Eqs. (13.34) and (13.35) reduces to one of finding the diagonal elements q_{kk}:

Problem 4. Find q_{kk} in order to minimize

$$\phi = \sum_{k=0}^{P-1} \frac{1}{1+\sigma_{h,k}^2 q_{kk}} \tag{13.41}$$

subject to the constraints

$$\sum_{k=0}^{P-1} q_{kk} = c \tag{13.42}$$

and

$$q_{kk} \geq 0, \quad 0 \leq k \leq P-1, \tag{13.43}$$

where $c > 0$ is fixed and $\sigma_{h,k}$ are fixed channel singular values. $\bigtriangledown \bigtriangledown \bigtriangledown$

Thus we have a constrained optimization problem, with one equality constraint and P inequality constraints. This problem is solved in Chap. 22, where we used it as an example to demonstrate the KKT theory of constrained optimization. The solution has the form

$$q_{kk} = \begin{cases} \dfrac{D}{\sigma_{h,k}} - \dfrac{1}{\sigma_{h,k}^2} & 0 \leq k \leq K-1 \\ 0 & \text{otherwise,} \end{cases} \tag{13.44}$$

assuming the ordering $\sigma_{h,0} \geq \sigma_{h,1} \geq \ldots$ Here $K \leq P$ is an appropriately chosen integer and D is a constant such that the equality constraint (13.42) is satisfied. We next elaborate on the choice of these constants.

13.5.1 Identification of the constants K and D

For any fixed K, the constant D is readily identified using the power constraint (13.42). Thus

$$c = \sum_{k=0}^{K-1} q_{kk} = D \sum_{k=0}^{K-1} \frac{1}{\sigma_{h,k}} - \sum_{k=0}^{K-1} \frac{1}{\sigma_{h,k}^2}$$

from which we obtain

$$D = \frac{c + \sum_{k=0}^{K-1} \frac{1}{\sigma_{h,k}^2}}{\sum_{k=0}^{K-1} \frac{1}{\sigma_{h,k}}} = \frac{\frac{p_0}{\sigma_q^2} + \sum_{k=0}^{K-1} \frac{1}{\sigma_{h,k}^2}}{\sum_{k=0}^{K-1} \frac{1}{\sigma_{h,k}}} \tag{13.45}$$

For fixed K, it also follows from the discussions of Sec. 22.4 (Chap. 22) that the channel singular values should be numbered such that

$$\sigma_{h,0} \geq \sigma_{h,1} \geq \ldots \geq \sigma_{h,P-1} \geq 0. \tag{13.46}$$

That is, the first K dominant singular values of the channel should be retained in order to minimize ϕ. Thus, in order to complete the solution, it only remains to identify the integer K in Eq. (13.44).

Choice of K. We now argue that it is best to choose K to be the largest integer subject to $K \leq M$ such that $q_{kk} \geq 0$ for all k. For this examine again the matrix $\mathbf{Q}$ defined in Eq. (13.31). Since $\mathbf{F}$ is $P \times M$, this matrix therefore has rank $\leq M$. Thus the quantity K in Eq. (13.44) has to be such that

$$K \leq M. \tag{13.47}$$

From Eq. (13.44) we see that the nonzero q_{kk} have the form

$$q_{kk} = \frac{1}{\sigma_{h,k}}\Big(D - \frac{1}{\sigma_{h,k}}\Big), \quad 0 \leq k \leq K-1. \tag{13.48}$$

As k increases in the range $0 \leq k \leq K-1$, the quantity $1/\sigma_{h,k}$ increases (see Eq. (13.46)), which shows that it is possible for q_{kk} to be negative for some k if K is too large. So the choice $K = M$ may not always be possible. If the power p_0 is large enough to make D in Eq. (13.45) sufficiently large, then we can ensure that $q_{kk} \geq 0$ for $0 \leq k \leq M-1$. But since we cannot always assume this to be the case, it is possible that $K < M$ in some cases. Notice here that D itself depends on K (from Eq. (13.45)). As explained in Sec. 22.4 of Chap. 22, it is best to choose K to be the largest integer (subject to $K \leq M$) such that $q_{kk} \geq 0$ in Eq. (13.48). $\bigtriangledown \bigtriangledown \bigtriangledown$

13.5.2 Expression for minimized mean square error

With the optimal q_{kk} given by Eq. (13.44) the minimized answer ϕ in the optimization problem (13.41)–(13.43) becomes

$$\phi_{min} = \sum_{k=0}^{P-1} \frac{1}{1+\sigma_{h,k}^2 q_{kk}} = \sum_{k=0}^{K-1} \frac{1}{1+\sigma_{h,k}^2 q_{kk}} + (P-K).$$

Substituting from Eq. (13.48) this simplifies to

$$\phi_{min} = \frac{1}{D}\sum_{k=0}^{K-1} \frac{1}{\sigma_{h,k}} + (P-K).$$

Substituting for D from Eq. (13.45), this yields

$$\phi_{min} = \frac{\left(\sum_{k=0}^{K-1} \frac{1}{\sigma_{h,k}}\right)^2}{\frac{p_0}{\sigma_q^2} + \sum_{k=0}^{K-1} \frac{1}{\sigma_{h,k}^2}} + (P-K). \tag{13.49}$$

The minimized mean square error for the transceiver can now be obtained from Eq. (13.30), which is reproduced below:

$$\operatorname{Tr}\mathbf{R_{ee}} = \sigma_s^2(M-P) + \sigma_s^2 \underbrace{\operatorname{Tr}\left(\mathbf{I}_P + \frac{\sigma_s^2}{\sigma_q^2}\mathbf{V}_h^\dagger \mathbf{F}\mathbf{F}^\dagger \mathbf{V}_h \mathbf{\Sigma}_h^2\right)^{-1}}_{\text{recall this is } \phi}.$$

Substituting the value of ϕ_{min} from Eq. (13.49), we therefore obtain

$$\mathcal{E}_{mmse} = \frac{\sigma_s^2\sigma_q^2\left(\sum_{k=0}^{K-1} \frac{1}{\sigma_{h,k}}\right)^2}{p_0 + \sigma_q^2 \sum_{k=0}^{K-1} \frac{1}{\sigma_{h,k}^2}} + \sigma_s^2(M-K). \tag{13.50}$$

It is interesting to compare this expression with the mean square error of the jointly optimal transceiver for a SISO channel given in Chap. 10 (Eq. (10.35)). The expression is (with $S_{ss}(e^{j\omega}) = \sigma_s^2$ and $S_{qq}(e^{j\omega}) = \sigma_q^2$):

$$\mathcal{E}_{mmse} = \frac{\sigma_s^2\sigma_q^2\left(\int_{\mathcal{F}} \frac{1}{|H(e^{j\omega})|}\frac{d\omega}{2\pi}\right)^2}{p_0 + \sigma_q^2 \int_{\mathcal{F}} \frac{1}{|H(e^{j\omega})|^2}\frac{d\omega}{2\pi}} + \sigma_s^2 \int_{\mathcal{F}^c} \frac{d\omega}{2\pi}. \tag{13.51}$$

Thus, if we replace $|H(e^{j\omega})|$ in the SISO system with $\sigma_{h,k}$ (singular values of the MIMO memoryless channel) and replace integrals with summations, we pass from Eq. (13.51) to Eq. (13.50).

13.6 Structure of the MMSE transceiver

Once the $P \times P$ diagonal matrix $\mathbf{Q}$ is identified as described in Sec. 13.5, the optimal $M \times P$ precoder $\mathbf{F}$ can be obtained from Eq. (13.31) which is reproduced below:

$$\mathbf{Q} = \alpha \mathbf{V}_h^\dagger \mathbf{F}\mathbf{F}^\dagger \mathbf{V}_h. \tag{13.52}$$

Since $\mathbf{V}_h$ is unitary, this yields

$$\mathbf{F}\mathbf{F}^\dagger = \frac{1}{\alpha}\mathbf{V}_h \mathbf{Q} \mathbf{V}_h^\dagger, \tag{13.53}$$

where $\alpha = \sigma_s^2/\sigma_q^2$. Recall also that $\mathbf{V}_h$ is the unitary matrix which diagonalizes $\mathbf{H}^\dagger\mathbf{H}$, where $\mathbf{H}$ is the channel:

$$\mathbf{H}^\dagger\mathbf{H} = \mathbf{V}_h \boldsymbol{\Sigma}_h^2 \mathbf{V}_h^\dagger, \quad \text{i.e.,} \quad \mathbf{V}_h^\dagger \mathbf{H}^\dagger \mathbf{H} \mathbf{V}_h = \boldsymbol{\Sigma}_h^2. \tag{13.54}$$

Only the first M diagonal elements of the $P \times P$ diagonal matrix $\mathbf{Q}$ are possibly nonzero. So, from Eq. (13.53), we can identify the $P \times M$ optimal precoder $\mathbf{F}$ as

$$\mathbf{F} = \frac{\sigma_q}{\sigma_s}\mathbf{V}_h \begin{bmatrix} \sqrt{q_{00}} & 0 & \ldots & 0 \\ 0 & \sqrt{q_{11}} & \ldots & 0 \\ \vdots & \vdots & \ddots & \vdots \\ 0 & 0 & \ldots & \sqrt{q_{M-1,M-1}} \\ 0 & 0 & \ldots & 0 \\ \vdots & \vdots & \ddots & \vdots \\ 0 & 0 & \ldots & 0 \end{bmatrix} = \mathbf{V}_h \begin{bmatrix} \boldsymbol{\Sigma}_f \\ \mathbf{0} \end{bmatrix}, \tag{13.55}$$

where

$$\boldsymbol{\Sigma}_f = \frac{\sigma_q}{\sigma_s} \begin{bmatrix} \sqrt{q_{00}} & 0 & \ldots & 0 \\ 0 & \sqrt{q_{11}} & \ldots & 0 \\ \vdots & \vdots & \ddots & \vdots \\ 0 & 0 & \ldots & \sqrt{q_{M-1,M-1}} \end{bmatrix}. \tag{13.56}$$

The optimal equalizer or receiver matrix $\mathbf{G}$ is given by Eq. (13.12) with the substitutions $\mathbf{R}_{\mathbf{ss}} = \sigma_s^2 \mathbf{I}_M$ and $\mathbf{R}_{\mathbf{qq}} = \sigma_q^2 \mathbf{I}_J$, that is,

$$\mathbf{G} = \mathbf{C}^\dagger \left(\mathbf{C}\mathbf{C}^\dagger + \frac{\sigma_q^2}{\sigma_s^2}\mathbf{I}_J \right)^{-1}, \tag{13.57}$$

where $\mathbf{C} = \mathbf{H}\mathbf{F}$. This can also be written as

$$\mathbf{G} = \left(\mathbf{C}^\dagger \mathbf{C} + \frac{\sigma_q^2}{\sigma_s^2}\mathbf{I}_M \right)^{-1} \mathbf{C}^\dagger. \tag{13.58}$$

Substituting $\mathbf{C} = \mathbf{H}\mathbf{F}$ and using Eqs. (13.54) and (13.55), this simplifies to

$$\mathbf{G} = \frac{\sigma_s}{\sigma_q} \begin{bmatrix} z_0 & 0 & \ldots & 0 \\ 0 & z_1 & \ldots & 0 \\ \vdots & \vdots & \ddots & \vdots \\ 0 & 0 & \ldots & z_{M-1} \end{bmatrix} [\mathbf{V}_h^\dagger]_{M \times P} \mathbf{H}^\dagger, \tag{13.59}$$

where $[\mathbf{V}_h^\dagger]_{M\times P}$ is the matrix obtained by retaining the first M rows of $\mathbf{V}_h^\dagger$, and

$$z_k = \frac{\sqrt{q_{kk}}}{1+q_{kk}\sigma_{h,k}^2} \tag{13.60}$$

With the channel SVD written in the form

$$\mathbf{H} = \underbrace{\mathbf{U}_h}_{J\times J}\,\underbrace{\mathbf{\Sigma}_h}_{J\times P}\,\underbrace{\mathbf{V}_h^\dagger}_{P\times P} \tag{13.61}$$

as usual, we have $\mathbf{V}_h^\dagger\mathbf{H}^\dagger = \mathbf{\Sigma}_h^\dagger\mathbf{U}_h^\dagger$, so that

$$\mathbf{G} = [\,\mathbf{\Sigma}_g \quad \mathbf{0}\,]\,\mathbf{U}_h^\dagger, \tag{13.62}$$

where

$$\mathbf{\Sigma}_g = \frac{\sigma_s}{\sigma_q}\begin{bmatrix} z_0 & 0 & \ldots & 0 \\ 0 & z_1 & \ldots & 0 \\ \vdots & \vdots & \ddots & \vdots \\ 0 & 0 & \ldots & z_{M-1}\end{bmatrix}\begin{bmatrix} \sigma_{h,0} & 0 & \ldots & 0 \\ 0 & \sigma_{h,1} & \ldots & 0 \\ \vdots & \vdots & \ddots & \vdots \\ 0 & 0 & \ldots & \sigma_{h,M-1}\end{bmatrix}. \tag{13.63}$$

The structure for the MMSE transceiver can therefore be drawn as in Fig. 13.3. We summarize the preceding results as a theorem:

♠**Theorem 13.1.** *Optimal MMSE transceiver.* The solution to the MMSE transceiver optimization problem described in Eqs. (13.8) and (13.9) can be summarized as follows:

1. ***Precoder.*** The optimal precoder has the form (13.55), where (a) $\mathbf{V}_h$ is the unitary matrix that occurs in the channel SVD Eq. (13.61) (i.e., it diagonalizes $\mathbf{H}^\dagger\mathbf{H}$ as in Eq. (13.54)), and (b) $\mathbf{\Sigma}_f$ is the diagonal matrix shown in Eq. (13.56). Here

$$q_{kk} = \begin{cases} \frac{D}{\sigma_{h,k}} - \frac{1}{\sigma_{h,k}^2} & 0 \le k \le K-1 \\ 0 & \text{otherwise,}\end{cases} \tag{13.64}$$

   where

$$D = \frac{\frac{p_0}{\sigma_q^2} + \sum_{k=0}^{K-1}\frac{1}{\sigma_{h,k}^2}}{\sum_{k=0}^{K-1}\frac{1}{\sigma_{h,k}}} \tag{13.65}$$

   In these expressions, the channel singular values are numbered such that $\sigma_{h,0} \ge \sigma_{h,1} \ge \ldots$, and the quantity K is the largest integer less than M which yields $q_{kk} \ge 0$ for all k in Eq. (13.64).

2. ***Equalizer.*** The optimal equalizer has the form (13.62), where $\mathbf{U}_h$ is the unitary matrix that occurs in the SVD of the channel Eq. (13.61), and $\mathbf{\Sigma}_g$ is the diagonal matrix shown in Eq. (13.63).

3. ***Minimized error.*** With the above choice of precoder and equalizer, the mean square error is minimized and has the value given by

$$\mathcal{E}_{mmse} = \sigma_s^2(M-K) + \frac{\sigma_s^2\sigma_q^2\left(\sum_{k=0}^{K-1}\frac{1}{\sigma_{h,k}}\right)^2}{p_0 + \sigma_q^2\sum_{k=0}^{K-1}\frac{1}{\sigma_{h,k}^2}}, \qquad (13.66)$$

as proved earlier in Eq. (13.50). ◇

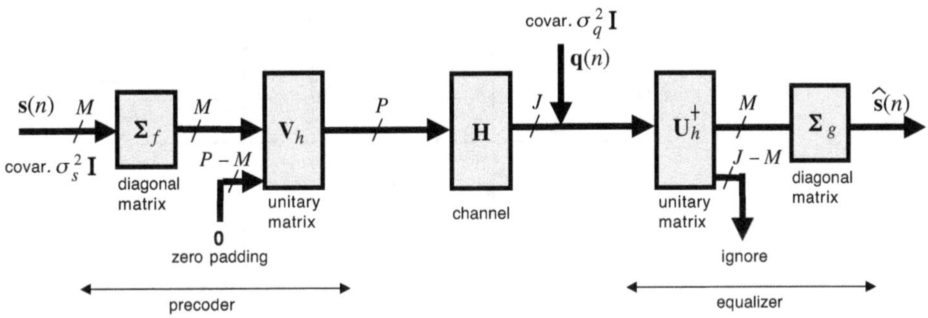

Figure 13.3. Structure of the optimal MMSE transceiver.

The integers P and J shown in Fig. 13.3 do not appear in this expression. Only the integers M and $K \leq M$ appear here.

Remarks

1. ***Similarity to zero-forcing solution.*** Note the similarity to the MMSE transceiver under the ZF constraint derived in Chap. 12 (see Fig. 12.12). The only difference is in the numerical values of the elements of the diagonal matrices $\boldsymbol{\Sigma}_f$ and $\boldsymbol{\Sigma}_g$, which are given in Eq. (13.56) and (13.63). In fact, as $p_0 \to \infty$, it can be verified that $K = M$, and the expression (13.66) reduces to the expression (12.56) for error in the zero-forcing case.

2. ***Orthogonality of precoder.*** The form (13.55) of the precoder shows that

$$\mathbf{F}^\dagger\mathbf{F} = \boldsymbol{\Sigma}_f^2,$$

that is, $\mathbf{F}^\dagger\mathbf{F}$ is diagonal. This means that the columns of the optimal precoder $\mathbf{F}$ can be assumed to be mutually orthogonal without loss of generality. Equivalently, the different users $s_k(n)$ can be assigned *mutually orthogonal codes.* This property is identical to what we observed for the ZF-MMSE transceiver.

3. ***Rank of the channel.*** Note finally that in the pure-MMSE problem there is no need to assume that the channel has rank $\geq M$. This is unlike in the ZF-MMSE case, where the zero-forcing property necessitated that the rank of the channel be at least equal to M.

4. ***Diagonal representation.*** Since the cascade of $\mathbf{V}_h, \mathbf{H}$, and $\mathbf{U}_h^\dagger$ in Fig. 13.3 is the diagonal matrix $\mathbf{\Sigma}_h$ of channel singular values, the optimal transceiver can be represented in diagonal form as shown in Fig. 13.4. The quantities $\sigma_{f,k}$ and $\sigma_{g,k}$, which depend only on $\sigma_{h,k}$ (and σ_s^2, σ_q^2, and p_0) can be calculated as described previously in this section.

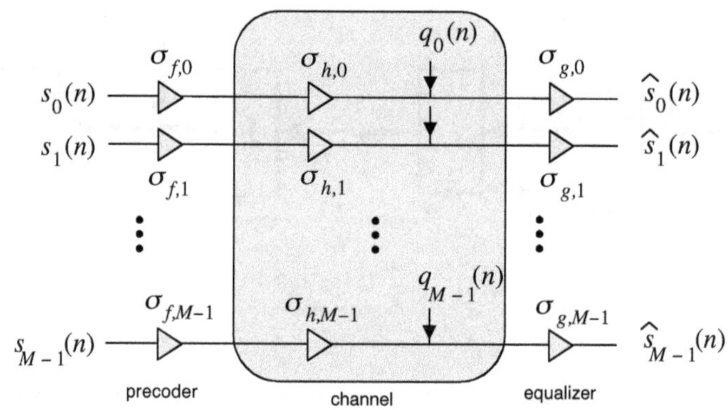

Figure 13.4. Diagonalized representation of the MMSE transceiver.

Example 13.1: Pure MMSE versus ZF-MMSE

Consider the case where $M = K$, which happens if the power is large enough to make $q_{kk} > 0$ for all k in Eq. (13.64). Assuming $\sigma_s^2 = 1$, the mean square error (13.66) for the pure-MMSE system becomes

$$\mathcal{E}_{pure} = \frac{\sigma_q^2}{p_0}\Big(\sum_{k=0}^{M-1}\frac{1}{\sigma_{h,k}}\Big)^2\left(\frac{1}{1+\frac{\sigma_q^2}{p_0}\sum_{k=0}^{M-1} 1/\sigma_{h,k}^2}\right). \qquad (13.67)$$

For the ZF-MMSE system of Chap. 12, the minimum error was given by

$$\mathcal{E}_{zf} = \frac{\sigma_q^2}{p_0}\Big(\sum_{k=0}^{M-1}\frac{1}{\sigma_{h,k}}\Big)^2. \qquad (13.68)$$

The gain obtained from giving up the ZF constraint is therefore

$$\mathcal{G} = 1 + \frac{\sigma_q^2}{p_0}\sum_{k=0}^{M-1} 1/\sigma_{h,k}^2. \qquad (13.69)$$

This depends on the channel singular values $\sigma_{h,k}$, power p_0, and noise variance σ_q^2. For example, let $M = 3$, $\sigma_q^2 = 0.01, p_0 = 1$, and

$$\sigma_{h,0} = 1, \ \sigma_{h,1} = 0.1, \ \text{and} \ \sigma_{h,2} = 0.01;$$

then the gain is about 102, which appears very impressive. In this example, the mean square errors per symbol are quite large:

$$\frac{\mathcal{E}_{zf}}{M} = 41.1, \quad \frac{\mathcal{E}_{pure}}{M} = 0.4026.$$

Compared to $\sigma_s^2 = 1$, these errors are unacceptably large. If we increase the power p_0 to make these errors smaller, then the gain $\mathcal{G}$ is also less impressive. For example, if $p_0 = 1000$ then

$$\frac{\mathcal{E}_{zf}}{M} = 0.0411, \quad \frac{\mathcal{E}_{pure}}{M} = 0.0373,$$

and the gain $\mathcal{G} = 1.1010$. In general, the benefit of using pure-MMSE over ZF-MMSE becomes negligible as p_0/σ_q^2 increases. In Chaps. 17 and 18 we shall see many examples demonstrating this fact. Notice finally that as $p_0 \to \infty$, the gain $\mathcal{G}$ in Eq. (13.69) approaches unity. Thus, for large powers, the difference between pure-MMSE and ZF-MMSE performances is negligible.

Example 13.2: Pure-MMSE transceiver

Consider again the 2×2 channel matrix of Ex. 12.2:

$$\mathbf{H} = \frac{1}{2\sqrt{2}} \begin{bmatrix} \sqrt{3} - (1/16) & 1 + (\sqrt{3}/16) \\ \sqrt{3} + (1/16) & 1 - (\sqrt{3}/16) \end{bmatrix}.$$

This has an SVD given by

$$\mathbf{H} = \underbrace{\frac{1}{\sqrt{2}} \begin{bmatrix} 1 & 1 \\ 1 & -1 \end{bmatrix}}_{\mathbf{U}_h} \times \underbrace{\begin{bmatrix} 1 & 0 \\ 0 & (1/16) \end{bmatrix}}_{\mathbf{\Sigma}_h} \times \underbrace{\frac{1}{2} \begin{bmatrix} \sqrt{3} & 1 \\ -1 & \sqrt{3} \end{bmatrix}}_{\mathbf{V}_h^\dagger}.$$

In this case the gain obtained from giving up the ZF constraint is therefore

$$\mathcal{G} = 1 + \frac{\sigma_q^2}{p_0} \sum_{k=0}^{M-1} 1/\sigma_{h,k}^2 = 3.57,$$

where we have used $\sigma_q^2 = 0.01$ and $p_0 = 1$, as in Ex. 13.1.

13.7 Concluding remarks

In this and the previous chapters we considered the minimization of mean square reconstruction error in transceivers. In the next chapter we shall consider the optimal allocation of bits among the transmitted components $s_k(n)$. We shall also consider the minimization of transmitted power for specified error probability and average bit rate. A one-page summary of the results of this chapter is included in Appendix I.

Appendix to Chapter 13

13.A A result on traces

Let $\mathbf{A}$ and $\mathbf{B}$ be two matrices of size $P \times M$ and $M \times P$, respectively, so that $\mathbf{AB}$ is $P \times P$ and $\mathbf{BA}$ is $M \times M$. We will prove the *trace identity* which says that

$$\mathrm{Tr}\,(\mathbf{I}_M + \mathbf{BA}) = M - P + \mathrm{Tr}\,(\mathbf{I}_P + \mathbf{AB}), \tag{13.70}$$

and furthermore

$$\mathrm{Tr}\,(\mathbf{I}_M + \mathbf{BA})^{-1} = M - P + \mathrm{Tr}\,(\mathbf{I}_P + \mathbf{AB})^{-1}, \tag{13.71}$$

as well. Note in particular that $\mathrm{Tr}\,(\mathbf{I}_M + \mathbf{BA})^{-1} = \mathrm{Tr}\,(\mathbf{I}_P + \mathbf{AB})^{-1}$ if and only if $M = P$, that is, if and only if $\mathbf{A}$ and $\mathbf{B}$ are square matrices.

Proof. Assume $M \geq P$ without loss of generality. It is shown in Horn and Johnson [1985] (p. 53, Theorem 1.3.20) that the M eigenvalues of $\mathbf{BA}$ are the same as the P eigenvalues of $\mathbf{AB}$ (with identical multiplicities), together with an additional set of $M - P$ eigenvalues equal to zero. Thus the characteristic polynomials are related as

$$\det(z\mathbf{I}_M - \mathbf{BA}) = cz^{M-P}\det(z\mathbf{I}_P - \mathbf{AB}),$$

where c is a nonzero constant. Since this is true for all z, we also have

$$\det\Big((y-1)\mathbf{I}_M - \mathbf{BA}\Big) = c(y-1)^{M-P}\det\Big((y-1)\mathbf{I}_P - \mathbf{AB}\Big)$$

for all y, that is,

$$\det\Big(y\mathbf{I}_M - (\mathbf{I}_M + \mathbf{BA})\Big) = c(y-1)^{M-P}\det\Big(y\mathbf{I}_P - (\mathbf{I}_P + \mathbf{AB})\Big)$$

Thus, if eigenvalues of $(\mathbf{I}_P + \mathbf{AB})$ are

$$\lambda_0, \lambda_1, \ldots, \lambda_{P-1},$$

then those of $(\mathbf{I}_M + \mathbf{BA})$ are

$$\lambda_0, \lambda_1, \ldots, \lambda_{P-1}, \underbrace{1, 1, \ldots, 1}_{M-P\ ones}$$

Since the trace is the sum of eigenvalues, this proves Eq. (13.70). Since the eigenvalues of the inverse (if it exists) are the reciprocals of the eigenvalues of the original matrix, the eigenvalues of $(\mathbf{I}_P+\mathbf{AB})^{-1}$ and those of $(\mathbf{I}_M+\mathbf{BA})^{-1}$ are

$$\frac{1}{\lambda_0}, \frac{1}{\lambda_1}, \ldots, \frac{1}{\lambda_{P-1}}$$

and

$$\frac{1}{\lambda_0}, \frac{1}{\lambda_1}, \ldots, \frac{1}{\lambda_{P-1}}, \underbrace{1, 1 \ldots, 1}_{M-P\ ones},$$

respectively. From this Eq. (13.71) follows immediately. $\bigtriangledown \bigtriangledown \bigtriangledown$

Problems

Note. Unless mentioned otherwise, $\sigma_s^2 = 1$.

13.1. Consider the case of a real scalar channel with real-valued precoder and equalizer. In this case, $\mathbf{F} = f, \mathbf{G} = g$, and $\mathbf{H} = h$ (scalar multipliers), and the mean square error (13.8) simplifies to

$$\mathcal{E}_{mse} = \sigma_s^2(ghf - 1)^2 + \sigma_q^2 g^2.$$

By setting $\partial \mathcal{E}_{mse}/\partial g = 0$, derive an expression for the optimal equalizer g for fixed f, h. Also find an expression for the mimimzed mean square error.

13.2. In Sec. 13.3 we showed that the MMSE equalizer matrix $\mathbf{G}$ has the form (13.12) and results in the mean square error (13.13). For the case where $\mathbf{R}_{\mathbf{ss}} = \sigma_s^2 \mathbf{I}$ and $\mathbf{R}_{\mathbf{qq}} = \sigma_q^2 \mathbf{I}$, we showed that this simplifies to

$$\mathcal{E}_{pure} = \sigma_q^2 \mathrm{Tr}\left(\mathbf{F}^\dagger \mathbf{H}^\dagger \mathbf{H}\mathbf{F} + \frac{\sigma_q^2}{\sigma_s^2}\mathbf{I}\right)^{-1}.$$

If the equalizer $\mathbf{G}$ is chosen to be the zero-forcing equalizer (instead of the MMSE equalizer) then (from Chap. 12) the mean square error is

$$\mathcal{E}_{zf} = \sigma_q^2 \mathrm{Tr}\left(\mathbf{F}^\dagger \mathbf{H}^\dagger \mathbf{H}\mathbf{F}\right)^{-1}.$$

Let σ_k denote the singular values of the product $\mathbf{HF}$.

1. Express $\mathcal{E}_{pure}$ and $\mathcal{E}_{zf}$ in terms of σ_k.
2. Consider an example where $\sigma_q^2/\sigma_s^2 = 0.01$, and $\mathbf{HF}$ is 2×2. If the singular values of $\mathbf{HF}$ are given by $\sigma_0 = 1, \sigma_1 = 0.1$, what is the ratio $\mathcal{E}_{zf}/\mathcal{E}_{pure}$? (This number represents the advantage of going from a zero-forcing system to a pure-MMSE system.)
3. Repeat for the case where $\sigma_1 = 0.001$.

In this problem we have compared the ZF-MMSE and pure-MMSE equalizers for fixed precoder $\mathbf{F}$. If we optimize the precoder, then $\mathbf{F}$ is different for the two cases, and the ratio $\mathcal{E}_{zf}/\mathcal{E}_{pure}$ is different from what we found above. This is addressed in the next few problems.

13.3. Consider the MIMO channel

$$\mathbf{H} = \begin{bmatrix} 3 & 3 \\ 3 & 2 \end{bmatrix}$$

with noise variance $\sigma_q^2 = 0.1$.

1. Compute the singular values of the channel.
2. With the transceiver $\{\mathbf{F}, \mathbf{G}\}$ designed to minimize the mean square error as summarized in Sec. 13.5, what is the mean square reconstruction error assuming the power is $p_0 = 2$?
3. Now assume $\{\mathbf{F}, \mathbf{G}\}$ is the MMSE solution with the ZF constraint as in Chap. 12. Then what is the mean square reconstuction error?
4. What is the ratio of the MSE values obtained in the above two methods? This is the gain $\mathcal{G}$ obtained by giving up the ZF constraint in the jointly optimal transceiver system.

13.4. Repeat Problem 13.3 for the channel

$$\mathbf{H} = \begin{bmatrix} 3 & 3 \\ 3 & 2.7 \end{bmatrix}.$$

Is the gain $\mathcal{G}$ higher or lower compared to Problem 13.3? Explain.

13.5. With all other quantities as in Problem 13.3, find an example of a channel $\mathbf{H}$ such that the gain $\mathcal{G}$ obtained (by giving up the ZF constraint) is at least 4.

13.6. Consider Eq. (13.64), which defines the optimal diagonal elements $\sigma_{f,k}$ of $\mathbf{\Sigma}_f$ in the precoder. Let $M = 3$, and let the dominant channel singular values be $\sigma_{h,0} = 1, \sigma_{h,1} = 0.1$, and $\sigma_{h,2} = 0.01$. The integer K represents the number of nonzero multipliers $\sigma_{f,k}$. If the power-to-noise ratio p_0/σ_q^2 is large enough, then we will have $K = 3$ (all three channel modes are used). If p_0/σ_q^2 is too small, then we will have $K = 1$ (only the dominant channel mode is used). Find three examples of p_0/σ_q^2 such that $K = 3, 2$, and 1, respectively.

13.7. The main results of this chapter have been derived assuming that $\mathbf{R_{ss}} = \sigma_s^2\mathbf{I}$. We now examine the possibility of making $\mathbf{R_{ss}}$ a more general diagonal matrix $\mathbf{\Lambda}_s$. In this case it makes more sense to minimize the sum of error-to-signal ratios (ESR)

$$ESR \triangleq \sum_{k=0}^{M-1} \frac{E|e_k(n)|^2}{E|s_k(n)|^2}$$

rather than the sum of errors. For, it is the error-to-signal ratios at the detectors that determine the performance of the receiver. We now show that the ESR minimization problem can readily be converted to the original problem (Eqs. (13.8) and (13.9)) which assumed $\mathbf{\Lambda}_s = \sigma_s^2\mathbf{I}$. Let $\mathbf{R_{ee}}$ denote the error covariance, and define $\mathbf{R}_{new} = \mathbf{\Lambda}_s^{-1/2}\mathbf{R_{ee}}\mathbf{\Lambda}_s^{-1/2}$. So we have to minimize $\text{Tr}\,(\mathbf{R}_{new})$. Note that the diagonal elements of $\mathbf{R_{ee}}$ and $\mathbf{\Lambda}_s$ represent $E|e_k(n)|^2$ and $E|s_k(n)|^2$, respectively.

1. Show that
$$\mathbf{R}_{new} = (\mathbf{G}_1\mathbf{H}\mathbf{F}_1 - \mathbf{I})(\mathbf{G}_1\mathbf{H}\mathbf{F}_1 - \mathbf{I})^\dagger + \sigma_q^2 \mathbf{G}_1\mathbf{G}_1^\dagger,$$
where $\mathbf{G}_1 = \mathbf{\Lambda}_s^{-1/2}\mathbf{G}$ and $\mathbf{F}_1 = \mathbf{F}\mathbf{\Lambda}_s^{1/2}$.
2. With p_0 denoting the power at the input of the channel as usual, express the power constraint in terms of $\mathbf{F}$ and $\mathbf{\Lambda}_s$, and show that this can be rewritten as $p_0 = \text{Tr}\,(\mathbf{F}_1\mathbf{F}_1^\dagger)$.

Thus, the problem reduces to finding $\mathbf{F}_1$ and $\mathbf{G}_1$ such that $\text{Tr}\,(\mathbf{R}_{new})$ is minimized subject to $p_0 = \text{Tr}\,(\mathbf{F}_1\mathbf{F}_1^\dagger)$. This is precisely the problem we solved in this chapter. Once $\mathbf{F}_1$ and $\mathbf{G}_1$ are found we obtain the original matrices readily using the inverse relations $\mathbf{G} = \mathbf{\Lambda}_s^{1/2}\mathbf{G}_1$ and $\mathbf{F} = \mathbf{F}_1\mathbf{\Lambda}_s^{-1/2}$.

13.8. We now show how to use matrix calculus (Chap. 20) to obtain a second derivation of the optimal equalizer of Sec. 13.3. The expression for $\mathcal{E}_{mse}$ is given by Eq. (13.7). Using the summary in Table 20.4 of Chap. 20, show that

$$\frac{\partial \mathcal{E}_{mse}}{\partial \mathbf{G}^*} = \mathbf{G}\mathbf{C}\mathbf{R}_{ss}\mathbf{C}^\dagger - \mathbf{R}_{ss}\mathbf{C}^\dagger + \mathbf{G}\mathbf{R}_{qq} = \mathbf{G}\Big(\mathbf{C}\mathbf{R}_{ss}\mathbf{C}^\dagger + \mathbf{R}_{qq}\Big) - \mathbf{R}_{ss}\mathbf{C}^\dagger,$$

where $\mathbf{C} = \mathbf{H}\mathbf{F}$. By setting this to zero we indeed obtain Eq. (13.12).

14

Bit allocation and power minimization

14.1 Introduction

Consider again the transceiver of Fig. 14.1 which was optimized in the preceding chapters for various criteria. In this chapter we consider a different objective function of significant importance. Namely, we minimize the transmitted power for a given performance constraint, by optimizing transceiver matrices and the bit allocation among the various transmitted symbol streams. More specifically we assign bits to $s_k(n)$ in an optimal way, for a fixed set of error probabilities (quality-of-service or QoS constraints) at the receiver. Such bit allocation is performed in discrete multitone, or **DMT**, systems used in DSL applications. We also optimize the precoder $\mathbf{F}$ such that the transmitted power is minimized for a given bit error rate. In this optimization the equalizer $\mathbf{G}$ will be assumed to have the zero-forcing property. The channel $\mathbf{H}$ is assumed to be a $J \times P$ matrix with

$$J \geq M \quad \text{and} \quad P \geq M.$$

It is also assumed that $\mathbf{H}$ has rank M so that the zero-forcing condition $\mathbf{GHF} = \mathbf{I}$ is possible. Throughout the chapter we assume that the channel noise $\mathbf{q}(n)$ has the diagonal covariance matrix

$$\mathbf{R_{qq}} = \sigma_q^2 \mathbf{I}. \tag{14.1}$$

The components $q_k(n)$ of the noise vector are assumed to be Gaussian (circular complex Gaussian in the complex case, Sec. 6.6) with variance σ_q^2.

The discrete multitone (DMT) transceiver with bit allocation, for DSL applications on telephone lines, was proposed in 1991 in the landmark paper by Chow *et al.* [1991]. The optimality results in this chapter are based mostly on

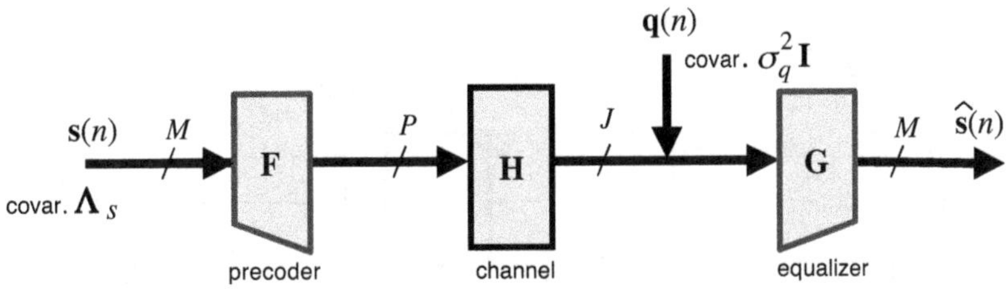

Figure 14.1. The transceiver with precoder **F**, equalizer **G** and a channel **H**.

the work of Lin and Phoong [2001b]. Other related work includes Pandharipande and Dasgupta [2003], Palomar and Barbarossa [2005], and Weng *et al.* [2010a].

14.2 Error probabilities, bit rates, and power

To understand how the problems of bit allocation and power minimization arise, we first examine the relationships between the error probabilities, bit rates, and user powers. Assume that the input signals $s_k(n)$ are zero-mean uncorrelated processes representing independent users (or independent components of a user obtained by blocking or parsing, Secs. 3.5 and 3.6), with powers P_k, so that the input covariance is

$$\mathbf{\Lambda}_s = \begin{bmatrix} P_0 & 0 & \dots & 0 \\ 0 & P_1 & \dots & 0 \\ \vdots & \vdots & \ddots & \vdots \\ 0 & 0 & \dots & P_{M-1} \end{bmatrix}.$$

Consider the situation where each user is represented by a different constellation size. Let us say that the kth user transmits b_k-bit symbols from a square QAM constellation with average power P_k. If the error $e_k(n)$ at the kth output of $\mathbf{G}$ (Fig. 14.2) has variance $\sigma_{e_k}^2$, then the probability of error for the kth symbol stream is given by (Sec. 2.3.2)

$$\mathcal{P}_e(k) \approx 4(1 - 2^{-b_k/2})\mathcal{Q}\left(\sqrt{\frac{3P_k}{(2^{b_k} - 1)\sigma_{e_k}^2}}\right). \tag{14.2}$$

If b_k is large enough this can be approximated by

$$\mathcal{P}_e(k) \approx 4\mathcal{Q}\left(\sqrt{\frac{3P_k}{(2^{b_k} - 1)\sigma_{e_k}^2}}\right), \tag{14.3}$$

so that the power P_k can be approximated as

$$P_k \approx \frac{(2^{b_k} - 1)\sigma_{e_k}^2}{3}\left(\mathcal{Q}^{-1}\left(\frac{\mathcal{P}_e(k)}{4}\right)\right)^2. \tag{14.4}$$

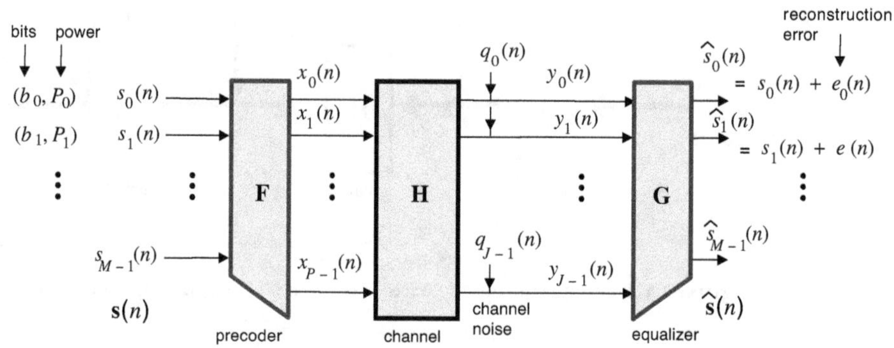

Figure 14.2. The transceiver with all signals shown in component form.

Note here that the approximate nature of Eq. (14.4) comes about for a number of reasons:

1. ***Low error assumption***. As we saw in Sec. 2.3.2, the probability of error for the QAM case is actually given by

$$2\mathcal{P}_{PAM} - \mathcal{P}_{PAM}^2,$$

   where $\mathcal{P}_{PAM}$ is the probability of error for an appropriately defined PAM constellation. Equation (14.2) makes the approximation that $\mathcal{P}_{PAM}^2 \approx 0$. This is certainly an excellent approximation for any practically acceptable value of $\mathcal{P}_{PAM}$.

2. ***High bit rate assumption***. The quantity $(1 - 2^{-b_k/2})$ has been replaced with unity. This approximation is valid only when b_k is reasonably large. Under this assumption, the further approximation $(2^{b_k} - 1) = 2^{b_k}$ is also reasonable, so that Eq. (14.4) is often replaced with

$$P_k \approx \frac{\sigma_{e_k}^2 2^{b_k}}{3} \left(\mathcal{Q}^{-1} \left(\frac{\mathcal{P}_e(k)}{4} \right) \right)^2. \tag{14.5}$$

This equation expresses the average power required for the kth user to operate at the probability of error $\mathcal{P}_e(k)$ with a b_k-bit QAM constellation, when the noise variance at the kth output of $\mathbf{G}$ (input of kth detector) is $\sigma_{e_k}^2$. Since the error probability $\mathcal{P}_e(k)$ is often regarded as the *quality of service* (QoS), the specification of the set of numbers $\{\mathcal{P}_e(k)\}$ is equivalent to the specification of QoS requirements.

DMT versus OFDM systems. In traditional OFDM systems the signals $s_k(n)$ are obtained from a scalar symbol stream $s(n)$ by the process of **blocking**. All the signals $s_k(n)$ then have identical power, that is, $P_k = P_0$ for all k. In such systems the error $\mathcal{P}_e(k)$ is fixed for all k, and no bit allocation is performed. While OFDM systems do not have bit allocation, discrete multitone, or **DMT**, systems used in

DSL applications allow us to perform bit allocation because the symbols $s_k(n)$ are obtained from a scalar symbol stream by **parsing** rather than blocking. See Secs. 3.5 and 3.6. $\bigtriangledown \bigtriangledown \bigtriangledown$

14.3 Minimization of transmitted power with bit allocation

The reader should notice that the quantity P_k in Eq. (14.5) does *not* represent the power at the input of the channel. Rather, it represents the kth *user* power, that is, the power in $s_k(n)$, before the precoder. The channel input power is the topic of this section. The total power transmitted on the channel can be written as

$$P_{trans} = \text{Tr}\left(\mathbf{F}\mathbf{\Lambda}_s\mathbf{F}^\dagger\right) = \text{Tr}\left(\mathbf{F}^\dagger\mathbf{F}\mathbf{\Lambda}_s\right) = \sum_{k=0}^{M-1} P_k[\mathbf{F}^\dagger\mathbf{F}]_{kk}. \tag{14.6}$$

Substituting from Eq. (14.5) we can rewrite this as

$$P_{trans} = \sum_{k=0}^{M-1} P_k[\mathbf{F}^\dagger\mathbf{F}]_{kk} = \frac{1}{3}\sum_{k=0}^{M-1} \sigma_{e_k}^2 2^{b_k}\left(\mathcal{Q}^{-1}\left(\frac{\mathcal{P}_e(k)}{4}\right)\right)^2 [\mathbf{F}^\dagger\mathbf{F}]_{kk}.$$

Since the channel noise has covariance $\sigma_q^2\mathbf{I}$ in Fig. 14.1, the error variance at the kth output of $\mathbf{G}$ is given by

$$\sigma_{e_k}^2 = \sigma_q^2[\mathbf{G}\mathbf{G}^\dagger]_{kk}. \tag{14.7}$$

The transmitted power can therefore be rewritten as

$$P_{trans} = \sum_{k=0}^{M-1} c_k 2^{b_k}[\mathbf{F}^\dagger\mathbf{F}]_{kk}[\mathbf{G}\mathbf{G}^\dagger]_{kk}, \tag{14.8}$$

where we have used the notation

$$c_k = \frac{\sigma_q^2}{3}\left(\mathcal{Q}^{-1}\left(\frac{\mathcal{P}_e(k)}{4}\right)\right)^2. \tag{14.9}$$

Since b_k denotes the number of bits for the kth user, the average number of bits is given by

$$b = \frac{1}{M}\sum_{k=0}^{M-1} b_k. \tag{14.10}$$

In Eq. (14.8) c_k are fixed constants. We have to optimize the precoder $\mathbf{F}$, equalizer $\mathbf{G}$, and the bits $\{b_k\}$ jointly, but we shall return to that later. For the present, we minimize the total power P_{trans} by optimizing the bit allocation $\{b_k\}$. The average number of bits b is assumed fixed, and so are $\mathbf{G}$, $\mathbf{F}$, and probabilities of error $P_e(k)$ (i.e., fixed set of c_k). Using the AM-GM inequality (see Appendix A) in Eq. (14.8) we see that

$$\begin{aligned} P_{trans} &= \sum_{k=0}^{M-1} c_k 2^{b_k} [\mathbf{F}^\dagger\mathbf{F}]_{kk}[\mathbf{G}\mathbf{G}^\dagger]_{kk} \\ &\geq M \prod_{k=0}^{M-1} \Big(c_k 2^{b_k} [\mathbf{F}^\dagger\mathbf{F}]_{kk}[\mathbf{G}\mathbf{G}^\dagger]_{kk}\Big)^{1/M} \\ &= c2^b \prod_{k=0}^{M-1} \Big([\mathbf{F}^\dagger\mathbf{F}]_{kk}[\mathbf{G}\mathbf{G}^\dagger]_{kk}\Big)^{1/M} \\ &= c2^b \Big(\prod_{k=0}^{M-1} [\mathbf{F}^\dagger\mathbf{F}]_{kk} \prod_{k=0}^{M-1} [\mathbf{G}\mathbf{G}^\dagger]_{kk}\Big)^{1/M}, \end{aligned}$$

where

$$c = M(\prod c_k)^{1/M},$$

and we have used the fact that $b = \sum b_k/M$. Equality can be achieved in the second line above if and only if the terms are identical for all k, that is,

$$c_k 2^{b_k} [\mathbf{F}^\dagger\mathbf{F}]_{kk}[\mathbf{G}\mathbf{G}^\dagger]_{kk} = A$$

for some constant A. Taking logarithms on both sides we get

$$\log_2 c_k + b_k + \log_2[\mathbf{F}^\dagger\mathbf{F}]_{kk} + \log_2[\mathbf{G}\mathbf{G}^\dagger]_{kk}, = \log_2 A$$

from which we obtain

$$b_k = D - \log_2 c_k - \log_2[\mathbf{F}^\dagger\mathbf{F}]_{kk} - \log_2[\mathbf{G}\mathbf{G}^\dagger]_{kk},$$

where D is a constant. Using the expression for c_k given in Eq. (14.9) we obtain

$$b_k = D - \log_2\left(\frac{\sigma_q^2}{3}\Big(\mathcal{Q}^{-1}\Big(\frac{\mathcal{P}_e(k)}{4}\Big)\Big)^2\right) - \log_2[\mathbf{F}^\dagger\mathbf{F}]_{kk} - \log_2[\mathbf{G}\mathbf{G}^\dagger]_{kk}. \quad (14.11)$$

This is called the *optimum bit allocation* or *bit loading* formula. The constant D is chosen such that $\sum b_k/M = b$. Note that if the average bit rate b is large enough then b_k computed from this formula will be non-negative and can be approximated well with integers. The formula (14.11) is similar to the water-filling formula for power allocation discussed in Chap. 22 (Eq. (22.26)).

For any fixed pair of $\mathbf{G}$ and $\mathbf{F}$, and a specified set of probabilities of error $\{\mathcal{P}_e(k)\}$, the bit allocation that minimizes the transmitted power is given by Eq. (14.11). With the bit allocation so chosen, the quantities P_k are computed from Eq. (14.5) where $\sigma_{e_k}^2$ is as in Eq. (14.7). With P_k so chosen, the specified probabilities of error are met, and the total power P_{trans} is minimized. This minimized power is given by

$$P_{trans} = c2^b \Big(\prod_{k=0}^{M-1} [\mathbf{F}^\dagger\mathbf{F}]_{kk} \prod_{k=0}^{M-1} [\mathbf{G}\mathbf{G}^\dagger]_{kk}\Big)^{1/M}. \quad (14.12)$$

14.4 Optimizing the precoder and equalizer

For any b, $\{\mathcal{P}_e(k)\}$, $\mathbf{G}$, and $\mathbf{F}$, Eq. (14.12) gives the minimum transmitted power with respect to $\{b_k\}$. The set $\{\mathcal{P}_e(k)\}$ is specified according to the desired QoS. The power (14.12) can further be minimized by optimizing the precoder $\mathbf{F}$ and equalizer $\mathbf{G}$. In this section we show how this optimization can be done.

14.4.1 Eliminating the precoder matrix F using the ZF constraint

We now use the zero-forcing constraint to eliminate one of the matrices $\mathbf{F}$ from further discussion. The problem will then reduce to that of optimizing the matrix $\mathbf{G}$. The precoder $\mathbf{F}$ satisfying the zero-forcing condition

$$\mathbf{GHF} = \mathbf{I} \tag{14.13}$$

is nothing but a right inverse of $\mathbf{GH}$. From Appendix C we know that the right inverse with minimum-norm (MNRI) is given by[1]

$$\mathbf{F} = \mathbf{H}^\dagger\mathbf{G}^\dagger\left(\mathbf{GHH}^\dagger\mathbf{G}^\dagger\right)^{-1}, \tag{14.14}$$

so that the expression for $\mathbf{F}^\dagger\mathbf{F}$ simplifies to

$$\mathbf{F}^\dagger\mathbf{F} = \left(\mathbf{GHH}^\dagger\mathbf{G}^\dagger\right)^{-1}. \tag{14.15}$$

Using Eq. (14.15), we can rewrite Eq. (14.12) as

$$P_{trans} = c2^b\Big(\prod_{k=0}^{M-1}[(\mathbf{GHH}^\dagger\mathbf{G}^\dagger)^{-1}]_{kk}\prod_{k=0}^{M-1}[\mathbf{GG}^\dagger]_{kk}\Big)^{1/M}. \tag{14.16}$$

So the optimal transceiver problem under the ZF constraint and fixed QoS constraints becomes one of solving for the $M \times J$ equalizer matrix $\mathbf{G}$ of rank M such that

$$\phi \triangleq \prod_{k=0}^{M-1}[(\mathbf{GHH}^\dagger\mathbf{G}^\dagger)^{-1}]_{kk}\prod_{k=0}^{M-1}[\mathbf{GG}^\dagger]_{kk} \tag{14.17}$$

is minimized, where $\mathbf{H}$ is the $J \times P$ channel matrix.

14.4.2 Finding the optimal equalizer G

To find the optimum $\mathbf{G}$ that minimizes Eq. (14.17), we first establish a very important result, namely that ϕ is lower bounded entirely in terms of the M dominant singular values of the channel $\mathbf{H}$. For this, let $\mathbf{U}_h$ denote the $J \times J$ unitary matrix diagonalizing the Hermitian matrix $\mathbf{HH}^\dagger$, that is,

$$\mathbf{U}_h^\dagger\mathbf{HH}^\dagger\mathbf{U}_h = \mathbf{\Lambda}_h^2, \tag{14.18}$$

[1]Since $\mathbf{GH}$ is rectangular, the right inverse is not unique. The choice of the minimum-norm right inverse, however, is not a loss of generality for the specific optimization problem to be solved in this section, as shown in Appendix 14.A at the end of the chapter.

where $\mathbf{\Lambda}_h$ is the diagonal matrix of J singular values of $\mathbf{H}$ in the order

$$\sigma_{h,0} \geq \sigma_{h,1} \geq \ldots \geq \sigma_{h,M-1} \geq \ldots. \tag{14.19}$$

We then have the following result:

♠**Lemma 14.1.** *Bound on ϕ.* For a given channel $\mathbf{H}$ the quantity ϕ in Eq. (14.17) is bounded as

$$\phi \geq \frac{1}{\prod_{k=0}^{M-1} \sigma_{h,k}^2} \tag{14.20}$$

The bound is achieved by choosing the $M \times J$ equalizer $\mathbf{G}$ to be

$$\mathbf{G} = [\mathbf{U}_h^\dagger]_{M\times J}, \tag{14.21}$$

where the right-hand side denotes the submatrix defined by the first M rows of $\mathbf{U}_h^\dagger$. ◊

Note that the assumption that $\mathbf{H}$ has rank $\geq M$ ensures that $\sigma_{h,k} \neq 0$ in the denominator of Eq. (14.20). To proceed with the proof of the lemma, it is convenient to represent $\mathbf{G}$ in terms of its SVD matrices:

$$\mathbf{G} = \mathbf{U}_g \left[\mathbf{\Sigma}_g \quad \mathbf{0} \right] \mathbf{V}_g^\dagger, \tag{14.22}$$

where $\mathbf{U}_g$ is $M \times M$ unitary, $\mathbf{V}_g$ is $J \times J$ unitary, and $\mathbf{\Sigma}_g$ is $M \times M$ diagonal with positive diagonal elements (singular values of $\mathbf{G}$).

Proof of Lemma 14.1. Using Hadamard's inequality for positive definite matrices (Appendix B) we have

$$\phi = \prod_{k=0}^{M-1} [(\mathbf{GHH}^\dagger\mathbf{G}^\dagger)^{-1}]_{kk} \prod_{k=0}^{M-1} [\mathbf{GG}^\dagger]_{kk} \geq \frac{\det(\mathbf{GG}^\dagger)}{\det(\mathbf{GHH}^\dagger\mathbf{G}^\dagger)} \tag{14.23}$$

with equality if and only if $(\mathbf{GG}^\dagger)$ and $[(\mathbf{GHH}^\dagger\mathbf{G}^\dagger)]$ are diagonal. From Eq. (14.22) we have

$$\mathbf{GG}^\dagger = \mathbf{U}_g \left[\mathbf{\Sigma}_g \quad \mathbf{0} \right] \mathbf{V}_g^\dagger \mathbf{V}_g \begin{bmatrix} \mathbf{\Sigma}_g^\dagger \\ \mathbf{0} \end{bmatrix} \mathbf{U}_g^\dagger = \mathbf{U}_g \left[\mathbf{\Sigma}_g \quad \mathbf{0} \right] \begin{bmatrix} \mathbf{\Sigma}_g^\dagger \\ \mathbf{0} \end{bmatrix} \mathbf{U}_g^\dagger$$

and

$$\mathbf{GHH}^\dagger\mathbf{G}^\dagger = \mathbf{U}_g \left[\mathbf{\Sigma}_g \quad \mathbf{0} \right] \mathbf{V}_g^\dagger \mathbf{HH}^\dagger \mathbf{V}_g \begin{bmatrix} \mathbf{\Sigma}_g^\dagger \\ \mathbf{0} \end{bmatrix} \mathbf{U}_g^\dagger.$$

So

$$\det(\mathbf{GG}^\dagger) = \det(\mathbf{\Sigma}_g\mathbf{\Sigma}_g^\dagger) \tag{14.24a}$$

and

$$\det(\mathbf{GHH}^\dagger\mathbf{G}^\dagger) = \det(\mathbf{\Sigma}_g\mathbf{\Sigma}_g^\dagger)\det(\mathbf{V}_g^\dagger\mathbf{HH}^\dagger\mathbf{V}_g)_M \tag{14.24b}$$

where we have used $(\det\mathbf{U}_g)(\det\mathbf{U}_g^\dagger) = 1$, and the notation $(.)_M$ denotes the $M \times M$ leading principal submatrix. Thus Eq. (14.23) becomes

$$\phi \geq \frac{1}{\det(\mathbf{V}_g^\dagger \mathbf{H}\mathbf{H}^\dagger \mathbf{V}_g)_M}. \tag{14.25}$$

Note that the diagonal matrix $\boldsymbol{\Sigma}_g$ has canceled off. Next, we know

$$\det(\mathbf{V}_g^\dagger \mathbf{H}\mathbf{H}^\dagger \mathbf{V}_g)_M = \prod_{k=0}^{M-1} \mu_k, \tag{14.26}$$

where $\mu_0 \geq \mu_1 \geq \ldots \geq \mu_{M-1}$ are the eigenvalues of $(\mathbf{V}_g^\dagger \mathbf{H}\mathbf{H}^\dagger \mathbf{V}_g)_M$. The $J \times J$ matrix $\mathbf{V}_g^\dagger \mathbf{H}\mathbf{H}^\dagger \mathbf{V}_g$ has eigenvalues $\sigma_{h,0}^2 \geq \sigma_{h,1}^2 \geq \ldots \geq \sigma_{h,J-1}^2$. So it follows from the interlacing property for Hermitian matrices (Sec. 12.4.2, see the proof following Lemma 12.2) that

$$\sigma_{h,0}^2 \geq \mu_0,\ \sigma_{h,1}^2 \geq \mu_1,\ \ldots,\ \sigma_{h,M-1}^2 \geq \mu_{M-1},$$

so that $\prod_{k=0}^{M-1} \mu_k \leq \prod_{k=0}^{M-1} \sigma_{h,k}^2$. Thus

$$\det\Big(\mathbf{V}_g^\dagger \mathbf{H}\mathbf{H}^\dagger \mathbf{V}_g\Big)_M = \prod_{k=0}^{M-1} \mu_k \leq \prod_{k=0}^{M-1} \sigma_{h,k}^2 \tag{14.27}$$

This, together with Eq. (14.25), immediately yields Eq. (14.20). With $\mathbf{G}$ chosen as in Eq. (14.21) we have

$$\mathbf{G}\mathbf{G}^\dagger = (\mathbf{U}_h^\dagger \mathbf{U}_h)_M = \mathbf{I}_M \quad \text{and} \quad \mathbf{G}\mathbf{H}\mathbf{H}^\dagger\mathbf{G}^\dagger = (\boldsymbol{\Lambda}_h^2)_M.$$

Since these matrices are both diagonal, equality holds in the Hadamard inequalities used to derive Eq. (14.23). That is,

$$\phi = \frac{\det(\mathbf{G}\mathbf{G}^\dagger)}{\det(\mathbf{G}\mathbf{H}\mathbf{H}^\dagger\mathbf{G}^\dagger)} = \frac{1}{\det\,(\boldsymbol{\Lambda}_h^2)_M} = \frac{1}{\prod_{k=0}^{M-1} \sigma_{h,k}^2},$$

and the bound (14.20) is achieved indeed. This is true regardless of the choice of $\boldsymbol{\Sigma}_g$ because it cancels off in Eq. (14.25). $\bigtriangledown\bigtriangledown\bigtriangledown$

The above lemma completely specifies how the optimal equalizer $\mathbf{G}$ should be chosen. The optimal precoder is found by substituting $\mathbf{G}$ into the zero-forcing equation (14.14). We have therefore proved the following:

♠**Theorem 14.1.** *Optimal transceiver summary.* The expression Eq. (14.12) for the transmitted power under optimal bit allocation (14.11) is minimized by the following choice of the transceiver matrices:

1. Choose the equalizer $\mathbf{G}$ as

$$\mathbf{G} = [\mathbf{U}_h^\dagger]_{M\times J}, \tag{14.28}$$

where $\mathbf{U}_h$ is the $J \times J$ unitary matrix diagonalizing $\mathbf{H}\mathbf{H}^\dagger$ as in Eq. (14.18) (with $\mathbf{H}$ denoting the $J \times P$ channel).

2. Choose the precoder $\mathbf{F}$ as the minimum-norm right inverse of $\mathbf{GH}$, that is,

$$\mathbf{F} = \mathbf{H}^{\dagger}\mathbf{G}^{\dagger}\Big(\mathbf{GHH}^{\dagger}\mathbf{G}^{\dagger}\Big)^{-1}. \tag{14.29}$$

With $\mathbf{G}$ and $\mathbf{F}$ so chosen, the minimized transmitted power is given by

$$P_{min} = c2^{b}\left(\frac{1}{\prod_{k=0}^{M-1}\sigma_{h,k}^{2}}\right)^{1/M}, \tag{14.30}$$

where $b = \sum_k b_k/M$ is the average number of bits. The constant $c = M\prod_k (c_k)^{1/M}$, where c_k is as in Eq. (14.9), that is,

$$c_k = \frac{\sigma_q^2}{3}\left(\mathcal{Q}^{-1}\Big(\frac{\mathcal{P}_e(k)}{4}\Big)\right)^2, \tag{14.31}$$

where σ_q^2 is the variance of the channel noise components and $\mathcal{P}_e(k)$ is the acceptable error probability specified for the kth user. ◇

With $\mathbf{G}$ and $\mathbf{F}$ chosen optimally as above, it can be shown (Sec. 14.6.2.B) that the optimal bit allocation formula (14.11) can be rewritten as

$$b_k = D - \log_2\left(\frac{\sigma_q^2}{3}\Big(\mathcal{Q}^{-1}\Big(\frac{\mathcal{P}_e(k)}{4}\Big)\Big)^2\right) + \log_2\sigma_{h,k}^2.$$

Thus, bit allocation is determined essentially by the channel singular values $\sigma_{h,k}$ and the desired error probabilties $\mathcal{P}_e(k)$. For completeness, note that the kth user power $P_k = E[|s_k(n)|^2]$ should be taken to be as in Eq. (14.5), so that the specified values of $\mathcal{P}_e(k)$ (QoS constraints) are satisfied. Since $\sigma_{e_k}^2 = \sigma_q^2[\mathbf{GG}^{\dagger}]_{kk} = \sigma_q^2$ (using $\mathbf{GG}^{\dagger} = \mathbf{I}$), this simplifies to

$$P_k \approx \frac{\sigma_q^2 2^{b_k}}{3}\left(\mathcal{Q}^{-1}\Big(\frac{\mathcal{P}_e(k)}{4}\Big)\right)^2 \tag{14.32}$$

for $0 \le k \le M-1$.

14.5 Optimal transceiver in terms of channel SVD

It is convenient to express the solution for the optimal transceiver in terms of the SVD of the $J \times P$ channel $\mathbf{H}$. Thus, let

$$\mathbf{H} = \underbrace{\mathbf{U}_h}_{J\times J}\underbrace{\mathbf{\Sigma}_h}_{J\times P}\underbrace{\mathbf{V}_h^{\dagger}}_{P\times P}, \tag{14.33}$$

where $\mathbf{U}_h$ and $\mathbf{V}_h$ are unitary and $\mathbf{\Sigma}_h$ is diagonal with diagonal elements $\sigma_{h,0} \geq \sigma_{h,1} \geq \ldots$ Then

$$\mathbf{H}\mathbf{H}^\dagger = \mathbf{U}_h\mathbf{\Sigma}_h\mathbf{V}_h^\dagger\mathbf{V}_h\mathbf{\Sigma}_h^\dagger\mathbf{U}_h^\dagger = \mathbf{U}_h\mathbf{\Lambda}_h^2\mathbf{U}_h^\dagger,$$

where $\mathbf{\Lambda}_h^2 = \mathbf{\Sigma}_h\mathbf{\Sigma}_h^\dagger$. Thus

$$\mathbf{U}_h^\dagger\mathbf{H}\mathbf{H}^\dagger\mathbf{U}_h = \mathbf{\Lambda}_h^2 \quad (J \times J \text{ diagonal matrix}). \tag{14.34}$$

Thus the receiver or equalizer matrix $\mathbf{G}$ in the optimal transceiver can be taken to be

$$\mathbf{G} = [\mathbf{U}_h^\dagger]_{M\times J}, \tag{14.35}$$

where the subscript $M \times J$ is used to denote the submatrix obtained by retaining the first M rows and first J columns. From Eq. (14.33) we have

$$\mathbf{U}_h^\dagger\mathbf{H} = \mathbf{\Sigma}_h\mathbf{V}_h^\dagger$$

so that

$$\mathbf{G}\mathbf{H} = [\mathbf{\Sigma}_h]_{M\times P}\mathbf{V}_h^\dagger = \left[\,[\mathbf{\Sigma}_h]_M \quad \mathbf{0}\,\right]\mathbf{V}_h^\dagger,$$

where $[\mathbf{\Sigma}_h]_M$ is the $M \times M$ leading principal submatrix of $\mathbf{\Sigma}_h$. Thus the optimal precoder $\mathbf{F}$ in Eq. (14.29) is nothing but

$$\mathbf{F} = \mathbf{V}_h \begin{bmatrix} [\mathbf{\Sigma}_h]_M^{-1} \\ \\ \mathbf{0} \end{bmatrix} = \mathbf{V}_h \begin{bmatrix} \frac{1}{\sigma_{h,0}} & 0 & \ldots & 0 \\ 0 & \frac{1}{\sigma_{h,1}} & \ldots & 0 \\ \vdots & \vdots & \ddots & \vdots \\ 0 & 0 & \ldots & \frac{1}{\sigma_{h,M-1}} \\ 0 & 0 & \ldots & 0 \\ \vdots & \vdots & \ddots & \vdots \\ 0 & 0 & \ldots & 0 \end{bmatrix}. \tag{14.36}$$

Thus we have shown the following:

♠**Theorem 14.2.** *Optimal transceiver in terms of channel SVD.* With the channel $\mathbf{H}$ expressed in SVD form (14.33), suppose the equalizer $\mathbf{G}$ and the precoder $\mathbf{F}$ are chosen as in Eqs. (14.35) and (14.36), respectively. Then the transceiver system is optimal when bit allocation is performed at the transmitter according to Eq. (14.11). The optimality is in the sense of minimizing the transmitted power (14.8) subject to the zero-forcing constraint. The minimized value of the power is given by Eq. (14.30), and the individual user powers are given by Eq. (14.32) so that the specified values of error probabilities $\mathcal{P}_e(k)$ (QoS constraints) are satisfied. ◇

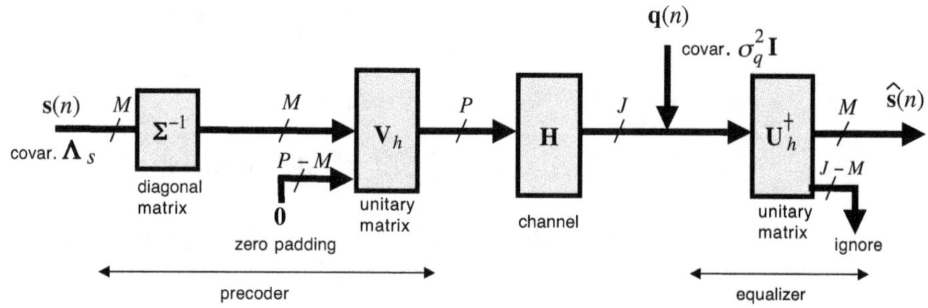

Figure 14.3. The optimal receiver structure expressed in terms of the SVD parameters of the channel. The diagonal matrix $\boldsymbol{\Sigma}$ has diagonal elements equal to the largest M singular values of $\mathbf{H}$.

Figure 14.3 shows the optimal receiver structure expressed in this form. Note that the *optimal solution is such that the channel is diagonalized*, because the cascade of $\mathbf{V}_h$, $\mathbf{H}$, and $\mathbf{U}_h^\dagger$ is the diagonal matrix $\boldsymbol{\Sigma}_h$, and this matrix is cancelled at the transmitter by the diagonal equalizer

$$\boldsymbol{\Sigma}^{-1} = [\boldsymbol{\Sigma}_h]_M^{-1} = \begin{bmatrix} \frac{1}{\sigma_{h,0}} & 0 & \cdots & 0 \\ 0 & \frac{1}{\sigma_{h,1}} & \cdots & 0 \\ \vdots & \vdots & \ddots & \vdots \\ 0 & 0 & \cdots & \frac{1}{\sigma_{h,M-1}} \end{bmatrix}. \tag{14.37}$$

The simpler notation $\boldsymbol{\Sigma}$ will be used for $[\boldsymbol{\Sigma}_h]_M$ from now on. The form of the solution is strikingly similar to the solutions for the ZF-MMSE and MMSE receivers described in Secs. 12.4.4 and 13.6, respectively.

14.5.1 Other equivalent forms of the optimal transceiver

Some readers might wonder why the equalizer $\boldsymbol{\Sigma}^{-1}$ is at the transmitter instead of at the receiver. It is possible to rearrange the optimum structure such that this diagonal equalizer is moved to the receiver as shown next.

With the optimal receiver matrix $\mathbf{G}$ written in the form (14.22), we found that the effect of $\boldsymbol{\Sigma}_g$ cancels in the expression for the transmitted power as seen from Eqs. (14.23) and (14.24b). So we eliminated it from further discussion by setting $\boldsymbol{\Sigma}_g = \mathbf{I}$. The presence of an arbitrary $\boldsymbol{\Sigma}_g$ changes $\mathbf{G}$ to

$$\mathbf{G} = [\boldsymbol{\Sigma}_g \quad \mathbf{0}]\,\mathbf{V}_g^\dagger = \boldsymbol{\Sigma}_g[\mathbf{U}_h^\dagger]_{M\times J}.$$

The corresponding receiver structure is shown in Fig. 14.4(a). Note that $\boldsymbol{\Sigma}_g^{-1}$ appears at the transmitter, so that the zero-forcing constraint is unaffected.

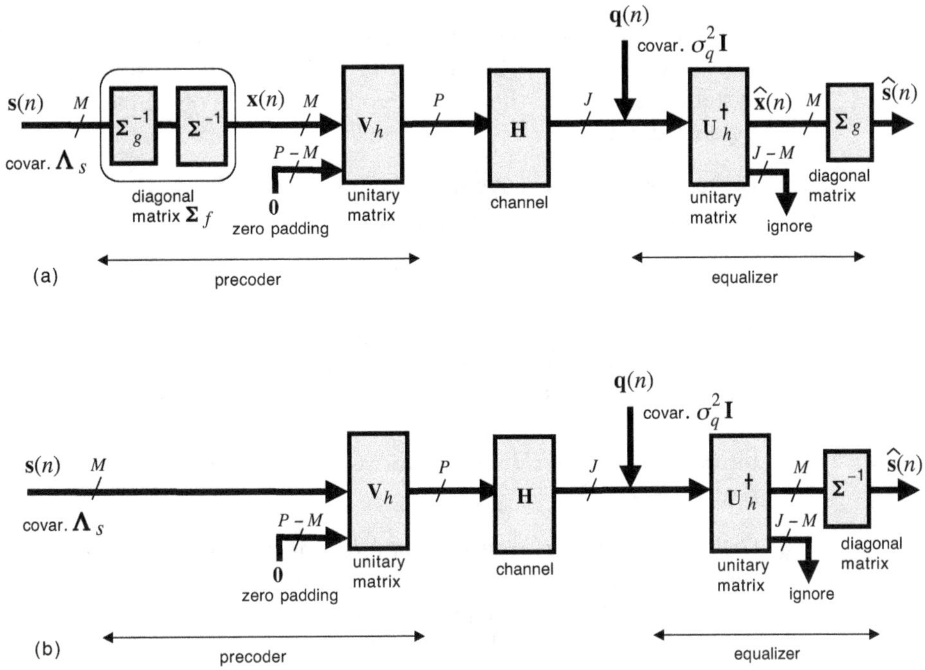

Figure 14.4. (a) Modification of the optimal transceiver by insertion of an arbitrary diagonal matrix $\mathbf{\Sigma}_g$ and its inverse. (b) The effect of choosing $\mathbf{\Sigma}_g = \mathbf{\Sigma}^{-1}$; this moves the diagonal equalizer $\mathbf{\Sigma}^{-1}$ to the receiver side.

A number of points should now be observed:

1. ***Movement of*** $\mathbf{\Sigma}_g$. Even though the rearrangement appears to change the transmitted power from the optimum value to something else, some reflection shows that all is well. For this notice first that $\mathbf{\Sigma}_g$ at the receiver scales the noise power at the kth output by $[\mathbf{\Sigma}_g]_{kk}$. In order to keep the probabilities of error $P_e(k)$ fixed, the signal power P_k also has to be scaled accordingly (because $P_e(k)$ is a function of the SNR at the input of the detector). So at the output of $\mathbf{\Sigma}_g^{-1}$ at the transmitter, the level of signal power is the same as before, so that the total transmitted power (at the input of $\mathbf{H}$) remains unchanged.

2. ***Equalization arbitrarily split.*** The freedom offered by $\mathbf{\Sigma}_g$ implies that the task of equalization can be divided between the transmitter and the receiver arbitrarily because of the arbitrary multipliers $\sigma_{g,k}$. This is unlike the *half and a half* property of the zero-forcing MMSE transceiver without bit allocation described in Chap. 12.

3. ***Orthogonality of precoder.*** From Eq. (14.36) we see that

$$\mathbf{F}^\dagger\mathbf{F} = \mathbf{\Sigma}^{-2},$$

which is diagonal. Thus the M columns of $\mathbf{F}$ are orthogonal. The same property was also observed for the ZF-MMSE transceiver and the MMSE transceiver described in Chaps. 12 and 13, respectively. So the optimal system is such that the M user codes (columns of $\mathbf{F}$) can be assumed to be orthogonal.

4. ***Orthonormal precoder system.*** The presence of $\boldsymbol{\Sigma}_g$ does give us some flexibility in choosing receiver structure. For example, suppose we choose

$$\boldsymbol{\Sigma}_g = \boldsymbol{\Sigma}^{-1}. \tag{14.38}$$

Then at the transmitter the inverses of $\boldsymbol{\Sigma}_g$ and $\boldsymbol{\Sigma}$ cancel each other and we get the equivalent optimal transceiver shown in Fig. 14.4(b). Thus the diagonal equalizer has been moved to the receiver side, which is more traditional. This is called an "orthonormal precoder" system, where the term orthonormal actually means that the precoder matrix $\mathbf{F}$ is unitary, that is, $\mathbf{F}^\dagger\mathbf{F} = \mathbf{I}_M$.

5. ***Orthonormality of precoder is not a loss of generality.*** Notice in particular that the optimal system can be assumed to have a unitary precoder without loss of generality. In Chaps. 12 and 13, where we optimized for MSE, the precoder was shown to be orthogonal but not unitary, that is, the columns of the precoder matrix did not have unit norm. In those chapters, the input covariance was of the form $\sigma_s^2\mathbf{I}$. In this chapter, however, additional flexibility is offered by the diagonal elements of the input covariance $\boldsymbol{\Lambda}_s$. The reader might wonder whether this explains intuitively why the columns of the optimal precoder are allowed to have unit norm because $\boldsymbol{\Lambda}_s$ makes up for this loss of generality. But it turns out that when optimal solution is chosen to have unitary $\mathbf{F}$, the diagonal matrix $\boldsymbol{\Lambda}_s$ (which depends on the specific optimum solution chosen) actually takes the trivial form $\boldsymbol{\Lambda}_s = \sigma_s^2\mathbf{I}$. This will be elaborated in Sec. 14.6.3 (Ex. 14.1, to be specific).

14.6 Further properties of the optimal solutions

In this chapter we optimized the transceiver $\{\mathbf{F}, \mathbf{G}\}$ to minimize the transmitted power under optimum bit allocation (and under the high bit rate assumption). Observe again that the solution is not unique; in fact, there are infinitely many solutions: given an optimal solution $\{\mathbf{F}, \mathbf{G}\}$, suppose we define

$$\mathbf{F}_1 = \mathbf{F}\boldsymbol{\Lambda}, \quad \mathbf{G}_1 = \boldsymbol{\Lambda}^{-1}\mathbf{G}, \tag{14.39}$$

where $\boldsymbol{\Lambda}$ is an arbitrary $M \times M$ diagonal matrix with diagonal elements $\lambda_k > 0$. Then $\{\mathbf{F}_1, \mathbf{G}_1\}$ continues to be an optimal solution. This follows from the fact that we are allowed to insert the arbitrary diagonal matrix $\boldsymbol{\Sigma}_g$ in Fig. 14.4(a). In fact, all solutions of the form in Fig. 14.4(a) can be obtained in this way.

14.6.1 Diagonal interpretation

Since the unitary matrices $\mathbf{V}_h$ and $\mathbf{U}_h$ in Fig. 14.4(a) are obtained from the channel SVD expression $\mathbf{H} = \mathbf{U}_h\mathbf{\Sigma}_h\mathbf{V}_h^\dagger$, it follows that

$$\mathbf{U}_h^\dagger\mathbf{H}\mathbf{V}_h = \mathbf{\Sigma}_h.$$

Since the matrix $\mathbf{V}_h$ in Fig. 14.4(a) has $P - M$ zero-inputs, and since the last $J-M$ outputs of $\mathbf{U}_h^\dagger$ are dropped, the transfer function from the signal $\mathbf{x}(n)$ to the signal $\widehat{\mathbf{x}}(n)$ in Fig. 14.4(a) is precisely the quantity

$$[\mathbf{U}_h^\dagger\mathbf{H}\mathbf{V}_h]_M = [\mathbf{\Sigma}_h]_M,$$

where the subscript M denotes the $M \times M$ submatrix represented by the first M rows and columns. The system can therefore be redrawn[2] as in Fig. 14.5(a). The special configuration of the optimal transceiver shown in Fig. 14.4(b) has the equivalent interpretation shown in Fig. 14.5(b). Thus, the equivalent channel is a set of parallel independent channels with gains $\sigma_{h,k}$ equal to the singular values. Since $\sigma_{h,k}^2$ is the kth **eigenvalue** of $\mathbf{H}^\dagger\mathbf{H}$, we also say that the kth path in the figure represents the kth **eigenmode** of the channel.

Thus, after all the dust clears, the optimal system is schematically equivalent to a diagonal system. The input signal $s_k(n)$ "effectively" feeds directly into $\sigma_{h,k}$. The equalizer effectively inverts the channel using $1/\sigma_{h,k}$. The only "sophisticated" thing that is involved here is bit allocation.

14.6.2 Some invariants of the optimal solutions

All the optimal solutions obtained by changing $\mathbf{\Lambda}$ in Eq. (14.39) obviously have the same transmitted power given by (14.30), because this is the objective function that has been optimized. There are a few other quantities which remain invariant as we change from one optimal solution to another.

14.6.2.A Invariance of $[\mathbf{G}\mathbf{G}^\dagger]_{kk}[\mathbf{F}^\dagger\mathbf{F}]_{kk}$

We now prove an interesting property pertaining to the product $[\mathbf{G}\mathbf{G}^\dagger]_{kk}[\mathbf{F}^\dagger\mathbf{F}]_{kk}$ which arises in the expression for total power (Eq. (14.8)). Refer again to Eq. (14.39) and consider the diagonal elements

$$[\mathbf{F}_1^\dagger\mathbf{F}_1]_{kk} = [\mathbf{\Lambda}\mathbf{F}^\dagger\mathbf{F}\mathbf{\Lambda}]_{kk} = \lambda_k^2[\mathbf{F}^\dagger\mathbf{F}]_{kk}$$

and

$$[\mathbf{G}_1\mathbf{G}_1^\dagger]_{kk} = [\mathbf{\Lambda}^{-1}\mathbf{G}\mathbf{G}^\dagger\mathbf{\Lambda}^{-1}]_{kk} = \frac{[\mathbf{G}\mathbf{G}^\dagger]_{kk}}{\lambda_k^2}.$$

Clearly

$$[\mathbf{G}_1\mathbf{G}_1^\dagger]_{kk}[\mathbf{F}_1^\dagger\mathbf{F}_1]_{kk} = [\mathbf{G}\mathbf{G}^\dagger]_{kk}[\mathbf{F}^\dagger\mathbf{F}]_{kk}.$$

---

[2] The noise components $q_k(n)$ are now the channel noise components filtered through $\mathbf{U}_h^\dagger$, and their covariance remains $\sigma_q^2\mathbf{I}$, since $\mathbf{U}_h^\dagger$ is unitary.

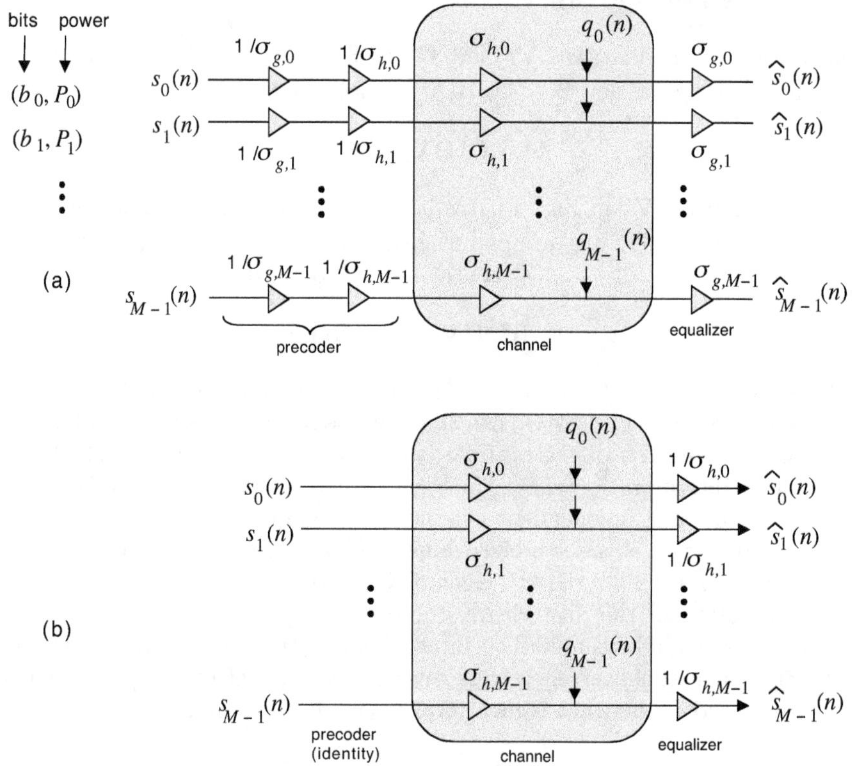

Figure 14.5. Equivalent diagonal representations. (a) The optimal transceiver of Fig. 14.4(a) redrawn in terms of the channel singular values. Here $\sigma_{h,k}$ are the largest M singular values of the channel $\mathbf{H}$. (b) The optimal transceiver of Fig. 14.4(b) redrawn in terms of the channel singular values.

For any fixed k, the value of this product is therefore the same for all the optimal solutions given by Fig. 14.4(a). To find what this value is, just consider any optimal solution, say the one in Fig. 14.4(b). This has the optimal $\mathbf{G}$ in the form

$$\mathbf{G} = [\,\boldsymbol{\Sigma}^{-1} \quad 0\,]\,\mathbf{U}_h^\dagger$$

so that $\mathbf{G}\mathbf{G}^\dagger = \boldsymbol{\Sigma}^{-2}$. Thus $[\mathbf{G}\mathbf{G}^\dagger]_{kk} = 1/\sigma_{h,k}^2$. The corresponding optimal $\mathbf{F}$ is

$$\mathbf{F} = \mathbf{V}_h \begin{bmatrix} \mathbf{I}_M \\ \mathbf{0} \end{bmatrix}$$

so that $\mathbf{F}^\dagger\mathbf{F} = \mathbf{I}_M$ and $[\mathbf{F}^\dagger\mathbf{F}]_{kk} = 1$. Thus

$$[\mathbf{G}\mathbf{G}^\dagger]_{kk}[\mathbf{F}^\dagger\mathbf{F}]_{kk} = \frac{1}{\sigma_{h,k}^2}, \quad 0 \le k \le M-1, \tag{14.40}$$

and it is independent of the specific optimal solution chosen.

14.6.2.B Invariance of bit allocation

From the preceding we can show that the bit allocation $\{b_k\}$ is the same for all optimal solutions. Recall that the bit allocation formula (14.11) is

$$\begin{aligned} b_k &= D - \log_2\left(\frac{\sigma_q^2}{3}\Big(\mathcal{Q}^{-1}\Big(\frac{\mathcal{P}_e(k)}{4}\Big)\Big)^2\right) - \log_2[\mathbf{F}^\dagger\mathbf{F}]_{kk} - \log_2[\mathbf{G}\mathbf{G}^\dagger]_{kk} \\ &= D - \log_2\left(\frac{\sigma_q^2}{3}\Big(\mathcal{Q}^{-1}\Big(\frac{\mathcal{P}_e(k)}{4}\Big)\Big)^2\right) - \log_2\Big([\mathbf{F}^\dagger\mathbf{F}]_{kk}[\mathbf{G}\mathbf{G}^\dagger]_{kk}\Big). \end{aligned}$$

This can be rewritten as

$$b_k = D - \log_2\left(\frac{\sigma_q^2}{3}\Big(\mathcal{Q}^{-1}\Big(\frac{\mathcal{P}_e(k)}{4}\Big)\Big)^2\right) + \log_2 \sigma_{h,k}^2. \tag{14.41}$$

A number of conclusions can now be drawn.

1. ***Bit allocation is invariant***. The last term in Eq. (14.41) depends only on the channel. The second term depends only on the QoS specifications. The first term is determined by the total number of bits, which is fixed. Thus, the optimal bit allocation $\{b_k\}$ is the same for all optimal solutions.

2. ***Good eigenmodes get more bits***. From Eq. (14.41) we see that if $\sigma_{h,k}$ is larger (large gain) then more bits are allocated to $s_k(n)$.

3. ***Dependence on QoS***. From the second term of Eq. (14.41) we see that if $\mathcal{P}_e(k)$ is larger then more bits are allocated.

4. ***Quantities independent of k***. From Eq. (14.41) it follows that

$$\frac{2^{b_k}\sigma_q^2}{3\sigma_{h,k}^2}\Big(\mathcal{Q}^{-1}\Big(\frac{\mathcal{P}_e(k)}{4}\Big)\Big)^2 = 2^D, \quad 0 \le k \le M-1. \tag{14.42}$$

   Thus, optimal bit allocation is such that the product on the left hand side is *independent of* k. This is true for *all* optimal solutions. This observation will be useful later.

14.6.3 Remarks on power allocation

Does the optimal solution have power allocation incorporated into it automatically? To address this question we have to understand what is meant by power allocation here. Referring again to the general form of the transceiver in Fig. 14.1, we know that the sum of the powers at the output terminals of $\mathbf{F}$ is the total transmitted power. But each of the individual power components at the output of $\mathbf{F}$ (i.e., the power transmitted from individual antennas) cannot be interpreted in a simple way; though the kth output of $\mathbf{F}$ is fed into the kth input of the channel, it affects all the outputs of $\mathbf{H}$ (because $\mathbf{H}$ is in general not diagonal).

The distribution of power among the symbols $s_k(n)$ (i.e., the diagonal elements of the covariance matrix $\mathbf{\Lambda}_s$), can be considered to be the power distribution "among the different users" if $s_k(n)$ can be regarded as the kth user (i.e., if the signals from various users are assumed to be collected in one place), but this has only restricted significance as we shall see (Sec. 14.6.3.B).

14.6.3.A Power allocated to eigenmodes of the channel

There is a different way to look at power allocation. Consider Fig. 14.5(a), which is an equivalent drawing of the optimal system of Fig. 14.4(a). The power inputs to the channel eigenmodes $\sigma_{h,k}$ can be regarded as the power allocation. The kth eigenmode has input power[3]

$$P_{eig,k} = \frac{P_k}{\sigma_{g,k}^2 \sigma_{h,k}^2}, \tag{14.43}$$

where

$$P_k = E|s_k(n)|^2;$$

P_k is the simpler notation for $[\mathbf{\Lambda}_s]_{kk}$ as before. Recall that in order to satisfy the QoS constraints, the kth user power should be taken as

$$P_k \approx \frac{\sigma_{e_k}^2 2^{b_k}}{3}\left(\mathcal{Q}^{-1}\left(\frac{\mathcal{P}_e(k)}{4}\right)\right)^2, \tag{14.44}$$

where $\sigma_{e_k}^2 = \sigma_q^2[\mathbf{G}\mathbf{G}^\dagger]_{kk}$. As the optimal $\mathbf{G}$ has the form $\mathbf{G} = [\,\mathbf{\Sigma}_g \quad \mathbf{0}\,]\,\mathbf{U}_h^\dagger$, it follows that $[\mathbf{G}\mathbf{G}^\dagger]_{kk} = \sigma_{g,k}^2$ so that

$$P_k \approx \frac{\sigma_q^2 \sigma_{g,k}^2 2^{b_k}}{3}\left(\mathcal{Q}^{-1}\left(\frac{\mathcal{P}_e(k)}{4}\right)\right)^2. \tag{14.45}$$

The input power to the kth eigenmode is therefore

$$P_{eig,k} \approx \frac{\sigma_q^2 2^{b_k}}{3\sigma_{h,k}^2}\left(\mathcal{Q}^{-1}\left(\frac{\mathcal{P}_e(k)}{4}\right)\right)^2 = 2^D \quad \text{for all } k, \tag{14.46}$$

where we have substituted from Eq. (14.42). This proves the following result:

♠**Theorem 14.3.** *Power in the eigenmodes.* The optimal transceiver solutions $\{\mathbf{F}, \mathbf{G}\}$ which minimize transmitted power under optimal bit allocation are such that all eigenmodes of the channel are allocated identical power. ◇

---

[3] Note that the power input to the eigenmode is not a "physical" power. It is conceptual.

14.6.3.B User power allocation

Equation (14.45) can be rewritten as

$$P_k = \sigma_{g,k}^2 \sigma_{h,k}^2 2^D, \tag{14.47}$$

where we have substituted from Eq. (14.42). This quantity is in general dependent on k. Thus the powers allocated to different signals $s_k(n)$ are in general different. But this allocation depends on the specific configuration chosen for the optimal solution. For example, it is not the same for Fig. 14.4(a) and 14.4(b).

Example 14.1: User power allocation with orthonormal precoding

Consider the specific optimal transceiver configuration where the free parameter $\sigma_{g,k}$ is chosen as $\sigma_{g,k} = 1/\sigma_{h,k}$. This corresponds to the optimal transceiver configuration shown in Fig. 14.4(b) (equivalently Fig. 14.5(b)), which has all equalization taking place at the receiver. In this case the precoder is *orthonormal*, that is,

$$\mathbf{F}^\dagger \mathbf{F} = \mathbf{I}_M, \tag{14.48}$$

because

$$\mathbf{F} = \mathbf{V}_h \begin{bmatrix} \mathbf{I}_M \\ \mathbf{0} \end{bmatrix}.$$

Since $\sigma_{g,k} = 1/\sigma_{h,k}$ we see from Eq. (14.47) that

$$P_k = 2^D = \text{constant for all } k \tag{14.49}$$

in this case. Thus, when the free parameters in the optimal transceiver solution are chosen such that the precoder is orthonormal, the input covariance matrix which satisfies the QoS requirements has the form

$$\mathbf{\Lambda}_s = \alpha \mathbf{I} \tag{14.50}$$

for constant α. The equivalent diagonal structure of Fig. 14.5(b) gives further insight as to why the user powers are identical. Since the powers fed into all the eigenmodes are identical for any optimal solution (Theorem 14.3), and since $s_k(n)$ feeds directly into the kth eigenmode $\sigma_{h,k}$ when the precoder is orthonormal, all the user powers P_k have to be identical.

Summarizing this section, we see that the only nontrivial thing that is happening in the optimal transceiver is optimal bit allocation. There is no **eigenmode power allocation** going into the channel. The largest eigenmode $\sigma_{h,0}$ and the smallest eigenmode $\sigma_{h,M-1}$ receive the same power no matter how different they are (they are nonzero because the channel is assumed to have rank M). The **user power allocation** $\{P_k\}$ changes with the specific configuration chosen for the nonunique optimal solution. In particular, when the optimal solution is chosen such that the precoder is orthonormal ($\mathbf{F}^\dagger\mathbf{F} = \mathbf{I}$), the user powers P_k are identical for all k because they feed directly into the channel eigenmodes (Fig. 14.4(b), equivalently Fig. 14.5(b)).

14.6.4 Comparison with MMSE transceivers

Figure 14.6(a) shows the general form of the optimal transceivers derived in Chaps. 12 and 13 for minimizing the mean square error under a power constraint. The diagonal matrices $\mathbf{\Sigma}_f$ and $\mathbf{\Sigma}_g$ depend on whether there is zero forcing or not. Compare this with the optimal transceiver configuration in Fig. 14.6(b), which minimizes the transmitted power under optimal bit allocation. The MMSE transceiver has input covariance $\sigma_s^2\mathbf{I}$ by assumption, and the power-minimizing transceiver also has input covariance $\sigma_s^2\mathbf{I}$ (as shown in Ex. 14.1). The MMSE transceiver has diagonal matrix $\mathbf{\Sigma}_f$ before the unitary matrix $\mathbf{V}_h$. This diagonal matrix performs part of the equalization, and the matrix $\mathbf{\Sigma}_g$ at the receiver performs part of the equalization. We cannot arbitrarily trade off the equalizers at the transmitter and receiver of the MMSE system. This is unlike the situation in the power minimizer of Fig. 14.4(a). Since the precoder of the MMSE transceiver has the form

$$\mathbf{F} = \mathbf{V}_h \begin{bmatrix} \mathbf{\Sigma}_f \\ \mathbf{0} \end{bmatrix}, \tag{14.51}$$

we have

$$\mathbf{F}^\dagger \mathbf{F} = \mathbf{\Sigma}_f^2. \tag{14.52}$$

That is, the columns of the precoder are orthogonal, though not orthonormal (i.e., the columns do not have unit norm). Compare this with the power minimizer configuration shown in Fig. 14.6(b), which has the precoder

$$\mathbf{F} = \mathbf{V}_h \begin{bmatrix} \mathbf{I}_M \\ \mathbf{0} \end{bmatrix}. \tag{14.53}$$

This precoder is orthonormal (or unitary) because it satisfies[4]

$$\mathbf{F}^\dagger \mathbf{F} = \mathbf{I}_M. \tag{14.54}$$

Thus, in the power-minimization problem the optimal solution can be assumed to have an orthonormal precoder without loss of generality. The same is not true for the MMSE problem.

For further clarity, Fig. 14.7 shows the diagonal equivalent structures for the two systems. Here we have used the fact that

$$\mathbf{U}_h^\dagger \mathbf{H} \mathbf{V}_h = \mathbf{\Sigma}_h. \tag{14.55}$$

The first M dominant singular values $\sigma_{h,k}$ of the channel are shown in the figure. We see that, in the MMSE system, the optimal multipliers $\sigma_{f,k}$ at the transmitter determine the power allocation into the different eigenmodes $\sigma_{h,k}$ of the channel. This power allocation is a crucial feature of the MMSE system, since the multipliers $\sigma_{f,k}$ and $\sigma_{g,k}$ cannot be traded arbitrarily in Fig. 14.7(a).

[4]Note that $\mathbf{F}\mathbf{F}^\dagger \neq \mathbf{I}_P$ unless $P = M$.

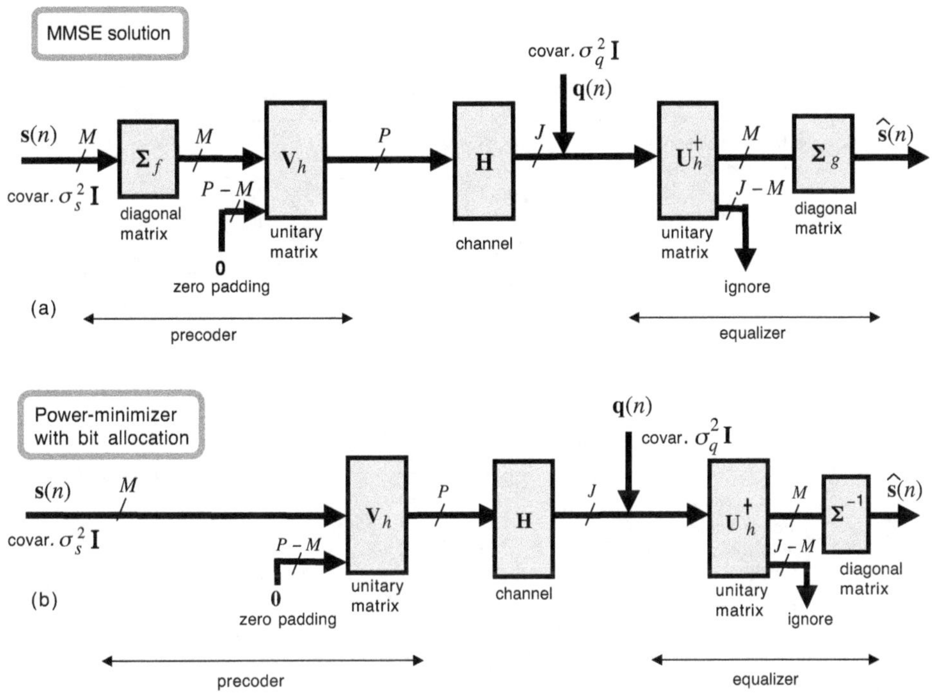

Figure 14.6. (a) General form of the optimal transceiver which minimizes MSE with or without zero forcing (ZF). (b) The optimal ZF transceiver which minimizes transmitted power under optimal bit allocation.

Compare this with the power minimizer shown in Fig. 14.7(b). In this system all eigenmodes receive identical power σ_s^2. The same was shown to be true for the more general configuration of the power minimizer (Theorem 14.3). Thus the power minimizer does not have power allocation; optimality is achieved essentially through bit allocation.

14.7 Coding gain due to bit allocation

If we take the optimal transceiver $\{\mathbf{F}, \mathbf{G}\}$ and forcefully assign identical bits to all the components $s_k(n)$ (i.e., set $b_k = b$), then the required channel power increases. Let P_{brute} denote this power. The coding gain due to bit allocation is defined as

$$\mathcal{G} = \frac{P_{brute}}{P_{min}}. \tag{14.56}$$

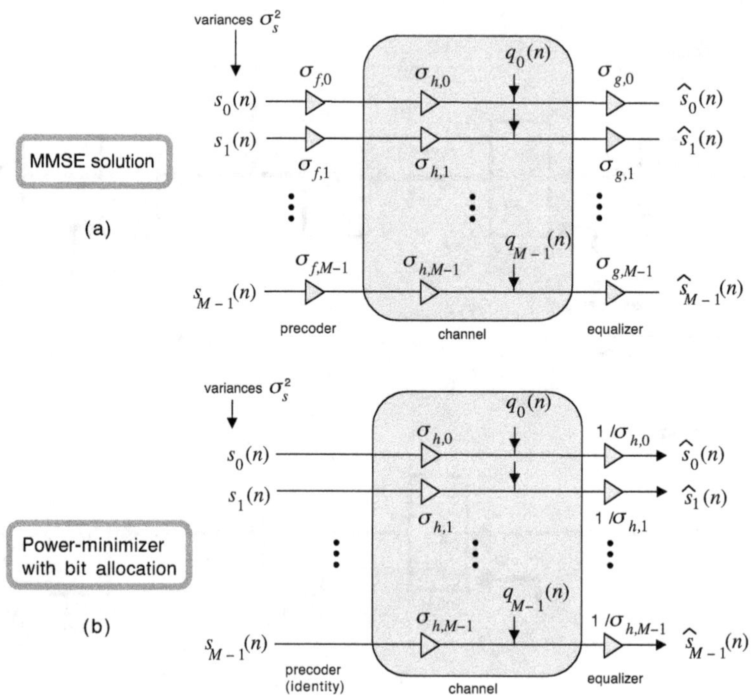

Figure 14.7. Diagonal representations of (a) the optimal transceiver which minimizes MSE with or without zero forcing, and (b) optimal ZF transceiver which minimizes transmitted power under optimal bit allocation.

Recall that the expression for the total transmitted power is given by Eq. (14.8), that is,

$$P_{trans} = \sum_{k=0}^{M-1} c_k 2^{b_k} [\mathbf{F}^\dagger \mathbf{F}]_{kk} [\mathbf{G}\mathbf{G}^\dagger]_{kk}, \tag{14.57}$$

where we have used the usual notation for c_k given in Eq. (14.9). Since the optimal transceiver satisfies Eq. (14.40) we have

$$P_{brute} = 2^b \sum_{k=0}^{M-1} \frac{c_k}{\sigma_{h,k}^2}. \tag{14.58}$$

Compare this with the expression for the optimal power (14.30), which can be rewritten as

$$P_{min} = M 2^b \Big(\prod_{k=0}^{M-1} \frac{c_k}{\sigma_{h,k}^2} \Big)^{1/M}. \tag{14.59}$$

The coding gain (14.56) is therefore given by

$$\mathcal{G} = \frac{\dfrac{1}{M}\displaystyle\sum_{k=0}^{M-1} \frac{c_k}{\sigma_{h,k}^2}}{\left(\displaystyle\prod_{k=0}^{M-1} \frac{c_k}{\sigma_{h,k}^2}\right)^{1/M}} = \frac{AM\left(\dfrac{c_k}{\sigma_{h,k}^2}\right)}{GM\left(\dfrac{c_k}{\sigma_{h,k}^2}\right)} \tag{14.60}$$

where AM and GM stand for arithmetic mean and geometric mean, respectively. For the special case where the required BER values $\mathcal{P}_e(k)$ are identical for all k, c_k is identical for all k. In this case the coding gain reduces to

$$\mathcal{G} = \frac{AM\left(\dfrac{1}{\sigma_{h,k}^2}\right)}{GM\left(\dfrac{1}{\sigma_{h,k}^2}\right)} \tag{14.61}$$

Recall here that $\sigma_{h,k}$ are the first M dominant singular values of the channel $\mathbf{H}$. For a channel with large variation in singular values, the coding gain is large whereas for a channel with $\sigma_{h,k}$ nearly identical for all k, the coding gain is quite small.

14.8 Concluding remarks

In the previous chapters we minimized the mean square reconstruction error in transceivers subject to the power constraint and possibly the zero-forcing constraint. In this chapter we showed how the transmitted power can be minimized for a given set of error probabilities by optimizing the transceiver matrices and bit allocation. For further interesting work on transceiver optimization with bit allocation, the reader should study Pandharipande and Dasgupta [2003], Palomar and Barbarossa [2005], Vemulapalli, Dasgupta, and Pandharipande [2006], and Weng *et al.*, [2010a]. In Chap. 15 we will consider a special class of transceivers which are restricted to have orthonormal precoders. Chapter 16 addresses the problem of minimizing the average symbol error probability. A one-page summary of the results of this chapter is included in Appendix I at the end of the book.

Appendix to Chapter 14

14.A Which right inverse is the best?

The precoder $\mathbf{F}$ in Eq. (14.14) represents the minimum-norm right inverse of $\mathbf{GH}$. We now show that this is better than using any other right inverse. For this it is convenient to use SVD representations for the matrices involved. The matrix $\mathbf{GH}$ is of size $M \times P$ and has the SVD

$$\mathbf{GH} = \mathbf{U}\,[\,\boldsymbol{\Sigma} \quad \mathbf{0}\,]\,\mathbf{V}^{\dagger}, \tag{14.62}$$

where $\mathbf{U}$ and $\mathbf{V}$ are unitary and $\boldsymbol{\Sigma}$ is $M \times M$. From Appendix C we know that any right inverse of $\mathbf{GH}$ can be expressed as

$$\mathbf{F} = \mathbf{V}\begin{bmatrix}\boldsymbol{\Sigma}^{-1}\\ \mathbf{X}\end{bmatrix}\mathbf{U}^{\dagger}, \tag{14.63}$$

where $\mathbf{X}$ is in general arbitrary, and $\mathbf{X} = 0$ for the minimum-norm solution. The transmitted power is given by

$$P_{trans} = \text{Tr}\left(\mathbf{F}\boldsymbol{\Lambda}_s\mathbf{F}^{\dagger}\right) = \text{Tr}\left(\mathbf{F}^{\dagger}\mathbf{F}\boldsymbol{\Lambda}_s\right).$$

Now,

$$\mathbf{F}^{\dagger}\mathbf{F} = \mathbf{U}\,[\,\boldsymbol{\Sigma}^{-1} \quad \mathbf{X}^{\dagger}\,]\,\mathbf{V}^{\dagger}\mathbf{V}\begin{bmatrix}\boldsymbol{\Sigma}^{-1}\\ \mathbf{X}\end{bmatrix}\mathbf{U}^{\dagger} = \mathbf{U}(\boldsymbol{\Sigma}^{2} + \mathbf{X}^{\dagger}\mathbf{X})\mathbf{U}^{\dagger}$$

so that

$$\begin{aligned} P_{trans} = \text{Tr}\left(\mathbf{F}^{\dagger}\mathbf{F}\boldsymbol{\Lambda}_s\right) &= \text{Tr}\left(\mathbf{U}(\boldsymbol{\Sigma}^{2} + \mathbf{X}^{\dagger}\mathbf{X})\mathbf{U}^{\dagger}\boldsymbol{\Lambda}_s\right) \\ &= \text{Tr}\left(\mathbf{U}\boldsymbol{\Sigma}^{2}\mathbf{U}^{\dagger}\boldsymbol{\Lambda}_s\right) + \text{Tr}\left(\mathbf{U}\mathbf{X}^{\dagger}\mathbf{X}\mathbf{U}^{\dagger}\boldsymbol{\Lambda}_s\right). \end{aligned}$$

Since $\boldsymbol{\Lambda}_s$ is a diagonal matrix with diagonal elements $P_k \geq 0$, the second term on the right-hand side is

$$\sum_k [\mathbf{U}\mathbf{X}^{\dagger}\mathbf{X}\mathbf{U}^{\dagger}]_{kk}P_k. \tag{14.64}$$

For any $\mathbf{X}$ whatsoever, the matrix $\mathbf{U}\mathbf{X}^{\dagger}\mathbf{X}\mathbf{U}^{\dagger}$ is positive semidefinite so that its diagonal elements are non-negative. So the preceding sum is always non-negative. This means that the arbitrary term $\mathbf{X}$ in the right inverse $\mathbf{F}$ can only increase the transmitted power without changing the received signal or noise in any way. The best strategy is therefore to set $\mathbf{X} = \mathbf{0}$, that is, to take $\mathbf{F}$ to be the minimum-norm right inverse.

Problems

14.1. Assume

$$\mathbf{G} = \begin{bmatrix} 1 & 1 \\ 1 & -1 \end{bmatrix}, \quad \mathbf{F} = \begin{bmatrix} 3 & 1 \\ 1 & 3 \end{bmatrix},$$

and a QAM constellation with desired error probabilities $\mathcal{P}_e(k) = 10^{-6}$ for all k. Assume an average number of bits $b = 8$, and $\sigma_q^2 = 0.01$.

1. Calculate the power (14.12) with optimal bit allocation.
2. Calculate the power (14.8) if the bits are uniformly allocated.
3. What is the gain due to bit allocation?

Note: The coding gain is the ratio of "P_{trans} with uniform bit allocation" over "P_{trans} with optimal bit allocation."

14.2. Assume all quantities are as in Problem 14.1 except that

$$\mathbf{F} = \begin{bmatrix} 3 & 0.1 \\ 3 & 0.2 \end{bmatrix}.$$

What is the gain due to bit allocation, and what makes it so different?

14.3. Assume all quantities are as in Problem 14.1 except $\mathbf{F}$. Find an $\mathbf{F}$ such that the gain due to bit allocation is greater than 20.

14.4. Consider the channel

$$\mathbf{H} = \begin{bmatrix} 4 & 4 & 4 \\ 4 & 3 & 4 \\ 4 & 2.9 & 3.9 \end{bmatrix}.$$

Assume a QAM constellation with $\mathcal{P}_e(k) = 10^{-7}$ for all k, $\sigma_q^2 = 0.01$, and $b = 6$ bits.

1. Calculate the minimized power P_{trans} in Eq. (14.30).
2. With the precoder and equalizer chosen optimally as in Eqs. (14.28) and (14.29), if we use uniform bit allocation instead of optimal allocation, what will the power P_{trans} be?

14.5. Consider the circulant channel

$$\mathbf{H} = \begin{bmatrix} 3.1 & 3 & 3 \\ 3 & 3.1 & 3 \\ 3 & 3 & 3.1 \end{bmatrix},$$

which happens to have positive eigenvalues.

1. Compute the precoder and equalizer matrices $\mathbf{F}$ and $\mathbf{G}$ in the optimal transceiver structure with optimal bit allocation.
2. With $\mathcal{P}_e(k) = 10^{-7}$ for all k, $\sigma_q^2 = 0.01$, and $b = 8$ bits, what is the optimal bit allocation? What is the coding gain due to bit allocation (Sec. 14.7).

Hint: Circulants are diagonalized by the DFT matrix.

14.6. Repeat Problem 14.5 for the circulant channel

$$\mathbf{H} = \begin{bmatrix} 4 & 1 & 1 \\ 1 & 4 & 1 \\ 1 & 1 & 4 \end{bmatrix},$$

which also has positive eigenvalues. Why is the coding gain so different from that in Problem 14.5?

14.7. For this problem assume $\mathcal{P}_e(k) = 10^{-7}$ for all k, and channel noise variance $\sigma_q^2 = 0.01$. Assume block size $M = 4$, and the average number of bits $b = 8$. The distribution of the bits b_k in Eq. (14.41) depends on the distribution of the channel singular values $\{\sigma_{h,k}\}$.

1. Find a set of singular values $\{\sigma_{h,k}\}$ such that all b_k calculated from Eq. (14.41) are positive.
2. Find a set of singular values $\{\sigma_{h,k}\}$ such that only one b_k calculated from Eq. (14.41) is positive.

15

Transceivers with orthonormal precoders

15.1 Introduction

In the preceding three chapters we addressed three different optimization problems for transceivers. Namely, MMSE transceivers under the zero-forcing (ZF) constraint (Chap. 12), MMSE transceivers without the ZF constraint (Chap. 13), and transceivers which minimize transmitted power under optimal bit allocation and the ZF constraint (Chap. 14). We found that the optimal precoder in each case turns out to be orthogonal, that is,

$$\mathbf{F}^\dagger\mathbf{F} = \mathbf{\Sigma}_f^2 \tag{15.1}$$

for some diagonal $\mathbf{\Sigma}_f$. Furthermore, for the transceiver which minimizes power under bit allocation, we found that the precoder can be assumed to be orthonormal without loss of generality:

$$\mathbf{F}^\dagger\mathbf{F} = \mathbf{I}_M. \tag{15.2}$$

Thus, for power minimization, orthonormality is not a loss of generality, but for the MMSE problem, orthonormality *is* a loss of generality. The reader should review Sec. 14.6.4 for a detailed comparison of these systems.

In this chapter we consider the MMSE problem with the precoders restricted to satisfy orthonormality (15.2). We call these *orthonormal precoders.* We also say that the $P \times M$ matrix $\mathbf{F}$ is "unitary" even though it should be remembered that when $P > M$, $\mathbf{F}\mathbf{F}^\dagger \neq \mathbf{I}_P$.

Scope and outline In Sec. 15.2 we consider the MMSE problem for the case of orthonormal precoders restricted to be square matrices. This is generalized in Sec. 15.3 for the case of rectangular matrix precoders. While orthonormality is a loss of generality it offers some simplifications in the design of the preocoder, as we shall see. One example is the case of OFDM systems which use the cyclic prefix. In this case the *optimal unitary precoder is independent of the channel* as we shall show. Thus the channel state information for equalization is required only at the receiver in these cases. The main results of the chapter are summarized in Theorems 15.1, 15.2, and 15.3.

Assumptions and reminders

1. ***Covariances***. We assume the signal and noise covariances to be

$$\mathbf{R_{ss}} = \sigma_s^2 \mathbf{I}_M, \quad \mathbf{R_{qq}} = \sigma_q^2 \mathbf{I}_J. \tag{15.3}$$

2. ***Channel SVD***. The channel $\mathbf{H}$ is a $J \times P$ matrix, and its SVD will be written in the form

$$\mathbf{H} = \underbrace{\mathbf{U}_h}_{J\times J} \underbrace{\mathbf{\Sigma}_h}_{J\times P} \underbrace{\mathbf{V}_h^\dagger}_{P\times P}, \tag{15.4}$$

where $\mathbf{U}_h$ and $\mathbf{V}_h$ are unitary.

3. ***Power constraint***. Since $\mathbf{F}^\dagger\mathbf{F} = \mathbf{I}_M$ we see that the transmitted power is

$$p_0 = \sigma_s^2 \text{Tr}\,(\mathbf{F}\mathbf{F}^\dagger) = \sigma_s^2 \text{Tr}\,(\mathbf{F}^\dagger\mathbf{F}) = M\sigma_s^2. \tag{15.5}$$

We assume that p_0 is fixed during all optimization.

15.2 Orthonormal precoders restricted to be square

We shall first consider the case where $P = M$, that is, the precoder is a square matrix with size $M \times M$. So we have a **non-redundant** precoder and a $J \times M$ channel $\mathbf{H}$.

15.2.1 ZF-MMSE transceivers

First we consider the case where the zero-forcing constraint $\mathbf{GHF} = \mathbf{I}$ is in place. (This automatically assumes the channel has rank M.) Recall from Sec. 12.3.1 that under the ZF constraint the MSE is

$$\mathcal{E}_{mse} = \sigma_q^2 \text{Tr}\,(\mathbf{F}^\dagger\mathbf{H}^\dagger\mathbf{H}\mathbf{F})^{-1}. \tag{15.6}$$

Since $\mathbf{F}$ is $M \times M$ unitary, we have $\mathbf{F}^\dagger\mathbf{F} = \mathbf{F}\mathbf{F}^\dagger = \mathbf{I}$, and the preceding equation can be simplified to

$$\mathcal{E}_{mse} = \sigma_q^2 \text{Tr}\left(\mathbf{F}^\dagger(\mathbf{H}^\dagger\mathbf{H})^{-1}\mathbf{F}\right) = \sigma_q^2 \text{Tr}\left(\mathbf{F}\mathbf{F}^\dagger(\mathbf{H}^\dagger\mathbf{H})^{-1}\right) = \sigma_q^2 \text{Tr}\,(\mathbf{H}^\dagger\mathbf{H})^{-1},$$

which is independent of the precoder. Thus, *any unitary precoder is as good as any other* if we restrict $\mathbf{F}$ to be $M \times M$ unitary.[1] For any chosen $\mathbf{F}$ we simply take the equalizer to be

$$\mathbf{G} = (\mathbf{HF})^{\#} \tag{15.7}$$

to satisfy the zero-forcing condition. Here the notation $\mathbf{A}^{\#}$ represents the (minimum-norm) left inverse of the matrix $\mathbf{A}$ (see Eq. (C.7b) in Appendix C). The reconstruction error then simplifies to

$$\mathcal{E}_{mse} = \sigma_q^2 \text{Tr}\,(\mathbf{H}^\dagger\mathbf{H})^{-1} = \sigma_q^2 \sum_{k=0}^{M-1} \frac{1}{\sigma_{h,k}^2} \tag{15.8}$$

where $\sigma_{h,k}$ are the M singular values of the channel $\mathbf{H}$ (all $\sigma_{h,k} > 0$ because the channel is assumed to have rank M).[2]

We now present three examples of the choice of unitary $\mathbf{F}$. Since any unitary $\mathbf{F}$ is as good as any other, all of these are "optimal" solutions.

1. ***Lazy precoder***. If $\mathbf{F} = \mathbf{I}$ we have $\mathbf{G} = \mathbf{H}^{\#}$. This transceiver is shown in Fig. 15.1(a). This system has the advantage that all "collaboration" between the different $s_k(n)$'s is done at the receiver. This is suitable for a multiuser system in multiple access mode (Sec. 4.5).

2. ***SVD precoder***. With the channel expressed in SVD form (15.4), if we decide to choose the unitary precoder $\mathbf{F} = \mathbf{V}_h$, then the equalizer is

$$\mathbf{G} = (\mathbf{HF})^{\#} = (\mathbf{U}_h\mathbf{\Sigma}_h)^{\#} = [\,[\mathbf{\Sigma}_h]_M^{-1} \quad \mathbf{0}\,]\,\mathbf{U}_h^\dagger, \tag{15.9}$$

   where $[\mathbf{\Sigma}_h]_M$ is the $M \times M$ diagonal matrix with diagonal elements equal to $\sigma_{h,k}$. This transceiver is shown in Fig. 15.1(b).

3. ***Cyclic-prefix precoder***. Recall that in the cyclic-prefix system the channel is made to look like an $M \times M$ circulant matrix. In this case

$$\mathbf{H} = \mathbf{W}^{-1}\mathbf{\Lambda}_c\mathbf{W}. \tag{15.10}$$

   Here $\mathbf{\Lambda}_c$ is a diagonal matrix containing the DFT coefficients of the channel, and $\mathbf{W}$ is the DFT matrix. Since $\mathbf{W}^\dagger\mathbf{W} = M\mathbf{I}$, the matrix $\mathbf{W}^\dagger/\sqrt{M}$ is unitary and we can take the unitary precoder and the zero-forcing equalizer to be

$$\mathbf{F} = \frac{\mathbf{W}^\dagger}{\sqrt{M}}, \quad \mathbf{G} = \frac{\mathbf{\Lambda}_c^{-1}\mathbf{W}}{\sqrt{M}}. \tag{15.11}$$

   This results in the familiar OFDM system described in Sec. 7.3. See Fig. 15.1(c). Since any unitary $\mathbf{F}$ is as good as any other, we can say that *the IDFT matrix* $\mathbf{W}^\dagger/\sqrt{M}$ *is also an "optimum unitary precoder" solution.*

---

[1] Note that we are only concerned about minimization of MSE. If we are interested in minimizing error probability, as in Chap. 16, then the situation is different.

[2] The second equality in Eq. (15.8) follows because (a) the trace is the sum of eignevalues, (b) the eigenvalues of $\mathbf{H}^\dagger\mathbf{H}$ are $\sigma_{h,k}^2$, and (c) the eigenvalues of $(\mathbf{H}^\dagger\mathbf{H})^{-1}$ are the reciprocals of those of $\mathbf{H}^\dagger\mathbf{H}$.

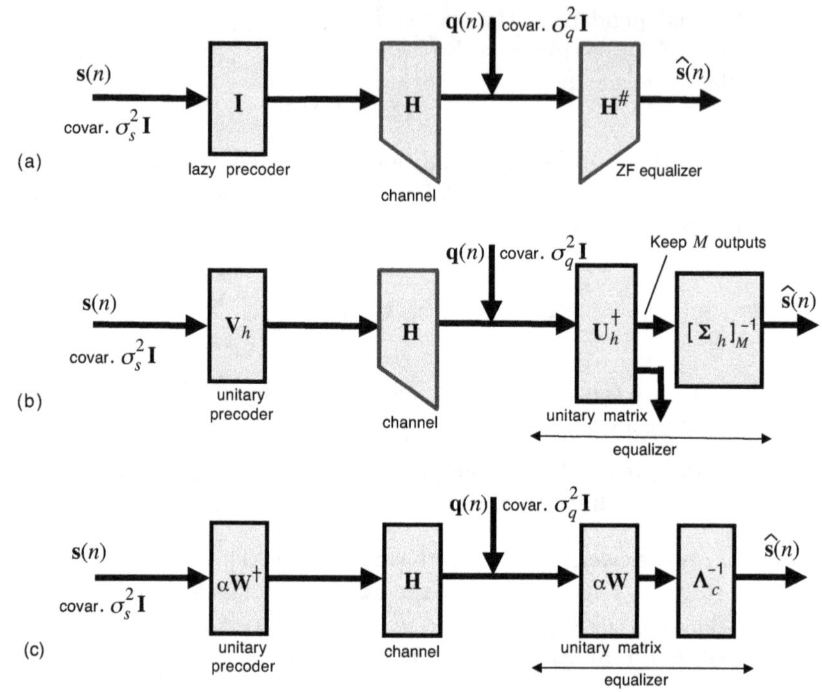

Figure 15.1. Examples of non-redundant ($P = M$) transceivers with orthonormal precoder and zero-forcing equalizer. (a) Identity (or "lazy") precoder, (b) precoder which "diagonalizes" the channel matrix, and (c) the IDFT precoder as used in cyclic prefix systems ($\alpha = 1/\sqrt{M}$). All of these are optimal in the ZF-MMSE sense.

15.2.1.A Comparison with general nonunitary precoder

We now compare the unitary precoder with the general precoder, assuming the channel input power to be the same for both systems. Recall that, when the precoder is unitary, the transmitted power (trace of the covariance at the input of the channel $\mathbf{H}$) is $p_0 = \sigma_s^2 M$. If we wish to have more control on p_0 (independent of σ_s^2), then we can insert a matrix $\alpha\mathbf{I}$ at the transmitter and a matrix $\alpha^{-1}\mathbf{I}$ at the receiver. This preserves the ZF condition. This is shown in Fig. 15.2 where the unitary precoder is denoted as $\mathbf{U}_f$, and its ZF equalizer is denoted as $\mathbf{G}_1$. The scalar variable $\alpha > 0$ is used for power control. The transmitted power is now

$$p_0 = \sigma_s^2 \alpha^2 M. \tag{15.12}$$

We already showed that when the precoder is unitary, the MMSE under the ZF constraint is as in Eq. (15.8). This corresponds to $\alpha = 1$. For arbitrary α we therefore have

$$\mathcal{E}_{mmse} = \frac{\sigma_q^2}{\alpha^2} \sum_{k=0}^{M-1} \frac{1}{\sigma_{h,k}^2}$$

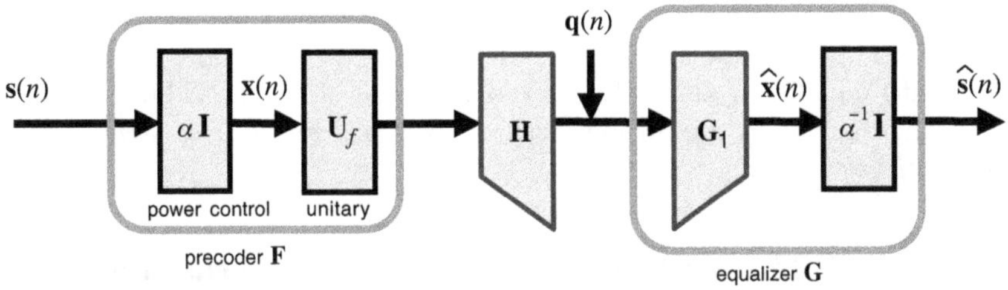

Figure 15.2. Inserting the multiplier α in the unitary precoder for power control.

We can eliminate α using Eq. (15.12) and write

$$\mathcal{E}_{mmse} = \frac{M\sigma_q^2\sigma_s^2}{p_0}\sum_{k=0}^{M-1}\frac{1}{\sigma_{h,k}^2} \tag{15.13}$$

For the case where the precoder is not restricted to be orthonormal (Sec. 12.4), the ZF-MMSE system has MSE

$$\mathcal{E}_{mmse} = \frac{\sigma_q^2\sigma_s^2}{p_0}\left(\sum_{k=0}^{M-1}\frac{1}{\sigma_{h,k}}\right)^2. \tag{15.14}$$

Summarizing, for the ZF-MMSE transceiver the minimized MSE is given by

$$\mathcal{E}_{mmse} = \begin{cases} \dfrac{M\sigma_q^2\sigma_s^2}{p_0}\displaystyle\sum_{k=0}^{M-1}\frac{1}{\sigma_{h,k}^2} & \text{(orthonormal precoder)} \\ \dfrac{\sigma_q^2\sigma_s^2}{p_0}\left(\displaystyle\sum_{k=0}^{M-1}\frac{1}{\sigma_{h,k}}\right)^2 & \text{(unrestricted precoder).} \end{cases} \tag{15.15}$$

Defining the vector

$$\mathbf{z} \stackrel{\Delta}{=} \begin{bmatrix} \frac{1}{\sigma_{h,0}} & \frac{1}{\sigma_{h,1}} & \cdots & \frac{1}{\sigma_{h,M-1}} \end{bmatrix}^T \tag{15.16}$$

we see that Eq. (15.15) can be written as

$$\mathcal{E}_{mmse} = \begin{cases} A\|\mathbf{z}\|_2^2 & \text{(orthonormal precoder)} \\ \frac{A}{M}\|\mathbf{z}\|_1^2 & \text{(unrestricted precoder),} \end{cases} \tag{15.17}$$

where A is a constant and $\|\mathbf{z}\|_p$ denotes the p-norm of the vector $\mathbf{z}$, that is,

$$\|\mathbf{z}\|_p = \left(\sum_{k=0}^{M-1}|z_k|^p\right)^{1/p}. \tag{15.18}$$

Using Cauchy-Schwartz inequality (see Appendix A) it follows readily that[3]

$$\frac{\|\mathbf{z}\|_1^2}{M} \leq \|\mathbf{z}\|_2^2 \tag{15.19}$$

with equality if and only if $\mathbf{z}$ has identical elements. Thus, unless the channel singular values are all identical, *the general precoder is strictly better than the unitary precoder.*

Example 15.1: Orthonormal versus non-orthonormal precoder

The gain obtained by going from an orthonormal precoder to an arbitrary precoder is, from Eq. (15.15),

$$\mathcal{G} = \frac{\mathcal{E}_{orth}}{\mathcal{E}_{non\text{-}orth}} = \frac{M\sum_{k=0}^{M-1} 1/\sigma_{h,k}^2}{\Big(\sum_{k=0}^{M-1} 1/\sigma_{h,k}\Big)^2} \tag{15.20}$$

This is precisely the gain we obtained by going from the lazy precoder to the ZF-MMSE precoder in Chap. 12. For example, if $M = 128$, $\sigma_{h,127} = 10^{-4}$, and $\sigma_{h,k} = 1$ for $k < 127$, then

$$\mathcal{G} = \frac{128(127 + 10^8)}{(127 + 10^4)^2} = 124.8,$$

which is quite significant.

15.2.2 Pure-MMSE transceivers

Consider again an $M \times M$ precoder and a $J \times M$ channel $\mathbf{H}$, but assume now that there is no zero-forcing constraint. To minimize the mean square error under these conditions recall first (Eq. (13.19)) that the MMSE equalizer for any precoder has the form

$$\mathbf{G} = \mathbf{C}^\dagger\Big(\mathbf{C}\mathbf{C}^\dagger + \frac{\sigma_q^2}{\sigma_s^2}\mathbf{I}\Big)^{-1}, \tag{15.21}$$

where $\mathbf{C} = \mathbf{HF}$. With this choice the mean square error can be expressed in terms of $\mathbf{F}$, as in Sec. 13.4.1, which is reproduced below (with $P = M$):

$$\mathcal{E}_{mse} = \sigma_s^2 \text{Tr}\Big(\mathbf{I}_J + \frac{\sigma_s^2}{\sigma_q^2}\mathbf{C}\mathbf{C}^\dagger\Big)^{-1} + (M - J)\sigma_s^2. \tag{15.22}$$

[3] Just write $\|\mathbf{z}\|_1^2 = (\sum_k |z_k|)^2 = (\sum_k |z_k| \times 1)^2 \leq \sum_k |z_k|^2 \sum_k 1^2 = M\sum_k |z_k|^2$. The inequality used here is the Cauchy-Schwartz inequaity.

Since $\mathbf{F}\mathbf{F}^\dagger = \mathbf{F}^\dagger\mathbf{F} = \mathbf{I}_M$, we have $\mathbf{C}\mathbf{C}^\dagger = \mathbf{H}\mathbf{F}\mathbf{F}^\dagger\mathbf{H}^\dagger = \mathbf{H}\mathbf{H}^\dagger$, so that

$$\mathbf{G} = \mathbf{F}^\dagger\mathbf{H}^\dagger\Big(\mathbf{H}\mathbf{H}^\dagger + \frac{\sigma_q^2}{\sigma_s^2}\mathbf{I}_J\Big)^{-1} \tag{15.23}$$

and

$$\mathcal{E}_{mse} = \sigma_s^2 \mathrm{Tr}\,\Big(\mathbf{I}_J + \frac{\sigma_s^2}{\sigma_q^2}\mathbf{H}\mathbf{H}^\dagger\Big)^{-1} + (M-J)\sigma_s^2, \tag{15.24}$$

which is independent of the precoder! *Thus the unitary $M \times M$ precoder $\mathbf{F}$ can again be chosen arbitrarily without affecting the reconstruction error.*

Here are three examples of unitary $\mathbf{F}$ with identical MSE (15.24). All these examples are similar to those in Fig. 15.1, with the difference that the equalizers are Wiener filters.

1. ***Lazy precoder***. The simplest choice is of course the identity matrix, $\mathbf{F} = \mathbf{I}$. In this case

$$\mathbf{F} = \mathbf{I}_M, \quad \mathbf{G} = \mathbf{H}^\dagger\Big(\mathbf{H}\mathbf{H}^\dagger + \frac{\sigma_q^2}{\sigma_s^2}\mathbf{I}_J\Big)^{-1}. \tag{15.25}$$

2. ***SVD precoder***. As a second example assume the channel is expressed in the SVD form $\mathbf{H} = \mathbf{U}_h\mathbf{\Sigma}_h\mathbf{V}_h^\dagger$. We can choose $\mathbf{F} = \mathbf{V}_h$ and then compute $\mathbf{G}$ from Eq. (15.23) to obtain the solution

$$\mathbf{F} = \mathbf{V}_h, \quad \mathbf{G} = [\,[\mathbf{\Sigma}_h]_M \;\; \mathbf{0}\,]\left(\begin{bmatrix}[\mathbf{\Sigma}_h]_M^2 & \mathbf{0}\\ \mathbf{0} & \mathbf{0}\end{bmatrix} + \frac{\sigma_q^2}{\sigma_s^2}\mathbf{I}_J\right)^{-1}\mathbf{U}_h^\dagger, \tag{15.26}$$

where $[\mathbf{\Sigma}_h]_M$ is the diagonal matrix with its M diagonal elements equal to the singular values $\sigma_{h,k}$ of the channel $\mathbf{H}$.

3. ***Cyclic-prefix precoder***. If $\mathbf{H}$ is circulant as in the case of a cyclic-prefix system, then, since $\mathbf{H} = \mathbf{W}^{-1}\mathbf{\Lambda}_c\mathbf{W}$, the choice $\mathbf{F} = \mathbf{W}^\dagger/\sqrt{M}$ yields

$$\mathbf{F} = \frac{\mathbf{W}^\dagger}{\sqrt{M}}, \qquad \mathbf{G} = \mathbf{\Lambda}_c^*\Big(\mathbf{\Lambda}_c\mathbf{\Lambda}_c^* + \frac{\sigma_q^2}{\sigma_s^2}\mathbf{I}\Big)^{-1}\frac{\mathbf{W}}{\sqrt{M}} \quad \approx \mathbf{\Lambda}_c^{-1}\frac{\mathbf{W}}{\sqrt{M}}, \tag{15.27}$$

where the approximation on the right is for large SNR. This results in the familiar OFDM system described in Sec. 7.5. Since any unitary $\mathbf{F}$ is as good as any other, we see that the IDFT matrix $\mathbf{W}^\dagger/\sqrt{M}$ is also an "optimum unitary precoder" solution.

15.2.2.A Comparison with general non-unitary precoder

The expression for the MSE given in Eq. (15.24) can be rewritten in terms of channel singular values $\sigma_{h,k}$. For this recall that the trace of a matrix is the sum of eigenvalues (Appendix B). Since the eigenvalues of $\mathbf{H}\mathbf{H}^\dagger$ are $\sigma_{h,k}^2$, the eigenvalues of the bracketted matrix in Eq. (15.24) are

$$\lambda_k = \begin{cases} 1 + \dfrac{\sigma_s^2\sigma_{h,k}^2}{\sigma_q^2} & \text{for } 0 \le k \le M-1 \\ 1 & \text{for } M \le k \le J-1. \end{cases}$$

The inverse of the bracketted matix has eigenvalues $1/\lambda_k$, so the trace in Eq. (15.24) is the sum $\sum_k 1/\lambda_k$. The mean square error can therefore be written as

$$\mathcal{E}_{mse} = \sum_{k=0}^{M-1} \frac{\sigma_s^2}{1 + \dfrac{\sigma_s^2}{\sigma_q^2}\sigma_{h,k}^2} + (J-M)\sigma_s^2 + (M-J)\sigma_s^2,$$

which simplifies to

$$\mathcal{E}_{mse} = \sum_{k=0}^{M-1} \frac{\sigma_q^2}{\sigma_{h,k}^2 + \dfrac{\sigma_q^2}{\sigma_s^2}} \tag{15.28}$$

Since the precoder is unitary, the transmitted power (the trace of the covariance at the input of the channel $\mathbf{H}$) is $\sigma_s^2 M$. To obtain independent control of the channel input power we insert the matrices $\alpha\mathbf{I}$ and $\alpha^{-1}\mathbf{I}$ in the transmitter and receiver as before (Fig. 15.2).

In the figure, the unitary precoder is denoted as $\mathbf{U}_f$, and its optimal equalizer (corresponding to the transmitted signal $\mathbf{x}(n)$) is denoted as $\mathbf{G}_1$. Since the input to the unitary precoder has power $\alpha^2\sigma_s^2$ rather than σ_s^2, the MSE between the signals $\mathbf{x}$ and $\widehat{\mathbf{x}}$ in the figure is given by

$$\mathcal{E}_{mmse,x} = \sum_{k=0}^{M-1} \frac{\alpha^2\sigma_s^2}{1 + \dfrac{\alpha^2\sigma_s^2}{\sigma_q^2}\sigma_{h,k}^2} \tag{15.29}$$

Since $\widehat{\mathbf{s}}(n) - \mathbf{s}(n) = (\widehat{\mathbf{x}}(n) - \mathbf{x}(n))/\alpha$, the MMSE between $\mathbf{s}(n)$ and $\widehat{\mathbf{s}}(n)$ is given by

$$\mathcal{E}_{mmse} = \frac{1}{\alpha^2}\sum_{k=0}^{M-1} \frac{\alpha^2\sigma_s^2}{1 + \dfrac{\alpha^2\sigma_s^2}{\sigma_q^2}\sigma_{h,k}^2} = \sum_{k=0}^{M-1} \frac{\sigma_s^2}{1 + \dfrac{\alpha^2\sigma_s^2}{\sigma_q^2}\sigma_{h,k}^2} \tag{15.30}$$

Since the transmitted power at the channel input is now $p_0 = M\alpha^2\sigma_s^2$, we can rewrite this as

$$\mathcal{E}_{mmse} = \sum_{k=0}^{M-1} \frac{\sigma_s^2}{1 + \dfrac{p_0}{M\sigma_q^2}\sigma_{h,k}^2} = \frac{M\sigma_s^2\sigma_q^2}{p_0}\sum_{k=0}^{M-1} \frac{1}{\dfrac{M\sigma_q^2}{p_0} + \sigma_{h,k}^2}$$

When σ_q^2 is sufficiently small, this reduces to the expression for the orthonormal precoder in Eq. (15.15). That is, the MMSE expression reduces to the ZF-MMSE expression for the low noise case. The general MMSE system with no orthonormality constraint on the precoder was studied in Chap. 13, and yielded the MMSE given by Eq. (13.66). We therefore finally arrive at the expressions

$$\mathcal{E}_{mmse} = \begin{cases} \frac{M\sigma_s^2\sigma_q^2}{p_0}\sum_{k=0}^{M-1}\left(\dfrac{1}{\frac{M\sigma_q^2}{p_0}+\sigma_{h,k}^2}\right) & \text{(orthonormal precoder)} \\ \dfrac{\sigma_s^2\sigma_q^2\Big(\sum\limits_{k=0}^{K-1} 1/\sigma_{h,k}\Big)^2}{p_0+\sigma_q^2\sum\limits_{k=0}^{K-1} 1/\sigma_{h,k}^2} + (M-K)\sigma_s^2 & \text{(unrestricted precoder),} \end{cases} \tag{15.31}$$

where K is the number of nonzero multipliers q_{kk} in the precoder (see Sec. 13.6).

A subtle point. If we insert the multiplier $\alpha\mathbf{I}$ at the transmitter, then for a ZF receiver it is clear that we have to insert $\alpha^{-1}\mathbf{I}$ at the receiver. For the MMSE case, is there a loss of generality in inserting $\alpha^{-1}\mathbf{I}$ at the receiver? It turns out that there is no such loss of generality. To see this let us solve for the optimal equalizer corresponding to the precoder $\mathbf{F} = \alpha\mathbf{U}_f$, where $\mathbf{U}_f$ is unitary. This equalizer is given by Eq. (15.21) with $\mathbf{C} = \mathbf{HF}$, that is,

$$\begin{aligned} \mathbf{G} &= \mathbf{F}^\dagger\mathbf{H}^\dagger\Big(\mathbf{HFF}^\dagger\mathbf{H}^\dagger + \frac{\sigma_q^2}{\sigma_s^2}\mathbf{I}\Big)^{-1} \\ &= \alpha\mathbf{U}_f^\dagger\mathbf{H}^\dagger\Big(\alpha^2\mathbf{HU}_f\mathbf{U}_f^\dagger\mathbf{H}^\dagger + \frac{\sigma_q^2}{\sigma_s^2}\mathbf{I}\Big)^{-1} = \alpha\mathbf{U}_f^\dagger\mathbf{H}^\dagger\Big(\alpha^2\mathbf{HH}^\dagger + \frac{\sigma_q^2}{\sigma_s^2}\mathbf{I}\Big)^{-1}, \end{aligned}$$

which can be rewritten as

$$\mathbf{G} = \frac{\mathbf{U}_f^\dagger\mathbf{H}^\dagger}{\alpha}\Big(\mathbf{HH}^\dagger + \frac{\sigma_q^2\mathbf{I}}{\alpha^2\sigma_s^2}\Big)^{-1}. \tag{15.32}$$

On the other hand, the optimal (MMSE) equalizer $\mathbf{G}_1$ corresponding to the precoder $\mathbf{U}_f$ and input $\mathbf{x}(n)$ in Fig. 15.2 is given by

$$\begin{aligned} \mathbf{G}_1 &= \mathbf{U}_f^\dagger\mathbf{H}^\dagger\Big(\mathbf{HU}_f\mathbf{U}_f^\dagger\mathbf{H}^\dagger + \frac{\sigma_q^2}{\sigma_x^2}\mathbf{I}\Big)^{-1} \\ &= \mathbf{U}_f^\dagger\mathbf{H}^\dagger\Big(\mathbf{HH}^\dagger + \frac{\sigma_q^2}{\alpha^2\sigma_s^2}\mathbf{I}\Big)^{-1} \end{aligned}$$

so that the effective equalizer in Fig. 15.2 is precisely the optimal equalizer (15.32) for the precoder $\alpha\mathbf{U}_f$ with input $\mathbf{s}(n)$. This shows that if we are given an MMSE system with unitary precoder, and we want to insert $\alpha\mathbf{I}$ at the transmitter for power control, the MMSE equalizer for the modified precoder is obtained simply by appending $\alpha^{-1}\mathbf{I}$ to the receiver. ▽▽▽

The results for $M \times M$ orthonormal precoders obtained so far, with and without zero forcing at the receiver, can be summarized as follows:

♠**Theorem 15.1.** *Optimal orthonormal precoders (non-redundant case).* For the case where the precoder is $M \times M$ and the channel is $J \times M$, assume that the precoder is restricted to be $\mathbf{F} = \alpha\mathbf{U}_f$, where $\mathbf{U}_f$ is unitary and α is a scalar for power control. Then the transceiver has the MMSE property for *any* choice of $\mathbf{U}_f$ provided the equalizer is chosen as follows:

$$\mathbf{G} = \begin{cases} \frac{1}{\alpha}\mathbf{U}_f^\dagger\mathbf{H}^\# & \text{(zero-forcing case)} \\ \frac{1}{\alpha}\mathbf{U}_f^\dagger\mathbf{H}^\dagger\Big(\mathbf{H}\mathbf{H}^\dagger + \frac{\sigma_q^2}{\sigma_s^2}\mathbf{I}_J\Big)^{-1} & \text{(non zero-forcing case).} \end{cases} \tag{15.33}$$

where $\mathbf{H}^\#$ is the minimum-norm left inverse of the channel $\mathbf{H}$. With channel input power equal to p_0 (realized by adjusting α), the minimized mean square errors are

$$\mathcal{E}_{mmse} = \begin{cases} \frac{M\sigma_q^2\sigma_s^2}{p_0}\sum_{k=0}^{M-1}\frac{1}{\sigma_{h,k}^2} & \text{(zero-forcing case)} \\ \frac{M\sigma_s^2\sigma_q^2}{p_0}\sum_{k=0}^{M-1}\left(\frac{1}{\frac{M\sigma_q^2}{p_0} + \sigma_{h,k}^2}\right) & \text{(non zero-forcing case),} \end{cases} \tag{15.34}$$

where $\sigma_{h,k}$ are the channel singular values, assumed nonzero in the zero-forcing case. $\diamondsuit$

15.3 Rectangular orthonormal precoder matrices

When the condition $P = M$ is not true, the optimization of the orthonormal precoder becomes more interesting and nontrivial. In this case it is not true any more that "any unitary choice is optimal" as we shall see. We shall assume $P, J \geq M$ so that it is possible for all matrices to have rank M.

15.3.1 ZF-MMSE transceivers

First consider the zero-forcing case and assume that $\mathbf{H}$ has rank $\geq M$ so that its M dominant singular values are nonzero. Even though $\mathbf{F}^\dagger\mathbf{F} = \mathbf{I}_M$, we have to remember that $\mathbf{F}\mathbf{F}^\dagger \neq \mathbf{I}_P$ because $P > M$. We first notice the following result:

♠**Lemma 15.1.** *Form of orthonormal precoder.* The $P \times M$ orthonormal precoder for the ZF-MMSE problem can be assumed, without loss of generality, to be of the form

$$\mathbf{F} = \mathbf{U}_f\begin{bmatrix}\mathbf{I}_M \\ \mathbf{0}\end{bmatrix}, \tag{15.35}$$

where $\mathbf{U}_f$ is $P \times P$ unitary. $\diamondsuit$

Proof. The unitary property

$$\mathbf{F}^\dagger\mathbf{F} = \mathbf{I} \tag{15.36}$$

simply says that the M columns of $\mathbf{F}$ have unit norm and are mutually orthogonal. We can therefore append $P - M$ new columns such that the resulting $P \times P$ matrix

$$\mathbf{U}_f \stackrel{\Delta}{=} [\mathbf{F} \quad \times] \tag{15.37}$$

is unitary. So the precoder $\mathbf{F}$ can be written as in Eq. (15.35). $\bigtriangledown \bigtriangledown \bigtriangledown$

Before proceeding further, some preliminary observations are useful. Let $\mu_0 \geq \mu_1 \geq \ldots \geq \mu_{M-1}$ denote the M eigenvalues of $(\mathbf{U}_f^\dagger \mathbf{H}^\dagger \mathbf{H} \mathbf{U}_f)_M$ as usual. Then the eigenvalues of the inverse are $1/\mu_k$, and since the trace is the sum of eigenvalues, we have from Eq. (15.37)

$$\mathcal{E}_{mse} = \sigma_q^2 \sum_{k=0}^{M-1} \frac{1}{\mu_k} \tag{15.38}$$

With $\sigma_{h,0} \geq \sigma_{h,1} \geq \ldots$ denoting the singular values of the channel $\mathbf{H}$, the P eigenvalues of the product $\mathbf{U}_f^\dagger \mathbf{H}^\dagger \mathbf{H} \mathbf{U}_f$ are

$$\sigma_{h,0}^2 \geq \sigma_{h,1}^2 \geq \ldots \geq \sigma_{h,P-1}^2. \tag{15.39}$$

With these preliminaries we are now ready to prove the main result:

♠**Theorem 15.2.** *ZF-MMSE transceiver with orthonormal precoder.* The optimum unitary percoder and the corresponding ZF equalizer can be taken to be

$$\mathbf{F} = \mathbf{V}_h \begin{bmatrix} \mathbf{I}_M \\ \mathbf{0} \end{bmatrix} \quad \text{and} \quad \mathbf{G} = [\,\mathbf{\Sigma}_g \quad \mathbf{0}\,]\,\mathbf{U}_h^\dagger, \tag{15.40}$$

where $\mathbf{\Sigma}_g = (\mathbf{\Sigma}_h)_M^{-1}$. The corresponding minimized mean square reconstruction error is

$$\mathcal{E}_{mse} = \sigma_q^2 \sum_{k=0}^{M-1} \frac{1}{\sigma_{h,k}^2}, \tag{15.41}$$

where the $\sigma_{h,k}$ are the M dominant singular values of the channel. $\diamondsuit$

This transceiver is shown in Fig. 15.3. Thus, unlike in the $P = M$ case (Sec. 15.2.1), the optimal $P \times M$ unitary precoder $\mathbf{F}$ is not arbitrary. However, the final expression (15.41) for the mean square error *is* identical to the case where $P = M$ (see Eq. (15.8)).

Note also that the solution given by Theorem 15.2 is **not unique**. If we insert an $M \times M$ unitary matrix $\mathbf{U}$ after $\mathbf{\Sigma}_g$ at the receiver, and its inverse $\mathbf{U}^\dagger$ between $\mathbf{s}(n)$ and $\mathbf{V}_h$ at the transmitter, then the transmitted power, the MSE, and the ZF property are unchanged (the proof is similar to that in Sec. 11.5.1). So the optimal solution is non unique.

Proof of Theorem 15.2. As explained in Sec. 12.4.2 we have $\sigma_{h,k}^2 \geq \mu_k^2$ for $0 \leq k \leq M-1$. Using this we see that

$$\mathcal{E}_{mse} = \sigma_q^2 \sum_{k=0}^{M-1} \frac{1}{\mu_k} \geq \sigma_q^2 \sum_{k=0}^{M-1} \frac{1}{\sigma_{h,k}^2} \tag{15.42}$$

The lower bound is achieved by choosing $\mathbf{U}_f = \mathbf{V}_h$ so that $\mathbf{U}_f^\dagger \mathbf{H}^\dagger \mathbf{H} \mathbf{U}_f = \boldsymbol{\Sigma}_h^\dagger \boldsymbol{\Sigma}_h$. Thus the optimal precoder $\mathbf{F}$ can be written in the form (15.40) indeed. The equalizer can be computed from the zero-forcing condition: $\mathbf{GHF} = \mathbf{I}$. Taking $\mathbf{G}$ to be the minimum-norm left inverse of $\mathbf{HF}$ we get

$$\mathbf{G} = \Big((\mathbf{HF})^\dagger \mathbf{HF}\Big)^{-1} (\mathbf{HF})^\dagger. \tag{15.43}$$

Since

$$\mathbf{HF} = \mathbf{U}_h \boldsymbol{\Sigma}_h \mathbf{V}_h^\dagger \mathbf{V}_h \begin{bmatrix} \mathbf{I}_M \\ \mathbf{0} \end{bmatrix} = \mathbf{U}_h \boldsymbol{\Sigma}_h \begin{bmatrix} \mathbf{I}_M \\ \mathbf{0} \end{bmatrix} = \mathbf{U}_h \begin{bmatrix} (\boldsymbol{\Sigma}_h)_M \\ \mathbf{0} \end{bmatrix} \tag{15.44}$$

we have

$$(\mathbf{HF})^\dagger \mathbf{HF} = [\,(\boldsymbol{\Sigma}_h)_M \quad \mathbf{0}\,]\, \mathbf{U}_h^\dagger \mathbf{U}_h \begin{bmatrix} \boldsymbol{\Sigma}_h)_M \\ \mathbf{0} \end{bmatrix} = (\boldsymbol{\Sigma}_h)_M^2,$$

where $(\boldsymbol{\Sigma}_h)_M$ is the $M \times M$ diagonal matrix containing the M dominant singular values of the channel. Substituting into Eq. (15.43), the equalizer $\mathbf{G}$ takes the final form in Eq. (15.40). The MSE expression then achieves the lower bound in Eq. (15.42), which proves Eq. (15.41). $\bigtriangledown \bigtriangledown \bigtriangledown$

15.3.2 Pure-MMSE transceivers

We now extend the results of Sec. 15.3.1 to the case where the zero-forcing condition is not imposed. From Sec. 13.3 we know that given an arbtirary precoder $\mathbf{F}$, the MMSE equalizer has the closed form expression

$$\mathbf{G} = \sigma_s^2 \mathbf{F}^\dagger \mathbf{H}^\dagger \Big(\sigma_s^2 \mathbf{HFF}^\dagger \mathbf{H}^\dagger + \sigma_q^2 \mathbf{I}\Big)^{-1}, \tag{15.45}$$

and the corresponding reconstruction error is

$$\mathcal{E}_{mse} = \sigma_s^2 (M-J) + \sigma_s^2 \mathrm{Tr}\,\Big(\mathbf{I}_J + \frac{\sigma_s^2}{\sigma_q^2} \mathbf{HFF}^\dagger \mathbf{H}^\dagger\Big)^{-1}. \tag{15.46}$$

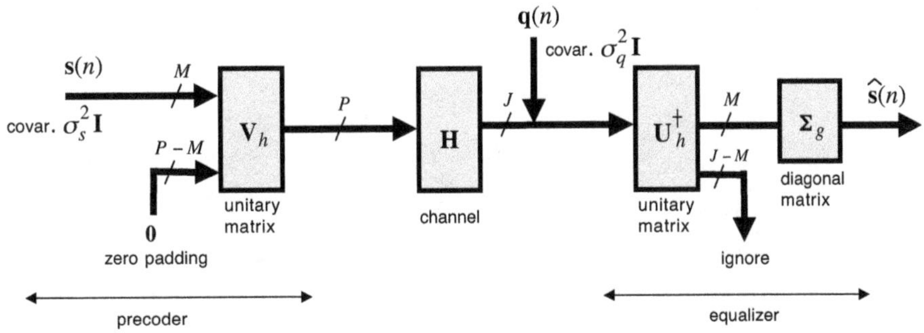

Figure 15.3. Transceiver with orthogonal precoder and minimum reconstruction error under the zero-forcing constraint. Here $\mathbf{\Sigma}_g = (\mathbf{\Sigma}_h)_M^{-1}$. The same form holds for the MMSE transceiver without the ZF constraint (Sec. 15.3.2), but the diagonal matrix $\mathbf{\Sigma}_g$ is different.

In the MMSE transceiver the matrix $\mathbf{F}$ is also optimized such that the above reconstruction error is minimized. Since the $P \times M$ matrix $\mathbf{F}$ is restricted to be such that $\mathbf{F}^\dagger\mathbf{F} = \mathbf{I}$, we write $\mathbf{F}$ in the form (15.35). We then have

$$\mathbf{HFF}^\dagger\mathbf{H}^\dagger = \mathbf{HU}_f \begin{bmatrix} \mathbf{I}_M & \mathbf{0} \\ \mathbf{0} & \mathbf{0} \end{bmatrix} \mathbf{U}_f^\dagger\mathbf{H}^\dagger = \underbrace{\mathbf{HU}_f \begin{bmatrix} \mathbf{I}_M & \mathbf{0} \\ \mathbf{0} & \mathbf{0} \end{bmatrix}}_{J\times P} \underbrace{\begin{bmatrix} \mathbf{I}_M & \mathbf{0} \\ \mathbf{0} & \mathbf{0} \end{bmatrix} \mathbf{U}_f^\dagger\mathbf{H}^\dagger}_{P\times J}.$$

Using the trace identity (Appendix 13.A of Chap. 13) we have:

$$\mathrm{Tr}\,(\mathbf{I}_J + \mathbf{BA})^{-1} = (J - P) + \mathrm{Tr}\,(\mathbf{I}_P + \mathbf{AB})^{-1}. \tag{15.47}$$

We can therefore rewrite the trace in the second term in Eq. (15.46) as

$$\begin{aligned}
\tau &= \mathrm{Tr}\left(\mathbf{I}_J + \frac{\sigma_s^2}{\sigma_q^2}\mathbf{HU}_f \begin{bmatrix} \mathbf{I}_M & \mathbf{0} \\ \mathbf{0} & \mathbf{0} \end{bmatrix}\begin{bmatrix} \mathbf{I}_M & \mathbf{0} \\ \mathbf{0} & \mathbf{0} \end{bmatrix} \mathbf{U}_f^\dagger\mathbf{H}^\dagger\right)^{-1} \\
&= (J - P) + \mathrm{Tr}\left(\mathbf{I}_P + \frac{\sigma_s^2}{\sigma_q^2}\begin{bmatrix} \mathbf{I}_M & \mathbf{0} \\ \mathbf{0} & \mathbf{0} \end{bmatrix} \mathbf{U}_f^\dagger\mathbf{H}^\dagger\mathbf{HU}_f \begin{bmatrix} \mathbf{I}_M & \mathbf{0} \\ \mathbf{0} & \mathbf{0} \end{bmatrix}\right)^{-1} \\
&= (J - P) + \mathrm{Tr}\left(\mathbf{I}_P + \frac{\sigma_s^2}{\sigma_q^2}\begin{bmatrix} (\mathbf{U}_f^\dagger\mathbf{H}^\dagger\mathbf{HU}_f)_M & \mathbf{0} \\ \mathbf{0} & \mathbf{0} \end{bmatrix}\right)^{-1}
\end{aligned}$$

where the notation $(.)_M$ denotes the $M \times M$ leading principal submatrix. With $\mu_0 \geq \mu_1 \geq \ldots \geq \mu_{M-1}$ denoting the M eigenvalues of $(\mathbf{U}_f^\dagger\mathbf{H}^\dagger\mathbf{HU}_f)_M$, the eigenvalues of the matrix inside the brackets are

$$\lambda_k = \begin{cases} 1 + \dfrac{\mu_k\sigma_s^2}{\sigma_q^2} & \text{for } 0 \leq k \leq M-1 \\ 1 & \text{for } M \leq k \leq P-1. \end{cases}$$

Since the eigenvalues of the inverse matrix are $1/\lambda_k$, the trace in the preceding equation becomes

$$\begin{aligned}\tau &= (J-P) + \sum_{k=0}^{P-1} \frac{1}{\lambda_k} \\ &= (J-P) + \sum_{k=0}^{M-1} \frac{1}{1+\frac{\mu_k \sigma_s^2}{\sigma_q^2}} + (P-M) \\ &= \sum_{k=0}^{M-1} \frac{1}{1+\left(\frac{\sigma_s^2}{\sigma_q^2}\right)\mu_k} + (J-M).\end{aligned}$$

The MSE in Eq. (15.46) therefore becomes

$$\mathcal{E}_{mse} = \sum_{k=0}^{M-1} \frac{\sigma_s^2}{1+\left(\frac{\sigma_s^2}{\sigma_q^2}\right)\mu_k} \qquad (15.48)$$

As in earlier sections let the channel SVD be written as follows:

$$\mathbf{H} = \underbrace{\mathbf{U}_h}_{J\times J} \underbrace{\mathbf{\Sigma}_h}_{J\times P} \underbrace{\mathbf{V}_h^\dagger}_{P\times P}. \qquad (15.49)$$

As usual let the matrix $(\mathbf{\Sigma}_h)_M$ denote the diagonal matrix of the dominant channel singular values $\sigma_{h,0} \geq \sigma_{h,1} \geq \ldots \geq \sigma_{h,M-1}$. We are now ready to prove the main result:

♠**Theorem 15.3.** *MMSE transceiver with orthonormal precoder.* Consider the transceiver with orthonormal precoder

$$\mathbf{F} = \mathbf{V}_h \begin{bmatrix} \mathbf{I}_M \\ \mathbf{0} \end{bmatrix} \qquad (15.50)$$

and equalizer

$$\mathbf{G} = [\,\mathbf{\Sigma}_g \quad \mathbf{0}\,]\,\mathbf{U}_h^\dagger, \qquad (15.51)$$

where $\mathbf{\Sigma}_g$ is the $M \times M$ diagonal matrix defined as

$$\mathbf{\Sigma}_g = (\mathbf{\Sigma}_h)_M^{-1} \left(\mathbf{I} + \frac{\sigma_q^2}{\sigma_s^2} (\mathbf{\Sigma}_h)_M^{-2}\right)^{-1}. \qquad (15.52)$$

This transceiver has the MMSE property. The minimized mean square error is given by

$$\mathcal{E}_{mse} = \sum_{k=0}^{M-1} \frac{\sigma_q^2}{\sigma_{h,k}^2 + \frac{\sigma_q^2}{\sigma_s^2}} \qquad (15.53)$$

where σ_s^2 and σ_q^2 are the signal and noise variances (Eq. (15.3)). $\diamond$

This is identical to Eq. (15.28) which was derived for $M\times M$ precoders. Only the M dominant singular values $\sigma_{h,k}$ of the channel are used in Eq. (15.53). The solution given by the theorem is not unique, because we can insert arbitrary $M \times M$ unitary matrices, as explained in the remarks following Theorem 15.2.

Proof of Theorem 15.3. The P eigenvalues of the product $\mathbf{U}_f^\dagger\mathbf{H}^\dagger\mathbf{H}\mathbf{U}_f$ are $\sigma_{h,0}^2 \geq \sigma_{h,1}^2 \ldots,$ and as explained in Sec. 12.4.2, we have $\sigma_{h,k}^2 \geq \mu_k^2$ for $0 \leq k \leq M-1$. Using this we see that

$$\mathcal{E}_{mse} = \sum_{k=0}^{M-1} \frac{\sigma_s^2}{1+\left(\frac{\sigma_s^2}{\sigma_q^2}\right)\mu_k} \geq \sum_{k=0}^{M-1} \frac{\sigma_s^2}{1+\left(\frac{\sigma_s^2}{\sigma_q^2}\right)\sigma_{h,k}^2} \tag{15.54}$$

This lower bound is therefore achieved by choosing $\mathbf{U}_f = \mathbf{V}_h$ so that $\mathbf{U}_f^\dagger\mathbf{H}^\dagger\mathbf{H}\mathbf{U}_f = \mathbf{\Sigma}_h^\dagger\mathbf{\Sigma}_h$. Thus the optimal precoder can be written in the form (15.50). The optimal equalizer (15.45) can now be simplified. For this observe that

$$\mathbf{F}^\dagger\mathbf{H}^\dagger = [\,\mathbf{I}_M \quad \mathbf{0}\,]\,\mathbf{V}_h^\dagger\mathbf{H}^\dagger = [\,\mathbf{I}_M \quad \mathbf{0}\,]\,\mathbf{V}_h^\dagger\mathbf{V}_h\mathbf{\Sigma}_h^\dagger\mathbf{U}_h^\dagger = [\,(\mathbf{\Sigma}_h)_M \quad \mathbf{0}\,]\,\mathbf{U}_h^\dagger$$

so that

$$\mathbf{H}\mathbf{F}\mathbf{F}^\dagger\mathbf{H}^\dagger = \mathbf{U}_h\begin{bmatrix}(\mathbf{\Sigma}_h)_M^2 & \mathbf{0}\\ \mathbf{0} & \mathbf{0}\end{bmatrix}\mathbf{U}_h^\dagger.$$

Substituting these into Eq. (15.45) we obtain

$$\mathbf{G} = [\,(\mathbf{\Sigma}_h)_M \quad \mathbf{0}\,]\left(\begin{bmatrix}(\mathbf{\Sigma}_h)_M^2 & \mathbf{0}\\ \mathbf{0} & \mathbf{0}\end{bmatrix} + \frac{\sigma_q^2}{\sigma_s^2}\mathbf{I}_P\right)^{-1}\mathbf{U}_h^\dagger. \tag{15.55}$$

Since the matrix inverse in the middle is a diagonal matrix, Eq. (15.51) follows from this. With this choice the lower bound in Eq. (15.54) is achieved, so the MMSE has the value shown in Eq. (15.53). $\bigtriangledown\bigtriangledown\bigtriangledown$

The transceiver with unitary precoder that minimizes the reconstruction error without the ZF constraint therefore has the same form as in Fig. 15.3 wherein the ZF constraint was imposed. The only difference is that the diagonal matrix $\mathbf{\Sigma}_g$ at the receiver is not $(\mathbf{\Sigma}_h)_M^{-1}$. Rather, it is a "polished version of $(\mathbf{\Sigma}_h)_M^{-1}$," as shown in Eq. (15.52). This is nothing but the Wiener filter expression, which takes into account the signal-to-noise ratio σ_s^2/σ_q^2. It is approximately equal to $(\mathbf{\Sigma}_h)_M^{-1}$ when noise is small, that is, when $\sigma_q^2 << \sigma_s^2$. We conclude this section with some remarks.

1. ***Witsenhausen's result.*** For the case where the precoder was not restricted to be unitary, we found in Sec. 13.5 that the derivation of the optimal

transceiver required a special result due to Witsenhausen. Notice that this result is not required in the derivation for the case where the precoder is restricted a priori to be unitary.

2. ***The integer K.*** In Chap. 13, where we solved the MMSE problem without imposing the unitary condition on the precoder, the precoder had rank $K \leq M$, where K is an appropriately chosen integer. In this section, with $\mathbf{F}^{\dagger}\mathbf{F} = \mathbf{I}_M$ the precoder has rank M and there is no K involved.

3. ***Laziness of the precoder.*** In all cases where the precoder is restricted to be unitary, the equalization (inversion of channel singular values $1/\sigma_{h,k}$) actually takes place at the receiver. For the $P \times M$ precoder the transmitter helps in the diagonalization of the channel. For the case where the precoder is $M \times M$ the transmitter need not even do this. It can just be the lazy precoder! This example arises in the zero-padding precoder (Sec. 7.2), where the precoder is a square matrix. Restricting the precoder to be unitary is as good as restricting it to be identity. In Chap. 16, where we minimize symbol error probabilities (rather than just minimize the MSE), we will show that the best unitary precoder cannot in general be lazy.

4. ***Power control.*** For power control we can insert $\alpha\mathbf{I}$ and $\alpha^{-1}\mathbf{I}$ as in Fig. 15.2. By using the same approach that was used to derive Eq. (15.34) we can verify that the minimized mean square errors for the case of rectangular precoders are still given by Eq. (15.34).

15.4 Concluding remarks

In this chapter we minimized the mean square error in transceivers under the constraint that the precoder be orthonormal. The zero-forcing transceiver and the non-ZF transceiver were both considered. In both cases we found that, if the precoder is square ($P = M$), then the unitary precoder can be chosen arbitrarily, with no loss of optimality. For $P > M$, the precoder should be chosen based on the SVD of the channel matrix. For all cases, we also derived the expression for the minimized MSE. In Chap. 16 we turn our attention to the case where the symbol error rate, rather than the MSE, will be the focus of our attention.

Problems

15.1. Consider the circulant channel

$$\mathbf{H} = \begin{bmatrix} 1 & -1 & 0.1 \\ 0.1 & 1 & -1 \\ -1 & 0.1 & 1 \end{bmatrix}.$$

Assume that the 3×3 precoder is orthonormal and that zero forcing is in place. Assume $\sigma_s^2 = 1, \sigma_q^2 = 0.01$, and $p_0 = 10$.

1. Find a pair of optimal precoder and equalizer matrices $\mathbf{F}$ and $\mathbf{G}$ to minimize the mean square reconstruction error.
2. Compute the mean square errors (per symbol $s_k(n)$) for the optimal system. What would this mean square error be for the optimal system where the precoder is not restricted to be orthonormal? What is the ratio $\mathcal{G}$ of these two errors? This ratio measures the loss due to orthonormality restriction.

15.2. Repeat Problem 15.1 for the circulant channel

$$\mathbf{H} = \begin{bmatrix} 4 & 1 & 1 \\ 1 & 4 & 1 \\ 1 & 1 & 4 \end{bmatrix}.$$

What is the ratio $\mathcal{G}$ of the two errors? How does it compare with the answer in Problem 15.1? Explain why it is so different.

15.3. Repeat Problem 15.1 for the case where the zero-forcing constraint is removed.

15.4. Repeat Problem 15.2 for the case where the zero-forcing constraint is removed.

16

Minimization of error probability in transceivers

16.1 Introduction

In the previous chapters we considered the minimization of the mean square error in linear transceivers. In this chapter we consider the problem of minimizing the symbol error probability, or bit error rate (BER), which is a practically important measure of performance. We will see that transceivers which are optimized for BER also have the minimum mean square error (MMSE) property, though not all MMSE transceivers have minimum BER.

In Sec. 16.2 we discuss the problem of minimizing BER under the zero-forcing (ZF) constraint. In Sec. 16.3 we discuss the issue of bias in the reconstruction error for transceivers which do not have a ZF constraint. This section is in preparation for Sec. 16.4, where we discuss the minimization of BER without the ZF constraint. In this chapter we do not discuss bit allocation. All the components of the transmitted vector $\mathbf{s}(n)$ are assumed to have the same number of bits. Appendices 16.A–16.D at the end of this chapter contain some important details that arise during the developments of the main results in the chapter. Some of the results on majorization and Schur convexity, reviewed in Chap. 21, are frequently used in our discussions in this chapter.

16.2 Minimizing error probability in ZF-transceivers

Consider Fig. 16.1, which shows the transceiver system considered in earlier chapters. Assume that $\mathbf{s}(n)$ and $\mathbf{q}(n)$ are jointly WSS and statistically independent with, zero mean and covariance matrices

$$E[\mathbf{s}(n)\mathbf{s}^\dagger(n)] = \sigma_s^2\mathbf{I}, \quad E[\mathbf{q}(n)\mathbf{q}^\dagger(n)] = \sigma_q^2\mathbf{I}, \tag{16.1}$$

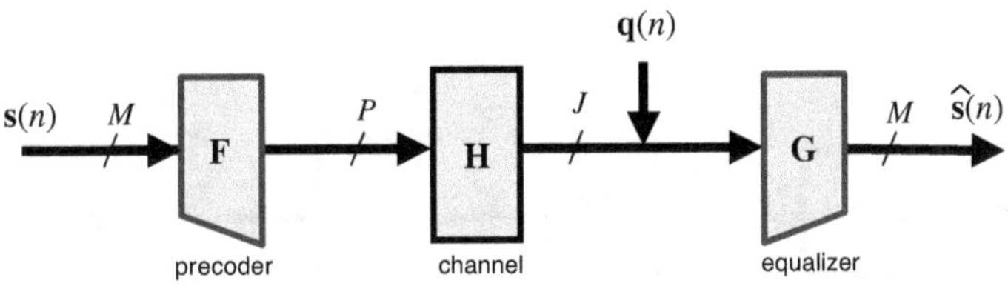

Figure 16.1. A transceiver with precoder **F** and equalizer **G**.

and that the components of $\mathbf{q}(n)$ are Gaussian.[1] In this section we consider the zero-forcing (ZF) transceiver, which satisfies

$$\mathbf{GHF} = \mathbf{I}. \tag{16.2}$$

In view of zero forcing, the reconstruction error

$$\mathbf{e}_s(n) = \widehat{\mathbf{s}}(n) - \mathbf{s}(n) \tag{16.3}$$

is nothing but $\mathbf{q}(n)$ filtered through $\mathbf{G}$. So, $\mathbf{e}_s(n)$ has zero mean, and the components of $\mathbf{e}_s(n)$ are Gaussian. The variances of these components are

$$\mathcal{E}_k = E[|\widehat{s}_k(n) - s_k(n)|^2]. \tag{16.4}$$

So the probability of error in the kth symbol is of the form (Sec. 11.5)

$$\mathcal{P}_e(k) = c\mathcal{Q}\Big(\frac{A}{\sqrt{\mathcal{E}_k}}\Big), \tag{16.5}$$

where the constants c and A depend on the type of symbol constellation used for $s_k(n)$ (e.g., PAM, QAM, and so forth). For example, in a b-bit QAM system,

$$c = 4(1 - 2^{-b/2}) \quad \text{and} \quad A = \sqrt{\frac{3\sigma_s^2}{2^b - 1}}. \tag{16.6}$$

The average symbol error probability therefore takes the form

$$\mathcal{P}_e = \frac{c}{M}\sum_{k=0}^{M-1} \mathcal{Q}\Big(\frac{A}{\sqrt{\mathcal{E}_k}}\Big). \tag{16.7}$$

16.2.1 Introducing the unitary matrix U

Now consider Fig. 16.2, where we have inserted a unitary matrix $\mathbf{U}$ (i.e., a matrix satisfying $\mathbf{U}^\dagger\mathbf{U} = \mathbf{I}$) at the receiver, and its inverse $\mathbf{U}^\dagger$ at the transmitter. Given any zero-forcing pair $\{\mathbf{F}, \mathbf{G}\}$, we will show how $\mathbf{U}$ should be chosen such that the average symbol error probability is minimized.

[1] For the PAM case the noise $\mathbf{q}(n)$ is assumed to be real and Gaussian, whereas for QAM we assume the noise is circularly symmetric Gaussian (Sec. 2.3.2).

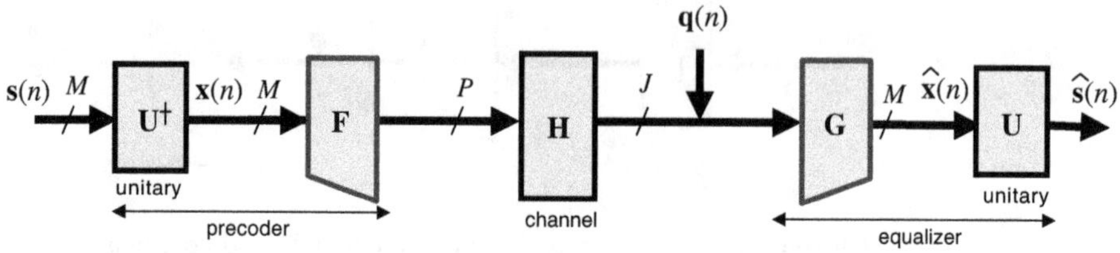

Figure 16.2. The modified transceiver with a unitary matrix $\mathbf{U}$ and its inverse $\mathbf{U}^\dagger$ inserted.

We first observe the following:

1. ***ZF-property unchanged.*** Since $\mathbf{GHF} = \mathbf{I}$, it follows that $\mathbf{UGHFU}^\dagger = \mathbf{UU}^\dagger = \mathbf{I}$, which shows that the modified system continues to have the zero-forcing property.

2. ***Transmitted power unchanged.*** The transmitted power is $\sigma_s^2 \text{Tr}\,(\mathbf{FU}^\dagger\mathbf{UF}^\dagger) = \sigma_s^2 \text{Tr}\,(\mathbf{FF}^\dagger)$, which is identical to the power before the unitary matrices were inserted.

3. ***MSE unchanged.*** Since $\mathbf{x}(n) = \mathbf{U}^\dagger\mathbf{s}(n)$ and $\widehat{\mathbf{s}}(n) = \mathbf{U}\widehat{\mathbf{x}}(n)$, it follows that $\widehat{\mathbf{s}}(n) - \mathbf{s}(n) = \mathbf{U}(\widehat{\mathbf{x}}(n) - \mathbf{x}(n))$. That is, the error signals with and without $\mathbf{U}$, defined as

$$\mathbf{e}_s(n) = \widehat{\mathbf{s}}(n) - \mathbf{s}(n) \quad \text{and} \quad \mathbf{e}_x = \widehat{\mathbf{x}}(n) - \mathbf{x}(n), \tag{16.8}$$

are related as follows:

$$\mathbf{e}_s(n) = \mathbf{U}\mathbf{e}_x(n). \tag{16.9}$$

The error covariances are therefore related as follows:

$$\mathbf{E}_s = \mathbf{U}\mathbf{E}_x\mathbf{U}^\dagger. \tag{16.10}$$

The total mean square error in Fig. 16.2 is then

$$\text{Tr}\,\mathbf{E}_s = \text{Tr}\left(\mathbf{U}\mathbf{E}_x\mathbf{U}^\dagger\right) = \text{Tr}\left(\mathbf{U}^\dagger\mathbf{U}\mathbf{E}_x\right) = \text{Tr}\,\mathbf{E}_x. \tag{16.11}$$

This shows that the average mean square error is unchanged by the insertion of the unitary matrices, that is,

$$\mathcal{E}_{mse} = \frac{1}{M}\sum_{k=0}^{M-1}\mathcal{E}_{s,k} = \frac{1}{M}\sum_{k=0}^{M-1}\mathcal{E}_{x,k}, \tag{16.12}$$

where

$$\mathcal{E}_{s,k} = [\mathbf{E}_s]_{kk}, \quad \mathcal{E}_{x,k} = [\mathbf{E}_x]_{kk}. \tag{16.13}$$

Thus, insertion of the unitary matrices does not affect the zero-forcing property, the transmitted power, or the average MSE. For example, if $\{\mathbf{F}, \mathbf{G}\}$ is an MMSE transceiver, then so is the modified transceiver $\{\mathbf{F}\mathbf{U}^\dagger, \mathbf{U}\mathbf{G}\}$. Although $\mathbf{U}$ does not change the MSE, we will see that we *can* reduce the average error probability (16.7) by optimizing $\mathbf{U}$.

16.2.2 Minimizing error probability by optimizing U

The average symbol error probability in Fig. 16.2 is, from Eq. (16.7),

$$\mathcal{P}_e(\mathbf{y}) = \frac{c}{M} \sum_{k=0}^{M-1} \mathcal{Q}\Big(\frac{A}{\sqrt{\mathcal{E}_{s,k}}}\Big), \tag{16.14}$$

where the argument vector $\mathbf{y}$ is defined as

$$\mathbf{y} \stackrel{\Delta}{=} [\, \mathcal{E}_{s,0} \quad \mathcal{E}_{s,1} \quad \ldots \quad \mathcal{E}_{s,M-1} \,]. \tag{16.15}$$

It is shown in Sec. 21.2.3 of Chap. 21 that $\mathcal{Q}(A/\sqrt{y})$ is a *convex* function of y as long as the signal-to-error ratio exceeds a certain threshold. This idea was also discussed in Chap. 11 (Sec. 11.5). More precisely

$$\mathcal{Q}\Big(\frac{A}{\sqrt{\mathcal{E}_{s,k}}}\Big) \text{ is } \begin{cases} \text{convex in } \mathcal{E}_{s,k} & \text{for } \mathcal{E}_{s,k} < A^2/3 \\ \text{concave in } \mathcal{E}_{s,k} & \text{for } \mathcal{E}_{s,k} > A^2/3. \end{cases} \tag{16.16}$$

Assuming that the errors $\mathcal{E}_{s,k}$ are sufficiently small, the convexity condition holds. In Sec. 11.5 of Chap. 11 we presented examples of typical signal-to-noise ratios for which convexity holds, and found that the assumption is often reasonable.

Assuming the convexity condition holds, we conclude from Chap. 21 (Sec. 21.4) that $\mathcal{P}_y(\mathbf{y})$ in Eq. (16.14) is a **Schur-convex** function of $\mathbf{y}$. By definition, what this means is that, if a vector $\mathbf{y}_1$ is majorized by $\mathbf{y}_2$ (for definitions see Sec. 21.3, Chap. 21), then

$$\mathcal{P}_e(\mathbf{y}_1) \leq \mathcal{P}_e(\mathbf{y}_2). \tag{16.17}$$

Now refer to Eq. (16.10). The freedom to change the unitary matrix $\mathbf{U}$ gives us the freedom to adjust the elements $[\mathbf{E}_s]_{kk}$, but only subject to the constraint that their sum be constant (from Eq. (16.12)). We now use a result from Chap. 21 (Lemma 21.1) which says that the vector

$$\alpha\,[1 \quad 1 \quad \ldots \quad 1], \quad \alpha > 0,$$

is majorized by any vector of the form

$$[\, y_0 \quad y_1 \quad \ldots \quad y_{M-1} \,], \quad y_k \geq 0,$$

with identical sum, that is, $\sum_k y_k = \alpha M$. Thus the unitary $\mathbf{U}$ that minimizes $\mathcal{P}_e(\mathbf{y})$ is the one that equalizes the diagonal elements of $\mathbf{E}_s$, that is,

$$\mathbf{U}\mathbf{E}_x\mathbf{U}^\dagger = \alpha \begin{bmatrix} 1 & \times & \ldots & \times \\ \times & 1 & \ldots & \times \\ \vdots & \vdots & \ddots & \vdots \\ \times & \times & \ldots & 1 \end{bmatrix} = \mathbf{E}_s. \tag{16.18}$$

Under this condition we have

$$\mathcal{E}_{s,0} = \mathcal{E}_{s,1} = \ldots = \mathcal{E}_{s,M-1} = \mathcal{E}_{mse}, \tag{16.19}$$

where $\mathcal{E}_{mse}$ is the average mean square error (16.12).

Finding U. Since $\mathbf{E}_x$ is Hermitian, such a unitary $\mathbf{U}$ is readily found. For this we first diagonalize $\mathbf{E}_x$ and then use a DFT matrix to equalize the diagonal elements. More specifically, define $\mathbf{U}$ as a product of two unitary matrices:

$$\mathbf{U} = \mathbf{U}_2\mathbf{U}_1. \tag{16.20}$$

Then

$$\mathbf{E}_s = \mathbf{U}_2\mathbf{U}_1\mathbf{E}_x\mathbf{U}_1^\dagger\mathbf{U}_2^\dagger. \tag{16.21}$$

We first choose $\mathbf{U}_1$ such that $\mathbf{U}_1\mathbf{E}_x\mathbf{U}_1^\dagger$ is diagonal. If we now choose $\mathbf{U}_2$ to be the DFT or IDFT matrix, then $\mathbf{E}_s$ becomes a circulant (Appendix D), which in particular has identical diagonal elements as desired. As explained in Sec. 21.5.1.B it is not necessary for $\mathbf{U}_2$ to be a DFT matrix. It can be *any* unitary matrix with all elements having identical magnitude:

$$\Big|[\mathbf{U}_2]_{km}\Big| = \frac{1}{\sqrt{M}}, \qquad 0 \le k,\, m \le M-1. \tag{16.22}$$

For example, we can choose it to be the Hadamard matrix. $\bigtriangledown \bigtriangledown \bigtriangledown$

With $\mathbf{U}$ chosen to equalize the mean square errors $\mathcal{E}_{s,k}$ of the individual components as in Eq. (16.19), the average symbol error probability (16.14) is minimized and has the value

$$\mathcal{P} = c\mathcal{Q}\bigg(\frac{A}{\sqrt{\mathcal{E}_{mse}}}\bigg). \tag{16.23}$$

The result of this subsection can be summarized as follows. It is assumed that the signal-to-noise ratios at the detector inputs are large enough to justify convexity (as explained above).

♠**Lemma 16.1.** *Optimal unitary matrix.* Given any precoder-equalizer pair $\{\mathbf{F}, \mathbf{G}\}$ as in Fig. 16.1 satisfying the zero-forcing constraint $\mathbf{GHF} = \mathbf{I}$, consider the modified system of Fig. 16.2 where $\mathbf{U}$ is unitary.

1. The insertion of $\mathbf{U}$ and $\mathbf{U}^\dagger$ does not change the transmitted power or the zero-forcing property. Nor does it change the average mean square error of reconstruction.

2. Among all unitary $\mathbf{U}$, the average symbol error probability (16.14) is minimized by choosing $\mathbf{U}$ such that the mean square errors $\mathcal{E}_{s,k}$ are equalized (see Eq. (16.19)).

3. With $\mathbf{U}$ so chosen, the average symbol error probability is given by Eq. (16.23) where $\mathcal{E}_{mse}$ is the average mean square error achieved by $\{\mathbf{F}, \mathbf{G}\}$ in Fig. 16.1 (which is identical to the average MSE in Fig. 16.2). $\diamond$

The reasoning based on majorization and Schur convexity can also be summarized directly without using **U** in the statement as follows:

♠**Lemma 16.2.** *Bound on average error probability.* Given any transceiver $\{\mathbf{F}, \mathbf{G}\}$ for a fixed channel and fixed transmitted power, let the mean square errors of the components be $\mathcal{E}_{s,k}$. Then the average symbol error probability (16.14) is lower bounded as follows:

$$\mathcal{P}_e \geq cQ\left(\frac{A}{\sqrt{\mathcal{E}_{mse}}}\right), \tag{16.24}$$

where $\mathcal{E}_{mse} = \sum_k \mathcal{E}_{s,k}/M$. Equality is achieved when $\mathcal{E}_{s,k} = \mathcal{E}_{mse}$ for all k. ♢

Proof. The vector $\mathcal{E}_{mse}\,[1 \quad 1 \quad \ldots \quad 1]$ is majorized by

$$\mathbf{y} = [\,\mathcal{E}_{s,0} \quad \mathcal{E}_{s,1} \quad \ldots \quad \mathcal{E}_{s,M-1}\,]$$

as shown in Lemma 21.1 (Chap. 21). Since Eq. (16.14) is Schur-convex in $\mathbf{y}$, the result follows immediately. ▽▽▽

16.2.3 ZF transceiver with minimum error probability

We are now ready for the main result. Recall again the assumption that the signal-to-noise ratio is large enough to satisfy the convexity condition in Eq. (16.16). The signal and noise statistics are as in Eq. (16.1).

♠**Theorem 16.1.** *Minimum error probability with zero-forcing.* For a fixed channel $\mathbf{H}$ and fixed transmitted power p_0, let $\mathcal{E}_{mmse}$ be the minimum achievable average mean square error among all zero-forcing transceivers $\{\mathbf{F}, \mathbf{G}\}$. Then the minimum possible average symbol error probability is

$$\mathcal{P}_{e,min} = cQ\left(\frac{A}{\sqrt{\mathcal{E}_{mmse}}}\right). \tag{16.25}$$

This minimum can be achieved as follows:

1. First design a ZF-MMSE transceiver $\{\mathbf{F}, \mathbf{G}\}$ by joint optimization of $\mathbf{F}$ and $\mathbf{G}$. This evidently has MSE equal to $\mathcal{E}_{mmse}$.
2. Insert unitary matrices $\mathbf{U}, \mathbf{U}^\dagger$ as in Fig. 16.2 so that the mean square errors $\mathcal{E}_{s,k}$ are equalized.

Then the resulting transceiver $\{\mathbf{F}\mathbf{U}^\dagger, \mathbf{U}\mathbf{G}\}$ achieves the minimum error probability (16.25). ♢

Proof. This follows from Lemmas 16.1 and 16.2. Lemma 16.2 says that, for any transceiver $\{\mathbf{F}, \mathbf{G}\}$, the average symbol error probability cannot be smaller than (16.24). The absolute lower bound is therefore given by Eq. (16.25) where $\mathcal{E}_{mmse}$ is the mean square error achieved by an MMSE transceiver. The bound (16.25) is indeed achievable by designing an MMSE transceiver and then inserting $\mathbf{U}$ and $\mathbf{U}^\dagger$ to equalize the errors $\mathcal{E}_k$ (Lemma 16.1). Such $\mathbf{U}$ and $\mathbf{U}^\dagger$ do not change the transmitted power, so the transceiver designed in this way indeed minimizes the average error probability while respecting the power constraint. $\bigtriangledown\bigtriangledown\bigtriangledown$

The unitary matrix $\mathbf{U}$ described in the theorem exists because it can be constructed as explained in the paragraph preceding Eq. (16.23).

16.3 Bias in the reconstruction error

In the next section we discuss the problem of minimizing error probability with the zero-forcing constraint removed. The approach will be similar to that given in Sec. 16.2 but there is an important difference. This arises owing to the fact that the reconstruction error is in general **biased** when the ZF condition is not imposed. The purpose of the present section is to explain this bias, and describe the so-called **bias removal** operation. For this section the reader might want to review portions from Kay [1998] as appropriate. A wealth of material on this topic can also be found in the lecture notes by John Cioffi on equalization.[2]

Example 16.1: Bias in MMSE transceivers for diagonal channels

Consider the simple MMSE system described in Sec. 11.3 (Chap. 11) for diagonal channels. We found that optimal multipliers α_k and β_k in Fig. 11.2 are such that

$$\alpha_k H_k \beta_k = \frac{|\alpha_k|^2}{|\alpha_k|^2 + \dfrac{\sigma_{q_k}^2}{|H_k|^2\sigma_s^2}} \tag{16.26}$$

This is the transfer function from $s_k(n)$ to $\widehat{s}_k(n)$, and is not unity unlike in a zero-forcing system (unless the noise is zero, that is, $\sigma_{q_k}^2 = 0$). Thus

$$\widehat{s}_k(n) = A_k s_k(n) + e_k(n),$$

where $A_k < 1$, and $e_k(n)$ is the effect of channel noise. Assuming that $s_k(n)$ and $e_k(n)$ are statistically independent and that $E[e_k(n)] = 0$, we have

$$E\big[\widehat{s}_k(n)\big|s_k(n)\big] = A_k s_k(n) \neq s_k(n) \tag{16.27}$$

because $A_k < 1$. That is, the expected value of $\widehat{s}_k(n)$, given $s_k(n)$, is not equal to $s_k(n)$. So we say that *the estimate of* $s_k(n)$ *is biased.* Similar analysis shows that the pure-MMSE estimates given in Chap. 13 for nondiagonal channels are also biased.

---

[2] http://www.stanford.edu/group/cioffi/ee379a/.

16.3.1 Bias removal

Consider Fig. 16.3, which shows a transceiver with linear precoder and equalizer. In general, the estimate $\widehat{s}_k(n)$ of $s_k(n)$ is biased in the sense that the expected value of $\widehat{s}_k(n)$, given $s_k(n)$, is different from $s_k(n)$:

$$E[\widehat{s}_k(n)\Big|s_k(n)] \neq s_k(n). \tag{16.28}$$

If a biased estimate is used at the input of the symbol detector in the receiver, then the detected symbol can suffer from serious errors (see Sec. 2.5.5). In practice therefore the bias is removed before symbol detection. In fact, bias removal is automatically done when the threshold lines of the detector are appropriately positioned with respect to the constellation symbols.

Let the transfer matrix from the input to the output of the transceiver before bias removal be denoted as $\mathbf{T}$. From Fig. 16.3 we have

$$\mathbf{T} = \mathbf{GHF}. \tag{16.29}$$

Then the kth output is given by

$$\widehat{s}_k = T_{kk}s_k + \sum_{m\neq k} T_{km}s_m + \eta_k, \tag{16.30}$$

where η_k is the kth component of the transformed noise vector $\mathbf{Gq}$. Here we have omitted the arguments (n) for simplicity. Since q_k are assumed to have zero mean, the quantities η_k also have zero mean. The symbols $s_k(n)$ come from the constellation. We can regard $s_m, m \neq k$, as zero-mean random variables statistically independent of s_k. We then have

$$E[\widehat{s}_k\Big|s_k] = T_{kk}s_k. \tag{16.31}$$

This is not equal to s_k unless $T_{kk} = 1$. So the estimate $\widehat{s}_k$ is biased. Note that for a zero-forcing system, since $\mathbf{GHF} = \mathbf{I}$, we have $T_{kk} = 1$, and there is no bias. But for MMSE systems without zero forcing, in general $T_{kk} \neq 1$, and this creates the bias.

We can remove the bias simply by dividing the estimate $\widehat{s}_k$ by the number T_{kk}. Thus the bias-removed estimate has the form

$$\widehat{s}_{k,br} = s_k + \sum_{m\neq k} T_{km}s_m/T_{kk} + \eta_k/T_{kk}. \tag{16.32}$$

Defining the multipliers

$$\gamma_k \stackrel{\Delta}{=} \frac{1}{T_{kk}}, \tag{16.33}$$

the receiver which incorporates bias removal can therefore be schematically represented as in Fig. 16.4. Our interest is in optimizing $\mathbf{G}$ and $\mathbf{H}$ in this bias-free system, for minimizing the symbol error rate.

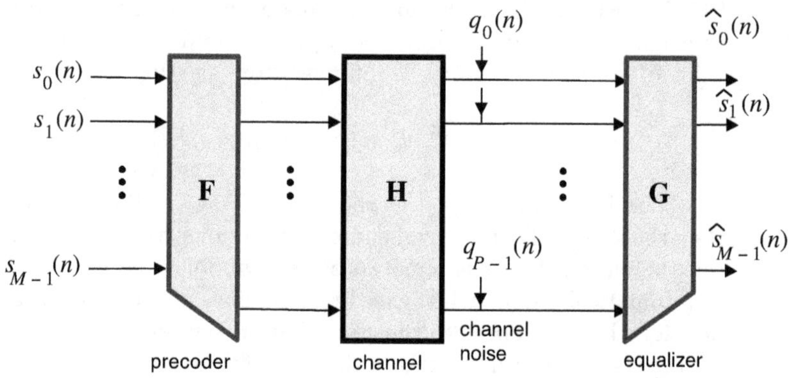

Figure 16.3. A linear transceiver with precoder **F** and equalizer **G**.

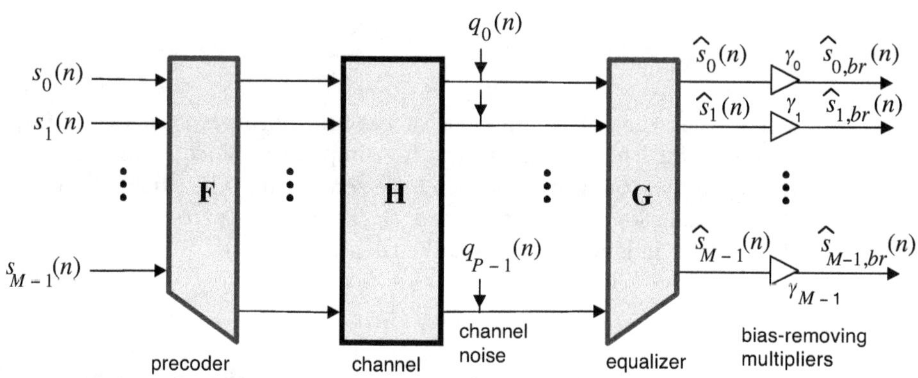

Figure 16.4. A linear transceiver with precoder **F**, equalizer **G**, and conceptual bias-removing multipliers γ_k.

16.3.2 Error probability after bias removal from an MMSE estimate

For the MMSE estimate, the reconstruction errors before and after bias removal are, respectively,

$$e_k = \widehat{s}_k - s_k \quad \text{and} \quad e_{k,br} = \widehat{s}_{k,br} - s_k. \tag{16.34}$$

Here the subscript "br" is a reminder for "bias-removed." Denote the corresponding mean square errors as $\mathcal{E}_k$ and $\mathcal{E}_{k,br}$ so that the signal-to-reconstruction error ratios (SER) before and after bias removal are given by

$$\beta_k = \frac{\sigma_s^2}{\mathcal{E}_k}, \quad \beta_{k,br} = \frac{\sigma_s^2}{\mathcal{E}_{k,br}} \tag{16.35}$$

If the receiver matrix $\mathbf{G}$ is chosen as the MMSE matrix (for a given precoder $\mathbf{F}$, channel $\mathbf{H}$, and noise statistics), then $\widehat{s}_k$ is an MMSE estimate of s_k. A fundamental result in this case is that the above two SERs are related as follows:

$$\beta_{k,br} = \beta_k - 1. \tag{16.36}$$

This is proved in Appendix 16.A (Lemma 16.6) at the end of the chapter. This relation can be rewritten as

$$\frac{1}{\mathcal{E}_{k,br}} = \frac{1}{\mathcal{E}_k} - \frac{1}{\sigma_s^2} \tag{16.37}$$

The kth symbol error probability after bias removal is therefore given by[3]

$$\mathcal{P}_{k,br} = c\mathcal{Q}\left(A\sqrt{\frac{1}{\mathcal{E}_{k,br}}}\right) = c\mathcal{Q}\left(A\sqrt{\frac{1}{\mathcal{E}_k} - \frac{1}{\sigma_s^2}}\right), \tag{16.38}$$

where c and A are the constants defined earlier (e.g., as in Eq. (16.6) for QAM). Thus the average error probability after bias removal is

$$\mathcal{P}_{br} = \frac{c}{M}\sum_{k=0}^{M-1} \mathcal{Q}\left(A\sqrt{\frac{1}{\mathcal{E}_k} - \frac{1}{\sigma_s^2}}\right). \tag{16.39}$$

16.3.3 Optimality of bias-removed MMSE estimate

For a fixed precoder $\mathbf{F}$ and channel $\mathbf{H}$, consider again the optimization of the equalizer $\mathbf{G}$. Since there is no constraint on $\mathbf{G}$ (e.g., no unitarity constraint, etc.), we can optimize one row at a time. For example, assume the kth row $\mathbf{g}_k$ is chosen such that its output is an MMSE estimate:[4]

$$\widehat{s}_\perp = \alpha s + \tau. \tag{16.40}$$

For simplicity, the argument (n) is omitted, and the subscript k is deleted temporarily. In Eq. (16.40), τ is a combination of noise and interference terms and will be assumed to be zero-mean, and statistically independent of s. The above estimate is used by the detector to identify the transmitted symbol s. In this process, the detector first removes the bias term automatically:

$$\widehat{s}_{\perp,br} = s + \frac{\tau}{\alpha}. \tag{16.41}$$

---

[3]This assumes that the error terms in Eq. (16.30), which come from both noise and interference, are Gaussian. The assumption is acceptable if M is large (in view of the central limit theorem).

[4]The subscript $\perp$ is a reminder that MMSE estimates satisfy the orthogonality condition (Sec. F.2.1 in Appendix F).

With bias removed, the accuracy of detection depends entirely upon the SINR (signa-to-interference-plus-noise ratio) given by

$$SINR_{\perp} = \frac{\alpha^2\sigma_s^2}{\mathcal{E}_\tau} \tag{16.42}$$

Notice here that the SINR is simply the mean square value of the signal term divided by the mean square value of the noise-and-interference terms in Eq. (16.40). Its value is the same whether we use the original equation (16.40) or the bias-removed equation (16.41).

Now, instead of obtaining the unbiased estimate from the MMSE solution as above, suppose we obtain another estimate of s (using a different value for the kth row $\mathbf{g}_k$) in some different way:

$$\widehat{s}_{other} = \beta s + \theta. \tag{16.43}$$

Here θ is a combination of noise and interference terms and can be assumed to be zero-mean, and statistically independent of s. The corresponding bias-removed version is

$$\widehat{s}_{other,br} = s + \frac{\theta}{\beta} \tag{16.44}$$

and the SINR is

$$SINR_{other} = \frac{\beta^2\sigma_s^2}{\mathcal{E}_\theta} \tag{16.45}$$

We now claim the following result:

♠**Lemma 16.3.** With quantities defined as above we have

$$SINR_{\perp} \geq SINR_{other}, \tag{16.46}$$

or equivalently

$$\frac{\alpha^2}{\mathcal{E}_\tau} \geq \frac{\beta^2}{\mathcal{E}_\theta} \tag{16.47}$$

That is, the linear MMSE estimate also has maximum SINR among all linear estimates. ◇

Proof. Lemma 16.5 in Appendix 16.A at the end of the chapter shows that the mean square error of a bias-free linear estimate (16.44) can never be better than the mean square error of the estimate (16.41) obtained by removing bias from the MMSE estimate. That is, $\mathcal{E}_\theta/\beta^2 \geq \mathcal{E}_\tau/\alpha^2$. This is the same as Eq. (16.47). ▽▽▽

Since the error probability depends only on the SINR at the input of the detector, the bias-removed MMSE estimate results in smaller error probability than any other linear bias-free estimate. This is the main conclusion of this subsection.

16.3.4 SINR, SER, and error probability

For completeness we would like to note here that one should carefully distinguish between the SINR and the signal-to-error ratio (SER). The SINR is the ratio of the mean square value of the signal term over the mean square value of the error term in expressions for the estimate such as Eqs. (16.40) and (16.43). Its value is unchanged by bias-removal operations. With $\mathcal{E}_k$ representing the mean square error in the estimate of s_k (which has mean square value σ_s^2), the quantity

$$\frac{\sigma_s^2}{\mathcal{E}_k} \tag{16.48}$$

can be regarded as the SER (signal-to-error ratio). It does not directly determine the error probability. In general, the SER and SINR are identical only when the bias is zero.

As shown in Sec. 16.3.2, the error probability after bias removal from the MMSE estimate is given by Eq. (16.38), and is reproduced below:

$$\mathcal{P}_{k,br} = c\mathcal{Q}\left(A\sqrt{\frac{1}{\mathcal{E}_k} - \frac{1}{\sigma_s^2}}\right) = c\mathcal{Q}\left(\frac{A}{\sigma_s}\sqrt{\frac{\sigma_s^2}{\mathcal{E}_k} - 1}\right), \tag{16.49}$$

where $\mathcal{E}_k$ is the MMSE (before bias removal). From Eq. (16.37) we have

$$\frac{\sigma_s^2}{\mathcal{E}_{k,br}} = \frac{\sigma_s^2}{\mathcal{E}_k} - 1 \tag{16.50}$$

so that

$$SINR_k = SER_{k,br} = SER_{k,\perp} - 1, \tag{16.51}$$

where the subscript *br* stands for "bias-removed" and $\perp$ stands for "mmse." So we can rewrite the kth error probability as

$$\mathcal{P}_{k,br} = c\mathcal{Q}\left(\frac{A}{\sigma_s}\sqrt{SINR_k}\right). \tag{16.52}$$

16.4 Minimizing error probability without ZF

In Sec. 16.3.3 we argued that the SINR of the bias-removed MMSE estimate is at least as good as the SINR of any bias-free linear estimate (such as, for example, the estimate with a zero-forcing constraint). Given that the detector in practice always removes the bias, it is therefore clear that, *for any precoder* $\mathbf{F}$ *and channel* $\mathbf{H}$*, the equalizer* $\mathbf{G}$ *should be chosen as the corresponding MMSE equalizer.* If $\mathbf{G}$ is followed by bias removal, as in Fig. 16.4, then the average error probability is as in Eq. (16.39), which is reproduced below:

$$\mathcal{P}_{br} = \frac{c}{M}\sum_{k=0}^{M-1} \mathcal{Q}\left(A\sqrt{\frac{1}{\mathcal{E}_k} - \frac{1}{\sigma_s^2}}\right). \tag{16.53}$$

In Appendix 16.B at the end of the chapter we show that

$$f(\mathcal{E}_k) \stackrel{\Delta}{=} \mathcal{Q}\left(A\sqrt{\frac{1}{\mathcal{E}_k} - \frac{1}{\sigma_s^2}}\right) \tag{16.54}$$

is a convex function of $\mathcal{E}_k$ (MSE before bias removal), as long as the error $\mathcal{E}_k$ is smaller than a threshold $\mathcal{E}_t$. In fact, for QPSK constellations, we will see that this is true[5] for all $\mathcal{E}_k \leq \sigma_s^2$. Thus the sum (16.53) is **Schur-convex** (Chap. 21) in the vector

$$\mathbf{y} = [\,\mathcal{E}_0 \quad \mathcal{E}_1 \quad \ldots \quad \mathcal{E}_{M-1}\,], \tag{16.55}$$

which is the vector of *mean square errors before bias removal.* Now consider Fig. 16.5, which shows the transceiver $\{\mathbf{F}, \mathbf{G}\}$, and the modified transceiver which has the unitary matrix $\mathbf{U}$ and its inverse $\mathbf{U}^\dagger$ inserted. Recall from Sec. 16.2.1 that this insertion of $\mathbf{U}$ and $\mathbf{U}^\dagger$ does not affect either the *transmitted power* or the *average mean square error*

$$\mathcal{E}_{mse} = \frac{1}{M}\sum_{k=0}^{M-1} \mathcal{E}_k. \tag{16.56}$$

Since $\mathcal{P}_{br}$ in Eq. (16.53) is a Schur-convex function of the vector $\mathbf{y}$, it follows that the best unitary $\mathbf{U}$ which minimizes $\mathcal{P}_{br}$ is such that the mean square errors $\mathcal{E}_k$ are equalized. The proof is precisely as in Sec. 16.2.2. Thus, for any precoder $\mathbf{F}$ and channel $\mathbf{H}$, the best $\mathbf{U}$ is such that

$$\mathcal{E}_0 = \mathcal{E}_1 = \ldots = \mathcal{E}_{M-1} = \mathcal{E}_{mse}, \tag{16.57}$$

where $\mathcal{E}_{mse}$ is the average MSE with or without $\mathbf{U}$. In Sec. 16.2.2 we explained how such an $\mathbf{U}$ can be designed (see the remarks after Eq. (16.19)). With $\mathbf{U}$ so chosen, the average error probability is clearly given by

$$\mathcal{P}_{e,min} = c\mathcal{Q}\left(A\sqrt{\frac{1}{\mathcal{E}_{mse}} - \frac{1}{\sigma_s^2}}\right). \tag{16.58}$$

From the preceding discussion it follows that if a transceiver $\{\mathbf{F}_o, \mathbf{G}_o\}$ minimizes the average error probability then the following should be true:

1. $\mathbf{G}_o$ is an MMSE equalizer for the precoder $\mathbf{F}_o$. As a result the expression for error probability is as in Eq. (16.53).
2. With no loss of generality, the mean square errors $\mathcal{E}_k$ can be assumed to be equal, as in Eq. (16.55), for otherwise we can append $\mathbf{U}$ and $\mathbf{U}^\dagger$ to achieve this without increasing the average error probability. So the expression for the average error probability actually reduces to the simpler form (16.58).
3. Since the function (16.58) decreases with decreasing $\mathcal{E}_{mse}$, the optimal precoder equalizer pair $\{\mathbf{F}_o, \mathbf{G}_o\}$ which minimizes the average error probability should be such that $\mathcal{E}_{mse}$ itself is minimized!

[5]Since the mean square error of an MMSE estimate of $s_k(n)$ cannot exceed the mean square value of $s_k(n)$, the condition $\mathcal{E}_k \leq \sigma_s^2$ is always satisfied!

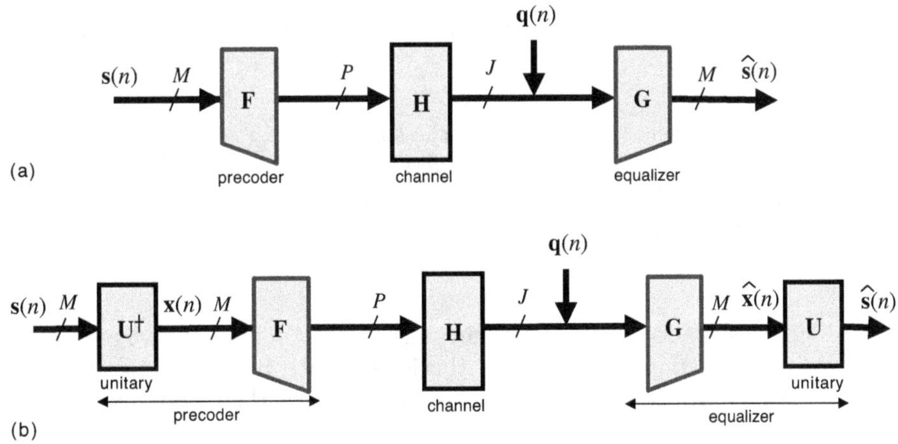

Figure 16.5. (a) A transceiver with precoder $\mathbf{F}$ and equalizer $\mathbf{G}$. (b) Insertion of unitary matrices $\mathbf{U}$ and $\mathbf{U}^\dagger$.

So, we first optimize $\mathbf{F}$ and $\mathbf{G}$ jointly so that we get an MMSE transceiver. We then insert $\mathbf{U}$ and $\mathbf{U}^\dagger$ such that Eq. (16.57) is achieved. The resulting system has precoder and equalizer given by

$$\mathbf{F}_o = \mathbf{F}\mathbf{U}^\dagger, \quad \mathbf{G}_o = \mathbf{U}\mathbf{G}, \tag{16.59}$$

and minimum average symbol error probability. This error probability is given by

$$\mathcal{P}_{e,min} = c\mathcal{Q}\left(A\sqrt{\frac{1}{\mathcal{E}_{mmse}} - \frac{1}{\sigma_s^2}}\right). \tag{16.60}$$

Note here that $\mathcal{E}_{mmse}$ is the average MSE (per scalar component $s_k(n)$). The result of this section can be summarized as follows. As a reminder, the underlying assumption here is that the mean square errors $\mathcal{E}_k$ are small enough to justify the convexity of the terms in Eq. (16.53). In practice this assumption is quite realistic.

♠**Theorem 16.2.** *Minimum error probability without zero-forcing.* For fixed channel and transmitted power, the transceiver that minimizes the average symbol error probability after bias removal can be designed as follows:

1. First design an MMSE transceiver $\{\mathbf{F}, \mathbf{G}\}$. Let $\mathcal{E}_{mmse}$ be the minimized average mean square error before bias removal.
2. Insert the unitary matrices $\mathbf{U}$ and $\mathbf{U}^\dagger$, as in Fig. 16.5(b), where $\mathbf{U}$ is such that the mean square errors of the components at its output are equalized.

The average mean square error continues to be $\mathcal{E}_{mmse}$, and the transmitted power p_0 is unchanged.

3. Insert bias-removal multipliers as in Fig. 16.4, and then use threshold detectors.

The resulting transceiver has the smallest possible average symbol error probability among all linear transceivers for the given channel with transmitted power p_0. The minimized error probability is given by Eq. (16.60), where σ_s^2 is the mean square value of $s_k(n)$. $\diamondsuit$

The systems described in Theorems 16.1 and 16.2 both have zero bias before symbol detection. In the former this is automatic because of zero forcing, and in the latter this is achieved by bias removal. Since the system described in Theorem 16.2 minimizes error probability without imposing the zero-forcing constraint, the average error probability (16.60) is at least as small as the error probability of the zero-forced optimal system in Eq. (16.25). A second proof of this is developed in Appendix 16.C at the end of the chapter, as it might bring additional insight to some readers. Also, the next section explains some interesting connections between the two systems.

16.5 Bias-removed MMSE versus ZF-MMSE

At the end of Sec. 16.4 we noted that bias-removed pure-MMSE systems are in general better than ZF-MMSE systems in the sense of having smaller average error probability. Under some conditions these two systems are actually identical. In this section we discuss how these two systems are related; this discussion can be skipped without loss of continuity.

16.5.1 Diagonal channels

First consider the diagonal channel shown in Fig. 16.6. Here we have a set of M scalar channels H_k. The precoder and equalizer are diagonal matrices, with diagonal elements α_k and β_k. The multipliers γ_k represent bias-removing multipliers. We have

$$\widehat{s}_k = H_k\alpha_k\beta_k s_k + \beta_k q_k, \tag{16.61}$$

If we divide this by $H_k\alpha_k\beta_k$ the result is

$$\widehat{s}_{k,br} = s_k + q_k/H_k\alpha_k, \tag{16.62}$$

which is unbiased. In fact the kth path now produces a *zero-forcing* estimate because the path from input s_k to output $\widehat{s}_k$ has transfer function equal to unity. In the pure-MMSE solution for Fig. 16.6, there are some paths for which $\alpha_k = 0$, depending on total available power. These paths are obviously not ZF-paths.

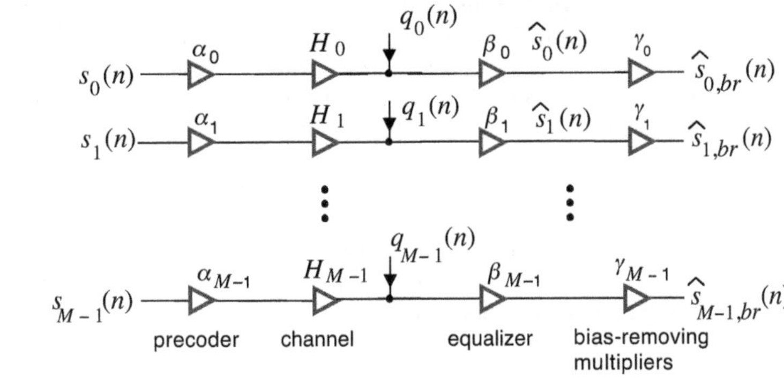

Figure 16.6. A diagonal channel with diagonal precoder and equalizer.

If we leave these zero paths unchanged, and replace the nonzero paths with the ZF versions as described above, then does the M-channel system have smaller MSE than the M-channel ZF-MMSE system? Not necessarily, because *the values of α_k are different in the two systems.*

For example, consider the case where $K = M$ (that is, there are no zero-paths). In this case if each MMSE path is bias-removed, the resulting ZF system is not necessarily the ZF-MMSE system – the precoder multipliers α_k in the pure-MMSE system are different from those in the ZF-MMSE system.[6] Since the ZF-MMSE system, by definition, is optimal under the ZF constraint, the *pure-MMSE system after bias removal is in general worse than the ZF-MMSE system* when $K = M$. More generally we have to distinguish between two types of scenarios:

1. If there are no zero-paths at all in the pure-MMSE solution, then the result of bias removal is a ZF system which can be worse than (but is no better than) the ZF-MMSE solution.

2. If there are some zero-paths in the MMSE solution, and we remove bias in the nonzero paths, the resulting system may or may not be better than the ZF-MMSE system. It will depend on the example under consideration.

For the second case above we can readily create an example where ZF-MMSE is worse than bias-removed pure MMSE: take $K = M - 1$. Then pure-MMSE has one path deleted. Now the ZF-MMSE expression is (Chap. 11)

$$\mathcal{E}_{ZF,mmse} = \frac{\sigma_s^2}{p_0}\left(\sum_{k=0}^{M-1} \frac{\sigma_{q_k}}{|H_k|}\right)^2, \tag{16.63}$$

and can be made arbitrarily large by constructing an example with $|H_{M-1}|$ arbitrarily small.

[6]The observations available at the receiver for estimating $s_k(n)$ are therefore different in the two cases.

How does unitary U make a difference? Why are things different when the unitary matrices $\mathbf{U}$ and $\mathbf{U}^\dagger$ are inserted as shown in Fig. 16.7? Recall that in this system we remove bias by inserting γ_k *after* the unitary $\mathbf{U}$, and the system *does not* reduce to a zero-forcing system. This is because the transfer matrix from $\mathbf{s}(n)$ to $\widehat{\mathbf{s}}(n)$ (before bias removal) has the form

$$\mathbf{T} = \mathbf{U}\boldsymbol{\Lambda}\mathbf{U}^\dagger,$$

where $\boldsymbol{\Lambda}$ is diagonal. The matrix $\mathbf{T}$ is in general not diagonal, so the insertion of γ_k does not make the overall transfer matrix identity. Bias removal forces only the diagonal elements to be unity:

$$[\boldsymbol{\Gamma}\mathbf{T}]_{kk} = 1,$$

where $\boldsymbol{\Gamma}$ is a diagonal matrix with γ_k on the diagonals. Furthermore, this transceiver with the unitary matrices inserted does not have any zero paths (i.e., $[\mathbf{T}]_{kk} \neq 0$ for any k in general), even though such paths may exist in the diagonal system $\boldsymbol{\Lambda}$ sandwiched between the matrices $\mathbf{U}$ and $\mathbf{U}^\dagger$. The effect of $\mathbf{U}$ will be discussed in greater detail below. $\bigtriangledown \bigtriangledown \bigtriangledown$

Note finally that for a *fixed* set of multipliers $\{\alpha_k\}$, the zero-forcing $\{\beta_k\}$ are identical to the MMSE solution $\{\beta_k\}$ followed by bias removal.

16.5.2 Non-diagonal channels

The preceding discussions can be generalized readily to non-diagonal channels. Let $\mathbf{H}$ be a possibly non-diagonal channel with singular value decomposition

$$\mathbf{H} = \mathbf{U}_h \boldsymbol{\Sigma}_h \mathbf{V}_h^\dagger, \tag{16.64}$$

where $\mathbf{U}_h$ and $\mathbf{V}_h$ are unitary and $\boldsymbol{\Sigma}_h$ diagonal. Thus the product

$$\mathbf{U}_h^\dagger \mathbf{H} \mathbf{V}_h = \boldsymbol{\Sigma}_h \tag{16.65}$$

is diagonal. If we now construct a precoder and an equalizer as in Fig. 16.8(a), then the system can be redrawn as in Fig. 16.8(b), which is a diagonal channel with diagonal precoder and equalizer. Thus all the discussions in the preceding paragraphs can be extended to non-diagonal channels.

1. For example, if we have an MMSE system and we remove the bias by inserting diagonal multipliers after the matrix $\boldsymbol{\Lambda}_\beta$ (Fig. 16.8(a)) then the resulting system may be better or worse than the ZF-MMSE system.
2. As a specific instance, for fixed precoder $\boldsymbol{\Lambda}_\alpha$ with no zero elements on the diagonal, if we compare a pure-MMSE equalizer with a ZF-MMSE equalizer, then the former reduces to the latter when bias is removed. Thus, **for a fixed precoder,** the pure-MMSE system (with bias removed) is identical to the ZF-MMSE system. In Chap. 17 we will see that multicarrier cyclic-prefix (MC-CP) systems provide an example falling under this category.

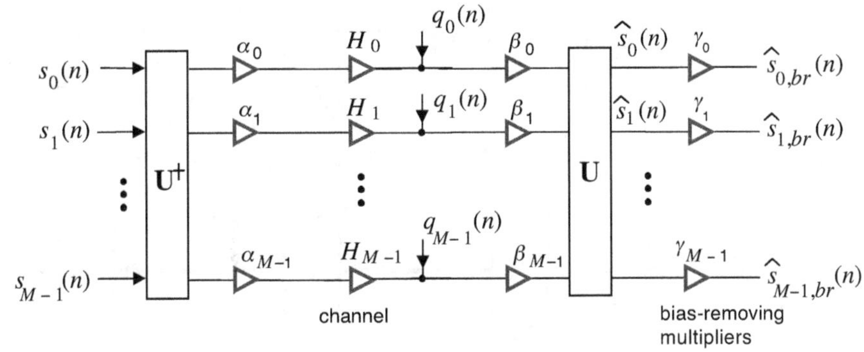

Figure 16.7. A diagonal channel with diagonal precoder and equalizer, with unitary matrices $\mathbf{U}$ and $\mathbf{U}^\dagger$ added to reduce average error probability.

3. Finally, for the nondiagonal channel we can insert the unitary matrices $\mathbf{U}$ and $\mathbf{U}^\dagger$ as before to minimize the average symbol error probability (Fig. 16.8(c)). The bias-removal multipliers γ_k are inserted after $\mathbf{U}$. The presence of $\mathbf{U}$ makes a difference: optimizing it makes the pure-MMSE system (followed by bias removal) superior to the ZF-MMSE system. Even though some of the diagonal elements of $\mathbf{\Lambda}_\alpha$ may be zero in the MMSE system, the presence of $\mathbf{U}$ and $\mathbf{U}^\dagger$ also ensures that none of the symbols $s_k(n)$ is left out, that is, the transfer function from $s_k(n)$ to $\widehat{s}_k(n)$ is nonzero for all k.

4. The optimal $\mathbf{U}$ is the same as described earlier, namely it is the matrix which forces the diagonal elements of the error covariance matrix to be equal. In Chap. 17 we will see that single-carrier cyclic-prefix (SC-CP) systems come under this category. They are therefore superior to OFDM systems because they take advantage of the presence of the matrix $\mathbf{U}$.

16.6 Concluding remarks

In this chapter we made use of the theory of convex and Schur-convex functions for the optimization of symbol error rates in linear transceivers. In the following few chapters we will consider the applications of these optimal transceiver results for the case of zero-padded and cyclic-prefix systems.

Based on the preparation obtained by reading this chapter (and Chap. 21), the reader can readily pursue advanced literature in this fascinating research area. While Schur-convex functions have been well known in the mathematics literature [Marshall and Olkin, 1979], their application in transceiver optimization was pioneered by Guess [2003] and Palomar *et al.* [2003].

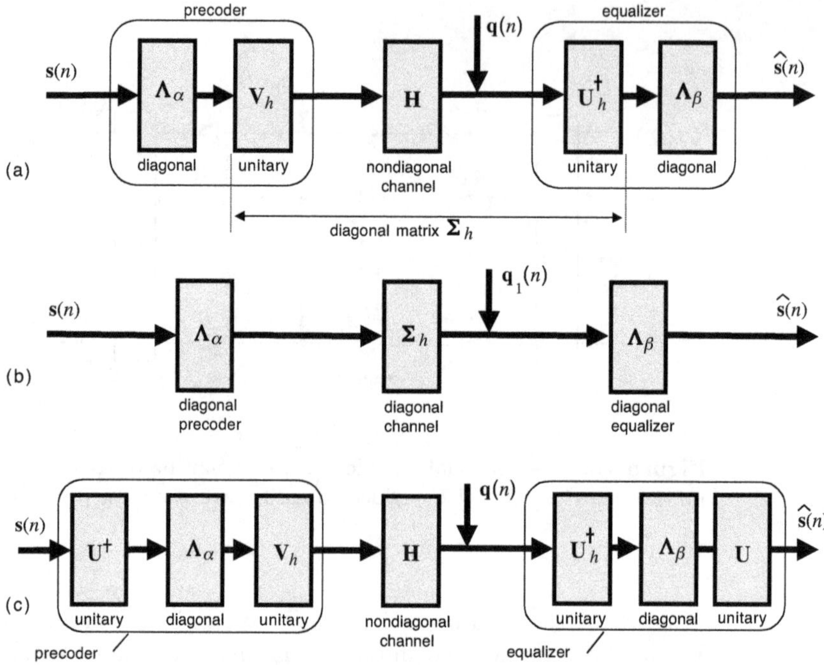

Figure 16.8. (a) A nondiagnal channel with precoder and equalizer chosen in a specific form, (b) equivalent diagonal transceiver system, and (c) insertion of extra unitary matrices $\mathbf{U}$ and $\mathbf{U}^\dagger$ in the original nondiagonal system.

Schur-convex functions and majorization theory have found many applications in transceiver optimization, such as optimization with QoS (quality-of-service constraints) [Pandharipande and Dasgupta, 2003], [Palomar *et al.*, 2004]. They have also found application in optimization of transceivers with nonlinear equalizers such as DFEs (decision feedback equalizers) [Jiang *et al.*, 2007], [Shenouda and Davidson, 2008], [Weng and Vaidyanathan, 2010a]. We shall give a detailed presentation of optimal DFE transceivers in Chap. 19. The results of this chapter are based on widely used approximations for the error probabilities in PAM and QAM systems. It will be interesting to extend these results using more accurate expressions that have been reported in recent years (see Cho and Yoon [2002] and references therein).

Appendices to Chapter 16

16.A Bias-removed MMSE estimates

Let $\widehat{x}$ be the MMSE estimate of a random variable x based on a linear combination of observations $\{y_k\}$:

$$\widehat{x} = \sum_{k=0}^{N-1} c_k y_k. \tag{16.66}$$

In the transceiver context, x could be a transmitted symbol, y_k the noisy measurements of the output of a linear time invariant channel, and $\widehat{x}$ the reconstructed version of x at the detector input. It is often possible to write the estimate in the form

$$\widehat{x} = \alpha x + \tau, \tag{16.67}$$

where τ is a zero-mean random variable related to the channel noise and other interfering symbols, but statistically independent of x. The independence implies in particular that

$$E[x\tau^*] = 0. \tag{16.68}$$

The estimation error is

$$e_{\perp} = \widehat{x} - x = (\alpha - 1)x + \tau. \tag{16.69}$$

In view of the MMSE property, this is orthogonal to the observations y_k (Sec. F.2.1 in Appendix F), and hence orthogonal to $\widehat{x}$:

$$E[e_{\perp}\widehat{x}^*] = 0. \tag{16.70}$$

We now prove a number of results pertaining to the form (16.67). One of these is that the quantity α is related to the mean square values of x and τ. In what follows, the notation $\mathcal{E}_y$ is used to denote the mean square value of a random variable y.

♠**Lemma 16.4.** Whenever the MMSE linear estimate is expressed as in Eq. (16.67), the mean square values of τ and x are necessarily related by

$$\mathcal{E}_{\tau} = \alpha(1 - \alpha)\mathcal{E}_x, \tag{16.71}$$

and the mean square estimation error can be written as

$$\mathcal{E}_{e_{\perp}} = \frac{\mathcal{E}_{\tau}}{\alpha} \tag{16.72}$$

Furthermore α itself is constrained by

$$\alpha = 1 - \frac{\mathcal{E}_{e_{\perp}}}{\mathcal{E}_x} \tag{16.73}$$

In particular, therefore, α is real and $0 \leq \alpha \leq 1$. ◇

Proof. The mean square error can be expressed as

$$\begin{aligned}\mathcal{E}_{e_\perp} = E[e_\perp e_\perp^*] = E[e_\perp(\widehat{x} - x)^*] &= -E[e_\perp x^*] \quad \text{(using Eq. (16.70))} \\ &= (1-\alpha)E|x|^2 \quad \text{(using Eq. (16.69))}\end{aligned}$$

so that

$$\mathcal{E}_{e_\perp} = (1-\alpha)\mathcal{E}_x, \tag{16.74}$$

which proves Eq. (16.73). But in view of Eq. (16.68), we also have from Eq. (16.69)

$$\mathcal{E}_{e_\perp} = (1-\alpha)^2\mathcal{E}_x + \mathcal{E}_\tau, \tag{16.75}$$

where we have used the fact that α is real. Equating the right-hand sides of Eqs. (16.74) and (16.75) we get Eq. (16.71). Dividing (16.74) by (16.71) yields Eq. (16.72). $\bigtriangledown \bigtriangledown \bigtriangledown$

16.A.1 Bias-removed estimates

From Eq. (16.67) we see that the expected value of the estimate $\widehat{x}$, given the transmitted symbol x, is given by

$$E[\widehat{x}|x] = \alpha x + E[\tau|x]. \tag{16.76}$$

Usually τ is a zero-mean random variable,

$$E[\tau] = 0, \tag{16.77}$$

and it is statistically independent of x. In particular, therefore, $E[\tau|x] = 0$, and we see that

$$E[\widehat{x}|x] = \alpha x \neq x. \tag{16.78}$$

Since the conditional expectation of $\widehat{x}$ is different from x, we say that the estimate is **biased**. An obvious way to remove the bias is to just "kick it out." That is, define the bias-removed estimate $\widehat{x}_{br} = \widehat{x}/\alpha$. This has the form

$$\widehat{x}_{br} = x + \frac{\tau}{\alpha} \tag{16.79}$$

where "*br*" is a reminder for "bias-removed." Clearly the mean squared error of the bias-removed estimate is $\mathcal{E}_{e_{br}} = \mathcal{E}_\tau/\alpha^2$, which simplifies to

$$\mathcal{E}_{e_{br}} = \frac{\mathcal{E}_{e_\perp}}{\alpha} \tag{16.80}$$

using Eq. (16.72). Since $\alpha < 1$ (unless the estimate is already unbiased), there is an *increase* in the MSE due to bias removal.

16.A.2 Optimum unbiased estimate

The preceding method to create an unbiased estimate might appear to be rather "ad hoc." But it turns out that no unbiased estimate can be better than this! To be more quantitative, assume that

$$\widehat{x}_{other} = x + \theta \tag{16.81}$$

is a linear estimate of x obtained from the same observations $\{y_k\}$ using some "other" procedure. This is clearly an unbiased estimate with mean squared error

$$\mathcal{E}_{other} = \mathcal{E}_{\theta}. \tag{16.82}$$

We now prove the following:

♠**Lemma 16.5.** Assume the error θ has zero mean, and that it is statistically independent of x (in particular, $E[x\theta^*] = 0$). Then the MSE $\mathcal{E}_{other}$ of the unbiased estimate (16.81) can never be better than that of the bias-removed MMSE estimate. In other words $\mathcal{E}_{other} \geq \mathcal{E}_{e_{br}}$. ◇

Proof. Assume the contrary, that is,

$$\begin{aligned}
\mathcal{E}_{\theta} = \mathcal{E}_{other} < \mathcal{E}_{e_{br}} &= \frac{\mathcal{E}_{e_{\perp}}}{\alpha} && \text{(from Eq. (16.80))} \\
&= \frac{\mathcal{E}_{e_{\perp}}}{1 - \frac{\mathcal{E}_{e_{\perp}}}{\mathcal{E}_x}} && \text{(from Eq. (16.73))} \\
&= \frac{1}{\frac{1}{\mathcal{E}_{e_{\perp}}} - \frac{1}{\mathcal{E}_x}}
\end{aligned}$$

This is readily rearranged as

$$\frac{\mathcal{E}_{\theta}\mathcal{E}_x}{\mathcal{E}_{\theta} + \mathcal{E}_x} < \mathcal{E}_{e_{\perp}}. \tag{16.83}$$

Now define a new linear estimate of x as follows:

$$\widehat{x}_{new} = \Big(\frac{\mathcal{E}_x}{\mathcal{E}_x + \mathcal{E}_{\theta}}\Big)\widehat{x}_{other} = \Big(\frac{\mathcal{E}_x}{\mathcal{E}_x + \mathcal{E}_{\theta}}\Big)(x + \theta). \tag{16.84}$$

The error is given by

$$e_{new} = \widehat{x}_{new} - x = \Big(\frac{\mathcal{E}_x}{\mathcal{E}_x + \mathcal{E}_{\theta}}\Big)\theta - \Big(\frac{\mathcal{E}_{\theta}}{\mathcal{E}_x + \mathcal{E}_{\theta}}\Big)x.$$

Since $E[x\theta^*] = 0$, this has mean square value

$$\mathcal{E}_{e,new} = \Big(\frac{\mathcal{E}_x}{\mathcal{E}_x + \mathcal{E}_{\theta}}\Big)^2\mathcal{E}_{\theta} + \Big(\frac{\mathcal{E}_{\theta}}{\mathcal{E}_x + \mathcal{E}_{\theta}}\Big)^2\mathcal{E}_x = \frac{\mathcal{E}_x\mathcal{E}_{\theta}}{\mathcal{E}_x + \mathcal{E}_{\theta}} < \mathcal{E}_{e_{\perp}} \tag{16.85}$$

using Eq. (16.83). Thus $\mathcal{E}_{e_{\perp}} > \mathcal{E}_{e,new}$, which contradicts the fact that $\mathcal{E}_{e_{\perp}}$ is the *minimum* mean squared error. ▽▽▽

16.A.3 Signal-to-error (SER) ratios

The signal-to-error (SER) ratios of the MMSE estimate and bias-removed estimate are given, respectively, by

$$\beta = \frac{\mathcal{E}_x}{\mathcal{E}_{e_\perp}} = \frac{\alpha\mathcal{E}_x}{\mathcal{E}_\tau} \quad \text{and} \quad \beta_{br} = \frac{\mathcal{E}_x}{\mathcal{E}_{e_{br}}} = \frac{\alpha^2\mathcal{E}_x}{\mathcal{E}_\tau} \tag{16.86}$$

using Eqs. (16.72) and (16.80). Thus $\beta - \beta_{br} = \alpha(1-\alpha)\mathcal{E}_x/\mathcal{E}_\tau$. Using Eq. (16.71) this equation simplifies to $\beta = \beta_{br} + 1$. This proves the following lemma:

♠**Lemma 16.6.** *Biased and unbiased SER.* The SER ratios of the MMSE estimate and bias-removed estimates are related by

$$\beta_{br} = \beta - 1. \tag{16.87}$$

Thus, bias removal from an MMSE estimate always decreases the SER ratio by unity. ◇

16.B Convexity proof

Consider the error probability expression

$$\mathcal{P}_{br}(\mathcal{E}) = c\mathcal{Q}\left(A\sqrt{\frac{1}{\mathcal{E}} - B}\right), \tag{16.88}$$

where $A, B > 0$. We have used the notation $B = 1/\sigma_s^2$ for convenience. We will show that the second derivative of this with respect to $\mathcal{E}$ is positive (i.e., $\mathcal{P}_{br}(\mathcal{E})$ is convex in $\mathcal{E}$) in a range

$$0 \leq \mathcal{E} \leq \mathcal{E}_t \tag{16.89}$$

for an appropriate threshold $\mathcal{E}_t$. That is, the error probability (16.88) *after* bias removal is convex in $\mathcal{E}$, where $\mathcal{E}$ is the MSE *before* bias removal. The only assumption we make is that

$$\frac{1}{\mathcal{E}} - B \geq 0,$$

which is necessary for the square root in Eq. (16.88) to be real. This assumption is automatically satisfied for an MMSE estimate because $\mathcal{E}/\sigma_s^2 \leq 1$.

Proof. It is shown in Sec. 21.2.3 that

$$\frac{d\,\text{erfc}(x)}{dx} = -\frac{2}{\sqrt{\pi}}e^{-x^2}. \tag{16.90}$$

Since $\mathcal{Q}(x) = 0.5\text{erfc}(x/\sqrt{2})$, we can write

$$\mathcal{P}_{br}(\mathcal{E}) = 0.5c \times \underbrace{\text{erfc}\left(\frac{A}{\sqrt{2}}\sqrt{\frac{1}{\mathcal{E}} - B}\right)}_{\text{call this } f(\mathcal{E})}.$$

Using Eq. (16.90) we have

$$\begin{aligned}\frac{df(\mathcal{E})}{d\mathcal{E}} &= -\frac{2}{\sqrt{\pi}}e^{-0.5A^2(\frac{1}{\mathcal{E}}-B)} \times \frac{A}{\sqrt{2}} \times 0.5 \times \left(\frac{1}{\mathcal{E}} - B\right)^{-1/2} \times \frac{-1}{\mathcal{E}^2} \\ &= \frac{A}{\sqrt{2\pi}} \underbrace{e^{-0.5A^2(\frac{1}{\mathcal{E}}-B)}\left(\frac{1}{\mathcal{E}} - B\right)^{-1/2}\frac{1}{\mathcal{E}^2}}_{\text{call this } g(\mathcal{E})}\end{aligned}$$

The sign of $d^2 f(\mathcal{E})/d\mathcal{E}^2$ is identical to that of $dg(\mathcal{E})/d\mathcal{E}$. Now,

$$\begin{aligned}\frac{dg(\mathcal{E})}{d\mathcal{E}} = &\quad \frac{0.5A^2}{\mathcal{E}^2} \times e^{-0.5A^2(\frac{1}{\mathcal{E}}-B)}\left(\frac{1}{\mathcal{E}} - B\right)^{-1/2}\frac{1}{\mathcal{E}^2} \\ &+ \; e^{-0.5A^2(\frac{1}{\mathcal{E}}-B)} \times \frac{0.5}{\mathcal{E}^2}\left(\frac{1}{\mathcal{E}} - B\right)^{-3/2}\frac{1}{\mathcal{E}^2} \\ &- \; 2e^{-0.5A^2(\frac{1}{\mathcal{E}}-B)}\left(\frac{1}{\mathcal{E}} - B\right)^{-1/2}\frac{1}{\mathcal{E}^3} \\ &= \; \frac{0.5}{\mathcal{E}^4}e^{-0.5A^2(\frac{1}{\mathcal{E}}-B)}\left(\frac{1}{\mathcal{E}} - B\right)^{-1/2}\left(A^2 + \left(\frac{1}{\mathcal{E}} - B\right)^{-1} - 4\mathcal{E}\right).\end{aligned}$$

This is nonnegative, or equivalently $\mathcal{P}_{br}(\mathcal{E})$ is convex, if and only if

$$A^2 + \left(\frac{1}{\mathcal{E}} - B\right)^{-1} - 4\mathcal{E} \geq 0,$$

which can be rewritten as

$$\mathcal{E}^2 - a_1\mathcal{E} + a_2 \geq 0, \tag{16.91}$$

where a_1 and a_2 are positive numbers given by

$$a_1 = \frac{3 + A^2B}{4B}, \quad a_2 = \frac{A^2}{4B}.$$

Figure 16.9 shows a plot of the left-hand side of Eq. (16.91). From the plot it is clear that there exists a range $0 \leq \mathcal{E} \leq \mathcal{E}_t$ such that Eq. (16.91) holds (since $a_2 > 0$). This concludes the proof. Notice that Eq. (16.91) also holds in the range $\mathcal{E} \geq \mathcal{E}_0$, where $\mathcal{E}_0$ is the second zero-crossing. ▽ ▽ ▽

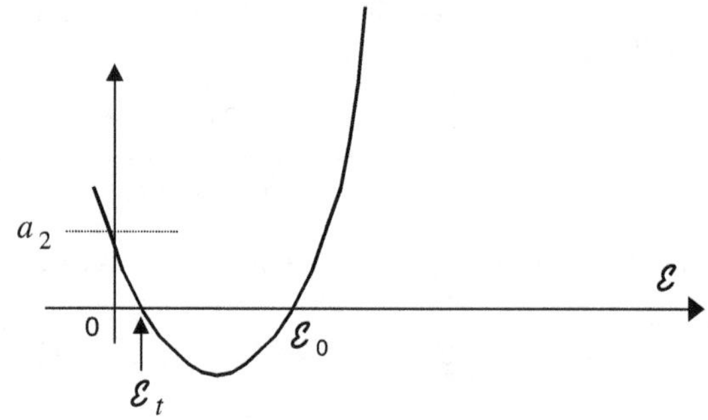

Figure 16.9. Plot of the left-hand side of Eq. (16.91).

The threshold $\mathcal{E}_t$ can be calculated from Eq. (16.91). For this we simply calculate the two roots of the quadratic equation $\mathcal{E}^2 - a_1\mathcal{E} + a_2 = 0$. The roots are positive and real (these are the zero crossings in Fig. 16.9), and the smaller root is equal to $\mathcal{E}_t$. To give a numerical example, consider a QAM constellation. We can compute c and A from Eq. (16.6), evaluate a_1 and a_2, and then compute the smaller root $\mathcal{E}_t$. Ths yields

$$\frac{\mathcal{E}_t}{\sigma_s^2} = \frac{3 \times 2^b}{8(2^b - 1)} - 0.5\sqrt{\left(\frac{3 \times 2^b}{4(2^b - 1)}\right)^2 - \frac{3}{2^b - 1}}$$

This ratio is shown in the table below for various choices of b (bits per symbol). This is the threshold error-to-signal ratio below which convexity holds. The table also shows the signal-to-error ratio $\sigma_s^2/\mathcal{E}_t$ in dB, and the symbol error probability if the detector were to have this SER at its input. Thus, if the symbol error probabilities exceed the numbers shown, then the convexity assumption fails. As long as the error probabilities in practice are smaller than these (rather large) values, the convexity assumption is therefore valid.

Number of bits b (QAM)	$\mathcal{E}_t/\sigma_s^2$ threshold	$\sigma_s^2/\mathcal{E}_t$ in dB	$\mathcal{P}_e$ per symbol
2	0.5000	03.0	0.317
4	0.0683	11.6	0.148
6	0.0160	18.0	0.152
8	0.0039	24.1	0.158

For the case of PAM constellations a similar table can be constructed and is shown below. The conclusions are similar. Note that b-bit PAM and $2b$-bit QAM have identical values for the constant A, so the threshold values are identical. However, the constant c in the error probability (16.88) has an extra factor of 2 in the QAM case (see Eqs. (11.41) and (11.42)), so that the threshold error probability for $2b$-bit QAM is twice as large the thereshold error probability for b-bit PAM, as seen from the last column of the table.

Number of bits b (PAM)	$\mathcal{E}_t/\sigma_s^2$ threshold	$\sigma_s^2/\mathcal{E}_t$ in dB	$\mathcal{P}_e$ per symbol
1	0.5000	03.0	0.159
2	0.0683	11.6	0.074
3	0.0160	18.0	0.076
4	0.0039	24.1	0.079
5	0.0010	30.1	0.081
6	0.0002	36.1	0.082

16.B.1 Special case of QPSK

For QPSK or 2-bit QAM we have

$$A = \sqrt{\frac{3\sigma_s^2}{2^b - 1}} = \sqrt{\frac{3\sigma_s^2}{3}} = \sigma_s$$

Since $B = 1/\sigma_s^2$ it then follows that

$$a_2 = \frac{A^2}{4B} = \frac{\sigma_s^4}{4}, \quad a_1 = \sigma_s^2.$$

So the left-hand side of Eq. (16.91) is

$$\mathcal{E}^2 - \sigma_s^2\mathcal{E} + \frac{\sigma_s^2}{4} = \left(\mathcal{E} - \frac{\sigma_s}{2}\right)^2, \tag{16.92}$$

which is always non-negative! See Fig. 16.10. So, for a QPSK system it is always true that the error probability (16.88) *after* bias removal is convex in $\mathcal{E}$, where $\mathcal{E}$ is the MMSE *before* bias removal. The threshold condition (16.89) is not required in this case. Since Eq. (16.92) is zero for $\mathcal{E}/\sigma_s^2 = 0.5$, the threshold is indicated as 0.5 in the QAM table above, but this threshold has no meaning since Eq. (16.92) is always ≥ 0.

The only assumption of course is that the quantity under the square root in Eq. (16.88) is positive, that is, $\sigma_s^2/\mathcal{E} \geq 1$. But this is always satisfied because $\mathcal{E}$ is the MMSE value. Interestingly enough, for the QPSK system it does not really matter whether the bias is removed or not. There are only four words in the constellation (located symmetrically with respect to the origin), so the decision boundaries are the X and Y axes. The decision of the threshold detector is therefore unaffected whether bias is removed or not, as long as the bias is created by a positive scale factor α as in Eq. (16.67). Such a scale factor simply brings all codewords closer to the origin, as shown in Fig. 16.11.

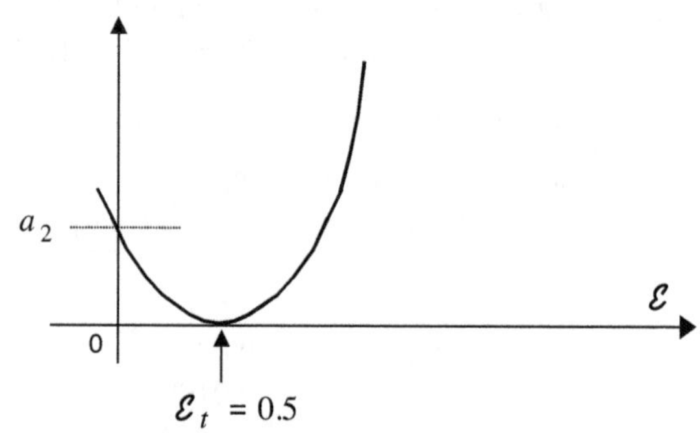

Figure 16.10. Plot of the left-hand side of Eq. (16.91) for the QPSK constellation.

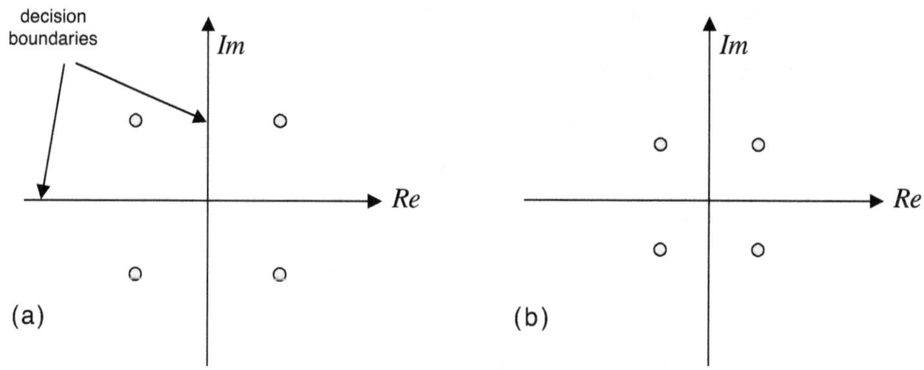

Figure 16.11. (a) The QPSK constellation, and (b) the constellation scaled by a positive scalar $\alpha < 1$.

16.C Pure MMSE with bias removal vs ZF-MMSE

Since the system described in Theorem 16.2 minimizes error probability without imposing the zero-forcing constraint, the average error probability (16.60) is at least as small as the error probability of the zero-forced optimal system in Eq. (16.25). We now develop a second proof of this, for additional insight. For the pure-MMSE system the optimized error probability is as in Eq. (16.60), that is,

$$\mathcal{P}_{br} = c\mathcal{Q}\left(A\sqrt{\frac{1}{\mathcal{E}_{pure}} - \frac{1}{\sigma_s^2}} \right), \tag{16.93}$$

where $\mathcal{E}_{pure}$ is the average MSE per symbol before bias removal, so

$$M\mathcal{E}_{pure} = \frac{\sigma_s^2(\sum_{\ell=0}^{K-1}\sigma_{q_\ell})^2}{p_0 + \sum_{\ell=0}^{K-1}\sigma_{q_\ell}^2} + (M-K)\sigma_s^2. \tag{16.94}$$

Note the slight simplification of notation. If we replace σ_{q_k} appropriately we get the MSE expressions for the cases of diagonal and non-diagonal channels (Chaps. 11 and 13, respectively). For the ZF-MMSE system the optimized error probability is as in Eq. (16.25):

$$\mathcal{P}_{ZF} = c\mathcal{Q}\left(A\sqrt{\frac{1}{\mathcal{E}_{ZF}}}\right), \tag{16.95}$$

where $\mathcal{E}_{ZF}$ is the average MSE per symbol, so

$$M\mathcal{E}_{ZF} = \frac{\sigma_s^2(\sum_{\ell=0}^{M-1}\sigma_{q_\ell})^2}{p_0} \tag{16.96}$$

Our goal is to prove

$$\mathcal{P}_{br} \leq \mathcal{P}_{ZF}. \tag{16.97}$$

Equivalently we have to prove

$$\frac{1}{\mathcal{E}_{pure}} - \frac{1}{\sigma_s^2} \geq \frac{1}{\mathcal{E}_{ZF}},$$

that is,

$$\frac{\sigma_s^2}{\mathcal{E}_{pure}} - \frac{\sigma_s^2}{\mathcal{E}_{ZF}} \geq 1. \tag{16.98}$$

That is, the signal-to-error ratios are related as follows:

$$SER_{pure} \geq SER_{ZF} + 1 \tag{16.99}$$

Substituting from Eqs. (16.94) and (16.96), this can be rewritten in the form

$$\frac{M\Big(p_0 + \sum_{\ell=0}^{K-1}\sigma_{q_\ell}^2\Big)}{\Big(\sum_{\ell=0}^{K-1}\sigma_{q_\ell}\Big)^2 + (M-K)\Big(p_0 + \sum_{\ell=0}^{K-1}\sigma_{q_\ell}^2\Big)} - \frac{Mp_0}{\Big(\sum_{\ell=0}^{M-1}\sigma_{q_\ell}\Big)^2} \geq 1. \tag{16.100}$$

In attempting to prove this, it should be remembered that the integer $K < M$ arises because the power p_0 is limited. The relation between p_0 and K can be found from Eq. (22.48) in Chap. 22. Thus, p_0 is such that

$$p_0 + \sum_{\ell=0}^{K-1}\sigma_{q_\ell}^2 \leq \sigma_{q_k}\sum_{\ell=0}^{K-1}\sigma_{q_\ell}, \quad K \leq k \leq M-1, \tag{16.101}$$

and

$$p_0 + \sum_{\ell=0}^{K-1}\sigma_{q_\ell}^2 > \sigma_{q_n}\sum_{\ell=0}^{K-1}\sigma_{q_\ell}, \quad 0 \leq n \leq K-1, \tag{16.102}$$

where the noise variances $\sigma_{q_k}^2$ are ordered such that

$$\sigma_{q_0}^2 \leq \sigma_{q_1}^2 \leq \ldots \leq \sigma_{q_{M-1}}^2 \tag{16.103}$$

Proof of Eq. (16.100). If $K = M$ then Eq. (16.100) reduces to

$$\frac{M\sum_{\ell=0}^{M-1}\sigma_{q_\ell}^2}{\left(\sum_{\ell=0}^{M-1}\sigma_{q_\ell}\right)^2} \geq 1. \tag{16.104}$$

Now, from Cauchy-Schwartz inequality we have

$$\left(\sum_{\ell=0}^{M-1}\sigma_{q_\ell}\right)^2 \leq \sum_{\ell=0}^{M-1}1^2\sum_{\ell=0}^{M-1}\sigma_{q_\ell}^2 = M\sum_{\ell=0}^{M-1}\sigma_{q_\ell}^2$$

which proves (16.104) indeed. Next assume that the power p_0 is such that $K < M$. That is, the noise variances are such that Eqs. (16.101) and (16.102) are true. For simplicity, we first assume that

$$\sigma_{q_K}^2 = \sigma_{q_{K+1}}^2 = \ldots = \sigma_{q_{M-1}}^2 \tag{16.105}$$

and that Eq. (16.101) holds with equality:

$$p_0 + \sum_{\ell=0}^{K-1}\sigma_{q_\ell}^2 = \sigma_{q_k}\sum_{\ell=0}^{K-1}\sigma_{q_\ell}, \quad K \leq k \leq M-1. \tag{16.106}$$

Substituting into Eq. (16.100) we can rewrite it as

$$\frac{M\left(\sigma_{q_K}\sum_{\ell=0}^{K-1}\sigma_{q_\ell}\right)}{\left(\sum_{\ell=0}^{K-1}\sigma_{q_\ell}\right)^2 + (M-K)\left(\sigma_{q_K}\sum_{\ell=0}^{K-1}\sigma_{q_\ell}\right)} - \frac{M\left(\sigma_{q_K}\sum_{\ell=0}^{K-1}\sigma_{q_\ell} - \sum_{\ell=0}^{K-1}\sigma_{q_\ell}^2\right)}{\left(\sum_{\ell=0}^{K-1}\sigma_{q_\ell} + (M-K)\sigma_{q_K}\right)^2} \geq 1,$$

which simplifies to

$$\frac{M\sigma_{q_K}}{\left(\sum_{\ell=0}^{K-1}\sigma_{q_\ell}\right) + (M-K)\sigma_{q_K}} - \frac{M\left(\sigma_{q_K}\sum_{\ell=0}^{K-1}\sigma_{q_\ell} - \sum_{\ell=0}^{K-1}\sigma_{q_\ell}^2\right)}{\left(\sum_{\ell=0}^{K-1}\sigma_{q_\ell} + (M-K)\sigma_{q_K}\right)^2} \geq 1.$$

This can be rewritten as

$$\frac{M(M-K)\sigma_{q_K}^2 + M\sum_{\ell=0}^{K-1}\sigma_{q_\ell}^2}{\left(\sum_{\ell=0}^{K-1}\sigma_{q_\ell} + (M-K)\sigma_{q_K}\right)^2} \geq 1,$$

which is nothing but a special case of Cauchy-Schwartz inequality (16.104) obtained by substituting Eq. (16.105). This proves Eq. (16.100) when Eq. (16.105) is true. Next assume Eq. (16.105) is not true, but only (16.103) holds. Then none of the terms in Eq. (16.100) is affected except the sum $\sum_{\ell=0}^{M-1}\sigma_{q_\ell}$ in the denominator of the second term. With (16.105) replaced by (16.103), $\sum_{\ell=0}^{M-1}\sigma_{q_\ell}$ can only get *larger*, increasing the left-hand-side of (16.100). Thus the inequality (16.100) continues to hold. $\bigtriangledown\bigtriangledown\bigtriangledown$

16.D SISO channel: ZF-MMSE vs bias removal

Consider the MMSE expressions with and without zero forcing given in Chap. 10 (Eq. (10.42)). We now give a direct proof that the bias-removed pure-MMSE system has smaller MSE than the ZF-MMSE system. For simplicity assume $S_{qq}(e^{j\omega}) = \sigma_q^2$ and $S_{ss}(e^{j\omega}) = \sigma_s^2$. The minimized mean square error without the ZF constraint is given by

$$\mathcal{E}_{mmse} = \frac{\sigma_q^2\sigma_s^2\Big(\int_{\mathcal{F}} \frac{1}{|H(e^{j\omega})|}\frac{d\omega}{2\pi}\Big)^2}{p_0 + \sigma_q^2\int_{\mathcal{F}} \frac{1}{|H(e^{j\omega})|^2}\frac{d\omega}{2\pi}} + \sigma_s^2\int_{\mathcal{F}^c}\frac{d\omega}{2\pi}. \tag{16.107}$$

The region $\mathcal{F}^c \in [0, 2\pi]$ is the part where $F(e^{j\omega})$ is zero. This is the frequency region in which the channel is excluded by the MMSE system. Substituting $S_{qq}(e^{j\omega}) = \sigma_q^2$ and $S_{ss}(e^{j\omega}) = \sigma_s^2$ into Eq. (10.33) we see that

$$\gamma - \frac{\sigma_q}{|H(e^{j\omega})|\sigma_s}\begin{cases} > 0 & \text{for } \omega \in \mathcal{F} \\ \leq 0 & \text{for } \omega \in \mathcal{F}^c. \end{cases} \tag{16.108}$$

The constant γ is given by Eq. (10.34), that is,

$$\gamma = \frac{p_0 + \sigma_q^2\int_{\mathcal{F}} \frac{1}{|H(e^{j\omega})|^2}\frac{d\omega}{2\pi}}{\sigma_q\sigma_s\int_{\mathcal{F}} \frac{1}{|H(e^{j\omega})|}\frac{d\omega}{2\pi}} \tag{16.109}$$

Let W be the total length of the region $\mathcal{F} \subset [0, 2\pi]$. Then the total length of the excluded part $\mathcal{F}^c$ (which need not be a single continuous band) is $2\pi - W$. Defining $\mu = W/2\pi$ we have

$$\mathcal{E}_{mmse} = \frac{\sigma_q^2\sigma_s^2\Big(\int_{\mathcal{F}} \frac{1}{|H(e^{j\omega})|}\frac{d\omega}{2\pi}\Big)^2}{p_0 + \sigma_q^2\int_{\mathcal{F}} \frac{1}{|H(e^{j\omega})|^2}\frac{d\omega}{2\pi}} + \Big(1 - \mu\Big)\sigma_s^2 \tag{16.110}$$

Next, from Eq. (10.41) we have

$$\mathcal{E}_{ZF} = \frac{\sigma_s^2 \sigma_q^2}{p_0} \left(\int_{-\pi}^{\pi} \frac{1}{|H(e^{j\omega})|} \frac{d\omega}{2\pi} \right)^2. \tag{16.111}$$

From Sec. 16.A.3 we know that, for the MMSE system with bias removed, the SNR at the detector input is given by

$$\frac{\sigma_s^2}{\mathcal{E}_{mmse,br}} = \frac{\sigma_s^2}{\mathcal{E}_{mmse}} - 1.$$

We will show that

$$\mathcal{E}_{mmse,br} \leq \mathcal{E}_{ZF}, \tag{16.112}$$

or equivalently

$$\frac{\sigma_s^2}{\mathcal{E}_{mmse}} - \frac{\sigma_s^2}{\mathcal{E}_{ZF}} \geq 1. \tag{16.113}$$

Proof of (16.113). From Eqs. (16.108) and (16.109) we obtain the inequalities

$$\frac{p_0 + \sigma_q^2 \displaystyle\int_{\mathcal{F}} \frac{1}{|H(e^{j\omega})|^2} \frac{d\omega}{2\pi}}{\displaystyle\int_{\mathcal{F}} \frac{1}{|H(e^{j\omega})|} \frac{d\omega}{2\pi}} - \frac{\sigma_q^2}{|H(e^{j\omega})|} \begin{cases} > 0 & \text{for } \omega \in \mathcal{F} \\ \leq 0 & \text{for } \omega \in \mathcal{F}^c. \end{cases}$$

The first term on the left-hand side is a constant, whereas the second term depends on frequency. If $|H(e^{j\omega})|$ drops below a threshold ϵ, the quantity $\sigma_q^2/|H(e^{j\omega})|$ becomes large enough to make the left-hand side negative. Thus, whenever the set $\mathcal{F}^c$ is non-empty, we have

$$\frac{p_0 + \sigma_q^2 \displaystyle\int_{\mathcal{F}} \frac{1}{|H(e^{j\omega})|^2} \frac{d\omega}{2\pi}}{\displaystyle\int_{\mathcal{F}} \frac{1}{|H(e^{j\omega})|} \frac{d\omega}{2\pi}} - \frac{\sigma_q^2}{|H(e^{j\omega})|} \leq 0 \tag{16.114}$$

for $\omega \in \mathcal{F}^c$. We will first consider a simple special case such that $|H(e^{j\omega})|$ is a constant in the excluded band $\mathcal{F}^c$, that is,

$$|H(e^{j\omega})| = \epsilon, \quad \omega \in \mathcal{F}^c, \tag{16.115}$$

and furthermore the left-hand side of Eq. (16.114) is exactly zero in $\mathcal{F}^c$, that is,

$$\frac{p_0 + \sigma_q^2 \displaystyle\int_{\mathcal{F}} \frac{1}{|H(e^{j\omega})|^2} \frac{d\omega}{2\pi}}{\displaystyle\int_{\mathcal{F}} \frac{1}{|H(e^{j\omega})|} \frac{d\omega}{2\pi}} = \frac{\sigma_q^2}{\epsilon} \tag{16.116}$$

Substituting this into Eq. (16.110) and simplifying, we get

$$\mathcal{E}_{mmse} = \epsilon\sigma_s^2 \int_{-\pi}^{\pi} \frac{1}{|H(e^{j\omega})|} \frac{d\omega}{2\pi} \tag{16.117}$$

Similarly, Eq. (16.111) can be rewritten by substituting from Eq. (16.116):

$$\mathcal{E}_{ZF} = \frac{\sigma_s^2 \left(\int_{-\pi}^{\pi} \frac{1}{|H(e^{j\omega})|} \frac{d\omega}{2\pi} \right)^2}{\frac{1}{\epsilon} \int_{\mathcal{F}} \frac{1}{|H(e^{j\omega})|} \frac{d\omega}{2\pi} - \int_{\mathcal{F}} \frac{1}{|H(e^{j\omega})|^2} \frac{d\omega}{2\pi}} \tag{16.118}$$

Substituting from Eqs. (16.117) and (16.118) into the left hand side of Eq. (16.113) and simplifying we get

$$\frac{\sigma_s^2}{\mathcal{E}_{mmse}} - \frac{\sigma_s^2}{\mathcal{E}_{ZF}} = \frac{\int_{-\pi}^{\pi} \frac{1}{|H(e^{j\omega})|^2} \frac{d\omega}{2\pi}}{\left(\int_{-\pi}^{\pi} \frac{1}{|H(e^{j\omega})|} \frac{d\omega}{2\pi} \right)^2} \tag{16.119}$$

Now, using Cauchy-Schwartz inequality (Appendix A) it readily follows that the above right-hand side is ≥ 1. This proves Eq. (16.113) or equivalently (16.112) under the assumption (16.115). Next consider the more general case where

$$|H(e^{j\omega})| \leq \epsilon, \quad \omega \in \mathcal{F}^c. \tag{16.120}$$

In the expression (16.110) for $\mathcal{E}_{mmse}$, only integrals over $\mathcal{F}$ are involved, so $\mathcal{E}_{mmse}$ is not affected by this. From the expression (16.111) for $\mathcal{E}_{ZF}$, it is clear that if we use Eq. (16.120) instead of Eq. (16.115), the result can only get larger because $1/|H(e^{j\omega})|$ gets larger in $\mathcal{F}_c$. Thus, Eq. (16.113) continues to hold. $\bigtriangledown \bigtriangledown \bigtriangledown$

Problems

16.1. Let the error covariance $\mathbf{E}_x$ defined in Sec. 16.2.1 be given by

$$\mathbf{E}_x = \begin{bmatrix} 2 & 1 \\ 1 & 3 \end{bmatrix}.$$

Find real unitary $\mathbf{U}$ such that $\mathbf{U}\mathbf{E}_x\mathbf{U}^\dagger$ has identical diagonal elements. What are the element values in the resulting matrix $\mathbf{U}\mathbf{E}_x\mathbf{U}^\dagger$?

16.2. Let the error covariance $\mathbf{E}_x$ be given by

$$\mathbf{E}_x = 10^{-3} \begin{bmatrix} 0.4 & 0 & 0 & 0 \\ 0 & 5 & 0 & 0 \\ 0 & 0 & 60 & 0 \\ 0 & 0 & 0 & 100 \end{bmatrix}.$$

1. Find a real unitary $\mathbf{U}$ such that $\mathbf{U}\mathbf{E}_x\mathbf{U}^\dagger$ has identical diagonal elements.
2. What are the element values in the resulting matrix $\mathbf{U}\mathbf{E}_x\mathbf{U}^\dagger$?
3. Assuming 2-bit PAM and $\sigma_s^2 = 1$, find the minimized error probability (16.23) in the zero-forcing case.
4. If the unitary matrix $\mathbf{U}$ were not used to equalize the errors, the average error probability would be as in Eq. (16.14). What is its value?

16.3. Let the error covariance $\mathbf{E}_x$ defined in Sec. 16.2.1 be given by

$$\mathbf{E}_x = \begin{bmatrix} 3 & 1 \\ 1 & 4 \end{bmatrix}.$$

Find a real unitary $\mathbf{U}$ such that the error probability (16.14) is *worse* than the case where no $\mathbf{U}$ is employed at all.

16.4. Let x and x_1 be random variables with identical mean m and cross correlation $r = E[xx_1^*]$. Let $\widehat{x}$ be the optimal linear estimate of x based on x_1, that is,

$$\widehat{x} = ax_1,$$

where a is such that $E[\,|\widehat{x} - x|^2]$ is minimized.

1. Find an expression for the mean value m_e of the error $e = \widehat{x} - x$.
2. Find specific numerical values for r and m and the mean square values of x and x_1 such that m_e is zero.

3. Find specific numerical values for r and m and the mean square values of x and x_1 such that m_e is nonzero.

16.5. Consider again the discussion centered around Fig. 16.6 in Sec. 16.5.1. Assume $M = 2$ so that K can be either 1 or 2.

1. Find a numerical example (i.e., choose p_0 and $\sigma_{q_k}/|H_k|$) such that $\mathcal{E}_{ZF,mmse}$ is smaller than the bias-removed MMSE.
2. Find another numerical example such that $\mathcal{E}_{ZF,mmse}$ is larger than the bias-removed MMSE.

17

Optimization of cyclic prefix transceivers

17.1 Introduction

Transceiver systems based on cyclic-prefix (CP) redundancy were introduced in Chap. 7. Two applications of this are OFDM systems and DMT systems, described in Secs. 7.5 and 7.6. In both systems the effect of the cyclic prefix is to convert a scalar FIR channel

$$C(z) = \sum_{n=0}^{L} c(n)z^{-n} \tag{17.1}$$

of order L to an $M \times M$ block channel $\mathbf{H}$, which is a circulant matrix (denoted as $\mathbf{C}$ in Chap. 7). Recall here that the DMT transceiver structure allows bit allocation, whereas in the OFDM case there is no bit allocation.

In this chapter we apply the optimality results derived in Chaps. 12–16 to the case of cyclic-prefix transceivers optimized for mean square error (Secs. 17.2–17.4), and for power (Sec. 17.5). We also consider the special case where the cyclic-prefix transceiver is optimized subject to the restriction that the precoder be unitary (Sec. 17.6). For this chapter the reader may want to review Sec. 7.3 where we introduced cyclic-prefix systems.

17.2 Optimal cyclic-prefix systems: preliminaries

Consider the transceiver shown in Fig. 17.1(a). In Chaps. 12 and 13 the precoder $\mathbf{F}$ and equalizer $\mathbf{G}$ were jointly optimized to mimimize the mean square error (MSE) with or without the zero-forcing (ZF) constraint. The optimal solutions for both of these cases have the form shown in Fig. 17.1(b), under the assumption

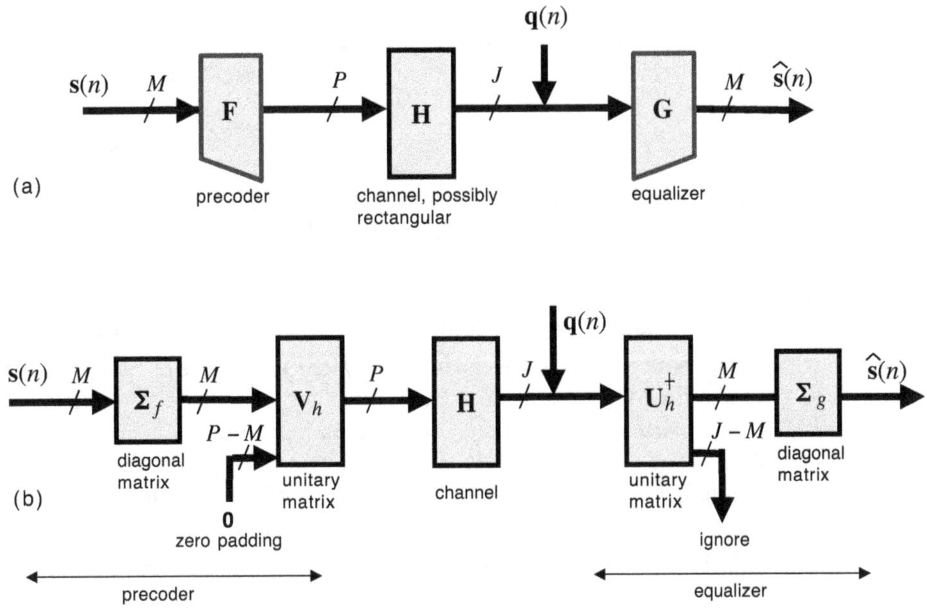

Figure 17.1. (a) General form of the transceiver. (b) The structure of the MMSE transceiver with or without the zero-forcing constraint. The optimal diagonal matrices $\mathbf{\Sigma}_f$ and $\mathbf{\Sigma}_g$ depend on whether the ZF constraint is imposed or not.

that the signal and noise covariance matrices are[1]

$$\mathbf{R}_{\mathbf{ss}} = \sigma_s^2 \mathbf{I}_M \quad \text{and} \quad \mathbf{R}_{\mathbf{qq}} = \sigma_q^2 \mathbf{I}_J. \tag{17.2}$$

The components $q_k(n)$ of the channel noise $\mathbf{q}(n)$ are assumed to be Gaussian with variance σ_q^2. More specifically, when the constellation is QAM, the channel can be complex and the noise components $q_k(n)$ are assumed to be complex circular Gaussian (Sec. 6.6). For PAM constellations and real channels, $q_k(n)$ is real and Gaussian. Here $\mathbf{U}_h^\dagger$ and $\mathbf{V}_h$ are the unitary matrices appearing in the SVD of the channel:

$$\mathbf{H} = \underbrace{\mathbf{U}_h}_{J\times J} \underbrace{\mathbf{\Sigma}_h}_{J\times P} \underbrace{\mathbf{V}_h^\dagger}_{P\times P}. \tag{17.3}$$

Besides $\mathbf{U}_h^\dagger$ and $\mathbf{V}_h$, the optimal transceiver contains only diagonal matrices, namely $\mathbf{\Sigma}_f$ and $\mathbf{\Sigma}_g$. These matrices depend on whether the ZF constraint is imposed or not. Thus the MMSE transceivers with or without the ZF constraint have identical form except for details of the diagonal matrices $\mathbf{\Sigma}_f$ and $\mathbf{\Sigma}_g$.

[1]When the channel noise $q(n)$ is colored, as is the case in practical DSL systems [Starr *et al.*, 1999], the techniques described in Chap. 12 should be used to obtain an equivalent system where the noise covariance has the form $\sigma_q^2\mathbf{I}$.

17.2.1 Optimal unitary matrices in the cyclic-prefix case

In a cyclic-prefix system a scalar FIR channel $H(z)$ is converted into an $M \times M$ circulant matrix $\mathbf{H}$. So, the equivalent vector representation has the form in Fig. 17.1(a) with

$$J = P = M,$$

where $\mathbf{H}$ is circulant (denoted as $\mathbf{C}$ in Sec. 7.4). The vector $\mathbf{s}(n)$ is the blocked version of an iid sequence $s(n)$ from a constellation such as the QAM constellation. Thus the vector $\mathbf{s}(n)$ indeed has the covariance $\sigma_s^2 \mathbf{I}_M$. Similarly the noise $\mathbf{q}(n)$ is the blocked version of a scalar white noise source $q(n)$, and the covariance of $\mathbf{q}(n)$ is of the form $\sigma_q^2 \mathbf{I}_M$. Thus the optimal transceiver (MMSE transceiver with or without ZF) has the form shown in Fig. 17.2(a). Since $\mathbf{U}_h^\dagger \mathbf{H} \mathbf{V}_h$ represents the diagonal matrix $\mathbf{\Sigma}_h$, the equivalent diagonal representation is as in Fig. 17.2(b), where $\sigma_{f,k}$ and $\sigma_{g,k}$ are diagonal elements of $\mathbf{\Sigma}_f$ and $\mathbf{\Sigma}_g$. It should be noticed that the components $q_k(n)$ of the vector $\mathbf{q}(n)$ in Fig. 17.2(a) are different from $q_k(n)$ in Fig. 17.2(b) because the latter is obtained from the former by using the transformation $\mathbf{U}_h^\dagger$. However, the noise covariance matrix $\sigma_q^2 \mathbf{I}$ is unchanged by this transformation, and we use the same notation $q_k(n)$ in both figures.

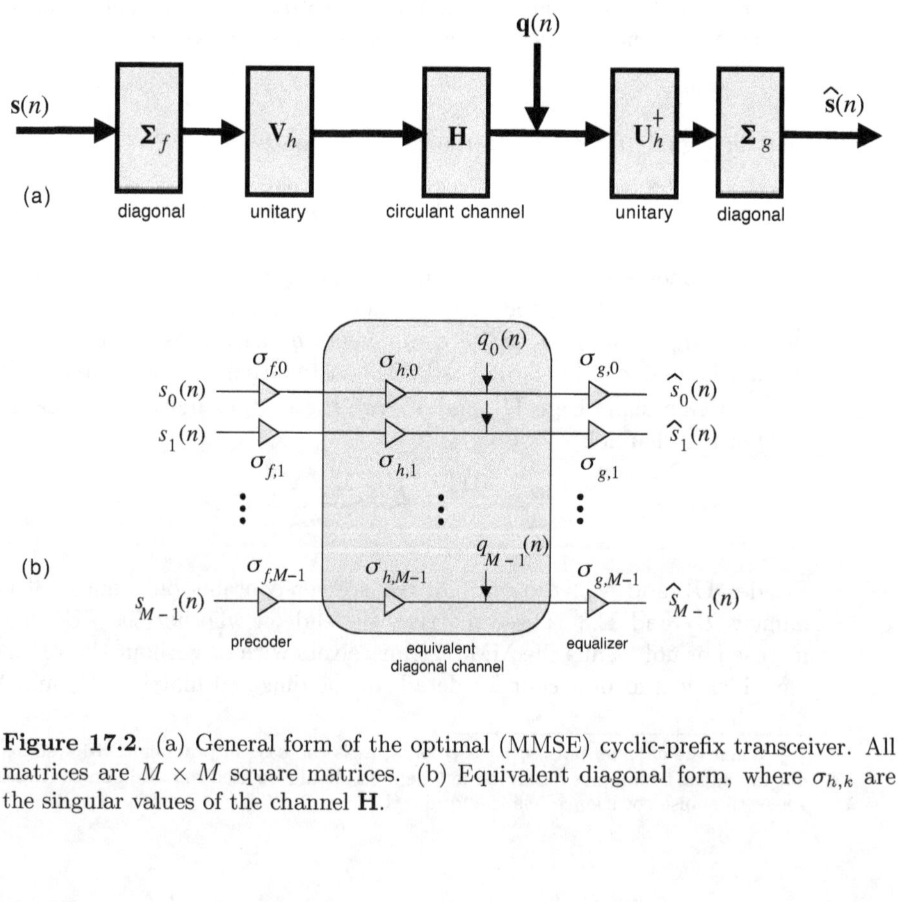

Figure 17.2. (a) General form of the optimal (MMSE) cyclic-prefix transceiver. All matrices are $M \times M$ square matrices. (b) Equivalent diagonal form, where $\sigma_{h,k}$ are the singular values of the channel $\mathbf{H}$.

17.2.2 SVD of a circulant channel

To identify the unitary matrices $\mathbf{U}_h$ and $\mathbf{V}_h$ in the SVD of a circulant channel, recall that circulants are diagonalized by the DFT matrix (Appendix D). More precisely,

$$\mathbf{W}\mathbf{H}\mathbf{W}^{-1} = \mathbf{\Lambda}_c, \tag{17.4}$$

where $\mathbf{W}$ is the $M \times M$ DFT matrix and $\mathbf{\Lambda}_c$ is a diagonal matrix with the channel DFT coefficients on the diagonal:

$$[\mathbf{\Lambda}_c]_{kk} = C[k] = \sum_{n=0}^{M-1} c(n)W^{nk}, \quad W = e^{-j2\pi/M} \tag{17.5}$$

Thus

$$\mathbf{H} = \mathbf{W}^{-1}\mathbf{\Lambda}_c\mathbf{W}. \tag{17.6}$$

Using $\mathbf{W}^{-1} = \mathbf{W}^\dagger/M$ this can be written as

$$\mathbf{H} = \frac{\mathbf{W}^\dagger}{\sqrt{M}}\mathbf{\Lambda}_c\frac{\mathbf{W}}{\sqrt{M}}. \tag{17.7}$$

Since $\mathbf{W}/\sqrt{M}$ is unitary, the preceding equation is "almost" the singular value decomposition

$$\mathbf{H} = \mathbf{U}_h\mathbf{\Sigma}_h\mathbf{V}_h^\dagger, \tag{17.8}$$

except for the fact that the diagonal matrix $\mathbf{\Lambda}_c$ can have complex entries (see Eq. (17.5)), whereas $\mathbf{\Sigma}_h$ is required to be real. But we can readily show that the magnitudes $|C[k]|$ are the singular values. Write

$$[\mathbf{\Lambda}_c]_{kk} = |C[k]|e^{j\theta_k} \tag{17.9}$$

and define $\mathbf{\Sigma}_h$ to be

$$\mathbf{\Sigma}_h = \mathbf{\Lambda}_\theta^{-1}\mathbf{\Lambda}_c, \tag{17.10}$$

where $\mathbf{\Lambda}_\theta$ is a diagonal matrix given by

$$\mathbf{\Lambda}_\theta = \text{diag}\left[\, e^{j\theta_0} \quad e^{j\theta_1} \quad \ldots \quad e^{j\theta_{M-1}} \,\right]. \tag{17.11}$$

Then the diagonal elements of $\mathbf{\Sigma}_h$ are nonnegative, and we can write

$$\begin{aligned}\mathbf{H} &= \frac{\mathbf{W}^\dagger}{\sqrt{M}}\mathbf{\Lambda}_c\frac{\mathbf{W}}{\sqrt{M}} \\ &= \underbrace{\left(\frac{\mathbf{W}^\dagger}{\sqrt{M}}\mathbf{\Lambda}_\theta\right)}_{\text{this is } \mathbf{U}_h} \mathbf{\Sigma}_h \underbrace{\left(\frac{\mathbf{W}}{\sqrt{M}}\right)}_{\text{this is } \mathbf{V}_h^\dagger}\end{aligned}$$

Thus, we can regard the preceding equation as the SVD of the circulant channel. In standard SVD notation (17.8) we have

$$\mathbf{U}_h = \frac{\mathbf{W}^\dagger\mathbf{\Lambda}_\theta}{\sqrt{M}} \quad \text{and} \quad \mathbf{V}_h^\dagger = \frac{\mathbf{W}}{\sqrt{M}}, \tag{17.12}$$

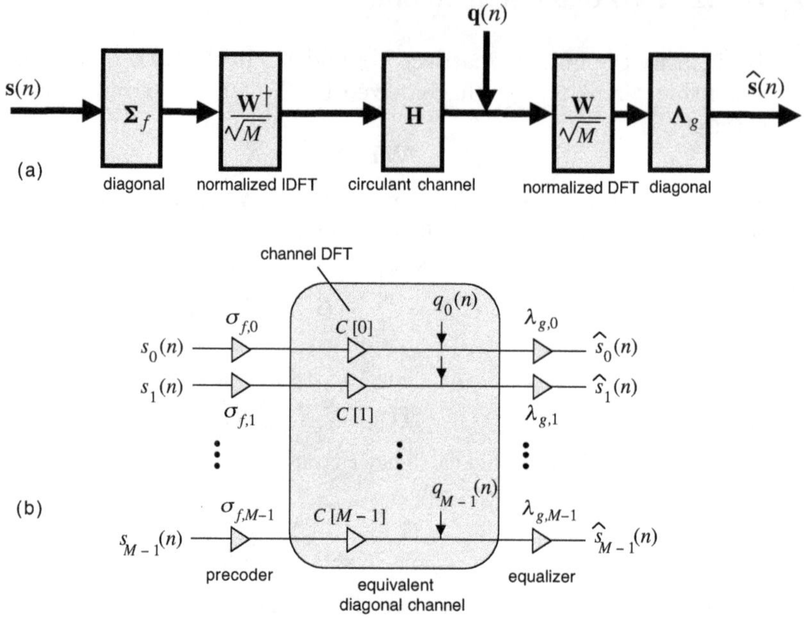

Figure 17.3. (a) Structure of the MMSE cyclic-prefix transceiver. All matrices are $M \times M$ square matrices. (b) Equivalent diagonal structure. In the pure-MMSE case (i.e., when there is no zero forcing), the diagonal elements of $\mathbf{\Sigma}_f$ and $\mathbf{\Lambda}_g$ may have to be in permuted order as explained later.

and the singular values are

$$\sigma_{h,k} = |C[k]|. \tag{17.13}$$

It should be noted that $C[k]$ are the samples of the channel frequency response $C(e^{j\omega})$ in natural order, that is,

$$C[k] = C(e^{j2\pi k/M}), \quad 0 \leq k \leq M-1. \tag{17.14}$$

Thus the singular values $\sigma_{h,k} = |C[k]|$ *do not*, in general, satisfy the ordering

$$\sigma_{h,0} \geq \sigma_{h,1} \geq \ldots \geq \sigma_{h,M-1}. \tag{17.15}$$

For the ZF-MMSE system this ordering is not important, but for the pure-MMSE system we have to use the ordered version to compute the matrices $\mathbf{\Sigma}_f$ and $\mathbf{\Sigma}_g$ in the optimal solution (see Sec. 13.5).

We shall take care of the ordering issue in Sec. 17.3.2. Ignoring it for the moment and using Eq. (17.12), the optimum transceiver of Fig. 17.2(a) therefore takes the form shown in Fig. 17.3(a). Here the diagonal matrix $\mathbf{\Lambda}_g$ is given by

$$\mathbf{\Lambda}_g \stackrel{\Delta}{=} \mathbf{\Sigma}_g \mathbf{\Lambda}_\theta^{-1} \tag{17.16}$$

where $\mathbf{\Lambda}_\theta$ is as in Eq. (17.11). The DFT and IDFT matrices (which are channel-independent constants) therefore play a crucial role in the optimal structure regardless of what the original scalar channel $C(z)$ is.

Equivalent diagonal form. From Eq. (17.7) we have

$$\frac{\mathbf{W}}{\sqrt{M}}\mathbf{H}\frac{\mathbf{W}^\dagger}{\sqrt{M}} = \mathbf{\Lambda}_c = \text{diag}\,[\,C[0] \quad C[1] \quad \ldots \quad C[M-1]\,]\,,$$

which shows that Fig. 17.3(a) can be redrawn in the equivalent diagonal form shown in Fig. 17.3(b). Here $\sigma_{f,k}$ are the diagonal elements of $\mathbf{\Sigma}_f$, and $\lambda_{g,k}$ are the diagonal elements of $\mathbf{\Lambda}_g$. This is nothing but a frequency-domain representation of the channel because the channel multipliers $C[k]$ are the DFT coefficients of $c(n)$. In this diagonal form, each DFT coefficient $C[k]$ is independently equalized by the precoder/equalizer pair $\{\sigma_{f,k}, \lambda_{g,k}\}$. $\bigtriangledown \bigtriangledown \bigtriangledown$

17.3 Cyclic-prefix systems optimized for MSE: details

We now apply the results of the preceding section to derive the detailed structure of the MMSE cyclic-prefix systems. We discuss the zero-forcing case and the non zero-forcing case separately for clarity.

17.3.1 ZF-MMSE transceiver

For the MMSE transceiver with zero-forcing constraint (ZF-MMSE transceiver) the diagonal matrices $\mathbf{\Sigma}_f$ and $\mathbf{\Sigma}_g$ were derived in Sec. 12.4.4. These are reproduced below:

$$\mathbf{\Sigma}_f = \alpha \begin{bmatrix} \sigma_{h,0}^{-1/2} & 0 & \ldots & 0 \\ 0 & \sigma_{h,1}^{-1/2} & \ldots & 0 \\ \vdots & \vdots & \ddots & \vdots \\ 0 & 0 & \ldots & \sigma_{h,M-1}^{-1/2} \end{bmatrix} = \alpha \mathbf{\Sigma}_h^{-1/2} \tag{17.17}$$

and

$$\mathbf{\Sigma}_g = \frac{1}{\alpha} \begin{bmatrix} \sigma_{h,0}^{-1/2} & 0 & \ldots & 0 \\ 0 & \sigma_{h,1}^{-1/2} & \ldots & 0 \\ \vdots & \vdots & \ddots & \vdots \\ 0 & 0 & \ldots & \sigma_{h,M-1}^{-1/2} \end{bmatrix} = \frac{1}{\alpha} \mathbf{\Sigma}_h^{-1/2}. \tag{17.18}$$

Note that the channel singular values $\sigma_{h,k}$ are assumed nonzero for all k. The constant α ensures that the power constraint is satisfied. The kth set of multipliers $\sigma_{f,k}$ and $\sigma_{g,k}$ depend only on the kth singular value $\sigma_{h,k}$. None of these multipliers is zero, and the ordering of the singular values $\sigma_{h,k}$ does not matter.

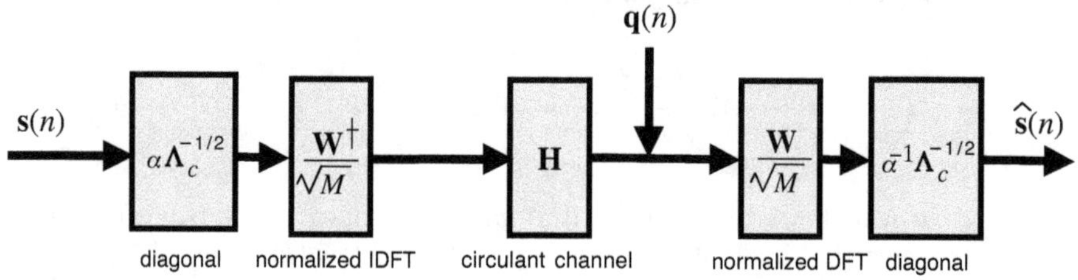

Figure 17.4. Structure of the zero-forcing MMSE cyclic prefix transceiver. The diagonal matric $\mathbf{\Lambda}_c$ has the channel DFT coefficients $C[k]$ on the diagonal.

We now prove:

♠**Theorem 17.1.** *ZF-MMSE cyclic-prefix system.* The cyclic-prefix transceiver optimized to have minimum MSE under the zero-forcing constraint has the form shown in Fig. 17.4. Here $\mathbf{W}$ is the $M \times M$ DFT matrix, and $\mathbf{\Lambda}_c$ is the diagonal matrix given by

$$\mathbf{\Lambda}_c = \begin{bmatrix} C[0] & 0 & \dots & 0 \\ 0 & C[1] & \dots & 0 \\ \vdots & \vdots & \ddots & \vdots \\ 0 & 0 & \dots & C[M-1] \end{bmatrix},$$

with $C[k]$ representing the DFT coefficients of the scalar channel as shown in Eq. (17.5). The constant α is chosen to satisfy the power constraint at the transmitter. The minimized mean square error is given by

$$\mathcal{E}_{mmse} = \frac{\sigma_s^2\sigma_q^2}{p_0}\Big(\sum_{k=0}^{M-1}\frac{1}{|C[k]|}\Big)^2, \tag{17.19}$$

where p_0 is the channel input power. ◇

Before proceeding to the proof, some remarks are in order.

1. The optimal system splits the equalizer $\mathbf{\Lambda}_c^{-1}$ equally between the transmitter and the receiver.
2. Since $\mathbf{s}(n)$ is the blocked version of a scalar symbol stream $s(n)$, the *average error per sample* of $s(n)$ is obtained by dividing Eq. (17.19) by M.
3. Even though the expression for the mean square error (17.19) depends on σ_q^2 and the power p_0, the optimal transceiver matrices do not depend on these quantities (except for the constant α which is chosen to satisfy the power constraint).

Proof of Theorem 17.1. The diagonal matrix $\mathbf{\Lambda}_g$ in the optimal structure of Fig. 17.3 is given by

$$\mathbf{\Lambda}_g = \mathbf{\Sigma}_g\mathbf{\Lambda}_\theta^{-1} \quad \text{(using Eq. (17.16))}$$

$$= \frac{1}{\alpha}\boldsymbol{\Sigma}_h^{-1/2}\boldsymbol{\Lambda}_\theta^{-1} \quad \text{(using Eq. (17.18))}$$

Substituting from Eq. (17.10) we therefore have

$$\boldsymbol{\Lambda}_g = \frac{1}{\alpha}\boldsymbol{\Lambda}_c^{-1/2}\boldsymbol{\Lambda}_\theta^{-1/2}. \tag{17.20}$$

Now, the diagonal matrix $\boldsymbol{\Lambda}_\theta^{-1/2}$ is unitary because the diagonal elements have the form $e^{-j\theta_k/2}$. Thus, if we replace $\boldsymbol{\Sigma}_f$ and $\boldsymbol{\Lambda}_g$ in Fig. 17.3(a) as follows:

$$\boldsymbol{\Sigma}_f \to \boldsymbol{\Sigma}_f \times \boldsymbol{\Lambda}_\theta^{-1/2} \quad \text{and} \quad \boldsymbol{\Lambda}_g \to \boldsymbol{\Lambda}_g \times \boldsymbol{\Lambda}_\theta^{1/2},$$

the system performance is unaffected, that is, neither the transmitted power nor the MSE is changed. From Eq. (17.20) we have

$$\boldsymbol{\Lambda}_g \times \boldsymbol{\Lambda}_\theta^{1/2} = \frac{1}{\alpha}\boldsymbol{\Lambda}_c^{-1/2},$$

whereas from Eqs. (17.17) and (17.10) we have

$$\boldsymbol{\Sigma}_f \times \boldsymbol{\Lambda}_\theta^{-1/2} = \alpha\boldsymbol{\Lambda}_c^{-1/2}.$$

This shows that the optimal system of Fig. 17.3(a) can be replaced with the system shown in Fig. 17.4. Equation (17.19) is obtained by replacing $\sigma_{h,k}$ with $|C[k]|$ in Eq. (12.66). $\bigtriangledown \bigtriangledown \bigtriangledown$

17.3.2 Pure-MMSE transceiver

For the MMSE transceiver without zero-forcing constraint the diagonal matrices $\boldsymbol{\Sigma}_f$ and $\boldsymbol{\Sigma}_g$ were derived in Sec. 13.6. These are reproduced below with the substitution $P = J = M$. First, the diagonal matrix $\boldsymbol{\Sigma}_f$ at the transmitter is

$$\boldsymbol{\Sigma}_f = \frac{\sigma_q}{\sigma_s}\begin{bmatrix} \sqrt{q_{00}} & 0 & \ldots & 0 \\ 0 & \sqrt{q_{11}} & \ldots & 0 \\ \vdots & \vdots & \ddots & \vdots \\ 0 & 0 & \ldots & \sqrt{q_{M-1,M-1}} \end{bmatrix}. \tag{17.21}$$

Here

$$q_{kk} = \begin{cases} \frac{D}{\sigma_{h,k}} - \frac{1}{\sigma_{h,k}^2} & 0 \le k \le K-1 \\ 0 & \text{otherwise,} \end{cases} \tag{17.22}$$

where

$$D = \frac{\dfrac{p_0}{\sigma_q^2} + \displaystyle\sum_{k=0}^{K-1}\frac{1}{\sigma_{h,k}^2}}{\displaystyle\sum_{k=0}^{K-1}\frac{1}{\sigma_{h,k}}} \tag{17.23}$$

In these expressions, the channel singular values are numbered such that

$$\sigma_{h,0} \geq \sigma_{h,1} \geq \ldots \geq \sigma_{h,M-1}, \tag{17.24}$$

and the quantity K is the largest integer $\leq M$ which yields $q_{kk} \geq 0$ for all k in Eq. (17.22). The diagonal matrix $\mathbf{\Sigma}_g$ in the receiver was given in Eq. (13.63), and with the substitutions $P = J = M$ it becomes

$$\mathbf{\Sigma}_g = \frac{\sigma_s}{\sigma_q} \begin{bmatrix} \frac{\sqrt{q_{00}}\,\sigma_{h,0}}{1+q_{00}\sigma_{h,0}^2} & 0 & \ldots & 0 \\ 0 & \frac{\sqrt{q_{11}}\,\sigma_{h,1}}{1+q_{11}\sigma_{h,1}^2} & \ldots & 0 \\ \vdots & \vdots & \ddots & \vdots \\ 0 & 0 & \ldots & \frac{\sqrt{q_{M-1,M-1}}\,\sigma_{h,M-1}}{1+q_{M-1,M-1}\sigma_{h,M-1}^2} \end{bmatrix}. \tag{17.25}$$

The quantities $\mathbf{U}_h, \mathbf{V}_h$, and $\sigma_{h,k}$ appearing in the SVD of the channel are given by Eqs. (17.12) and (17.13). But since Eq. (17.24) has to be satisfied, we have to reorder the DFT coefficients $C[k]$ before computing $\mathbf{\Sigma}_f$ and $\mathbf{\Lambda}_g$. Thus, define the singular values $\sigma_{h,k}$ to be

$$\sigma_{h,k} = |\widehat{C}[k]|, \quad 0 \leq k \leq M-1,$$

where $\widehat{C}[k]$ is the reordered version satisfying

$$|\widehat{C}[0]| \geq |\widehat{C}[1]| \geq \ldots \geq |\widehat{C}[M-1]|. \tag{17.26}$$

We then compute the multiplier coefficients $\sigma_{f,k}$ and $\sigma_{g,k}$ (elements of the diagonal matrices $\mathbf{\Sigma}_f$ and $\mathbf{\Sigma}_g$ as described in Eq. (17.21)–(17.25)). The elements $\lambda_{g,k}$ in the diagonal matrix $\mathbf{\Lambda}_g$ are then given by Eq. (17.16), that is,

$$\lambda_{g,k} = \sigma_{g,k} e^{-j\widehat{\theta}_k},$$

where $\widehat{C}[k] = |\widehat{C}[k]|e^{j\widehat{\theta}_k}$. Once this is done, we assign the multiplier pair $\{\sigma_{g,i}, \lambda_{g,i}\}$ in Fig. 17.3 to the channel which has the ith largest $|C[k]|$. The optimal transceiver structure therefore takes the form shown in Fig. 17.5, where $\mathbf{\Sigma}_p$ and $\mathbf{\Lambda}_p$ are diagonal matrices whose diagonal elements are appropriately *permuted* versions of the diagonal elements of $\mathbf{\Sigma}_f$ and $\mathbf{\Lambda}_g$. The minimized mean square error is given by

$$\mathcal{E}_{mmse} = \sigma_s^2(M-K) + \frac{\sigma_s^2\sigma_q^2\Big(\sum_{k=0}^{K-1}\frac{1}{|\widehat{C}[k]|}\Big)^2}{p_0 + \sigma_q^2\sum_{k=0}^{K-1}\frac{1}{|\widehat{C}[k]|^2}} \tag{17.27}$$

as seen by replacing $\sigma_{h,k}$ with $|\widehat{C}[k]|$ in Eq. (13.66). Thus the K largest values of the DFT magnitudes $|C[k]|$ are involved in this expression. Since $\mathbf{s}(n)$ is the blocked version of a scalar symbol stream $s(n)$, the error per sample is obtained by dividing (17.27) with M.

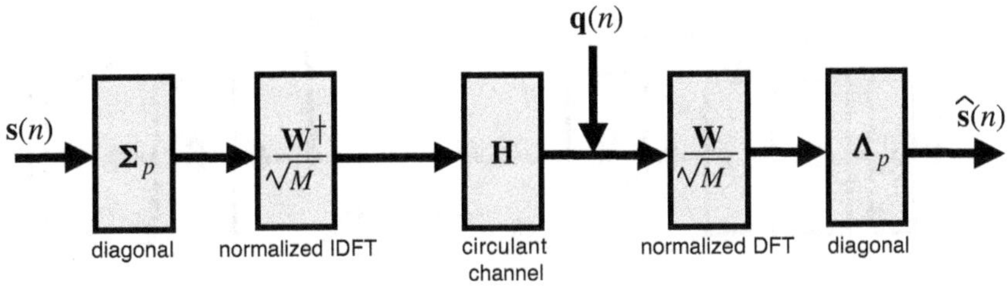

Figure 17.5. Structure of the pure-MMSE cyclic-prefix transceiver. All matrices are square matrices. $\mathbf{\Sigma}_p$ and $\mathbf{\Lambda}_p$ are diagonal matrices whose diagonal elements are permuted versions of the diagonals of $\mathbf{\Sigma}_f$ and $\mathbf{\Lambda}_g$.

Permutation matrix notation. In Sec. B.5.1 (Appendix B) we describe permutation matrices. We show that if two diagonal matrices are such that their diagonal elements are permuted versions of each other, then they can be related using a permutation matrix. Thus, the diagonal matrix $\mathbf{\Sigma}_f$ is related to $\mathbf{\Sigma}_p$ as follows:

$$\mathbf{\Sigma}_p = \mathbf{P}\mathbf{\Sigma}_f\mathbf{P}^T,$$

for an appropriate permutation matrix $\mathbf{P}$. Similarly, $\mathbf{\Lambda}_p = \mathbf{P}\mathbf{\Lambda}_g\mathbf{P}^T$. Thus, the pure-MMSE transceiver in Fig. 17.5 can be redrawn in terms of the original matrices $\mathbf{\Sigma}_f$ and $\mathbf{\Lambda}_g$ as shown in Fig. 17.6(a). Since the permutation matrix $\mathbf{P}$ is unitary, we can eliminate the matrices $\mathbf{P}^T$ and $\mathbf{P}$ in the leftmost and rightmost ends of the figure to obtain Fig. 17.6(b) – this system has identical transmitted power and MSE as the system in part (a). $\triangledown\triangledown\triangledown$

For convenience, the explicit expressions for the diagonal matrices $\mathbf{\Sigma}_f$ and $\mathbf{\Lambda}_g$ in Fig. 17.6 are given below, in terms of $\widehat{C}[k]$, which is the ordered version (17.26) of the channel DFT coefficients:

$$\mathbf{\Sigma}_f = \frac{\sigma_q}{\sigma_s}\begin{bmatrix} \sqrt{q_{00}} & 0 & \ldots & 0 \\ 0 & \sqrt{q_{11}} & \ldots & 0 \\ \vdots & \vdots & \ddots & \vdots \\ 0 & 0 & \ldots & \sqrt{q_{M-1,M-1}} \end{bmatrix}. \tag{17.28}$$

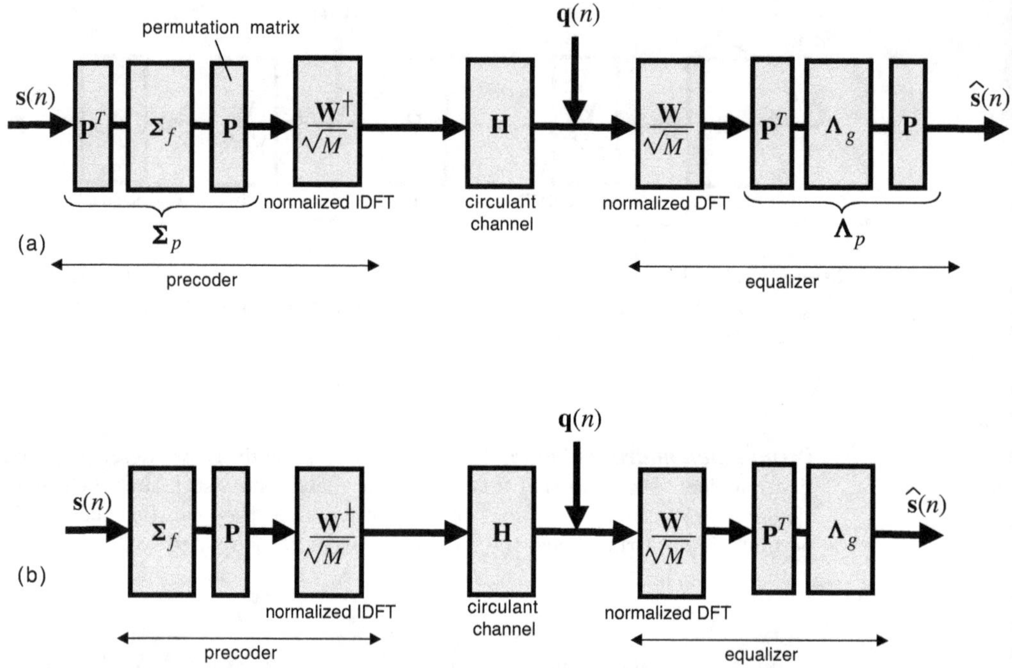

Figure 17.6. (a) Structure of the MMSE cyclic-prefix transceiver in Fig. 17.5, redrawn using a permutation matrix $\mathbf{P}$. (b) A variation with identical transmitted power and MSE.

Here

$$q_{kk} = \begin{cases} \frac{D}{|\widehat{C}[k]|} - \frac{1}{|\widehat{C}[k]|^2} & 0 \le k \le K-1 \\ 0 & \text{otherwise,} \end{cases} \tag{17.29}$$

with D defined as

$$D = \frac{\dfrac{p_0}{\sigma_q^2} + \displaystyle\sum_{k=0}^{K-1} \frac{1}{|\widehat{C}[k]|^2}}{\displaystyle\sum_{k=0}^{K-1} \frac{1}{|\widehat{C}[k]|}} \tag{17.30}$$

and K is the largest integer $\le M$, which yields $q_{kk} \ge 0$ for all k in Eq. (17.29). The diagonal matrix $\mathbf{\Lambda}_g$ is given by

$$\mathbf{\Lambda}_g = \frac{\sigma_s}{\sigma_q} \times \begin{bmatrix} \frac{\sqrt{q_{00}}\,|\widehat{C}[0]|\,e^{-j\widehat{\theta}_0}}{1+q_{00}|\widehat{C}[0]|^2} & \cdots & 0 \\ \vdots & \ddots & \vdots \\ 0 & \cdots & \frac{\sqrt{q_{M-1,M-1}}\,|\widehat{C}[M-1]|\,e^{-j\widehat{\theta}_{M-1}}}{1+q_{M-1,M-1}|\widehat{C}[M-1]|^2} \end{bmatrix}. \quad (17.31)$$

Here the angles $\widehat{\theta}_k$ are the phases of the reordered channel DFT coefficients, that is, $\widehat{C}[k] = |\widehat{C}[k]|e^{j\widehat{\theta}_k}$.

17.4 CP systems with minimum error probability

The cyclic-prefix transceivers with MMSE property are summarized in Fig. 17.7(a) with fresh notations for the diagonal matrices at the transmitter and receiver. These diagonal matrices are now indicated as $\mathbf{\Lambda}_a$ and $\mathbf{\Lambda}_b$. The matrices are chosen as follows: for the ZF-MMSE system choose

$$\mathbf{\Lambda}_a = \alpha\mathbf{\Lambda}_c^{-1/2}, \quad \mathbf{\Lambda}_b = \alpha^{-1}\mathbf{\Lambda}_c^{-1/2} \quad \text{(ZF-MMSE)}, \quad (17.32)$$

where α is chosen to satisfy the power constraint. And for the pure-MMSE system choose

$$\mathbf{\Lambda}_a = \mathbf{\Sigma}_p, \quad \mathbf{\Lambda}_b = \mathbf{\Lambda}_p \quad \text{(pure-MMSE)}, \quad (17.33)$$

where $\mathbf{\Sigma}_p$ and $\mathbf{\Lambda}_p$ are computed as described in Sec. 17.3.2. In Chap. 16 we showed that systems with minimum symbol error probability can be obtained from systems with minimum MSE (with the power constraint and possibly the ZF constraint) by using a unitary matrix $\mathbf{U}$ at the receiver and its inverse $\mathbf{U}^\dagger$ at the transmitter.[2] With such matrices inserted the MMSE system of Fig. 17.7(a) takes the form shown in Fig. 17.7(b). The matrix $\mathbf{U}$ does not change the transmitted power or the MSE. The purpose of $\mathbf{U}$ is to equalize the mean square errors of the M component signals at its output (review Sec. 16.2.2 here).

17.4.1 Choice of the unitary matrix U

We begin by proving an important result on error covariances. For this, first recall that the MMSE cyclic-prefix system shown in Fig. 17.7(a) assumes that the signal $\mathbf{s}(n)$ and noise $\mathbf{q}(n)$ are zero-mean uncorrelated processes with covariances $\sigma_s^2\mathbf{I}$ and $\sigma_q^2\mathbf{I}$, respectively. The components $q_k(n)$ of the noise $\mathbf{q}(n)$ are assumed to be Gaussian with variance σ_q^2.

[2]Such systems are also called minimum BER systems, since the bit error rate is directly related to the symbol error rate for Gray-coded systems (Sec. 2.3.3).

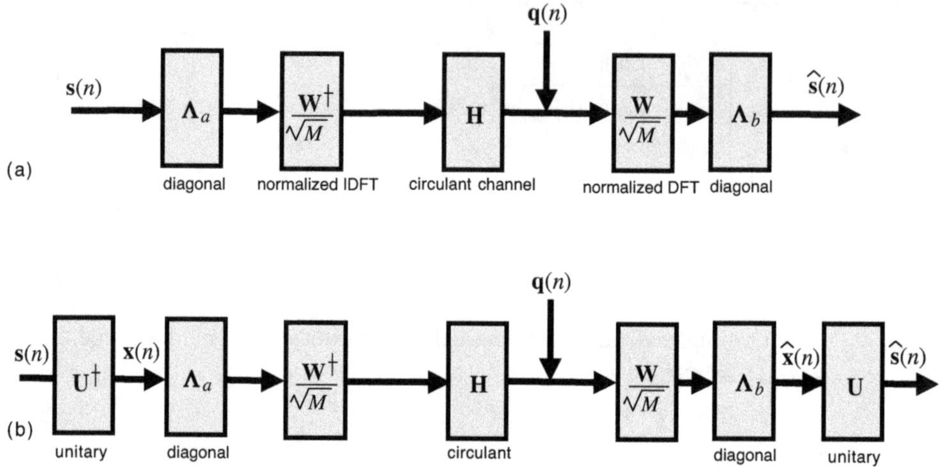

Figure 17.7. (a) The MMSE cyclic-prefix transceiver, and (b) the modified version with unitary matrices $\mathbf{U}$ and $\mathbf{U}^\dagger$ inserted.

♠**Lemma 17.1.** *Diagonal structure of error covariance.* For the MMSE cyclic-prefix system shown in Fig. 17.7(a) (with or without zero forcing), the error covariance matrix is diagonal. $\diamondsuit$

Proof. The reconstructed signal in Fig. 17.7(a) is given by

$$\widehat{\mathbf{s}}(n) = \frac{\mathbf{\Lambda}_b\mathbf{W}\mathbf{H}\mathbf{W}^\dagger\mathbf{\Lambda}_a}{M}\mathbf{s}(n) + \frac{\mathbf{\Lambda}_b\mathbf{W}}{\sqrt{M}}\mathbf{q}(n).$$

Since $\mathbf{H}$ is circulant, $\mathbf{W}\mathbf{H}\mathbf{W}^\dagger$ is diagonal (Appendix D). So

$$\widehat{\mathbf{s}}(n) = \mathbf{\Lambda}_0\mathbf{s}(n) + \mathbf{\Lambda}_1\mathbf{W}\mathbf{q}(n)$$

for some diagonal matrices $\mathbf{\Lambda}_0$ and $\mathbf{\Lambda}_1$. The reconstruction error is therefore

$$\mathbf{e}(n) = \widehat{\mathbf{s}}(n) - \mathbf{s}(n) = (\mathbf{\Lambda}_0 - \mathbf{I})\mathbf{s}(n) + \mathbf{\Lambda}_1\mathbf{W}\mathbf{q}(n) = \mathbf{\Lambda}_2\mathbf{s}(n) + \mathbf{\Lambda}_1\mathbf{W}\mathbf{q}(n),$$

where $\mathbf{\Lambda}_2$ is also diagonal. Since $\mathbf{s}(n)$ and $\mathbf{q}(n)$ are zero-mean uncorrelated processes with covariances $\sigma_s^2\mathbf{I}$ and $\sigma_q^2\mathbf{I}$, respectively, the covariance of the error is

$$\mathbf{R}_{\mathbf{ee}} = \sigma_s^2\mathbf{\Lambda}_2\mathbf{\Lambda}_2^\dagger + \sigma_q^2\mathbf{\Lambda}_1\mathbf{W}\mathbf{W}^\dagger\mathbf{\Lambda}_1^\dagger = \sigma_s^2\mathbf{\Lambda}_2\mathbf{\Lambda}_2^\dagger + M\sigma_q^2\mathbf{\Lambda}_1\mathbf{\Lambda}_1^\dagger,$$

which is diagonal indeed. $\bigtriangledown \bigtriangledown \bigtriangledown$

As explained in detail in Sec. 16.3, the MMSE system without zero forcing creates a bias in the reconstruction error. This **bias should be removed** before the threshold detector is used at the receiver to identify the transmitted symbols. This was explained in detail in Sec. 16.3.1. The following result implicitly assumes that in the non zero-forcing case the bias is removed.

♠**Theorem 17.2.** *Minimum error probability in cyclic-prefix systems.* The MMSE cyclic-prefix system shown in Fig. 17.7(b) has minimum average error probability (with or without zero forcing) for any unitary matrix $\mathbf{U}$ satisfying the property

$$|U_{km}| = \frac{1}{\sqrt{M}} \tag{17.34}$$

for all k, m. In particular, therefore, the normalized IDFT matrix

$$\mathbf{U} = \frac{\mathbf{W}^\dagger}{\sqrt{M}} \tag{17.35}$$

also provides a solution with minimum error probability. ◇

Proof. Since $\mathbf{U}^\dagger$ is unitary the covariance of $\mathbf{x}(n)$ is the same as that of $\mathbf{s}(n)$ (namely $\sigma_s^2\mathbf{I}$). It therefore follows from application of Lemma 17.1 that the error $\widehat{\mathbf{x}}(n) - \mathbf{x}(n)$ has a diagonal covariance matrix. Next, since $\widehat{\mathbf{s}}(n) - \mathbf{s}(n) = \mathbf{U}[\widehat{\mathbf{x}}(n) - \mathbf{x}(n)]$ it follows that the covariances of the errors $\widehat{\mathbf{s}}(n) - \mathbf{s}(n)$ and $\widehat{\mathbf{x}}(n) - \mathbf{x}(n)$ are related as

$$\mathbf{R}_{\mathbf{ee,s}} = \mathbf{U}\mathbf{R}_{\mathbf{ee,x}}\mathbf{U}^\dagger.$$

Since $\mathbf{R}_{\mathbf{ee,x}}$ is diagonal according to Lemma 17.1, it follows that if we choose $\mathbf{U}$ to satisfy Eq. (17.34) then $\mathbf{R}_{\mathbf{ee,s}}$ has identical diagonal elements (Sec. 21.5.1.B). This ensures that the system in Fig. 17.7(b) has minimum BER (for reasons explained in detail in Chap. 16). Since $\mathbf{W}$ has elements with unit magnitude, the last part of the theorem follows readily. ▽▽▽

The unitary property of $\mathbf{U}$ ensures that the MMSE property continues to hold for any choice of $\mathbf{U}$ and furthermore the transmitted power remains unchanged. Thus the minimum-BER system continues to be an MMSE system. With $\mathbf{U} = \mathbf{W}^\dagger/\sqrt{M}$ as in the theorem statement, the minimum-BER cyclic-prefix system takes the specific form shown in Fig. 17.8. With $\mathcal{E}_{mmse}$ denoting the minimized MSE, the average MSE per symbol is $\mathcal{E}_{ave} = \mathcal{E}_{mmse}/M$. For the zero-forcing system, with $\mathcal{E}_{mmse}$ given by Eq. (17.19), the minimized average symbol error probability is given by

$$\mathcal{P}_{e,min} = c\mathcal{Q}\Big(\frac{A}{\sqrt{\mathcal{E}_{ave}}}\Big) \quad \text{(ZF-MMSE case)}. \tag{17.36}$$

See Sec. 16.2. Here the constants c and A depend on the constellation used. For example, in a b-bit PAM system,

$$c = 2(1 - 2^{-b}) \quad \text{and} \quad A = \sqrt{\frac{3\sigma_s^2}{2^{2b} - 1}}. \tag{17.37}$$

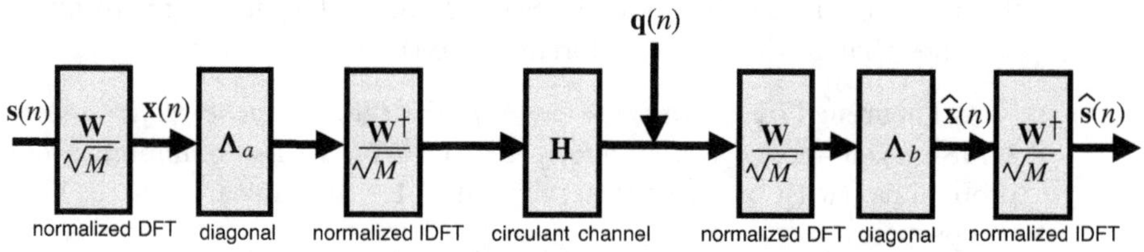

Figure 17.8. The MMSE cyclic-prefix transceiver optimized for error probability. The diagonal matrices $\mathbf{\Lambda}_a$ and $\mathbf{\Lambda}_b$ depend on whether zero forcing is enforced or not.

For the system without zero forcing, with $\mathcal{E}_{mmse}$ denoting the minimized mean square error (17.27), the minimized symbol error probability is given by (see Sec. 16.4)

$$\mathcal{P}_{e,min} = c\mathcal{Q}\Big(A\sqrt{\frac{1}{\mathcal{E}_{ave}} - \frac{1}{\sigma_s^2}}\Big) \quad \text{(pure-MMSE case)}, \tag{17.38}$$

where $\mathcal{E}_{ave} = \mathcal{E}_{mmse}/M$.

17.4.2 Circulant form of minimum-BER cyclic prefix transceivers

The minimum-BER cyclic-prefix system shown in Fig. 17.8 can be redrawn as in Fig. 17.9(a), where

$$\mathbf{H}_a^{-1} = \frac{1}{M}\mathbf{W}^\dagger\mathbf{\Lambda}_a\mathbf{W}, \quad \mathbf{H}_b^{-1} = \frac{1}{M}\mathbf{W}^\dagger\mathbf{\Lambda}_b\mathbf{W}. \tag{17.39}$$

Since $\mathbf{\Lambda}_a$ and $\mathbf{\Lambda}_b$ are diagonal matrices, we see that $\mathbf{H}_a^{-1}$ and $\mathbf{H}_b^{-1}$ are circulants. These matrices share the task of equalization. For the special case where zero forcing is in effect, we have

$$\mathbf{\Lambda}_a = \alpha\mathbf{\Lambda}_c^{-1/2}, \qquad \mathbf{\Lambda}_b = \alpha^{-1}\mathbf{\Lambda}_c^{-1/2}, \tag{17.40}$$

where $\mathbf{\Lambda}_c$ is the diagonal matrix of the channel DFT coefficients $C[k]$. Thus

$$\mathbf{H}_a^{-1} = \frac{\alpha}{M}\mathbf{W}^\dagger\mathbf{\Lambda}_c^{-1/2}\mathbf{W}, \quad \mathbf{H}_b^{-1} = \frac{1}{\alpha M}\mathbf{W}^\dagger\mathbf{\Lambda}_c^{-1/2}\mathbf{W}. \tag{17.41}$$

We can write this in the form

$$\mathbf{H}_a^{-1} = \alpha\mathbf{H}_{1/2}^{-1}, \quad \mathbf{H}_b^{-1} = \frac{1}{\alpha}\mathbf{H}_{1/2}^{-1} \quad \text{(ZF-MMSE case)}, \tag{17.42}$$

where

$$\mathbf{H}_{1/2}^{-1} \stackrel{\Delta}{=} \frac{1}{M}\mathbf{W}^\dagger\mathbf{\Lambda}_c^{-1/2}\mathbf{W}. \tag{17.43}$$

The zero-forcing case is shown in Fig. 17.9(b). Since $\mathbf{H} = \mathbf{W}^{-1}\mathbf{\Lambda}_c\mathbf{W}$, we have

$$\mathbf{H}_{1/2}^{-1}\mathbf{H}_{1/2}^{-1} = \mathbf{H}^{-1}. \tag{17.44}$$

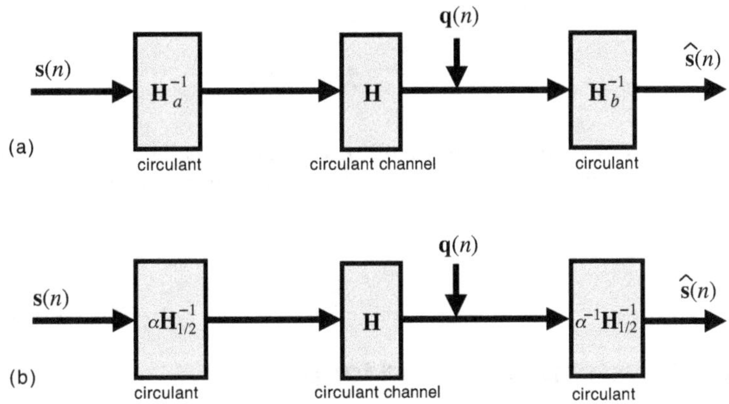

Figure 17.9. (a) Circulant representation of the cyclic-prefix transceiver optimized for error probability. (b) Special case where zero forcing is in place.

Thus the equalization in the zero-forcing case is *split evenly* between the transmitter and receiver. In the PAM case it is possible that the discrete-time equivalent channel is real as explained in Sec. 2.A.2 (Appendix of Chap. 2). For the case of real channels, it is clear that the circulant channel matrix $\mathbf{H}$ is real. In this case it can be verified that the precoder $\mathbf{H}_a^{-1}$ is real (see below). This implies that a real input $\mathbf{s}(n)$ produces a real precoder output, that is, the channel input remains real. The signal which comes into the receiver is therefore also real.

Realness. To see why $\mathbf{H}_a^{-1}$ is real for real channels, recall from Eq. (17.39) that

$$\mathbf{H}_a^{-1} = \frac{1}{M}\mathbf{W}^\dagger \boldsymbol{\Lambda}_a \mathbf{W}.$$

The diagonal elements of $\boldsymbol{\Lambda}_a$ are essentially the DFT coefficients of (a permutation of) the top row of $\mathbf{H}_a^{-1}$. So, to show that this matrix is real, we only have to argue that $\boldsymbol{\Lambda}_a$ has conjugate symmetry [Oppenheim and Schafer, 1999], that is,

$$[\boldsymbol{\Lambda}_a]_{kk} = [\boldsymbol{\Lambda}_a]^*_{M-k,M-k}$$

First consider the zero-forcing case, where we have, from Eq. (17.40),

$$\boldsymbol{\Lambda}_a = \alpha \boldsymbol{\Lambda}_c^{-1/2},$$

with $\boldsymbol{\Lambda}_c$ denoting a diagonal matrix whose diagonal elements are the DFT coefficients of the channel. If the channel is real, these DFT coefficients have conjugate symmetry, so the diagonal elements of $\boldsymbol{\Lambda}_c^{-1/2}$ have the same symmetry, and the elements of $\mathbf{H}_a^{-1}$ are real. Next consider the non-ZF MMSE case. In this case $\boldsymbol{\Lambda}_a$ is a permuted version of the real matrix $\boldsymbol{\Sigma}_f$ shown in Eq. (17.21). If the channel is real then its DFT coefficients are conjugate symmetric. This means that the singular values $\sigma_{h,k}$ (absolute values of the DFT coefficients) satisfy the magnitude symmetry

$$\sigma_{h,k} = \sigma_{h,M-k}$$

before they are ordered as in Eq. (17.24). Thus the real matrix $\mathbf{\Lambda}_a$ can always be reordered such that it has the same magnitude symmetry. In this case $\mathbf{H}_a^{-1}$ will indeed be real. $\bigtriangledown \bigtriangledown \bigtriangledown$

17.5 DMT systems optimized for power

In Chap. 14 we considered the problem of optimum bit allocation in a transceiver system which has a specified set of error probabilities $\mathcal{P}_e(k)$ (quality-of-service or QoS constraints) for the M signals $s_k(n)$. We showed how the transmitted power can be minimized (under the ZF constraint) by choosing the precoder and equalizer appropriately and allocating the bits among the M signals. The optimal system (minimum-power system subject to zero forcing) takes the form shown in Fig. 17.10(a) for the special case where all matrices are square (i.e., $P = J = M$ in the notation of Chap. 14).

In this figure $\mathbf{U}_h, \mathbf{\Sigma}_h$, and $\mathbf{V}_h$ are the matrices arising in the SVD of $\mathbf{H}$, that is $\mathbf{H} = \mathbf{U}_h\mathbf{\Sigma}_h\mathbf{V}_h^\dagger$. The matrix $\mathbf{\Sigma}_g$ is a diagonal matrix that can be chosen arbitrarily. For the case of the DMT system the matrix $\mathbf{H}$ is a circulant, so it is diagonalized by the DFT matrix $\mathbf{W}$, that is, $\mathbf{H} = \mathbf{W}^{-1}\mathbf{\Lambda}_c\mathbf{W}$, where $\mathbf{\Lambda}_c$ is a diagonal matrix whose diagonal elements are the DFT coefficients of the scalar channel:

$$\mathbf{\Lambda}_c = \begin{bmatrix} C[0] & 0 & \ldots & 0 \\ 0 & C[1] & \ldots & 0 \\ \vdots & \vdots & \ddots & \vdots \\ 0 & 0 & \ldots & C[M-1] \end{bmatrix}. \tag{17.45}$$

Thus, following the line of arguments in Eqs. (17.6)–(17.12) we conclude again that $\mathbf{U}_h, \mathbf{\Sigma}_h$, and $\mathbf{V}_h^\dagger$ are as in Eqs. (17.10) and (17.12), so that:

$$\mathbf{\Sigma}_h = \mathbf{\Lambda}_c\mathbf{\Lambda}_\theta^{-1}, \quad \mathbf{U}_h^\dagger = \frac{\mathbf{\Lambda}_\theta^{-1}\mathbf{W}}{\sqrt{M}}, \quad \text{and} \quad \mathbf{V}_h = \frac{\mathbf{W}^\dagger}{\sqrt{M}}. \tag{17.46}$$

This results in the structure shown in Fig. 17.10(b) for the optimal DMT system. If we choose the free diagonal matrix $\mathbf{\Sigma}_g$ to be $\mathbf{\Sigma}_g = \mathbf{\Sigma}_h^{-1}$, then the optimal transceiver simplifies to the one shown in Fig. 17.11 in view of the relation $\mathbf{\Sigma}_h\mathbf{\Lambda}_\theta = \mathbf{\Lambda}_c$.

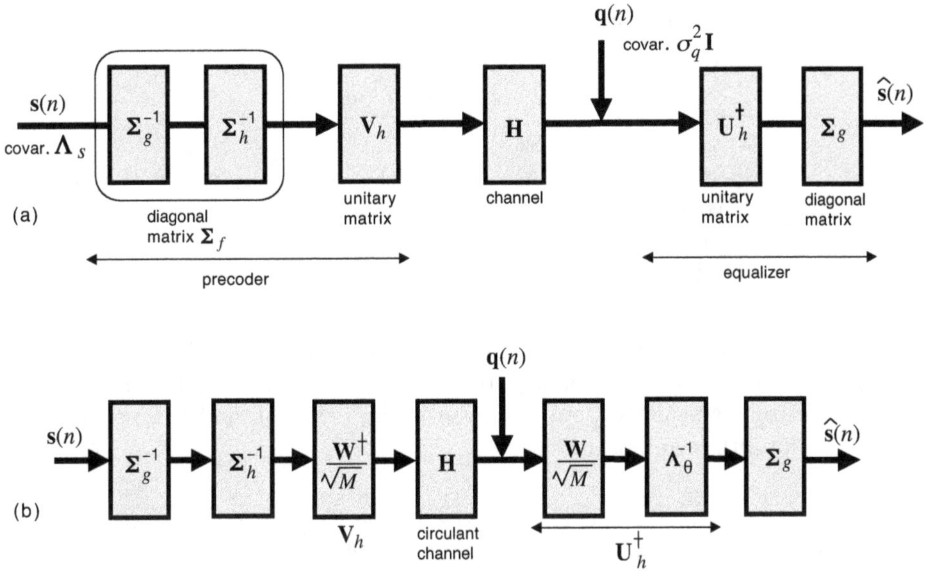

Figure 17.10. (a) General form of the transceiver optimized for minimum transmitted power using bit allocation. Zero forcing is assumed. (b) Special case of the DMT system with cyclic prefix where the channel **H** is circulant. All matrices in both figures are $M \times M$.

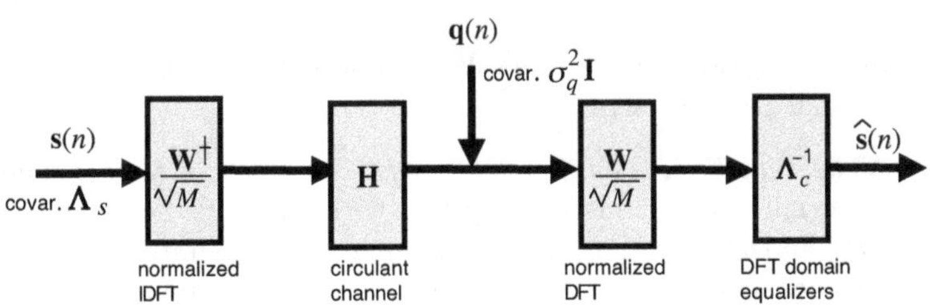

Figure 17.11. The DMT system optimized to minimize transmitted power. Zero forcing and optimal bit allocation are assumed.

Thus the optimal DMT system (which minimizes power subject to optimum bit allocation and zero forcing) is nothing but the "traditional DMT system" introduced in Sec. 7.6, with the DFT and IDFT matrices [Lin and Phoong, 2001b]. The optimal bit allocation formula is as in Eq. (14.41), with $\sigma_{h,k} = |C[k]|$. Thus

$$b_k = D_0 - \log_2\left(\frac{\sigma_q^2}{3}\Big(\mathcal{Q}^{-1}\Big(\frac{\mathcal{P}_e(k)}{4}\Big)\Big)^2\right) + \log_2 |C[k]|^2. \tag{17.47}$$

It is assumed here that the right-hand side of Eq. (17.47) yields sufficiently large values for all k so that it can be truncated to positive integer values of b_k

without compromising optimality seriously. In Eq. (17.47) the quantity $\mathcal{P}_e(k)$ is the specified error probability for the kth symbol stream $s_k(n)$. The constant D_0 is such that the bit rate constraint

$$\frac{1}{M}\sum_{k=0}^{M-1} b_k = b \tag{17.48}$$

is satisfied. In Fig. 17.11 the kth signal power $[\mathbf{\Lambda}_s]_{kk}$ is such that the probability of error is kept at the specified level $\mathcal{P}_e(k)$. As shown in Ex. 14.1 of Chap. 14, the appropriate value of $[\mathbf{\Lambda}_s]_{kk}$ (also denoted as P_k in Chap. 14) is independent of k, that is,

$$[\mathbf{\Lambda}_s]_{kk} = 2^{D_0}. \tag{17.49}$$

This is a consequence of the fact that the free matrix $\mathbf{\Sigma}_g$ was chosen to be such that the precoder is orthonormal (review Ex. 14.1 here). The minimized value of the transmitted power is given by Eq. (14.30), and is reproduced here with appropriate substitutions:

$$P_{min} = c2^b\Big(\frac{1}{\prod_{k=0}^{M-1}|C[k]|^2}\Big)^{1/M}, \tag{17.50}$$

where $c = M(\prod_k c_k)^{1/M}$ with

$$c_k = \frac{\sigma_q^2}{3}\left(\mathcal{Q}^{-1}\Big(\frac{\mathcal{P}_e(k)}{4}\Big)\right)^2, \tag{17.51}$$

assuming a QAM constellation. We are now ready to summarize the main results on optimal DMT systems:

♠**Theorem 17.3.** *Optimal DMT system with zero-forcing constraint.* Consider a cyclic-prefix transceiver system with a specified set of error probabilities $\{\mathcal{P}_e(k)\}$ and average bit rate b. Assume the transmitter allows bit allocation among the signals $s_k(n)$, as in the DMT system. Assume the precoder $\mathbf{F}$, equalizer $\mathbf{G}$, and bit allocation are optimized under the zero-forcing constraint such that the transmitted power is minimized subject to the specified error probability constraints $\{\mathcal{P}_e(k)\}$. This optimized system has the same transmitted power as the DFT-based DMT system shown in Fig. 17.11. Thus the DMT system shown in Fig. 17.11 is optimal for power. This system is designed as follows:

1. The bit allocation is chosen as in Eq. (17.47), where D_0 is a constant such that the average bit rate constraint (17.48) is satisfied.
2. The diagonal elements of $\mathbf{\Lambda}_s$ are chosen as in Eq. (17.49), so that the specified $\mathcal{P}_e(k)$ is satisfied.

The minimized value of the transmitted power is given by Eq. (17.50). ◇

A number of results in this section were first presented in [Lin and Phoong, 2001b]. As a final remark, it should be noticed that the results of this section require modification when the channel noise $q(n)$ is colored, as is the case in practical DSL systems [Starr *et al.*, 1999]. In such situations the covariance $\mathbf{R_{qq}}$ of the noise vector $\mathbf{q}(n)$ is not diagonal. The general techniques described in Chap. 12 should be used to obtain an equivalent system where the noise covariance has the form $\sigma_q^2\mathbf{I}$. The final form of the transceiver will not be entirely DFT-based, as it depends also on $\mathbf{R_{qq}}$.

17.6 The cyclic-prefix system with unitary precoder

In Chap. 15 we studied the special case of transceivers with unitary or orthonormal precoders. In particular we considered in Sec. 15.2 orthonormal precoders where all matrices ($\mathbf{F}, \mathbf{H}$, and $\mathbf{G}$) are square. We showed that if such transceivers are optimized to minimize the MSE under the power constraint (with or without the zero-forcing constraint), then the precoder $\mathbf{F}$ can be assumed to be *any unitary matrix* without loss of optimality.

In this section we consider the special case of cyclic-prefix systems with precoder restricted to be unitary. Such systems will be referred to as *orthonormal cyclic-prefix systems*, even though the receiver matrix is not unitary (it has to perform equalization).[3] All the results of Sec. 15.2 can be applied, with the additional feature that the channel matrix $\mathbf{H}$ is circulant. Since the precoder is restricted to be unitary, all equalization in the optimized sytem will take place at the receiver (instead of being split evenly between the transmitter and the receiver).

17.6.1 Single-carrier and multicarrier cyclic-prefix systems

Since the unitary precoder $\mathbf{F}$ can be assumed to be any unitary matrix without loss of optimality, we first consider the simplest example, where the precoder is chosen to be identity:

$$\mathbf{F} = \mathbf{I} \tag{17.52}$$

This is called the **lazy precoder**. In this case the transceiver is as in Fig. 17.12(a). From Sec. 15.2 in Chap. 15 we then conclude that the optimal equalizer $\mathbf{G}$ (for MMSE property) is given by

$$\mathbf{G} = \begin{cases} \mathbf{H}^{-1} & \text{ZF-MMSE case} \\ \mathbf{H}^{\dagger}\left(\mathbf{H}\mathbf{H}^{\dagger} + \dfrac{\sigma_q^2}{\sigma_s^2}\mathbf{I}\right)^{-1} & \text{pure-MMSE case.} \end{cases} \tag{17.53}$$

[3] In OFDM (orthogonal frequency division multiplex) systems the columns of the precoder are mutually orthogonal but do not necessarily have unit norm. Thus, the *O* in *OFDM* signifies *orthogonal* rather than *orthonormal.*

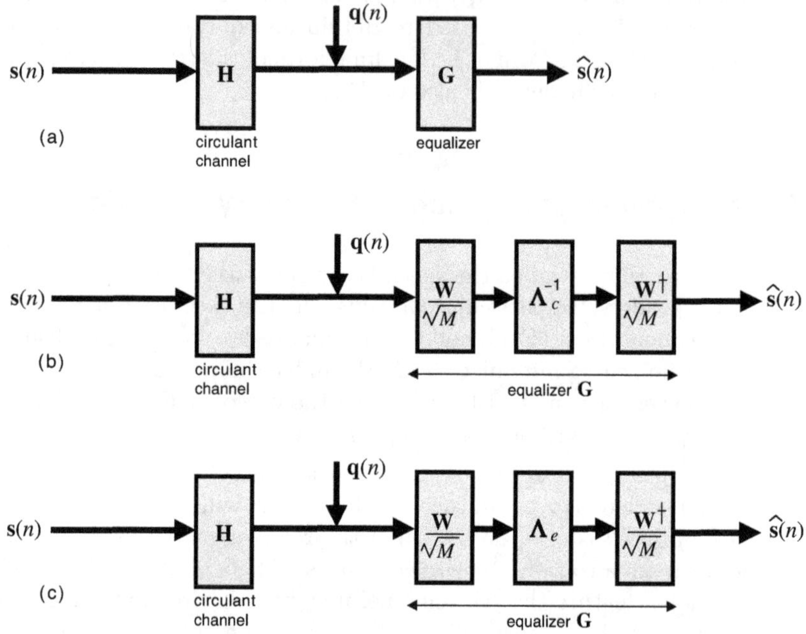

Figure 17.12. (a) Structure of the MMSE-optimized cyclic-prefix transceiver with unitary precoder chosen as identity. (b) Special case when the zero-forcing constraint is in place, and (c) case when ZF constraint is removed. The input $\mathbf{s}(n)$ and noise $\mathbf{q}(n)$ are assumed to have covariances $\sigma_s^2\mathbf{I}$ and $\sigma_q^2\mathbf{I}$, respectively. Parts (b) and (c) are also called single-carrier cyclic-prefix (SC-CP) systems. All matrices have size $M \times M$.

Since $\mathbf{H}$ is circulant, it can be written as $\mathbf{H} = \mathbf{W}^\dagger\boldsymbol{\Lambda}_c\mathbf{W}/M$, where $\mathbf{W}$ is the DFT matrix. From this it follows that

$$\mathbf{G} = \begin{cases} \frac{1}{M}\mathbf{W}^\dagger\boldsymbol{\Lambda}_c^{-1}\mathbf{W} & \text{ZF-MMSE case,} \\ \frac{1}{M}\mathbf{W}^\dagger\boldsymbol{\Lambda}_c^*\Big(\boldsymbol{\Lambda}_c\boldsymbol{\Lambda}_c^* + \frac{\sigma_q^2}{\sigma_s^2}\mathbf{I}\Big)^{-1}\mathbf{W} & \text{pure-MMSE case.} \end{cases} \tag{17.54}$$

This is shown in Figs. 17.12(b) and 17.12(c), where

$$\boldsymbol{\Lambda}_e \stackrel{\Delta}{=} \boldsymbol{\Lambda}_c^*\Big(\boldsymbol{\Lambda}_c\boldsymbol{\Lambda}_c^* + \frac{\sigma_q^2}{\sigma_s^2}\mathbf{I}\Big)^{-1}. \tag{17.55}$$

The reader will recall from Sec. 7.5 that this is nothing but the **single-carrier** cyclic-prefix system, abbreviated as the **SC-CP** system. Thus, the SC-CP system is also an optimal system among the class of orthonormal cyclic-prefix systems. Instead of choosing the precoder to be the identity matrix, we could have chosen it to be the normalized IDFT matrix, without any increase of MSE. In this case we have

$$\mathbf{F} = \mathbf{W}^{\dagger}/\sqrt{M}, \tag{17.56}$$

and the optimal equalizer is (Sec. 15.2)

$$\mathbf{G} = \begin{cases} \dfrac{\mathbf{\Lambda}_c^{-1}\mathbf{W}}{\sqrt{M}} & \text{ZF-MMSE case,} \\ \underbrace{\mathbf{\Lambda}_c^*\Big(\mathbf{\Lambda}_c\mathbf{\Lambda}_c^* + \dfrac{\sigma_q^2}{\sigma_s^2}\mathbf{I}\Big)^{-1}}_{\mathbf{\Lambda}_e} \dfrac{\mathbf{W}}{\sqrt{M}} & \text{pure-MMSE case.} \end{cases} \tag{17.57}$$

This example is shown in Fig. 17.13.

The reader will recall from Sec. 7.5 that this is nothing but the multicarrier system or the OFDM system. We also use the term **MC-CP** (multicarrier cyclic-prefix) to describe this. Note that the SC-CP system and the MC-CP system can be schematically related as shown in Fig. 17.14. The mean square errors of the SC-CP and MC-CP MMSE systems can be obtained from Eq. (15.34) in Chap. 15, by substituting $\sigma_{h,k} = |C[k]|$. Thus, for the optimal cyclic-prefix transceiver with unitary precoder,

$$\mathcal{E}_{mmse} = \begin{cases} \dfrac{M\sigma_q^2\sigma_s^2}{p_0}\displaystyle\sum_{k=0}^{M-1}\dfrac{1}{|C[k]|^2} & \text{(ZF-MMSE)} \\ \dfrac{M\sigma_s^2\sigma_q^2}{p_0}\displaystyle\sum_{k=0}^{M-1}\left(\dfrac{1}{\frac{M\sigma_q^2}{p_0} + |C[k]|^2}\right) & \text{(pure-MMSE).} \end{cases} \tag{17.58}$$

The average MSE per symbol can be obtained by dividing these by M. We summarize the results derived so far in the following lemma. Here "orthonormal transceivers" simply means that the precoders are restricted to be unitary.

♠**Theorem 17.4.** *Optimality of cyclic-prefix systems.* The single-carrier (SC) system and the multicarrier (MC) system are both optimal systems among the class of orthonormal transceivers for cyclic-prefix systems. The optimality is in the sense of minimizing the mean square error (with or without the zero-forcing constraint). The minimized errors are given by Eq. (17.58), where $C[k]$ denote the channel DFT coefficients. ◇

17.6.2 Error covariances

The error covariance matrices in the MC-CP and SC-CP cases have some structural differences. These are important when we address the problem of minimizing the error probability. We first prove the following result:

♠**Lemma 17.2.** *Structure of error covariance.* For the MC-CP system the error covariance is diagonal, whereas for the SC-CP system it is circulant. This property holds whether there is zero forcing or not. ◇

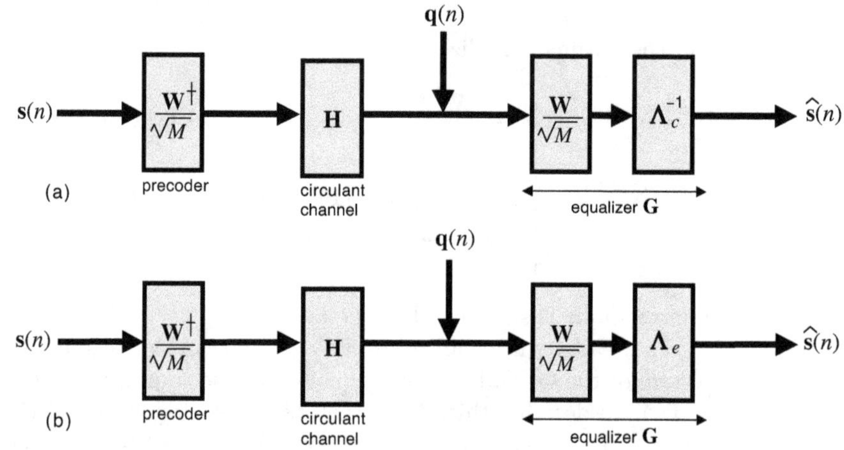

Figure 17.13. Another variation of the MMSE-optimized cyclic-prefix transceiver, where the unitary precoders are chosen as $\mathbf{W}^\dagger/\sqrt{M}$. (a) Case when the zero-forcing constraint is in place, and (b) case when the ZF constraint is removed. The input $\mathbf{s}(n)$ and noise $\mathbf{q}(n)$ are assumed to have covariances $\sigma_s^2\mathbf{I}$ and $\sigma_q^2\mathbf{I}$, respectively. This is also called a multicarrier or MC-CP system.

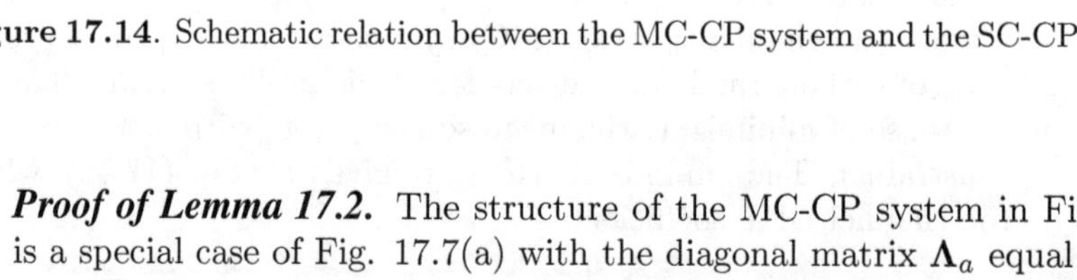

Figure 17.14. Schematic relation between the MC-CP system and the SC-CP systems.

Proof of Lemma 17.2. The structure of the MC-CP system in Fig. 17.13 is a special case of Fig. 17.7(a) with the diagonal matrix $\mathbf{\Lambda}_a$ equal to identity. From Lemma 17.1 it therefore follows that the error covariance is diagonal. Now, the SC-CP system is schematically related to the MC-CP system as in Fig. 17.14. In this figure we have $\widehat{\mathbf{s}}(n) = \mathbf{W}^\dagger\widehat{\mathbf{x}}(n)/\sqrt{M}$ and $\mathbf{s}(n) = \mathbf{W}^\dagger\mathbf{x}(n)/\sqrt{M}$. The reconstruction errors $\mathbf{e}_{SC}(n) = \widehat{\mathbf{s}}(n) - \mathbf{s}(n)$, and $\mathbf{e}_{MC}(n) = \widehat{\mathbf{x}}(n) - \mathbf{x}(n)$ are therefore related by $\mathbf{e}_{SC}(n) = \mathbf{W}^\dagger\mathbf{e}_{MC}(n)/\sqrt{M}$. So their covariance matrices are related by

$$\mathbf{C}_{SC} = \frac{\mathbf{W}^\dagger\mathbf{C}_{MC}\mathbf{W}}{M} \tag{17.59}$$

Since $\mathbf{C}_{MC}$ is diagonal and $\mathbf{W}$ is the DFT, $\mathbf{C}_{SC}$ is indeed circulant. $\bigtriangledown\bigtriangledown\bigtriangledown$

17.6.3 Minimum-error-probability property of SC-CP systems

We have shown that both the SC-CP and MC-CP systems are optimal (MMSE) orthonormal cyclic-prefix systems. We now show that one of these is better than the other if we also wish to have minimum symbol error probability. We shall refer to these as minimum-BER systems (because for Gray-coded systems BER is proportional to symbol error probability, see Sec. 2.3.3). Some of the results of this section were first observed in Ding *et al.* [2003] and Lin and Phoong [2003].

Figure 17.15 shows the minimum-MSE cyclic-prefix systems with a unitary matrix $\mathbf{U}$ and its inverse $\mathbf{U}^\dagger$ inserted, in order to minimize the average BER. As explained in Chap. 16, the optimal $\mathbf{U}$ is the matrix that produces an error covariance with identical diagonal elements (e.g., as in a circulant). From Lemma 17.2 we know that, for the SC-CP system, the error covariance already satisfies this property. So we simply have to set $\mathbf{U} = \mathbf{I}$. Thus the SC-CP system with MMSE property (with or without the ZF constraint) also has *minimum BER among all orthonormal cyclic-prefix transceivers.* According to Lemma 17.2 the MC-CP system with the MMSE property shown in Fig. 17.15(b) has a *diagonal* error covariance matrix, and we therefore need to choose $\mathbf{U}$ to make it a circulant. This is readily achieved by choosing $\mathbf{U} = \mathbf{W}$, which merely results in the SC-CP system again.

Summarizing, if we want the cyclic-prefix system with the MMSE property (with or without the ZF constraint) to enjoy the minimum-BER property as well, then we have to choose the single-carrier (SC-CP) system. A few remarks are in order before we conclude this section:

1. ***MC-CP worst for BER***. It is interesting to note that, among all the cyclic prefix MMSE transceivers with orthonormal precoders (with or without the ZF constraint), the multicarrier system (OFDM system) is the *worst* from the view-point of BER. It has maximum BER because the diagonal nature of the covariance *maximizes* the BER function rather than minimizing it . This follows from the discussion on Schur-convex functions and majorization given in Sec. 21.5.1.

2. ***Low SNR case***. The preceding discussion assumes that the noise-to-signal ratios (NSRs) at the inputs of the detectors are reasonably low. If this is not the case, and all NSRs are higher than the threshold described in Sec. 11.5.2, then the MC-CP system becomes the minimum-BER system because the average error probability becomes Schur-concave rather than Schur-convex.

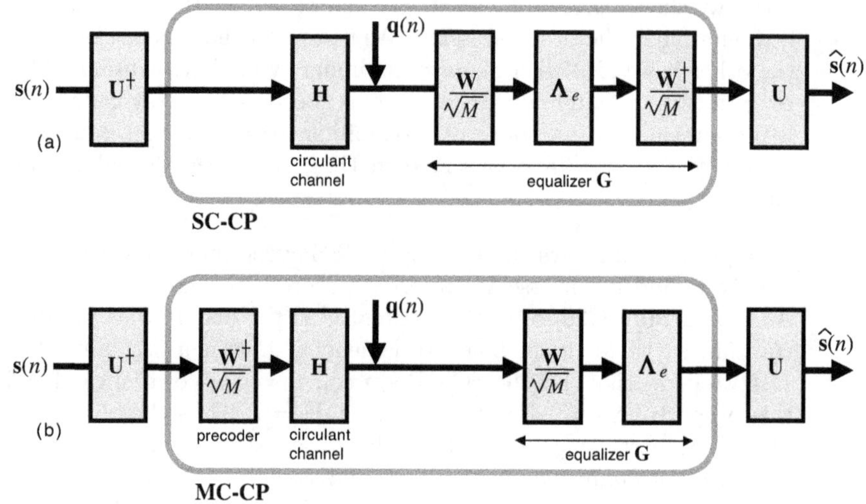

Figure 17.15. (a), (b) The SC-CP system and MC-CP system, with unitary matrices $\mathbf{U}$ and $\mathbf{U}^\dagger$ attached to them.

Discussions on bias removal. At this point it is insightful to relate the above discussions to Sec. 16.3, where we had a detailed discussion of bias removal in MMSE receivers. If the cascade of the channel and precoder is a fixed diagonal matrix, and if the equalizer is also restricted to be diagonal, then removing bias from the MMSE equalizer simply yields the ZF equalizer (Sec. 16.5). The bias-removal multipliers are defined by a diagonal matrix that immediately follows $\mathbf{\Lambda}_e$. For the MC-CP system, the diagonal equalizer $\mathbf{\Lambda}_e$ in Fig. 17.13(b) indeed sees a diagonal precoder-channel cascade, and the MMSE equalizer (with bias removal) is no better than a ZF equalizer. Compare this with an SC-CP system (Fig. 17.12(c)). In this case, the equalizer $\mathbf{G}$ is not diagonal, and furthermore the bias-removal multipliers are not inserted right after the diagonal part $\mathbf{\Lambda}_e$; they are inserted after the IDFT matrix. So, in this case, the bias-removed MMSE system is different from the ZF equalizer. Thus, unlike in the MC-CP case, the SC-CP MMSE system with bias removal does not reduce to a ZF system. $\bigtriangledown\bigtriangledown\bigtriangledown$

17.7 Cyclic-prefix optimization examples

We now present numerical examples which demonstrate the performance of optimal cyclic-prefix systems. We will compare MMSE systems with and without zero-forcing equalizers. For each of these we will consider both systems with orthonormal precoders and general precoders. The MMSE expressions for these cases were summarized in Sec. 15.2 of Chap. 15 and are reproduced below for convenience:

MMSE systems with zero-forcing constraint:

$$\mathcal{E}_{mmse} = \begin{cases} \dfrac{M\sigma_q^2\sigma_s^2}{p_0} \displaystyle\sum_{k=0}^{M-1} \frac{1}{|C[k]|^2} & \text{(orthonormal precoder)} \\ \\ \dfrac{\sigma_q^2\sigma_s^2}{p_0} \left(\displaystyle\sum_{k=0}^{M-1} \frac{1}{|C[k]|}\right)^2 & \text{(unrestricted precoder);} \end{cases} \tag{17.60}$$

MMSE systems without zero-forcing constraint:

$$\mathcal{E}_{mmse} = \begin{cases} \frac{M\sigma_s^2\sigma_q^2}{p_0} \sum_{k=0}^{M-1} \left(\dfrac{1}{\frac{M\sigma_q^2}{p_0} + |C[k]|^2} \right) & \text{(orthonormal precoder)} \\ \\ \dfrac{\sigma_s^2\sigma_q^2 \left(\displaystyle\sum_{k=0}^{K-1} 1/|\widehat{C}[k]|\right)^2}{p_0 + \sigma_q^2 \displaystyle\sum_{k=0}^{K-1} 1/|\widehat{C}[k]|^2} + (M-K)\sigma_s^2 & \text{(unrestricted precoder),} \end{cases} \tag{17.61}$$

where $\widehat{C}[k]$ are the channel DFT coefficients $C[k]$ reordered as in Eq. (17.26), and K is the number of nonzero multipliers q_{kk} in the precoder (see Sec. 13.5). In all these cases we compute the symbol error probability using Eq. (17.36) for the zero-forced system and Eq. (17.38) for the non-zero-forced system. These formulas represent the *minimized* average error probability per symbol (among all orthonormal precoders), achieved by using an extra unitary matrix, which makes the cyclic-prefix system correspond to the *single-carrier* system (see Sec. 17.6.3).

In the examples we assume a real channel and a PAM input. As explained in earlier sections, the minimum-BER system has a real precoder in this case (Sec. 17.4.2). For the special case where the precoder is restricted to be orthonormal, we showed in Sec. 17.6 that the single-carrier system (with identity precoder) is optimal. So the channel input and output are real in all these cases.

Example 17.1.

Consider the 9th order channel with transfer function

$$C(z) = 0.0331 + 0.0894z^{-1} + 0.1407z^{-2} + 0.1522z^{-3} + 0.1050z^{-4} + 0.0827z^{-5}$$
$$+0.0882z^{-6} + 0.1389z^{-7} + 0.1099z^{-8} + 0.0598z^{-9}.$$

We call this *Channel 1.* This has the frequency response magnitude shown in Fig. 17.16 (top and middle). In this example we use 2-bit PAM (i.e., 4-PAM) for the components of the signal $\mathbf{s}(n)$. As explained at the end of Sec. 17.4 the precoder output and channel output can be assumed to be real. The signal and noise variances (real Gaussian noise) are assumed to be

$$\sigma_s^2 = 1 \quad \text{and} \quad \sigma_q^2 = 0.01.$$

The system designed for this channel has $M = 16$ and $L = 9$, so the bandwidth expansion ratio is

$$\gamma = \frac{M+L}{M} = \frac{25}{16} \approx 1.56.$$

Figure 17.17 shows the performance of the four optimal systems described above. The top plot shows the error probability curves as a function of the transmitted power p_0. Actually we plot it against $p_0/M\sigma_q^2$, i.e., the transmitted-power-to-noise ratio per symbol). A number of points should be noticed about this example:

1. The channel frequency response has some deep nulls, as seen from the middle plot.

2. The pure-MMSE system with precoder not constrained to be orthonormal is clearly the best (smallest error probability for fixed power p_0).

3. The system with ZF constraint and orthonormal precoder constraint is the worst. The other two systems have intermediate performaces as expected.

4. The ZF systems approach the pure-MMSE systems as p_0/σ_q^2 increases. For *large powers*, the *ZF constraint* is not a serious loss of generality.

5. The orthonormality constraint makes a difference only at large values of p_0/σ_q^2. For *small powers*, *orthonormality* is not a serious loss of generality.

6. The mean square error (MSE) plots are shown in Fig. 17.17 (middle). The MSE shown is the average error per symbol, that is,

$$\frac{1}{M} E[\, \|\widehat{\mathbf{s}}(n) - \mathbf{s}(n)\|^2],$$

   where M is the size of the vector $\mathbf{s}(n)$. For the ZF-MMSE case note that the MSE is inversely proportional to $p_0/M\sigma_q^2$, as seen from Eq. (17.60). The log-log plot in Fig. 17.17 (middle) is therefore a straight line. For the pure-MMSE case, the MSE is smaller, and it is a more complicated function of $p_0/M\sigma_q^2$ as seen from Eq. (17.61). Note that the MSE of the pure-MMSE system never exceeds σ_s^2, but the MSE of the ZF-MMSE system is not bounded like that. It takes on very large values for low power p_0, as seen from the plots. The error probability of course does not become arbitrarily large (even though the MMSE can be arbitrarily large) because the error probability, by definition, is bounded by unity! Thus, even though the ZF-MMSE system has much larger MSE than the pure-MMSE system for low p_0, both of these systems have unacceptably low error probability for low p_0.

7. The bottom plot in Fig. 17.17 shows the error probability versus channel-output SNR. Note that the ZF-MMSE system appears to be better than the pure-MMSE system for some values of the SNR. This apparent contradiction is explained in Appendix 17.A at the end of this chapter, where we argue that a plot with respect to channel-output SNR can be misleading.

Example 17.2

This example has all parameters identical to Ex. 17.1, except that the channel is taken to be

$$C(z) = 0.1472 + 0.0928z^{-1} + 0.0451z^{-2} + 0.0274z^{-3} + 0.0977z^{-4} + 0.1684z^{-5}$$

$$+0.1667z^{-6} + 0.1288z^{-7} + 0.0390z^{-8} + 0.0870z^{-9}.$$

From the frequency response of this channel (Fig. 17.18) we see that there are nulls in frequency, but these are not as bad as in Ex. 17.1. The error probability and SNR plots of Fig. 17.17 are now replaced with Fig. 17.19. Note that the error probability plots are significantly better. Note also that, at $\omega = 0$, we have $C(e^{j\omega}) = 1$ in this as well as the previous examples. The channel energy is nearly the same in both examples (0.1245 in this example and 0.1126 in the previous one).

Comparsion with distortionless channel. The error probability plots for the channels in the preceding two examples are summarized in the top two plots of Fig. 17.20. The bottom plot shows the performance when $C(z) = 1$. In this case the only effect of the channel is to introduce the additive noise with variance σ_q^2. There is no other distortion that needs to be equalized. The performances of the four different methods in this ideal case are nearly identical. These plots can be regarded as upper bounds on the performance that we can obtain for arbitrary $C(z)$. $\triangledown\triangledown\triangledown$

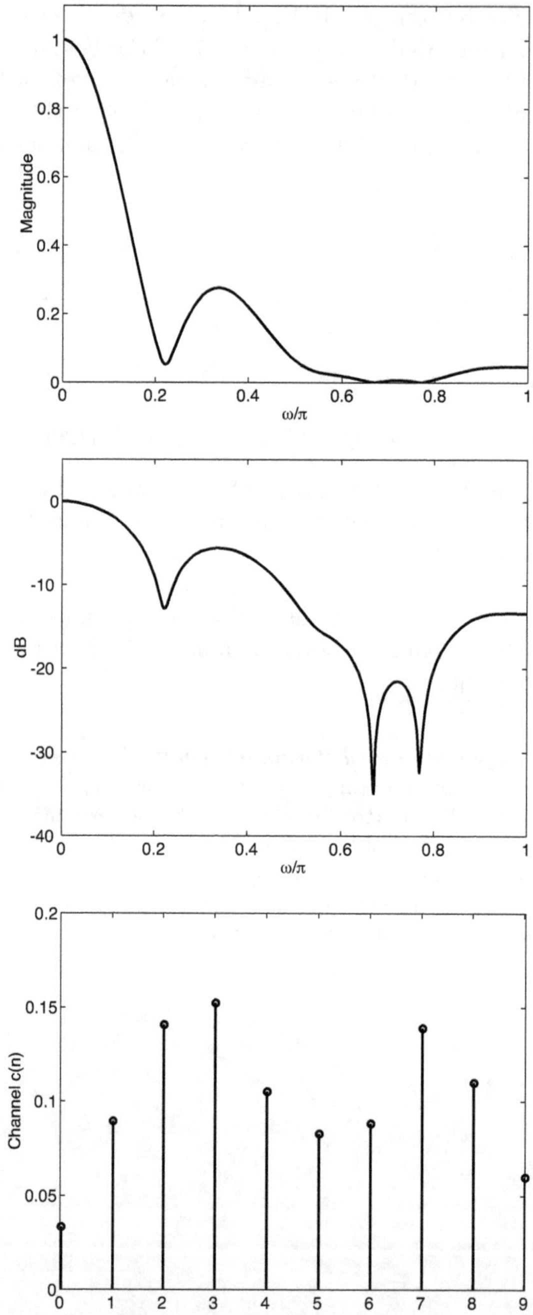

Figure 17.16. Responses of the channel used in Ex. 17.1. Frequency response magnitude (top), frequency response in dB (middle), and impulse response (bottom).

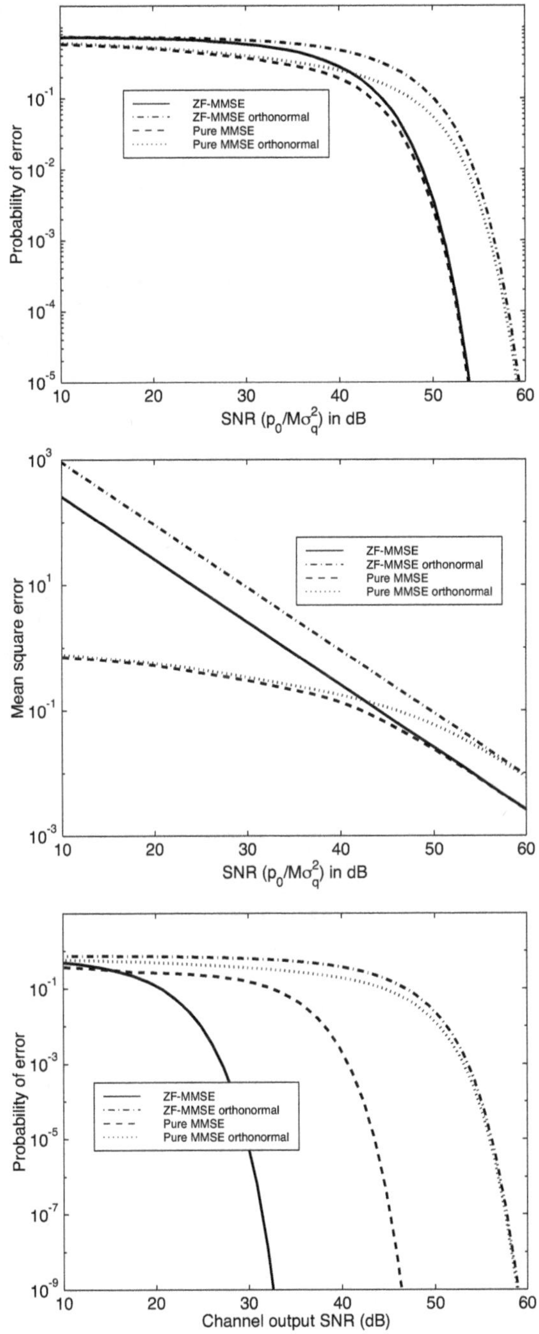

Figure 17.17. Performances of various transceivers in Example 17.1. Error probability plots (top), MSE plots (middle), and error probability verus channel-output SNR (bottom).

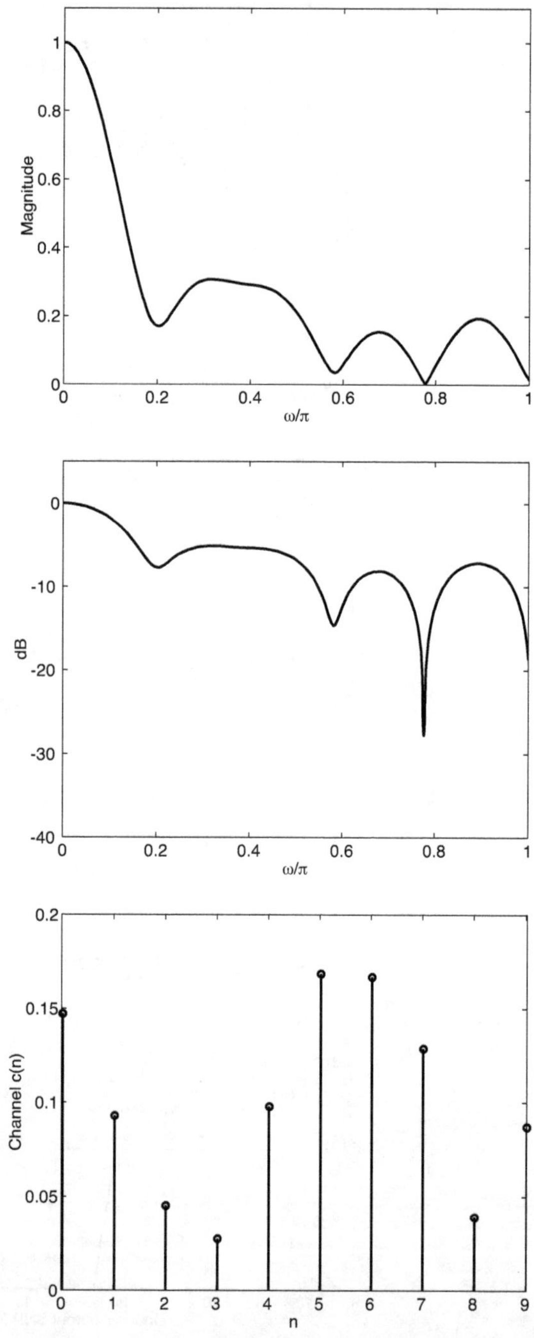

Figure 17.18. Responses of the channel used in Ex. 17.2. Frequency response magnitude (top), frequency response in dB (middle), and impulse response (bottom).

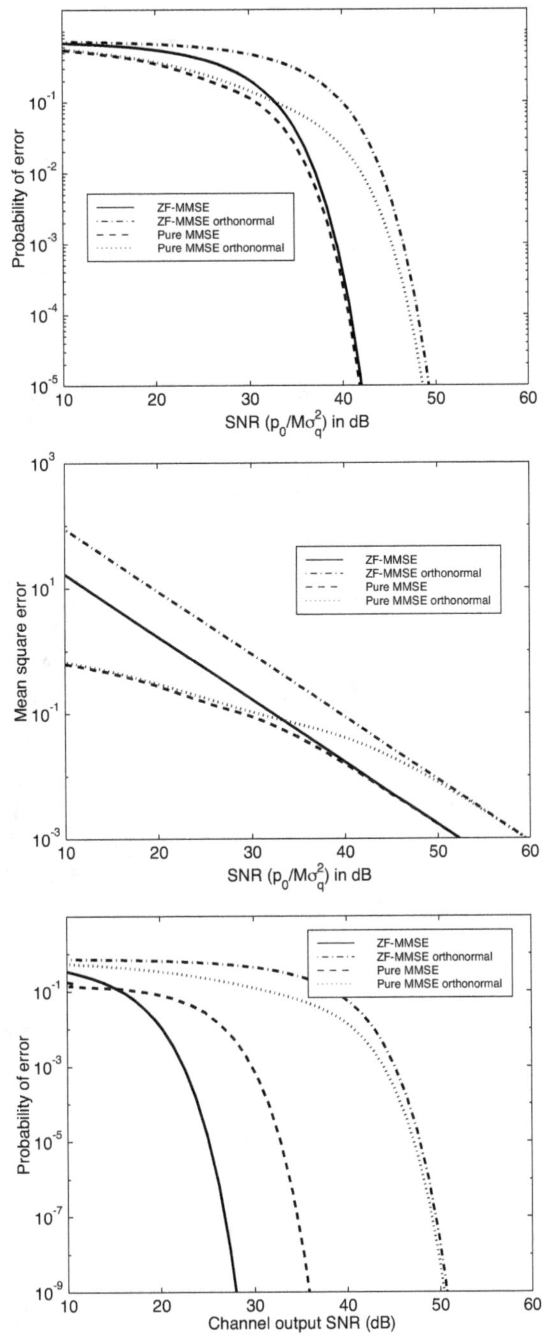

Figure 17.19. Performances of various transceivers in Example 17.2. Error probability plots (top), MSE plots (middle), and error probability verus channel-output SNR (bottom).

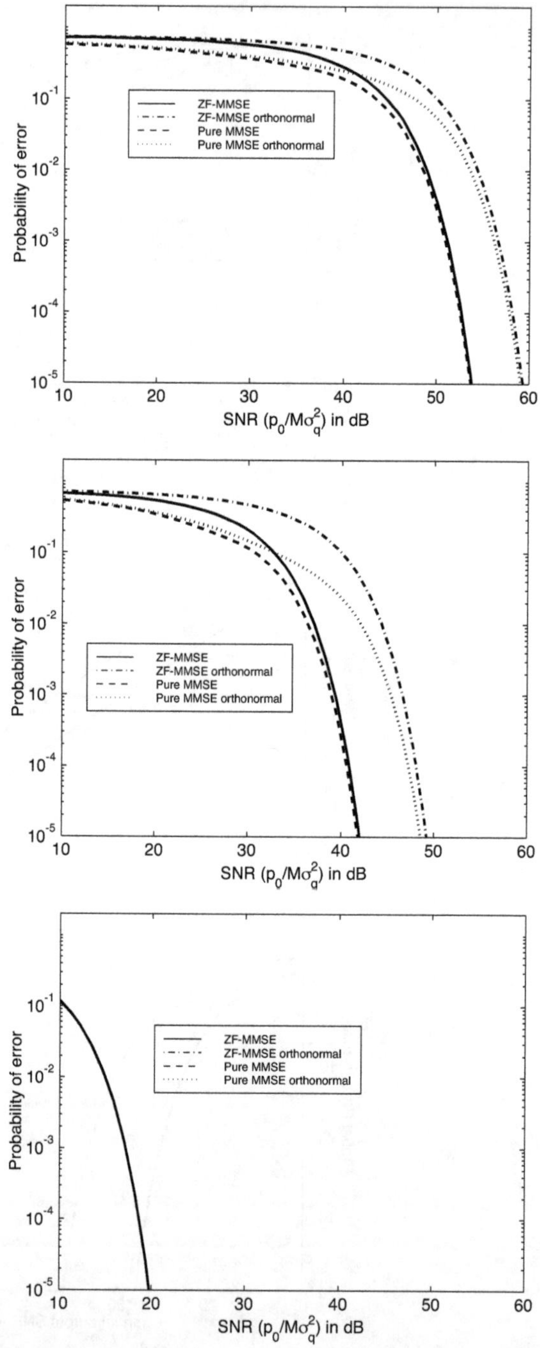

Figure 17.20. Error probability plots for various channels. The channel in Ex. 17.1 (top), the channel in Ex. 17.2 (middle) and the ideal channel $C(z) = 1$ (bottom).

17.8 Increasing the block-size in cyclic-prefix systems

For the channel in Ex. 17.1 we now consider the effect of varying the block size M. The value $M = 16$ was used in Ex. 17.1. Figure 17.21 shows the error probability plots for $M = 17, 18$, and 19. As M increases from 17 to 18 there is a dramatic degradation in performance. When M increases to 19 the performance improves dramatically again. The performances for $M = 16, 17$, and 19 are about the same, whereas the performance for $M = 18$ is very poor.

The reason for this awkward behavior is that the performance of the cyclic-prefix transceiver depends crucially on the magnitude of the M DFT coefficients $C[k]$. These numbers are samples of $C(e^{j\omega})$ in $0 \leq \omega < 2\pi$. If one of the samples gets very close to zero, then there is severe degradation in performance, even for the pure-MMSE equalizers. To demonstrate this effect of channel nulls, Fig. 17.22 shows the plot of $|C[k]|$ on a dB scale for $M = 17, 18$, and 19. Only the first half of the samples are shown, corresponding to $0 \leq \omega \leq \pi$, because of magnitude symmetry (arising from realness of $c(n)$). We see that there are deep nulls for all M, but the null for $M = 18$ is very much deeper (in fact, the sample $C[6] = 0$). This is the reason for the poor performance when $M = 18$. The example shows that the performance, even when the transceiver is optimized using MMSE or optimal error probability criterion without zero forcing, can be very poor if the choice of M is such that the DFT sample location accidently hits a channel null exactly. In practice, of course, the probability of $C(e^{j\omega})$ being exactly zero for any frequency is zero, and this situation arises rarely. But it is not uncommon for $C(e^{j\omega})$ to be very close to zero at some points.

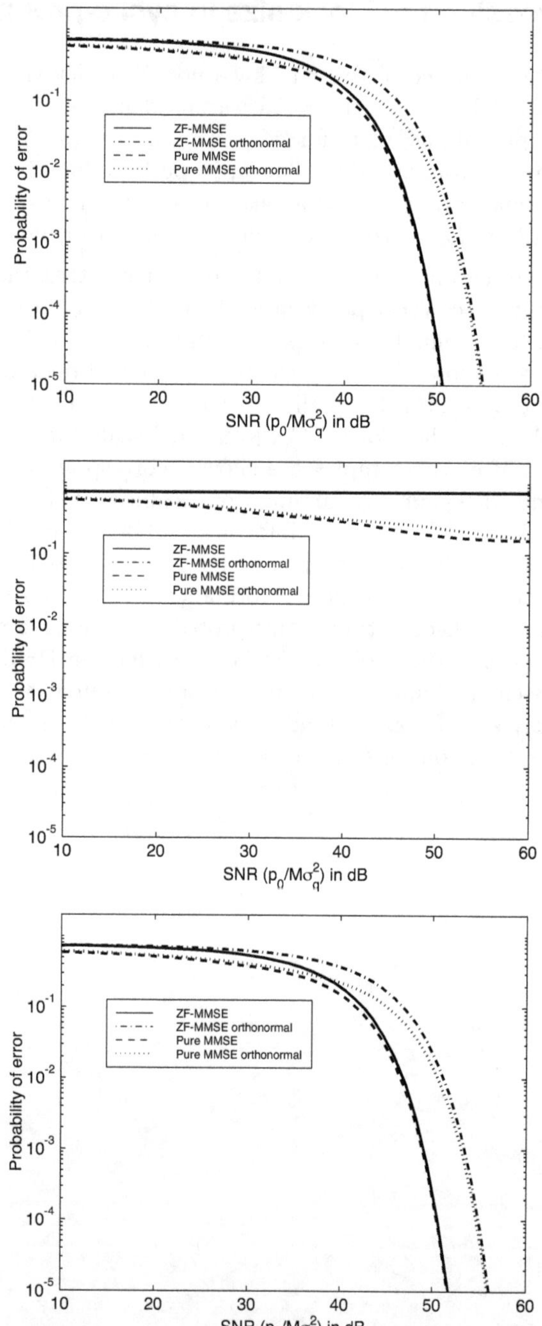

Figure 17.21. Error probability plots for the system in Ex. 17.1 as the block size M is varied. $M = 17$ (top), $M = 18$ (middle), and $M = 19$ (bottom).

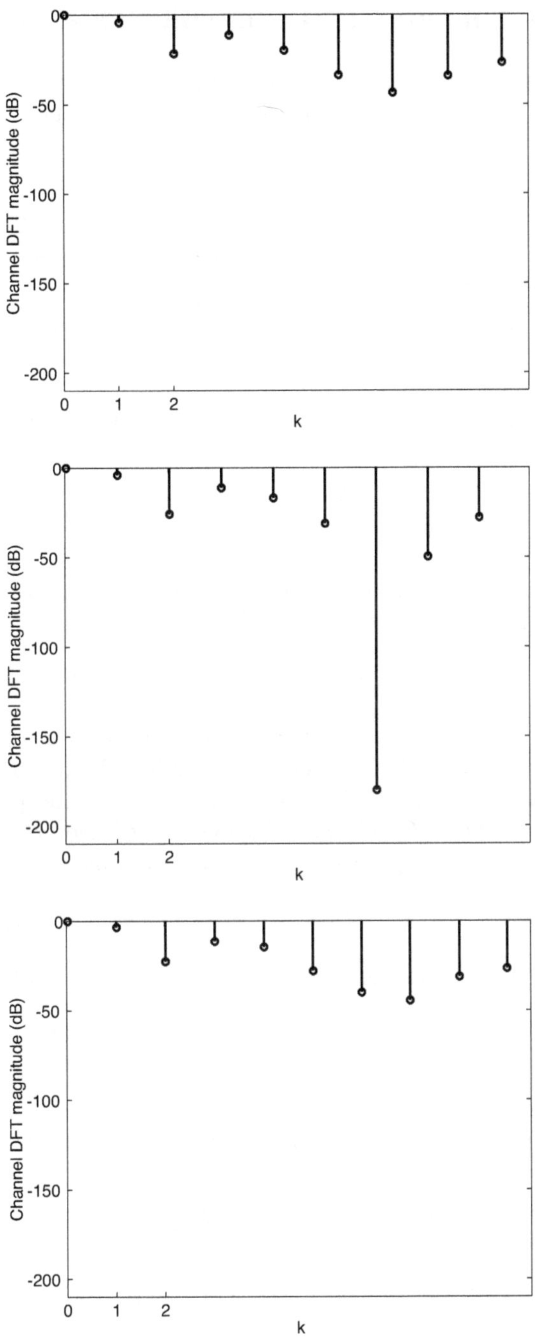

Figure 17.22. Plot of DFT magnitude $|C[k]|$ in dB for the channel of Ex. 17.1, as the block size M is varied. $M = 17$ (top), $M = 18$ (middle), and $M = 19$ (bottom).

17.9 Power minimization using bit allocation

We now demonstrate the optimal DMT design with optimal bit allocation (Sec. 17.5). Consider the channel with a complex coefficient

$$C(z) = 1 - 0.9jz^{-1}$$

with a zero at $0.9j$, rather close to the unit circle. We choose the block size $M = 4$ (any $M > L = 1$ will work). We use a square QAM constellation. For this example we assume $\sigma_q^2 = 0.01$ (complex, circular Gaussian noise), and $b = 8$. The required error probability is assumed to be $\mathcal{P}_e(k) = 10^{-5}$ for all k. With optimal bit allocation the DMT system has symbol power $P_{min}/M = 30.32$. If we do not perform bit allocation and assume $b_k = b$ for all k then the symbol power is $P_{brute}/M = 450.7$. So the **coding gain**, defined in Sec. 14.7 to be the ratio P_{brute}/P_{min}, is given by

$$\frac{P_{brute}}{P_{min}} = 14.86,$$

which corresponds to a gain of 11.7 dB. The optimal bit allocation obtained from the theoretical formula is as follows:

$$b_0 = 9.6260, \; b_1 = 2.1261, \; b_2 = 9.6260, \; b_3 = 10.6220.$$

Since the numbers are not integers, we have to truncate them to integers. Truncation ensures that the error probability is at least as small as the specified values (in this case 10^{-5} for all k), for the same power $P_{min} = 30.32$. In fact, since a square QAM constellation is assumed, we have to truncate b_k to the nearest *even* integer value. So in our example the final allocation is

$$b_0 = 8, \quad b_1 = 2, \quad b_2 = 8, \quad b_3 = 10.$$

With this the average bit rate is

$$\frac{8+2+8+10}{4} = 7,$$

which is smaller than the desired average rate $b = 8$. Thus the process of truncation to even integers has resulted in a loss of bit rate.

Example 17.3: Channel with deep nulls

In this and the following examples we have $b = 8, M = 16$, and $\sigma_q^2 = 0.01$ (complex, circular Gaussian noise). First consider the channel $C(z) = C_0(ze^{-j\omega_0})$, where

$$C_0(z) = 0.1653 + 0.1549z^{-1} + 0.1700z^{-2} + 0.1714z^{-3} + 0.1593z^{-4} + 0.1790z^{-5}$$

and $\omega_0 = 6\pi/M$. The channel $C(z)$ has a complex impulse response, and its frequency response magnitude is shown in Fig. 17.23 (top and middle) for $0 \leq \omega < 2\pi$. The channel has rather deep nulls at several frequencies. The bottom plot in the figure shows the channel input power per symbol as a function of the symbol error probability, which is specified to be identical for all subchannels (i.e., $\mathcal{P}_e(k)$ is identical for all k). The plot shows the minimum power P_{min} of the optimized system as well as the power P_{brute} for the system with no bit allocation ($b_k = b$ for all k), both divided by M to get the per-symbol value. In this example the coding gain is

$$\frac{P_{brute}}{P_{min}} = 19.17,$$

which corresponds to 12.8 dB for any symbol error probability. Thus the separation between the two curves in the bottom plot (which has a log-log axis) is identical for all symbol error probabilities.

Example 17.4: Channel with deeper nulls

In this example all quantities are as in the preceding example, but the channel is different. We take $C(z) = C_0(ze^{-j\omega_0})$, where $\omega_0 = 6\pi/M$ as before, and

$$C_0(z) = 0.1734 + 0.1634z^{-1} + 0.1664z^{-2} + 0.1651z^{-3} + 0.1621z^{-4} + 0.1696z^{-5}.$$

Figure 17.24 shows the channel response and the required channel input power as a function of error probability. In this example the coding gain is

$$\frac{P_{brute}}{P_{min}} = 127,$$

which corresponds to 21 dB. The higher coding gain is a consequence of the fact that the channel nulls in this example are deeper than in Example 17.3. As a result some of the DFT coefficients $C[k]$ have very small values, and the set of numbers $1/|C[k]|^2$ have large variation. This makes their arithmetic to geometric mean ratio (AM/GM) quite large. Recall here that the coding gain due to bit allocation is precisely the AM/GM ratio (Sec. 14.7):

$$\mathcal{G} = \frac{P_{brute}}{P_{min}} = \frac{\dfrac{1}{M}\displaystyle\sum_{k=0}^{M-1} \frac{1}{|C[k]|^2}}{\left(\displaystyle\prod_{k=0}^{M-1} \frac{1}{|C[k]|^2}\right)^{1/M}}$$

The aforementioned variation of $C[k]$ can be seen qualitatively from the following table. Indeed, for $k = 11$ the second channel has a much smaller value of $C[k]$, though for other values of k the two channels have comparable magnitudes.

k	$\|C[k]\|$ Ex. 17.3	$\|C[k]\|$ Ex. 17.4
0	0.1154	0.1243
1	0.3044	0.2996
2	0.7875	0.7863
3	0.9999	1.0000
4	0.7875	0.7863
5	0.3044	0.2996
6	0.1154	0.1243
7	0.2243	0.2383
8	0.0526	0.0682
9	0.1586	0.1430
10	0.1784	0.1690
11	**0.0107**	**0.0038**
12	0.1784	0.1690
13	0.1586	0.1430
14	0.0526	0.0682
15	0.2243	0.2383

Dependence on block size M. For Ex. 17.3, the transmitted power per symbol is shown in Fig. 17.25 as a function of the block size M (with $b = 8$ and $\mathcal{P}_e(k) = 10^{-6}$). It is seen that the power can increase or decrease as M increases. There is no fixed pattern. Notice that the power per symbol is less sensitive to the block size M when there is optimal bit allocation. Thus, if we seek to reduce the **bandwidth expansion factor** owing to cyclic-prefix redundancy by increasing M, this does not seriously increase the power. Figure 17.26 shows the coding gains for the two examples as a function of block size M. Again, the coding gain can increase or decrease with M. Figure 17.27 shows a similar plot, with the value of M restricted to be powers of two.

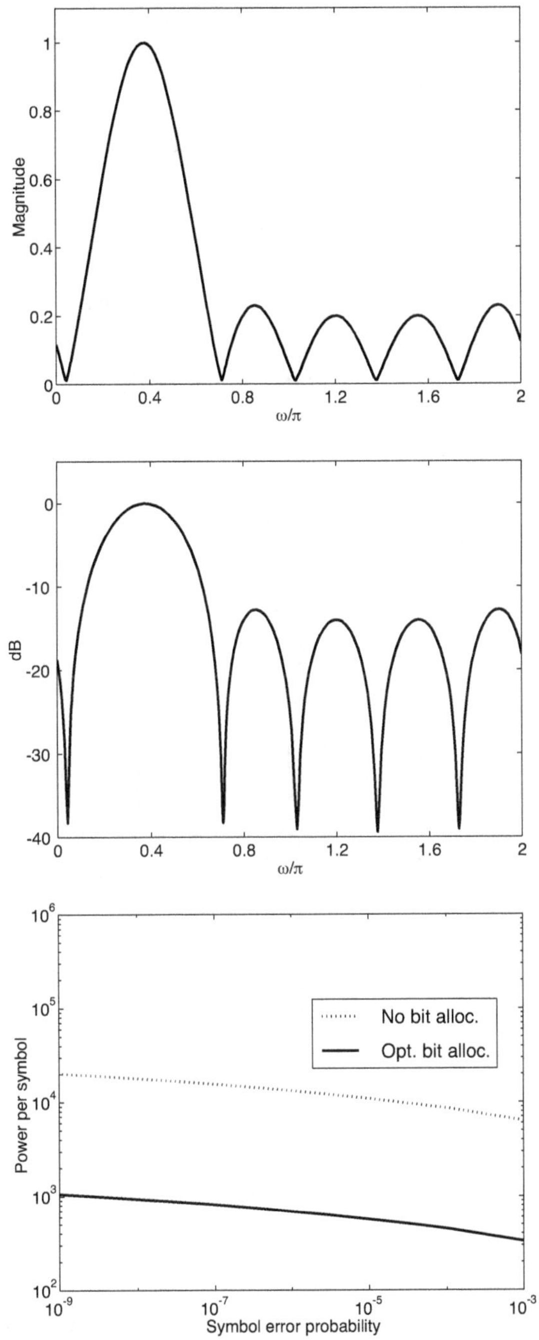

Figure 17.23. Example 17.3. Power minimization using bit allocation in DMT systems. Channel frequency response magnitude (top), channel frequency response in dB (middle), and power per symbol for various bit error rates, for minimum-power design and design with no bit allocation (bottom).

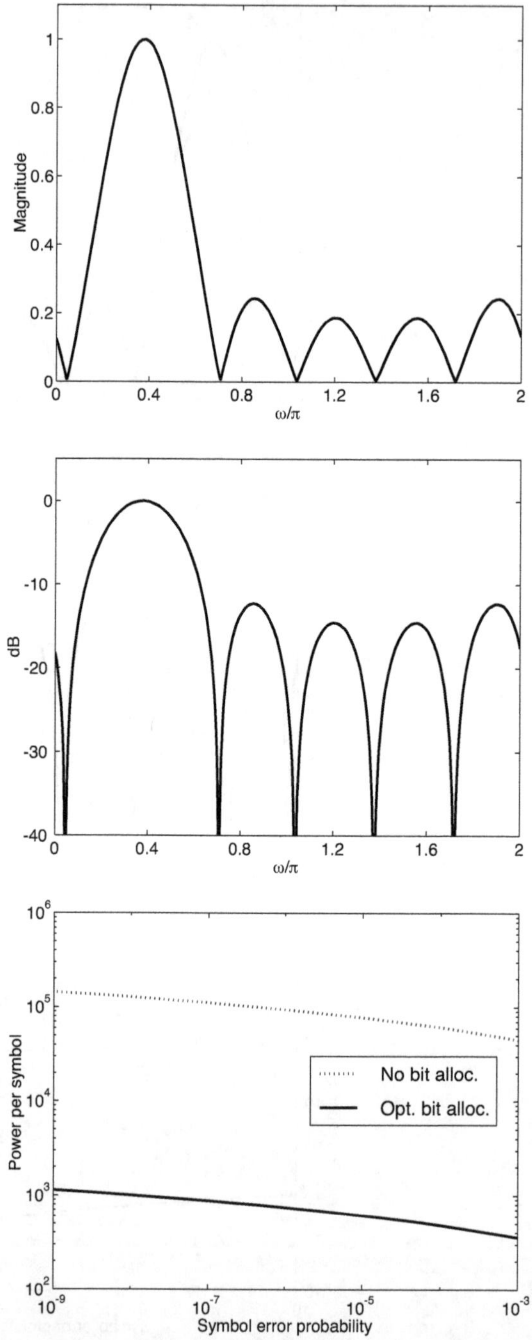

Figure 17.24. Example 17.4. Power minimization using bit allocation in DMT systems. Channel frequency response magnitude (top), channel frequency response in dB (middle), and power per symbol for various bit error rates, for minimum-power design and design with no bit allocation (bottom).

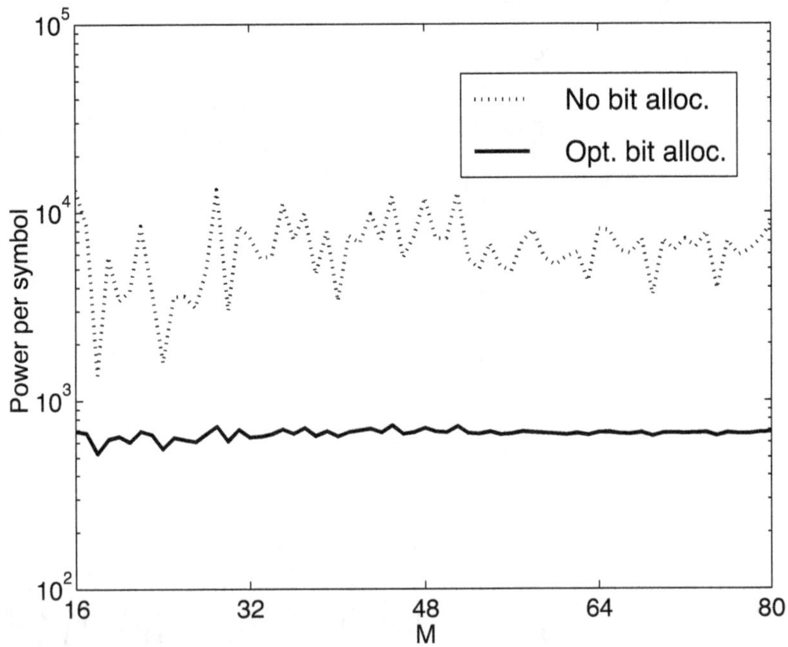

Figure 17.25. Example 17.3. Power minimization using bit allocation in DMT systems. Symbol power for minimum-power design, and design with no bit allocation, for various block sizes M. Note that the power per symbol is less sensitive to the block size M when there is optimal bit allocation.

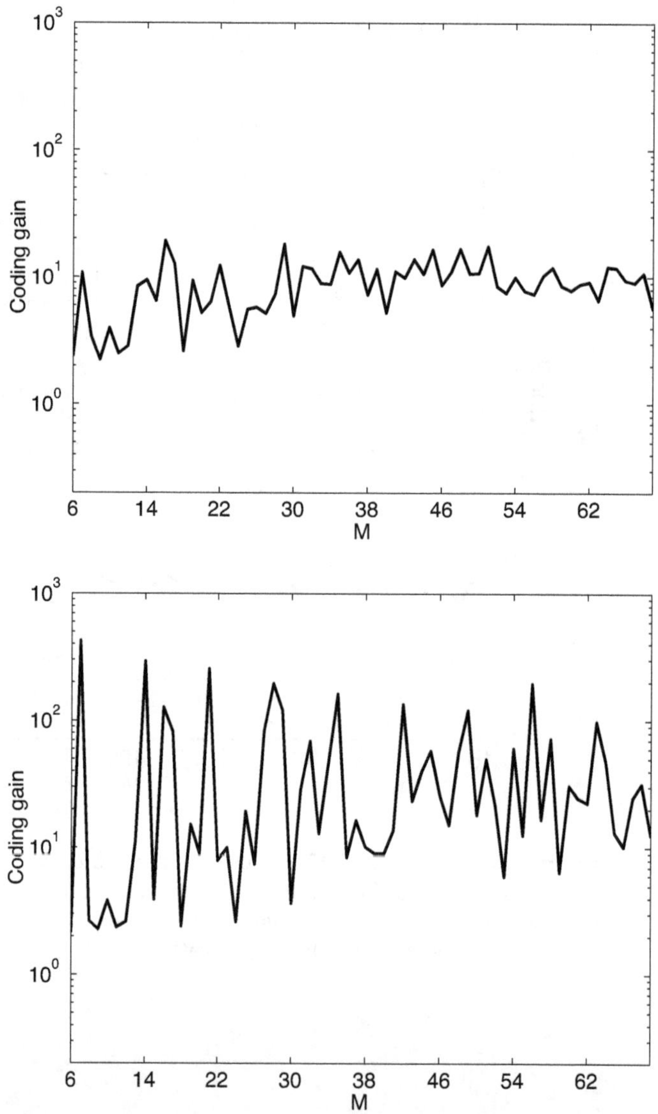

Figure 17.26. Coding gain due to optimal bit allocation for the channel in Ex. 17.3 (top), and the channel in Ex. 17.4 (bottom) as a function of block size M.

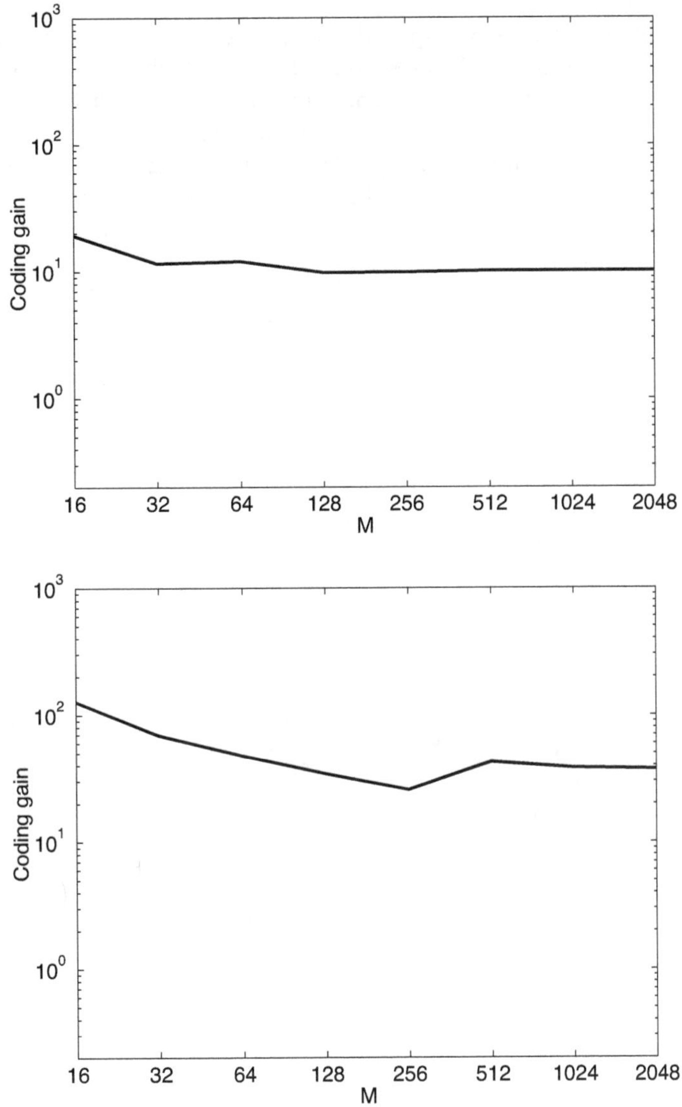

Figure 17.27. Coding gain due to optimal bit allocation for the channel in Ex. 17.3 (top), and the channel in Ex. 17.4 (bottom) as a function of block size M. In this plot M is a power of two.

17.10 Concluding remarks

In this chapter we applied the results of earlier chapters for the case of cyclic-prefix systems. We presented CP systems with minimum MSE and showed how to modify them to obtain minimum error probability. Many simulation examples were presented to demonstrate various features of the optimal systems. In Chap. 18 we will present similar derivations for the case of zero-padded systems. A few one-page summaries in Appendix I are dedicated to summarizing the main results of this chapter.

Appendix to Chapter 17

17.A Error probability versus channel-output SNR

The bottom plot in Fig. 17.17 shows the error probability as a function of channel output SNR. Even though such a plot is sometimes included in discussions it should be warned that comparisons based on this plot have only limited value. Some conclusions can be misleading. For example, the error probability for the pure-MMSE system is "worse" than that of the ZF-MMSE system in some parts of the figure!

The reason is that, for fixed p_0 (channel input power) the channel output power is not the same for all four methods. This is because, the covariance of the signal at the channel input is not identical in all these methods, even though the trace of this covariance is fixed at p_0. Even though the covariance of the input $\mathbf{s}(n)$ is $\sigma_s^2\mathbf{I}$, the above differences in covariance arise because the precoders are different in these systems. To analyze this, consider again the optimal (MMSE) cyclic-prefix system in Fig. 17.7(a). The diagonal matrices $\mathbf{\Lambda}_a$ and $\mathbf{\Lambda}_b$ depend on whether the MMSE system has the zero-forcing property or not. Since $\mathbf{s}(n)$ is assumed to have covariance $\sigma_s^2\mathbf{I}$, the covariance of the channel input is

$$\mathbf{R}_{in} = \frac{\sigma_s^2}{M}\mathbf{W}^\dagger\mathbf{\Lambda}_a\mathbf{\Lambda}_a^*\mathbf{W}.$$

The channel input power, which is the trace of this, is given by

$$p_0 = \frac{\sigma_s^2}{M}\text{Tr}\left(\mathbf{W}^\dagger\mathbf{\Lambda}_a\mathbf{\Lambda}_a^*\mathbf{W}\right) = \frac{\sigma_s^2}{M}\text{Tr}\left(\mathbf{W}\mathbf{W}^\dagger\mathbf{\Lambda}_a\mathbf{\Lambda}_a^*\right) = \sigma_s^2\text{Tr}\left(\mathbf{\Lambda}_a\mathbf{\Lambda}_a^*\right),$$

where we have used $\text{Tr}\,(\mathbf{AB}) = \text{Tr}\,(\mathbf{BA})$, and the fact that $\mathbf{W}\mathbf{W}^\dagger = M\mathbf{I}$. The channel output has covariance $\mathbf{R}_{out} = \mathbf{H}\mathbf{R}_{in}\mathbf{H}^\dagger$, so the channel-output power is

$$p_{out} = \frac{\sigma_s^2}{M}\text{Tr}\left(\mathbf{H}\mathbf{W}^\dagger\mathbf{\Lambda}_a\mathbf{\Lambda}_a^*\mathbf{W}\mathbf{H}^\dagger\right).$$

To simplify this, recall that in cyclic-prefix systems $\mathbf{H}$ is circulant so that

$$\mathbf{H} = \frac{\mathbf{W}^\dagger\mathbf{\Lambda}_c\mathbf{W}}{M}$$

where $\mathbf{\Lambda}_c$ is diagonal. Using $\mathbf{W}^\dagger\mathbf{W} = \mathbf{W}\mathbf{W}^\dagger = M\mathbf{I}$, we therefore obtain $\mathbf{H}\mathbf{W}^\dagger = \mathbf{W}^\dagger\mathbf{\Lambda}_c$, so that

$$\begin{aligned} p_{out} &= \frac{\sigma_s^2}{M}\text{Tr}\left(\mathbf{W}^\dagger\mathbf{\Lambda}_c\mathbf{\Lambda}_a\mathbf{\Lambda}_a^*\mathbf{\Lambda}_c^*\mathbf{W}\right) \\ &= \frac{\sigma_s^2}{M}\text{Tr}\left(\mathbf{W}\mathbf{W}^\dagger\mathbf{\Lambda}_c\mathbf{\Lambda}_a\mathbf{\Lambda}_a^*\mathbf{\Lambda}_c^*\right) = \sigma_s^2\text{Tr}\left(\mathbf{\Lambda}_c\mathbf{\Lambda}_a\mathbf{\Lambda}_a^*\mathbf{\Lambda}_c^*\right). \end{aligned}$$

Summarizing, we have the channel-input power

$$p_0 = \sigma_s^2\text{Tr}\left(\mathbf{\Lambda}_a\mathbf{\Lambda}_a^*\right) = \sigma_s^2\sum_{k=0}^{M-1}|\lambda_{a,k}|^2, \qquad (17.62)$$

and channel-output power

$$p_{out} = \sigma_s^2 \text{Tr}\left(\mathbf{\Lambda}_c \mathbf{\Lambda}_a \mathbf{\Lambda}_a^* \mathbf{\Lambda}_c^*\right) = \sigma_s^2 \sum_{k=0}^{M-1} |\lambda_{a,k}|^2 |C[k]|^2 \tag{17.63}$$

where $\lambda_{a,k}$ are the diagonal elements of $\mathbf{\Lambda}_a$, and $C[k]$ are the diagonal elements of $\mathbf{\Lambda}_c$ (DFT coefficients of the channel $c(n)$). The quantity $\lambda_{a,k}$ in Eqs. (17.62) and (17.63) depends on whether the zero-forcing constraint is used or not. We now consider these special cases:

1. ***MMSE cyclic-prefix system with zero-forcing constraint.*** From Sec. 17.3.1 we know that when ZF is enforced the diagonal matrix $\mathbf{\Lambda}_a$ of the optimal precoder can be taken as

$$\mathbf{\Lambda}_a = \alpha \mathbf{\Lambda}_c^{-1/2}, \tag{17.64}$$

where α is a scale factor used to satisfy the power constraint p_0. Thus

$$p_0 = \alpha^2 \sigma_s^2 \sum_{k=0}^{M-1} \frac{1}{|C[k]|}, \quad p_{out} = \alpha^2 \sigma_s^2 \sum_{k=0}^{M-1} |C[k]|.$$

We can eliminate α^2 in the second expression using the first expression. The result is

$$p_{out} = \frac{p_0 \sum_{k=0}^{M-1} |C[k]|}{\sum_{k=0}^{M-1} \frac{1}{|C[k]|}} \tag{17.65}$$

So the SNR at the channel-output is

$$SNR_{out} = \frac{p_0}{\sigma_q^2} \left(\frac{\sum_{k=0}^{M-1} |C[k]|}{\sum_{k=0}^{M-1} \frac{1}{|C[k]|}} \right). \tag{17.66}$$

One consequence of this expression is that if $C[k]$ is very small for some k then $1/|C[k]|$ in the denominator is very large, so the channel-output power is very small. For example let $C[k_0]$ be such that

$$|C[k_0]| << |C[k]|, \;\; k \neq k_0$$

Then

$$SNR_{out} \approx \frac{p_0}{\sigma_q^2} |C[k_0]| \sum_{k=0}^{M-1} |C[k]|.$$

Thus, even if all other channel DFT coefficients are close to unity, *the presence of one DFT coefficient very close to zero makes the channel output SNR very small.*

2. ***MMSE cyclic-prefix without zero-forcing constraint.*** When there is no ZF constraint the MMSE cyclic prefix-system has

$$\mathbf{\Lambda}_a = \frac{\sigma_q}{\sigma_s} \mathbf{P} \mathbf{\Lambda}_q^{1/2} \mathbf{P}^T,$$

where $\mathbf{P}$ is a permutation matrix, and $\boldsymbol{\Lambda}_q$ has the diagonal elements q_{kk} defined in Eq. (17.22). Thus Eqs. (17.62) and (17.63) become

$$p_0 = \sigma_q^2 \sum_{k=0}^{M-1} |q_{kk}| \quad \text{and} \quad p_{out} = \sigma_q^2 \sum_{k=0}^{M-1} |q_{kk}|\, |\widehat{C}[k]|^2,$$

where $\widehat{C}[k]$ are the permuted versions of the DFT coefficients $C[k]$, and satisfy Eq. (17.26). So the channel output power can be written as

$$p_{out} = \frac{p_0 \sum_{k=0}^{M-1} |q_{kk}|\, |\widehat{C}[k]|^2}{\sum_{k=0}^{M-1} |q_{kk}|}, \tag{17.67}$$

and the channel output SNR is

$$SNR_{out} = \frac{p_0}{\sigma_q^2}\left(\frac{\sum_{k=0}^{M-1} |q_{kk}|\, |\widehat{C}[k]|^2}{\sum_{k=0}^{M-1} |q_{kk}|}\right). \tag{17.68}$$

When some of the $\widehat{C}[k]$ are very small, the quantity q_{kk} will be forced to be zero by the MSE system (see Eq. (17.29)). As a result, the quantity SNR_{out} does not become arbitrarily small as in the ZF-MMSE case.

Summarizing, for a given channel-input power, the channel-output power depends on whether we use the ZF-MMSE system or the pure MMSE system. In the former, the output power can be arbitrarily small if some of the $|C[k]|$ are very small. This shows why the error probability plots in the bottom of Fig. 17.17 give the impression that the ZF-MMSE system is "better" in some parts. In view of these discussions, the error probability versus channel-output SNR plots will be omitted from most discussions.

Orthonormal precoder case. If a precoder $\mathbf{F}$ is orthonormal except for a scale factor (i.e., $\mathbf{F}\mathbf{F}^\dagger = \alpha^2\mathbf{I}$), then the covariance martrix at its output is $\sigma_s^2\mathbf{F}\mathbf{F}^\dagger = \alpha^2\sigma_s^2\mathbf{I}$. Thus the channel-input power, which is the trace of this matrix, is given by

$$p_0 = M\alpha^2\sigma_s^2$$

The channel-output covariance is $\alpha^2\sigma_s^2\mathbf{H}\mathbf{H}^\dagger$, and its trace is the channel-output power:

$$p_{out} = \alpha^2\sigma_s^2 \sum_{k=0}^{M-1} \sigma_{h,k}^2.$$

Thus

$$p_{out} = \frac{p_0}{M} \sum_{k=0}^{M-1} \sigma_{h,k}^2.$$

Details of the unitary matrix $\mathbf{F}$ do not enter this equation. Thus, any transceiver with an orthonormal precoder has the same channel-output power (for fixed channel and fixed input power) regardless of other details of the precoder. So a comparison of error probability versus channel-output SNR is equivalent to a comparison of error probability versus p_0/σ_q^2. $\bigtriangledown \bigtriangledown \bigtriangledown$

Problems

Note. Unless mentioned otherwise assume, $\sigma_s^2 = 1$.

17.1. Consider the fourth order channel

$$C(z) = 0.26 + 0.1962z^{-1} + 0.1738z^{-2} + 0.1738z^{-3} + 0.1962z^{-4}$$

and assume we perform cyclic prefixing with block size $M = 5$. Assume $\sigma_q^2 = 0.01$, and assume the power p_0 is twice the minimum value required to make the transition from $K = M-1$ to $K = M$ in the MMSE solution (Sec. 17.3.2). Compute the MMSE of the optimized cyclic-prefix transceiver for the case where there is no zero forcing. For the same p_0, what is the MMSE if the optimal system has the zero-forcing constraint?

17.2. Consider Problem 17.1 again, but with the channel replaced with

$$C(z) = 0.244 + 0.2091z^{-1} + 0.1689z^{-2} + 0.1689z^{-3} + 0.2091z^{-4}.$$

Assume p_0 is chosen to be the same numerical value as in Problem 17.1. How does the MMSE compare with that in Problem 17.1? (Repeat this with and without zero forcing.)

17.3. Repeat Problem 17.1 for the case where the precoder $\mathbf{F}$ is restricted to be orthonormal (with appropriate scale factor to achieve the same power p_0 as in Problem 17.1).

17.4. In Problem 17.1 the optimal solutions have the form shown in Fig. 17.2(a). What are the diagonal matrices $\mathbf{\Sigma}_f$ and $\mathbf{\Lambda}_g$ in the example under consideration?

17.5. For the channel $C(z)$ in Problem 17.1 consider now the problem of minimizing power as in Sec. 17.5 (optimal transceiver with bit allocation). Assume $b = 4, \mathcal{P}_e(k) = 10^{-6}$, and $\sigma_q^2 = 0.01$. What is the minimized power? What is the minimized power if the channel is the one used in Problem 17.2? Use $L = 4, M = 5$.

17.6. In Fig. 17.21 we plotted the error probabilities for three values of the block size M for a certain channel. For this example, suppose we fix the SNR $p_0/M\sigma_q^2$ at a specific value of 45 dB. Plot the error probability as a function of M for $16 \leq M \leq 32$. Repeat this exercise for SNR $p_0/M\sigma_q^2$ equal to 50 dB and 55 dB. All unspecified parameters are as in Sec. 17.8 (i.e., as in Ex. 17.1). What are your conclusions based on this plot? Is there a specific pattern in these functional dependencies on M?

18

Optimization of zero-padded systems

18.1 Introduction

The zero-padded (ZP) transceiver was discussed at length in Sec. 7.2. In this chapter we consider numerical examples of optimal ZP systems. Performance plots are shown in Sec. 18.2 for the ZP system with minimum symbol error probability (with and without the zero-forcing constraint). We also discuss the effect of the block size M on performance of optimal transceivers (Sec. 18.3). It will be found, unlike in cyclic-prefix systems, that the performance of optimal ZP systems degrades as the block size M is increased. The interest in increasing M arises from the fact that it reduces the bandwidth expansion factor (Sec. 7.2). But in the zero-padded case, there is a clear prize paid in terms of performance as we shall see.

18.2 Zero-padded optimal transceivers

The transceiver with zero-padded precoding is reproduced in Fig. 18.1. Throughout the chapter the channel is assumed to be FIR with order $\leq L$, that is,

$$C(z) = \sum_{n=0}^{L} c(n) z^{-n}. \tag{18.1}$$

With L zeros inserted per block as indicated in the figure, the system shown in Fig. 18.1 is equivalent to Fig. 18.2, where $\mathbf{A}$ is a full-banded Toeplitz matrix:

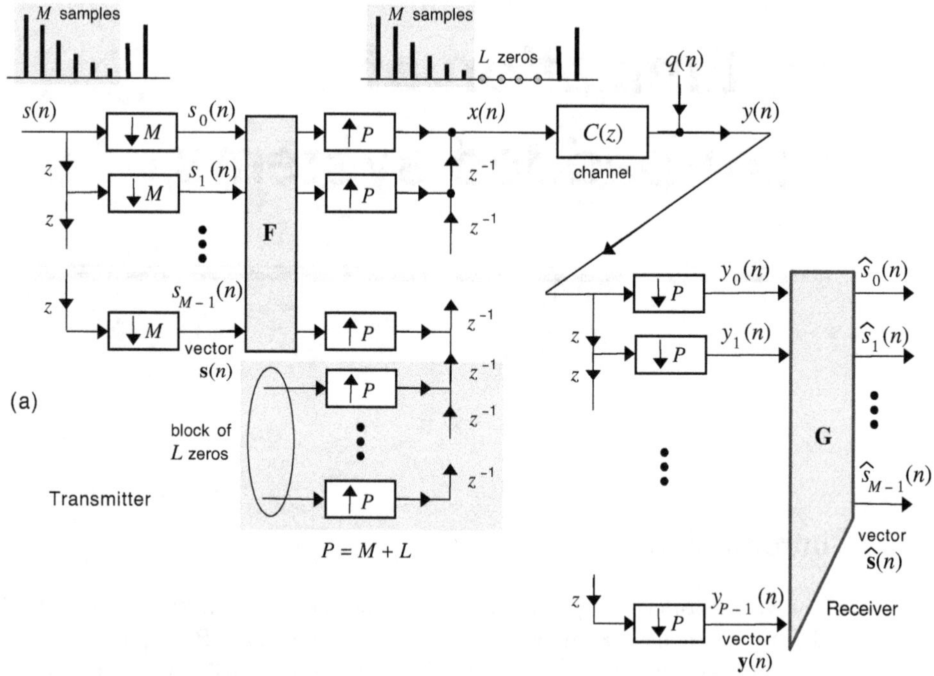

Figure 18.1. The zero-padded transceiver system with a precoder matrix **F** and an equalizer matrix **G**.

$$\mathbf{A} = \begin{bmatrix} c(0) & 0 & \dots & 0 \\ c(1) & c(0) & \dots & 0 \\ \vdots & \vdots & \ddots & \vdots \\ c(L) & & & \\ 0 & c(L) & & \\ \vdots & & \ddots & \vdots \\ 0 & 0 & \dots & c(L) \end{bmatrix}. \tag{18.2}$$

Thus the precoder is an $M \times M$ matrix and the channel is $P \times M$ with $P = M + L$. Given the channel $\mathbf{A}$ we can directly apply the results of Chap. 12 and Chap. 13 to obtain the optimal transceiver matrices $\mathbf{F}$ and $\mathbf{G}$ with or without the zero-forcing constraint. We simply take the channel matrix $\mathbf{H}$ in those chapters to be the matrix $\mathbf{A}$ given above.[1]

[1]The notation $\mathbf{A}$ rather than $\mathbf{H}$ serves as a reminder that this matrix has the banded-

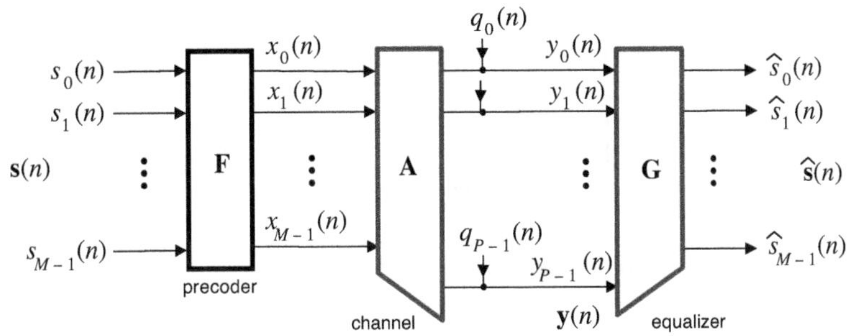

Figure 18.2. The zero-padded transceiver system redrawn in matrix-vector notation.

The expressions for MMSE and symbol error probability derived in earlier chapters can also be applied here directly. We will consider optimal zero-padded precoders of four types. These are

1. Pure MMSE systems,
2. MMSE systems with zero-forcing (ZF-MMSE systems),
3. Pure MMSE systems with orthonormal precoder, and
4. ZF-MMSE systems with orthonormal precoder.

In the orthonormal case, a scalar multiplier α is used at the transmitter to satisfy the power constraint as explained in Chap. 15. It is assumed that extra unitary matrices $\mathbf{U}$ and $\mathbf{U}^\dagger$ are inserted to minimize error probability as in Chap. 16. So the expression for the symbol error probability has the form

$$\mathcal{P}_{e,min} = \begin{cases} c\mathcal{Q}\Big(\dfrac{A}{\sqrt{\mathcal{E}_{ave}}}\Big) & \text{(ZF-MMSE case)} \\ \\ c\mathcal{Q}\left(A\sqrt{\dfrac{1}{\mathcal{E}_{ave}} - \dfrac{1}{\sigma_s^2}}\right) & \text{(pure-MMSE case),} \end{cases} \tag{18.3}$$

where $\mathcal{E}_{ave} = \mathcal{E}_{mmse}/M$. The second expression holds under the assumption that bias has been removed before symbol detection (Sec. 16.3). Here the constants c and A depend on the constellation used. For example, in a b-bit PAM system

$$c = 2(1 - 2^{-b}), \quad \text{and} \quad A = \sqrt{\frac{3\sigma_s^2}{2^{2b} - 1}} \tag{18.4}$$

The expression for the error $\mathcal{E}_{mmse}$ depends on whether we have the zero-forcing constraint or not, and whether the precoder is constrained to be orthonormal or

Toeplitz structure.

not. Repeating the expressions summarized in Chap. 15 we have

$$\mathcal{E}_{mmse} = \begin{cases} \dfrac{M\sigma_q^2\sigma_s^2}{p_0}\displaystyle\sum_{k=0}^{M-1}\frac{1}{\sigma_{h,k}^2} & \text{(orthonormal precoder)} \\[2ex] \dfrac{\sigma_q^2\sigma_s^2}{p_0}\left(\displaystyle\sum_{k=0}^{M-1}\frac{1}{\sigma_{h,k}}\right)^2 & \text{(unrestricted precoder),} \end{cases} \tag{18.5}$$

for the zero-forcing case, and

$$\mathcal{E}_{mmse} = \begin{cases} \frac{M\sigma_s^2\sigma_q^2}{p_0}\sum_{k=0}^{M-1}\left(\dfrac{1}{\frac{M\sigma_q^2}{p_0}+\sigma_{h,k}^2}\right) & \text{(orthonormal precoder)} \\[2ex] \dfrac{\sigma_s^2\sigma_q^2\left(\displaystyle\sum_{k=0}^{K-1}1/\sigma_{h,k}\right)^2}{p_0+\sigma_q^2\displaystyle\sum_{k=0}^{K-1}1/\sigma_{h,k}^2}+(M-K)\sigma_s^2 & \text{(unrestricted precoder),} \end{cases} \tag{18.6}$$

for the non-zero-forcing case. In the last expression K is the number of nonzero multipliers q_{kk} in the precoder (see Sec. 13.5). In all expressions $\sigma_{h,k}$ are the M nonzero singular values of the Toeplitz matrix $\mathbf{A}$ in Eq. (18.2), ordered such that $\sigma_{h,0} \geq \sigma_{h,1} \geq \ldots \geq \sigma_{h,M-1}$.

In the examples we assume that the input is PAM, and that the channel has real coefficients. Even though the equivalent baseband channel is usually complex, one can use a time-domain equalizer first to get a shorter real equivalent channel in the PAM case as mentioned in Sec. 2.A.2 (Appendix of Chap. 2).

Example 18.1:

Consider the channel

$$C(z) = 0.0331 + 0.0894z^{-1} + 0.1407z^{-2} + 0.1522z^{-3} + 0.1050z^{-4} + 0.0827z^{-5}$$
$$+0.0882z^{-6} + 0.1389z^{-7} + 0.1099z^{-8} + 0.0598z^{-9}.$$

The order is $L = 9$. This channel was also used in the cyclic-prefix example in Sec. 17.7. Its frequency response is reproduced in Fig. 18.3. As in the cyclic prefix example we assume 2-bit PAM (i.e., 4-PAM) for the samples $s(n)$. Since the channel and the input are real, all the signals and matrices in Fig. 18.2 are real. The signal and noise variances are assumed to be $\sigma_s^2 = 1$ and $\sigma_q^2 = 0.01$. For each of the four types of optimal zero-padded transceivers, Fig. 18.4 shows the symbol error probability. Three block sizes are shown in the figure, namely $M = 16, 4$, and 1. A number of observations can be made from these plots:

1. Small block sizes yield better performances.[2] The advantage of increasing the block size is that the bandwidth expansion factor gets closer to unity. But this is obtained only at the price of performance. In Sec. 18.3 we will present more detailed examples to further demonstrate this point.

2. The performance of the ZF-MMSE transceiver approaches that of the pure-MMSE transceiver for large power p_0. Thus zero-forcing is not a serious loss of optimality for large powers. This behavior is similar to what we observed in Chap. 17 for optimal cyclic-prefix systems.

3. On the other hand, transceivers without orthonormality constraint on the precoder are significantly better than those with orthonormality, only for large power. This behavior is similar to that of optimal cyclic-prefix systems (Sec. 17.7).

Example 18.2:

The reader will notice that the difference between ZF-MMSE and pure-MMSE systems in Fig. 18.4 is rather negligible. This difference depends on the channel, the block size M, and on the SNR. For example, consider the channel

$$C(z) = \alpha(1 + 3z^{-1} + 3z^{-2} + z^{-3}),$$

where α is such that the channel energy is unity. This channel has three zeros on the unit circle. The error probability plots for $M = 32$ are shown in Fig. 18.5 (2-bit PAM, $\sigma_s^2 = 1$, and $\sigma_q^2 = 0.01$, as before). The systems with orthonormal precoders do not perform well at all. For the systems with unrestricted precoders the difference between ZF-MMSE and pure-MMSE is rather significant, especially for small SNR.

[2]In cyclic-prefix systems the minimum acceptable value of M is the channel order L. But in zero-padded systems, M can be as small as unity. See Ex. 7.1. In fact, even in cyclic-prefix systems, there are some clever ways to reduce M.

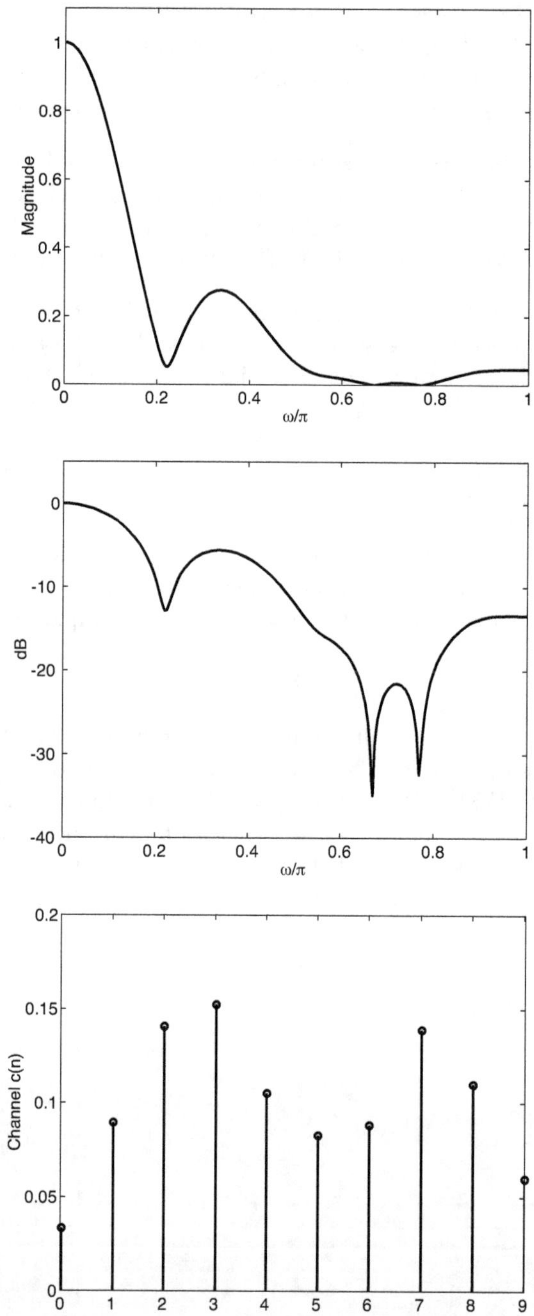

Figure 18.3. Responses of the channel used in Ex. 18.1. Frequency response magnitude (top), frequency response in dB (middle), and impulse response (bottom).

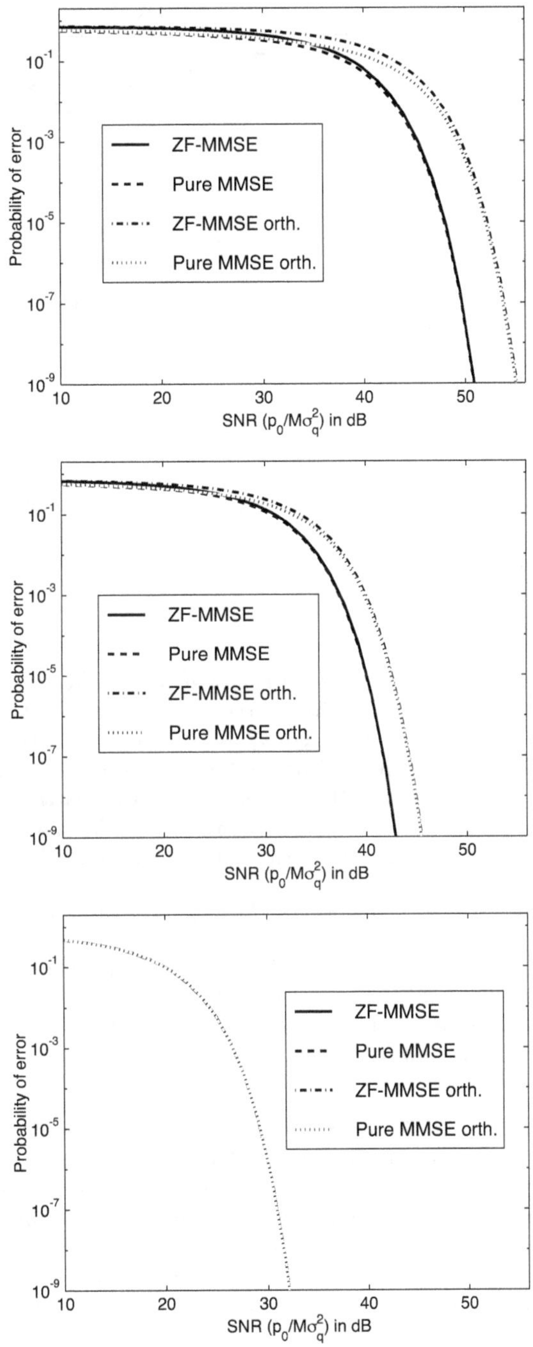

Figure 18.4. Symbol error probability for the channel in Ex. 18.1 with optimal zero padding transceivers, as the block size M is varied. $M = 16$ (top), $M = 4$ (middle), and $M = 1$ (bottom).

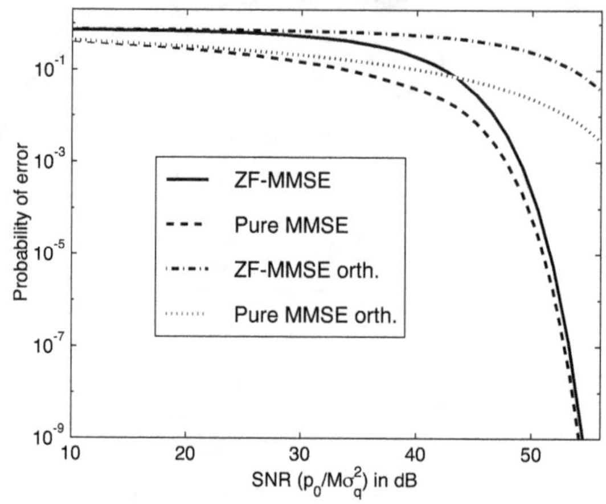

Figure 18.5. Symbol error probability for the channel $C(z) = \alpha(1+3z^{-1}+3z^{-2}+z^{-3})$ in Ex. 18.2. Notice the significant gap between the ZF-MMSE and pure-MMSE systems for small SNR.

Zero-padded optimal systems with block-size $M = 1$. From Fig. 18.4 (bottom plot) we see that the four curves merge into one when the block size $M = 1$. There is no difference between ZF MMSE and pure MMSE, nor is there a difference between orthonormal and arbitrary precoders. To understand why this is the case, observe first that when $M = 1$ the matrix $\mathbf{F}$ reduces to a scalar α whose only role is to control the power. This explains why there is no difference between orthonormal and non-orthonormal precoders. Next, we have

$$P = M + L = 1 + L,$$

so that the transceiver reduces to the form shown in Fig. 18.6. Thus the channel matrix $\mathbf{A}$ is a column vector denoted as $\mathbf{a}$, and the equalizer $\mathbf{G}$ is a row vector denoted as $\mathbf{b}^\dagger$. In the zero-forcing case the transfer function from $s(n)$ to $\widehat{s}(n)$ should be unity, that is,

$$\alpha \mathbf{b}^\dagger \mathbf{a} = 1. \tag{18.7}$$

In the non-ZF case, we insert a bias-removal multiplier before $\widehat{s}(n)$, but it can be absorbed into $\mathbf{b}$. It merely enforces the property $\alpha \mathbf{b}^\dagger \mathbf{a} = 1$ as in the zero-forcing case. Thus the best choice of $\mathbf{b}$ which minimizes the MSE is the same whether we have the ZF constraint or not. More explicitly, the pure-MMSE solution followed by bias removal reduces to the ZF-MMSE solution, for the case $M = 1$ (similar to what we saw in Sec. 16.5). The symbol error rate is therefore identical in these two cases. $\bigtriangledown \bigtriangledown \bigtriangledown$

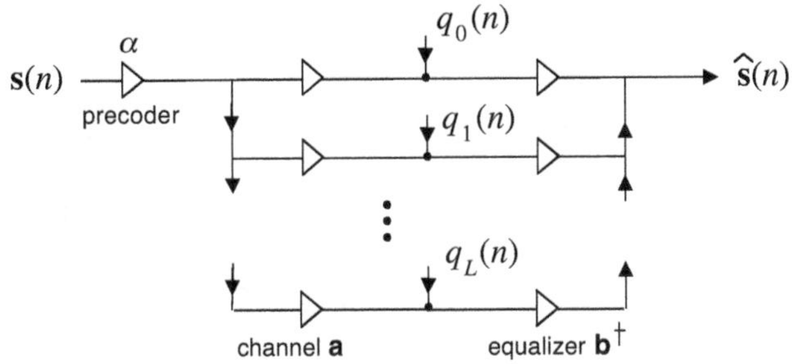

Figure 18.6. The zero-padded transceiver system with $M = 1$.

Example 18.3: Zero padding versus cyclic prefixing

For the channel in Ex. 18.1 we now compare the zero-padded BER-optimal pure MMSE system (with no orthonormal restriction) to the corresponding optimal cyclic-prefix system (Chap. 17). The error probability plots are shown in Fig. 18.7 for three values of the block size M. As in Ex. 18.1 we have chosen 2-bit PAM with $\sigma_s^2 = 1$ and $\sigma_q^2 = 0.01$. For small M we see that ZP system is significantly better, but as M increases the ZP system slowly deteriorates and comes close to the CP system. In the next section we will study in greater detail the performance of the ZP system as a function of the block size.

18.3 Effect of increasing M in zero-padded systems

We now plot symbol error probabilities as a function of block size M for optimal zero-padded transceivers. This is shown in Fig. 18.8 for three examples of channels. All unspecified quantities are as in Ex. 18.1. In all cases we observe that the performance gets worse as M increases. This is true for pure-MMSE as well as ZF-MMSE systems, whether the precoder is constrained to be orthonormal or not. This behavior is unlike the case of optimal cyclic-prefix systems where the error probability may increase or decrease with M depending on where the channel zeros are relative to the DFT zeros (Sec. 17.8). The three channels used for the three plots in Fig. 18.8 are given below. In each case the scale factor β is chosen such that the channel has energy $\sum |c(n)|^2 = 4$. The transmitted power per symbol was fixed: $p_0/M = 1$ in all cases.

1. *Top:* Channel transfer function as in Ex. 18.1:

$$\begin{aligned} C(z) \quad = \quad & \beta(0.0331 + 0.0894z^{-1} + 0.1407z^{-2} + 0.1522z^{-3} + 0.1050z^{-4} \\ & + 0.0827z^{-5} + 0.0882z^{-6} + 0.1389z^{-7} + 0.1099z^{-8} + 0.0598z^{-9}). \end{aligned}$$

   This channel has deep nulls on the unit circle, as seen from the plot of Fig. 18.3 (middle). The error probability deteriorates to an unacceptable value even for moderate values of M.

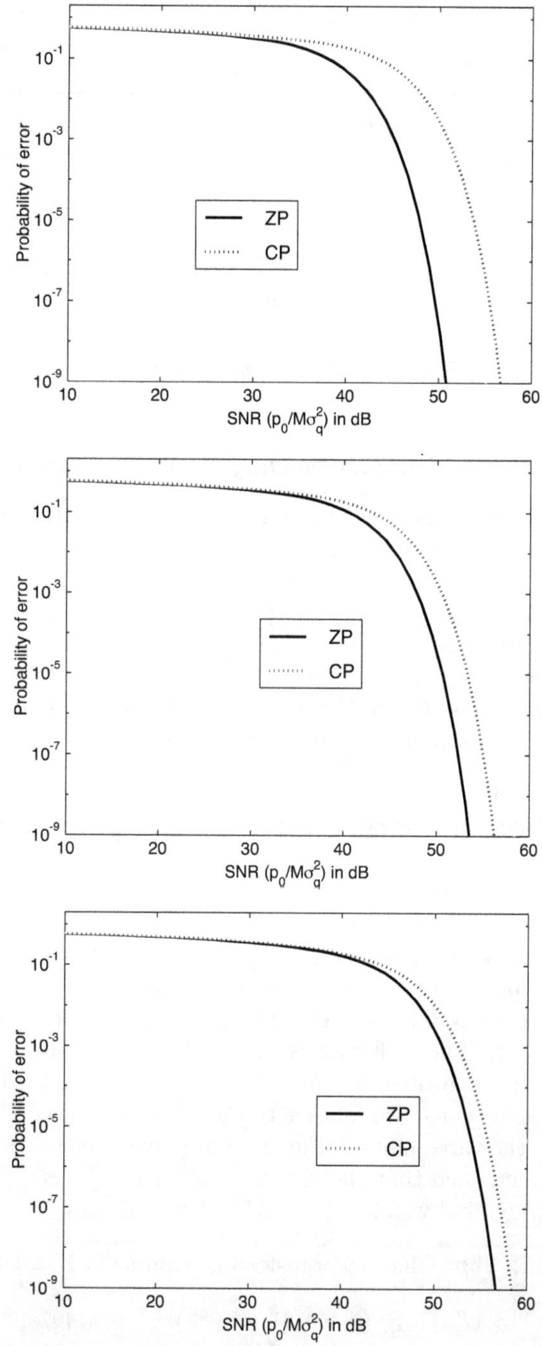

Figure 18.7. Example 18.3. Symbol error probability for the channel in Ex. 18.1 wit optimal zero padding transceiver and optimal cyclic-prefix transceiver, as the block siz M is varied. $M = 16$ (top), $M = 32$ (middle), and $M = 100$ (bottom).

2. *Middle.* $C(z) = \beta(5 + 4z^{-1} + 3z^{-2} + 2z^{-3} + z^{-4})$. This has all its zeros inside the unit circle. The error probability performance is much better than in the top plot.

3. *Bottom.* $C(z) = \beta(1 + 2z^{-1} + 3z^{-2} + 4z^{-3} + 5z^{-4})$. This is the time reversed version of the middle channel, so its zeros are outside the unit circle (two complex conjugate pairs with magnitudes 1.44 and 1.55). The error probability plot is identical to that in the middle plot.

We see that the zero-padded transceiver is not sensitive to whether the zeros are inside or outside the unit circle. This is similar to the behavior of the lazy precoder system described in Chap. 8 (see Sec. 8.5).[3] This behavior is a consequence of the fact that $\mathcal{E}_{mmse}$ depends only on the singular values of the full-banded channel matrix $\mathbf{A}$, which in turn is determined by the magnitude $|C(e^{j\omega})|$, and does not depend on the phase of $C(e^{j\omega})$ (review Sec. 8.5). Notice also that zeros on or close to the unit circle (deep nulls of $C(e^{j\omega})$) result in poor performance. To emphasize the effect of unit circle zeros, the next set of plots (Fig. 18.9) shows the error probability performance as a function of block size M for three channels with unit-circle zeros:

1. *Top:* $C(z) = \beta(1 + z^{-1})$ (single zero at $\omega = \pi$).
2. *Middle:* $C(z) = \beta(1 + z^{-1})^2$ (double zero at $\omega = \pi$).
3. *Bottom:* $C(z) = \beta(1 + z^{-1})^3$ (triple zero at $\omega = \pi$).

The number of padded zeros per block is $L = 1, 2$, and 3, respectively. In each case the scale factor β is chosen such that the channel has energy $\sum |c(n)|^2 = 6$. The transmitted power per symbol was $p_0/M = 1$ in each case. It is clear from the plots that the performance gets worse as we allow more zeros on the unit circle, even for the pure-MMSE system. The performance deteriorates rapidly with M for all the four types of optimal designs.

[3]For cyclic-prefix systems also the same is true, because only the channel DFT coefficients matter. If we replace a channel zero z_k with $1/z_k^*$, the magnitude of the DFT coefficients are unchanged.

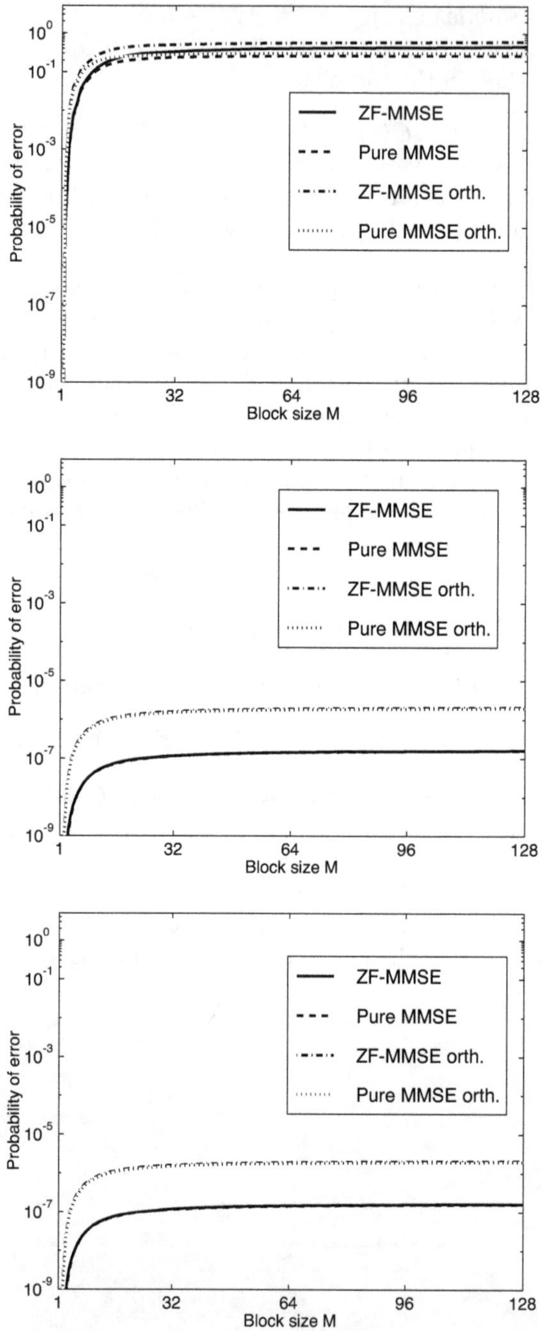

Figure 18.8. Symbol error probability of optimal zero-padded transceivers as a function of block size M for three channels. Channel with deep nulls (top), channel with all zeros inside the unit circle (middle), and channel with all zeros outside the unit circle (bottom).

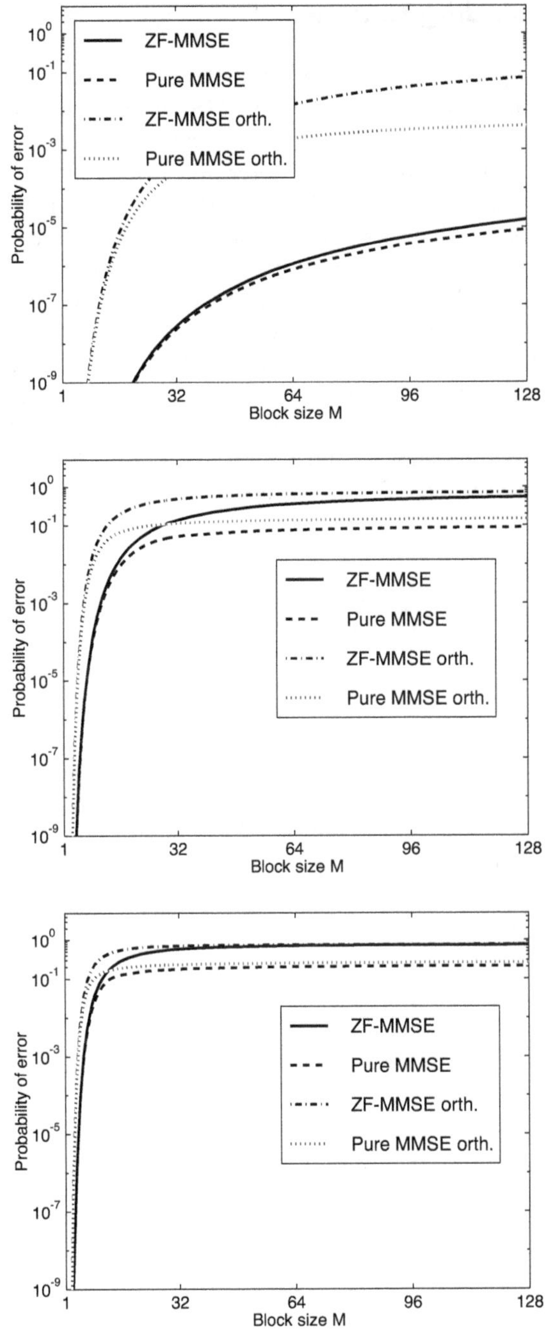

Figure 18.9. Symbol error probability of optimal zero-padded transceivers as a function of block size M for three channels with unit circle zeros. Channel transfer function $C(z) = \beta(1 + z^{-1})$ (top), $C(z) = \beta(1 + z^{-1})^2$ (middle), and $C(z) = \beta(1 + z^{-1})^3$ (bottom).

18.4 Concluding remarks

In Chap. 8, where we discussed the lazy precoder with a zero-forcing equalizer, we proved theoretically that the MSE with a zero-forcing equalizer increases with the block size M. This proof was based on intricate properties of the Toeplitz matrix that represents the deterministic autocorrelation of the channel (see Appendix 8.A at the end of Chap. 8). In this chapter we have demonstrated with examples that the symbol error probability increases with block size M for all four types of optimal systems, namely MMSE transceivers with and without zero forcing, and with and without orthonormality constraint on the precoder. This behavior is unlike the behavior observed for optimal cyclic-prefix systems. For those systems we showed in Sec. 17.8 that the error probability of the optimal system may increase or decrease with M. The monotone decrease in the performance of the zero-padded optimal transceivers (all four types) with increasing block size has not yet been theoretically explained. Except for the special case discussed in Sec. 8.6, the proof remains an interesting open problem.

Problems

Note. Unless mentioned otherwise, assume $\sigma_s^2 = 1$.

18.1. Consider again the channel in Problem 17.1, but assume that we do zero padding instead of cyclic prefixing. With $L = 4$ and $M = 5$, and with all parameters (including the numerical value of power p_0) as in Problem 17.1, compute the MMSE values (Sec. 18.2) with and without zero forcing. How do these values compare with what you obtain for the cyclic-prefix system of Problem 17.1? Assume the precoder is not restricted to be orthonormal.

18.2. Repeat Problem 18.1 for the channel in Problem 17.2.

18.3. Repeat Problem 18.1 for the case where the zero-padded system has its precoder restricted to be orthonormal (but scaled appropriately to satisfy the power constraint).

18.4. Figure 18.9 showed the error probability as a function of block size for three examples of channels which use zero padding. With all parameters as described in Sec. 18.3 for Fig. 18.9, repeat these plots for the modified channels $C(z) = \beta(1 + \rho z^{-1})$, $C(z) = \beta(1 + \rho z^{-1})^2$, and $C(z) = \beta(1 + \rho z^{-1})^3$, for $\rho = 0.99$. Repeat for $\rho = 0.9$. In each case assume β is such that the channel energy $\sum |c(n)|^2 = 6$ as in Sec. 18.3. Explain qualitatively how and why the behavior is different from Fig. 18.9.

18.5. Repeat Problem 18.4 for $C(z) = \beta(1 + \rho z^{-1})$, where $\rho = 1/0.9$ and β such that $\sum |c(n)|^2 = 6$. How does this compare with the case $\rho = 0.9$ (with β adjusted again such that $\sum |c(n)|^2 = 6$)?

19

Transceivers with decision feedback equalizers

19.1 Introduction

In Sec. 5.7 we described decision feedback equalizers (DFEs). These equalizers use past decisions on symbols to refine the decision on the present symbol. In this chapter we extend this idea to decision feedback equalization in memoryless MIMO channels. Here past decisions within a block are used to improve future decisions in that block, at a given block-time instant. The basic idea is introduced in Sec. 19.2. The decision feedback transceiver has a precoder $\mathbf{F}$, feedforward equalizer $\mathbf{G}$, and a feedback matrix $\mathbf{B}$. In Sec. 19.3 we show how these matrices can be jointly optimized to minimize the mean square reconstruction error, under the zero-forcing constraint. We will see that in DFE systems the minimized mean square error depends on the geometric mean of $1/\sigma_{h,k}$ rather than on their arithmetic mean (where $\sigma_{h,k}$ are the channel singular values). This transition from AM to GM is a fundamental property of DFE systems, and results in substantial decrease of the mean square error. In Sec. 19.4 we extend the joint optimization to the case where there is no zero forcing. DFE systems with minimum symbol error rate are presented in Sec. 19.5. Error probability plots are presented in Sec. 19.6, showing the superiority of DFE transceivers over linear transceivers. It turns out that optimal DFE systems are related to systems optimizing mutual information; this is elaborated in Sec. 19.7. A brief description of DFE receivers based on QR decomposition, and the VBLAST receiver are given in Sec. 19.8. This chapter is based mostly on the work of Zhang *et al.* [2005], Jiang *et al.* [2005b, 2005c], and Xu *et al.* [2006]. A history of DFE systems was given at the end of Sec. 5.7 which the reader may want to review at this time.

19.2 Fundamentals of decision feedback equalizers

First let us review the basic idea behind time-domain decision feedback. Figure

19.1 shows the digital communication system described earlier in Chap. 2, with one modification. The detector output $s_{est}(n)$, which is the decision about the transmitted symbol $s(n)$, is delayed and fed back through a digital filter to modify the value of the detector input $\widehat{s}(n)$. As explained in Sec. 5.7, this receiver is called a decision feedback receiver, or a *decision feedback equalizer*, abbreviated as **DFE**. Here the filter $B(z)$ has the form

$$B(z) = b_1 z^{-1} + b_2 z^{-2} + \ldots + b_N z^{-N}. \tag{19.1}$$

Since $b_0 = 0$, there is no delay-free loop. That is, the detector input $\widehat{s}(n)$ at time n is affected only by the **past** detected symbols or past decisions

$$s_{est}(n-1), s_{est}(n-2), \ldots, s_{est}(n-N). \tag{19.2}$$

Figure 19.2 shows the detailed structure of the DFE system for the case of a third order feedback filter $B(z)$.

The basic idea behind DFE is that if the past symbols have been identified with low probability of error, then those decisions can be successfully used to improve the detection performance of the present symbol. This idea works when there is a sufficient amount of statistical dependence between the received signal samples $r(n)$. Even though the transmitted symbol stream $s(n)$ may have independent samples, such a dependence can arise in $r(n)$ owing to the filtering effect of the cascaded system $G(j\omega)H(j\omega)F(j\omega)$. Since the DFE system involves a nonlinear device (the detector) in the feedback loop, the equalizer is **nonlinear**, and we say that we have a nonlinear transceiver. The transceivers considered in earlier sections, which do not contain DFEs, are called linear transceivers; more specifically, they have linear equalizers.

The performance of the DFE system depends on the mean square error at the detector input, namely

$$\mathcal{E} = E|\widehat{s}(n) - s(n)|^2. \tag{19.3}$$

For fixed $F(j\omega), H(j\omega)$, and $G(j\omega)$, this error depends on how the feedback filter $B(z)$ is chosen. The mean square error can be estimated theoretically under the assumption that the past decisions have been perfect, that is, the decisions fed back via $B(z)$ are correct. More specifically, one assumes that

$$s_{est}(n-k) = s(n-k), \quad 1 \le k \le N, \tag{19.4}$$

and proceeds with the theoretical analysis to compute $\mathcal{E}$. Under the preceding assumption of **perfect past decisions**, it is in fact possible to optimize the prefilter $F(j\omega)$, equalizer $G(j\omega)$, and feedback filter $B(z)$ such that the mean square error $\mathcal{E}$ is minimized. The details of such analysis can be found in Proakis [1995] and references therein (also see the end of Sec. 5.7 for historical remarks). When the error probabilities are small, the "perfect past decisions" assumption is quite reasonable, but when the receiver operates in the range of larger error probabilities, such assumptions tend to be unsatisfactory. The error in the decision $s_{est}(n)$ tends to "propagate" via the feedback loop, and many future decisions can be affected this way.

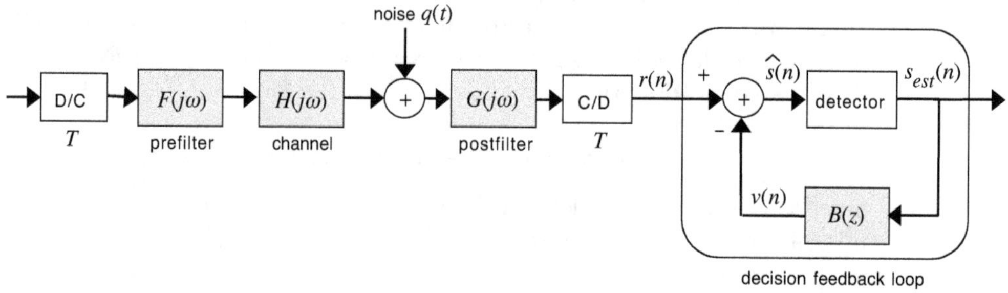

Figure 19.1. Transceiver with a time-domain decision feedback equalizer at the receiver.

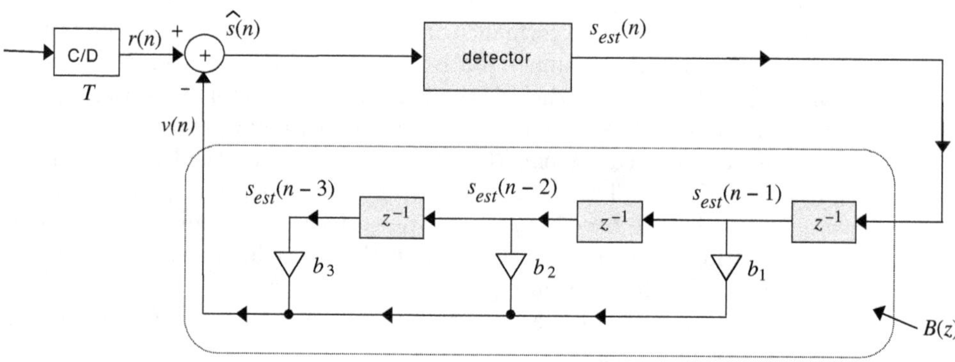

Figure 19.2. Detailed structure of the decision feedback equalizer for third order feedback filter $B(z)$.

It has been amply demonstrated that the use of DFE results in considerable reduction in the mean square error $\mathcal{E}$ compared to transceivers based on linear equalization only. For sufficiently large order N of the feedback filter, the performance of the DFE is much better than linear MMSE systems, and sometimes comes close to that of the maximum likelihood detector (MLD, see Sec. 5.5). Thus the DFE system is a very practical system which can perform significantly better than linear MMSE systems, and has much lower complexity than the theoretically optimal MLD system. For further details the reader is referred to Proakis [1995] and references therein.

19.2.1 Decision feedback within a block

Historically, the original form of decision feedback makes use of the past estimated symbols in time to correct the present estimate. A similar idea can be implemented sequentially within a block in a block-based communication system such as, for example, the zero-padded or the cyclic-prefix system. The general form of the system is shown in Fig. 19.3. Comparing this with the linear transceiver of Fig. 16.1, we see that the only new element here is the matrix $\mathbf{B}$, called the **feedback matrix**, which provides a feedback path from the detected symbols $\mathbf{s}_{est}(n)$ to the *input* of the detector. The matrix $\mathbf{G}$ will henceforth be called the **feedforward matrix.** The matrix $\mathbf{B}$ is chosen to be *strictly upper triangular*, that is, upper triangular with zeros on the main diagonal, so that the signal $\mathbf{v}$ is related to $\mathbf{s}_{est}$ as follows:

$$\underbrace{\begin{bmatrix} v_0 \\ v_1 \\ \vdots \\ v_{M-2} \\ v_{M-1} \end{bmatrix}}_{\mathbf{v}} = \begin{bmatrix} 0 & b_{01} & b_{02} & \cdots & b_{0,M-1} \\ 0 & 0 & b_{12} & \cdots & b_{1,M-1} \\ \vdots & \vdots & \vdots & \ddots & \vdots \\ 0 & 0 & 0 & \cdots & b_{M-2,M-1} \\ 0 & 0 & 0 & \cdots & 0 \end{bmatrix} \underbrace{\begin{bmatrix} s_{0,est} \\ s_{1,est} \\ \vdots \\ s_{M-2,est} \\ s_{M-1,est} \end{bmatrix}}_{\mathbf{s}_{est}}. \qquad (19.5)$$

For example if $M = 3$ we have

$$\begin{bmatrix} v_0 \\ v_1 \\ v_2 \end{bmatrix} = \begin{bmatrix} 0 & b_{01} & b_{02} \\ 0 & 0 & b_{12} \\ 0 & 0 & 0 \end{bmatrix} \begin{bmatrix} s_{0,est} \\ s_{1,est} \\ s_{2,est} \end{bmatrix}, \qquad (19.6)$$

and the details of the feedback part are shown in Fig. 19.4. In this case $v_2 = 0$. The generation of v_1 and v_0 are detailed in the figure. Though it is not indicated in the figure, the operation of the structure should be interpreted **sequentially**, from bottom to top, as follows:

1. First, the bottom-most detector input $\widehat{s}_2(n)$ is used to obtain the estimate $s_{2,est}(n)$ of $s_2(n)$ directly.

2. This estimate $s_{2,est}(n)$ is then used to correct the input $\widehat{s}_1(n)$ to the next detector, and the estimate $s_{1,est}(n)$ of $s_1(n)$ is produced.

3. The estimates $s_{2,est}(n)$ and $s_{1,est}(n)$ are then used to correct the input $\widehat{s}_0(n)$ to the next detector (topmost in this case). This detector then produces the estimate $s_{0,est}(n)$ of $s_0(n)$.

Summarizing, we start from the bottom-most signal and proceed to the top. When a symbol estimate $s_{k,est}(n)$ is produced, it is used (along with the estimates $s_{k+1,est}(n), \ldots, s_{M-1,est}(n)$) to correct the signal $\widehat{s}_{k-1}(n)$ and produce the estimate $s_{k-1,est}(n)$. Thus, even though the feedback structure in Figs. 19.3 and 19.4 appear to have delay-free loops, the actual operation (M steps described above) follows sequentially within one block-processing interval. It is clear that the strict upper triangular nature of the matrix $\mathbf{B}$ is crucial for the successful operation of this receiver structure.

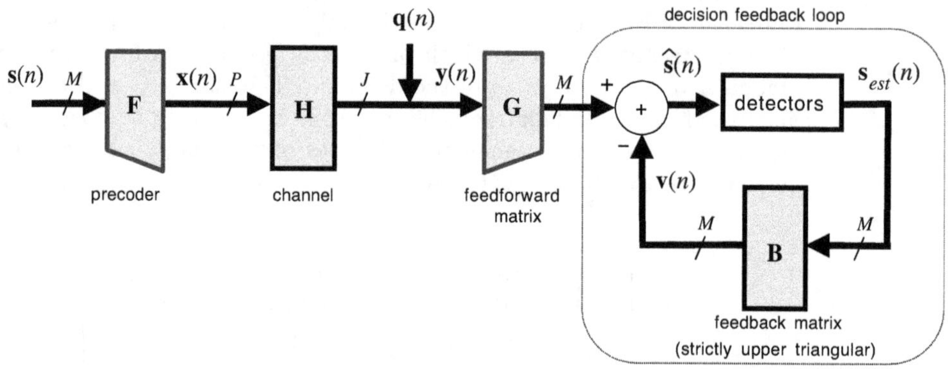

Figure 19.3. Transceiver with intrablock decision feedback equalization.

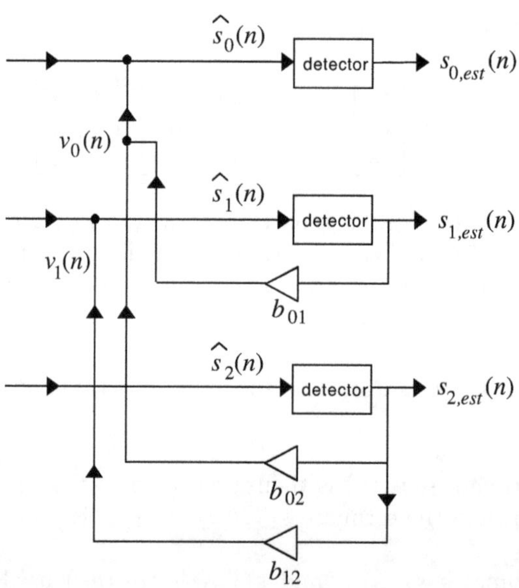

Figure 19.4. Structure implementing the intrablock decision feedback system described in Eq. (19.6).

Decision feedback performed within a block can be regarded as **vertical** decision feedback, or **space-domain** decision feedback (because the symbols in a block are shown vertically in figures), or **intrablock decision feedback** to distinguish from Fig. 19.1, which shows **time-domain** decision feedback. Some sophisticated variations of the space-domain DFE include the Bell Labs V-BLAST system and the so-called QRS decomposition system, to be described later. As in the case of time-domain DFEs, the blockwise DFE also suffers from error propagation when the receiver operates in the region of medium or large error rates. However, the error propagation is limited to one block, and at most $M-1$ symbols in that block are affected. Notice that the argument n in Fig. 19.3 refers to a block-processing instant of time.

19.2.2 Equivalent structure under no error propagation

Figure 19.5 shows an equivalent schematic drawing of the DFE system under the assumption that there is no error propagation. This figure is obtained simply by setting $\mathbf{s}_{est} = \mathbf{s}$ in Fig. 19.3. The input to the detector is therefore given by

$$\widehat{\mathbf{s}}(n) = \mathbf{GHFs}(n) - \mathbf{Bs}(n) + \mathbf{Gq}(n). \tag{19.7}$$

The reconstruction error at the input of the detector is

$$\mathbf{e}(n) = \widehat{\mathbf{s}}(n) - \mathbf{s}(n), \tag{19.8}$$

so that

$$\mathbf{e}(n) = \mathbf{GHFs}(n) - \mathbf{Bs}(n) - \mathbf{s}(n) + \mathbf{Gq}(n). \tag{19.9}$$

This can be rewritten as

$$\mathbf{e}(n) = \Big(\mathbf{GHF} - \mathbf{B} - \mathbf{I}\Big)\mathbf{s}(n) + \mathbf{Gq}(n). \tag{19.10}$$

Assuming that the input symbol stream $\mathbf{s}(n)$ and the noise $\mathbf{q}(n)$ are zero-mean and uncorrelated, the error covariance matrix is therefore

$$\mathbf{R_{ee}} = \sigma_s^2\Big(\mathbf{GHF} - \mathbf{B} - \mathbf{I}\Big)\Big(\mathbf{GHF} - \mathbf{B} - \mathbf{I}\Big)^{\dagger} + \sigma_q^2\mathbf{GG}^{\dagger}, \tag{19.11}$$

where it is assumed that the signal and noise covariances are

$$\mathbf{R_{ss}} = \sigma_s^2\mathbf{I}_M, \quad \mathbf{R_{qq}} = \sigma_q^2\mathbf{I}_J, \tag{19.12}$$

respectively.

19.3 The optimal DFE system with zero forcing

We now consider the optimization of the DFE system with zero forcing. From Eq. (19.7) we see that the transfer function from $\mathbf{s}(n)$ to $\widehat{\mathbf{s}}(n)$ in absence of noise $\mathbf{q}(n)$ is

$$\mathbf{GHF} - \mathbf{B},$$

and the zero-forcing (ZF) condition means that this should be forced to identity:

$$\mathbf{GHF} = \mathbf{I}_M + \mathbf{B}. \tag{19.13}$$

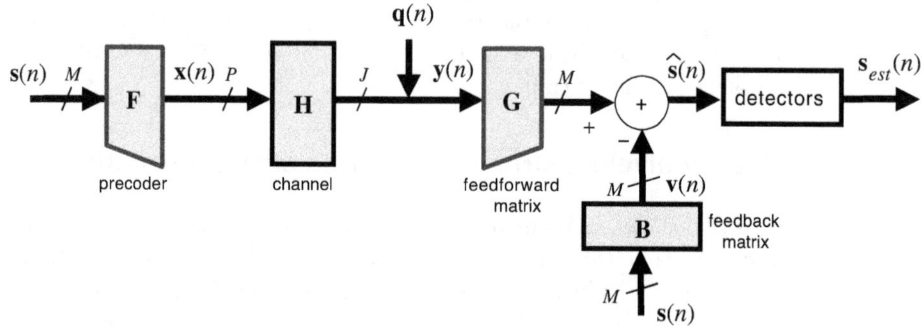

Figure 19.5. Equivalent schematic drawing of the DFE system under the assumption of no error propagation.

This equation imposes certain **rank conditions** on the matrices. Since the matrix $\mathbf{I}_M + \mathbf{B}$ is upper triangular with unit diagonal elements, it has full rank M which shows that the $P \times M$ matrix $\mathbf{F}$ is required to have rank M, and so is the $J \times M$ matrix $\mathbf{HF}$. Similarly $\mathbf{G}$ is required to have rank M and the product $\mathbf{GH}$ has to have rank M.

19.3.1 Optimal feedforward matrix G for fixed F and B

Under the ZF condition the reconstruction error (19.10) is simply

$$\mathbf{e}(n) = \mathbf{Gq}(n), \tag{19.14}$$

so that the error covariance is

$$\mathbf{R_{ee}} = \sigma_q^2 \mathbf{GG}^\dagger. \tag{19.15}$$

The objective function to be minimized is the mean square reconstruction error

$$\mathcal{E}_{mse} = \sigma_q^2 \text{Tr}\Big(\mathbf{GG}^\dagger\Big) \tag{19.16}$$

under the usual power constraint

$$\sigma_s^2 \text{Tr}\Big(\mathbf{FF}^\dagger\Big) = p_0. \tag{19.17}$$

With $(\mathbf{HF})^{\#}$ denoting a left inverse of $\mathbf{HF}$, if we choose

$$\mathbf{G} = (\mathbf{I}_M + \mathbf{B})(\mathbf{HF})^{\#}, \tag{19.18}$$

then $\mathbf{GHF} = (\mathbf{I}_M + B)(\mathbf{HF})^{\#}\mathbf{HF} = \mathbf{I}_M + \mathbf{B}$ as desired. Choosing the left inverse to be the minimum-norm left inverse (Appendix C), that is,[1]

$$(\mathbf{HF})^{\#} = \Big((\mathbf{HF})^\dagger \mathbf{HF}\Big)^{-1} (\mathbf{HF})^\dagger, \tag{19.19}$$

[1] From Eq. (19.15) we see that the total error is $\text{Tr}\,\mathbf{R_{ee}} = \sigma_q^2 \|\mathbf{G}\|^2$, where $\|\mathbf{G}\|$ is the (Frobenius) norm of $\mathbf{G}$ (see Sec. 8.2 in Chap. 8). The choice of minimum-norm inverse therefore minimizes the total error.

the receiver matrix for zero forcing becomes

$$\mathbf{G} = \Big(\mathbf{I} + \mathbf{B}\Big)\Big((\mathbf{HF})^{\dagger}\mathbf{HF}\Big)^{-1}(\mathbf{HF})^{\dagger}. \tag{19.20}$$

The error covariance (19.15) therefore becomes

$$\mathbf{R_{ee}} = \sigma_q^2\Big(\mathbf{I} + \mathbf{B}\Big)\Big((\mathbf{HF})^{\dagger}\mathbf{HF}\Big)^{-1}\Big(\mathbf{I} + \mathbf{B}\Big)^{\dagger}, \tag{19.21}$$

and the mean square error is

$$\mathcal{E}_{mse} = \sigma_q^2 \text{Tr}\Big[\Big(\mathbf{I} + \mathbf{B}\Big)\Big(\mathbf{F}^{\dagger}\mathbf{H}^{\dagger}\mathbf{HF}\Big)^{-1}\Big(\mathbf{I} + \mathbf{B}\Big)^{\dagger}\Big]. \tag{19.22}$$

We have to find $\mathbf{B}$ and $\mathbf{F}$ such that this is minimized under the power constraint (19.17). Our strategy will be in two steps. First we derive a lower bound on (19.22) which depends only on the channel $\mathbf{H}$, the power p_0, and the noise σ_q^2. Our second step is to show how to choose $\mathbf{F}$ and $\mathbf{B}$ so that this bound is achieved.

19.3.2 Lower bound on the MSE

Given a square matrix, we know that the trace is the sum of eigenvalues λ_k. Since the matrix in square brackets in Eq. (19.22) is positive definite, it has positive eigenvalues λ_k, and we can apply the AM-GM inequality (Appendix A) to obtain

$$\begin{aligned}\mathcal{E}_{mse} &= \sigma_q^2 \textstyle\sum_{k=0}^{M-1} \lambda_k \geq M\sigma_q^2\Big(\prod_{k=0}^{M-1}\lambda_k\Big)^{1/M} \\ &= M\sigma_q^2\left(\det\Big[\Big(\mathbf{I} + \mathbf{B}\Big)\Big(\mathbf{F}^{\dagger}\mathbf{H}^{\dagger}\mathbf{HF}\Big)^{-1}\Big(\mathbf{I} + \mathbf{B}\Big)^{\dagger}\Big]\right)^{1/M},\end{aligned} \tag{19.23}$$

where we have used the fact that the product of the eigenvalues is equal to the determinant of the matrix. Now the matrix $\mathbf{I} + \mathbf{B}$ is upper triangular with diagonal elements equal to unity. Its determinant is therefore unity. Using the fact that the determinant of a product of square matrices is equal to the product of the determinants, we therefore obtain

$$\mathcal{E}_{mse} \geq M\sigma_q^2\Big[\det\Big(\mathbf{F}^{\dagger}\mathbf{H}^{\dagger}\mathbf{HF}\Big)\Big]^{-1/M}. \tag{19.24}$$

Starting from this we will first show that

$$\mathcal{E}_{mse} \geq \frac{M^2\sigma_s^2\sigma_q^2}{p_0}\Big(\prod_{k=0}^{M-1}\sigma_{h,k}^2\Big)^{-1/M}, \tag{19.25}$$

where $\sigma_{h,k}$ are the first M dominant singular values of the channel $\mathbf{H}$. We will also show that the bound in Eq. (19.25) can actually be achieved by choosing the precoder matrix **F** and the DFE matrix **B** appropriately. The minimized MSE under the zero-forcing constraint and the power constraint is therefore

$$\mathcal{E}_{mmse} = \frac{M^2\sigma_s^2\sigma_q^2}{p_0}\Big(\prod_{k=0}^{M-1}\frac{1}{\sigma_{h,k}}\Big)^{2/M}. \tag{19.26}$$

Compare this with the minimum MSE for the case where there is no DFE, which was derived in Chap. 12 under the zero-forcing constraint and the power constraint:

$$\mathcal{E}_{no\,dfe} = \frac{\sigma_s^2\sigma_q^2}{p_0}\Big(\sum_{k=0}^{M-1}\frac{1}{\sigma_{h,k}}\Big)^2. \tag{19.27}$$

Thus

$$\frac{\mathcal{E}_{no\,dfe}}{\mathcal{E}_{mmse}} = \frac{\left(\dfrac{1}{M}\displaystyle\sum_{k=0}^{M-1}\frac{1}{\sigma_{h,k}}\right)^2}{\Big(\displaystyle\prod_{k=0}^{M-1}\frac{1}{\sigma_{h,k}}\Big)^{2/M}} \tag{19.28}$$

The benefit of going from a linear transceiver to a transceiver with DFE is therefore measured by the square of the ratio of the arithmetic mean to geometric mean of the reciprocal of the M dominant channel singular values. The ratio (19.28) will be referred to as the *gain due to DFE in the zero forcing case:*

$$\mathcal{G}_{DFE,zf} = \left(\frac{AM\Big\{\dfrac{1}{\sigma_{h,k}}\Big\}}{GM\Big\{\dfrac{1}{\sigma_{h,k}}\Big\}}\right)^2. \tag{19.29}$$

For the case of scalar channels with time-domain DFE, a similar formula has been observed in the pioneering work of Price [1972]. See p. 729 of the review by Lucky [1973].

It is interesting to recall here a result from the theory of optimal bit allocation (Chap. 14): the "coding gain" due to bit allocation in a zero-forced system was shown to be an AM/GM ratio (Sec. 14.7). In that AM/GM ratio, the numbers $1/\sigma_{h,k}^2$ appeared instead of $1/\sigma_{h,k}$.

19.3.3 Proof of Eq. (19.25)

We will find it convenient to express the $P\times M$ precoder **F** in terms of its SVD:

$$\mathbf{F} = \underbrace{\mathbf{U}_f}_{P\times P}\underbrace{\begin{bmatrix}\mathbf{\Sigma}_f\\ \mathbf{0}\end{bmatrix}}_{P\times M}\underbrace{\mathbf{V}_f^\dagger}_{M\times M} \tag{19.30}$$

Here $\mathbf{U}_f$ and $\mathbf{V}_f$ are unitary matrices, and $\mathbf{\Sigma}_f$ is diagonal with nonnegative diagonal elements. Since **F** has rank M as explained after Eq. (19.13), the

$M \times M$ diagonal matrix $\mathbf{\Sigma}_f$ has positive diagonal elements. Recall also that the channel SVD is

$$\mathbf{H} = \underbrace{\mathbf{U}_h}_{J\times J}\underbrace{\mathbf{\Sigma}_h}_{J\times P}\underbrace{\mathbf{V}_h^\dagger}_{P\times P} \tag{19.31}$$

Since this has to have rank $\geq M$ (in view of Eq. (19.13)), we can assume the first M dominant singular values are positive:

$$\sigma_{h,0} \geq \sigma_{h,1} \geq \ldots \geq \sigma_{h,M-1} > 0.$$

Thus, the determinant in Eq. (19.24) can be written as

$$\begin{aligned}\det\left(\mathbf{F}^\dagger\mathbf{H}^\dagger\mathbf{H}\mathbf{F}\right) &= \det\left(\mathbf{V}_f\,[\mathbf{\Sigma}_f \quad \mathbf{0}]\,\mathbf{U}_f^\dagger\mathbf{V}_h\mathbf{\Sigma}_h^\dagger\mathbf{\Sigma}_h\mathbf{V}_h^\dagger\mathbf{U}_f\begin{bmatrix}\mathbf{\Sigma}_f\\ \mathbf{0}\end{bmatrix}\mathbf{V}_f^\dagger\right)\\ &= \det\left([\mathbf{\Sigma}_f \quad \mathbf{0}]\,\mathbf{U}_f^\dagger\mathbf{V}_h\mathbf{\Sigma}_h^\dagger\mathbf{\Sigma}_h\mathbf{V}_h^\dagger\mathbf{U}_f\begin{bmatrix}\mathbf{\Sigma}_f\\ \mathbf{0}\end{bmatrix}\right),\end{aligned}$$

where we have used the fact that $\det[\mathbf{V}_f\mathbf{B}\mathbf{V}_f^\dagger] = \det\mathbf{B}$ because $\mathbf{V}_f$ is a square unitary matrix. From the preceding we finally obtain the expression

$$\det\left(\mathbf{F}^\dagger\mathbf{H}^\dagger\mathbf{H}\mathbf{F}\right) = \det\left(\mathbf{\Sigma}_f^2\left[\mathbf{U}_f^\dagger\mathbf{V}_h\mathbf{\Sigma}_h^\dagger\mathbf{\Sigma}_h\mathbf{V}_h^\dagger\mathbf{U}_f\right]_M\right), \tag{19.32}$$

where $[\mathbf{A}]_M$ indicates the $M \times M$ leading principal submatrix of

$$\mathbf{A} \stackrel{\Delta}{=} \mathbf{U}_f^\dagger\mathbf{V}_h\mathbf{\Sigma}_h^\dagger\mathbf{\Sigma}_h\mathbf{V}_h^\dagger\mathbf{U}_f. \tag{19.33}$$

We will establish the desired lower bound on the right-hand side of Eq. (19.24) by establishing an upper bound on Eq. (19.32). Since $\mathbf{U}_f^\dagger\mathbf{V}_h$ is unitary, the eigenvalues of the $P \times P$ matrix $\mathbf{A}$ are the diagonal elements of $\mathbf{\Sigma}_h^\dagger\mathbf{\Sigma}_h$, namely

$$\sigma_{h,0}^2 \geq \sigma_{h,1}^2 \geq \ldots \geq \sigma_{h,M-1}^2 \geq \ldots \geq \sigma_{h,P-1}^2. \tag{19.34}$$

Let

$$\eta_0 \geq \eta_1 \geq \ldots \geq \eta_{M-1} > 0 \tag{19.35}$$

be the M eigenvalues of $[\mathbf{A}]_M$ (evidently positive since this matrix has rank $\geq M$). We now use the same technique that we used in Sec. 12.4.2, namely the eigenvalue interlace property described in Fig. 12.4. Thus

$$\sigma_{h,k}^2 \geq \eta_k, \quad 0 \leq k \leq M-1. \tag{19.36}$$

Since the determinant is the product of eigenvalues, we therefore have

$$\det\left(\mathbf{F}^\dagger\mathbf{H}^\dagger\mathbf{H}\mathbf{F}\right) = \prod_{k=0}^{M-1}\sigma_{f,k}^2\eta_k \leq \prod_{k=0}^{M-1}\sigma_{f,k}^2\sigma_{h,k}^2, \tag{19.37}$$

where

$$\sigma_{f,k} = [\boldsymbol{\Sigma}_f]_{kk}$$

The right-hand side of Eq. (19.37) can be further simplified by using the power constraint (19.17). From (19.30) we have

$$\mathbf{F}\mathbf{F}^\dagger = \mathbf{U}_f \begin{bmatrix} \boldsymbol{\Sigma}_f \\ \\ \mathbf{0} \end{bmatrix} \mathbf{V}_f^\dagger \mathbf{V}_f [\boldsymbol{\Sigma}_f \quad \mathbf{0}]\, \mathbf{U}_f^\dagger = \mathbf{U}_f \begin{bmatrix} \boldsymbol{\Sigma}_f^2 & \mathbf{0} \\ \mathbf{0} & \mathbf{0} \end{bmatrix} \mathbf{U}_f^\dagger,$$

where we have used the unitarity of $\mathbf{V}_f$. Taking traces we have

$$\begin{aligned} \mathrm{Tr}\,(\mathbf{F}\mathbf{F}^\dagger) &= \mathrm{Tr}\left(\mathbf{U}_f \begin{bmatrix} \boldsymbol{\Sigma}_f^2 & \mathbf{0} \\ \mathbf{0} & \mathbf{0} \end{bmatrix} \mathbf{U}_f^\dagger\right) \\ &= \mathrm{Tr}\left(\mathbf{U}_f^\dagger \mathbf{U}_f \begin{bmatrix} \boldsymbol{\Sigma}_f^2 & \mathbf{0} \\ \mathbf{0} & \mathbf{0} \end{bmatrix}\right) \\ &= \mathrm{Tr}\,\boldsymbol{\Sigma}_f^2 = \sum_{k=0}^{M-1} \sigma_{f,k}^2, \end{aligned}$$

where we have used the facts that $\mathrm{Tr}\,(\mathbf{AB}) = \mathrm{Tr}\,(\mathbf{BA})$ and $\mathbf{U}_f^\dagger \mathbf{U}_f = \mathbf{I}$. Thus the power constraint (19.17) can be rewritten as

$$\sum_{k=0}^{M-1} \sigma_{f,k}^2 = \frac{p_0}{\sigma_s^2} \tag{19.38}$$

Now, from the AM-GM inequality we know that

$$\left(\prod_{k=0}^{M-1} \sigma_{f,k}^2\right)^{1/M} \leq \frac{1}{M} \sum_{k=0}^{M-1} \sigma_{f,k}^2 \tag{19.39}$$

with equality if and only if $\sigma_{f,k}$ are identical. From Eq. (19.37) we therefore obtain

$$\det\left(\mathbf{F}^\dagger \mathbf{H}^\dagger \mathbf{H} \mathbf{F}\right) \leq \prod_{k=0}^{M-1} \sigma_{f,k}^2 \sigma_{h,k}^2 \leq \left(\frac{1}{M} \sum_{k=0}^{M-1} \sigma_{f,k}^2\right)^M \prod_{k=0}^{M-1} \sigma_{h,k}^2. \tag{19.40}$$

In view of Eq. (19.38) this finally yields

$$\det\left(\mathbf{F}^\dagger \mathbf{H}^\dagger \mathbf{H} \mathbf{F}\right) \leq \left(\frac{p_0}{M\sigma_s^2}\right)^M \prod_{k=0}^{M-1} \sigma_{h,k}^2. \tag{19.41}$$

Substituting into Eq. (19.24), the inequality (19.25) follows immediately.

19.3.4 Achieving the lower bound (19.25)

Since the bound (19.25) was derived using several inequalities, we can achieve the bound by satisfying each of these inequalities with equality. We now summarize the inequalities that have been used:

1. In Eq. (19.23) we used the AM-GM inequality on the eigenvalues λ_k of the positive definite matrix $(\mathbf{I}+\mathbf{B})(\mathbf{F}^\dagger\mathbf{H}^\dagger\mathbf{H}\mathbf{F})^{-1}(\mathbf{I}+\mathbf{B})^\dagger$. To satisfy equality, these eigenvalues should therefore be all identical. Since the matrix is Hermitian this is possible if and only if

$$\Big(\mathbf{I}+\mathbf{B}\Big)\Big(\mathbf{F}^\dagger\mathbf{H}^\dagger\mathbf{H}\mathbf{F}\Big)^{-1}\Big(\mathbf{I}+\mathbf{B}\Big)^\dagger = c\mathbf{I} \tag{19.42}$$

for some scalar $c > 0$. From Eq. (19.21) it then follows that the error covariance is

$$\mathbf{R}_{\mathbf{ee}} = \sigma_e^2\mathbf{I}. \tag{19.43}$$

Thus the optimal system is such that the error components at the input of the detector are uncorrelated and have identical mean square values.

2. Another inequality we used was (19.36). This is satisfied with equality if the $P\times P$ matrix $\mathbf{A}$ in Eq. (19.33) is diagonal. This is readily accomplished by choosing the precoder's unitary factor $\mathbf{U}_f$ as

$$\mathbf{U}_f = \mathbf{V}_h, \tag{19.44}$$

where $\mathbf{V}_h$ is one of the unitary factors in the channel SVD (see Eqs. (19.30) and (19.31)). Thus the precoder should be restricted to be of the form

$$\mathbf{F} = \mathbf{V}_h\begin{bmatrix}\boldsymbol{\Sigma}_f\\ \mathbf{0}\end{bmatrix}\mathbf{V}_f^\dagger. \tag{19.45}$$

3. The last inequality we used was (19.39). This can be satisfied with equality if and only if $\sigma_{f,k}$ is identical for all k. That is, the diagonal component of the precoder matrix $\mathbf{F}$ should be $\boldsymbol{\Sigma}_f = \alpha\mathbf{I}_M$. Combining with the preceding expression (19.45) for $\mathbf{F}$ we therefore obtain

$$\mathbf{F} = \alpha\mathbf{V}_h\begin{bmatrix}\mathbf{I}_M\\ \\ \mathbf{0}\end{bmatrix}\mathbf{V}_f^\dagger, \tag{19.46}$$

where the constant $\alpha > 0$ can be identified by using the power constraint.

It remains to identify the unitary factor $\mathbf{V}_f$ in the precoder matrix (19.46) and the upper triangular feedback matrix $\mathbf{B}$. These remaining degrees of freedom must be chosen such that Eq. (19.42) is satisfied for some $c > 0$. Now, with $\mathbf{F}$ chosen as in Eq. (19.46), we have

$$\mathbf{F}^\dagger\mathbf{H}^\dagger\mathbf{H}\mathbf{F} = \alpha^2\mathbf{V}_f\,[\mathbf{I}_M\quad \mathbf{0}]\,\mathbf{V}_h^\dagger\mathbf{V}_h\boldsymbol{\Sigma}_h^\dagger\boldsymbol{\Sigma}_h\mathbf{V}_h^\dagger\mathbf{V}_h\begin{bmatrix}\mathbf{I}_M\\ \\ \mathbf{0}\end{bmatrix}\mathbf{V}_f^\dagger,$$

which simplifies to

$$\mathbf{F}^\dagger\mathbf{H}^\dagger\mathbf{H}\mathbf{F} = \alpha^2\mathbf{V}_f[\boldsymbol{\Sigma}_h]_M^2\mathbf{V}_f^\dagger,$$

where

$$[\boldsymbol{\Sigma}_h]_M^2 = \text{diag}\,[\sigma_{h,0}^2\quad \sigma_{h,1}^2\quad \ldots\sigma_{h,M-1}^2]$$

Thus the requirement (19.42) takes the form

$$\Big(\mathbf{I}+\mathbf{B}\Big)\mathbf{V}_f[\mathbf{\Sigma}_h]_M^{-2}\mathbf{V}_f^\dagger\Big(\mathbf{I}+\mathbf{B}\Big)^\dagger = \beta^2\mathbf{I}_M, \tag{19.47}$$

where $\beta^2 = c\alpha^2$. The preceding equation simply says that

$$\Big(\mathbf{I}+\mathbf{B}\Big)\mathbf{V}_f[\mathbf{\Sigma}_h]_M^{-1} = \beta\mathbf{U} \tag{19.48}$$

for some unitary matrix $\mathbf{U}$. This can be rewritten as

$$\mathbf{U}^\dagger\Big(\mathbf{I}+\mathbf{B}\Big)\mathbf{V}_f = \beta[\mathbf{\Sigma}_h]_M. \tag{19.49}$$

Given the diagonal matrix $[\mathbf{\Sigma}_h]_M$ of the M dominant channel singular values $\sigma_{h,k}$, we can always find unitary matrices $\mathbf{U}$ and $\mathbf{V}_f$ such that the preceding equation is satisfied for an appropriately chosen, strictly upper triangular, $\mathbf{B}$ matrix. This is a non-obvious result, a direct consequence of the **QRS** decomposition or the **GMD** (geometric mean decomposition) introduced in the pioneering papers of Jiang *et al.* [2005a, 2005b], and Zhang *et al.* [2005]. It is stated in detail in Appendix 19.A and proved in Appendix 19.B (both at the end of this chapter). In what follows we use this decomposition, and summarize the results of this section.

19.3.4.A Summary: Optimal DFE transceiver with zero forcing

The optimal zero-forcing DFE transceiver which minimizes the mean square error at the input of the detector in Fig. 19.3 can be constructed as follows. Let the channel SVD be given by

$$\mathbf{H} = \underbrace{\mathbf{U}_h}_{J\times J}\underbrace{\mathbf{\Sigma}_h}_{J\times P}\underbrace{\mathbf{V}_h^\dagger}_{P\times P}, \tag{19.50}$$

where the singular values are ordered as $\sigma_{h,0} \geq \sigma_{h,1} \geq \ldots$ Let the $M\times M$ leading principal matrix of $\mathbf{\Sigma}_h$ containing the M dominant singular values be

$$[\mathbf{\Sigma}_h]_M = \begin{bmatrix} \sigma_{h,0} & 0 & \ldots & 0 \\ 0 & \sigma_{h,1} & \ldots & 0 \\ \vdots & \vdots & \ddots & \vdots \\ 0 & 0 & \ldots & \sigma_{h,M-1} \end{bmatrix}. \tag{19.51}$$

The channel is assumed to have rank $\geq M$ so that

$$\sigma_{h,0} \geq \sigma_{h,1} \geq \ldots \geq \sigma_{h,M-1} > 0. \tag{19.52}$$

Compute the QRS decomposition (i.e., the GMD) of $[\mathbf{\Sigma}_h]_M$ (see Appendix 19.A at the end of this chapter):

$$[\mathbf{\Sigma}_h]_M = \sigma\mathbf{QRS}^\dagger, \tag{19.53}$$

where $\mathbf{Q}$ and $\mathbf{S}$ are $M \times M$ unitary, $\mathbf{R}$ is upper triangular with diagonal elements $R_{kk} = 1$, and σ is the geometric mean of the dominant singular values:

$$\sigma = \Big(\prod_{k=0}^{M-1} \sigma_{h,k}\Big)^{1/M}. \tag{19.54}$$

Comparing Eq. (19.53) with Eq. (19.49) reproduced below:

$$[\mathbf{\Sigma}_h]_M = \beta^{-1}\mathbf{U}^\dagger\Big(\mathbf{I}+\mathbf{B}\Big)\mathbf{V}_f,$$

we see that $\mathbf{V}_f = \mathbf{S}^\dagger$ and $\mathbf{R} = \mathbf{I}+\mathbf{B}$. Thus the matrix $\mathbf{R}$ in the QRS decompostion determines the decision feedback matrix $\mathbf{B}$:

$$\mathbf{B} = \mathbf{R} - \mathbf{I}_M. \tag{19.55}$$

Since $\mathbf{R}$ is upper triangular with $[\mathbf{R}]_{kk} = 1$, the matrix $\mathbf{B}$ is upper triangular with $[\mathbf{B}]_{kk} = 0$, as desired. The optimal $P \times M$ precoder is given by

$$\mathbf{F} = \alpha \mathbf{V}_h \begin{bmatrix} \mathbf{I}_M \\ \mathbf{0} \end{bmatrix} \mathbf{S}, \tag{19.56}$$

where α is chosen to satisfy the power constraint (19.17), that is,

$$\alpha = \sqrt{p_0/(M\sigma_s^2)}. \tag{19.57}$$

The optimal feedforward matrix $\mathbf{G}$ in the receiver is computed using the zero-forcing constraint, and has the closed form (19.20). Using the preceding forms of $\mathbf{H}$ and $\mathbf{F}$ we can show (see below) that this simplifies to

$$\mathbf{G} = \Big(\frac{\mathbf{I}+\mathbf{B}}{\alpha}\Big)\mathbf{S}^\dagger\Big([\mathbf{\Sigma}_h]_M\Big)^{-1}\Big([\mathbf{U}_h]_{J\times M}\Big)^\dagger, \tag{19.58}$$

where $[\mathbf{U}_h]_{J\times M}$ is the submatrix of $\mathbf{U}_h$ (defined in Eq. (19.50)), obtained by retaining the first J rows and M columns. With $\mathbf{F}, \mathbf{G}$, and $\mathbf{B}$ chosen as above, the mean square reconstruction error (at the input of the detector in Fig. 19.3) is minimized, that is, it attains the lower bound given in Eq. (19.25). Thus the minimized MSE (at the detector input) under the zero-forcing constraint and the power constraint is given by

$$\mathcal{E}_{mmse} = \frac{M^2\sigma_s^2\sigma_q^2}{p_0}\Big(\prod_{k=0}^{M-1}\frac{1}{\sigma_{h,k}}\Big)^{2/M}. \tag{19.59}$$

The optimal system is such that the error covariance at the detector input is given by

$$\mathbf{R}_{\mathbf{ee}} = \frac{\mathcal{E}_{mmse}\mathbf{I}_M}{M}. \tag{19.60}$$

A number of points should now be noted.

1. From Eq. (19.60) we find that the error components are *uncorrelated and have identical mean square values.* Recall that for optimum linear transceivers (i.e., without DFE) the error covariance is not diagonal, although the diagonal elements are equalized in minimum error probability solutions (Chap. 16). Note also that the minimized error (19.59) in a DFE system depends on the *geometric mean* rather than the arithmetic mean of $1/\sigma_{h,k}$.

2. The optimal solution can be schematically shown in Fig. 19.6. Since $\mathbf{H} = \mathbf{U}_h\boldsymbol{\Sigma}_h\mathbf{V}_h^\dagger$, the matrices $\mathbf{V}_h$ in the precoder and $\mathbf{U}_h^\dagger$ in the receiver essentially diagonalize the channel. The equivalent system can therefore be drawn as in Fig. 19.7. Here the diagonal matrix $[\boldsymbol{\Sigma}_h]_M^{-1}$ simply performs zero-forcing equalization of the diagonal channel $[\boldsymbol{\Sigma}_h]_M$. The unitary matrix $\mathbf{S}$ and triangular matrix $\mathbf{B}$ play crucial roles in minimizing the reconstruction error.

3. Note that the matrix $\mathbf{Q}$ which arises in the QRS decomposition does not appear in the implementation.

4. From Eq. (19.56) we see that $\mathbf{F}^\dagger\mathbf{F} = \mathbf{I}_M$, that is, the columns of the optimal precoder $\mathbf{F}$ are orthonormal.

Justification of Eq. (19.58). We conclude this section by deriving Eq. (19.58). From Eqs. (19.56) and (19.50) we have

$$\mathbf{HF} = \alpha\mathbf{U}_h\boldsymbol{\Sigma}_h\mathbf{V}_h^\dagger\mathbf{V}_h\begin{bmatrix}\mathbf{I}_M\\ \mathbf{0}\end{bmatrix}\mathbf{S} = \alpha\mathbf{U}_h\boldsymbol{\Sigma}_h\begin{bmatrix}\mathbf{I}_M\\ \mathbf{0}\end{bmatrix}\mathbf{S} = \alpha[\mathbf{U}_h]_{J\times M}[\boldsymbol{\Sigma}_h]_M\mathbf{S},$$

so that

$$\Big((\mathbf{HF})^\dagger\mathbf{HF}\Big) = \alpha^2\mathbf{S}^\dagger[\boldsymbol{\Sigma}_h]_M\Big([\mathbf{U}_h]_{J\times M}\Big)^\dagger[\mathbf{U}_h]_{J\times M}[\boldsymbol{\Sigma}_h]_M\mathbf{S} = \alpha^2\mathbf{S}^\dagger\Big([\boldsymbol{\Sigma}_h]_M\Big)^2\mathbf{S},$$

where we have used the fact that the columns of $[\mathbf{U}_h]_{J\times M}$ are orthonormal. Thus

$$\Big((\mathbf{HF})^\dagger\mathbf{HF}\Big)^{-1}(\mathbf{HF})^\dagger = \alpha^{-1}\mathbf{S}^\dagger\Big([\boldsymbol{\Sigma}_h]_M\Big)^{-1}\Big([\mathbf{U}_h]_{J\times M}\Big)^\dagger.$$

Substituting this into Eq. (19.20), we obtain Eq. (19.58) indeed. ▽ ▽ ▽

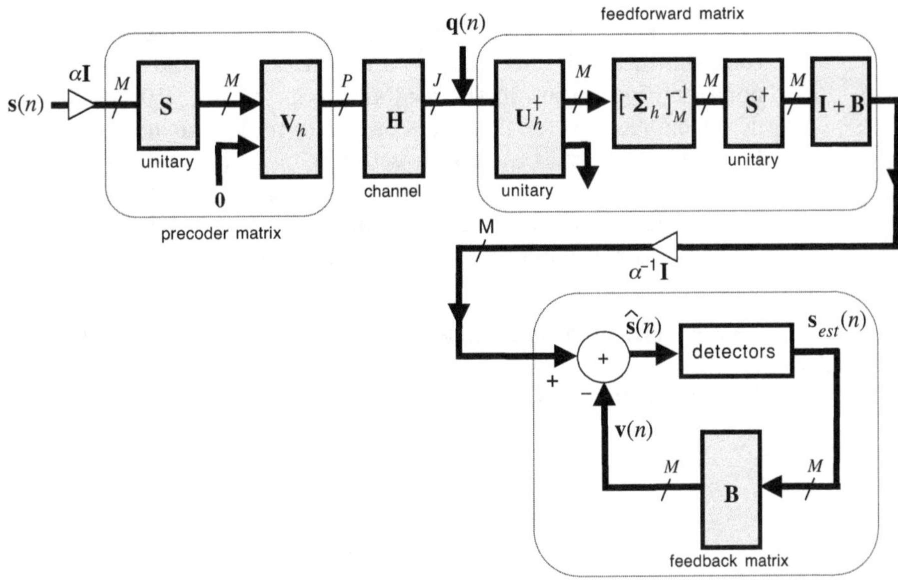

Figure 19.6. The optimal DFE system with zero-forcing constraint.

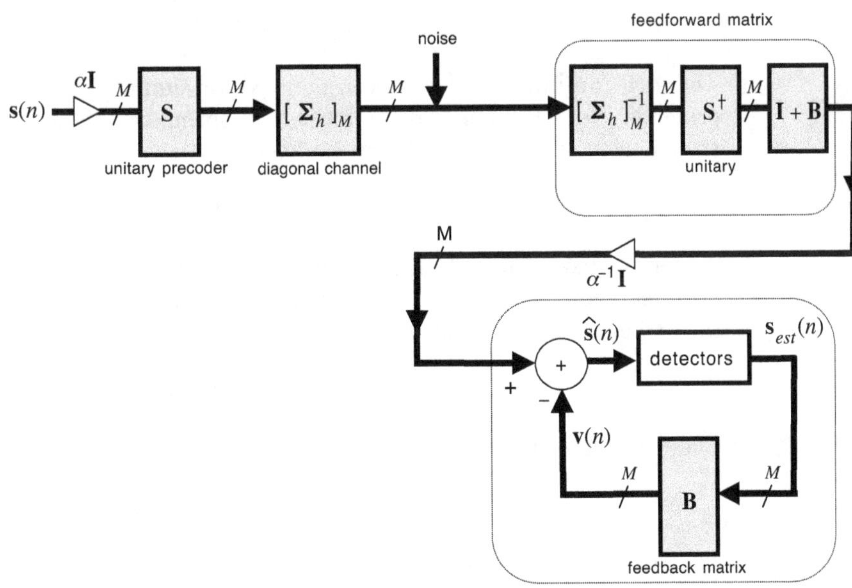

Figure 19.7. Equivalent diagonal form of the optimal DFE system with zero-forcing constraint. The dominant diagonal part of the channel $[\mathbf{\Sigma}_h]_M$ is equalized by its inverse. The matrices $\mathbf{S}$ and $\mathbf{B}$ are chosen using the QRS decomposition.

19.4 Optimal DFE system without zero forcing

Consider again the equivalent schematic for the transceiver with decision feedback shown in Fig. 19.8. This representation is valid under the assumption of no error propagation in the feedback loop. In Sec. 19.3 we used this figure to derive expressions for the error covariance under the assumption of zero forcing. We now assume that there is no zero forcing and proceed with a similar analysis. The goal now is to optimize the matrices $\mathbf{F}, \mathbf{G}$, and $\mathbf{B}$ jointly to minimize the total mean square value of the error at the detector input. First we derive an expression for the optimal feedforward matrix $\mathbf{G}$ for fixed $\mathbf{F}$ and $\mathbf{B}$.

19.4.1 Optimal feedforward matrix G for fixed F and B

The reconstruction error at the input of the detector is

$$\mathbf{e}(n) = \widehat{\mathbf{s}}(n) - \mathbf{s}(n).$$

But

$$\widehat{\mathbf{s}}(n) = \mathbf{G}\mathbf{y}(n) - \mathbf{B}\mathbf{s}_{est}(n). \tag{19.61}$$

Under the "no error propagation" assumption, this can be rewritten as

$$\widehat{\mathbf{s}}(n) = \mathbf{G}\mathbf{y}(n) - \mathbf{B}\mathbf{s}(n)$$

so that the reconstruction error at the input of the detector becomes

$$\mathbf{e} = \mathbf{G}\mathbf{y} - (\mathbf{B} + \mathbf{I})\mathbf{s}, \tag{19.62}$$

where the argument (n) has been omitted for simplicity. By the orthogonality principle (Sec. F.2.1 in Appendix F), the optimal $\mathbf{G}$ should be such that this error is orthogonal to the observation vector $\mathbf{y}$, that is,

$$E[\mathbf{e}\mathbf{y}^\dagger] = \mathbf{0}. \tag{19.63}$$

So, for any fixed pair of $\mathbf{F}$ and $\mathbf{B}$ matrices, we can use this condition to obtain an expression for the optimal feedforward matrix $\mathbf{G}$. Substituting from Eq. (19.62), the orthogonality condition becomes

$$E[\mathbf{e}\mathbf{y}^\dagger] = \mathbf{G}\mathbf{R}_{\mathbf{yy}} - (\mathbf{B} + \mathbf{I})\mathbf{R}_{\mathbf{sy}} = \mathbf{0}, \tag{19.64}$$

where $\mathbf{R}_{\mathbf{yy}}$ and $\mathbf{R}_{\mathbf{sy}}$ are appropriate correlation matrices:

$$\mathbf{R}_{\mathbf{yy}} = E[\mathbf{y}\mathbf{y}^\dagger], \quad \mathbf{R}_{\mathbf{sy}} = E[\mathbf{s}\mathbf{y}^\dagger]. \tag{19.65}$$

Thus the optimal feedforward matrix is

$$\mathbf{G} = (\mathbf{B} + \mathbf{I})\mathbf{R}_{\mathbf{sy}}\mathbf{R}_{\mathbf{yy}}^{-1}. \tag{19.66}$$

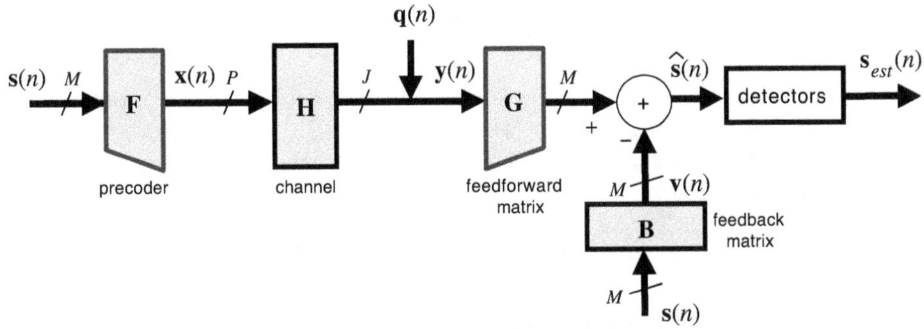

Figure 19.8. Equivalent schematic drawing of the DFE system under the assumption of no error propagation.

19.4.2 Error covariance

With $\mathbf{G}$ chosen as above, we now derive the error covariance matrix. This expression will be useful in optimizing the precoder $\mathbf{F}$ and the feedback matrix $\mathbf{B}$. First we find explicit expressions for the correlation matrices $\mathbf{R_{yy}}$ and $\mathbf{R_{sy}}$ and use Eq. (19.66) to write $\mathbf{G}$ more explicitly. For this observe that

$$\mathbf{y} = \mathbf{HFs} + \mathbf{q}. \tag{19.67}$$

We assume that the signal and noise are zero-mean uncorrelated processes with covariances

$$\mathbf{R_{ss}} = \sigma_s^2 \mathbf{I}_M, \quad \mathbf{R_{qq}} = \sigma_q^2 \mathbf{I}_J, \tag{19.68}$$

respectively. In particular we have $\mathbf{R_{sq}} = \mathbf{0}$, so that

$$\mathbf{R_{yy}} = \sigma_s^2 \mathbf{HFF}^\dagger \mathbf{H}^\dagger + \sigma_q^2 \mathbf{I}.$$

Similarly

$$\mathbf{R_{sy}} = E[\mathbf{sy}^\dagger] = E[\mathbf{s}(\mathbf{HFs} + \mathbf{q})^\dagger] = \sigma_s^2 \mathbf{F}^\dagger \mathbf{H}^\dagger.$$

Thus the optimal feedforward matrix becomes

$$\mathbf{G} = (\mathbf{B} + \mathbf{I})\mathbf{F}^\dagger \mathbf{H}^\dagger \Big(\mathbf{HFF}^\dagger \mathbf{H}^\dagger + \frac{\sigma_q^2}{\sigma_s^2} \mathbf{I}\Big)^{-1}. \tag{19.69}$$

With $\mathbf{G}$ chosen as above we now compute the error covariance matrix. Substituting Eq. (19.67) into Eq. (19.62) we have

$$\mathbf{e} = \Big(\mathbf{GHF} - (\mathbf{B} + \mathbf{I})\Big)\mathbf{s} + \mathbf{Gq}. \tag{19.70}$$

Using $\mathbf{R_{sq}} = \mathbf{0}$ we find that the error covariance is given by

$$\mathbf{R_{ee}} = \sigma_s^2 \Big(\mathbf{GHF} - (\mathbf{B} + \mathbf{I})\Big)\Big(\mathbf{GHF} - (\mathbf{B} + \mathbf{I})\Big)^\dagger + \sigma_q^2 \mathbf{GG}^\dagger. \tag{19.71}$$

In Appendix 19.C at the end of the chapter we show that when $\mathbf{G}$ is as in Eq. (19.69), the preceding expression for $\mathbf{R_{ee}}$ can be rewritten in the following form:

$$\mathbf{R_{ee}} = \sigma_s^2(\mathbf{B}+\mathbf{I})\Big(\mathbf{I}+\frac{\sigma_s^2}{\sigma_q^2}\mathbf{F}^\dagger\mathbf{H}^\dagger\mathbf{H}\mathbf{F}\Big)^{-1}(\mathbf{B}+\mathbf{I})^\dagger, \tag{19.72}$$

which is more useful for optimization. The goal is now to optimize the matrices $\mathbf{F}$ and $\mathbf{B}$ to minimize the trace of Eq. (19.72) subject to the power constraint and the constraint that $\mathbf{B}$ be strictly upper triangular. Thus the objective to be minimized is

$$\mathcal{E}_{mse} = \sigma_s^2\text{Tr}\Big[(\mathbf{B}+\mathbf{I})\Big(\mathbf{I}+\frac{\sigma_s^2}{\sigma_q^2}\mathbf{F}^\dagger\mathbf{H}^\dagger\mathbf{H}\mathbf{F}\Big)^{-1}(\mathbf{B}+\mathbf{I})^\dagger\Big], \tag{19.73}$$

subject to the power constraint

$$\sigma_s^2\text{Tr}\Big(\mathbf{F}\mathbf{F}^\dagger\Big) = p_0 \tag{19.74}$$

and the upper triangular constraint

$$b_{km} = 0, \quad k \geq m. \tag{19.75}$$

19.4.3 Bound on the error

We now find a lower bound on the reconstruction error $\mathcal{E}_{mse}$ in terms of quantities that depend only on $\mathbf{F}$ and $\mathbf{H}$. For the reader who is eager for a preview, this bound is shown in Eq. (19.88), where $\sigma_{f,k}$ are the singular values of the precoder $\mathbf{F}$, and $\sigma_{h,k}$ are the first M dominant singular values of the channel $\mathbf{H}$. Depending on the available power, this bound can further be refined as in Eq. (19.98), where K and λ are quantities that depend on power p_0. Finally we show in Sec. 19.4.5 how this bound can be achieved by choice of $\mathbf{F}, \mathbf{G}$, and $\mathbf{B}$.

Now for the details. The trace is the sum of the eigenvalues λ_k of $\mathbf{R_{ee}}$, whereas the determinant is the product of λ_k. Since $\mathbf{R_{ee}}$ is positive definite, we have $\lambda_k > 0$, and the AM-GM inequality can be applied to obtain

$$\frac{\text{Tr}\,\mathbf{R_{ee}}}{M} \geq (\det\mathbf{R_{ee}})^{1/M}. \tag{19.76}$$

That is,

$$\begin{aligned}\frac{\mathcal{E}_{mse}}{\sigma_s^2} &\geq M\det\Big[(\mathbf{B}+\mathbf{I})\Big(\mathbf{I}+\frac{\sigma_s^2}{\sigma_q^2}\mathbf{F}^\dagger\mathbf{H}^\dagger\mathbf{H}\mathbf{F}\Big)^{-1}(\mathbf{B}+\mathbf{I})^\dagger\Big]^{1/M} \\ &= M\det\Big(\mathbf{I}+\frac{\sigma_s^2}{\sigma_q^2}\mathbf{F}^\dagger\mathbf{H}^\dagger\mathbf{H}\mathbf{F}\Big)^{-1/M},\end{aligned}$$

where we have used the fact that $\mathbf{B}+\mathbf{I}$ is upper triangular with diagonal elements equal to unity. So

$$\mathcal{E}_{mse} \geq M\sigma_s^2\det\Big(\mathbf{I}+\frac{\sigma_s^2}{\sigma_q^2}\mathbf{F}^\dagger\mathbf{H}^\dagger\mathbf{H}\mathbf{F}\Big)^{-1/M}. \tag{19.77}$$

We will now find an *upper bound* on the quantity

$$\phi = \det\left(\mathbf{I} + \frac{\sigma_s^2}{\sigma_q^2}\mathbf{F}^\dagger\mathbf{H}^\dagger\mathbf{H}\mathbf{F}\right) \tag{19.78}$$

and show that the upper bound depends only on the channel singular values and the power p_0. This will establish a *lower bound* on $\mathcal{E}_{mse}$. We will then argue that this lower bound can actually be attained by appropriate choice of the matrices $\mathbf{F}$ and $\mathbf{B}$. With $\mathbf{F}$ expressed in terms of its SVD:

$$\mathbf{F} = \underbrace{\mathbf{U}_f}_{P\times P}\underbrace{\begin{bmatrix}\boldsymbol{\Sigma}_f \\ \mathbf{0}\end{bmatrix}}_{P\times M}\underbrace{\mathbf{V}_f^\dagger}_{M\times M}, \tag{19.79}$$

we have

$$\begin{aligned}\phi &= \det\left(\mathbf{I} + \frac{\sigma_s^2}{\sigma_q^2}\mathbf{V}_f\boldsymbol{\Sigma}_f(\mathbf{U}_f^\dagger\mathbf{H}^\dagger\mathbf{H}\mathbf{U}_f)_M\boldsymbol{\Sigma}_f\mathbf{V}_f^\dagger\right)\\ &= \det\mathbf{V}_f\left(\mathbf{I} + \frac{\sigma_s^2}{\sigma_q^2}\boldsymbol{\Sigma}_f(\mathbf{U}_f^\dagger\mathbf{H}^\dagger\mathbf{H}\mathbf{U}_f)_M\boldsymbol{\Sigma}_f\right)\mathbf{V}_f^\dagger\\ &= \det\left(\mathbf{I} + \frac{\sigma_s^2}{\sigma_q^2}\boldsymbol{\Sigma}_f(\mathbf{U}_f^\dagger\mathbf{H}^\dagger\mathbf{H}\mathbf{U}_f)_M\boldsymbol{\Sigma}_f\right).\end{aligned}$$

Here we have used $(\mathbf{A})_M$ to denote the $M\times M$ leading principal submatrix of $\mathbf{A}$. In the second and third lines we have used the fact that $\mathbf{V}_f$ is unitary. We now use the fact that, given an $N\times M$ matrix $\mathbf{A}$ and an $M\times N$ matrix $\mathbf{B}$, the products $\mathbf{AB}$ and $\mathbf{BA}$ have the same set of nonzero eigenvalues including multiplicity [Horn and Johnson, 1985, p. 53]. Using this we conclude that

$$\det(\mathbf{I}_N + \mathbf{AB}) = \det(\mathbf{I}_M + \mathbf{BA}). \tag{19.80}$$

Using this (with $N = M$) we find that ϕ can be written as

$$\phi = \det\left(\mathbf{I} + \frac{\sigma_s^2}{\sigma_q^2}\boldsymbol{\Sigma}_f^2(\mathbf{U}_f^\dagger\mathbf{H}^\dagger\mathbf{H}\mathbf{U}_f)_M\right). \tag{19.81}$$

Next, given two $M\times M$ positive semidefinite matrices $\mathbf{A}$ and $\mathbf{B}$ with eigenvalues

$$\lambda_0 \geq \lambda_1 \geq \ldots \geq \lambda_{M-1}, \quad \text{and} \quad \mu_0 \geq \mu_1 \geq \ldots \geq \mu_{M-1},$$

it can be shown that

$$\det(\mathbf{I} + \mathbf{AB}) \leq \prod_{k=0}^{M-1}(1 + \lambda_k\mu_k). \tag{19.82}$$

This follows from a result called **Fiedler's inequality** [Fiedler, 1971], [Witsenhausen, 1975]. Using Eq. (19.82) in Eq. (19.81) we see that

$$\phi = \det\left(\mathbf{I} + \frac{\sigma_s^2}{\sigma_q^2}\boldsymbol{\Sigma}_f^2(\mathbf{U}_f^\dagger\mathbf{H}^\dagger\mathbf{H}\mathbf{U}_f)_M\right) \leq \prod_{k=0}^{M-1}\left(1 + \frac{\sigma_s^2}{\sigma_q^2}\sigma_{f,k}^2\mu_k\right), \tag{19.83}$$

where μ_k are the eigenvalues of $(\mathbf{U}_f^\dagger \mathbf{H}^\dagger \mathbf{H} \mathbf{U}_f)_M$. Denoting the M dominant singular values of the channel as

$$\sigma_{0,h} \geq \sigma_{1,h} \geq \ldots \geq \sigma_{h,M-1} \tag{19.84}$$

we can show, as we did in Sec. 12.4.2, that

$$\sigma_{h,0}^2 \geq \mu_0, \quad \sigma_{h,1}^2 \geq \mu_1, \quad \sigma_{h,M-1}^2 \geq \mu_{M-1}. \tag{19.85}$$

Thus,

$$\phi = \det\left(\mathbf{I} + \frac{\sigma_s^2}{\sigma_q^2} \mathbf{\Sigma}_f^2 (\mathbf{U}_f^\dagger \mathbf{H}^\dagger \mathbf{H} \mathbf{U}_f)_M\right) \leq \prod_{k=0}^{M-1} (1 + \frac{\sigma_s^2}{\sigma_q^2} \sigma_{f,k}^2 \sigma_{h,k}^2) \tag{19.86}$$

with equality achieved whenever $\mathbf{U}_f$ is chosen such that $\mathbf{U}_f^\dagger \mathbf{H}^\dagger \mathbf{H} \mathbf{U}_f$ is diagonalized, with the top M diagonal elements equal to the M dominant values of $\sigma_{h,k}^2$. This is readily achieved by choosing

$$\mathbf{U}_f = \mathbf{V}_h, \tag{19.87}$$

where $\mathbf{V}_h$ is the unitary matrix appearing in the channel SVD (Eq. (19.31)). Summarizing, we have proved that

$$\mathcal{E}_{mse} \geq M\sigma_s^2 \prod_{k=0}^{M-1} \left(1 + \frac{\sigma_s^2}{\sigma_q^2} \sigma_{f,k}^2 \sigma_{h,k}^2\right)^{-1/M}. \tag{19.88}$$

The bound in Eq. (19.88) depends on the elements $\sigma_{f,k} \geq 0$ of the precoder matrix. These can be optimized subject to the power constraint to minimize the right-hand side of Eq. (19.88) as shown next. After this we will show how the minimized bound can actually be achieved by choosing the remaining free parameters (feedback matrix $\mathbf{B}$, and the unitary part $\mathbf{V}_f$ of the precoder).

19.4.4 Minimizing the bound (19.88)

For simplicity of notation let

$$\alpha_k = \sigma_{f,k}^2 \geq 0, \quad \beta_k = \frac{\sigma_s^2}{\sigma_q^2} \sigma_{h,k}^2. \tag{19.89}$$

But the power constraint can be rewritten as Eq. (19.38), that is,

$$\sum_{k=0}^{M-1} \sigma_{f,k}^2 = \frac{p_0}{\sigma_s^2}$$

To minimize the right-hand side of Eq. (19.88) we have to maximize

$$\prod_{k=0}^{M-1} (1 + \beta_k \alpha_k) \tag{19.90}$$

subject to the constraint

$$\sum_{k=0}^{M-1} \alpha_k = p_0/\sigma_s^2. \tag{19.91}$$

For convenience we will instead maximize the logarithm of the above product (19.90). Since the logarithm is a real-valued monotone increasing function for positive values of its argument this achieves the same end. So, our problem is to optimize α_k (for fixed β_k) to maximize

$$\psi = \sum_{k=0}^{M-1} \ln(1 + \beta_k \alpha_k) \tag{19.92}$$

subject to the constraint (19.91). To solve this we define the Lagrangian

$$\mathcal{L} = \sum_{k=0}^{M-1} \ln(1 + \beta_k \alpha_k) - \lambda\Big(\sum_{k=0}^{M-1} \alpha_k - p_0/\sigma_s^2\Big). \tag{19.93}$$

Setting $\partial\mathcal{L}/\partial\alpha_k = 0$ we get the following necessary conditions for the optimum:

$$\frac{\beta_k}{1 + \beta_k \alpha_k} = \lambda, \quad 0 \le k \le M - 1,$$

which can be rewritten as

$$\alpha_k = \frac{1}{\lambda} - \frac{1}{\beta_k}$$

We have ignored the positivity constraint $\alpha_k \ge 0$ in the above derivation. The correct way to take this into account is to use the KKT theory described in Chap. 22. If this is done then the result will be

$$\alpha_k = \begin{cases} \dfrac{1}{\lambda} - \dfrac{1}{\beta_k} & \text{if this is non-negative} \\ 0 & \text{otherwise.} \end{cases} \tag{19.94}$$

Substituting for the notations α_k and β_k from Eq. (19.89), we obtain

$$\sigma_{f,k}^2 = \begin{cases} \dfrac{1}{\lambda} - \dfrac{\sigma_q^2}{\sigma_s^2 \sigma_{h,k}^2} & \text{if this is non-negative} \\ 0 & \text{otherwise.} \end{cases} \tag{19.95}$$

This expression for the optimal choice of precoder multipliers is similar to the **water-filling** allocation explained in Sec. 22.3 in connection with channel capacity maximization. Since $\sigma_{h,k}$ are ordered as in Eq. (19.84), we can assume that the first K values of $\sigma_{f,k}^2$ are nonzero for some $K \le M$. Thus

$$\sigma_{f,k}^2 = \begin{cases} \dfrac{1}{\lambda} - \dfrac{\sigma_q^2}{\sigma_s^2 \sigma_{h,k}^2} & 0 \le k \le K - 1 \\ 0 & \text{otherwise.} \end{cases} \tag{19.96}$$

The constant λ is determined by using the power constraint (19.91):

$$p_0 = \sigma_s^2 \sum_{k=0}^{K-1} \sigma_{f,k}^2 = \frac{K\sigma_s^2}{\lambda} - \sigma_q^2 \sum_{k=0}^{K-1} \frac{1}{\sigma_{h,k}^2},$$

which yields

$$\frac{1}{\lambda} = \frac{1}{K\sigma_s^2}\left(p_0 + \sigma_q^2 \sum_{k=0}^{K-1} \frac{1}{\sigma_{h,k}^2}\right). \tag{19.97}$$

Substituting the optimal values (19.96) into the right-hand side of Eq. (19.88) we obtain

$$\mathcal{E}_{mse} \geq M\sigma_s^2 \Big(\frac{\lambda\sigma_q^2}{\sigma_s^2}\Big)^{K/M} \prod_{k=0}^{K-1} \Big(\frac{1}{\sigma_{h,k}}\Big)^{2/M}. \tag{19.98}$$

When the noise $\sigma_q^2 << p_0$ we have $K = M$, and from Eq. (19.97),

$$\frac{1}{\lambda} \approx \frac{p_0}{K\sigma_s^2} = \frac{p_0}{M\sigma_s^2}$$

Thus Eq. (19.98) reduces to

$$\mathcal{E}_{mse} \geq \frac{M^2\sigma_s^2\sigma_q^2}{p_0} \prod_{k=0}^{M-1} \Big(\frac{1}{\sigma_{h,k}}\Big)^{2/M}, \tag{19.99}$$

which is identical to the expression (19.59) derived in the zero-forcing case.

19.4.5 Achieving the minimized bound (19.98)

We now explain how the lower bound on the mean square error given by Eq. (19.98) can be achieved by appropriate choice of the precoder $\mathbf{F}$ and the feedback matrix $\mathbf{B}$. The feedforward matrix $\mathbf{G}$ is then determined by Eq. (19.69).

1. First consider Fiedler's inequality given by Eq. (19.82). To satisfy it with equality, all we have to do is to choose $\mathbf{U}_f$ (unitary matrix appearing in the SVD of $\mathbf{F}$) to be $\mathbf{U}_f = \mathbf{V}_h$. With the channel singular values ordered as in Eq. (19.84), the bound (19.88) is achieved.

2. With the diagonal elements $\sigma_{f,k}$ in the precoder chosen to satisfy Eq. (19.96), the final bound (19.98) is achieved. This determines the matrix $\boldsymbol{\Sigma}_f$ in the precoder (19.79). Only the unitary matrix $\mathbf{V}_f$ in the precoder remains to be determined.

3. Next consider the inequality (19.77), which is obtained from the determinant-trace inequality (19.76). Equality can be achieved here if and only if the eigenvalues of the Hermitian matrix $\mathbf{R_{ee}}$ are identical, that is,

$$\mathbf{R_{ee}} = \sigma_e^2\mathbf{I}.$$

Using the expression (19.72) for $\mathbf{R_{ee}}$, this condition can be written as

$$\sigma_s^2(\mathbf{B}+\mathbf{I})\Big(\mathbf{I}+\frac{\sigma_s^2}{\sigma_q^2}\mathbf{F}^\dagger\mathbf{H}^\dagger\mathbf{H}\mathbf{F}\Big)^{-1}(\mathbf{B}+\mathbf{I})^\dagger=\sigma_e^2\mathbf{I}. \tag{19.100}$$

To simplify this, observe that

$$\begin{aligned}\mathbf{F}^\dagger\mathbf{H}^\dagger\mathbf{H}\mathbf{F} &= \mathbf{V}_f\,[\mathbf{\Sigma}_f\quad \mathbf{0}]\,\mathbf{U}_f^\dagger\mathbf{H}^\dagger\mathbf{H}\mathbf{U}_f\begin{bmatrix}\mathbf{\Sigma}_f\\ \mathbf{0}\end{bmatrix}\mathbf{V}_f^\dagger\\ &= \mathbf{V}_f\,[\mathbf{\Sigma}_f\quad \mathbf{0}]\,\mathbf{V}_h^\dagger\mathbf{H}^\dagger\mathbf{H}\mathbf{V}_h\begin{bmatrix}\mathbf{\Sigma}_f\\ \mathbf{0}\end{bmatrix}\mathbf{V}_f^\dagger\\ &= \mathbf{V}_f\,[\mathbf{\Sigma}_f\quad \mathbf{0}]\,\mathbf{\Sigma}_h^\dagger\mathbf{\Sigma}_h\begin{bmatrix}\mathbf{\Sigma}_f\\ \mathbf{0}\end{bmatrix}\mathbf{V}_f^\dagger\\ &= \mathbf{V}_f\mathbf{\Sigma}_f[\mathbf{\Sigma}_h]_M^2\mathbf{\Sigma}_f\mathbf{V}_f^\dagger,\end{aligned}$$

so that

$$\begin{aligned}\Big(\mathbf{I}+\frac{\sigma_s^2}{\sigma_q^2}\mathbf{F}^\dagger\mathbf{H}^\dagger\mathbf{H}\mathbf{F}\Big) &= \Big(\mathbf{I}+\frac{\sigma_s^2}{\sigma_q^2}\mathbf{V}_f\mathbf{\Sigma}_f[\mathbf{\Sigma}_h]_M^2\mathbf{\Sigma}_f\mathbf{V}_f^\dagger\Big)\\ &= \mathbf{V}_f\Big(\mathbf{I}+\frac{\sigma_s^2}{\sigma_q^2}\mathbf{\Sigma}_f^2[\mathbf{\Sigma}_h]_M^2\Big)\mathbf{V}_f^\dagger.\end{aligned}$$

Thus Eq. (19.100) can be rewritten as

$$\sigma_s^2(\mathbf{B}+\mathbf{I})\mathbf{V}_f\Big(\mathbf{I}+\frac{\sigma_s^2}{\sigma_q^2}\mathbf{\Sigma}_f^2[\mathbf{\Sigma}_h]_M^2\Big)^{-1}\mathbf{V}_f^\dagger(\mathbf{B}+\mathbf{I})^\dagger=\sigma_e^2\mathbf{I}. \tag{19.101}$$

Since the left-hand side is a Hermitian positive definite matrix, we can rewrite Eq. (19.101) in the form

$$\beta(\mathbf{B}+\mathbf{I})\mathbf{V}_f\Big(\mathbf{I}+\frac{\sigma_s^2}{\sigma_q^2}\mathbf{\Sigma}_f^2[\mathbf{\Sigma}_h]_M^2\Big)^{-1/2}=\mathbf{U}, \tag{19.102}$$

where $\mathbf{U}$ is unitary and $\beta=\sigma_s/\sigma_e$. That is,

$$\beta\mathbf{U}^\dagger(\mathbf{B}+\mathbf{I})\mathbf{V}_f=\mathbf{\Sigma}, \tag{19.103}$$

where $\mathbf{\Sigma}$ is the diagonal matrix with positive diagonal elements, given by

$$\mathbf{\Sigma}=\Big(\mathbf{I}+\frac{\sigma_s^2}{\sigma_q^2}\mathbf{\Sigma}_f^2[\mathbf{\Sigma}_h]_M^2\Big)^{1/2}. \tag{19.104}$$

Since $\mathbf{\Sigma}_f$ and $\mathbf{\Sigma}_h$ are known, this matrix is known. We can compute a QRS decomposition of this matrix and identify $\mathbf{V}_f$ and $\mathbf{B}$ as we did in Sec. 19.3.4. This completely identifies all the matrices in the optimal DFE transceiver.

19.4.5.A Summary: Optimal DFE transceiver without zero forcing

The optimal (non zero-forcing) DFE transceiver which minimizes the mean square error at the input of the detector in Fig. 19.3 can be constructed as follows. Let the channel SVD be given by

$$\mathbf{H} = \underbrace{\mathbf{U}_h}_{J\times J} \underbrace{\mathbf{\Sigma}_h}_{J\times P} \underbrace{\mathbf{V}_h^\dagger}_{P\times P}, \tag{19.105}$$

where the singular values are ordered as $\sigma_{h,0} \geq \sigma_{h,1} \geq \ldots$ The $M \times M$ leading principal matrix of $\mathbf{\Sigma}_h$ contains the M dominant singular values:

$$[\mathbf{\Sigma}_h]_M = \text{diag}\,[\,\sigma_{h,0} \quad \sigma_{h,1} \quad \ldots \quad \sigma_{h,M-1}\,]. \tag{19.106}$$

Then the optimal precoder has the form

$$\mathbf{F} = \underbrace{\mathbf{V}_h}_{P\times P} \underbrace{\begin{bmatrix} \mathbf{\Sigma}_f \\ \mathbf{0} \end{bmatrix}}_{P\times M} \underbrace{\mathbf{V}_f^\dagger}_{M\times M}, \tag{19.107}$$

where $\mathbf{\Sigma}_f$ is an $M \times M$ diagonal matrix with diagonal elements $\sigma_{f,k}$ given by the **water-filling** formula

$$\sigma_{f,k}^2 = \begin{cases} \dfrac{1}{\lambda} - \dfrac{\sigma_q^2}{\sigma_s^2 \sigma_{h,k}^2} & 0 \leq k \leq K-1 \\ 0 & \text{otherwise.} \end{cases} \tag{19.108}$$

Here λ and K depend on the power p_0, and are detemined using the procedure described following Eq. (19.96). The unitary matrix $\mathbf{V}_f$ in the precoder, and the diagonal matrix $\mathbf{B}$ in the decision feedback loop, are determined by performing a QRS decomposition (Appendix 19.A) on the $M \times M$ matrix $\mathbf{\Sigma}$ defined in Eq. (19.104):

$$\mathbf{\Sigma} = \sigma \mathbf{QRS}^\dagger, \tag{19.109}$$

where σ is the geometric mean of the diagonal elements of $\mathbf{\Sigma}$, and the matrix $\mathbf{R}$ is upper triangular with diagonal elements $R_{kk} = 1$. Comparing with Eq. (19.103), that is,

$$\mathbf{\Sigma} = \beta \mathbf{U}^\dagger (\mathbf{B} + \mathbf{I}) \mathbf{V}_f,$$

we see that $\mathbf{V}_f$ should be chosen as

$$\mathbf{V}_f = \mathbf{S}^\dagger \tag{19.110}$$

and that $\mathbf{B}$ should be chosen as

$$\mathbf{B} = \mathbf{R} - \mathbf{I}. \tag{19.111}$$

Since the upper triangular matrix $\mathbf{R}$ has $\mathbf{R}_{kk} = 1$, the upper triangular matrix $\mathbf{B}$ has $B_{kk} = 0$ as desired. This determines the precoder $\mathbf{F}$ and the feedback matrix $\mathbf{B}$ completely. The feedforward part of the equalizer is then calculated from Eq. (19.69) and takes the following form:

$$\mathbf{G} = (\mathbf{B}+\mathbf{I})\mathbf{S}^\dagger \mathbf{\Lambda}\Big([\mathbf{U}_h]_{J\times M}\Big)^\dagger, \tag{19.112}$$

where

$$\mathbf{\Lambda} = \mathbf{\Sigma}_f(\mathbf{\Sigma}_h)_M\left(\mathbf{\Sigma}_f^2[\mathbf{\Sigma}_h]_M^2 + \frac{\sigma_q^2}{\sigma_s^2}\mathbf{I}\right)^{-1}.$$

With the above choices of $\mathbf{F}, \mathbf{G}$, and $\mathbf{B}$, the minimized total mean square error at the input of the detector is given by

$$\mathcal{E}_{mmse} = M\sigma_s^2\Big(\frac{\lambda\sigma_q^2}{\sigma_s^2}\Big)^{K/M}\prod_{k=0}^{K-1}\Big(\frac{1}{\sigma_{h,k}}\Big)^{2/M}. \tag{19.113}$$

The optimal system is such that the error covariance at the detector input is given by

$$\mathbf{R_{ee}} = \frac{\mathcal{E}_{mmse}\mathbf{I}_M}{M} \tag{19.114}$$

A number of points should be noted about this optimal solution.

1. The error components are uncorrelated and have identical mean square values, as in the zero-forcing case. In an optimized DFE system the mean square errors of the components are therefore identical. In particular, therefore, we can say that the **geometric mean** of errors has been minimized (since it is equal to the arithmetic mean, which was the original objective function).

2. Note also that the minimized error (19.113) in a DFE system depends on the geometric rather than the arithmetic mean of $1/\sigma_{h,k}$.

3. The optimal solution can be schematically shown as in Fig. 19.9. Since $\mathbf{H} = \mathbf{U}_h\mathbf{\Sigma}_h\mathbf{V}_h^\dagger$, the matrices $\mathbf{V}_h$ in the precoder and $\mathbf{U}_h^\dagger$ in the receiver essentially diagonalize the channel. The equivalent system can therefore be drawn as in Fig. 19.10.

4. The main difference from the zero-forcing system is that the diagonal matrix $\mathbf{\Sigma}_f$ is not proportional to identity, and the diagonal equalizer $\mathbf{\Lambda}$ is not $[\mathbf{\Sigma}_h]_M^{-1}$. These diagonal matrices do reduce to the simple forms taken in the zero-forcing case, when $\sigma_q^2/\sigma_s^2 << 1$. In this case the expression for $\mathcal{E}_{mmse}$ reduces to the corresponding expression (19.59) of the zero-forcing system.

5. Note that the matrix $\mathbf{Q}$ which arises in the QRS decomposition *does not* appear in the implementation.

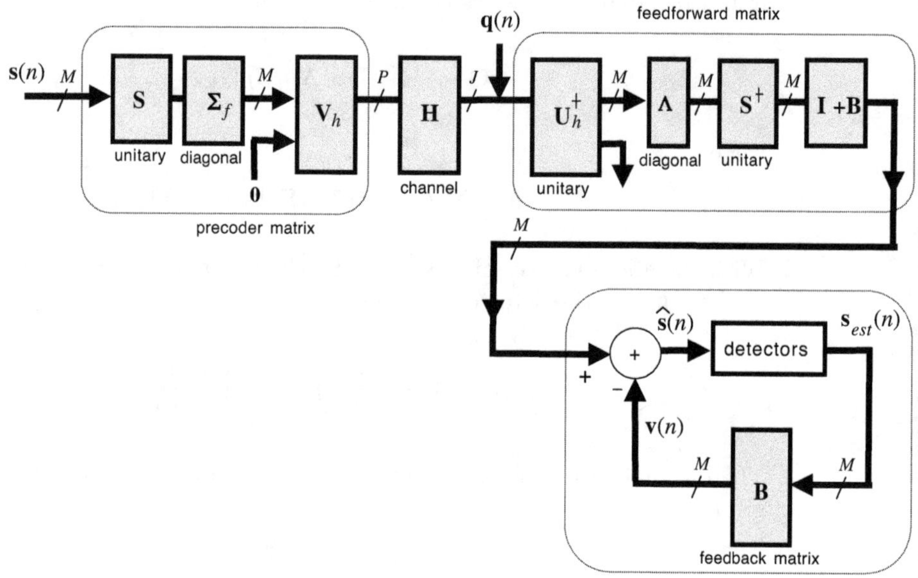

Figure 19.9. The optimal DFE system without zero-forcing constraint.

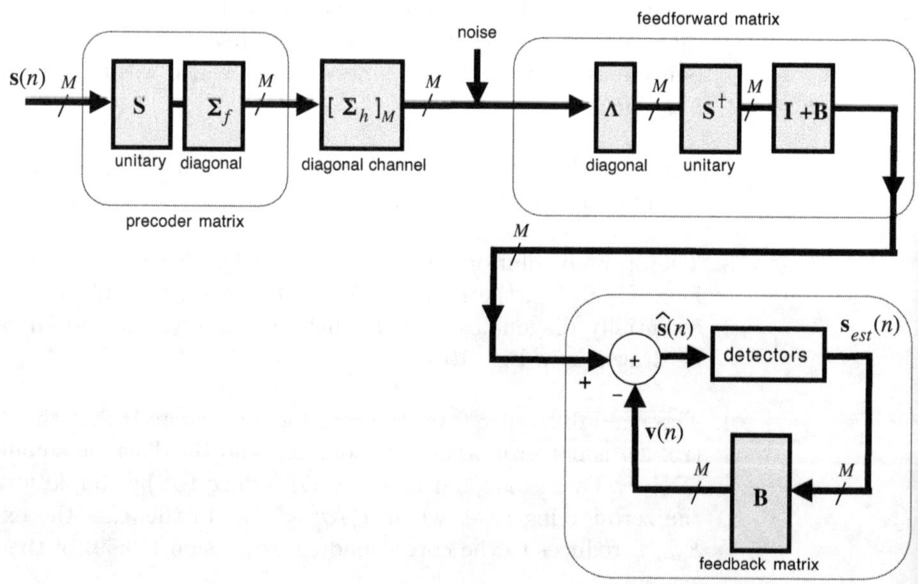

Figure 19.10. Equivalent diagonal form of the optimal DFE system without zero-forcing constraint.

19.5 Minimizing error probability in DFE transceivers

We now show that the optimal DFE transceivers derived in Sec. 19.3 (with zero forcing) and in Sec. 19.4 (without zero forcing) are also optimal in the sense of minimizing the average symbol error probability. This is related to the fact that the mean square error components $\mathcal{E}_k$ at the input of the detector are equalized in the optimal solutions (because $\mathbf{R_{ee}} = \sigma_e^2\mathbf{I}$, as shown by Eqs. (19.60) and (19.114)).

For completeness, we shall provide a formal argument to justify the minimum error probability property when $\mathbf{s}(n)$ comes from a PAM or a QAM constellation. The reconstruction error at the input of the detector is $\mathbf{e}(n) = \widehat{\mathbf{s}}(n) - \mathbf{s}(n)$. Letting $\mathcal{E}_k$ denote the mean square value of the kth component, the average error probability in the ZF case is given by

$$\mathcal{P}_{ZF} = \frac{c}{M}\sum_{k=0}^{M-1} \mathcal{Q}\left(A\sqrt{\frac{1}{\mathcal{E}_k}}\right) \quad \text{(ZF case).} \tag{19.115}$$

In the non-ZF case (assuming bias removal as explained in Sec. 16.3), the average error probability is

$$\mathcal{P}_{no\,ZF} = \frac{c}{M}\sum_{k=0}^{M-1} \mathcal{Q}\left(A\sqrt{\frac{1}{\mathcal{E}_k} - \frac{1}{\sigma_s^2}}\right) \quad \text{(non-ZF case).} \tag{19.116}$$

Here c and A are constants that depend on the number of bits in the symbol streams. For any fixed set of matrices $\mathbf{F}$ and $\mathbf{B}$ it is therefore clear that $\mathbf{G}$ should be such that $\mathcal{E}_k$ is minimized for each k. Indeed, this is how $\mathbf{G}$ was chosen in Secs. 19.3 and 19.4.

We will now argue that if a system minimizes the average error probability, then the errors $\mathcal{E}_k$ should necessarily be identical for all k. This situation is similar to what we found in Sec. 16.4. To make the argument formally, consider the modified system shown in Fig. 19.11(a), where a unitary matrix $\mathbf{U}$ and its inverse $\mathbf{U}^\dagger$ are inserted at certain places, essentially replacing the matrices $\mathbf{F}, \mathbf{G}$, and $\mathbf{B}$ as follows:

$$\mathbf{F}_1 = \mathbf{F}\mathbf{U}^\dagger, \quad \mathbf{G}_1 = \mathbf{U}\mathbf{G}, \quad \mathbf{B}_1 = \mathbf{U}\mathbf{B}\mathbf{U}^\dagger.$$

Even though it appears that this system is unrealizable because $\mathbf{U}\mathbf{B}\mathbf{U}^\dagger$ is not strictly upper triangular, the system can readily be redrawn by using the equivalent form shown in Fig. 19.11(b) where $\mathbf{B}$ is strictly upper triangular. We now observe the following:

1. ***Power unchanged.*** The power constraint (19.74) continues to be satisfied since $\mathbf{F}_1\mathbf{F}_1^\dagger = \mathbf{F}\mathbf{F}^\dagger$.

2. ***ZF-property unchanged.*** The quantity $[\mathbf{GHF} - (\mathbf{B}+\mathbf{I})]$ which arises in the zero-forcing constraint (19.13) is replaced with

$$\mathbf{G}_1\mathbf{H}\mathbf{F}_1 - (\mathbf{B}_1 + \mathbf{I}) = \mathbf{U}[\mathbf{GHF} - (\mathbf{B}+\mathbf{I})]\mathbf{U}^\dagger.$$

   Thus, if the original system satisfies the zero-forcing condition $[\mathbf{GHF} - (\mathbf{B}+\mathbf{I})] = \mathbf{0}$, the modified system also satisfies zero forcing.

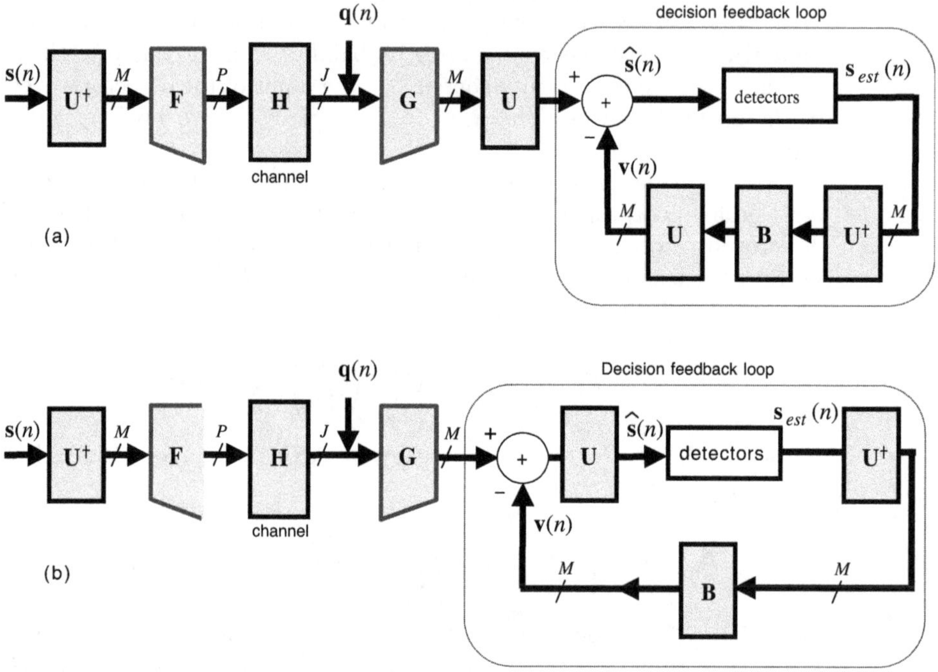

Figure 19.11. (a) The decision feedback transceiver with extra unitary matrices $\mathbf{U}$ and $\mathbf{U}^\dagger$, and (b) an equivalent rearrangement.

3. ***Total error unchanged.*** The reconstruction error (19.70) is replaced with

$$\mathbf{e} = \mathbf{U}\Big(\mathbf{GHF} - (\mathbf{B}+\mathbf{I})\Big)\mathbf{U}^\dagger\mathbf{s} + \mathbf{UGq}.$$

Thus the error covariance (19.71) is replaced with

$$\mathbf{R_{ee}} = \sigma_s^2\mathbf{U}\Big(\mathbf{GHF} - (\mathbf{B}+\mathbf{I})\Big)\Big(\mathbf{GHF} - (\mathbf{B}+\mathbf{I})\Big)^\dagger\mathbf{U}^\dagger + \sigma_q^2\mathbf{UGG}^\dagger\mathbf{U}^\dagger. \tag{19.117}$$

The trace of this quantity is unaffected by the presence of $\mathbf{U}$ because

$$\mathrm{Tr}\,(\mathbf{UAU}^\dagger) = \mathrm{Tr}\,(\mathbf{U}^\dagger\mathbf{UA}) = \mathrm{Tr}\,\mathbf{A}$$

for any $\mathbf{A}$. Thus, the total reconstruction error $\mathcal{E}_{mse}$ is unchanged. This is true whether there is zero forcing or not. If there is zero forcing, the only difference is that the first term on the right-hand side of Eq. (19.117) is absent.

Summarizing, the insertion of $\mathbf{U}$ and $\mathbf{U}^\dagger$ does not affect the power, the total mean square error

$$\mathcal{E}_{mse} = \sum_{k=0}^{M-1} \mathcal{E}_k, \tag{19.118}$$

or the zero-forcing constraint (if there is one). The only effect of $\mathbf{U}$ is to redistribute the values of the components $\mathcal{E}_k$. Next, for a fixed sum (19.118), we know that the vector

$$[\,\mathcal{E}_0 \quad \mathcal{E}_1 \quad \ldots \quad \mathcal{E}_{M-1}\,] \tag{19.119}$$

majorizes

$$(\mathcal{E}_{mse}/M)\,[\,1 \quad 1 \quad \ldots \quad 1\,] \tag{19.120}$$

(see Lemma 21.1). But since Eqs. (19.115) and (19.116) are Schur-convex in the vector (19.119) (for sufficiently large SNRs; see Sec. 11.5.2), it follows that the equal-error distribution (19.120) yields a smaller error probability than any other distribution. Thus, given any triplet $\{\mathbf{F}, \mathbf{G}, \mathbf{B}\}$ with unequal values for $\mathcal{E}_k$, we can always improve the average error probability by inserting $\mathbf{U}$ and $\mathbf{U}^\dagger$ to equalize $\mathcal{E}_k$. This shows that the system that minimizes average error probability should be such that

$$\mathcal{E}_0 = \mathcal{E}_1 = \ldots = \mathcal{E}_{M-1} = \frac{\mathcal{E}_{mse}}{M} \tag{19.121}$$

So the minimum error probability has the forms

$$\mathcal{P}_{ZF} = c\mathcal{Q}\left(A\sqrt{\frac{1}{(\mathcal{E}_{mse}/M)}} \right) \tag{19.122}$$

and

$$\mathcal{P}_{no\,ZF} = c\mathcal{Q}\left(A\sqrt{\frac{1}{(\mathcal{E}_{mse}/M)} - \frac{1}{\sigma_s^2}} \right). \tag{19.123}$$

Since the above probabilities are monotone increasing functions of $\mathcal{E}_{mse}$ it follows that the optimal system should also minimize the mean square error.

Summarizing, any system which minimizes the total error $\mathcal{E}_{mse}$, subject to the condition that the components $\mathcal{E}_k$ are identical, is also a system that minimizes the average error probability. But the optimal DFE systems derived in Sec. 19.3 (with zero forcing) and in Sec. 19.4 (without zero forcing) already have the property (19.121), because the error covariance is $\mathbf{R_{ee}} = \sigma_e^2\mathbf{I}$. Furthermore they are designed to have the MMSE property. So it follows that these systems are also optimal in the sense of minimizing the average error probability (assuming that the SNRs at the detector inputs are large enough to satisfy Schur convexity). *The matrix* $\mathbf{U}$ *is therefore not necessary.* The minimized error probability is given by Eq. (19.122) or Eq. (19.123) with the minimum error $\mathcal{E}_{mmse}$ substituted for $\mathcal{E}_{mse}$, namely

$$\mathcal{E}_{mmse} = \begin{cases} \dfrac{M^2\sigma_s^2\sigma_q^2}{p_0}\Big(\prod_{k=0}^{M-1}\dfrac{1}{\sigma_{h,k}}\Big)^{2/M} & \text{(ZF case)} \\ \\ M\sigma_s^2\Big(\dfrac{\lambda\sigma_q^2}{\sigma_s^2}\Big)^{K/M}\prod_{k=0}^{K-1}\Big(\dfrac{1}{\sigma_{h,k}}\Big)^{2/M} & \text{(no ZF).} \end{cases} \quad (19.124)$$

where λ and K are as described in Sec. 19.4.4.

19.6 Examples of optimal transceivers with DFE

In this section we present a number of examples which compare jointly optimal DFE transceivers with and without zero forcing with jointly optimal linear transceivers studied in earlier chapters. The superior performance of DFE transceivers will be amply demonstrated by these examples. In the examples we assume that the input is PAM, and that the channel has real coefficients. Even though the equivalent baseband channel is usually complex, one can use a time-domain equalizer first to get a shorter real equivalent channel in the PAM case, as mentioned in Sec. 2.A.2 (Appendix of Chap. 2).

Example 19.1: DFE gain

Consider the scalar channel

$$C_1(z) = \beta(0.0331 + 0.0894z^{-1} + 0.1407z^{-2} + 0.1522z^{-3} + 0.1050z^{-4} + 0.0827z^{-5}$$
$$+0.0882z^{-6} + 0.1389z^{-7} + 0.1099z^{-8} + 0.0598z^{-9}),$$

which was also considered in the examples of Chaps. 17 and 18. The frequency response plot and impulse response plot for this channel were given in Fig. 17.16. Assume that we have a zero-padded precoder for this channel. The channel order is $L = 9$, so we pad L zeros per block of size M, and the extended block size is $J = M + L$. Then the equivalent MIMO channel is a $J \times M$ full-banded Toeplitz matrix $\mathbf{H}$. In the notation of Fig. 19.3 this means $P = M$ and $J = M + L$. Taking this $\mathbf{H}$ to be our MIMO channel we can design an optimal DFE system with precoder $\mathbf{F}$, feedforward equalizer $\mathbf{G}$, and feedback equalizer $\mathbf{B}$. We can do this with or without zero forcing. We will plot the DFE gain as a function of block size M. Recall that the DFE gain (Sec. 19.3.2) is defined by the ratio

$$\mathcal{G}_{DFE} = \frac{\mathcal{E}_{nodfe}}{\mathcal{E}_{dfe}}$$

where $\mathcal{E}_{nodfe}$ is the minimized mean square error for an optimal linear transceiver (no DFE, see Chaps. 12 and 13) and $\mathcal{E}_{dfe}$ denotes the minimized mean square error for an optimal DFE transceiver (Secs. 19.3 and 19.4). The value of this

DFE gain depends on whether we have zero forcing or not. In the zero-forcing case it can be calculated from the AM-GM ratio (19.29), reproduced below:

$$\mathcal{G}_{DFE,zf} = \left(\frac{AM\left\{ \dfrac{1}{\sigma_{h,k}} \right\}}{GM\left\{ \dfrac{1}{\sigma_{h,k}} \right\}} \right)^2.$$

In the non-zero-forcing case the DFE gain can be calculated by using the expressions for the errors (19.113) and Eq. (13.66) in Chap. 13. The DFE gain for the zero-forcing case is plotted in Fig. 19.12 for the above channel $C_1(z)$ (top) and for the channel

$$C_2(z) = \beta(1 + 2z^{-1} + z^{-2})$$

(bottom plot), as a function of the block size M. Both channels are normalized to have unit energy, by choosing β appropriately. For the same two channels the DFE gain without zero forcing is plotted in Fig. 19.13. In these plots we have assumed symbol power $p_0/M = 10$, $\sigma_s^2 = 1$, and $\sigma_q^2 = 0.01$.

The DFE gain is indeed impressive in all these plots. For the zero-forcing case with channel $C_2(z)$ (bottom plot, Fig. 19.12) the gain is indeed very large for even modestly large block size M; this happens because channel $C_2(z)$ has a double zero on unit-circle (at $z = -1$).

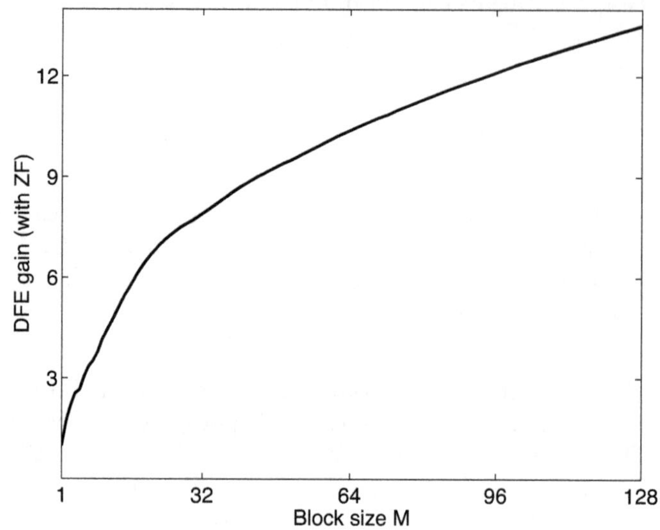

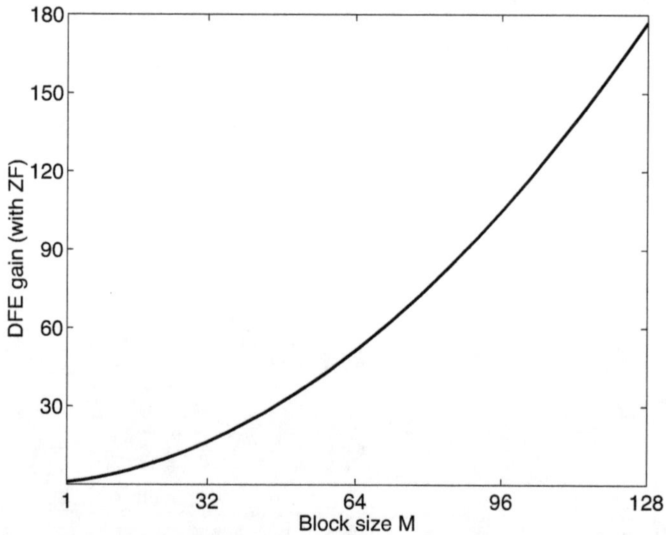

Figure 19.12. Example 19.1. Plot of DFE gain as a function of block size M for optimal zero-forcing DFE transceivers. Channel $C_1(z)$ (top) and channel $C_2(z)$ (bottom).

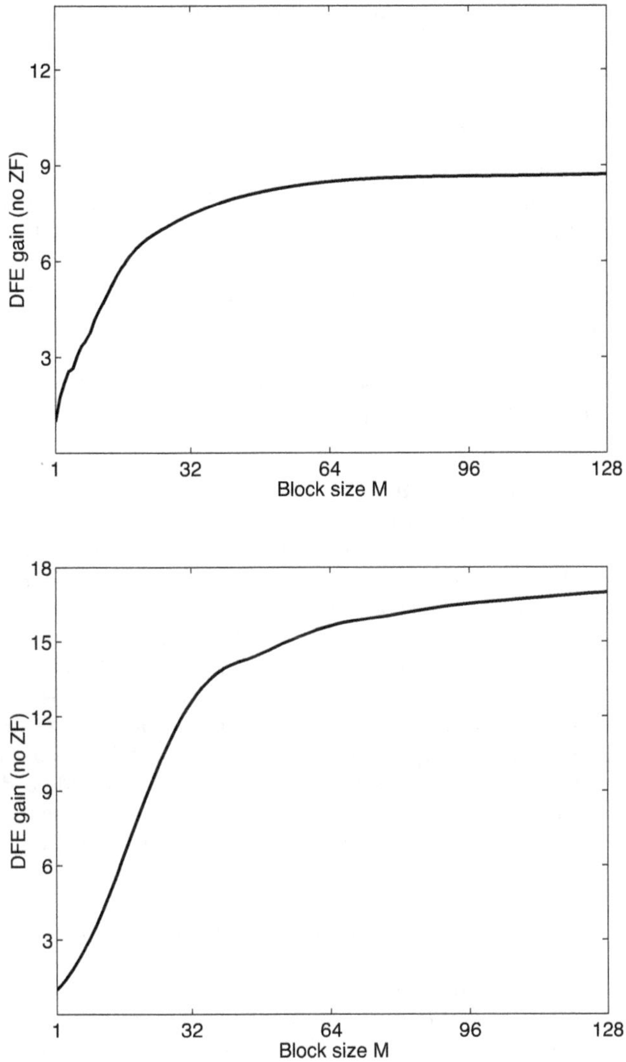

Figure 19.13. Example 19.1 (continued). Plot of DFE gain as a function of block size M for optimal *non zero-forcing* DFE transceivers. Channel $C_1(z)$ (top) and channel $C_2(z)$ (bottom).

Example 19.2: Symbol error probability with DFE (zero-padded systems)

For the zero-padded optimal DFE system considered in Ex. 19.1 we now plot the symbol error probabilities as a function of the quantity $p_0/M\sigma_q^2$ (signal power per symbol $s_k(n)$ over channel noise variance). In this example we consider 2-bit PAM, and set $\sigma_s^2 = 1$ and $\sigma_q^2 = 0.01$. The block size is fixed at $M = 16$. First we consider channel $C_1(z)$ given in Ex. 19.1. The top plot in Fig. 19.14 shows the symbol error probabilities for the optimal linear transceiver and the optimal DFE transceivers, both with zero forcing. We see that the improvement as we go from linear to DFE systems is substantial. The same holds if we compare linear and DFE transceivers, both without zero forcing (i.e., pure-MMSE transceivers), as seen from Fig. 19.14 (middle). From the bottom plot in the figure which shows all four graphs together, we see that as we move from zero-forced systems to pure-MMSE systems (with or without DFE) the improvement is not that substantial. Figure 19.15 shows symbol error probabilities for the channel

$$C_3(z) = 1 + 3z^{-1} + 3z^{-2} + z^{-3},$$

which has a triple zero on the unit circle (at $z = -1$). The conclusions for this channel are similar. Both channels are normalized to have unit energy before making all computations.

Simulation results. The plots shown in Figs. 19.14 and 19.15 use the closed form expressions for the error probabilities given in Sec. 19.5. Instead of using these formulas, one can actually *simulate* the transceiver using Monte Carlo techniques. If we do this, then the error probabilities for the DFE system for channel $C_1(z)$ have the values shown by the circles in Fig. 19.16, where the plots from theoretical formulas are also shown (dashed lines). Note that the agreement with theory is not very good for low values of the SNR but improves as the SNR increases. The theoretical formula for the linear transceiver is also shown in the figure. In the region of large error probabilities the linear transceiver is found to be even better than the DFE system! The disagreement between the theoretical formula and simulation results for the DFE system at low SNR is due to *error propagation* in the DFE loop. The reader may recall that the theoretical development assumed that there is no error propagation (Sec. 19.2.2). In fact, it is possible to perform a simulation in which the exact symbols are fed back as in Fig. 19.5. Such simulation with *no error propagation* yields the results shown in Fig. 19.17. Indeed the theoretical formula and simulation results agree very well for all SNR in this case.

For completeness, Fig. 19.18 shows the comparison between the theoretical formula and Monte Carlo simulations for optimal linear transceivers.[2] Once again, channel $C_1(z)$ is used. The theoretical formula is shown by the solid line, and the circles represent Monte Carlo simulations. Again, as the SNR increases, the agreement between theory and simulation improves significantly. For insightful discussions on the accuracy of BER formulas, see Verdu [1998]. ▽ ▽ ▽

[2]The authors are thankful to Ching-Chih Weng, graduate student at the California Institute of Technology, for performing these simulations.

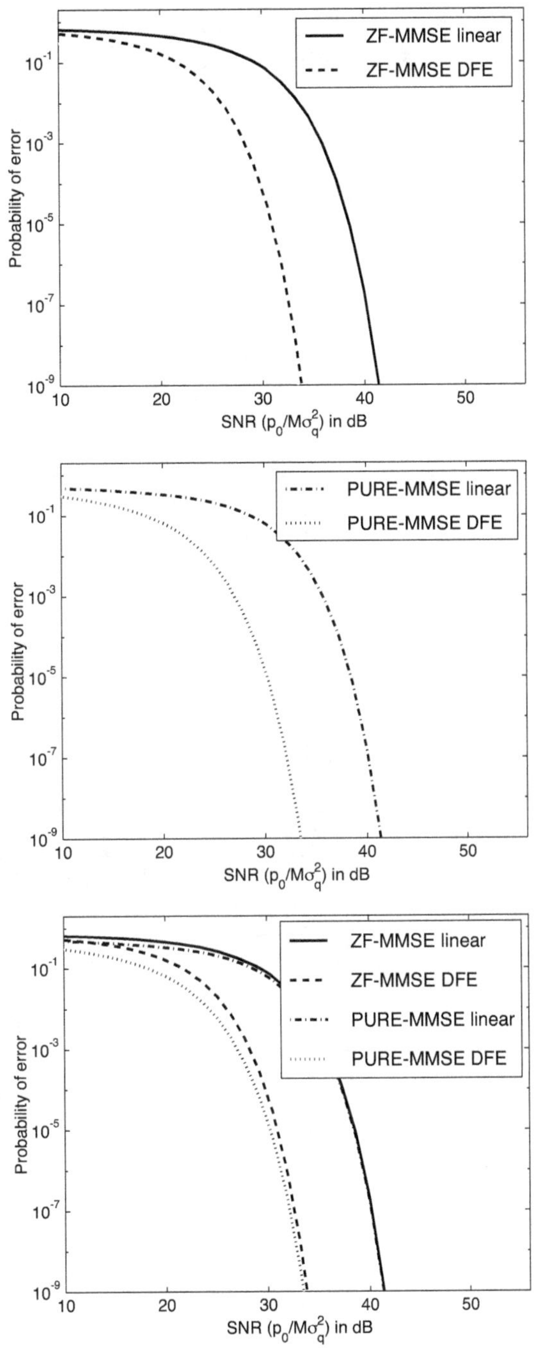

Figure 19.14. Example 19.2. Symbol error probabilities of optimal zero-padded DFE transceivers compared with those of optimal zero-padded linear transceivers, for channel $C_1(z)$. Top plot shows zero-forced systems, and middle plot shows systems without zero forcing. All four plots are overlaid in the bottom.

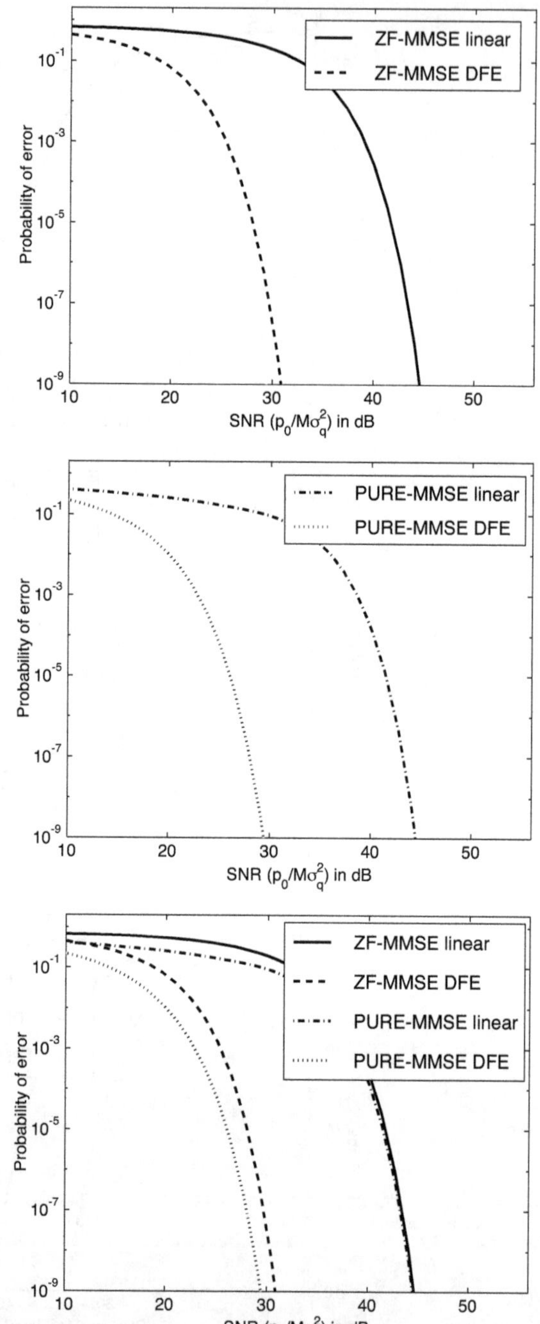

Figure 19.15. Example 19.2 (continued). Symbol error probabilities of optimal zero-padded DFE transceivers compared with those of optimal zero-padded linear transceivers, for channel $C_3(z)$. Top plot shows zero-forced systems, and middle plot shows systems without zero forcing. All four plots are overlaid in the bottom.

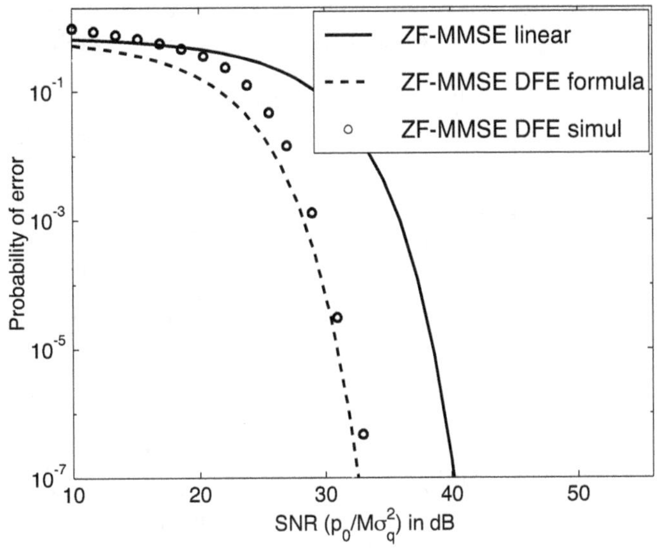

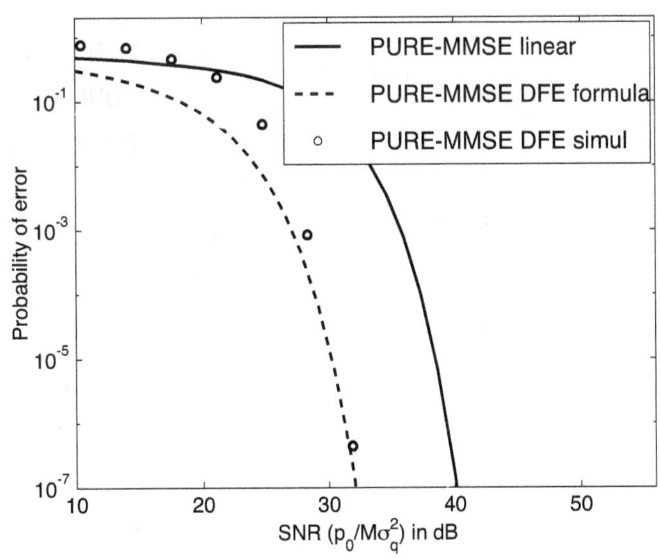

Figure 19.16. Example 19.2 (continued). Simulation versus theory. Symbol error probabilities of optimal zero-padded DFE transceivers for channel $C_1(z)$. The dashed lines are from the closed-form formula for the error probability, whereas the circles are obtained from actual Monte Carlo simulation. Top plot shows zero-forced systems, and bottom plot shows MMSE systems without zero forcing. The error probabilities for linear transceivers are shown as solid lines for reference.

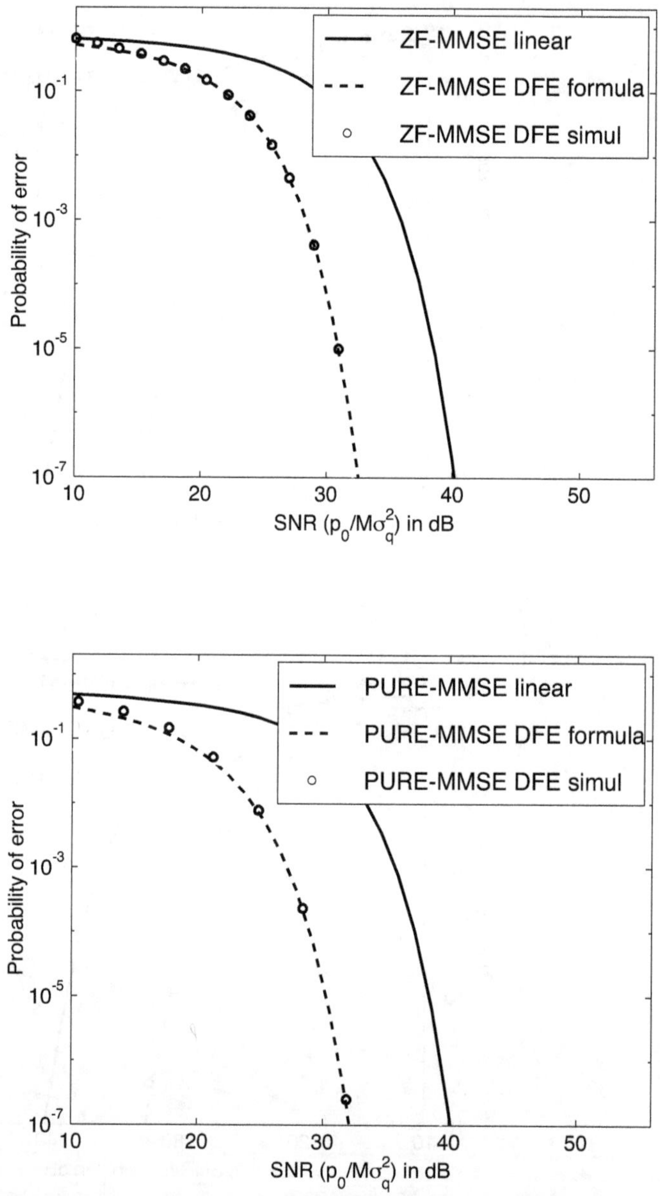

Figure 19.17. Example 19.2 (continued). Simulation with no error propagation. Repetition of the simulations in Fig. 19.16 with the difference that the symbol error propagation in the DFE loop is artificially curbed by feeding back the actual symbol values as in Fig. 19.5.

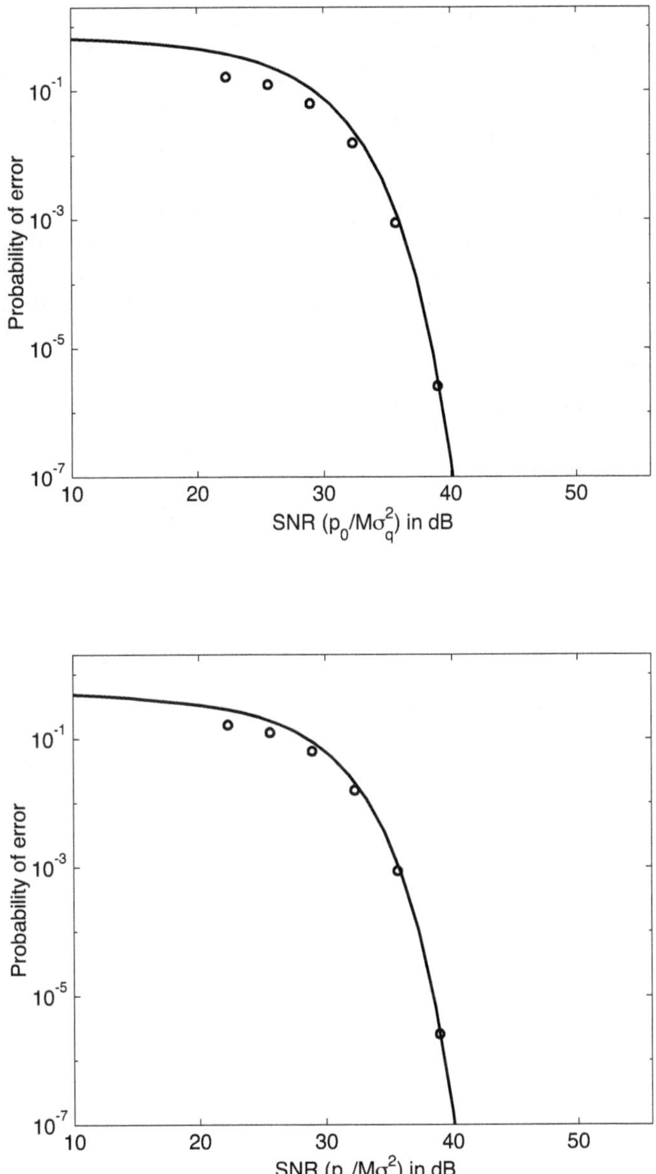

Figure 19.18. Example 19.2 (continued). Formula versus simulation for linear transceivers. Channel $C_1(z)$ is used. Top: Zero-forced optimal linear transceiver, and bottom: optimal linear transceiver without zero forcing. The theoretical formula is shown by the solid line, and the little circles represent Monte Carlo simulations.

Example 19.3: Symbol error probability with DFE (cyclic-prefix systems)

We now repeat Ex. 19.2 with cyclic prefixing instead of zero padding. Channel $C_1(z)$ from Ex. 19.1 is used again (with energy normalized to unity). The block size is taken to be $M = 16$. A 2-bit PAM constellation was used. In this example $\sigma_s^2 = 1$ as usual, and $\sigma_q^2 = 0.01$. With cyclic prefixing, the equivalent matrix channel $\mathbf{H}$ is a circulant matrix, and it is diagonalized by the DFT matrix. We can therefore take the matrices $\mathbf{V}_h$ and $\mathbf{U}_h$ in Figs. 19.6 and 19.9 to be the IDFT and DFT matrices (with minor adjustments to account for the fact that the singular values $\sigma_{h,k}$ are the magnitudes of the channel DFT coefficients, in decreasing order). The unitary matrix $\mathbf{S}$ in the precoder and the upper triangular matrix $\mathbf{B}$ in the decision feedback loop do not have additional structure because they are obtained from a QRS decomposition of a general diagonal matrix as in previous examples.

From the plots of the error probability in Fig. 19.19 we see once again that DFE leads to substantial improvement in performance. This improvement due to DFE is much more significant than the improvement that arises from replacing zero-forced systems with non-zero-forced (pure-MMSE) systems.

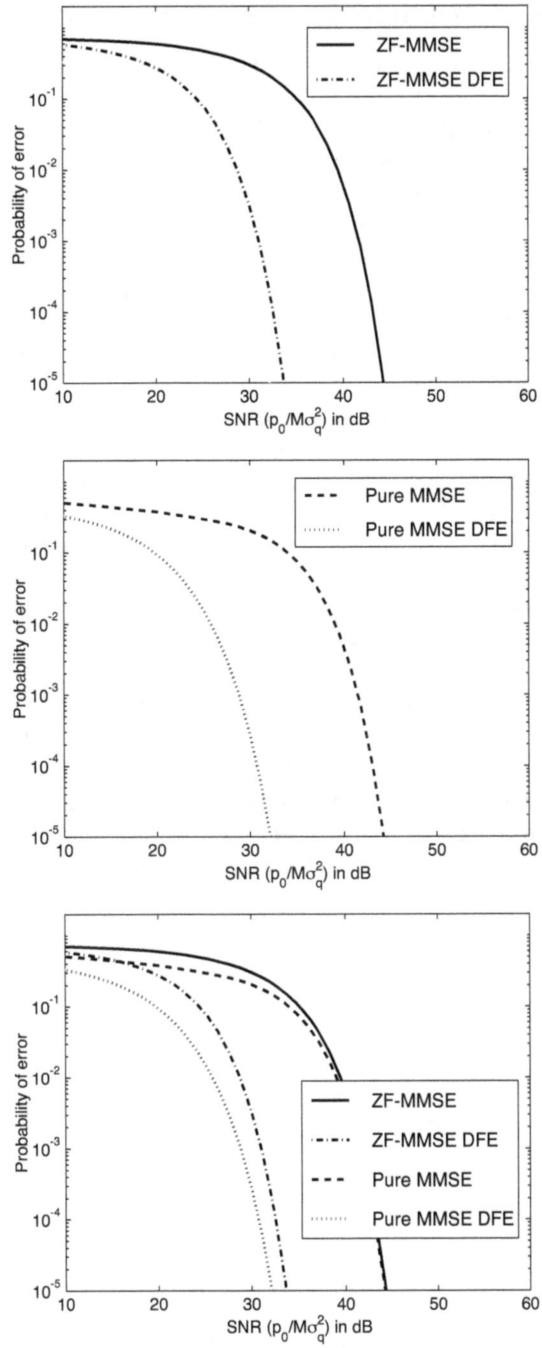

Figure 19.19. Example 19.3. Symbol error probabilities of optimal cyclic-prefixed DFE transceivers compared with those of optimal cyclic-prefixed linear transceivers, for channel $C_1(z)$. Top plot shows zero-forced systems, and middle plot shows systems without zero-forcing. All four plots are overlaid in the bottom.

Example 19.4: Effect of increasing the block size

In Chap. 18 we observed that zero-padded optimal linear transceivers have the property that the symbol error probability increases with block size M. A similar behavior is also observed in the case of zero-padded optimized DFE transceivers with or without zero forcing. This is demonstrated in Fig. 19.20 for channel $C_1(z)$ used in earlier examples. To obtain these plots we have used 2-bit PAM with $\sigma_s^2 = 1$ and channel noise $\sigma_q^2 = 0.01$. The channel input power per symbol was $p_0/M = 5$, and the channel energy is normalized to 4 in this example. Figure 19.21 shows similar plots for the channel $C_2(z) = 1 + 2z^{-1} + z^{-2}$ with energy normalized to 4 and power per symbol $p_0/M = 1$. All other parameters are as above. In these examples, the error probabilities for the linear transceivers are very large, even for moderately large block sizes.

Note that the error probability increases monotonically with block sizes for all four jointly optimal transceivers – linear as well as DFE transceivers, with or without zero forcing. No theoretical proof of this monotonic behavior is known. In Chap. 8 where we considered lazy zero-padded precoders with zero-forcing linear equalizers, we formally proved that the error probability increases monotonically with block size M (see Appendix 8.A of Chap. 8). For the case of jointly optimized transceivers (linear or DFE), such as the ones shown in Fig. 19.20, such a formal proof has not been reported in the literature, though counter examples have not been known either.

Note finally that, even though the symbol error probability increases monotonically with block size, for the case of DFE systems it tends to saturate at a much lower value than for linear transceivers. Thus, even with $C_2(z)$, which has a double zero on the unit circle, the error probabiliy for the DFE system without zero forcing continues to be excellent as M grows. For the linear transceivers, however, the error probability becomes impractically large in this example.

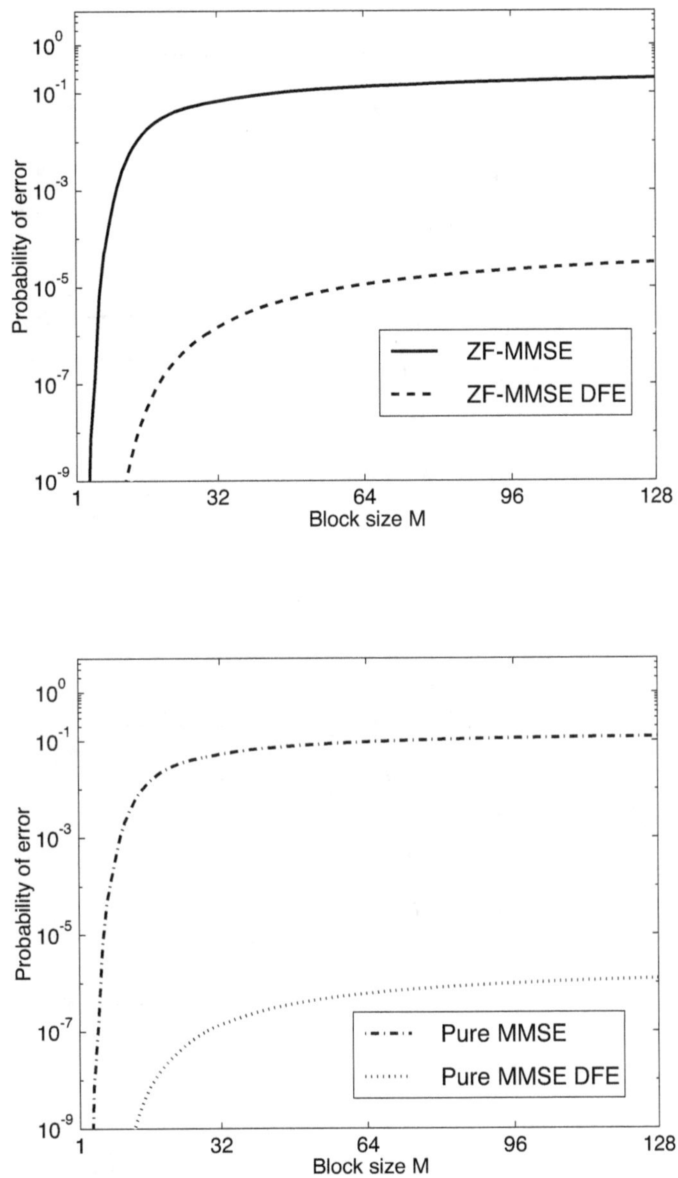

Figure 19.20. Example 19.4. Symbol error probabilities of optimal zero-padded transceivers as a function of block size M. Channel $C_1(z)$ is used, with energy normalized to 4.

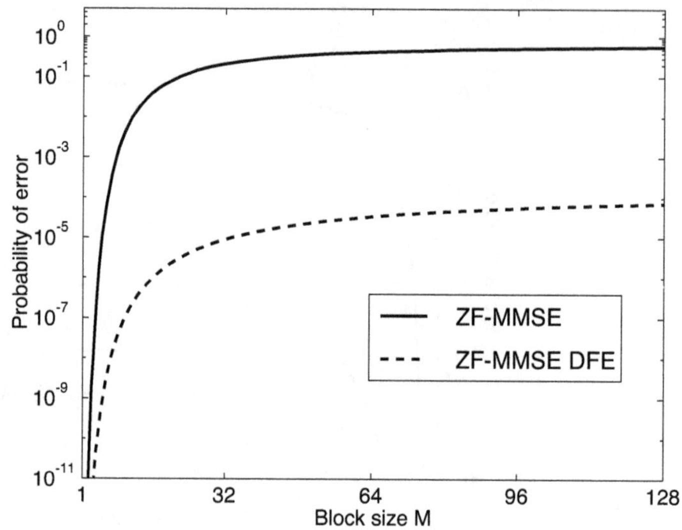

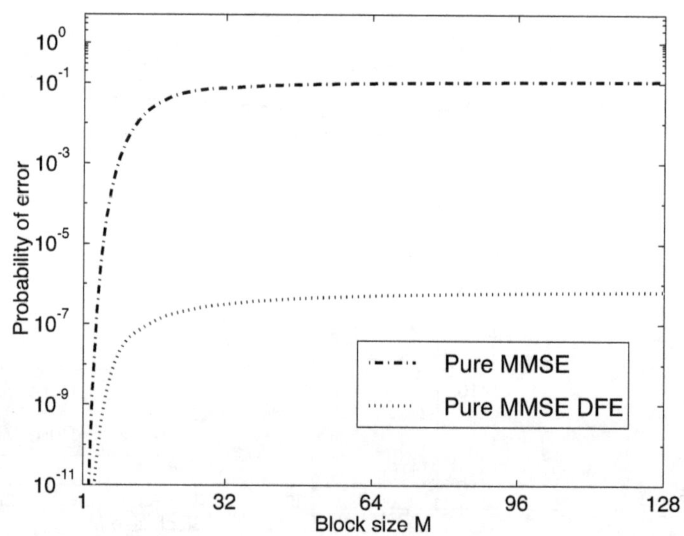

Figure 19.21. Example 19.4 (continued). Symbol error probabilities of optimal zero-padded transceivers as a function of block size M. Channel $C_2(z) = 1 + 2z^{-1} + z^{-2}$ is used, with energy normalized to 4.

19.7 DFE optimization and mutual information

In Sec. 6.7 we provided a brief discussion of mutual information between the input and output of a channel. Based on that discussion it can be seen that the maximization of Eq. (19.78) is the same problem as maximizing mutual information. More precisely, assuming that $\mathbf{s}(n)$ and $\mathbf{q}(n)$ are zero-mean statistically independent circularly symmetric Gaussian processes (Sec. 6.6), with covariances $\sigma_s^2\mathbf{I}$ and $\sigma_q^2\mathbf{I}$, the mutual information between $\mathbf{x}$ and $\mathbf{y}$ in Fig. 19.22 is given by

$$\mathcal{I}(\mathbf{x};\mathbf{y}) = \log\det\Big(\mathbf{I}_J + \frac{\sigma_s^2}{\sigma_q^2}\mathbf{H}\mathbf{F}\mathbf{F}^\dagger\mathbf{H}^\dagger\Big). \tag{19.125}$$

Using Eq. (19.80) this can also be written as

$$\mathcal{I}(\mathbf{x};\mathbf{y}) = \log\det\Big(\mathbf{I}_M + \frac{\sigma_s^2}{\sigma_q^2}\mathbf{F}^\dagger\mathbf{H}^\dagger\mathbf{H}\mathbf{F}\Big). \tag{19.126}$$

Now consider the quantity ϕ defined in Eq. (19.78) and reproduced below:

$$\phi = \det\Big(\mathbf{I}_M + \frac{\sigma_s^2}{\sigma_q^2}\mathbf{F}^\dagger\mathbf{H}^\dagger\mathbf{H}\mathbf{F}\Big). \tag{19.127}$$

In Sec. 19.4 we essentially chose $\mathbf{F}$ to maximize this quantity. Since $\mathcal{I}(\mathbf{x};\mathbf{y}) = \log\phi$ we see that the maximization of ϕ is equivalent to maximizing[3] the mutual information $\mathcal{I}(\mathbf{x};\mathbf{y})$. The MMSE DFE problem (without zero forcing) therefore has a precoder $\mathbf{F}$ which also maximizes mutual information between the channel input $\mathbf{x}$ and noisy output $\mathbf{y}$ under a power constraint at the input of $\mathbf{H}$. Now, since the minimized MSE achieves the bound in Eq. (19.77), we conclude that

$$\mathcal{E}_{mmse} = M\sigma_s^2\det\Big(\mathbf{I} + \frac{\sigma_s^2}{\sigma_q^2}\mathbf{F}^\dagger\mathbf{H}^\dagger\mathbf{H}\mathbf{F}\Big)^{-1/M} = M\sigma_s^2\phi^{-1/M}. \tag{19.128}$$

Since the optimized system has identical MSE for each of the signal components, the error $E[|\widehat{s}_k - s_k|^2]$ is given by

$$\mathcal{E}_k = \sigma_s^2\det\Big(\mathbf{I} + \frac{\sigma_s^2}{\sigma_q^2}\mathbf{F}^\dagger\mathbf{H}^\dagger\mathbf{H}\mathbf{F}\Big)^{-1/M} = \sigma_s^2\phi^{-1/M} \tag{19.129}$$

for each k so that

$$\mathcal{I}(\mathbf{x};\mathbf{y}) = \log_2\phi = M\log_2\Big(\frac{\sigma_s^2}{\mathcal{E}_{ave}}\Big), \tag{19.130}$$

where

$$\mathcal{E}_{ave} = \mathcal{E}_{mmse}/M \tag{19.131}$$

for all k.

---

[3] Since $\mathbf{F}$ is allowed to be rectangular, this is the maximum mutual information subject to the additional constraint that the rank of $\mathbf{R}_{\mathbf{xx}}$ be at most M. For the special case where $M = P$, this rank constraint on $\mathbf{R}_{\mathbf{xx}}$ goes away.

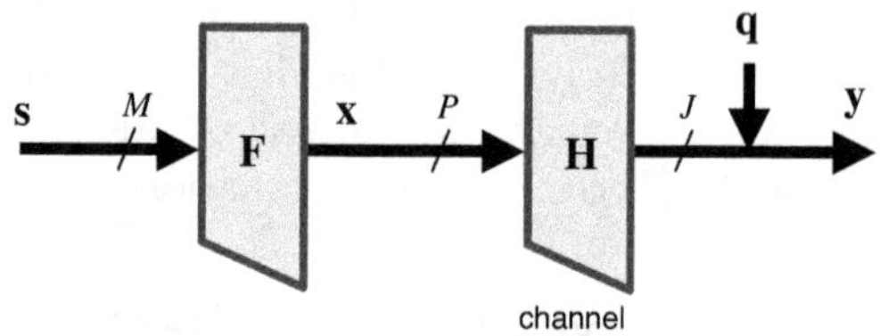

Figure 19.22. A MIMO channel with a linear precoder.

Recall now that

$$\frac{1}{\mathcal{E}_k} = \frac{1}{\mathcal{E}_{k,br}} + \frac{1}{\sigma_s^2} \tag{19.132}$$

where $\mathcal{E}_{k,br}$ is the mean square error after bias removal (Sec. 16.3.2). In our case, $\mathcal{E}_k = \mathcal{E}_{ave}$ for all k, so $\mathcal{E}_{k,br} = \mathcal{E}_{ave,br}$ for all k, where

$$\frac{1}{\mathcal{E}_{ave}} = \frac{1}{\mathcal{E}_{ave,br}} + \frac{1}{\sigma_s^2}$$

Thus Eq. (19.130) yields

$$\mathcal{I}(\mathbf{x};\mathbf{y}) = \sum_{k=0}^{M-1} \log_2\Big(1 + \frac{\sigma_s^2}{\mathcal{E}_{ave,br}}\Big). \tag{19.133}$$

This is the mutual information between the input and output of the channel. To understand this expression in the proper light, observe that the reconstructed vector $\widehat{\mathbf{s}}$ (at the detector input, after bias removal) can be written as

$$\widehat{\mathbf{s}} = \mathbf{s} + \mathbf{e}, \tag{19.134}$$

where the error $\mathbf{e}$ is zero-mean with covariance

$$\mathbf{R_{ee}} = \mathcal{E}_{ave,br}\mathbf{I}. \tag{19.135}$$

So, in the optimal system we can regard the path from $\mathbf{s}$ to $\widehat{\mathbf{s}}$ as an (approximately) AWGN channel with transfer matrix $\mathbf{I}$, and noise covariance $\mathbf{R_{ee}} = \mathcal{E}_{ave,br}\mathbf{I}$. This is demonstrated in Fig. 19.23. This is a set of identical, parallel, independent channels. This explains in an independent way that the mutual information between $\mathbf{s}$ and $\widehat{\mathbf{s}}$ (with $\mathbf{s}$ chosen to be zero-mean circularly symmetric Gaussian with covariance $\sigma_s^2\mathbf{I}$) is indeed of the form (19.133).

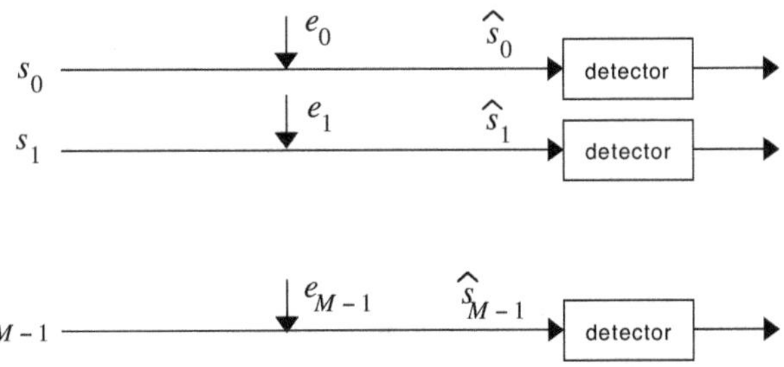

Figure 19.23. The diagonal channel model for the optimal DFE system represented by Eq. (19.134). The errors e_k are zero-mean and independent random variables with identical variance.

Discussions

1. ***Zero-forcing case.*** We should notice carefully that the MMSE problem is equivalent to mutual information maximization only for the *DFE system with the precoder and equalizer matrices jointly optimized without the zero-forcing constraint*, and under the assumption that there is no error propagation in the feedback loop. If we impose zero forcing, for example, then the solution is different. The model (19.134) continues to hold with error covariance

$$\mathbf{R_{ee}} = \mathcal{E}_{ZF}\mathbf{I}, \tag{19.136}$$

but since $\mathcal{E}_{ZF} > \mathcal{E}_{ave,br}$ in general (see Appendix 19.D at the end of this chapter), the optimized system has a smaller mutual information (and larger mean square error).

2. ***Water filling.*** For the DFE system without zero forcing, the diagonal elements $\sigma_{f,k}^2$ of the optimal precoder are chosen using the water-filling allocation (19.95), which has the general form

$$\sigma_{f,k}^2 = \begin{cases} c\Big(A - \dfrac{1}{\sigma_{h,k}^2}\Big) & \text{if this is non-negative} \\ 0 & \text{otherwise.} \end{cases} \tag{19.137}$$

This is not surprising because this is the form which also arises in the maximization of mutual information in Sec. 22.3 (see Eq. (22.26)). By contrast, the formula for $\sigma_{f,k}^2$ in the zero-forcing case is totally different, namely $\sigma_{f,k}$ was constant for all k (see Eq. (19.56)).

3. ***Linear MMSE case.*** In Chaps. 12 and 13 we optimized linear transceivers (without DFE) with and without zero-forcing constraints. The solutions

in these cases do not maximize mutual information. Indeed the formulas for $\sigma_{f,k}^2$ were very different from the water-filling formula. For example, the linear MMSE transceiver without zero forcing used a formula of the form

$$\sigma_{f,k}^2 = \begin{cases} \dfrac{c}{\sigma_{h,k}}\Big(A - \dfrac{1}{\sigma_{h,k}}\Big) & \text{if this is non-negative} \\ 0 & \text{otherwise.} \end{cases} \tag{19.138}$$

The bracketted term

$$\Big(A - \frac{1}{\sigma_{h,k}}\Big)$$

does indeed look like a water-filling formula, but it is multiplied by $c/\sigma_{h,k}$, which makes it different. So Eq. (19.138) is only a "pseudo" water-filling equation. For the optimal linear transceivers of Chaps. 12 and 13, the reconstructed signal at the detector input can still be written as in Eq. (19.134), but the covariance of the error is not proportional to identity as in Eq. (19.135). We used an extra unitary matrix $\mathbf{U}$ to minimize the error probability. This matrix equalized the mean square errors, so $[\mathbf{R_{ee}}]_{kk}$ were identical for all k, but $\mathbf{R_{ee}}$ was not diagonal.

4. ***Losslessness of mutual information.*** Since the MMSE DFE system without ZF maximizes the mutual information, it will be called a *mutual-information lossless* system. In the literature it has been referred to as a *capacity lossless* system [Jiang *et al.*, 2005b]. Technically, since the ZF constraint does not make the capacity smaller (because capacity cannot be decreased by insertion of invertible precoders at the transmitter), we prefer the term *mutual-information lossless.* Since the system from $\mathbf{s}$ (transmitted symbol) to $\widehat{\mathbf{s}}$ (detector input) is characterized by Eq. (19.134) we can conceptually represent it as in Fig. 19.23. Here s_k are statistically independent with identical power σ_s^2. The fact that the mutual information has been maximized means that we can in principle use an identical **channel coding** scheme for each individual symbol stream $s_k(n)$ and get arbitrarily close to capacity [Gallager, 1968] (while making error probabilities arbitarily small). By contrast, in any of the other systems (e.g., DFE with ZF, or any linear transceiver), since there is loss of mutual information, we cannot independently code $s_k(n)$ to get close to the capacity of the channel.

19.8 Other algorithms related to decision feedback

The use of space-domain decision feedback within the block has been prevalent owing to its importance in wireless MIMO channels and in multiuser detection theory see Zhang *et al.* [2005] and references therein). In most of the early versions, the precoder was identity, that is, $\mathbf{F} = \mathbf{I}$ (the so-called lazy precoder). The focus of this chapter so far, on the other hand, has been to optimize a general precoder matrix jointly with the receiver matrices (Secs. 19.3–19.5). In this section we describe some of the earlier lazy-precoder DFE systems, in order to place the main results of this chapter in proper historical context. Recall at this point that systems with a lazy precoder do not have collaboration between the components $s_k(n)$ at the transmitter, and are applicable to multiuser systems operating in the multiple access mode (Sec. 4.5).

19.8.1 DFE based on QR decomposition of the channel

Consider a $J \times M$ MIMO channel $\mathbf{H}$ with additive noise $\mathbf{q}$. With precoder $\mathbf{F} = \mathbf{I}$ the received signal is given by

$$\mathbf{y} = \mathbf{H}\mathbf{s} + \mathbf{q}. \tag{19.139}$$

With the channel $\mathbf{H}$ represented using QR decomposition (Appendix 19.A at the end of this chapter), we have

$$\mathbf{H} = \mathbf{Q}\mathbf{R}, \tag{19.140}$$

where $\mathbf{Q}$ has orthonormal columns, that is,

$$\mathbf{Q}^\dagger\mathbf{Q} = \mathbf{I}_M,$$

and $\mathbf{R}$ is upper triangular. Thus Eq. (19.139) becomes

$$\mathbf{y} = \mathbf{Q}\mathbf{R}\mathbf{s} + \mathbf{q},$$

so that

$$\mathbf{v} \stackrel{\Delta}{=} \mathbf{Q}^\dagger\mathbf{y} = \mathbf{R}\mathbf{s} + \mathbf{Q}^\dagger\mathbf{q}. \tag{19.141}$$

Thus, if we design a receiver with $\mathbf{Q}^\dagger$ in the front, the $\mathbf{Q}$ part is cancelled, and the receiver sees a triangular channel $\mathbf{R}$. Writing Eq. (19.141) out explicitly, we have

$$\begin{bmatrix} v_0 \\ v_1 \\ \vdots \\ v_{M-1} \end{bmatrix} = \begin{bmatrix} r_{00} & r_{01} & r_{02} & \ldots & r_{0,M-1} \\ 0 & r_{11} & r_{12} & \ldots & r_{1,M-1} \\ \vdots & \vdots & \vdots & \ddots & \vdots \\ 0 & 0 & 0 & \ldots & r_{M-1,M-1} \end{bmatrix} \begin{bmatrix} s_0 \\ s_1 \\ \vdots \\ s_{M-1} \end{bmatrix} + \mathbf{Q}^\dagger\mathbf{q}. \tag{19.142}$$

A linear zero-forcing receiver would estimate $\mathbf{s}$ by inverting $\mathbf{R}$ and then using a decision device. Instead of this approach, we can detect one symbol at a time and use it in future decisions by proceeding as follows: from the last equation in Eq. (19.142) we see that the received noisy signal v_{M-1} depends only on one transmitted symbol, namely s_{M-1}. So we estimate this first:

$$\widehat{s}_{M-1} = \frac{v_{M-1}}{r_{M-1,M-1}} \tag{19.143}$$

This is similar to zero forcing. An MMSE version can also be readily developed. The symbol s_{M-1} is now estimated using the decision device $\mathcal{D}(.)$:

$$s_{M-1,est} = \mathcal{D}(\widehat{s}_{M-1}).$$

We then substitute this estimated symbol back into Eq. (19.142) in the place of s_{M-1}. Then the second-to-last equation in (19.142) becomes

$$v_{M-2} = r_{M-2,M-2}s_{M-2} + r_{M-2,M-1}s_{M-1,est}.$$

Using this we compute

$$\widehat{s}_{M-2} = \frac{v_{M-2} - r_{M-2,M-1}s_{M-1,est}}{r_{M-2,M-2}} \tag{19.144}$$

and estimate the next symbol:

$$s_{M-2,est} = \mathcal{D}(\widehat{s}_{M-2}).$$

The steps are then repeated. Thus, after s_k has been estimated, the estimate is substituted back into Eq. (19.142) along with the estimates of $s_{k+1}, \ldots, s_{M-1}$. The next symbol s_{k-1} is then estimated as described above. Thus past decisions on the symbols are utilized in making future decisions, where "past" and "future" refer to the spatial (vertical) domain.

Discussions

1. The DFE receiver described above based on the QR decomposition suffers from some deficiencies. The diagonal elements of $\mathbf{R}$ appear in the denominators of the estimates as shown by Eqs. (19.143) and (19.144). If these are small, then there is *severe noise amplification.* By itself, this is not a serious problem as it can be overcome by using an MMSE version of this algorithm. Note however that the order in which the symbols are estimated (bottom-up, that is, starting with s_{M-1} and ending in s_0) is rather arbitrary. It seems intuitively, that the symbol to be estimated first should be the one for which the noise amplification is least, because this symbol estimate is used for estimating all the remaining $M-1$ symbols.

2. In general, the *optimal ordering* of the symbol estimates is a nontrivial question. Instead of a QR decomposition of the channel $\mathbf{H}$ suppose we decompose the product $\mathbf{HP}$ where $\mathbf{P}$ is some permutation matrix:

   $$\mathbf{HP} = \mathbf{QR}. \tag{19.145}$$

   We can write the received signal as

   $$\mathbf{y} = \mathbf{HPP}^{-1}\mathbf{s} + \mathbf{q} = \mathbf{QRP}^{-1}\mathbf{s} + \mathbf{q}. \tag{19.146}$$

   This is equivalent to inserting a unitary precoder (permutation) $\mathbf{P}^{-1}$ at the transmitter. We then have

   $$\mathbf{v} \stackrel{\Delta}{=} \mathbf{Q}^{\dagger}\mathbf{y} = \mathbf{RP}^{-1}\mathbf{s} + \mathbf{Q}^{\dagger}\mathbf{q}. \tag{19.147}$$

   Note that $\mathbf{P}^{-1}\mathbf{s}$ is a permuted version of the vector $\mathbf{s}$. If we now use the above algorithm to estimate the symbols bottom up, this is equivalent to estimating the original symbols in a permuted order. This adds generality to the algorithm, although the optimal permutation $\mathbf{P}$ is not obvious in this formulation.

3. A further generalization would be to replace $\mathbf{P}$ with a general unitary matrix $\mathbf{S}$:
$$\mathbf{HS} = \mathbf{QR}. \tag{19.148}$$
This is equivalent to decomposing $\mathbf{H}$ in the form $\mathbf{H} = \mathbf{QRS}^\dagger$. This is a major generalization. As explained in Appendices 19.A and 19.B at the end of this chapter, this can be done in such a way that all diagonal elements of $\mathbf{R}$ are identical. In this case we call it the QRS decomposition or the GMD. As shown in detail in Secs. 19.3 and 19.4 the QRS decomposition allows us to obtain the jointly optimal precoder and receiver matrices in the general DFE transceiver. From the details in the preceding sections the reader should notice that the QRS decomposition was not performed directly on $\mathbf{H}$ but on the diagonal matrix of singular values of the channel $\mathbf{H}$.

19.8.2 The VBLAST algorithm

The VBLAST system (Vertical Bell Labs Layered Space Time system) is a well-known transceiver system with space-domain or vertical decision feedback within each block. It incorporates a way to optimize the order in which the symbols are detected and fed back. In this system the precoder is assumed to be identity ($\mathbf{F} = \mathbf{I}$), and all computation is performed at the receiver. VBLAST was introduced by Wolniansky *et al.* [1998] and further developed by Golden *et al.* [1999]. The original VBLAST algorithm was computationally intensive. Later, a clever algorithm was developed by Hassibi [2000] based on the QR decomposition. This algorithm was not only much faster, it also offered numerical robustness. In this section we briefly mention the basic idea behind VBLAST, but omit details about the fast algorithm.

Consider a $J \times M$ MIMO channel $\mathbf{H}$ with additive noise $\mathbf{q}$. With precoder $\mathbf{F} = \mathbf{I}$, the received signal is

$$\mathbf{y} = \mathbf{Hs} + \mathbf{q}. \tag{19.149}$$

Assume as usual that $\mathbf{s}$ and $\mathbf{q}$ are zero-mean and uncorrelated with covariances $\mathbf{R_{qq}} = \sigma_q^2\mathbf{I}, \mathbf{R_{ss}} = \sigma_s^2\mathbf{I}$, and $\mathbf{R_{sq}} = \mathbf{0}$. Then the linear MMSE equalizer $\mathbf{G}$ is given by the expression (13.25) derived in Chap. 13:

$$\mathbf{G} = \Big(\mathbf{H}^\dagger\mathbf{H} + \alpha\mathbf{I}\Big)^{-1}\mathbf{H}^\dagger. \tag{19.150}$$

Here $\alpha = \sigma_q^2/\sigma_s^2$. The reconstructed signal is

$$\widehat{\mathbf{s}} = \Big(\mathbf{H}^\dagger\mathbf{H} + \alpha\mathbf{I}\Big)^{-1}\mathbf{H}^\dagger\mathbf{y}. \tag{19.151}$$

From Sec. 13.4.1 in Chap. 13 the error covariance is

$$\mathbf{R_{ee}} = \sigma_q^2\Big(\mathbf{H}^\dagger\mathbf{H} + \alpha\mathbf{I}\Big)^{-1}. \tag{19.152}$$

The VBLAST system is a nonlinear receiver which detects one symbol at a time *using a certain ordering scheme*, and uses the detected symbols for the detection of the remaining symbols. Recall that the kth diagonal element of $\mathbf{R_{ee}}$ represents the mean square error $E|\widehat{s}_k - s_k|^2$. Let $[\mathbf{R_{ee}}]_{L,L}$ be the smallest diagonal element. Then the Lth symbol is first estimated. From Eq. (19.151)

$$\widehat{s}_L = \left[\left(\mathbf{H}^\dagger\mathbf{H} + \alpha\mathbf{I}\right)^{-1}\right]_{row\,L} \times \mathbf{H}^\dagger\mathbf{y}. \tag{19.153}$$

This is used to detect the Lth symbol:

$$s_{L,est} = \mathcal{D}(\widehat{s}_L), \tag{19.154}$$

where $\mathcal{D}(.)$ represents the detector function. Once the Lth symbol is detected, the received signal $\mathbf{y}$ is modified as follows:

$$\mathbf{y}_{new} = \mathbf{y} - s_{L,est}[\mathbf{H}]_{col\,L}. \tag{19.155}$$

That is, the effect of the already detected symbol is removed from $\mathbf{y}$. Next, the Lth column from the $J \times M$ channel $\mathbf{H}$ is removed to define a "deflated" $J \times (M-1)$ channel $\mathbf{H}_{new}$. The deflated channel $\mathbf{H}_{new}$ and the modified received signal $\mathbf{y}_{new}$ are used to repeat the entire process. This results in the detection of one more symbol. This process is repeated until all the symbols are detected. This is the essence of the VBLAST algorithm.

Discussions

1. ***High complexity.*** Implemented as described above, the algorithm requires the computation of the matrix inverse

$$\left(\mathbf{H}_k^\dagger\mathbf{H}_k + \alpha\mathbf{I}\right)^{-1}$$

for the original channel $\mathbf{H}_k = \mathbf{H}$, and for each of the deflated channels at each step. At each step, the algorithm also needs to identify the smallest diagonal element of this matrix, and compute the next estimated symbol, and then proceed to the next step. Each step is therefore computationally intensive.

2. ***Fast algorithm.*** However, there is plenty of structure in the equations that has not been exploited in the above description. Hassibi [2000] placed this structure in evidence by performing a QR decomposition of the augmented matrix

$$\mathbf{H}_a \stackrel{\Delta}{=} \begin{bmatrix} \mathbf{H} \\ \alpha\mathbf{I} \end{bmatrix}, \tag{19.156}$$

where $\mathbf{H}$ is the original channel matrix. Such a decomposition reveals a great deal of information about the matrices that occur in the successive stages of the VBLAST algorithm, and this is the key to the development of the fast algorithm.

3. ***Generalizations.*** The relation between the VBLAST algorithm and generalized DFE is also explained in Ginis and Cioffi [2001] and Zhang *et al.* [2005]. The QRS method developed in Zhang *et al.* [2005] and Xu *et al.* [2006], and the GMD based methods of Jiang *et al.* [2005a, 2005b] are generalizations of the VBLAST in a major way. Namely, they show how the transmitter (precoder) can be optimized jointly with the matrices $\mathbf{G}$ and $\mathbf{B}$ in the DFE receiver. A knowledge of the channel $\mathbf{H}$ is therefore required at the transmitter, unlike the VBLAST system which requires a knowledge of the channel only at the receiver.

19.9 Concluding remarks

In this chapter we have shown that the use of a decision feedback matrix can improve the performance of a MIMO transceiver significantly. We showed how to design the jointly optimal DFE transceiver with and without zero forcing. We also showed that these optimal transceivers have minimum average symbol error probability. All derivations were under the assumption that there is no error propagation in the feedback loop. The climax point in all derivations was based on the algebraic tool called the GMD or the QRS decomposition.

It is often said that the DFE system minimizes the geometric mean rather than the arithmetic mean of the mean square errors at the input of the decision device. The fact is that in an optimized DFE system the AM and GM of the errors are identical because the error covariance matrix at the detector input is diagonal with identical diagonal elements. More important is the fact that in an optimized DFE system the mean square error depends on the geometric mean rather than the arithmetic mean of $\{1/\sigma_{h,k}\}$ (reciprocals of the channel singular values). This is what makes the performance of the DFE transceiver so much better than that of a linear transceiver.

The feedback loop involved in DFE systems can be moved to the transmitter side by using ideas similar to Tomlinson-Harashima precoding (Sec. 5.8). Optimal nonlinear transceivers can be developed based on this idea. For further details the reader should study Jiang *et al.* [2005b] and Shenouda and Davidson [2008]. As in the case of linear transceivers with bit allocation (Chap. 14) one can design transceivers with decision feedback equalizers to minimize power under bit allocation. A generalized version of the GMD, called the GTD (generalized triangular decomposition), introduced by Jiang, Hager, and Li [2008] (Appendix 19.A) plays a crucial role in such systems. It can be shown that with bit allocation optimized, the linear transceiver performs as well as a DFE system [Weng *et al.*, 2010a]. The advantage of a DFE system using bit allocation is that the GTD offers extra flexibility which often allows one to restrain the bits to be integers and still obtain a global optimum. See Weng *et al.* [2010a] for details.

A one-page summary of the results of this chapter is included at the end of the book in Appendix I.

Appendices to Chapter 19

19.A Matrix triangularization results

We summarize here some matrix factorization results which play a crucial role in decision feedback equalizers (DFE systems). The general idea of these factorizations is to convert an arbitrary matrix into an upper triangular matrix with certain additional properties. This conversion is done by multiplication of the given matrix with appropriate unitary matrices.

19.A.1 The QR decomposition

Given any $N \times M$ matrix $\mathbf{A}$, possibly complex, with $N \geq M$ we can always factor it as

$$\mathbf{A} = \underbrace{\mathbf{Q}}_{N\times M} \underbrace{\mathbf{R}}_{M\times M}, \tag{19.157}$$

where $\mathbf{Q}$ has orthonormal columns (i.e., $\mathbf{Q}^\dagger\mathbf{Q} = \mathbf{I}_M$) and $\mathbf{R}$ is upper triangular. This is called the **QR decomposition** Horn and Johnson [1985].

1. If $M = N$ and $\mathbf{A}$ is nonsingular, then $\mathbf{R}$ can be chosen such that all its diagonal elements satisfy $R_{kk} > 0$. With R_{kk} so restricted, $\mathbf{Q}$ and $\mathbf{R}$ are unique.
2. If the $N \times M$ matrix $\mathbf{A}$ is real, then $\mathbf{Q}$ and $\mathbf{R}$ can be taken to be real.

19.A.2 The QRS decomposition

A generalization of the QR decomposition, called the **QRS decomposition**, or the **GMD** (geometric mean decomposition) was proved in Jiang *et al.* [2005a], and Zhang *et al.* [2005]. The result can be traced back to a work by Zhang *et al.* [2002] on the design of unitary precoders. Early results in the mathematics literature related to these decompositions can be found in Horn [1954]. The QRS decomposition can be stated as follows: Given an $N \times M$ matrix $\mathbf{A}$, possibly complex, with rank r, we can always factor it as

$$\mathbf{A} = \underbrace{\mathbf{Q}}_{N\times r} \underbrace{\mathbf{R}}_{r\times r} \underbrace{\mathbf{S}^\dagger}_{r\times M}, \tag{19.158}$$

where

1. $\mathbf{Q}^\dagger\mathbf{Q} = \mathbf{I}_r$ (the columns of $\mathbf{Q}$ are orthonormal),
2. $\mathbf{S}^\dagger\mathbf{S} = \mathbf{I}_r$ (the columns of $\mathbf{S}$ are orthonormal), and
3. $\mathbf{R}$ is upper triangular with identical diagonal elements

$$R_{kk} = (\sigma_0\sigma_1 \ldots \sigma_{r-1})^{1/r} > 0, \quad 0 \leq k \leq r-1,$$

where σ_k are the r positive singular values of $\mathbf{A}$.

Thus the diagonal elements of $\mathbf{R}$ are equal to the **geometric mean** of the r positive singular values of $\mathbf{A}$. This decomposition is also referred to as the equal-diagonal QRS decomposition. It is also clear from the construction (Appendix 19.B) that all the other elements of the triangular matrix $\mathbf{R}$ can be restricted to be real without loss of generality. As seen in Secs. 19.3 and 19.4 this decomposition plays a crucial role in the optimization of transceivers with decision feedback matrices [Jiang *et al.*, 2005a, 2005b], [Xu *et al.*, 2006].

The generalized triangular decomposition (GTD). A further generalization was developed in Jiang *et al.* [2008]. For related early results in the mathematics literature see Weyl [1949] and Horn [1954]. With all notations as above and r denoting the rank of $\mathbf{A}$ as before, suppose we are given a set r of possibly complex numbers r_k such that

$$\left[\, |r_0| \quad |r_1| \quad \ldots \quad |r_{r-1}| \,\right] \prec_{\times} \left[\, \sigma_0 \quad \sigma_1 \ldots \sigma_{r-1} \,\right]. \tag{19.159}$$

Here the notation $\prec_{\times}$ means that the vector on the left is *multiplicatively majorized* by the vector on the right (see Sec. 21.6). Then we can find a decomposition of the form (19.158), where $\mathbf{Q}^{\dagger}\mathbf{Q} = \mathbf{S}^{\dagger}\mathbf{S} = \mathbf{I}_r$ as before, and the upper triangular matrix $\mathbf{R}$ has the specified diagonal entries r_k, that is,

$$R_{kk} = r_k, \quad 0 \leq k \leq r-1.$$

Thus, subject only to the restriction (19.159), the diagonal elements R_{kk} in the triangular decomposition can be chosen freely! This decomposition is called *generalized triangular decomposition* or the *GTD*. It plays an important role in the optimization of transceivers with decision feedback matrices, including DFE systems involving bit allocation at the transmitter. We shall not elaborate on this further, but the interested reader should refer to Jiang *et al.* [2007, 2008], Shenouda and Davidson [2008], and Weng *et al.* [2010a]. $\bigtriangledown \bigtriangledown \bigtriangledown$

19.B Proof of the GMD or QRS decomposition

Our proof of the decomposition (19.158) will be patterned after the proof given by Jiang *et al.* [2005a], which provides a direct constructive approach. The essence of this proof can be traced back to Horn [1954]. We will first prove it for 2×2 diagonal matrices and then for arbitrary diagonal matrices, with positive diagonal elements. The proof for arbitrary nondiagonal matrices will then follow based on the well known singular value decomposition.

19.B.1 Case of two-by-two diagonal matrices

First consider a 2×2 diagonal matrix

$$\begin{bmatrix} \sigma_0 & 0 \\ 0 & \sigma_1 \end{bmatrix}$$

with

$$\sigma_0 > \sigma_1 > 0. \tag{19.160}$$

Let σ be any number such that

$$\sigma_0 > \sigma > \sigma_1. \tag{19.161}$$

We will show how to find real unitary matrices $\mathbf{U}_1$ and $\mathbf{U}_2$ such that

$$\mathbf{U}_1 \begin{bmatrix} \sigma_0 & 0 \\ 0 & \sigma_1 \end{bmatrix} \mathbf{U}_2 = \begin{bmatrix} \sigma & \times \\ 0 & \sigma_0\sigma_1/\sigma \end{bmatrix}. \tag{19.162}$$

Thus, the product of the diagonal elements is unchanged by the premultiplication with $\mathbf{U}_1$ and postmultiplication with $\mathbf{U}_2$. Since $\mathbf{U}_1$ and $\mathbf{U}_2$ are real and unitary, they can both be regarded as **planar rotation** operators [Horn and Johnson, 1985]. Note that the case where σ is the geometric mean $\sqrt{\sigma_0\sigma_1}$ is a special case which still satisfies Eq. (19.161); the diagonal elements on the right hand side of Eq. (19.162) become identical in this case.

Construction of $\mathbf{U}_1$ ***and*** $\mathbf{U}_2$**.** Choose

$$\mathbf{U}_2 = \begin{bmatrix} c & -s \\ s & c \end{bmatrix}, \quad \mathbf{U}_1 = \frac{1}{\sigma} \begin{bmatrix} c\sigma_0 & s\sigma_1 \\ -s\sigma_1 & c\sigma_0 \end{bmatrix}, \tag{19.163}$$

where $c = \cos\theta$ and $s = \sin\theta$. We will show how to choose $\cos\theta$ such that all the claims are satisfied. We have

$$\mathbf{U}_1 \begin{bmatrix} \sigma_0 & 0 \\ 0 & \sigma_1 \end{bmatrix} \mathbf{U}_2 = \frac{1}{\sigma} \begin{bmatrix} c^2\sigma_0^2 + s^2\sigma_1^2 & sc(\sigma_1^2 - \sigma_0^2) \\ 0 & \sigma_0\sigma_1 \end{bmatrix}.$$

So we have to choose $\cos\theta$ such that

$$c^2\sigma_0^2 + s^2\sigma_1^2 = \sigma^2, \tag{19.164}$$

that is,

$$c^2 = \frac{\sigma^2 - \sigma_1^2}{\sigma_0^2 - \sigma_1^2}$$

The right-hand side of the above expression is in the range $0 < x < 1$ as required (because of Eqs. (19.160) and (19.161)). With c chosen to satisfy Eq. (19.164) we have

$$\mathbf{U}_1 \begin{bmatrix} \sigma_0 & 0 \\ 0 & \sigma_1 \end{bmatrix} \mathbf{U}_2 = \begin{bmatrix} \sigma & sc(\sigma_1^2 - \sigma_0^2)/\sigma \\ 0 & \sigma_0\sigma_1/\sigma \end{bmatrix}$$

as claimed. Note that $\mathbf{U}_2$ is unitary, and with c chosen to satisfy Eq. (19.164) $\mathbf{U}_1$ is unitary as well. $\triangledown \triangledown \triangledown$

19.B.2 Diagonal matrices of arbitrary size

Given an $r \times r$ diagonal matrix with diagonal elements $\sigma_k > 0$ we will now show how to construct $r \times r$ real unitary matrices $\mathbf{U}_1$ and $\mathbf{U}_2$ such that

$$\mathbf{U}_1 \begin{bmatrix} \sigma_0 & 0 & \dots & 0 \\ 0 & \sigma_1 & \dots & 0 \\ \vdots & \vdots & \ddots & \vdots \\ 0 & 0 & \dots & \sigma_{r-1} \end{bmatrix} \mathbf{U}_2 = \begin{bmatrix} \sigma & \times & \dots & \times \\ 0 & \sigma & \dots & \times \\ \vdots & \vdots & \ddots & \vdots \\ 0 & 0 & \dots & \sigma \end{bmatrix}, \tag{19.165}$$

where

$$\sigma = \left(\prod_{k=0}^{r-1} \sigma_k \right)^{1/r}, \tag{19.166}$$

and the symbol $\times$ denotes (possibly nonzero) entries that we do not care about. The matrix on the right is upper triangular with identical diagonal elements σ, equal to the **geometric mean** of the numbers σ_k.

Construction of Eq. (19.165)

If σ_k are identical for all k, we simply take $\mathbf{U}_1 = \mathbf{U}_2 = \mathbf{I}$, and the proof is complete. Assuming this is not the case, there exist σ_k and σ_m such that

$$\sigma_m > \sigma > \sigma_k.$$

By using a permutation matrix $\mathbf{P}$ (Appendix B, Sec. B.5.1) we can reorder the diagonal elements as follows:

$$\mathbf{P} \begin{bmatrix} \sigma_0 & 0 & \dots & 0 \\ 0 & \sigma_1 & \dots & 0 \\ \vdots & \vdots & \ddots & \vdots \\ 0 & 0 & \dots & \sigma_{r-1} \end{bmatrix} \mathbf{P}^T = \begin{bmatrix} \sigma_m & 0 & \dots & 0 \\ 0 & \sigma_k & \dots & 0 \\ \vdots & \vdots & \ddots & \vdots \\ 0 & 0 & \dots & \sigma_{r-1} \end{bmatrix}. \tag{19.167}$$

We can now apply planar rotation matrices on the 2×2 top-left submatrix to obtain:

$$\underbrace{\begin{bmatrix} e_1 & f_1 & 0 & \dots & 0 \\ g_1 & h_1 & 0 & \dots & 0 \\ 0 & 0 & 1 & \dots & 0 \\ \vdots & \vdots & \vdots & \ddots & \vdots \\ 0 & 0 & 0 & \dots & 1 \end{bmatrix}}_{\mathbf{V}_1} \underbrace{\begin{bmatrix} \sigma_m & 0 & 0 & \dots & 0 \\ 0 & \sigma_k & 0 & \dots & 0 \\ 0 & 0 & \sigma_\ell & \dots & 0 \\ \vdots & \vdots & \vdots & \ddots & \vdots \\ 0 & 0 & 0 & \dots & \sigma_n \end{bmatrix}}_{\mathbf{D}_1} \underbrace{\begin{bmatrix} e_2 & f_2 & 0 & \dots & 0 \\ g_2 & h_2 & 0 & \dots & 0 \\ 0 & 0 & 1 & \dots & 0 \\ \vdots & \vdots & \vdots & \ddots & \vdots \\ 0 & 0 & 0 & \dots & 1 \end{bmatrix}}_{\mathbf{V}_2}$$

$$= \underbrace{\begin{bmatrix} \sigma & \times & 0 & \dots & 0 \\ 0 & \eta & 0 & \dots & 0 \\ 0 & 0 & \sigma_\ell & \dots & 0 \\ \vdots & \vdots & \vdots & \ddots & \vdots \\ 0 & 0 & 0 & \dots & \sigma_n \end{bmatrix}}_{\mathbf{D}_2}. \tag{19.168}$$

Here

$$\begin{bmatrix} e_1 & f_1 \\ g_1 & h_1 \end{bmatrix}, \quad \begin{bmatrix} e_2 & f_2 \\ g_2 & h_2 \end{bmatrix} \tag{19.169}$$

play the role of the matrices $\mathbf{U}_1$ and $\mathbf{U}_2$ in Eq. (19.163). The elements of these matrices are chosen precisely as described in Sec. 19.B.1; from that construction we see that the product of the diagonal elements of $\mathbf{D}_1$ and those of $\mathbf{D}_2$ are identical. We also see that only the top-left 2×2 submatrix of $\mathbf{D}_1$ is different from that of $\mathbf{D}_2$ and there are no other differences. Now, using Eq. (19.167) we can rewrite Eq. (19.168) as

$$\mathbf{W}_1 \begin{bmatrix} \sigma_0 & 0 & \ldots & 0 \\ 0 & \sigma_1 & \ldots & 0 \\ \vdots & \vdots & \ddots & \vdots \\ 0 & 0 & \ldots & \sigma_{r-1} \end{bmatrix} \mathbf{Z}_1 = \underbrace{\begin{bmatrix} \sigma & \times & \ldots & 0 \\ 0 & \eta & \ldots & 0 \\ \vdots & \vdots & \ddots & \vdots \\ 0 & 0 & \ldots & \sigma_n \end{bmatrix}}_{\mathbf{D}_2}, \tag{19.170}$$

where $\mathbf{W}_1$ and $\mathbf{Z}_1$ are real unitary (because $\mathbf{V}_1, \mathbf{V}_2$, and $\mathbf{P}$ are). We can repeat this process until the form shown on the right-hand side of Eq. (19.165) is obtained. This is the essence of the construction.

Details of the proof

For readers who like to see details, we describe some of the steps involved. Consider a matrix of the form

$$\begin{bmatrix} \mathbf{B}_1 & \mathbf{B}_2 \\ \mathbf{0} & \mathbf{B}_3 \end{bmatrix} =$$

$$\begin{array}{c} \\ k \\ \\ r-k \end{array} \begin{pmatrix} \overbrace{\begin{array}{cccc} \sigma & \times & \ldots & \times \\ 0 & \sigma & \ldots & \times \\ \vdots & \vdots & \ddots & \vdots \\ 0 & 0 & \ldots & \sigma \end{array}}^{k} & \vline & \overbrace{\begin{array}{cccc} \times & \times & \ldots & \times \\ \times & \times & \ldots & \times \\ \vdots & \vdots & \ddots & \vdots \\ \times & \times & \ldots & \times \end{array}}^{r-k} \\ \hline \begin{array}{cccc} 0 & 0 & \ldots & 0 \\ 0 & 0 & \ldots & 0 \\ \vdots & \vdots & \ddots & \vdots \\ 0 & 0 & \ldots & 0 \end{array} & \vline & \begin{array}{cccc} \eta_k & 0 & \ldots & 0 \\ 0 & \eta_{k+1} & \ldots & 0 \\ \vdots & \vdots & \ddots & 0 \\ 0 & 0 & \ldots & \eta_{r-1} \end{array} \end{pmatrix}. \tag{19.171}$$

This matrix has the following properties: the first k diagonal elements are equal to σ and there are $r-k$ diagonal elements remaining. The $(r-k) \times (r-k)$ submatrix $\mathbf{B}_3$ is diagonal. Assuming the geometric mean of its $r-k$ diagonal elements is σ, we now show how to find unitary $\mathbf{W}_{k+1}$ and $\mathbf{Z}_{k+1}$ such that

$$\mathbf{W}_{k+1}\begin{bmatrix}\mathbf{B}_1 & \mathbf{B}_2\\ \mathbf{0} & \mathbf{B}_3\end{bmatrix}\mathbf{Z}_{k+1} =$$

$$\begin{array}{c} \\ \\ k+1 \\ \\ \\ \\ r-k-1 \\ \\ \\ \end{array}\left(\begin{array}{cccc|cccc} \multicolumn{4}{c}{k+1} & \multicolumn{4}{c}{r-k-1}\\ \sigma & \times & \ldots & \times & \times & \times & \ldots & \times\\ 0 & \sigma & \ldots & \times & \times & \times & \ldots & \times\\ \vdots & \vdots & \ddots & \vdots & \vdots & \vdots & \ddots & \vdots\\ 0 & 0 & \ldots & \sigma & \times & \times & \ldots & \times\\ \hline 0 & 0 & \ldots & 0 & \mu_{k+1} & 0 & \ldots & 0\\ 0 & 0 & \ldots & 0 & 0 & \mu_{k+2} & \ldots & 0\\ \vdots & \vdots & \ddots & \vdots & \vdots & \vdots & \ddots & 0\\ 0 & 0 & \ldots & 0 & 0 & 0 & \ldots & \mu_{r-1} \end{array}\right), \tag{19.172}$$

where the $r-k-1$ diagonal elements μ_k still have geometric mean σ.

1. First apply a permutation $\mathbf{P}$ to the submatrix $\mathbf{B}_3$:

$$\begin{bmatrix}\mathbf{I} & \mathbf{0}\\ \mathbf{0} & \mathbf{P}\end{bmatrix}\begin{bmatrix}\mathbf{B}_1 & \mathbf{B}_2\\ \mathbf{0} & \mathbf{B}_3\end{bmatrix}\begin{bmatrix}\mathbf{I} & \mathbf{0}\\ \mathbf{0} & \mathbf{P}^T\end{bmatrix} = \begin{bmatrix}\mathbf{B}_1 & \mathbf{B}_2\mathbf{P}^T\\ \mathbf{0} & \mathbf{P}\mathbf{B}_3\mathbf{P}^T\end{bmatrix}. \tag{19.173}$$

   Note that $\mathbf{B}_3$ is diagonal. The matrix $\mathbf{P}$ is chosen such that the first two diagonal elements of $\mathbf{P}\mathbf{B}_3\mathbf{P}^T$, denoted σ_0 and σ_1 temporarily, satisfy $\sigma_0 > \sigma > \sigma_1$. (This is always possible unless all the diagonal elements of $\mathbf{B}_3$ are already equal to σ.)

2. By using a process similar to the planar rotation step in Eq. (19.168), we then convert the matrix $\mathbf{P}\mathbf{B}_3\mathbf{P}^T$ to the form

$$\mathbf{V}_1\left(\mathbf{P}\mathbf{B}_3\mathbf{P}^T\right)\mathbf{V}_2 = \begin{bmatrix}\sigma & \times & \ldots & 0\\ 0 & \eta & \ldots & 0\\ \vdots & \vdots & \ddots & \vdots\\ 0 & 0 & \ldots & \sigma_n\end{bmatrix}, \tag{19.174}$$

   where $\mathbf{V}_1$ and $\mathbf{V}_2$ are square unitary matrices with $r-k$ rows. The complete process when going from step k to $k+1$ is

$$\underbrace{\begin{bmatrix}\mathbf{I} & \mathbf{0}\\ \mathbf{0} & \mathbf{V}_1\end{bmatrix}\begin{bmatrix}\mathbf{I} & \mathbf{0}\\ \mathbf{0} & \mathbf{P}\end{bmatrix}}_{\mathbf{W}_{k+1}}\begin{bmatrix}\mathbf{B}_1 & \mathbf{B}_2\\ \mathbf{0} & \mathbf{B}_3\end{bmatrix}\underbrace{\begin{bmatrix}\mathbf{I} & \mathbf{0}\\ \mathbf{0} & \mathbf{P}^T\end{bmatrix}\begin{bmatrix}\mathbf{I} & \mathbf{0}\\ \mathbf{0} & \mathbf{V}_2\end{bmatrix}}_{\mathbf{Z}_{k+1}}$$

$$= \begin{bmatrix}\mathbf{B}_1 & \mathbf{B}_2\mathbf{P}^T\mathbf{V}_2\\ \mathbf{0} & \mathbf{V}_1\mathbf{P}\mathbf{B}_3\mathbf{P}^T\mathbf{V}_2\end{bmatrix}. \tag{19.175}$$

   Note that the submatrix $\mathbf{B}_1$ remains unaffected in the whole process. This means that the diagonal elements, which were forced to be σ in earlier steps, remain unchanged.

The unitary matrices $\mathbf{V}_1$ and $\mathbf{V}_2$ have determinant equal to unity (because the planar rotations in Eq. (19.163) have this property). The matrices $\mathbf{P}$ and $\mathbf{P}^T$ have identical determinants (± 1). So the determinant of Eq. (19.171) is identical to that of Eq. (19.175). Since the determinant of a triangular matrix is the product of its diagonal elements, this implies that the geometric mean of the $r-k-1$ elements μ_k continues to be σ.

The preceding construction shows how we can go from Eq. (19.171) to Eq. (19.172) for some $k>0$. For the initialization step, we already know how the initial diagonal matrix can be converted to the form (19.168) (which has $k=2$). So we can repeat the process, and eventually obtain an upper triangular matrix whose diagonal elements are all equal to σ.

19.B.3 Case of arbitrary matrices

Now consider an $N\times M$ matrix $\mathbf{A}$, possibly complex, with rank r. Using singular value decomposition (Appendix C) we can always write this as

$$\mathbf{A}=\mathbf{U}\boldsymbol{\Sigma}\mathbf{V}^\dagger, \tag{19.176}$$

where $\mathbf{U}$ and $\mathbf{V}$ are square unitary matrices and $\boldsymbol{\Sigma}$ is an $N\times M$ diagonal matrix of the form

$$\boldsymbol{\Sigma}=\begin{bmatrix}\boldsymbol{\Sigma}_a & \mathbf{0}\\ \mathbf{0} & \mathbf{0}\end{bmatrix}. \tag{19.177}$$

Here $\boldsymbol{\Sigma}_a$ is an $r\times r$ diagonal matrix with positive diagonal elements (nonzero singular values of $\mathbf{A}$). Thus $\mathbf{A}$ can be rewritten as

$$\mathbf{A}=\mathbf{U}_a\boldsymbol{\Sigma}_a\mathbf{V}_a^\dagger, \tag{19.178}$$

where $\mathbf{U}_a$ is $N\times r$ and $\mathbf{V}_a^\dagger$ is $r\times M$, and

$$\mathbf{U}_a^\dagger\mathbf{U}_a=\mathbf{V}_a^\dagger\mathbf{V}_a=\mathbf{I}_r. \tag{19.179}$$

That is, the columns of $\mathbf{U}_a$ are orthonormal and so are the columns of $\mathbf{V}_a$. From Sec. 19.B.2 we know that we can find $r\times r$ unitary matrices $\mathbf{U}_1$ and $\mathbf{U}_2$ such that $\boldsymbol{\Sigma}_a$ can be converted into the lower triangular form shown in Eq. (19.165). That is,

$$\mathbf{U}_1\boldsymbol{\Sigma}_a\mathbf{U}_2=\mathbf{R}, \quad \text{or} \quad \boldsymbol{\Sigma}_a=\mathbf{U}_1^\dagger\mathbf{R}\mathbf{U}_2^\dagger, \tag{19.180}$$

where

$$\mathbf{R}=\begin{bmatrix}\sigma & \times & \dots & \times\\ 0 & \sigma & \dots & \times\\ \vdots & \vdots & \ddots & \vdots\\ 0 & 0 & \dots & \sigma\end{bmatrix}. \tag{19.181}$$

Thus

$$\mathbf{A}=\mathbf{U}_a\boldsymbol{\Sigma}_a\mathbf{V}_a^\dagger=\underbrace{\mathbf{U}_a\mathbf{U}_1^\dagger}_{\mathbf{Q}}\mathbf{R}\underbrace{\mathbf{U}_2^\dagger\mathbf{V}_a^\dagger}_{\mathbf{S}^\dagger}=\mathbf{Q}\mathbf{R}\mathbf{S}^\dagger. \tag{19.182}$$

Using Eq. (19.179) and the unitarity of $\mathbf{U}_1$ and $\mathbf{U}_2$, we see that

$$\mathbf{Q}^\dagger\mathbf{Q}=\mathbf{U}_1\mathbf{U}_a^\dagger\mathbf{U}_a\mathbf{U}_1^\dagger=\mathbf{I}_r \quad \text{and} \quad \mathbf{S}^\dagger\mathbf{S}=\mathbf{U}_2^\dagger\mathbf{V}_a^\dagger\mathbf{V}_a\mathbf{U}_2=\mathbf{I}_r. \tag{19.183}$$

That is, the columns of the $N \times r$ matrix $\mathbf{Q}$ are orthonormal and so are the columns of the $M \times r$ matrix $\mathbf{S}$. This proves the QRS decomposition (19.158) in its general form.

19.C Derivation of Eq. (19.72)

For convenience let $\mathbf{C} = \mathbf{HF}$ and $\alpha = \sigma_q^2/\sigma_s^2$. Then Eqs. (19.69) and (19.71) become

$$\mathbf{G} = (\mathbf{B}+\mathbf{I})\mathbf{C}^\dagger\Big(\mathbf{C}\mathbf{C}^\dagger + \alpha\mathbf{I}\Big)^{-1} \tag{19.184}$$

and

$$\mathbf{R_{ee}} = \sigma_s^2\Big(\mathbf{GC} - (\mathbf{B}+\mathbf{I})\Big)\Big(\mathbf{GC} - (\mathbf{B}+\mathbf{I})\Big)^\dagger + \sigma_q^2\mathbf{G}\mathbf{G}^\dagger. \tag{19.185}$$

Details of arriving at Eq. (19.72). Using the identity $(\mathbf{C}\mathbf{C}^\dagger + \alpha\mathbf{I})\mathbf{C} = \mathbf{C}(\mathbf{C}^\dagger\mathbf{C} + \alpha\mathbf{I})$, we have

$$\mathbf{C}^\dagger(\mathbf{C}\mathbf{C}^\dagger + \alpha\mathbf{I})^{-1} = (\mathbf{C}^\dagger\mathbf{C} + \alpha\mathbf{I})^{-1}\mathbf{C}^\dagger. \tag{19.186}$$

So $\mathbf{GC} - (\mathbf{B}+\mathbf{I}) = (\mathbf{B}+\mathbf{I})[(\mathbf{B}+\mathbf{I})^{-1}\mathbf{GC} - \mathbf{I}]$ can be written as

$$\begin{aligned}
& (\mathbf{B}+\mathbf{I})\Big(\mathbf{C}^\dagger(\mathbf{C}\mathbf{C}^\dagger + \alpha\mathbf{I})^{-1}\mathbf{C} - \mathbf{I}\Big) \quad \text{(use Eq. (19.184))} \\
= \ & (\mathbf{B}+\mathbf{I})\Big((\mathbf{C}^\dagger\mathbf{C} + \alpha\mathbf{I})^{-1}\mathbf{C}^\dagger\mathbf{C} - \mathbf{I}\Big) \quad \text{(use Eq. (19.186))} \\
= \ & (\mathbf{B}+\mathbf{I})(\mathbf{C}^\dagger\mathbf{C} + \alpha\mathbf{I})^{-1}\Big(\mathbf{C}^\dagger\mathbf{C} - (\mathbf{C}^\dagger\mathbf{C} + \alpha\mathbf{I})\Big) \\
= \ & -\alpha(\mathbf{B}+\mathbf{I})(\mathbf{C}^\dagger\mathbf{C} + \alpha\mathbf{I})^{-1}.
\end{aligned}$$

The first term in Eq. (19.185) therefore reduces to

$$\sigma_s^2\alpha^2(\mathbf{B}+\mathbf{I})(\mathbf{C}^\dagger\mathbf{C} + \alpha\mathbf{I})^{-2}(\mathbf{B}+\mathbf{I})^\dagger. \tag{19.187}$$

Now, from Eq. (19.184) we have (again using Eq. (19.186))

$$\mathbf{G} = (\mathbf{B}+\mathbf{I})\mathbf{C}^\dagger\Big(\mathbf{C}\mathbf{C}^\dagger + \alpha\mathbf{I}\Big)^{-1} = (\mathbf{B}+\mathbf{I})\Big(\mathbf{C}^\dagger\mathbf{C} + \alpha\mathbf{I}\Big)^{-1}\mathbf{C}^\dagger, \tag{19.188}$$

so that the second term in Eq. (19.185) becomes

$$\sigma_q^2\mathbf{G}\mathbf{G}^\dagger = \sigma_q^2(\mathbf{B}+\mathbf{I})\Big(\mathbf{C}^\dagger\mathbf{C} + \alpha\mathbf{I}\Big)^{-1}\mathbf{C}^\dagger\mathbf{C}\Big(\mathbf{C}^\dagger\mathbf{C} + \alpha\mathbf{I}\Big)^{-1}(\mathbf{B}+\mathbf{I})^\dagger. \tag{19.189}$$

Thus $\mathbf{R_{ee}}$ is the sum of Eqs. (19.187) and (19.189), and simplifies as follows:

$$\begin{aligned}
& (\mathbf{B}+\mathbf{I})(\mathbf{C}^\dagger\mathbf{C} + \alpha\mathbf{I})^{-1}\Big(\sigma_s^2\alpha^2\mathbf{I} + \sigma_q^2\mathbf{C}^\dagger\mathbf{C}\Big)(\mathbf{C}^\dagger\mathbf{C} + \alpha\mathbf{I})^{-1}(\mathbf{B}+\mathbf{I})^\dagger \\
= \ & \sigma_q^2(\mathbf{B}+\mathbf{I})(\mathbf{C}^\dagger\mathbf{C} + \alpha\mathbf{I})^{-1}(\mathbf{C}^\dagger\mathbf{C} + \alpha\mathbf{I})(\mathbf{C}^\dagger\mathbf{C} + \alpha\mathbf{I})^{-1}(\mathbf{B}+\mathbf{I})^\dagger \\
= \ & \sigma_q^2(\mathbf{B}+\mathbf{I})(\mathbf{C}^\dagger\mathbf{C} + \alpha\mathbf{I})^{-1}(\mathbf{B}+\mathbf{I})^\dagger = \sigma_s^2(\mathbf{B}+\mathbf{I})\Big(\mathbf{I} + \frac{\sigma_s^2}{\sigma_q^2}\mathbf{C}^\dagger\mathbf{C}\Big)^{-1}(\mathbf{B}+\mathbf{I})^\dagger.
\end{aligned}$$

Resubstituting $\mathbf{C} = \mathbf{HF}$ this reduces to Eq. (19.72) indeed. $\bigtriangledown\bigtriangledown\bigtriangledown$

19.D Bias-removed MMSE is better than ZF MMSE

In this section we will use the simple notations $\mathcal{E}_{ZF}$ and $\mathcal{E}_{pure}$ to denote the average minimized mean square error per scalar symbol for the DFE system with and without the zero-forcing constraint. Dividing the expressions in Eq. (19.124) by M (to get per-symbol values) we obtain

$$\mathcal{E}_{ZF} = \frac{M\sigma_s^2\sigma_q^2}{p_0}\Big(\prod_{k=0}^{M-1}\frac{1}{\sigma_{h,k}^2}\Big)^{1/M} \tag{19.190}$$

and

$$\mathcal{E}_{pure} = \sigma_s^2\Big(\frac{\lambda\sigma_q^2}{\sigma_s^2}\Big)^{K/M}\prod_{k=0}^{K-1}\Big(\frac{1}{\sigma_{h,k}^2}\Big)^{1/M}. \tag{19.191}$$

Substituting for λ from Eq. (19.97), we have

$$\mathcal{E}_{pure} = \sigma_s^2\left(\frac{K}{\frac{p_0}{\sigma_q^2}+\sum_{k=0}^{K-1}\frac{1}{\sigma_{h,k}^2}}\right)^{K/M}\prod_{k=0}^{K-1}\Big(\frac{1}{\sigma_{h,k}^2}\Big)^{1/M}. \tag{19.192}$$

Since the optimal DFE system equalizes the mean square errors of the M individual components in the block, these expressions represent the errors in the individual symbols. For the case where there is no zero forcing, we have to perform bias removal, as explained in Sec. 16.3. Denoting the mean square error per symbol after bias removal by $\mathcal{E}_{br}$, we have

$$\frac{\sigma_s^2}{\mathcal{E}_{br}} = \frac{\sigma_s^2}{\mathcal{E}_{pure}} - 1 \tag{19.193}$$

(review Sec. 16.3.2). Our goal now is to prove that

$$\mathcal{E}_{br} \le \mathcal{E}_{ZF}. \tag{19.194}$$

Even though it is obvious that $\mathcal{E}_{pure} \le \mathcal{E}_{ZF}$, it takes more effort to verify Eq. (19.194). For a given K, the error $\mathcal{E}_{pure}$ (hence $\mathcal{E}_{br}$) does not depend on $\sigma_{h,k}$ with $k \ge K$. But since $\mathcal{E}_{ZF}$ does depend on all the components $\sigma_{h,k}$, it is necessary (and sufficient) to prove $\mathcal{E}_{br} \le \mathcal{E}_{ZF}$ for the extreme case where $\sigma_{h,K}, \ldots, \sigma_{h,M-1}$ are as large as possible (i.e., $\mathcal{E}_{ZF}$ as small as possible). Now recall from Eq. (19.96) that K is such that $(1/\lambda) - (\sigma_q^2/\sigma_s^2\sigma_{h,k}^2) \le 0$ for $k \ge K$. That is,

$$\sigma_{h,k}^2 \le \frac{\lambda\sigma_q^2}{\sigma_s^2}, \quad K \le k \le M-1. \tag{19.195}$$

So, it is necessary and sufficient to prove $\mathcal{E}_{br} \le \mathcal{E}_{ZF}$ for the case where

$$\sigma_{h,K}^2 = \sigma_{h,K+1}^2 = \ldots = \sigma_{h,M-1}^2 = \frac{\lambda\sigma_q^2}{\sigma_s^2} \tag{19.196}$$

The details of this proof are given below. Substituting from Eq. (19.196), Eq. (19.191) can be rewritten as

$$\begin{aligned}
\mathcal{E}_{pure} &= \sigma_s^2(\sigma_{h,K}^2)^{K/M}\prod_{k=0}^{K-1}\Big(\frac{1}{\sigma_{h,k}^2}\Big)^{1/M} \\
&= \sigma_s^2\Big(\frac{1}{\sigma_{h,K}^2}\Big)^{-K/M}\prod_{k=0}^{K-1}\Big(\frac{1}{\sigma_{h,k}^2}\Big)^{1/M} \\
&= \sigma_s^2\sigma_{h,K}^2\Big(\frac{1}{\sigma_{h,K}^2}\Big)^{(M-K)/M}\prod_{k=0}^{K-1}\Big(\frac{1}{\sigma_{h,k}^2}\Big)^{1/M} \\
&= \sigma_s^2\sigma_{h,K}^2\prod_{k=0}^{M-1}\Big(\frac{1}{\sigma_{h,k}^2}\Big)^{1/M} \quad \text{(using Eq. (19.196))}
\end{aligned}$$

so that, from Eq. (19.193), we get

$$\frac{\sigma_s^2}{\mathcal{E}_{br}} = \frac{1}{\sigma_{h,K}^2}\prod_{k=0}^{M-1}\Big(\sigma_{h,k}^2\Big)^{1/M} - 1. \tag{19.197}$$

Next, from Eq. (19.190) we have

$$\frac{\sigma_s^2}{\mathcal{E}_{ZF}} = \frac{p_0}{M\sigma_q^2}\Big(\prod_{k=0}^{M-1}\sigma_{h,k}^2\Big)^{1/M}. \tag{19.198}$$

Using the expression for the power p_0 given in the equation following Eq. (19.96), we get

$$\begin{aligned}
\frac{p_0}{\sigma_q^2} &= \frac{K\sigma_s^2}{\sigma_q^2\lambda} - \sum_{k=0}^{K-1}\frac{1}{\sigma_{h,k}^2} \\
&= \frac{K}{\sigma_{h,K}^2} - \sum_{k=0}^{K-1}\frac{1}{\sigma_{h,k}^2} \quad \text{(from Eq. (19.196))} \\
&= \frac{M}{\sigma_{h,K}^2} - \sum_{k=0}^{K-1}\frac{1}{\sigma_{h,k}^2} - \frac{M-K}{\sigma_{h,K}^2} \\
&= \frac{M}{\sigma_{h,K}^2} - \sum_{k=0}^{M-1}\frac{1}{\sigma_{h,k}^2} \quad \text{(again from Eq. (19.196))}
\end{aligned}$$

Substituting this into Eq. (19.198) we get

$$\begin{aligned}
\frac{\sigma_s^2}{\mathcal{E}_{ZF}} &= \Big(\frac{1}{\sigma_{h,K}^2} - \frac{1}{M}\sum_{k=0}^{M-1}\frac{1}{\sigma_{h,k}^2}\Big)\Big(\prod_{k=0}^{M-1}\sigma_{h,k}^2\Big)^{1/M} \\
&= \frac{1}{\sigma_{h,K}^2}\Big(\prod_{k=0}^{M-1}\sigma_{h,k}^2\Big)^{1/M} - \Big(\frac{1}{M}\sum_{k=0}^{M-1}\frac{1}{\sigma_{h,k}^2}\Big)\Big(\prod_{k=0}^{M-1}\sigma_{h,k}^2\Big)^{1/M}.
\end{aligned}$$

This can be rewritten as

$$\frac{\sigma_s^2}{\mathcal{E}_{ZF}} = \frac{1}{\sigma_{h,K}^2}\Big(\prod_{k=0}^{M-1}\sigma_{h,k}^2\Big)^{1/M} - \frac{AM\{1/\sigma_{h,k}^2)\}}{GM\{1/\sigma_{h,k}^2)\}}, \tag{19.199}$$

where $AM\{1/\sigma_{h,k}^2)\}$ and $GM\{1/\sigma_{h,k}^2)\}$ represent the arithmetic and geometric means of the set of positive numbers $1/\sigma_{h,k}^2$. Subtracting the preceding expression from Eq. (19.197) we have

$$\frac{\sigma_s^2}{\mathcal{E}_{br}} - \frac{\sigma_s^2}{\mathcal{E}_{ZF}} = \frac{AM\{1/\sigma_{h,k}^2)\}}{GM\{1/\sigma_{h,k}^2)\}} - 1 \geq 0, \tag{19.200}$$

since AM/GM is ≥ 1 for any set of positive numbers. This proves $\mathcal{E}_{br} \leq \mathcal{E}_{ZF}$.

Problems

Note. Unless mentioned otherwise, assume $\sigma_s^2 = 1$.

19.1. In Sec. 19.3.2 we presented an expression for the DFE gain $\mathcal{G}_{DFE,zf}$ in the zero-forcing case. This depends entirely on the channel singular values $\sigma_{h,k}$. Assume $\sigma_{h,k} = \rho^{-k}, 0 \leq k \leq M-1$, where $0 < \rho < 1$. Find a closed-form expression for $\mathcal{G}_{DFE,zf}$, and compute its values for $\rho = 0.01$ and $\rho = 0.99$. What are its limiting values as $\rho \to 0$ and $\rho \to 1$?

19.2. One of the crucial inequalities we used to derive the MMSE bound in Sec. 19.4.3 was Fiedler's inequality. Verify this inequality when the two matrices are

$$\mathbf{A} = \begin{bmatrix} 3 & 1 \\ 1 & 3 \end{bmatrix}, \quad \mathbf{B} = \begin{bmatrix} 2 & 1 \\ 1 & 2 \end{bmatrix}.$$

Repeat for

$$\mathbf{A} = \begin{bmatrix} 3 & -1 \\ -1 & 2 \end{bmatrix}, \quad \mathbf{B} = \begin{bmatrix} 2 & 1 \\ 1 & 1 \end{bmatrix}.$$

19.3. For the diagonal matrix

$$\boldsymbol{\Sigma} = \begin{bmatrix} 3 & 0 & 0 \\ 0 & 2 & 0 \\ 0 & 0 & 1 \end{bmatrix}$$

perform the QRS decomposition and identify the matrices $\mathbf{Q}$, $\mathbf{R}$, and $\mathbf{S}$.

19.4. For the following 2×2 matrix:

$$\mathbf{H} = \begin{bmatrix} 3 & 1 \\ 1 & 3 \end{bmatrix}$$

perform the QRS decomposition and identify the matrices $\mathbf{Q}$, $\mathbf{R}$, and $\mathbf{S}$.

19.5. Consider the following 2×2 channel

$$\mathbf{H} = \begin{bmatrix} 3 & 2.9 \\ 2.9 & 3 \end{bmatrix}.$$

Find the matrices $\mathbf{F}$ (up to scale), $\mathbf{G}$, and $\mathbf{B}$ in the optimal zero-forcing DFE transceiver.

19.6. Consider again the channel in Problem 19.5.

1. What is the minimized mean square error if the channel noise variance is $\sigma_q^2 = 0.01$ and $p_0 = 1$?
2. If we use an optimal linear zero-forcing transceiver as in Chap. 12, what would be the minimized mean square error?

3. What is the DFE gain when we go from the linear to the DFE transceiver?
4. In the above optimal designs, assume a 2-bit QAM constellation. Then what are the error probabilities for the DFE transceiver? And what are the error probabilities for the linear transceiver (with unitary $\mathbf{U}$ inserted to minimize error probabilities as in Chap. 16)?
5. Repeat the above for 4-bit QAM.

19.7. Consider the following 2×2 channel

$$\mathbf{H} = \begin{bmatrix} 3 & 1 \\ 1 & 3 \end{bmatrix}.$$

For this channel, suppose we design the optimal DFE transceiver and an optimal linear transceiver (Chap. 12), both with zero forcing. What is the DFE gain? How does it compare with the DFE gain in Problem 19.6? Explain why there is such a difference.

19.8. Consider a 3×3 channel with singular values $\{1, 0.1, 0.01\}$. Assume $\sigma_s^2 = 1$, and $\sigma_q^2 = 0.01$. For the optimal DFE system without zero forcing, the precoder singular values $\sigma_{f,k}$ can be obtained using the water-filling formula (19.95). The results depend on the power p_0. Choose three different values of the power, and compute $\sigma_{f,k}$. The three cases should be such that (a) only one $\sigma_{f,k}$ is nonzero, (b) two of the $\sigma_{f,k}$ values are nonzero, and (c) all three $\sigma_{f,k}$ values are nonzero.

19.9. Repeat Problem 19.5 for the case where the transceivers do not have the zero-forcing constraint. Assume $\sigma_q^2 = 0.01$ and $p_0 = 1$.

19.10. Repeat Problem 19.6 for the case where the transceivers do not have the zero-forcing constraint.

19.11. For the optimal DFE system, the ratio of the minimized errors with and without the zero-forcing constraint represents the benefit obtained from giving up the ZF constraint. Derive an expression for this ratio for the simple case where $K = M$ in the water-filling formula (19.108).

19.12. For the optimal DFE system in Problem 19.9 compute the mutual information $\mathcal{I}(\mathbf{x}; \mathbf{y})$.

19.13. Consider a scalar channel $H(z) = 1 - 2z^{-1} + 0.95z^{-2}$. Suppose we use a cyclic-prefix system as in Sec. 7.3 with $M = 3$ and $L = 2$ to convert this into a matrix channel.

1. Write down the circulant matrix $\mathbf{H}$ representing the matrix channel.
2. What are the singular values of this matrix?

3. Now assume that we design an optimal cyclic-prefix DFE transceiver. That is, we design the optimal transceiver matrices $\mathbf{F}, \mathbf{G}$, and $\mathbf{B}$ for the channel $\mathbf{H}$, with the zero-forcing constraint. We also design a linear MMSE transceiver with zero forcing as in Chap. 12. What is the DFE gain (i.e., the ratio of the minimized mean square errors in the two designs)?

19.14. Repeat Problem 19.13 for the case where $H(z) = 1 + 2z^{-1} + 3z^{-2}$. Explain why there is so much difference between the DFE gains in these two problems.

19.15. In Problem 19.13 assume that, instead of cyclic prefixing, we use a zero-padding system to convert the channel into a matrix channel $\mathbf{H}$. Clearly this is a 5×3 full-banded Toeplitz matrix. In this case what is the DFE gain (assuming zero-forcing designs)?

Part 3 Mathematical background

20 Matrix differentiation

20.1 Introduction

In this chapter we review matrix calculus, more specifically matrix derivatives. We explain how functions of matrices can be differentiated with respect to matrices. These functions can be matrices (e.g., inverses) or scalars (determinant, trace). We begin with the real case; excellent references for this include Graham [1981], Moon and Stirling [2000], and Van Trees [2002]. Appendix E of Moon and Stirling [2000] is especially thorough and lucid from a signal processor's viewpoint, and we follow their conventions and notations for the most part. For the complex case we first explain Brandwood's conjugate derivatives and their role in optimization [Brandwood, 1983]. We then extend the discussion of matrix calculus to the complex case, which is also briefly covered in Haykin [2002], Sayed [2003], and in somewhat greater detail in Moon and Stirling [2000]. Several applications from the area of transceiver optimization are presented throughout this book.

Notation reminder. Superscripts as in $\mathbf{A}^T, \mathbf{A}^*$, and $\mathbf{A}^\dagger$ denote, respectively, the transpose, conjugate, and transpose-conjugate of the matrix $\mathbf{A}$. $\bigtriangledown \bigtriangledown \bigtriangledown$

20.2 Real matrices and functions

Let $\mathbf{X}$ be a possibly rectangular matrix, and let $\mathbf{Y}(\mathbf{X})$ be a function of $\mathbf{X}$. Note that $\mathbf{Y}$ is also a possibly rectangular matrix, possibly of different size than $\mathbf{X}$. Assume both $\mathbf{X}$ and $\mathbf{Y}$ are real. The notation $\partial \mathbf{Y}/\partial x_{rs}$ represents the following matrix:

$$\frac{\partial \mathbf{Y}}{\partial x_{rs}} = \begin{bmatrix} \dfrac{\partial y_{00}}{\partial x_{rs}} & \dfrac{\partial y_{01}}{\partial x_{rs}} & \dfrac{\partial y_{02}}{\partial x_{rs}} & \cdots \\ \dfrac{\partial y_{10}}{\partial x_{rs}} & \dfrac{\partial y_{11}}{\partial x_{rs}} & \dfrac{\partial y_{12}}{\partial x_{rs}} & \cdots \\ \dfrac{\partial y_{20}}{\partial x_{rs}} & \dfrac{\partial y_{21}}{\partial x_{rs}} & \dfrac{\partial y_{22}}{\partial x_{rs}} & \cdots \\ \vdots & \vdots & \vdots & \ddots \end{bmatrix}.$$

Clearly the size of $\partial \mathbf{Y}/\partial x_{rs}$ is the same as that of $\mathbf{Y}$. Similarly the notation $\partial y_{mn}/\partial \mathbf{X}$ represents the following matrix:

$$\frac{\partial y_{mn}}{\partial \mathbf{X}} = \begin{bmatrix} \dfrac{\partial y_{mn}}{\partial x_{00}} & \dfrac{\partial y_{mn}}{\partial x_{01}} & \dfrac{\partial y_{mn}}{\partial x_{02}} & \cdots \\ \dfrac{\partial y_{mn}}{\partial x_{10}} & \dfrac{\partial y_{mn}}{\partial x_{11}} & \dfrac{\partial y_{mn}}{\partial x_{12}} & \cdots \\ \dfrac{\partial y_{mn}}{\partial x_{20}} & \dfrac{\partial y_{mn}}{\partial x_{21}} & \dfrac{\partial y_{mn}}{\partial x_{22}} & \cdots \\ \vdots & \vdots & \vdots & \ddots \end{bmatrix}.$$

This matrix has the same size as $\mathbf{X}$. Note that it is possible to define $\partial \mathbf{Y}/\partial \mathbf{X}$ by stacking up the elements $\partial y_{mn}/\partial x_{rs}$ in four dimensions. We shall not, however, require it in this chapter.

The notation $\mathcal{I}_{rs}$. When working with examples based on these definitions, it is convenient to define a matrix $\mathcal{I}_{rs}$ whose elements are all zero, except the (r, s)-element, which is unity. Note that the size of $\mathcal{I}_{rs}$ is not indicated in the notation, and is usually clear from the context. For example, assuming the size is 5×4, the matrix $\mathcal{I}_{2,3}$ is

$$\mathcal{I}_{2,3} = \begin{matrix} & \begin{matrix} 0 & 1 & 2 & 3 \end{matrix} \\ \begin{matrix} 0 \\ 1 \\ 2 \\ 3 \\ 4 \end{matrix} & \begin{pmatrix} 0 & 0 & 0 & 0 \\ 0 & 0 & 0 & 0 \\ 0 & 0 & 0 & 1 \\ 0 & 0 & 0 & 0 \\ 0 & 0 & 0 & 0 \end{pmatrix} \end{matrix}. \qquad (20.1)$$

Note carefully that rows and columns are numbered from zero as in earlier chapters. Note also that $\mathcal{I}_{rs}$ is a rank-one matrix and can be written as

$$\mathcal{I}_{rs} = r \text{ rows} \begin{pmatrix} 0 \\ \vdots \\ 0 \\ 1 \\ 0 \\ \vdots \\ 0 \end{pmatrix} \overbrace{\begin{pmatrix} 0 & \ldots & 0 \end{pmatrix}}^{s \text{ columns}} \!\!\!\!\!\! \begin{pmatrix} & 1 & 0 & \ldots & 0 \end{pmatrix} = \mathbf{e}_r \mathbf{e}_s^T, \qquad (20.2)$$

where $\mathbf{e}_k$ is the unit column vector with "1" in the kth row and "0" in all other rows.
$\bigtriangledown\bigtriangledown\bigtriangledown$

As a simple application of these notations, observe that

$$\left(\frac{\partial \mathbf{X}}{\partial x_{rs}}\right) = \mathcal{I}_{rs}. \tag{20.3}$$

That is, all elements of the derivative are zero except the (r, s)-element. Similarly

$$\left(\frac{\partial \mathbf{X}^T}{\partial x_{rs}}\right) = \mathcal{I}_{rs}^T = \mathcal{I}_{sr}. \tag{20.4}$$

These examples assume that there are no special relationships among the elements of $\mathbf{X}$ such as, for example, symmetry. If there is such a relation, then the derivative matrix $\partial\mathbf{X}/\partial x_{rs}$ would be nonzero in more than one location. *Unless mentioned otherwise, we shall assume that matrices do not have such special relationships between their elements.*

Example 20.1.

Let $\mathbf{Y} = \mathbf{AXB}$, then $y_{mn} = \sum_k \sum_i a_{mk}x_{ki}b_{in}$ so that

$$\frac{\partial y_{mn}}{\partial x_{rs}} = a_{mr}b_{sn}.$$

To write this in a more convenient form for future use, note that

$$\mathbf{A}\mathcal{I}_{rs}\mathbf{B} = \mathbf{A}\mathbf{e}_r\mathbf{e}_s^T\mathbf{B} = [\mathbf{A}]_{r\text{th } col.}[\mathbf{B}]_{s\text{th } row}.$$

Thus $[\mathbf{A}\mathcal{I}_{rs}\mathbf{B}]_{mn} = a_{mr}b_{sn}$, and we can rewrite

$$\frac{\partial y_{mn}}{\partial x_{rs}} = [\mathbf{A}\mathcal{I}_{rs}\mathbf{B}]_{mn}$$

so that

$$\frac{\partial \mathbf{Y}}{\partial x_{rs}} = \mathbf{A}\mathcal{I}_{rs}\mathbf{B}.$$

Since

$$\partial y_{mn}/\partial x_{rs} = a_{mr}b_{sn} = [\mathbf{A}^T]_{rm}[\mathbf{B}^T]_{ns} = [\mathbf{A}^T\mathcal{I}_{mn}\mathbf{B}^T]_{rs}$$

it then follows that

$$\frac{\partial y_{mn}}{\partial \mathbf{X}} = \mathbf{A}^T\mathcal{I}_{mn}\mathbf{B}^T.$$

Similarly, if $\mathbf{Y} = \mathbf{AX}^T\mathbf{B}$ it follows that

$$\frac{\partial \mathbf{Y}}{\partial x_{rs}} = \mathbf{A}\mathcal{I}_{rs}^T\mathbf{B} \quad \text{and} \quad \frac{\partial y_{mn}}{\partial \mathbf{X}} = \mathbf{B}\mathcal{I}_{mn}^T\mathbf{A}.$$

The transformation principle. Moon and Stirling [2000] observe that even when $\mathbf{A}$ and $\mathbf{B}$ depend on $\mathbf{X}$ one can show that

$$\frac{\partial \mathbf{Y}}{\partial x_{rs}} = \mathbf{A}\mathcal{I}_{rs}\mathbf{B} \quad \text{if and only if} \quad \frac{\partial y_{mn}}{\partial \mathbf{X}} = \mathbf{A}^T\mathcal{I}_{mn}\mathbf{B}^T$$

and

$$\frac{\partial \mathbf{Y}}{\partial x_{rs}} = \mathbf{A}\mathcal{I}_{rs}^T\mathbf{B} \quad \text{if and only if} \quad \frac{\partial y_{mn}}{\partial \mathbf{X}} = \mathbf{B}\mathcal{I}_{mn}^T\mathbf{A}.$$

This is called the *transformation principle.* ▽ ▽ ▽

20.2.1 Product rule

Consider the matrix product

$$\mathbf{Y}(\mathbf{X}) = \mathbf{U}(\mathbf{X})\mathbf{V}(\mathbf{X}).$$

Then

$$\frac{\partial \mathbf{Y}}{\partial x_{rs}} = \Big(\frac{\partial \mathbf{U}}{\partial x_{rs}}\Big)\mathbf{V} + \mathbf{U}\Big(\frac{\partial \mathbf{V}}{\partial x_{rs}}\Big),$$

where the argument ($\mathbf{X}$) has been omitted for brevity. This is the product rule of matrix calculus, and can be proved easily by writing $y_{mn} = \sum_k u_{mk}v_{kn}$, and using the conventional product rule for differentiation.

Example 20.2.

Let $\mathbf{Y} = \mathbf{X}^T\mathbf{A}\mathbf{X}$. Using the product rule we find

$$\frac{\partial \mathbf{Y}}{\partial x_{rs}} = \Big(\frac{\partial \mathbf{X}^T}{\partial x_{rs}}\Big)\mathbf{A}\mathbf{X} \; + \; \mathbf{X}^T\mathbf{A}\Big(\frac{\partial \mathbf{X}}{\partial x_{rs}}\Big) = \mathcal{I}_{rs}^T\mathbf{A}\mathbf{X} + \mathbf{X}^T\mathbf{A}\mathcal{I}_{rs}.$$

Example 20.3.

Let $\mathbf{Y} = \mathbf{X}^{-1}$. To calculate $\partial\mathbf{Y}/\partial x_{rs}$, observe that $\mathbf{X}\mathbf{X}^{-1} = \mathbf{I}$. Differentiating both sides and using the product rule, we therefore obtain

$$\Big(\frac{\partial \mathbf{X}}{\partial x_{rs}}\Big)\mathbf{X}^{-1} + \mathbf{X}\Big(\frac{\partial \mathbf{X}^{-1}}{\partial x_{rs}}\Big) = \mathbf{0}.$$

But since $\Big(\partial\mathbf{X}/\partial x_{rs}\Big) = \mathcal{I}_{rs}$ for arbitrary $\mathbf{X}$, it follows from the above that

$$\frac{\partial \mathbf{X}^{-1}}{\partial x_{rs}} = -\mathbf{X}^{-1}\mathcal{I}_{rs}\mathbf{X}^{-1}. \tag{20.5}$$

If we have an expression like $\mathbf{Y} = \mathbf{A}\mathbf{X}^{-1}\mathbf{B}$ then using the product rule and remembering that $\mathbf{A}$ and $\mathbf{B}$ are constants we get

$$\frac{\partial(\mathbf{A}\mathbf{X}^{-1}\mathbf{B})}{\partial x_{rs}} = -\mathbf{A}\mathbf{X}^{-1}\mathcal{I}_{rs}\mathbf{X}^{-1}\mathbf{B}.$$

Example 20.4.

Let $\mathbf{Y} = \mathbf{f}(\mathbf{X})$. Suppose we are interested in calculating $\partial \mathbf{Y}^{-1}/\partial x_{rs}$. Since $\mathbf{Y}\mathbf{Y}^{-1} = \mathbf{I}$ we have

$$\Big(\frac{\partial \mathbf{Y}}{\partial x_{rs}}\Big)\mathbf{Y}^{-1} + \mathbf{Y}\Big(\frac{\partial \mathbf{Y}^{-1}}{\partial x_{rs}}\Big) = \mathbf{0}$$

so that

$$\frac{\partial \mathbf{Y}^{-1}}{\partial x_{rs}} = -\mathbf{Y}^{-1}\Big(\frac{\partial \mathbf{Y}}{\partial x_{rs}}\Big)\mathbf{Y}^{-1}. \qquad (20.6)$$

20.2.2 Scalar functions of vectors

We now consider the special case where $\mathbf{Y}$ is a scalar (denoted ϕ) and $\mathbf{X}$ a column vector (denoted $\mathbf{x}$). For example, suppose $\phi = \mathbf{c}^T\mathbf{x}$. Then $\partial\phi/\partial\mathbf{x}$ is a column vector with elements $\partial\phi/\partial x_r$. Since $\partial\phi/\partial x_r = c_r$ it follows that

$$\frac{\partial(\mathbf{c}^T\mathbf{x})}{\partial \mathbf{x}} = \mathbf{c}. \qquad (20.7)$$

As another example let $\phi = \mathbf{x}^T\mathbf{R}\mathbf{x}$, where $\mathbf{R}$ is a square matrix and $\mathbf{x}$ is a column vector. To compute $\partial\phi/\partial\mathbf{x}$ just observe that

$$\frac{\partial(\mathbf{x}^T\mathbf{R}\mathbf{x})}{\partial x_r} = \Big(\frac{\partial \mathbf{x}^T}{\partial x_r}\Big)\mathbf{R}\mathbf{x} + \mathbf{x}^T\mathbf{R}\Big(\frac{\partial \mathbf{x}}{\partial x_r}\Big) = \mathbf{e}_r^T\mathbf{R}\mathbf{x} + \mathbf{x}^T\mathbf{R}\mathbf{e}_r,$$

where $\mathbf{e}_k$ is the unit column vector with "1" in the kth row. The preceding clearly simplifies to

$$\frac{\partial(\mathbf{x}^T\mathbf{R}\mathbf{x})}{\partial x_r} = (\mathbf{R}\mathbf{x})_r + (\mathbf{x}^T\mathbf{R})_r = (\mathbf{R}\mathbf{x})_r + (\mathbf{R}^T\mathbf{x})_r.$$

This shows that

$$\frac{\partial(\mathbf{x}^T\mathbf{R}\mathbf{x})}{\partial \mathbf{x}} = \mathbf{R}\mathbf{x} + \mathbf{R}^T\mathbf{x}. \qquad (20.8)$$

For example, when $\mathbf{R} = \mathbf{I}$,

$$\frac{\partial(\mathbf{x}^T\mathbf{x})}{\partial \mathbf{x}} = 2\mathbf{x}. \qquad (20.9)$$

20.2.3 Differentiating the trace of a matrix

The trace is a scalar function of a matrix. This arises frequently in optimization problems in signal processing. Recall that the trace of a square matrix is the sum of its diagonal elements. Let $\mathbf{Y} = f(\mathbf{X})$. Then $\text{Tr}(\mathbf{Y})$ is a function of the elements y_{kk}. Assuming the derivative $\partial\,\text{Tr}(\mathbf{Y})/\partial x_{rs}$ is well defined, we can stack up the numbers $\partial\,\text{Tr}(\mathbf{Y})/\partial x_{rs}$ for all (r, s) combinations, and define the matrix

$$\frac{\partial\,\mathrm{Tr}(\mathbf{Y})}{\partial \mathbf{X}} = \begin{bmatrix} \frac{\partial\,\mathrm{Tr}(\mathbf{Y})}{\partial x_{00}} & \frac{\partial\,\mathrm{Tr}(\mathbf{Y})}{\partial x_{01}} & \frac{\partial\,\mathrm{Tr}(\mathbf{Y})}{\partial x_{02}} & \cdots \\ \frac{\partial\,\mathrm{Tr}(\mathbf{Y})}{\partial x_{10}} & \frac{\partial\,\mathrm{Tr}(\mathbf{Y})}{\partial x_{11}} & \frac{\partial\,\mathrm{Tr}(\mathbf{Y})}{\partial x_{12}} & \cdots \\ \frac{\partial\,\mathrm{Tr}(\mathbf{Y})}{\partial x_{20}} & \frac{\partial\,\mathrm{Tr}(\mathbf{Y})}{\partial x_{21}} & \frac{\partial\,\mathrm{Tr}(\mathbf{Y})}{\partial x_{22}} & \cdots \\ \vdots & \vdots & \vdots & \ddots \end{bmatrix}.$$

Since

$$\mathrm{Tr}(\mathbf{X}) = x_{00} + x_{11} + x_{22} + \ldots$$

it follows that

$$\frac{\partial\,\mathrm{Tr}(\mathbf{X})}{\partial x_{rs}} = \delta(r-s)$$

so that

$$\frac{\partial\,\mathrm{Tr}(\mathbf{X})}{\partial \mathbf{X}} = \mathbf{I}. \tag{20.10}$$

In some of the examples to follow, we will find the following identity very useful:

$$\mathrm{Tr}(\mathcal{I}_{rs}\mathbf{X}) = x_{sr}. \tag{20.11}$$

Proof of Eq. (20.11). The only nonzero row of $\mathcal{I}_{rs}\mathbf{X}$ is the rth row. This has the form

$$[0\,0\ldots\,0\,1\,0\ldots 0]\mathbf{X} = [x_{s0} \quad x_{s1} \quad x_{s2} \quad \ldots]$$

because the "1" is in the sth place. So the matrix $\mathcal{I}_{rs}\mathbf{X}$ has only one (possibly) nonzero diagonal element, and that is in the rth row. Its value is clearly x_{sr}. This justifies Eq. (20.11). $\bigtriangledown\bigtriangledown\bigtriangledown$

In some of the following examples it is also useful to remember the identity

$$\frac{\partial\,\mathrm{Tr}(\mathbf{Y})}{\partial x_{rs}} = \mathrm{Tr}\Big(\frac{\partial \mathbf{Y}}{\partial x_{rs}}\Big). \tag{20.12}$$

This follows from the fact that the derivative of a finite sum is the sum of the derivatives of individual terms. It is also very useful to remember that

$$\mathrm{Tr}(\mathbf{AB}) = \mathrm{Tr}(\mathbf{BA}), \tag{20.13}$$

where $\mathbf{A}$ is $P \times M$ and $\mathbf{B}$ is $M \times P$ so that both $\mathbf{AB}$ and $\mathbf{BA}$ are defined.

Example 20.5.

Let $\mathbf{Y} = \mathbf{AX}$. To find $\partial\,\mathrm{Tr}(\mathbf{Y})/\partial\mathbf{X}$ observe first that

$$\begin{aligned}
\frac{\partial\,\mathrm{Tr}(\mathbf{AX})}{\partial x_{rs}} &= \mathrm{Tr}\Big(\frac{\partial \mathbf{AX}}{\partial x_{rs}}\Big) \\
&= \mathrm{Tr}\Big(\mathbf{A}\frac{\partial \mathbf{X}}{\partial x_{rs}}\Big) \quad \text{(using the product rule)} \\
&= \mathrm{Tr}\Big(\mathbf{A}\mathcal{I}_{rs}\Big) \quad \text{(using Eq. (20.3))} \\
&= \mathrm{Tr}\Big(\mathcal{I}_{rs}\mathbf{A}\Big) \quad \text{(using Eq. (20.13))} \\
&= a_{sr} \quad \text{(using Eq. (20.11))}
\end{aligned}$$

This proves that

$$\frac{\partial\,\mathrm{Tr}(\mathbf{AX})}{\partial \mathbf{X}} = \mathbf{A}^T.$$

In a similar way we can prove that

$$\frac{\partial\,\mathrm{Tr}(\mathbf{AXB})}{\partial \mathbf{X}} = \mathbf{A}^T\mathbf{B}^T.$$

As mentioned at the beginning, we have assumed that $\mathbf{X}$ does not have a special structure, such as symmetry.

Example 20.6.

Let $\mathbf{Y} = \mathbf{X}^T\mathbf{AX}$. To find $\partial\,\mathrm{Tr}(\mathbf{Y})/\partial\mathbf{X}$ we proceed as follows:

$$\begin{aligned}
\frac{\partial\,\mathrm{Tr}(\mathbf{X}^T\mathbf{AX})}{\partial x_{rs}} &= \mathrm{Tr}\Big(\frac{\partial(\mathbf{X}^T\mathbf{AX})}{\partial x_{rs}}\Big) \\
&= \mathrm{Tr}\Big(\mathbf{X}^T\mathbf{A}\frac{\partial \mathbf{X}}{\partial x_{rs}} + \frac{\partial \mathbf{X}^T}{\partial x_{rs}}\mathbf{AX}\Big) \\
&= \mathrm{Tr}\Big(\mathbf{X}^T\mathbf{A}\mathcal{I}_{rs}\Big) + \mathrm{Tr}\Big(\mathcal{I}_{sr}\mathbf{AX}\Big) \\
&= \mathrm{Tr}\Big(\mathcal{I}_{rs}\mathbf{X}^T\mathbf{A}\Big) + \mathrm{Tr}\Big(\mathcal{I}_{sr}\mathbf{AX}\Big) \\
&= (\mathbf{X}^T\mathbf{A})_{sr} + (\mathbf{AX})_{rs},
\end{aligned}$$

where we have used Eqs. (20.3) and (20.4) in the 3rd line and Eq. (20.11) in the last line. This proves that

$$\frac{\partial\,\mathrm{Tr}(\mathbf{X}^T\mathbf{AX})}{\partial \mathbf{X}} = \mathbf{A}^T\mathbf{X} + \mathbf{AX}.$$

For the special case $\mathbf{A} = \mathbf{I}$ we therefore have $\partial\,\mathrm{Tr}(\mathbf{X}^T\mathbf{X})/\partial\mathbf{X} = 2\mathbf{X}$. Using $\mathrm{Tr}(\mathbf{XX}^T) = \mathrm{Tr}(\mathbf{X}^T\mathbf{X})$, we get

$$\frac{\partial\,\mathrm{Tr}(\mathbf{XX}^T)}{\partial \mathbf{X}} = \frac{\partial\,\mathrm{Tr}(\mathbf{X}^T\mathbf{X})}{\partial \mathbf{X}} = 2\mathbf{X}.$$

It can similarly be shown that

$$\frac{\partial\, \mathrm{Tr}(\mathbf{XAX}^T)}{\partial \mathbf{X}} = \mathbf{XA} + \mathbf{XA}^T.$$

By a slight modification of the steps in the preceding example we can show that

$$\frac{\partial\, \mathrm{Tr}(\mathbf{X}^T\mathbf{AXB})}{\partial \mathbf{X}} = \mathbf{A}^T\mathbf{XB}^T + \mathbf{AXB}.$$

Example 20.7.

Let $\mathbf{Y} = \mathbf{AX}^{-1}\mathbf{B}$. To find $\partial(\,\mathrm{Tr}\mathbf{Y})/\partial\mathbf{X}$ observe that

$$\begin{aligned}\frac{\partial\, \mathrm{Tr}(\mathbf{AX}^{-1}\mathbf{B})}{\partial x_{rs}} &= \mathrm{Tr}\Big(\frac{\partial(\mathbf{AX}^{-1}\mathbf{B})}{\partial x_{rs}}\Big)\\ &= \mathrm{Tr}\Big(\mathbf{A}\frac{\partial(\mathbf{X}^{-1})}{\partial x_{rs}}\mathbf{B}\Big)\\ &= -\,\mathrm{Tr}\Big(\mathbf{AX}^{-1}\mathcal{I}_{rs}\mathbf{X}^{-1}\mathbf{B}\Big) \quad \text{(from Eq. (20.5))}\\ &= -\,\mathrm{Tr}\Big(\mathcal{I}_{rs}\mathbf{X}^{-1}\mathbf{BAX}^{-1}\Big) \quad \text{(from Eq. (20.13))}\\ &= -\Big(\mathbf{X}^{-1}\mathbf{BAX}^{-1}\Big)_{sr} \quad \text{(from Eq. (20.11))},\end{aligned}$$

which shows that

$$\frac{\partial\, \mathrm{Tr}(\mathbf{AX}^{-1}\mathbf{B})}{\partial \mathbf{X}} = -\Big(\mathbf{X}^{-1}\mathbf{BAX}^{-1}\Big)^T = -\mathbf{X}^{-T}\mathbf{A}^T\mathbf{B}^T\mathbf{X}^{-T}. \tag{20.14}$$

Example 20.8.

Let $\mathbf{Y} = (\mathbf{XAX}^T)^{-1}$. To compute $\partial(\,\mathrm{Tr}\mathbf{Y})/\partial\mathbf{X}$ notice first that

$$\frac{\partial\, \mathrm{Tr}((\mathbf{XAX}^T)^{-1})}{\partial x_{rs}} = \mathrm{Tr}\Big(\frac{\partial(\mathbf{XAX}^T)^{-1}}{\partial x_{rs}}\Big). \tag{20.15}$$

Using the product rule we write

$$\frac{\partial(\mathbf{XAX}^T)}{\partial x_{rs}} = \Big(\frac{\partial \mathbf{X}}{\partial x_{rs}}\Big)\mathbf{AX}^T + \mathbf{XA}\Big(\frac{\partial \mathbf{X}^T}{\partial x_{rs}}\Big) = \mathcal{I}_{rs}\mathbf{AX}^T + \mathbf{XA}\mathcal{I}_{rs}^T.$$

Using Eq. (20.6) with $\mathbf{Y} = (\mathbf{XAX}^T)$ we therefore obtain

$$\begin{aligned}\frac{\partial(\mathbf{XAX}^T)^{-1}}{\partial x_{rs}} &= -(\mathbf{XAX}^T)^{-1}\frac{\partial(\mathbf{XAX}^T)}{\partial x_{rs}}(\mathbf{XAX}^T)^{-1}\\ &= -(\mathbf{XAX}^T)^{-1}\Big(\mathcal{I}_{rs}\mathbf{AX}^T + \mathbf{XA}\mathcal{I}_{rs}^T\Big)(\mathbf{XAX}^T)^{-1}.\end{aligned}$$

Substituting this into Eq. (20.15), $\partial \operatorname{Tr}(\mathbf{XAX}^T)^{-1}/\partial x_{rs}$ simplifies to

$$\begin{aligned}
& -\operatorname{Tr}\left((\mathbf{XAX}^T)^{-1}\left(\mathcal{I}_{rs}\mathbf{AX}^T+\mathbf{XA}\mathcal{I}_{rs}^T\right)(\mathbf{XAX}^T)^{-1}\right)\\
= & -\operatorname{Tr}\left(\left(\mathcal{I}_{rs}\mathbf{AX}^T+\mathbf{XA}\mathcal{I}_{rs}^T\right)(\mathbf{XAX}^T)^{-2}\right)\\
= & -\operatorname{Tr}\left(\mathcal{I}_{rs}\mathbf{AX}^T(\mathbf{XAX}^T)^{-2}\right)-\operatorname{Tr}\left(\mathcal{I}_{rs}^T(\mathbf{XAX}^T)^{-2}\mathbf{XA}\right)\\
= & -\left(\mathbf{AX}^T(\mathbf{XAX}^T)^{-2}\right)_{sr}-\left((\mathbf{XAX}^T)^{-2}\mathbf{XA}\right)_{rs},
\end{aligned}$$

where we have used Eqs. (20.13) and (20.11). This proves that

$$\frac{\partial \operatorname{Tr}(\mathbf{XAX}^T)^{-1}}{\partial \mathbf{X}}=-(\mathbf{XA}^T\mathbf{X}^T)^{-2}\mathbf{XA}^T-(\mathbf{XAX}^T)^{-2}\mathbf{XA}. \tag{20.16}$$

A slight variation of this derivation can be used to show that

$$\begin{aligned}
\frac{\partial \operatorname{Tr}(\mathbf{B}(\mathbf{XAX}^T)^{-1})}{\partial \mathbf{X}} &= -(\mathbf{XA}^T\mathbf{X}^T)^{-1}\mathbf{B}^T(\mathbf{XA}^T\mathbf{X}^T)^{-1}\mathbf{XA}^T\\
&\quad -(\mathbf{XAX}^T)^{-1}\mathbf{B}(\mathbf{XAX}^T)^{-1}\mathbf{XA}.
\end{aligned} \tag{20.17}$$

Tables 20.1 and 20.2 at the end of this chapter summarize the important definitions and formulas discussed in this section.

20.3 Complex gradient operators

Derivatives and gradients have been used in optimization theory for many years. Most of the theory was initially developed for the case of functions of real variables. But in digital communications, array processing, and adaptive signal processing, **real functions of complex variables** occur frequently. Even though the real and imaginary parts can be separated out and regarded as independent variables, it is more elegant and economical to introduce complex differentiation operators. This is complicated by the fact that the objective functions are not usually analytic functions of the complex variables involved. The conventional definition of derivatives from analytic function theory is therefore not useful.

In 1983, D. H. Brandwood introduced the idea of *complex gradient operators* in the array processing literature. Since its introduction, this has found wide application in all areas of signal processing where optimization of complex parameters is involved. This topic has been briefly covered in the appendices of some signal processing books [Moon and Stirling, 2000], [Haykin, 2002], [Sayed, 2003]. Readers interested in a detailed and clear exposition should also read the original article [Brandwood, 1983].

20.3.1 Definitions

To motivate the idea consider the example $h(z) = |z|^2$. Since $|z|^2 = zz^*$ this is not analytic in the variable z (the presence of z^* makes it non-analytic [Churchill and Brown, 1984]). So $dh(z)/dz$ does not exist. However, imagine for a moment that z and z^* denote two separate, independent, variables and define the two-variable function

$$g(z, z^*) = zz^*.$$

We can differentiate with respect to z by holding z^* constant, and vice versa. That is, we can define the partial *derivatives*

$$\frac{\partial g(z,z^*)}{\partial z} = z^* \qquad \text{and} \qquad \frac{\partial g(z,z^*)}{\partial z^*} = z.$$

A subtle point about notation. We know that z and z^* are not independent; knowing one we can find the other. However, given the original function $h(z)$, if we replace all occurrence of z with the notation z_1 and all occurrence of z^* with z_2, then the two-variable function $g(z_1, z_2)$ can certainly be differentiated with respect to each variable separately. Instead of using the new notations z_1 and z_2 we simply carry on with z and z^*. The fact that any $h(z)$ can be expressed uniquely as $g(z, z^*)$ is explained in a footnote in Sec. 20.3.2. $\triangledown\triangledown\triangledown$

The usefulness of this "brave" definition will become clear as soon as some of its properties are established. For now, just consider another example: let $h(z) = 2\text{Re}[z]$. This is not analytic because $h(z) = z + z^*$. So $dh(z)/dz$ does not exist. Define the two-variable function

$$g(z, z^*) = z + z^*.$$

Then

$$\frac{\partial g(z,z^*)}{\partial z} = 1 \qquad \text{and} \qquad \frac{\partial g(z,z^*)}{\partial z^*} = 1.$$

Finally consider the simple example $h(z) = z$. In this case $g(z, z^*) = z$ so that $\partial g/\partial z = 1$ and $\partial g/\partial z^* = 0$. Thus

$$\frac{\partial z}{\partial z} = 1 \qquad \text{and} \qquad \frac{\partial z}{\partial z^*} = 0. \tag{20.18}$$

Similarly,

$$\frac{\partial z^*}{\partial z} = 0 \qquad \text{and} \qquad \frac{\partial z^*}{\partial z^*} = 1. \tag{20.19}$$

More generally,

$$\text{if } g(z, z^*) \text{ is free from } z^* \text{ then } \partial g/\partial z^* = 0, \tag{20.20}$$

$$\text{if } g(z, z^*) \text{ is free from } z \text{ then } \partial g/\partial z = 0. \tag{20.21}$$

In complex variable theory the derivative $\partial z^*/\partial z$ is undefined because z^* is not analytic. Observe however that the meaning of the notation here is different.

20.3.2 Functions of real and imaginary parts

Let $z = x + jy$ so that x and y denote the real and imaginary parts of z. Let $f(x, y)$ denote the function $g(z, z^*)$ regarded as a function of the two real variables x and y. Thus the original function $h(z)$ now has three different notations:[1]

$$h(z) = g(z, z^*) = f(x, y). \tag{20.22}$$

Now, how are the partial derivatives $\partial f(x, y)/\partial x$ and $\partial f(x, y)/\partial y$ related to the partial derivatives $\partial g(z, z^*)/\partial z$ and $\partial g(z, z^*)/\partial z^*$? The answer is (see proof below):

$$\frac{\partial g(z, z^*)}{\partial z} = 0.5\Big(\frac{\partial f(x, y)}{\partial x} - j\frac{\partial f(x, y)}{\partial y}\Big) \text{ and } \frac{\partial g(z, z^*)}{\partial z^*} = 0.5\Big(\frac{\partial f(x, y)}{\partial x} + j\frac{\partial f(x, y)}{\partial y}\Big). \tag{20.23}$$

By adding and subtracting these two equations we obtain

$$\frac{\partial f(x, y)}{\partial x} = \Big(\frac{\partial g(z, z^*)}{\partial z} + \frac{\partial g(z, z^*)}{\partial z^*}\Big) \quad \text{and} \quad \frac{\partial f(x, y)}{\partial y} = j\Big(\frac{\partial g(z, z^*)}{\partial z} - \frac{\partial g(z, z^*)}{\partial z^*}\Big). \tag{20.24}$$

The distinction between the notations h, g, and f is usually clear from the context, and we often simplify by saying $\partial g/\partial x = \partial g/\partial z + \partial g/\partial z^*$, and so forth. The relation (20.23) is often summarized in operator form as shown below:

$$\frac{\partial}{\partial z} = 0.5\Big(\frac{\partial}{\partial x} - j\frac{\partial}{\partial y}\Big) \quad \text{and} \quad \frac{\partial}{\partial z^*} = 0.5\Big(\frac{\partial}{\partial x} + j\frac{\partial}{\partial y}\Big). \tag{20.25}$$

Proof of Eq. (20.23). From $g(z, z^*) = f(x, y)$ we see that

$$\frac{\partial g}{\partial z} = \frac{\partial f}{\partial x}\frac{\partial x}{\partial z} + \frac{\partial f}{\partial y}\frac{\partial y}{\partial z} \quad \text{and} \quad \frac{\partial g}{\partial z^*} = \frac{\partial f}{\partial x}\frac{\partial x}{\partial z^*} + \frac{\partial f}{\partial y}\frac{\partial y}{\partial z^*}$$

Substituting

$$\frac{\partial x}{\partial z} = 0.5, \quad \frac{\partial x}{\partial z^*} = 0.5, \quad \frac{\partial y}{\partial z} = -0.5j, \quad \frac{\partial y}{\partial z^*} = 0.5j,$$

(which follows from $x = (z + z^*)/2$ and $y = (z - z^*)/2j$) we therefore obtain

$$\frac{\partial g}{\partial z} = 0.5\Big(\frac{\partial f}{\partial x} - j\frac{\partial f}{\partial y}\Big) \quad \text{and} \quad \frac{\partial g}{\partial z^*} = 0.5\Big(\frac{\partial f}{\partial x} + j\frac{\partial f}{\partial y}\Big)$$

This proves Eq. (20.23). ▽ ▽ ▽

[1] Given $h(z)$, we can substitute $z = x + jy$ and obtain the form $h(z) = f(x, y)$. In this expression if we substitute $x = (z + z^*)/2$ and $y = (z - z^*)/2j$, then we get the form $h(z) = g(z, z^*)$.

Readers who are still uncomfortable regarding z and z^* as "independent variables" can regard the two equations in (20.23) as the definitions of $\partial g/\partial z$ and $\partial g/\partial z^*$. This should cause no difficulty, as $\partial f/\partial x$ and $\partial f/\partial y$ are familiar partial derivatives from elementary calculus. Please also see Sec. 20.6 for remarks on the correct interpretations of these equations.

20.3.3 Cauchy-Riemann equations

If the expression $g(z, z^*)$ does not have any z^* in it, then $\partial g(z, z^*)/\partial z^* = 0$. From Eqs. (20.23) it then follows that

$$\frac{\partial f(x,y)}{\partial x} + j\frac{\partial f(x,y)}{\partial y} = 0.$$

Writing $f(x, y) = u(x, y) + jv(x, y)$, where $u(x.y)$ and $v(x, y)$ are real, and separating out the real and imaginary parts of the preceding equation, we obtain

$$\frac{\partial u(x,y)}{\partial x} = \frac{\partial v(x,y)}{\partial y} \quad \text{and} \quad \frac{\partial u(x,y)}{\partial y} = -\frac{\partial v(x,y)}{\partial x}.$$

These are nothing but the Cauchy-Riemann equations from complex variable theory [Churchill and Brown, 1984]. Thus, when the expression $g(z, z^*)$ is independent of z^*, the Cauchy-Riemann equations are satisfied.[2]

20.3.4 Extension to the case of vector and matrix variables

If g is a function of a complex vector $\mathbf{z}$ or a complex matrix $\mathbf{Z}$, we define the complex gradient operators in an obvious way element by element. Thus

$$\frac{\partial g(\mathbf{z},\mathbf{z}^*)}{\partial \mathbf{z}} = \begin{bmatrix} \dfrac{\partial g(\mathbf{z},\mathbf{z}^*)}{\partial z_0} \\ \dfrac{\partial g(\mathbf{z},\mathbf{z}^*)}{\partial z_1} \\ \dfrac{\partial g(\mathbf{z},\mathbf{z}^*)}{\partial z_2} \\ \vdots \end{bmatrix} \quad \text{and} \quad \frac{\partial g(\mathbf{z},\mathbf{z}^*)}{\partial \mathbf{z}^*} = \begin{bmatrix} \dfrac{\partial g(\mathbf{z},\mathbf{z}^*)}{\partial z_0^*} \\ \dfrac{\partial g(\mathbf{z},\mathbf{z}^*)}{\partial z_1^*} \\ \dfrac{\partial g(\mathbf{z},\mathbf{z}^*)}{\partial z_2^*} \\ \vdots \end{bmatrix}$$

and similarly

$$\frac{\partial g(\mathbf{Z},\mathbf{Z}^*)}{\partial \mathbf{Z}} = \begin{bmatrix} \dfrac{\partial g(\mathbf{Z},\mathbf{Z}^*)}{\partial z_{00}} & \dfrac{\partial g(\mathbf{Z},\mathbf{Z}^*)}{\partial z_{01}} & \cdots \\ \dfrac{\partial g(\mathbf{Z},\mathbf{Z}^*)}{\partial z_{10}} & \dfrac{\partial g(\mathbf{Z},\mathbf{Z}^*)}{\partial z_{11}} & \cdots \\ \vdots & \vdots & \ddots \end{bmatrix}$$

[2]The reader will recall that these equations are necessary but not sufficient for analyticity [Churchill and Brown, 1984].

and

$$\frac{\partial g(\mathbf{Z},\mathbf{Z}^*)}{\partial \mathbf{Z}^*} = \begin{bmatrix} \frac{\partial g(\mathbf{Z},\mathbf{Z}^*)}{\partial z_{00}^*} & \frac{\partial g(\mathbf{Z},\mathbf{Z}^*)}{\partial z_{01}^*} & \cdots \\ \frac{\partial g(\mathbf{Z},\mathbf{Z}^*)}{\partial z_{10}^*} & \frac{\partial g(\mathbf{Z},\mathbf{Z}^*)}{\partial z_{11}^*} & \cdots \\ \vdots & \vdots & \ddots \end{bmatrix}$$

The properties of complex gradients outlined for the scalar case can be applied to each element in the matrix or the vector to arrive at the properties of complex vector gradients and complex matrix gradients. For example, if $g(\mathbf{Z},\mathbf{Z}^*)$ does not have any $\mathbf{Z}^*$ in it, then

$$\frac{\partial g(\mathbf{Z},\mathbf{Z}^*)}{\partial \mathbf{Z}^*} = \mathbf{0}.$$

Also, Eqs. (20.23) can be applied elementwise to obtain

$$\frac{\partial g(\mathbf{Z},\mathbf{Z}^*)}{\partial \mathbf{Z}} = 0.5\Big(\frac{\partial f(\mathbf{X},\mathbf{Y})}{\partial \mathbf{X}} - j\frac{\partial f(\mathbf{X},\mathbf{Y})}{\partial \mathbf{Y}}\Big) \tag{20.26}$$

and

$$\frac{\partial g(\mathbf{Z},\mathbf{Z}^*)}{\partial \mathbf{Z}^*} = 0.5\Big(\frac{\partial f(\mathbf{X},\mathbf{Y})}{\partial \mathbf{X}} + j\frac{\partial f(\mathbf{X},\mathbf{Y})}{\partial \mathbf{Y}}\Big), \tag{20.27}$$

where $\mathbf{Z} = \mathbf{X} + j\mathbf{Y}$ and $g(\mathbf{Z},\mathbf{Z}^*) = f(\mathbf{X},\mathbf{Y})$.

20.3.5 Stationary points and optimization

Let $g(\mathbf{Z},\mathbf{Z}^*)$ be a **real scalar function** and write $g(\mathbf{Z},\mathbf{Z}^*) = f(\mathbf{X},\mathbf{Y})$ as usual. By definition, the stationary point of g is such that

$$\frac{\partial f(\mathbf{X},\mathbf{Y})}{\partial \mathbf{X}} = \mathbf{0} \quad \text{and} \quad \frac{\partial f(\mathbf{X},\mathbf{Y})}{\partial \mathbf{Y}} = \mathbf{0}.$$

Since g (and hence f) is real, it follows from Eq. (20.26) that a point is a stationary point if and only if

$$\frac{\partial g(\mathbf{Z},\mathbf{Z}^*)}{\partial \mathbf{Z}} = \mathbf{0} \tag{20.28}$$

at that point. We also see that a point is a stationary point if and only if

$$\frac{\partial g(\mathbf{Z},\mathbf{Z}^*)}{\partial \mathbf{Z}^*} = \mathbf{0} \tag{20.29}$$

at that point. So either of the above conditions can be used to test stationarity. We conclude with the following remarks:

1. Equations (20.28) and (20.29) are fundamental in applications involving **optimization**. At a **local interior extremum** of the real function $g(\mathbf{Z},\mathbf{Z}^*)$, the gradient $\partial g(\mathbf{Z},\mathbf{Z}^*)/\partial \mathbf{Z}$ (equivalently $\partial g(\mathbf{Z},\mathbf{Z}^*)/\partial \mathbf{Z}^*$) is set to zero to obtain a **necessary** condition for optimality.

2. Writing out the above necessary condition yields equations from which it is often possible to figure out the optimum matrix $\mathbf{Z}$ completely. One such example arises in Chap. 12.

3. As we shall see through examples, Eq. (20.29) is often more convenient than Eq. (20.28) in these applications.

4. For real $g(\mathbf{z}, \mathbf{z}^*)$, the relation between complex gradients and the "direction of *maximum rate of change* of $g(\mathbf{z}, \mathbf{z}^*)$" is explained in detail in Brandwood [1983].

20.4 Complex matrices and derivatives

We now consider a number of examples and applications of complex gradients with respect to matrix variables. Observe first that if $\mathbf{Z}$ is a complex matrix then

$$\frac{\partial \mathbf{Z}}{\partial z_{rs}} = \mathcal{I}_{rs} \quad \text{and} \quad \frac{\partial \mathbf{Z}}{\partial z_{rs}^*} = \mathbf{0}. \tag{20.30}$$

Similarly,

$$\frac{\partial \mathbf{Z}^\dagger}{\partial z_{rs}} = \mathbf{0} \quad \text{and} \quad \frac{\partial \mathbf{Z}^\dagger}{\partial z_{rs}^*} = \mathcal{I}_{rs}^T = \mathcal{I}_{sr}, \tag{20.31}$$

If we have an expression like $\mathbf{AZB}$ then we can use the "product rule" to obtain

$$\frac{\partial(\mathbf{AZB})}{\partial z_{rs}} = \mathbf{A}\Big(\frac{\partial \mathbf{Z}}{\partial z_{rs}}\Big)\mathbf{B} = \mathbf{A}\mathcal{I}_{rs}\mathbf{B}.$$

Example 20.9.

Let $g(\mathbf{z}, \mathbf{z}^*) = \mathbf{z}^\dagger \mathbf{a}$, where $\mathbf{z}$ is a column vector. Writing

$$g(\mathbf{z}, \mathbf{z}^*) = \sum_r z_r^* a_r$$

we see that $\partial g(\mathbf{z}, \mathbf{z}^*)/\partial z_r^* = a_r$ and $\partial g(\mathbf{z}, \mathbf{z}^*)/\partial z_r = 0$. This shows that

$$\frac{\partial(\mathbf{z}^\dagger \mathbf{a})}{\partial \mathbf{z}^*} = \mathbf{a} \qquad \frac{\partial(\mathbf{z}^\dagger \mathbf{a})}{\partial \mathbf{z}} = \mathbf{0}.$$

Similarly we can show that

$$\frac{\partial(\mathbf{a}^\dagger \mathbf{z})}{\partial \mathbf{z}^*} = \mathbf{0} \qquad \frac{\partial(\mathbf{a}^\dagger \mathbf{z})}{\partial \mathbf{z}} = \mathbf{a}^*.$$

What if $g(\mathbf{z}, \mathbf{z}^*) = \mathbf{z}^T \mathbf{b}$? Since there is no conjugation of $\mathbf{z}$ we see right away that $\partial g(\mathbf{z}, \mathbf{z}^*)/\partial \mathbf{z}^* = \mathbf{0}$. Simple calculation also shows that $\partial g(\mathbf{z}, \mathbf{z}^*)/\partial \mathbf{z} = \mathbf{b}$. Thus,

$$\frac{\partial(\mathbf{z}^T \mathbf{b})}{\partial \mathbf{z}^*} = \mathbf{0}, \qquad \frac{\partial(\mathbf{z}^T \mathbf{b})}{\partial \mathbf{z}} = \frac{\partial(\mathbf{b}^T \mathbf{z})}{\partial \mathbf{z}} = \mathbf{b}.$$

Example 20.10.

Let $g(\mathbf{z},\mathbf{z}^*) = \mathbf{z}^\dagger\mathbf{A}\mathbf{z}$, where $\mathbf{z}$ is a column vector. Writing this in expanded form

$$g(\mathbf{z},\mathbf{z}^*) = \sum_r\sum_s z_r^* A_{rs} z_s,$$

we see that

$$\frac{\partial g(\mathbf{z},\mathbf{z}^*)}{\partial z_r^*} = \sum_s A_{rs}z_s = (\mathbf{A}\mathbf{z})_{r\text{th row}}, \qquad \frac{\partial g(\mathbf{z},\mathbf{z}^*)}{\partial z_s} = \sum_r z_r^* A_{rs} = (\mathbf{z}^\dagger\mathbf{A})_{s\text{th col}}$$

Thus

$$\frac{\partial(\mathbf{z}^\dagger\mathbf{A}\mathbf{z})}{\partial\mathbf{z}^*} = \mathbf{A}\mathbf{z}, \qquad \frac{\partial(\mathbf{z}^\dagger\mathbf{A}\mathbf{z})}{\partial\mathbf{z}} = \mathbf{A}^T\mathbf{z}^*$$

As a special case note that

$$\frac{\partial(\mathbf{z}^\dagger\mathbf{z})}{\partial\mathbf{z}^*} = \mathbf{z}, \qquad \frac{\partial(\mathbf{z}^\dagger\mathbf{z})}{\partial\mathbf{z}} = \mathbf{z}^*. \tag{20.32}$$

If $g(\mathbf{z},\mathbf{z}^*) = \mathbf{z}^T\mathbf{z}$ then it is a bit different. First, since there is no conjugation of $\mathbf{z}$, $\partial g(\mathbf{z},\mathbf{z}^*)/\partial\mathbf{z}^* = \mathbf{0}$. Next, when we differentiate with respect to $\mathbf{z}$ we get one contribution from the $\mathbf{z}^T$ part and one from the $\mathbf{z}$ part, so the answer is doubled up: $\partial(\mathbf{z}^T\mathbf{z})/\partial\mathbf{z} = 2\mathbf{z}$. To see this simply observe that

$$\mathbf{z}^T\mathbf{z} = \sum_r z_r^2.$$

So $\partial(\mathbf{z}^T\mathbf{z})/\partial z_r = 2z_r$, from which the result follows. Thus

$$\frac{\partial(\mathbf{z}^T\mathbf{z})}{\partial\mathbf{z}^*} = \mathbf{0}, \qquad \frac{\partial(\mathbf{z}^T\mathbf{z})}{\partial\mathbf{z}} = 2\mathbf{z}. \tag{20.33}$$

For the special case when $\mathbf{z} = \mathbf{x}$ is real we know from Eq. (20.9) that $\partial(\mathbf{x}^T\mathbf{x})/\partial\mathbf{x} = 2\mathbf{x}$. This special case can be obtained by setting $\mathbf{z} = \mathbf{x}$ in Eq. (20.33) but cannot be obtained by setting $\mathbf{z} = \mathbf{x}$ in Eq. (20.32) which lacks the factor 2. See comments in Sec. 20.6.

Example 20.11.

Let $\mathbf{Y}(\mathbf{Z},\mathbf{Z}^*) = \mathbf{Z}^\dagger\mathbf{A}\mathbf{Z}$. We have

$$\frac{\partial(\mathbf{Z}^\dagger\mathbf{A}\mathbf{Z})}{\partial z_{rs}} = \Big(\frac{\partial\mathbf{Z}^\dagger}{\partial z_{rs}}\Big)\mathbf{A}\mathbf{Z} + \mathbf{Z}^\dagger\mathbf{A}\Big(\frac{\partial\mathbf{Z}}{\partial z_{rs}}\Big).$$

Since $\partial\mathbf{Z}^\dagger/\partial z_{rs} = \mathbf{0}$ and $\partial\mathbf{Z}/\partial z_{rs} = \mathcal{I}_{rs}$, we obtain

$$\frac{\partial(\mathbf{Z}^\dagger\mathbf{A}\mathbf{Z})}{\partial z_{rs}} = \mathbf{Z}^\dagger\mathbf{A}\mathcal{I}_{rs}. \tag{20.34}$$

Similarly, using $\partial \mathbf{Z}^\dagger/\partial z_{rs}^* = \mathcal{I}_{rs}^T = \mathcal{I}_{sr}$ and $\partial \mathbf{Z}/\partial z_{rs}^* = \mathbf{0}$, we get

$$\frac{\partial(\mathbf{Z}^\dagger \mathbf{A}\mathbf{Z})}{\partial z_{rs}^*} = \mathcal{I}_{rs}^T \mathbf{A}\mathbf{Z} = \mathcal{I}_{sr}\mathbf{A}\mathbf{Z}. \tag{20.35}$$

These yield the special cases

$$\frac{\partial(\mathbf{Z}^\dagger \mathbf{Z})}{\partial z_{rs}} = \mathbf{Z}^\dagger \mathcal{I}_{rs} \quad \text{and} \quad \frac{\partial(\mathbf{Z}^\dagger \mathbf{Z})}{\partial z_{rs}^*} = \mathcal{I}_{rs}^T \mathbf{Z} = \mathcal{I}_{sr}\mathbf{Z}.$$

Similarly we can show

$$\frac{\partial(\mathbf{Z}\mathbf{Z}^\dagger)}{\partial z_{rs}} = \mathcal{I}_{rs}\mathbf{Z}^\dagger, \qquad \frac{\partial(\mathbf{Z}\mathbf{Z}^\dagger)}{\partial z_{rs}^*} = \mathbf{Z}\mathcal{I}_{rs}^T = \mathbf{Z}\mathcal{I}_{sr}.$$

Example 20.12. Symmetric and Hermitian matrices

For a real symmetric matrix $\mathbf{X}$, since $x_{sr} = x_{rs}$, $\partial \mathbf{X}/\partial x_{rs}$ is unity at the (r, s) position as well as at the (s, r) position. So we get $\partial \mathbf{X}/\partial x_{rs} = \mathcal{I}_{rs} + \mathcal{I}_{sr}$ for $r \neq s$, and for $r = s$ we have $\partial \mathbf{X}/\partial x_{rr} = \mathcal{I}_{rr}$. So for real symmetric $\mathbf{X}$,

$$\frac{\partial \mathbf{X}}{\partial x_{rs}} = \begin{cases} \mathcal{I}_{rs} + \mathcal{I}_{sr}, & r \neq s \\ \mathcal{I}_{rr}, & r = s. \end{cases}$$

Next consider a complex matrix $\mathbf{Z}$. If $\mathbf{Z}$ is arbitrary with no structure we know

$$\frac{\partial \mathbf{Z}}{\partial z_{rs}^*} = \mathbf{0}.$$

But if $\mathbf{Z}$ is constrained to be Hermitian then z_{rs}^* happens to be the element in the (s, r) position. Thus for complex Hermitian $\mathbf{Z}$,

$$\frac{\partial \mathbf{Z}}{\partial z_{rs}^*} = \mathcal{I}_{sr},$$

and, of course, $\partial \mathbf{Z}/\partial z_{rs} = \mathcal{I}_{rs}$.

20.4.1 Differentiating matrix functions with traces in them

Traces of matrices arise frequently in signal processing problems. Consider the example $\mathbf{Y} = \mathbf{Z}^\dagger \mathbf{A}\mathbf{Z}\mathbf{B}$, where $\mathbf{Z}$ is a complex matrix. To compute $\partial\, \text{Tr}\mathbf{Y}/\partial z_{rs}^*$ we proceed as follows:

$$\frac{\partial\, \text{Tr}(\mathbf{Z}^\dagger \mathbf{A}\mathbf{Z}\mathbf{B})}{\partial z_{rs}^*} = \text{Tr}\left(\frac{\partial \mathbf{Z}^\dagger}{\partial z_{rs}^*}\mathbf{A}\mathbf{Z}\mathbf{B}\right) + \text{Tr}\left(\mathbf{Z}^\dagger \mathbf{A}\frac{\partial \mathbf{Z}}{\partial z_{rs}^*}\mathbf{B}\right).$$

Here we have used the fact that the trace of a derivative is equal to the derivative of the trace. This is followed by the application of the product rule for derivatives. In view of Eqs. (20.30) and (20.31) this simplifies to

$$\frac{\partial\, \text{Tr}(\mathbf{Z}^\dagger \mathbf{A}\mathbf{Z}\mathbf{B})}{\partial z_{rs}^*} = \text{Tr}\left(\mathcal{I}_{rs}^T \mathbf{A}\mathbf{Z}\mathbf{B}\right) = (\mathbf{A}\mathbf{Z}\mathbf{B})_{rs},$$

where the last equality above follows from Eq. (20.11). The derivative with respect to z_{rs} can be handled similarly. Thus we have the results

$$\frac{\partial\, \mathrm{Tr}(\mathbf{Z}^\dagger\mathbf{AZB})}{\partial\mathbf{Z}^*} = \mathbf{AZB} \quad \text{and} \quad \frac{\partial\, \mathrm{Tr}(\mathbf{Z}^\dagger\mathbf{AZB})}{\partial\mathbf{Z}} = \mathbf{A}^T\mathbf{Z}^*\mathbf{B}^T. \tag{20.36}$$

From this observe the following special cases:

$$\frac{\partial\, \mathrm{Tr}(\mathbf{Z}^\dagger\mathbf{Z})}{\partial\mathbf{Z}^*} = \mathbf{Z} \quad \text{and} \quad \frac{\partial\, \mathrm{Tr}(\mathbf{Z}^\dagger\mathbf{Z})}{\partial\mathbf{Z}} = \mathbf{Z}^*. \tag{20.37}$$

Note that $\partial/\partial\mathbf{Z}^*$ is usually cleaner than $\partial/\partial\mathbf{Z}$. For this reason the former is more popularly used in optimization problems than the latter.

Example 20.13.

For complex $\mathbf{Z}$ we now evaluate $\partial\mathrm{Tr}(\mathbf{AZ}^\dagger)/\partial\mathbf{Z}^*$. We have

$$\begin{aligned}
\frac{\partial\mathrm{Tr}(\mathbf{AZ}^\dagger)}{\partial z_{rs}^*} &= \mathrm{Tr}\frac{\partial(\mathbf{AZ}^\dagger)}{\partial z_{rs}^*} \\
&= \mathrm{Tr}\Big(\mathbf{A}\frac{\partial\mathbf{Z}^\dagger}{\partial z_{rs}^*}\Big) \\
&= \mathrm{Tr}(\mathbf{A}\mathcal{I}_{sr}) = \mathrm{Tr}(\mathcal{I}_{sr}\mathbf{A}) = [\mathbf{A}]_{rs},
\end{aligned}$$

which shows that $\partial\mathrm{Tr}(\mathbf{AZ}^\dagger)/\partial\mathbf{Z}^* = \mathbf{A}$.

Example 20.14.

For complex $\mathbf{Z}$ we now consider derivatives involving $(\mathbf{ZAZ}^\dagger)$. First we have

$$\frac{\partial(\mathbf{ZAZ}^\dagger)}{\partial z_{rs}} = \frac{\partial\mathbf{Z}}{\partial z_{rs}}\mathbf{AZ}^\dagger = \mathcal{I}_{rs}\mathbf{AZ}^\dagger$$

because $\mathbf{Z}^\dagger$ only has elements z_{rs}^*. Similarly

$$\frac{\partial(\mathbf{ZAZ}^\dagger)}{\partial z_{rs}^*} = \mathbf{ZA}\frac{\partial\mathbf{Z}^\dagger}{\partial z_{rs}^*} = \mathbf{ZA}\mathcal{I}_{rs}^T$$

because the element z_{rs}^* occurs as the (s,r)th element in $\mathbf{Z}^\dagger$. Next

$$\frac{\partial\, \mathrm{Tr}(\mathbf{ZAZ}^\dagger)}{\partial z_{rs}} = \mathrm{Tr}\frac{\partial(\mathbf{ZAZ}^\dagger)}{\partial z_{rs}} = \mathrm{Tr}\Big(\mathcal{I}_{rs}\mathbf{AZ}^\dagger\Big) = (\mathbf{AZ}^\dagger)_{sr}$$

so that

$$\frac{\partial\, \mathrm{Tr}(\mathbf{ZAZ}^\dagger)}{\partial\mathbf{Z}} = (\mathbf{AZ}^\dagger)^T.$$

Similarly

$$\frac{\partial\, \mathrm{Tr}(\mathbf{ZAZ}^\dagger)}{\partial\mathbf{Z}^*} = \mathbf{ZA}.$$

20.4.2 Differentiating matrix functions with inverses in them

The inverses of complex matrix functions often arise in signal processing problems. When optimizing matrices it often becomes necessary to know how to differentiate such functions with respect to the matrix parameters.

Example 20.15.

Consider $\mathbf{Y} = \mathbf{f}(\mathbf{Z})$ where $\mathbf{Z}$ is complex. In exactly the same way that we obtained (20.6), we can show that

$$\frac{\partial \mathbf{Y}^{-1}}{\partial z_{rs}^*} = -\mathbf{Y}^{-1}\Big(\frac{\partial \mathbf{Y}}{\partial z_{rs}^*}\Big)\mathbf{Y}^{-1} \quad \text{and} \quad \frac{\partial \mathbf{Y}^{-1}}{\partial z_{rs}} = -\mathbf{Y}^{-1}\Big(\frac{\partial \mathbf{Y}}{\partial z_{rs}}\Big)\mathbf{Y}^{-1}. \tag{20.38}$$

Setting $\mathbf{Y} = \mathbf{Z}$ and remembering that $\partial \mathbf{Z}/\partial z_{rs} = \mathcal{I}_{rs}$ and $\partial \mathbf{Z}/\partial z_{rs}^* = \mathbf{0}$, we get the following special case:

$$\frac{\partial \mathbf{Z}^{-1}}{\partial z_{rs}^*} = \mathbf{0} \qquad \text{and} \qquad \frac{\partial \mathbf{Z}^{-1}}{\partial z_{rs}} = -\mathbf{Z}^{-1}\mathcal{I}_{rs}\mathbf{Z}^{-1}. \tag{20.39}$$

For a nontrivial special case, consider $\mathbf{Y} = \mathbf{Z}^\dagger\mathbf{A}\mathbf{Z}$. From Eq. (20.38) we see that

$$\begin{aligned}\frac{\partial (\mathbf{Z}^\dagger\mathbf{A}\mathbf{Z})^{-1}}{\partial z_{rs}^*} &= -(\mathbf{Z}^\dagger\mathbf{A}\mathbf{Z})^{-1}\Big(\frac{\partial (\mathbf{Z}^\dagger\mathbf{A}\mathbf{Z})}{\partial z_{rs}^*}\Big)(\mathbf{Z}^\dagger\mathbf{A}\mathbf{Z})^{-1} \\ &= -(\mathbf{Z}^\dagger\mathbf{A}\mathbf{Z})^{-1}\Big(\frac{\partial \mathbf{Z}^\dagger}{\partial z_{rs}^*}\mathbf{A}\mathbf{Z}\Big)(\mathbf{Z}^\dagger\mathbf{A}\mathbf{Z})^{-1}.\end{aligned}$$

Using $\partial \mathbf{Z}^\dagger/\partial z_{rs}^* = \mathcal{I}_{rs}^T$, this simplifies to

$$\frac{\partial (\mathbf{Z}^\dagger\mathbf{A}\mathbf{Z})^{-1}}{\partial z_{rs}^*} = -(\mathbf{Z}^\dagger\mathbf{A}\mathbf{Z})^{-1}\mathcal{I}_{rs}^T\mathbf{A}\mathbf{Z}(\mathbf{Z}^\dagger\mathbf{A}\mathbf{Z})^{-1}. \tag{20.40}$$

The derivative $\partial(\mathbf{Z}^\dagger\mathbf{A}\mathbf{Z})^{-1}/\partial z_{rs}$ can be similarly handled, and the result is

$$\frac{\partial (\mathbf{Z}^\dagger\mathbf{A}\mathbf{Z})^{-1}}{\partial z_{rs}} = -(\mathbf{Z}^\dagger\mathbf{A}\mathbf{Z})^{-1}\mathbf{Z}^\dagger\mathbf{A}\mathcal{I}_{rs}(\mathbf{Z}^\dagger\mathbf{A}\mathbf{Z})^{-1}. \tag{20.41}$$

20.4.3 Differentiating traces of inverses of complex matrices

Let $\mathbf{Y} = (\mathbf{Z}^\dagger\mathbf{A}\mathbf{Z})^{-1}$, where $\mathbf{Z}$ is a complex matrix. To compute the derivative of the trace, observe that

$$\frac{\partial \operatorname{Tr}(\mathbf{Z}^\dagger\mathbf{A}\mathbf{Z})^{-1}}{\partial z_{rs}^*} = \operatorname{Tr}\Big(\frac{\partial (\mathbf{Z}^\dagger\mathbf{A}\mathbf{Z})^{-1}}{\partial z_{rs}^*}\Big) = -\operatorname{Tr}\Big((\mathbf{Z}^\dagger\mathbf{A}\mathbf{Z})^{-1}\mathcal{I}_{rs}^T\mathbf{A}\mathbf{Z}(\mathbf{Z}^\dagger\mathbf{A}\mathbf{Z})^{-1}\Big)$$

from Eq. (20.40). By using commutativity of trace (20.13), we can simplify this as

$$\frac{\partial \operatorname{Tr}(\mathbf{Z}^\dagger\mathbf{A}\mathbf{Z})^{-1}}{\partial z_{rs}^*} = -\operatorname{Tr}\Big(\mathcal{I}_{rs}^T\mathbf{A}\mathbf{Z}(\mathbf{Z}^\dagger\mathbf{A}\mathbf{Z})^{-2}\Big) = -\Big(\mathbf{A}\mathbf{Z}(\mathbf{Z}^\dagger\mathbf{A}\mathbf{Z})^{-2}\Big)_{rs},$$

where the last equality follows from Eq. (20.11); $\partial \operatorname{Tr}(\mathbf{Z}^\dagger \mathbf{A}\mathbf{Z})^{-1}/\partial z_{rs}$ can be obtained similarly. The final results are

$$\frac{\partial \operatorname{Tr}(\mathbf{Z}^\dagger \mathbf{A}\mathbf{Z})^{-1}}{\partial \mathbf{Z}^*} = -\mathbf{A}\mathbf{Z}(\mathbf{Z}^\dagger \mathbf{A}\mathbf{Z})^{-2} \text{ and } \frac{\partial \operatorname{Tr}(\mathbf{Z}^\dagger \mathbf{A}\mathbf{Z})^{-1}}{\partial \mathbf{Z}} = -\left((\mathbf{Z}^\dagger \mathbf{A}\mathbf{Z})^{-2}\mathbf{Z}^\dagger \mathbf{A}\right)^T.$$

Tables 20.3 and 20.4 summarize many of the formulas derived in this and the previous sections for the complex case. Some of the entries in the tables are left as exercises at the end of the chapter.

20.5 Optimization examples

We now present two examples which demonstrate the usefuless of complex matrix differentiation in optimization.

Example 20.16. Beamforming

Let $\mathbf{R}$ be a Hermitian positive definite matrix ($\mathbf{R} = \mathbf{R}^\dagger > \mathbf{0}$) and consider the function

$$\phi = \mathbf{z}^\dagger \mathbf{R}\mathbf{z}, \tag{20.42}$$

where $\mathbf{z}$ is a complex vector. Since

$$\phi^* = \phi^\dagger = \mathbf{z}^\dagger \mathbf{R}^\dagger \mathbf{z} = \mathbf{z}^\dagger \mathbf{R}\mathbf{z} = \phi,$$

it follows that ϕ is real. Suppose we wish to optimize $\mathbf{z}$ such that ϕ is minimized under the constraint

$$\mathbf{z}^\dagger \mathbf{a} = 1. \tag{20.43}$$

This problem arises in beamforming applications [Haykin, 2002] and results in the so-called **Capon** beamformer. To solve this problem we can use the Largrange multiplier method and formulate a composite objective function ψ [Luenberger, 1969], [Brandwood, 1983], [Boyd and Vandenberghe, 2004]. Setting $\partial\psi/\partial\mathbf{z}^* = 0$ would then give us a *necessary* condition for a stationary point, and hence optimality. As we shall see, in this example this condition is also sufficient, and yields a global minimum solution.

We have to make sure that **the Lagrangian ψ is real**, so that the conditions of Sec. 20.3.5 can be applied. For this we first rewrite the constraint (20.43) in the form of two real constraints

$$\mathbf{z}^\dagger \mathbf{a} + \mathbf{a}^\dagger \mathbf{z} = 2 \tag{20.44}$$

and

$$\frac{\mathbf{z}^\dagger \mathbf{a} - \mathbf{a}^\dagger \mathbf{z}}{j} = 0. \tag{20.45}$$

The first constraint says that the real part of $\mathbf{z}^\dagger \mathbf{a}$ should be unity, and the second constraint says that the imaginary part of $\mathbf{z}^\dagger \mathbf{a}$ should be zero. We now construct the Lagrangian

$$\psi(\mathbf{z}, \mathbf{z}^*) = \mathbf{z}^\dagger \mathbf{R}\mathbf{z} + \lambda_1(\mathbf{z}^\dagger \mathbf{a} + \mathbf{a}^\dagger \mathbf{z} - 2) + \lambda_2(\mathbf{z}^\dagger \mathbf{a} - \mathbf{a}^\dagger \mathbf{z})/j, \tag{20.46}$$

where λ_1 and λ_2 are real-valued Lagrange multipliers. Since ψ is real by definition, at a stationary point of ψ we should have $\partial\psi/\partial\mathbf{z}^* = \mathbf{0}$ (Sec. 20.3.5). We already showed (Ex. 20.9 and Ex. 20.10) that

$$\frac{\partial(\mathbf{z}^\dagger\mathbf{a})}{\partial\mathbf{z}^*} = \mathbf{a}, \qquad \frac{\partial(\mathbf{z}^\dagger\mathbf{R}\mathbf{z})}{\partial\mathbf{z}^*} = \mathbf{R}\mathbf{z},$$

and furthermore that $\partial(\mathbf{a}^\dagger\mathbf{z})/\partial\mathbf{z}^* = \mathbf{0}$. So it follows that

$$\frac{\partial\psi(\mathbf{z},\mathbf{z}^*)}{\partial\mathbf{z}^*} = \mathbf{R}\mathbf{z} + (\lambda_1 - j\lambda_2)\mathbf{a}.$$

Setting this to zero we obtain

$$\mathbf{z} = c\mathbf{R}^{-1}\mathbf{a},$$

where $c = -(\lambda_1 - j\lambda_2)$. This is the optimum vector $\mathbf{z}$. The constant c can be found by substituting $\mathbf{z} = c\mathbf{R}^{-1}\mathbf{a}$ into the constraint (20.43). The result is $c = 1/(\mathbf{a}^\dagger\mathbf{R}^{-1}\mathbf{a})$. (Since this is real we automatically have $\lambda_2 = 0$.) Thus the stationarity condition yields a unique value for $\mathbf{z}$, which is therefore a global optimal point:

$$\mathbf{z}_{opt} = \frac{\mathbf{R}^{-1}\mathbf{a}}{\mathbf{a}^\dagger\mathbf{R}^{-1}\mathbf{a}} \tag{20.47}$$

The minimized objective function is given by

$$\phi_{opt} = \frac{1}{\mathbf{a}^\dagger\mathbf{R}^{-1}\mathbf{a}} \tag{20.48}$$

To verify that the preceding solution is a *minimum* rather than a maximum, one computes the so-called Hessian matrix and verifies that it is positive semidefinite [Moon and Stirling, 2000]. Problem 20.13 outlines a different approach.

Example 20.17. Rayleigh-Ritz principle

A variation of Ex. 20.16 arises frequently when the same objective function $\phi = \mathbf{z}^\dagger\mathbf{R}\mathbf{z}$ has to be minimized (for Hermitian $\mathbf{R}$) under a different constraint, namely the unit energy constraint

$$\mathbf{z}^\dagger\mathbf{z} = 1. \tag{20.49}$$

In this case we define the Lagrangian as

$$\psi = \mathbf{z}^\dagger\mathbf{R}\mathbf{z} - \lambda(\mathbf{z}^\dagger\mathbf{z} - 1).$$

where $-\lambda$ is a real Lagrange multiplier. So ψ is real, and a necessary condition for optimality is $\partial\psi(\mathbf{z},\mathbf{z}^*)/\partial\mathbf{z}^* = \mathbf{0}$. Thus

$$\frac{\partial\psi(\mathbf{z},\mathbf{z}^*)}{\partial\mathbf{z}^*} = \mathbf{R}\mathbf{z} - \lambda\mathbf{z} = \mathbf{0}$$

should be satisfied for optimality. Equivalently,

$$\mathbf{R}\mathbf{z} = \lambda\mathbf{z}.$$

Thus at *any* extremum of ϕ, subject to the constraint (20.49), the vector $\mathbf{z}$ must be an eigenvector of $\mathbf{R}$, and λ is the corresponding eigenvalue. The objective function now becomes

$$\phi = \mathbf{z}^\dagger \mathbf{R} \mathbf{z} = \lambda \mathbf{z}^\dagger \mathbf{z} = \lambda.$$

If $\mathbf{z}$ is an eigenvector corresponding to the minimum eigenvalue, then ϕ is indeed a global minimum. Interestingly, if $\mathbf{z}$ is chosen as an eigenvector corresponding to the maximum eigenvalue of $\mathbf{R}$, then ϕ is maximized. Any other eigenvector of $\mathbf{R}$ yields an extremum which is neither a global minimum nor a global maximum. The results of this problem are special cases of the Rayleigh-Ritz principle, well known in the theory of matrices [Horn and Johnson, 1985].

Example 20.18. A problem with a diagonal solution

Let $\mathbf{\Lambda}$ be a fixed diagonal matrix with $\lambda_k \geq 0$. Consider the problem of optimizing the positive definite matrix $\mathbf{Q}$ (possibly complex) such that

$$\phi = \text{Tr}\Big(\mathbf{I} + \mathbf{Q}\mathbf{\Lambda}\Big)^{-1} \tag{20.50}$$

is minimized, subject to $\text{Tr}(\mathbf{Q}) = c > 0$. Since $\mathbf{Q}$ is Hermitian and positive definite, we can write $\mathbf{Q} = \mathbf{T}\mathbf{T}^\dagger$ for some nonsingular $\mathbf{T}$. So we will find $\mathbf{T}$ such that $\phi = \text{Tr}(\mathbf{I} + \mathbf{T}\mathbf{T}^\dagger\mathbf{\Lambda})^{-1}$ is minimized subject to $\text{Tr}(\mathbf{T}\mathbf{T}^\dagger) = c$. For this define the Lagrangian

$$\psi = \text{Tr}\Big(\mathbf{I} + \mathbf{T}\mathbf{T}^\dagger\mathbf{\Lambda}\Big)^{-1} + \mu\Big(\text{Tr}(\mathbf{T}\mathbf{T}^\dagger) - c\Big).$$

Setting $\partial\psi/\partial\mathbf{T}^* = \mathbf{0}$ we can obtain the necessary conditions for an interior optimum. From Table 20.4 (entry 11) we have

$$\frac{\partial \text{Tr}(\mathbf{T}\mathbf{T}^\dagger)}{\partial \mathbf{T}^*} = \mathbf{T}. \tag{20.51}$$

We now show that

$$\frac{\partial \text{Tr}(\mathbf{I} + \mathbf{T}\mathbf{T}^\dagger\mathbf{\Lambda})^{-1}}{\partial \mathbf{T}^*} = -\mathbf{\Lambda}(\mathbf{I} + \mathbf{T}\mathbf{T}^\dagger\mathbf{\Lambda})^{-2}\mathbf{T}. \tag{20.52}$$

Proof. We have

$$\begin{aligned}
\frac{\partial \text{Tr}(\mathbf{I} + \mathbf{T}\mathbf{T}^\dagger\mathbf{\Lambda})^{-1}}{\partial t^*_{rs}} &= \text{Tr}\Big(\frac{\partial (\mathbf{I} + \mathbf{T}\mathbf{T}^\dagger\mathbf{\Lambda})^{-1}}{\partial t^*_{rs}}\Big) \\
&= -\text{Tr}\Big[(\mathbf{I} + \mathbf{T}\mathbf{T}^\dagger\mathbf{\Lambda})^{-1}\Big(\frac{\partial (\mathbf{I} + \mathbf{T}\mathbf{T}^\dagger\mathbf{\Lambda})}{\partial t^*_{rs}}\Big)(\mathbf{I} + \mathbf{T}\mathbf{T}^\dagger\mathbf{\Lambda})^{-1}\Big] \\
&= -\text{Tr}\Big[(\mathbf{I} + \mathbf{T}\mathbf{T}^\dagger\mathbf{\Lambda})^{-1}\mathbf{T}\mathcal{I}_{sr}\mathbf{\Lambda}(\mathbf{I} + \mathbf{T}\mathbf{T}^\dagger\mathbf{\Lambda})^{-1}\Big] \\
&= -\text{Tr}[\mathcal{I}_{sr}\mathbf{\Lambda}(\mathbf{I} + \mathbf{T}\mathbf{T}^\dagger\mathbf{\Lambda})^{-2}\mathbf{T}] \\
&= -\Big[\mathbf{\Lambda}(\mathbf{I} + \mathbf{T}\mathbf{T}^\dagger\mathbf{\Lambda})^{-2}\mathbf{T}\Big]_{rs},
\end{aligned}$$

which proves Eq. (20.52) indeed. In the preceding derivation the second equality follows from Eq. (20.38), the third from Table 20.3 (entry 9), the fourth from $\text{Tr}[\mathbf{AB}] = \text{Tr}[\mathbf{BA}]$, and the last line from Table 20.2. $\bigtriangledown\bigtriangledown\bigtriangledown$

Thus the stationarity condition $\partial\psi/\partial\mathbf{T}^* = \mathbf{0}$ yields

$$\mathbf{\Lambda}(\mathbf{I} + \mathbf{T}\mathbf{T}^\dagger\mathbf{\Lambda})^{-2}\mathbf{T} = \mu\mathbf{T}. \tag{20.53}$$

Since $\mathbf{T}$ is nonsingular, this yields $\mathbf{\Lambda}(\mathbf{I} + \mathbf{T}\mathbf{T}^\dagger\mathbf{\Lambda})^{-2} = \mu\mathbf{I}$, which shows that

$$\mathbf{I} + \mathbf{T}\mathbf{T}^\dagger\mathbf{\Lambda} = \left(\frac{\mathbf{\Lambda}}{\mu}\right)^{1/2}.$$

The solution $\mathbf{Q} = \mathbf{T}\mathbf{T}^\dagger$ therefore takes the form

$$\mathbf{Q} = \left(\frac{\mathbf{\Lambda}}{\mu}\right)^{-1/2} - \mathbf{\Lambda}^{-1}, \tag{20.54}$$

which is a diagonal matrix. Thus there is no loss of generality in restricting $\mathbf{Q}$ to be a diagonal matrix in the above problem. The constant μ should be chosen such that the condition $\text{Tr}(\mathbf{Q}) = c$ is satisfied.

The careful reader will notice that the solution $\mathbf{Q}$ calculated above is not guaranteed to be positive definite, since the diagonal elements on the right-hand side of Eq. (20.54) are not guaranteed to be positive. This problem can be solved by a more careful formulation which uses inequality constraints (positivity) as in Chap. 22. A more elegant approach to this problem based on Schur-convex functions was presented by Witsenhausen (see Salz [1985]). This is given in Chap. 13 (Sec. 13.5), and is applicable even when the optimum point is not an interior point; it is also applicable when $\mathbf{Q}$ is only positive semidefinite rather than positive definite.

Example 20.19. Optimum noise canceller

Figure 20.1 shows a signal $\mathbf{y}(n)$ with additive noise $\mathbf{e}_1(n)$. There is another pure noise source $\mathbf{e}_2(n)$ also shown in the figure. We wish to add an appropriately transformed version of $\mathbf{e}_2(n)$ to the noisy signal, so that the result $\widehat{\mathbf{y}}(n)$ is as close to $\mathbf{y}(n)$ as possible in the mean square sense. This system is called the noise canceller.

If the pure noise component $\mathbf{e}_2(n)$ has some correlation with $\mathbf{e}_1(n)$, then indeed such a cancellation is possible to some extent. As an extreme example, if $\mathbf{e}_2(n) = \mathbf{e}_1(n)$, then we only have to choose $\mathbf{A} = -\mathbf{I}$, and the noise $\mathbf{e}_1(n)$ is completely cancelled, that is, $\widehat{\mathbf{y}}(n) = \mathbf{y}(n)$. Similarly if $\mathbf{e}_2(n) = \mathbf{T}\mathbf{e}_1(n)$, where $\mathbf{T}$ has a left inverse $\mathbf{T}^\#$, then choosing $\mathbf{A} = -\mathbf{T}^\#$ makes $\widehat{\mathbf{y}}(n) = \mathbf{y}(n)$. These are examples where $\mathbf{e}_2(n)$ is completely correlated to $\mathbf{e}_1(n)$, that is, we can estimate the noise $\mathbf{e}_1(n)$ with no error based on measurement of $\mathbf{e}_2(n)$. In practice, there are situations when a separate measurement of correlated noise is available, although the correlation will only be partial. For example, $\mathbf{y}(n) + \mathbf{e}_1(n)$ might be the noisy speech from a microphone and $\mathbf{e}_2(n)$ might be the speech-free noise measured with another microphone far away from the speaker.

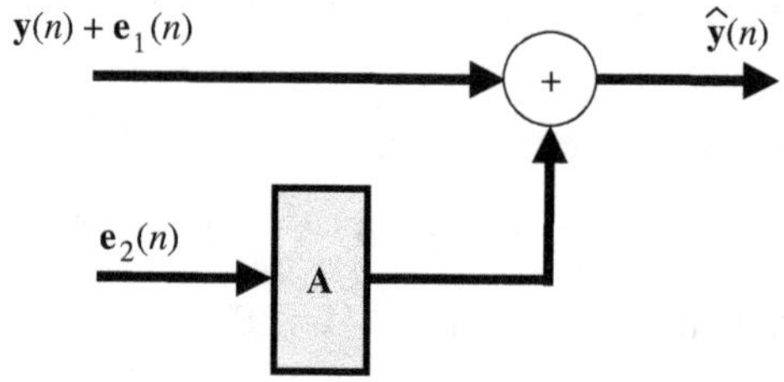

Figure 20.1. The noise cancellation system.

At this point we assume that all signals and noise components are jointly wide sense stationary. We use the notation $\mathbf{R}_{km}$ to denote the correlation matrix between the two noise sources, that is,

$$\mathbf{R}_{km} = E[\mathbf{e}_k(n)\mathbf{e}_m^\dagger(n)].$$

Clearly, $\mathbf{R}_{11}$ and $\mathbf{R}_{22}$ denote the autocorrelations of the individual noise vectors. Since the correlation matrices and the transform $\mathbf{A}$ are independent of time, all time arguments (n) will henceforth be dropped. The reconstruction error is $\mathbf{e} = \widehat{\mathbf{y}} - \mathbf{y} = \mathbf{y} + \mathbf{e}_1 + \mathbf{A}\mathbf{e}_2 - \mathbf{y}$, which simplifies to

$$\mathbf{e} = \mathbf{e}_1 + \mathbf{A}\mathbf{e}_2. \tag{20.55}$$

Its correlation matrix is therefore given by

$$E[\mathbf{c}\mathbf{c}^\dagger] = \mathbf{R}_{11} + \mathbf{A}\mathbf{R}_{22}\mathbf{A}^\dagger + \mathbf{R}_{12}\mathbf{A}^\dagger + \mathbf{A}\mathbf{R}_{21}. \tag{20.56}$$

The mean square reconstruction error is defined as $E[\mathbf{e}^\dagger\mathbf{e}]$. This can be expressed as

$$\mathcal{E} = E[\mathbf{e}^\dagger\mathbf{e}] = \text{Tr}E[\mathbf{e}\mathbf{e}^\dagger] = \text{Tr}\Big(\mathbf{R}_{11} + \mathbf{A}\mathbf{R}_{22}\mathbf{A}^\dagger + \mathbf{R}_{12}\mathbf{A}^\dagger + \mathbf{A}\mathbf{R}_{21}\Big). \tag{20.57}$$

The optimal choice of $\mathbf{A}$ which minimizes this trace can be obtained from matrix calculus. Thus, a necessary condition for optimality is that the *complex gradient* of $\mathcal{E}$ with respect to $\mathbf{A}^*$ be zero (Sec. 20.3.5), that is,

$$\frac{\partial \mathcal{E}}{\partial \mathbf{A}^*} = \mathbf{0}. \tag{20.58}$$

To evaluate the left-hand side, we first note (using entries from Table 20.4) that

$$\frac{\partial \text{Tr}(\mathbf{A}\mathbf{R}_{22}\mathbf{A}^\dagger)}{\partial \mathbf{A}^*} = \mathbf{A}\mathbf{R}_{22}, \quad \frac{\partial \text{Tr}(\mathbf{R}_{12}\mathbf{A}^\dagger)}{\partial \mathbf{A}^*} = \mathbf{R}_{12}, \quad \text{and} \quad \frac{\partial \text{Tr}(\mathbf{A}\mathbf{R}_{12})}{\partial \mathbf{A}^*} = \mathbf{0}, \tag{20.59}$$

so that Eq. (20.58) becomes

$$\frac{\partial \mathcal{E}}{\partial \mathbf{A}^*} = \mathbf{A}\mathbf{R}_{22} + \mathbf{R}_{12} = \mathbf{0}. \tag{20.60}$$

From this we see that the optimum noise-cancelling transform is

$$\mathbf{A}_{opt} = -\mathbf{R}_{12}\mathbf{R}_{22}^{-1} \tag{20.61}$$

Substituting into Eq. (20.57) we see that the minimized error is

$$\mathcal{E}_{opt} = \mathrm{Tr}\Big(\mathbf{R}_{11} - \mathbf{R}_{12}\mathbf{R}_{22}^{-1}\mathbf{R}_{12}^{\dagger}\Big). \tag{20.62}$$

This completes the solution to the problem. Let us consider two extreme special cases. First, if the noise components are identical ($\mathbf{e}_1 = \mathbf{e}_2$) then $\mathbf{R}_{12} = \mathbf{R}_{11} = \mathbf{R}_{22}$, whence

$$\mathbf{A}_{opt} = -\mathbf{I} \quad \text{and} \quad \mathcal{E}_{opt} = 0,$$

as expected. This is complete noise cancellation. At the other extreme, if the noise components are orthogonal, that is, $\mathbf{R}_{12} = \mathbf{0}$, then

$$\mathbf{A}_{opt} = \mathbf{0} \quad \text{and} \quad \mathcal{E}_{opt} = \mathrm{Tr}\,\mathbf{R}_{11}.$$

That is, knowledge of $\mathbf{e}_2$ does not help to reduce the effect of the noise $\mathbf{e}_1$. The reconstruction error is $\mathbf{e}_1$ itself, so the mean square error is the trace of its autocorrelation $\mathbf{R}_{11}$.

This example is included here only to demonstrate another application of matrix calculus. In this particular example the use of the orthogonality principle leads to the same result rather quickly. See Problem 12.6.

20.6 Being careful with interpretations ...

The main purpose of this section is to emphasize the need for careful attention in the application of notations such as $\partial/\partial\mathbf{Z}$ and $\partial/\partial\mathbf{Z}^*$. For example, even though some of the real examples summarized in Table 20.1 appear to be special cases of the complex examples in Table 20.3, the formulas in the latter table must be correctly interpreted to get those in the former. Consider for example the formulas

$$\frac{\partial(\mathbf{Z}^\dagger\mathbf{A}\mathbf{Z})}{\partial z_{rs}} = \mathbf{Z}^\dagger\mathbf{A}\mathcal{I}_{rs}, \qquad \frac{\partial(\mathbf{Z}^\dagger\mathbf{A}\mathbf{Z})}{\partial z_{rs}^*} = \mathcal{I}_{rs}^T\mathbf{A}\mathbf{Z} \tag{20.63}$$

in Table 20.3. Suppose we want to recover the special case when everything is real, that is, find $\partial(\mathbf{X}^T\mathbf{A}\mathbf{X})/\partial x_{rs}$. If we let $\mathbf{Z} = \mathbf{X}$ (i.e., set $\mathbf{Y} = \mathbf{0}$ in $\mathbf{Z} = \mathbf{X} + \jmath\mathbf{Y}$), and similarly let $z_{rs} = x_{rs}$, then Eqs. (20.63) reduce to

$$\frac{\partial(\mathbf{X}^T\mathbf{A}\mathbf{X})}{\partial x_{rs}} = \mathbf{X}^T\mathbf{A}\mathcal{I}_{rs}, \qquad \frac{\partial(\mathbf{X}^T\mathbf{A}\mathbf{X})}{\partial x_{rs}} = \mathcal{I}_{rs}^T\mathbf{A}\mathbf{X} \tag{20.64}$$

That is, we get two different answers for the same real derivative! Something is clearly wrong. Similar contradictions can be found even in the first entry of Table 20.3, which has $\partial z/\partial z = 1$, $\partial z/\partial z^* = 0$. If we set $z = x = real$ "everywhere" we get $\partial x/\partial x = 1$, $\partial x/\partial x = 0$, which again is a contradiction. So we have to be careful in order to derive the real-valued special cases from the complex formulas of Tables 20.3 and 20.4. Here are the points to be noted carefully, in order to make correct use of the complex gradient operators:

1. ***Operator***. First, $\partial/\partial\mathbf{Z}$ and $\partial/\partial\mathbf{Z}^*$ are just notations for operators, and we should not be replacing $\mathbf{Z}$ in $\partial\mathbf{Z}$ or $\partial\mathbf{Z}^*$ with anything. These operators are actually equivalent to a linear combination of the familiar partial derivatives $\partial/\partial\mathbf{X}$ and $\partial/\partial\mathbf{Y}$, as shown in Sec. 20.3:

$$\frac{\partial}{\partial\mathbf{Z}} = 0.5\Big(\frac{\partial}{\partial\mathbf{X}} - j\frac{\partial}{\partial\mathbf{Y}}\Big) \quad \text{and} \quad \frac{\partial}{\partial\mathbf{Z}^*} = 0.5\Big(\frac{\partial}{\partial\mathbf{X}} + j\frac{\partial}{\partial\mathbf{Y}}\Big). \tag{20.65}$$

   Inverting these relations we obtain

$$\frac{\partial}{\partial\mathbf{X}} = \Big(\frac{\partial}{\partial\mathbf{Z}} + \frac{\partial}{\partial\mathbf{Z}^*}\Big) \quad \text{and} \quad \frac{\partial}{\partial\mathbf{Y}} = j\Big(\frac{\partial}{\partial\mathbf{Z}} - \frac{\partial}{\partial\mathbf{Z}^*}\Big). \tag{20.66}$$

2. ***Operand***. Second, the operand $\mathbf{g}$ in the operators $\partial\mathbf{g}/\partial\mathbf{Z}$ and $\partial\mathbf{g}/\partial\mathbf{Z}^*$ should be written in terms of $\mathbf{Z}$ and $\mathbf{Z}^*$. Given an arbitrary operand $\mathbf{h}(\mathbf{Z})$ which is a function of $\mathbf{Z}$, we know it can always be written as a function $\mathbf{f}(\mathbf{X},\mathbf{Y})$. By replacing $\mathbf{X}$ and $\mathbf{Y}$ everywhere in this expression with $(\mathbf{Z}+\mathbf{Z}^*)/2$ and $(\mathbf{Z}-\mathbf{Z}^*)/2j$, respectively, we get the correct form $\mathbf{g}(\mathbf{Z},\mathbf{Z}^*)$ of the operand for use in $\partial\mathbf{g}/\partial\mathbf{Z}$ and $\partial\mathbf{g}/\partial\mathbf{Z}^*$.

Consider the toy example where we wish to obtain the formula $\partial x^2/x = 2x$ as a special case from the complex gradients $\partial x^2/z$ and $\partial x^2/z^*$. We first write

$$x^2 = \Big(\frac{z+z^*}{2}\Big)^2 = \frac{z^2+(z^*)^2+2zz^*}{4}$$

Then

$$\frac{\partial x^2}{\partial z} = \frac{2z+2z^*}{4} = \frac{z+z^*}{2}, \quad \frac{\partial x^2}{\partial z^*} = \frac{2z^*+2z}{4} = \frac{z+z^*}{2}$$

Since $(z+z^*)/2 = x$ it then follows that

$$\frac{\partial x^2}{\partial z} = \frac{\partial x^2}{\partial z^*} = x,$$

so that, from Eq. (20.66), $\partial x^2/\partial x = 2x$ as expected.

Matrix exercise. We conclude this section by showing the correct derivation for recovering the expression

$$\frac{\partial(\mathbf{X}^T\mathbf{A}\mathbf{X})}{\partial x_{rs}} = \mathcal{I}_{rs}^T\mathbf{A}\mathbf{X} + \mathbf{X}^T\mathbf{A}\mathcal{I}_{rs} \tag{20.67}$$

in Table 20.1 as a special case of the formulas we have in the complex case. First we substitute $\mathbf{X} = (\mathbf{Z}+\mathbf{Z}^*)/2$ and write

$$\mathbf{X}^T\mathbf{A}\mathbf{X} = \frac{\mathbf{Z}^T\mathbf{A}\mathbf{Z} + \mathbf{Z}^\dagger\mathbf{A}\mathbf{Z} + \mathbf{Z}^T\mathbf{A}\mathbf{Z}^* + \mathbf{Z}^\dagger\mathbf{A}\mathbf{Z}^*}{4}.$$

The idea is to compute $\partial(\mathbf{X}^T\mathbf{A}\mathbf{X})/\partial z_{rs}$ and $\partial(\mathbf{X}^T\mathbf{A}\mathbf{X})/\partial z_{rs}^*$ from the above expression, and add the results to get $\partial(\mathbf{X}^T\mathbf{A}\mathbf{X})/\partial x_{rs}$. For $\partial(\mathbf{X}^T\mathbf{A}\mathbf{X})/\partial z_{rs}$ the quantity z_{rs}^* is regarded as constant. Thus, using the product rule and expressions such as $\partial \mathbf{Z}/\partial z_{rs} = \mathcal{I}_{rs}$ and $\partial \mathbf{Z}^T/\partial z_{rs} = \mathcal{I}_{rs}^T$, we obtain

$$\frac{\partial(\mathbf{X}^T\mathbf{A}\mathbf{X})}{\partial z_{rs}} = \frac{1}{4}\Big(\mathcal{I}_{rs}^T\mathbf{A}\mathbf{Z} + \mathbf{Z}^T\mathbf{A}\mathcal{I}_{rs} + \mathbf{Z}^\dagger\mathbf{A}\mathcal{I}_{rs} + \mathcal{I}_{rs}^T\mathbf{A}\mathbf{Z}^*\Big).$$

Similarly

$$\frac{\partial(\mathbf{X}^T\mathbf{A}\mathbf{X})}{\partial z_{rs}^*} = \frac{1}{4}\Big(\mathcal{I}_{rs}^T\mathbf{A}\mathbf{Z} + \mathbf{Z}^T\mathbf{A}\mathcal{I}_{rs} + \mathcal{I}_{rs}^T\mathbf{A}\mathbf{Z}^* + \mathbf{Z}^\dagger\mathbf{A}\mathcal{I}_{rs}\Big).$$

Adding these and simplifying we can obtain $\partial(\mathbf{X}^T\mathbf{A}\mathbf{X})/\partial x_{rs}$. After simplification,

$$\frac{\partial(\mathbf{X}^T\mathbf{A}\mathbf{X})}{\partial x_{rs}} = \frac{1}{2}\Big(\mathcal{I}_{rs}^T\mathbf{A}(\mathbf{Z} + \mathbf{Z}^*) + (\mathbf{Z}^T + \mathbf{Z}^\dagger)\mathbf{A}\mathcal{I}_{rs}\Big).$$

Since $(\mathbf{Z} + \mathbf{Z}^*) = 2\mathbf{X}$, this simplifies to Eq. (20.67) indeed. $\bigtriangledown \bigtriangledown \bigtriangledown$

The preceding example also shows that it is more economical to derive the special cases of real matrices directly, rather than obtain them from formulas for the complex case. Tables which show real entries and complex entries therefore have their own separate usefulness.

20.7 Summary and conclusions

In this chapter we have given an extensive review of matrix calculus methods as applied to optimization. These methods sometimes lead to quick derivations of optimality conditions. There also exist other elegant ways to find gradients based on perturbation techniques (for example, see Sec. A.4 of Boyd and Vandenberghe [2004]). These are often very convenient. For some simpler problems, such as MMSE filtering, a direct use of the orthogonality principle (instead of using a gradient approach) is the best way to find optimal filter matrices (e.g., the examples in Appendix F at the end of the book). The main definitions and formulas discussed in the chapter are summarized in Tables 20.1–20.5.

Table 20.1. Derivatives of real matrices

The notation $\partial\mathbf{Y}/\partial x_{rs}$ is defined by

$$\frac{\partial\mathbf{Y}}{\partial x_{rs}} = \begin{bmatrix} \frac{\partial y_{00}}{\partial x_{rs}} & \frac{\partial y_{01}}{\partial x_{rs}} & \frac{\partial y_{02}}{\partial x_{rs}} & \cdots \\ \frac{\partial y_{10}}{\partial x_{rs}} & \frac{\partial y_{11}}{\partial x_{rs}} & \frac{\partial y_{12}}{\partial x_{rs}} & \cdots \\ \frac{\partial y_{20}}{\partial x_{rs}} & \frac{\partial y_{21}}{\partial x_{rs}} & \frac{\partial y_{22}}{\partial x_{rs}} & \cdots \\ \vdots & \vdots & \vdots & \ddots \end{bmatrix}.$$

The notation $\mathcal{I}_{rs}$ represents a matrix of appropriate size with all elements equal to zero except the (r, s)th element, which is unity. Unless mentioned otherwise, it is assumed that the matrix $\mathbf{X}$ has no special structure that imposes any relationship between elements (such as symmetry, Toeplitz property, and so forth). Some useful facts are summarized below:

1. $\frac{\partial\mathbf{X}}{\partial x_{rs}} = \mathcal{I}_{rs}, \quad \frac{\partial\mathbf{X}^T}{\partial x_{rs}} = \mathcal{I}_{rs}^T.$

2. For real symmetric $\mathbf{X}$, $\frac{\partial\mathbf{X}}{\partial x_{rs}} = \begin{cases} \mathcal{I}_{rs} + \mathcal{I}_{sr}, & r \neq s \\ \mathcal{I}_{rr}, & r = s. \end{cases}$

3. $\frac{\partial(\mathbf{AXB})}{\partial x_{rs}} = \mathbf{A}\mathcal{I}_{rs}\mathbf{B}, \quad \frac{\partial(\mathbf{AX}^T\mathbf{B})}{\partial x_{rs}} = \mathbf{A}\mathcal{I}_{rs}^T\mathbf{B}.$

4. $\frac{\partial(\mathbf{UV})}{\partial x_{rs}} = \left(\frac{\partial\mathbf{U}}{\partial x_{rs}}\right)\mathbf{V} + \mathbf{U}\left(\frac{\partial\mathbf{V}}{\partial x_{rs}}\right)$ (product rule).

5. $\frac{\partial(\mathbf{X}^T\mathbf{AX})}{\partial x_{rs}} = \mathcal{I}_{rs}^T\mathbf{AX} + \mathbf{X}^T\mathbf{A}\mathcal{I}_{rs}.$

6. $\frac{\partial(\mathbf{X}^{-1})}{\partial x_{rs}} = -\mathbf{X}^{-1}\mathcal{I}_{rs}\mathbf{X}^{-1}.$

7. $\frac{\partial(\mathbf{AX}^{-1}\mathbf{B})}{\partial x_{rs}} = -\mathbf{AX}^{-1}\mathcal{I}_{rs}\mathbf{X}^{-1}\mathbf{B}.$

8. $\frac{\partial\mathbf{Y}^{-1}}{\partial x_{rs}} = -\mathbf{Y}^{-1}\left(\frac{\partial\mathbf{Y}}{\partial x_{rs}}\right)\mathbf{Y}^{-1}.$

Table 20.2. Derivatives of scalar functions of real matrices

It is assumed that the matrix $\mathbf{X}$ has no special structure that imposes any relationship between elements (such as symmetry, Toeplitz property, and so forth). The notation $\mathrm{Tr}(\mathbf{X})$ stands for the trace of $\mathbf{X}$.

1. $\dfrac{\partial(\mathbf{c}^T\mathbf{x})}{\partial\mathbf{x}} = \mathbf{c} \quad (\mathbf{x} \text{ a column vector}).$

2. $\dfrac{\partial(\mathbf{x}^T\mathbf{x})}{\partial\mathbf{x}} = 2\mathbf{x} \quad (\mathbf{x} \text{ a column vector}).$

3. $\dfrac{\partial(\mathbf{x}^T\mathbf{R}\mathbf{x})}{\partial\mathbf{x}} = \mathbf{R}\mathbf{x} + \mathbf{R}^T\mathbf{x} \quad (\mathbf{x} \text{ a column vector}).$

4. $\dfrac{\partial\,\mathrm{Tr}(\mathbf{X})}{\partial x_{rs}} = \delta(r-s).$

5. $\dfrac{\partial\,\mathrm{Tr}(\mathbf{X})}{\partial\mathbf{X}} = \mathbf{I}.$

6. $\dfrac{\partial\,\mathrm{Tr}(\mathbf{AXB})}{\partial\mathbf{X}} = \mathbf{A}^T\mathbf{B}^T.$

7. $\dfrac{\partial\,\mathrm{Tr}(\mathbf{X}^T\mathbf{X})}{\partial\mathbf{X}} = \dfrac{\partial\,\mathrm{Tr}(\mathbf{X}\mathbf{X}^T)}{\partial\mathbf{X}} = 2\mathbf{X}.$

8. $\dfrac{\partial\,\mathrm{Tr}(\mathbf{X}^T\mathbf{AX})}{\partial\mathbf{X}} = \mathbf{A}^T\mathbf{X} + \mathbf{AX}.$

9. $\dfrac{\partial\,\mathrm{Tr}(\mathbf{XAX}^T)}{\partial\mathbf{X}} = \mathbf{XA} + \mathbf{XA}^T.$

10. $\dfrac{\partial\,\mathrm{Tr}(\mathbf{X}^T\mathbf{AXB})}{\partial\mathbf{X}} = \mathbf{A}^T\mathbf{XB}^T + \mathbf{AXB}.$

11. $\dfrac{\partial\,\mathrm{Tr}(\mathbf{AX}^{-1}\mathbf{B})}{\partial\mathbf{X}} = -\left(\mathbf{X}^{-1}\mathbf{BAX}^{-1}\right)^T = -\mathbf{X}^{-T}\mathbf{A}^T\mathbf{B}^T\mathbf{X}^{-T}.$

12. $\dfrac{\partial\,\mathrm{Tr}(\mathbf{XAX}^T)^{-1}}{\partial\mathbf{X}} = -(\mathbf{XA}^T\mathbf{X}^T)^{-2}\mathbf{XA}^T - (\mathbf{XAX}^T)^{-2}\mathbf{XA}.$

Other useful relations:

1. $\mathrm{Tr}(\mathcal{I}_{rs}\mathbf{X}) = x_{sr}.$

2. $\mathrm{Tr}(\mathbf{AB}) = \mathrm{Tr}(\mathbf{BA}).$

3. $\dfrac{\partial\,\mathrm{Tr}(\mathbf{f}(\mathbf{X}))}{\partial x_{rs}} = \mathrm{Tr}\left(\dfrac{\partial(\mathbf{f}(\mathbf{X}))}{\partial x_{rs}}\right).$

Table 20.3. Derivatives with respect to complex matrices

See Secs. 20.3.1 and 20.3.4 for definitions of $\partial/\partial z$, $\partial/\partial z^*$, $\partial/\partial \mathbf{Z}$, and $\partial/\partial \mathbf{Z}^*$. The real and imaginary parts of $\mathbf{Z}$ are indicated as $\mathbf{X}$ and $\mathbf{Y}$, as in $\mathbf{Z} = \mathbf{X} + j\mathbf{Y}$. The notation $\mathcal{I}_{rs}$ represents a matrix of appropriate size with all elements equal to zero, except the (r, s)th element, which is unity. Superscripts as in $\mathbf{A}^T$, $\mathbf{A}^*$, and $\mathbf{A}^\dagger$ denote, respectively, the transpose, conjugate, and transpose-conjugate of the matrix $\mathbf{A}$.

1. $\dfrac{\partial z}{\partial z} = 1, \quad \dfrac{\partial z}{\partial z^*} = 0, \quad \dfrac{\partial z^*}{\partial z} = 0, \quad \dfrac{\partial z^*}{\partial z^*} = 1.$

2. If $g(\mathbf{Z}, \mathbf{Z}^*)$ has no $\mathbf{Z}^*$, then $\dfrac{\partial g}{\partial \mathbf{Z}^*} = \mathbf{0}$. If $g(\mathbf{Z}, \mathbf{Z}^*)$ has no $\mathbf{Z}$, then $\dfrac{\partial g}{\partial \mathbf{Z}} = \mathbf{0}$.

3. $\dfrac{\partial}{\partial \mathbf{Z}} = 0.5\Big(\dfrac{\partial}{\partial \mathbf{X}} - j\dfrac{\partial}{\partial \mathbf{Y}}\Big)$ and $\dfrac{\partial}{\partial \mathbf{Z}^*} = 0.5\Big(\dfrac{\partial}{\partial \mathbf{X}} + j\dfrac{\partial}{\partial \mathbf{Y}}\Big).$

4. *Cauchy-Riemann equations*: hold if $g(\mathbf{Z}, \mathbf{Z}^*)$ has no $\mathbf{Z}^*$ (Sec. 20.3.3).

5. *Stationarity*: $\dfrac{\partial g(\mathbf{Z}, \mathbf{Z}^*)}{\partial \mathbf{Z}} = \mathbf{0}$, equivalently $\dfrac{\partial g(\mathbf{Z}, \mathbf{Z}^*)}{\partial \mathbf{Z}^*} = \mathbf{0}$ (real scalar g).

6. $\dfrac{\partial \mathbf{Z}}{\partial z_{rs}} = \mathcal{I}_{rs}, \quad \dfrac{\partial \mathbf{Z}}{\partial z^*_{rs}} = \mathbf{0}, \quad \dfrac{\partial \mathbf{Z}^\dagger}{\partial z_{rs}} = \mathbf{0}, \quad \dfrac{\partial \mathbf{Z}^\dagger}{\partial z^*_{rs}} = \mathcal{I}^T_{rs} = \mathcal{I}_{sr}.$

7. If $\mathbf{Z}$ is complex Hermitian, $\partial \mathbf{Z}/\partial z^*_{rs} = \mathcal{I}_{sr}$.

8. $\dfrac{\partial (\mathbf{AZB})}{\partial z_{rs}} = \mathbf{A}\mathcal{I}_{rs}\mathbf{B}.$

9. $\dfrac{\partial (\mathbf{Z}^\dagger \mathbf{Z})}{\partial z_{rs}} = \mathbf{Z}^\dagger \mathcal{I}_{rs}, \quad \dfrac{\partial (\mathbf{Z}^\dagger \mathbf{Z})}{\partial z^*_{rs}} = \mathcal{I}^T_{rs}\mathbf{Z} = \mathcal{I}_{sr}\mathbf{Z}.$

10. $\dfrac{\partial (\mathbf{Z}\mathbf{Z}^\dagger)}{\partial z_{rs}} = \mathcal{I}_{rs}\mathbf{Z}^\dagger, \quad \dfrac{\partial (\mathbf{Z}\mathbf{Z}^\dagger)}{\partial z^*_{rs}} = \mathbf{Z}\mathcal{I}^T_{rs} = \mathbf{Z}\mathcal{I}_{sr}.$

11. $\dfrac{\partial (\mathbf{Z}^\dagger \mathbf{A}\mathbf{Z})}{\partial z_{rs}} = \mathbf{Z}^\dagger \mathbf{A}\mathcal{I}_{rs}, \quad \dfrac{\partial (\mathbf{Z}^\dagger \mathbf{A}\mathbf{Z})}{\partial z^*_{rs}} = \mathcal{I}^T_{rs}\mathbf{A}\mathbf{Z} = \mathcal{I}_{sr}\mathbf{A}\mathbf{Z}.$

12. $\dfrac{\partial (\mathbf{Z}\mathbf{A}\mathbf{Z}^\dagger)}{\partial z_{rs}} = \mathcal{I}_{rs}\mathbf{A}\mathbf{Z}^\dagger, \quad \dfrac{\partial (\mathbf{Z}\mathbf{A}\mathbf{Z}^\dagger)}{\partial z^*_{rs}} = \mathbf{Z}\mathbf{A}\mathcal{I}^T_{rs} = \mathbf{Z}\mathbf{A}\mathcal{I}_{sr}.$

13. $\dfrac{\partial \mathbf{Y}^{-1}}{\partial z^*_{rs}} = -\mathbf{Y}^{-1}\Big(\dfrac{\partial \mathbf{Y}}{\partial z^*_{rs}}\Big)\mathbf{Y}^{-1},$ and $\dfrac{\partial \mathbf{Y}^{-1}}{\partial z_{rs}} = -\mathbf{Y}^{-1}\Big(\dfrac{\partial \mathbf{Y}}{\partial z_{rs}}\Big)\mathbf{Y}^{-1}.$

14. $\dfrac{\partial \mathbf{Z}^{-1}}{\partial z_{rs}} = -\mathbf{Z}^{-1}\mathcal{I}_{rs}\mathbf{Z}^{-1}, \quad \dfrac{\partial \mathbf{Z}^{-1}}{\partial z^*_{rs}} = \mathbf{0}.$

15. $\dfrac{\partial (\mathbf{Z}^\dagger \mathbf{A}\mathbf{Z})^{-1}}{\partial z_{rs}} = -(\mathbf{Z}^\dagger \mathbf{A}\mathbf{Z})^{-1}\mathbf{Z}^\dagger \mathbf{A}\mathcal{I}_{rs}(\mathbf{Z}^\dagger \mathbf{A}\mathbf{Z})^{-1}.$

Table 20.4. Derivatives of scalar functions of complex matrices

See Secs. 20.3.1 and 20.3.4 for definitions of $\partial/\partial z$, $\partial/\partial z^*$, $\partial/\partial \mathbf{Z}$, and $\partial/\partial \mathbf{Z}^*$. The real and imaginary parts of $\mathbf{Z}$ are indicated as $\mathbf{X}$ and $\mathbf{Y}$, as in $\mathbf{Z} = \mathbf{X} + j\mathbf{Y}$. The notation $\mathcal{I}_{rs}$ represents a matrix of appropriate size with all elements equal to zero, except the (r,s)th element which is unity. Superscripts as in $\mathbf{A}^T, \mathbf{A}^*$, and $\mathbf{A}^\dagger$ denote, respectively, the transpose, conjugate, and transpose-conjugate of the matrix $\mathbf{A}$. The notation $\text{Tr}(\mathbf{X})$ stands for the trace of $\mathbf{X}$.

1. $\dfrac{\partial(\mathbf{z}^\dagger\mathbf{a})}{\partial\mathbf{z}} = \mathbf{0}, \quad \dfrac{\partial(\mathbf{z}^\dagger\mathbf{a})}{\partial\mathbf{z}^*} = \mathbf{a}, \quad \dfrac{\partial(\mathbf{a}^\dagger\mathbf{z})}{\partial\mathbf{z}} = \mathbf{a}^*, \quad \dfrac{\partial(\mathbf{a}^\dagger\mathbf{z})}{\partial\mathbf{z}^*} = \mathbf{0} \quad (\mathbf{z} = \text{column vector}).$

2. $\dfrac{\partial(\mathbf{z}^T\mathbf{b})}{\partial\mathbf{z}} = \dfrac{\partial(\mathbf{b}^T\mathbf{z})}{\partial\mathbf{z}} = \mathbf{b}, \quad \dfrac{\partial(\mathbf{z}^T\mathbf{b})}{\partial\mathbf{z}^*} = \mathbf{0} \quad (\mathbf{z} = \text{column vector}).$

3. $\dfrac{\partial(\mathbf{z}^\dagger\mathbf{z})}{\partial\mathbf{z}} = \mathbf{z}^*, \quad \dfrac{\partial(\mathbf{z}^\dagger\mathbf{z})}{\partial\mathbf{z}^*} = \mathbf{z} \quad (\mathbf{z} = \text{column vector}).$

4. $\dfrac{\partial(\mathbf{z}^\dagger\mathbf{A}\mathbf{z})}{\partial\mathbf{z}} = \mathbf{A}^T\mathbf{z}^*, \quad \dfrac{\partial(\mathbf{z}^\dagger\mathbf{A}\mathbf{z})}{\partial\mathbf{z}^*} = \mathbf{A}\mathbf{z} \quad (\mathbf{z} = \text{column vector}).$

5. $\dfrac{\partial(\mathbf{z}^T\mathbf{z})}{\partial\mathbf{z}} = 2\mathbf{z}, \quad \dfrac{\partial(\mathbf{z}^T\mathbf{z})}{\partial\mathbf{z}^*} = \mathbf{0}, \quad (\mathbf{z} = \text{column vector}).$

6. $\dfrac{\partial\,\text{Tr}(\mathbf{Z})}{\partial\mathbf{Z}} = \mathbf{I}, \quad \dfrac{\partial\,\text{Tr}(\mathbf{Z})}{\partial\mathbf{Z}^*} = \mathbf{0}.$

7. $\dfrac{\partial\,\text{Tr}(\mathbf{Z}^\dagger)}{\partial\mathbf{Z}} = \mathbf{0}, \quad \dfrac{\partial\,\text{Tr}(\mathbf{Z}^\dagger)}{\partial\mathbf{Z}^*} = \mathbf{I}.$

8. $\dfrac{\partial\,\text{Tr}(\mathbf{AZB})}{\partial\mathbf{Z}} = (\mathbf{BA})^T, \quad \dfrac{\partial\,\text{Tr}(\mathbf{AZB})}{\partial\mathbf{Z}^*} = \mathbf{0} \quad (\text{Problem 20.2}).$

9. $\dfrac{\partial\,\text{Tr}(\mathbf{Z}^\dagger\mathbf{AZB})}{\partial\mathbf{Z}} = \mathbf{A}^T\mathbf{Z}^*\mathbf{B}^T, \quad \dfrac{\partial\,\text{Tr}(\mathbf{Z}^\dagger\mathbf{AZB})}{\partial\mathbf{Z}^*} = \mathbf{AZB}.$

10. $\dfrac{\partial\,\text{Tr}(\mathbf{Z}^\dagger\mathbf{Z})}{\partial\mathbf{Z}} = \mathbf{Z}^*, \quad \dfrac{\partial\,\text{Tr}(\mathbf{Z}^\dagger\mathbf{Z})}{\partial\mathbf{Z}^*} = \mathbf{Z}.$

11. $\dfrac{\partial\,\text{Tr}(\mathbf{ZAZ}^\dagger)}{\partial\mathbf{Z}} = (\mathbf{AZ}^\dagger)^T, \quad \dfrac{\partial\,\text{Tr}(\mathbf{ZAZ}^\dagger)}{\partial\mathbf{Z}^*} = \mathbf{ZA}.$

12. $\partial\,\text{Tr}(\mathbf{AZ}^\dagger)/\partial\mathbf{Z}^* = \mathbf{A}.$

13. $\dfrac{\partial\,\text{Tr}(\mathbf{AZ}^{-1}\mathbf{B})}{\partial\mathbf{Z}} = -(\mathbf{Z}^{-1}\mathbf{BAZ}^{-1})^T, \quad \dfrac{\partial\,\text{Tr}(\mathbf{AZ}^{-1}\mathbf{B})}{\partial\mathbf{Z}^*} = \mathbf{0} \quad (\text{Problem 20.3}).$

14. $\dfrac{\partial\,\text{Tr}(\mathbf{Z}^\dagger\mathbf{AZ})^{-1}}{\partial\mathbf{Z}} = -\Big((\mathbf{Z}^\dagger\mathbf{AZ})^{-2}\mathbf{Z}^\dagger\mathbf{A}\Big)^T.$

15. $\dfrac{\partial\,\text{Tr}(\mathbf{Z}^\dagger\mathbf{AZ})^{-1}}{\partial\mathbf{Z}^*} = -\mathbf{AZ}(\mathbf{Z}^\dagger\mathbf{AZ})^{-2}.$

Table 20.5. Derivatives of more complicated functions of matrices

The derivatives in this table involve determinants, logarithms, powers, and exponents. The proofs are requested in the homework problems at the end of the chapter. The notation $\det(\mathbf{A})$ stands for the determinant of the square matrix $\mathbf{A}$. In the table $\mathbf{X}$ denotes a real matrix and $\mathbf{Z}$ denotes a complex matrix. The quantity $e^{\mathbf{Z}}$ is defined by $e^{\mathbf{Z}} = \mathbf{I} + \mathbf{Z} + \frac{\mathbf{Z}^2}{2!} + \frac{\mathbf{Z}^3}{3!} + \dots$

1. $\dfrac{\partial \det(\mathbf{X})}{\partial \mathbf{X}} = \det(\mathbf{X})\, \mathbf{X}^{-T} \quad$ (unstructured $\mathbf{X}$).

2. $\dfrac{\partial \det(\mathbf{X})}{\partial \mathbf{X}} = \det(\mathbf{X}) \left(2\mathbf{X}^{-T} - \mathrm{diag}(\mathbf{X}^{-T})\right) \quad (\mathbf{X} = \mathbf{X}^T)$.

3. $\dfrac{\partial \det(\mathbf{Z})}{\partial \mathbf{Z}} = \det(\mathbf{Z})\, \mathbf{Z}^{-T}, \quad \dfrac{\partial \det(\mathbf{Z})}{\partial \mathbf{Z}^*} = \mathbf{0} \quad$ (unstructured $\mathbf{Z}$).

4. $\dfrac{\partial \det(\mathbf{Z})}{\partial \mathbf{Z}} = \det(\mathbf{Z})(\mathbf{Z}^{-1})^*, \quad \dfrac{\partial \det(\mathbf{Z})}{\partial \mathbf{Z}^*} = \det(\mathbf{Z})\mathbf{Z}^{-1} \quad (\mathbf{Z} = \mathbf{Z}^\dagger)$.

5. $\dfrac{\partial [\ln \det(\mathbf{X})]}{\partial \mathbf{X}} = \begin{cases} \mathbf{X}^{-T} & \text{(unstructured } \mathbf{X}) \\ 2\mathbf{X}^{-T} - \mathrm{diag}(\mathbf{X}^{-T}) & (\mathbf{X} = \mathbf{X}^T). \end{cases}$

6. $\dfrac{\partial [\ln \det(\mathbf{Z})]}{\partial \mathbf{Z}} = \mathbf{Z}^{-T}, \quad \dfrac{\partial [\ln \det(\mathbf{Z})]}{\partial \mathbf{Z}^*} = \mathbf{0} \quad$ (unstructured $\mathbf{Z}$).

7. $\dfrac{\partial [\ln \det(\mathbf{Z})]}{\partial \mathbf{Z}} = \mathbf{Z}^{-T} = (\mathbf{Z}^{-1})^*, \quad \dfrac{\partial [\ln \det(\mathbf{Z})]}{\partial \mathbf{Z}^*} = \mathbf{Z}^{-1} \quad (\mathbf{Z} = \mathbf{Z}^\dagger)$.

8. $\dfrac{\partial \mathbf{X}^n}{\partial x_{rs}} = \displaystyle\sum_{k=0}^{n-1} \mathbf{X}^k \mathcal{I}_{rs} \mathbf{X}^{n-k-1}$.

9. $\dfrac{\partial \mathrm{Tr}\,(\mathbf{X}^n)}{\partial \mathbf{X}} = n(\mathbf{X}^{n-1})^T$.

10. $\dfrac{\partial \mathbf{Z}^n}{\partial z_{rs}} = \displaystyle\sum_{k=0}^{n-1} \mathbf{Z}^k \mathcal{I}_{rs} \mathbf{Z}^{n-k-1}, \quad \dfrac{\partial \mathbf{Z}^n}{\partial z^*_{rs}} = 0 \quad$ (unstructured $\mathbf{Z}$).

11. $\dfrac{\partial \mathbf{Z}^n}{\partial z_{rs}} = \displaystyle\sum_{k=0}^{n-1} \mathbf{Z}^k \mathcal{I}_{rs} \mathbf{Z}^{n-k-1}, \quad \dfrac{\partial \mathbf{Z}^n}{\partial z^*_{rs}} = \displaystyle\sum_{k=0}^{n-1} \mathbf{Z}^k \mathcal{I}^T_{rs} \mathbf{Z}^{n-k-1} \quad (\mathbf{Z} = \mathbf{Z}^\dagger)$.

12. $\dfrac{\partial\, \mathrm{Tr}\,(\mathbf{Z}^n)}{\partial \mathbf{Z}} = n(\mathbf{Z}^{n-1})^T, \quad \dfrac{\partial\, \mathrm{Tr}\,(\mathbf{Z}^n)}{\partial \mathbf{Z}^*} = \mathbf{0} \quad$ (unstructured $\mathbf{Z}$).

13. $\dfrac{\partial\, \mathrm{Tr}\,(\mathbf{Z}^n)}{\partial \mathbf{Z}} = n(\mathbf{Z}^{n-1})^*, \quad \dfrac{\partial\, \mathrm{Tr}\,(\mathbf{Z}^n)}{\partial \mathbf{Z}^*} = n(\mathbf{Z}^{n-1}) \quad (\mathbf{Z} = \mathbf{Z}^\dagger)$.

14. $\dfrac{\partial \mathrm{Tr}\,(e^{\mathbf{X}})}{\partial \mathbf{X}} = e^{\mathbf{X}^T}$.

15. $\dfrac{\partial\, \mathrm{Tr}\,(e^{\mathbf{Z}})}{\partial \mathbf{Z}} = \mathbf{e}^{\mathbf{Z}^T}$ and $\dfrac{\partial\, \mathrm{Tr}\,(e^{\mathbf{Z}})}{\partial \mathbf{Z}^*} = \mathbf{0}$ (unstructured $\mathbf{Z}$).

16. $\dfrac{\partial\, \mathrm{Tr}\,(e^{\mathbf{Z}})}{\partial \mathbf{Z}} = \mathbf{e}^{\mathbf{Z}^*}$ and $\dfrac{\partial\, \mathrm{Tr}\,(e^{\mathbf{Z}})}{\partial \mathbf{Z}^*} = \mathbf{e}^{\mathbf{Z}} \quad (\mathbf{Z} = \mathbf{Z}^\dagger)$.

Problems

Unless mentioned otherwise, all results assume that the matrices $\mathbf{X}$ and $\mathbf{Z}$ are unstructured (i.e., they are not symmetric, or Hermitian, or Toepltiz, or any other type which imposes relationships among elements).

20.1. For real $\mathbf{X}$ find an expression for $\partial\,\mathrm{Tr}(\mathbf{BXAX}^T)^{-1}/\partial\mathbf{X}$.

20.2. For complex $\mathbf{Z}$ show that

$$\frac{\partial\,\mathrm{Tr}(\mathbf{AZB})}{\partial\mathbf{Z}} = (\mathbf{BA})^T \quad \text{and} \quad \frac{\partial\,\mathrm{Tr}(\mathbf{AZB})}{\partial\mathbf{Z}^*} = \mathbf{0}. \tag{P20.2}$$

20.3. For complex $\mathbf{Z}$ show that

$$\frac{\partial\,\mathrm{Tr}(\mathbf{AZ}^{-1}\mathbf{B})}{\partial\mathbf{Z}} = -(\mathbf{Z}^{-1}\mathbf{BAZ}^{-1})^T \quad \text{and} \quad \frac{\partial\,\mathrm{Tr}(\mathbf{AZ}^{-1}\mathbf{B})}{\partial\mathbf{Z}^*} = \mathbf{0}. \tag{P20.3}$$

20.4. For complex $\mathbf{Z}$ find expressions for $\partial\mathrm{Tr}(\mathbf{AZ}^\dagger\mathbf{B})/\partial\mathbf{Z}$, $\partial\mathrm{Tr}(\mathbf{AZ}^\dagger\mathbf{B})/\partial\mathbf{Z}^*$, $\partial\mathrm{Tr}(\mathbf{AZZ}^\dagger\mathbf{B})/\partial\mathbf{Z}$, and $\partial\mathrm{Tr}(\mathbf{AZZ}^\dagger\mathbf{B})/\partial\mathbf{Z}^*$.

20.5. *Derivatives of determinants.* In this problem we find an expression for $\partial[\det(\mathbf{X})]/\partial\mathbf{X}$, where $\mathbf{X}$ is a real square matrix. For this observe that

$$\det(\mathbf{X}) = x_{r0}C_{r0} + x_{r1}C_{r1} + \ldots$$

where C_{rk} is the cofactor of x_{rk}. Using this and the fact that $\mathbf{X}^{-1} = \mathbf{C}^T/\det(\mathbf{X})$, where $\mathbf{C}$ is the matrix of cofactors, show that

$$\frac{\partial\,\det(\mathbf{X})}{\partial\mathbf{X}} = \det(\mathbf{X})\,\mathbf{X}^{-T}, \tag{P20.5a}$$

assuming that $\mathbf{X}$ has no structure. For the case where $\mathbf{X}$ has symmetric structure ($\mathbf{X} = \mathbf{X}^T$), show that

$$\frac{\partial\,\det(\mathbf{X})}{\partial\mathbf{X}} = \det(\mathbf{X})\left(2\mathbf{X}^{-T} - \mathrm{diag}(\mathbf{X}^{-T})\right), \tag{P20.5b}$$

where $\mathrm{diag}(\mathbf{A})$ is a diagonal matrix whose diagonal elements are a_{ii}.

20.6. *Derivatives of determinants, complex case.* Let $\mathbf{Z}$ be a square, complex, matrix. Show that

$$\frac{\partial\det(\mathbf{Z})}{\partial\mathbf{Z}} = \det(\mathbf{Z})\,\mathbf{Z}^{-T} \quad \text{and} \quad \frac{\partial\det(\mathbf{Z})}{\partial\mathbf{Z}^*} = \mathbf{0} \tag{P20.6a}$$

for unstructured $\mathbf{Z}$. If $\mathbf{Z}$ is Hermitian rather than unstructured, show that the above result should be modified to

$$\frac{\partial\det(\mathbf{Z})}{\partial\mathbf{Z}} = \det(\mathbf{Z})(\mathbf{Z}^{-1})^* \quad \text{and} \quad \frac{\partial\det(\mathbf{Z})}{\partial\mathbf{Z}^*} = \det(\mathbf{Z})\mathbf{Z}^{-1}. \tag{P20.6b}$$

20.7. *Derivatives of logarithms of determinants.* For a real square matrix $\mathbf{X}$, the quantity $\partial[\ln\det(\mathbf{X})]/\partial x_{rs}$ can be calculated by using the standard chain rule of calculus. Using this idea show that

$$\frac{\partial[\ln\det(\mathbf{X})]}{\partial\mathbf{X}} = \begin{cases} \mathbf{X}^{-T} & \text{for unstructured } \mathbf{X} \\ 2\mathbf{X}^{-T} - \mathrm{diag}(\mathbf{X}^{-T}) & \text{for symmetric } \mathbf{X}. \end{cases} \tag{P20.7}$$

20.8. *Derivatives of logarithms of determinants, complex case.* By extending the methods of the preceding problem to complex square matrices $\mathbf{Z}$, show that

$$\frac{\partial[\ln\det(\mathbf{Z})]}{\partial\mathbf{Z}} = \mathbf{Z}^{-T} \quad \text{and} \quad \frac{\partial[\ln\det(\mathbf{Z})]}{\partial\mathbf{Z}^*} = \mathbf{0} \tag{P20.8a}$$

if $\mathbf{Z}$ is unstructured, and

$$\frac{\partial[\ln\det(\mathbf{Z})]}{\partial\mathbf{Z}} = \mathbf{Z}^{-T} = (\mathbf{Z}^{-1})^* \quad \text{and} \quad \frac{\partial[\ln\det(\mathbf{Z})]}{\partial\mathbf{Z}^*} = \mathbf{Z}^{-1} \tag{P20.8b}$$

when $\mathbf{Z}$ is Hermitian.

20.9. *Derivatives of powers of matrices.* For a real square matrix $\mathbf{X}$ show that

$$\frac{\partial\mathbf{X}^n}{\partial x_{rs}} = \sum_{k=0}^{n-1} \mathbf{X}^k \mathcal{I}_{rs} \mathbf{X}^{n-k-1}. \tag{P20.9a}$$

For example, $\partial\mathbf{X}^2/\partial x_{rs} = \mathcal{I}_{rs}\mathbf{X} + \mathbf{X}\mathcal{I}_{rs}$. From the preceding show that

$$\frac{\partial \mathrm{Tr}\,(\mathbf{X}^n)}{\partial\mathbf{X}} = n(\mathbf{X}^{n-1})^T. \tag{P20.9b}$$

20.10. *Derivatives of powers of complex matrices.* For a square complex matrix $\mathbf{Z}$ show that

$$\frac{\partial\mathbf{Z}^n}{\partial z_{rs}} = \sum_{k=0}^{n-1} \mathbf{Z}^k \mathcal{I}_{rs} \mathbf{Z}^{n-k-1} \quad \text{and} \quad \frac{\partial\mathbf{Z}^n}{\partial z_{rs}^*} = 0 \tag{P20.10a}$$

when $\mathbf{Z}$ is unstructured. Show also that

$$\frac{\partial\mathbf{Z}^n}{\partial z_{rs}} = \sum_{k=0}^{n-1} \mathbf{Z}^k \mathcal{I}_{rs} \mathbf{Z}^{n-k-1} \quad \text{and} \quad \frac{\partial\mathbf{Z}^n}{\partial z_{rs}^*} = \sum_{k=0}^{n-1} \mathbf{Z}^k \mathcal{I}_{rs}^T \mathbf{Z}^{n-k-1} \tag{P20.10b}$$

when $\mathbf{Z}$ is Hermitian. Taking traces and simplifying show that

$$\frac{\partial\,\mathrm{Tr}\,(\mathbf{Z}^n)}{\partial\mathbf{Z}} = n(\mathbf{Z}^{n-1})^T \quad \text{and} \quad \frac{\partial\,\mathrm{Tr}\,(\mathbf{Z}^n)}{\partial\mathbf{Z}^*} = \mathbf{0} \tag{P20.10c}$$

for unstructured $\mathbf{Z}$, and

$$\frac{\partial\,\mathrm{Tr}\,(\mathbf{Z}^n)}{\partial\mathbf{Z}} = n(\mathbf{Z}^{n-1})^* \quad \text{and} \quad \frac{\partial\,\mathrm{Tr}\,(\mathbf{Z}^n)}{\partial\mathbf{Z}^*} = n(\mathbf{Z}^{n-1}) \tag{P20.10d}$$

for Hermitian $\mathbf{Z}$.

20.11. *Derivatives of exponentials of real matrices.* Given a square matrix $\mathbf{X}$, the exponential matrix $e^{\mathbf{X}}$ is defined as

$$e^{\mathbf{X}} = \mathbf{I} + \mathbf{X} + \frac{\mathbf{X}^2}{2!} + \frac{\mathbf{X}^3}{3!} + \ldots \tag{P20.11a}$$

By using the result of Problem 20.9 show that

$$\frac{\partial \text{Tr}\,(e^{\mathbf{X}})}{\partial \mathbf{X}} = e^{\mathbf{X}^T}. \tag{P20.11b}$$

20.12. *Derivatives of exponentials of complex matrices.* For the case where $\mathbf{Z}$ is complex show that

$$\frac{\partial\ \text{Tr}\,(e^{\mathbf{Z}})}{\partial \mathbf{Z}} = \mathbf{e}^{\mathbf{Z}^T} \quad \text{and} \quad \frac{\partial\ \text{Tr}\,(e^{\mathbf{Z}})}{\partial \mathbf{Z}^*} = \mathbf{0} \tag{P20.12a}$$

for unstructured $\mathbf{Z}$, and

$$\frac{\partial\ \text{Tr}\,(e^{\mathbf{Z}})}{\partial \mathbf{Z}} = \mathbf{e}^{\mathbf{Z}^*} \quad \text{and} \quad \frac{\partial\ \text{Tr}\,(e^{\mathbf{Z}})}{\partial \mathbf{Z}^*} = \mathbf{e}^{\mathbf{Z}} \tag{P20.12b}$$

for Hermitian $\mathbf{Z}$.

20.13. *Mimimum or maximum?* Consider again the optimization problem in Ex. 20.16, where the optimum vector $\mathbf{z}$ and the objective function were shown to be as in Eqs. (20.47) and (20.48), respectively. Let $\mathbf{b}$ be any vector orthogonal to $\mathbf{a}$, that is, $\mathbf{b}^\dagger \mathbf{a} = \mathbf{0}$. Then the vector

$$\mathbf{z} = \mathbf{z}_{opt} + \mathbf{b}$$

still satisfies the constraint $\mathbf{z}^\dagger \mathbf{a} = 1$. Show that when $\mathbf{z}_{opt}$ is replaced with $\mathbf{z}$, the objective function ϕ_{opt} changes to

$$\phi = \phi_{opt} + \mathbf{b}^\dagger \mathbf{R} \mathbf{b},$$

where $\mathbf{R}$ is the Hermitian positive definite matrix given in the optimization problem formulation. Based on this, argue that Eq. (20.48) indeed represents a minimum rather than a maximum.

21

Convexity, Schur convexity, and majorization theory

21.1 Introduction

In this chapter we give a brief overview of convex functions, majorization theory, and Schur-convex functions. The idea of majorization is very fundamental to many problems in linear algebra and optimization. One of the earliest references on this topic is the book by the mathematicians Hardy, Littlewood, and Pólya [1952]. More recent references include Marshall and Olkin [1979], Horn and Johnson [1985], and Pečarić, Proschan, and Tong [1992]. A good reference for convex functions is the book by Marcus and Minc [1964].

Only a small subset of results reviewed in this chapter are actually used in this book. However, the material in the chapter is valuable for readers wishing to pursue some of the advanced research articles published in the areas of transceiver and filter bank optimization.

21.2 Review of convex functions

Before proceeding further with the property of Schur-convex functions, it is convenient at this point to review the notion of convexity, which is more familiar to engineers.

♠**Definition 21.1.** *Convex sets and convex combinations.* A linear combination of vectors $\mathbf{x}_k$ in the form $\sum_{k=0}^{P-1} \alpha_k \mathbf{x}_k$ is said to be a *convex combination* if $\alpha_k \geq 0$ and $\sum_k \alpha_k = 1$. A set of real vectors $\mathcal{S}$ is said to be a convex set if any convex combination of vectors in $\mathcal{S}$ is still a vector belonging to the set $\mathcal{S}$. ◇

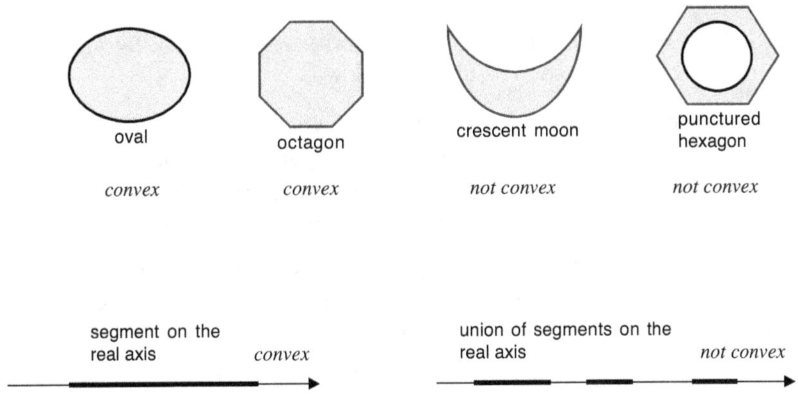

Figure 21.1. Examples of convex and non-convex sets.

Figure 21.1 shows examples of convex and non convex sets. Geometrically speaking, if we draw a chord from one point to another in a convex set then all points on this chord are still contained in the set.

♠**Definition 21.2.** *Convex functions.* Let $f(\mathbf{x})$ be a real function defined on a convex set $\mathcal{S}$. We say $f(\mathbf{x})$ is a convex function if

$$f(\alpha\mathbf{x}_1 + (1-\alpha)\mathbf{x}_2) \leq \alpha f(\mathbf{x}_1) + (1-\alpha)f(\mathbf{x}_2) \tag{21.1}$$

for any α in $0 \leq \alpha \leq 1$ and any $\mathbf{x}_1, \mathbf{x}_2 \in \mathcal{S}$. A function $f(\mathbf{x})$ is said to be *concave* if $-f(\mathbf{x})$ is convex. ♢

By repeated application of the definition we can show that $f(\mathbf{x})$ is convex if and only if

$$f\Big(\sum_{k=0}^{P-1} \alpha_k \mathbf{x}_k\Big) \leq \sum_{k=0}^{P-1} \alpha_k f(\mathbf{x}_k) \tag{21.2}$$

for any set of non-negative numbers α_k such that $\sum \alpha_k = 1$. A function is said to be *strictly* convex if equality holds in Eq. (21.1) only for the trivial choices $\alpha = 0$ and $\alpha = 1$. A function $f(\mathbf{x})$ is *strictly* concave if $-f(\mathbf{x})$ is strictly convex.

Why is convexity important? In engineering sciences, one often encounters the need to minimize an objective function $f(\mathbf{x})$ subject to some constraint on the variables x_i (components of $\mathbf{x}$). If an objective function is convex and the constraint set (i.e., the set $\mathcal{X}$ from which $\mathbf{x}$ is allowed to be drawn) is a convex set, then the optimization problem is said to be a convex optimization problem. In this case a local minimum is also global. Furthermore, first-order necessary conditions for optimality based on derivatives are also sufficient, that is, there is no need to look at second-order derivative information such as Hessian matrices [Chong and Żak, 2001]. Examples of convex optimization includes linear programming problems and problems where the objective functions are quadratic and constraints are linear. Many excellent references exist for this topic; for example, see Chong and Żak [2001], Boyd and Vandenberghe [2004], and Antoniou and Lu [2007]. ▽ ▽ ▽

21.2.1 Testing convexity

It is often possible to test for convexity or concavity by applying some simple rules. Two of these are outlined below.

1. ***The chord test.*** For functions of *one variable*, convexity pictorially means this: if we draw a chord from one point to another in the plot of $f(x)$, then all points on the plot lie below (or possibly on) the chord. For concave functions, it is the other way. Figure 21.2 shows several examples.

2. ***The second-derivative test.*** If $f(\mathbf{x})$ is twice differentiable, then it is easy to test convexity. We compute second partial derivatives $\partial^2 f(\mathbf{x})/\partial x_k \partial x_m$ for all k, m and form the Hessian matrix

$$\mathbf{F} = \begin{bmatrix} \frac{\partial^2 f(\mathbf{x})}{\partial x_0 \partial x_0} & \frac{\partial^2 f(\mathbf{x})}{\partial x_0 \partial x_1} & \cdots \\ \frac{\partial^2 f(\mathbf{x})}{\partial x_1 \partial x_0} & \frac{\partial^2 f(\mathbf{x})}{\partial x_1 \partial x_1} & \cdots \\ \vdots & \vdots & \ddots \end{bmatrix}. \tag{21.3}$$

This is a Hermitian matrix, since $\partial^2 f(\mathbf{x})/\partial x_k \partial x_m = \partial^2 f(\mathbf{x})/\partial x_m \partial x_k$. It can be shown (see Horn and Johnson [1985], p. 535) that $f(\mathbf{x})$ is convex if and only if $\mathbf{F}$ is positive semidefinite, and *strictly* convex if and only if $\mathbf{F}$ is positive definite. A similar result holds for concave functions (replace "positive" with "negative"). For the special case of scalar functions which can be differentiated twice, convexity is therefore equivalent to the condition

$$\frac{d^2 f(x)}{dx^2} \geq 0, \tag{21.4}$$

with strict convexity holding if and only if

$$\frac{d^2 f(x)}{dx^2} > 0. \tag{21.5}$$

21.2.2 Examples

We now present a number of insightful examples of convex and concave functions.

1. The absolute value function

$$f_{abs}(x) = |x|$$

is convex as seen from the chord test (Fig. 21.3(a)).

2. The *affine function* defined as

$$f_{aff}(x) = A + Bx$$

is *both convex and concave* because the second derivative is zero. The plot of this is simply a straight line (Fig. 21.3(b)). So the chord that we draw in the "chord test" always lies *on* this line.

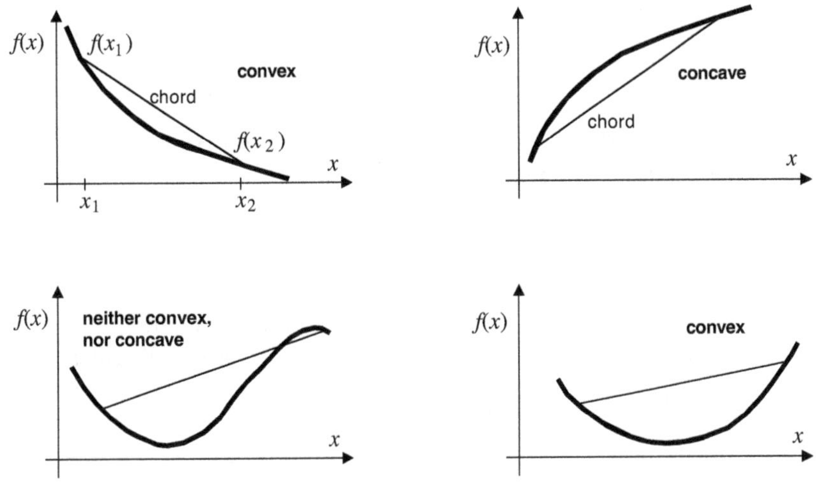

Figure 21.2. Examples of convex and concave functions.

3. By examining second derivatives we can readily verify that monomials like x, x^2, and so forth are convex for $x \geq 0$. For positive integer n, x^{2n} is convex in $-\infty < x < \infty$.

4. The two-variable function $f_1(x_0, x_1) = A + Bx_0 + Cx_1$ is both concave and convex. This is because the second derivatives in the Hessian matrix are all zero, so the Hessian is zero (trivially both positive- and negative-semi definite).

5. Consider the scalar function

$$f_2(x) = 1/x^a, \quad x > 0, \tag{21.6}$$

for some positive a. Since

$$\frac{d^2 f_2(x)}{dx^2} = \frac{a(a+1)}{x^{a+2}} > 0, \quad \text{in } 0 < x < \infty,$$

it follows that $f_2(x)$ is convex in $0 < x < \infty$. For example, $1/\sqrt{x}$ is convex in $0 < x < \infty$, and so is $1/x$.

6. Next consider the logarithm

$$f_3(x) = \ln x, \quad 0 < x < \infty. \tag{21.7}$$

We have $d^2 f_3(x)/dx^2 = -1/x^2 < 0$ for $x > 0$, so it follows that $\ln x$ is strictly *concave* (i.e., $-\ln x$ is strictly convex) in $0 < x < \infty$.

7. Note that *shifted versions* of convex functions are also convex with the domain appropriately shifted. Thus $1/\sqrt{1+x}$ is convex for $1+x>0$, that is, for $x>-1$, and so is $1/(x+1)$. Similarly $\ln(1+x)$ is strictly concave for $x>-1$.

8. For the function
$$f_4(x) = x\ln x, \quad 0 < x < \infty, \tag{21.8}$$
we have $d^2 f_4(x)/dx^2 = 1/x > 0$, so $x\ln x$ is *convex* in $x>0$. Figure 21.4 shows the plots of $\ln x$ and $x\ln x$.

9. Similarly it is readily verified that
$$f_5(x) = e^{\alpha x}, \quad \alpha \text{ real}, \tag{21.9}$$
is convex in $-\infty < x < \infty$. This is because $d^2 f_5(x)/dx^2 = \alpha^2 e^{\alpha x} \geq 0$ for all real x. Thus e^x and e^{-x} are both convex for all real x.

10. Next consider the Gaussian pulse
$$f_6(x) = e^{-x^2}, \ -\infty < x < \infty, \tag{21.10}$$
which represents the Gaussian pdf except for a scale factor. We have
$$\frac{d^2 f_6(x)}{dx^2} = 2(2x^2-1)e^{-x^2} \begin{cases} \leq 0 & \text{for } |x| \leq \frac{1}{\sqrt{2}} \\ \geq 0 & \text{for } |x| \geq \frac{1}{\sqrt{2}} \end{cases} \tag{21.11}$$
so that the Gaussian is concave in $|x| \leq 1/\sqrt{2}$, convex in $-\infty < x \leq -1/\sqrt{2}$, and convex in $1/\sqrt{2} \leq x < \infty$. See Fig. 21.5.

21.2.3 The complementary error function

From Sec. 16.2 we know that the probability of error at the output of the receiver in many digital communications systems is approximately proportional to
$$\mathcal{P}(x) = \text{erfc}\Big(\frac{K}{\sqrt{x}}\Big) = 2\mathcal{Q}\Big(\frac{\sqrt{2}K}{\sqrt{x}}\Big), \tag{21.12}$$
where K is a positive constant whose value depends on the modulation scheme, and x is the *noise-to-signal ratio* at the detector input. Here $\text{erfc}(x)$ is the *complementary error function*, defined as the integral
$$\text{erfc}(x) = \frac{2}{\sqrt{\pi}} \int_x^{\infty} e^{-y^2} dy. \tag{21.13}$$
This is related to the $\mathcal{Q}$ function by $2\mathcal{Q}(\sqrt{2}\,x) = \text{erfc}(x)$, or equivalently $\mathcal{Q}(x) = 0.5\text{erfc}(x/\sqrt{2})$. To study the convexity properties of $\mathcal{P}(x)$ we first compute the derivative of $\text{erfc}(x)$. Let the indefinite integral $2\int e^{-y^2}dy/\sqrt{\pi}$ be denoted as $\eta(y)$ so that $d\eta(y)/dy = 2e^{-y^2}/\sqrt{\pi}$. Then
$$\text{erfc}(x) = \eta(\infty) - \eta(x),$$

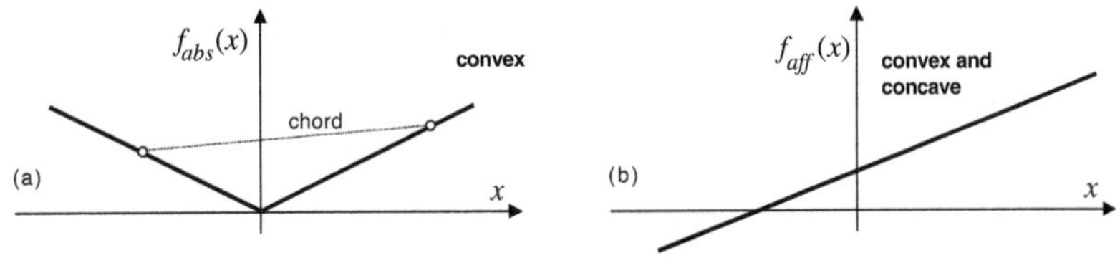

Figure 21.3. (a) The absolute-value function is convex because the chord lies above it. (b) The affine function is both convex and concave.

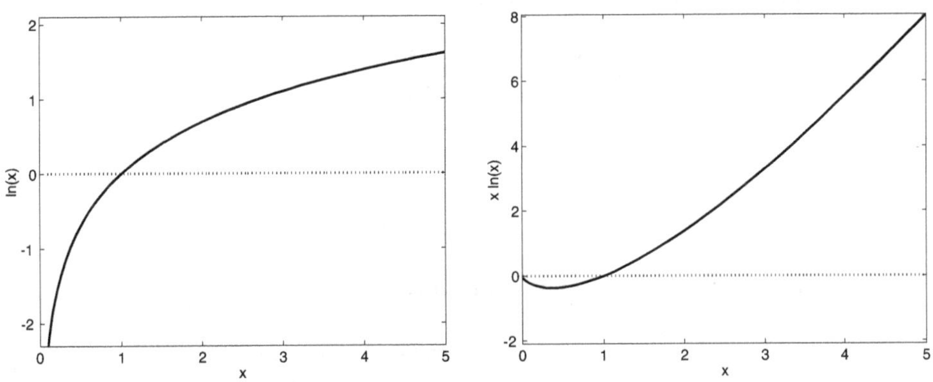

Figure 21.4. Plots of the logarithm $\ln x$ (left), and the function $x \ln x$ (right), for $x > 0$. The former is concave whereas the latter is convex.

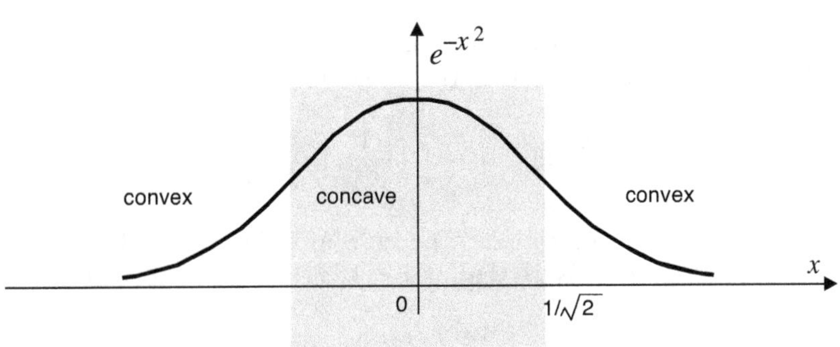

Figure 21.5. Plot of the Gaussian pulse e^{-x^2}. This function is concave in the middle and convex in the outer parts as indicated.

so that

$$\frac{d\,\text{erfc}(x)}{dx} = \frac{-d\eta(x)}{dx} = -\frac{2}{\sqrt{\pi}}e^{-x^2}, \tag{21.14}$$

from which we verify that

$$\frac{d^2\,\text{erfc}(x)}{dx^2} = \frac{4xe^{-x^2}}{\sqrt{\pi}} \geq 0, \quad \text{for } x \geq 0. \tag{21.15}$$

This shows that $\text{erfc}(x)$ and $\mathcal{Q}(x)$ are convex for $x \geq 0$. Figure 21.6 shows a plot of $\mathcal{Q}(x)$, from which we see indeed that it is convex for $x \geq 0$. From Eq. (21.12) we also see that

$$\frac{d\mathcal{P}(x)}{dx} = \frac{K\,x^{-3/2}e^{-K^2/x}}{\sqrt{\pi}}$$

Differentiating one more time, we obtain

$$\frac{d^2\mathcal{P}(x)}{dx^2} = \frac{K\,x^{-7/2}e^{-K^2/x}}{2\sqrt{\pi}}(2K^2 - 3x). \tag{21.16}$$

So

$$\frac{d^2\mathcal{P}(x)}{dx^2} \begin{cases} \geq 0 & \text{for } x \leq 2K^2/3 \\ \leq 0 & \text{for } x \geq 2K^2/3. \end{cases} \tag{21.17}$$

$\mathcal{P}(x)$ is therefore first convex and then concave:

$$\mathcal{P}(x) \text{ is } \begin{cases} \text{convex} & \text{in } x \leq 2K^2/3 \\ \text{concave} & \text{in } x \geq 2K^2/3. \end{cases} \tag{21.18}$$

Since the $\mathcal{Q}$ function is related to $\mathcal{P}$ as in Eq. (21.12) it follows therefore that

$$\mathcal{Q}\Big(\frac{A}{\sqrt{x}}\Big) \text{ is } \begin{cases} \text{convex} & \text{in } x \leq A^2/3 \\ \text{concave} & \text{in } x \geq A^2/3. \end{cases} \tag{21.19}$$

As an application, consider binary modulation (one bit PAM) which has two symbols (1 and -1). In this case the probability of error is (Sec. 2.3.1)

$$\mathcal{P}_e(x) = 0.5\,\text{erfc}\Big(\frac{1}{\sqrt{2x}}\Big) = \mathcal{Q}\Big(\frac{1}{\sqrt{x}}\Big). \tag{21.20}$$

Using Eq. (21.19) we see that $\mathcal{P}_e(x)$ is convex for $x < 1/3$ and concave for $x > 1/3$. See Fig. 21.7. Note that at $x = 1/3$ the error probability is

$$\mathcal{P}_e(x) \approx 0.042,$$

which is quite large. If the application is such that the error probabilities are required to be smaller than this (which is often the case), it is reasonable indeed to assume that $\mathcal{P}_e(x)$ is convex.

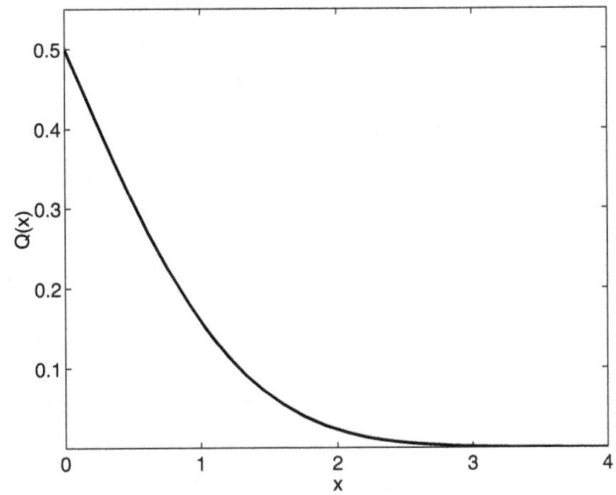

Figure 21.6. Convexity of the $\mathcal{Q}(x)$ function for $x \geq 0$.

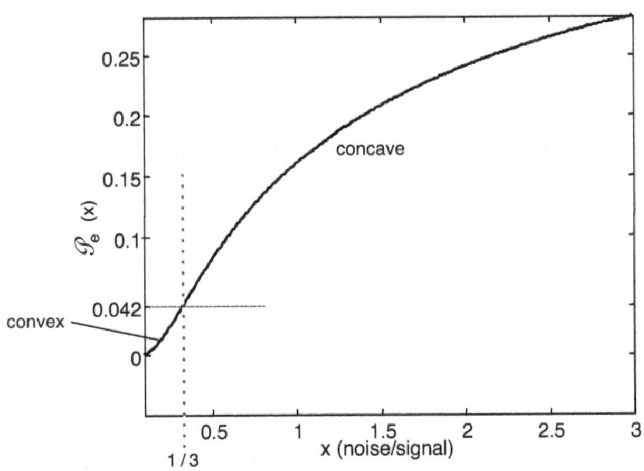

Figure 21.7. The probability of error for binary modulation as a function of the noise-to-signal ratio x. The plot is convex for $x \leq 1/3$ and concave for $x \geq 1/3$.

21.2.4 Convex and concave functions of two variables

We now consider examples of two-variable functions.

1. First consider $g_1(x_0, x_1) = x_0 x_1$. We have

$$\frac{\partial^2 g_1}{\partial x_0^2} = \frac{\partial^2 g_1}{\partial x_1^2} = 0 \quad \text{and} \quad \frac{\partial^2 g_1}{\partial x_0 \partial x_1} = \frac{\partial^2 g_1}{\partial x_1 \partial x_0} = 1,$$

which shows that the Hessian matrix is

$$\mathbf{G} = \begin{bmatrix} 0 & 1 \\ 1 & 0 \end{bmatrix}.$$

This is Hermitian, so the eigenvalues are real: $\lambda_0 = 1$ and $\lambda_1 = -1$ (with corresponding eigenvectors $[1 \quad 1]^T$ and $[1 \quad -1]^T$). So it is neither positive (semi)definite nor negative (semi)definite. Since the Hessian is an indefinite matrix, *the product x_0x_1 is neither convex nor concave.*

2. If we repeat this exercise for the *geometric-mean* function

$$g_2(x_0, x_1) = \sqrt{x_0x_1},$$

where $0 < x_i < \infty$, we will find that the Hessian is

$$\mathbf{G} = \frac{1}{4\sqrt{x_0x_1}} \begin{bmatrix} -x_1/x_0 & 1 \\ 1 & -x_0/x_1 \end{bmatrix}.$$

Once again this is Hermitian, so the eigenvalues are real. Since the determinant (product of the eigenvalues) is zero, at least one of the eigenvaules is zero. The trace (sum of the diagonal elements, which equals the sum of the eigenvalues) is clearly negative for $x_0, x_1 > 0$, showing that the other eigenvalue is negative, so the Hessian is negative semidefinite. The *geometric mean is therefore concave in* $x_0, x_1 > 0$. The same is true for the geometric mean of P numbers, as shown later.

3. Next consider

$$g_3(x_0, x_1) = \ln(x_0x_1) = \ln x_0 + \ln x_1.$$

Now the Hessian is

$$\mathbf{G} = \begin{bmatrix} -1/x_0^2 & 0 \\ 0 & -1/x_1^2 \end{bmatrix},$$

which is negative definite for $x_0, x_1 > 0$. Thus $\ln(x_0x_1)$ is concave when $x_0, x_1 > 0$. Since $\ln\sqrt{x_0x_1} = 0.5\ln(x_0x_1)$, this is also concave.

4. Finally, let $g_4(x_0, x_1, x_2) = x_0 + x_1 + x_2$. Then $\partial g_4(x_0, x_1, x_2)/\partial x_k = 1$, for all k, so

$$\frac{\partial^2 g_4(x_0, x_1, x_2)}{\partial x_k \partial x_m} = 0$$

for all k, m. So the Hessian $\mathbf{G} = \mathbf{0}$, which is both positive semidefinite, and negative semidefinite. Thus $g_4(x_0, x_1, x_2)$ is both convex and concave.

The preceding examples show that even though a function (such as x_0x_1) may not be concave, a monotone function of this function (such as $\sqrt{x_0x_1}$ or $\ln(x_0x_1)$) can be concave. Such results are useful in optimization because optimizing an objective function is the same thing as optimizing a monotone function of the objective function.

Concavity of the geometric mean. In the preceding discussions we found that the geometric mean of two positive numbers is concave in the positive orthant $x_i > 0$. Similarly it is true that the geometric mean of P numbers

$$g(x_0, x_1, \ldots, x_{P-1}) = \Big(\prod_{k=0}^{P-1} x_k\Big)^{1/P} \tag{21.21}$$

is concave in the positive orthant $x_k > 0$. To see this we first compute the Hessian matrix, which turns out to be $\mathbf{G} = c\mathbf{H}$, where $c = (\prod_{k=0}^{P-1} x_k)^{1/P}/P^2 > 0$, and

$$\mathbf{H} = \mathbf{y}\mathbf{y}^T - P\begin{bmatrix} y_0^2 & 0 & \ldots & 0 \\ 0 & y_1^2 & \ldots & 0 \\ \vdots & \vdots & \ddots & \vdots \\ 0 & 0 & \ldots & y_{P-1}^2 \end{bmatrix},$$

where $\mathbf{y} = [\,y_0 \quad y_1 \quad \ldots \quad y_{P-1}\,]^T$, is a real column vector with $y_k = 1/x_k$. To prove concavity of the geometric mean, it therefore suffices to prove that the real Hermitian matrix $\mathbf{H}$ is negative semidefinite, that is, $\mathbf{v}^T\mathbf{H}\mathbf{v} \leq 0$ for all real $\mathbf{v}$. Now,

$$\mathbf{v}^T\mathbf{H}\mathbf{v} = (\mathbf{v}^T\mathbf{y})(\mathbf{y}^T\mathbf{v}) - P\mathbf{v}^T\begin{bmatrix} y_0^2 & 0 & \ldots & 0 \\ 0 & y_1^2 & \ldots & 0 \\ \vdots & \vdots & \ddots & \vdots \\ 0 & 0 & \ldots & y_{P-1}^2 \end{bmatrix}\mathbf{v}$$

which can be rewritten as

$$\mathbf{v}^T\mathbf{H}\mathbf{v} = \Big(\sum_k v_k y_k\Big)^2 - P\sum_k v_k^2 y_k^2 \tag{21.22}$$

Defining the real vectors $\mathbf{u} = [\,v_0 y_0 \quad \ldots \quad v_{P-1}y_{P-1}\,]^T$, $\mathbf{z} = [\,1 \quad 1 \quad \ldots \quad 1\,]^T$, and applying the Cauchy-Schwartz inequality, we get, $(\mathbf{u}^T\mathbf{z})^2 \leq (\mathbf{u}^T\mathbf{u})(\mathbf{z}^T\mathbf{z})$, that is,

$$\Big(\sum_k v_k y_k\Big)^2 \leq P\sum_k v_k^2 y_k^2.$$

Using this in Eq. (21.22) shows that $\mathbf{v}^T\mathbf{H}\mathbf{v} \leq 0$ for all real $\mathbf{v}$. So $\mathbf{H}$ is negative semidefinite, proving that the geometric mean is a concave function in the positive orthant $x_k > 0$. $\bigtriangledown\bigtriangledown\bigtriangledown$

21.2.5 Composite functions

If $f(x)$ and $h(x)$ are convex, can we say $h(f(x))$ is convex as well? If $h(x)$ is an *increasing* function, this is indeed true. More precisely we have the following (see Marcus and Minc [1964] p. 102):

♠**Theorem 21.1.** *Convex functions of convex functions.* Assume $f(\mathbf{x})$ is convex on the set $\mathcal{A}$ and $a \leq f(\mathbf{x}) \leq b$ on $\mathcal{A}$. If $h(y)$ is an increasing convex function on $a \leq y \leq b$, then

$$g(\mathbf{x}) \stackrel{\Delta}{=} h(f(\mathbf{x})) \tag{21.23}$$

is convex on $\mathcal{A}$. ◇

Example 21.1: Composite functions and convexity

We consider three combinations of $f(x)$ and $h(y)$ to demonstrate how composite functions work. All of these are simple examples which can be addressed directly without the above theorem, but the examples serve to demonstrate the theroem.

- ***Combination 1.*** We know $f(x) = x^2$ is convex everywhere, and $h(y) = e^y$ increasing convex everywhere. According to Theorem 21.1, the composite function $g(x) = h(f(x)) = e^{x^2}$ should therefore be convex everywhere. Indeed, the second derivative is $d^2g(x)/dx^2 = 2e^{x^2}(1 + 2x^2) > 0$ verifying this assertion.
- ***Combination 2.*** Consider next the example where $f(x) = x^2$ (convex everywhere), and $h(y) = -y^2$, which is decreasing and concave in $y > 0$. Since $-h(y)$ is increasing and convex, it follows from Theorem 21.1 that $-h(f(x)) = x^4$ is convex, from which we conclude $g(x) = h(f(x)) = -x^4$ is concave. Indeed, the second derivative is given by $d^2g(x)/dx^2 = -12x^2 < 0$, showing the concavity directly.
- ***Combination 3.*** Consider finally the following combination: $f(x) = x^2$ (convex everywhere), and $h(y) = e^{-y}$, which is *decreasing convex* everywhere. In this case Theorem 21.1 cannot be applied. Now $g(x) = h(f(x)) = e^{-x^2}$, which is neither convex nor concave, as we saw earlier. More precisely this function is convex in some parts and concave in some parts (Fig. 21.5).

21.2.6 Jensen's inequality

We know that $f(\mathbf{x})$ is convex if and only if

$$f\Big(\sum_{k=0}^{P-1} \alpha_k \mathbf{x}_k\Big) \leq \sum_{k=0}^{P-1} \alpha_k f(\mathbf{x}_k) \tag{21.24}$$

for any set of non-negative numbers α_k such that $\sum \alpha_k = 1$. Such a set of numbers $\{\alpha_k\}$ can be interpreted as probabilites for a discrete random variable X, as demonstrated in Fig. 21.8. The sum $\sum_{k=0}^{P-1} \alpha_k x_k$ can therefore be regarded as the expected value of the random variable X, which can take the value x_k

with probability α_k. Similarly, $\sum_{k=0}^{P-1} \alpha_k f(x_k)$ is the expected value of $f(X)$. Thus, if $f(x)$ is a convex function of a real scalar then

$$f(E[X]) \leq E[f(X)] \tag{21.25}$$

for any discrete random variable X. This inequality is called Jensen's inequality, and can also be extended to the case where X is a continuous random variable [Cover and Thomas, 1991]. Furthermore, if $f(x)$ is strictly convex then equality holds in Eq. (21.25) if and only if $X = E[X]$ (i.e., X is just a constant rather than a random variable, in which case Fig. 21.8 becomes an impulse concentrated at one point).

Consider the special case where $f(x) = -\ln x$, which is strictly convex for $x > 0$ (Sec. 21.2.2). This yields an interesting result. Suppose we set $\alpha_k = 1/P$ for all k, then, according to Jensen's inequality,

$$-\ln \frac{1}{P}\sum_{k=0}^{P-1} x_k < -\frac{1}{P}\sum_{k=0}^{P-1} \ln x_k$$

(unless $P = 1$), which can readily be rewritten as

$$\frac{1}{P}\sum_{k=0}^{P-1} x_k > \left(\prod_{k=0}^{P-1} x_k\right)^{\frac{1}{P}}, \tag{21.26}$$

where $x_k > 0$. This is nothing but the familiar AM-GM inequality (Appendix A) which is therefore a special case of Jensen's inequality. We know that equality in the AM-GM inequality holds if and only if all x_k are identical. The reason why we have strict inequality in Eq. (21.26) is because the x_k in Fig. 21.8 are distinct.

21.2.7 Further properties

We conclude by mentioning a few other useful properties of convex functions. More details can be found in several references, including Marcus and Minc [1964].

1. ***The max function.*** If $f_k(\mathbf{x}), 0 \leq k \leq N-1$, are convex on the set $\mathcal{A}$ then the maximum function
$$f(\mathbf{x}) = \max_k f_k(\mathbf{x})$$
is convex on $\mathcal{A}$.

2. ***Subset where function is bounded.*** If $f(\mathbf{x})$ is convex on $\mathcal{A}$ and $\mathcal{A}_1 \subset \mathcal{A}$ is the subset of all $\mathbf{x}$ such that $f(\mathbf{x}) \leq K$ in $\mathcal{A}_1$ (for some fixed K), then $\mathcal{A}_1$ is a convex set.

3. ***Sum of convex functions.*** If the functions $f_i(x)$ are convex on the interval $a_i \leq x \leq b_i$, then $g(\mathbf{x}) = \sum_{i=0}^{N-1} f_i(x_i)$ is convex for all $\mathbf{x}$ such that the ith component satisfies $a_i \leq x_i \leq b_i$.

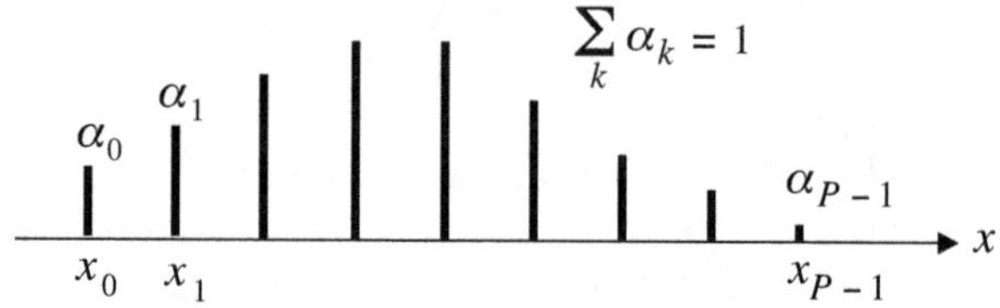

Figure 21.8. Example of a discrete probability density function.

4. ***The ℓ_p norm.*** The p-norm of a vector $\mathbf{x} = [x_0 \quad x_1 \quad \ldots \quad x_{N-1}]^T$, defined as

$$\|\mathbf{x}\|_p = (\sum_{k=0}^{N-1} |x_k|^p)^{1/p}, \quad p > 1,$$

is convex for all real $\mathbf{x}$. A proof can be found in Marcus and Minc [1964], p. 105. For example, ℓ_1 and ℓ_2 norms are convex.

21.3 Schur-convex functions

To explain what Schur-convex functions are, we first define the idea of majorization. Given a sequence of real numbers $x_k, 0 \leq k \leq P-1$, the notation $x_{[k]}$ refers to the permuted sequence such that

$$x_{[0]} \geq x_{[1]} \geq x_{[2]} \ldots$$

This notation will be used throughout the chapter.

♠**Definition 21.3.** *Majorization.* Given two real vectors

$$\mathbf{x} = [x_0 \quad x_1 \quad \ldots \quad x_{P-1}]^T, \quad \mathbf{y} = [y_0 \quad y_1 \quad \ldots \quad y_{P-1}]^T,$$

we say that $\mathbf{y}$ majorizes $\mathbf{x}$ if the following two conditions are satisfied. First, the sum of the elements is identical:

$$\sum_{k=0}^{P-1} y_k = \sum_{k=0}^{P-1} x_k,$$

and second, any partial sum of the ordered sequence $y_{[k]}$ is at least as large as the corresponding partial sum of $x_{[k]}$, that is,

$$\sum_{k=0}^{n} y_{[k]} \geq \sum_{k=0}^{n} x_{[k]}, \qquad 0 \leq n \leq P-2. \tag{21.27}$$

When $\mathbf{y}$ majorizes $\mathbf{x}$, this is denoted as

$$\mathbf{y} \succ \mathbf{x}$$

or equivalently $\mathbf{x} \prec \mathbf{y}$. $\diamondsuit$

Other equivalent statements are (a) the sequence y_k majorizes x_k, and (b) $\mathbf{x}$ is majorized by $\mathbf{y}$. The statement

$$\mathbf{y} \succ \mathbf{x} \quad \text{on } \mathcal{S}$$

means that $\mathbf{x}$ and $\mathbf{y}$ both belong to a subset $\mathcal{S}$ of real vectors (e.g., subset with non-negative elements), and that $\mathbf{y}$ majorizes $\mathbf{x}$ in that subset. A closely related concept is the idea of *multiplicative* majorization, mentioned briefly in Sec. 21.6.

Relation to convex functions

There is a simple relation between the concept of majorization and the concept of convex functions. (Convex functions are reviewed in Sec. 21.2.) The following relation was observed by Hardy, Littlewood, and Pólya [1952]:

♠**Theorem 21.2.** *Convex functions and majorization.* Given two vectors $\mathbf{x}$ and $\mathbf{y}$, we have $\mathbf{y} \succ \mathbf{x}$ if and only if

$$\sum_{k=0}^{P-1} g(y_k) \geq \sum_{k=0}^{P-1} g(x_k)$$

for all continuous convex functions $g(x)$. $\diamondsuit$

A proof of this beautiful result can be found on p. 108 of Marshall and Olkin [1979].

Example 21.2: Majorization

We can readily verify that $[4 \;\; 2 \;\; 1] \succ [3 \;\; 2 \;\; 2]$. More nontrivially, consider the vectors

$$\mathbf{x} = [\, \tfrac{1}{P} \;\; \tfrac{1}{P} \;\; \ldots \;\; \tfrac{1}{P} \,]^T \quad \text{and} \quad \mathbf{z} = [\, 1 \;\; 0 \;\; \ldots \;\; 0 \,]^T,$$

where P is the number of elements in each vector. Then $\sum x_i = \sum z_i = 1$. Furthermore it is obvious that

$$\sum_{k=0}^{n} z_{[k]} \geq \sum_{k=0}^{n} x_{[k]}, \qquad 0 \leq n \leq P-2.$$

So $\mathbf{z} \succ \mathbf{x}$. Given any vector $\mathbf{y} = [\, y_0 \;\; y_1 \;\; \ldots \;\; y_{P-1} \,]^T$ such that $y_k \geq 0$ and $\sum_k y_k = 1$, we will in fact show that

$$\mathbf{z} \succ \mathbf{y} \succ \mathbf{x}. \tag{21.28}$$

Assume without loss of generality that $y_0 \geq y_1 \geq \ldots \geq y_{P-1}$ because it is only the ordered sets that should undergo the test of Eq. (21.27). Since $\mathbf{z} \succ \mathbf{y}$ is obvious, it only remains to show $\mathbf{y} \succ \mathbf{x}$. Assume the contrary. Since $y_0 \geq x_0$, this assumption means that there exists $m < P - 1$ such that

$$\sum_{k=0}^{m-1} y_k \geq \frac{m}{P} \quad \text{and} \quad \sum_{k=0}^{m} y_k < \frac{m+1}{P}$$

These two inequalities are together possible only if $y_m < 1/P$, which implies $y_n < 1/P$ for $m \leq n \leq P-1$ (because of the ordering $y_0 \geq y_1 \geq \ldots \geq y_{P-1}$). Thus

$$\sum_{k=0}^{P-1} y_k = \underbrace{\sum_{k=0}^{m} y_k}_{\text{less than } (m+1)/P} + \underbrace{\sum_{k=m+1}^{P-1} y_k}_{\text{less than } (P-m-1)/P}$$

This shows that $\sum_{k=0}^{P-1} y_k < 1$, which contradicts $\sum_{k=0}^{P-1} y_k = 1$. So we have proved $\mathbf{y} \succ \mathbf{x}$. This completes the proof of Eq. (21.28).

The outcome of Ex. 21.2 is important enough to be summarized in the following form:

♠**Lemma 21.1.** *The two extreme vectors.* Given a P-vector $\mathbf{y}$ whose components satisfy $y_k \geq 0$ and $\sum_k y_k = 1$, we have

$$[1 \;\; 0 \;\; \ldots \;\; 0]^T \succ [y_0 \;\; y_1 \;\; \ldots \;\; y_{P-1}]^T \succ [\tfrac{1}{P} \;\; \tfrac{1}{P} \;\; \ldots \;\; \tfrac{1}{P}]^T \qquad (21.29)$$

Thus, as far as ranking by majorization goes, concentration of all the energy in one component is one extreme, and equal distribution of energy among the components is the other extreme. ◇

We are now ready to define Schur-convex functions. These were introduced and studied by Schur as early as 1923. Since then they have been studied by many mathematicians, and their importance in optimization appropriately recognized.

♠**Definition 21.4.** *Schur convexity.* Let $f(\mathbf{x})$ be a real-valued function of a real vector $\mathbf{x}$. We say that $f(\mathbf{x})$ is Schur-convex if

$$\mathbf{x}_1 \succ \mathbf{x}_2 \quad \text{implies} \quad f(\mathbf{x}_1) \geq f(\mathbf{x}_2),$$

and Schur-concave if

$$\mathbf{x}_1 \succ \mathbf{x}_2 \quad \text{implies} \quad f(\mathbf{x}_1) \leq f(\mathbf{x}_2).$$

The term *strictly Schur-convex* is used if

$$\mathbf{x}_1 \succ \mathbf{x}_2 \quad \text{implies} \quad f(\mathbf{x}_1) > f(\mathbf{x}_2),$$

whenever $\mathbf{x}_2$ is not a permutation of $\mathbf{x}_1$. $\diamondsuit$

Clearly $f(\mathbf{x})$ is Schur-convex if and only if $-f(\mathbf{x})$ is Schur-concave. When we say that $f(\mathbf{x})$ is Schur-convex in a subset $\mathcal{S}$ of real vectors, we mean that the argument $\mathbf{x}$ is constrained to be in the domain $\mathcal{S}$.

Example 21.3: Schur-convex functions

Perhaps the simplest example of a Schur-convex function is

$$f(\mathbf{x}) = \max_k \{x_k\} = x_{[0]}. \tag{21.30}$$

If $\mathbf{x} \succ \mathbf{y}$ then clearly $x_{[0]} \geq y_{[0]}$, which proves that $f(\mathbf{x}) \geq f(\mathbf{y})$ indeed. For a less trivial example consider

$$f(\mathbf{x}) = \frac{1}{x_0} + \frac{1}{x_1}$$

We will show later (see remarks around Eq. (21.38)) that this is Schur-convex in $0 < x_k < \infty$. Here we will evaluate this function for the two vectors

$$\mathbf{x}_1 = [4 \quad 2], \quad \mathbf{x}_2 = [3 \quad 3].$$

Clearly $\mathbf{x}_1 \succ \mathbf{x}_2$. We have

$$f(\mathbf{x}_1) = \frac{1}{4} + \frac{1}{2} = 0.75 \quad \text{and} \quad f(\mathbf{x}_2) = \frac{1}{3} + \frac{1}{3} \approx 0.667,$$

so that $f(\mathbf{x}_1) > f(\mathbf{x}_2)$ indeed.

21.4 Examples of Schur-convex functions

There are many beautiful examples of Schur-concave/convex functions which arise in optimization problems in signal processing and communications. Some of these examples follow from the theorems to be presented in this section.

The first theorem shows the relation between convex functions and Schur-convex functions (see p. 64 of Marshall and Olkin [1979]). In what follows, the symbol $\mathcal{I}$ denotes an interval on the real line (e.g., $0.5 < x \leq \infty$, $-2 \leq x \leq 3$, and so forth), and $\mathcal{I}^P$ denotes the set of P-vectors whose components x_k are in $\mathcal{I}$.

♠**Theorem 21.3.** *From convex to Schur-convex functions.* Let $g(x)$ be convex on some interval $\mathcal{I}$ of the real line. Then the function $f(\mathbf{x}) = \sum_{k=0}^{P-1} g(x_k)$ is Schur-convex on $\mathcal{I}^P$. $\diamondsuit$

Example 21.4: Schur-convex functions

From Sec. 21.2.2 we know that $1/x^a$, $a > 0$, is convex in $x > 0$. Thus the function

$$f(\mathbf{x}) = \sum_{k=0}^{P-1} \frac{1}{x_k^a}, \quad a > 0,$$

is Schur-convex when $x_k > 0$. For example, the following special cases

$$\sum_{k=0}^{P-1} \frac{1}{\sqrt{x_k}}, \quad \sum_{k=0}^{P-1} \frac{1}{x_k}, \quad \text{and} \quad \sum_{k=0}^{P-1} \frac{1}{x_k^2}$$

are Schur-convex when $x_k > 0$. Similarly

$$\sum_{k=0}^{P-1} \frac{1}{1+x_k}$$

is Schur-convex for $x_k > 0$ because $1/(1+x)$ is convex for $x > 0$ (its second derivative is $2/(1+x)^3 > 0$ for $x > 0$, in fact for $x > -1$). In Sec. 21.2.2 we also showed that $-\ln x$ and $x \ln x$ are convex in $x > 0$. Thus

$$-\sum_{k=0}^{P-1} \ln(x_k) \quad \text{and} \quad \sum_{k=0}^{P-1} x_k \ln(x_k)$$

are Schur-convex in $x_k > 0$. Finally consider the Gaussian pulse e^{-x^2}, which is convex in some parts and concave in some parts as indicated in Fig. 21.5. Thus the sum of Gaussian pulses given by

$$\sum_{k=0}^{P-1} e^{-x_k^2},$$

is Schur-concave if $|x_k| < 1/\sqrt{2}$ for all k, Schur-convex if $-\infty < x_k < -1/\sqrt{2}$ for all k, and Schur-convex if $1/\sqrt{2} < x_k < \infty$ for all k.

Example 21.5: The average probability of error

In Sec. 21.2.3 we showed that the probability of error in the case of a binary constellation is convex for $0 < x < 2/3$ and concave for $x > 2/3$. In MIMO systems one often deals with the average probability of error of several symbol streams. For example, if we have P symbol streams using binary constellations, the average error probability is given by

$$\mathcal{P}_e(\mathbf{x}) = \frac{1}{2P} \sum_{k=0}^{P-1} \text{erfc}\Big(\frac{1}{\sqrt{x_k}}\Big), \tag{21.31}$$

where x_k is the noise-to-signal ratio at the input of the kth detector. Since each term is convex in $0 < x_k < 2/3$, it follows that the average probability of error is Schur-convex in the noise-to-signal vector $\mathbf{x} = [x_0 \ldots x_{P-1}]^T$.

Assume the sum of the noise-to-signal ratios $s = \sum_k x_k$ is fixed for some reason. By using the result of Lemma 21.1 and the fact that $\mathcal{P}_e(\mathbf{x})$ is Schur-convex it then follows that the distribution of x_k that has the smallest error probability $\mathcal{P}_e(\mathbf{x})$ is the equipartition distribution

$$\mathbf{x} = \frac{s}{P}[1 \quad 1 \quad \ldots \quad 1]$$

This is a result that has been elaborated further in Sec. 11.5.

In the following discussions, $\mathcal{D}$ denotes the set of vectors $\mathbf{x}$ whose components satisfy

$$x_0 \geq x_1 \geq \ldots \geq x_{P-1} \quad (\text{set } \mathcal{D}), \tag{21.32}$$

whereas $\mathcal{D}_+$ denotes the set of vectors $\mathbf{x}$ satisfying

$$x_0 \geq x_1 \geq \ldots \geq x_{P-1} \geq 0 \quad (\text{set } \mathcal{D}_+). \tag{21.33}$$

Similarly, $\mathcal{D}_{++}$ denoes

$$x_0 \geq x_1 \geq \ldots \geq x_{P-1} > 0 \quad (\text{set } \mathcal{D}_{++}). \tag{21.34}$$

The following result (p. 92 of Marshall and Olkin [1979]) allows us to generate many Schur-convex examples.

♠**Theorem 21.4.** *Schur-convex sum.* Let

$$f(\mathbf{x}) = \sum_{k=0}^{P-1} g_k(x_k), \tag{21.35}$$

where x_k is the kth component of $\mathbf{x}$. Assume $g_k(x)$ is differentiable in the scalar argument x. Then $f(\mathbf{x})$ is Schur-convex in $\mathcal{D}$ *if and only if* the derivatives of $g_k(x)$ satisfy

$$\frac{dg_k(a)}{dx} \geq \frac{dg_{k+1}(b)}{dx} \quad \text{whenever} \quad a \geq b,$$

for $0 \leq k \leq P-2$. $\diamond$

We now prove the following corollary which is especially convenient to generate nice examples.

♠**Corollary 21.1.** *Schur-convex linear combination.* Let

$$f(\mathbf{x}) = \sum_{k=0}^{P-1} a_k g(x_k) \tag{21.36}$$

and assume the following three conditions:

1. $0 \leq a_0 \leq a_1 \leq \ldots \leq a_{P-1}$;

2. $\frac{dg(x)}{dx} \leq 0$ ($g(x)$ monotone non increasing);

3. $\frac{d^2g(x)}{dx^2} \geq 0$ ($g(x)$ convex).

Then $f(\mathbf{x})$ is Schur-convex on $\mathcal{D}$ (i.e., for $x_0 \geq x_1 \geq \ldots \geq x_{P-1}$). ◇

Proof. Throughout the proof remember that the condition $x_0 \geq x_1 \geq \ldots \geq x_{P-1}$ is assumed. The notation $dg(x_k)/dx$ denotes the derivative of $g(x)$ evaluated at x_k. Since $d^2g(x)/dx^2 \geq 0$, we have

$$\frac{dg(x_0)}{dx} \geq \frac{dg(x_1)}{dx} \geq \ldots \geq \frac{dg(x_{P-1})}{dx}.$$

Using $dg(x)/dx \leq 0$ we therefore get

$$0 \leq -\frac{dg(x_0)}{dx} \leq -\frac{dg(x_1)}{dx} \leq \ldots \leq -\frac{dg(x_{P-1})}{dx}.$$

We now combine this with the first condition of the corollary:

$$0 \leq a_0 \leq a_1 \leq \ldots \leq a_{P-1}.$$

Since a_k and $-dg(x_k)/dx$ are non-negative we conclude from the preceding two sets of inequalities that

$$0 \leq -a_0\frac{dg(x_0)}{dx} \leq -a_1\frac{dg(x_1)}{dx} \leq \ldots \leq -a_{P-1}\frac{dg(x_{P-1})}{dx}$$

which is equivalent to

$$0 \geq a_0\frac{dg(x_0)}{dx} \geq a_1\frac{dg(x_1)}{dx} \geq \ldots \geq a_{P-1}\frac{dg(x_{P-1})}{dx}$$

The function (21.36) therefore has the form (21.35) (just set $g_k(x) = a_k g(x)$) and satisfies all the conditions of Theorem 21.4. It is therefore a Schur-convex function in $\mathcal{D}$. ▽▽▽

For example, let

$$g(x) = \frac{1}{x^p}, \quad p > 0.$$

Then, for $x > 0$,

$$\frac{dg(x)}{dx} = \frac{-p}{x^{p+1}} < 0 \quad \text{and} \quad \frac{d^2g(x)}{dx^2} = \frac{p(p+1)}{x^{p+2}} > 0.$$

Applying Corollay 21.1 it therefore follows that

$$f(\mathbf{x}) = \sum_{k=0}^{P-1} \frac{a_k}{x_k^p}, \quad p > 0 \tag{21.37}$$

is Schur-convex in $x_0 \geq x_1 \geq \ldots \geq x_{P-1} > 0$ for $0 \leq a_0 \leq a_1 \leq \ldots \leq a_{P-1}$. Here are specific examples of such Schur convex functions:

$$\sum_{k=0}^{P-1} \frac{a_k}{\sqrt{x_k}}, \quad \sum_{k=0}^{P-1} \frac{a_k}{x_k}, \quad \sum_{k=0}^{P-1} \frac{a_k}{x_k^2} \tag{21.38}$$

and so forth. Similar reasoning shows that the following functions are Schur-convex in $x_0 \geq x_1 \geq \ldots \geq x_{P-1} > 0$ for $0 \leq a_0 \leq a_1 \leq \ldots \leq a_{P-1}$.

$$\sum_{k=0}^{P-1} \frac{a_k}{1+x_k}, \quad \sum_{k=0}^{P-1} a_k e^{-\alpha x_k} \ (\alpha > 0), \quad \text{and} \quad -\sum_{k=0}^{P-1} a_k \ln x_k. \tag{21.39}$$

Since an increasing function of a Schur-concave function is Schur-concave (Sec. 21.4.2), it follows that $e^{f(\mathbf{x})}$ is Schur-concave whenever $f(\mathbf{x})$ is. For example, since $\sum_{k=0}^{P-1} a_k \ln x_k$ is Schur-concave (from the preceding list) it follows that

$$\exp\Big(\sum_{k=0}^{P-1} a_k \ln x_k\Big)$$

is Schur-concave under the stated conditions. Simplifying, we see that

$$g(\mathbf{x}) = \prod_{k=0}^{P-1} x_k^{a_k} \tag{21.40}$$

is Schur-concave in $x_0 \geq x_1 \geq \ldots \geq x_{P-1} > 0$, if $0 \leq a_0 \leq a_1 \leq \ldots \leq a_{P-1}$.

Example 21.6: Non-Schur-convex example

Even though

$$\sum_{k=0}^{P-1} a_k e^{-\alpha x_k} \tag{21.41}$$

is Schur-convex in $x_0 \geq x_1 \geq \ldots \geq x_{P-1} > 0$ for $0 \leq a_0 \leq a_1 \leq \ldots a_{P-1}$, the function

$$f(\mathbf{x}) = \sum_{k=0}^{P-1} a_k e^{\alpha x_k} \ (\alpha > 0) \tag{21.42}$$

is not, because $dg(x)/dx = \alpha e^{\alpha x}$ and $d^2 g(x)/dx^2 = \alpha^2 e^{\alpha x}$ have identical signs, and Corollary 21.1 cannot be applied. For example, consider the functions

$$f_1(\mathbf{x}) = 0.5e^{x_0} + e^{x_1}, \quad f_2(\mathbf{x}) = 0.25e^{x_0} + e^{x_1}.$$

Let us evaluate these for $\mathbf{x}_a = [3 \ \ 1]$ and $\mathbf{x}_b = [2 \ \ 2]$. Note that $\mathbf{x}_a \succ \mathbf{x}_b$. It is readily verified that

$$f_1(\mathbf{x}_a) = 12.76, \ f_1(\mathbf{x}_b) = 11.08, \ f_2(\mathbf{x}_a) = 7.74, \ f_2(\mathbf{x}_b) = 9.24,$$

which shows that

$$f_1(\mathbf{x}_a) > f_1(\mathbf{x}_b), \quad \text{whereas} \quad f_2(\mathbf{x}_a) < f_2(\mathbf{x}_b).$$

Thus the conditions $x_0 \geq x_1 \geq \ldots \geq x_{P-1} > 0$ and $0 \leq a_0 \leq a_1 \leq \ldots a_{P-1}$ are not enough to ensure that Eq. (21.42) is Schur-convex or Schur-concave.

Example 21.7: Example which is Schur-convex for restricted *x*

A slightly different example is offered by the function

$$g(x) = \frac{1}{1+x^2}$$

which has

$$\frac{dg(x)}{dx} = \frac{-2x}{(1+x^2)^2}, \quad \frac{d^2g(x)}{dx^2} = \frac{2(3x^2-1)}{(1+x^2)^3}$$

Thus $dg(x)/dx \leq 0$ for $x \geq 0$ and

$$\frac{d^2g(x)}{dx^2} \begin{cases} \geq 0 & \text{for } x \geq 1/\sqrt{3} \\ < 0 & \text{for } x < 1/\sqrt{3}. \end{cases}$$

This shows that

$$f(\mathbf{x}) = \sum_{k=0}^{P-1} \frac{a_k}{(1+x_k)^2} \tag{21.43}$$

is Schur-convex in $x_0 \geq x_1 \geq \ldots \geq x_{P-1} \geq 1/\sqrt{3}$, for $0 \leq a_0 \leq a_1 \leq \ldots \leq a_{P-1}$. Note that the function $g(x)$ in this example is concave at the beginning and then turns convex (Fig. 21.9).

21.4.1 More examples

Corollary 21.1 is very powerful and can be used to generate many examples. Here are some. Most of these have been found to arise in transceiver optimization problems [Palomar *et al.*, 2003].

1. Consider $f(\mathbf{x}) = \sum_{k=0}^{P-1} a_k x_k$. Here $g(x) = x$ so that $dg/dx = 1 > 0$ and $d^2g/dx^2 = 0$. From Corollary 21.1 we conclude that $-f(\mathbf{x})$ is Schur-convex, that is,

$$f(\mathbf{x}) = \sum_{k=0}^{P-1} a_k x_k \tag{21.44}$$

is Schur-concave in $x_0 \geq x_1 \geq \ldots \geq x_{P-1}$, if $0 \leq a_0 \leq a_1 \leq \ldots \leq a_{P-1}$.

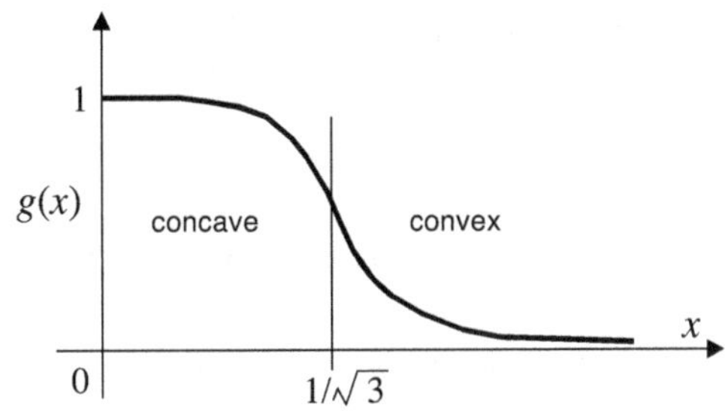

Figure 21.9. The function $g(x) = 1/(1 + x^2)$, which is convex in some parts and concave in some parts.

2. Defining $g(x) = (1 - x)/x$ we find that

$$\frac{dg(x)}{dx} = -\frac{1}{x^2} < 0 \quad \text{and} \quad \frac{d^2g(x)}{dx^2} = \frac{2}{x^3} > 0$$

for $x > 0$. Using Corollary 21.1 we therefore conclude that the function

$$f(\mathbf{x}) = \sum_{k=0}^{P-1} a_k \Big(\frac{1 - x_k}{x_k}\Big), \quad x_k > 0,$$

is Schur-convex in $x_0 \geq x_1 \geq \ldots \geq x_{P-1} > 0$, if $0 \leq a_0 \leq a_1 \leq \ldots \leq a_{P-1}$.

3. The function

$$f(\mathbf{x}) = \sum_{k=0}^{P-1} a_k \ln\Big(\frac{1 - x_k}{x_k}\Big), \quad 0 < x_k \leq 0.5, \tag{21.45}$$

is Schur-convex in $x_0 \geq x_1 \geq \ldots \geq x_{P-1}$, if $0 \leq a_0 \leq a_1 \leq \ldots \leq a_{P-1}$. To see this let $g(x) = \ln[(1 - x)/x]$. Then

$$\frac{dg(x)}{dx} = \frac{1}{x(x-1)} \quad \text{and} \quad \frac{d^2g(x)}{dx^2} = \frac{1 - 2x}{x^2(x-1)^2}$$

so that $dg(x)/dx \leq 0$ and $d^2g(x)/dx^2 \geq 0$ in $0 < x \leq 0.5$. The rest follows from Corollary 21.1.

4. Since an increasing function of a Schur-convex function is Schur-convex (Sec. 21.4.2), it follows from the preceding example that $e^{f(\mathbf{x})}$ is Schur-convex, that is,

$$\prod_{k=0}^{P-1} \Big(\frac{1 - x_k}{x_k}\Big)^{a_k}, \quad 0 < x_k \leq 0.5, \tag{21.46}$$

is Schur-convex in $x_0 \geq x_1 \geq \ldots \geq x_{P-1}$, if $0 \leq a_0 \leq a_1 \leq \ldots \leq a_{P-1}$.

A typo

The result H.2.b on p. 92 of Marshall and Olkin [1979] appears to have a typo. First we repeat the statement as it is:

♠**Assertion H.2.b.** (From Marshall and Olkin [1979]). Let

$$f(\mathbf{x}) = \sum_{k=0}^{P-1} a_k g(x_k) \tag{21.47}$$

and assume the following three conditions:

1. $a_0 \geq a_1 \geq \ldots \geq a_{P-1} \geq 0$;
2. $g(x)$ is increasing (decreasing);
3. $g(x)$ is convex.

Then $f(\mathbf{x})$ is Schur-convex (Schur-concave) on $\mathcal{D}$ (i.e., for $x_0 \geq x_1 \geq \ldots \geq x_{P-1}$). ◊

Example 21.8: A counter example

Here is a counter example to disprove the above assertion. Let $g(x) = 1/x$ which is decreasing and convex for $x > 0$. Let

$$f_1(\mathbf{x}) = \frac{5}{x_0} + \frac{4}{x_1}, \quad f_2(\mathbf{x}) = \frac{3}{x_0} + \frac{1}{x_1}$$

Let $\mathbf{x}_a = [4 \;\; 2]$ and $\mathbf{x}_b = [3 \;\; 3]$ so that $\mathbf{x}_a \succ \mathbf{x}_b$. Then

$$f_1(\mathbf{x}_a) = 3.25, \; f_1(\mathbf{x}_b) = 3.00, \quad f_2(\mathbf{x}_a) = 1.25, \; f_2(\mathbf{x}_b) = 1.33.$$

Thus $f_1(\mathbf{x}_a) > f_1(\mathbf{x}_b)$, whereas $f_2(\mathbf{x}_a) < f_2(\mathbf{x}_b)$. This shows that even though both $f_1(\mathbf{x})$ and $f_2(\mathbf{x})$ satisfy the conditions of H.2.b, $f_1(\mathbf{x})$ violates Schur concavity and $f_2(\mathbf{x})$ violates Schur convexity. (According to H.2.b both of these functions should have been Schur-concave.)

21.4.2 Schur convexity of compositions of functions

If $g(\mathbf{x})$ is Schur-convex then $f(\mathbf{x}) = h(g(\mathbf{x}))$ is Schur-convex, as long as $h(y)$ is a non-decreasing real function of its argument (p. 61 of Marshall and Olkin [1979]). More exhaustively,

$$\begin{aligned}
g(\mathbf{x}) \text{ is Schur-convex and } h(y) \text{ non-decreasing} &\Rightarrow f(\mathbf{x}) \text{ Schur-convex;}\\
g(\mathbf{x}) \text{ is Schur-convex and } h(y) \text{ non-increasing} &\Rightarrow f(\mathbf{x}) \text{ Schur-concave;}\\
g(\mathbf{x}) \text{ is Schur-concave and } h(y) \text{ non-decreasing} &\Rightarrow f(\mathbf{x}) \text{ Schur-concave;}\\
g(\mathbf{x}) \text{ is Schur-concave and } h(y) \text{ non-increasing} &\Rightarrow f(\mathbf{x}) \text{ Schur-convex.}
\end{aligned}$$

For example,

$$e^{\left(\frac{1}{x_0}+\frac{1}{x_1}\right)} \quad \text{and} \quad e^{-\left(\frac{1}{x_0}+\frac{1}{x_1}\right)}$$

are, respectively, Schur-convex and Schur-concave in $x_k > 0$.

21.4.3 Permutation invariance

Let $\mathcal{S}$ be a set of real vectors of the form $[x_0 \; x_1 \; \ldots \; x_{P-1}]^T$. We say that $\mathcal{S}$ is *permutation invariant* if every permutation of a vector in $\mathcal{S}$ is still in $\mathcal{S}$. Here are some interesting points pertaining to Schur convexity and permutation invariance.[1]

1. If $f(\mathbf{x})$ is Schur-convex in a permutation invariant set $\mathcal{S}$, then $f(\mathbf{Px}) = f(\mathbf{x})$ for any permutation matrix $\mathbf{P}$, that is, $f(\mathbf{x})$ is a *permutation invariant function* on $\mathcal{S}$. This is because we know, trivially from the definition of majorization, that $\mathbf{x} \succ \mathbf{Px}$ and $\mathbf{Px} \succ \mathbf{x}$. By Schur convexity we then have

$$f(\mathbf{x}) \geq f(\mathbf{Px}) \quad \text{and} \quad f(\mathbf{Px}) \geq f(\mathbf{x}),$$

   which proves that $f(\mathbf{x}) = f(\mathbf{Px})$.

2. Conversely, suppose $f(\mathbf{x})$ is permutation invariant in a permutation invariant set $\mathcal{S}$. If $f(\mathbf{x})$ is also convex in $\mathcal{S}$ then it has been shown that it is Schur-convex in $\mathcal{S}$ as well (see p. 67-68 of Marshall and Olkin [1979]).

3. Finally, assume $f(\mathbf{x})$ is permutation invariant on a permutation invariant set $\mathcal{S}$ and Schur-convex on the subset $\mathcal{D} \cap \mathcal{S}$, where $\mathcal{D}$ is the set satisfying $x_0 \geq x_1 \geq \ldots \geq x_{P-1}$. Then $f(\mathbf{x})$ is also Schur-convex on the bigger set $\mathcal{S}$ (see p. 54 of Marshall and Olkin [1979]).

21.5 Relation to matrix theory

A number of interesting results have been developed connecting matrix theory to majorization [Marshall and Olkin, 1979], [Horn and Johnson, 1985]. In this section we mention some of these. For proofs the reader should consult the above excellent references.

21.5.1 Hermitian matrices and majorization

Schur proved a crucial theorem in 1923 that relates the diagonal elements of a Hermitian matrix to its eigenvalues (see p. 218 of Marshall and Olkin [1979]). Before we review this, recall first that for any square matrix the sum of its eigenvalues is equal to the sum of diagonal elements (trace). Recall also that for a Hermitian matrix the diagonal elements a_{kk} are real, and so are the eigenvalues λ_k.

[1] In the literature permutation invariance has often been just referred to as "symmetry."

♠**Theorem 21.5.** *Diagonal elements and eigenvalues.* Let $\mathbf{A}$ be a $P \times P$ Hermitian matrix. Define the vectors

$$\mathbf{a}_{eigen} = [\lambda_0 \quad \lambda_1 \quad \ldots \quad \lambda_{P-1}]^T, \quad \mathbf{a}_{diag} = [a_{00} \quad a_{11} \quad \ldots \quad a_{P-1,P-1}]^T$$

Then

$$\mathbf{a}_{eigen} \succ \mathbf{a}_{diag}. \tag{21.48}$$

That is, for a Hermitian matrix, the vector of eigenvalues majorizes the vector of diagonal elements. ◇

For example, suppose

$$\mathbf{A} = \begin{bmatrix} 3 & 1 \\ 1 & 3 \end{bmatrix}.$$

Then the eigenvalues are $\lambda_0 = 4$ and $\lambda_1 = 2$ so that $\mathbf{a}_{eigen} = [4 \quad 2]^T$, and $\mathbf{a}_{diag} = [3 \quad 3]^T$. Clearly $\mathbf{a}_{eigen} \succ \mathbf{a}_{diag}$.

Example 21.9: Witsenhausen's observation

A nice application of the inequality (21.48) is the following: we know from Eq. (21.39) that

$$f(\mathbf{x}) = \sum_{k=0}^{P-1} \frac{1}{1+x_k}$$

is Schur-convex in $x_k \geq 0$. From Eq. (21.48), and from the very definition of Schur convexity, it then follows that

$$\sum_{k=0}^{P-1} \frac{1}{1+\lambda_k} \geq \sum_{k=0}^{P-1} \frac{1}{1+a_{kk}} \tag{21.49}$$

for any Hermitian matrix. The importance of this inequality arises in the design of MMSE transceivers (Sec. 13.5), and was first pointed out by Witsenhausen in Salz [1985].

The following theorem is due to Fan. A proof can be found on p. 241 of Marshall and Olkin [1979]. In the theorem $\lambda(\mathbf{A})$ denotes the **vector** of eigenvalues of the Hermitian matrix $\mathbf{A}$, arranged in decreasing order: $\lambda_0(\mathbf{A}) \geq \lambda_1(\mathbf{A}) \geq \ldots$

♠**Theorem 21.6.** *Sum of Hermitian matrices.* Let $\mathbf{A}_1$ and $\mathbf{A}_2$ be $P \times P$ Hermitian matrices. Then

$$\lambda(\mathbf{A}_1) + \lambda(\mathbf{A}_2) \succ \lambda(\mathbf{A}_1 + \mathbf{A}_2).$$

Thus the *sum of eigenvalues* majorizes the *eigenvalues of the sum.* ◇

21.5.1.A Example of a maximizing unitary matrix

Consider a function of the form

$$\phi(\mathbf{U}) = \sum_{k=0}^{P-1} \frac{1}{1+[\mathbf{U}^\dagger\mathbf{A}\mathbf{U}]_{kk}} \tag{21.50}$$

where $\mathbf{A}$ is a fixed $P \times P$ Hermitian positive semidefinite matrix, and $\mathbf{U}$ is a unitary matrix to be chosen such that $\phi(\mathbf{U})$ is maximized. Here the quantities

$$d_k = [\mathbf{U}^\dagger\mathbf{A}\mathbf{U}]_{kk}$$

denote the diagonal elements of $\mathbf{U}^\dagger\mathbf{A}\mathbf{U}$. First observe that

$$\sum_k d_k = \mathrm{Tr}[\mathbf{U}^\dagger\mathbf{A}\mathbf{U}] = \mathrm{Tr}[\mathbf{U}\mathbf{U}^\dagger\mathbf{A}] = \mathrm{Tr}[\mathbf{A}], \tag{21.51}$$

which is fixed, independent of $\mathbf{U}$. Moreover $d_k \geq 0$ because $\mathbf{A}$ is positive definite. If $\mathbf{U}$ is chosen to diagonalize $\mathbf{A}$ then $[\mathbf{U}^\dagger\mathbf{A}\mathbf{U}]_{kk} = \lambda_k$ (eigenvalues of $\mathbf{A}$). From Ex. 21.9 we know that

$$\sum_k \frac{1}{1+\lambda_k} \geq \sum_k \frac{1}{1+d_k}$$

Summarizing, the unitary matrix $\mathbf{U}$ that maximizes $\phi(\mathbf{U})$ is the one that diagonalizes $\mathbf{A}$. The maximized objective function is given by

$$\phi(\mathbf{U}) = \sum_k \frac{1}{1+\lambda_k}$$

21.5.1.B Example of a minimizing unitary matrix

Now imagine that our goal is to *minimize* rather than maximize $\phi(\mathbf{U})$ in Eq. (21.50) by choice of the unitary matrix $\mathbf{U}$. Then what is the best $\mathbf{U}$? Since the average value $c = \sum_k d_k/P$ is independent of $\mathbf{U}$ (by Eq. (21.51)) it follows from Lemma 21.1 that

$$\begin{bmatrix} d_0 & d_1 & \ldots & d_{P-1} \end{bmatrix}^T \succ c\begin{bmatrix} 1 & 1 & \ldots & 1 \end{bmatrix}^T$$

no matter how $\mathbf{U}$ is chosen. Since

$$f(\mathbf{x}) = \sum_{i=0}^{P-1} \frac{1}{1+x_i}$$

is Schur-convex in $\mathbf{x}$ for $x_i \geq 0$ (from Eq. (21.39)), it follows that

$$\sum_{k=0}^{P-1} \frac{1}{1+[\mathbf{U}^\dagger\mathbf{A}\mathbf{U}]_{kk}} \geq \sum_{k=0}^{P-1} \frac{1}{1+c} = \frac{P}{1+c}$$

Moreover, there is a particular choice of $\mathbf{U}$ that achives $d_k = c$ for all k, that is,

$$\mathbf{U}^\dagger \mathbf{A}\mathbf{U} = \begin{bmatrix} c & \times & \dots & \times \\ \times & c & \dots & \times \\ \vdots & \vdots & \ddots & \vdots \\ \times & \times & \dots & c \end{bmatrix}. \tag{21.52}$$

So the unitary $\mathbf{U}$ which minimizes $\phi(\mathbf{U})$ is the one that makes the diagonal elements of $\mathbf{U}^\dagger\mathbf{A}\mathbf{U}$ identical.

Proof of existence. To prove the existence of a unitary $\mathbf{U}$ which achieves Eq. (21.52), observe that if $\mathbf{V}$ is *any* unitary matrix with identical magnitudes for all its elements (i.e., $|V_{km}| = 1/\sqrt{P}$ for all (k, m)), then, for *any* diagonal matrix $\mathbf{\Lambda}$,

$$[\mathbf{V}^\dagger \mathbf{\Lambda}\mathbf{V}]_{kk} = \sum_{m=0}^{P-1} [\mathbf{V}^\dagger]_{km}\lambda_m[\mathbf{V}]_{mk} = \sum_{m=0}^{P-1} \lambda_m |V_{mk}|^2 = \frac{1}{P}\sum_m \lambda_m = c,$$

and this is independent of k. Thus, given arbitrary Hermitian $\mathbf{A}$, if we choose $\mathbf{U} = \mathbf{T}\mathbf{V}$, where $\mathbf{T}$ diagonalizes $\mathbf{A}$, and $\mathbf{V}$ is as above, then

$$[\mathbf{U}^\dagger\mathbf{A}\mathbf{U}]_{kk} = [\mathbf{V}^\dagger\mathbf{T}^\dagger\mathbf{A}\mathbf{T}\mathbf{V}]_{kk} = [\mathbf{V}^\dagger\mathbf{\Lambda}\mathbf{V}]_{kk} = c.$$

Such a $\mathbf{U}$ therefore minimizes $\phi(\mathbf{U})$. There are many examples of unitary matrices $\mathbf{V}$ which satisfy

$$|V_{km}| = \frac{1}{\sqrt{P}} \quad \text{for all } (k, m).$$

One example is the normalized DFT matrix $\mathbf{W}$, which has

$$[\mathbf{W}]_{km} = \frac{e^{-j2\pi km/P}}{\sqrt{P}}$$

Another example is the Hadamard matrix, which is a real orthogonal matrix with elements ± 1 (divided by $\sqrt{P}$ for normalization). Hadamard matrices exist for certain values of P, e.g., when P is a power of two [Moon and Stirling, 2000]. $\triangledown\triangledown\triangledown$

Here is the summary of what we have shown:

♠**Theorem 21.7.** *A minimization problem with unitary* $\mathbf{U}$. Consider the function

$$\phi(\mathbf{U}) = \sum_{k=0}^{P-1} \frac{1}{1 + [\mathbf{U}^\dagger\mathbf{A}\mathbf{U}]_{kk}} \tag{21.53}$$

where $\mathbf{A}$ is a fixed $P \times P$ Hermitian positive semidefinite matrix. Then the minimum value of $\phi(\mathbf{U})$ as $\mathbf{U}$ varies over the set of all unitary matrices is given by

$$\phi(\mathbf{U}) = \frac{P}{1 + c}$$

where $c = \text{Tr}(\mathbf{A})/P = \sum_m \lambda_m/P$, with λ_m representing the P eigenvalues of $\mathbf{A}$. Furthermore the following are true:

1. This minimum is achieved by choosing $\mathbf{U}$ such that the diagonal elements of $\mathbf{U}^\dagger\mathbf{A}\mathbf{U}$ are identical.
2. Any unitary $\mathbf{U}$ of the form $\mathbf{U} = \mathbf{T}\mathbf{V}$ achieves this, where (a) $\mathbf{T}$ is any unitary matrix that diagonalizes $\mathbf{A}$ (i.e., $\mathbf{T}^\dagger\mathbf{A}\mathbf{T}$ is diagonal), and (b) $\mathbf{V}$ is any unitary matrix with all elements having identical magnitude $1/\sqrt{P}$.
3. Examples of the above kind of $\mathbf{V}$ include the normalized DFT matrix, and the Hadamard matrix. ◊

21.5.2 Doubly stochastic matrices, and majorization

A class of matrices called doubly stochastic matrices plays a fundamental role in the theory of majorization, and therefore in the theory of Schur-convex functions. In this section we briefly review the key ideas.

♠**Definition 21.5.** *Stochastic matrices.* A $P \times P$ matrix $\mathbf{A}$ is said to be *doubly stochastic* if $a_{mn} \geq 0$ and $\sum_n a_{mn} = \sum_m a_{mn} = 1$. That is, elements of each row add up to unity, and so do the elements of each column. A $P \times P$ matrix $\mathbf{A}$ is said to be *orthostochastic* if it is constructed from a unitary matrix $\mathbf{U}$ by defining $a_{mn} = |u_{mn}|^2$. Orthostochastic matrices are also doubly stochastic (because unitary matrices have unit-norm columns and rows). ◊

The term doubly stochastic was apparently first introduced by William Feller in his classic book on probability theory [Feller, 1968]. The relation between these matrices and majorization was observed by Hardy and colleagues in 1929; see Hardy *et. al.* [1952]. Here are some simple properties of these matrices:

1. ***Eigenvectors.*** The doubly stochastic property means that $\mathbf{Ae} = \mathbf{e}$ and $\mathbf{e}^T\mathbf{A} = \mathbf{e}^T$, where $\mathbf{e} = [1 \quad 1 \quad \ldots \quad 1]^T$. This is just another way to say that the elements in each row and each column add up to unity. So $\mathbf{e}$ is an eigenvector of both $\mathbf{A}$ and $\mathbf{A}^T$, with eigenvalue equal to unity. Combined with the condition $a_{mn} \geq 0$, these are also sufficient to make $\mathbf{A}$ doubly stochastic.
2. ***Products.*** If $\mathbf{A}_1$ and $\mathbf{A}_2$ are $P \times P$ doubly stochastic matrices then so is the product $\mathbf{A}_1\mathbf{A}_2$. To see this observe first that, since the elements of $\mathbf{A}_1$ and $\mathbf{A}_2$ are non-negative, so are the elements of the product. Second,
$$\mathbf{A}_1\mathbf{A}_2\mathbf{e} = \mathbf{A}_1\mathbf{e} = \mathbf{e},$$
and similarly $\mathbf{e}^T\mathbf{A}_1\mathbf{A}_2 = \mathbf{e}^T$, so that $\mathbf{A}_1\mathbf{A}_2$ is doubly stochastic indeed.
3. ***Inverses.*** If a doubly stochastic matrix is nonsingular then its inverse clearly satisfies $\mathbf{A}^{-1}\mathbf{e} = \mathbf{e}$ and $\mathbf{e}^T\mathbf{A}^{-1} = \mathbf{e}^T$. However, it may not be doubly

stochastic (because some elements may be negative). In fact, the *inverse of a doubly stochastic matrix is doubly stochastic if and only if it is a permutation matrix* (see p. 48 of Marshall and Olkin [1979]).

Example 21.10: Stochastic matrices

The following matrices are doubly stochastic as can be verified by inspection.

$$\mathbf{A}_1 = \begin{bmatrix} \cos^2\theta & \sin^2\theta \\ \sin^2\theta & \cos^2\theta \end{bmatrix}, \quad \mathbf{A}_2 = \frac{1}{3}\begin{bmatrix} 1 & 1 & 1 \\ 1 & 1 & 1 \\ 1 & 1 & 1 \end{bmatrix}, \quad \mathbf{A}_3 = \frac{1}{2}\begin{bmatrix} 1 & 1 & 0 \\ 1 & 0 & 1 \\ 0 & 1 & 1 \end{bmatrix}.$$

The first two matrices are also orthostochastic, though the third one is not. Thus $\mathbf{A}_1$ can be generated from the unitary matrix

$$\begin{bmatrix} \cos\theta & \sin\theta \\ -\sin\theta & \cos\theta \end{bmatrix}$$

by squaring its elements, and $\mathbf{A}_2$ can be generated from the (normalized) DFT matrix whose (m, n) element is $e^{-j2\pi mn/3}/\sqrt{3}$. The third matrix is not orthostochastic because there cannot be a unitary matrix of the form

$$\begin{bmatrix} a & b & 0 \\ c & 0 & d \\ 0 & e & f \end{bmatrix},$$

where $a, b, c, d, e,$ and f are nonzero. For example, the inner product of the first two columns is a^*b, which cannot be zero, as required for unitarity. Another family of examples is obtained from circulant matrices (Appendix D), e.g.,

$$\mathbf{A} = \begin{bmatrix} c_0 & c_1 & c_2 & c_3 \\ c_3 & c_0 & c_1 & c_2 \\ c_2 & c_3 & c_0 & c_1 \\ c_1 & c_2 & c_3 & c_0 \end{bmatrix}.$$

Circulants are doubly stochastic for any set of $c_k \geq 0$ such that $\sum_k c_k = 1$.

It can be shown (see p. 527 of Horn and Johnson [1985]) that $\mathbf{A}$ is doubly stochastic if and only if it is a convex combination of permutation matrices, that is,

$$\mathbf{A} = \sum_{k=0}^{K} \alpha_k \mathbf{P}_k,$$

where $\mathbf{P}_k$ are permutation matrices, and $\alpha_k \geq 0$ with $\sum_k \alpha_k = 1$. This result is called **Birkhoff's theorem.** The connection between majorization and stochastic matrices is given by the following beautiful theorems [Marshall and Olkin, 1979]:

♠**Theorem 21.8.** *Majorization and stochastic matrices.* $\mathbf{A}$ is doubly stochastic if and only if $\mathbf{y}$ majorizes $\mathbf{Ay}$ for every real vector $\mathbf{y}$. ♢

♠**Theorem 21.9.** *Majorization, and existence of a stochastic matrix.* Given two real vectors $\mathbf{x}$ and $\mathbf{y}$, we have $\mathbf{y} \succ \mathbf{x}$ if and only if $\mathbf{x} = \mathbf{Ay}$ for some doubly stochastic matrix $\mathbf{A}$. ♢

♠**Theorem 21.10.** *Majorization and orthostochastic matrices.* Given two real vectors $\mathbf{x}$ and $\mathbf{y}$, we have $\mathbf{y} \succ \mathbf{x}$ if and only if $\mathbf{x} = \mathbf{Ay}$ for some orthostochastic $\mathbf{A}$. ♢

Example 21.11: T-transforms

A special case of doubly stochastic matrices called T-transforms (see p. 21 of Marshall and Olkin [1979]) has considerable importance in the theory of majorization. A T-transform is a matrix of the form

$$\mathbf{T} = \alpha \mathbf{I} + (1-\alpha)\mathbf{J}, \tag{21.54}$$

where $0 \leq \alpha \leq 1$, $\mathbf{I}$ is the identity, and $\mathbf{J}$ is the identity with *exactly one pair of columns interchanged.* For example, the $\mathbf{J}$ matrices

$$\mathbf{J}_1 = \begin{bmatrix} 0 & 1 & 0 \\ 1 & 0 & 0 \\ 0 & 0 & 1 \end{bmatrix} \quad \text{and} \quad \mathbf{J}_2 = \begin{bmatrix} 1 & 0 & 0 \\ 0 & 0 & 1 \\ 0 & 1 & 0 \end{bmatrix}$$

generate, respectively, the T-transforms

$$\mathbf{T}_1 = \begin{bmatrix} \alpha & 1-\alpha & 0 \\ 1-\alpha & \alpha & 0 \\ 0 & 0 & 1 \end{bmatrix} \quad \text{and} \quad \mathbf{T}_2 = \begin{bmatrix} 1 & 0 & 0 \\ 0 & \alpha & 1-\alpha \\ 0 & 1-\alpha & \alpha \end{bmatrix}.$$

From the definition and the above examples it is amply clear that a T-transform is *doubly stochastic* as well as *symmetric* ($\mathbf{T}^T = \mathbf{T}$). Note that

$$\mathbf{T}_1 \begin{bmatrix} x_0 \\ x_1 \\ x_2 \end{bmatrix} = \begin{bmatrix} \alpha x_0 + (1-\alpha)x_1 \\ \alpha x_1 + (1-\alpha)x_0 \\ x_2 \end{bmatrix} \quad \text{and} \quad \mathbf{T}_2 \begin{bmatrix} x_0 \\ x_1 \\ x_2 \end{bmatrix} = \begin{bmatrix} x_0 \\ \alpha x_1 + (1-\alpha)x_2 \\ \alpha x_2 + (1-\alpha)x_1 \end{bmatrix}.$$

Thus, a T-transform operating on a vector simply replaces a pair of components x_k and x_m with their convex combinations $\alpha x_k + (1-\alpha)x_m$ and $\alpha x_m + (1-\alpha)x_k$. All other components are unchanged.

A beautiful result which historically preceded Theorem 21.9 is the fact that $\mathbf{y} = [y_0 \quad y_1 \quad \ldots \quad y_{P-1}]^T$ majorizes $\mathbf{x} = [x_0 \quad x_1 \quad \ldots \quad x_{P-1}]^T$ if and only if we can go from $\mathbf{y}$ to $\mathbf{x}$ by using a sequence of *no more than* $(P-1)$ T-transforms [Marshall and Olkin, 1979], that is,

$$\mathbf{y} \succ \mathbf{x} \quad \text{if and only if} \quad \mathbf{x} = (\mathbf{T}_1\mathbf{T}_2 \ldots \mathbf{T}_{P-1})\mathbf{y} \tag{21.55}$$

for some sequence of T-transforms $\mathbf{T}_k$.

21.6 Multiplicative majorization

In this section we briefly introduce an idea called multiplicative majorization. This finds applications in optimization of transceivers with DFE and bit allocation [Guess, 2003], [Jiang, *et al.*, 2007], [Weng, *et al.*, 2010a].

♠**Definition 21.6.** *Multiplicative majorization.* Given two real vectors with positive elements,

$$\mathbf{x} = [\,x_0 \quad x_1 \quad \dots \quad x_{P-1}\,]^T, \quad \mathbf{y} = [\,y_0 \quad y_1 \quad \dots \quad y_{P-1}\,]^T$$

we say that $\mathbf{y}$ majorizes $\mathbf{x}$ multiplicatively if the following two conditions are satisfied. First, the product of the elements is identical:

$$\prod_{k=0}^{P-1} y_k = \prod_{k=0}^{P-1} x_k, \tag{21.56}$$

and second, any partial product of the ordered sequence $y_{[k]}$ is at least as large as the corresponding partial product of $x_{[k]}$, that is,

$$\prod_{k=0}^{n} y_{[k]} \geq \prod_{k=0}^{n} x_{[k]}, \qquad 0 \leq n \leq P-2. \tag{21.57}$$

When $\mathbf{y}$ multiplicatively majorizes $\mathbf{x}$, we denote it as

$$\mathbf{y} \succ_{\times} \mathbf{x}, \tag{21.58}$$

or equivalently $\mathbf{x} \prec_{\times} \mathbf{y}$. ◇

For clarity, the majorization introduced in Sec. 21.3 will be referred to as **additive majorization**, and denoted by the symbol $\succ_{+}$. The symbol $\succ$ without qualification always refers to additive majorization. Let $\widehat{\mathbf{x}}$ and $\widehat{\mathbf{y}}$ denote vectors whose elements are logarithms of the elements of $\mathbf{x}$ and $\mathbf{y}$. Then we see that $\mathbf{y} \succ_{\times} \mathbf{x}$ if and only if $\widehat{\mathbf{y}} \succ_{+} \widehat{\mathbf{x}}$. A number of properties pertaining to multiplicative majorization are summarized next [Marshall and Olkin, 1979], [Horn and Johnson, 1991]:

1. Let $\mathbf{y} = [\,y_0 \quad y_1 \quad \dots \quad y_{P-1}\,]^T$ where $y_k > 0$, and $\mathbf{x} = [\,\sigma \quad \sigma \quad \dots \quad \sigma\,]^T$ where $\sigma = (\prod_{k=0}^{P-1} y_k)^{1/P}$ = geometric mean of y_k. Then $\mathbf{y} \succ_{\times} \mathbf{x}$.

2. Let $\mathbf{A}$ be $P \times P$ nonsingular, and let $\{\lambda_k\}$ and $\{\sigma_k\}$ denote the eigenvalues (possibly complex) and singular values, respectively. Then

$$[\,\sigma_0^2 \quad \sigma_1^2 \quad \dots \quad \sigma_{P-1}^2\,]^T \succ_{\times} [\,|\lambda_0|^2 \quad |\lambda_1|^2 \quad \dots \quad |\lambda_{P-1}|^2\,]^T. \tag{21.59}$$

   So, the singular values multiplicatively majorize the eigenvalues in the above sense.

3. Given two sets of numbers $\{\lambda_k\}$ (possibly complex) and $\{\sigma_k\}$ (positive) satisfying the multiplicative majorization property (21.59), there exists a $P \times P$ matrix whose eigenvalues are $\{\lambda_k\}$ and singular values are $\{\sigma_k\}$.

The first property is similar to the property described in Ex. 21.2 for additive majorization. The last two properties are very similar to the relation between additive majorization and Hermitian matrices summarized in Sec. 21.5.

21.7 Summary and conclusions

Schur-convex functions and the theory of majorization have been applied increasingly in many problems in digital communications. This chapter has provided an extensive review of these mathematical tools, to enable the reader to study some of the advanced literature in this area. In the next few pages we summarize the key definitions and examples in the form of tables.

Table 21.1. Convex functions: examples and properties

Examples of convex functions

1. $|x|$, for all real x.
2. $A + Bx$ (A, B real), for all real x. *This is concave and convex.*
3. x^k (k positive integer), for $x \geq 0$.
4. x^{2k} (k positive integer), for $-\infty < x < \infty$.
5. $1/x^a$ ($a > 0$) for $x > 0$. Examples: $1/\sqrt{x}, 1/x, 1/x^2, \ldots$
6. $-\ln x$, for $x > 0$.
7. $x \ln x$, for $x > 0$.
8. $-\ln(1+x)$, $(1+x)\ln(1+x)$, $1/\sqrt{1+x}, 1/(1+x), \ldots$, for $x > -1$.
9. $e^{\alpha x}$ (α real) for all real x.
10. e^{-x^2} for $|x| \geq 1/\sqrt{2}$, and $-e^{-x^2}$ for $|x| \leq 1/\sqrt{2}$.
11. $\text{erfc}(K/\sqrt{x})$ in $0 < x \leq 2K^2/3$ and $-\text{erfc}(K/\sqrt{x})$ in $x \geq 2K^2/3$.
12. $\mathcal{Q}(A/\sqrt{x})$ in $0 < x \leq A^2/3$ and $-\mathcal{Q}(A/\sqrt{x})$ in $x \geq A^2/3$.
13. $A + Bx_0 + Cx_1$ (A, B, C real), for real x_0, x_1. *This is concave and convex.*
14. $-\ln(x_0 x_1)$, for $x_k > 0$. *Note:* The product $x_0 x_1$ is neither convex nor concave.
15. $-\prod_{k=0}^{P-1}(x_k)^{1/P}$ for $x_k > 0$ (geometric mean is *concave*; Sec. 21.2.4).
16. ℓ_p norm $\|\mathbf{x}\|_p \stackrel{\Delta}{=} (\sum_{k=0}^{P-1} |x_k|^p)^{1/p}$, $p > 1$ for all real $\mathbf{x}$.

Properties of convex functions

1. $f(\mathbf{x})$ is convex $\Leftrightarrow -f(\mathbf{x})$ is concave.
2. $f(x)$ convex $\Leftrightarrow d^2 f(x)/dx^2 \geq 0$ (assuming $f(x)$ is twice differentiable).
3. $f(\mathbf{x})$ convex $\Leftrightarrow$ Hessian $\geq \mathbf{0}$ (assuming Hessian exists; Sec. 21.2.1).
4. $f(\mathbf{x})$ convex $\Leftrightarrow f(\sum_{k=0}^{P-1} \alpha_k \mathbf{x}_k) \leq \sum_{k=0}^{P-1} \alpha_k f(\mathbf{x}_k)$ (see Eq. (21.2)).
5. $f(x)$ convex $\Rightarrow f(E[X]) \leq E[f(X)]$ (Jensen's inequality, Eq. (21.25)).
6. $f_k(\mathbf{x})$ convex on $\mathcal{A} \Rightarrow f(\mathbf{x}) \stackrel{\Delta}{=} \max_k f_k(\mathbf{x})$ is convex on $\mathcal{A}$ (Sec. 21.2.7).
7. $f_k(x)$ convex on $a_k \leq x \leq b_k \Rightarrow g(\mathbf{x}) = \sum_{k=0}^{P-1} f_k(x_k)$ convex on $a_k \leq x_k \leq b_k$ (Sec. 21.2.7).
8. Increasing convex functions of convex functions are convex (Theorem 21.1).

Note: All functions, arguments, and constants are *real-valued.*

Table 21.2. Majorization: summary

Some of the key points about majorization are summarized here; P denotes the size of the vectors.

1. $\mathbf{y}$ majorizes $\mathbf{x}$ ($\mathbf{y} \succ \mathbf{x}$) if $\sum_{k=0}^{n} y_{[k]} \geq \sum_{k=0}^{n} x_{[k]}$, for $0 \leq n \leq P-2$, and $\sum_{k=0}^{P-1} y_k = \sum_{k=0}^{P-1} x_k$ (Definition 21.3).
2. $[1 \;\; 0 \;\; \ldots \;\; 0]^T \succ [y_0 \;\; y_1 \;\; \ldots \;\; y_{P-1}]^T \succ [\frac{1}{P} \;\; \frac{1}{P} \;\; \ldots \;\; \frac{1}{P}]^T$ for $y_k \geq 0$, and $\sum_k y_k = 1$ (Lemma 21.1).
3. $\mathbf{y} \succ \mathbf{x}$ if and only if $\sum_{k=0}^{P-1} g(y_k) \geq \sum_{k=0}^{P-1} g(x_k)$ for all continuous convex functions $g(x)$ (Theorem 21.2).
4. $\mathbf{y} \succ \mathbf{x}$ if and only if $\mathbf{x} = (\mathbf{T}_1\mathbf{T}_2 \ldots \mathbf{T}_{P-1})\mathbf{y}$ for some sequence of T-transforms $\mathbf{T}_k$ (end of Sec. 21.5.2).
5. $\mathbf{y} \succ \mathbf{x}$ if and only if there exists a doubly stochastic matrix $\mathbf{A}$ such that $\mathbf{x} = \mathbf{Ay}$ (Theorem 21.9).
6. $\mathbf{y} \succ \mathbf{x}$ if and only if there exists an orthostochastic matrix $\mathbf{A}$ such that $\mathbf{x} = \mathbf{Ay}$ (Theorem 21.10).
7. $\mathbf{A}$ is doubly stochastic if and only if $\mathbf{y}$ majorizes $\mathbf{Ay}$ for every real vector $\mathbf{y}$ (Theorem 21.8).
8. For Hermitian $\mathbf{A}$, the vector of eigenvalues (λ_k) majorizes the vector of diagonal elements (a_{kk}) (Theorem 21.5).
9. For Hermitian $\mathbf{A}$, $\sum_{k=0}^{P-1} 1/(1+\lambda_k) \geq \sum_{k=0}^{P-1} 1/(1+a_{kk})$ (Ex. 21.9).
10. For Hermitian matrices $\mathbf{A}_1$ and $\mathbf{A}_2$, the *sum of eigenvalues* majorizes the *eigenvalues of the sum,* i.e., $\lambda(\mathbf{A}_1) + \lambda(\mathbf{A}_2) \succ \lambda(\mathbf{A}_1 + \mathbf{A}_2)$ (Theorem 21.6).

Note: All functions, arguments, and constants are *real-valued.*

Table 21.3. Schur-convex functions: summary

Some of the key points about Schur-convex functions are summarized here. Set $\mathcal{D}$ represents $x_0 \geq x_1 \geq \ldots \geq x_{P-1}$.

1. $f(\mathbf{x})$ is Schur-convex $\Leftrightarrow$ $\mathbf{x}_1 \succ \mathbf{x}_2$ implies $f(\mathbf{x}_1) \geq f(\mathbf{x}_2)$ (Definition 21.4).
2. $g(x)$ convex $\Rightarrow f(\mathbf{x}) = \sum_{k=0}^{P-1} g(x_k)$ Schur-convex (Theorem 21.3).
3. $f(\mathbf{x}) = \sum_{k=0}^{P-1} g_k(x_k)$ ($g_k(x)$ differentiable) is Schur-convex on $\mathcal{D}$ *if and only if* $dg_k(a)/dx \geq dg_{k+1}(b)/dx$ whenever $a \geq b$ (Theorem 21.4).
4. $f(\mathbf{x}) = \sum_{k=0}^{P-1} a_k g(x_k)$ is Schur-convex on $\mathcal{D}$ if (a) $0 \leq a_0 \leq a_1 \leq \ldots$, (b) $dg(x)/dx \leq 0$, and (c) $d^2g(x)/dx^2 \geq 0$ (Corollary 21.1).
5. *An increasing function of a Schur-convex function is Schur-convex* (Sec. 21.4.2).
6. If $f(\mathbf{x})$ is Schur convex in a *permutation invariant* set $\mathcal{S}$, then $f(\mathbf{Px}) = f(\mathbf{x})$ for any *permutation matrix* $\mathbf{P}$ (Sec. 21.4.3).

Note: All functions, arguments, and constants are *real-valued.*

Table 21.4. Schur-convex functions: examples

Set $\mathcal{D}$ is $x_0 \geq x_1 \geq \ldots \geq x_{P-1}$, set $\mathcal{D}_+$ is $x_0 \geq x_1 \geq \ldots \geq x_{P-1} \geq 0$, and set $\mathcal{D}_{++}$ is $x_0 \geq x_1 \geq \ldots \geq x_{P-1} > 0$. It is assumed throughout that $0 \leq a_0 \leq a_1 \leq \ldots \leq a_{P-1}$. Here are some examples of Schur-convex functions.

1. $\exp(\sum_{k=0}^{P-1} 1/x_k)$ for $x_k > 0$.
2. $\sum_{k=0}^{P-1} e^{-x_k^2}$ in $|x_k| \geq 1/\sqrt{2}$ and $-\sum_{k=0}^{P-1} e^{-x_k^2}$ in $|x_k| \leq 1/\sqrt{2}$.
3. $\sum_{k=0}^{P-1} \text{erfc}(1/\sqrt{x_k})$ in $0 < x_k \leq 2/3$.
4. $\max_k\{x_k\}$ for any real $\mathbf{x}$.
5. $\sum_{k=0}^{P-1} a_k/x_k^p$ $(p > 0)$ in $\mathcal{D}_{++}$
6. Examples of above: $\sum_{k=0}^{P-1} a_k/\sqrt{x_k}$, $\sum_{k=0}^{P-1} a_k/x_k$, $\sum_{k=0}^{P-1} a_k/x_k^2$ in $\mathcal{D}_+$
7. $\sum_{k=0}^{P-1} a_k/(1+x_k)$ in $\mathcal{D}_+$
8. $\sum_{k=0}^{P-1} a_k e^{-\alpha x_k}$ $(\alpha > 0)$ in $\mathcal{D}_+$
9. $-\sum_{k=0}^{P-1} a_k \ln x_k$ in $\mathcal{D}_{++}$
10. $-\prod_{k=0}^{P-1} x_k^{a_k}$ in $\mathcal{D}_{++}$
11. $\sum_{k=0}^{P-1} a_k/(1+x_k)^2$ is Schur convex in $x_0 \geq x_1 \geq \ldots \geq x_{P-1} \geq 1/\sqrt{3}$.
12. $-\sum_{k=0}^{P-1} a_k x_k$ in $\mathcal{D}$.
13. $\sum_{k=0}^{P-1} a_k(1-x_k)/x_k$ in $\mathcal{D}_{++}$
14. $\sum_{k=0}^{P-1} a_k \ln[(1-x_k)/x_k]$ in $\mathcal{D}_{++}$ if $x_k \leq 0.5$.
15. $\prod_{k=0}^{P-1} [(1-x_k)/x_k]^{a_k}$ in $\mathcal{D}_{++}$ if $x_k \leq 0.5$.

Note: All functions, arguments, and constants are *real-valued.*

22

Optimization with equality and inequality constraints

22.1 Introduction

If we have an objective function to be optimized under an equality constraint, we use the Lagrange multiplier method and set up a Lagrangian which is then differentiated and set to zero to obtain a set of necessary conditions for an extremum. If there are inequality constraints as well, then a further modification is necessary. The resulting necessary conditions for optimality are called the Karush–Kuhn–Tucker (KKT) conditions. Essentially we add another term to the Lagrangian involving what is called a *KKT multiplier*. We explain the idea briefly here. There are several references dedicated to a detailed discussion of this topic; for example, see Luenberger [1969], Chong and Żak [2001], Boyd and Vandenberghe [2004], or Antoniou and Lu [2007].

22.2 Setting up the problem

The optimization problem to be addressed here is as follows:

$$\text{minimize} \quad f(\mathbf{x}) \triangleq f(x_0, x_1, \ldots, x_{N-1}), \tag{22.1}$$

subject to M equality constraints,

$$\mathbf{h}(\mathbf{x}) = \mathbf{0}, \tag{22.2}$$

and P inequality constraints,

$$\mathbf{g}(\mathbf{x}) \le \mathbf{0}, \tag{22.3}$$

where the notation means that each component satisfies $g_k(\mathbf{x}) \le 0$. So the constraints can be written more explicitly as follows:

$$h_k(\mathbf{x}) = 0, \quad 0 \le k \le M-1 \quad \text{and} \quad g_k(\mathbf{x}) \le 0, \quad 0 \le k \le P-1. \tag{22.4}$$

We assume that $\mathbf{x}$ is real. The functions $f(\mathbf{x})$, $\mathbf{h}(\mathbf{x})$, and $\mathbf{g}(\mathbf{x})$ are obviously real too.

22.2.1 Definitions and conventions

We use the convention that the constraint vectors are column vectors:

$$\mathbf{h}(\mathbf{x}) = \begin{bmatrix} h_0(\mathbf{x}) \\ h_1(\mathbf{x}) \\ \vdots \\ h_{M-1}(\mathbf{x}) \end{bmatrix}, \quad \mathbf{g}(\mathbf{x}) = \begin{bmatrix} g_0(\mathbf{x}) \\ g_1(\mathbf{x}) \\ \vdots \\ g_{P-1}(\mathbf{x}) \end{bmatrix}. \tag{22.5}$$

For the scalar function $f(\mathbf{x})$ we use the gradient notation

$$\nabla f(\mathbf{x}) = \begin{bmatrix} \frac{\partial f(\mathbf{x})}{\partial x_0} & \frac{\partial f(\mathbf{x})}{\partial x_1} & \cdots & \frac{\partial f(\mathbf{x})}{\partial x_{N-1}} \end{bmatrix}. \tag{22.6}$$

For the constraint vectors the gradients are therefore matrices:

$$\nabla \mathbf{h}(\mathbf{x}) = \begin{bmatrix} \frac{\partial h_0(\mathbf{x})}{\partial x_0} & \frac{\partial h_0(\mathbf{x})}{\partial x_1} & \cdots & \frac{\partial h_0(\mathbf{x})}{\partial x_{N-1}} \\ \frac{\partial h_1(\mathbf{x})}{\partial x_0} & \frac{\partial h_1(\mathbf{x})}{\partial x_1} & \cdots & \frac{\partial h_1(\mathbf{x})}{\partial x_{N-1}} \\ \vdots & \vdots & \ddots & \vdots \\ \frac{\partial h_{M-1}(\mathbf{x})}{\partial x_0} & \frac{\partial h_{M-1}(\mathbf{x})}{\partial x_1} & \cdots & \frac{\partial h_{M-1}(\mathbf{x})}{\partial x_{N-1}} \end{bmatrix},$$

$$\nabla \mathbf{g}(\mathbf{x}) = \begin{bmatrix} \frac{\partial g_0(\mathbf{x})}{\partial x_0} & \frac{\partial g_0(\mathbf{x})}{\partial x_1} & \cdots & \frac{\partial g_0(\mathbf{x})}{\partial x_{N-1}} \\ \frac{\partial g_1(\mathbf{x})}{\partial x_0} & \frac{\partial g_1(\mathbf{x})}{\partial x_1} & \cdots & \frac{\partial g_1(\mathbf{x})}{\partial x_{N-1}} \\ \vdots & \vdots & \ddots & \vdots \\ \frac{\partial g_{P-1}(\mathbf{x})}{\partial x_0} & \frac{\partial g_{P-1}(\mathbf{x})}{\partial x_1} & \cdots & \frac{\partial g_{P-1}(\mathbf{x})}{\partial x_{N-1}} \end{bmatrix}. \tag{22.7}$$

As there are M equality constraints, we define M **Lagrange multipliers** λ_k. For the P inequality constraints we define P multipliers μ_k called the **KKT multipliers**. Thus

$$\mathcal{L} = \begin{bmatrix} \lambda_0 \\ \lambda_1 \\ \vdots \\ \lambda_{M-1} \end{bmatrix}, \quad \mathcal{M} = \begin{bmatrix} \mu_0 \\ \mu_1 \\ \vdots \\ \mu_{P-1} \end{bmatrix}. \tag{22.8}$$

Given a point $\widehat{\mathbf{x}}$ which satisfies the constraints

$$\mathbf{h}(\widehat{\mathbf{x}}) = \mathbf{0}, \quad \mathbf{g}(\widehat{\mathbf{x}}) \le \mathbf{0},$$

it is possible that some of the inequality constraints are satisfied with equality:

$$g_{k_0}(\widehat{\mathbf{x}}) = 0,\ g_{k_1}(\widehat{\mathbf{x}}) = 0,\ \ldots\ , g_{k_{L-1}}(\widehat{\mathbf{x}}) = 0, \tag{22.9}$$

and the remaining $P-L$ constraints are satisfied as strict inequalities. We then say that the above L constraints are **active constraints** at $\widehat{\mathbf{x}}$, and remaining $P-L$ constraints are **inactive**. Now consider the $M+L$ gradient vectors

$$\nabla h_i(\widehat{\mathbf{x}}),\ 0 \le i \le M-1, \quad \nabla g_{k_i}(\widehat{\mathbf{x}}),\ 0 \le i \le L-1, \tag{22.10}$$

defined from the M equality constraints and the L active inequality constraints. If these $M+L$ vectors are linearly independent then we say that $\widehat{\mathbf{x}}$ is a **regular point** for the optimization problem under consideration.

22.2.2 Karush–Kuhn–Tucker (KKT) conditions

For the optimization problem (22.1)–(22.3), let $\widehat{\mathbf{x}}$ be a regular point and a local minimum of $f(\mathbf{x})$. Then there exists a real Lagrange vector $\mathcal{L}$ and a real KKT vector $\mathcal{M}$ such that the following three conditions are satisfied:

1. ***Non-negativity***. $\mathcal{M} \ge \mathbf{0}$;
2. ***Stationarity***. $\nabla f(\widehat{\mathbf{x}}) + \mathcal{L}^T \nabla \mathbf{h}(\widehat{\mathbf{x}}) + \mathcal{M}^T \nabla \mathbf{g}(\widehat{\mathbf{x}}) = \mathbf{0}$;
3. ***Orthogonality***. $\mathcal{M}^T \mathbf{g}(\widehat{\mathbf{x}}) = 0$.

Thus, the above conditions are *necessary* for a local minimum. Some discussions on sufficiency are presented in Sec. 22.5. Written out explicitly in terms of components, the KKT conditions take the form

$$\mu_\ell \ge 0, \quad 0 \le \ell \le P-1 \quad \text{(non-negativity)}; \tag{22.11}$$

$$\frac{\partial f(\widehat{\mathbf{x}})}{\partial x_k} + \sum_{\ell=0}^{M-1} \lambda_\ell \frac{\partial h_\ell(\widehat{\mathbf{x}})}{\partial x_k} + \sum_{\ell=0}^{P-1} \mu_\ell \frac{\partial g_\ell(\widehat{\mathbf{x}})}{\partial x_k} = 0, \quad 0 \le k \le N-1 \quad \text{(stationarity)}; \tag{22.12}$$

and

$$\sum_{\ell=0}^{P-1} \mu_\ell g_\ell(\widehat{\mathbf{x}}) = 0 \quad \text{(orthogonality)}. \tag{22.13}$$

The first condition says that the KKT multipliers are non-negative. The second condition is similar to the idea in introductory calculus that the derivative be zero for a local minimum. The third condition says that the KKT multiplier vector is orthogonal to the inequality constraint vector $\mathbf{g}$ evaluated at $\widehat{\mathbf{x}}$. Note that if the optimization problem does not have inequality constraints then the first and third equations do not arise. Only the stationarity condition remains:

$$\frac{\partial f(\widehat{\mathbf{x}})}{\partial x_k} + \sum_{\ell=0}^{M-1} \lambda_\ell \frac{\partial h_\ell(\widehat{\mathbf{x}})}{\partial x_k} = 0, \quad 0 \le k \le N-1 \quad \text{(stationarity)}. \tag{22.14}$$

This is nothing but the traditional Lagrangian stationarity.

22.2.3 Non-negative variables

In many optimization problems that arise in signal processing, the variables x_k are required to be non-negative. For example, x_k can represent a resistor in a passive circuit, or it can represent the power allocated to the kth user in a multiuser communication system. The functions $g_k(\mathbf{x})$ can be appropriately defined to incorporate such constraints. For example, if we define

$$g_k(\mathbf{x}) = -x_k, \quad 0 \leq k \leq N-1,$$

then the constraints $\mathbf{g}(\mathbf{x}) \leq \mathbf{0}$ are equivalent to $x_k \geq 0$ for all k. Since

$$\frac{\partial g_\ell(\mathbf{x})}{\partial x_k} = \begin{cases} -1 & k = \ell \\ 0 & k \neq \ell, \end{cases}$$

the stationarity condition now takes the form

$$\frac{\partial f(\widehat{\mathbf{x}})}{\partial x_k} + \sum_{\ell=0}^{M-1} \lambda_\ell \frac{\partial h_\ell(\widehat{\mathbf{x}})}{\partial x_k} - \mu_k = 0, \quad 0 \leq k \leq N-1 \quad \text{(stationarity)}. \tag{22.15}$$

The orthogonality condition (22.13) is now equivalent to

$$\sum_{\ell=0}^{N-1} \mu_\ell \widehat{x}_\ell = 0 \quad \text{(orthogonality)}. \tag{22.16}$$

That is, the KKT vector $\mathcal{M}$ is orthogonal to the optimal vector $\widehat{\mathbf{x}}$. Since μ_ℓ and $\widehat{x}_\ell$ are both non-negative at the optimum point, this implies in particular that

$$\mu_\ell \widehat{x}_\ell = 0 \quad \text{for all } \ell. \tag{22.17}$$

That is, for each ℓ, either $\widehat{x}_\ell = 0$ or $\mu_\ell = 0$. If $\widehat{x}_\ell = 0$ then the ℓth constraint is active, whereas if $\widehat{x}_\ell \neq 0$ then the ℓth constraint is inactive, in which case we must have $\mu_\ell = 0$. The significance of this is that if there are N_1 nonzero components

$$\widehat{x}_{k_m}, \quad 0 \leq m \leq N_1 - 1,$$

then there are N_1 equations of the form

$$\frac{\partial f(\widehat{\mathbf{x}})}{\partial x_{k_m}} + \sum_{\ell=0}^{M-1} \lambda_\ell \frac{\partial h_\ell(\widehat{\mathbf{x}})}{\partial x_{k_m}} = 0, \quad 0 \leq m \leq N_1 - 1 \quad \text{(stationarity)}. \tag{22.18}$$

This is the conventional stationarity condition with Lagrange multipliers. The N_1 nonzero variables x_{k_m} can be obtained by solving these N_1 equations.

A further special case of the above situation with non-negative variables x_k arises in many examples. In this special case, each equation of the form (22.18) gives a unique answer for the corresponding variable x_{k_m}. That is, the equations get decoupled. If the answer x_{k_m} turns out to be negative for some m, we immediately know that the assumption which leads to Eq. (22.18) is not correct. The only assumption required to write down Eq. (22.18) is $\mu_{k_m} = 0$. If this is not true then $x_{k_m} = 0$ (as seen from Eq. (22.17)). This gives us the following procedure to solve for the optimal $\mathbf{x}$: first solve for each x_k using the simple Lagrangian stationary condition

$$\frac{\partial f(\widehat{\mathbf{x}})}{\partial x_k} + \sum_{\ell=0}^{M-1} \lambda_\ell \frac{\partial h_\ell(\widehat{\mathbf{x}})}{\partial x_k} = 0, \quad 0 \leq k \leq N-1, \tag{22.19}$$

with all the μ_k ignored. This yields the correct solution for x_k if $x_k \geq 0$. For some k, if we find $x_k < 0$ we simply drop that answer and take $x_k = 0$. The resulting $\widehat{\mathbf{x}}$ then satisfies the KKT conditions for optimality.

Example 22.1: Optimum resistor for power transfer

In the electrical circuit shown in Fig. 22.1 the power in the load resistor of resistance R_l is

$$\mathcal{P} = i^2 R_l = \left(\frac{v}{R_s + R_l}\right)^2 R_l. \tag{22.20}$$

For fixed voltage v and source resistance R_s, what is the value of R_l that maximizes the power $\mathcal{P}$? If we differentiate the preceding expression with respect to R_l and set it to zero we get the solution

$$R_l = R_s. \tag{22.21}$$

So the *load should be matched to the source resistor* in order for the source to transfer maximum power. In obtaining this well known solution we ignored the constraint $R_l \geq 0$, but it did not matter because the answer turned out to be positive anyway. But for the sake of illustration of the KKT theory, let us formally incorporate the constraint and see how the KKT conditions yield the same solution. We can define the objective function $f(R_l) = -\mathcal{P}$ and minimize it. The only constraint in this problem is the inequality constraint $R_l \geq 0$. So we define the constraint function $g(R_l) = -R_l$. There is one KKT multiplier μ and no Lagrange multipliers (because there are no equality constraints). The three KKT conditions for this optimization are

$$\mu \geq 0, \tag{22.22}$$

$$\frac{\partial f(R_l)}{\partial R_l} + \mu \frac{\partial g(R_l)}{\partial R_l} = 0, \tag{22.23}$$

and

$$\mu R_l = 0, \tag{22.24}$$

respectively. Substituting the values of $f(R_l)$ and $g(R_l)$ this yields

$$\mu = -\left(\frac{v}{R_s + R_l}\right)^2 + \frac{2v^2 R_l}{(R_s + R_l)^3}. \tag{22.25}$$

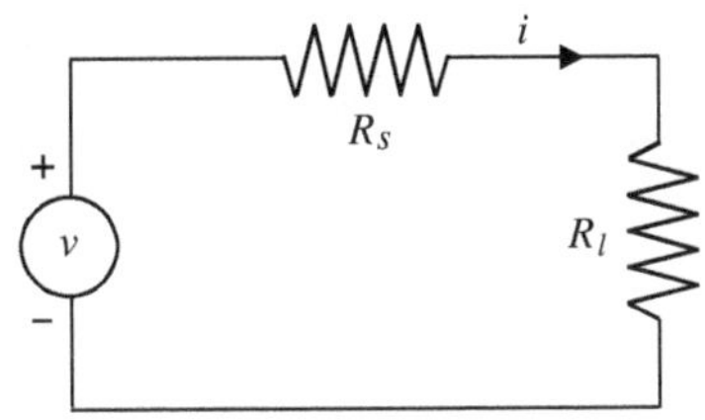

Figure 22.1. An electrical circuit with a voltage source v, source resistance R_s, and load resistance R_l.

If $\mu \neq 0$ we have to have $R_l = 0$ (because of Eq. (22.24)), which means $\mu < 0$ from the preceding equation. Since this contradicts the first KKT condition (22.22), we cannot have $\mu \neq 0$. Setting $\mu = 0$ in Eq. (22.25) yields the solution (22.21) indeed. Thus, from a formal view point, the KKT condition in this example yields $\mu = 0$, which is equivalent to ignoring the constraint $R_l \geq 0$.

Maximixing power delivered to R_s**.** An interesting modification of the problem is this: what is the optimal load resistance R_l such that the power delivered to the fixed *source resistance* R_s (assumed nonzero) is maximized? The power in R_s is given by

$$\mathcal{P}_s = \left(\frac{v}{R_s + R_l}\right)^2 R_s.$$

Proceeding as before we now find that the stationary equation of the *KKT* conditions yields

$$\frac{2v^2 R_s}{(R_s + R_l)^3} - \mu = 0.$$

This shows that μ cannot be zero (since $R_s \neq 0$). Since $\mu R_l = 0$ (third KKT condition), it follows that R_l has to be zero. Thus, maximum power is delivered to R_s when the load R_l is zero, as one would expect! We say that *the solution is at the boundary of the constraint set,* because R_l cannot be any smaller. $\bigtriangledown \bigtriangledown \bigtriangledown$

22.3 Maximizing channel capacity

In many communication systems we are faced with maximizing the quantity

$$\mathcal{C} = 0.5 \sum_{i=0}^{N-1} \log_2(1 + \frac{x_k}{Q_k}).$$

This quantity is proportional to the capacity of a parallel set of N independent channels [Cover and Thomas, 1991] with input powers x_k and zero-mean additive Gaussian noise components with powers Q_k, as shown in Fig. 22.2.

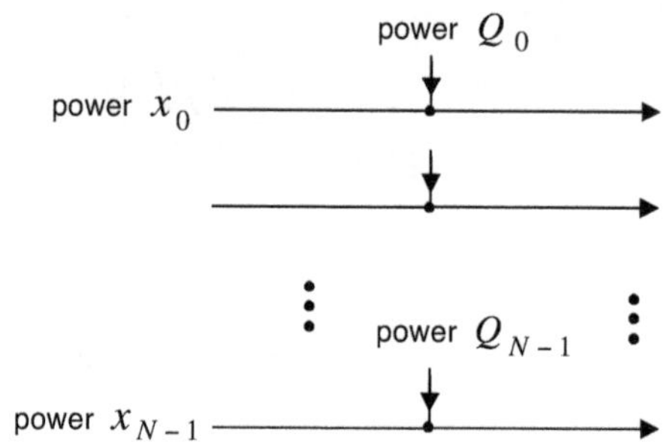

Figure 22.2. A parallel set of N independent channels, with input powers x_k and noise powers Q_k.

With total power fixed by the equality constraint

$$\sum_{i=0}^{N-1} x_i = p_0 \quad \text{(power constraint)},$$

what is the distribution of powers $\{x_k\}$ that maximizes the capacity $\mathcal{C}$? Since the power can only be non-negative, we also have to take into account the N inequality constraints

$$-x_k \leq 0, \quad 0 \leq i \leq N-1.$$

So, in terms of KKT notations, we have $g_k(\mathbf{x}) = -x_k$ in this problem. This problem has one Lagrange multiplier λ for the equality constraint and N separate KKT multipliers μ_k for the N inequality constraints. The first and the third KKT conditions are of course

$$\mu_k \geq 0, \quad 0 \leq k \leq N-1,$$

and $\sum \mu_k g_k(\mathbf{x}) = 0$, that is,

$$\sum_{k=0}^{N-1} \mu_k x_k = 0$$

because $g_k(x) = -x_k$. Since the objective function to be minimized is $f(\mathbf{x}) = -\mathcal{C}$, the second KKT condition is

$$-\frac{\partial \mathcal{C}}{\partial x_k} + \lambda \frac{\partial(\sum_{i=0}^{N-1} x_i - p_0)}{\partial x_k} - \mu_k = 0,$$

which yields

$$\frac{-0.5}{x_k + Q_k} + \lambda - \mu_k = 0.$$

For any k, we know that x_k and μ_k cannot both be nonzero. If $x_k \neq 0$ then $\mu_k = 0$, and we get from the preceding equation

$$x_k = \eta - Q_k,$$

where $\eta = 0.5/\lambda$. So we can write the capacity-maximizing power allocation in the form

$$x_k = \begin{cases} \eta - Q_k & \text{if this is } \geq 0 \\ 0 & \text{otherwise.} \end{cases} \tag{22.26}$$

Water-pouring interpretation. Equation (22.26) is called the *water-pouring solution* or water-filling solution. Whenever the noise Q_k is large we pour less power x_k into that channel, and vice versa, as demonstrated in Fig. 22.3. It is like a water tank whose bottom level is not a constant but varies according to the noise power Q_k. The power x_k allocated to each subchannel is such that the water height $x_k + Q_k$ is constant at the value η. At some point, if the bottom of the tank is too high (large noise) then no power goes into that channel at all. If the available power p_0 is large enough then all channels get some power. If available power is too small then the best channel gets all the power. $\bigtriangledown \bigtriangledown \bigtriangledown$

The reader might wonder how the water-pouring solution Eq. (22.26) can be implemented in practice without knowing η first. We now describe a simple procedure for implementing Eq. (22.26) which automatically determines η uniquely. For convenience first renumber the channels so that the noise variances satisfy

$$Q_0 \leq Q_1 \leq \ldots \leq Q_{N-1}.$$

We start increasing the power x_0 in the 0th channel until $x_0 + Q_0 = Q_1$. At this point we start increasing the powers x_0 and x_1 *equally*, so that x_0+Q_0 and x_1+Q_1 remain equal. We do this until we reach the point $x_0 + Q_0 = x_1 + Q_1 = Q_2$. We then start adding power to the first three channels, that is, increment x_0, x_1, and x_2 in equal amounts until their common value reaches Q_3. We proceed in this way until $\sum_k x_k = p_0$, the total power. At this point the sum $x_k + Q_k$ is identical for all channels which received nonzero power (and this sum is equal to η). Thus, a unique solution $\{x_k\}$ satisfying all the constraints and the KKT conditions (which are necessary conditions for optimality) has been found. Even though the KKT conditions are not sufficient for optimality, the fact that the above solution represents a *global maximum* is justified in Sec. 22.5.

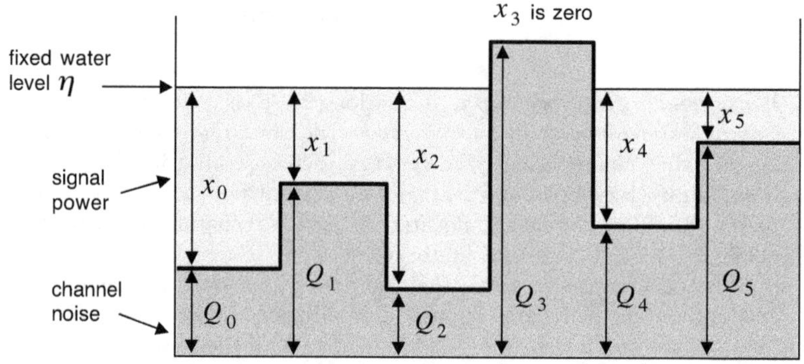

Figure 22.3. Optimizing capacity by power allocation. This is called the water-pouring solution.

Expression for optimized capacity. The water-filling method assigns power according to Eq. (22.26). Let K denote the number of channels which have been allocated nonzero power (K depends on the total available power p_0). Assume that the channel ordering is such that these are the first K channels. Thus the capacity can be obtained as follows:

$$\begin{aligned} 2C_{water} &= \sum_{k=0}^{K-1} \log_2(1+\frac{x_k}{Q_k}) \\ &= \sum_{k=0}^{K-1} \log_2(\frac{Q_k+x_k}{Q_k}) \\ &= \sum_{k=0}^{K-1} \log_2 \frac{\eta}{Q_k} = K\log_2\eta - \sum_{k=0}^{K-1}\log_2 Q_k. \end{aligned}$$

To proceed further we have to know the value of the constant η. For this observe from Eq. (22.26) that $\sum_{k=0}^{K-1} x_k = K\eta - \sum_{k=0}^{K-1} Q_k$. Using $p_0 = \sum_k x_k$ it then follows that

$$\eta = \frac{p_0 + \sum_{k=0}^{K-1} Q_k}{K} \tag{22.27}$$

Thus

$$C_{water} = 0.5\log_2\left(\frac{p_0+\sum_{k=0}^{K-1} Q_k}{K}\right)^K - 0.5\sum_{k=0}^{K-1}\log_2 Q_k. \tag{22.28}$$

We shall have occasion to use this expression in Sec. 22.5.1. $\bigtriangledown \bigtriangledown \bigtriangledown$

22.4 MMSE transceiver

We now consider a slightly different aspect of the parallel independent channel problem. Consider Fig. 22.4, where N independent symbol streams $s_k(n)$ are transmitted through N independent channels which are described by additive noise sources $q_k(n)$. The transmitters have the premultipliers α_k and the receivers have the postmultipliers β_k. Assume further that all processes are stationary so the arguments (n) can be dropped from the discussion. We assume all signals and noise sources to have zero mean, and

$$\sigma_s^2 = E[|s_k|^2], \quad \sigma_{q_k}^2 = E[|q_k|^2], \quad E[q_k s_m^*] = 0. \tag{22.29}$$

Let us consider the problem of optimally choosing the multipliers α_k and β_k such that the reconstruction error is minimized in the mean square sense, that is,

$$\mathcal{E} = \sum_{k=0}^{N-1} E\Big[|\widehat{s}_k - s_k|^2\Big] \tag{22.30}$$

is minimized. This is the *MMSE transceiver design problem* for diagonal channels (i.e., channels whose transfer matrices $\mathbf{H}$ are diagonal). The channel input power is the sum of $|\alpha_k|^2\sigma_s^2$. Assume this is constrained as follows:

$$\sigma_s^2 \sum_{k=0}^{N-1} |\alpha_k|^2 = p_0. \tag{22.31}$$

We will adopt the convention that the channels are numbered such that

$$\sigma_{q_0}^2 \leq \sigma_{q_1}^2 \leq \ldots \leq \sigma_{q_{N-1}}^2. \tag{22.32}$$

The reconstruction error in the kth channel is given by

$$e_k = \widehat{s}_k - s_k = \beta_k\alpha_k s_k + \beta_k q_k - s_k = (\beta_k\alpha_k - 1)s_k + \beta_k q_k. \tag{22.33}$$

For a fixed set of multipliers α_k, the best set of receiver multipliers β_k can be found by using the standard Wiener filter theory, which uses the orthogonality principle (Sec. F.2.1 in Appendix F at the end of the book). According to this principle β_k should be such that the error e_k is orthogonal to the observation

$$y_k \stackrel{\Delta}{=} \alpha_k s_k + q_k, \tag{22.34}$$

that is,

$$E[e_k y_k^*] = 0. \tag{22.35}$$

Substituting from the preceding equations we therefore get

$$\alpha_k^*(\beta_k\alpha_k - 1)\sigma_s^2 + \beta_k\sigma_{q_k}^2 = 0$$

from which the optimal value of the receiver multiplier (Wiener multiplier) is found to be

$$\beta_k = \frac{\alpha_k^*}{|\alpha_k|^2 + \frac{\sigma_{q_k}^2}{\sigma_s^2}} \tag{22.36}$$

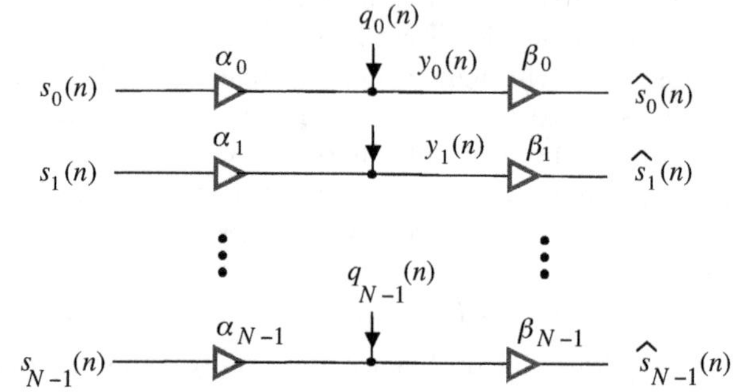

Figure 22.4. A parallel connection of N independent channels.

Observe that if the noise-to-signal ratio $\sigma_{q_k}^2/\sigma_s^2$ is small then this reduces to $\beta_k \approx 1/\alpha_k$ as expected. The preceding expression holds for any fixed set of transmitting multipliers α_k. With this choice the mean square error depends only on α_k, which can be optimized to minimize the MSE further. For this we first show that the expression for the MSE with the above choice of β_k is

$$\mathcal{E}_{mse} = \sum_{k=0}^{N-1} \frac{\sigma_{q_k}^2}{|\alpha_k|^2 + \frac{\sigma_{q_k}^2}{\sigma_s^2}} \tag{22.37}$$

Proof. The mean square error in the kth channel with optimal β_k is

$$\begin{aligned}
E[e_k^* e_k] &= E[e_k^*(\widehat{s}_k - s_k)] \\
&= E[e_k^*(y_k\beta_k - s_k)] \\
&= -E[e_k^* s_k] \quad \text{(from Eq. (22.35))} \\
&= -E\Big[\Big((\beta_k\alpha_k - 1)s_k + \beta_k q_k\Big)^* s_k\Big] \quad \text{(from Eq. (22.33))} \\
&= \sigma_s^2(1 - \beta_k\alpha_k)^* \quad \text{(from Eq. (22.29))}
\end{aligned}$$

Substituting the optimal value of β_k from Eq. (22.36) and adding for all k, Eq. (22.37) follows immediately. $\bigtriangledown\bigtriangledown\bigtriangledown$

Now consider the problem of optimizing the distribution of multipliers $\{\alpha_k\}$ such that Eq. (22.37) is minimized. For simplicity we write Eq. (22.37) in the form

$$\mathcal{E}_{mse} = \sum_{k=0}^{N-1} \frac{A_k}{x_k + B_k} \tag{22.38}$$

where $x_k = |\alpha_k|^2$. The problem therefore is to minimize this quantity under the equality constraint

$$\sum_{k=0}^{N-1} x_k = c \tag{22.39}$$

and the inequality constraints

$$x_k \geq 0. \tag{22.40}$$

where A_k, B_k, and c are positive constants. We can then solve for α_k as the non-negative square root of x_k (as the sign or phase of α_k contributes nothing). We can apply the three KKT conditions for this problem again. The stationary condition is again of the form

$$\frac{\partial \mathcal{E}_{mse}}{\partial x_k} + \lambda \frac{\partial (\sum_{\ell=0}^{N-1} x_\ell - c)}{\partial x_k} - \mu_k = 0, \quad 0 \leq k \leq N-1,$$

that is,

$$-\frac{A_k}{(x_k + B_k)^2} + \lambda - \mu_k = 0, \quad 0 \leq k \leq N-1. \tag{22.41}$$

The first and third KKT conditions are again $\mu_k \geq 0$ and $\sum_k \mu_k x_k = 0$, which implies that either μ_k or x_k is zero for each k. The nonzero x_k can therefore be calculated from Eq. (22.41) by setting $\mu_k = 0$:

$$x_k = \sqrt{\frac{A_k}{\lambda}} - B_k, \tag{22.42}$$

that is,

$$|\alpha_k|^2 = D\sigma_{q_k} - \frac{\sigma_{q_k}^2}{\sigma_s^2}, \tag{22.43}$$

where the constant D should be adjusted so that the power constraint (22.31) is satisfied. Since α_k can be taken as the positive real square root of the above expression, we can drop the absolute value sign and write

$$\alpha_k^2 = \begin{cases} D\sigma_{q_k} - \frac{\sigma_{q_k}^2}{\sigma_s^2} & \text{if this is non-negative} \\ 0 & \text{otherwise.} \end{cases} \tag{22.44}$$

Since α_k is real, the optimum receiver multipliers β_k in Eq. (22.36) take the form

$$\beta_k = \frac{\alpha_k}{\alpha_k^2 + \frac{\sigma_{q_k}^2}{\sigma_s^2}} \tag{22.45}$$

Recall that the channels are numbered such that

$$\sigma_{q_0}^2 \leq \sigma_{q_1}^2 \leq \ldots \leq \sigma_{q_{N-1}}^2. \tag{22.46}$$

Now,

$$D\sigma_{q_k} - \frac{\sigma_{q_k}^2}{\sigma_s^2} = \sigma_{q_k}\Big(D - \frac{\sigma_{q_k}}{\sigma_s^2}\Big),$$

and it is clear that, if D is not large enough, then

$$D - \frac{\sigma_{q_k}}{\sigma_s^2}$$

may be negative for all integers $k \geq K$ for some K. So, in view of the ordering convention (22.46), we can assume in general that the first K values of α_k are nonzero, for some $K \leq N$.

22.4.1 Final form of the solution

From Eq. (22.44) we have

$$\sum_{k=0}^{K-1} \alpha_k^2 = D \sum_{k=0}^{K-1} \sigma_{q_k} - \frac{1}{\sigma_s^2} \sum_{k=0}^{K-1} \sigma_{q_k}^2$$

where K is the number of nonzero values of α_k. Substituting the preceding equation into the power constraint (22.31) we obtain

$$D = \frac{\dfrac{p_0}{\sigma_s^2} + \dfrac{1}{\sigma_s^2} \displaystyle\sum_{\ell=0}^{K-1} \sigma_{q_\ell}^2}{\displaystyle\sum_{\ell=0}^{K-1} \sigma_{q_\ell}} \tag{22.47}$$

Thus

$$\alpha_k^2 = \begin{cases} \left(\dfrac{p_0 + \sum_{\ell=0}^{K-1} \sigma_{q_\ell}^2}{\sum_{\ell=0}^{K-1} \sigma_{q_\ell}} - \sigma_{q_k} \right) \dfrac{\sigma_{q_k}}{\sigma_s^2} & 0 \leq k \leq K-1 \\ 0 & \text{otherwise.} \end{cases} \tag{22.48}$$

The expression for the minimized MSE can be obtained by substituting this into Eq. (22.37):

$$\mathcal{E}_{mmse} = \frac{\sigma_s^2 \left(\sum_{\ell=0}^{K-1} \sigma_{q_\ell} \right)^2}{p_0 + \sum_{\ell=0}^{K-1} \sigma_{q_\ell}^2} + (N-K)\sigma_s^2. \tag{22.49}$$

The second term $(N-K)\sigma_s^2$ arises because whenever $\alpha_k = 0$, the kth term in Eq. (22.37) reduces to σ_s^2.

22.4.2 Choice of K

We have to figure out the value of the integer K in Eq. (22.48). For sufficiently large power p_0 we can assume $K = N$, but in general this is not so. To appreciate this, recall that the ordering convention in Eq. (22.32) implies that the quantity

$$\left(\frac{p_0 + \sum_{\ell=0}^{K-1} \sigma_{q_\ell}^2}{\sum_{\ell=0}^{K-1} \sigma_{q_\ell}} - \sigma_{q_k} \right)$$

is a decreasing function of k. Suppose we choose $K = N$. Then α_{N-1}^2 computed from Eq. (22.48) might turn out to be negative (for a given power p_0). If this is the case we take $K = N-1$ and try it again. We proceed like this by successively decreasing K, until we reach a K for which all the computed values for α_k^2 from Eq. (22.48) are non-negative. This is the right value of K to use.

22.4.2.A Optimality of K

How do we know that the integer K chosen as above is optimal? That is, how can we be sure that there is no smaller integer $L < K$ that will yield a smaller value of the error $\mathcal{E}_{mse}$? An intuitive explanation is as follows: suppose we have found a K using the described procedure. If we use $L < K$ channels, that would be like imposing an additional constraint that only L channels be used. So using $L < K$ cannot be better. For the reader who prefers a formal argument, we can proceed as follows. Let $A_k = \sigma_{q_k}^2, B_k = \sigma_{q_k}^2/\sigma_s^2$, as before, and let $x_k = \alpha_k^2$. For the chosen K we have

$$x_k = \begin{cases} \sqrt{\dfrac{A_k}{\lambda_x}} - B_k & 0 \le k \le K-1 \\ 0 & \text{otherwise,} \end{cases} \tag{22.50}$$

where λ_x is the Lagrange multiplier such that the power constraint $\sum_{k=0}^{K-1} x_k = c$ is satisfied. Now assume that there is a smaller integer $L < K$ yielding the solution

$$y_k = \begin{cases} \sqrt{\dfrac{A_k}{\lambda_y}} - B_k & 0 \le k \le L-1 \\ 0 & \text{otherwise,} \end{cases} \tag{22.51}$$

where λ_y is the new Lagrange multiplier to satisfy the power constraint $\sum_{k=0}^{L-1} x_k = c$. We first prove that

$$1/\lambda_y \ge 1/\lambda_x \tag{22.52}$$

Proof. In view of the power constraint we have $\sum_{k=0}^{L-1} y_k = \sum_{k=0}^{K-1} x_k = c$ so that

$$\sum_{k=0}^{L-1} y_k = \sum_{k=0}^{L-1} x_k + \sum_{k=L}^{K-1} x_k.$$

This shows that $\sum_{k=0}^{L-1} y_k - \sum_{k=0}^{L-1} x_k \ge 0$. In view of Eqs. (22.50) and (22.51) this yields

$$\Big(\frac{1}{\sqrt{\lambda_y}} - \frac{1}{\sqrt{\lambda_x}}\Big) \sum_{k=0}^{L-1} \sqrt{A_k} \ge 0.$$

It follows therefore that $1/\lambda_y \ge 1/\lambda_x$. $\bigtriangledown \bigtriangledown \bigtriangledown$

Observe now that in Eq. (22.50) we can assume

$$x_{K-1} > 0, \tag{22.53}$$

for otherwise we can define K to be the next smaller integer. So it follows that

$$\sqrt{\frac{A_{K-1}}{\lambda_x}} - B_{K-1} > 0,$$

and, since $1/\lambda_y \geq 1/\lambda_x$, we also have

$$\sqrt{\frac{A_{K-1}}{\lambda_y}} - B_{K-1} > 0. \tag{22.54}$$

This means that the solution (22.51) is such that the variable y_{K-1} is forced to be zero, even though the expression (22.54) would have yielded a positive value. We will now show that such "forced zeros" violate optimality!

Forced zeros and optimality. The KKT condition (22.41) becomes, for $k = K-1$,

$$-\frac{A_{K-1}}{(y_{K-1} + B_{K-1})^2} + \lambda_y - \mu_{K-1} = 0. \tag{22.55}$$

If we force $y_{K-1} = 0$, the KKT multiplier μ_{K-1} has to be

$$\mu_{K-1} = \lambda_y - \frac{A_{K-1}}{B_{K-1}^2} < 0, \tag{22.56}$$

where the inequality follows from Eq. (22.54). But since the KKT multipliers are required to satisfy $\mu_k \geq 0$, Eq. (22.56) violates optimality. $\triangledown\triangledown\triangledown$

Summarizing, whenever the expression

$$\sqrt{\frac{A_k}{\lambda_y}} - B_k$$

is positive, we have to take y_k to be equal to this expression instead of forcing it to be zero. Otherwise KKT conditions, which are necessary for optimality, will be violated. This shows that when there exists an optimal set $\{x_k\}$ for some K as described in Eq. (22.50), we *cannot find a smaller optimal set* $\{y_k\}$ as in Eq. (22.51).

22.4.3 How does power in a channel depend on noise?

When we optimized the capacity of a set of independent channels by power allocation we found that the optimal solution allocates more power to less noisy channels. How about the MMSE solution discussed in this section? Does this also allocate more power to less noisy channels? From Eq. (22.48) we see that the power allocated to the kth channel $(0 \leq k \leq K-1)$ is

$$\epsilon(k) = \alpha_k^2\sigma_s^2 = A\sigma_{q_k} - \sigma_{q_k}^2 = (A - \sigma_{q_k})\sigma_{q_k},$$

where $A = (p_0 + \sum_{\ell=0}^{K-1} \sigma_{q_\ell}^2)/\sum_{\ell=0}^{K-1} \sigma_{q_\ell}$. This is positive for $\sigma_{q_k} \leq A$, and the power is taken to be zero for larger σ_{q_k}. Note that $\epsilon(k)$ first increases and then decreases with σ_{q_k}, as shown in Fig. 22.5. Since the derivative

$$\frac{\partial \epsilon(k)}{\partial \sigma_{q_k}} = A - 2\sigma_{q_k}$$

is zero for $\sigma_{q_k} = A/2$, the maximum value $A^2/4$ occurs at $\sigma_{q_k} = A/2$. Thus, with

$$\sigma_{q_0} \leq \sigma_{q_1} \leq \ldots \leq \sigma_{q_{K-1}}$$

we see that the power need not increase monotonically with k. Typically it increases first and then decreases, as shown in the example of Fig. 22.6.

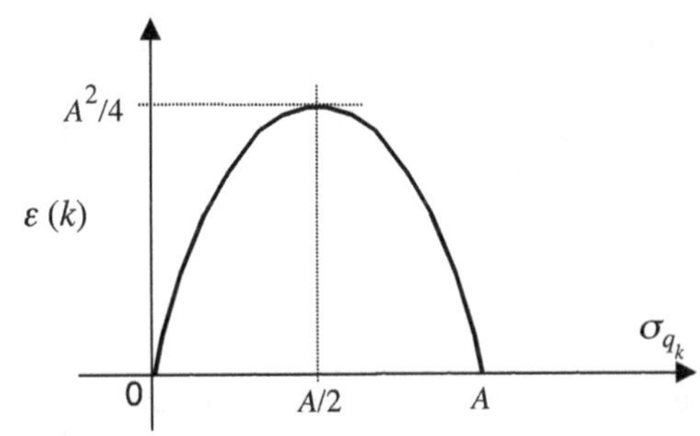

Figure 22.5. Dependence of allocated power on noise variance (MMSE design).

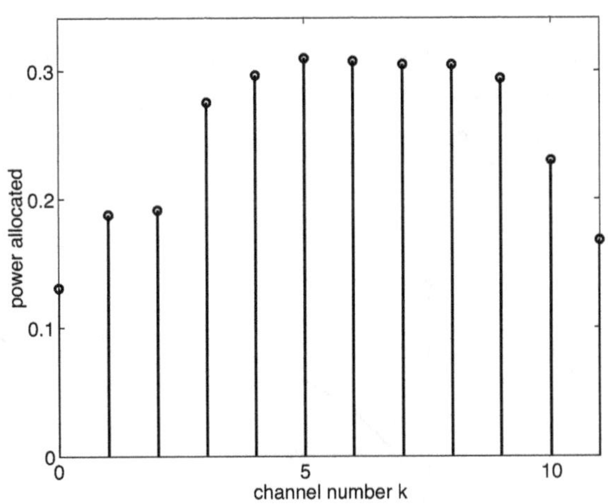

Figure 22.6. Example of allocated power as a function of channel number (MMSE design). Here $N = 12$, $p_0 = 3$, and $A = 1.1134$, so that $A^2/4 = 0.3099$.

Example 22.2: Equal power by accident

Let $N = 2$ and assume the total power allowed is

$$p_0 = 2\sigma_{q_0}\sigma_{q_1}.$$

We will find that the choice $K = 2$ yields positive answers for α_0^2 and α_1^2, so the best choice in this example is $K = 2$. To see this observe that

$$\frac{p_0 + \sigma_{q_0}^2 + \sigma_{q_1}^2}{\sigma_{q_0} + \sigma_{q_1}} = \frac{2\sigma_{q_0}\sigma_{q_1} + \sigma_{q_0}^2 + \sigma_{q_1}^2}{\sigma_{q_0} + \sigma_{q_1}} = \sigma_{q_0} + \sigma_{q_1},$$

so that Eq. (22.48) yields

$$\alpha_0^2 = \alpha_1^2 = \frac{\sigma_{q_1}\sigma_{q_0}}{\sigma_s^2}$$

This example shows that it is possible for the optimal solution to be such that $\alpha_0^2 = \alpha_1^2$, even though the channel noise variances may be different. The mean square errors corresponding to the two channels are

$$\frac{\sigma_{q_0}^2}{\alpha^2 + \dfrac{\sigma_{q_0}^2}{\sigma_s^2}} \quad \text{and} \quad \frac{\sigma_{q_1}^2}{\alpha^2 + \dfrac{\sigma_{q_1}^2}{\sigma_s^2}},$$

where $\alpha = \alpha_0 = \alpha_1$. Since the quantity

$$\psi(x) \stackrel{\Delta}{=} \frac{x}{\alpha^2 + \dfrac{x}{\sigma_s^2}} \tag{22.57}$$

increases monotonically as x increases from 0 to ∞ (Fig. 22.7), it follows that, even though $\alpha_0^2 = \alpha_1^2$, the two MMSE error components are *not equal* unless $\sigma_{q_0}^2 = \sigma_{q_1}^2$.

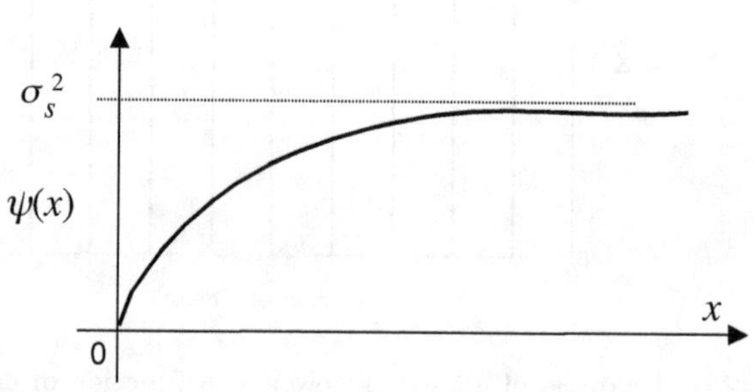

Figure 22.7. Plot of the MMSE component $\psi(x)$ in Eq. (22.57).

22.5 KKT conditions are only necessary conditions

The KKT conditions are necessary conditions for a local optimum. In many problems the KKT conditions can be solved to find a **unique** solution, in which case we know that the local optimum is also the global solution. One question that often arises is whether the solution to the stationarity equation yields a minimum or a maximum. In the calculus of one variable, one examines the sign of the second derivative to resolve this. A similar procedure is involved in the case of a vector variable $\mathbf{x}$. But it is more complicated, and involves Hessian matrices [Chong and Żak, 2001]. It is often more convenient to use simple physical considerations to figure out whether an extremum is a minimum or a maximum. For example, the solution (22.21) could not have *minimized* the load power because $R_l = 0$ obviously yields the smallest power. We will now make similar arguments for the other examples presented in this chapter.

22.5.1 Capacity is maximized

For the example where we used the KKT conditions to optimize channel capacity we now show that the result is indeed a maximum and not a minimum. This is done by constructing an example which has a smaller capacity under the same power constraint. Recall that the unique solution satisfying the KKT conditions was Eq. (22.26). Instead of using the power allocation given by Eq. (22.26) suppose we assign equal power to the same K channels:

$$x_k = \begin{cases} \frac{p_0}{K} & 0 \leq k \leq K-1 \\ 0 & \text{otherwise.} \end{cases} \tag{22.58}$$

This is called the *flat allocation.* We will show that this allocation is necessarily worse than the one given by Eq. (22.26). This proves that Eq. (22.26) represents a local maximum. Since Eq. (22.26) is the *unique solution* satisfying the stationary KKT conditions, the conclusion indeed is that it represents a *global maximum.*

Details. From Eq. (22.28) the capacity of the water-filling allocation is

$$C_{water} = 0.5 \log \left(\frac{p_0 + \sum_{k=0}^{K-1} Q_k}{K} \right)^K - 0.5 \sum_{k=0}^{K-1} \log Q_k. \tag{22.59}$$

The capacity of the flat allocation (22.58) is

$$C_{flat} = 0.5 \sum_{k=0}^{N-1} \log \left(\frac{x_k + Q_k}{Q_k} \right) = 0.5 \sum_{k=0}^{K-1} \log \left(\frac{p_0}{K} + Q_k \right) - 0.5 \sum_{k=0}^{K-1} \log Q_k. \tag{22.60}$$

To compare C_{water} and C_{flat} it is therefore sufficient to compare

$$T_{water} \triangleq \log \left(\frac{p_0 + \sum_{k=0}^{K-1} Q_k}{K} \right)^K$$

and

$$T_{flat} \triangleq \sum_{k=0}^{K-1} \log\left(\frac{p_0}{K} + Q_k\right) = \log \prod_{k=0}^{K-1} \left(\frac{p_0}{K} + Q_k\right)$$

occurring in the capacity expressions. Rewrite these equations as follows:

$$\frac{T_{water}}{K} = \log\left(\frac{p_0 + \sum_{k=0}^{K-1} Q_k}{K}\right) \quad \text{and} \quad \frac{T_{flat}}{K} = \log \prod_{k=0}^{K-1} \left(\frac{p_0}{K} + Q_k\right)^{1/K}.$$

Now, the quantity

$$\prod_{k=0}^{K-1} \left(\frac{p_0}{K} + Q_k\right)^{1/K}$$

is nothing but the geometric mean (GM) of the K positive numbers $(p_0/K) + Q_k$, whereas $(p_0 + \sum_{k=0}^{K-1} Q_k)/K$ is the arithmetic mean (AM) of these numbers. Since $AM \geq GM$, and since log(.) is a monotone increasing function for positive arguments, it follows that $T_{water} \geq T_{flat}$. Thus

$$C_{water} \geq C_{flat}$$

indeed! $\bigtriangledown \bigtriangledown \bigtriangledown$

22.5.2 MSE is minimized

For the example where we used the KKT conditions to optimize the mean square error, we now show that the result is indeed a minimum and not a maximum. This is done by constructing an example which has a larger MSE under the same power constraint. Thus, let $N = 2$, $\sigma_s^2 = 1$, and $\sigma_{q_1}^2 > \sigma_{q_0}^2$. Assume the noise variances and total power are such that $K = 1$. Then the theory in Sec. 22.4 yields the optimum solution with $\alpha_0^2 = p_0 > 0$ and $\alpha_1^2 = 0$. The corresponding mean square error is

$$\mathcal{E}_{mmse} = \frac{\sigma_{q_0}^2}{p_0 + \sigma_{q_0}^2} + 1.$$

On the other hand, the arbitrary choice $\alpha_0^2 = 0$ and $\alpha_1^2 = p_0 > 0$, which also satisfies the power constraint, yields

$$\mathcal{E}_{arb} = \frac{\sigma_{q_1}^2}{p_0 + \sigma_{q_1}^2} + 1.$$

Since $\sigma_{q_1}^2 > \sigma_{q_0}^2$, it follows that $\mathcal{E}_{arb}$ is larger than $\mathcal{E}_{mmse}$ (owing to the monotone behavior in Fig. 22.7). Thus we have demonstrated a solution which has higher reconstruction error than the solution obtained using the stationarity conditions. The solution obtained from the stationary KKT conditions is therefore a minimum rather than a maximum. Furthermore, since we have been able to obtain a unique solution satisfying the KKT conditions, it is the *unique* minimum.

22.6 Concluding remarks

Optimization problems with equality and inequality constraints frequently arise in signal processing and communications. A review of KKT conditions for such optimization was provided in this chapter. There are many excellent references that the interested reader can pursue for a deeper study; see, for example, Luenberger [1969], Chong and Żak [2001], Boyd and Vandenberghe [2004], and Antoniou and Lu [2007].

Part 4 Appendices

Appendix A

Inner products, norms, and inequalities

A.1 Inner products and norms

The inner product between two column vectors

$$\mathbf{a} = \begin{bmatrix} a_0 & a_1 \ldots a_{M-1} \end{bmatrix}^T, \quad \text{and} \quad \mathbf{b} = \begin{bmatrix} b_0 & b_1 \ldots b_{M-1} \end{bmatrix}^T,$$

is defined to be $\mathbf{b}^\dagger \mathbf{a}$, and is denoted as $< \mathbf{a}, \mathbf{b} >$. Thus

$$< \mathbf{a}, \mathbf{b} > = \mathbf{b}^\dagger \mathbf{a} = \sum_{k=0}^{M-1} a_k b_k^*. \tag{A.1}$$

The vectors are said to be **orthogonal** if the inner product is zero, that is, $\mathbf{b}^\dagger \mathbf{a} = 0$. The ℓ_2**-norm** of the vector $\mathbf{a}$, denoted as $\|\mathbf{a}\|_2$, is defined by

$$\|\mathbf{a}\|_2 = (\mathbf{a}^\dagger \mathbf{a})^{1/2} = \Big(\sum_{k=0}^{M-1} |a_k|^2 \Big)^{1/2}. \tag{A.2}$$

The quantity $\|\mathbf{a}\|_2^2$ is also called the energy in the vector, and $\|\mathbf{a}\|_2$ is sometimes called the length of the vector. More generally the ℓ_p**-norm** $\|\mathbf{a}\|_p$ is defined by

$$\|\mathbf{a}\|_p = \Big(\sum_{k=0}^{M-1} |a_k|^p \Big)^{1/p} \tag{A.3}$$

The definitions also extend to sequences of infinite length such as $a_k, -\infty \leq k \leq \infty$. If the ℓ_p-norm of a sequence is finite we say that it is an ℓ_p**-sequence**. The definitions readily extend to functions of continuous argument. For example, the inner product of two functions $f(t)$ and $g(t)$ defined in $0 \leq t \leq T$ is given by

$$< f, g >= \int_0^T f(t)g^*(t)dt. \tag{A.4}$$

The $L_2[0,T]$-norm, or simply L_2-norm $\|f(t)\|_2$, is defined by the equation

$$\|f(t)\|_2^2 = \int_0^T |f(t)|^2 dt. \tag{A.5}$$

More generally the L_p-norm is defined by

$$\|f(t)\|_p = \left(\int_0^T |f(t)|^p dt\right)^{1/p}. \tag{A.6}$$

For functions supported on infinite intervals such as $-\infty < t < \infty$ or $0 \leq t < \infty$, the definitions extend naturally with all integrals appropriately replaced.

The set of all ℓ_p-sequences forms a linear **vector space** called the ℓ_p-space. This means in particular that any finite linear combination of ℓ_p sequences is ℓ_p. Similarly the set of all L_p-functions on a given interval $a \leq t \leq b$ forms a linear vector space called the $L_p[a,b]$-space. When we mention inner products we implicitly assume the objects are from an ℓ_2- or an L_2-space as appropriate.

A.2 Cauchy-Schwartz inequality

Given two sets of numbers a_k and b_k, $0 \leq k \leq M-1$, the Cauchy-Schwartz (CS) inequality says that

$$\Big|\sum_{k=0}^{M-1} a_k b_k^*\Big|^2 \leq \sum_{k=0}^{M-1} |a_k|^2 \sum_{k=0}^{M-1} |b_k|^2, \tag{A.7}$$

with equality if and only if $a_k = cb_k$ for all k, for some constant c. The result also extends to integrals of functions. Thus

$$\Big|\int_{-\infty}^{\infty} f(t)g^*(t)dt\Big|^2 \leq \int_{-\infty}^{\infty} |f(t)|^2 dt \int_{-\infty}^{\infty} |g(t)|^2 dt, \tag{A.8}$$

with equality if and only if $f(t) = cg(t)$ for all t, for some constant c. The result is also true for finite integrals. For example, in the case of discrete-time Fourier transforms,

$$\Big|\int_{-\pi}^{\pi} F(e^{j\omega})G^*(e^{j\omega})\frac{d\omega}{2\pi}\Big|^2 \leq \int_{-\pi}^{\pi} |F(e^{j\omega})|^2 \frac{d\omega}{2\pi} \int_{-\pi}^{\pi} |G(e^{j\omega})|^2 \frac{d\omega}{2\pi}, \tag{A.9}$$

with equality if and only if $F(e^{j\omega}) = cG(e^{j\omega})$ for all ω, for some constant c. Note that in the notations of inner products and norms, Cauchy-Schwartz inequality can be expressed as

$$| < \mathbf{a}, \mathbf{b} > |^2 \leq \|\mathbf{a}\|_2^2 \|\mathbf{b}\|_2^2 \tag{A.10}$$

and similarly, for functions of continuous argument,

$$| < f(t), g(t) > |^2 \leq \|f(t)\|_2^2 \|g(t)\|_2^2. \tag{A.11}$$

A.3 The AM-GM inequality

For a set of positive numbers a_k, $0 \leq k \leq N-1$, the arithmetic mean (AM) and the geometric mean (GM), defined as

$$AM = \frac{1}{N} \sum_{k=0}^{N-1} a_k$$

and

$$GM = \left(\prod_{k=0}^{N-1} a_k \right)^{1/N},$$

satisfy the inequality

$$AM \geq GM. \tag{A.12}$$

There is equality if and only if all the numbers a_k are identical. For a guided proof please see Problem C.2 in Vaidyanathan [1993]. Other proofs can be found in Bellman [1960].

Appendix B

Matrices: a brief overview

B.1 Introduction

In this appendix we review basic concepts from matrices which are frequently used in this book. More details can be found in several standard texts, e.g., Franklin [1968], Horn and Johnson [1985], and Golub and Van Loan [1989]. A matrix with N rows and M columns is referred to as an $N \times M$ matrix. For example,

$$\mathbf{P} = \begin{bmatrix} 1 & 2+j \\ 1 & 3 \\ 2 & 4-j \end{bmatrix} = \begin{bmatrix} p_{00} & p_{01} \\ p_{10} & p_{11} \\ p_{20} & p_{21} \end{bmatrix}$$

is a 3×2 matrix. Note that the elements are denoted by double subscripts, as in p_{km}, sometimes with a comma for clarity as in $p_{k,m}$. Sometimes we use upper case to denote the elements, as in P_{km}. In some situations we also use $[\mathbf{P}]_{km}$ to denote p_{km}.

The notations $\mathbf{P}^T, \mathbf{P}^*$, and $\mathbf{P}^\dagger$ indicate the transpose, conjugate, and transpose-conjugate of a matrix, respectively. For the matrix $\mathbf{P}$ given above,

$$\mathbf{P}^T = \begin{bmatrix} 1 & 1 & 2 \\ 2+j & 3 & 4-j \end{bmatrix}, \quad \mathbf{P}^* = \begin{bmatrix} 1 & 2-j \\ 1 & 3 \\ 2 & 4+j \end{bmatrix}, \quad \mathbf{P}^\dagger = \begin{bmatrix} 1 & 1 & 2 \\ 2-j & 3 & 4+j \end{bmatrix}.$$

Some other notations were mentioned in Sec. 1.6, which the reader may want to review at this time. When $M = N$ we say that $\mathbf{P}$ is a **square matrix.** A $1 \times M$ matrix is said to be a **row vector**, and an $N \times 1$ matrix is said to be a **column vector**. A matrix with $p_{km} = 0$ for $k \neq m$ is called a **diagonal** matrix. For example,

$$\mathbf{P} = \begin{bmatrix} 1 & 0 & 0 \\ 0 & 4 & 0 \\ 0 & 0 & -2 \end{bmatrix}$$

is a diagonal matrix. An $M \times M$ diagonal matrix with all diagonal elements equal to unity is called the **identity** matrix, and is indicated as $\mathbf{I}_M$ or simply $\mathbf{I}$. A matrix with $p_{km} = 0$ for $k > m$ is called an **upper triangular** matrix:

$$\mathbf{P} = \begin{bmatrix} 1 & 2 & 3 \\ 0 & 1 & 4 \\ 0 & 0 & 5 \end{bmatrix}.$$

If $p_{km} = 0$ for $k \geq m$ we say it is *strictly* upper triangular:

$$\mathbf{P} = \begin{bmatrix} 0 & 2 & 3 \\ 0 & 0 & 4 \\ 0 & 0 & 0 \end{bmatrix}.$$

Lower triangular matrices are defined similarly. Note that the diagonal matrix is both upper and lower triangular. Given two matrices $\mathbf{P}$ and $\mathbf{Q}$ of identical size, the sum $\mathbf{P} + \mathbf{Q}$ has elements $p_{km} + q_{km}$. If $\mathbf{P}$ is $N \times M$ and $\mathbf{Q}$ is $M \times K$, then the product $\mathbf{PQ}$ is well defined. It is an $N \times K$ matrix with elements

$$[\mathbf{PQ}]_{nk} = \sum_{m=0}^{M-1} p_{nm} q_{mk}.$$

Observe that in general $\mathbf{PQ} \neq \mathbf{QP}$ (even if both products exist and have the same size). If $\mathbf{PQ} = \mathbf{QP}$ we say that the matrices $\mathbf{P}$ and $\mathbf{Q}$ **commute** in multiplication. For this to happen it is necessary (but not sufficient) that both matrices be square.

B.2 Determinant and trace

The **trace** of a square matrix $\mathbf{P}$ is defined to be the sum of the diagonal elements:

$$\text{Tr}(\mathbf{P}) = \sum_k p_{kk}. \tag{B.1}$$

It can be shown that

$$\text{Tr}(\mathbf{PQ}) = \text{Tr}(\mathbf{QP}) \tag{B.2}$$

provided both products are square (even if each matrix may not be square). The **determinant** of an $M \times M$ matrix $\mathbf{P}$ is defined recursively as follows:

$$\det \mathbf{P} = \sum_{k=0}^{M-1} (-1)^{k+m} p_{km} M_{km}, \tag{B.3}$$

where m is a fixed integer in $0 \leq m \leq M-1$. Here M_{km} is the determinant of the $(M-1) \times (M-1)$ submatrix obtained by deleting the kth row and mth column of $\mathbf{P}$. The quantity M_{km} is said to be the **minor** of the element p_{km}, and $(-1)^{k+m} M_{km}$ is said to be the **cofactor** of p_{km}. The determinant can also be defined in terms of the mth *row* as $\det \mathbf{P} = \sum_{k=0}^{M-1}(-1)^{k+m} p_{mk} M_{mk}$. In the 2×2 case the determinant has a closed form expression:

$$\begin{vmatrix} a & b \\ c & d \end{vmatrix} = ad - bc. \tag{B.4}$$

A square matrix is said to be **singular** if $[\det \mathbf{P}] = 0$, and **nonsingular** if $[\det \mathbf{P}] \neq 0$.

B.2.1 Principal minors

A **submatrix** of $\mathbf{P}$ is any matrix obtained by deleting some rows and some columns. The determinant of any square submatrix of $\mathbf{P}$ is said to be a minor of $\mathbf{P}$. A *principal submatrix* is any submatrix whose diagonal elements are also diagonal elements of $\mathbf{P}$. A *principal minor* is the determinant of a principal submatrix. For example, consider a 3×3 matrix $\mathbf{P}$. The principal minors are given by

$$p_{00}, p_{11}, p_{22}, \begin{vmatrix} p_{00} & p_{01} \\ p_{10} & p_{11} \end{vmatrix}, \begin{vmatrix} p_{00} & p_{02} \\ p_{20} & p_{22} \end{vmatrix}, \begin{vmatrix} p_{11} & p_{12} \\ p_{21} & p_{22} \end{vmatrix}, \text{and } [\det \mathbf{P}].$$

A *leading principal submatrix* of an $M \times M$ matrix $\mathbf{P}$ is a principal submatrix obtained by retaining the first K rows and columns of $\mathbf{P}$ for some $K \leq M$. The determinant of such a submatrix is said to be a leading principal minor. Thus, the leading principal minors for a 3×3 matrix are

$$p_{00}, \begin{vmatrix} p_{00} & a_{01} \\ p_{10} & p_{11} \end{vmatrix}, \text{ and } [\det \mathbf{P}].$$

B.2.2 Properties of determinants

We now summarize a number of properties relating to the determinant.

1. The determinant of a diagonal matrix is the product of its diagonal elements.

2. The determinant of a lower or upper triangular matrix is the product of its diagonal elements.

3. For an $M \times M$ matrix $\mathbf{P}$, $[\det c\mathbf{P}] = c^M [\det \mathbf{P}]$, for any scalar c.

4. $[\det \mathbf{PQ}] = [\det \mathbf{P}][\det \mathbf{Q}]$, assuming $\mathbf{P}$ and $\mathbf{Q}$ are both square.

5. If any row (or column) is a scalar multiple of another row (column), the determinant is zero. If any row (or column) is zero, the determinant is zero.

6. If $\mathbf{Q}$ is obtained from $\mathbf{P}$ by exchanging two rows then $[\det \mathbf{Q}] = -[\det \mathbf{P}]$. The same holds if columns are exchanged.

7. $[\det \mathbf{P}^T] = [\det \mathbf{P}]$.

8. A matrix of the form
$$\mathbf{A} = \begin{bmatrix} \mathbf{P} & \mathbf{0} \\ \mathbf{0} & \mathbf{Q} \end{bmatrix}, \tag{B.5}$$
where $\mathbf{P}$ and $\mathbf{Q}$ are arbitrary, is called a **block-diagonal** matrix. When $\mathbf{P}$ and $\mathbf{Q}$ are square matrices, we can show that $[\det \mathbf{A}] = [\det \mathbf{P}][\det \mathbf{Q}]$.

B.3 Rank

A set of nonzero vectors $\mathbf{v}_k$, $0 \leq k \leq M-1$ (row or column vectors), is said to be **linearly independent** if the equality

$$\sum_{k=0}^{M-1} c_k \mathbf{v}_k = \mathbf{0}$$

implies $c_k = 0$ for all k. The rank of a matrix $\mathbf{P}$ is equal to the number of linearly independent rows in the matrix. It can be shown that this is also equal to the number of linearly independent columns. A number of properties of rank are listed below.

1. An $M \times M$ matrix is nonsingular (i.e., its determinant is nonzero) if and only if it has full rank, that is, its rank is M.

2. Let $\mathbf{A}$ and $\mathbf{B}$ be $M \times N$ and $N \times K$ matrices with ranks ρ_a and ρ_b. Let ρ be the rank of $\mathbf{AB}$. Then,
$$\rho_a + \rho_b - N \leq \rho \leq \min(\rho_a, \rho_b).$$
This is called **Sylvester's inequality.**

3. Given two square matrices $\mathbf{A}$ and $\mathbf{B}$, the matrices $\mathbf{I} - \mathbf{AB}$ and $\mathbf{I} - \mathbf{BA}$ have identical rank. However, the products $\mathbf{AB}$ and $\mathbf{BA}$ may not have the same rank in general.

4. Given a $p \times r$ matrix $\mathbf{P}$, the space of all vectors of the form $\mathbf{Px}$ is called the **range space** or column space of $\mathbf{P}$. The **dimension** of the range space (i.e., the number of linearly independent vectors in that space) is equal to the rank of $\mathbf{P}$. The **null space** of $\mathbf{P}$ is the set of all vectors $\mathbf{y}$ such that $\mathbf{Py} = \mathbf{0}$. It turns out that the set of all linear combinations from the range space of $\mathbf{P}$ and the null space of $\mathbf{P}^\dagger$ is equal to the complete space of all vectors of size p.

B.3.1 Inverse of a matrix

Given an $N \times M$ matrix $\mathbf{P}$, if there exists a matrix $\mathbf{Q}$ such that $\mathbf{QP} = \mathbf{I}_M$, we say that $\mathbf{Q}$ is a *left inverse* of $\mathbf{P}$. A left inverse exists if and only if $\mathbf{P}$ has rank M. Similarly, if $\mathbf{PR} = \mathbf{I}_N$ then $\mathbf{R}$ is a *right inverse* of $\mathbf{P}$; this exists if and only if $\mathbf{P}$ has rank N. If $\mathbf{P}$ is rectangular, then at most one of these inverses can exist. If $\mathbf{P}$ is square then the inverse exists if and only if $\det \mathbf{P} \neq \mathbf{0}$, that is, $\mathbf{P}$ is nonsingular. And in this case the left and right inverses are identical, and furthermore the inverse can be written in closed form:

$$\mathbf{P}^{-1} = \frac{\text{Adj } \mathbf{P}}{\det \mathbf{P}} \tag{B.6}$$

Here [Adj $\mathbf{P}$] denotes the **adjugate** of $\mathbf{P}$ (sometimes known as the adjoint), defined as

$$[\text{Adj } \mathbf{P}]_{km} = \text{cofactor of } P_{mk}. \tag{B.7}$$

In other words, the (k, m) element of the adjugate is equal to the cofactor of the (m, k) element of $\mathbf{P}$. Note that the inverse of a square matrix, if it exists, is unique. For a product $\mathbf{P} = \mathbf{AB}$ the inverse is given by

$$\mathbf{P}^{-1} = \mathbf{B}^{-1}\mathbf{A}^{-1}, \tag{B.8}$$

assuming all inverses exist. Thus the order of the matrices in a product is reversed in the inversion. The same holds for products of rectangular matrices which have left (or right) inverses.

B.4 Eigenvalues and eigenvectors

We say that the nonzero vector $\mathbf{v}$ is an **eigenvector** of the $M \times M$ matrix $\mathbf{P}$ if

$$\mathbf{Pv} = \lambda\mathbf{v} \tag{B.9}$$

for some scalar λ. The scalar λ is the **eigenvalue** corresponding to this eigenvector $\mathbf{v}$. A number of properties of eigenvalues and eigenvectors are listed next.

1. The eigenvalues are solutions to the so-called **characteristic equation**

   $$\det [s\mathbf{I} - \mathbf{P}] = 0. \tag{B.10}$$

   The left-hand side can be shown to be a degree-M polynomial, and is called the *characteristic polynomial.* Equation (B.10) has M solutions. If these are distinct, then there are M distinct eigenvalues; otherwise the values are repeated. For example,in the 3×3 case, if the three solutions to Eq. (B.10) are $1, 4$, and 4 then the eigenvalue 4 has multiplicity two, and the eigenvalue 1 has multiplicity one.

2. If $\mathbf{v}_1$ and $\mathbf{v}_2$ are eigenvectors of $\mathbf{P}$ corresponding to eigenvalues $\lambda_1 \neq \lambda_2$, then $\mathbf{v}_1$ and $\mathbf{v}_2$ are linearly independent, that is, $\mathbf{v}_1 \neq \alpha\mathbf{v}_2$ for any α.

3. If the M eigenvalues are distinct, then the corresponding eigenvectors $\mathbf{v}_k$, $0 \leq k \leq M - 1$, are *linearly independent.* Also, each eigenvector $\mathbf{v}_k$ is unique (except for a scale factor, since $c\mathbf{v}_k$ is also an eigenvector). If $\mathbf{P}$ has fewer than M distinct eigenvalues, then there may or may not exist a set of M linearly independent eigenvectors.

4. $\mathbf{P}$ has an eigenvalue equal to zero if and only if it is singular (determinant equal to zero).

5. $\mathbf{P}$ and $\mathbf{P}^T$ have the same set of eigenvalues including multiplicity.

6. For a (lower or upper) triangular matrix, the eigenvalues are equal to the diagonal elements. Diagonal matrices also have this property.

7. Even if $\mathbf{P} \neq \mathbf{0}$, it is possible for all eigenvalues to be zero. For example,

$$\mathbf{P} = \begin{bmatrix} 0 & 1 & 2 \\ 0 & 0 & 4 \\ 0 & 0 & 0 \end{bmatrix} \tag{B.11}$$

   has all eigenvalues equal to zero.

8. The determinant and trace of an $M \times M$ matrix $\mathbf{P}$ are related to its M eigenvalues λ_k as follows:

$$\det \mathbf{P} = \prod_{k=0}^{M-1} \lambda_k \quad \text{and} \quad \text{Tr}\,(\mathbf{P}) = \sum_{k=0}^{M-1} \lambda_k. \tag{B.12}$$

9. For nonsingular $\mathbf{P}$, the eigenvalues of $\mathbf{P}^{-1}$ are reciprocals of those of $\mathbf{P}$.

10. If λ_k are the eigenvalues of $\mathbf{P}$, the eigenvalues of $\mathbf{P} + \sigma\mathbf{I}$ are $\lambda_k + \sigma$.

11. The matrix $\mathbf{T}^{-1}\mathbf{PT}$ has the same set of eigenvalues as $\mathbf{P}$ (including multiplicity). This is true for any nonsingular $\mathbf{T}$. The matrix $\mathbf{T}^{-1}\mathbf{PT}$ is said to be a **similarity transformation** of $\mathbf{P}$.

B.4.1 Invertible square matrices

From the preceding discussions we see that for an $M \times M$ matrix $\mathbf{P}$ the following statements are equivalent:

1. $\mathbf{P}^{-1}$ exists.

2. $\mathbf{P}$ is nonsingular.

3. $[\det \mathbf{P}] \neq 0$.

4. All eigenvalues of $\mathbf{P}$ are nonzero.

5. There is no nonzero vector $\mathbf{v}$ that annihilates $\mathbf{P}$ (i.e., makes $\mathbf{Pv} = \mathbf{0}$).

6. The rank of $\mathbf{P}$ is M.

7. The M columns of $\mathbf{P}$ are linearly independent (and so are the rows).

B.4.2 Matrix inversion lemma

Let $\mathbf{P}$ and $\mathbf{R}$ be square and nonsingular. Then, according to the matrix inversion lemma, we have

$$(\mathbf{P}+\mathbf{QRS})^{-1}=\mathbf{P}^{-1}-\mathbf{P}^{-1}\mathbf{Q}(\mathbf{SP}^{-1}\mathbf{Q}+\mathbf{R}^{-1})^{-1}\mathbf{SP}^{-1}. \tag{B.13}$$

This holds even if $\mathbf{Q}$ and $\mathbf{S}$ are not square. To prove Eq. (B.13) just observe that

$$\begin{aligned}
&\Big(\mathbf{P}^{-1}-\mathbf{P}^{-1}\mathbf{Q}(\mathbf{SP}^{-1}\mathbf{Q}+\mathbf{R}^{-1})^{-1}\mathbf{SP}^{-1}\Big)(\mathbf{P}+\mathbf{QRS})\\
&= \mathbf{I}-\mathbf{P}^{-1}\mathbf{Q}(\mathbf{SP}^{-1}\mathbf{Q}+\mathbf{R}^{-1})^{-1}\mathbf{S}+\mathbf{P}^{-1}\mathbf{Q}\Big(\mathbf{I}-(\mathbf{SP}^{-1}\mathbf{Q}+\mathbf{R}^{-1})^{-1}\mathbf{SP}^{-1}\mathbf{Q}\Big)\mathbf{RS}\\
&= \mathbf{I}-\mathbf{P}^{-1}\mathbf{Q}(\mathbf{SP}^{-1}\mathbf{Q}+\mathbf{R}^{-1})^{-1}\mathbf{S}\\
&\quad+\mathbf{P}^{-1}\mathbf{Q}(\mathbf{SP}^{-1}\mathbf{Q}+\mathbf{R}^{-1})^{-1}\Big(\mathbf{SP}^{-1}\mathbf{Q}+\mathbf{R}^{-1}-\mathbf{SP}^{-1}\mathbf{Q}\Big)\mathbf{RS}\\
&= \mathbf{I}-\mathbf{P}^{-1}\mathbf{Q}(\mathbf{SP}^{-1}\mathbf{Q}+\mathbf{R}^{-1})^{-1}\mathbf{S}+\mathbf{P}^{-1}\mathbf{Q}(\mathbf{SP}^{-1}\mathbf{Q}+\mathbf{R}^{-1})^{-1}\mathbf{S}=\mathbf{I}
\end{aligned}$$

indeed.

B.4.3 Partitioned matrices

Consider a matrix in the form

$$\begin{bmatrix}\mathbf{A} & \mathbf{B}\\ \mathbf{C} & \mathbf{D}\end{bmatrix}. \tag{B.14}$$

Assuming that $\mathbf{A}$ is nonsingular, we can verify by explicit multiplication that

$$\begin{bmatrix}\mathbf{A} & \mathbf{B}\\ \mathbf{C} & \mathbf{D}\end{bmatrix}=\begin{bmatrix}\mathbf{I} & \mathbf{0}\\ \mathbf{CA}^{-1} & \mathbf{I}\end{bmatrix}\begin{bmatrix}\mathbf{A} & \mathbf{0}\\ \mathbf{0} & \mathbf{D}-\mathbf{CA}^{-1}\mathbf{B}\end{bmatrix}\begin{bmatrix}\mathbf{I} & \mathbf{A}^{-1}\mathbf{B}\\ \mathbf{0} & \mathbf{I}\end{bmatrix}. \tag{B.15}$$

By recognizing that

$$\begin{bmatrix}\mathbf{I} & \mathbf{0}\\ \mathbf{CA}^{-1} & \mathbf{I}\end{bmatrix}^{-1}=\begin{bmatrix}\mathbf{I} & \mathbf{0}\\ -\mathbf{CA}^{-1} & \mathbf{I}\end{bmatrix} \quad \text{and} \quad \begin{bmatrix}\mathbf{I} & \mathbf{A}^{-1}\mathbf{B}\\ \mathbf{0} & \mathbf{I}\end{bmatrix}^{-1}=\begin{bmatrix}\mathbf{I} & -\mathbf{A}^{-1}\mathbf{B}\\ \mathbf{0} & \mathbf{I}\end{bmatrix}$$

we therefore obtain

$$\begin{bmatrix}\mathbf{A} & \mathbf{B}\\ \mathbf{C} & \mathbf{D}\end{bmatrix}^{-1}=\begin{bmatrix}\mathbf{I} & -\mathbf{A}^{-1}\mathbf{B}\\ \mathbf{0} & \mathbf{I}\end{bmatrix}\begin{bmatrix}\mathbf{A}^{-1} & \mathbf{0}\\ \mathbf{0} & (\mathbf{D}-\mathbf{CA}^{-1}\mathbf{B})^{-1}\end{bmatrix}\begin{bmatrix}\mathbf{I} & \mathbf{0}\\ -\mathbf{CA}^{-1} & \mathbf{I}\end{bmatrix}, \tag{B.16}$$

where it is assumed that $\mathbf{D}-\mathbf{CA}^{-1}\mathbf{B}$ is nonsingular as well. Similarly, for the case where $\mathbf{D}$ is nonsingular, we can write

$$\begin{bmatrix}\mathbf{A} & \mathbf{B}\\ \mathbf{C} & \mathbf{D}\end{bmatrix}=\begin{bmatrix}\mathbf{I} & \mathbf{BD}^{-1}\\ \mathbf{0} & \mathbf{I}\end{bmatrix}\begin{bmatrix}\mathbf{A}-\mathbf{BD}^{-1}\mathbf{C} & \mathbf{0}\\ \mathbf{0} & \mathbf{D}\end{bmatrix}\begin{bmatrix}\mathbf{I} & \mathbf{0}\\ \mathbf{D}^{-1}\mathbf{C} & \mathbf{I}\end{bmatrix}, \tag{B.17}$$

which yields

$$\begin{bmatrix}\mathbf{A} & \mathbf{B}\\ \mathbf{C} & \mathbf{D}\end{bmatrix}^{-1}=\begin{bmatrix}\mathbf{I} & \mathbf{0}\\ -\mathbf{D}^{-1}\mathbf{C} & \mathbf{I}\end{bmatrix}\begin{bmatrix}(\mathbf{A}-\mathbf{BD}^{-1}\mathbf{C})^{-1} & \mathbf{0}\\ \mathbf{0} & \mathbf{D}^{-1}\end{bmatrix}\begin{bmatrix}\mathbf{I} & -\mathbf{BD}^{-1}\\ \mathbf{0} & \mathbf{I}\end{bmatrix}. \tag{B.18}$$

This inversion formula assumes that $\mathbf{D}$ and $(\mathbf{A}-\mathbf{BD}^{-1}\mathbf{C})$ are nonsingular. From Eq. (B.15) we have

$$\det \begin{bmatrix} \mathbf{A} & \mathbf{B} \\ \mathbf{C} & \mathbf{D} \end{bmatrix} = \det \mathbf{A} \times \det \left(\mathbf{D} - \mathbf{CA}^{-1}\mathbf{B} \right). \tag{B.19}$$

Similarly from Eq. (B.17) we get

$$\det \begin{bmatrix} \mathbf{A} & \mathbf{B} \\ \mathbf{C} & \mathbf{D} \end{bmatrix} = \det \mathbf{D} \times \left(\mathbf{A} - \mathbf{BD}^{-1}\mathbf{C} \right). \tag{B.20}$$

B.4.4 Diagonalizable matrices

Suppose $\mathbf{P}$ is $N \times N$, and assume that it has N linearly independent eigenvectors $\mathbf{t}_k$:

$$\mathbf{Pt}_k = \lambda_k \mathbf{t}_k, \quad 0 \le k \le N-1. \tag{B.21}$$

Defining

$$\mathbf{T} = [\, \mathbf{t}_0 \quad \mathbf{t}_1 \quad \ldots \quad \mathbf{t}_{N-1} \,]$$

we can write $\mathbf{PT} = \mathbf{T\Lambda}$, where $\mathbf{\Lambda}$ is an $N \times N$ diagonal matrix with kth diagonal element equal to λ_k. Note that $\mathbf{T}$ is nonsingular because $\mathbf{t}_k$ are independent. So we can express Eq. (B.21) as

$$\mathbf{T}^{-1}\mathbf{PT} = \mathbf{\Lambda}. \tag{B.22}$$

Thus, if there exist N linearly independent eigenvectors, we can diagonalize $\mathbf{P}$ by using a similarity transform. An $N \times N$ matrix $\mathbf{P}$ is *diagonalizable* if it can be written as in Eq. (B.22) for some diagonal $\mathbf{\Lambda}$ and nonsingular $\mathbf{T}$. Every $N \times N$ matrix with N distinct eigenvalues is diagonalizable (because the N eigenvectors are linearly independent). Two remarks are in order:

1. There is a class of matrices called **normal matrices**, which are diagonalizable even if the eigenvalues are not distinct (see below).

2. There are examples of matrices which *cannot be diagonalized.* For example, consider the matrix in Eq. (B.11). Since the matrix is triangular, the eigenvalues are the diagonal elements, and are all equal to zero. If it were diagonalizable then $\mathbf{\Lambda}$ would be zero in Eq. (B.22), which would imply $\mathbf{P} = \mathbf{0}$, a contradiction.

Even though not all matrices can be diagonalized, we can **triangularize** any square matrix. More precisely, if $\mathbf{P}$ is $N \times N$, we can always write

$$\mathbf{P} = \mathbf{U\Delta U}^{\dagger}, \tag{B.23}$$

where $\mathbf{U}$ is $N \times N$ unitary and $\mathbf{\Delta}$ is (lower or upper) triangular. This is called the **Schur decomposition**. In the preceding equation the diagonal elements of $\mathbf{\Delta}$ are the eigenvalues of $\mathbf{P}$. But the columns of $\mathbf{U}$ are not necessarily the eigenvectors of $\mathbf{P}$, unlike in diagonalization.

B.5 Matrices with special properties

In this section we summarize a number of matrices with special properties which arise frequently in our discussions.

1. A matrix $\mathbf{H}$ is said to be **Hermitian** if $\mathbf{H}^\dagger = \mathbf{H}$. Equivalently, the elements are such that $H_{km} = H_{mk}^*$. Clearly $\mathbf{H}$ has to be square for this. A real Hermitian matrix is said to be **symmetric** ($\mathbf{H}^T = \mathbf{H}$). Other related types include skew-Hermitian matrices ($\mathbf{H}^\dagger = -\mathbf{H}$), and antisymmetric matrices ($\mathbf{H}^T = -\mathbf{H}$). If $\mathbf{H}$ is Hermitian, then *all its eigenvalues are real*, and moreover $\mathbf{v}^\dagger\mathbf{H}\mathbf{v}$ is real for all $\mathbf{v}$.

2. A matrix $\mathbf{U}$ is said to be **unitary** if $\mathbf{U}^\dagger\mathbf{U} = \mathbf{I}$. Denoting the columns of $\mathbf{U}$ by $\mathbf{u}_0, \mathbf{u}_1, \ldots$, we see that unitarity implies

$$\mathbf{u}_k^\dagger\mathbf{u}_m = \delta(k-m).$$

   That is, the columns are **orthogonal**. Moreover, each column has unit norm. Note that a unitary matrix need not be square; it can be $N \times M$ with $N \geq M$. If $N = M$, then $\mathbf{U}^\dagger$ is also unitary, that is, $\mathbf{U}\mathbf{U}^\dagger = \mathbf{I}$. If $\mathbf{U}$ is square and unitary then *all its eigenvalues have unit magnitude.*

3. The $N \times N$ DFT (discrete Fourier transform) matrix has the elements

$$[\mathbf{W}]_{km} = [W^{km}], \tag{B.24}$$

   where $W = e^{-j2\pi/N}$. This is a symmetric (but complex) matrix. The matrix $\mathbf{W}/\sqrt{N}$ can be verified to be unitary, that is,

$$\mathbf{W}^\dagger\mathbf{W} = N\mathbf{I}. \tag{B.25}$$

4. A matrix $\mathbf{P}$ is said to be **Toeplitz** if the elements P_{km} are determined completely by the difference $k - m$. For example,

$$\mathbf{P} = \begin{bmatrix} p_0 & p_1 & p_2 \\ p_3 & p_0 & p_1 \\ p_4 & p_3 & p_0 \end{bmatrix} \tag{B.26}$$

   is Toeplitz. Thus, all elements on a line parallel to the diagonal are identical. The matrix need not be square.

5. An $N \times N$ matrix, each of whose rows has the form

$$[1 \quad a_m \quad a_m^2 \quad \ldots \quad a_m^{N-1}], \tag{B.27}$$

   is called a **Vandermonde** matrix. An example is the DFT matrix described above. The determinant of a Vandermonde matrix $\mathbf{V}$ is given by

$$\det \mathbf{V} = \prod_{m>n} (a_m - a_n). \tag{B.28}$$

   A Vandermonde matrix is nonsingular (determinant nonzero) if and only if the a_m's are distinct. A vector of the form (B.27) is called a Vandermonde vector.

6. A square matrix is **right-circulant** if each row is obtained by a right-circular shift of the previous row as in the following 3×3 example:

$$\mathbf{C} = \begin{bmatrix} c_0 & c_1 & c_2 \\ c_2 & c_0 & c_1 \\ c_1 & c_2 & c_0 \end{bmatrix}. \tag{B.29}$$

For a left-circulant matrix, each row is obtained by a left-circular shift of the previous row. A generalization of the circulant is the **pseudocirculant** matrix, which we discuss in Appendix D. Note that circulants are also Toeplitz.

7. If $\mathbf{C}$ is $N \times N$ circulant, then $\mathbf{C} = \mathbf{W}^{-1}\mathbf{\Lambda}_c\mathbf{W}$, where $\mathbf{W}$ is the $N \times N$ DFT matrix and $\mathbf{\Lambda}_c$ is diagonal. So the columns of $\mathbf{W}^{-1}$ are eigenvectors (equivalently the columns of $\mathbf{W}$, which are the columns of $\mathbf{W}^{-1}$ renumbered and scaled by a constant). See Appendix D for more details.

8. A matrix $\mathbf{P}$ is said to be **normal** if $\mathbf{PP}^\dagger = \mathbf{P}^\dagger\mathbf{P}$. Clearly $\mathbf{P}$ has to be a square matrix. It can be shown that $\mathbf{P}$ is normal if and only if it can be diagonalized by a unitary matrix, that is,

$$\mathbf{U}^\dagger\mathbf{PU} = \mathbf{\Lambda} \tag{B.30}$$

for diagonal $\mathbf{\Lambda}$ and unitary $\mathbf{U}$. Since the columns of $\mathbf{U}$ are eigenvectors, we see that normal matrices can also be defined to be those for which there exists a complete set of mutually orthogonal eigenvectors. It can be shown that the following are examples of normal matrices:

   (a) Hermitian matrices;
   (b) skew-Hermitian matrices;
   (c) unitary matrices;
   (d) circulants.

So these can be diagonalized by unitary matrices, even if the eigenvalues may not all be distinct.

B.5.1 Permutations

An $N \times N$ permutation matrix $\mathbf{P}$ has all entries equal to zero except N unit entries, which are distributed such that in each row and in each column there is precisely one unit element. Here are some examples of permutation matrices:

$$\begin{bmatrix} 0 & 1 & 0 \\ 1 & 0 & 0 \\ 0 & 0 & 1 \end{bmatrix}, \begin{bmatrix} 0 & 0 & 1 \\ 1 & 0 & 0 \\ 0 & 1 & 0 \end{bmatrix}, \begin{bmatrix} 0 & 0 & 1 \\ 0 & 1 & 0 \\ 1 & 0 & 0 \end{bmatrix}, \text{ and } \begin{bmatrix} 1 & 0 & 0 \\ 0 & 0 & 1 \\ 0 & 1 & 0 \end{bmatrix}. \tag{B.31}$$

In the ith row, let n_i denote the location of the unit element. So we see that the set of N integers n_i is a permutation of the integers $0 \leq i \leq N-1$. It is easily verified that $\mathbf{P}^T\mathbf{P} = \mathbf{I}$, so the permutation matrix is unitary. Given any $N \times N$ matrix $\mathbf{A}$, consider now the product

$$\mathbf{B} = \mathbf{P}\mathbf{A}\mathbf{P}^T. \tag{B.32}$$

The ith diagonal element is

$$b_{ii} = \sum_k \sum_m p_{i,k} a_{k,m} [\mathbf{P}^T]_{mi} = \sum_k \sum_m p_{i,k} a_{k,m} p_{i,m}. \tag{B.33}$$

All terms in the summation are zero except the terms which have $k = n_i$ and $m = n_i$. Thus

$$b_{ii} = p_{i,n_i} a_{n_i,n_i} p_{i,n_i} = a_{n_i,n_i}.$$

That is, the diagonal elements of $\mathbf{B}$ are permuted versions of the diagonal elements of $\mathbf{A}$. In fact the diagonal elements of $\mathbf{B}$ are related to the diagonal elements of $\mathbf{A}$ by the same permutation that relates i to n_i.

Exchange matrices. Let $\mathbf{P}_e$ be a permutation matrix obtained from the identity matrix $\mathbf{I}$ by interchanging two rows. For example, let

$$\mathbf{P}_e = \begin{bmatrix} 1 & 0 & 0 & 0 & 0 \\ 0 & 0 & 1 & 0 & 0 \\ 0 & 1 & 0 & 0 & 0 \\ 0 & 0 & 0 & 1 & 0 \\ 0 & 0 & 0 & 0 & 1 \end{bmatrix}. \tag{B.34}$$

This is equivalent to interchanging the corresponding columns of $\mathbf{I}$, and the matrix $\mathbf{P}_e$ is symmetric (i.e., $\mathbf{P}_e^T = \mathbf{P}_e$). Permutation matrices of the above form are called **exchange matrices**. The reason is, if we premultiply a matrix $\mathbf{A}$ with $\mathbf{P}_e$ it is equivalent to interchanging rows k and m of the matrix $\mathbf{A}$. No other rows are affected. Similarly if we postmultiply $\mathbf{A}$ with $\mathbf{P}_e$ it is equivalent to interchanging columns k and m of $\mathbf{A}$. It is readily verified from the properties of $\mathbf{P}_e$ that the matrix $\mathbf{P}_e\mathbf{A}\mathbf{P}_e$ $(= \mathbf{P}_e\mathbf{A}\mathbf{P}_e^T)$ has the kth and mth diagonal elements of $\mathbf{A}$ exchanged, with all other diagonal elements unaffected.

B.6 Positive definite matrices

For any $N \times N$ matrix $\mathbf{P}$, the scalar

$$\phi = \mathbf{v}^\dagger \mathbf{P} \mathbf{v} \tag{B.35}$$

is said to be a **quadratic form**, where $\mathbf{v}$ is a column vector. When $\mathbf{P}$ is Hermitian, $\mathbf{v}^\dagger\mathbf{P}\mathbf{v}$ is guaranteed to be real. If the Hermitian matrix $\mathbf{P}$ is such that $\mathbf{v}^\dagger\mathbf{P}\mathbf{v} > 0$ for $\mathbf{v} \neq \mathbf{0}$, we say that $\mathbf{P}$ is positive definite. If $\mathbf{v}^\dagger\mathbf{P}\mathbf{v} \geq 0$ for all $\mathbf{v}$, then $\mathbf{P}$ is positive semidefinite. Negative definiteness and semidefiniteness are similarly defined. If $\mathbf{P}$ is positive definite, we write it as

$$\mathbf{P} > \mathbf{0}, \tag{B.36}$$

and we write $\mathbf{P} \geq \mathbf{0}$ when $\mathbf{P}$ is positive semidefinite. If $\mathbf{P}$ is positive (semi)definite, then $-\mathbf{P}$ is *negative* (semi)definite. The notation

$$\mathbf{P} > \mathbf{Q} \tag{B.37}$$

is a short hand for the statement "$\mathbf{P} - \mathbf{Q}$ is positive definite" ($\mathbf{P} \geq \mathbf{Q}$ for semidefinite). For two Hermitian matrices $\mathbf{P}$ and $\mathbf{Q}$, it is possibe for the difference $\mathbf{P} - \mathbf{Q}$ to be neither positive nor negative semidefinite. For example, try

$$\mathbf{P} = \begin{bmatrix} 1 & 0 \\ 0 & 1 \end{bmatrix}, \quad \mathbf{Q} = \begin{bmatrix} 2 & 0 \\ 0 & 0.5 \end{bmatrix}. \tag{B.38}$$

In such cases we say that $\mathbf{P} - \mathbf{Q}$ is **indefinite**. Here are some properties of positive (semi)definite matrices.

1. ***Relation to eigenvalues.*** The Hermitian matrix $\mathbf{P}$ is positive definite (semidefinite) if and only if all the eigenvalues are positive (non-negative).

2. ***Relation to minors.*** The Hermitian matrix $\mathbf{P}$ is positive definite if and only if all *leading principal minors* of $\mathbf{P}$ are positive, and positive semidefinite if and only if all *principal minors* are non-negative. In particular, therefore, all diagonal elements of a positive definite (semidefinite) matrix are positive (non-negative).

3. ***Square root factorization.*** It can be shown that any $N \times N$ positive semidefinite $\mathbf{P}$ with rank $\rho \leq N$ can be factorized as

$$\mathbf{P} = \mathbf{Q}^\dagger \mathbf{Q}, \tag{B.39}$$

   where $\mathbf{Q}$ is $\rho \times N$. The factor $\mathbf{Q}$ is called a **square root** of $\mathbf{P}$. One technique to find such a factor $\mathbf{Q}$ is called *Cholesky decomposition* [Golub and Van Loan, 1989], which produces a lower triangular square root. When $\mathbf{P}$ is positive definite, it has full rank ($\rho = N$), and the square root $\mathbf{Q}$ is square and nonsingular. Conversely, a product of the form $\mathbf{Q}^\dagger \mathbf{Q}$ is positive semidefinite for any $\mathbf{Q}$, and positive definite if $\mathbf{Q}$ has linearly independent columns (e.g., when $\mathbf{Q}$ is square and nonsingular).

4. ***Determinant and diagonal elements.*** Let $\mathbf{P}$ be $N \times N$ Hermitian positive definite, and let P_{ii} denote its diagonal elements. Then

$$\det \mathbf{P} \leq \prod_{i=0}^{N-1} P_{ii}, \tag{B.40}$$

   with equality if and only if $\mathbf{P}$ is diagonal. This is called the **Hadamard inequality**.

B.7 Rayleigh-Ritz principle

Let $\mathbf{P}$ be $N \times N$ Hermitian, so that all its eigenvalues are real. Let λ_{min} and λ_{max} denote the smallest and largest eigenvalues, respectively. Then the maximum value of $\mathbf{v}^\dagger \mathbf{P} \mathbf{v}$ under the constraint that the vector $\mathbf{v}$ have unit norm is given by λ_{max}. That is,

$$\max_{\mathbf{v}^\dagger \mathbf{v}=1} \mathbf{v}^\dagger \mathbf{P} \mathbf{v} = \lambda_{max}. \tag{B.41}$$

This maximum is achieved if and only if $\mathbf{v}$ is an eigenvector of $\mathbf{P}$ corresponding to the maximum eigenvalue. This is called the Rayleigh-Ritz principle. Note that λ_{max} may have multiplicity > 1, in which case the optimizing eigenvector is not unique. Similar to Eq. (B.41) we also have

$$\min_{\mathbf{v}^\dagger \mathbf{v}=1} \mathbf{v}^\dagger \mathbf{P} \mathbf{v} = \lambda_{min}. \tag{B.42}$$

The minimum is achieved if and only if $\mathbf{v}$ is an eigenvector corresponding to the minimum eigenvalue. If the Hermitian matrix $\mathbf{P}$ also happens to be positive definite, then λ_{max} and λ_{min} are positive.

Appendix C

Singular value decomposition

C.1 Introduction

In this appendix we review the singular value decomposition of matrices, and its application in computing inverses of rectangular matrices. The matrix norm called the Frobenius norm is also reviewed, and its connection to the singular values indicated. More detailed treatments can be found in Horn and Johnson [1985]. The singular value decomposition was introduced by Autonne [1902], and further developments were made by Eckart and Young [1939]. Further historical remarks can be found in Haykin [1996].

Given any $P \times M$ matrix, one can show that it can be written in the form

$$\mathbf{A} = \underbrace{\mathbf{U}}_{P\times P}\underbrace{\mathbf{S}}_{P\times M}\underbrace{\mathbf{V}^{\dagger}}_{M\times M}, \tag{C.1}$$

where $\mathbf{U}$ and $\mathbf{V}$ are unitary matrices, that is,

$$\mathbf{U}^{\dagger}\mathbf{U} = \mathbf{U}\mathbf{U}^{\dagger} = \mathbf{I}_P, \qquad \mathbf{V}\mathbf{V}^{\dagger} = \mathbf{V}^{\dagger}\mathbf{V} = \mathbf{I}_M, \tag{C.2}$$

and $\mathbf{S}$ is a diagonal matrix with real non-negative diagonal elements $\sigma_k \geq 0$. This factorization is called the singular value decomposition, abbreviated as **SVD**, of the matrix $\mathbf{A}$. The diagonal elements σ_k are called the **singular values** of $\mathbf{A}$. If ρ is the rank of $\mathbf{A}$ then there are precisely ρ nonzero singular values, and we can write

$$\mathbf{A} = \mathbf{U}\begin{bmatrix} \mathbf{\Sigma}_\rho & \mathbf{0} \\ \mathbf{0} & \mathbf{0} \end{bmatrix}\mathbf{V}^{\dagger}, \tag{C.3}$$

where $\mathbf{\Sigma}_\rho$ is a diagonal matrix with positive diagonal elements:

$$\mathbf{\Sigma}_\rho = \begin{bmatrix} \sigma_0 & 0 & \ldots & 0 \\ 0 & \sigma_1 & \ldots & 0 \\ \vdots & \vdots & \ddots & \vdots \\ 0 & 0 & \ldots & \sigma_{\rho-1} \end{bmatrix}. \tag{C.4}$$

This is often written as $\mathbf{\Sigma}$ with the subscript ρ deleted, when the meaning is clear from the context. For example, when $P = M + L$ and $\mathbf{A}$ has full rank M, we can rewrite

$$\mathbf{A} = \mathbf{U}\begin{bmatrix} \mathbf{\Sigma} \\ \mathbf{0} \end{bmatrix}\mathbf{V}^\dagger, \tag{C.5}$$

where $\mathbf{\Sigma}$ is an $M \times M$ nonsingular diagonal matrix. Similarly, for $P < M$ if $\mathbf{A}$ has full rank P we can write

$$\mathbf{A} = \mathbf{U}[\,\mathbf{\Sigma} \quad \mathbf{0}\,]\mathbf{V}^\dagger, \tag{C.6}$$

where $\mathbf{\Sigma}$ is a $P \times P$ nonsingular diagonal matrix.

C.2 Left inverse computed from SVD

For the rest of the section we assume $P \geq M$ and that $\mathbf{A}$ has rank M. In this case a left inverse (i.e., a matrix $\mathbf{B}$ such that $\mathbf{BA} = \mathbf{I}$) exists. Left inverses of channel matrices often arise in communication systems. The left inverse can be expressed elegantly in terms of the SVD of $\mathbf{A}$. Since $\mathbf{A}$ has rank M, the diagonal matrix $\mathbf{\Sigma}$ in Eq. (C.5) is nonsingular. Define the $M \times P$ matrix

$$\mathbf{A}^\# = \mathbf{V}[\,\mathbf{\Sigma}^{-1} \quad \mathbf{0}\,]\mathbf{U}^\dagger. \tag{C.7a}$$

Then

$$\begin{aligned} \mathbf{A}^\#\mathbf{A} &= \mathbf{V}[\,\mathbf{\Sigma}^{-1} \quad \mathbf{0}\,]\mathbf{U}^\dagger\mathbf{U}\begin{bmatrix} \mathbf{\Sigma} \\ \mathbf{0} \end{bmatrix}\mathbf{V}^\dagger \\ &= \mathbf{V}[\,\mathbf{\Sigma}^{-1} \quad \mathbf{0}\,]\begin{bmatrix} \mathbf{\Sigma} \\ \mathbf{0} \end{bmatrix}\mathbf{V}^\dagger \quad (\text{since } \mathbf{U}^\dagger\mathbf{U} = \mathbf{I}) \\ &= \mathbf{V}\mathbf{\Sigma}^{-1}\mathbf{\Sigma}\mathbf{V}^\dagger \\ &= \mathbf{V}\mathbf{V}^\dagger = \mathbf{I}. \end{aligned}$$

This shows that $\mathbf{A}^\#$ is a left inverse of $\mathbf{A}$. We can readily verify that Eq. (C.7a) can be rewritten as

$$\mathbf{A}^\# = (\mathbf{A}^\dagger\mathbf{A})^{-1}\mathbf{A}^\dagger \tag{C.7b}$$

So this left inverse can be calculated directly from $\mathbf{A}$ without computing the SVD. Similarly, when the $P \times M$ matrix $\mathbf{A}$ has rank P (which requires $P \leq M$) the *right inverse* computed from the SVD takes the form

$$\mathbf{A}^\# = \mathbf{A}^\dagger(\mathbf{A}\mathbf{A}^\dagger)^{-1}.$$

Pseudoinverse. If $\mathbf{A}$ has rank $\rho < M$ then the left inverse does not exist. In this case we can still write $\mathbf{A}$ as in Eq. (C.3), where $\boldsymbol{\Sigma}_\rho$ is a nonsingular diagonal matrix. The *pseudoinverse* of $\mathbf{A}$ is then defined to be the matrix

$$\mathbf{A}^{\#} = \mathbf{V}\begin{bmatrix}\boldsymbol{\Sigma}_\rho^{-1} & \mathbf{0}\\ \mathbf{0} & \mathbf{0}\end{bmatrix}\mathbf{U}^\dagger. \tag{C.8}$$

Note that

$$\mathbf{A}^{\#}\mathbf{A} = \mathbf{V}\begin{bmatrix}\mathbf{I}_\rho & \mathbf{0}\\ \mathbf{0} & \mathbf{0}\end{bmatrix}\mathbf{V}^\dagger,$$

which reduces to $\mathbf{I}_M$ when the rank $\rho = M$. Thus the pseudoinverse reduces to the left inverse when $\mathbf{A}$ has full rank M. Pseudoinverses are also known as Moore-Penrose pseudoinverses in recognition of the authors who have studied many of their properties. The pseudoinverse of a matrix can be found by using the Matlab command *pinv*. ▽ ▽ ▽

C.3 Frobenius norm and SVD

The Frobenius norm $\|\mathbf{A}\|$ of a matrix $\mathbf{A}$ is defined as follows:

$$\|\mathbf{A}\|^2 = \sum_{k=0}^{P-1}\sum_{m=0}^{M-1}|a_{km}|^2. \tag{C.9}$$

From the definition it follows that the transpose-conjugate $\mathbf{A}^\dagger$ has the same Frobenius norm as $\mathbf{A}$. The right-hand side of Eq. (C.9) is nothing but the total energy in all the elements of $\mathbf{A}$. Note that the ℓ_2-norm of a vector is defined in precisely the same way (square root of the energy of the vector). We shall refer to the Frobenius norm as simply the *norm*. From the definition of the norm it follows that

$$\|\mathbf{A}\|^2 = \text{Tr}(\mathbf{A}^\dagger\mathbf{A}) = \text{Tr}(\mathbf{A}\mathbf{A}^\dagger), \tag{C.10}$$

where Tr denotes the trace of a matrix (sum of diagonal elements). To see this note that the diagonal element $[\mathbf{A}^\dagger\mathbf{A}]_{kk}$ is nothing but the energy in the kth column of $\mathbf{A}$, and adding up these energies we get $\|\mathbf{A}\|^2$ indeed. The identity $\|\mathbf{A}\|^2 = \text{Tr}(\mathbf{A}\mathbf{A}^\dagger)$ follows similarly.

We now show that the Frobenius norm of $\mathbf{A}$ is related to its singular values σ_k as follows:

$$\|\mathbf{A}\|^2 = \sum_{k=0}^{M-1}\sigma_k^2. \tag{C.11}$$

In other words, the energy in the elements a_{km} is equal to the energy in the singular values σ_k.

Proof of Eq. (C.11). From Eq. (C.5) we have

$$\mathbf{A}^\dagger\mathbf{A} = \mathbf{V}[\boldsymbol{\Sigma}\quad \mathbf{0}]\mathbf{U}^\dagger\mathbf{U}\begin{bmatrix}\boldsymbol{\Sigma}\\ \mathbf{0}\end{bmatrix}\mathbf{V}^\dagger$$

$$
\begin{aligned}
&= \mathbf{V}\,[\,\boldsymbol{\Sigma} \quad \mathbf{0}\,]\begin{bmatrix}\boldsymbol{\Sigma}\\ \mathbf{0}\end{bmatrix}\mathbf{V}^\dagger \\
&= \mathbf{V}\boldsymbol{\Sigma}^2\mathbf{V}^\dagger.
\end{aligned}
$$

From this we obtain

$$
\begin{aligned}
\text{Tr}\,(\mathbf{A}^\dagger\mathbf{A}) &= \text{Tr}\,(\mathbf{V}\boldsymbol{\Sigma}^2\mathbf{V}^\dagger) \\
&= \text{Tr}\,(\mathbf{V}^\dagger\mathbf{V}\boldsymbol{\Sigma}^2) \\
&= \text{Tr}\,(\boldsymbol{\Sigma}^2) = \sum_k \sigma_k^2\,,
\end{aligned}
$$

which proves the claim. In the second equality above we have used the trace identity, which says that, $\text{Tr}\,(\mathbf{PQ}) = \text{Tr}\,(\mathbf{QP})$ whenever $\mathbf{PQ}$ and $\mathbf{QP}$ are both defined. $\bigtriangledown\bigtriangledown\bigtriangledown$

C.4 Frobenius norm of the left inverse

For the $P \times M$ matrix $\mathbf{A}$ with rank M the SVD is

$$\mathbf{A} = \mathbf{U}\begin{bmatrix}\boldsymbol{\Sigma}\\ \mathbf{0}\end{bmatrix}\mathbf{V}^\dagger, \tag{C.12}$$

and a valid left inverse is

$$\mathbf{A}^{\#} = \mathbf{V}\,[\,\boldsymbol{\Sigma}^{-1} \quad \mathbf{0}\,]\,\mathbf{U}^\dagger. \tag{C.13}$$

The Frobenius norm of the left inverse $\mathbf{A}^{\#}$ can be calculated by observing that

$$\|\mathbf{A}^{\#}\|^2 = \text{Tr}\left(\mathbf{A}^{\#}[\mathbf{A}^{\#}]^\dagger\right). \tag{C.14}$$

Proceeding as in the proof of Eq. (C.11) this simplifies to $\|\mathbf{A}^{\#}\|^2 = \sum_{k=0}^{M-1} 1/\sigma_k^2$. Summarizing, the Frobenius norms of $\mathbf{A}$ and its left inverse $\mathbf{A}^{\#}$ are given by

$$\|\mathbf{A}\|^2 = \sum_{k=0}^{M-1} \sigma_k^2, \qquad \|\mathbf{A}^{\#}\|^2 = \sum_{k=0}^{M-1} \frac{1}{\sigma_k^2}. \tag{C.15}$$

Rank-deficient case. For the case where $\mathbf{A}$ does not have full rank we can use the representation (C.3) for $\mathbf{A}$ and the representation (C.8) for the pseudoinverse $\mathbf{A}^{\#}$, where ρ is the rank of $\mathbf{A}$ (and of $\mathbf{A}^{\#}$). In this case the expressions (C.15) continue to hold with the modification that both the summations have the range $0 \leq k \leq \rho - 1$. $\bigtriangledown\bigtriangledown\bigtriangledown$

Minimum-norm left inverse

If $\mathbf{A}$ is $P \times M$ with rank M, and if $P > M$, then there are infinitely many left inverses, and $\mathbf{A}^{\#}$ is just one of them. To see this, observe that there are $P - M$ linearly independent nonzero vectors $\mathbf{v}$ such that

$$\mathbf{v}^\dagger\mathbf{A} = \mathbf{0}.$$

Adding $\alpha\mathbf{v}^\dagger$ to any row of $\mathbf{A}^\#$ (for arbitrary α) does not therefore alter the left-inverse property. So there are infinitely many left inverses. We now show that the *left inverse defined in Eq. (C.13) has the smallest possible Frobenius norm.* So it is called the *minimum-norm left inverse (MNLI).* Since Eq. (C.13) can also be written as Eq. (C.7b), it follows that the minimum-norm left inverse can be written as

$$\mathbf{A}^\# = \mathbf{V}\,[\,\boldsymbol{\Sigma}^{-1} \quad \mathbf{0}\,]\,\mathbf{U}^\dagger = (\mathbf{A}^\dagger\mathbf{A})^{-1}\mathbf{A}^\dagger. \tag{C.16}$$

Proof that* $\mathbf{A}^\#$ *is the MNLI. Let $\mathbf{B}$ be another left inverse, so that $\mathbf{BA} = \mathbf{I}_M$. Without loss of generality we can write

$$\mathbf{B} = \mathbf{VCU}^\dagger$$

for appropriate $\mathbf{C}$ (just define $\mathbf{C} = \mathbf{V}^\dagger\mathbf{BU}$). Thus $\mathbf{BA} = \mathbf{I}_M$ implies

$$\mathbf{VCU}^\dagger\mathbf{U}\begin{bmatrix}\boldsymbol{\Sigma} \\ \mathbf{0}\end{bmatrix}\mathbf{V}^\dagger = \mathbf{I}_M,$$

which shows that

$$\mathbf{C}\begin{bmatrix}\boldsymbol{\Sigma} \\ \mathbf{0}\end{bmatrix} = \mathbf{I}.$$

That is, $\mathbf{C}$ can be written in the form $\mathbf{C} = [\,\boldsymbol{\Sigma}^{-1} \quad \mathbf{D}\,]$ for arbitrary $\mathbf{D}$. So all the non-uniqueness of the left inverse is captured by the arbitrariness of $\mathbf{D}$. We have just shown that any left inverse of $\mathbf{A}$ has the form

$$\mathbf{B} = \mathbf{V}\,[\,\boldsymbol{\Sigma}^{-1} \quad \mathbf{D}\,]\,\mathbf{U}^\dagger.$$

It follows from the preceding expression that

$$\begin{aligned}
\|\mathbf{B}\|^2 &= \mathrm{Tr}\,(\mathbf{BB}^\dagger) \\
&= \mathrm{Tr}\left(\mathbf{V}\,[\,\boldsymbol{\Sigma}^{-1} \quad \mathbf{D}\,]\,\mathbf{U}^\dagger\mathbf{U}\begin{bmatrix}\boldsymbol{\Sigma}^{-1} \\ \mathbf{D}^\dagger\end{bmatrix}\mathbf{V}^\dagger\right) \\
&= \mathrm{Tr}\left(\mathbf{V}\,[\,\boldsymbol{\Sigma}^{-1} \quad \mathbf{D}\,]\begin{bmatrix}\boldsymbol{\Sigma}^{-1} \\ \mathbf{D}^\dagger\end{bmatrix}\mathbf{V}^\dagger\right) \\
&= \mathrm{Tr}\left(\mathbf{V}^\dagger\mathbf{V}[\boldsymbol{\Sigma}^{-2} + \mathbf{DD}^\dagger]\right) \\
&= \mathrm{Tr}\,(\boldsymbol{\Sigma}^{-2}) + \mathrm{Tr}\,(\mathbf{DD}^\dagger) = \|\mathbf{A}^\#\|^2 + \|\mathbf{D}\|^2.
\end{aligned}$$

In the fourth line we have used the trace identity $\mathrm{Tr}\,(\mathbf{PQ}) = \mathrm{Tr}\,(\mathbf{QP})$. The quantity $\|\mathbf{D}\|^2$ is positive unless $\mathbf{D} = \mathbf{0}$. Thus

$$\|\mathbf{B}\|^2 \geq \|\mathbf{A}^\#\|^2$$

with equality if and only if $\mathbf{D} = \mathbf{0}$, that is if and only if $\mathbf{B} = \mathbf{A}^\#$. $\bigtriangledown\bigtriangledown\bigtriangledown$

Appendix D

Properties of pseudocirculant matrices

D.1 Introduction

Pseudocirculant matrices arise in filter bank precoder theory (Sec. 3.9.2). We also find them arising in the context of alias cancellation and in the description of blocked versions of transfer functions [Vaidyanathan and Mitra, 1988], [Vaidyanathan, 1993]. In this section we mention a few mathematical properties of pseudocirculants which have been found to be useful in research, e.g., in the use of filter banks in digital communications.

D.2 Circulant matrices

We begin by reviewing circulant matrices, and then we move on to pseudocirculants. To define a circulant matrix first consider an example:

$$\mathbf{C} = \begin{bmatrix} c(0) & c(4) & c(3) & c(2) & c(1) \\ c(1) & c(0) & c(4) & c(3) & c(2) \\ c(2) & c(1) & c(0) & c(4) & c(3) \\ c(3) & c(2) & c(1) & c(0) & c(4) \\ c(4) & c(3) & c(2) & c(1) & c(0) \end{bmatrix}. \tag{D.1}$$

Note that the leftmost column is arbitrary, but the other columns are obtained by shifting *down* the previous column and recirculating the element that spills over. This matrix is therefore called a circulant, or more specifically a **down circulant** (to distinguish from an up circulant, which is defined in an obvious way). In this text the term *circulant* always refers to down circulants. Notice that in a down-circulant matrix, any row is obtained from the preceding row by performing a right shift followed by recirculation of the element that spills over. The terms down circulant and **right circulant** therefore refer to the same property.

An important property of circulants is the fact that they can be **diagonalized** by the DFT matrix [Papoulis, 1977]. To understand this, first consider the product $\mathbf{CW}^*$, where $\mathbf{W}$ is the $P \times P$ DFT matrix and the superscript $*$ denotes conjugation. The kth column of this product has the form demonstrated below for $P = 5$:

$$[\mathbf{CW}^*]_k = \begin{bmatrix} c(0) & c(4) & c(3) & c(2) & c(1) \\ c(1) & c(0) & c(4) & c(3) & c(2) \\ c(2) & c(1) & c(0) & c(4) & c(3) \\ c(3) & c(2) & c(1) & c(0) & c(4) \\ c(4) & c(3) & c(2) & c(1) & c(0) \end{bmatrix} \begin{bmatrix} 1 \\ W^{-k} \\ W^{-2k} \\ W^{-3k} \\ W^{-4k} \end{bmatrix}, \tag{D.2}$$

where $W = e^{-j2\pi/P}$. The top row of this product is $\sum_{n=0}^{P-1} c((P-n))W^{-nk}$, where the argument $((P-n))$ is interpreted modulo P. Using $W^P = 1$ this can be simplified to

$$\sum_{n=0}^{P-1} c((P-n))W^{-nk} = \sum_{\ell=0}^{P-1} c(\ell)W^{-(P-\ell)k} = \sum_{\ell=0}^{P-1} c(\ell)W^{\ell k} = C[k],$$

where $C[k]$ is the kth point in the DFT of $c(n)$:

$$C[k] = \sum_{n=0}^{P-1} c(n)W^{nk}. \tag{D.3}$$

The mth row of the product (D.2) can be written as

$$\sum_{n=0}^{P-1} c((P-n+m))W^{-nk} = W^{-mk} \sum_{\ell=0}^{P-1} c(\ell)W^{\ell k} = W^{-mk}C[k].$$

Thus, in the example with $P = 5$ we have

$$\begin{bmatrix} c(0) & c(4) & c(3) & c(2) & c(1) \\ c(1) & c(0) & c(4) & c(3) & c(2) \\ c(2) & c(1) & c(0) & c(4) & c(3) \\ c(3) & c(2) & c(1) & c(0) & c(4) \\ c(4) & c(3) & c(2) & c(1) & c(0) \end{bmatrix} \begin{bmatrix} 1 \\ W^{-k} \\ W^{-2k} \\ W^{-3k} \\ W^{-4k} \end{bmatrix} = C[k] \begin{bmatrix} 1 \\ W^{-k} \\ W^{-2k} \\ W^{-3k} \\ W^{-4k} \end{bmatrix}. \tag{D.4}$$

That is, the DFT coefficients $C[k]$ of the leftmost column of $\mathbf{C}$ are the eigenvalues, and the corresponding eigenvectors are of the form

$$[1 \; W^{-k} \; \ldots \; W^{-(P-1)k}]^T.$$

Writing Eq. (D.4) for all k in $0 \le k \le P-1$ we therefore get the identity

$$\mathbf{CW}^* = \mathbf{W}^* \mathbf{\Lambda}_c, \tag{D.5}$$

where $\mathbf{\Lambda}_c$ is the diagonal matrix

$$\mathbf{\Lambda}_c = \begin{bmatrix} C[0] & 0 & \dots & 0 \\ 0 & C[1] & \dots & 0 \\ \vdots & \vdots & \ddots & \vdots \\ 0 & 0 & \dots & C[P-1] \end{bmatrix}. \tag{D.6}$$

Using the fact that $\mathbf{W}^*\mathbf{W} = P\mathbf{I}$ we then get $\mathbf{C}\mathbf{W}^{-1} = \mathbf{W}^{-1}\mathbf{\Lambda}_c$, that is,

$$\mathbf{W}\mathbf{C}\mathbf{W}^{-1} = \mathbf{\Lambda}_c, \tag{D.7}$$

or equivalently

$$\mathbf{C} = \mathbf{W}^{-1}\mathbf{\Lambda}_c\mathbf{W}. \tag{D.8}$$

Conversely, we can start from the preceding equation and easily prove that $\mathbf{C}$ is a circulant. Summarizing, we have proved the following:

♠**Theorem D.1.** *Diagonalization of circulants.* If $\mathbf{C}$ is a circulant matrix as demonstrated in Eq. (D.1), it can be written as in Eq. (D.8), where $\mathbf{\Lambda}_c$ is the diagonal matrix (D.6) of the DFT coefficients $C[k]$ defined in Eq. (D.3), and $\mathbf{W}$ is the $P \times P$ DFT matrix. *Conversely,* a matrix $\mathbf{C}$ expressed as in Eq. (D.8) is necessarily a circulant, and its leftmost column is $[c(0)\ c(1) \dots c(P-1)]^T$, where $c(n)$ are the inverse DFT coefficients of the diagonal elements $C[k]$ shown in Eq. (D.6). ◇

D.3 Diagonalization of pseudocirculants

Pseudocirculants were defined in Sec. 3.9.2. Consider the example of a 3×3 pseudocirculant

$$\mathbf{C}(z) = \begin{bmatrix} C_0(z) & z^{-1}C_2(z) & z^{-1}C_1(z) \\ C_1(z) & C_0(z) & z^{-1}C_2(z) \\ C_2(z) & C_1(z) & C_0(z) \end{bmatrix},$$

and note that the product

$$\begin{bmatrix} 1 & 0 & 0 \\ 0 & z^{-1} & 0 \\ 0 & 0 & z^{-2} \end{bmatrix} \begin{bmatrix} C_0(z^3) & z^{-3}C_2(z^3) & z^{-3}C_1(z^3) \\ C_1(z^3) & C_0(z^3) & z^{-3}C_2(z^3) \\ C_2(z^3) & C_1(z^3) & C_0(z^3) \end{bmatrix} \begin{bmatrix} 1 & 0 & 0 \\ 0 & z & 0 \\ 0 & 0 & z^2 \end{bmatrix}$$

is the circulant matrix

$$\begin{bmatrix} C_0(z^3) & z^{-2}C_2(z^3) & z^{-1}C_1(z^3) \\ z^{-1}C_1(z^3) & C_0(z^3) & z^{-2}C_2(z^3) \\ z^{-2}C_2(z^3) & z^{-1}C_1(z^3) & C_0(z^3) \end{bmatrix}.$$

More generally, we can readily verify the following relation between pseudocirculants and circulants [Vaidyanathan and Mitra, 1988]:

♠**Theorem D.2.** *Pseudocirculants and circulants.* Let $\mathbf{C}(z)$ be a $P \times P$ pseudocirculant with leftmost column equal to

$$\begin{bmatrix} C_0(z) & C_1(z) & \dots & C_{P-1}(z) \end{bmatrix}^T. \tag{D.9}$$

Then we can write

$$\mathbf{\Lambda}(z)\mathbf{C}(z^P)\mathbf{\Lambda}(z^{-1}) = \mathbf{C}_{circ}(z), \tag{D.10}$$

where $\mathbf{C}_{circ}(z)$ is circulant with the leftmost column equal to

$$\begin{bmatrix} C_0(z^P) & z^{-1}C_1(z^P) & \dots & z^{-(P-1)}C_{P-1}(z^P) \end{bmatrix}^T, \tag{D.11}$$

and $\mathbf{\Lambda}(z)$ is a diagonal matrix of delay elements:

$$\mathbf{\Lambda}(z) = \begin{bmatrix} 1 & 0 & \dots & 0 \\ 0 & z^{-1} & \dots & 0 \\ \vdots & \vdots & \ddots & \vdots \\ 0 & 0 & \dots & z^{-(P-1)} \end{bmatrix}. \tag{D.12}$$

Conversely if Eq. (D.10) is true, where $\mathbf{C}_{circ}$ is a circulant with leftmost column as in Eq. (D.11), then $\mathbf{C}(z)$ is a pseudocirculant with leftmost column as in Eq. (D.9). $\diamondsuit$

The theorem also asserts the converse, which the reader can readily verify. Since a circulant can be diagonalized with the DFT matrix (Theorem D.1), we can write

$$\mathbf{C}_{circ}(z) = \mathbf{W}^{-1}\mathbf{\Lambda}_c(z)\mathbf{W}, \tag{D.13}$$

where $\mathbf{\Lambda}_c(z)$ is a diagonal matrix whose diagonal elements are the DFT coefficients of the leftmost column of $\mathbf{C}_{circ}(z)$, that is,

$$[\mathbf{\Lambda}_c(z)]_{kk} = \sum_{n=0}^{P-1} [\mathbf{C}_{circ}(z)]_{n,0} W^{kn}, \tag{D.14}$$

where $W = e^{-j2\pi/P}$. Since the leftmost column of $\mathbf{C}_{circ}(z)$ is as in Eq. (D.11) this can be rewritten as

$$[\mathbf{\Lambda}_c(z)]_{kk} = \sum_{n=0}^{P-1} z^{-n} C_n(z^P) W^{kn}. \tag{D.15}$$

That is,

$$[\mathbf{\Lambda}_c(z)]_{kk} = C(zW^{-k}), \tag{D.16}$$

where

$$C(z) = \sum_{n=0}^{P-1} z^{-n} C_n(z^P). \tag{D.17}$$

Substituting Eq. (D.13) into Eq. (D.10) we find $\mathbf{C}(z^P) = \mathbf{\Lambda}(z^{-1})\mathbf{W}^{-1}\mathbf{\Lambda}_c(z)\mathbf{W}\mathbf{\Lambda}(z)$. Summarizing, we have proved the following:

♠**Theorem D.3.** *Diagonalization of pseudocirculants.* Let $\mathbf{C}(z)$ be a $P \times P$ pseudocirculant with leftmost column equal to

$$\begin{bmatrix} C_0(z) & C_1(z) & \dots & C_{P-1}(z) \end{bmatrix}^T. \tag{D.18}$$

Then we can write

$$\mathbf{C}(z^P) = \mathbf{\Lambda}(z^{-1})\mathbf{W}^{-1}\mathbf{\Lambda}_c(z)\mathbf{W}\mathbf{\Lambda}(z), \tag{D.19}$$

where $\mathbf{\Lambda}(z)$ is the diagonal matrix of delay elements (D.12), $\mathbf{\Lambda}_c(z)$ is a diagonal matrix with elements as defined in Eq. (D.15) or equivalently Eq. (D.16), and $\mathbf{W}$ is the $P \times P$ DFT matrix. Conversely, assume Eq. (D.19) is true, where $\mathbf{W}$ and $\mathbf{\Lambda}(z)$ are as described above and $\mathbf{\Lambda}_c(z)$ is a diagonal matrix with elements as defined in Eq. (D.16) for some $C(z)$. Then $\mathbf{C}(z)$ is a pseudocirculant. Its leftmost column has elements $C_k(z)$, which are the P polyphase components of $[\mathbf{\Lambda}_c(z)]_{00}$ (which is $C(z)$ given in Eq. (D.17)). $\diamondsuit$

The diagonalizing matrix $\mathbf{W}\mathbf{\Lambda}(e^{j\omega})$ is **unitary** (up to a scale factor $\sqrt{P}$). Thus, Eq. (D.19) establishes a **unitary diagonalization** for pseudocirculants.

D.4 Further properties of pseudocirculants

We now present a number of further properties of pseudocirculants, some of which are based on Theorem D.3 on diagonalization.

1. ***Blocked versions***. From Sec. 3.9 we know that, given a transfer function $C(z)$ as in Eq. (D.17), its $P \times P$ blocked version $\mathbf{C}(z)$ is a pseudocirculant whose leftmost column is precisely as in Eq. (D.18). Theorem D.3 shows that $\mathbf{C}(z^P)$ can be diagonalized by using the matrix $\mathbf{W}\mathbf{\Lambda}(e^{j\omega})$. The diagonal elements of the diagonalized version $\mathbf{\Lambda}_c(z)$ are *uniformly shifted versions* $C(zW^{-k})$ of the original transfer function $C(z)$.

2. ***Determinant of the blocked version.*** Taking determinants on both sides of Eq. (D.19) we arrive at

$$\det \mathbf{C}(z^P) = \det \mathbf{\Lambda}_c(z).$$

Since $\mathbf{\Lambda}_c(z)$ is a diagonal matrix with diagonal elements as in Eq. (D.16) it follows that

$$\det \mathbf{C}(z^P) = \prod_{k=0}^{P-1} C(zW^{-k}). \tag{D.20}$$

This therefore is the relation between a scalar transfer function $C(z)$ and the determinant of its blocked version $\mathbf{C}(z)$.

3. ***Bracelet of zeros.*** Suppose for example that $C(z)$ is FIR with a zero at z_0 so that there is a factor $(1 - z_0z^{-1})$. Suppose we replace this zero with the uniformly shifted set of zeros

$$z_0,\ z_0W,\ z_0W^2,\ \ldots,\ z_0W^{P-1}.$$

This is demonstrated in Fig. D.1 for a real zero z_0 and for a complex zero z_1. The set of P zeros resulting from a single zero z_k like this lies on a circle, and is called a **bracelet** of zeros generated by z_k. If each zero of $C(z)$ is replaced like this, then each factor $(1 - z^{-1}z_0)$ is replaced with

$$\prod_{k=0}^{P-1}(1 - z_0 W^k z^{-1}),$$

and the new FIR filter is precisely the righthand side of Eq. (D.20). Equation (D.20) shows that the determinant of the blocked version $\mathbf{C}(z)$ (or rather $\mathbf{C}(z^P)$) is precisely this bracelet polynomial.

4. ***Eigenvalues and eigenvectors.*** From Eq. (D.19) we have

$$\mathbf{C}(z^P)\mathbf{T}_z = \mathbf{T}_z \boldsymbol{\Lambda}_c(z),$$

where $\mathbf{T}_z = \boldsymbol{\Lambda}(z^{-1})\mathbf{W}^{-1}$. This shows that the diagonal elements of $\boldsymbol{\Lambda}_c(z)$ are the eigenvalues of $\mathbf{C}(z^P)$. The corresponding eigenvectors are given by the columns of $\mathbf{T}_z$. Summarizing, $\mathbf{C}(z^P)$ has *eigenvalues*

$$C(z), C(zW^{-1}), \ldots, C(zW^{-(P-1)}),$$

and the *eigenvector* corresponding to $C(zW^{-k})$ is

$$[1 \quad \alpha_k \quad \alpha_k^2 \quad \ldots \quad \alpha_k^{P-1}]^T,$$

where $\alpha_k = zW^{-k}$. The *eigenvectors are universal*, in the sense that they do not depend on $C(z)$. Only the *eigenvalues* depend on $C(z)$.

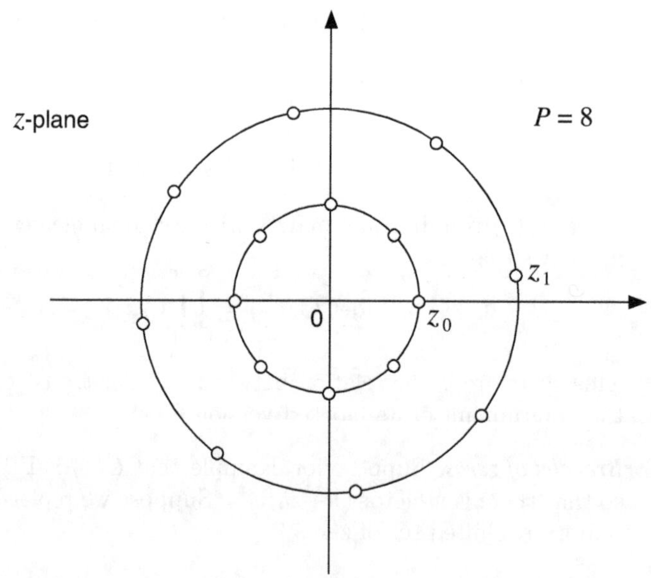

Figure D.1. The bracelet of zeros created from a real zero z_0, and from a complex zero z_1.

5. ***Existence of inverses.*** Note that the eigenvectors are polynomial functions of z. If $C(z)$ is an FIR filter, then the blocked version $\mathbf{C}(z)$, the eigenvalues $C(zW^k)$, and the determinant of $\mathbf{C}(z^P)$ are all FIR functions. Except in the trivial case where $C(z)$ is identically zero, the determinant (D.20) cannot be identically zero. So the inverse of $\mathbf{C}(z^P)$ *always* exists as a rational function. If the determinant has no unit circle zeros this inverse is **stable**; if the determinant has all zeros inside the unit circle then there is a **causal stable inverse**. Similar statements follow when $C(z)$ is a rational IIR transfer function.

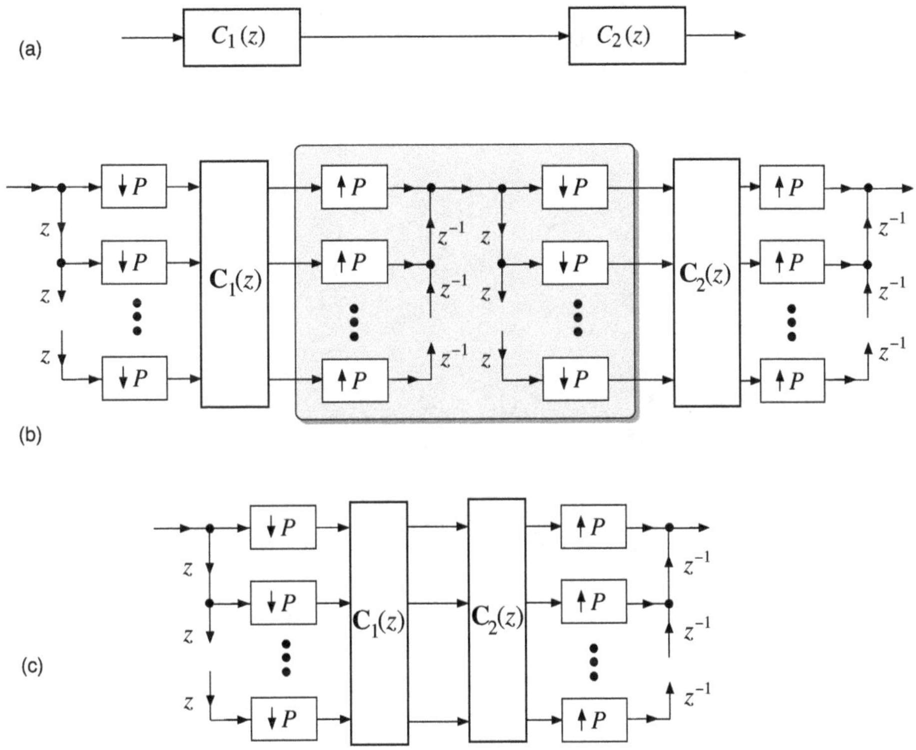

Figure D.2. (a) A cascade of two LTI systems, (b) the blocked versions, and (c) the simplified system.

6. ***Cascade of pseudocirculants.*** Next consider a cascade of two LTI systems $C_1(z)$ and $C_2(z)$ as shown in Fig. D.2(a). If we block these two systems (Sec. 3.9), the result is as shown in part (b). But the system shown in the gray box is the identity system (Fig. 3.17) so we obtain the simplified version of Fig. D.2(c). This means that the blocked version of the cascade $C_2(z)C_1(z)$ is the cascade of blocked versions $\mathbf{C}_2(z)\mathbf{C}_1(z)$. But since $C_1(z)$

and $C_2(z)$ commute in cascade, it follows entirely from this logic that

$$\mathbf{C}_2(z)\mathbf{C}_1(z) = \mathbf{C}_1(z)\mathbf{C}_2(z).$$

Thus, *two pseudocirculants in a product can be interchanged without changing the product, and the product itself is a pseudocirculant* because it is the blocked version of a scalar LTI system. Note also that the sum of two pseudocirculants is trivially a pseudocirculant as well.

7. ***Inverse of a pseudocirculant.*** In Fig. D.2(a) suppose $C_2(z) = 1/C_1(z)$. Then we have an identity system. But this system is also equivalent to Fig. D.2(c), which shows that the pseudocirculant $\mathbf{C}_2(z)$ is equal to $\mathbf{C}_1^{-1}(z)$. But since $C_1(z)$ is arbitrary, this allows us to argue that $\mathbf{C}_1^{-1}(z)$ is a pseudocirculant! This proves that *the inverse of any pseudocirculant is a pseudocirculant.*

Appendix E

Random processes

E.1 Introduction

In this appendix we provide a brief summary of wide sense stationary (WSS) random processes and their characterization. We begin with scalar processes and then consider vector processes. Optimal linear filtering of WSS processes is also reviewed. Detailed discussions can be found in Papoulis [1965], Peebles [1987], and Therrien [1992]. We assume the reader has basic familiarity with ideas from probability and random variable theory [Feller, 1968], [Therrien and Tummala, 2004].

E.2 Wide sense stationary processes

A discrete-time random process is a sequence of random variables $x(n)$, where n usually denotes time. In general the mean value $E[x(n)]$ depends on the time index n, and the autocorrelation function

$$E[x(n)x^*(n-k)]$$

depends both on the time n and the distance between samples k. We say that the process is **wide sense stationary** (abbreviated WSS) if two things are true. First, the mean is independent of time:

$$m_x = E[x(n)],$$

and second, the autocorrelation depends only on the time difference k, and not on n. In this case the **autocorrelation sequence** is given by

$$R_{xx}(k) = E[x(n)x^*(n-k)]. \tag{E.1}$$

The argument k is said to be the "lag variable." The subscript "xx" is deleted whenever there is no confusion as to which process we are discussing. Note that

$R_{xx}(0) = E|x(n)|^2$ is the mean square value, and is independent of n. It can be shown that

$$R_{xx}(0) \geq |R_{xx}(k)| \tag{E.2}$$

for all k. Furthermore the autocorrelation enjoys the following symmetry property:

$$R_{xx}(-k) = R^*_{xx}(k). \tag{E.3}$$

The **covariance** $C_{xx}(k)$ of the WSS process is the autocorrelation of $x(n) - m_x$, which is a zero-mean process. It can be shown that

$$R_{xx}(k) = C_{xx}(k) + |m_x|^2. \tag{E.4}$$

In most situations we consider only zero-mean processes, in which case the covariance sequence $C_{xx}(k)$ is identical to the autocorrelation sequence $R_{xx}(k)$. The Fourier transform of $R_{xx}(k)$ is called the **power spectrum** or **PSD** (power spectral density) of the WSS process, and is denoted as $S_{xx}(e^{j\omega})$. Thus

$$S_{xx}(e^{j\omega}) = \sum_{k=-\infty}^{\infty} R_{xx}(k)e^{-j\omega k}. \tag{E.5}$$

In view of the symmetry in Eq. (E.3), the Fourier transform $S_{xx}(e^{j\omega})$ is real-valued. In fact it turns out that

$$S_{xx}(e^{j\omega}) \geq 0, \quad \text{for all } \omega, \tag{E.6}$$

justifying the name *power* spectrum. If the process has nonzero mean, then the presence of the constant term in Eq. (E.4) contributes to the impulse component $2\pi|m_x|^2\delta(\omega)$ in the power spectrum.

E.2.1 Vector processes

A vector random process is a sequence of column vectors, where each sample $\mathbf{x}(n)$ is an $M \times 1$ random vector. The autocorrelation of this process is defined as

$$E[\mathbf{x}(n)\mathbf{x}^\dagger(n-k)].$$

This is an $M \times M$ matrix, and its value depends in general on n and k. We say that the process is wide sense stationary (WSS) if the mean value is a constant and the autocorrelation depends only on k, and not on n. Thus

$$\mathbf{m}_x = E[\mathbf{x}(n)]$$

and

$$\mathbf{R}_{\mathbf{xx}}(k) = E[\mathbf{x}(n)\mathbf{x}^\dagger(n-k)]. \tag{E.7}$$

The **autocorrelation matrix** $\mathbf{R}_{\mathbf{xx}}(k)$ is a sequence of $M \times M$ matrices. Note that

$$\mathbf{R}_{\mathbf{xx}}(0) = E[\mathbf{x}(n)\mathbf{x}^\dagger(n)] \tag{E.8}$$

and is a positive semidefinite matrix. As in the scalar case, the covariance sequence $\mathbf{C}_{\mathbf{xx}}(k)$ is the autocorrelation of $\mathbf{x}(n) - \mathbf{m}_{\mathbf{x}}$, which is a zero-mean process. The autocorrelation of a WSS process satisfies the property

$$\mathbf{R}_{\mathbf{xx}}(-k) = \mathbf{R}^{\dagger}_{\mathbf{xx}}(k). \tag{E.9}$$

The Fourier transform of $\mathbf{R}_{\mathbf{xx}}(k)$ is called the power spectrum or **PSD matrix**:

$$\mathbf{S}_{\mathbf{xx}}(e^{j\omega}) = \sum_{k=-\infty}^{\infty} \mathbf{R}_{\mathbf{xx}}(k)e^{-j\omega k}. \tag{E.10}$$

This is an $M \times M$ matrix function of ω. It can be shown that this is a *positive semidefinite* matrix for all ω.

E.2.2 Joint stationarity and cross power spectra

Two random processes $\mathbf{x}(n)$ and $\mathbf{y}(n)$ are said to be jointly WSS if each of them is WSS and, in addition, the cross correlation defined by $E[\mathbf{x}(n)\mathbf{y}^{\dagger}(n-k)]$ is independent of n. In this case we denote the cross correlation by

$$\mathbf{R}_{\mathbf{xy}}(k) = E[\mathbf{x}(n)\mathbf{y}^{\dagger}(n-k)]. \tag{E.11}$$

Note that this is an $N \times M$ matrix sequence, where $\mathbf{x}(n)$ is $N \times 1$ and $\mathbf{y}(n)$ is $M \times 1$. The cross power spectrum of $\mathbf{x}(n)$ with $\mathbf{y}(n)$ is given by the Fourier transform of $\mathbf{R}_{\mathbf{xy}}(k)$:

$$\mathbf{S}_{\mathbf{xy}}(e^{j\omega}) = \sum_{k=-\infty}^{\infty} \mathbf{R}_{\mathbf{xy}}(k)e^{-j\omega k}. \tag{E.12}$$

It is readily verified that

$$\mathbf{R}_{\mathbf{yx}}(k) = \mathbf{R}^{\dagger}_{\mathbf{xy}}(-k) \tag{E.13}$$

and

$$\mathbf{S}_{\mathbf{yx}}(e^{j\omega}) = \mathbf{S}^{\dagger}_{\mathbf{xy}}(e^{j\omega}).$$

For example, in the scalar case

$$R_{yx}(k) = R^*_{xy}(-k), \qquad S_{yx}(e^{j\omega}) = S^*_{xy}(e^{j\omega}). \tag{E.14}$$

It can be shown that the joint WSS condition is equivalent to the condition that the vector process

$$\mathbf{v}(n) = \begin{bmatrix} \mathbf{x}(n) \\ \mathbf{y}(n) \end{bmatrix} \tag{E.15}$$

be WSS. In this case the power spectrum of $\mathbf{v}(n)$ is

$$\mathbf{S}_{\mathbf{vv}}(e^{j\omega}) = \begin{bmatrix} \mathbf{S}_{\mathbf{xx}}(e^{j\omega}) & \mathbf{S}_{\mathbf{xy}}(e^{j\omega}) \\ \mathbf{S}_{\mathbf{yx}}(e^{j\omega}) & \mathbf{S}_{\mathbf{yy}}(e^{j\omega}) \end{bmatrix}. \tag{E.16}$$

Here are a few other properties which follow directly from the definition and from linearity of the expectation operator:

$$\mathbf{R}_{(\mathbf{x}+\mathbf{y})\mathbf{z}}(k) = \mathbf{R}_{\mathbf{xz}}(k) + \mathbf{R}_{\mathbf{yz}}(k), \tag{E.17}$$

$$\mathbf{R}_{\mathbf{x}(\mathbf{y}+\mathbf{z})}(k) = \mathbf{R}_{\mathbf{xy}}(k) + \mathbf{R}_{\mathbf{xz}}(k). \tag{E.18}$$

Similarly for the cross power spectra:

$$\mathbf{S}_{(\mathbf{x}+\mathbf{y})\mathbf{z}}(e^{j\omega}) = \mathbf{S}_{\mathbf{xz}}(e^{j\omega}) + \mathbf{S}_{\mathbf{yz}}(e^{j\omega}), \tag{E.19}$$

$$\mathbf{S}_{\mathbf{x}(\mathbf{y}+\mathbf{z})}(e^{j\omega}) = \mathbf{S}_{\mathbf{xy}}(e^{j\omega}) + \mathbf{S}_{\mathbf{xz}}(e^{j\omega}). \tag{E.20}$$

E.2.3 Uncorrelatedness and orthogonality

Two random vectors $\mathbf{x}$ and $\mathbf{y}$ are said to be uncorrelated if

$$E[\mathbf{x}\mathbf{y}^\dagger] = E[\mathbf{x}]E[\mathbf{y}^\dagger] \tag{E.21}$$

and orthogonal if

$$E[\mathbf{x}\mathbf{y}^\dagger] = \mathbf{0}. \tag{E.22}$$

In terms of the individual components of $\mathbf{x}$ and $\mathbf{y}$, orthogonality means that

$$E[x_k y_m^*] = 0, \tag{E.23}$$

for all k, m. That is, every component of $\mathbf{x}$ is orthogonal to every component of $\mathbf{y}$. Note that *uncorrelatedness is equivalent to orthogonality if and only if at least one of the two vectors has zero mean.* Next, two random processes $\mathbf{x}(n)$ and $\mathbf{y}(n)$ are uncorrelated if

$$E[\mathbf{x}(n)\mathbf{y}^\dagger(m)] = E[\mathbf{x}(n)]E[\mathbf{y}^\dagger(m)], \tag{E.24}$$

for all n, m and orthogonal if

$$E[\mathbf{x}(n)\mathbf{y}^\dagger(m)] = \mathbf{0}, \tag{E.25}$$

for all n, m. In the jointly WSS case therefore, the processes are uncorrelated if

$$\mathbf{R}_{\mathbf{xy}}(k) = \mathbf{m}_{\mathbf{x}}\mathbf{m}_{\mathbf{y}}^\dagger, \tag{E.26}$$

for all k, where $\mathbf{m}_{\mathbf{x}}$ and $\mathbf{m}_{\mathbf{y}}$ are the mean values. The jointly WSS processes are orthogonal if

$$\mathbf{R}_{\mathbf{xy}}(k) = \mathbf{0}, \tag{E.27}$$

for all k. In the case of jointly Gaussian random vectors, i.e., when the composite vector

$$\mathbf{v} = \begin{bmatrix} \mathbf{x} \\ \mathbf{y} \end{bmatrix}$$

is Gaussian, uncorrelatedness also implies that $\mathbf{x}$ and $\mathbf{y}$ are independent. Further discussions on Gaussian vectors, and the so-called circularly symmetric random vectors, can be found in Sec. 6.6.

E.2.4 Linear filtering of a WSS process

Let a WSS process $\mathbf{x}(n)$ be input to an LTI system with impulse response $\mathbf{h}(n)$ and transfer function $\mathbf{H}(z)$. Then the output is

$$\mathbf{y}(n) = \sum_{m=-\infty}^{\infty} \mathbf{h}(m)\mathbf{x}(n-m). \tag{E.28}$$

It can be shown that this is also a WSS process. Its mean value is given by

$$\mathbf{m_y} = \Big[\sum_m \mathbf{h}(m)\Big]\mathbf{m_x}, \tag{E.29}$$

where $\mathbf{m_x}$ is the mean value of the input process. Defining the output autocorrelation,

$$\mathbf{R_{yy}}(k) = E[\mathbf{y}(n)\mathbf{y}^\dagger(n-k)] \tag{E.30}$$

and taking its Fourier transform, we get the power spectrum of $\mathbf{y}(n)$. This can be shown to be

$$\mathbf{S_{yy}}(e^{j\omega}) = \mathbf{H}(e^{j\omega})\mathbf{S_{xx}}(e^{j\omega})\mathbf{H}^\dagger(e^{j\omega}). \tag{E.31}$$

It is sometimes of interest to know the cross correlation between the input $\mathbf{x}(n)$ and the output $\mathbf{y}(n)$. This can be calculated as follows:

$$\mathbf{R_{yx}}(k) = E[\mathbf{y}(n)\mathbf{x}^\dagger(n-k)] = \sum_{m=-\infty}^{\infty} \mathbf{h}(m)E[\mathbf{x}(n-m)\mathbf{x}^\dagger(n-k)],$$

which simplifies to

$$\mathbf{R_{yx}}(k) = \sum_{m=-\infty}^{\infty} \mathbf{h}(m)\mathbf{R_{xx}}(k-m). \tag{E.32}$$

Thus the cross correlation sequence can be visualized as the output of $\mathbf{H}(z)$ in response to an input sequence equal to $\mathbf{R_{xx}}(k)$. Taking Fourier transforms, Eq. (E.32) yields an expression for the cross power spectrum:

$$\mathbf{S_{yx}}(e^{j\omega}) = \mathbf{H}(e^{j\omega})\mathbf{S_{xx}}(e^{j\omega}). \tag{E.33}$$

For the special case where $\mathbf{H}(z)$ is memoryless, $\mathbf{H}(z) = \mathbf{H} = \mathbf{h}(0)$, and so

$$\mathbf{R_{yx}}(k) = \mathbf{H}\mathbf{R_{xx}}(k). \tag{E.34}$$

Similarly, from (E.31), we have, for memoryless LTI systems,

$$\mathbf{S_{yy}}(e^{j\omega}) = \mathbf{H}\mathbf{S_{xx}}(e^{j\omega})\mathbf{H}^\dagger. \tag{E.35}$$

In particular therefore,

$$\mathbf{R_{yy}}(0) = \mathbf{H}\mathbf{R_{xx}}(0)\mathbf{H}^\dagger. \tag{E.36}$$

From time-domain description to power-spectral description. Here is an example of how the preceding concepts are applied in communication systems. In digital communications, the received sequence can often be represented by the equation

$$\mathbf{y}(n) = \sum_{m=-\infty}^{\infty} \mathbf{c}(m)\mathbf{s}(n-m) + \mathbf{q}(n), \tag{E.37}$$

where $\mathbf{s}(n)$ is a transmitted vector sequence, $\mathbf{c}(n)$ is the impulse response of a MIMO channel, and $\mathbf{q}(n)$ is additive noise. It is typical to assume that $\mathbf{s}(n)$ and $\mathbf{q}(n)$ are jointly wide sense stationary processes. Given any other process $\mathbf{u}(n)$, we can write

$$\mathbf{y}(n)\mathbf{u}^\dagger(n-k) = \sum_{m=-\infty}^{\infty} \mathbf{c}(m)\mathbf{s}(n-m)\mathbf{u}^\dagger(n-k) + \mathbf{q}(n)\mathbf{u}^\dagger(n-k).$$

If $\mathbf{u}(n)$ is jointly WSS with all other processes under discussion, then we can take expectations on both sides to get

$$\mathbf{R}_{\mathbf{yu}}(k) = \sum_{m=-\infty}^{\infty} \mathbf{c}(m)\mathbf{R}_{\mathbf{su}}(k-m) + \mathbf{R}_{\mathbf{qu}}(k). \tag{E.38}$$

Taking Fourier transforms, we get

$$\mathbf{S}_{\mathbf{yu}}(e^{j\omega}) = \mathbf{C}(e^{j\omega})\mathbf{S}_{\mathbf{su}}(e^{j\omega}) + \mathbf{S}_{\mathbf{qu}}(e^{j\omega}). \tag{E.39}$$

Using z-transform notation, this is sometimes written as

$$\mathbf{S}_{\mathbf{yu}}(z) = \mathbf{C}(z)\mathbf{S}_{\mathbf{su}}(z) + \mathbf{S}_{\mathbf{qu}}(z). \tag{E.40}$$

Such equations serve as starting points for finding optimal prefilters and equalizers that minimize the reconstruction error. Many such examples can be found in this book. $\bigtriangledown \bigtriangledown \bigtriangledown$

E.3 Cyclo WSS processes

Imagine we convert a wide sense stationary (WSS) discrete-time signal $s(n)$ into a continuous-time signal $x(t)$ as in Fig. E.1. Then $x(t)$ is a "cyclo WSS process" with period T, rather than a WSS process. In this section we shall explain what this term means. For such processes the power spectrum can only be defined in a time-average sense, as elaborated below. With the power spectrum appropriately defined, we will see that the **average power spectrum** of $x(t)$ is given by

$$S_{xx}(j\omega) = \frac{1}{T}|F(j\omega)|^2 S_{ss}(e^{j\omega T}), \tag{E.41}$$

where $S_{ss}(e^{j\omega})$ is the power spectrum of $s(n)$. Thus the **power** at the channel input takes the form

$$p_0 = \frac{1}{T}\int_{-\infty}^{\infty} |F(j\omega)|^2 S_{ss}(e^{j\omega T})\frac{d\omega}{2\pi} \tag{E.42}$$

Now for the details. The process $x(t)$ given in Fig. E.1 can be expressed as

$$x(t) = \sum_{n=-\infty}^{\infty} s(n)f(t - nT). \tag{E.43}$$

Assume that $s(n)$ is a WSS random process with autocorrelation

$$R_{ss}(k) = E[s(n)s^*(n-k)]. \tag{E.44}$$

The mean value of $x(t)$ is given by

$$E[x(t)] = E[s(n)] \sum_{n=-\infty}^{\infty} f(t - nT), \tag{E.45}$$

which is clearly periodic in t with period T. Thus the mean value is in general not a constant. Rather, it is a periodic function of time. The correlation between $x(t)$ and $x(t-\tau)$ is given by

$$\begin{aligned} E[x(t)x^*(t-\tau)] &= \sum_{n=-\infty}^{\infty} \sum_{m=-\infty}^{\infty} E[s(n)s^*(m)]f(t-nT)f^*(t-\tau-mT) \\ &= \sum_{n=-\infty}^{\infty} \sum_{m=-\infty}^{\infty} R_{ss}(n-m)f(t-nT)f^*(t-\tau-mT), \end{aligned}$$

so that

$$E[x(t)x^*(t-\tau)] = \sum_{k=-\infty}^{\infty} \sum_{n=-\infty}^{\infty} R_{ss}(k)f(t-nT)f^*(t-\tau+kT-nT). \tag{E.46}$$

For a WSS process, $E[x(t)x^*(t-\tau)]$ is independent of t. But since Eq. (E.46) has dependence on both t and τ, $x(t)$ is not WSS in general. Note however that the dependence on t is periodic with period T because replacing t with $t+T$ does not change the right-hand side.

♠**Definition E.1.** *Cyclo WSS processes.* A random process whose mean $E[x(t)]$ and autocorrelation $E[x(t)x^*(t-\tau)]$ are periodic functions of t with period T is said to be a cyclo WSS process with period T, and is abbreviated as a CWSS(T) process. ◇

For a CWSS(T) process we can define an autocorrelation function by taking the average of $E[x(t)x^*(t-\tau)]$ for one period:

$$R_{xx}(\tau) \stackrel{\Delta}{=} \frac{1}{T}\int_0^T E[x(t)x^*(t-\tau)]dt. \tag{E.47}$$

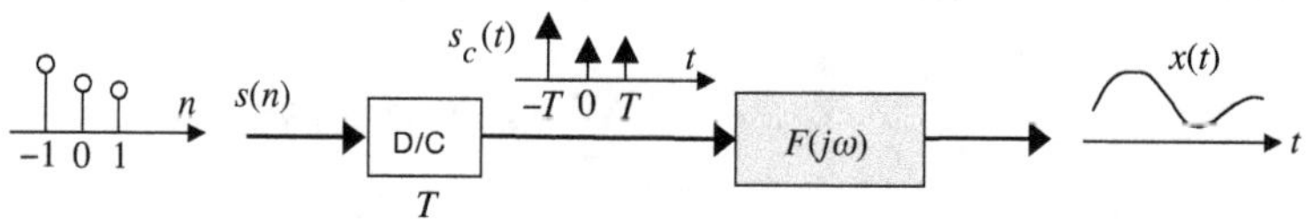

Figure E.1. The continuous-time signal created by D/C conversion followed by filtering.

Using Eq. (E.46) this becomes

$$\begin{aligned} R_{xx}(\tau) &= \frac{1}{T}\int_0^T \sum_{k=-\infty}^{\infty}\sum_{n=-\infty}^{\infty} R_{ss}(k)f(t-nT)f^*(t-\tau+kT-nT)dt \\ &= \frac{1}{T}\sum_{k=-\infty}^{\infty} R_{ss}(k)\sum_{n=-\infty}^{\infty}\int_0^T f(t-nT)f^*(t-\tau+kT-nT)dt \\ &= \frac{1}{T}\sum_{k=-\infty}^{\infty} R_{ss}(k)\int_{-\infty}^{\infty} f(t)f^*(t-\tau+kT)dt \\ &= \frac{1}{T}\sum_{k=-\infty}^{\infty} R_{ss}(k)R_{ff}(\tau-kT), \end{aligned}$$

where

$$R_{ff}(\tau) \triangleq \int_{-\infty}^{\infty} f(t)f^*(t-\tau)dt \tag{E.48}$$

is the deterministic autocorrelation of the impulse response $f(t)$. The Fourier transform of $R_{xx}(\tau)$ is regarded as the power spectrum of the CWSS process $x(t)$. Thus

$$\begin{aligned} S_{xx}(j\omega) &= \frac{1}{T}\sum_{k=-\infty}^{\infty} R_{ss}(k)\int_\tau R_{ff}(\tau-kT)e^{-j\omega\tau}d\tau \\ &= \frac{1}{T}\sum_{k=-\infty}^{\infty} e^{-j\omega kT}R_{ss}(k)\int_\tau R_{ff}(\tau)e^{-j\omega\tau}d\tau \\ &= \frac{1}{T}S_{ss}(e^{j\omega T})|F(j\omega)|^2, \end{aligned}$$

where $S_{ss}(e^{j\omega})$ is the power spectrum of the discrete process $s(n)$ and $F(j\omega)$ is the Fourier transform of $f(t)$.

Summary. If $s(n)$ is a discrete-time WSS process with autocorrelation $R_{ss}(k)$, the process described by

$$x(t) = \sum_{n=-\infty}^{\infty} s(n)f(t-nT) \tag{E.49}$$

is CWSS(T), and its autocorrelation (averaged over a duration T) is given by

$$R_{xx}(\tau) = \frac{1}{T} \sum_{k=-\infty}^{\infty} R_{ss}(k) R_{ff}(\tau - kT), \tag{E.50}$$

where $R_{ff}(\tau)$ is the deterministic autocorrelation of $f(t)$. The power spectrum of $x(t)$ is the Fourier transform of $R_{xx}(\tau)$, and is given by

$$S_{xx}(j\omega) = \frac{1}{T} S_{ss}(e^{j\omega T}) |F(j\omega)|^2, \tag{E.51}$$

where $S_{ss}(e^{j\omega})$ is the power spectrum of $s(n)$ and $F(j\omega)$ is the Fourier transform of $f(t)$.

E.4 Linear combinations of random variables

Linear combinations of random variables arise frequently in communication systems. For example, the output samples of an LTI system in response to a random process input are linear combinations of random variables. A few important properties of such linear combinations, not mentioned in earlier sections, are reviewed in this section. The simplest linear combination of two random variables x and y is the sum

$$z = x + y. \tag{E.52}$$

Assume x and y have a joint probability density function (pdf) $f(x, y)$. For fixed z, given any x we can find y such that $y = z - x$. So the pdf of z is given by

$$f_Z(z) = \int_{-\infty}^{\infty} f(x, z - x) dx. \tag{E.53}$$

For a more detailed derivation see Papoulis [1965] or Therrien and Tummala [2004]. For the special case where x and y are statistically independent, we have $f(x, y) = f_X(x) f_Y(y)$, so that

$$f_Z(z) = \int_{-\infty}^{\infty} f_X(x) f_Y(z - x) dx. \tag{E.54}$$

Thus the pdf of the sum of two *independent* random variables is the convolution of the individual pdfs. If the random variables are not indepedent we have to use Eq. (E.53) in general.

It should be mentioned that there are some hypothetical examples where x and y are not independent, but the pdf is still given by the convolution (E.54). One such example, due to Robbins, can be found on p. 98 of Feller [1965]. Referring to Fig. E.2(a), the joint pdf of x and y in this example is

$$f(x, y) = \begin{cases} 2 & \text{in the shaded area} \\ 0 & \text{otherwise.} \end{cases} \tag{E.55}$$

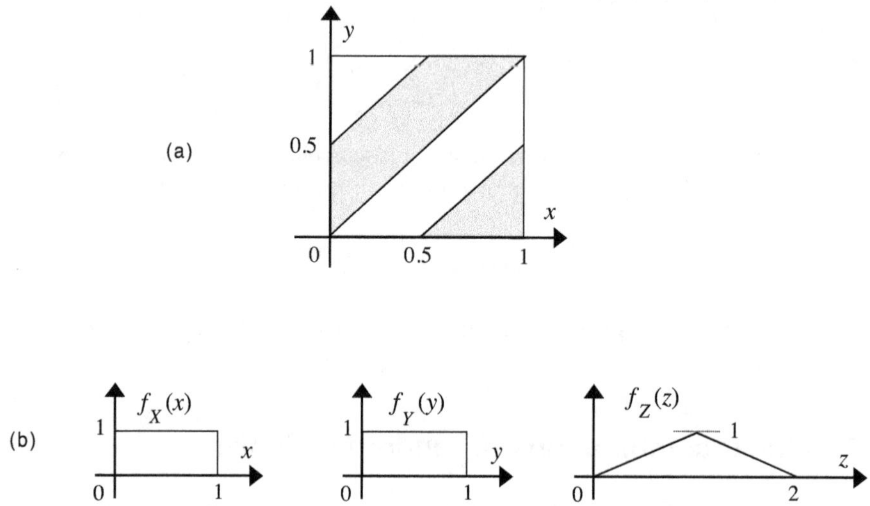

Figure E.2. (a) Defining the pdf $f(x, y)$ for the example in Eq. (E.55), and (b) various one-dimensional pdfs involved in the discussion. See text.

It can be shown that the marginals (defined as $f_X(x) = \int f(x, y)dy$ and $f_Y(y) = \int f(x, y)dx$) are uniform pdfs with support in the region $[0, 1]$. Furthermore, the pdf of $z = x + y$ is indeed the convolution of these uniform pdfs, that is, it is a triangle supported in $[0, 2]$. All these pdfs are shown in Fig. E.2(b). Clearly $f(x, y) \neq f_X(x)f_Y(y)$ in this case, so x and y are not independent. But still, the pdf of $x + y$ is the convolution of individual pdfs.

E.4.1 Linear transformation of a Gaussian vector

A real random vector $\mathbf{x}$ with mean $\mathbf{m_x}$ and covariance $\mathbf{C_{xx}}$ is said to be a **Gaussian** random vector if its pdf has the form

$$f(\mathbf{x}) = \frac{1}{\sqrt{\det(2\pi\mathbf{C_{xx}})}} e^{-\frac{1}{2}(\mathbf{x}-\mathbf{m_x})^T\mathbf{C_{xx}^{-1}}(\mathbf{x}-\mathbf{m_x})}. \tag{E.56}$$

Indicating the individual components of $\mathbf{x}$ as x_k, that is,

$$\mathbf{x} = \begin{bmatrix} x_0 & x_1 & \ldots, x_{N-1} \end{bmatrix}^T,$$

we also denote $f(\mathbf{x})$ as $f(x_0, x_1, \ldots, x_{N-1})$. The integral

$$f_{X_0}(x_0) = \int_{x_1} \ldots \int_{x_{N-1}} f(\mathbf{x})dx_1 \ldots dx_{N-1} \tag{E.57}$$

is said to be the **marginal** pdf of x_0. Similarly, $f_{X_k}(x_k)$ can be defined by integrating over all variables except x_k. We now mention some properties of Gaussians.

1. When $f(\mathbf{x})$ is Gaussian, it can be verified that each of the marginals $f_{X_k}(x_k)$ is Gaussian.

2. Defining the **linearly transformed** vector $\mathbf{y} = \mathbf{Ax}$, where $\mathbf{A}$ is a possibly rectangular real matrix, it can be shown [Feller, 1965], [Therrien and Tummala, 2004] that $\mathbf{y}$ is Gaussian. It has mean $\mathbf{m_y} = \mathbf{Am_x}$ and covariance $\mathbf{C_{yy}} = \mathbf{AC_{xx}A}^T$. Since the correlation matrix has the form $\mathbf{R_{yy}} = \mathbf{C_{yy}} + \mathbf{m_y m_y}^T$, we also have $\mathbf{R_{yy}} = \mathbf{AR_{xx}A}^T$.

3. A **complex** random vector $\mathbf{x} = \mathbf{x}_{re} + j\mathbf{x}_{im}$ is said to be Gaussian if the real vector
$$\mathbf{u} = \begin{bmatrix} \mathbf{x}_{re} \\ \mathbf{x}_{im} \end{bmatrix}$$
is Gaussian. The linearly transformed version $\mathbf{y} = \mathbf{Ax}$ is also Gaussian for any (possibly complex and rectangular) $\mathbf{A}$. The means and covariances are related as $\mathbf{m_y} = \mathbf{Am_x}$ and $\mathbf{C_{yy}} = \mathbf{AC_{xx}A}^\dagger$. A complex Gaussian vector is said to be *circularly symmetric* if it satisfies some additional properties as described in Sec. 6.6.

4. Note in particular that, when $\mathbf{x}$ is Gaussian, the sum of its individual components
$$y = x_0 + x_1 + \ldots + x_{N-1} \tag{E.58}$$
is Gaussian. More generally, any linear combination of x_k's is Gaussian, as this has the form $\mathbf{y} = \mathbf{Ax}$, where $\mathbf{A}$ is a row vector.

E.4.2 Jointly Gaussian random variables

Let x_0 and x_1 be two real random variables. We say that they are jointly Gaussian if the vector
$$\mathbf{x} = \begin{bmatrix} x_0 \\ x_1 \end{bmatrix} \tag{E.59}$$
has Gaussian pdf $f(\mathbf{x})$ (also denoted as $f(x_0, x_1)$). In this case the marginal pdfs $f_{X_0}(x_0)$ and $f_{X_1}(x_1)$ are also Gaussian, as mentioned above. This raises the following question: is it possible to have two random variables x_0 and x_1 that are individually Gaussian, but not jointly Gaussain? In other words, is it possible that $f_{X_k}(x_k)$ are Gaussian but that $f(\mathbf{x})$ does not take the form (E.56)? This is indeed possible! A number of examples can be found in Feller [1965]. An amusing example due to E. Nelson is given below.

Example E.1: Gaussian variables that are not jointly Gaussian

Let $g(y)$ be the Gaussian pdf
$$g(y) = \frac{e^{-y^2/2}}{\sqrt{2\pi}},$$

and let $h(y)$ be an antisymmetric function ($h(y) = -h(-y)$) so that

$$\int_{-\infty}^{\infty} g(y)dy = 1 \quad \text{and} \quad \int_{-\infty}^{\infty} h(y)dy = 0. \tag{E.60}$$

Define a two-variable function

$$f(x_0, x_1) = g(x_0)g(x_1) + h(x_0)h(x_1). \tag{E.61}$$

Then, in view of Eqs. (E.60) we have

$$\int_{-\infty}^{\infty}\int_{-\infty}^{\infty} f(x_0, x_1)dx_0 dx_1 = 1.$$

Furthermore, if we constrain the antisymmetric function $h(y)$ such that

$$h(y) = 0, \quad |y| \geq 1,$$

and upper bound it as follows:

$$|h(y)| < \frac{1}{\sqrt{2\pi e}}, \tag{E.62}$$

then $f(x_0, x_1) \geq 0$. Thus $f(x_0, x_1)$ becomes a valid pdf in this case. Furthermore,

$$\int_{-\infty}^{\infty} f(x_0, x_1)dx_0 = g(x_1), \qquad \int_{-\infty}^{\infty} f(x_0, x_1)dx_1 = g(x_0), \tag{E.63}$$

so that the marginals of $f(x_0, x_1)$ are Gaussian. However, $f(x_0, x_1)$ is not Gaussian, since the second term in Eq. (E.61) spoils it! Note that the function $h(y)$ can be *any* real antisymmetric function supported in $[-1, 1]$ and bounded by Eq. (E.62). So there are infinitely many examples like this, where $f(x_0, x_1)$ is not Gaussian though the marginals are. Note by the way that since

$$f(x_0, x_1) \neq g(x_0)g(x_1), \tag{E.64}$$

that is, the joint pdf is not the product of marginal pdfs, the random variables x_0 and x_1 in this example are *not* independent.

Sum of Gaussians may not be Gaussian. Now let

$$z = x_0 + x_1, \tag{E.65}$$

where x_0 and x_1 are the Gaussian random variables defined in Ex. E.1. We will now show that z is not Gaussian even though it is a sum of two Gaussians. The pdf of z is, from Eq. (E.53),

$$f_Z(z) = \int_{-\infty}^{\infty} f(x_0, z - x_0)dx_0 = \int_{-\infty}^{\infty} g(x_0)g(z - x_0)dx_0 + \int_{-\infty}^{\infty} h(x_0)h(z - x_0)dx_0.$$

The first term on the right-hand side is a convolution of the Gaussian pdf with itself and is therefore a Gaussian. The second term is the convolution of $h(x_0)$ with itself and is in general nonzero. By choosing $h(.)$ subject to the mild constraints mentioned in Ex. E.1 we can get infinitely many nonzero shapes for the second term. In particular, the pdf of z is not Gaussian even though it is a sum of two Gaussian random variables! While this appears to violate what we mentioned around Eq. (E.58), it happens because x_0 and x_1 are not *jointly* Gaussian, that is, the vector defined in Eq. (E.59) is not a Gaussian vector.

Appendix F

Wiener filtering

F.1 Introduction

In this appendix we review the theory of Wiener filtering, which is fundamental to many of the optimal signal processing methods involved in digital communications [Wiener, 1949], [Wiener and Masani, 1957, 1958]. First we review the fundamentals of statistically optimal filtering (Sec. F.2). We then consider the important special case where the signal and noise are uncorrelated (Sec. F.3). In Chap. 4 there are examples of applications of Wiener filter theory in the equalization of noisy communication channels (Sec. 4.10). In fact the topics discussed in this appendix are used throughout the book.

F.2 Theory of statistically optimal filtering

In this section we discuss the theory of optimal filtering of a signal buried in noise. Readers not familiar with random process theory should review Appendix E before reading this section.

F.2.1 The optimal linear estimation problem

First we need a fundamental result from linear estimation theory. Let $\mathbf{x}$ be an $M \times 1$ random vector, possibly complex, which we wish to estimate from observation of another random vector $\mathbf{y}$, called the observation vector. The vector $\mathbf{y}$ can be of different size, say $K \times 1$. We would like to obtain a **linear estimate**, that is, an estimate of the form

$$\widehat{\mathbf{x}} = \mathbf{A}\mathbf{y}. \tag{F.1}$$

Our aim is to find the matrix $\mathbf{A}$ based on our knowledge of the joint statistics of $\mathbf{x}$ and $\mathbf{y}$, such that the estimate is the best in the mean square sense. More precisely, define the estimation error

$$\mathbf{e} = \widehat{\mathbf{x}} - \mathbf{x} \tag{F.2}$$

and its mean square value

$$\mathcal{E} = E[\mathbf{e}^\dagger \mathbf{e}]. \tag{F.3}$$

We wish to find $\mathbf{A}$ such that $\mathcal{E}$ is minimized. Such an estimate $\widehat{\mathbf{x}}$ is called a linear *minimum mean square error estimate*, abbreviated as **linear MMSE estimate**. The fundamental theorem which governs the solution to this problem is the following.

♠**Lemma F.1.** *The orthogonality principle.* The estimate $\widehat{\mathbf{x}}$ in the preceding discussion is an MMSE estimate if and only if

$$E[\mathbf{e}\mathbf{y}^\dagger] = \mathbf{0}, \tag{F.4}$$

that is, if and only if the estimation error vector $\mathbf{e}$ is *orthogonal* to the observation vector $\mathbf{y}$. ◊

Proof. Let $\widehat{\mathbf{x}}_\perp$ be a linear estimate of $\mathbf{x}$ satisfying the orthogonality condition. So the error $\mathbf{e}_\perp = \widehat{\mathbf{x}}_\perp - \mathbf{x}$ satisfies

$$E[\mathbf{e}_\perp \mathbf{y}^\dagger] = \mathbf{0}.$$

Let $\widehat{\mathbf{x}}$ denote some arbitrary linear estimate of $\mathbf{x}$. Then

$$\mathbf{e} = \widehat{\mathbf{x}} - \mathbf{x} = \widehat{\mathbf{x}} - \widehat{\mathbf{x}}_\perp + \widehat{\mathbf{x}}_\perp - \mathbf{x} = \widehat{\mathbf{x}} - \widehat{\mathbf{x}}_\perp + \mathbf{e}_\perp. \tag{F.5a}$$

Since $\mathbf{e}_\perp$ is orthogonal to $\mathbf{y}$ it is also orthogonal to $\widehat{\mathbf{x}}$ and $\widehat{\mathbf{x}}_\perp$ (because these estimates have the linear form Eq. (F.1)). Thus the mean square error is

$$E[\mathbf{e}^\dagger \mathbf{e}] = E\Big[(\widehat{\mathbf{x}} - \widehat{\mathbf{x}}_\perp)^\dagger (\widehat{\mathbf{x}} - \widehat{\mathbf{x}}_\perp)\Big] + E[\mathbf{e}_\perp^\dagger \mathbf{e}_\perp]. \tag{F.5b}$$

Thus the mean square error $E[\mathbf{e}^\dagger \mathbf{e}] \geq E[\mathbf{e}_\perp^\dagger \mathbf{e}_\perp]$, with equality if and only if $\widehat{\mathbf{x}}_\perp = \widehat{\mathbf{x}}$. This proves the claimed result. ▽▽▽

F.2.1.A Properties of the linear MMSE estimator

1. ***Optimal estimator matrix.*** Using the condition that the error $\mathbf{e}$ is orthogonal to $\mathbf{y}$ we obtain $E[(\mathbf{x} - \widehat{\mathbf{x}})\mathbf{y}^\dagger] = \mathbf{0}$, that is, $E[(\mathbf{x} - \mathbf{A}\mathbf{y})\mathbf{y}^\dagger] = \mathbf{0}$. Thus

$$\mathbf{A} = E[\mathbf{x}\mathbf{y}^\dagger]\Big\{E[\mathbf{y}\mathbf{y}^\dagger]\Big\}^{-1} = \mathbf{R}_{\mathbf{xy}} \mathbf{R}_{\mathbf{yy}}^{-1},$$

   where $\mathbf{R}_{\mathbf{xy}}$ is the cross correlation between $\mathbf{x}$ and $\mathbf{y}$ and $\mathbf{R}_{\mathbf{yy}}$ is the correlation matrix for $\mathbf{y}$.

2. ***Error correlation matrix***. By using the orthogonality property we see that the correlation matrix of the random vector $\mathbf{e}$ is

$$E[\mathbf{e}\mathbf{e}^\dagger] = E[\mathbf{e}_\perp \mathbf{e}_\perp^\dagger] + E\Big[(\widehat{\mathbf{x}} - \widehat{\mathbf{x}}_\perp)(\widehat{\mathbf{x}} - \widehat{\mathbf{x}}_\perp)^\dagger\Big]. \tag{F.6}$$

Evidently the second term on the right-hand side of Eq. (F.6) is Hermitian and positive semidefinite (all eigenvalues ≥ 0). So the error correlation matrices satisfy[1]

$$E[\mathbf{e}\mathbf{e}^\dagger] \geq E[\mathbf{e}_\perp \mathbf{e}_\perp^\dagger]. \tag{F.7}$$

Thus the error correlation matrix for an arbitrary linear estimate is "at least as large as" the error correlation matrix of the optimum estimate. This is something we may not have guessed from the property $E[\mathbf{e}^\dagger\mathbf{e}] \geq E[\mathbf{e}_\perp^\dagger \mathbf{e}_\perp]$ of the estimator (because $\mathbf{u}^\dagger\mathbf{u} \geq \mathbf{v}^\dagger\mathbf{v}$ does not in general imply $\mathbf{u}\mathbf{u}^\dagger \geq \mathbf{v}\mathbf{v}^\dagger$; for example, try $\mathbf{u} = [1\ 0]^T$ and $\mathbf{v} = [0\ 1]^T$).

3. ***Orthogonality***. From Eq. (F.4) we know that for optimality of the estimate $\widehat{\mathbf{x}}_\perp$, every component of the error $\mathbf{e}_\perp$ must be orthogonal to every component of the observation $\mathbf{y}$, that is,

$$E[(\mathbf{e}_\perp)_i y_k^*] = 0, \quad \text{for all } i \text{ and } k.$$

Orthogonality therefore implies that

$$E[\mathbf{e}_\perp \widehat{\mathbf{x}}_\perp^\dagger] = E[\mathbf{e}_\perp \mathbf{y}^\dagger \mathbf{A}^\dagger] = \mathbf{0}.$$

Thus the error is orthogonal to the estimate itself. It then follows that

$$E[\mathbf{e}_\perp^\dagger \widehat{\mathbf{x}}_\perp] = 0.$$

4. ***Right-triangle analog.*** From $\mathbf{x} = \widehat{\mathbf{x}}_\perp - \mathbf{e}_\perp$ we can verify using orthogonality that the correlation matrices for the optimal estimator satisfy

$$E[\mathbf{x}\mathbf{x}^\dagger] = E[\widehat{\mathbf{x}}_\perp \widehat{\mathbf{x}}_\perp^\dagger] + E[\mathbf{e}_\perp \mathbf{e}_\perp^\dagger]. \tag{F.8}$$

Taking the trace on both sides, this implies in particular that

$$E[\mathbf{x}^\dagger\mathbf{x}] = E[\widehat{\mathbf{x}}_\perp^\dagger \widehat{\mathbf{x}}_\perp] + E[\mathbf{e}_\perp^\dagger \mathbf{e}_\perp]. \tag{F.9}$$

Thus the mean square value of the estimate and that of the error add up to the mean square value of the original variable $\mathbf{x}$. The same is true of the corresponding correlation matrices. It is as if $\mathbf{x}$ is the hypotenuse of a right triangle whose base is $\widehat{\mathbf{x}}_\perp$ and height is $\mathbf{e}_\perp$. See Fig. F.1.

[1] The notation $\mathbf{A} \geq \mathbf{B}$ for two Hermitian matrices $\mathbf{A}$ and $\mathbf{B}$ simply means that $\mathbf{A} - \mathbf{B}$ is positive semidefinite.

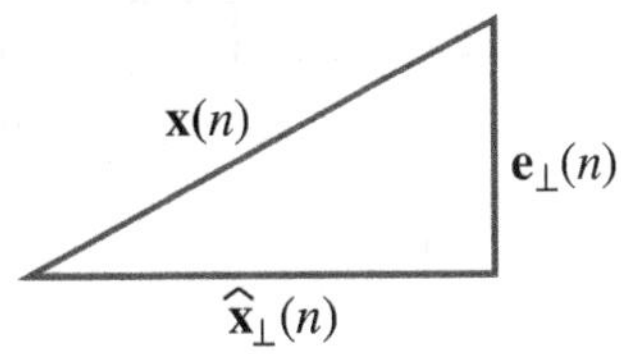

Figure F.1. The right-triangle analogy. Note that the error vector is orthogonal to the estimate. See text.

5. ***A generalization.*** In some applications, the linear estimation problem is written in a slightly different form: let $\mathbf{y}_1, \mathbf{y}_2, \ldots, \mathbf{y}_N$ be a set of N random vectors, possibly complex, all of these having common size $L \times 1$. Suppose we wish to estimate another $M \times 1$ random vector $\mathbf{x}$ by a linear combination of the vectors $\mathbf{y}_i$. The estimate of $\mathbf{x}$ has the form
$$\widehat{\mathbf{x}} = \sum_{i=1}^{N} \mathbf{A}_i \mathbf{y}_i,$$
where $\mathbf{A}_i$ are $M \times L$ matrices. This can be reformulated as in Eq. (F.1) simply by defining $\mathbf{A} = [\mathbf{A}_1 \quad \ldots \quad \mathbf{A}_N]$ and defining $\mathbf{y}$ in an obvious way. The optimum choice of $\mathbf{A}_i$ is such that the estimation error $\widehat{\mathbf{x}} - \mathbf{x}$ is orthogonal to each of the vectors $\mathbf{y}_i$.

F.2.2 The Wiener filter

Figure F.2 shows an $M \times 1$ vector sequence $\mathbf{x}(n)$ contaminated with additive noise $\mathbf{w}(n)$. The noisy signal
$$\mathbf{y}(n) = \mathbf{x}(n) + \mathbf{w}(n)$$
is filtered by a linear time invariant filter with transfer matrix $\mathbf{G}(z)$. The goal is to choose the filter such that the output $\widehat{\mathbf{x}}(n)$ is as close to $\mathbf{x}(n)$ as possible. In order to formulate this problem mathematically we assume that $\mathbf{x}(n)$ and $\mathbf{w}(n)$ are wide sense stationary (WSS) random processes with power spectra $\mathbf{S_{xx}}(z)$ and $\mathbf{S_{ww}}(z)$ (Appendix E).

Assume further that $\mathbf{x}(n)$ and $\mathbf{w}(n)$ are **jointly WSS** (Appendix E) so that their cross power spectrum $\mathbf{S_{xw}}(z)$ is well defined.[2] Then $\widehat{\mathbf{x}}(n)$ is also WSS, and so is the error
$$\mathbf{e}(n) = \widehat{\mathbf{x}}(n) - \mathbf{x}(n). \tag{F.10}$$
Define the mean square error
$$\mathcal{E} = E[\mathbf{e}^\dagger(n)\mathbf{e}(n)]. \tag{F.11}$$

[2] As elaborated in Appendix E, the notation $\mathbf{R_{xw}}(k)$ stands for the cross correlation matrix between two jointly WSS processes $\mathbf{x}(n)$ and $\mathbf{w}(n)$. The z-transform of $\mathbf{R_{xw}}(k)$ is denoted $\mathbf{S_{xw}}(z)$. This (evaluated for $z = e^{j\omega}$) is the cross power spectrum.

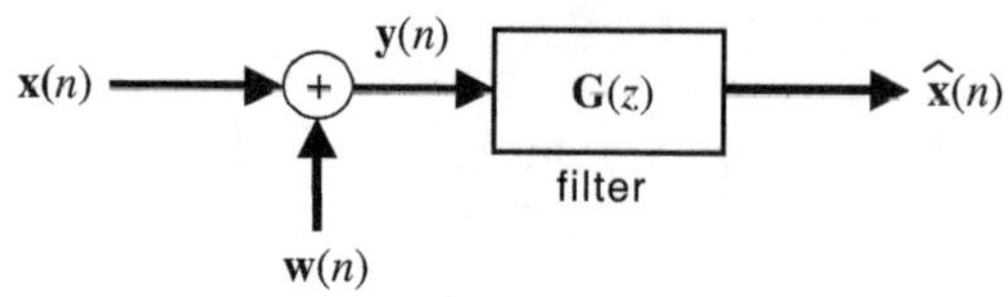

Figure F.2. Filtering of a signal contaminated by additive noise. Here $\mathbf{x}(n)$ is an $M \times 1$ vector sequence and so is the output $\widehat{\mathbf{x}}(n)$.

The filter $\mathbf{G}(z)$ which minimizes this quantity is called the **Wiener filter** for this pair $\{\mathbf{x}(n), \mathbf{w}(n)\}$. We will see that this filter has a simple closed form expression and can be expressed in terms of the various power spectra.

♠**Theorem F.1.** *The Wiener filter.* In the above setting, the optimal filter $\mathbf{G}(z)$ which minimizes the mean square error $\mathcal{E}$ is given by

$$\mathbf{G}(z) = \mathbf{S_{xy}}(z)\mathbf{S}^{-1}_{\mathbf{yy}}(z), \tag{F.12}$$

assuming $[\det \mathbf{S_{yy}}(e^{j\omega})]$ is nonzero for all ω. The expressions

$$\mathbf{G}(z) = [\mathbf{S_{xx}}(z) + \mathbf{S_{xw}}(z)]\mathbf{S}^{-1}_{\mathbf{yy}}(z) \tag{F.13}$$

and

$$\mathbf{G}(z) = [\mathbf{S_{yy}}(z) - \mathbf{S_{wy}}(z)]\mathbf{S}^{-1}_{\mathbf{yy}}(z) \tag{F.14}$$

are equivalent forms of Eq. (F.12). ◇

Proof. The filter output is given by the matrix vector convolution

$$\widehat{\mathbf{x}}(n) = \sum_m \mathbf{g}(m)\mathbf{y}(n-m), \tag{F.15}$$

where $\mathbf{g}(m)$ is the impulse response matrix, that is $\mathbf{G}(z) = \sum_m \mathbf{g}(m)z^{-m}$. Applying the orthogonality principle (Lemma F.1) we find that the best filter should satisfy the condition

$$E[\mathbf{e}(n)\mathbf{y}^\dagger(n-k)] = \mathbf{0}, \quad \text{for all } k, \tag{F.16}$$

where $\mathbf{e}(n) = \widehat{\mathbf{x}}(n) - \mathbf{x}(n)$. That is,

$$E[\mathbf{x}(n)\mathbf{y}^\dagger(n-k)] = E[\widehat{\mathbf{x}}(n)\mathbf{y}^\dagger(n-k)], \quad \text{for all } k. \tag{F.17}$$

By joint wide sense stationarity, the above expectations are independent of n. Thus the left-hand side is $\mathbf{R_{xy}}(k)$. Substituting from Eq. (F.15) the equation simplifies to

$$\mathbf{R_{xy}}(k) = \sum_m \mathbf{g}(m)\mathbf{R_{yy}}(k-m). \tag{F.18}$$

Taking z-transforms on both sides, we obtain $\mathbf{S}_{\mathbf{xy}}(z) = \mathbf{G}(z)\mathbf{S}_{\mathbf{yy}}(z)$, from which the stated form (F.12) of $\mathbf{G}(z)$ follows. The alternative forms in Eqs. (F.13) and (F.14) follow from identities such as $\mathbf{S}_{(\mathbf{x}_1+\mathbf{x}_2),\mathbf{y}}(z) = \mathbf{S}_{\mathbf{x}_1,\mathbf{y}}(z) + \mathbf{S}_{\mathbf{x}_2,\mathbf{y}}(z)$ and so forth (Appendix E). $\bigtriangledown\bigtriangledown\bigtriangledown$

F.2.2.A Special Cases

1. ***When is the Wiener filter identity?*** Under what conditions do we have $\mathbf{G}(z) = \mathbf{I}$? We see from Eq. (F.14) that this happens if and only if $\mathbf{S}_{\mathbf{wy}}(z) = \mathbf{0}$, or equivalently $\mathbf{R}_{\mathbf{wy}}(k) = \mathbf{0}$. In other words, the *noise* $\mathbf{w}(n)$ and the *noisy signal* $\mathbf{y}(m)$ must be orthogonal for all choices of n, m. This is expressed by saying that the *processes* $\mathbf{w}(.)$ and $\mathbf{y}(.)$ are orthogonal. (This is stronger than requiring that the samples $\mathbf{w}(n)$ and $\mathbf{y}(n)$ be orthogonal for each n.) Such a property indeed arises in the literature on optimal vector quantization where the quantization error is orthogonal to the quantized result [Gersho and Gray, 1992]. But we do not usually encounter this situation in communication systems.

2. ***Case when noise is orthogonal to signal.*** What happens when the signal and noise components $\mathbf{x}(.)$ and $\mathbf{w}(.)$ are orthogonal (so that $\mathbf{S}_{\mathbf{xw}}(z) = 0$)? In this case Eq. (F.13) shows that $\mathbf{G}(z) = \mathbf{S}_{\mathbf{xx}}(z)\mathbf{S}_{\mathbf{yy}}^{-1}(z)$, which further simplifies to
$$\mathbf{G}(z) = \mathbf{S}_{\mathbf{xx}}(z)[\mathbf{S}_{\mathbf{xx}}(z) + \mathbf{S}_{\mathbf{ww}}(z)]^{-1}. \tag{F.19}$$
This is a very important expression. We make repeated use of it in many chapters.

3. ***Memoryless case.*** In Fig. F.2 suppose we are interested in finding a constant matrix $\mathbf{T}$ rather than a filter $\mathbf{G}(z)$. In this case, the optimal $\mathbf{T}$ can be obtained by modifying the steps in the proof of Theorem F.1 as follows: Equation (F.16) has to hold only for $k = 0$, and Eq. (F.18) is replaced with
$$\mathbf{R}_{\mathbf{xy}}(0) = \mathbf{T}\mathbf{R}_{\mathbf{yy}}(0), \tag{F.20}$$
so that the optimal solution is
$$\mathbf{T} = \mathbf{R}_{\mathbf{xy}}(0)\mathbf{R}_{\mathbf{yy}}^{-1}(0). \tag{F.21a}$$
This assumes nonsingularity of $\mathbf{R}_{\mathbf{yy}}(0)$.[3] We can write
$$\mathbf{R}_{\mathbf{xy}}(0) = \mathbf{R}_{\mathbf{xx}}(0) + \mathbf{R}_{\mathbf{xw}}(0) = \mathbf{R}_{\mathbf{yy}}(0) - \mathbf{R}_{\mathbf{wy}}(0)$$
using $\mathbf{y}(n) = \mathbf{x}(n) + \mathbf{w}(n)$. Thus the optimal $\mathbf{T}$ can also be written in the forms
$$\mathbf{T} = \Big(\mathbf{R}_{\mathbf{xx}}(0) + \mathbf{R}_{\mathbf{xw}}(0)\Big)\mathbf{R}_{\mathbf{yy}}^{-1}(0) \tag{F.21b}$$

[3]If this is not true then the components of $\mathbf{y}(n)$ are linearly dependent. So we can find $\mathbf{v} \neq \mathbf{0}$ such that $\mathbf{v}^\dagger\mathbf{y}(n) = 0$ for all n and thereby reduce the size of the vector process $\mathbf{y}(n)$ [Therrien, 1992], [Vaidyanathan, 2008].

and

$$\mathbf{T} = \Big(\mathbf{R_{yy}}(0) - \mathbf{R_{wy}}(0)\Big)\mathbf{R}_{\mathbf{yy}}^{-1}(0). \tag{F.21c}$$

From this we can verify the following:

(a) $\mathbf{T} = \mathbf{I}$ if and only if $\mathbf{R_{wy}}(0) = \mathbf{0}$, that is, the *noise* $\mathbf{w}(n)$ and the *noisy signal* $\mathbf{y}(n)$ are orthogonal for each n. This is weaker than the requirement for $\mathbf{G}(z) = \mathbf{I}$ in remark 1 above.

(b) When the noise $\mathbf{w}(n)$ and the signal $\mathbf{x}(n)$ are orthogonal for each n, that is, $\mathbf{R_{xw}}(0) = \mathbf{0}$, then

$$\mathbf{T} = \mathbf{R_{xx}}(0)[\mathbf{R_{yy}}(0)]^{-1} = \mathbf{R_{xx}}(0)[\mathbf{R_{xx}}(0) + \mathbf{R_{ww}}(0)]^{-1}. \tag{F.22}$$

This is the counterpart of Eq. (F.19) for the memoryless case.

A very important practical case arises when the Wiener filter is constrained to be an FIR matrix of the form $\mathbf{G}(z) = \sum_{n=0}^{N} \mathbf{g}(n)z^{-n}$. In this case it is possible to develop a time-domain formulation of the optimal filtering problem and solve for the optimal impulse response matrices $\mathbf{g}(n)$ directly.

F.3 Wiener filter for zero-mean uncorrelated noise

We now consider the special case of the Wiener filtering problem where the additive noise $\mathbf{w}(n)$ has zero mean, that is,

$$E[\mathbf{w}(n)] = \mathbf{0}, \tag{F.23}$$

and is uncorrelated to the signal, that is,

$$E[\mathbf{x}(n)\mathbf{w}^{\dagger}(m)] = \mathbf{0}, \tag{F.24}$$

for all m, n. This is a common assumption made in the analysis of communication systems. By definition, the condition (F.24) says that the random processes are orthogonal, but the term *uncorrelated* is equally appropriate because of the zero-mean property (F.23). Since Eq. (F.24) implies that $\mathbf{S_{xw}}(z) = \mathbf{0}$, the Wiener filter reduces to Eq. (F.19), that is,

$$\mathbf{G}(z) = \mathbf{S_{xx}}(z)[\mathbf{S_{xx}}(z) + \mathbf{S_{ww}}(z)]^{-1}. \tag{F.25}$$

Since this is a popular special case arising in communication applications, we shall derive an expression for the mean square value of the error

$$\mathbf{e}(n) = \widehat{\mathbf{x}}(n) - \mathbf{x}(n).$$

In the following discussions, z is an abbreviation for $e^{j\omega}$ and $\widetilde{\mathbf{G}}(z)$ is transpose conjugation on the unit circle. (Recall from Sec. 1.6 that $\widetilde{\mathbf{G}}(z) = [\mathbf{G}(1/z^*)]^{\dagger}$.)

♠**Theorem F.2.** *Power spectrum of error.* Assume the noise $\mathbf{w}(n)$ is zero-mean and uncorrelated to the signal $\mathbf{x}(n)$, so that the Wiener filter is as in Eq. (F.25). Then the error $\mathbf{e}(n)$ has the power spectrum

$$\mathbf{S_{ee}}(e^{j\omega}) = \left(\mathbf{S}_{\mathbf{xx}}^{-1}(e^{j\omega}) + \mathbf{S}_{\mathbf{ww}}^{-1}(e^{j\omega})\right)^{-1}, \tag{F.26}$$

assuming that $\mathbf{S_{ww}}(e^{j\omega})$ and $\mathbf{S_{xx}}(e^{j\omega})$ are invertible for all ω. Thus the error spectrum is the *inverse of the sum of inverses* of the spectra of the signal and the noise. ◇

Proof. Since $\widehat{\mathbf{x}}(n) = [\mathbf{g} * (\mathbf{x}+\mathbf{w})](n)$, where $\mathbf{g}(n)$ is the impulse response of $\mathbf{G}(z)$ and $*$ denotes convolution, we see that

$$\mathbf{e}(n) = [\mathbf{g} * \mathbf{x}](n) \ + \ [\mathbf{g} * \mathbf{w}](n) \ - \ \mathbf{x}(n). \tag{F.27}$$

The orthogonality principle dictates that $\mathbf{e}(n)$ be orthogonal to $\mathbf{y}(m)$ for all n, m. Since $\widehat{\mathbf{x}}(n)$ is a linear combination of samples of $\mathbf{y}(.)$, it then follows that $\widehat{\mathbf{x}}(n)$ is orthogonal to $\mathbf{e}(m)$ for all n, m. Thus

$$E[\mathbf{e}(n)\widehat{\mathbf{x}}^\dagger(n-k)] = \mathbf{0},$$

for all k, which implies that $\mathbf{S}_{\mathbf{e}\widehat{\mathbf{x}}}(z) = \mathbf{0}$. Thus

$$\mathbf{S_{ee}}(z) = \mathbf{S}_{\mathbf{e},(\widehat{\mathbf{x}}-\mathbf{x})}(z) = \mathbf{S}_{\mathbf{e}\widehat{\mathbf{x}}}(z) - \mathbf{S_{ex}}(z) = -\mathbf{S_{ex}}(z).$$

From Eq. (F.27) we therefore obtain

$$\mathbf{S_{ee}}(z) = -\mathbf{S_{ex}}(z) = -\mathbf{G}(z)\mathbf{S_{xx}}(z) - \mathbf{G}(z)\mathbf{S_{wx}}(z) + \mathbf{S_{xx}}(z).$$

In view of the assumption that $\mathbf{S_{wx}}(z) = \mathbf{0}$, it therefore follows that

$$\mathbf{S_{ee}}(z) = (\mathbf{I} - \mathbf{G}(z))\mathbf{S_{xx}}(z). \tag{F.28}$$

Substituting for $\mathbf{G}(z)$ from the Wiener filter expression (F.25) we have

$$\begin{aligned}
\mathbf{I} - \mathbf{G}(z) &= \mathbf{I} - \mathbf{S_{xx}}(z)[\mathbf{S_{xx}}(z) + \mathbf{S_{ww}}(z)]^{-1} \\
&= [\mathbf{S_{xx}}(z) + \mathbf{S_{ww}}(z) - \mathbf{S_{xx}}(z)][\mathbf{S_{xx}}(z) + \mathbf{S_{ww}}(z)]^{-1} \\
&= \mathbf{S_{ww}}(z)[\mathbf{S_{xx}}(z) + \mathbf{S_{ww}}(z)]^{-1}.
\end{aligned}$$

Using this in Eq. (F.28) we have

$$\mathbf{S_{ee}}(z) = \mathbf{S_{ww}}(z)[\mathbf{S_{xx}}(z) + \mathbf{S_{ww}}(z)]^{-1}\mathbf{S_{xx}}(z),$$

which indeed can be rewritten as Eq. (F.26). ▽ ▽ ▽

The error covariance martrix is simply the integral of the power spectral matrix, that is,

$$\mathbf{R_{ee}}(0) = E[\mathbf{e}(n)\mathbf{e}^\dagger(n)], = \frac{1}{2\pi}\int_0^{2\pi} \mathbf{S_{ee}}(e^{j\omega})d\omega,$$

and the MMSE

$$E[\mathbf{e}^\dagger(n)\mathbf{e}(n)] = \mathrm{Tr}[\mathbf{R}_{\mathbf{ee}}(0)], \tag{F.29}$$

can be evaluated from here.

Memoryless case with zero-mean uncorrelated noise. For the case where we restrict the Wiener filter to be a memoryless matrix $\mathbf{T}$ rather than a filter $\mathbf{G}(z)$, Eqs. (F.25), (F.26) should be replaced with

$$\mathbf{T} = \mathbf{R}_{\mathbf{xx}}(0)[\mathbf{R}_{\mathbf{xx}}(0) + \mathbf{R}_{\mathbf{ww}}(0)]^{-1} \tag{F.30}$$

and

$$\mathbf{R}_{\mathbf{ee}}(0) = [\mathbf{R}_{\mathbf{xx}}^{-1}(0) + \mathbf{R}_{\mathbf{ww}}^{-1}(0)]^{-1} \tag{F.31}$$

where $\mathbf{R}_{\mathbf{xx}}^{-1}(0)$ denotes the inverse of $\mathbf{R}_{\mathbf{xx}}(0)$, and so on.

F.3.1 Special case of scalar Wiener filter

Consider the special case of scalar signals. Assume that the noise $w(.)$ is zero-mean and uncorrelated to the signal $x(.)$. From the preceding results we conclude that the Wiener filter is

$$G(z) = \frac{S_{xx}(z)}{S_{xx}(z) + S_{ww}(z)} \tag{F.32}$$

and that the power spectrum of the minimized error is

$$S_{ee}(z) = \frac{1}{\frac{1}{S_{xx}(z)} + \frac{1}{S_{ww}(z)}} = \frac{S_{xx}(z)S_{ww}(z)}{S_{xx}(z) + S_{ww}(z)} \tag{F.33}$$

Recall now that the harmonic mean $\mathcal{H}$ of two nonzero numbers a and b is defined as the reciprocal of the arithmetic mean of their reciprocals, that is,

$$\mathcal{H} = \frac{1}{0.5(\frac{1}{a} + \frac{1}{b})} \tag{F.34}$$

Thus the error spectrum is half the **harmonic mean** of the signal and noise spectra. For the case where $G(z)$ is restricted to be a constant multiplier α we can verify from Eqs. (F.30) and (F.31) that the optimum (Wiener) multiplier is

$$\alpha = \frac{\sigma_x^2}{\sigma_x^2 + \sigma_w^2} \tag{F.35}$$

and that the minimized mean square error is

$$\mathcal{E}_e = \frac{\sigma_x^2\sigma_w^2}{\sigma_x^2 + \sigma_w^2} \tag{F.36}$$

which can be rewritten in the following forms

$$\mathcal{E}_e = \sigma_x^2\Big(\frac{\sigma_w^2}{\sigma_x^2 + \sigma_w^2}\Big) = \sigma_w^2\Big(\frac{\sigma_x^2}{\sigma_x^2 + \sigma_w^2}\Big) = \frac{1}{\frac{1}{\sigma_x^2} + \frac{1}{\sigma_w^2}} \tag{F.37}$$

This shows in particular that the error $\mathcal{E}_e$ is smaller than both σ_x^2 and σ_w^2.

F.4 Concluding remarks

In transceiver design the primary use of Wiener filter theory is in the design of the receiver filter for fixed transmitting filter, channel matrix, and noise. This is the first step in any attempt to optimize the transmitter and receiver jointly, for a fixed channel. All the filters involved are, in general, MIMO systems, and the results of this appendix are directly applicable.

Appendix G

Review of concepts from sampling theory

G.1 Introduction

In this appendix we review some concepts relating to sampling theory. If a continuous-time signal $x_c(t)$ is uniformly sampled with sample spacing T seconds, then the sampling rate is denoted as

$$f_s = \frac{1}{T} \quad \text{Hz}, \tag{G.1}$$

or equivalently as

$$\omega_s = \frac{2\pi}{T} \quad \text{radians/s}. \tag{G.2}$$

The sampled signal is

$$x_s(t) = \sum_{n=-\infty}^{\infty} x_c(t)\delta_c(t - nT), \tag{G.3}$$

where $\delta_c(t)$ is the continuous-time Dirac delta function. The Fourier transform of $x_s(t)$ is given by

$$X_s(j\omega) = \frac{1}{T}\sum_{k=-\infty}^{\infty} X_c(j(\omega + k\omega_s)) = \frac{1}{T}\sum_{k=-\infty}^{\infty} X_c\Big(j(\omega + \frac{2\pi k}{T})\Big). \tag{G.4}$$

This is a periodic function with period $2\pi/T$. Defining the discrete-time signal

$$x_d(n) \stackrel{\Delta}{=} x_c(nT),$$

we have

$$X_d(e^{j\omega}) = \frac{1}{T}\sum_k X_c\Big(j\Big(\frac{\omega + 2\pi k}{T}\Big)\Big). \tag{G.5}$$

This is a periodic function with period 2π. We use the notation $[X_c(j\omega)]_{\downarrow T}$ to denote this. Thus

$$X_d(e^{j\omega}) = [X_c(j\omega)]_{\downarrow T} = \frac{1}{T}\sum_k X_c\Big(j\Big(\frac{\omega + 2\pi k}{T}\Big)\Big).$$

G.2 Noble identities for C/D and D/C converters

We now derive some simple identities pertaining to the C/D and D/C operators defined in Sec. 4.2. These are shown in Figs. G.1 and G.2, and are called the **noble identities**. These are quite easily proved as follows:

Proof. First consider Fig. G.1. Since

$$X_d(e^{j\omega}) = \frac{1}{T}\sum_k X_c\Big(j\Big(\frac{\omega + 2\pi k}{T}\Big)\Big)$$

it follows that

$$Y(e^{j\omega}) = \frac{P(e^{j\omega})}{T}\sum_k X_c\Big(j\Big(\frac{\omega + 2\pi k}{T}\Big)\Big)$$

for the system on the left. For the system on the right $Y_c(j\omega) = X_c(j\omega)P(e^{j\omega T})$ so that

$$\begin{aligned} Y(e^{j\omega}) &= \frac{1}{T}\sum_k X_c\Big(j\Big(\frac{\omega + 2\pi k}{T}\Big)\Big)P\Big(e^{j\left(\frac{\omega+2\pi k}{T}\right)T}\Big) \\ &= \frac{P(e^{j\omega})}{T}\sum_k X_c\Big(j\Big(\frac{\omega + 2\pi k}{T}\Big)\Big), \end{aligned}$$

which proves the equivalence of the two systems. Next consider Fig. G.2. For the system on the left, we have

$$Y_c(j\omega) = P(e^{j\omega T})X_c(j\omega) = P(e^{j\omega T})X(e^{j\omega T}),$$

whereas for the system on the right $Y_c(j\omega) = Y(e^{j\omega T}) = P(e^{j\omega T})X(e^{j\omega T})$, which proves that the two systems are equivalent. $\triangledown\triangledown\triangledown$

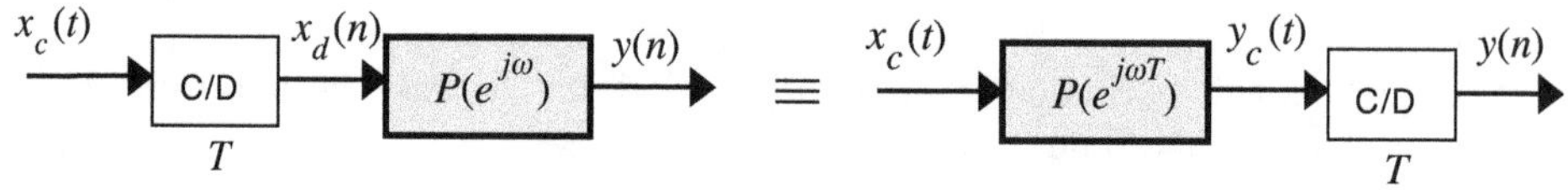

Figure G.1. The noble identity for C/D converters.

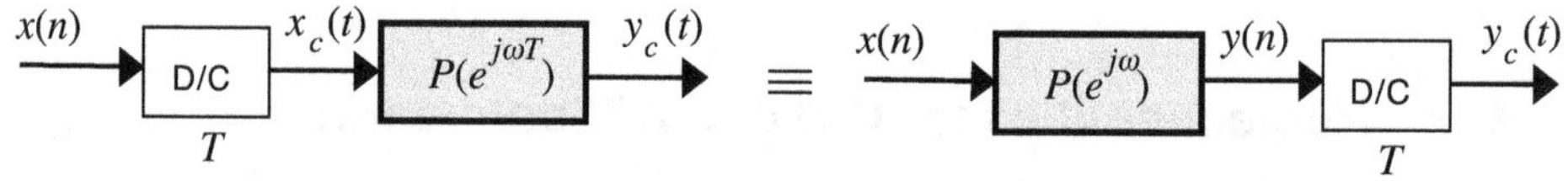

Figure G.2. The noble identity for D/C converters.

Note that the first noble identity can also be written in the form of an equation, namely

$$P(e^{j\omega})\Big[X_c(j\omega)\Big]_{\downarrow T} = \Big[X_c(j\omega)P(e^{j\omega T})\Big]_{\downarrow T}. \qquad (G.6)$$

G.3 The generalized alias-free(T) band

We know that if a signal $x(t)$ is bandlimited to the interval $-\sigma < \omega < \sigma$, then it can be sampled at the rate $\omega_s = 2\sigma$ without aliasing. This is because the terms in

$$X_s(j\omega) = \frac{1}{T}\sum_{k=-\infty}^{\infty} X(j(\omega + k\omega_s))$$

do not overlap for any two values of k. Equivalently, for any fixed ω, we do not have more than one nonzero term in the summation. We say that the interval $-\sigma < \omega < \sigma$ represents an alias-free(T) band for the sampling rate ω_s. Here T represents the sample spacing

$$T = \frac{2\pi}{\omega_s}$$

as usual. The lowpass band $-\sigma < \omega < \sigma$ shown in Fig. G.3(a) is not the only possible band of this kind. For example, the bandpass region

$$-\sigma + c < \omega < \sigma + c$$

is a valid alias-free(T) band for any constant c. See Fig. G.3(b). A third example is shown in part (c) of the figure, where the band of length 2σ has been split into two halves. If the individual halves are appropriately positioned, then the copies of the left half shifted in multiples of 2σ do not overlap with similar shifted copies of the right half. This happens if and only if σ_1 (equivalently σ_2) is a multiple of σ. In this case the combination of the two bands is an alias-free(T)

band. More generally the alias-free(T) band can be a union of non overlapping subbands

$$\omega_{\ell,k} < \omega < \omega_{r,k}, \tag{G.7}$$

where $\omega_{\ell,k}$ and $\omega_{r,k}$ are the left and right edges of the kth subband. Figure G.3(d) shows how to create such an example: start from a traditional lowpass alias-free(T) band, cut out an abritrary portion, and move it by a multiple of ω_s. We can repeat this any number of times and create multiple pieces, the sum of whose lengths is still 2σ. When we sample a signal bandlimited to such a region with sampling rate $\omega_s = 2\sigma$, there is no aliasing.

♠**Definition G.1.** *Alias-free(T) bands.* We say that a frequency band $\mathcal{A}$ (which can be made of many individual pieces) is an alias-free(T) band (or alias-free($2\pi/\omega_s$) band where $\omega_s = 2\pi/T$), if the regions

$$\mathcal{A} + k\omega_s$$

do not overlap for any two distinct integers k, and if the total length of $\mathcal{A}$ (the sum of the bandwidths of individual pieces) is precisely equal to ω_s. The signal $x(t)$ is said to be alias-free(T) if $X(j\omega)$ is restricted to some alias-free(T) region so that no two terms in the summation

$$X_s(j\omega) = \frac{1}{T} \sum_{k=-\infty}^{\infty} X(j(\omega + k\omega_s)) \tag{G.8}$$

overlap. In this case the sampled version $x(nT)$ suffers no aliasing. ◇

Remarks

1. The union of the sets
$$\cup_{k=-\infty}^{\infty}(\mathcal{A} + k\omega_s)$$
is the set of all frequencies in $-\infty < \omega < \infty$. Since the sets $\mathcal{A} + k\omega_s$ are disjoint for different k, these sets form a partition of the real line into subbands. Thus, given any frequency ω in $-\infty < \omega < \infty$, there is precisely one integer k such that $\omega + k\omega_s$ is in $\mathcal{A}$.

2. An alias-free(T) signal $x(t)$ can be reconstructed from the samples $x(nT)$ by using an ideal reconstruction filter which is bandlimited to the alias-free(T) region, that is,
$$H_{reco}(j\omega) = \begin{cases} T & \text{if } \omega \in \text{alias-free}(T) \\ 0 & \text{otherwise.} \end{cases}$$
The scale factor T is needed to cancel the factor $1/T$ in Eq. (G.8). Thus we have
$$x(t) = \sum_n x(nT) h_{reco}(t - nT).$$

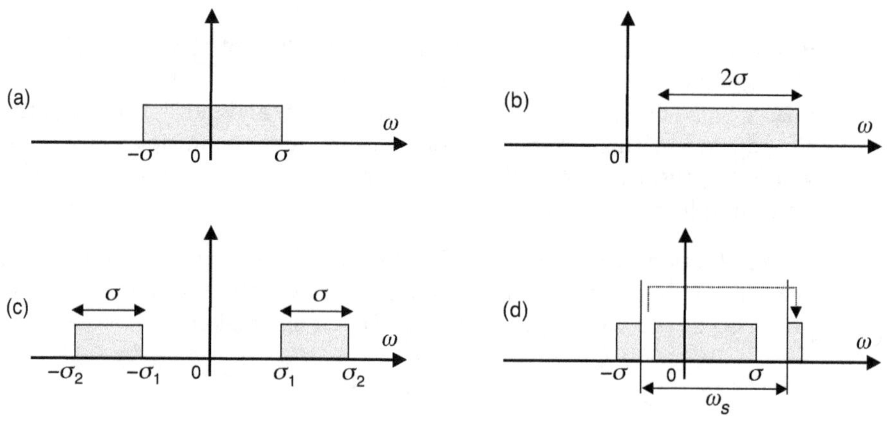

Figure G.3. (a)–(d). Examples of alias-free(T) bands.

G.4 Alias-free(T) signals with identical samples

We conclude this section with a few interesting results about the sampled versions of alias-free(T) signals:

♠**Lemma G.1.** Given an arbitrary Fourier transform $Y(j\omega)$, suppose we define the alias-free(T) Fourier transform

$$Y_a(j\omega) = \begin{cases} \sum_k Y(j(\omega + k\omega_s)) & \omega \in \mathcal{A} \\ 0 & \text{otherwise,} \end{cases} \tag{G.9}$$

for some alias-free($2\pi/\omega_s$) region $\mathcal{A}$. Then

$$\sum_\ell Y_a(j(\omega + \ell\omega_s)) = \sum_k Y(j(\omega + k\omega_s)), \tag{G.10}$$

for all ω. That is, the sampled versions $y_a(nT)$ and $y(nT)$ are identical. $\diamond$

Proof. Pick a frequency ω_0. Since $\mathcal{A}$ is an alias-free($2\pi/\omega_s$) region, the quantity $\omega_0 + \ell\omega_s$ belongs to $\mathcal{A}$ precisely for one value of the integer ℓ, say $\ell = \ell_0$. From the definition (G.9) we then have

$$\begin{aligned} \sum_\ell Y_a(j(\omega_0 + \ell\omega_s)) &= Y_a(j(\omega_0 + \ell_0\omega_s)) \\ &= \sum_k Y(j(\omega_0 + \ell_0\omega_s + k\omega_s)) \\ &= \sum_k Y(j(\omega_0 + k\omega_s)) \end{aligned}$$

because the sum is over all k. This proves Eq. (G.10) for any ω_0. $\bigtriangledown \bigtriangledown \bigtriangledown$

Reflections of alias-free(T) regions

Let $\mathcal{A}_1$ and $\mathcal{A}_2$ be two alias-free($2\pi/\omega_s$) regions. Given any frequency ω_1 in $\mathcal{A}_1$, there is precisely one integer k such that

$$\omega_2 \stackrel{\Delta}{=} \omega_1 + k\omega_s$$

is in $\mathcal{A}_2$. If we have a signal $X_{\mathcal{A}_1}(j\omega)$ which is bandlimited to $\mathcal{A}_1$ we can therefore define a unique signal $X_{\mathcal{A}_2}(j\omega)$ bandlimited to $\mathcal{A}_2$ such that

$$X_{\mathcal{A}_2}(j\omega_2) = X_{\mathcal{A}_1}(j\omega_1).$$

We say that $X_{\mathcal{A}_2}(j\omega)$ is the *reflection* of $X_{\mathcal{A}_1}(j\omega)$ onto $\mathcal{A}_2$. Similarly, $X_{\mathcal{A}_1}(j\omega)$ is the reflection of $X_{\mathcal{A}_2}(j\omega)$ onto $\mathcal{A}_1$.

♠**Lemma G.2.** With $X_{\mathcal{A}_1}(j\omega)$ and $X_{\mathcal{A}_2}(j\omega)$ defined as above, the sampled versions of $X_{\mathcal{A}_1}(j\omega)$ and $X_{\mathcal{A}_2}(j\omega)$ (with sampling rate ω_s) are identical, that is,

$$x_{\mathcal{A}_1}(nT) = x_{\mathcal{A}_2}(nT),$$

for all n, or equivalently

$$\sum_k X_{\mathcal{A}_1}(j(\omega + k\omega_s)) = \sum_k X_{\mathcal{A}_2}(j(\omega + k\omega_s)), \tag{G.11}$$

for all ω. $\diamondsuit$

Proof. Given any frequency ω_0 there is at most one nonzero term on the left-hand side of Eq. (G.11). Thus

$$\begin{aligned}
\sum_k X_{\mathcal{A}_1}(j(\omega_0 + k\omega_s)) &= X_{\mathcal{A}_1}(j(\omega_0 + k_1\omega_s)) \quad \text{(for some } k_1\text{)} \\
&= X_{\mathcal{A}_2}(j(\omega_0 + k_2\omega_s)) \quad \text{(for some } k_2\text{)} \\
&= \sum_k X_{\mathcal{A}_2}(j(\omega_0 + k\omega_s))
\end{aligned}$$

because the right-hand side is zero for $k \neq k_2$ anyway. $\bigtriangledown \bigtriangledown \bigtriangledown$

Appendix H

Euclid's algorithm

In this appendix we review one of Euclid's fundamental results concerning polynomials. This can be stated as follows:

♠**Theorem H.1.** *Euclid's theorem.* Consider two polynomials with degrees N_0 and N_1, given by

$$R_0(x) = \sum_{n=0}^{N_0} r_0(n)x^n, \quad R_1(x) = \sum_{n=0}^{N_1} r_1(n)x^n. \tag{H.1}$$

Assuming that these do not have any common factors with degree > 0, there exists a unique pair of polynomials

$$E_0(x) = \sum_{n=0}^{N_1-1} e_0(n)x^n, \quad E_1(x) = \sum_{n=0}^{N_0-1} e_1(n)x^n \tag{H.2}$$

such that

$$R_0(x)E_0(x) + R_1(x)E_1(x) = 1 \tag{H.3}$$

for all x. ◇

A number of points should first be noted.

1. Note the use of x instead of z^{-1}. This is mostly for convenience.
2. Since the theorem requires that there be no common factors, in particular x cannot be a common factor. So $r_0(0)$ and $r_1(0)$ cannot *both* be zero.
3. Observe carefully that the theorem asserts that there exist polynomials $E_0(x)$ and $E_1(x)$ with *smaller* degrees, that is,

   $$\deg E_1(x) < \deg R_0(x), \quad \deg E_0(x) < \deg R_1(x) \tag{H.4}$$

   satisfying Eq. (H.3). If the polynomials $E_0(x)$ and $E_1(x)$ are allowed to have higher degree, then there exist infinitely many solutions satisfying Eq. (H.3) as shown in Sec. 4.8.4.

Proof of Theorem H.1. First let us rewrite Eq. (H.3) in terms of matrix vector multiplications. The product $R_k(x)E_k(x)$ is a polynomial whose coefficients are obtained by convolving the coefficients of $R_k(x)$ and $E_k(x)$. Thus Eq. (H.3) can be rewritten as

$$\mathbf{R}_0\mathbf{e}_0 + \mathbf{R}_1\mathbf{e}_1 = \begin{bmatrix} 1 \\ 0 \\ \vdots \\ 0 \end{bmatrix}, \tag{H.5}$$

where $\mathbf{e}_0$ is a column vector with the N_1 elements of $E_0(x)$, $\mathbf{e}_1$ is a column vector with the N_0 elements of $E_1(x)$, and the right-hand side is the unit vector with $N_0 + N_1$ elements (number of coefficients in $R_k(x)E_k(x)$). The matrices $\mathbf{R}_0$ and $\mathbf{R}_1$ have elements which are coefficients of the polynomials $R_0(x)$ and $R_1(x)$. The sizes are $(N_0+N_1)\times N_1$ for $R_0(x)$ and $(N_0+N_1)\times N_0$ for $R_1(x)$. For example, if $N_0 = 2$ and $N_1 = 3$, Eq. (H.5) takes the form

$$\begin{bmatrix} r_0(0) & 0 & 0 \\ r_0(1) & r_0(0) & 0 \\ r_0(2) & r_0(1) & r_0(0) \\ 0 & r_0(2) & r_0(1) \\ 0 & 0 & r_0(2) \end{bmatrix} \begin{bmatrix} e_0(0) \\ e_0(1) \\ e_0(2) \end{bmatrix} + \begin{bmatrix} r_1(0) & 0 \\ r_1(1) & r_1(0) \\ r_1(2) & r_1(1) \\ r_1(3) & r_1(2) \\ 0 & r_1(3) \end{bmatrix} \begin{bmatrix} e_1(0) \\ e_1(1) \end{bmatrix} = \begin{bmatrix} 1 \\ 0 \\ 0 \\ 0 \\ 0 \end{bmatrix}. \tag{H.6}$$

The matrices $\mathbf{R}_0$ and $\mathbf{R}_1$ turn out to be lower triangular and Toeplitz because the product $R_k(x)E_k(x)$ represents a convolution of the polynomial coefficients. It is more economical to write Eq. (H.5) in the form

$$\underbrace{\begin{bmatrix} \mathbf{R}_0 & \mathbf{R}_1 \end{bmatrix}}_{\text{call this } \mathbf{R}} \underbrace{\begin{bmatrix} \mathbf{e}_0 \\ \mathbf{e}_1 \end{bmatrix}}_{\text{call this } \mathbf{e}} = \begin{bmatrix} 1 \\ 0 \\ \vdots \\ 0 \end{bmatrix}, \tag{H.7}$$

where $\mathbf{R}$ is a square matrix with (N_0+N_1) rows and columns. For example, Eq. (H.6) becomes

$$\underbrace{\begin{bmatrix} r_0(0) & 0 & 0 & r_1(0) & 0 \\ r_0(1) & r_0(0) & 0 & r_1(1) & r_1(0) \\ r_0(2) & r_0(1) & r_0(0) & r_1(2) & r_1(1) \\ 0 & r_0(2) & r_0(1) & r_1(3) & r_1(2) \\ 0 & 0 & r_0(2) & 0 & r_1(3) \end{bmatrix}}_{\mathbf{R}} \underbrace{\begin{bmatrix} e_0(0) \\ e_0(1) \\ e_0(2) \\ e_1(0) \\ e_1(1) \end{bmatrix}}_{\mathbf{e}} = \begin{bmatrix} 1 \\ 0 \\ 0 \\ 0 \\ 0 \end{bmatrix}. \tag{H.8}$$

Our goal is to prove that under the conditions stated in the theorem there exists a unique vector $\mathbf{e}$ satisfying Eq. (H.7). This is equivalent to proving that $\mathbf{R}$ is nonsingular. Assume this is *not* the case. Then there exists a nonzero vector $\mathbf{e}$ such that $\mathbf{Re} = \mathbf{0}$, that is, there exist $E_0(x)$ and $E_1(x)$, not both zero, and with degrees satisfying Eq. (H.4) such that $R_0(x)E_0(x) + R_1(x)E_1(x) = 0$, that is,

$$R_0(x)E_0(x) = -R_1(x)E_1(x).$$

Since $R_0(x)$ and $R_1(x)$ have no common factors, this implies that $E_1(x)$ has all the factors of $R_0(x)$, which is not possible because of the degree constraint (H.4). This shows that $\mathbf{R}$ cannot be singular. Equation (H.7) therefore has a unique solution $\mathbf{e}$ which defines $E_0(x)$ and $E_1(x)$ uniquely subject to the constraint (H.4). This proves the theorem. $\bigtriangledown\bigtriangledown\bigtriangledown$

Remarks

1. ***Euclid's algorithm.*** Note that the proof is constructive. That is, it allows us to find the polynomials $E_0(x)$ and $E_1(x)$ by solving for the vector $\mathbf{e}$ from Eq. (H.7). This is called the modified Euclid's algorithm (the plain Euclid's algorithm is a method to compute the greatest common divisor of polynomials; see Knuth [1998]).

2. ***What if there is a common factor?*** Note from Eq. (H.3) that if $R_0(x)$ and $R_1(x)$ have a common factor $G(x)$ then it is also a factor of the left-hand side, which shows that the right-hand side has to have this factor. In short, Eq. (H.3) cannot be true unless the greatest common divisor (gcd) of $R_0(x)$ and $R_1(x)$ is unity. So we can say that there exist polynomials $E_0(x)$ and $E_1(x)$ satisfying Eq. (H.3) **if and only if** the gcd of $R_0(x)$ and $R_1(x)$ is unity (i.e., they are coprime). More generally, if

$$G(x) = \text{gcd}\,\{R_0(x), R_1(x)\} \tag{H.9}$$

then we can factor out the gcd and apply the theorem to show that there exist polynomials $E_0(x)$ and $E_1(x)$ such that

$$R_0(x)E_0(x) + R_1(x)E_1(x) = G(x) \tag{H.10}$$

For example, if x^K is a common factor then the right-hand side has x^K instead of unity.

3. ***Case of more polynomials.*** Suppose $R_k(x)$, $0 \leq k \leq L-1$, are polynomials in x such that

$$\text{gcd}\,\{R_k(x)\} = G(x).$$

It can then be shown that there exist polynomials $E_k(x)$ such that

$$\sum_k R_k(x)E_k(x) = G(x).$$

For example, if there is no common factor (i.e., $G(x) = 1$) then

$$R_0(x)E_0(x) + R_1(x)E_1(x) + \ldots + R_{L-1}(x)E_{L-1}(x) = 1. \tag{H.11}$$

To understand why this result is true, consider the case where $L = 3$. Let $G_1(x)$ be the gcd of $R_0(x)$ and $R_1(x)$. Then we can find polynomials $A_0(x)$ and $A_1(x)$ such that

$$R_0(x)A_0(x) + R_1(x)A_1(x) = G_1(x). \tag{H.12}$$

If there is no factor common to all three $R_k(x)$'s, the gcd of $G_1(x)$ and $R_2(x)$ is unity, so there exist polynomials $B_1(x)$ and $B_2(x)$ such that

$$G_1(x)B_1(x) + R_2(x)B_2(x) = 1.$$

Substituting from Eq. (H.12) this yields

$$R_0(x)E_0(x) + R_1(x)E_1(x) + R_2(x)E_2(x) = 1 \tag{H.13}$$

for appropriately defined polynomials $E_k(x)$. The proof for arbitrary L and $G(x)$ can readily be developed from here.

4. ***History.*** Euclid's algorithm can be traced back to *Euclid's* work in 300 BC. Later extensions, which gave rise to the present form, came from a number of sources including *Āryabhaṭa* in AD 499. An impressive account of Euclid's algorithm and his history can be found in Knuth's classic book on seminumerical algorithms [Knuth, 1998].

Example H.1: Euclid's algorithm

Let

$$\begin{aligned} R_0(x) &= (1-x)(1-2x) = 1 - 3x + 2x^2, \\ R_1(x) &= (1+x)(1+2x)(1-3x) = 1 - 7x^2 - 6x^3. \end{aligned}$$

These have no common factors, so the theorem can be applied. In this example $N_0 = 2$ and $N_1 = 3$, so Eq. (H.7) becomes

$$\begin{bmatrix} 1 & 0 & 0 & 1 & 0 \\ -3 & 1 & 0 & 0 & 1 \\ 2 & -3 & 1 & -7 & 0 \\ 0 & 2 & -3 & -6 & -7 \\ 0 & 0 & 2 & 0 & -6 \end{bmatrix} \begin{bmatrix} e_0(0) \\ e_0(1) \\ e_0(2) \\ e_1(0) \\ e_1(1) \end{bmatrix} = \begin{bmatrix} 1 \\ 0 \\ 0 \\ 0 \\ 0 \end{bmatrix}.$$

Inverting the matrix, we can solve for the coefficients $e_k(n)$. The result is

$$e_0(0) = \frac{9}{4}, \quad e_0(1) = \frac{67}{12}, \quad e_0(2) = \frac{7}{2}, \quad e_1(0) = \frac{-5}{4}, \quad e_1(1) = \frac{7}{6}$$

From this we obtain the polynomials

$$E_0(x) = \frac{9}{4} + \frac{67x}{12} + \frac{7x^2}{2}, \quad E_1(x) = \frac{-5}{4} + \frac{7x}{6}$$

From this we can indeed verify by explicit computation that $R_0(x)E_0(x) + R_1(x)E_1(x) = 1$.

Appendix I

Transceiver optimization: summary and tables

This appendix provides a summary of each of the main optimization results presented in Part 2 of this book. Each main result is summarized in one page. For further details see the appropriate sections of the chapters which are cited below. The general form of the memoryless MIMO transceiver is shown in Fig. I.1. Set $\mathbf{B} = \mathbf{0}$ for a linear transceiver. For nonzero $\mathbf{B}$ (strictly upper triangular) we have a DFE system. Note that ZF-MMSE stands for the MMSE system with zero forcing and pure MMSE stands for the MMSE system without zero forcing.

General notations and assumptions

1. The transmitted power is denoted as p_0. The term "power constraint" means that the total power at the output of the precoder is constrained to be p_0.
2. The mean square error is defined as $E[\,\|\widehat{\mathbf{s}}(n)-\mathbf{s}(n)\|^2]$, and the mean square error per symbol is this quantity divided by M, where M is the number of elements in $\mathbf{s}(n)$.
3. The notation $[\mathbf{A}]_M$ denotes the $M \times M$ leading principal submatrix of $\mathbf{A}$ (i.e., the matrix obtained by retaining the first M rows and columns of the matrix $\mathbf{A}$).
4. We assume $\mathbf{q}(n)$ and $\mathbf{s}(n)$ are zero-mean, jointly WSS processes with $\mathbf{R}_{\mathbf{qq}} = \sigma_q^2\mathbf{I}$. Furthermore, $\mathbf{R}_{\mathbf{ss}} = \sigma_s^2\mathbf{I}$, except in the bit allocation problem.
5. The channel $\mathbf{H}$ has singular value decomposition (SVD) of the form

$$\mathbf{H} = \mathbf{U}_h\mathbf{\Sigma}_h\mathbf{V}_h^\dagger,$$

where $\mathbf{U}_h$ and $\mathbf{V}_h$ are unitary matrices and $\boldsymbol{\Sigma}_h$ is diagonal with nonnegative diagonal elements (singular values of $\mathbf{H}$). The precoder $\mathbf{F}$ and equalizer $\mathbf{G}$ have similar singular value decompositions.

6. Channel singular values (diagonal elements of $\boldsymbol{\Sigma}_h$) are assumed to be ordered such that $\sigma_{h,0} \geq \sigma_{h,1} \geq \ldots$ In the zero-forcing (ZF) case $\mathbf{H}$ has to have rank $\geq M$, so the first M singular values $\sigma_{h,k}$ are nonzero.

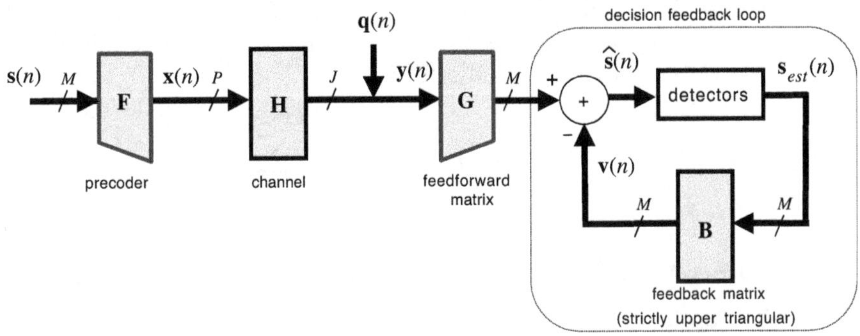

Figure I.1. The general form of transceiver with decision feedback.

Table I.1. Linear MIMO transceivers (Chaps. 12, 13)

ZF-MMSE linear transceiver (Chap. 12; detailed summary in Sec. 12.4.4)

The optimal solution can be taken as

$$\mathbf{F} = \mathbf{V}_h \begin{bmatrix} \mathbf{\Sigma}_f \\ \mathbf{0} \end{bmatrix} \quad \text{and} \quad \mathbf{G} = [\,\mathbf{\Sigma}_g \quad \mathbf{0}\,]\,\mathbf{U}_h^\dagger, \tag{I.1}$$

where $\mathbf{\Sigma}_f = c[\mathbf{\Sigma}_h]_M^{-1/2}$ and $\mathbf{\Sigma}_g = c^{-1}[\mathbf{\Sigma}_h]_M^{-1/2}$ (equalization shared equally between transmitter and receiver). The scalar $c > 0$ is determined from the power constraint. The minimized MSE is

$$\mathcal{E}_{mmse} = \frac{\sigma_s^2\sigma_q^2}{p_0}\Big(\sum_{k=0}^{M-1}\frac{1}{\sigma_{h,k}}\Big)^2 \tag{I.2}$$

Pure-MMSE linear transceiver (Chap. 13; detailed summary in Sec. 13.6)

The optimal solution can again be taken as

$$\mathbf{F} = \mathbf{V}_h \begin{bmatrix} \mathbf{\Sigma}_f \\ \mathbf{0} \end{bmatrix} \quad \text{and} \quad \mathbf{G} = [\,\mathbf{\Sigma}_g \quad \mathbf{0}\,]\,\mathbf{U}_h^\dagger, \tag{I.3}$$

but the diagonal matrices $\mathbf{\Sigma}_f$ and $\mathbf{\Sigma}_g$ are now given by

$$\mathbf{\Sigma}_f = \frac{\sigma_q}{\sigma_s}\begin{bmatrix} \sqrt{q_{00}} & 0 & \ldots & 0 \\ 0 & \sqrt{q_{11}} & \ldots & 0 \\ \vdots & \vdots & \ddots & \vdots \\ 0 & 0 & \ldots & \sqrt{q_{M-1,M-1}} \end{bmatrix}, \tag{I.4}$$

$$\mathbf{\Sigma}_g = \frac{\sigma_s}{\sigma_q}\begin{bmatrix} z_0 & 0 & \ldots & 0 \\ 0 & z_1 & \ldots & 0 \\ \vdots & \vdots & \ddots & \vdots \\ 0 & 0 & \ldots & z_{M-1} \end{bmatrix}\begin{bmatrix} \sigma_{h,0} & 0 & \ldots & 0 \\ 0 & \sigma_{h,1} & \ldots & 0 \\ \vdots & \vdots & \ddots & \vdots \\ 0 & 0 & \ldots & \sigma_{h,M-1} \end{bmatrix} \tag{I.5}$$

where

$$q_{kk} = \begin{cases} \frac{D}{\sigma_{h,k}} - \frac{1}{\sigma_{h,k}^2} & 0 \le k \le K-1 \\ 0 & \text{otherwise,} \end{cases} \tag{I.6}$$

and $z_k = \sqrt{q_{kk}}/(1+q_{kk}\sigma_{h,k}^2)$. Note that Eq. (I.6) is *not* a water-filling formula. The transmitted power p_0 determines the constant $D > 0$ and integer K. The minimized MSE is given by

$$\mathcal{E}_{mmse} = \sigma_s^2(M-K) + \frac{\sigma_s^2\sigma_q^2\Big(\sum_{k=0}^{K-1}\frac{1}{\sigma_{h,k}}\Big)^2}{p_0 + \sigma_q^2\sum_{k=0}^{K-1}\frac{1}{\sigma_{h,k}^2}} \tag{I.7}$$

Table I.2. Linear MIMO system with orthonormal precoder (Chap. 15)

ZF-MMSE linear transceiver (Chap. 15, Sec. 15.3.1)

The optimal solution is such that

$$\mathbf{F} = \alpha \mathbf{V}_h \begin{bmatrix} \mathbf{I}_M \\ \mathbf{0} \end{bmatrix} \quad \text{and} \quad \mathbf{G} = \alpha^{-1} [\, \boldsymbol{\Sigma}_g \quad \mathbf{0}\,] \, \mathbf{U}_h^{\dagger}, \tag{I.8}$$

where $\boldsymbol{\Sigma}_g = ([\boldsymbol{\Sigma}_h]_M)^{-1}$ and α is determined by the power p_0. The minimized MSE is given by

$$\mathcal{E}_{mmse} = \frac{M\sigma_q^2\sigma_s^2}{p_0} \sum_{k=0}^{M-1} \frac{1}{\sigma_{h,k}^2} \tag{I.9}$$

Pure-MMSE linear transceiver (Chap. 15, Sec. 15.3.2)

The optimal solution is such that

$$\mathbf{F} = \alpha \mathbf{V}_h \begin{bmatrix} \mathbf{I}_M \\ \mathbf{0} \end{bmatrix} \quad \text{and} \quad \mathbf{G} = \alpha^{-1} [\, \boldsymbol{\Sigma}_g \quad \mathbf{0}\,] \, \mathbf{U}_h^{\dagger}, \tag{I.10}$$

where

$$\boldsymbol{\Sigma}_g = (\boldsymbol{\Sigma}_h)_M^{-1} \left(\mathbf{I} + \frac{\sigma_q^2}{\sigma_s^2} (\boldsymbol{\Sigma}_h)_M^{-2} \right)^{-1} \tag{I.11}$$

and α is determined by the power p_0. The minimized MSE is given by

$$\mathcal{E}_{mmse} = \frac{M\sigma_s^2\sigma_q^2}{p_0} \sum_{k=0}^{M-1} \left(\frac{1}{\frac{M\sigma_q^2}{p_0} + \sigma_{h,k}^2} \right). \tag{I.12}$$

Table I.3. MIMO transceivers with DFE (Chap. 19)

ZF-MMSE DFE transceiver (Chap. 19; detailed summary in Sec. 19.3.4)

The optimal solution is based on the QRS decomposition of $[\mathbf{\Sigma}_h]_M$ given by

$$[\mathbf{\Sigma}_h]_M = \sigma \mathbf{QRS}^\dagger, \tag{I.13}$$

where $\mathbf{Q}$ and $\mathbf{S}$ are $M \times M$ unitary, $\mathbf{R}$ is upper triangular with diagonal elements $R_{kk} = 1$, and σ is the geometric mean $\sigma = (\prod_{k=0}^{M-1} \sigma_{h,k})^{1/M}$. In the optimal system the precoder $\mathbf{F}$, feedforward matrix $\mathbf{G}$, and feedback matrix $\mathbf{B}$ are given by

$$\mathbf{F} = \alpha \mathbf{V}_h \begin{bmatrix} \mathbf{I}_M \\ \mathbf{0} \end{bmatrix} \mathbf{S}, \quad \mathbf{G} = \Big(\frac{\mathbf{I}+\mathbf{B}}{\alpha}\Big)\mathbf{S}^\dagger \Big([\mathbf{\Sigma}_h]_M\Big)^{-1}\Big([\mathbf{U}_h]_{J\times M}\Big)^\dagger, \tag{I.14}$$

$$\text{and} \quad \mathbf{B} = \mathbf{R} - \mathbf{I}_M. \tag{I.15}$$

The minimized MSE is

$$\mathcal{E}_{mmse} = \frac{M^2 \sigma_s^2 \sigma_q^2}{p_0} \Big(\prod_{k=0}^{M-1} \frac{1}{\sigma_{h,k}}\Big)^{2/M}.$$

Pure-MMSE DFE transceiver (Chap. 19; detailed summary in Sec. 19.4.5)

The optimal precoder has the form

$$\mathbf{F} = \mathbf{V}_h \begin{bmatrix} \mathbf{\Sigma}_f \\ \mathbf{0} \end{bmatrix} \mathbf{S}, \tag{I.16}$$

where $\mathbf{\Sigma}_f$ is an $M \times M$ diagonal matrix with diagonal elements

$$\sigma_{f,k}^2 = \begin{cases} \dfrac{1}{\lambda} - \dfrac{\sigma_q^2}{\sigma_s^2 \sigma_{h,k}^2} & 0 \le k \le K-1 \\ 0 & \text{otherwise.} \end{cases} \tag{I.17}$$

This is the *water-filling* formula. Here λ and $K \le M$ depend on the power p_0. The matrix $\mathbf{S}$ comes from the QRS decomposition of the diagonal matrix

$$\mathbf{\Sigma} = \Big(\mathbf{I} + \frac{\sigma_s^2}{\sigma_q^2}\mathbf{\Sigma}_f^2 [\mathbf{\Sigma}_h]_M^2\Big)^{1/2}. \tag{I.18}$$

Thus $\mathbf{\Sigma} = \sigma \mathbf{QRS}^\dagger$, where $\mathbf{Q}$ and $\mathbf{S}$ are $M \times M$ unitary, $\mathbf{R}$ is upper triangular with diagonal elements $R_{kk} = 1$, and σ is the geometric mean of the diagonal elements of $\mathbf{\Sigma}$. The optimal feedforward matrix $\mathbf{G}$ and feedback matrix $\mathbf{B}$ are given by

$$\mathbf{G} = \mathbf{RS}^\dagger \mathbf{\Lambda}\Big([\mathbf{U}_h]_{J\times M}\Big)^\dagger \quad \text{and} \quad \mathbf{B} = \mathbf{R} - \mathbf{I}, \tag{I.19}$$

where $\mathbf{\Lambda} = \mathbf{\Sigma}_f (\mathbf{\Sigma}_h)_M \Big(\mathbf{\Sigma}_f^2 [\mathbf{\Sigma}_h]_M^2 + \frac{\sigma_q^2}{\sigma_s^2}\mathbf{I}\Big)^{-1}$. The minimized MSE is

$$\mathcal{E}_{mmse} = M\sigma_s^2 \Big(\frac{\lambda \sigma_q^2}{\sigma_s^2}\Big)^{K/M} \prod_{k=0}^{K-1} \Big(\frac{1}{\sigma_{h,k}}\Big)^{2/M}. \tag{I.20}$$

The pure-MMSE DFE system also maximizes *mutual information* between the channel input and output (Sec. 19.7).

Table I.4. Minimum error probability systems (Chap. 16)

MIMO transceivers which are optimized to minimize mean square error also minimize the symbol error probability if the MSE components $E[|\widehat{s}_k(n)-s_k(n)|^2]$ are equalized. This is achieved by using a unitary matrix at the transmitter and its inverse at the receiver (Sec. 16.2.1). The derivation of such a matrix in the case of linear transceivers is described in Sec. 11.5 and revisited in Secs. 16.2 and 16.4. The optimal DFE transceivers derived in Sec. 19.3 (with zero forcing) and in Sec. 19.4 (without zero forcing), both summarized in Table I.3, are automatically optimal in the sense of minimizing the average symbol error probability (Sec. 19.5). There is no need for extra unitary matrices. For PAM and QAM constellations, the minimized average symbol error probability per component of $\mathbf{s}(n)$ is given by

$$\mathcal{P}_{e,min} = \begin{cases} cQ\left(A\sqrt{\frac{1}{\mathcal{E}_{ave}}}\right) & \text{(ZF-MMSE case)} \\ \\ cQ\left(A\sqrt{\frac{1}{\mathcal{E}_{ave}} - \frac{1}{\sigma_s^2}}\right) & \text{(pure-MMSE case),} \end{cases} \tag{I.21}$$

where $\mathcal{E}_{ave}$ is the average MSE (per scalar component $s_k(n)$). The second term $-1/\sigma_s^2$ arises in the pure-MMSE case because of the need to remove bias before using the symbol detectors (Sec. 16.3). In spite of this term, it can be shown that $\mathcal{P}_{e,min}$ without zero forcing is smaller than $\mathcal{P}_{e,min}$ with zero forcing (Appendix 16.C). The constants c and A depend on the constellation:

$$c = 2(1-2^{-b}), \quad A = \sqrt{\frac{3\sigma_s^2}{2^{2b}-1}} \quad \text{(PAM),} \tag{I.22}$$

$$c = 4(1-2^{-b/2}), \quad A = \sqrt{\frac{3\sigma_s^2}{2^{b}-1}} \quad \text{(QAM),} \tag{I.23}$$

where b is the number of bits used for the constellation.

Table I.5. Linear transceivers with ZF and bit allocation (Chap. 14)

In the transceiver below assume $\mathbf{R_{qq}} = \sigma_q^2\mathbf{I}$ and that the components $s_k(n)$ of $\mathbf{s}(n)$ are zero-mean uncorrelated processes representing independent users with powers $[\mathbf{\Lambda}_s]_{kk} = P_k$. Assume the kth user transmits b_k-bit symbols, and the desired error probability is $\mathcal{P}_e(k)$. Assume that zero forcing is in effect, that is $\mathbf{GHF} = \mathbf{I}$.

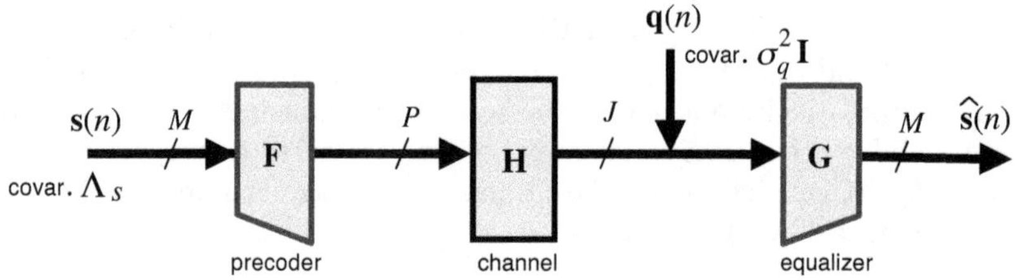

Under the high-bit rate assumption (Sec. 14.2), the user power required to achieve the error probability $\leq \mathcal{P}_e(k)$ is $P_k \approx c_k 2^{b_k}[\mathbf{GG}]^\dagger_{kk}$, where $c_k = (\sigma_q^2/3)[\mathcal{Q}^{-1}(\mathcal{P}_e(k)/4)]^2$ (assuming QAM). The total transmitted power is given by

$$P_{trans} = \sum_{k=0}^{M-1} c_k 2^{b_k}[\mathbf{F}^\dagger\mathbf{F}]_{kk}[\mathbf{GG}^\dagger]_{kk}. \tag{I.24}$$

For fixed $b = \sum_k b_k/M$, the jointly optimal choices of bit allocation $\{b_k\}$, precoder $\mathbf{F}$, and equalizer $\mathbf{G}$ which minimize this power are as follows (Sec. 14.4):

$$b_k = D - \log_2\left(\frac{\sigma_q^2}{3}\left(\mathcal{Q}^{-1}\left(\frac{\mathcal{P}_e(k)}{4}\right)\right)^2\right) + \log_2 \sigma_{h,k}^2, \tag{I.25}$$

$$\mathbf{G} = [\mathbf{U}_h^\dagger]_{M\times J}, \quad \text{and} \quad \mathbf{F} = \mathbf{V}_h\begin{bmatrix}[\mathbf{\Sigma}_h]_M^{-1} \\ \mathbf{0}\end{bmatrix}, \tag{I.26}$$

where $\sigma_{h,k}$ are the M dominant channel singular values (nonzero owing to the ZF assumption). The unitary matrices above come from the channel SVD, that is, $\mathbf{H} = \mathbf{U}_h\mathbf{\Sigma}_h\mathbf{V}_h^\dagger$. The notation $[\mathbf{A}]_{M\times J}$ denotes the top-left $M\times J$ submatrix of $\mathbf{A}$, and $\mathbf{A}_M$ denotes the top-left $M \times M$ submatrix. The minimized transmitted power is given by

$$P_{min} = c2^b\left(\frac{1}{\prod_{k=0}^{M-1}\sigma_{h,k}^2}\right)^{1/M}. \tag{I.27}$$

If the above $\{\mathbf{F},\mathbf{G}\}$ pair is replaced with $\{\mathbf{F}\mathbf{\Sigma}_g^{-1}, \mathbf{\Sigma}_g\mathbf{G}\}$ for arbitrary nonsingular diagonal matrix $\mathbf{\Sigma}_g$, the system continues to be optimal with the above bit allocation (Sec. 14.5.1). The optimal solution is such that the channel eigenmodes receive the same power for any diagonal $\mathbf{\Sigma}_g$ (Sec. 14.6.3.A). In particular, if we choose $[\mathbf{\Sigma}_g]_{kk} = 1/\sigma_{h,k}$ then the precoder has orthonormal columns (that is, $\mathbf{F}^\dagger\mathbf{F} = \mathbf{I}$), all equalization is at the receiver, and P_k is identical for all users ($\mathbf{\Lambda}_s = \alpha\mathbf{I}$; see Ex. 14.1).

Table I.6. Linear MMSE transceiver for circulant channel (Chap. 17)

MIMO circulant channels arise when a SISO channel is converted into MIMO by using the cyclic prefix (Secs. 7.3, 7.4). Assuming $\mathbf{R_{ss}} = \sigma_s^2\mathbf{I}$ and $\mathbf{R_{qq}} = \sigma_q^2\mathbf{I}$, the MMSE linear transceiver for a circulant channel has the form shown below. Here $\mathbf{W}$ is the $M \times M$ DFT matrix. Circulants admit the diagonalization $\mathbf{H} = \mathbf{W}^{-1}\boldsymbol{\Lambda}_c\mathbf{W}$. In what follows, $C[k]$ is the kth diagonal element of $\boldsymbol{\Lambda}_c$.

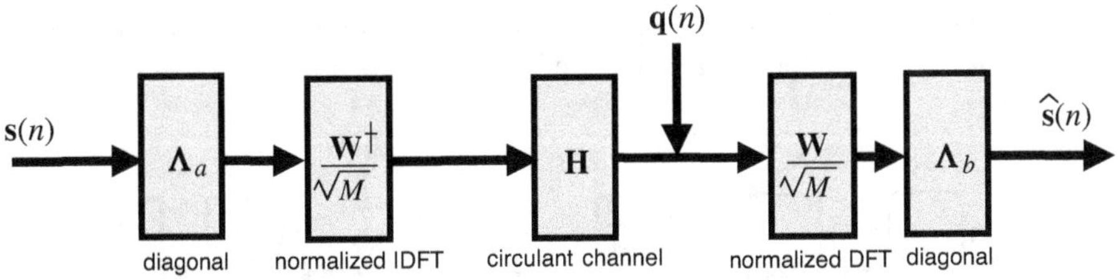

For the ZF-MMSE system,

$$\boldsymbol{\Lambda}_a = \alpha\boldsymbol{\Lambda}_c^{-1/2} \quad \text{and} \quad \boldsymbol{\Lambda}_b = \alpha^{-1}\boldsymbol{\Lambda}_c^{-1/2}, \tag{I.28}$$

where α is chosen to make the total power $= p_0$. For the pure-MMSE system

$$\boldsymbol{\Lambda}_a = \mathbf{P}\boldsymbol{\Sigma}_f\mathbf{P}^T \quad \text{and} \quad \boldsymbol{\Lambda}_b = \mathbf{P}\boldsymbol{\Lambda}_g\mathbf{P}^T, \tag{I.29}$$

where $\boldsymbol{\Sigma}_f$ and $\boldsymbol{\Lambda}_g$ are the diagonal matrices given in Eqs. (17.28) and (17.31). The quantity $\widehat{C}[k] = |\widehat{C}[k]|e^{j\widehat{\theta}_k}$ in those equations is a permuted version of $C[k]$ so that $|\widehat{C}[0]| \geq |\widehat{C}[1]| \ldots$ The matrix $\mathbf{P}$ is merely a permutation that reorders the diagonal elements of $\boldsymbol{\Sigma}_f$ and $\boldsymbol{\Lambda}_g$ in natural order before using the diagonal matrices with the DFT and IDFT matrices in the figure. The minimized mean square error is given by

$$\mathcal{E}_{mmse} = \begin{cases} \dfrac{\sigma_s^2\sigma_q^2}{p_0}\Big(\displaystyle\sum_{k=0}^{M-1}\frac{1}{|C[k]|}\Big)^2 & \text{(ZF-MMSE case)} \\[3ex] \sigma_s^2(M-K) + \dfrac{\sigma_s^2\sigma_q^2\Big(\sum_{k=0}^{K-1}\frac{1}{|\widehat{C}[k]|}\Big)^2}{p_0 + \sigma_q^2\sum_{k=0}^{K-1}\frac{1}{|\widehat{C}[k]|^2}} & \text{(pure-MMSE case),} \end{cases} \tag{I.30}$$

where K is the number of nonzero diagonal elements in $\boldsymbol{\Sigma}_f$ (Eq. (17.28)) and depends on the power p_0. For a detailed summary see Sec. 17.3.

Table I.7. Minimum error probability for circulant channel (Chap. 17)

The MMSE transceiver in Table I.6 can be converted to a transceiver with minimum symbol error probability by inserting additional DFT and IDFT matrices as shown in the figure below (Sec. 17.4). The MSE (with or without zero forcing) and the channel input power for this system are the same as in Table I.6. So the system below has minimum MSE as well as minimum error probability. This assumes that the SNRs are large enough to validate the convexity assumptions of Sec. 16.2.2. Since the product $\mathbf{W}^\dagger \boldsymbol{\Lambda}_a \mathbf{W}$ is circulant for any diagonal $\boldsymbol{\Lambda}_a$, the optimal precoder is circulant, and so is the optimal equalizer!

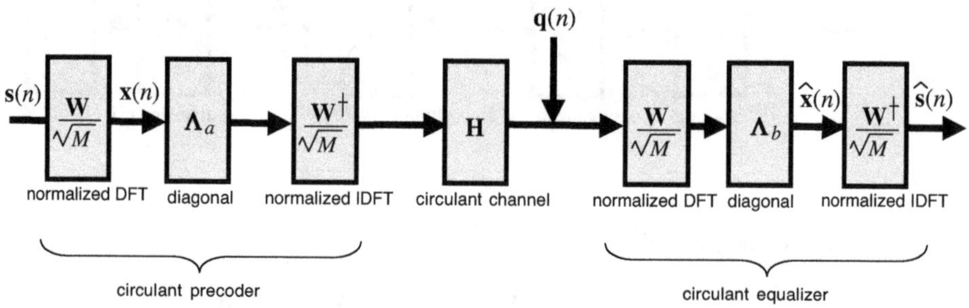

For PAM and QAM constellations, the minimized average symbol error probability per component of $\mathbf{s}(n)$ is given by

$$\mathcal{P}_{e,min} = \begin{cases} cQ\left(A\sqrt{\frac{1}{\mathcal{E}_{ave}}}\right) & \text{(ZF-MMSE case)} \\ \\ cQ\left(A\sqrt{\frac{1}{\mathcal{E}_{ave}} - \frac{1}{\sigma_s^2}}\right) & \text{(pure-MMSE case),} \end{cases} \tag{I.31}$$

where $\mathcal{E}_{ave}$ is the average MSE (per scalar component $s_k(n)$). The constants c and A depend on the constellation, as summarized in Eqs. (I.22) and (I.23), where b is the number of bits used for the constellation.

Table I.8. Circulant channel with orthonormal precoder (Chap. 17)

Consider the linear transceiver for a circulant channel under the constraint that the precoder be orthonormal (except for a scale factor for power control). Assume $\mathbf{R_{qq}} = \sigma_q^2\mathbf{I}$ and $\mathbf{R_s} = \sigma_s^2\mathbf{I}$. If we now minimize the MSE (with or without ZF) then the precoder can be taken as *any* unitary matrix (Sec. 17.6). The special case where the precoder is identity is called the single-carrier cyclic-prefix system (SC-CP), and the special case where the precoder is the IDFT matrix is called the multicarrier cyclic-prefix (MC-CP) system. With such precoders, the equalizers achieving the MMSE property are shown in the figure below.

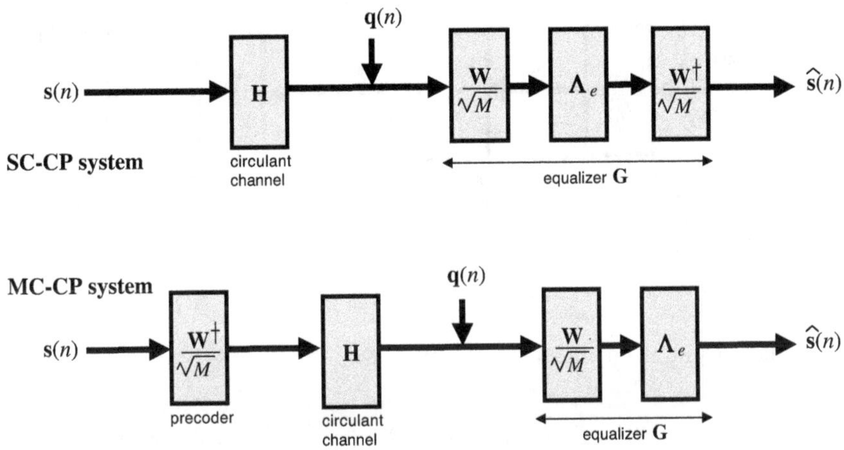

Here $\mathbf{\Lambda}_e$ is a diagonal matrix of equalizer coefficients, given by

$$\mathbf{\Lambda}_e = \begin{cases} \mathbf{\Lambda}_c^{-1} & \text{(ZF-MMSE)} \\ \mathbf{\Lambda}_c^*\left(\mathbf{\Lambda}_c\mathbf{\Lambda}_c^* + \dfrac{\sigma_q^2}{\sigma_s^2}\mathbf{I}\right)^{-1} & \text{(pure MMSE).} \end{cases} \tag{I.32}$$

The SC-CP and MC-CP systems have total MSE given by

$$\mathcal{E}_{mmse} = \begin{cases} \dfrac{M\sigma_q^2\sigma_s^2}{p_0}\displaystyle\sum_{k=0}^{M-1}\frac{1}{|C[k]|^2} & \text{(ZF-MMSE)} \\ \dfrac{M\sigma_s^2\sigma_q^2}{p_0}\displaystyle\sum_{k=0}^{M-1}\left(\frac{1}{\frac{M\sigma_q^2}{p_0} + |C[k]|^2}\right) & \text{(pure-MMSE).} \end{cases} \tag{I.33}$$

Among all MMSE systems with unitary precoder, the average *error probability is minimized by the SC-CP system*, under the convexity assumptions of Sec. 16.2.2. With $\mathcal{E}_{ave} = \mathcal{E}_{mmse}/M$ (MSE per component), this error probability per symbol is given by Eq. (I.21). The MC-CP system, unfortunately, has the maximum error probability (among all MMSE systems with orthonormal precoder).

Table I.9. Circulant channel with bit allocation (Chap. 17)

The circulant channel with bit allocation at the transmitter arises in DMT systems. The optimal linear transceiver which minimizes transmitted power with bit allocation under the *zero-forcing* condition was derived in Sec. 17.5. The optimum solution is shown in the figure below, where $\mathbf{W}$ is the DFT matrix. Thus the optimal precoder is automatically orthonormal in this problem set-up! Since $\mathbf{H}$ is circulant, we have $\mathbf{H} = \mathbf{W}^{-1}\mathbf{\Lambda}_c\mathbf{W}$. The inverse $\mathbf{\Lambda}_c^{-1}$ appears in the optimal receiver. In what follows $C[k]$ denotes the kth diagonal element of $\mathbf{\Lambda}_c$.

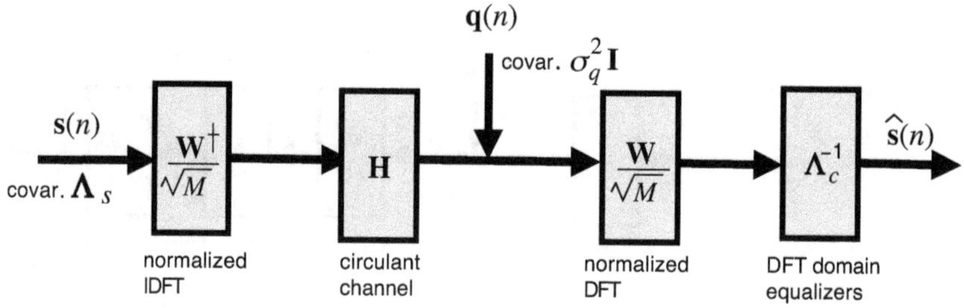

The optimum bit allocation formula is

$$b_k = D_0 - \log_2\left(\frac{\sigma_q^2}{3}\left(\mathcal{Q}^{-1}\left(\frac{\mathcal{P}_e(k)}{4}\right)\right)^2\right) + \log_2|C[k]|^2, \tag{I.34}$$

where $\mathcal{P}_e(k)$ is the specified error probability for the kth component of $\mathbf{s}(n)$. D_0 is chosen such that $\sum_{k=0}^{M-1} b_k/M = b$ for fixed b. The minimized power is

$$P_{min} = c2^b\Big(\frac{1}{\prod_{k=0}^{M-1}|C[k]|^2}\Big)^{1/M}, \tag{I.35}$$

where $c = M(\prod_k c_k)^{1/M}$, with $c_k = (\sigma_q^2/3)[\mathcal{Q}^{-1}(\mathcal{P}_e(k)/4)]^2$ assuming a QAM constellation. The optimal choice of user powers $[\mathbf{\Lambda}_s]_{kk}$ happens to be identical for all k, and in fact $\mathbf{\Lambda}_s = 2^{D_0}\mathbf{I}$. Thus the precoder can be taken to be orthonormal without loss of optimality, and furthermore $\mathbf{\Lambda}_s$ is proportional to identity.

Table I.10. Continuous-time SISO channel with ZF (Chap. 10)

The digital communication system with a continuous-time channel is shown below. This is a SISO linear transceiver. In what follows $S_{qq}(j\omega)$ is the power spectrum of noise, and σ_s^2 is the variance of $s(n)$, which is assumed to be zero-mean and white. The filters $F(j\omega)$ and $G(j\omega)$ optimized for MMSE are specified below.

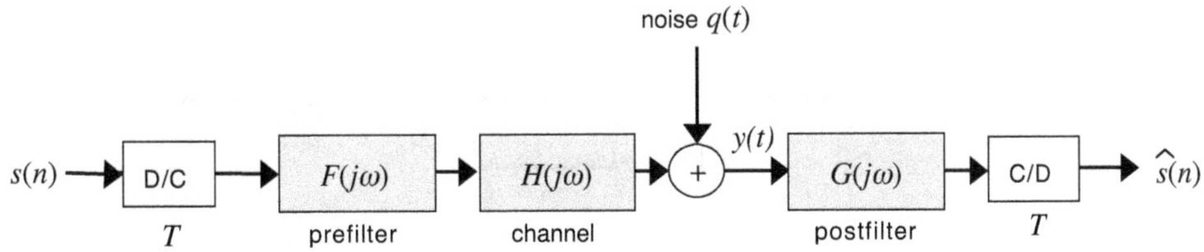

Under the constraint

$$G(j\omega)H(j\omega)F(j\omega) = H_c(j\omega), \tag{I.36}$$

where $H_c(j\omega)$ is a specified Nyquist(T) function (i.e., $h_c(nT) = \delta(n)$), the *zero-forcing* property is satisfied. Then the filters for the MMSE solution are obtained from

$$F(j\omega) = \beta e^{j\theta_f(\omega)} \left| \frac{H_c(j\omega)}{H(j\omega)} \right|^{1/2} S_{qq}^{1/4}(j\omega) \tag{I.37}$$

and

$$G(j\omega) = \frac{e^{-j[\theta(\omega)+\theta_f(\omega)]}}{\beta} \left| \frac{H_c(j\omega)}{H(j\omega)} \right|^{1/2} S_{qq}^{-1/4}(j\omega). \tag{I.38}$$

Here $\theta_f(\omega)$ is arbitrary, $\theta(\omega)$ is the phase of $H(j\omega)/H_c(j\omega)$, and β is computed using the power constraint

$$p_0 = \frac{\sigma_s^2}{T} \int_{-\infty}^{\infty} |F(j\omega)|^2 \frac{d\omega}{2\pi} \tag{I.39}$$

The minimized mean square error is given by

$$\mathcal{E}_{mmse} = \frac{\sigma_s^2}{p_0 T} \left(\int_{-\infty}^{\infty} \frac{S_{qq}^{1/2}(j\omega)}{|H(j\omega)|} |H_c(j\omega)| \frac{d\omega}{2\pi} \right)^2. \tag{I.40}$$

The optimum choice of the Nyquist(T) product filter $H_c(j\omega)$ is just the *optimum compaction filter* (Sec. 10.2.2) for the function $H_{eff}(j\omega) = H(j\omega)/S_{qq}^{1/2}(j\omega)$. See Sec. 10.2 for details.

Table I.11. All-digital SISO channel (Chap. 10)

The all-digital equivalent of the linear SISO transceiver is reproduced below. In what follows $S_{ss}(e^{j\omega})$ and $S_{qq}(e^{j\omega})$ are the power spectra of the input and the noise, which are assumed to be zero-mean and uncorrelated processes. The filters $F(e^{j\omega})$ and $G(e^{j\omega})$ optimized for MMSE are specified below; when magnitude squares are specified, the filters are taken to be stable spectral factors.

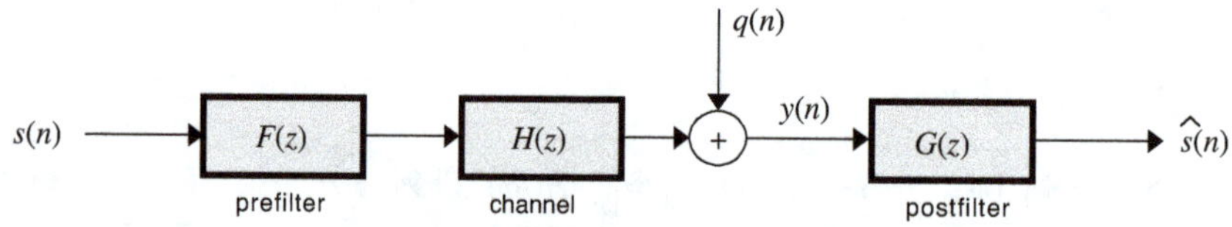

ZF-MMSE case (Sec. 10.3.2)

The optimal precoder is obtained as a spectral factor of

$$|F(e^{j\omega})|^2 = \gamma\sqrt{\frac{S_{qq}(e^{j\omega})}{|H(e^{j\omega})|^2 S_{ss}(e^{j\omega})}}, \tag{I.41}$$

and the zero-forcing equalizer is $G(e^{j\omega}) = 1/[F(e^{j\omega})H(e^{j\omega})]$. Here γ is determined from the power constraint. The minimized MSE is given by

$$\mathcal{E}_{mmse} = \frac{1}{p_0}\left(\int_{-\pi}^{\pi}\sqrt{\frac{S_{qq}(e^{j\omega})S_{ss}(e^{j\omega})}{|H(e^{j\omega})|^2}}\frac{d\omega}{2\pi}\right)^2. \tag{I.42}$$

Pure-MMSE case (Sec. 10.3.1)

The optimal prefilter is computed from the expression

$$|F(e^{j\omega})|^2 = \begin{cases} \gamma\sqrt{\dfrac{S_{qq}(e^{j\omega})}{|H(e^{j\omega})|^2 S_{ss}(e^{j\omega})}} - \dfrac{S_{qq}(e^{j\omega})}{|H(e^{j\omega})|^2 S_{ss}(e^{j\omega})} & \text{if this is} \geq 0 \\ 0 & \text{otherwise.} \end{cases} \tag{I.43}$$

The constant γ is computed from the power constraint. The equalizer is

$$G(e^{j\omega}) = \frac{H^*(e^{j\omega})F^*(e^{j\omega})S_{ss}(e^{j\omega})}{S_{ss}(e^{j\omega})|H(e^{j\omega})F(e^{j\omega})|^2 + S_{qq}(e^{j\omega})} \tag{I.44}$$

The minimized MSE is given by

$$\mathcal{E}_{mmse} = \frac{\left(\displaystyle\int_{\mathcal{F}}\sqrt{\frac{S_{qq}(e^{j\omega})S_{ss}(e^{j\omega})}{|H(e^{j\omega})|^2}}\frac{d\omega}{2\pi}\right)^2}{p_0 + \displaystyle\int_{\mathcal{F}}\frac{S_{qq}(e^{j\omega})}{|H(e^{j\omega})|^2}\frac{d\omega}{2\pi}} + \int_{\mathcal{F}^c} S_{ss}(e^{j\omega})\frac{d\omega}{2\pi} \tag{I.45}$$

where p_0 is the transmitted power, $\mathcal{F}$ is the set of frequencies in $[-\pi, \pi]$ for which $|F(e^{j\omega})|^2$ is nonzero, and $\mathcal{F}^c$ is the complementary set.

Glossary

1. Bold-faced quantities denote matrices and vectors.
2. $|a|$ denotes the absolute value.
3. $\det \mathbf{A}$ denotes the determinant of $\mathbf{A}$.
4. $\mathrm{Tr}(\mathbf{A})$ denotes the trace of $\mathbf{A}$.
5. $\mathbf{A}^T$ denotes the transpose of $\mathbf{A}$.
6. $\mathbf{A}^\dagger$ denotes the transpose-conjugate of $\mathbf{A}$.
7. $\mathbf{A}^*$ denotes the conjugate of $\mathbf{A}$.
8. $\mathbf{A}^{-1}$ denotes the inverse of $\mathbf{A}$.
9. $\mathbf{A}^{-T}$ denotes the inverse of the transpose of $\mathbf{A}$.
10. $\mathbf{A}^{-\dagger}$ denotes the inverse of the transpose-conjugate of $\mathbf{A}$.
11. $\widetilde{\mathbf{H}}(z) = \mathbf{H}^\dagger(1/z^*)$, and $\widetilde{\mathbf{H}}(e^{j\omega}) = \mathbf{H}^\dagger(e^{j\omega})$.
12. $\|\mathbf{a}\|$ denotes the ℓ_2-norm of the vector $\mathbf{a}$.
13. $W_M = e^{-j2\pi/M}$; subscript M is often omitted.
14. $\mathbf{W}$ denotes the $M \times M$ DFT matrix with $[\mathbf{W}]_{km} = W^{km}$.
15. $\delta(n)$ denotes the unit pulse or impulse function; $\delta_c(t)$ denotes the Dirac delta function or impulse function.
16. $\stackrel{\Delta}{=}$ denotes "defined as."

Acronyms

- CP: cyclic prefixing;
- DFE: decision feedback equalizer;
- DFT: discrete Fourier transform;
- DMT: discrete multitone;
- DSL: digital subscriber loop;
- FFT: fast Fourier transform;
- FIR: finite impulse response;
- FSE: fractionally spaced equalizer;
- IIR: infinite impulse response;
- MIMO: multi-input multi-output;
- OFDM: orthogonal frequency division multiplexing;
- SISO: single-input single-output;
- SSE: symbol spaced equalizer;
- ZP: zero padding;

Further details on notations can be found in Chap. 1, Sec. 1.6. For further abbreviations and acronyms please refer to the index.

References

1. Abend, K. and Fritchman, B. D., "Statistical detection for communication channels with intersymbol interference," *Proc. IEEE*, vol. **58**, no. 5, pp. 779–785, May 1970.

2. Abramowitz, M. and Stegun. I. A., *Handbook of Mathematical Functions*, New York: Dover Publications, Inc., 1965.

3. Akansu, A. N., Duhamel, P., Lin, X., and de Courville, M., "Orthogonal transmultiplexers in communication: a review," *IEEE Trans. Signal Proc.*, vol. **46**, no. 4, pp. 979–995, April 1998.

4. Amitay, N. and Salz, J., "Linear equalization theory in digital data transmission over dually polarized fading radio channels," *AT&T Tech. J.*, vol. **63**, no. 10, pp. 2215–2259, Dec. 1984.

5. Andersen, IB. N., "Sample-whitened matched filters," *IEEE Trans. Info. Theory*, vol. **19**, no. 5, pp. 653–660, Sept. 1973.

6. Ansari, R. and Liu, B., "Transmultiplexer design using allpass filters," *IEEE Trans. Commun.*, vol. **30**, pp. 1569–1574, July 1982.

7. Antoniou, A., *Digital Signal Processing: Signals, Systems, and Filters*, New York: McGraw Hill Co., Inc., 2006.

8. Antoniou, A. and Lu, W.-S., *Practical Optimization: Algorithms and Engineering Applications*, New York: Springer, 2007.

9. Arslan, G., Evans, B. L., and Kiaei, S., "Equalization for discrete multitone transceivers to maximize bit rate," *IEEE Trans. Signal Proc.*, vol. **49**, no. 12, pp. 3123–3135, Dec. 2001.

10. Assalini, A. and Tonello, A. M., "Improved Nyquist pulses", *IEEE Communications Letters*, vol. **8**, no. 2, pp. 87–89, Feb. 2004.

11. Austin, M. E., *Decision Feedback Equalization for Digital Communication Over Dispersive Channels*, Tech. Report, MIT Lincoln Labs., August 11, 1967.

12. Autonne, L., "Sur les groupes linéaires, réels et orthogonaux," *Bull. Soc. Math. France*, vol. **30**, pp. 121–133, 1902.

13. Beaulieu, N. C., Tan, C. C., and Damen, M. O., "A 'better than' Nyquist pulse," *IEEE Communications Lett.*, vol. **5**, pp. 367–368, Sept. 2001.

14. Becker, F. K., Holzman, L. N., Lucky, R. W., and Port, E., "Automatic equalization for digital communication," *Proc. IEEE Letters*, pp. 96–98, Jan. 1965.

15. Bell, M. R., "Information theory and radar waveform design," *IEEE Trans. Info. Theory*, vol. **39**, no. 5, pp. 1578–1597, Sept. 1993.

16. Bellanger, M. and Daguet, J. L., "TDM-FDM transmultiplexer: digital polyphase and FFT," *IEEE Trans. Commun.*, vol. **22**, pp. 1199–1205, Sept. 1974.

17. Bellanger, M., Bonnerot, G., and Coudreuse, M., "Digital filtering by polyphase network: application to sample rate alteration and filter banks," *IEEE Trans. Acoust. Speech & Signal Proc.*, vol. **24**, pp. 109–114, April 1976.

18. Bellanger, M., "On computational complexity in digital transmultiplexer filters," *IEEE Trans. Commun.*, vol. **30**, pp. 1461–1465, July 1982.

19. Bellman, R., *Introduction to Matrix Analysis,* New York: McGraw-Hill Book Co., Inc., 1960.

20. Berger, T. and Tufts, D. W., "Optimum pulse amplitude modulation Part I: Transmitter-receiver design and bounds from information theory," *IEEE Trans. Info. Theory,* vol. **13**, no. 2, pp. 196–208, April 1967.

21. Bingham, J. A. C., "Multicarrier modulation for data transmission: an idea whose time has come," *IEEE Comm. Mag.*, pp. 5–14, May 1990.

22. Blahut, R. E., *Digital Transmission of Information*, New York: Addison Wesley, 1990.

23. Bliss, D. W. and Forsythe, K. W., "Multiple-input multiple-output (MIMO) radar and imaging: degrees of freedom and resolution," *Proc. 37th IEEE Asilomar Conf. on Signals, Systems, and Computers*, Monterey, CA, pp. 54–59, Nov. 2003.

24. Bölcskei, H., "A necessary and sufficient condition for dual Weyl-Heisenberg frames to be compactly supported," *J. Fourier Analysis & Applic.*, vol. **5**, no. 5, pp. 409–419, 1999.

25. Bölcskei, H., Duhamel, P., and Hleiss, R., "Design of pulse shaping OFDM/OQAM systems for high data-rate transmission over wireless channels," *Proc. IEEE Int. Conf. on Commun.*, Vancouver, BC, Canada, pp. 559–564, June 1999.

26. Borah, D. K., Kennedy, R. A., Ding. Z., and Fijalkow, I., "Sampling and prefiltering effects on blind equalizer design," *IEEE Trans. Signal Proc.*, vol. **49**, no. 1, pp. 209–218, Jan. 2001.

27. Boyd, S. and Vandenberghe, L., *Convex Optimization,* Cambridge: Cambridge University Press, 2004.

28. Brandenburg, L. H. and Wyner, A. D., "Capacity of the Gaussian channel with memory: the multivariable case," *Bell Sys. Tech. J.*, pp. 745–778, May–June 1974.

29. Brandwood, D. H., "A complex gradient operator and its application in adaptive array theory," *IEE Proc.* vol. **130**, parts F and H, no. 1, pp. 11–16, Feb. 1983.

30. Bruce, R. A., "Optimum pre-emphasis and de-emphasis networks for transmission of television by PCM," *IEEE Trans. on Commun. Tech.*, pp. 91–96, Sept. 1964.

31. Burrus, C. S., Gopinath, R. A., and Guo, H., *Introduction to Wavelets and Wavelet Transforms,* Upper Saddle River, NJ: Prentice Hall, 1998.

32. Chan, D. and Donaldson, R. W., "Optimum pre- and postfiltering of sampled signals with application to pulse modulation and data compression systems," *IEEE Trans. on Commun. Tech.*, vol. **19**, no. 2, pp. 141–157, April 1971.

33. Chang, R. W., "Synthesis of bandlimited orthogonal signals for multichannel data transmission," *Bell Syst. Tech. J.*, vol. **45**, pp. 1775–1796, Dec. 1966.

34. Chang, R. W., *Orthogonal Frequency Multiplex Data Transmission Systems*, US Patent no. 3,488,445, Jan. 6, 1970.

35. Chang, R. W. and Hancock, J. C., "On receiver structures for channels having memory," *IEEE Trans. Info. Theory,* vol. **12**, no. 4, pp. 463–468, Oct. 1966.

36. Chen, C.-Y. and Vaidyanathan, P. P., "Precoded FIR and redundant V-BLAST systems for frequency-selective MIMO channels", *IEEE Trans. Signal Proc.*, vol. **55**, no. 7, pp. 3390–3404, July 2007a.

37. Chen, C.-Y. and Vaidyanathan, P. P., "MIMO radar ambiguity optimization using frequency-hopping waveforms," *Proc. 41st IEEE Asilomar Conference on Signals, Systems, and Computers*, Monterey, CA, Nov. 2007b.

38. Chen, C.-Y. and Vaidyanathan, P. P., "MIMO radar space-time adaptive processing using prolate spheroidal wave functions," *IEEE Trans. Signal Proc.*, vol. **56**, no. 2, pp. 623–635, Feb. 2008.

39. Chen, C.-Y. and Vaidyanathan, P. P., "MIMO radar space-time adaptive processing and signal design," in *MIMO Radar Signal Processing,* eds., J. Li and P. Stoica, New York: John Wiley and Sons, Inc., 2009.

40. Chevillat, P. R. and Ungerboeck, G., "Optimum FIR transmitter and receiver filters for data transmission over band-limited channels," *IEEE Trans. on Commun.*, vol. **30**, no. 8, pp. 1909–1915, Aug. 1982.

41. Cho, K. and Yoon, D., "On the general BER expression of one- and two-dimensional amplitude modulations," *IEEE Trans. Commun.*, vol. **50**, no. 7, pp. 1074–1080, July 2002.

42. Chong, E. K. P. and Żak, S. H., *An Introduction to Optimization*, New York: John Wiley & Sons, Inc., 2001.

43. Chow, J. S., Tu, J. C., and Cioffi, J. M., "A discrete multitone transceiver system for HDSL applications," *IEEE J. Selected Areas in Commun.*, vol. **9**, no. 6, pp. 895–908, Aug. 1991.

44. Churchill, R. V. and Brown, J. W., *Introduction to Complex Variables and Applications,* New York: McGraw-Hill Book Co., 1984.

45. Cimini, Jr., L. J., "Analysis and simulation of a digital mobile channel using orthogonal frequency division multiplexing," *IEEE Trans. Commun.*, vol. **33**, no. 7, pp. 665–675, July 1985.

46. Costas, J. P., "Coding with linear systems," *Proc. IRE,* pp. 1101–1103, Sept. 1952.

47. Cover, T. M. and Thomas, J. A., *Elements of Information Theory,* New York: John Wiley & Sons, Inc., 1991.

48. Crochiere, R. E. and Rabiner, L. R., *Multirate Digital Signal Processing,* Englewood Cliffs, NJ: Prentice Hall, 1983.

49. Dasgupta, S. and Pandharipande, A., "Complete characterization of channel-resistant DMT with cyclic prefix," *IEEE Signal Proc. Lett.*, vol. **10**, no. 6, pp. 161–163, June 2003.

50. Daubechies, I., *Ten Lectures on Wavelets,* Philadelphia, PA: SIAM, 1992.

51. Daubechies, I., Landau, H. J., and Landau, Z., "Gabor time-frequency lattices and the Wexler-Raz identity," *J. Fourier Analysis and Appl.*, vol. **1**, no. 4, pp. 437–478, 1995.

52. Ding, Y., Davidson, T. N., Luo, Z-Q., and Wong, K. M., "Minimum BER block precoders for zero-forcing equalization," *IEEE Trans. Signal Proc.*, vol. **51**, no. 9, pp. 2410–2423, Sept. 2003.

53. Ding, Z. and Li, Y., *Blind Equalization and Identification (Signal Processing and Communications),* New York: CRC Press, 2001.

54. Eckart, C. and Young, G., "A principal axis transformation for non-Hermitian matrices," *Bull. Amer. Math. Soc.*, vol. **45**, pp. 118–121, 1939.

55. Ericson, T., "Structure of optimum receiving filters in data transmission systems," *IEEE Trans. Info. Theory,* pp. 352–353, May 1971.

56. Ericson, T., "Optimum PAM filters are always band limited," *IEEE Trans. on Info. Theory,* pp. 570–573, July 1973.

57. Falconer, D. and Foschini, G. J., "Theory of minimum mean square error QAM systems employing decision feedback equalization," *Bell Syst. Tech. J.*, vol. **52**, no. 10, pp. 1821–1849, Dec. 1973.

58. Falconer, D., Ariyavisitakul, S. L., Benyamin-Seeyar, A., and Eidson, B., "Frequency domain equalization for single-carrier broadband wireless systems," *IEEE Commun. Mag.*, vol. **40**, no. 4, pp. 58–66, April 2002.

59. Farhang-Boroujeny, B. and Lin, L., "Analysis of post-combiner equalizers in cosine-modulated filterbank-based transmultiplexer systems," *IEEE Trans. Signal Proc.*, vol. **51**, no. 12, pp. 3249–3262, Dec. 2003.

60. Feller, W., *An Introduction to Probability Theory and its Applications, vol. I,* New York: John Wiley & Sons, Inc., 1968.

61. Feller, W., *An Introduction to Probability Theory and its Applications, vol. II,* New York: John Wiley & Sons, Inc., 1965.

62. Fiedler, M., "Bounds for the determinant of the sum of Hermitian matrices," *Proc. Amer. Math. Soc.*, vol. **30**, pp. 27–31, 1971.

63. Fliege, N. J., *Multirate Digital Signal Processing: Multirate Systems, Filter Banks, Wavelets,* New York: John Wiley & Sons, Inc., 1994.

64. Forney, G. D., Jr., "Maximum-likelihood sequence estimation of digital sequences in the presence of intersymbol interference," *IEEE Trans. Info. Theory,* vol. **18**, no. 3, pp. 363–378, May 1972.

65. Forney, G. D., Jr., "The Viterbi algorithm," *Proc. IEEE,* vol. **61**, no. 3, pp. 268–278, March 1973.

66. Franklin, J. N., *Matrix Theory,* Englewood Cliffs, NJ: Prentice Hall, Inc., 1968.

67. Gallager, R. G., *Information Theory and Reliable Communication,* New York: John Wiley and Sons, Inc., 1968.

68. Gerrish, A. M. and Howson, R. D., "Multilevel partial response signaling," *Rec. Intl. Commun. Conf.*, 1967.

69. Gersho, A. and Gray, R. M., *Vector Quantization and Signal Compression,* Boston, MA: Kluwer Academic Publishers, 1992.

70. Giannakis, G. B., "Filter banks for blind channel identification and equalization," *IEEE Signal Proc. Lett.*, vol. **4**, no. 6, pp. 184–187, June 1997.

71. Giannakis, G. B., Wang, Z., Scaglione, A., and Barbarossa, S., "AMOUR–generalized multicarrier transceivers for blind CDMA regardless of multipath," *IEEE Trans. Commun.*, vol. **48**, no. 12, pp. 2064–2076, Dec. 2000.

72. Giannakis, G. B., Hua, Y., Stoica, P., and Tong, L., *Signal Processing Advances in Wireless and Mobile Communications,* vol. 1., Upper Saddle River, NJ: Prentice Hall, 2001.

73. Ginis, G. and Cioffi, J. M., "On the relation between V-BLAST and the GDFE," *IEEE Commun. Lett.*, vol. **5**, no. 9, pp. 364–366, Sept. 2001.

74. Gitlin, R. D. and Ho, E. Y., "The performance of staggered quadrature amplitude modulation in the presence of jitter," *IEEE Trans. Commun.*, vol. **23**, no. 3, pp. 348–352, March 1975.

75. Gitlin, R. D. and Weinstein, S. B., "Fractionally spaced equalization: an improved digital transversal equalizer," *Bell System Tech. J.*, vol. **60**, no. 2, pp. 275–296, Feb. 1981.

76. Goldmsith, A., *Wireless Communications*, Cambridge: Cambridge University Press, 2005.

77. Golden, G. D., Foschini, G. J., Valenzuela, R. A., and Wolniansky, P. W., "Detection algorithm and initial laboratory results using V-BLAST space-time communication architecture," *Electron. Lett.*, vol. **35**, no. 1, pp. 14–16, Jan. 1999.

78. Golub, G. H. and Van Loan, C. F., *Matrix Computations,* Baltimore, MD: The Johns Hopkins Univeristy Press, 1989.

79. Graham, A., *Kronecker Products and Matrix Calculus,* New York: Halsted Press, 1981.

80. Guess, T., "Optimal sequences for CDMA with decision-feedback receivers," *IEEE Trans. Info. Theory,* vol. **49**, no. 4, pp. 886–900, April 2003.

81. Haas, R. and Belfiore, J.-C., "A time-frequency well-localized pulse for multiple carrier transmission," *Wireless Personal Commun.*, vol. **5**, pp. 1–18, 1997.

82. Harashima, H. and Miyakawa, H., "Matched-transmission technique for channels with intersymbol interference," *IEEE Trans. Commun.*, vol. **20**, no. 4, pp. 774–780, August 1972.

83. Hardy, G., Littlewood, J. E., and Pólya, G., *Inequalities*, Cambridge: Cambridge University Press, 1952.

84. Harris, F. J., *Multirate Signal Processing for Communication Systems,* Englewood Cliffs, NJ: Prentice Hall, 2004.

85. Hassibi, B., "An efficient square-root algorithm for BLAST," *Proc. 34th IEEE Asilomar Conf. on Signals, Systems, and Computers*, Monterey, CA, pp. 737–740, Nov. 2000.

86. Haykin, S., *Adaptive Filter Theory,* Englewood Cliffs, NJ: Prentice Hall, Inc., 1996 and 2002.

87. Haykin, S., *Communication Systems,* New York: John Wiley, and Sons, Inc., 2001.

88. Haykin, S., *Digital Communications,* New York: John Wiley and Sons, Inc., 1988.

89. Haykin, S. and Moher, M., *Modern Wireless Communications,* Upper Saddle River, NJ: Prentice Hall, Inc., 2005.

90. Heil, C. E. and Walnut, D. F., "Continuous and discrete wavelet transforms," *SIAM Rev.*, vol. **31**, pp. 628–666, Dec. 1989.

91. Helstrom, C. W., *Statistical Theory of Detection*, New York: Pergamon Press, 1968.

92. Hirosaki, B., "An analysis of automatic equalizers for orthogonally multiplexed QAM systems," *IEEE Trans. Commun.*, vol. **28**, no. 1, pp. 73–83, Jan. 1980.

93. Hirosaki, B., "An orthogonally multiplexed QAM system using the discrete Fourier transform," *IEEE Trans. Commun.*, vol. **29**, no. 7, pp. 982–989, July 1981.

94. Hjϕrungnes, A., Diniz, P. S. R., and de Campos, M. L. R., "Jointly minimum BER transmitter and receiver FIR MIMO filters for binary signal vectors," *IEEE Trans. Signal Proc.*, vol. **52**, no. 4, pp. 1021–1036, April 2004.

95. Horn, A., "On the eigenvalues of a matrix with prescribed singular values," *Proc. Amer. Math. Soc.*, vol. **5**, no. 1, pp. 4–7, Feb. 1954.

96. Horn, R. A. and Johnson, C. R., *Matrix Analysis,* Cambridge: Cambridge University Press, 1985.

97. Howson, R. D., "An analysis of the capabilities of polybinary data transmission", *IEEE Trans. Commun. Tech.*, vol. **13**, no. 3, pp. 312–319, Sept. 1965.

98. Janssen, A. J. E. M., "From continuous to discrete Weyl-Heisenberg frames through sampling," *J. Fourier Analysis and Appl.*, vol. **3**, no. 5, pp. 583–596, 1997.

99. Jayant, N. S. and Noll, P., *Digital Coding of Waveforms,* Englewood Cliffs, NJ: Prentice Hall, Inc., 1984.

100. Jiang, Y., Hager, W. W., and Li, J., "The geometric mean decomposition," *Linear Algebra and its Applications*, pp. 373–384, 2005a.

101. Jiang, Y., Hager, W. W., and Li, J., "The generalized triangular decomposition," *Mathematics of Computation*, vol. **77**, no. 262, 1037–1056, April 2008.

102. Jiang, Y., Li, J., and Hager, W. W., "Joint transceiver design for MIMO communications using geometric mean decomposition," *IEEE Trans. Signal Proc.*, vol. **53**, no. 10, pp. 3791–3803, Oct. 2005b.

103. Jiang, Y., Li, J., and Hager, W. W., "Uniform channel decomposition for MIMO communication", *IEEE Trans. Signal Proc.*, vol. **53**, no. 11, pp. 4283–4294, Nov. 2005c.

104. Jiang, Y., Palomar, D. P., and Varanasi, M. K., "Precoder optimization for nonlinear MIMO transceiver based on arbitrary cost function", *Proc. Conf. Info. Sci. and Sys.*, Baltimore, MD, March 2007.

105. Jones, D. L., "Peak power reduction in OFDM and DMT via active channel modification," *Proc. 33rd IEEE Asilomar Conf. on Signals, Systems, and Computers*, Monterey, CA, pp. 1076-1079, Nov. 1999.

106. Kalet, I., "The multitone channel," *IEEE Trans. Commun.*, vol. **37**, no. 2, pp. 119–124, Feb. 1989.

107. Kay, S. M., *Fundamentals of Statistical Signal Processing, Vol. II: Detection theory*, Englewood Cliffs, NJ: Prentice Hall, 1998.

108. Kirac, A. and Vaidyanathan, P. P., "Theory and design of optimum FIR compaction filters," *IEEE Trans. Signal Proc.*, vol. **46**, no. 4, pp. 903–919, April 1998.

109. Kirk, D. E., *Optimal Control Theory,* Englewood Cliffs, NJ: Prentice Hall, Inc., 1970.

110. Klauder, J. R., Price, A. C., Darington, S., and Albersheim, W. J., "The theory and design of chirp radars," *Bell Syst. Tech. J.*, vol. **XXXIX**, no. 4., pp. 745–808, July 1960.

111. Knuth, D. E., *The Art of Computer Programming, vol. 2., Seminumerical Algorithms,* Reading, MA: Addison Wesley, 1998.

112. Koilpillai, R. D., Nguyen, T. Q., and Vaidyanathan, P. P. "Some results in the theory of cross-talk free transmultiplexers," *IEEE Trans. Acoustics, Speech & Signal Proc.*, vol. **39**, pp. 2174–2183, Oct. 1991.

113. Kou, Y. J., Lu, W.-S., and Antoniou, A., "An improved peak-to-average power-ratio reduction algorithm for multicarrier communications," *Proc. IEEE Intl. Symp. Circuits and Systems*, pp. I-5–I-9, Arizona, May 2002.

114. Kretzmer, E. R., "Generalization of a technique for binary data communication," *IEEE Trans. on Commun. Tech.*, pp. 67–68, Feb. 1966.

115. Lathi, B. P., *Modern Digital and Analog Communication Systems,* New York: Oxford University Press, 1998.

116. Le Floch, B., Alard, M., and Berrou, C., "Coded orthogonal frequency division multiplex," *Proc. IEEE,* vol. **83**, no. 6, pp. 982–996, June 1995.

117. Lee, K.-H. and Petersen, D. P., "Optimum linear coding for vector channels," *IEEE Trans. on Commun.*, vol. **24**, no. 12, pp. 1283–1290, Dec. 1976.

118. Lee, S. J. and Beaulieu, N. C., "Performance improvement of non-data-aided feedforward symbol timing estimation using the "better than" raised-cosine pulse," *IEEE Trans. commun.*, vol. **56**, no. 4, pp. 542–544, April 2008.

119. Lender, A., "The duobinary technique for high speed data transmission," *IEEE Trans. on Commun. & Electron.*, vol. **82**, pp. 214–218, May 1963.

120. Lender, A., "Correlative digital communication techniques", *IEEE Trans. Commun. Tech.*, pp. 128–135, Dec. 1964.

121. Levanon, N., and Mozeson, E., *Radar Signals,* New York: Wiley Interscience, 2004.

122. Li, J. and Stoica, P., "MIMO radar with colocated antennas," *IEEE Signal Proc. Mag.*, pp. 106–114, Sept. 2007.

123. Li, T. and Ding, Z., "Joint transmitter-receiver optimization for partial response channels based on nonmaximally decimated filter bank precoding technique", *IEEE Trans. Signal Proc.*, vol. **47**, no. 9, pp. 2407–2414, Sept. 1999.

124. Liang, J. and Tran, T. D., "A partial DFT based OFDM system resistant to channel nulls," *Proc. 36th IEEE Asilomar Conf. on Signals, Systems, and Computers*, Monterey, CA, Nov. 2002.

125. Lin, H., Siohan, P., Tanguy, P., and Javaudin, J.-P., "An analysis of the EIC method for OFDM/OQAM systems," *J. of Commun.*, vol. **4**, no. 1, pp. 52–60, Feb. 2009.

126. Lin, Y.-P. and Phoong, S.-M., "Perfect discrete multitone modulation with optimal transceivers," *IEEE Trans. Signal Proc.*, vol. **48**, no. 6, pp. 1702–1711, June 2000.

127. Lin, Y.-P. and Phoong, S.-M., "ISI free FIR filter bank transceivers for frequency selective channels," *IEEE Trans. Signal Proc.*, vol. **49**, no. 11, pp. 2648–2658, Nov. 2001a.

128. Lin, Y.-P. and Phoong, S.-M., "Optimal ISI free DMT transceivers for distorted channels with colored noise," *IEEE Trans. Signal Proc.*, vol. **49**, no. 11, pp. 2702–2712, Nov. 2001b.

129. Lin, Y.-P. and Phoong, S.-M., "Minimum redundancy for ISI free FIR filter bank transceivers," *IEEE Trans. on Signal Proc.*, vol. **50**, no. 4, pp. 842–853, April 2002.

130. Lin, Y.-P. and Phoong, S.-M., "BER minimized OFDM systems with channel independent precoders," *IEEE Trans. Signal Proc.*, vol. **51**, no. 9, pp. 2369–2380, Sep. 2003.

131. Lin, Y.-P., Vaidyanathan, P. P., Akkarakaran, S., and Phoong, S.-M., "On the duality of optimal DMT systems and biorthogonal subband coders," *Proc. IEEE Intl. Conf. Acoustics, Speech, and Signal Processing,* Istanbul, June 2000.

132. Lucky, R. W., "Automatic equalization for digital communication," *Bell Syst. Tech. J.*, pp. 547–588, April 1965.

133. Lucky, R. W., "Techniques for adaptive equalization of digital communication systems," *Bell Syst. Tech. J.*, pp. 255–286, Feb. 1966.

134. Lucky, R. W. and Rudin, H. R., "Generalized automatic equalization for communication channels," *Proc. IEEE Lett.*, pp. 439–440, March 1966.

135. Lucky, R. W., Salz, J., and Weldon, Jr., E. J., *Principles of Data Communication*, New York: Mc-Graw Hill Inc., 1968.

136. Lucky, R. W., "A survey of communication theory literature: 1968-1973," *IEEE Trans. Info. Theory*, vol. **19**, no. 5, pp. 725–739, Nov. 1973.

137. Luenberger, D. G., *Optimization By Vector Space Methods,* New York: John Wiley & Sons, Inc., 1969.

138. MacColl, L. R. A., *Signaling Method and Apparatus*, US Patent no. 2,056, 284, Oct. 6, 1936.

139. McEliece, R. J., *The Theory of Information and Coding,* Reading, MA: Addison-Wesley, 1977.

140. McEliece, R. J., Ash, R. B., and Ash, C., *Introduction to Discrete Mathematics,* New York: Random House, Inc., 1989.

141. Mallat, S., *A Wavelet Tour of Signal Processing,* San Diego, CA : Academic Press, 1998.

142. Malvar, H. S. and Staelin, D. H., "Optimal FIR pre- and postfilters for decimation and interpolation of random signals," *IEEE Trans. Commun.*, vol. **36**, no. 1, pp. 67–74, Jan. 1988a.

143. Malvar, H. S. and Staelin, D. H., "Optimal pre- and postfilters for multichannel signal processing", *IEEE Trans. Acoust., Speech, & Signal Proc.*, vol. **36**, no. 2, pp. 287–289, Feb. 1988b.

144. Marcus, M. and Minc, H., *A Survey of Matrix Theory and Matrix Inequalities,* New York: Dover Publications, 1964.

145. Marshall, A. W. and Olkin, I., *Inequalities: Theory of Majorization and its Applications*, New York: Academic Press, Inc., 1979.

146. Mertins, A., "MMSE design of redundant FIR precoders for arbitrary channel lengths," *IEEE Trans. Signal Proc.*, vol. **51**, no. 9, pp. 2402–2409, Sept. 2003.

147. Milnor, J. W., *Correction Circuit for Ocean Cable Relays*, US Patent no. 1,717,116, June 11, 1929.

148. Mitra, S. K., *Digital Signal Processing: A Computer-Based Approach*, New York: McGraw Hill Co., Inc., 2001.

149. Monsen, P., "Feedback equalization for fading dispersive channels," *IEEE Trans. on Info. Theory*, vol. **17**, no. 1, pp. 56–64, Jan. 1971.

150. Moon, T. K. and Stirling, W. C., *Mathematical Methods and Algorithms for Signal Processing,* Englewood Cliffs, NJ: Prentice Hall, Inc., 2000.

151. Müller, S. H. and Huber, J. B., "A comparison of peak power reduction schemes for OFDM," *Proc. IEEE Globecom*, pp. 1–5, Phoenix, AZ, Nov. 1997.

152. Narasimha, M. J. and Peterson, A. M. "Design of a 24-channel transmultiplexer," *IEEE Trans. Acoustics., Speech, & Signal Proc.*, vol. **27**, Dec. 1979.

153. North, D. O., "Analysis of the factors which determine signal/noise discrimination in radar," *Proc. IRE*, vol. **51**, pp. 1016–1028, July 1963 (first appeared as an *RCA Tech. Rep.*, June 1943).

154. Nyquist, H., "Certain topics in telegraphic transmission theory", *Trans. AIEE*, pp. 617–644, Feb. 1928 (reprinted in *Proc. IEEE*, vol. **90**, no. 2, pp. 280–305, Feb. 2002).

155. Ohno, S., "Performance of single-carrier block transmissions over multipath fading channels with linear equalization," *IEEE Trans. Signal Proc.*, vol. **54**, pp. 3678–3687, Oct. 2006.

156. Oppenheim, A. V. and Schafer, R. W., *Discrete-Time Signal Processing,* Englewood Cliffs, NJ: Prentice Hall, Inc., 1999.

157. Oppenheim, A. V. and Willsky, A. S., *Signals and Systems,* Englewood Cliffs, NJ: Prentice Hall, Inc., 1997.

158. Palomar, D. P. and Barbarossa, S., "Designing MIMO communications systems: constellation choice and linear transceiver design," *IEEE Trans. Signal Proc.*, vol. **53**, no. 10, pp. 3804–3818, Oct. 2005.

159. Palomar, D. P., Cioffi, J. M., and Lagunas, M. A., "Joint Tx-Rx beamforming design for multicarrier MIMO channels: a unified framework for convex optimization," *IEEE Trans. Signal Proc.*, vol. **51**, no. 9, pp. 2381–2401, Sept. 2003.

160. Palomar, D. P. and Jiang, Y., *MIMO Transceiver Design Via Majorization Theory,* Foundations and Trends in Communications and Information Theory, vol. **3**, no. 4-5, Now Publishers, pp. 331–551, 2006.

161. Palomar, D. P., Lagunas, M. A., and Cioffi, J. M., "Optimum linear joint transmit-receive processing for MIMO channels with QoS constraints," *IEEE Trans. Signal Proc.*, vol. **52**, no. 5, pp. 1179–1197, May 2004.

162. Pandharipande, A. and Dasgupta, S., "Optimum DMT-based transceivers for multiuser communications," *IEEE Trans. Commun.*, vol. **51**, no. 12, pp. 2038–2046, Dec. 2003.

163. Papoulis, A., *Probability, Random Variables, and Stochastic Processes,* New York: McGraw-Hill, Inc., 1965.

164. Papoulis, A., *Signal Analysis,* New York: McGraw-Hill, Inc., 1977.

165. Pečarić, J. E., Proschan, F., and Tong, Y. L., *Convex Functions, Partial Orderings, and Statistical Applications,* San Diego, CA: Academic Press, Inc., 1992.

166. Peebles, Jr., P. Z., *Probability, Random Variables, and Random Signal Principles,* New York: McGraw-Hill, Inc., 1987.

167. Peled, A. and Ruiz A., "Frequency domain data transmission using reduced computational complexity algorithms," *Proc. IEEE Intl. Conf. On Acoustics, Speech, and Signal Proc.*, pp. 964–967, 1980.

168. Phoong, S.-M., Chang, Y., and Chen C.-Y., "DFT-modulated filterbank transceivers for multipath fading channels," *IEEE Trans. Signal Proc.*, vol. **53**, no. 1, pp. 182–192, Jan. 2005.

169. Picinbono, B., "On circularity," *IEEE Trans. Signal Proc.*, vol. **42**, no. 12, pp. 3473–3482, Dec. 1994.

170. Price, R., "Nonlinearly feedback-equalized PAM versus capacity for noisy filter channels," *Rec. Int. Conf. Commun.*, 1972.

171. Proakis, J. G., *Digital Communications,* New York: McGraw Hill, 1995.

172. Proakis, J. G. and Salehi, M., *Digital Communications,* New York: McGraw Hill, 2008.

173. Rappaport, T. S., *Wireless Communications: Principle and Practice*, Upper Saddle River, NJ: Prentice Hall, 1996.

174. Ribeiro, C. B., de Campos, M. L. R, and Diniz, P. S. R., "FIR equalizers with minimum redundancy," *Proc. IEEE Intl. Conf. On Acoustics, Speech, and Signal Proc.*, vol. **III**, pp. 2673–2676, May 2002.

175. Rizos, A. D., Proakis, J. G., and Nguyen, T. Q., "Comparison of DFT and cosine modulated filter banks in multicarrier modulation," *Proc. of Globecom*, pp. 687–691, Nov. 1994.

176. Robey, F. C., Coutts, S., Weikle, D., McHarg, J. C., and Cuomo, K., "MIMO radar theory and experimental results," *Proc. 38th IEEE Asilomar Conf. on Signals, Systems, and Computers*, Monterey, CA, pp. 300–304, Nov. 2004.

177. Saltzberg, B. R., "Performance of an efficient parallel data transmission system," *IEEE Trans. Commun. Tech.*, vol. **15**, no. 6, pp. 805–811, Dec. 1967.

178. Salz, J., "Optimum mean-square decision feedback equalization," *Bell Sys. Tech. J.*, vol. **52**, no. 8, pp. 1341–1373, Oct. 1973.

179. Salz, J., "Digital transmission over cross-coupled linear channels", *AT&T Tech. J.*, vol. **64**, no. 6, pp. 1147–1159, July–Aug. 1985.

180. San Antonio, G., Fuhrmann, D. R., and Robey, F. C., "MIMO radar mmbiguity functions," *Proc. 40th IEEE Asilomar Conf. on Signals, Systems, and Computers*, Monterey, CA, pp. 36–40, Nov. 2006.

181. Sandberg, S. D., and Tzannes, M. A., "Overlapped discrete multitone modulation for high speed copper wire communications," *IEEE J. Selected Areas Commun.*, vol. **13**, no. 9, pp. 1571–1585, Dec. 1995.

182. Sari, H., Karam, G., and Jeanclaude, I., "An analysis of orthogonal frequency-division multiplexing for mobile radio applications," *IEEE Vehicular Technology Conference*, pp. 1635–1639, June 1994.

183. Sayed, A. H., *Fundamentals of Adaptive Filtering,* New York: John Wiley & Sons, Inc., 2003.

184. Scaglione, A., Giannakis, G. B., and Barbarossa, S., "Redundant filter bank precoders and equalizers Part I: Unification and optimal designs", *IEEE Trans. Signal Proc.*, vol. **47**, no. 7, pp. 1988–2006, July 1999a.

185. Scaglione, A., Giannakis, G. B., and Barbarossa, S., "Redundant filter bank precoders and equalizers Part II: Synchronization and direct equalization," *IEEE Trans. Signal Proc.*, vol. **47**, no. 7, pp. 2007–2022, July 1999b.

186. Scaglione, A., Giannakis, G. B., and Barbarossa, S., "Lagrange/ Vandermonde MUI eliminating user codes for quasi-synchronous CDMA in unknown multipath", *IEEE Trans. Signal Proc.*, vol. **48**, no. 7, pp. 2057–2073, July 2000.

187. Serbetli, S. and Yener, A., "Transceiver optimization for multiuser MIMO systems," *IEEE Trans. on Signal Proc.*, vol. **52**, no. 1, pp. 214–226, Jan. 2004.

188. Shannon, C. E., "A mathematical theory of communication," *Bell Syst. Tech. J.*, vol. **27**, pp. 379–423, 1948.

189. Shannon, C. E., "Communication in the presence of noise," *Proc. of the IRE*, vol. **37**, pp. 10–21, 1949.

190. Shenouda, M. B. and Davidson, T. N., "A framework for designing MIMO systems with decision feedback equalization or Tomlinson-Harashima precoding," *IEEE J. Selected areas in Commun.*, vol. **26**, no. 2, pp. 401–411, Feb. 2008.

191. Siohan, P., Siclet, C., and Lacaille, N., "Analysis and design of OFDM/ OQAM systems based on filter bank theory," *IEEE Trans. on Signal Proc.*, vol. **50**, no. 5, pp. 1170–1183, May 2002.

192. Skolnik, M. I., *Introduction to Radar Systems,* New York: Mc-Graw Hill, 2001.

193. Slepian, D., "Prolate spheroidal wave functions, Fourier analysis, and uncertainty – V: The discrete case," *Bell Sys. Tech. J.*, vol. **57**, pp. 1371–1430, May-June 1978.

194. Smith, J. W., "The joint optimization of transmitted signal and receiving filter for data transmission systems," *Bell Sys. Tech. J.*, pp. 2363–2392, Dec. 1965.

195. Starr, T., Cioffi, J. M., and Silverman, P. J., *Understanding DSL Technology,* Englewood Cliffs, NJ: Prentice Hall, Inc., 1999.

196. Strang, G. and Nguyen, T., *Wavelets and Filter Banks,* Wellesley, MA: Wellesley-Cambridge Press, 1996.

197. Su, B. and Vaidyanathan, P. P., "A generalized algorithm for blind channel identification with linear redundant precoders," *EURASIP J. Appl. Signal Proc.*, 2007a.

198. Su, B. and Vaidyanathan, P. P., "Subspace-based blind channel identification for cyclic prefix systems using very few received blocks," *IEEE Trans. Signal Pro.*, vol. **55**, no. 10, pp. 4979–4993, Oct. 2007b.

199. Tan, P. and Beaulieu, N. C., "Reduced ICI in OFDM systems using "better than" raised-cosine pulse," *IEEE Commun. Lett.*, vol. **8**, no. 3, pp. 135–137, March 2004.

200. Telatar, I. E., "Capacity of multi-antenna Gaussian channels," *European Trans. on Telecomm.*, vol. **10**, pp. 585–595, Nov.–Dec. 1999.

201. Telatar, I. E. and Tse, D., "Capacity and mutual information of wideband multipath fading channels," *IEEE Trans. Info. theory*, vol. **46**, no. 4 , pp. 1384–1400, July 2000.

202. Tepedelenlioĝlu, C. and Challagulla, R., "Low-complexity multipath diversity through fractional sampling in OFDM", *IEEE Trans. Signal Proc.*, vol. **52**, no. 11, pp. 3104–3116, Nov. 2004.

203. Therrien, C. W., *Discrete Random Signals and Statistical Signal Procesing,* Upper Saddle River, NJ: Prentice Hall, Inc., 1992.

204. Therrien, C. W. and Tummala, M., *Probability for Electrical and Computer Engineers,* New York: CRC Press, 2004.

205. Tomlinson, M., "New automatic equaliser employing modulo arithmetic," *Electron. Lett.*, vol. **7**, no. 5–6, pp. 138–139, Feb. 1971.

206. Tong, L., Xu, G., and Kailath, T., "Blind identification and equalization based on second order statistics: a time domain approach," *IEEE Trans. on Info. Theory,* vol. **40**, no. 2, pp. 340-349, March 1994.

207. Tong, L., Xu, G., Hassibi, B., and Kailath, T., "Blind channel identification based on second order statistics: a frequency domain approach," *IEEE Trans. on Info. Theory,* vol. **41**, no. 1, pp. 329–334, Jan. 1995.

208. Treichler, J. R., Fijalkow, I., and Johnson, C. R., Jr., "Fractionally spaced equalizers: how long should they really be?," *IEEE Signal Processing Mag.*, pp. 65–81, May 1996.

209. Tsatsanis, M. K. and Giannakis, G. B., "Principal component filter banks for optimal multiresolution analysis", *IEEE Trans. Signal Proc.*, vol. **43**, no. 8, pp. 1766–1777, Aug. 1995.

210. Tse, D. and Viswanath, P., *Fundamentals of Wireless Communication,* Cambridge: Cambridge University Press, 2005.

211. Tu, J. C., "Optimum MMSE equalization for staggered modulation," *Proc. 27th IEEE Asilomar Conf. on Signals, Systems, and Computers*, pp. Monterey, CA, 1401–1406, Nov. 1993.

212. Tufts, D. W., *Design Problems in Pulse Transmission*, Tech. Report 368, Research Laboratory of Electronics, MIT, Cambridge, MA, July 1960.

213. Tufts, D. W., "Nyquist's problem – the joint optimization of transmitter and receiver in pulse amplitude modulation," *Proc. IEEE*, pp. 248–259, March 1965.

214. Tzannes, M. A., Tzannes, M. C., Proakis, J. G., and Heller, P. N., "DMT systems, DWMT systems, and digital filter banks," *Proc. Intl. Conf. Commun.*, pp. 311–315, 1994.

215. Unser, M., "On the optimality of ideal filters for pyramid and wavelet signal approximation," *IEEE Trans. Signal Proc.*, vol. **41**, no. 12, pp. 3591–3596, Dec. 1993.

216. Vahlin, A. and Holte, N., "Optimal finite duration pulses for OFDM," *IEEE Trans. on Commun.*, vol. **44**, no. 1, pp. 10–14, Jan. 1996.

217. Vaidyanathan, P. P., "Multirate digital filters, filter banks, polyphase networks, and applications: a tutorial," *Proc. IEEE*, vol. **78**, pp. 56–93, Jan. 1990.

218. Vaidyanathan, P. P., *Multirate Systems and Filter Banks,* Englewood Cliffs, NJ: Prentice Hall, Inc., 1993.

219. Vaidyanathan, P. P., "Theory of optimal orthonormal subband coders," *IEEE Trans. on Signal Proc.*, vol. **46**, no. 6, pp. 1528–1543, June 1998.

220. Vaidyanathan, P. P., "Filter banks in digital communications," *IEEE CAS Mag.*, vol. **1**, pp. 4–25, 2001.

221. Vaidyanathan, P. P., "On equalization of channels with ZP precoders," *Proc. IEEE Int. Symp. Circuits and Systems*, pp. 1329–1332, New Orleans, LA, May 2007.

222. Vaidyanathan, P. P., *The Theory of Linear Prediction,* Morgan and Claypool Publishers, 2008.

223. Vaidyanathan, P. P. and Djokovic, I., "Wavelet transforms", in *Mathematics for Circuits and Filters*, ed. W.-K. Chen, New York: CRC Press, 2000.

224. Vaidyanathan, P. P., Lin, Y.-P., Akkarakaran, S., and Phoong, S.-M., "Discrete multitone modulation with principal component filter banks," *IEEE Trans. Circ. and Syst.-I*, vol. **49**, pp. 1397–1412, Oct. 2002.

225. Vaidyanathan, P. P. and Mitra, S. K., "Polyphase networks, block digital filtering, LPTV systems, and alias-free QMF banks: a unified approach based on pseudocirculants," *IEEE Trans. Acoust., Speech, & Signal Proc.*, vol. **36**, pp. 381–391, March 1988.

226. Vaidyanathan, P. P. and Vrcelj, B., "Biorthogonal partners and applications," *IEEE Trans. Signal Proc.*, vol. **49**, pp. 1013–1027, May 2001.

227. Vaidyanathan, P. P. and Vrcelj, B., "Theory of fractionally spaced cyclic-prefix equalizers", *Proc. IEEE Int. Conf. Acoust. Speech, and Signal Proc.*, Orlando, FL, May 2002.

228. Vaidyanathan, P. P. and Vrcelj, B., "Transmultiplexers as precoders in modern digital communication: a tutorial review," *IEEE Int. Symp. on Circuits and Systems*, pp. V-405–V-412, Vancouver, Canada, 2004.

229. Van Trees, H. L., *Radar and Sonar Signal Processing: Part III of Detection Estimation and Modulation Theory*, New York: John Wiley & Sons, Inc., 2001.

230. Van Trees, H. L., *Optimum Array Processing: Part IV of Detection Estimation and Modulation Theory*, New York: Wiley Interscience, 2002.

231. Vandendorpe, L., Cuvelier, L., Deryck, F., Louveaux, J., and van de Wiel, O., "Fractionally spaced linear and decision feedback detectors for transmultiplexers," *IEEE Trans. on Signal Proc.*, vol. **46**, no. 4, pp. 996–1011, April 1998.

232. Vangelista, L. and Laurenti, N., "Efficient implementations and alternative architectures for OFDM-OQAM systems," ÎEEE Trans. on Commun., vol. **49**, no. 4, pp. 664–675, April 2001.

233. Vemulapalli, M., Dasgupta, S., and Pandharipande, A., "A new algorithm for optimum bit loading with a general cost," *Proc. IEEE Intl. Symp. On Circuits and Systems*, pp. 5403–5406, Kos, Greece, May 2006.

234. Verdu, S., *Multiuser Detection*, Cambridge: Cambridge University Press, 1998.

235. Vetterli, M. "Perfect transmultiplexers," *Proc. IEEE Intl. Conf. Acoustics, Speech, and Signal Proc.*, pp. 2567–2570, Japan, April 1986.

236. Vetterli, M. and Kovačević, J., *Wavelets and Subband Coding*, Englewood cliffs, NJ: Prentice Hall, Inc., 1995.

237. Vijaya Krishna, A. and Hari, K. V. S., "Filter bank precoding for FIR equalization in high-rate MIMO communications," *IEEE Trans. Signal Proc.*, vol. **54**, no. 5, pp. 1645–1652, May 2006.

238. Viterbi, A. J., "Error bounds for convolutional codes and an asymptotically optimum decoding algorithm," *IEEE Trans. on Info. Theory*, vol. **13**, no. 2, pp. 260–269, April 1967.

239. Vrcelj, B. and Vaidyanathan, P. P., "MIMO biorthogonal partners and applications," *IEEE Trans. Signal Proc.*, vol. **50**, pp. 528–542, March 2002.

240. Vrcelj, B. and Vaidyanathan, P. P., "Fractional biorthogonal partners in channel equalization and signal interpolation," *IEEE Trans. Signal Proc.*, vol. **51**, pp. 1928–1940, July 2003.

241. Wang, X., Lu, W.-S., and Antoniou, A., "Constrained minimum-BER multiuser detection", *IEEE Trans. Signal Proc.*, vol. **48**, no. 10, pp. 2903–2909, Oct. 2000.

242. Wang, Z., Ma, X., and Giannakis, G. B., "OFDM or single-carrier block transmissions?" *IEEE Trans. on Commun.*, vol. **52**, no. 3, pp. 380–394, March 2004.

243. Weng, C.-C., Chen, C.-Y., and Vaidyanathan, P. P., "MIMO transceivers with decision feedback and bit loading: theory and optimization," *IEEE Trans. Signal Proc.*, vol. **58**, 2010a.

244. Weng, C.-C., Chen, C.-Y., and Vaidyanathan, P. P., "Generalized triangular decomposition in transform coding," *IEEE Trans. Signal Proc.*, vol. **58**, 2010b.

245. Weng C.-C. and Vaidyanathan, P. P., "Joint optimization of transceivers with fractionally spaced equalizers", *Proc. IEEE Int. Conf. Acoustics Speech, and Signal Processing*, Las Vegas, March–April 2008.

246. Weng, C.-C. and Vaidyanathan, P. P., "MIMO transceiver optimization with linear constraints on transmitted signal covariance components," *IEEE Trans. Signal Proc.*, vol. **58**, 2010.

247. Weyl, H., "Inequalities between the two kinds of eigenvalues of a linear transformation,", *Proc. Natl. Acad. Sci., USA*, vol. **35** pp. 408–411, 1949.

248. Wiener, N., *Extrapolation, Interpolation, and Smoothing of Stationary Time Series,* New York: Wiley, 1949.

249. Wiener, N. and Masani, P., "The prediction theory of multivariate stochastic processes: Part 1," *Acta Mathematica*, vol. **98**, pp. 111–150, 1957.

250. Wiener, N. and Masani, P., "The prediction theory of multivariate stochastic processes: Part 2," *Acta Mathematica*, vol. **99**, pp. 93–137, 1958.

251. Witsenhausen, H. S., "A determinant maximization problem occurring in the theory of data communication," *SIAM. J. Appl. Math.*, pp. 515–522, Nov. 1975.

252. Wolniansky, P. W., Foschini, G. J., Golden, G. D., and Valenzuela, R. A., "V-BLAST: an architecture for realizing very high data rates over the rich-scattering wireless channel," *Proc. URSI Int. Symp. Signals, Systems, and Electronics*, pp. 295–300, 1998.

253. Xia, X-G., "New precoding for intersymbol interference cancellation using nonmaximally decimated multirate filter banks with ideal FIR equalizers," *IEEE Trans. Signal Proc.*, vol. **45**, no. 10, pp. 2431–2441, Oct. 1997.

254. Xia, X-G., *Modulated Coding for Intersymbol Interference Channels*, New York: Marcel Dekker, Inc., 2001.

255. Xu, F., Davidson, T. N., Zhang, J-K., and Wong, K. M., "Design of block transceivers with decision feedback detection," *IEEE Trans. Signal Proc.*, vol. **54**, no. 3, pp. 965–978, March 2006.

256. Yang. J. and Roy, S., "Joint transmitter-receiver optimization for multiple-input-multiple-output systems with decision feedback," *IEEE Trans. Info. Theory*, vol. **40**, no. 5, pp. 1334–1347, Sept. 1994a.

257. Yang. J. and Roy, S., "On joint transmitter and receiver optimization for multiple-input-multiple-output (MIMO) transmission systems," *IEEE Trans. on Commun.*, vol. **42**, no. 12, pp. 3221–3231, Dec. 1994b.

258. Yasotharan, A., "Multirate zero-forcing Tx-Rx design for MIMO channel under BER constraints," *IEEE Trans. Signal Proc.*, vol. **54**, no. 6, pp. 2288–2301, June 2006.

259. Zhang, J-K., Kavčić, A., Ma, X., and Wong, K. M., "Design of unitary precoders for ISI channels," *Proc. IEEE Intl. Conf. Acoustics, Speech, and Sig. Proc.*, pp. III-2265–III-2268, Orlando, FL, April 2002.

260. Zhang, J-K., Kavčić, A., and Wong, K. M., "Equal-diagonal QR decomposition and its application to precoder design for successive-cancellation detection," *IEEE Trans. Info. Theory*, vol. **51**, no. 1, pp. 154–172, Jan. 2005.

Index

www.ingramcontent.com/pod-product-compliance
Lightning Source LLC
LaVergne TN
LVHW062201200926
848484LV00006B/1296
* 9 7 8 0 5 2 1 7 6 0 7 9 9 *